CLYMER
HONDA
GL1800 GOLD WING • 2001-2005

CLYMER
P.O. Box 12901, Overland Park, Kansas 66282-2901

Copyright ©2006 Prism Business Media Inc.

FIRST EDITION
First Printing April, 2004

SECOND EDITION
First Printing May, 2006

Printed in U.S.A.

CLYMER and colophon are registered trademarks of Prism Business Media Inc.

ISBN: 1-59969-043-8

Library of Congress: 2006903380

AUTHOR: Ron Wright.

TECHNICAL PHOTOGRAPHY: Ron Wright. Assistance from Clawson Motorsports, Fresno, California. Special thanks to Service Manager Josh Ray and Service Technician Steve Friebe for their technical assistance.

TECHNICAL ILLUSTRATIONS: Steve Amos.

WIRING DIAGRAMS: Bob Meyer and Lee Buell.

EDITOR: James Grooms.

PRODUCTION: Susan Hartington.

TOOLS AND EQUIPMENT: K&L Supply Co. at www.klsupply.com.

COVER: Mark Clifford Photography at www.markclifford.com. 2001 GL1800 courtesy of David Jay, Santa Clarita, California.

All rights reserved. Reproduction or use, without express permission, of editorial or pictorial content, in any manner, is prohibited. No patent liability is assumed with respect to the use of the information contained herein. While every precaution has been taken in the preparation of this book, the publisher assumes no responsibility for errors or omissions. Neither is any liability assumed for damages resulting from use of the information contained herein. Publication of the servicing information in this manual does not imply approval of the manufacturers of the products covered.

All instructions and diagrams have been checked for accuracy and ease of application; however, success and safety in working with tools depend to a great extent upon individual accuracy, skill and caution. For this reason, the publishers are not able to guarantee the result of any procedure contained herein. Nor can they assume responsibility for any damage to property or injury to persons occasioned from the procedures. Persons engaging in the procedure do so entirely at their own risk.

CONTENTS

QUICK REFERENCE DATA XI

CHAPTER ONE
GENERAL INFORMATION 1
- Manual organization
- Warnings, cautions and notes
- Safety
- Serial numbers
- Fasteners
- Shop supplies
- Basic tools
- Precision measuring tools
- Electrical system fundamentals
- Basic service methods
- Storage
- Specifications

CHAPTER TWO
TROUBLESHOOTING 32
- Operating requirements
- Starting the engine
- Engine will not start
- Engine performance
- Fuel system
- Engine
- Engine lubrication
- Cylinder leakdown test
- Clutch
- Gearshift linkage
- Transmission
- Electrical testing
- Final drive
- Front suspension and steering
- Brake System
- Cruise control

CHAPTER THREE
LUBRICATION, MAINTENANCE AND TUNE-UP 56
- Fuel
- Tune-up
- Cylinder identification and firing order
- Air filter
- Engine compression test
- Spark plugs
- Ignition timing
- Valve clearance
- Engine idle speed inspection
- Fuel hose inspection
- Throttle cable lubrication
- Throttle cable operation and adjustment
- Reverse operation
- Clutch
- Engine oil and filter
- Cooling system
- Crankcase breather inspection
- Emission control systems
- Final drive oil
- Tires and wheels
- Brakes
- Lights and horn inspection
- Sidestand switch and ignition cut-off system
- Steering head bearing inspection
- Front suspension inspection
- Rear suspension inspection
- Fasteners
- Unscheduled maintenance
- Specifications

CHAPTER FOUR
ENGINE TOP END 89
- Cylinder head covers
- Camshafts
- Cam chains and timing sprocket
- Cam chain tensioner
- Cylinder head
- Valves and valve components
- Specifications

CHAPTER FIVE
ENGINE LOWER END 123
- Engine service
- Engine
- Rear crankcase cover
- Primary gears/output shaft
- Output shaft
- Reverse shift arm
- Reverse shifter and shift drum lock arm
- Crankcase
- Pistons and connecting rods
- Crankshaft
- Oil strainer and oil pressure relief valve
- Oil pump
- Break-in procedure
- Specifications

CHAPTER SIX
CLUTCH 187
- Brake fluid
- Preventing brake fluid damage
- Clutch fluid draining
- Clutch system bleeding
- Clutch master cylinder
- Clutch slave cylinder
- Clutch cover
- Clutch
- Specifications

CHAPTER SEVEN
SHIFT MECHANISM AND TRANSMISSION 211
- External shift linkage
- Transmission assembly
- Mainshaft
- Gearshift spindle assembly
- Shift forks and shift drum
- Countershaft
- Transmission shifting check
- Specifications

CHAPTER EIGHT
FUEL INJECTION AND
EMISSION CONTROL SYSTEMS 237

Fuel injection system precautions
Fuel and vacuum hose identification
Depressurizing the fuel system
Fuel pressure test
Fuel flow test
Fuel tank
Air filter housing
Fuel injectors
Fuel pressure reulator
Throttle body
Fuel pump
Fuel pump relay
Baro and map sensor
Idle air control (IAC) valve
Throttle position (TP) sensor

ECM Test Harness
MIL Lamp Circuit Test
Cam pulse generator
Knock sensors
Oxygen (O_2) sensors
Vehicle speed sensor (VSS)
Engine control module (ECM)
Intake air temperature (IAT) sensor
Engine idle speed adjustment
Emission control system labels
Secondary air supply system
Evaporative emission control system
Intake manifold
Malfunction indicator lamp (MIL)
Specifications

CHAPTER NINE
ELECTRICAL SYSTEM 327

Electrical component replacement
Electrical connectors
Battery
Charging system
Alternator
Ignition system troubleshooting
Ignition coil
Ignition pulse generator
Ignition pulse generator rotor
PGM-FI ignition relay
Bank angle sensor
Engine control module
Ignition timing
Starter/reverse system troubleshooting
Starter relay A
Starter relay B
Reverse switch
Reverse relay
Reverse shift relays
Reverse shift actutor
Reverse regulator
Reverse resistor assembly
Power control relays
Speed limiter relay
Starter/reverse motor service
Headlight
Headlight adjusters
Headlight adjusting switch
Headlight adjuster relay
Front turn signal light
Trunk brake/taillight
License plate light
Saddlebag combination light
Combination meter

Speedometer
Tachometer
Coolant temperature gauge
ECT sensor
Fuel gauge (low fuel indicator)
Fuel level sensor
Oil pressure indicator and oil pressure switch
Neutral indicator
Turn signal relay
Turn signal cancel unit
Position light relay
Open air temperature sensor
Multi-display control switch
Left panel switch
Hazard switch
Hazard switch diode
Handlebar switches
Ignition switch
Cruise control switches
Clutch switch
Front brake light switch
Rear brake light switch
Gear position switch
Sidestand switch
Horns
Relay box
Fuses
Audio system
Audio unit
Audio switch
Antenna
Headset junction cable
Speakers
Specifications

CHAPTER TEN
COOLING SYSTEM 413
- Temperature warning system
- Cooling system inspection
- Cooling system hoses
- water pump mechanical seal inspection
- Pressure test
- Coolant reserve tank
- Radiators
- Cooling fans
- Fan relay testing
- Water pump/thermostat
- Specifications

CHAPTER ELEVEN
WHEELS AND TIRES 431
- Bike lift
- Front wheel
- Rear wheel
- Wheel runout and balance
- Front hub
- Tire safety
- Tire changing
- Tire repairs
- Specifications

CHAPTER TWELVE
FRONT SUSPENSION AND STEERING 449
- Handlebars
- Front fork
- Steering head and stem
- Steering bearing preload
- Steering head bearing races
- Specifications

CHAPTER THIRTEEN
REAR SUSPENSION AND FINAL DRIVE 489
- Shock absorber
- Suspension level relays
- Suspension linkage
- Swing arm
- Final drive housing
- Specifications

CHAPTER FOURTEEN
BRAKES 520
- Linked brake system (LBS)
- Brake fluid selection
- Preventing brake fluid damage
- Brake service
- Brake pads
- Brake calipers
- Brake caliper overhaul
- Front master cylinder
- Rear master cylinder and rear brake pedal
- Secondary master cylinder
- Delay valve
- Proportional control valve
- Brake hose and brake pipe replacement
- Brake disc
- Brake bleeding service tips and tools
- Brake fluid draining
- Flushing the brake system
- Bleeding the system
- Specifications

CHAPTER FIFTEEN
ANTI-LOCK BRAKE SYSTEM 560
- ABS service precautions
- ABS troubleshooting
- Wheel speed sensors
- Pulser ring
- ABS modulators
- ABS electronic control unit (ECU)
- Specifications

CHAPTER SIXTEEN
CRUISE CONTROL SYSTEM 597
- Cruise control system components
- Cruise control troubleshooting
- Cruise/reverse control unit
- Cruise control relays
- Cruise actuator
- Cruise control switches

CHAPTER SEVENTEEN
BODY COMPONENTS AND EXHAUST SYSTEM 621
Color label
Engine side covers
Side covers
Seat
Passenger seat back
Front fender assembly
Front fender top cover
Front fender B
Front lower fairings
Meter panel
Mirrors and front turn signals
Windshield assembly
Inner fairings
Front fairing and fairing molding
Fairing pockets
Top shelter
Rear fender A
Trunk lower cover and molding strips
Trunk
Saddlebags
Rear fender B
Saddlebag/trunk stay
Remote transmitter
Remote transmitter and circuit troubleshooting
Trunk lock control unit
Trunk actuator test
Saddlebag open switch test
Open switch diode test
Exhaust system
Body repair
Specifications

INDEX 661

WIRING DIAGRAMS 667
Starting and reverse systems
Ignition, charging and cooling systems
Fuel injection
ABS system
Lighting system
Turn signal and horn systems
Combination meter
Cruise control system
Accessory and trunk lock control system
Audio system
Suspension level control system
Fuse box

QUICK REFERENCE DATA

MOTORCYCLE INFORMATION

MODEL: _____ YEAR: _____
VIN NUMBER: _____
ENGINE SERIAL NUMBER: _____
THROTTLE BODY NUMBER: _____
COLOR CODE: _____

TIRE INFLATION PRESSURE*

	Front kPa (psi)	Rear kPa (psi)
Rider	250 (36)	280 (41)
Maximum weight capacity	250 (36)	280 (41)

*The tire inflation pressures are for original equipment tires. Aftermarket tires may require different inflation pressure. Refer to the tire manufacturer's specifications.

RECOMMENDED LUBRICANTS AND FUEL

Brake fluid	DOT 4 brake fluid
Clutch fluid	DOT 4 brake fluid
Control cables	Cable lubricant
Cooling system	Honda HP Coolant or equivalent[1]
Engine oil[2]	
Classification	
JASO T 903 standard rating	MA
API classification	SG or later[3]
Viscosity rating	SAE 10W-40
Final drive unit	Hypoid gear oil, SAE 80
Fork oil	Pro Honda Suspension Fluid SS-8 or equivalent 10 wt fork oil
Fuel	Unleaded gasoline with a pump octane number of 86 or higher[4]

1. Coolant must not contain silicate inhibitors.
2. Do not use oil with molybdenum additives.
3. Use API SG or later classified oils not specified as ENERGY CONSERVING. See text for additional information.
4. Leaded gasoline will damage the catalytic converter.

ENGINE OIL CAPACITY

	Liters	U.S. qt.
At oil change only	3.6	3.8
Oil and filter change	3.7	3.9
After engine disassembly	4.6	4.9

ENGINE COOLANT CAPACITY

	Liters	U.S. qt.
Radiator and engine	3.53	3.73
Reserve tank	0.65	0.69

FINAL DRIVE OIL CAPACITY

	ml	U.S. oz.
At oil change	120	4.1
After final drive disassembly	150	5.1

FORK OIL CAPACITY

Fork oil capacity	
Left fork tube	526.5-531.5 ml (17.82-17.98 U.S. oz.)
Right fork tube	482.5-487.5 ml (16.32-16.48 U.S. oz.)
Fork oil level	128 mm (5.0 in.)
Fork oil type	Pro-Honda Suspension Fluid SS-8 or equivalent 10 wt. fork oil

SERVICE TORQUE SPECIFICATIONS

	N•m	in.-lb.	ft.-lb.
Clutch reservoir stopper plate screw	2	18	—
Cylinder head cover bolt	12	106	—
Cylinder head side cover bolt	10	88	—
Engine oil drain bolt	34	—	25
Final drive oil drain plug	20	—	15
Final drive oil filler plug	12	106	—
Front brake reservoir cap screw	2	18	—
Oil filter cartridge[1]	26	—	19
Oil pressure switch[2]	12	106	—
Spark plugs	18	—	13
Timing hole cap[3]	18	—	13

1. Lubricate rubber seal and threads with engine oil.
2. Apply RTV sealant to switch threads as described in text.
3. Lubricate threads with grease.

TUNE-UP SPECIFICATIONS

Engine compression	1383 kPa (201 psi) at 300 rpm
Engine firing order	1-4-5-2-3-6
Engine idle speed	700 ± 70 rpm
Oil pressure	530 kPa (77 psi) at 5000 rpm
Spark plug type	
NGK	
Standard	BKR6E-11
Cold climate*	BKR5E-11
Extended high speed riding	BKR7E-11
Denso	
Standard	K20PR-U11
Cold climate*	K16PR-U11
Extended high speed riding	K22PR-U11
Spark plug gap	1.00-1.10 mm (0.039-0.043 in.)
Throttle lever free play	2-6 mm (0.08-0.24 in.)
Valve clearance	
Intake	0.12-0.18 mm (0.005-0.007 in.)
Exhaust	0.19-0.25 mm (0.008-0.010 in.)

*Cold climate conditions are when the ambient temperature is below 5° C (41° F).

CHAPTER ONE

GENERAL INFORMATION

This detailed and comprehensive manual covers the Honda GL1800 Gold Wing from 2001-2005.

The text provides complete information on maintenance, tune-up, repair and overhaul. Hundreds of photographs and illustrations created during the complete disassembly of the motorcycle guide the reader through every job. All procedures are in step-by-step format and designed for the reader who may be working on the motorcycle for the first time.

MANUAL ORGANIZATION

A shop manual is a tool and, as in all Clymer manuals, the chapters are thumb tabbed for easy reference. Main headings are listed in the table of contents and the index. Frequently used specifications and capacities from the tables at the end of each individual chapter are listed in the *Quick Reference Data* section at the front of the manual. Specifications and capacities are provided in U.S. standard and metric units of measure.

During some of the procedures there will be references to headings in other chapters or sections of the manual. When a specific heading is called out in a step it will be *italicized* as it appears in the manual. If a sub-heading is indicated as being "in this section" it is located within the same main heading. For example, the sub-heading *Handling Gasoline Safely* is located within the main heading **SAFETY**.

This chapter provides general information on shop safety, tools and their usage, service fundamentals and shop supplies. **Tables 1-8**, at the end of the chapter, list the following:

Table 1 lists engine and frame serial numbers.
Table 2 lists motorcycle dimensions.
Table 3 lists motorcycle weight.
Table 4 lists metric, inch and fractional equivalents.
Table 5 lists conversion formulas.
Table 6 lists general torque specifications.
Table 7 lists technical abbreviations.
Table 8 lists metric tap and drill sizes.

Chapter Two provides methods for quick and accurate diagnosis of problems. Troubleshooting procedures present typical symptoms and logical methods to pinpoint and repair the problem.

Chapter Three explains all routine maintenance necessary to keep the motorcycle running well. Chapter Three also includes recommended tune-up procedures, eliminating the need to constantly consult the chapters on the various assemblies.

Subsequent chapters describe specific systems such as engine, transmission, clutch, drive system, fuel system, suspension, brakes, cruise control, fairing and exhaust system. Each disassembly, repair and assembly procedure is discussed in step-by-step form.

WARNINGS, CAUTIONS AND NOTES

The terms WARNING, CAUTION and NOTE have specific meanings in this manual.

A WARNING emphasizes areas where injury or even death could result from negligence. Mechanical damage may also occur. WARNINGS *are to be taken seriously*.

A CAUTION emphasizes areas where equipment damage could result. Disregarding a CAUTION could cause permanent mechanical damage, though injury is unlikely.

A NOTE provides additional information to make a step or procedure easier or clearer. Disregarding a NOTE could cause inconvenience, but would not cause equipment damage or injury.

SAFETY

Professional mechanics can work for years and never sustain a serious injury or mishap. Follow these guidelines and practice common sense to safely service the motorcycle.

1. Do not operate the motorcycle in an enclosed area. The exhaust gasses contain carbon monoxide, an odorless, colorless and tasteless poisonous gas. Carbon monoxide levels build quickly in small enclosed areas and can cause unconsciousness and death in a short time. Make sure to properly ventilate the work area or operate the motorcycle outside.
2. *Never* use gasoline or any extremely flammable liquid to clean parts. Refer to *Cleaning Parts* and *Handling Gasoline Safely* in this section.
3. *Never* smoke or use a torch in the vicinity of flammable liquids, such as gasoline or cleaning solvent.
4. If welding or brazing on the motorcycle, remove the fuel tank to a safe distance at least 50 ft. (15 m) away.
5. Use the correct type and size of tools to avoid damaging fasteners.
6. Keep tools clean and in good condition. Replace or repair worn or damaged equipment.
7. When loosening a tight fastener, be guided by what would happen if the tool slips.
8. When replacing fasteners, make sure the new fasteners are the same size and strength as the original ones.
9. Keep the work area clean and organized.
10. Wear eye protection *anytime* the safety of the eyes is in question. This includes procedures that involve drilling, grinding, hammering, compressed air and chemicals.
11. Wear the correct clothing for the job. Tie up or cover long hair so it does not get caught in moving equipment.
12. Do not carry sharp tools in clothing pockets.
13. Always have an approved fire extinguisher available. Make sure it is rated for gasoline (Class B) and electrical (Class C) fires.
14. Do not use compressed air to clean clothes, the motorcycle or the work area. Debris may be blown into the eyes or skin. *Never* direct compressed air at anyone. Do not allow children to use or play with any compressed air equipment.
15. When using compressed air to dry rotating parts, hold the part so it does not rotate. Do not allow the force of the air to spin the part. The air jet is capable of rotating parts at extreme speed. The part may disintegrate or become damaged, causing serious injury.
16. Do not inhale the dust created by brake pad and clutch wear. These particles may contain asbestos. In addition, some types of insulating materials and gaskets may contain asbestos. Inhaling asbestos particles is hazardous to health.
17. Never work on the motorcycle while someone is working under it.
18. When placing the motorcycle on a stand, make sure it is secure before walking away.

Handling Gasoline Safely

Gasoline is a volatile flammable liquid and is one of the most dangerous items in the shop. Because gasoline is used so often, many people forget it is hazardous. Only use gasoline as fuel for gasoline internal combustion engines. Keep in mind when working on the machine, gasoline is always present in the fuel tank, fuel line and throttle body. To avoid a disastrous accident when working

GENERAL INFORMATION

around the fuel system, carefully observe the following precautions:

1. *Never* use gasoline to clean parts. Refer to *Cleaning Parts* in this section.
2. When working on the fuel system, work outside or in a well-ventilated area.
3. Do not add fuel to the fuel tank or service the fuel system while the motorcycle is near open flames, sparks or where someone is smoking. Gasoline vapor is heavier than air; it collects in low areas and is more easily ignited than liquid gasoline.
4. Allow the engine to cool completely before working on any fuel system component.
5. Do not store gasoline in glass containers. If the glass breaks, a serious explosion or fire may occur.
6. Immediately wipe up spilled gasoline with rags. Store the rags in a metal container with a lid until they can be properly disposed of, or place them outside in a safe place for the fuel to evaporate.
7. Do not pour water onto a gasoline fire. Water spreads the fire and makes it more difficult to put out. Use a class B, BC or ABC fire extinguisher to extinguish the fire.
8. Always turn off the engine before refueling. Do not spill fuel onto the engine or exhaust system. Do not overfill the fuel tank. Leave an air space at the top of the tank to allow room for the fuel to expand due to temperature fluctuations.

Cleaning Parts

Cleaning parts is one of the more tedious and difficult service jobs performed in the home garage. Many types of chemical cleaners and solvents are available for shop use. Most are poisonous and extremely flammable. To prevent chemical exposure, vapor buildup, fire and serious injury, observe each product warning label and note the following:

1. Read and observe the entire product label before using any chemical. Always know what type of chemical is being used and whether it is poisonous and/or flammable.
2. Do not use more than one type of cleaning solvent at a time. If mixing chemicals is required, measure the proper amounts according to the manufacturer.
3. Work in a well-ventilated area.
4. Wear chemical-resistant gloves.
5. Wear safety glasses.
6. Wear a vapor respirator if the instructions call for it.
7. Wash hands and arms thoroughly after cleaning parts.
8. Keep chemical products away from children and pets.
9. Thoroughly clean all oil, grease and cleaner residue from any part that must be heated.
10. Use a nylon brush when cleaning parts. Metal brushes may cause a spark.
11. When using a parts washer, only use the solvent recommended by the manufacturer. Make sure the parts washer is equipped with a metal lid that will lower in case of fire.

Warning Labels

Most manufacturers attach information and warning labels to the motorcycle. These labels contain instructions that are important to personal safety when operating, servicing, transporting and storing the motorcycle. Refer to the owner's manual for the description and location of labels. Order replacement labels from the manufacturer if they are missing or damaged.

SERIAL NUMBERS

Serial numbers are stamped on various locations on the frame, engine, transmission and throttle body. Record these numbers in the *Quick Reference Data* section in the front of the manual. Have these numbers available when ordering parts.

The VIN number label (**Figure 1**) is located on the left side of the frame, next to the steering head.

The frame serial number (**Figure 2**) is stamped on the right side of the steering head.

The engine serial number (**Figure 3**) is stamped on a raised pad on the right crankcase.

The fuel injection throttle body serial number (**Figure 4**) is located on the rear side of the throttle body.

The color label (**Figure 5**) is located on the inside of the fuel compartment fill lid.

FASTENERS

Proper fastener selection and installation is important to ensure the motorcycle operates as designed and can be serviced efficiently. The choice of original equipment fasteners is not arrived at by chance. Make sure replacement fasteners meet all the same requirements as the originals.

Threaded Fasteners

Threaded fasteners secure most of the components on the motorcycle. Most are tightened by turning them clockwise (right-hand threads). If the normal rotation of the component being tightened would loosen the fastener, it may have left-hand threads. If a left-hand threaded fastener is used, it is noted in the text.

Two dimensions are required to match the thread size of the fastener: the number of threads in a given distance and the outside diameter of the threads.

The two systems currently used to specify threaded fastener dimensions are the U.S. standard system and the metric system (**Figure 6**). Pay particular attention when working with unidentified fasteners; mismatching thread types can damage threads.

NOTE
To ensure that the fastener threads are not mismatched or cross-threaded, start all fasteners by hand. If a fastener is hard to start or turn, determine the cause before tightening with a wrench.

The length (L, **Figure 7**), diameter (D) and distance between thread crests (pitch) (T) classify metric screws and bolts. A typical bolt may be identified by the numbers, 8—1.25 × 130. This indicates the bolt has a diameter of 8 mm, the distance between thread crests is 1.25 mm and the length is 130 mm. Always measure bolt length as shown in L, **Figure 7** to avoid purchasing replacements of the wrong length.

The numbers on the top of the fastener (**Figure 7**) indicate the strength of metric screws and bolts. The higher the number, the stronger the fastener is. Typically, unnumbered fasteners are the weakest.

Many screws, bolts and studs are combined with nuts to secure particular components. To indicate the size of a nut, manufacturers specify the internal diameter and the thread pitch.

The measurement across two flats on a nut or bolt indicates the wrench size.

WARNING
Do not install fasteners with a strength classification lower than what was originally installed by the manufacturer. Doing so may cause equipment failure and/or damage.

Torque Specifications

The materials used in the manufacturing of the motorcycle may be subjected to uneven stresses if the fasteners of the various subassemblies are not installed and tightened correctly. Fasteners that are improperly installed or

GENERAL INFORMATION

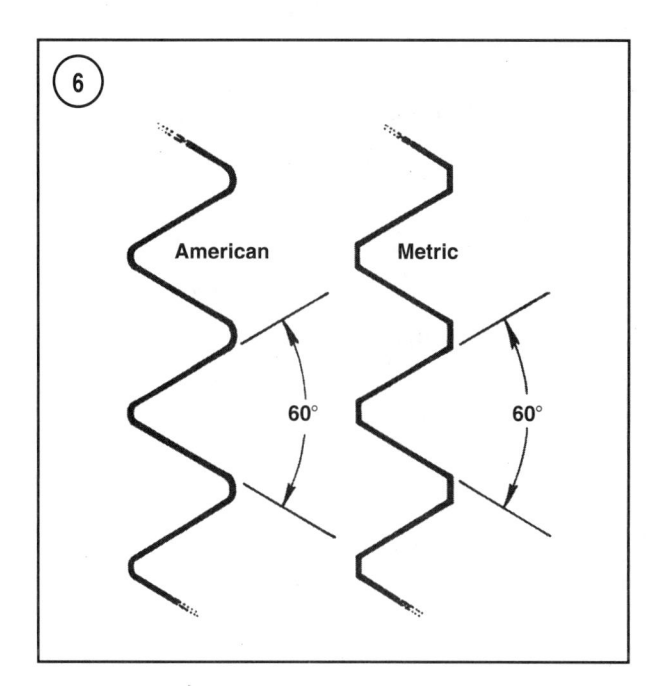

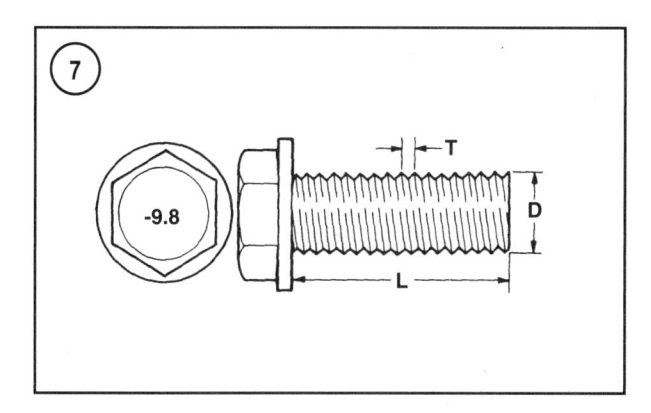

work loose can cause extensive damage. It is essential to use an accurate torque wrench as described in this chapter.

Specifications for torque are provided in Newton-meters (N•m), foot-pounds (ft.-lb.) and inch-pounds (in.-lb.). Refer to **Table 6** for general torque specifications. To determine the torque requirement, first determine the size of the fastener as described in *Threaded Fasteners* in this section. Torque specifications for specific components are at the end of the appropriate chapters. Torque wrenches are covered in *Basic Tools* in this chapter.

Self-Locking Fasteners

Several types of bolts, screws and nuts incorporate a system that creates interference between the two fasteners. Interference is achieved in various ways. The most common types are the nylon insert nut and a dry adhesive coating on the threads of a bolt.

Self-locking fasteners offer greater holding strength than standard fasteners, which improves their resistance to vibration. All self-locking fasteners cannot be reused. The materials used to form the lock become distorted after the initial installation and removal. Discard and replace self-locking fasteners after removing them. Do not replace self-locking fasteners with standard fasteners.

Washers

The two basic types of washers are flat washers and lockwashers. Flat washers are simple discs with a hole to fit a screw or bolt. Lockwashers are used to prevent a fastener from working loose. Washers can be used as spacers and seals, or can help distribute fastener load and prevent the fastener from damaging the component.

As with fasteners, when replacing washers make sure the replacement washers are of the same design and quality.

Cotter Pins

A cotter pin is a split metal pin inserted into a hole or slot to prevent a fastener from loosening. In certain applications, such as the rear axle on an ATV or motorcycle, the fastener must be secured in this way. For these applications, a cotter pin and castellated (slotted) nut is used.

To use a cotter pin, first make sure the diameter is correct for the hole in the fastener. After correctly tightening the fastener and aligning the holes, insert the cotter pin through the hole and bend the ends over the fastener (**Figure 8**). Unless instructed to do so, never loosen a tightened fastener to align the holes. If the holes do not align, tighten the fastener enough to achieve alignment.

Cotter pins are available in various diameters and lengths. Measure the length from the bottom of the head to the tip of the shortest pin.

Snap Rings and E-clips

Snap rings (**Figure 9**) are circular-shaped metal retaining clips. They are required to secure parts and gears in place on parts such as shafts, pins or rods. External type snap rings are used to retain items on shafts. Internal type snap rings secure parts within housing bores. In some applications, in addition to securing the component(s), snap rings of varying thicknesses also determine endplay. These are usually called selective snap rings.

Figure 8: Correct installation of cotter pin

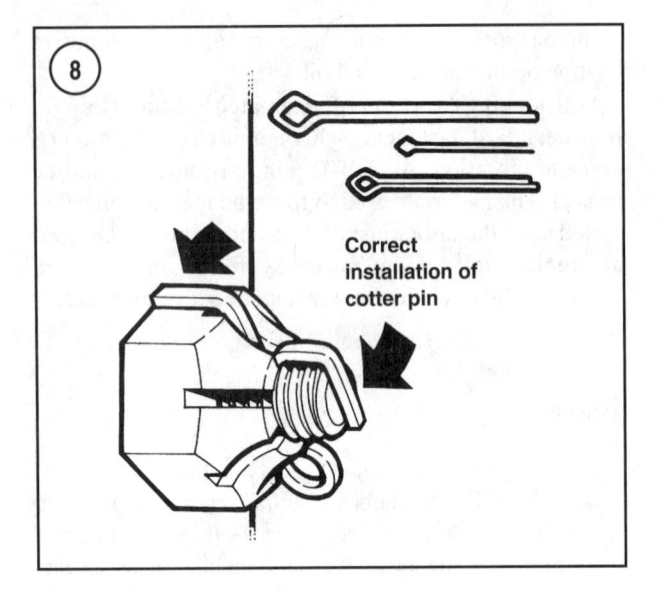

Figure 9: Internal snap ring, Plain clip, External snap ring, E-clip

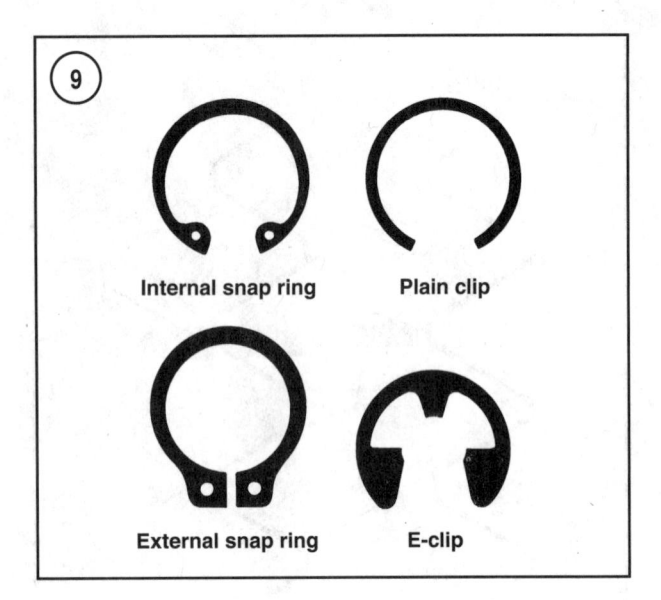

The two basic types of snap rings are machined and stamped snap rings. Machined snap rings (**Figure 10**) can be installed in either direction, because both faces have sharp edges. Stamped snap rings (**Figure 11**) are manufactured with a sharp and a round edge. When installing a stamped snap ring in a thrust application, install the sharp edge facing away from the part producing the thrust.

E-clips are used when it is not practical to use a snap ring. Remove E-clips with a flat blade screwdriver by prying between the shaft and E-clip. To install an E-clip, center it over the shaft groove and push or tap it into place.

Observe the following when installing snap rings:
1. Remove and install snap rings with snap ring pliers. Refer to *Basic Tools* in this chapter.
2. In some applications, it may be necessary to replace snap rings after removing them.
3. Compress or expand snap rings only enough to install them. If overly expanded, they lose their retaining ability.
4. After installing a snap ring, make sure it seats completely.
5. Wear eye protection when removing and installing snap rings.

SHOP SUPPLIES

Lubricants and Fluids

Periodic lubrication helps ensure a long service life for any type of equipment. Using the correct type of lubricant is as important as performing the lubrication service, although in an emergency the wrong type is better than not using one. The following section describes the types of lubricants most often required. Make sure to follow the manufacturer's recommendations for lubricant types.

Engine oils

Engine oil for four-stroke motorcycle engine use is classified by three standards: the American Petroleum Institute (API) service classification, the Society of Automotive Engineers (SAE) viscosity rating and the Japanese Automobile Standards Organization (JASO) T 903 Standard rating.

The API and SAE information is on all oil container labels. The JASO information is found on oil containers sold by the oil manufacturer specifically for motorcycle use. Two letters indicate the API service classification. The number or sequence of numbers and letter (10W-40 for example) is the oil's viscosity rating. The API service classification and the SAE viscosity index are not indications of oil quality.

The API service classification indicates that the oil meets specific lubrication standards. The first letter in the classification *S* indicates that the oil is for gasoline engines. The second letter indicates the standard the oil satisfies.

The JASO certification label identifies two separate oil classifications and a registration number to ensure the oil has passed all JASO certification standards for use in four-stroke motorcycle engines. The classifications are: MA (high friction applications) and MB (low friction applications). Only oil that has passed JASO standards can carry the JASO certification label.

NOTE
*Refer to **Engine Oil and Filter** in Chapter Three for further information on API, SAE and JASO ratings.*

GENERAL INFORMATION

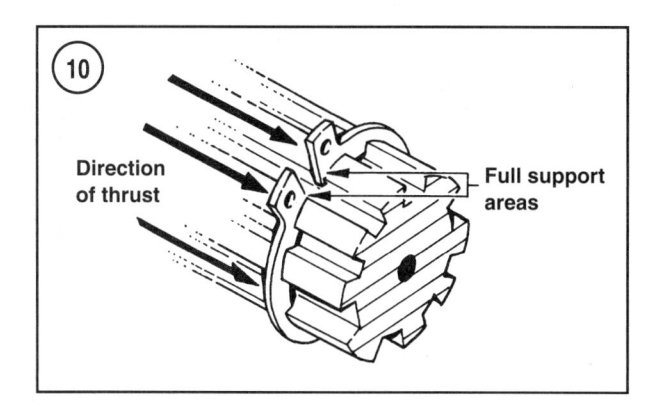

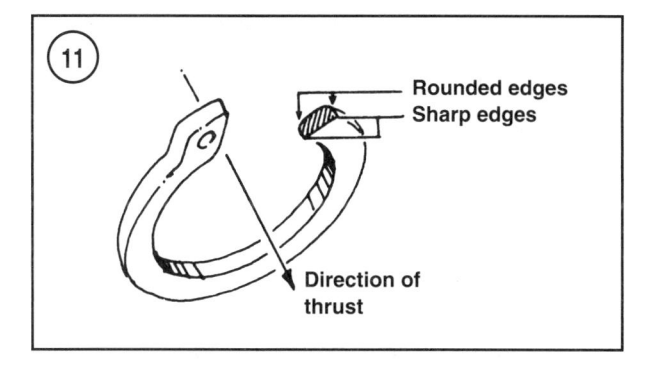

Figure 10: Direction of thrust; Full support areas

Figure 11: Rounded edges; Sharp edges; Direction of thrust

Always use an oil with a classification recommended by the manufacturer. Using an oil with a different classification can cause engine damage.

Viscosity is an indication of the oil's thickness. Thin oils have a lower number while thick oils have a higher number. Engine oils fall into the 5- to 50-weight range for single-grade oils.

Most manufacturers recommend multi-grade oil. These oils perform efficiently across a wide range of operating conditions. Multi-grade oils are identified by a *W* after the first number, which indicates the low-temperature viscosity.

Engine oils are most commonly mineral (petroleum) based, but synthetic and semi-synthetic types are used more frequently. When selecting engine oil, follow the manufacturer's recommendation for type, classification and viscosity when selecting engine oil.

Greases

Grease is lubricating oil with thickening agents added to it. The National Lubricating Grease Institute (NLGI) grades grease. Grades range from No. 000 to No. 6, with No. 6 being the thickest. Typical multipurpose grease is NLGI No. 2. For specific applications, manufacturers may recommend water-resistant type grease or one with an additive such as molybdenum disulfide (MoS_2).

Brake fluid

Brake fluid is the hydraulic fluid used to transmit hydraulic pressure (force) to the wheel brakes. Brake fluid is classified by the Department of Transportation (DOT). Current designations for brake fluid are DOT 3, DOT 4 and DOT 5. This classification appears on the fluid container.

Each type of brake fluid has its own definite characteristics. Do not intermix different types of brake fluid as this may cause brake system failure. DOT 5 brake fluid is silicone based. DOT 5 is not compatible with other brake fluids or in systems for which it was not designed. Mixing DOT 5 fluid with other fluids may cause brake system failure. When adding brake fluid, *only* use the fluid recommended by the manufacturer.

Brake fluid will damage any plastic, painted or plated surface it contacts. Use extreme care when working with brake fluid and remove any spills immediately with soap and water.

Hydraulic brake systems require clean and moisture free brake fluid. Never reuse brake fluid. Keep containers and reservoirs properly sealed.

> **WARNING**
> *Never put a mineral-based (petroleum) oil into the brake system. Mineral oil causes rubber parts in the system to swell and break apart, causing complete brake failure.*

Coolant

Coolant is a mixture of water and antifreeze used to dissipate engine heat. Ethylene glycol is the most common form of antifreeze. Check the motorcycle manufacturer's recommendations when selecting antifreeze. Most require one specifically designed for use in aluminum engines. These types of antifreeze have additives that inhibit corrosion.

Only mix antifreeze with distilled water. Impurities in tap water may damage internal cooling system passages.

Cleaners, Degreasers and Solvents

Many chemicals are available to remove oil, grease and other residue from the motorcycle. Before using cleaning solvents, consider how they will be used and disposed of, particularly if they are not water-soluble. Local ordinances may require special procedures for the disposal of many types of cleaning chemicals. Refer to *Safety* in this chapter.

Use brake parts cleaner to clean brake system components. Brake parts cleaner leaves no residue. Use electrical contact cleaner to clean electrical connections and components without leaving any residue. Carburetor cleaner is a powerful solvent used to remove fuel deposits and varnish from fuel system components. Use this cleaner carefully, as it may damage finishes.

Generally, degreasers are strong cleaners used to remove heavy accumulations of grease from engine and frame components.

Most solvents are designed to be used with a parts washing cabinet for individual component cleaning. For safety, use only nonflammable or high flash point solvents.

Gasket Sealant

Sealant is used in combination with a gasket or seal. In other applications, such as between crankcase halves, only a sealant is used. Follow the manufacturer's recommendation when using a sealant. Use extreme care when choosing a sealant different from the type originally recommended. Choose sealant based on its resistance to heat, various fluids and its sealing capabilities.

A common sealant is room temperature vulcanization sealant, or RTV. This sealant cures at room temperature over a specific time period. This allows the repositioning of components without damaging gaskets.

Moisture in the air causes the RTV sealant to cure. Always install the tube cap as soon as possible after applying RTV sealant. RTV sealant has a limited shelf life and will not cure properly if the shelf life has expired. Keep partial tubes sealed and discard them if they have surpassed the expiration date.

Applying RTV sealant

Clean all old gasket residue from the mating surfaces. Remove all gasket material from blind threaded holes to avoid inaccurate bolt torque. Spray the mating surfaces with aerosol parts cleaner and then wipe with a lint-free cloth. The area must be clean for the sealant to adhere.

Apply RTV sealant in a continuous bead 2-3 mm (0.08-0.12 in.) thick. Circle all the fastener holes unless otherwise specified. Do not allow any sealant to enter these holes. Assemble and tighten the fasteners to the specified torque within the time frame recommended by the sealant manufacturer.

Gasket Remover

Aerosol gasket remover can help remove stubborn gaskets. This product can speed up the removal process and prevent damage to the mating surface that may be caused by using a scraping tool. Most of these types of products are very caustic. Follow the gasket remover manufacturer's instructions for use.

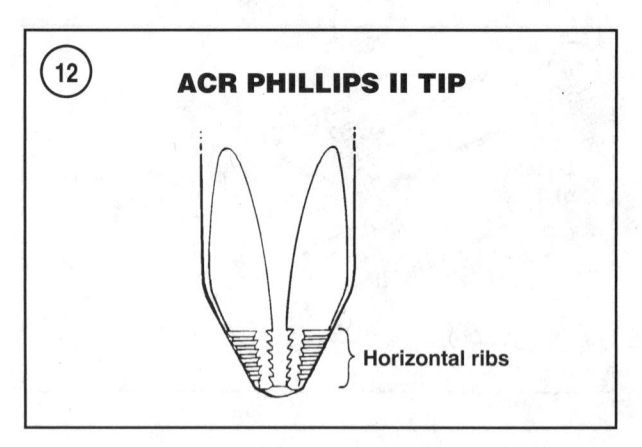

Threadlocking Compound

A threadlocking compound is a fluid applied to the threads of fasteners. After tightening the fastener, the fluid dries and becomes a solid filler between the threads. This makes it difficult for the fastener to work loose from vibration or heat expansion and contraction. Some threadlocking compounds also provide a seal against fluid leaks.

Before applying a threadlocking compound, remove any old compound from both thread areas and clean them with aerosol parts cleaner. Use the compound sparingly. Excess fluid can run into adjoining parts.

CAUTION
Threadlocking compounds are anaerobic and will stress, crack and attack most plastics. Use caution when using these products in areas where there are plastic components.

Threadlocking compounds are available in a wide range of compounds for various strength, temperature and repair applications. Follow the manufacturer's recommendations regarding compound selection.

BASIC TOOLS

Most of the procedures in this manual can be carried out with simple hand tools and test equipment familiar to the home mechanic. Always use the correct tools for the job at hand. Keep tools organized and clean. Store them in a tool chest with related tools organized together.

Quality tools are essential. The best are constructed of high-strength alloy steel. These tools are light, easy to use and resistant to wear. Their working surface is devoid of sharp edges and carefully polished. They have an

GENERAL INFORMATION

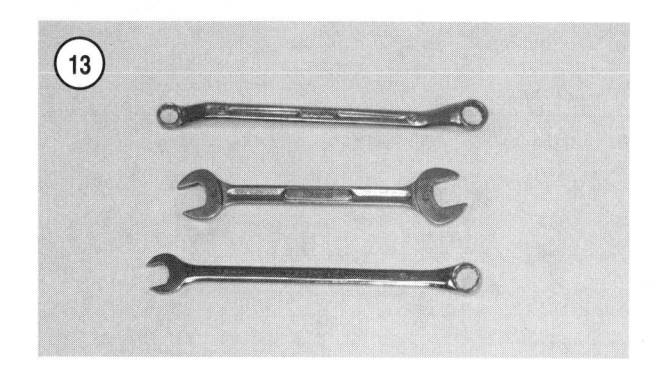

easy-to-clean finish and are comfortable to use. Quality tools are a good investment.

Some of the procedures in this manual specify special tools. In many cases the tool is illustrated in use. Those with a large tool kit may be able to use a suitable substitute or fabricate a suitable replacement. However, in some cases, the specialized equipment or expertise may make it impractical for the home mechanic to attempt the procedure. When necessary, such operations come with the recommendation to have a dealership or specialist perform the task. It may be less expensive to have a professional perform these jobs, especially when considering the cost of equipment.

When purchasing tools to perform the procedures covered in this manual, consider the tool's potential frequency of use. If a tool kit is just now being started, consider purchasing a basic tool set from a quality tool supplier. These sets are available in many tool combinations and offer substantial savings when compared to individually purchased tools. As work experience grows and tasks become more complicated, specialized tools can be added.

Screwdrivers

Screwdrivers of various lengths and types are mandatory for the simplest tool kit. The two basic types are the slotted tip (flat blade) and the Phillips tip. These are available in sets that often include an assortment of tip sizes and shaft lengths.

As with all tools, use a screwdriver designed for the job. Make sure the size of the tip conforms to the size and shape of the fastener. Use them only for driving screws. Never use a screwdriver for prying or chiseling metal. Repair or replace worn or damaged screwdrivers. A worn tip may damage the fastener, making it difficult to remove.

Phillips-head screws are often damaged by incorrectly fitting screwdrivers. Quality Phillips screwdrivers are manufactured with their crosshead tip machined to Phillips Screw Company specifications. Poor quality or damaged Phillips screwdrivers can back out (camout) and round over the screw head. In addition, weak or soft screw materials can make removal difficult.

The best type of screwdriver to use on Phillips screws is the ACR Phillips II screwdriver, patented by the Phillips Screw Company. ACR stands for the horizontal anti-camout ribs found on the driving faces or flutes of the screwdriver's tip (**Figure 12**). ACR Phillips II screwdrivers were designed as part of a manufacturing drive system to be used with ACR Phillips II screws, but they work well on all common Phillips screws. A number of tool companies offer ACR Phillips II screwdrivers in different tip sizes and interchangeable bits to fit screwdriver bit holders.

NOTE
Another way to prevent camout and to increase the grip of a Phillips screwdriver is to apply valve grinding compound or Permatex Screw & Socket Gripper onto the screwdriver tip. After loosening/tightening the screw, clean the screw recess to prevent engine oil contamination.

Wrenches

Open-end, box-end and combination wrenches (**Figure 13**) are available in a variety of types and sizes.

The number stamped on the wrench refers to the distance between the work areas. This size must match the size of the fastener head.

The box-end wrench is an excellent tool because it grips the fastener on all sides. This reduces the chance of the tool slipping. The box-end wrench is designed with either a 6 or 12-point opening. For stubborn or damaged fasteners, the 6-point provides superior holding because it contacts the fastener across a wider area at all six edges. For general use, the 12-point works well. It allows the wrench to be removed and reinstalled without moving the handle over such a wide arc.

An open-end wrench is fast and works best in areas with limited overhead access. It contacts the fastener at only two points and is subject to slipping if under heavy force, or if the tool or fastener is worn. A box-end wrench is preferred in most instances, especially when breaking loose and applying the final tightness to a fastener.

The combination wrench has a box-end on one end and an open-end on the other. This combination makes it a convenient tool.

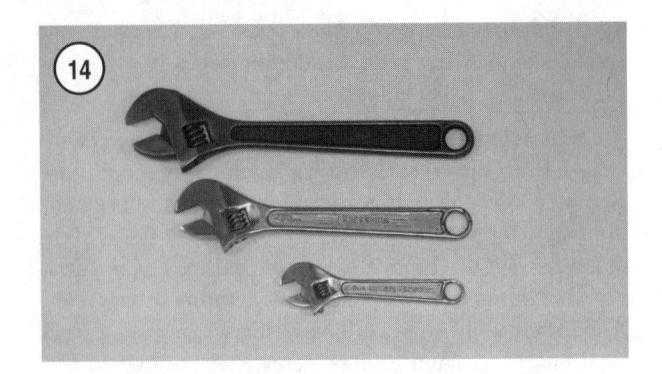

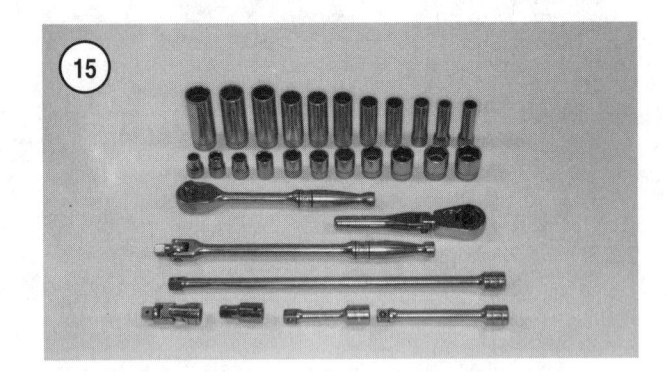

Adjustable Wrenches

An adjustable wrench or Crescent wrench (**Figure 14**) can fit nearly any nut or bolt head that has clear access around its entire perimeter. An adjustable wrench is best used as a backup wrench to keep a large nut or bolt from turning while the other end is being loosened or tightened with a box-end or socket wrench.

Adjustable wrenches contact the fastener at only two points, which makes them more subject to slipping off the fastener. Because one jaw is adjustable and may become loose, this shortcoming is aggravated. Make certain the solid jaw is the one transmitting the force.

Socket Wrenches, Ratchets and Handles

Sockets that attach to a ratchet handle (**Figure 15**) are available with 6-point or 12-point openings (**Figure 16**) and different drive sizes. The drive size indicates the size of the square hole that accepts the ratchet handle. The number stamped on the socket is the size of the work area and must match the fastener head.

As with wrenches, a 6-point socket provides superior-holding ability, while a 12-point socket needs to be moved only half as far to reposition it on the fastener.

Sockets are designated for either hand or impact use. Impact sockets are made of thicker material for more durability. Compare the size and wall thickness of a 19-mm hand socket (A, **Figure 17**) and the 19-mm impact socket (B). Use impact sockets when using an impact driver or air tools. Use hand sockets with hand-driven attachments.

WARNING
Do not use hand sockets with air or impact tools because they may shatter and cause injury. Always wear eye protection when using impact or air tools.

Various handles are available for sockets. Use the speed handle for fast operation. Flexible ratchet heads in varying lengths allow the socket to be turned with varying force and at odd angles. Extension bars allow the socket setup to reach difficult areas. The ratchet is the most versatile. It allows the user to install or remove the nut without removing the socket.

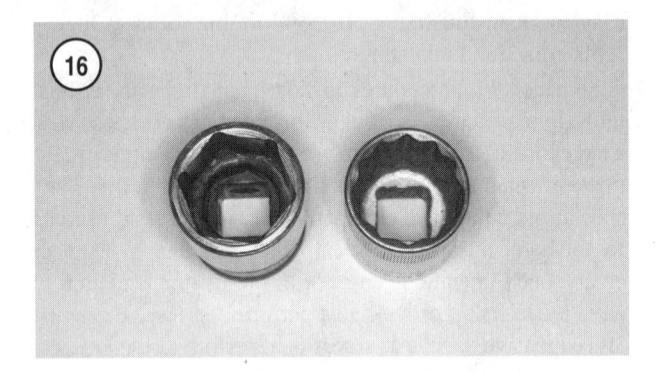

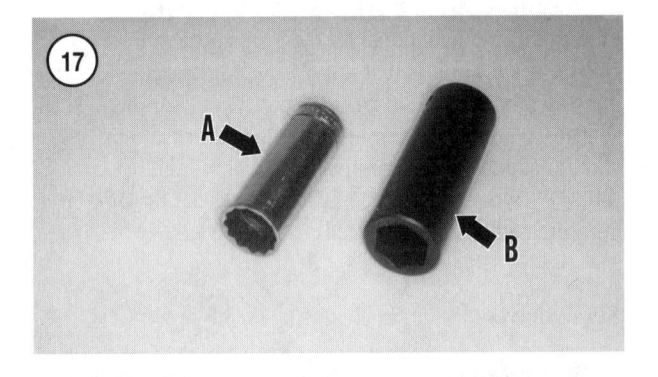

Sockets combined with any number of drivers make them undoubtedly the fastest, safest and most convenient tool for fastener removal and installation.

Impact Drivers

An impact driver provides extra force for removing fasteners by converting the impact of a hammer into a turning motion. This makes it possible to remove stubborn

GENERAL INFORMATION

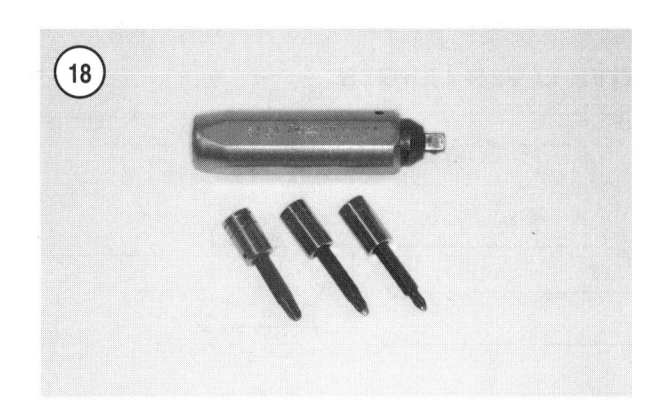

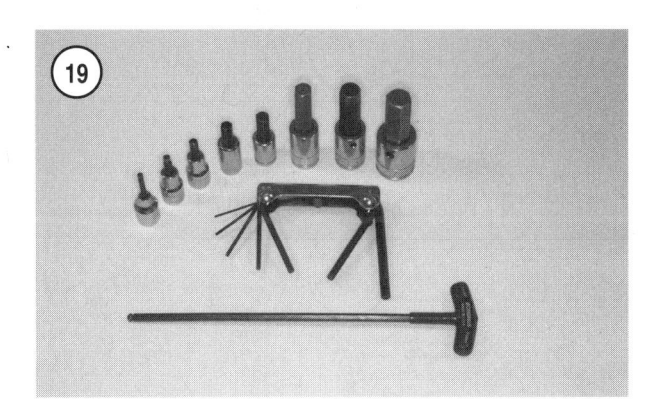

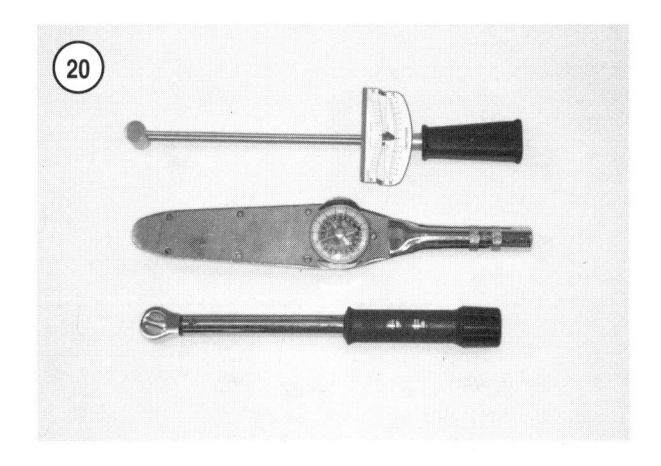

fasteners without damaging them. Impact drivers and interchangeable bits (**Figure 18**) are available from most tool suppliers. When using a socket with an impact driver, make sure the socket is designed for impact use. Refer to *Socket Wrenches, Ratchets and Handles* in this section.

> *WARNING*
> *Do not use hand sockets with air or impact tools because they may shatter and cause injury. Always wear eye protection when using impact or air tools.*

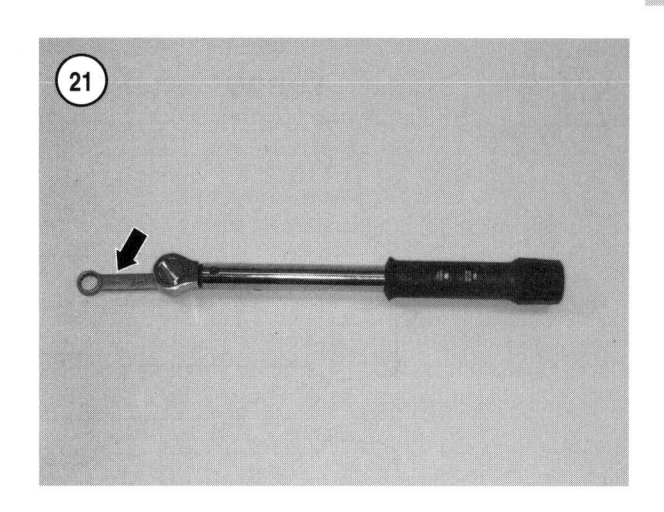

Allen Wrenches

Use Allen or setscrew wrenches (**Figure 19**) on fasteners with hexagonal recesses in the fastener head. These wrenches are available in L-shaped bar, socket and T-handle types. A metric set is required when working on most motorcycles. Allen bolts are sometimes called socket bolts.

Torque Wrenches

Use a torque wrench with a socket, torque adapter or similar extension to tighten a fastener to a measured torque. Torque wrenches come in several drive sizes (1/4, 3/8, 1/2 and 3/4) and have various methods of reading the torque value. The drive size indicates the size of the square drive that accepts the socket, adapter or extension. Common methods of reading the torque value are the deflecting beam, the dial indicator and the audible click (**Figure 20**).

When choosing a torque wrench, consider the torque range, drive size and accuracy. The torque specifications in this manual provide an indication of the range required.

A torque wrench is a precision tool that must be properly cared for to remain accurate. Store torque wrenches in cases or separate padded drawers within a toolbox. Follow the manufacturer's instructions for their care and calibration.

Torque Adapters

Torque adapters or extensions extend or reduce the reach of a torque wrench. The torque adapter shown in **Figure 21** is used to tighten a fastener that cannot be reached because of the size of the torque wrench head, drive, and socket. If a torque adapter changes the effective

22   TORQUE WRENCH EFFECTIVE LEVER LENGTH

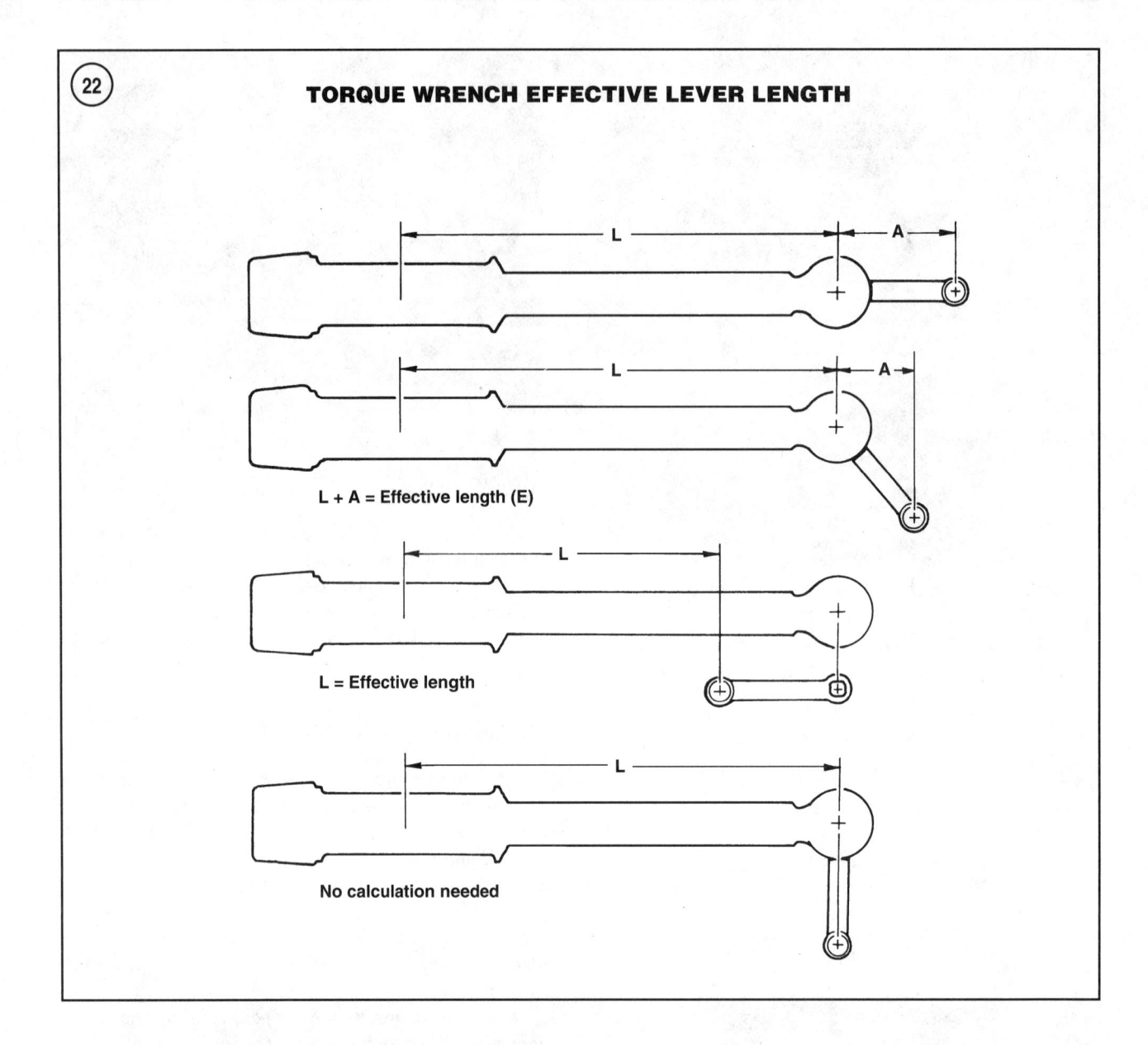

L + A = Effective length (E)

L = Effective length

No calculation needed

lever length (**Figure 22**), the torque reading on the wrench will not equal the actual torque applied to the fastener. It is necessary to recalibrate the torque setting on the wrench to compensate for the change of lever length. When using a torque adapter at a right angle to the drive head, calibration is not required, because the effective length has not changed.

To recalculate a torque reading when using a torque adapter, use the following formula and refer to **Figure 22**:

$$TW = \frac{TA \times L}{L + A}$$

TW is the torque setting or dial reading on the wrench.

TA is the torque specification and the actual amount of torque that is applied to the fastener.

A is the amount that the adapter increases (or in some cases reduces) the effective lever length as measured along the centerline of the torque wrench.

L is the lever length of the wrench as measured from the center of the drive to the center of the grip.

The effective length is the sum of *L* and *A*.

Example:
TA = 20 ft.-lb.
A = 3 in.
L = 14 in.

$$TW = \frac{20 \times 14}{14 + 3} = \frac{280}{17} = 16.5 \text{ ft. lb.}$$

In this example, the torque wrench would be set to the recalculated torque value (TW = 16.5 ft.-lb.). When using

GENERAL INFORMATION

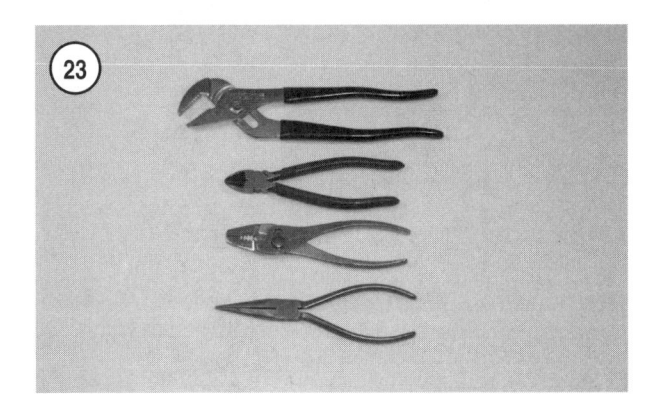

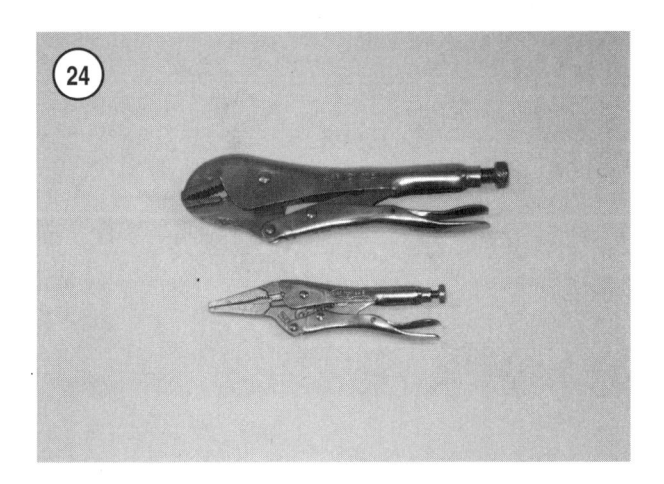

a beam-type wrench, tighten the fastener until the pointer aligns with 16.5 ft.-lb. In this example, although the torque wrench is pre set to 16.5 ft.-lb., the actual torque is 20 ft.-lb.

Pliers

Pliers come in a wide range of types and sizes. Pliers are useful for holding, cutting, bending, and crimping. Do not use them to turn fasteners. **Figure 23** and **Figure 24** show several types of useful pliers. Each design has a specialized function. Slip-joint pliers are general-purpose pliers used for gripping and bending. Diagonal cutting pliers are needed to cut wire and can be used to remove cotter pins. Use needlenose pliers to hold or bend small objects. Locking pliers (**Figure 24**), sometimes called Vise-Grips, are used to hold objects very tightly. They have many uses ranging from holding two parts together, to gripping the end of a broken stud. Use caution when using locking pliers, as the sharp jaws will damage the objects they hold.

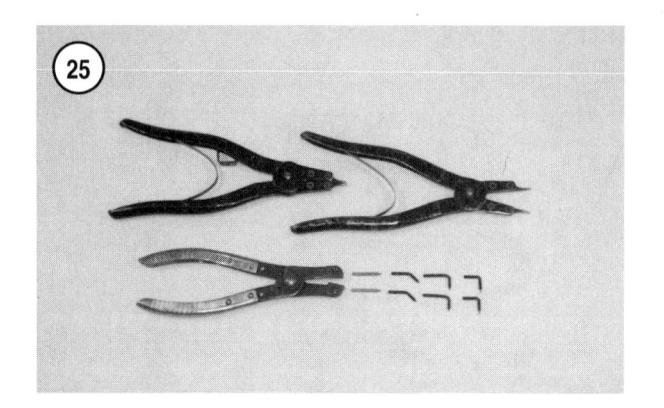

Snap Ring Pliers

Snap ring pliers are specialized pliers with tips that fit into the ends of snap rings to remove and install them.

Snap ring pliers (**Figure 25**) are available with a fixed action (either internal or external) or convertible (one tool works on both internal and external snap rings). They may have fixed tips or interchangeable ones of various sizes and angles. For general use, select a convertible type pliers with interchangeable tips (**Figure 25**).

WARNING
Snap rings can slip and fly off when removing and installing them. Also, the snap ring pliers' tips may break. Always wear eye protection when using snap ring pliers.

Hammers

Various types of hammers are available to fit a number of applications. Use a ball-peen hammer to strike another tool, such as a punch or chisel. Use soft-faced hammers when a metal object must be struck without damaging it. *Never* use a metal-faced hammer on engine and suspension components because damage occurs in most cases.

Always wear eye protection when using hammers. Make sure the hammer face is in good condition and the handle is not cracked. Select the correct hammer for the job and make sure to strike the object squarely. Do not use the handle or the side of the hammer to strike an object.

PRECISION MEASURING TOOLS

The ability to accurately measure components is essential to perform many of the procedures described in this manual. Equipment is manufactured to close tolerances, and obtaining consistently accurate measurements is essential to determine which components require replacement or further service.

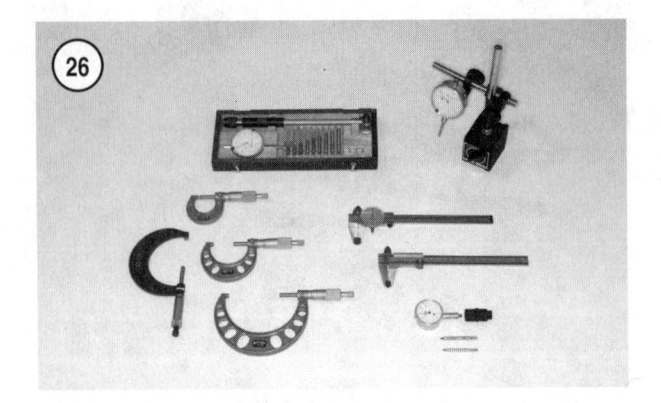

Each type of measuring instrument (**Figure 26**) is designed to measure a dimension with a certain degree of accuracy and within a certain range. When selecting the measuring tool, make sure it is applicable to the task.

As with all tools, measuring tools provide the best results if cared for properly. Improper use can damage the tool and cause inaccurate results. If any measurement is questionable, verify the measurement using another tool. A standard gauge is usually provided with micrometers to check accuracy and calibrate the tool if necessary.

Precision measurements can vary according to the experience of the person performing the procedure. Accurate results are only possible if the mechanic possesses a feel for using the tool. Heavy-handed use of measuring tools produces less accurate results. Hold the tool gently by the fingertips to easily feel the point at which the tool contacts the object. This feel for the equipment produces more accurate measurements and reduces the risk of damaging the tool or component. Refer to the following sections for specific measuring tools.

Feeler Gauge

Use feeler or thickness gauges (**Figure 27**) for measuring the distance between two surfaces.

A feeler gauge set consists of an assortment of steel strips of graduated thickness. Each blade is marked with its thickness. Blades can be of various lengths and angles for different procedures.

A common use for a feeler gauge is to measure valve clearance. Use wire (round) type gauges to measure spark plug gap.

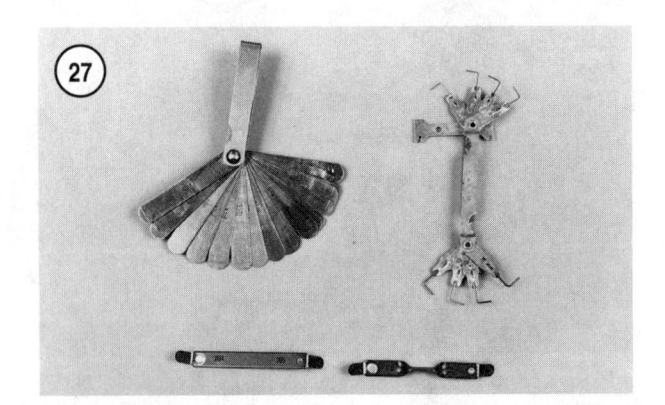

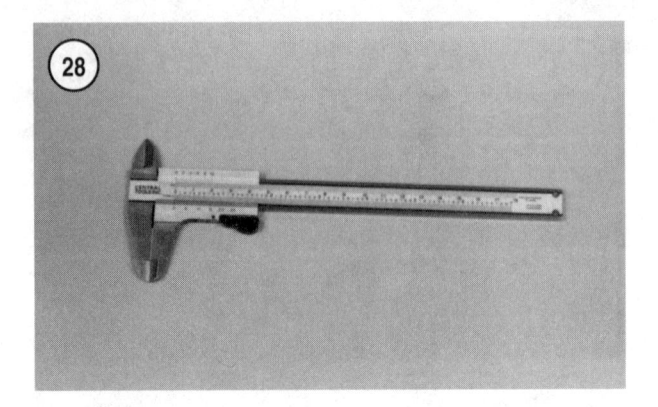

Calipers

Calipers (**Figure 28**) are excellent tools for obtaining inside, outside and depth measurements. Although not as precise as a micrometer, they allow reasonable precision, typically to within 0.05 mm (0.001 in.). Most calipers have a range up to 150 mm (6 in.).

Calipers are available in dial, vernier or digital versions. Dial calipers have a dial readout that provides convenient reading. Vernier calipers have marked scales that must be compared to determine the measurement. The digital caliper uses a liquid-crystal display (LCD) to show the measurement.

Properly maintain the measuring surfaces of the caliper. There must not be any dirt or burrs between the tool and the object being measured. Never force the caliper to close around an object. Close the caliper around the highest point so it can be removed with a slight drag. Some calipers require calibration. Always refer to the manufacturer's instructions when using a new or unfamiliar caliper.

To read a vernier caliper refer to **Figure 29**. The fixed scale is marked in 1-mm increments. Ten individual lines on the fixed scale equal 1 cm. The movable scale is marked in 0.05 mm (hundredth) increments. To obtain a reading, establish the first number by the location of the 0

GENERAL INFORMATION 15

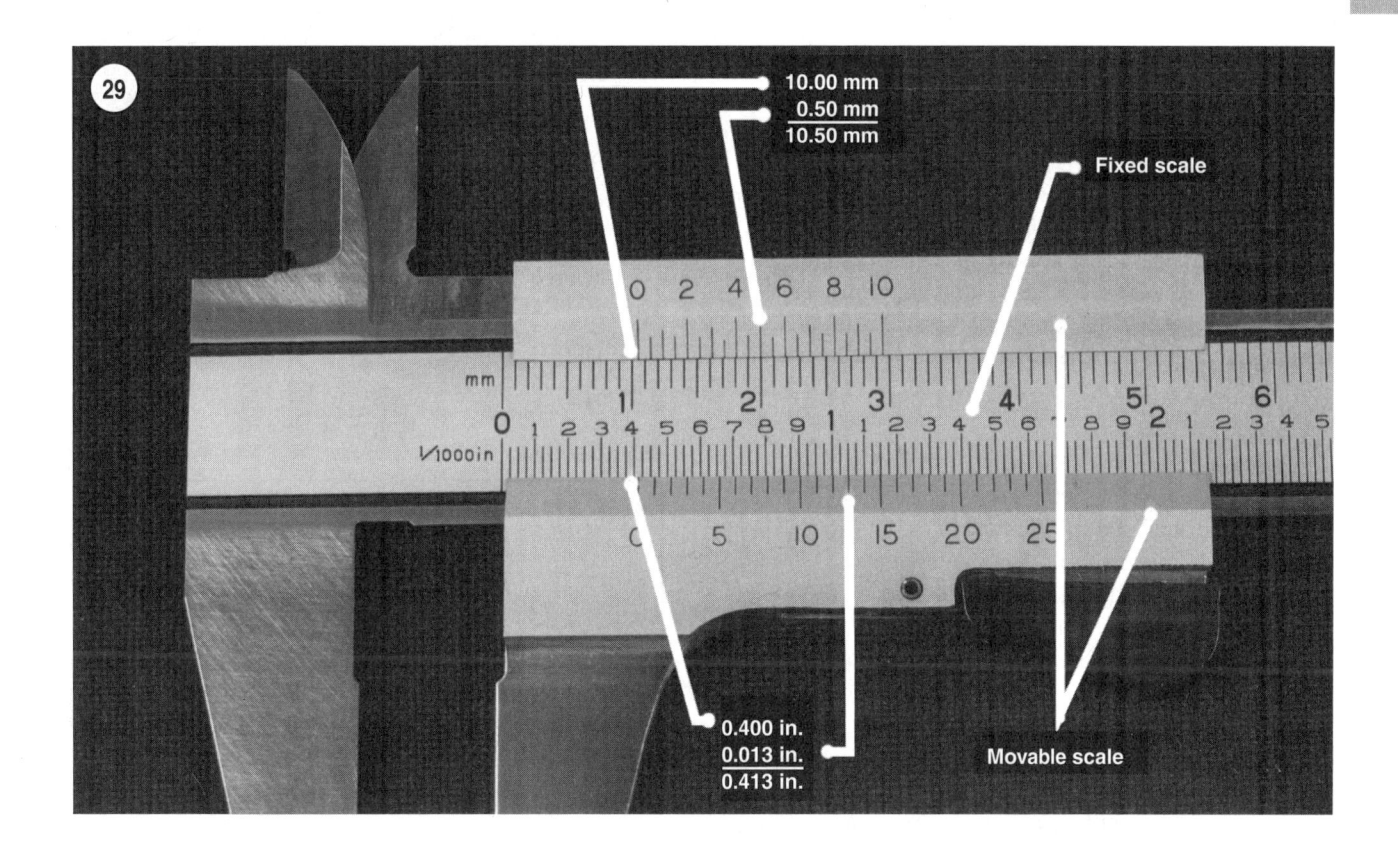

29

- 10.00 mm
- 0.50 mm
- 10.50 mm
- Fixed scale
- 0.400 in.
- 0.013 in.
- 0.413 in.
- Movable scale

30

DECIMAL PLACE VALUES*

0.1	Indicates 1/10 (one tenth of an inch or millimeter)
0.010	Indicates 1/100 (one one-hundreth of an inch or millimeter)
0.001	Indicates 1/1000 (one one-thousandth of an inch or millimeter)

*This chart represents the values of figures placed to the right of the decimal point. Use it when reading decimals from one-tenth to one one-thousandth of an inch or millimeter. It is not a conversion chart (for example: 0.001 in. is not equal to 0.001 mm).

line on the movable scale in relation to the first line to the left on the fixed scale. In this example, the number is 10 mm. To determine the next number, note which of the lines on the movable scale align with a mark on the fixed scale. A number of lines will seem close, but only one will align exactly. In this case, 0.50 mm is the reading to add to the first number. Adding 10 mm and 0.50 mm equals a measurement of 10.50 mm.

Micrometers

A micrometer is an instrument designed for linear measurement using the decimal divisions of the inch or meter (**Figure 30**). While there are many types and styles of micrometers, most of the procedures in this manual call for an outside micrometer. Use the outside micrometer to measure the outside diameter of cylindrical forms and the thickness of materials.

A micrometer's size indicates the minimum and maximum size of a part that it can measure. The usual sizes (**Figure 31**) are 0-25 mm (0-1 in.), 25-50 mm (1-2 in.), 50-75 mm (2-3 in.) and 75-100 mm (3-4 in.).

Micrometers that cover a wider range of measurements are available. These use a large frame with interchangeable anvils of various lengths. This type of micrometer offers a cost savings, but its overall size may make it less convenient.

Adjustment

Before using a micrometer, check its adjustment as follows:
1. Clean the anvil and spindle faces.
2A. To check a 0-1 in. or 0-25 mm micrometer:
 a. Turn the thimble until the spindle contacts the anvil. If the micrometer has a ratchet stop, use it to ensure that the proper amount of pressure is applied.
 b. If the adjustment is correct, the 0 mark on the thimble will align exactly with the 0 mark on the sleeve line. If the marks do not align, the micrometer is out of adjustment.
 c. Follow the manufacturer's instructions to adjust the micrometer.
2B. To check a micrometer larger than 1 in. or 25 mm use the standard gauge supplied by the manufacturer. A standard gauge is a steel block, disc or rod that is machined to an exact size.
 a. Place the standard gauge between the spindle and anvil, and measure its outside diameter or length. If the micrometer has a ratchet stop, use it to ensure that the proper amount of pressure is applied.
 b. If the adjustment is correct, the 0 mark on the thimble will align exactly with the 0 mark on the sleeve line. If the marks do not align, the micrometer is out of adjustment.
 c. Follow the manufacturer's instructions to adjust the micrometer.

Care

Micrometers are precision instruments. They must be used and maintained with great care. Note the following:
1. Store micrometers in protective cases or separate padded drawers in a toolbox.
2. When in storage, make sure the spindle and anvil faces do not contact each other or an other object. If they do, temperature changes and corrosion may damage the contact faces.
3. Do not clean a micrometer with compressed air. Dirt forced into the tool will cause wear.

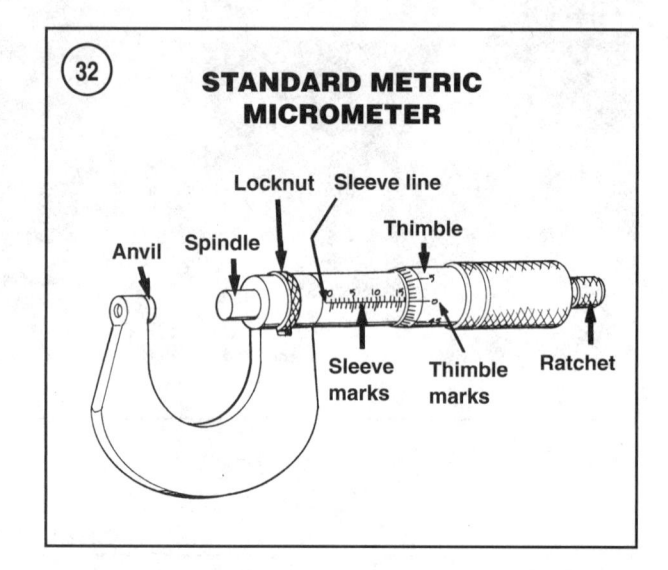

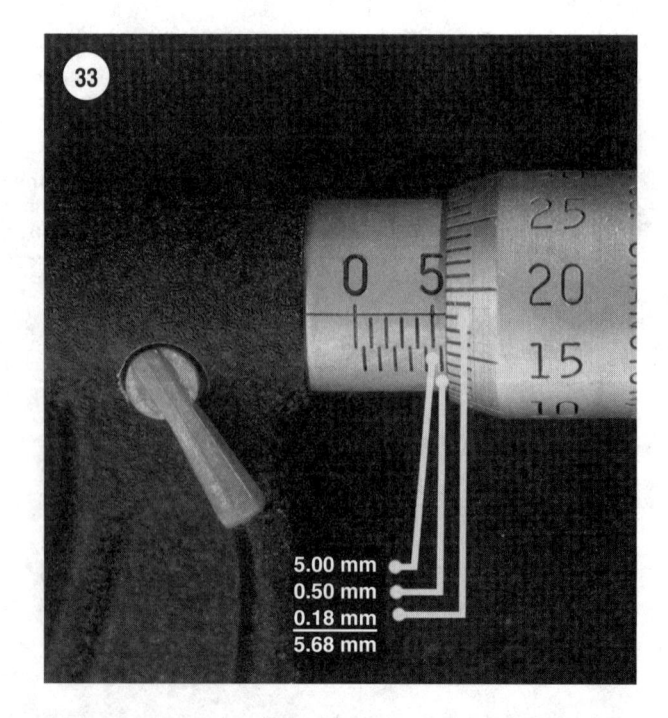

GENERAL INFORMATION

34 STANDARD INCH MICROMETER

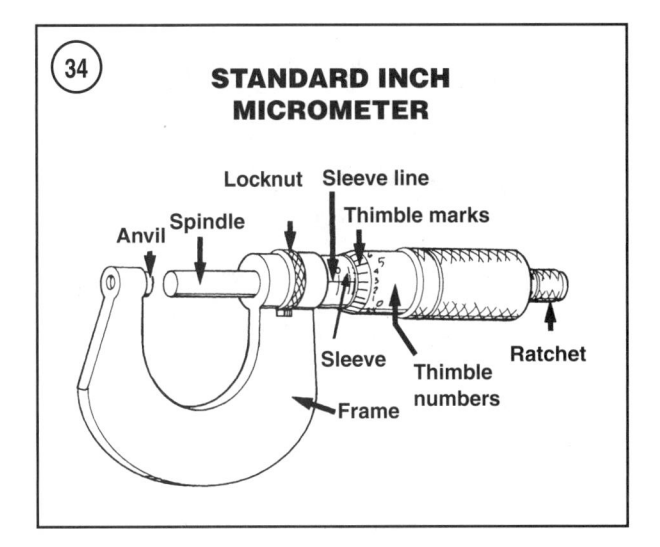

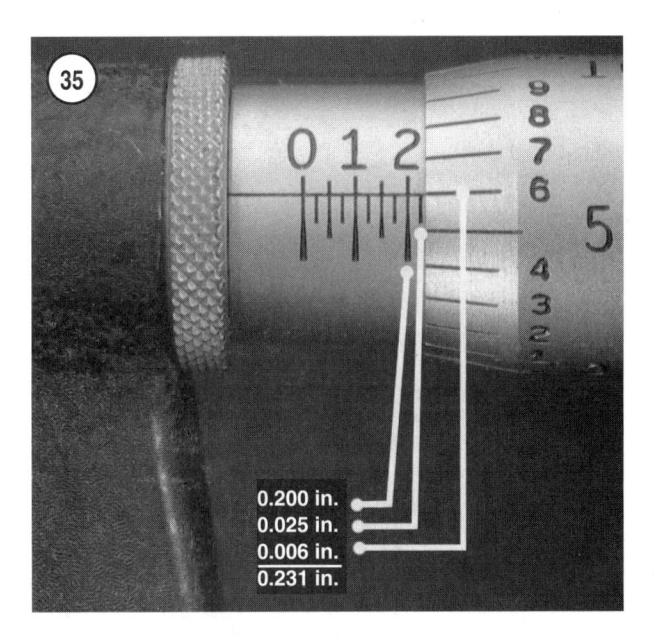

Anvil, Spindle, Locknut, Sleeve line, Thimble marks, Sleeve, Thimble numbers, Frame, Ratchet

35
- 0.200 in.
- 0.025 in.
- 0.006 in.
- 0.231 in.

4. Lubricate micrometers with WD-40 to prevent corrosion.

Reading

When reading a micrometer, numbers are taken from different scales and added together. The following sections describe how to read the measurements of various types of outside micrometers.

For accurate results, properly maintain the measuring surfaces of the micrometer. There cannot be any dirt or burrs between the tool and the measured object. Never force the micrometer to close around an object. Close the micrometer around the highest point so it can be removed with a slight drag.

Metric micrometer

The standard metric micrometer (**Figure 32**) is accurate to one one-hundredth of a millimeter (0.01 mm). The sleeve line is graduated in millimeter and half millimeter increments. The marks on the upper half of the sleeve line equal 1.00 mm. Each fifth mark above the sleeve line is identified with a number. The number sequence depends on the size of the micrometer. A 0-25 mm micrometer, for example, will have sleeve marks numbered 0 through 25 in 5 mm increments. This numbering sequence continues with larger micrometers. On all metric micrometers, each mark on the lower half of the sleeve equals 0.50 mm.

The tapered end of the thimble has 50 lines marked around it. Each mark equals 0.01 mm. One complete turn of the thimble aligns its 0 mark with the first line on the lower half of the sleeve line or 0.50 mm.

When reading a metric micrometer, add the number of millimeters and half-millimeters on the sleeve line to the number of one one-hundredth millimeters on the thimble. Perform the following steps while referring to **Figure 33**.

1. Read the upper half of the sleeve line and count the number of lines visible. Each upper line equals 1 mm.
2. See if the half-millimeter line is visible on the lower sleeve line. If so, add 0.50 mm to the reading in Step 1.
3. Read the thimble mark that aligns with the sleeve line. Each thimble mark equals 0.01 mm.

> *NOTE*
> *If a thimble mark does not align exactly with the sleeve line, estimate the amount between the lines. For accurate readings in two-thousandths of a millimeter (0.002 mm), use a metric vernier micrometer.*

4. Add the readings from Steps 1-3.

Standard Inch Micrometer

The standard inch micrometer (**Figure 34**) is accurate to one-thousandth of an inch or 0.001. The sleeve is marked in 0.025 in. increments. Every fourth sleeve mark is numbered 1, 2, 3, 4, 5, 6, 7, 8, 9. These numbers indicate 0.100, 0.200, 0.300, and so on.

The tapered end of the thimble has 25 lines marked around it. Each mark equals 0.001 in. One complete turn of the thimble will align its zero mark with the first mark on the sleeve or 0.025 in.

To read a standard inch micrometer, perform the following steps and refer to **Figure 35**.

1. Read the sleeve and find the largest number visible. Each sleeve number equals 0.100 in.

2. Count the number of lines between the numbered sleeve mark and the edge of the thimble. Each sleeve mark equals 0.025 in.

3. Read the thimble mark that aligns with the sleeve line. Each thimble mark equals 0.001 in.

NOTE
If a thimble mark does not align exactly with the sleeve line, estimate the amount between the lines. For accurate readings in ten-thousandths of an inch (0.0001 in.), use a vernier inch micrometer.

4. Add the readings from Steps 1-3.

Telescoping and Small Bore Gauges

Use telescoping gauges (**Figure 36**) and small bore gauges (**Figure 37**) to measure bores. Neither gauge has a scale for direct readings. Use an outside micrometer to determine the reading.

To use a telescoping gauge, select the correct size gauge for the bore. Compress the movable post and carefully insert the gauge into the bore. Carefully move the gauge in the bore to make sure it is centered. Tighten the knurled end of the gauge to hold the movable post in position. Remove the gauge and measure the length of the posts. Telescoping gauges are typically used to measure cylinder bores.

To use a small bore gauge, select the correct size gauge for the bore. Carefully insert the gauge into the bore. Tighten the knurled end of the gauge to carefully expand the gauge fingers to the limit within the bore. Do not overtighten the gauge because there is no built-in release. Excessive tightening can damage the bore surface and damage the tool. Remove the gauge and measure the outside dimension (**Figure 38**). Small bore gauges are typically used to measure valve guides.

Dial Indicator

A dial indicator (A, **Figure 39**) is a gauge with a dial face and needle used to measure variations in dimensions and movements. Measuring brake rotor runout is a typical use for a dial indicator.

Dial indicators are available in various ranges and graduations and with three basic types of mounting bases: magnetic (B, **Figure 39**), clamp, or screw-in stud. When purchasing a dial indicator, select one with a continuous dial (A, **Figure 39**).

GENERAL INFORMATION

Compression Gauge

A compression gauge (**Figure 41**) measures combustion chamber (cylinder) pressure, usually in psi or kg/cm^2. The gauge adapter is either inserted or screwed into the spark plug hole to obtain the reading. Disable the engine so it does not start and hold the throttle in the wide-open position when performing a compression test. An engine that does not have adequate compression cannot be properly tuned. Refer to Chapter Three.

Multimeter

A multimeter (**Figure 42**) is an essential tool for electrical system diagnosis. The voltage function indicates the voltage applied or available to various electrical components. The ohmmeter function tests circuits for continuity, or lack of continuity, and measures the resistance of a circuit.

Some manufacturers' specifications for electrical components are based on results using a specific test meter. Results may vary if using a meter not recommended by the manufacturer is used. Such requirements are noted when applicable.

Ohmmeter (analog) calibration

Each time an analog ohmmeter is used or if the scale is changed, the ohmmeter must be calibrated.

Digital ohmmeters do not require calibration.

1. Make sure the meter battery is in good condition.
2. Make sure the meter probes are in good condition.
3. Touch the two probes together and observe the needle location on the ohms scale. The needle must align with the 0 mark to obtain accurate measurements.
4. If necessary, rotate the meter ohms adjust knob until the needle and 0 mark align.

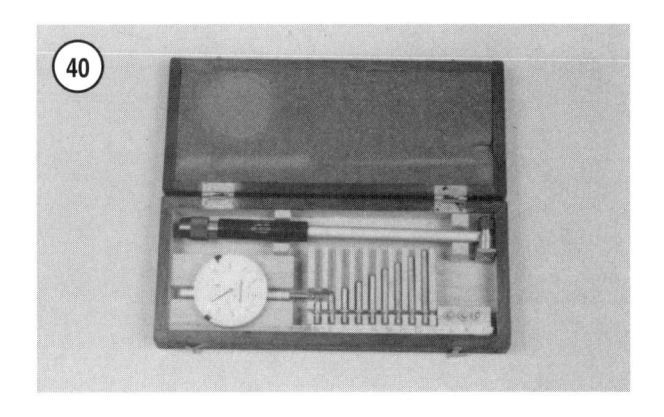

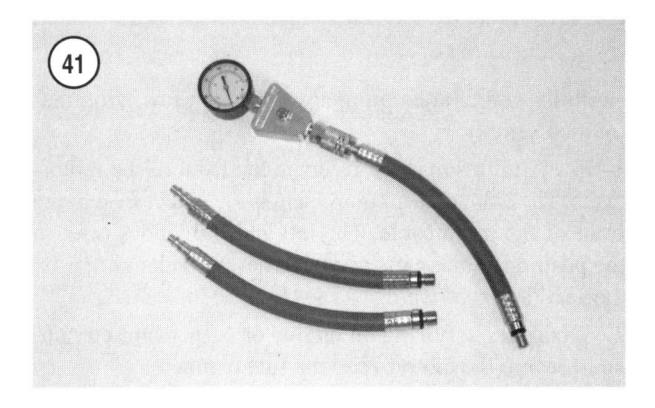

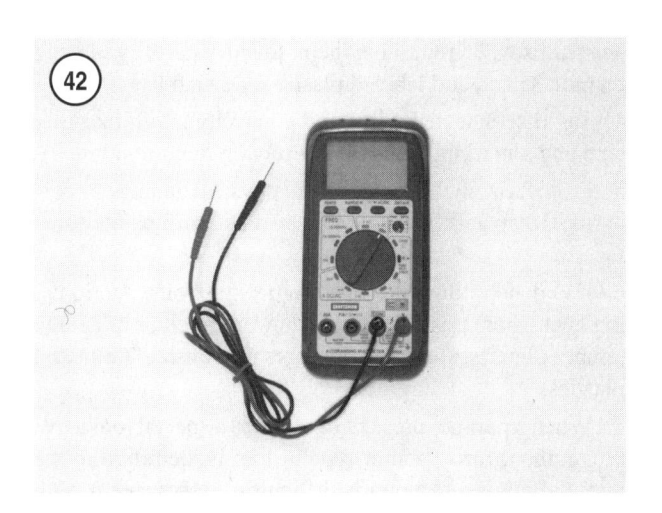

Cylinder Bore Gauge

A cylinder bore gauge is similar to a dial indicator. The gauge set shown in **Figure 40** consists of a dial indicator, handle, and different length adapters (anvils) to fit the gauge to various bore sizes. The bore gauge is used to measure bore size, taper and out-of-round. When using a bore gauge, follow the manufacturer's instructions.

ELECTRICAL SYSTEM FUNDAMENTALS

A thorough study of the many types of electrical systems used in today's motorcycles is beyond the scope of this manual. However, a basic understanding of electrical basics is necessary to perform simple diagnostic tests.

Refer to *Electrical Testing* in Chapter Two for typical test procedures and equipment. Refer to Chapter Nine for specific system test procedures.

Voltage

Voltage is the electrical potential or pressure in an electrical circuit and is expressed in volts. The more pressure (voltage) in a circuit, the more work can be performed.

Direct current (DC) voltage means the electricity flows in one direction. All circuits powered by a battery are DC circuits.

Alternating current (AC) means the electricity flows in one direction momentarily and then switches to the opposite direction. Alternator output is an example of AC voltage. This voltage must be changed or rectified to direct current to operate in a battery powered system.

Resistance

Resistance is the opposition to the flow of electricity within a circuit or component and is measured in ohms. Resistance causes a reduction in available current and voltage.

Resistance is measured in an inactive circuit with an ohmmeter. The ohmmeter sends a small amount of current into the circuit and measures how difficult it is to push the current through the circuit.

An ohmmeter, although useful, is not always a good indicator of a circuit's actual ability under operating conditions. This is because of the low voltage (6-9 volts) the meter uses to test the circuit. The voltage in an ignition coil secondary winding can be several thousand volts. Such high voltage can cause the coil to malfunction, even though it tests acceptable during a resistance test.

Resistance generally increases with temperature. Perform all testing with the component or circuit at room temperature. Resistance tests performed at high temperatures may indicate high resistance readings and cause unnecessary replacement of a component.

Amperage

Amperage is the unit of measurement for the amount of current within a circuit. Current is the actual flow of electricity. The higher the current, the more work can be performed up to a given point. If the current flow exceeds the circuit or component capacity, it will damage the system.

BASIC SERVICE METHODS

Most of the procedures in this manual are straightforward and can be performed by anyone reasonably competent with tools. However, consider personal capabilities carefully before attempting any operation involving major disassembly.

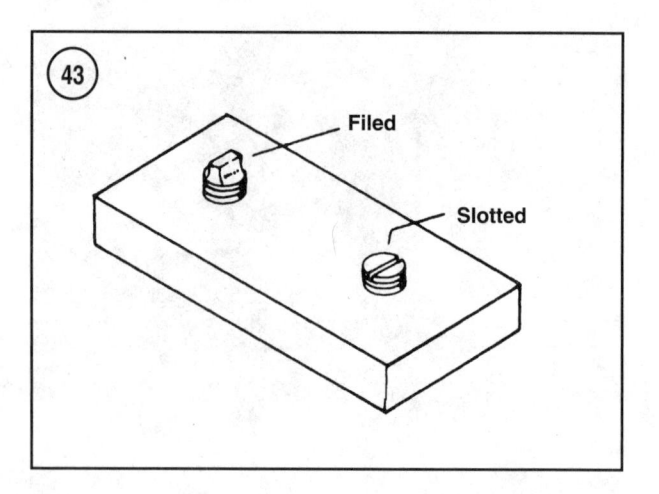

1. *Front*, in this manual, refers to the front of the motorcycle. The front of any component is the end closest to the front of the motorcycle. The left and right sides refer to the position of the parts as viewed by the rider sitting on the seat facing forward.

2. Whenever servicing an engine or suspension component, secure the motorcycle in a safe manner.

3. Tag all similar parts for location and mark all mating parts for position. Record the number and thickness of any shims when removing them. Identify parts by placing them in sealed and labeled plastic sandwich bags.

4. Tag disconnected wires and connectors with masking tape and a marking pen. Do not rely on memory alone.

5. Protect finished surfaces from physical damage or corrosion. Keep gasoline and other chemicals off painted surfaces.

6. Use penetrating oil on frozen or tight bolts. Avoid using heat where possible. Heat can warp, melt or affect the temper of parts. Heat also damages the finish of paint and plastics.

7. When a part is a press fit or requires a special tool to remove, the information or type of tool is identified in the text. Otherwise, if a part is difficult to remove or install, determine the cause before proceeding.

8. To prevent objects or debris from falling into the engine, cover all openings.

9. Read each procedure thoroughly and compare the illustrations to the actual components before starting the procedure. Perform the procedure in sequence.

10. Recommendations are occasionally made to refer service to a dealership or specialist. In these cases, the work can be performed more economically by the specialist than by the home mechanic.

GENERAL INFORMATION

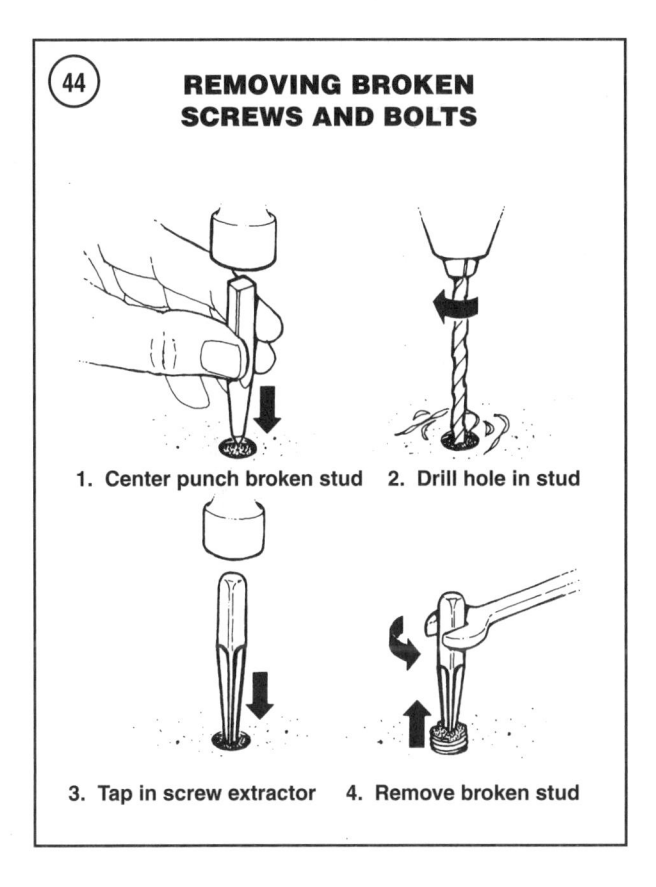

REMOVING BROKEN SCREWS AND BOLTS (Figure 44)

1. Center punch broken stud
2. Drill hole in stud
3. Tap in screw extractor
4. Remove broken stud

11. The term *replace* means to discard a defective part and replace it with a new part. *Overhaul* means to remove, disassemble, inspect, measure, repair and/or replace parts as required to recondition an assembly.

12. Some operations require using a hydraulic press. If a press is not available, have these operations performed by a shop equipped with the necessary equipment. Do not use makeshift equipment that may damage the motorcycle.

13. Repairs are much faster and easier if the motorcycle is clean before starting work. Degrease the motorcycle with a commercial degreaser; follow the directions on the container for the best results. Clean all parts with cleaning solvent when removing them.

CAUTION
Do not direct high-pressure water at steering bearings, fuel hoses, wheel bearings, suspension and electrical components. Water may force grease out of the bearings and possibly damage the seals.

14. If special tools are required, have them available before starting the procedure. When special tools are required, they are described at the beginning of the procedure.

15. Make diagrams of similar-appearing parts. For instance, crankcase bolts are often not the same lengths. Do not rely on memory alone. Carefully laid out parts can become disturbed, making it difficult to reassemble the components correctly.

16. Make sure all shims and washers are reinstalled in the same location and position.

17. Whenever rotating parts contact a stationary part, look for a shim or washer.

18. Use new gaskets if there is any doubt about the condition of old ones.

19. If using self-locking fasteners, replace them with new ones. Do not install standard fasteners in place of self-locking ones.

20. Use grease to hold small parts in place if they tend to fall out during assembly. Do not apply grease to electrical or brake components.

Removing Frozen Fasteners

If a fastener cannot be removed, several methods may be used to loosen it. First, apply penetrating oil such as Liquid Wrench or WD-40. Apply it liberally and let it penetrate for 10-15 minutes. Rap the fastener several times with a small hammer. Do not hit it hard enough to cause damage. Reapply the penetrating oil if necessary.

For frozen screws, apply penetrating oil as described, then insert a screwdriver in the slot and rap the top of the screwdriver with a hammer. This loosens the rust so the screw can be removed in the normal way. If the screw head is too damaged to use this method, grip the head with locking pliers and twist the screw out.

Avoid applying heat unless specifically instructed. Heat may melt, warp or remove the temper from parts.

Removing Broken Fasteners

If the head breaks off a screw or bolt, several methods are available for removing the remaining portion. If a large portion of the remainder projects out, try gripping it with locking pliers. If the projecting portion is too small, file it to fit a wrench or cut a slot in it to fit a screwdriver (**Figure 43**).

If the head breaks off flush, use a screw extractor. To do this, centerpunch the exact center of the remaining portion of the screw or bolt. Drill a small hole in the screw and tap the extractor into the hole. Back the screw out with a wrench on the extractor (**Figure 44**).

Repairing Damaged Threads

Occasionally, threads are stripped through carelessness or impact damage. Often the threads can be repaired by running a tap (for internal threads on nuts) or die (for external threads on bolts) through the threads (**Figure 45**). To clean or repair spark plug threads, use a spark plug tap.

If an internal thread is damaged, it may be necessary to install a Helicoil or some other type of thread insert. Follow the manufacturer's instructions when installing their insert.

If it is necessary to drill and tap a hole, refer to **Table 8** for metric tap and drill sizes.

Stud Removal/Installation

A stud removal tool (**Figure 46**) is available from most tool suppliers. This tool makes the removal and installation of studs easier. If one is not available, thread two nuts onto the stud and tighten them against each other. Remove the stud by turning the lower nut (**Figure 47**).

1. Measure the height of the stud above the surface.
2. Thread the stud removal tool onto the stud and tighten it, or thread two nuts onto the stud.
3. Remove the stud by turning the stud remover or the lower nut.
4. Remove any threadlocking compound from the threaded hole. Clean the threads with an aerosol parts cleaner.
5. Install the stud removal tool onto the new stud or thread two nuts onto the stud.
6. Apply threadlocking compound to the threads of the stud.
7. Install the stud and tighten with the stud removal tool or the top nut.
8. Install the stud to the height noted in Step 1 or its torque specification.
9. Remove the stud removal tool or the two nuts.

Removing Hoses

When removing stubborn hoses, do not exert excessive force on the hose or fitting. Remove the hose clamp and carefully insert a small screwdriver or pick tool between the fitting and hose. Apply a spray lubricant under the hose and carefully twist the hose off the fitting. Clean the fitting of any corrosion or rubber hose material with a wire brush. Clean the inside of the hose thoroughly. Do not use any lubricant when installing the hose (new or old). The lubricant may allow the hose to come off the fitting, even with the clamp secure.

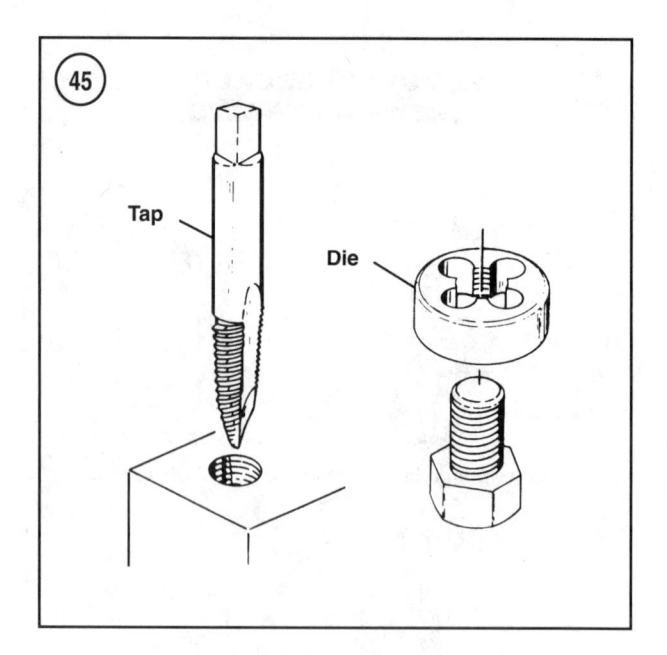

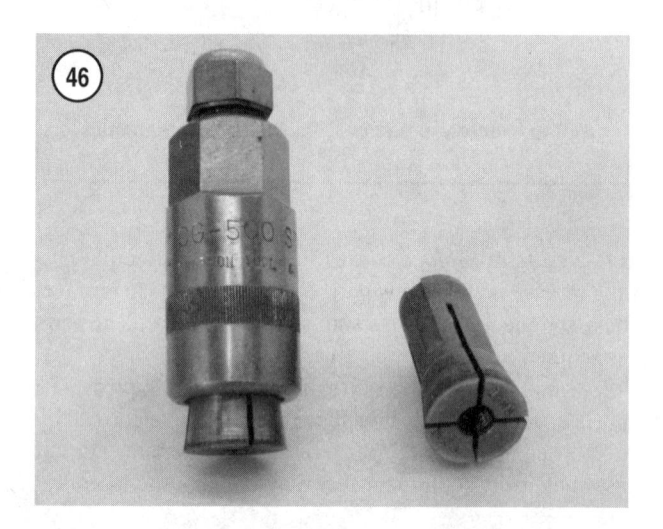

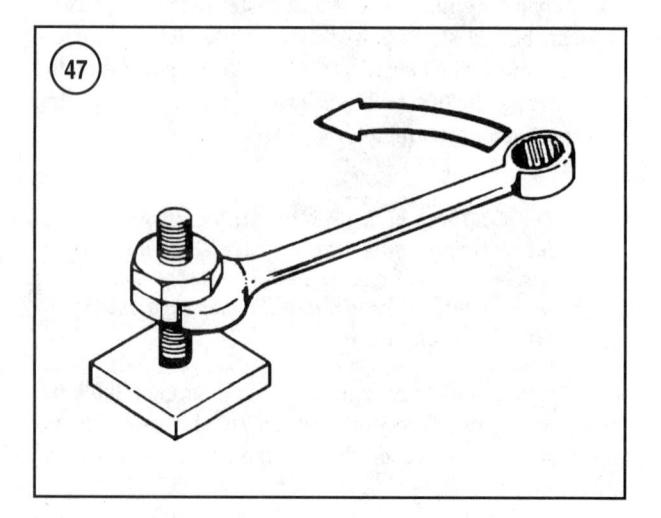

GENERAL INFORMATION

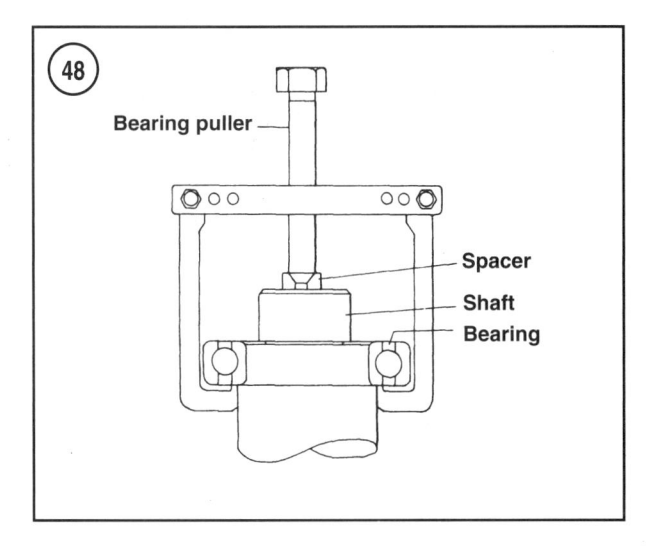

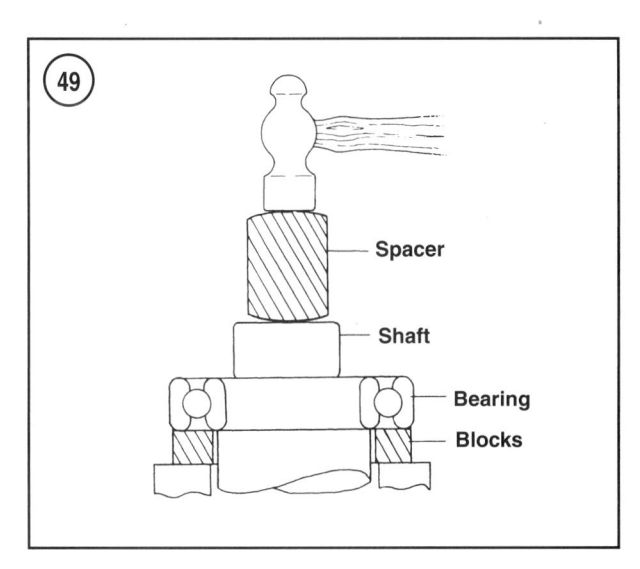

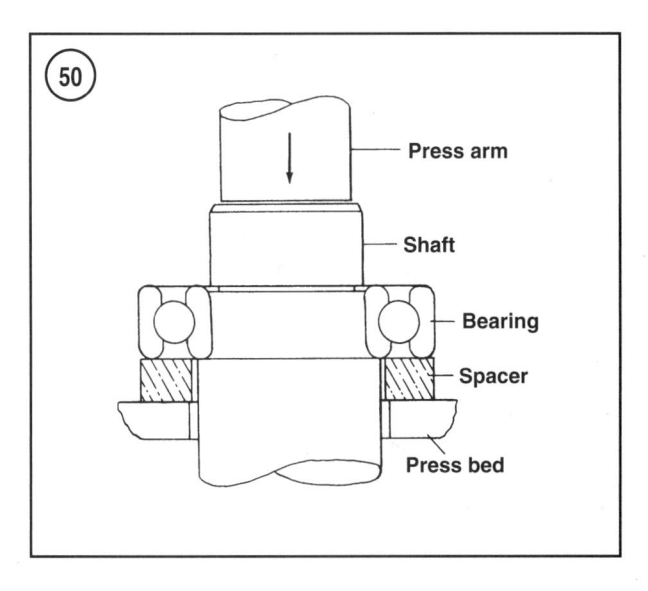

Bearings

Bearings are used in the engine and transmission assembly to reduce power loss, heat and noise resulting from friction. Because bearings are precision parts, they must be maintained with proper lubrication and maintenance. If a bearing is damaged, replace it immediately. When installing a new bearing, take care to prevent damaging it. Bearing replacement procedures are included in the individual chapters where applicable; however, use the following sections as a guideline.

NOTE
Unless otherwise specified, install bearings with the manufacturer's mark or number facing outward.

Removal

While bearings are normally removed only when damaged, there may be times when it is necessary to remove a bearing that is in good condition. However, improper bearing removal will damage the bearing and possibly the shaft or case. Note the following when removing bearings:

1. When using a puller to remove a bearing from a shaft, take care that the shaft is not damaged. Always place a piece of metal between the end of the shaft and the puller screw. In addition, place the puller arms next to the inner bearing race. See **Figure 48**.

2. When using a hammer to remove a bearing from a shaft, do not strike the hammer directly against the shaft. Instead, use a brass or aluminum rod between the hammer and shaft (**Figure 49**) and make sure to support both bearing races with wooden blocks as shown.

3. The ideal method of bearing removal is with a hydraulic press. Note the following when using a press:

   a. Always support the inner and outer bearing races with a suitable size wooden or aluminum spacer (**Figure 50**). If only the outer race is supported, pressure applied against the balls and/or the inner race will damage them.

   b. Always make sure the press arm (**Figure 50**) aligns with the center of the shaft. If the arm is not centered, it may damage the bearing and/or shaft.

   c. The moment the shaft is free of the bearing, it drops to the floor. Secure or hold the shaft to prevent it from falling.

Installation

1. When installing a bearing in a housing, apply pressure to the *outer* bearing race (**Figure 51**). When installing a bearing on a shaft, apply pressure to the *inner* bearing race (**Figure 52**).

2. When installing a bearing as described in Step 1, some type of driver is required. Never strike the bearing directly with a hammer or it will damage the bearing. When installing a bearing, use a piece of pipe or a driver with a diameter that matches the bearing inner race. **Figure 53** shows the correct way to use a driver and hammer to install a bearing.

3. Step 1 describes how to install a bearing in a case half or over a shaft. However, when installing a bearing over a shaft and into the housing at the same time, a tight fit is required for both outer and inner bearing races. In this situation, install a spacer underneath the driver tool so that pressure is applied evenly across both races. See **Figure 54**. If the outer race is not supported as shown, the balls will push against the outer bearing race and damage it.

Interference fit

1. Follow this procedure when installing a bearing over a shaft. When a tight fit is required, the bearing inside diameter is smaller than the shaft. In this case, driving the bearing on the shaft using normal methods may cause bearing damage. Instead, heat the bearing before installation. Note the following:
    a. Secure the shaft so it is ready for bearing installation.
    b. Clean all residues from the bearing surface of the shaft. Remove burrs with a file or sandpaper.
    c. Fill a suitable pot or beaker with clean mineral oil. Place a thermometer rated above 120° C (248° F) in the oil. Support the thermometer so it does not rest on the bottom or side of the pot.
    d. Remove the bearing from its wrapper and secure it with a piece of heavy wire bent to hold it in the pot. Hang the bearing in the pot so it does not touch the bottom or sides of the pot.
    e. Turn the heat on and monitor the thermometer. When the oil temperature rises to approximately 120° C (248° F), remove the bearing from the pot and quickly install it. If necessary, place a socket on the inner bearing race and tap the bearing into place. As the bearing chills, it will tighten on the shaft, so installation must be done quickly. Make sure the bearing is installed completely.
2. Follow this step when installing a bearing in a housing. Bearings are generally installed in a housing with a slight interference fit. Driving the bearing into the housing using normal methods may damage the housing or cause bearing damage. Instead, heat the housing before the bearing is installed. Note the following:

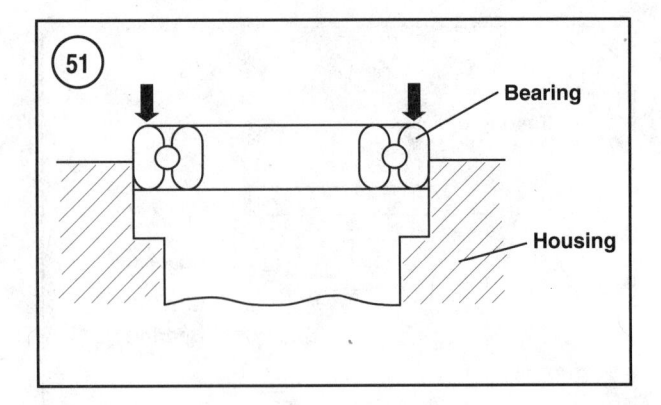

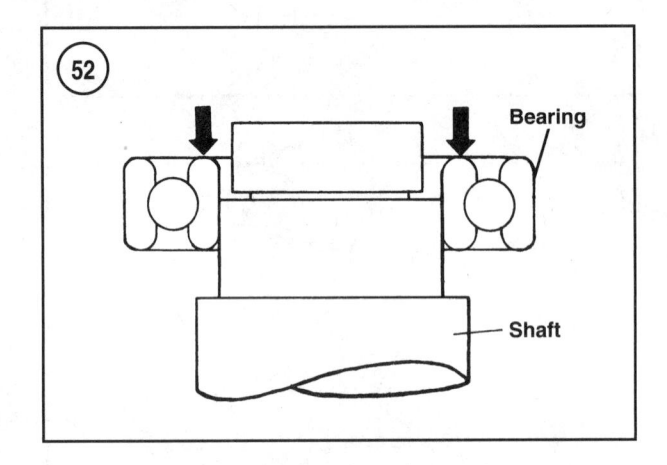

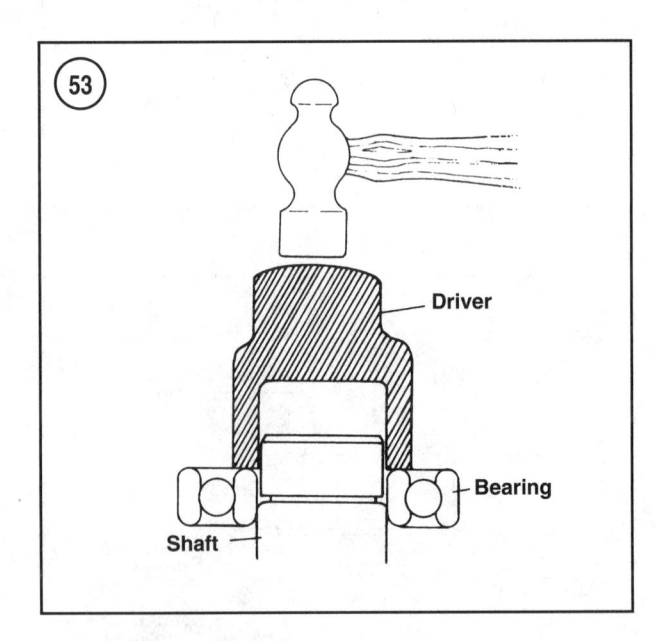

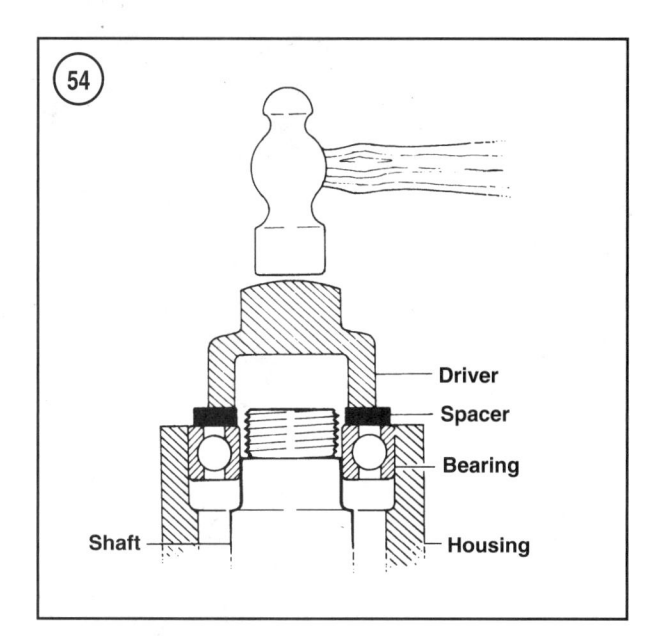

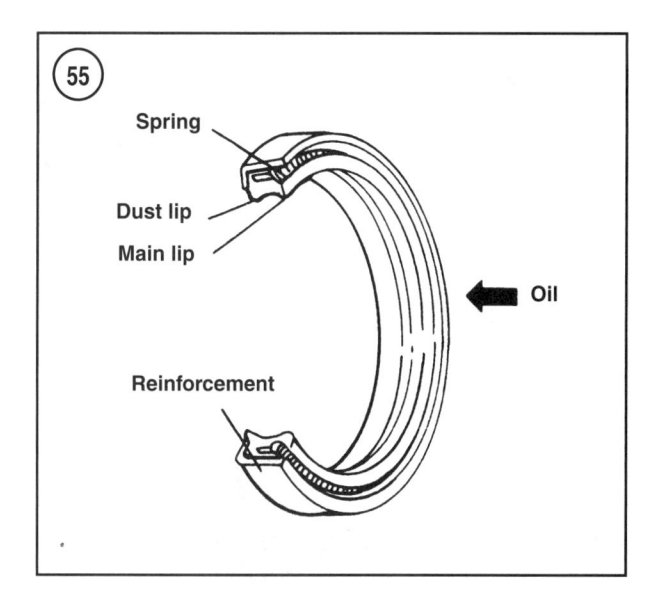

GENERAL INFORMATION

CAUTION
Do not heat the housing with a propane or acetylene torch. Never bring a flame into contact with the bearing or housing. The direct heat will destroy the case hardening of the bearing and will likely warp the housing.

b. Remove the housing from the oven or hot plate, and hold onto the housing with welding gloves. It is hot!

NOTE
Remove and install the bearings with a suitable size socket and extension.

c. Hold the housing with the bearing side down and tap the bearing out. Repeat for all bearings in the housing.
d. Before heating the bearing housing, place the new bearing in a freezer if possible. Chilling a bearing slightly reduces its outside diameter while the heated bearing housing assembly is slightly larger due to heat expansion. This makes bearing installation easier.

NOTE
Always install bearings with the manufacturer's mark or number facing outward.

e. While the housing is still hot, install the new bearing(s) into the housing. Install the bearings by hand, if possible. If necessary, lightly tap the bearing(s) into the housing with a driver placed on the outer-bearing race (**Figure 51**). Do not install new bearings by driving on the inner-bearing race. Install the bearing(s) until it seats completely.

Seal Replacement

Seals (**Figure 55**) contain oil, water, grease or combustion gasses in a housing or shaft. Improperly removing a seal can damage the housing or shaft. Improperly installing the seal can damage the seal. Note the following:

1. Prying is generally the easiest and most effective method of removing a seal from the housing. However, always place a rag underneath the pry tool (**Figure 56**) to prevent damage to the housing. Note the seal's installed depth or if it is installed flush.
2. Pack waterproof grease in the seal lips before the seal is installed.
3. In most cases, install seals with the manufacturer's numbers or marks facing out.
4. Install seals with a socket or driver placed on the outside of the seal as shown in **Figure 57**. Drive the seal

CAUTION
Before heating the housing in this procedure, wash the housing thoroughly with detergent and water. Rinse and rewash the cases as required to remove all traces of oil and other chemical deposits.

a. Heat the housing to approximately 100° C (212° F) in an oven or on a hot plate. An easy way to check that it is the proper temperature is to place tiny drops of water on the housing; if they sizzle and evaporate immediately, the temperature is correct. Heat only one housing at a time.

squarely into the housing until it is to the correct depth or flush (**Figure 58**) as noted during removal. Never install a seal by hitting against the top of it with a hammer.

STORAGE

Several months of non-use can cause a general deterioration of the motorcycle. This is especially true in areas of extreme temperature variations. This deterioration can be minimized with careful preparation for storage. A properly stored motorcycle is much easier to return to service.

Storage Area Selection

When selecting a storage area, consider the following:
1. The storage area must be dry. A heated area is best, but not necessary. It should be insulated to minimize extreme temperature variations.
2. If the building has large window areas, mask them to keep sunlight off the motorcycle.
3. Avoid buildings in industrial areas where corrosive emissions may be present. Avoid areas close to saltwater.
4. Consider the area's risk of fire, theft or vandalism. Check with an insurer regarding motorcycle coverage while in storage.

Preparing the Motorcycle for Storage

The amount of preparation a motorcycle should undergo before storage depends on the expected length of non-use, storage area conditions and personal preference. Consider the following list the minimum requirement:
1. Wash the motorcycle thoroughly. Make sure all dirt, mud and road debris are removed.
2. Start the engine and allow it to reach operating temperature. Drain the engine oil regardless of the riding time since the last service. Fill the engine with the recommended type of oil.
3. Fill the fuel tank completely. There is no need to try to empty the fuel delivery or return lines since they are not vented to the atmosphere.
4. Remove the spark plugs and pour a teaspoon (15-20 ml) of engine oil into the cylinders. Place a rag over the openings and slowly turn the engine over to distribute the oil. Reinstall the spark plugs.
5. Remove the battery. Store the battery in a cool and dry location. Charge the battery once a month.
6. Cover the exhaust and intake openings.

7. Apply a protective substance to the plastic and rubber components. Make sure to follow the manufacturer's instructions for each type of product being used.

8. Place the motorcycle on its centerstand. Rotate the front tire periodically to prevent a flat spot from developing and damaging the tire.

9. Cover the motorcycle with old bed sheets or something similar. Do not cover it with any plastic material that will trap moisture.

GENERAL INFORMATION

Returning the Motorcycle to Service

The amount of service required when returning a motorcycle to service after storage depends on the length of non-use and storage conditions. In addition to performing the reverse of the above procedure, make sure the brakes, clutch, throttle and engine stop switch work properly before operating the motorcycle. Refer to Chapter Three and evaluate the service intervals to determine which areas require service.

Table 1 ENGINE AND FRAME SERIAL NUMBERS

Model*	Frame serial number	Engine serial number
2001		
GL1800	1HFSC470*1A000001-on	SC47E-2000101-on
GL1800A	1HFSC474*1A000001-on	SC47E-2000101-on
2002		
GL1800	1HFSC470*2A100001-on	SC47E-2100101-on
GL1800A	1HFSC474*2A100001-on	SC47E-2100101-on
2003		
GL1800	1HFSC470*3A200001-on	SC47E-2200011-on
GL1800A	1HFSC474*3A200001-on	SC47E-2200011-on
2004		
GL1800	1HFSC470*4A300001-on	SC47E-2300001-on
GL1800A	1HFSC474*4A300001-on	SC47E-2300001-on
2005		
GL1800	1HFSC470*5A400001-on	SC47E-2400001-on
GL1800A	1HFSC474*5A400001-on	SC47E-2400001-on

*The GL1800A model is equipped with ABS.

Table 2 GENERAL DIMENSIONS

	mm	in.
Footpeg height	251	9.9
Ground clearance	125	4.9
Overall height	1455	57.3
Overall length	2635	103.7
Overall width	945	37.2
Seat height	740	29.1
Wheelbase	1690	66.5

Table 3 WEIGHT SPECIFICATIONS

	kg	lb.
Curb weight		
GL1800A	402	886
GL1800	399	880
Dry weight		
GL1800A	362	798
GL1800	359	791
Maximum weight capacity		
Canada	193	425
U.S.	189	417

Table 4 METRIC, INCH AND FRACTIONAL EQUIVALENTS

mm	in.	Nearest fraction	mm	in.	Nearest fraction
1	0.0394	1/32	26	1.0236	1 1/32
2	0.0787	3/32	27	1.0630	1 1/16
3	0.1181	1/8	28	1.1024	1 3/32
4	0.1575	5/32	29	1.1417	1 5/32
5	0.1969	3/16	30	1.1811	1 3/16
6	0.2362	1/4	31	1.2205	1 7/32
7	0.2756	9/32	32	1.2598	1 1/4
8	0.3150	5/16	33	1.2992	1 5/16
9	0.3543	11/32	34	1.3386	1 11/32
10	0.3937	13/32	35	1.3780	1 3/8
11	0.4331	7/16	36	1.4173	1 13/32
12	0.4724	15/32	37	1.4567	1 15/32
13	0.5118	1/2	38	1.4961	1 1/2
14	0.5512	9/16	39	1.5354	1 17/32
15	0.5906	19/32	40	1.5748	1 9/16
16	0.6299	5/8	41	1.6142	1 5/8
17	0.6693	21/32	42	1.6535	1 21/32
18	0.7087	23/32	43	1.6929	1 11/16
19	0.7480	3/4	44	1.7323	1 23/32
20	0.7874	25/32	45	1.7717	1 25/32
21	0.8268	13/16	46	1.8110	1 13/16
22	0.8661	7/8	47	1.8504	1 27/32
23	0.9055	29/32	48	1.8898	1 7/8
24	0.9449	15/16	49	1.9291	1 15/16
25	0.9843	31/32	50	1.9685	1 31/32

Table 5 CONVERSION FORMULAS

Multiply:	By:	To get the equivalent of:
Length		
Inches	25.4	Millimeter
Inches	2.54	Centimeter
Miles	1.609	Kilometer
Feet	0.3048	Meter
Millimeter	0.03937	Inches
Centimeter	0.3937	Inches
Kilometer	0.6214	Mile
Meter	3.281	Feet
Fluid volume		
U.S. quarts	0.9463	Liters
U.S. gallons	3.785	Liters
U.S. ounces	29.573529	Milliliters
Imperial gallons	4.54609	Liters
Imperial quarts	1.1365	Liters
Liters	0.2641721	U.S. gallons
Liters	1.0566882	U.S. quarts
Liters	33.814023	U.S. ounces
Liters	0.22	Imperial gallons
Liters	0.8799	Imperial quarts
Milliliters	0.033814	U.S. ounces
Milliliters	1.0	Cubic centimeters
Milliliters	0.001	Liters

(continued)

GENERAL INFORMATION

Table 5 CONVERSION FORMULAS (continued)

Multiply:	By:	To get the equivalent of:
Torque		
Foot-pounds	1.3558	Newton-meters
Foot-pounds	0.138255	Meters-kilograms
Inch-pounds	0.11299	Newton-meters
Newton-meters	0.7375622	Foot-pounds
Newton-meters	8.8507	Inch-pounds
Meters-kilograms	7.2330139	Foot-pounds
Volume		
Cubic inches	16.387064	Cubic centimeters
Cubic centimeters	0.0610237	Cubic inches
Temperature		
Fahrenheit	(°F − 32) × 0.556	Centigrade
Centigrade	(°C × 1.8) + 32	Fahrenheit
Weight		
Ounces	28.3495	Grams
Pounds	0.4535924	Kilograms
Grams	0.035274	Ounces
Kilograms	2.2046224	Pounds
Pressure		
Pounds per square inch	0.070307	Kilograms per square centimeter
Kilograms per square centimeter	14.223343	Pounds per square inch
Kilopascals	0.1450	Pounds per square inch
Pounds per square inch	6.895	Kilopascals
Speed		
Miles per hour	1.609344	Kilometers per hour
Kilometers per hour	0.6213712	Miles per hour

Table 6 GENERAL TORQUE SPECIFICATIONS

Fastener size or type	N•m	in.-lb.	ft.-lb.
5 mm screw	4	35	–
5 mm bolt and nut	5	44	–
6 mm screw	9	80	–
6 mm bolt and nut	10	88	–
6 mm flange bolt (8 mm head, small flange)	9	80	–
6 mm flange bolt (10 mm head) and nut	12	106	–
8 mm bolt and nut	22	–	16
8 mm flange bolt and nut	27	–	20
10 mm bolt and nut	35	–	25
10 mm flange bolt and nut	40	–	29
12 mm bolt and nut	55	–	40

Table 7 TECHNICAL ABBREVIATIONS

ABDC	After bottom dead center
ABS	Anti-lock brake system
ATDC	After top dead center
AVC	Auto volume control
BBDC	Before bottom dead center
BDC	Bottom dead center
BTDC	Before top dead center
BARO	Barometric pressure sensor
C	Celsius (centigrade)
cc	Cubic centimeters
cid	Cubic inch displacement
CDI	Capacitor discharge ignition
cu. in.	Cubic inches
DAI	Direct air induction
DLC	Data link connector
DTC	Diagnostic trouble code
ECM	Engine control module
ECT	Engine coolant temperature sensor
EFI	Electronic fuel injection
F	Fahrenheit
ft.	Feet
ft.-lb.	Foot-pounds
gal.	Gallons
H/A	High altitude
hp	Horsepower
IAC	Idle air control valve
IAT	Intake air temperature sensor
ICM	Ignition control module
in.	Inches
in.-lb.	Inch-pounds
I.D.	Inside diameter
kg	Kilograms
kgm	Kilogram meters
km	Kilometer
kPa	Kilopascals
L	Liter
m	Meter
MIL	Malfunction indicator lamp
MAP	Manifold absolute pressure sensor
MAG	Magneto
ml	Milliliter
mm	Millimeter
N•m	Newton-meters
O_2	Oxygen sensor
O.D.	Outside diameter
oz.	Ounces
PAIR	Pulse secondary air injection system
PGM-FI	Programmed fuel injection
psi	Pounds per square inch
PTO	Power take off
pt.	Pint
qt.	Quart
rpm	Revolutions per minute
RTV	Room temperature vulcanization
TP	Throttle position sensor
VSS	Vehicle speed sensor

GENERAL INFORMATION

Table 8 METRIC TAP AND DRILL SIZES

Metric size	Drill equivalent	Decimal fraction	Nearest fraction
3 × 0.50	No. 39	0.0995	3/32
3 × 0.60	3/32	0.0937	3/32
4 × 0.70	No. 30	0.1285	1/8
4 × 0.75	1/8	0.125	1/8
5 × 0.80	No. 19	0.166	11/64
5 × 0.90	No. 20	0.161	5/32
6 × 1.00	No. 9	0.196	13/64
7 × 1.00	16/64	0.234	15/64
8 × 1.00	J	0.277	9/32
8 × 1.25	17/64	0.265	17/64
9 × 1.00	5/16	0.3125	5/16
9 × 1.25	5/16	0.3125	5/16
10 × 1.25	11/32	0.3437	11/32
10 × 1.50	R	0.339	11/32
11 × 1.50	3/8	0.375	3/8
12 × 1.50	13/32	0.406	13/32
12 × 1.75	13/32	0.406	13/32

CHAPTER TWO

TROUBLESHOOTING

The troubleshooting procedures described in this chapter provide typical symptoms and logical methods for isolating the cause(s). There may be several ways to solve a problem, but only a systematic approach is successful in avoiding wasted time and possibly unnecessary parts replacement.

Gather as much information as possible to aid in diagnosis. Never assume anything and do not overlook the obvious. Make sure the start switch is in the RUN position and there is fuel in the tank. Learning to recognize symptoms will make troubleshooting easier. In most cases, expensive and complicated test equipment is not needed to determine whether repairs can be performed at home. On the other hand, be realistic and do not start procedures that are beyond the experience and equipment available. If the motorcycle does require the attention of a professional, describe symptoms and conditions accurately and fully. The more information a technician has available, the easier it is to diagnose the problem.

Proper lubrication, maintenance and periodic tune-ups reduce the chance that problems will occur. However, even with the best of care the motorcycle may require troubleshooting.

OPERATING REQUIREMENTS

An engine needs three basics to run properly: correct air/fuel mixture, compression and a spark (**Figure 1**) at the right time. If one basic requirement is missing, the engine will not run.

A four-stroke engine performs these functions as shown in **Figure 2**.

STARTING THE ENGINE

When experiencing engine-starting troubles, it is easy to work out of sequence and forget basic starting procedures. The following sections describe the recommended starting procedures for the GL1800.

Starting System Operation

1. The position of the sidestand can affect engine starting. Note the following:
   a. The engine cannot turn over when the sidestand is down and the transmission is in gear.

TROUBLESHOOTING

①

FUEL

↓

COMPRESSION

↓

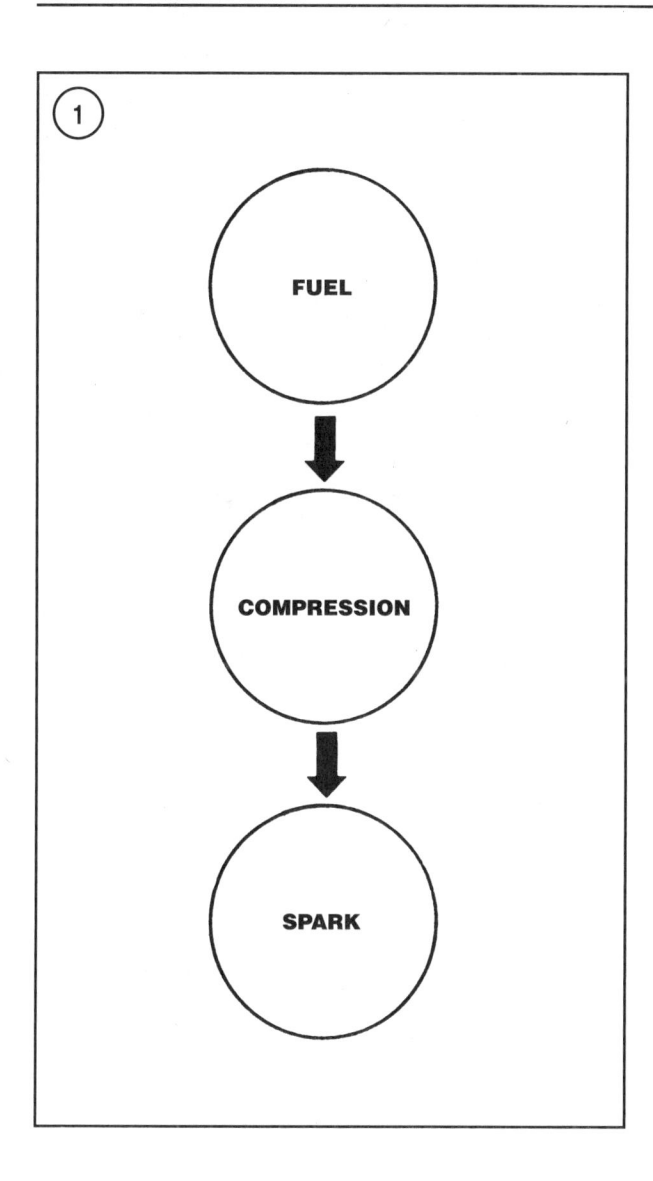

SPARK

b. The engine can turn over when the sidestand is down and the transmission is in neutral. The engine will stop when the transmission is shifted into gear with the sidestand down.

c. The engine can turn over when the sidestand is up and the transmission is in neutral, or in gear with the clutch lever pulled in.

2. Before starting the engine, shift the transmission into neutral and turn the engine stop switch to RUN (A, **Figure 3**).

3. Confirm that the reverse (RVS) switch is off (B, **Figure 3**).

4. Turn the ignition switch on and confirm the following:
   a. The neutral indicator light is on (when transmission is in neutral).
   b. The low oil pressure warning light is on. The warning light should go off a few seconds after the en-

gine starts. If the light stays on, turn the engine off and check the oil level (Chapter Three).

c. The malfunction indicator lamp (MIL) illuminates for a few seconds and then goes off (**Figure 4**). If the indicator stays on, or comes on when the motorcycle is being operated, there is a problem in the programmed fuel injection (PGM-FI) system. Refer to Chapter Eight to retrieve diagnostic trouble codes (DTC) and troubleshoot the system.

d. The low fuel indicator illuminates for a few seconds and then goes off.

e. On GL1800A models, the ABS indicator illuminates and stays on until the motorcycle starts moving. When the engine is started, the ABS control unit (ECU) performs self-diagnosis and checks the operating conditions of the ABS components. Self-diagnosis starts when the ignition switch is turned on and ends when the motorcycle speed reaches 10 km/h (6 mph). If the system is operational, the ABS indicator turns off. Observe this check each time the motorcycle is ridden.

NOTE
The low oil pressure, MIL, low fuel and ABS indicators come on when the ignition switch is turned on to show that the indicators are working correctly, and then turn off (except the ABS indicator). This is considered a lamp circuit test. If an indicator does not light after turning on the ignition switch, test the appropriate indicator circuit as described in Chapter Eight or Chapter Nine.

5. The engine is now ready to start. Refer to the correct *Starting Procedure* in this section.

Bank angle sensor ignition cut-off system

The GL1800 is equipped with a bank angle (lean angle) sensor system that turns the engine and fuel pump off if the motorcycle falls on its side with the engine running. After the motorcycle is returned to the upright position, the ignition switch must be turned off, then turned back on before the engine will restart.

Starting Procedures

NOTE
Do not operate the starter for more than five seconds at a time. Wait approximately 10 seconds between starting attempts.

FOUR-STROKE OPERATING PRINCIPLES

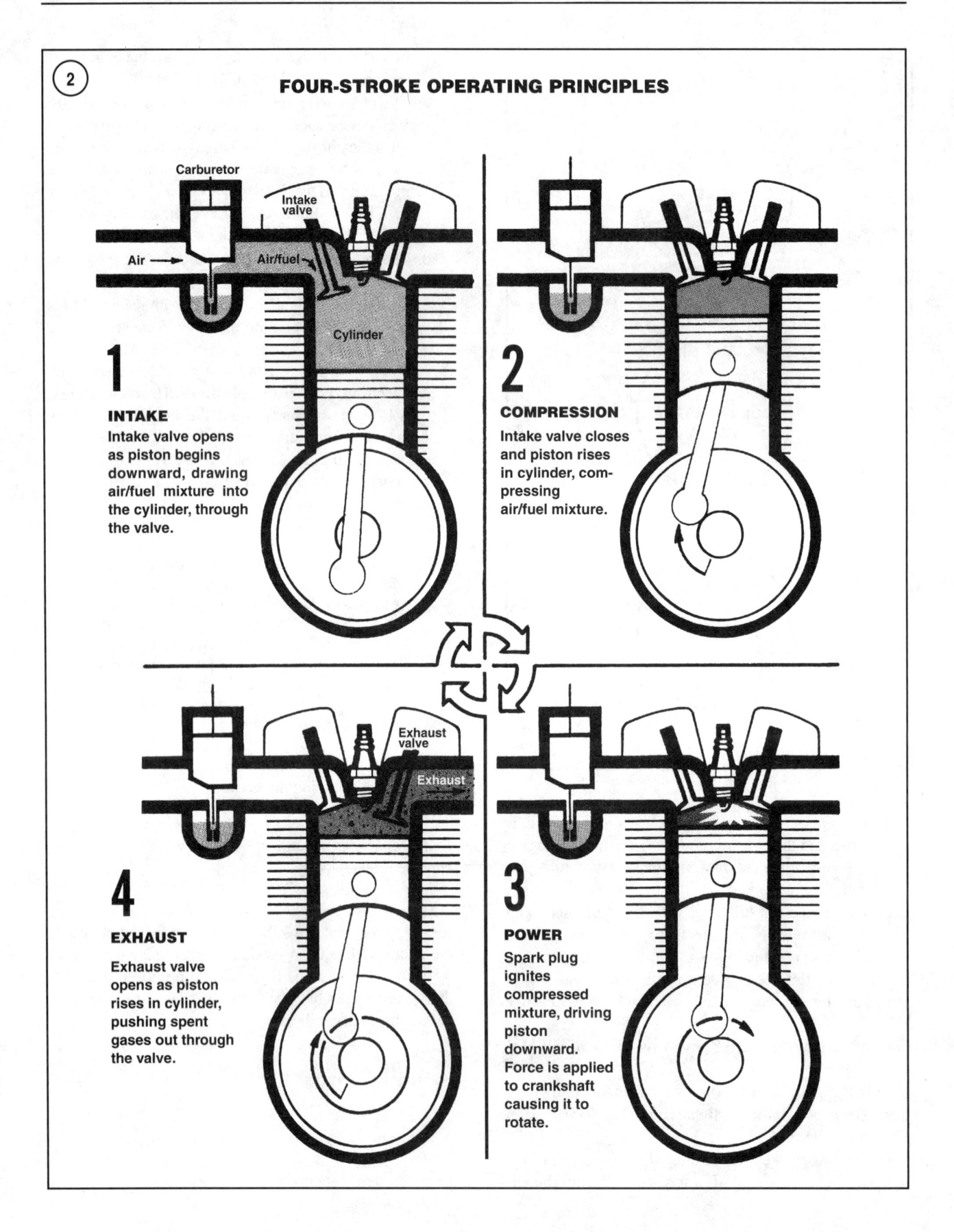

1 INTAKE
Intake valve opens as piston begins downward, drawing air/fuel mixture into the cylinder, through the valve.

2 COMPRESSION
Intake valve closes and piston rises in cylinder, compressing air/fuel mixture.

3 POWER
Spark plug ignites compressed mixture, driving piston downward. Force is applied to crankshaft causing it to rotate.

4 EXHAUST
Exhaust valve opens as piston rises in cylinder, pushing spent gases out through the valve.

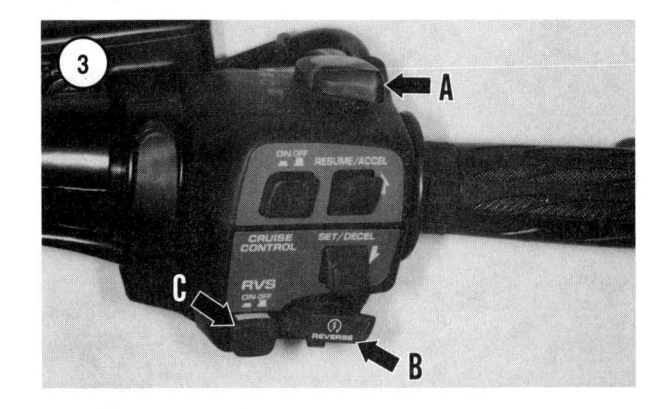

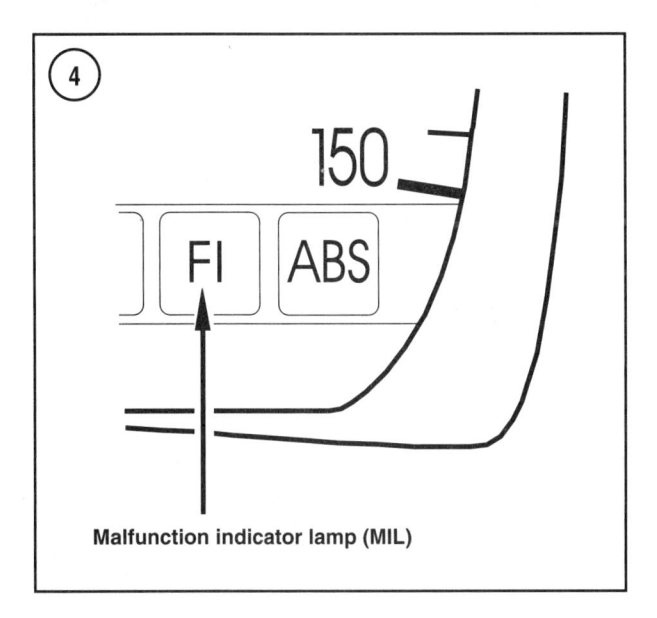

TROUBLESHOOTING

Flooded engine

If the engine does not start after a few attempts and there is a strong gasoline smell, it may be flooded. To start the engine when flooded:

1. Turn the engine stop switch to RUN (A, **Figure 3**).
2. Open the throttle fully.
3. Turn the ignition switch on and operate the starter button (C, **Figure 3**) for five seconds. Release the starter button and close the throttle.
4. Follow the normal starting procedure listed under *Any air temperature* in this section. Note the following:
    a. If the engine starts but idles roughly, vary the throttle position slightly until the engine idles and responds smoothly.
    b. If the engine does not start, turn the ignition switch off and wait approximately ten seconds. Then repeat Steps 1-4 again. If the engine still does not start, refer to *Engine Will Not Start* in this chapter.

ENGINE WILL NOT START

Malfunction Indicator Lamp (MIL)

If the malfunction indicator lamp (MIL) stays on when the engine is running, there is a problem in the PGM-FI system (**Figure 4**). Refer to Chapter Eight to retrieve diagnostic trouble codes (DTC) and troubleshoot the system. If the MIL indicator is not on, refer to *Identifying the Problem* in this section to isolate the problem to the fuel, ignition or compression system.

Any air temperature

1. Refer to *Starting System Operation* in this section.
2. Install the ignition key and turn the ignition switch on.
3. Turn the engine stop switch to RUN (A, **Figure 3**).
4. Depress the starter button (C, **Figure 3**) and start the engine. Do *not* open the throttle when pressing the starter button.

NOTE
To prevent the engine from starting with the throttle in the wide-open position, the electronic control module (ECM) interrupts the fuel supply if the throttle is in this position while the engine is cranking. The only time it would be necessary to open the throttle all the way is when attempting to start a flooded engine. See **Flooded engine** *in this section.*

Identifying the Problem

The first step in troubleshooting a no start condition is to narrow the possibilities by following a specific troubleshooting procedure—while never overlooking the obvious. Many times, a starting problem is simple, such as a disconnected or damaged wire.

If the engine does not start, perform the following steps in order while remembering the *Operating Requirements* described in this chapter. If the engine fails to start after performing these checks, refer to the troubleshooting procedures indicated in the steps. If the engine starts, but idles or runs roughly, refer to *Engine Performance* in this chapter.

NOTE
The GL1800 is equipped with a bank angle (lean angle) sensor system that turns the engine and fuel pump off if the motorcycle falls on its side when the engine is running. After

positioning the motorcycle upright, the ignition switch must be turned off, then turned back on before the engine will restart.

1. Refer to *Starting the Engine* in this chapter to make sure all switch positions and starting procedures are correct.
2. If the starter does not operate, or turns over slowly, refer to the appropriate starter procedure in this section.
3. If the starter turns over, and the engine seems flooded, refer to *Flooded Engine* under *Starting The Engine* in this chapter. If the engine is not flooded, continue with Step 4.
4. Turn the ignition switch on and check the fuel gauge. If the low fuel indicator at the bottom of the fuel gauge is flashing, the fuel level in the tank is low. The amount of fuel remaining in the tank when the low fuel indicator *first* starts to flash is approximately 4.4 liters (1.16 U.S. Gal.). When the fuel gauge needle enters the red band, there is approximately 3.0 liters (0.79 U.S. Gal.) of fuel remaining in the tank.
5. If there is sufficient fuel in the fuel tank, remove one of the spark plugs immediately after attempting to start the engine. The plug's insulator should be wet, indicating that fuel is reaching the engine. If the plug tip is dry, fuel is not reaching the engine. Confirm this condition by checking another spark plug. A faulty fuel flow problem will cause this condition. Refer to *Fuel System* in this chapter and to *Fuel Flow Test* in Chapter Eight. If there is fuel on each spark plug and the engine will not start, the engine may not have adequate spark. Continue with Step 6.

NOTE
When examining the spark plug caps in the following steps, check for the presence of water in the plug caps.

6. Make sure each spark plug wire is secure. Push the spark plug caps and slightly rotate them to clean the electrical connection between the plug and the connector. If the engine does not start, continue with Step 7.

NOTE
Cracked or damaged spark plug caps and cables can cause intermittent problems that are hard to diagnose. If the engine occasionally misfires or cuts out, use a spray bottle to wet the spark plug cables and caps while the engine is running. Water that enters a damaged cap or cable will cause an arc through the insulating material, causing an engine misfire.

7. Perform the *Spark Test* in this section. If there is a strong spark at each plug, perform Step 8. If there is no spark or if the spark is very weak, refer to *Ignition System Troubleshooting* in Chapter Nine.

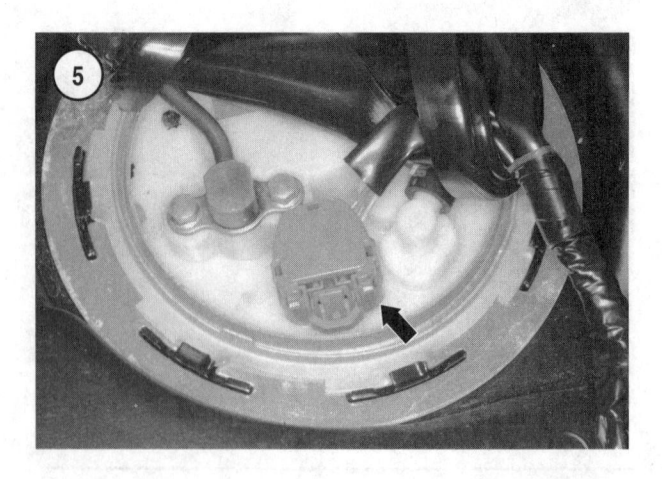

NOTE
Performing a spark test is the quickest way to isolate an ignition or fuel system problem. If a spark is recorded at each plug wire, and the fuel and mechanical systems are working correctly, the engine should start.

8. If the fuel and ignition systems are working correctly, perform a leakdown test (this chapter) and cylinder compression test (Chapter Three). If the leakdown test indicates a problem with a cylinder(s), or the compression is low, refer to *Low Engine Compression* under *Engine* in this chapter.

Spark Test

Perform a spark test to determine if the ignition system is producing adequate spark. This test should be performed with a spark tester (Motion Pro part No. 08-0122). A spark tester looks like a spark plug with a large adjustable gap between the center electrode and grounded base. Because the voltage required to jump the spark tester gap is sufficiently larger than that of a normally gapped spark plug, the test results are more accurate than with a spark plug. Do not assume that because a spark jumped across a spark plug gap, the ignition system is working correctly.

Perform this test on a cold and hot engine. If the test results are positive for each test, the ignition system is working correctly.

CAUTION
After removing the spark plug caps and before removing the spark plugs in Step 1, clean the area around each spark plug with

TROUBLESHOOTING

compressed air. Dirt that falls into the cylinder causes rapid engine wear.

1. Remove the seat. Refer to *Seat* in Chapter Seventeen.

   WARNING
   Step 2 must be performed to disable the fuel injection system. Otherwise, fuel will be injected into the cylinders when the engine is turned over during the spark test, flooding the cylinders and creating explosive fuel vapors.

2. Disconnect the fuel pump 5P connector (**Figure 5**).

   NOTE
   When removing the spark plug caps in Step 3, check for water in the plug caps.

3. Remove the spark plugs as described in Chapter Three.

   NOTE
   After removing each spark plug, shake it to see if the insulator slides down over the center electrode. Insulators can be broken when the spark plugs are mishandled or when a plug wrench is used incorrectly.

4. Connect a spark tester to one of the spark plug leads. Touch the spark tester base to a screw threaded into the cylinder head (**Figure 6**). The screw helps to position the spark tester or spark plug firing tip away from the open spark plug hole. Position the spark tester so the electrodes are visible.

   WARNING
   Mount the spark tester and spark plugs away from the spark plug holes in the cylinder head so the spark plugs or tester cannot ignite the gasoline vapors in the cylinder. If the engine is flooded, do not perform this test. The firing of the spark plugs or spark tester can ignite fuel that is ejected through the spark plug holes.

5. With the transmission in neutral, turn the ignition system on and the engine stop switch to RUN (A, **Figure 3**).

   WARNING
   Do not hold the spark tester, spark plug or connector because a serious electrical shock may result.

6. Push the starter button to turn the engine over. There must be a fat blue spark between the spark tester terminals. Repeat for each cylinder.

7. If there is a spark at each plug wire, the ignition system is functioning properly. Check for one or more of the following possible malfunctions:
   a. Faulty fuel system component.
   b. Flooded engine.
   c. Engine damage (low compression).

8. If the spark was weak or if there was no spark at one or more plugs, note the following:
   a. If there is no spark at all six plugs, check for a damaged PGM-FI ignition relay as described in Chapter Nine. If the relay is good, perform the peak voltage tests described under *Ignition System Troubleshooting* in Chapter Nine. If the ignition coil peak voltage signal is correct, but there is still no spark, perform the *Power/Ground Circuit Check* described under *Ignition Coil* in Chapter Nine.

   NOTE
   An accidentally triggered anti-theft device can cut off power to the ignition system (or starter or fuel system, depending on how it is wired). If such a device is installed, check its operation for a short circuit.

   b. If there is no spark at one spark plug only, and the plug is good, there is a problem with the spark plug wire or plug cap. Tighten the spark plug cap and repeat the test.
   c. If there is no spark with one ignition group (two spark plugs, same ignition coil), switch the ignition coils and retest. If there is now spark (both spark plugs), the ignition coil is faulty.

   NOTE
   The three ignition coils are grouped as follows: Cylinders 1-2, 3-4, and 5-6. The three ignition coils are identical (same part number).

   d. If the problem cannot be found, refer to *Ignition System Troubleshooting* in Chapter Nine.

Starter Does Not Turn Over

If the engine does not turn over, the battery or starting system is usually at fault. Check the following steps in order:

CAUTION
Jump starting is not recommended, especially if an automotive battery is used. If the jump start is being performed with the car running, the large amperage output of the battery can damage the electrical system.

1. Refer to *Starting the Engine* in this chapter for proper switch and sidestand operation.
2. Check the condition of the following fuses as described in Chapter Nine:
    a. Starter kill fuse.
    b. RVS start fuse.
    c. RVS fuse B.
3. Loose, contaminated or damaged battery cables. Loose connections at battery.
4. Discharged or damaged battery. Check the battery and battery cables as described in Chapter Nine.
5. Damaged sidestand switch. Refer to Chapter Nine.
6. Damaged starter, starter solenoids or starter switch. Test the starting circuit as described in Chapter Nine.
7. Ignition system failure. Perform the peak voltage tests in Chapter Nine.
8. Engine damage.

Starter Turns Over Slowly

For the starter to work correctly, the battery must be 75 percent charged and the battery cables clean and in good condition. Inspect and test the battery as described in Chapter Nine.

Starter Turns Over Correctly, But Engine Will Not Start

If the starter turns over correctly, the battery and starting circuit are working correctly. Perform the *Spark Test* in this chapter to isolate the problem to the fuel or ignition system. If the ignition and fuel systems are working correctly, the engine may not have enough compression to start. Refer to *Low Engine Compression* under *Engine* in this chapter.

Starter Reverse System

If the starter reverse system does not operate, refer to the appropriate troubleshooting procedures under *Starter/Reverse System Troubleshooting* in Chapter Nine.

ENGINE PERFORMANCE

If the engine runs, but performance or drivability is unsatisfactory, refer to the following procedure(s) that best describes the symptom(s).

Malfunction Indicator Lamp (MIL)

If the malfunction indicator lamp (MIL) stays on when the engine is running, there is a problem in the PGM-FI system (**Figure 4**). Refer to Chapter Eight to retrieve the diagnostic trouble code (DTC) and troubleshoot the system. If the MIL indicator is not on, refer to the procedure in this section that best describes the performance complaint.

Engine Starts But Stalls and is Hard to Restart

Check for the following:
1. The engine idle speed is out of adjustment. Check for a diagnostic trouble code (DTC) as described in Chapter Eight. If there is no DTC code, refer to *Engine Idle Speed Adjustment* in Chapter Eight to reset the idle speed.

NOTE
The engine idle speed is controlled by the ECM.

2. Plugged fuel tank breather tube.
3. Plugged fuel feed hose.
4. Contaminated or stale fuel.
5. Intake air leak.
6. Damaged idle air control valve (IAC), mounted on the throttle body. Refer to *Throttle Body* in Chapter Eight.
7. Faulty ECM or ignition pulse generator.

Engine Backfires, Cuts Out or Misfires During Acceleration

A backfire occurs when fuel is burned or ignited in the exhaust system.
1. A lean air/fuel mixture can cause these engine performance problems. Check for the following conditions:
    a. Low fuel pressure.
    b. Clogged fuel injectors.
    c. Vacuum leak.

TROUBLESHOOTING

2. Loose exhaust pipe-to-cylinder head connection.
3. Leaks in the intake system.
4. Incorrect ignition timing or a damaged ignition system can cause these conditions. Perform the peak voltage tests in Chapter Nine to isolate the damaged ignition system component. Check the ignition timing as described in Chapter Three.

NOTE
The ignition timing is controlled by the ECM and cannot be adjusted. Checking the ignition timing is used as a diagnostic aid.

5. Check the following engine components:
   a. Broken valve springs.
   b. Stuck or leaking valves.
   c. Worn or damaged camshaft lobes.
   d. Incorrect valve timing due to an incorrect camshaft installation or a mechanical failure.

Engine Backfires on Deceleration

If the engine backfires when the throttle is released, check the following:
1. Damaged pulse secondary air injection (PAIR) system:

NOTE
The PAIR system injects fresh air into the exhaust port.

   a. Clogged, damaged or disconnected PAIR system hoses.
   b. Damaged PAIR check valve.
   c. Damaged PAIR control solenoid valve.
2. Damaged ignition system. Because the PAIR solenoid control valve is controlled by the fuel injection system, a damaged ECM unit may be at fault.
3. Check the following engine components:
   a. Broken valve springs.
   b. Stuck or leaking valves.
   c. Worn or damaged camshaft lobes.
   d. Incorrect valve timing due to an incorrect camshaft installation or a mechanical failure.

Poor Fuel Mileage

1. Damaged pressure regulator.
2. Clogged fuel feed hose.
3. Faulty thermostat.
4. Dirty or clogged air filter.
5. Incorrect ignition timing.
6. Vacuum leak.

Engine Will Not Idle or Idles Roughly

1. Clogged air filter element.
2. Clogged fuel feed hose.
3. Damaged idle air control valve (IAC).
4. Obstructed or defective fuel injector.
5. Poor fuel flow.
6. Contaminated or stale fuel.
7. Incorrect idle adjustment.
8. Leaking head gasket(s) or vacuum leak.
9. Intake air leak.
10. Incorrect ignition timing (defective ECM or ignition pulse generator).
11. Low engine compression.

Low Engine Power

1. Support the bike on its centerstand with the rear wheel off the ground and then spin the rear wheel by hand. If the wheel spins freely, perform Step 2. If the wheel does not spin freely, check for the following conditions:
   a. Dragging brakes.

NOTE
After riding the motorcycle, come to a stop on a level surface (in a safe area away from all traffic). Turn the engine off and shift the transmission into neutral. Walk or push the motorcycle forward. If the motorcycle is harder to push than normal, check for dragging brakes.

   b. Damaged final gear bearings. Excessive noise from the final gear housing may indicate bearing or gear damage.

NOTE
The motorcycle can seem to lose power when riding into a strong head wind due to the large frontal area (windshield and fairing assembly). Always consider the current riding conditions and motorcycle load when troubleshooting low engine power.

2. Test ride the bike and accelerate quickly from first to second gear. If the engine speed increased according to throttle position, perform Step 3. If the engine speed did not increase, check for one or more of the following problems:
   a. Slipping clutch.
   b. Warped clutch plates/discs.
   c. Worn clutch plates/discs.
   d. Weak or damaged clutch spring.
   e. Faulty clutch hydraulic system.

3. Test ride the bike and accelerate lightly. If the engine speed increased according to throttle position, perform Step 4. If the engine speed did not increase, check for one or more of the following problems:
   a. Clogged air filter.
   b. Restricted fuel flow.
   c. Pinched fuel tank breather hose.
   d. Clogged or damaged muffler. Tap the muffler with a rubber mallet and check for loose or broken baffles.

NOTE
A clogged muffler or exhaust system will prevent some of the burned exhaust gasses from exiting the exhaust port at the end of the exhaust stroke. This condition affects the incoming air/fuel mixture on the intake stroke and reduces engine power.

4. Check for retarded ignition timing as described in Chapter Three. A decrease in power results when the plugs fire later than normal.
5. Check for one or more of the following problems:
   a. Low engine compression.
   b. Worn spark plugs.
   c. Fouled spark plug(s).
   d. Incorrect spark plug heat range.
   e. Weak ignition coil(s).
   f. Clogged or defective fuel injector(s).
   g. Incorrect ignition timing (defective ECM or ignition pulse generator).
   h. Incorrect oil level (too high or too low).
   i. Contaminated oil.
   j. Worn or damaged valve train assembly.
   k. Engine overheating. Refer to *Engine* in this chapter.
6. If the engine knocks when it is accelerated or when running at high speed, check for one or more of the following possible malfunctions:
   a. Incorrect type of fuel.
   b. Lean fuel mixture.
   c. Advanced ignition timing (defective ECM).

NOTE
Other signs of advanced ignition timing are engine overheating and hard or uneven engine starting.

   d. Excessive carbon buildup in combustion chamber.
   e. Worn pistons and/or cylinder bores.

Poor Idle or Low Speed Performance

1. Check for damaged fuel body insulators or a loose throttle body and air filter housing hose clamps. These conditions will cause an air leak.
2. Perform the spark test described in this chapter. Note the following:
   a. If the spark is good, go to Step 3.
   b. If the spark is weak, refer to the *Ignition System Troubleshooting* described in Chapter Nine.
3. Check the ignition timing as described in Chapter Three. If the ignition timing is correct, perform Step 4. If the timing is incorrect, refer to *Ignition System Troubleshooting*.
4. Check the fuel system as described in this chapter.

Poor High Speed Performance

1. Check ignition timing as described in Chapter Three. If ignition timing is correct, perform Step 2. If the timing is incorrect, perform the peak voltage tests in Chapter Nine.
2. Check the fuel system as described in this chapter.
3. Check the valve clearance as described in Chapter Three. Note the following:
   a. If the valve clearance is correct, perform Step 4.
   b. If the clearance is incorrect, readjust the valves.
4. Incorrect valve timing and worn or damaged valve springs can cause poor high-speed performance. If the camshafts were timed just prior to the bike experiencing this type of problem, the cam timing may be incorrect. If the cam timing was not set or changed, and all of the other inspection procedures in this section failed to locate the problem, inspect the camshafts and valve assembly.

FUEL SYSTEM

The following section isolates common fuel system problems under specific complaints.

If the starter turns over and there is spark at each spark plug, low fuel pressure may prevent fuel from being supplied to the spark plugs. Troubleshoot fuel flow as follows:

1. After attempting to start the engine, remove one of the spark plugs (Chapter Three) and check for fuel on the plug tip. Note the following:
   a. If there is no fuel visible on the plug, remove another spark plug. If there is no fuel on this plug, perform Step 2.
   b. If there is fuel on the plug tip, and the engine has spark at all of the spark plugs, check for an intake air leak or contaminated or stale fuel.

TROUBLESHOOTING

NOTE
If the motorcycle was not used for some time, and was not properly stored, the fuel may have gone stale, where lighter parts of the fuel have evaporated. Depending on the condition of the fuel, a no-start condition can result.

   c. If there is an excessive amount of fuel on the plug, check for a clogged, plugged air filter or stuck open fuel injector(s).
2. Perform the *Fuel Pressure Check* in Chapter Eight. Note the following:
   a. If the fuel pressure reading is correct, go to Step 3.
   b. If the fuel pressure reading is too low or too high, follow the inspection procedures listed in the *Fuel Pressure Check* section.
3. Perform the *Fuel Flow Test* in Chapter Eight. Note the following:
   a. If the fuel flow is normal, go to Step 4.
   b. If the fuel flow volume is less than specified, follow the inspection procedures listed in the *Fuel Flow Test* section.
4. Perform the *Fuel Pump Operational Test* in Chapter Eight. Note the following:
   a. If the fuel pump operation is correct, go to Step 5.
   b. If the fuel pump operation is faulty, replace the fuel pump and retest the fuel system.
5. Check the fuel injectors as described in Chapter Eight.
6. Check for a stuck idle air control valve (IAC), mounted on the throttle body. A stuck open IAC valve lowers cranking vacuum and makes it difficult for the engine to start. Refer to *Throttle Body* in Chapter Eight.

NOTE
If the IAC valve or its wires or connectors malfunctions, a diagnostic trouble code (DTC) may be displayed on the malfunction indicator lamp (MIL) indicating a problem in the PGM-FI system. See Chapter Eight.

ENGINE

Exhaust Smoke

The color of the exhaust smoke can help diagnose engine problems or operating conditions.

Black smoke

Black smoke is an indication of a rich air/fuel mixture where an excessive amount of fuel is being burned in the combustion chamber. Check for a leaking fuel injector(s) or a damaged pressure regulator as described in Chapter Eight.

Blue smoke

Blue smoke indicates that the engine is burning oil in the combustion chamber as it leaks past worn valve stem seals and piston rings. Excessive oil consumption is another indicator of an engine that is burning oil. Perform a compression test (Chapter Three) to isolate the problem.

White smoke or steam

It is normal to see white smoke or steam from the exhaust after first starting the engine in cold weather. This is actually condensed steam formed by the engine during combustion. If the motorcycle is ridden far enough, the water cannot build up in the crankcase and should not be a problem. Once the engine heats up to normal operating temperature, the water evaporates and exits the engine through the crankcase vent system. However, if the motorcycle is ridden for short trips or repeatedly started and stopped and allowed to cool off without the engine getting warm enough, water will start to collect in the crankcase. With each short run of the engine, more water collects. As this water mixes with the oil in the crankcase, sludge is produced. Sludge can eventually cause engine damage as it circulates through the lubrication system and blocks off oil passages. Water draining from drain holes in exhaust pipes indicates water buildup.

Large amounts of steam can also be caused by a cracked cylinder head or cylinder block surface that allows antifreeze to leak into the combustion chamber. Perform a coolant pressure test as described in Chapter Ten.

Low Engine Compression

Problems with the engine top end will affect engine performance and drivability. When the engine is suspect, perform a leakdown test (this chapter) and a compression test (Chapter Three). Interpret the results as described in each procedure to troubleshoot the suspect area. An engine can lose compression through the following areas:
1. Valves:
   a. Incorrect valve adjustment.
   b. Incorrect valve timing.
   c. Worn or damaged valve seats (valve and/or cylinder head).
   d. Bent valves.
   e. Weak or broken valve springs.

2. Cylinder head:
   a. Loose spark plug or damaged spark plug hole.
   b. Damaged cylinder head gasket.
   c. Warped or cracked cylinder head.

Engine Overheating (Cooling System)

WARNING
Do not remove the radiator cap, coolant drain plug or disconnect any coolant hose immediately after or during engine operation. Scalding fluid and steam may be blown out under pressure and cause serious injury. When the engine has been operated, the coolant is very hot and under pressure. Attempting to remove these items when the engine is hot can cause the coolant to spray violently from the radiator, water pump or hose, causing severe burns and injury on contact.

Test the electrical circuit of the cooling system as described in Chapter Ten.
1. Low coolant level.
2. Air in cooling system.
3. Clogged radiator, hose or engine coolant passages.
4. Thermostat stuck closed.
5. Worn or damaged radiator cap.
6. Open or short circuit in the cooling system wiring harness.
7. Damaged water pump.
8. Defective engine coolant temperature (ECT) sensor.
9. Damaged temperature gauge.
10. Radiator fans inoperative.

Engine Overheating (Engine)

1. Improper spark plug heat range.
2. Low oil level.
3. Oil not circulating properly.
4. Valves leaking.
5. Heavy engine carbon deposits in combustion chamber.
6. Dragging brake(s).
7. Clutch slipping.

Engine Temperature Too Low

1. Thermostat stuck open.
2. Defective engine coolant temperature (ECT) sensor.
3. Damaged temperature gauge.

Preignition

Preignition is the premature burning of fuel and is caused by hot spots in the combustion chambers. Glowing deposits in the combustion chambers, inadequate cooling or an overheated spark plug(s) can all cause preignition. This is first noticed as a power loss but eventually causes damage to the internal parts of the engine because of higher combustion chamber temperatures.

Detonation

Commonly called spark knock or fuel knock, detonation is the violent explosion of fuel in the combustion chamber before the proper time of ignition. Severe damage can result. Using low octane gasoline is a common cause of detonation.

Even when using a high octane gasoline, detonation can still occur. Other causes are over-advanced ignition timing, lean air/fuel mixture at or near full throttle, inadequate engine cooling, or the excessive accumulation of carbon deposits in the combustion chamber (cylinder head and piston crowns).

Power Loss

Refer to *Engine Performance* in this chapter.

Engine Noises

Unusual noises are often the first indication of a developing problem. Investigate any new noises as soon as possible. Something that may be a minor problem, if corrected, could prevent the possibility of more extensive damage later on.

Use a mechanic's stethoscope or a small section of hose held near your ear (not directly on your ear) with the other end close to the source of the noise to isolate the location. Determining the exact cause of a noise can be difficult. If this is the case, consult with a professional mechanic to determine the cause. Do not disassemble major components until all other possibilities have been eliminated.

Consider the following when troubleshooting engine noises:

1. A knocking or pinging during acceleration can be caused by using a lower octane fuel than recommended. May also be caused by poor fuel. Pinging can also be caused by an incorrect spark plug heat range or carbon buildup in the combustion chamber.

TROUBLESHOOTING

2. A slapping or rattling noise at low speed or during acceleration may be caused by excessive piston-to-cylinder wall clearance (piston slap).

NOTE
Piston slap is easier to detect when the engine is cold and before the pistons have expanded. Once the engine has warmed up, piston expansion reduces piston-to-cylinder clearance.

3. A knocking or rapping while decelerating is usually caused by excessive rod bearing clearance.
4. A persistent knocking and vibration occurring every crankshaft rotation is usually caused by a worn rod or main bearing(s). It can also be caused by broken piston rings or damaged piston pins.
5. A rapid on-off squeal may indicate a compression leak around cylinder head gasket or spark plug(s).
6. For valve train noises, check for the following:
   a. Excessive valve clearance.
   b. Worn or damaged camshaft.
   c. Damaged camshaft.
   d. Worn or damaged valve train components.
   e. Damaged valve lifter bore(s).
   f. Valve sticking in guide.
   g. Broken valve spring.
   h. Low oil pressure.
   i. Clogged cylinder oil hole or oil passage.

ENGINE LUBRICATION

An improperly operating engine lubrication system will quickly lead to engine seizure. Check the engine oil level and oil pressure as described in Chapter Three. Oil pump service is described in Chapter Five.

High Oil Consumption or Excessive Exhaust Smoke

1. Worn valve guides.
2. Worn or damaged piston rings.

Low Oil Pressure

1. Low oil level.
2. Worn or damaged oil pump.
3. Clogged oil strainer screen.
4. Clogged oil filter.
5. Internal oil leakage.
6. Incorrect type of engine oil.

High Oil Pressure

1. Oil pressure relief valve stuck closed.
2. Clogged oil filter.
3. Clogged oil gallery or metering orifices.

No Oil Pressure

1. Low oil level.
2. Oil pressure relief valve stuck closed.
3. Damaged oil pump.
4. Damaged oil pump sprocket(s) or chain.
5. Incorrect oil pump installation.
6. Internal oil leak.

Oil Pressure Indicator Stays On

1. Low oil pressure.
2. No oil pressure.
3. Damaged oil pressure switch.
4. Short circuit in warning indicator circuit.

Low Oil Level

1. Oil level not maintained at correct level.
2. Worn piston rings.
3. Worn cylinder.
4. Worn valve guides.
5. Worn valve stem seals.
6. Piston rings incorrectly installed during engine overhaul.
7. External oil leakage.
8. Oil leaking into the cooling system.

Oil Contamination

1. Head gasket allowing coolant to leak into the engine. Pressure test the cooling system and eliminate any external leaks, and/or perform the *Cylinder Leakdown Test* described in this chapter.
2. Oil and filter not changed at specified intervals or when operating conditions demand more frequent changes.

CYLINDER LEAKDOWN TEST

A cylinder leakdown test can locate engine problems from leaking valves, blown head gaskets or broken, worn or stuck piston rings. Performed this test by applying compressed air to the cylinder and then measuring the

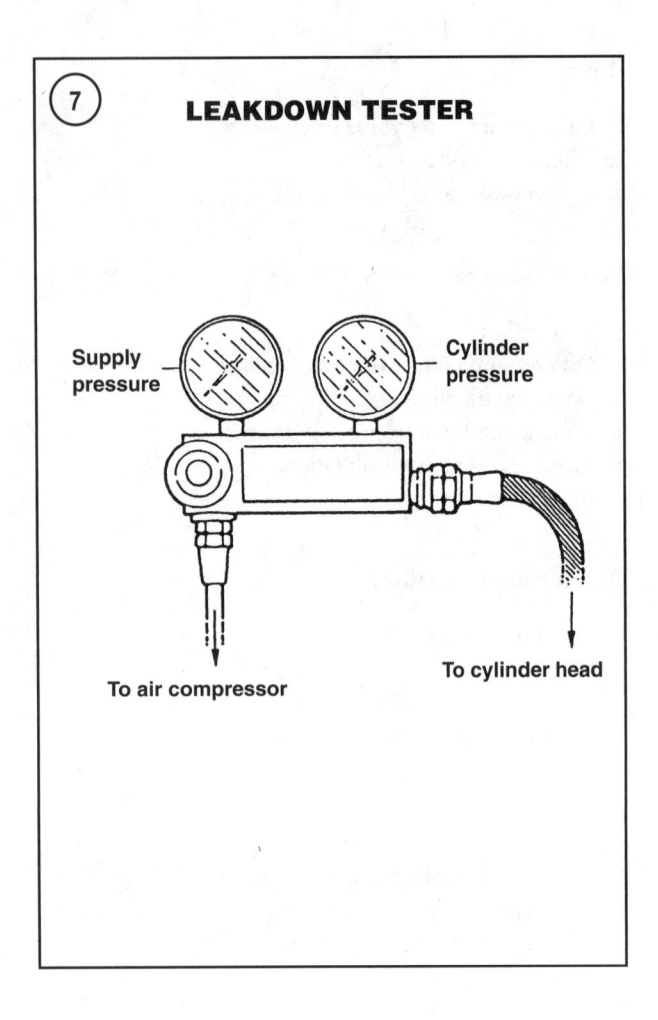

Figure 7 LEAKDOWN TESTER
- Supply pressure
- Cylinder pressure
- To air compressor
- To cylinder head

percent of leakage. Use a cylinder leakdown tester (**Figure 7**) and an air compressor to perform this test.

Follow the manufacturer's directions along with the following information when performing a cylinder leakdown test.

1. Run the engine until it is warm. Turn the engine off.

2. Remove the No. 1 cylinder spark plug as described in *Spark Plugs* in Chapter Three.

3. Set the No. 1 piston to TDC on its compression stroke. Refer to *Valve Clearance* in Chapter Three.

> **WARNING**
> *The crankshaft may rotate when compressed air is applied to the cylinder. Remove any tools attached to the end of the crankshaft. To prevent the engine from turning over as compressed air is applied to the cylinder, shift the transmission into fifth gear and have an assistant apply the rear brake.*

4. Thread the test adapter into the No. 1 spark plug hole. Connect the air compressor hose to the tester (**Figure 8**) following the manufacturer's instructions.

5. Apply compressed air to the tester and perform the test following the manufacturer's instructions. Read the leak rate percentage on the gauge, following the manufacturer's instructions. Note the following:
   a. For a new or rebuilt engine, a leakage rate of 0 to 5 percent per cylinder is desired. A leakage rate of 6 to 14 percent is acceptable and means the engine is in good condition.
   b. If testing a used engine, the critical rate is not the percent of leakage for each cylinder, but instead, the difference between the cylinders. On a used engine, a leakage rate of 10 percent or less between cylinders is satisfactory.
   c. A leakage rate exceeding 10 percent between cylinders points to an engine that is in poor condition and requires further inspection and possible repair.

6. After checking the leak rate percentage, and with air pressure still applied to the combustion chamber, listen for air escaping from the following areas. If necessary, use a mechanic's stethoscope to pinpoint the source.
   a. Air leaking through the exhaust pipe indicates a leaking exhaust valve.
   b. Air leaking through the throttle body indicates a leaking intake valve.
   c. Air leaking through the crankcase breather tube indicates worn piston rings or a worn cylinder bore.
   d. Air leaking into the cooling system causes coolant to bubble in the radiator. If this condition exists, check for damaged cylinder head gaskets and warped cylinder head or cylinder block surfaces.

7. Remove the tester and repeat these steps for each cylinder. Set each cylinder at TDC as described in *Valve Clearance* in Chapter Three.

8. After testing each cylinder, reinstall the spark plug.

CLUTCH

Basic clutch troubleshooting is listed in this section. Clutch service is covered in Chapter Six.

Clutch Lever Feels Soft or Spongy

1. Air in clutch hydraulic system.
2. Loose banjo bolts.
3. Damaged hoses.
4. Leaking clutch fluid.
5. Low clutch fluid level.

TROUBLESHOOTING

Clutch Lever Hard to Pull In

1. Sticking master cylinder piston.
2. Sticking slave cylinder piston.
3. Incorrect clutch assembly.
4. Damaged clutch lifter mechanism.
5. Damaged clutch lifter plate bearing.
6. Clogged clutch hydraulic system.

Rough Clutch Operation

1. Worn, grooved or damaged clutch hub and clutch housing slots.

Clutch Slippage

If the engine speed increases without an increase in motorcycle speed, the clutch is probably slipping. The main causes of clutch slippage are:
1. Sticking clutch hydraulic system.
2. Worn clutch plates.
3. Weak clutch spring.
4. Clutch plates contaminated by engine oil additive.
5. Clutch regulator valve stuck open.

Clutch Drag

If the clutch will not disengage or if the bike creeps with the transmission in gear and the clutch disengaged, the clutch is dragging. Some main causes of clutch drag are:
1. Air in hydraulic system.

NOTE
If different handlebars were installed, an extreme bar angle may affect the brake fluid level in the clutch master cylinder reservoir. Check the fluid level with the handlebar in both left and right lock positions.

2. Clogged hydraulic system.
3. Low clutch fluid level.
4. Warped clutch plates.
5. Damaged clutch lifter assembly.
6. Loose clutch housing locknut.
7. High oil level.
8. Incorrect oil viscosity.
9. Engine oil additive being used.
10. Damaged clutch hub and clutch housing splines.
11. Incorrect clutch lifter installation.

GEARSHIFT LINKAGE

The gearshift linkage assembly connects the shift pedal (external shift mechanism) to the shift drum (internal shift mechanism). See Chapter Seven to identify the components called out in this section.

Transmission Jumps Out of Gear

1. Damaged stopper arm.
2. Damaged stopper arm spring.
3. Worn or damaged shift drum cam.
4. Damaged gearshift arm spring.
5. Loose or damaged shift drum.
6. Bent shift fork shaft(s).
7. Bent or damaged shift fork(s).
8. Worn gear dogs or slots.

Difficult Shifting

1. Incorrect clutch operation.
2. Incorrect oil viscosity.
3. Loose or damaged stopper arm assembly.
4. Bent shift fork shaft(s).
5. Bent or damaged shift fork(s).
6. Worn gear dogs or slots.
7. Damaged shift drum grooves.
8. Damaged gearshift spindle.
9. Damaged joint arm.
10. Weak or damaged gearshift spindle B return spring.
11. Incorrect gearshift linkage installation.

Shift Pedal Does Not Return

1. Bent gearshift spindle.
2. Bent gearshift arm.

3. Weak or damaged gearshift arm return spring.
4. Shift shaft incorrectly installed (return spring incorrectly indexed around pin).

Excessive Engine/Transmission Noise

1. Damaged primary drive and driven gears or bearing.
2. Damaged alternator drive and driven gears or bearing.
3. Damaged final drive and driven gears or bearing.
4. Damaged transmission bearings or gears.

TRANSMISSION

Transmission symptoms are sometimes hard to distinguish from clutch symptoms. Basic transmission troubleshooting is listed below. Refer to Chapter Seven for transmission service procedures. Prior to working on the transmission, make sure the clutch and gearshift linkage assembly are not causing the problem.

Difficult Shifting

1. Incorrect clutch operation.
2. Bent shift fork(s).
3. Damaged shift fork guide pin(s).
4. Bent shift fork shaft(s).
5. Bent gearshift spindle.
6. Damaged shift drum grooves.

Jumps Out of Gear

1. Loose or damaged shift drum stopper arm.
2. Bent or damaged shift fork(s).
3. Bent shift fork shaft(s).
4. Damaged shift drum grooves.
5. Worn gear dogs or slots.
6. Weak or damaged gearshift arm return spring.

Incorrect Shift Lever Operation

1. Bent shift pedal or linkage.
2. Stripped shift pedal splines.
3. Damaged shift linkage.
4. Damaged gearshift spindle.
5. Damaged joint arm.

Excessive Gear Noise

1. Worn or damaged transmission bearings.
2. Worn or damaged gears.

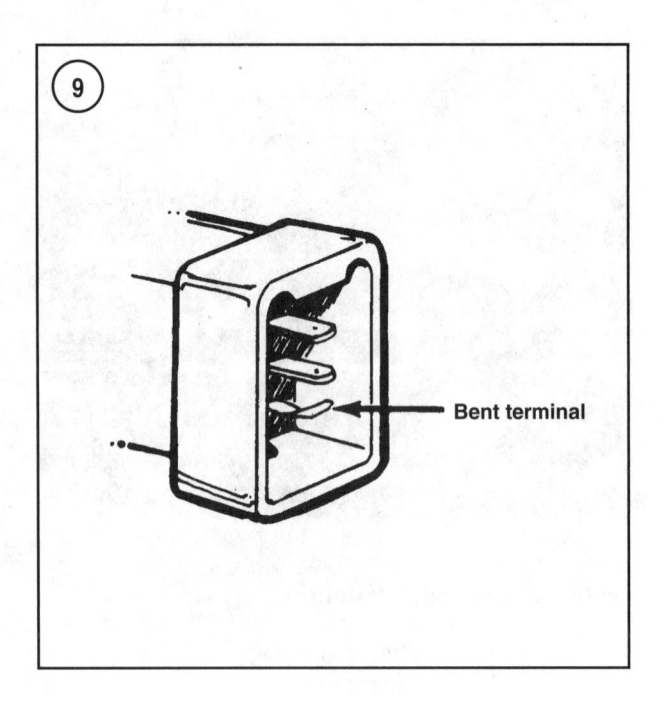

3. Excessive gear backlash.

ELECTRICAL TESTING

This section describes basic electrical troubleshooting and the use of test equipment. Refer to *Electrical System Fundamentals* in Chapter One. Refer to Chapter Nine for specific system test procedures.

Electrical troubleshooting can be very time consuming and frustrating without a plan. Refer to the color wiring diagrams at the end of the manual for component and connector identification. Use the wiring diagrams to determine how the circuit should work by tracing the current flow from the power source through the components to ground. Also, check any circuits that share the same fuse, ground or switch. If the other circuits work properly and the shared wiring is good, the cause must be in the wiring used only by the suspect circuit. If all related circuits are faulty at the same time, the probable cause is a poor ground connection or a blown fuse(s).

As with all troubleshooting procedures, analyze typical symptoms in a systematic manner. Never assume anything and do not overlook the obvious like a blown fuse or an electrical connector that has separated. Test the simplest and most obvious items first and try to make tests at easily accessible points on the bike.

TROUBLESHOOTING

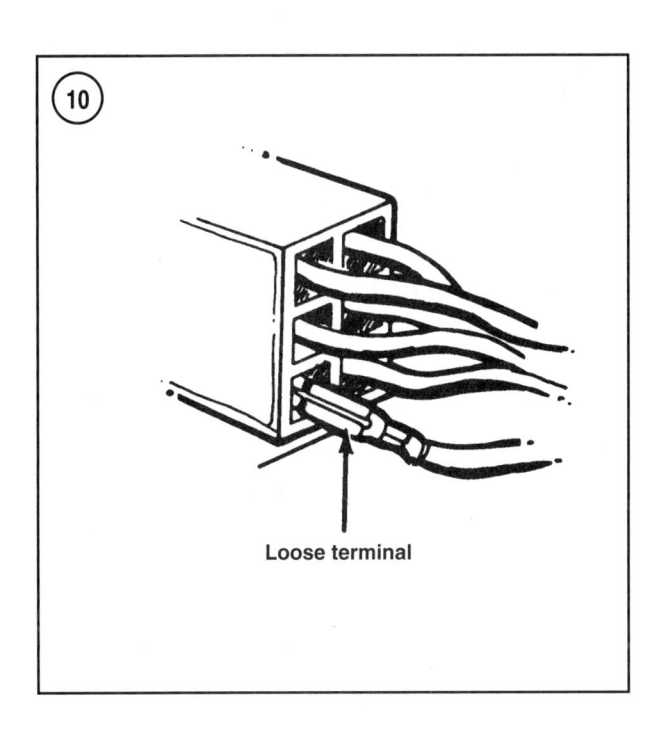

Loose terminal

Preliminary Checks and Precautions

Prior to starting any electrical troubleshooting, perform the following:

1. Check the main fuse (Chapter Nine). If the fuse is blown, replace it.
2. Check the individual fuses mounted in the fuse box (Chapter Nine). Inspect the suspected fuse, and replace it if blown.
3. Inspect the battery. Make sure it is fully charged, and that the battery leads are clean and securely attached to the battery terminals. Refer to *Battery* Chapter Nine.
4. Disconnect each electrical connector in the suspect circuit and make sure there are no bent terminals in the connector (**Figure 9**). A bent terminal will not connect to its mate, causing an open circuit.
5. Make sure the terminals are pushed all the way into the plastic housing (**Figure 10**). If not, carefully push them in with a narrow blade screwdriver.
6. Check the wires where they enter connector housings.
7. Make sure all electrical terminals are clean and free of corrosion. Clean them, if necessary, and pack the connectors with dielectric grease.
8. Push the connector halves together. Make sure the connectors are fully engaged and locked together.
9. Never pull the wires when disconnecting a connector—pull only on the connector housing.

NOTE
Always consider electrical connectors the weak link in the electrical system. Dirty, loose fitting and corroded connectors cause numerous electrical related problems, especially on high-mileage motorcycles. When troubleshooting an electrical problem, carefully inspect the connectors and wiring harness.

10. Never use a self-powered test light on circuits that contain solid-state devices. The solid-state devices may be damaged.

Intermittent Problems

To locate and repair intermittent problems, simulate the condition in the shop when testing the components. Note the following:

1. Vibration—This is a common problem with loose or damaged electrical connectors.
   a. Perform a continuity test as described in the appropriate service procedure or refer to *Continuity Testing* in this section.
   b. Lightly pull or wiggle the connectors while repeating the test. Do the same when checking the wiring harness and individual components, especially where the wires enter a housing or connector.
   c. A change in meter readings indicates a poor connection. Find and repair the problem or replace the part. Check for wires with cracked or broken insulation.

NOTE
An analog ohmmeter is useful when making this type of test. Slight needle movements are visibly apparent, and they indicate a loose connection.

2. Heat—This is a common problem with connectors or joints that have loose or poor connections. As these connections heat up, the connection or joint expands and separates, causing an open circuit. Other heat related problems occur when a component creates its own heat as it starts to fail.
   a. To check a connector, perform a continuity test as described in the appropriate service procedure, or refer to *Continuity Testing* in this section. Then repeat the test while heating the connector with a hair dryer. If the meter reading was normal (continuity) when the connector was cold, then fluctuated or read infinity when heat was applied, the connection is bad.
   b. To check a component, allow the engine to cool, then start and run the engine. Note operational differences when the engine is cold and hot.

c. If the engine does not start, isolate and remove the component. First test it at room temperature, and then after heating it with a hair dryer. A change in meter readings indicates a temperature problem.

CAUTION
A hair dryer will quickly raise the heat of the component being tested. Do not apply heat directly to the ECM or use heat in excess of 60° C (140° F) on any electrical component.

3. Water—When a problem occurs when riding in wet conditions, or in areas with high humidity, run the engine in a dry area. Then, with the engine running, spray water onto the suspected component. Often times, water related problems clear up after the component becomes hot enough to evaporate the moisture.

Test Light or Voltmeter

A test light can be constructed from a 12-volt light bulb with a pair of test leads soldered to the bulb. To check for battery voltage in a circuit, attach one lead to ground and the other lead to various points along the circuit. The bulb lights when battery voltage is present.

A voltmeter is used in the same manner as the test light to find out if battery voltage is present in any given circuit. The voltmeter, unlike the test light, also indicates how much voltage is present at each test point. When using a voltmeter, attach the positive lead to the component or wire to be checked and the negative lead to a good ground (**Figure 11**).

Ammeter

An ammeter measures the flow of current (amps) in a circuit (**Figure 12**). When connected in series in a circuit, the ammeter determines if current is flowing through the circuit and if that current flow is excessive because of a short in the circuit. Current flow is often referred to as current draw. Comparing actual current draw in the circuit or component to the manufacturer's specified current draw provides useful diagnostic information.

Self-Powered Test Light

A self-powered test light can be constructed from a 12-volt light bulb, a pair of test leads and a 12-volt battery. When the test leads are touched together, the light bulb should go on.

Use a self-powered test light as follows:

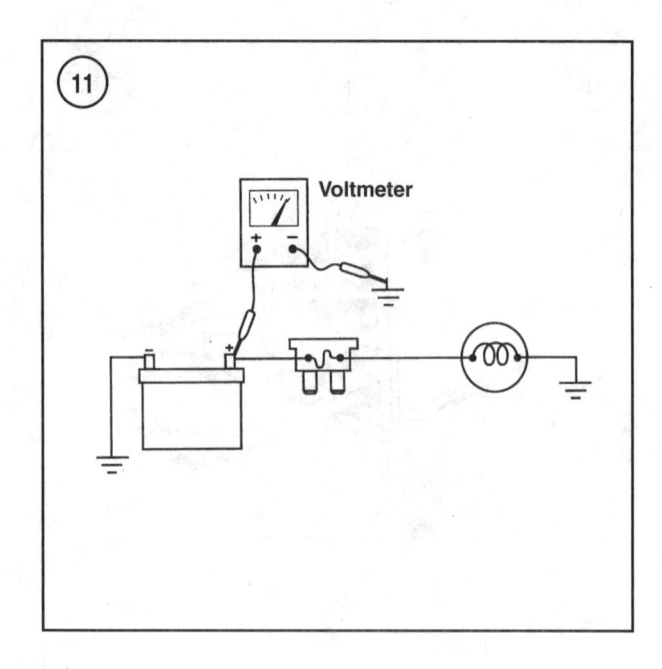

1. Touch the test leads together to make sure the light bulb goes on. If not, correct the problem before using it in a test procedure.
2. Disconnect the motorcycle's battery or remove the fuse(s) that protects the circuit to be tested.
3. Select two points within the circuit where there should be continuity.
4. Attach one lead of the self-powered test light to each point.
5. If there is continuity, the self-powered test light bulb will come on.
6. If there is no continuity, the self-powered test light bulb will not come on, indicating an open circuit.

Ohmmeter

An ohmmeter measures the resistance (in ohms) to current flow in a circuit or component. Like the self-powered test light, an ohmmeter contains its own power source and should not be connected to a live circuit.

Ohmmeters may be analog type (needle scale) or digital type (LCD or LED readout). Both types of ohmmeters have a switch that allows the user to select different ranges of resistance for accurate readings. The analog ohmmeter also has a set-adjust control which is used to zero or calibrate the meter (digital ohmmeters do not require calibration).

An ohmmeter is used by connecting its test leads to the terminals or leads of the circuit or component to be tested (**Figure 13**). If using an analog meter, calibrate by touching the test leads together and turning the set-adjust knob

TROUBLESHOOTING

resistance in the circuit or component being tested. If the meter needle falls between these two ends of the scale, this indicates the actual resistance to current flow that is present. To determine the resistance, multiply the meter reading by the ohmmeter scale. For example, a meter reading of 5 multiplied by the R × 1000 scale is 5000 ohms of resistance.

CAUTION
Never connect an ohmmeter to a circuit which has power applied to it. Always disconnect the battery negative lead before using an ohmmeter.

Jumper Wire

A jumper wire is a simple way to bypass a potential problem and isolate it to a particular point in a circuit. If a faulty circuit works properly with a jumper wire installed, an open circuit exists between the two jumper points in the circuit.

To troubleshoot with a jumper wire, first use the wire to determine if the problem is on the ground side or the load side of a device. Test the ground by connecting a jumper between the lamp and a good ground. If the lamp comes on, the problem is the connection between the lamp and ground. If the lamp does not come on with the jumper installed, the lamp's connection to ground is good so the problem is between the lamp and the power source.

To isolate the problem, connect the jumper between the battery and the lamp. If it comes on, the problem is between these two points. Next, connect the jumper between the battery and the fuse side of the switch. If the lamp comes on, the switch is good. By successively moving the jumper from one point to another, the problem can be isolated to a particular place in the circuit.

Pay attention to the following when using a jumper wire:

1. Make sure the jumper wire gauge (thickness) is the same as that used in the tested circuit. A smaller gauge wire could overheat and melt.
2. Install insulated boots over alligator clips. This prevents accidental grounding, sparks or possible shock when working in cramped quarters.
3. Jumper wires are temporary test measures only. Do not leave a jumper wire installed as a permanent solution. This creates a severe fire hazard that could easily lead to complete loss of the motorcycle.
4. When using a jumper wire, always install an inline fuse/fuse holder (available at most auto supply stores or electronic supply stores) to the jumper wire. Never use a jumper wire across any load (a component that is con-

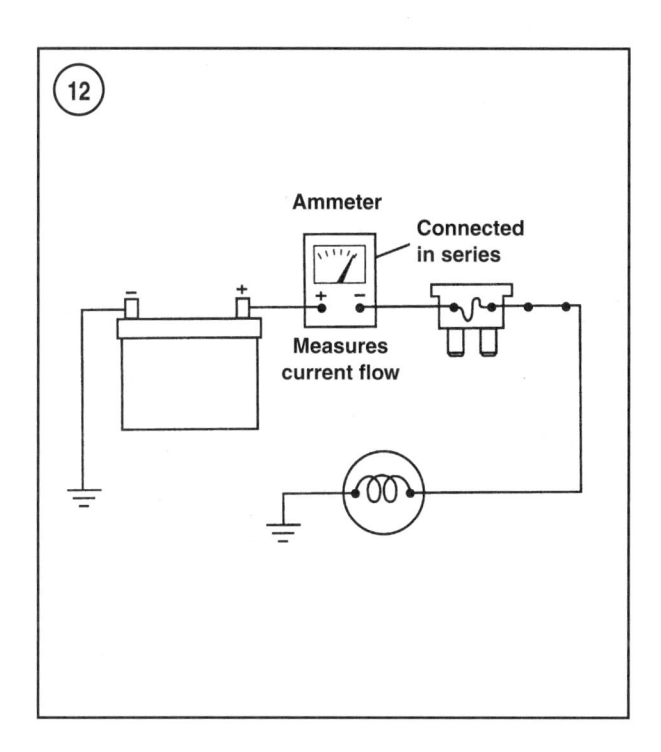

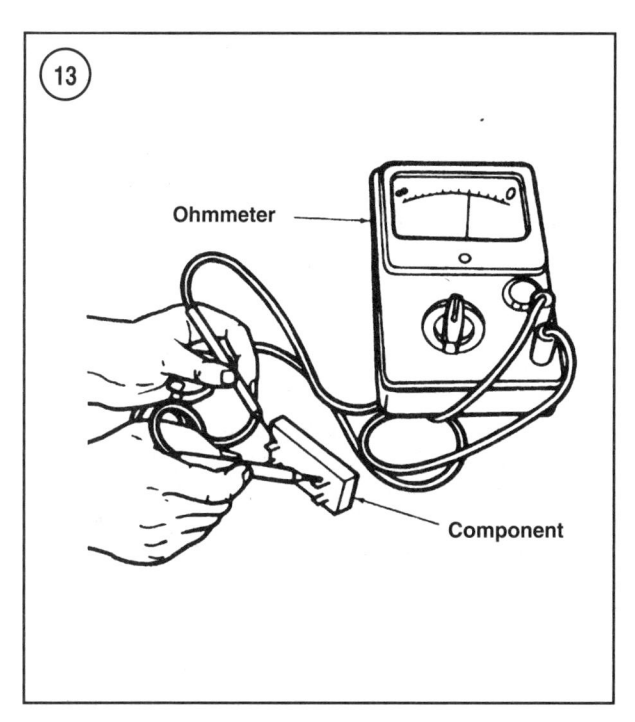

until the meter needle reads zero. When the leads are uncrossed, the needle should move to the other end of the scale indicating infinite resistance.

During a continuity test, a reading of infinity indicates that there is an open in the circuit or component. A reading of zero indicates continuity, that is, there is no measurable nected and turned on). This would result in a direct short and will blow the fuse(s).

Voltage Testing

Unless otherwise specified, make all voltage tests with the electrical connectors still connected. Insert the test leads into the backside of the connector and make sure the test lead touches the electrical wire or metal terminal within the connector housing. If the test lead only touches the wire insulation, there will be a false reading.

Always check both sides of the connector as one side may be loose or corroded thus preventing electrical flow through the connector. This type of test can be performed with a test light or a voltmeter. A voltmeter gives the best results.

NOTE
If using a test light, it does not make any difference which test lead is attached to ground.

1. Attach the voltmeter negative test lead to a good ground (bare metal). Make sure the part used for ground is not insulated with a rubber gasket or rubber grommet.
2. Attach the voltmeter positive test lead to the point (electrical connector, etc.) to be tested (**Figure 11**).
3. Turn the ignition switch on. If using a test light, the test light will come on if voltage is present. If using a voltmeter, note the voltage reading. The reading should be within 1 volt of battery voltage. If the voltage is less there is a problem in the circuit.

Voltage Drop Testing

The wires, cables, connectors and switches in the electrical circuit are designed to carry current with low resistance. This ensures that current can flow through the circuit with a minimum loss of voltage. Voltage drop indicates where there is resistance in a circuit. A higher than normal amount of resistance in a circuit decreases the flow of current and causes the voltage to drop between the source and destination in the circuit.

Because resistance causes voltage to drop, a voltmeter is used to measure voltage drop when current is running through the circuit. If the circuit has no resistance, there is no voltage drop so the voltmeter indicates 0 volts. The greater the resistance in a circuit, the greater the voltage drop reading.

To perform a voltage drop:
1. Connect the positive meter test lead to the electrical source (where electricity is coming from).

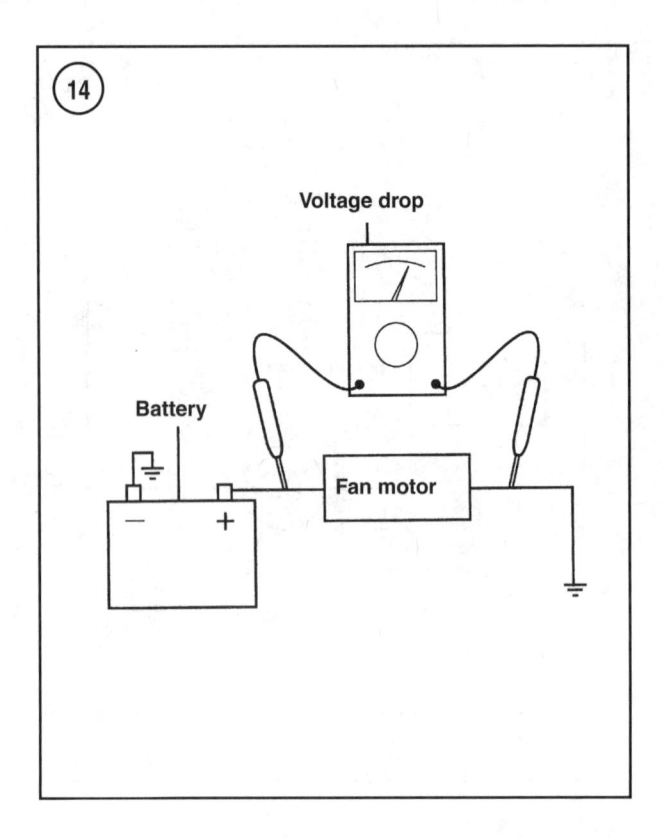

2. Connect the voltmeter negative test lead to the electrical load (where the electricity is going). See **Figure 14**.

3. If necessary, activate the component(s) in the circuit.

4. Read the voltage drop (difference in voltage between the source and destination) on the voltmeter. Note the following:

   a. The voltmeter should indicate 0 volts. If there is a drop of 1 volt or more, there is a problem within the circuit. A voltage drop reading of 12 volts indicates an open in the circuit.

   b. A voltage drop of 1 or more volts indicates that a circuit has excessive resistance.

   c. For example, consider a starting problem where the battery is fully charged but the starter turns over slowly. Voltage drop would be the difference in the voltage at the battery (source) and the voltage at the starter (destination) as the engine is being cranked (current is flowing through the battery cables). A corroded battery cable would cause a high voltage drop (high resistance) and slow engine cranking.

   d. Common sources of voltage drop are loose or contaminated connectors and poor ground connections.

TROUBLESHOOTING

Peak Voltage Testing

Peak voltage tests check the voltage output of the ignition coil and ignition pulse generator at normal cranking speed. These tests make it possible to identify ignition system problems quickly and accurately.

Peak voltage tests require a peak voltage adapter or tester. Refer to *Ignition System Troubleshooting* in Chapter Nine.

Continuity Testing

A continuity test is used to determine the integrity of a circuit, wire or component. A circuit has continuity if it forms a complete circuit; that is there are no opens in either the electrical wires or components within the circuit. An open circuit, on the other hand, has no continuity.

This type of test can be performed with a self-powered test light or an ohmmeter. An ohmmeter gives the best results. If using an analog ohmmeter, calibrate the meter by touching the leads together and turning the calibration knob until the meter reads zero.

1. Disconnect the negative battery cable.
2. Attach one test lead (test light or ohmmeter) to one end of the part of the circuit to be tested.
3. Attach the other test lead to the other end of the part or the circuit to be tested.
4. The self-powered test light comes on if there is continuity. An ohmmeter reads 0 or very low resistance if there is continuity. A reading of infinite resistance indicates no continuity; the circuit is open.

Testing For a Short with a Self-Powered Test Light or Ohmmeter

1. Disconnect the negative battery cable.
2. Remove the blown fuse from the fuse panel.
3. Connect one test lead of the test light or ohmmeter to the load side (battery side) of the fuse terminal in the fuse panel.
4. Connect the other test lead to a good ground (bare metal). Make sure the part used for a ground is not insulated with a rubber gasket or rubber grommet.
5. With the self-powered test light or ohmmeter attached to the fuse terminal and ground, wiggle the wiring harness relating to the suspect circuit at 6 in. (15.2 cm) intervals. Start next to the fuse panel and work systematically away from the fuse panel. Watch the self-powered test light or ohmmeter while progressing along the harness.
6. If the test light blinks or the needle on the ohmmeter moves, there is a short-to-ground at that point in the harness.

Testing For a Short with a Test Light or Voltmeter

1. Remove the blown fuse from the fuse panel.
2. Connect the test light or voltmeter across the fuse terminals in the fuse panel. Turn the ignition switch on and check for battery voltage.
3. With the test light or voltmeter attached to the fuse terminals, wiggle the wiring harness relating to the suspect circuit at 6 in. (15.2 cm) intervals. Start next to the fuse panel and work systematically away from the panel. Watch the test light or voltmeter while progressing along the harness.
4. If the test light blinks or if the needle on the voltmeter moves, there is a short-to-ground at that point in the harness.

FINAL DRIVE

Oil Leaks

1. Clogged breather.
2. Oil level too high.
3. Loose or missing case cover mounting bolts.
4. Damaged oil seal(s).

Excessive Noise

1. Oil level too low.

NOTE
Check the oil level as described in Chapter Three. If the oil level is low, check the assembly for an oil leak.

2. Worn or damaged pinion and ring gears.
3. Worn or scored pinion and splines.
4. Excessive backlash between ring and pinion gears.

FRONT SUSPENSION AND STEERING

Steering is Sluggish

1. Incorrect steering stem adjustment (too tight).
2. Damaged steering head bearings.
3. Low tire pressure.

Bike Steers to One Side

1. Bent axle.
2. Bent frame.
3. Worn or damaged wheel bearings.

4. Worn or damaged swing arm pivot bearings.
5. Damaged steering head bearings.
6. Bent swing arm.
7. Incorrectly installed wheels.
8. Front and rear wheels are not aligned.
9. Front fork legs positioned unevenly in steering stem.
10. Damaged tire.
11. Loose or damaged fairing mounts.

Front Suspension Noise

1. Loose mounting fasteners.
2. Damaged fork leg(s) or rear shock absorber.
3. Low fork oil capacity.
4. Loose or damaged fairing mounts.

Front Wheel Wobble/Vibration

1. Loose front wheel axle.
2. Loose or damaged wheel bearing(s).
3. Damaged wheel rim(s).
4. Damaged tire(s).
5. Flat spot on tire.

NOTE
If the motorcycle is put in storage for a considerable amount of time, the weight placed on the front tire can cause the tire to flatten on the spot resting against the floor.

6. Unbalanced tire and wheel assembly.
7. Loose or damaged front faring mounts.

Hard Suspension
(Front Fork)

1. Excessive tire pressure.
2. Damaged steering head bearings.
3. Incorrect steering head bearing adjustment.
4. Bent fork tubes.
5. Binding slider.

NOTE
If a fork brace was installed onto the fork tubes, make sure it was installed correctly.

6. Incorrect weight fork oil.
7. Fork oil level too high.
8. Plugged fork oil passage.
9. Plugged anti-dive orifice.

Hard Suspension
(Rear Shock Absorber)

1. Excessive rear tire pressure.
2. Bent damper rod.
3. Incorrect shock adjustment.
4. Damaged shock absorber bushing(s).
5. Damaged shock absorber bearing.
6. Damaged swing arm pivot bearings.

Soft Suspension
(Front Fork)

1. Insufficient tire pressure.
2. Insufficient fork oil level or fluid capacity.
3. Incorrect oil viscosity.
4. Weak or damaged fork springs.

Soft Suspension
(Rear Shock Absorber)

1. Insufficient rear tire pressure.
2. Weak or damaged shock absorber spring.
3. Damaged shock absorber.
4. Incorrect shock absorber adjustment.
5. Leaking damper unit.

BRAKE SYSTEM

The GL1800 (standard model) is equipped with a linked brake system (LBS). This system is designed to operate both the front and rear brakes when either the front brake lever or the rear brake pedal is applied. The LBS system is a hydraulic system—no electronic controls are used. See Chapter Fourteen for information on servicing the LBS.

The GL1800A (ABS) model is equipped with a linked brake system (LBS) and anti-lock brake system (ABS). The ABS system is an electronically controlled hydraulic system designed to prevent wheel lockup during hard braking or when braking on slippery and loose road surfaces. When the wheel is about to lock, the ABS modulates the hydraulic pressure in the system by reducing pressure at the brake calipers. When the system senses that the wheel lock condition is reduced, full hydraulic pressure to the calipers is restored. Hydraulic pressure is regulated continuously. See Chapter Fifteen for information on servicing the ABS.

On the GL1800A, the LBS and ABS are integrated systems. That is, the same hydraulic system components (master cylinders, brake calipers and ABS hydraulic brake lines) are used for both systems. A number of

TROUBLESHOOTING

ABS indicator

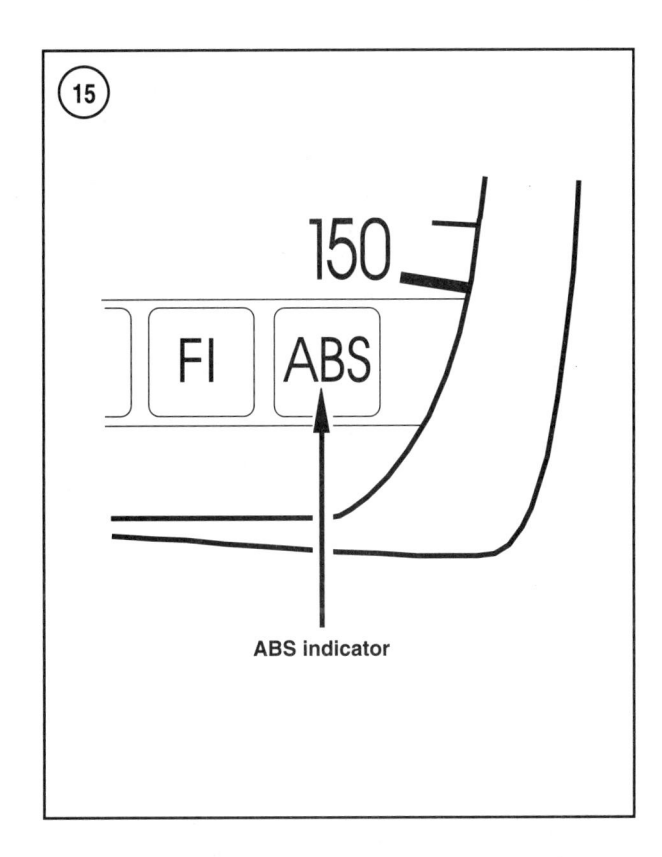

anti-lock components are installed in the brake system to provide anti-lock braking. During normal braking operations, the LBS system provides the braking and operates like the LBS system used on the standard GL1800 model.

The front and rear brake units are critical to riding performance and safety. Inspect the front and rear brakes frequently and repair any problem immediately. When replacing or refilling the disc brake fluid, use only DOT 4 brake fluid from a closed container. See Chapter Three for additional information on brake fluid selection and routine brake inspection and service.

When checking brake pad wear, check that the brake pads in each caliper contact the disc squarely. If one of the brake pads is wearing unevenly, a warped or bent brake disc or damaged caliper could be the cause.

Always check the brake operation before riding the motorcycle.

ABS Indicator Light

The ABS system is programmed with a self-diagnostic capability. Self-diagnosis starts when the ignition switch is turned on. It is indicated when the red ABS indicator on the speedometer panel comes on, and ends when the motorcycle speed reaches 10 kilometers per hour (6 mph), which is indicated by the ABS indicator turning off.

When the ABS control unit (ECU) detects a fault in the ABS system, the ABS indicator (**Figure 15**) will flash or stay on to alert the rider. A diagnostic trouble code (DTC) is also set in the ECU memory. When the ABS indicator is flashing or stays on, the ABS function is disabled. However, even when the ABS is disabled, the LBS system (both front and rear brakes) still operates normally.

NOTE
*If the ABS indicator does not turn on when the ignition switch is turned on, perform the **Power/Ground Circuit Test** under **Combination Meter** in Chapter Nine.*

The ABS indicator may flash or stay on when the motorcycle is used or ridden under the following conditions:

1. The ABS indicator may blink if the rear wheel is turned when the motorcycle is supported on its centerstand with the ignition switch on.

2. The motorcycle is ridden continuously on a stretch of rough or bumpy roads.

3. When the motorcycle is ridden through an area with a strong electromagnetic interference, the ECU may set a DTC and disable the ABS. When the ABS is disabled in this manner, erase the DTC and perform the self-diagnosis as described in Chapter Fifteen. If the ABS indicator goes out during the self-diagnosis, the ABS function has returned and the system is operating correctly.

4. If the ABS indicator comes on when riding, perform the following:

   a. Ride the motorcycle to a safe area away from all traffic and turn the ignition switch OFF.

   b. Restart the engine. The ABS indicator should light and stay on until the motorcycle is ridden. If the ABS indicator flashes or stays on, there is a problem with the ABS and the system has been disabled. Retrieve the DTC and service the ABS system as described in Chapter Fifteen.

Troubleshooting (Linked Brake System)

The following sections troubleshoot common hydraulic and mechanical problems with the linked brake system (LBS) used on the GL1800 and GL1800A models. If the ABS indicator (**Figure 15**) flashes or stays on (GL1800A models), there is a problem with the ABS and the system has been disabled. Retrieve the DTC and service the ABS system as described in Chapter Fifteen.

Soft or Spongy Brake Lever or Pedal

Quickly operate the front brake lever or rear brake pedal repeatedly and check to see if the lever/pedal travel distance increases. If the lever/pedal travel does increase while being operated, or feels soft or spongy, there may be air in the brake lines. In this condition, the brake system is not capable of producing sufficient brake force. When there is an increase in lever/pedal travel or when the brake feels soft or spongy, check the following possible causes:

1. Air in system.

NOTE
If the brake level in the reservoir drops too low, air can enter the hydraulic system through the master cylinder. Air can also enter the system from loose or damaged hose fittings. Air in the hydraulic system causes a soft or spongy brake lever or pedal action. This condition is noticeable and reduces brake performance. When it is suspected that air has entered the hydraulic system, flush the brake system and bleed the brakes as described in Chapter Fourteen.

NOTE
If different handlebars were installed, an extreme bar angle may affect the brake fluid level in the reservoir. Check the fluid level with the handlebar in both left and right lock positions.

2. Low brake fluid level.

NOTE
As the brake pads wear, the brake fluid level in the master cylinder reservoir drops. Whenever adding brake fluid to the reservoirs, visually check the brake pads for wear. If it does not appear that there is an increase in pad wear, check the brake hoses, lines and banjo bolts for leaks.

3. Leak in the brake system.
4. Contaminated brake fluid.
5. Plugged brake fluid passages.
6. Damaged brake lever or pedal assembly.
7. Worn or damaged brake pads.
8. Worn or damaged brake disc.
9. Warped brake disc.
10. Contaminated brake pads and disc.

NOTE
A leaking fork seal can allow oil to contaminate the brake pads and disc.

11. Worn or damaged master cylinder cups and/or cylinder bore.
12. Worn or damaged brake caliper piston seals.
13. Contaminated master cylinder assembly.
14. Contaminated brake caliper assembly.
15. Brake caliper not sliding correctly on slide pins.
16. Sticking master cylinder piston assembly.
17. Sticking brake caliper pistons.

NOTE
If the brake system does not operate correctly and nothing was discovered during the previous checks, carefully test ride the bike and determine if the motorcycle front-end dive under braking is excessive.

Brake Drag

When the brakes drag, the brake pads are not capable of moving away from the brake disc when the brake lever or pedal is released. Any of the following causes, if they occur, would prevent correct brake pad movement and cause brake drag.

1. Warped or damaged brake disc.
2. Brake caliper not sliding correctly on slide pins.
3. Sticking or damaged brake caliper pistons.
4. Contaminated brake pads and disc.
5. Plugged master cylinder port.
6. Contaminated brake fluid and hydraulic passages.
7. Restricted brake hose joint.
8. Loose brake disc mounting bolts.
9. Damaged or misaligned wheel.
10. Incorrect wheel alignment.
11. Incorrectly installed brake caliper.
12. Damaged front wheel.
13. Damaged proportional control valve (rear brake).
14. Incorrect secondary master cylinder push rod length (rear brake).

Hard Brake Lever or Pedal Operation

When applying the brakes and there is sufficient brake performance but the operation of the brake lever or pedal feels excessively hard, check for the following possible causes:

1. Clogged brake hydraulic system.
2. Sticking caliper piston.
3. Sticking master cylinder piston.
4. Glazed or worn brake pads.
5. Mismatched brake pads.
6. Damaged front brake lever.
7. Damaged rear brake pedal.

TROUBLESHOOTING

8. Brake caliper not sliding correctly on slide pins.
9. Worn or damaged brake caliper seals.

Brakes Grab

1. Damaged brake pad pin bolt. Look for steps or cracks along the pad pin bolt surface.
2. Contaminated brake pads and disc.
3. Incorrect wheel alignment.
4. Warped brake disc.
5. Loose brake disc mounting bolts.
6. Brake caliper not sliding correctly on slide pins.
7. Mismatched brake pads.
8. Damaged wheel bearings.

Brake Squeal or Chatter

1. Contaminated brake pads and disc.
2. Incorrectly installed brake caliper.
3. Warped brake disc.
4. Incorrect wheel alignment.
5. Mismatched brake pads.
6. Incorrectly installed brake pads.

Leaking Brake Caliper

1. Damaged dust and piston seals.
2. Damaged cylinder bore.
3. Loose caliper body bolts.
4. Loose banjo bolt.
5. Damaged banjo bolt washers.
6. Damaged banjo bolt threads in caliper body.

Leaking Master Cylinder

1. Damaged piston secondary seal.
2. Damaged piston snap ring or snap ring groove.
3. Worn or damaged master cylinder bore.
4. Loose banjo bolt.
5. Damaged banjo bolt washers.
6. Damaged banjo bolt threads in master cylinder body.
7. Loose or damaged reservoir cap.

Excessive Fork Drive

1. Faulty secondary master cylinder or related system component.

CRUISE CONTROL

Refer to Chapter Sixteen.

CHAPTER THREE

LUBRICATION, MAINTENANCE AND TUNE-UP

This chapter describes lubrication, maintenance and tune-up procedures required for the Honda GL1800.

To maximize the service life of the motorcycle and gain the utmost in safety and performance, it is necessary to perform periodic inspections and maintenance. Minor problems found during routine service can be corrected before they develop into major ones. A neglected motorcycle will be unreliable and may be dangerous to ride.

Table 1 lists the recommended lubrication, maintenance and tune-up intervals. When operating the motorcycle in extreme conditions, it may be appropriate to reduce the time interval between some maintenance items.

For convenience, most of the services listed in **Table 1** are described in this chapter. Procedures that require more than minor disassembly or adjustment are covered in the appropriate chapter.

Before servicing the motorcycle, make sure the procedures and the required skills are thoroughly understood. If your experience and equipment are limited, start by performing basic procedures. Perform more involved tasks after gaining further experience and acquiring the necessary tools.

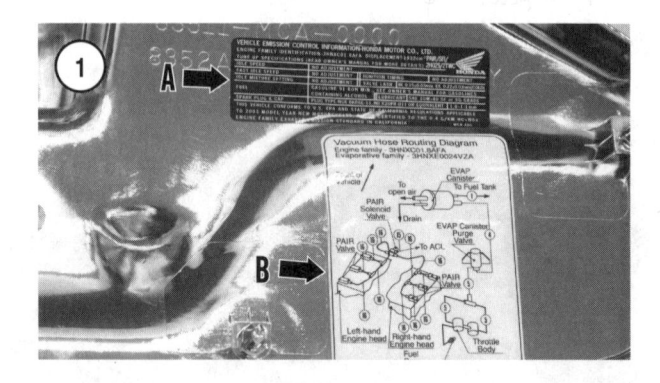

FUEL

The engine requires gasoline with a pump octane number of 86 or higher. Using a gasoline with a lower octane number can cause pinging or spark knock, leading to engine damage.

When choosing gasoline and filling the fuel tank, note the following:

1. When filling the tank, do not overfill it. There should be no fuel in the filler neck (tube located between the fuel cap and tank).

LUBRICATION, MAINTENANCE AND TUNE-UP

CYLINDER IDENTIFICATION

Firing order: 1-4-5-2-3-6

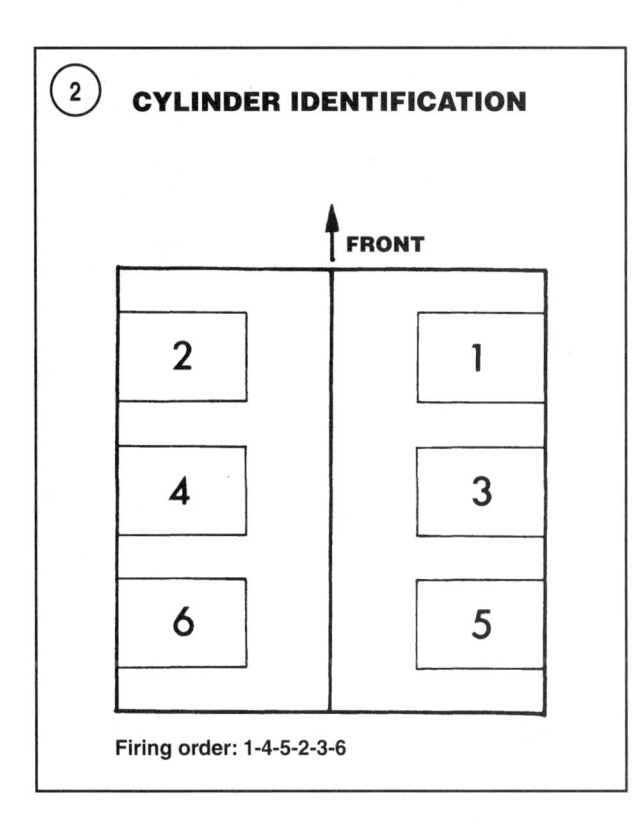

2. Because oxygenated fuels can damage plastic and paint, make sure not to spill fuel onto the top shelter during filling. Wipe up spills with a soft cloth. If using oxygenated fuel, make sure it meets the minimum octane requirements.

3. An ethanol (ethyl or grain alcohol) gasoline that contains more than 10 percent ethanol by volume may cause engine starting and performance related problems.

4. A methanol (methyl or wood alcohol) gasoline that contains more than 5 percent methanol by volume may cause engine starting and performance related problems. Gasoline that contains methanol must have corrosion inhibitors to protect the metal, plastic and rubber parts in the fuel system from damage.

TUNE-UP

A complete tune-up restores performance lost due to normal wear and deterioration of engine parts. Perform the engine tune-up procedures at the mileage intervals specified in **Table 1**. Note that some items require service every two years or at the indicated mileage, whichever comes first. More frequent tune-ups may be required if the bike is operated primarily in stop-and-go traffic or in areas were there is a large amount of blowing dirt and dust.

The Vehicle Emission Control Information label attached to the inside of the right engine side cover (A, **Figure 1**) lists tune-up information. Refer to **Table 2** for tune-up specifications. The vacuum hose routing diagram (B, **Figure 1**) identifies emission control system hoses.

NOTE
*If the specifications on the Vehicle Emission Control Information label differ from those in **Table 2**, use those on the label.*

To perform a tune-up, service the following items as described in this chapter:
1. Air filter.
2. Spark plugs.
3. Engine compression.
4. Ignition timing.
5. Valve clearance.
6. Engine oil and filter.
7. Cooling system.
8. Final drive unit.
9. Brake system.
10. Tires.
11. Suspension components.
12. Fasteners.

CYLINDER IDENTIFICATION AND FIRING ORDER

Refer to **Figure 2** for the cylinder number identification. The cylinder firing order is 1-4-5-2-3-6.

AIR FILTER

The air filter removes dust and abrasive particles from the air before the air enters the engine. A clogged air filter decreases the efficiency and life of the engine. With a damaged air filter, very fine particles could enter the engine and cause rapid wear of the piston rings, cylinder and bearings. Never run the bike without the air filter element installed.

Replace the air filter element at the service intervals specified in **Table 1**.

NOTE
*The service intervals specified in **Table 1** should be followed with general use. However, replace the air filter more often if dusty areas are frequently encountered.*

Replacement

1. Remove the top shelter. Refer to *Top Shelter* in Chapter Seventeen.

2. Remove the screw and the left air duct (**Figure 3**).
3. Remove the screw and the right air duct (**Figure 4**).

NOTE
*To disconnect the connectors in Step 4 and Step 5, press the locking tab on the bottom of each connector and pull on the connector. **Figure 5** shows the locking tab on the gray 22-pin ECM harness connector.*

4. Disconnect the black 26-pin (A, **Figure 6**) and gray 26-pin (B) harness connectors from the cruise/reverse control module.
5. Disconnect the following harness connectors at the ECM:
   a. Gray 22-pin (A, **Figure 7**).
   b. Black 6-pin (B, **Figure 7**).
   c. Black 22-pin (C, **Figure 7**).
6. Disconnect the BARO sensor harness connector (A, **Figure 8**). Press the tab on top of the connector to disconnect.
7. Disconnect the black 6-pin cruise actuator harness connector (B, **Figure 8**). Press the tab on bottom of the connector to disconnect.
8. Disconnect the connector at the ignition switch (**Figure 9**) by inserting a small screwdriver between the front side of the metal connector bracket and the connector. Pry

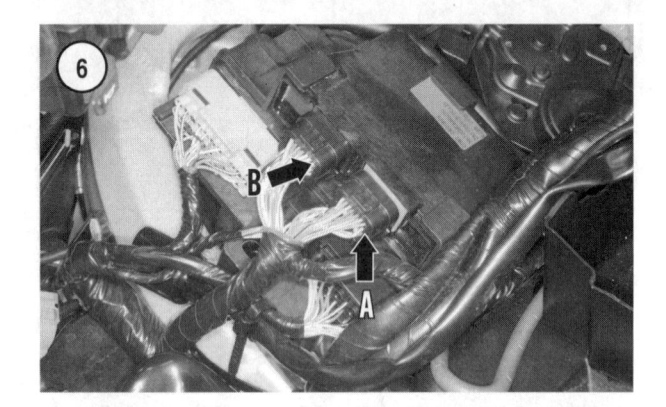

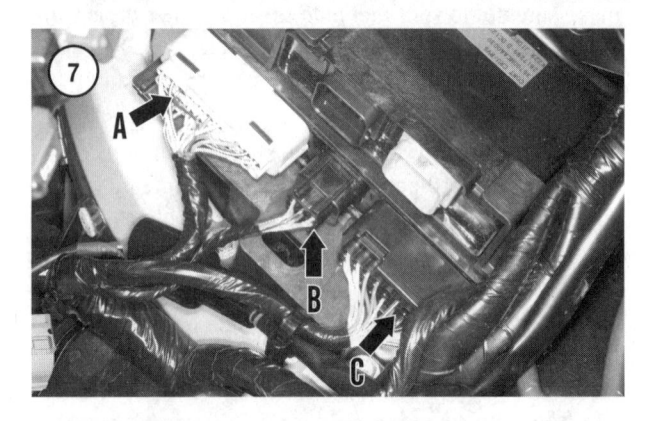

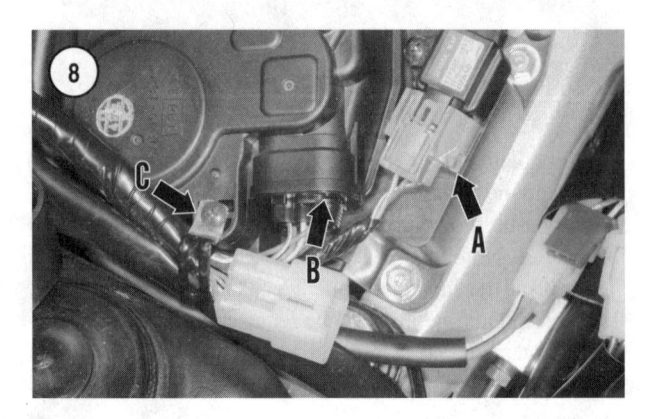

LUBRICATION, MAINTENANCE AND TUNE-UP

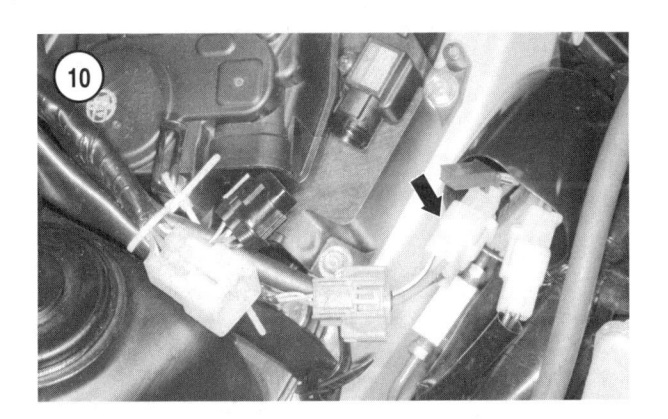

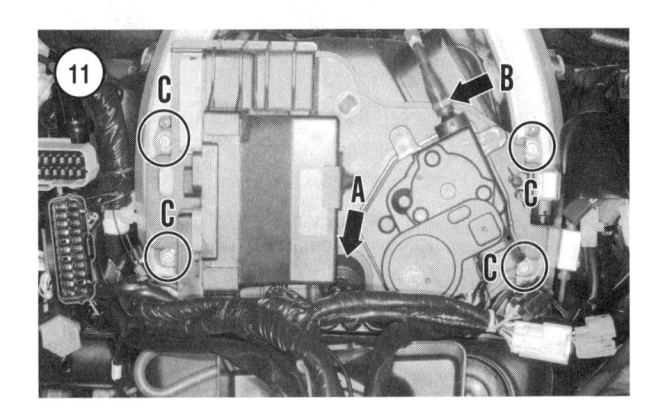

the screwdriver forward to unlock and remove the connector. Move the wiring harness assembly over the fuel tank.

9. Raise the harness guide (C, **Figure 8**) and position the wire harness group out of the way. Wrap the group with a colored plastic tie to help identify them for reassembly.

10. Disconnect the ignition switch white 4-pin connector (**Figure 10**).

11. Disconnect the black 2-pin intake air temperature (IAT) sensor harness connector (A, **Figure 11**). Press the lock tab to disconnect.

CAUTION
*The cruise cable identified in B, **Figure 11** is permanently fixed to the cruise actuator. Do not attempt to disconnect it.*

CAUTION
Handle the ECU carefully when removing it in Step 12.

12. Remove the four mounting bolts (C, **Figure 11**) and remove the control module holder. Set the holder aside. See **Figure 12**. Because of the cruise cable, the holder cannot be removed from the motorcycle.

CAUTION
The pins in the control modules are exposed. Handle the assembly carefully to avoid damaging them.

NOTE
A lip at the top of each filter cover screw bore prevents the screws from falling out.

13. Loosen the screws and remove the air filter cover (**Figure 13**).

14. Remove the air filter (A, **Figure 14**).

NOTE
*The air filter element contains a dust adhesive. Do not clean the filter element (**Figure 15**)*

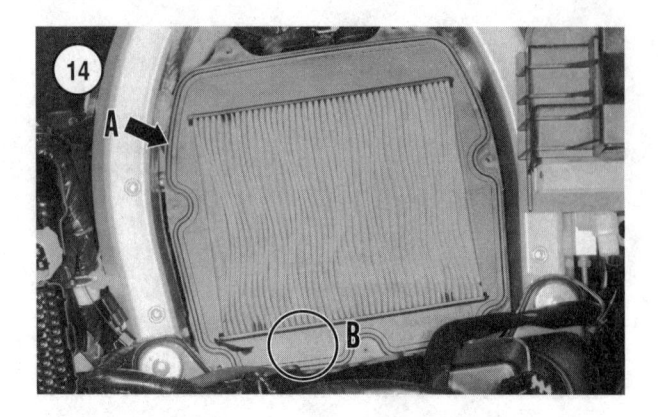

with air or any type of chemical cleaner or water.

15. Inspect the air filter element for excessive dirt, debris and possible damage. Do not run the motorcycle with a damaged air filter element or attempt to clean it. It may allow dirt to enter the engine. If the element is in good condition and it has not been in use longer than the indicated service interval, return it to service.

16. Inspect the air box for dirt and debris that may have passed through the element. Wipe the inside of the air box with a clean cloth.

NOTE
The seals used in the air box cover and air box grooves are not available separately. Handle the seals carefully to avoid damaging them.

17. Make sure the rubber seal seats fully in the air box cover (**Figure 16**) and air box grooves.
18. Install the air filter with the paper side facing up. Check that the filter seats flush in the air box with its UP mark (B, **Figure 14**) facing up.
19. Install the air box cover (**Figure 13**) and secure with the mounting screws.
20. Install the module unit holder (**Figure 12**) over the air filter cover. Do not twist the cruise control cable (B, **Figure 11**) when repositioning the control unit holder. The ECU connector terminals must face toward the left side. Install the four shoulder bolts (C, **Figure 11**) and tighten securely.
21. Reverse Steps 1-11 to reconnect the connectors and complete installation. Record the service interval in the maintenance log at the end of this manual.

ENGINE COMPRESSION TEST

A cranking compression test is one of the quickest ways to check the internal condition of the engine (piston rings, pistons, head gasket, valves and cylinders). It is a good idea to check compression at each tune-up, record it and compare it with the reading obtained at the next tune-up.

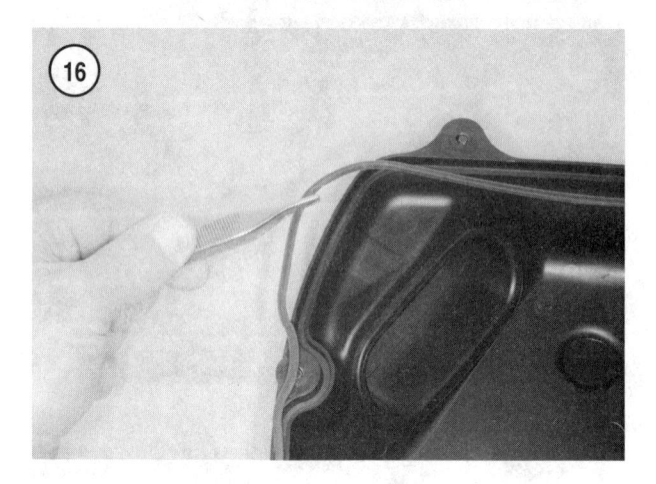

Use the spark plug tool included in the bike's tool kit and a screw-in type compression gauge with a flexible adapter (**Figure 17**). Before using the gauge, check that the rubber gasket on the end of the adapter is not cracked or damaged; this gasket seals the cylinder to ensure accurate compression readings.

1. Make sure the battery is fully charged to ensure proper engine cranking speed (300 rpm).

2. Run the engine until it reaches normal operating temperature, then turn it off.

3. Remove the seat. Refer to *Seat* in Chapter Seventeen.

4. Pull the cover part way off the fuel pump and disconnect the gray 5-pin fuel pump connector (**Figure 18**).

WARNING
Step 4 must be performed to disable the fuel injection system. Otherwise, fuel will be injected into the cylinders when the engine is turned over during the compression test,

LUBRICATION, MAINTENANCE AND TUNE-UP

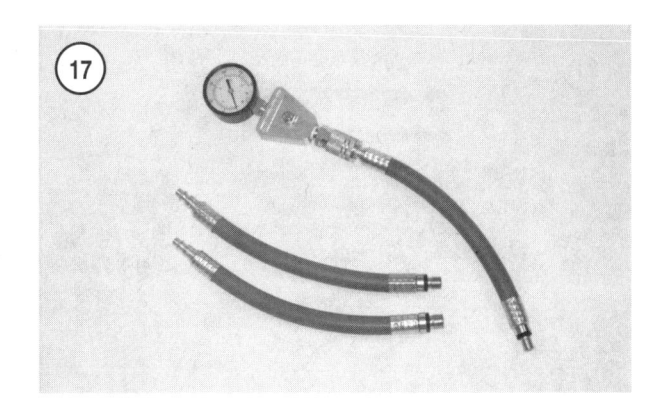

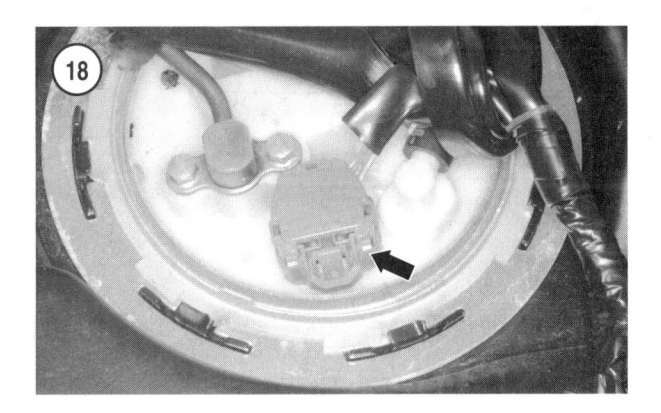

flooding the cylinders and creating explosive fuel vapors.

5. Remove the spark plugs as described under *Spark Plugs* in this chapter.

6. Lubricate the threads of the compression gauge adapter with a *small* amount of antiseize compound and carefully thread the gauge into one of the spark plug holes (**Figure 19**). Tighten the hose by hand to form a good seal.

CAUTION
Do not crank the engine more than necessary. When spark plug leads are disconnected, the electronic ignition produces the highest voltage possible and the coils may overheat and become damaged. Ground each plug cap with a grounding tool.

7. Turn the engine stop switch to RUN, then turn the ignition switch on. *Open the throttle completely* and using the starter, crank the engine over while reading the compression gauge until there is no further rise in pressure. The compression reading should increase on each stroke. Maximum pressure is should be reached within 4-7 seconds of engine cranking. Record the reading.

NOTE
If a cylinder requires a longer cranking time to reach its maximum compression reading, there is a problem with that cylinder.

8. Repeat for each cylinder.

9. When interpreting the results, note the individual readings and the difference between the readings. Standard compression pressure is specified in **Table 2**. Low compression indicates worn or broken rings, leaking or sticky valves, a blown head gasket or a combination of all three. Readings that are lower than normal, but are relatively even among all the cylinders indicates piston, ring and cylinder wear. Note the following:

   a. If the compression readings do not differ between cylinders by more than 10 percent, the rings and valves are in good condition.
   b. If a low reading (10 percent or more) is obtained on one of the cylinders, it indicates valve or ring trouble. To determine which, perform a wet compression test. Pour about a teaspoon of engine oil into the spark plug hole. Repeat the compression test and record the reading. If the compression increases significantly, the valves are good but the rings are defective on that cylinder. If compression does not increase, the valves require servicing.

NOTE
If the compression is low, the engine cannot be tuned to maximum performance.

10. Reverse Steps 1-6 to complete installation. Reinstall the spark plugs and caps as described under *Spark Plugs* in this chapter.

SPARK PLUGS

Inspect and replace the spark plugs at the service intervals specified in **Table 1**.

Removal

When properly read, a spark plug can reveal the operating condition of its cylinder. When removing the spark plugs, label each one with its cylinder number (**Figure 2**).
1. Remove the rubber plugs, bolts and cylinder head side covers (**Figure 20**).
2. Carefully disconnect the spark plug caps from the spark plugs (**Figure 21**). The plug cap forms a tight seal on the spark plug. Grasp the plug cap and twist it to break it loose from the plug.
3. Clean the spark plugs and the area around the plugs with compressed air.

CAUTION
Whenever a spark plug is removed, dirt around it can fall into the plug hole. This can cause serious engine damage.

4. Install the spark plug socket onto the spark plug. Make sure it is correctly seated on the plug, then loosen and remove the spark plug.
5. Inspect the spark plugs for a broken center porcelain, excessively eroded electrodes and excessive carbon or oil fouling. Replace such plugs. Refer to *Inspection* in this section.
6. Inspect the spark plug caps and spark plugs wires for cracks, hardness and other damage.

Gap

Carefully gap new and used plugs to ensure a reliable, consistent spark. To do this, use a spark plug gapping tool with a wire gauge.
1. Remove the new plugs from the box. If removed, install and tighten the terminal nut (A, **Figure 22**) on the end of the plug.
2. Insert a round feeler gauge between the center and the side electrode of the plug (**Figure 23**). The correct gap is listed in **Table 2**. If the gap is correct, there will be a slight drag as the wire is pulled through. If there is no drag, or if the wire will not pass through, bend the side electrode with a gapping tool (**Figure 24**) and set the gap to specification.
3. Repeat for each plug.

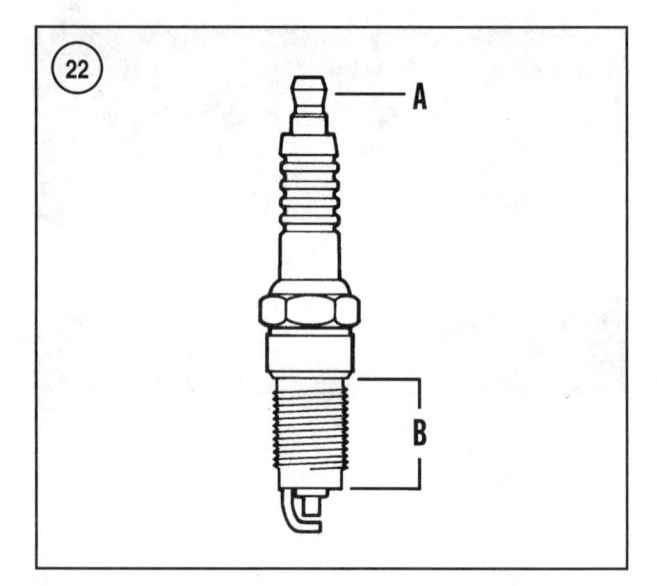

Installation

1. Apply a *light* coat of antiseize compound onto the threads of the spark plug before installing it. Remove any compound that contacts the plug's firing tip. Do not use engine oil on the plug threads.

CAUTION
The cylinder head is aluminum. The spark plug threads can be easily damaged by cross threading the spark plug.

LUBRICATION, MAINTENANCE AND TUNE-UP

Heat Range

Spark plugs are available in various heat ranges, hotter or colder than the plugs originally installed by the manufacturer.

Select a plug with a heat range designed for the loads and conditions under which the bike will be operated. A plug with an incorrect heat range can foul, overheat and cause piston damage.

In general, use a hot plug for low speeds and low temperatures. Use a cold plug for high speeds, high engine loads and high temperatures. The plug should operate hot enough to burn off unwanted deposits, but not so hot that it becomes damaged or causes preignition. To determine if the plug heat range is correct, remove each spark plug and examine the insulator.

Do not change the spark plug heat range to compensate for adverse engine or air/fuel mixture conditions.

When replacing plugs, make sure the reach (B, **Figure 22**) is correct. A shorter than standard plug will reduce engine performance and allow carbon to build on the exposed cylinder head threads, interfering with the installation of the correct reach plugs.

Refer to **Table 2** for recommended spark plugs.

Inspection

Reading the spark plugs can provide a significant amount of information regarding engine performance. Reading plugs that have been used will give an indication of spark plug operation, air/fuel mixture composition and engine condition (oil consumption, pistons, etc.). Before checking new spark plugs, operate the motorcycle under a medium load for approximately 10 km (6 miles). Avoid prolonged idling before shutting off the engine. Remove the spark plugs as described in this chapter. Examine each plug and compare it to those in **Figure 25**.

Normal condition

If the plug has a light tan- or gray-colored deposit and no abnormal gap wear or erosion, good engine, fuel system and ignition condition are indicated. The plug is the proper heat range and may be serviced and returned to use.

Carbon fouled

Soft, dry, sooty deposits covering the entire firing end of the plug are evidence of incomplete combustion. Even though the firing end of the plug is dry, the plug's insula-

2. Screw the spark plug in by hand until it seats. Very little effort is required. If force is necessary, the plug may be cross-threaded. Unscrew it and try again.

3. Tighten the spark plug to 18 N•m (13 ft.-lb.). If a torque wrench is not available, tighten it 1/2 turn after the gasket contacts the head.

CAUTION
Do not overtighten the spark plug. This will crush the gasket and destroy its sealing ability. It may also damage the spark plug threads in the cylinder head.

NOTE
The original equipment spark plug leads are marked with a number indicating their respective cylinder number.

4. Install the spark plug cap onto the correct spark plug. Push the cap down hard so it seats onto the spark plug.

5. Repeat for each spark plug.

6. Reinstall the cylinder head side covers (**Figure 20**).

25 SPARK PLUG CONDITIONS

Normal | Carbon fouled | Oil fouled
Gap bridged | Overheated | Sustained preignition

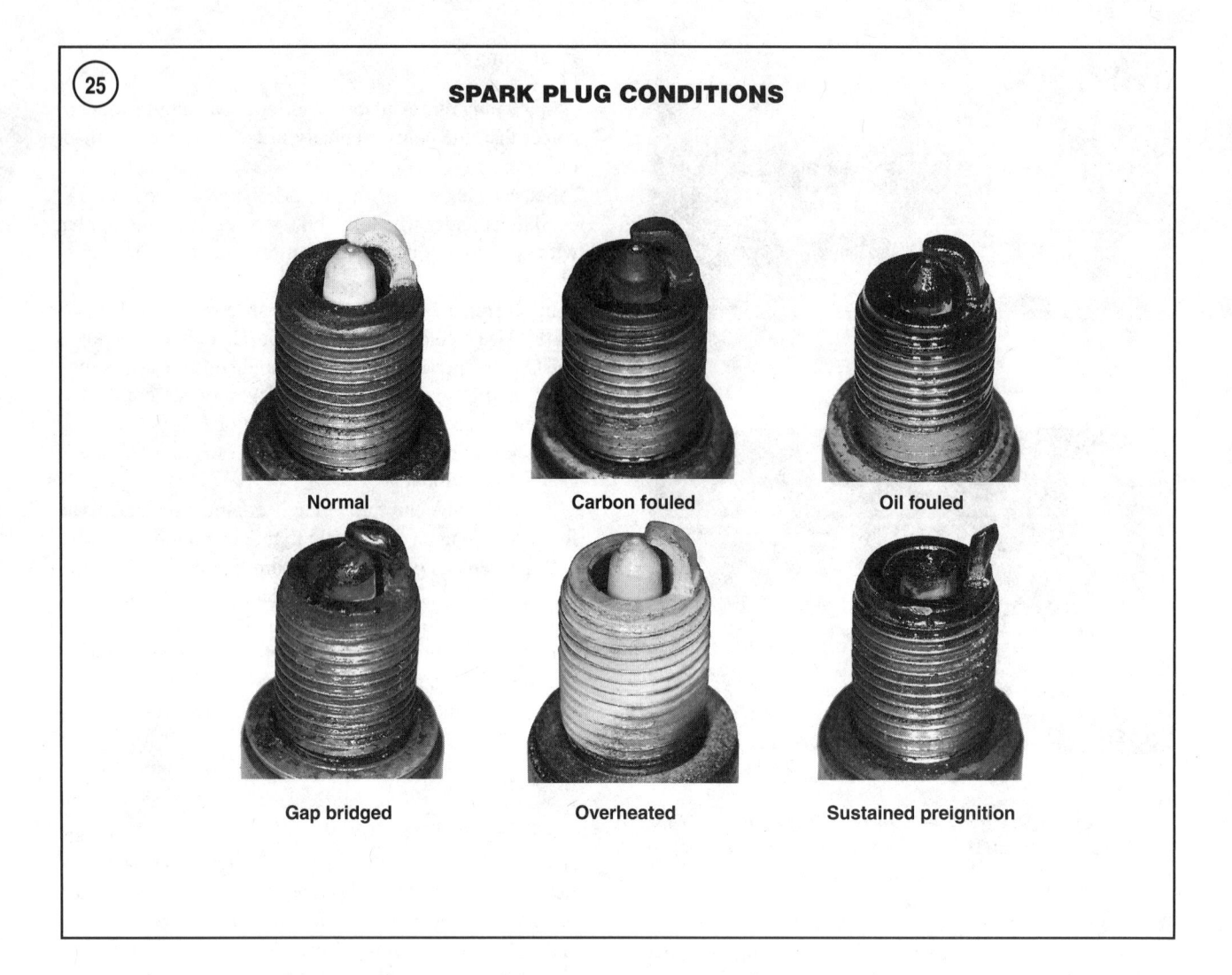

tion decreases when in this condition. The carbon forms an electrical path that bypasses the spark plug electrodes causing a misfire condition. One or more of the following can cause carbon fouling:
1. Fuel mixture too rich.
2. Spark plug heat range too cold.
3. Clogged air filter.
4. Improperly operating ignition component.
5. Ignition component failure.
6. Low engine compression.
7. Prolonged idling.

Oil fouled

The tip of an oil fouled plug has a black insulator tip, a damp oily film over the firing end and a carbon layer over the entire nose. The electrodes are not worn. Common causes for this condition are:

1. Faulty fuel injection system.
2. Low idle speed or prolonged idling.
3. Ignition component failure.
4. Spark plug heat range too cold.
5. Engine still being broken in.
6. Valve guides worn.
7. Piston rings worn or broken.

It is important to correct the cause of fouling before returning the engine to service.

Gap bridging

Plugs with this condition exhibit gaps shorted out by combustion deposits between the electrodes. If this condition is encountered, check for an improper oil type or excessive carbon in the combustion chamber. Make sure to locate and correct the cause of this condition.

LUBRICATION, MAINTENANCE AND TUNE-UP

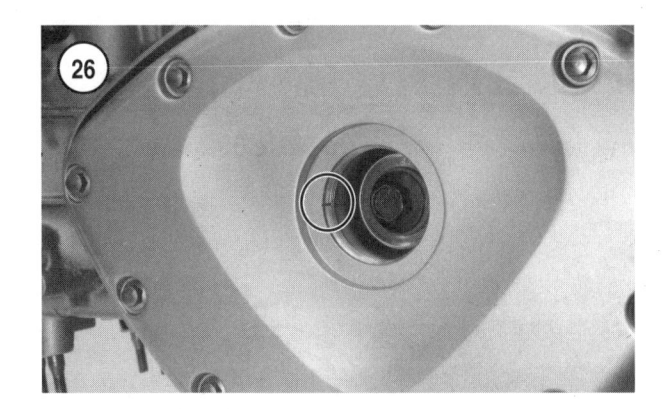

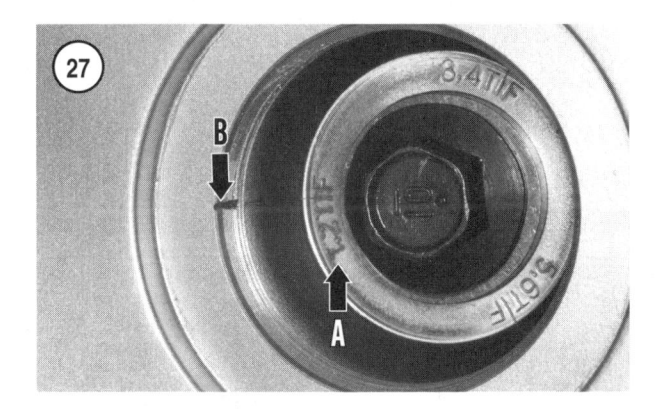

Overheating

Badly worn electrodes and premature gap wear are signs of overheating, along with a gray or white blistered porcelain insulator surface. The most common cause for this condition is using a spark plug of the wrong heat range (too hot). If the spark plug has not recently been changed to a hotter one, but the plug is overheated, consider the following causes:

1. Faulty fuel injection operation.
2. Improperly operating ignition component.
3. Engine lubrication system malfunction.
4. Engine air leak.
5. Improper spark plug installation (overtightening).
6. No spark plug gasket.

Worn out

This happens when corrosive gasses formed by combustion and high voltage sparks have eroded the electrodes. A worn out spark plug requires more voltage to fire under hard acceleration. Replace with a new spark plug.

Preignition

If the electrodes are melted, preignition is usually the cause. Check for throttle body mounting or intake manifold leaks and advanced ignition timing. It is also possible that a plug of the wrong heat range (too hot) is being used. Find the cause of the preignition before returning the engine into service.

IGNITION TIMING

The engine is equipped with a fully transistorized ignition system. Ignition timing is controlled by the ECM and is not adjustable. However, the ignition timing can be checked to make sure all components are operating correctly. Because of the solid-state design, problems with the ignition system are rare. If an ignition related problem is suspected, check the ignition timing to confirm the condition of the ignition system.

Incorrect ignition timing can cause a drastic loss of engine performance. It may also cause overheating.

WARNING
Do not start and run the motorcycle in an enclosed area. The exhaust gasses contain carbon monoxide, a colorless, odorless, poisonous gas. Carbon monoxide levels build quickly in an enclosed area and can cause unconsciousness and death in a short time.

1. Start the engine and let it reach normal operating temperature. Shut the engine off.
2. Remove the rubber plugs, bolts and either cylinder head side cover (**Figure 20**).
3. Remove the crankshaft hole cap and its O-ring to access the index mark (**Figure 26**).
4. Connect a timing light to the No. 1 or No. 2 spark plug wire (**Figure 2**) following the manufacturer's instructions.
5. Start the engine and allow it to run at idle speed.
6. Check the ignition timing as follows:
    a. Aim the timing light at the timing hole and pull the trigger. The ignition timing is correct if the ignition pulse generator rotor 1.2 T|F mark (A, **Figure 27**) aligns with the index mark on the front crankcase cover (B). See **Figure 28**. Turn the engine off.
    b. Connect the timing light to the No. 3 or No. 4 spark plug wire (**Figure 2**). Start the engine and recheck the ignition timing. The ignition timing is correct if the ignition pulse generator rotor 3.4 T|F mark (**Figure 28**) aligns with the index mark on the front crankcase rotor cover. Turn the engine off.

TIMING MARKS

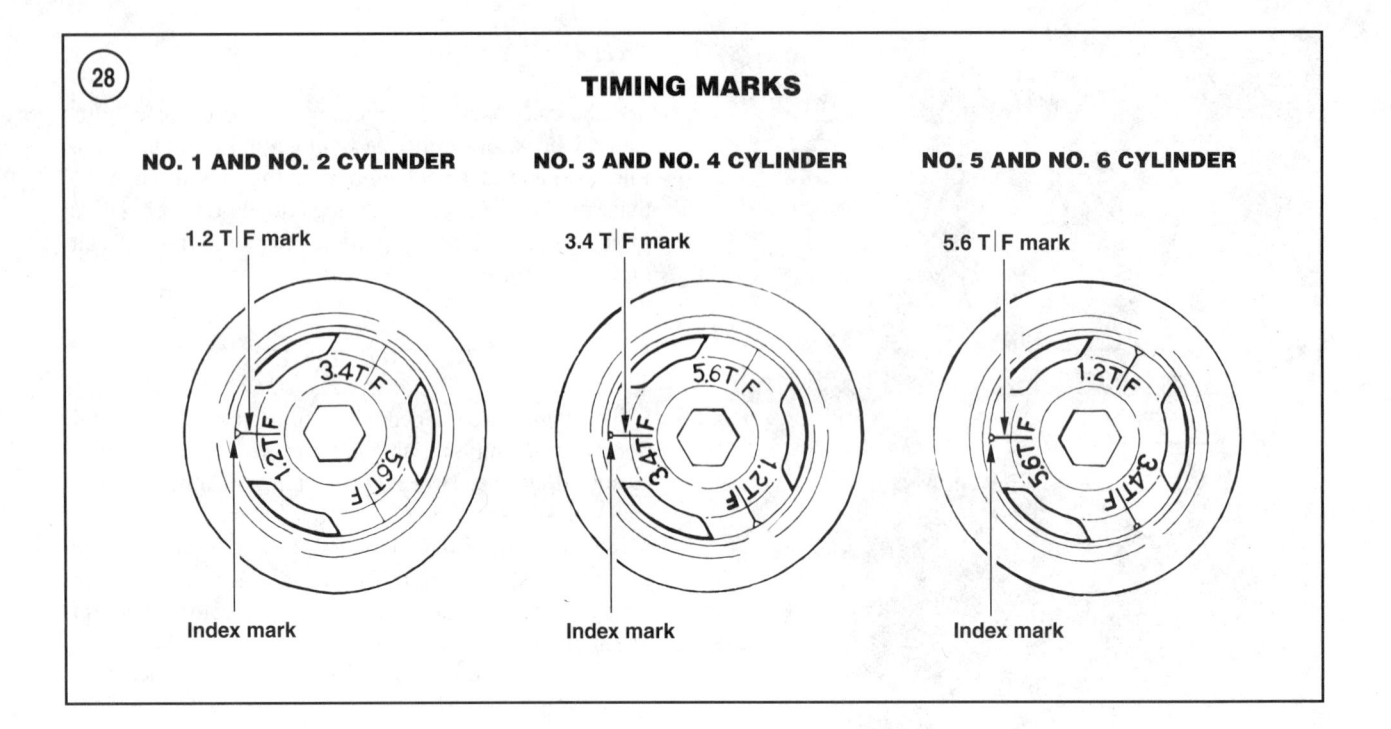

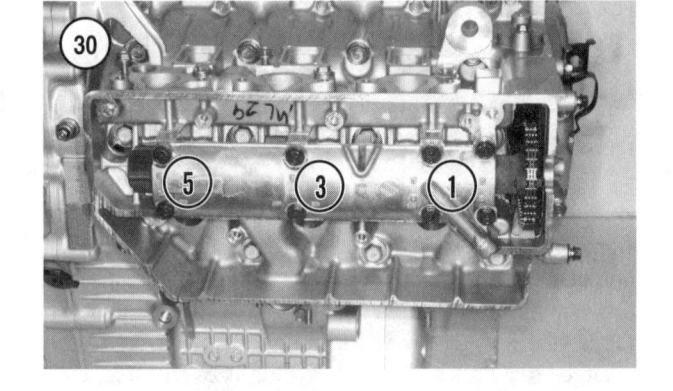

NO. 1 AND NO. 2 CYLINDER — 1.2 T|F mark, Index mark

NO. 3 AND NO. 4 CYLINDER — 3.4 T|F mark, Index mark

NO. 5 AND NO. 6 CYLINDER — 5.6 T|F mark, Index mark

c. Connect the timing light to the No. 5 or No. 6 spark plug wire (**Figure 2**). Start the engine and recheck the ignition timing. The ignition timing is correct if the ignition pulse generator rotor 5.6 T|F mark (**Figure 28**) aligns with the index mark on the front crankcase cover.

7. Turn the engine off and disconnect the timing light.

8. If the timing is incorrect, there is a problem with one or more ignition system components; refer to *Ignition System Troubleshooting* in Chapter Nine. There is no method of adjusting ignition timing.

9. Lubricate the timing hole cap threads and O-ring with grease and tighten to 18 N•m (13 ft.-lb.).

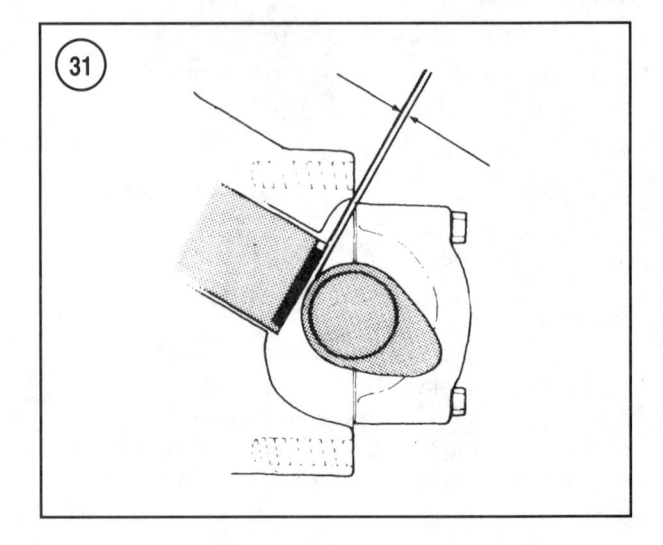

LUBRICATION, MAINTENANCE AND TUNE-UP

(32) CAMSHAFT SHIM CHART

← FRONT

RIGHT CYLINDER HEAD						
	Cylinder No. 1		Cylinder No. 3		Cylinder No. 5	
	Intake	Exhaust	Intake	Exhaust	Intake	Exhaust
Clearance						
Shim No.						

LEFT CYLINDER HEAD						
	Cylinder No. 2		Cylinder No. 4		Cylinder No. 6	
	Intake	Exhaust	Intake	Exhaust	Intake	Exhaust
Clearance						
Shim No.						

VALVE CLEARANCE

Measurement

The correct valve clearance for all models is listed in **Table 2**.

Adjust the valves in the firing order: 1-4-5-2-3-6.

NOTE
Valve clearance measurement and adjustment must be performed with the engine cold (below 35° C [95° F]).

1. Remove the cylinder head covers. Refer to *Cylinder Head Covers* in Chapter Four.
2. Loosen and remove the timing hole cap and O-ring (**Figure 26**).
3. Remove all the spark plugs as described in this chapter. This makes it easier to turn the engine by hand.
4. Loosen the cam chain tensioners as described under *Camshaft, Removal* in Chapter Four. This reduces the force the cam chains apply to the camshafts and ensures accurate measurements.
5. Turn the crankshaft (**Figure 29**) counterclockwise and align the 1.2T | F mark on the ignition pulse generator rotor (A, **Figure 27**) with the index mark on the front crankcase cover (B). See **Figure 28**. Confirm that the No.1 cylinder (**Figure 30**) camshaft lobes are facing out. If the No. 1 camshaft lobes are facing inward, rotate the crankshaft counterclockwise 360° (1 full turn) and realign the 1.2T | F mark.
6. With the engine in this position, check the clearance of the No. 1 cylinder intake and exhaust valves. Check the clearance by inserting a flat feeler gauge between the cam lobe and the valve lifter (**Figure 31**). When the clearance is correct, there will be a slight drag on the feeler gauge when it is inserted and withdrawn. Record the clearance and cylinder number and whether it is an intake or exhaust valve (**Figure 32**). The clearance dimensions will be used during the adjustment procedure if valve adjustment is necessary.
7. Turn the crankshaft counterclockwise 120° and align the 3.4T | F mark on the ignition pulse generator rotor with the index mark on the front crankcase cover (**Figure 28**). Check the clearance of the No. 4 cylinder (**Figure 33**) intake and exhaust valves. Record the clearance (**Figure 32**).

8. Turn the crankshaft counterclockwise 120° and align the 5.6T|F mark on the ignition pulse generator rotor with the index mark on the front crankcase cover (**Figure 28**). Check the clearance of the No. 5 cylinder (**Figure 30**) intake and exhaust valves. Record the clearance (**Figure 32**).
9. Turn the crankshaft counterclockwise 120° and align the 1.2T|F mark on the ignition pulse generator rotor with the index mark on the front crankcase cover (**Figure 28**). Check the clearance of the No. 2 cylinder (**Figure 33**) intake and exhaust valves. Record the clearance (**Figure 32**).
10. Turn the crankshaft counterclockwise 120° and align the 3.4T|F mark on the ignition pulse generator rotor with the index mark on the front crankcase cover (**Figure 28**). Check the clearance of the No. 3 cylinder (**Figure 30**) intake and exhaust valves. Record the clearance (**Figure 32**).
11. Turn the crankshaft counterclockwise 120° and align the 5.6T|F mark on the ignition pulse generator rotor with the index mark on the front crankcase cover (**Figure 28**). Check the clearance of the No. 6 cylinder (**Figure 33**) intake and exhaust valves. Record the clearance (**Figure 32**).
12. If any of the valves require adjustment, follow the adjustment procedure described below.
13. If all the valve clearances are correct, continue with Step 14.
14. Remove the tensioner tools or the lock plate as described under *Camshaft, Installation* in Chapter Four.
15. Reinstall the spark plugs as described in this chapter.
16. Install the cylinder head covers as described in Chapter Four.
17. Lubricate the timing hole cap threads and O-ring with grease and tighten to 18 N•m (13 ft.-lb.).

Adjustment

To adjust the valve clearance, replace the shim located under the valve lifter with a shim of a different thickness. The camshaft(s) must be removed to gain access to the shims. The shims are available from Honda dealerships in thickness increments of 0.025 mm that range from 1.200 to 2.800 mm in thickness.
1. Remove the camshafts. Refer to *Camshafts* in Chapter Four.
2. Before removing the valve lifters and shims, note the following:
    a. Identify and store the valve lifters and shims so they can be installed in their original locations. This step is critical to ensure correct valve clearance adjustment and reassembly.
    b. The shims are located under the valve lifters and may stick to the valve lifters when removing the lifter. Remove the lifter carefully to avoid dropping the shim. Use a magnet to remove the lifter and shims.
    c. Clean the valve lifters and shims in solvent and dry with compressed air.
    d. Use the chart in **Figure 32** to keep track of the shim numbers and measured valve clearances.
3. Remove the valve lifter (**Figure 34**) and shim (**Figure 35**) for each valve to be adjusted. Label and store each lifter and shim assembly in a marked container.

NOTE
*Always measure the thickness of the old shim with a micrometer to confirm the exact thickness. If the shim is worn to less than the marked indicated thickness (**Figure 36**), the calculations for a new shim will be inaccurate.*

4. Measure the thickness of the old shim (**Figure 37**) with a micrometer.
5. Using the measured valve clearance, the specified valve clearance (**Table 2**) and the old shim thickness, determine the new shim thickness with the following equation:
$a = (b - c) + d$
Where:
a is the new shim thickness.

LUBRICATION, MAINTENANCE AND TUNE-UP

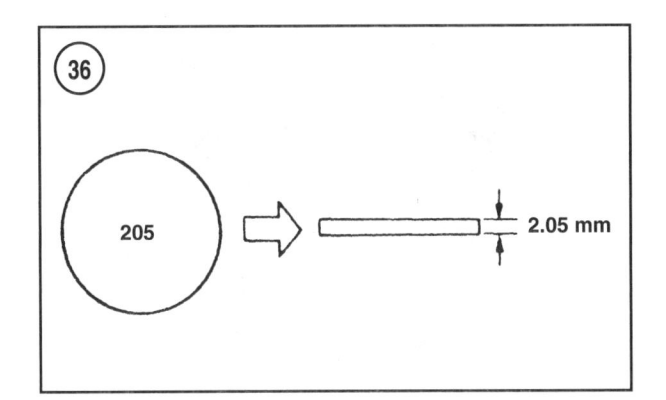

b is the measured valve clearance.

c is the specified valve clearance.

d is the old shim thickness.

For *example*: If the measured valve clearance is 0.26 mm, the old shim thickness is 1.870 mm and the specified valve clearance is 0.18 mm, then: a = (0.26 - 0.18) + 1.870, a = 1.95 (new shim thickness).

NOTE
If the required shim thickness exceeds 2.800 mm, the valve seat is heavily carboned and must be cleaned or refaced.

6. Apply clean engine oil to both sides of the new shim.
7. Install the new shim and the valve lifter (**Figure 34**).
8. Repeat for each valve to be adjusted.
9. Install the camshafts as described in Chapter Four.
10. Rotate the crankshaft counterclockwise several times to turn the camshafts and help seat the new shims.
11. Recheck the valve clearances as described in the preceding procedure. Repeat this procedure until all valve clearances are correct.
12. Reinstall the spark plugs as described in this chapter.
13. Install the cylinder head cover as described in Chapter Four.

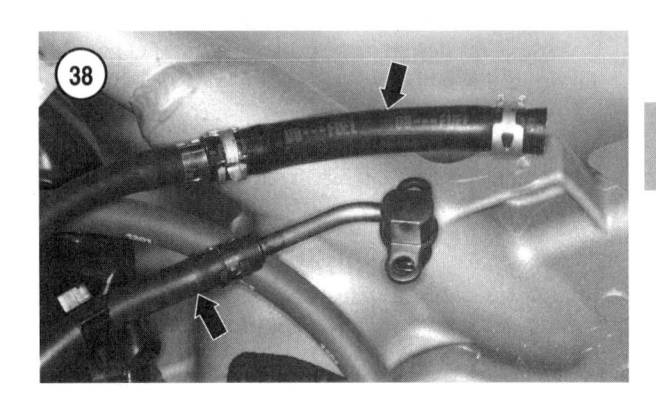

14. Lubricate the timing hole cap threads and O-ring with grease and tighten to 18 N•m (13 ft.-lb.).

ENGINE IDLE SPEED INSPECTION

Routine adjustment of the engine idle speed is not required. However, it is important to periodically check the engine idle speed to ensure proper operation of the IAC valve and the idle air control system.

If the idle speed is not correct, refer to *Engine Idle Speed Adjustment* in Chapter Eight.

FUEL HOSE INSPECTION

Inspect the fuel hoses at the intervals specified in **Table 1**.

1. Remove the fuel tank and air filter housing as described in Chapter Eight.
2. Inspect the fuel hoses for leakage, hardness, age deterioration or other damage. **Figure 38** shows the fuel hose connections at the fuel tank. Follow the length of these hoses to inspect them.
3. Replace damaged fuel hoses and weak or damaged hose clamps as required.
4. Install the air filter and fuel tank (Chapter Eight).

THROTTLE CABLE LUBRICATION

Lubricate the throttle cables at the intervals specified in **Table 1**, or whenever the throttle becomes stiff and sluggish and fails to snap back after releasing it.

The main cause of cable breakage or stiffness is improper lubrication. Periodic lubrication assures long service life. Inspect the cables for fraying and check the sheath for chafing. Replace any defective cables.

CHAPTER THREE

Because of the design of the cable ends, it is not possible to lubricate the cables with a cable lubricant tool while mounted on the motorcycle. Instead, use a can of graphite with a thin hollow tube or an aerosol cable lubricant with an attached tube that can be inserted into the upper end of the cables.

CAUTION
When servicing nylon-lined and other aftermarket cables, follow the cable manufacturer's instructions.

NOTE
The throttle housing can be removed without having to remove the handlebar. The handlebar is removed in this procedure to better illustrate the steps.

1. Loosen the return cable nut (A, **Figure 39**) at the throttle housing.

NOTE
*The pull cable (B, **Figure 39**) upper end threads into the throttle housing. This cable end cannot be separated from the throttle grip or removed from the throttle housing unless the cable is disconnected at its lower end, or the throttle assembly is removed from the handlebar. The pull cable end will be lubricated through its cable adjuster (C, **Figure 39**).*

2. Loosen the pull cable adjuster locknut (C, **Figure 39**).
3. Remove the screws (A, **Figure 40**) from the lower switch housing half and lift it away from the handlebar (B).
4. Remove the screw (A, **Figure 41**) and setting plate (B).
5. Disconnect the return throttle cable from the throttle grip.

NOTE
When using an aerosol type lubricant, cover the area around the nozzle and tube with a plastic bag.

6. Insert the lubricator tube into the return throttle cable end and lubricate the cable.
7. Lubricate the pull cable as follows:
   a. Loosen the locknut (C, **Figure 39**) and turn the upper adjuster (D) to separate the adjuster halves and expose the cable.
   b. Insert the lubricator tube (**Figure 42**) into the lower cable end and lubricate the cable. 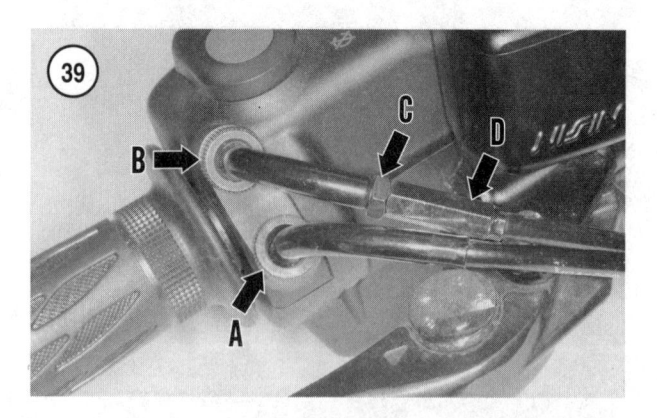 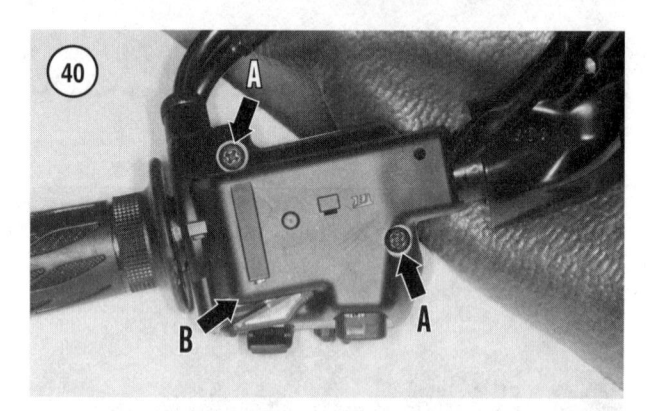 
   c. Reconnect the adjuster halves and turn the adjuster all the way to obtain as much cable slack as possible.
8. Installation is the reverse of removal. Note the following:
   a. Lightly lubricate the cable ends with grease.
   b. Align the pin in the upper switch half with the hole in the handlebar and install the upper switch half. Make sure the switch half seats over the throttle grip.

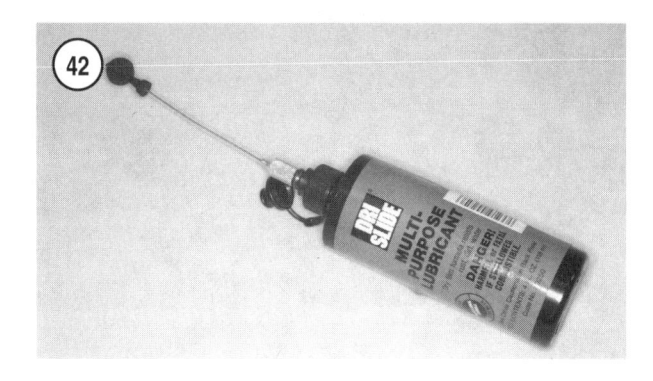

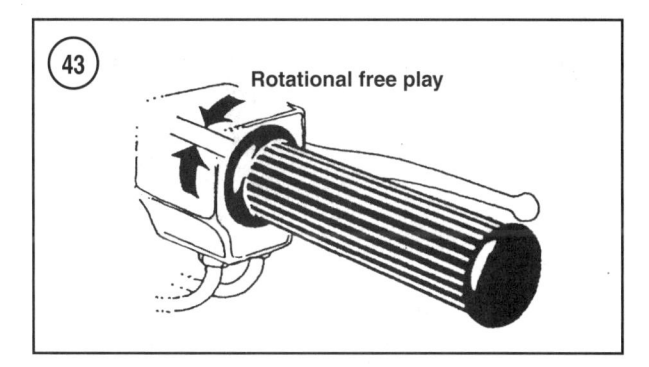

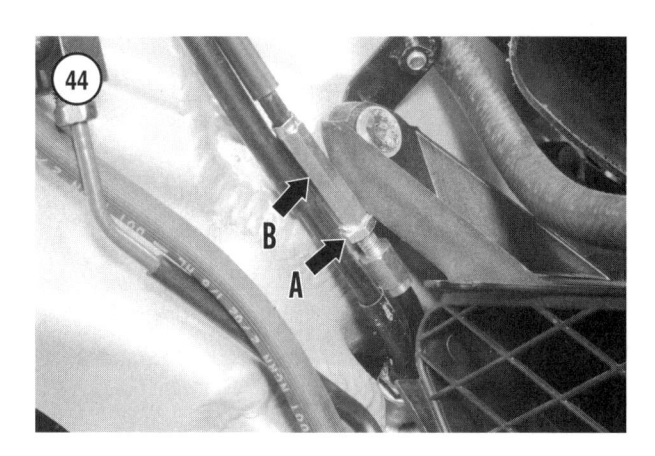

LUBRICATION, MAINTENANCE AND TUNE-UP

the motorcycle until the throttle grip is correctly installed and snaps back when released.

THROTTLE CABLE OPERATION AND ADJUSTMENT

Throttle Lever Operation

Check the throttle operation at the intervals specified in **Table 1**.

Check for smooth throttle operation from the fully closed to fully open positions. Check at various steering positions. The throttle lever must return to the fully closed position without any hesitation.

Check the throttle cables for damage, wear or deterioration. Make sure the throttle cables are not kinked at any place.

If the throttle lever does not return to the fully closed position smoothly and the cables do not appear to be damaged, lubricate the throttle cables as described in this chapter. If the throttle still does not return properly, the cables are probably kinked or routed incorrectly. Replace damaged throttle cables.

Check free play at the throttle grip flange (**Figure 43**). The free play specification is 2-6 mm (0.08-0.24 in.). If adjustment is required perform the following procedure.

Throttle Cable Adjustment

WARNING
If idle speed increases when turning the handlebar, check the throttle cable routing. Correct this problem immediately. Do not ride the bike in this unsafe condition.

1. If minor adjustment is necessary, perform the following at the throttle grip:
   a. Loosen the locknut (C, **Figure 39**) and turn the pull cable upper adjuster (D) in or out to achieve proper free play at the throttle grip (**Figure 43**).
   b. Tighten the locknut and recheck the adjustment. If necessary, continue with Step 2.
2. If major adjustment is necessary, perform the following:
   a. Remove the top shelter, as described in Chapter Seventeen.
   b. Loosen the locknut (A, **Figure 44**) and turn the pull cable lower adjuster (B) to achieve proper free play at the throttle grip (**Figure 43**).

c. Hook the setting plate (B, **Figure 41**) into the switch and secure with the screw (A).
d. Install the lower switch half (B, **Figure 40**) and secure with the lower mounting screws (A). Tighten the front screw first, then the rear screw.
e. Adjust the throttle cables as described in this chapter.

WARNING
An improperly installed throttle grip assembly may cause the throttle to stick open. Failure to properly assemble and adjust the throttle cables and throttle grip could cause a loss of steering control. Do not start or ride

c. Tighten the locknut securely and recheck free play. If necessary, readjust the upper cable adjuster as described in Step 1.

3. Operate the throttle a few times. The throttle grip should now be adjusted correctly. If not, the throttle cables may be stretched. Replace cables in this condition.
4. Reinstall all parts previously removed.
5. Open and release the throttle grip. Make sure it opens and closes (snaps back) without any binding or roughness. Then support the bike on its centerstand and turn the handlebar from side to side, checking throttle operation at both steering lock positions.
6. Sit on the seat and start the engine with the transmission in neutral. Turn the handlebars from lock-to-lock to check for idle speed variances due to improper cable adjustment, routing or damage.

WARNING
If idle speed increases when the handlebar is turned, recheck the throttle cable adjustment and routing. Do not ride the motorcycle in this unsafe condition.

7. Test ride the bike, slowly at first, to make sure the throttle cables are operating correctly. Readjust if necessary.

REVERSE OPERATION

1. Sit on the seat, raise the sidestand and start the engine with the transmission in neutral.
2. Push the reverse switch (A, **Figure 45**). The neutral indicator should turn *off* and the reverse indicator should turn *on*. Push the starter/reverse switch (B, **Figure 45**). The motorcycle should move rearward.
3. If the reverse system does not work correctly, refer to *Starter/Reverse System Troubleshooting* in Chapter Nine.
4. Turn the ignition switch off.

CLUTCH

Clutch Lever Adjustment

The distance between the clutch lever and grip can be adjusted to suit rider preference.

Turn the adjuster (**Figure 46**) and align one of the index marks on the adjuster with the arrow on the clutch lever.

Clutch Fluid Selection

Use DOT 4 brake fluid in the clutch master cylinder reservoir.

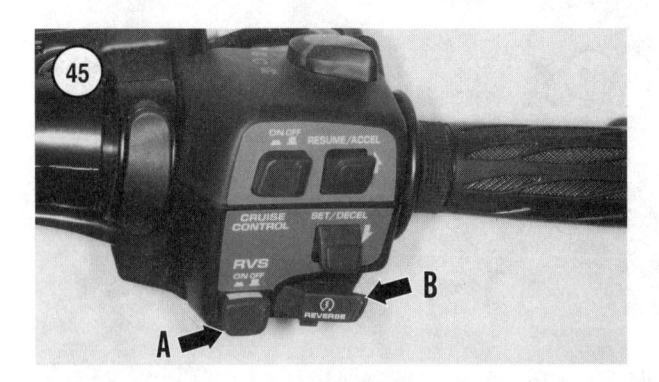

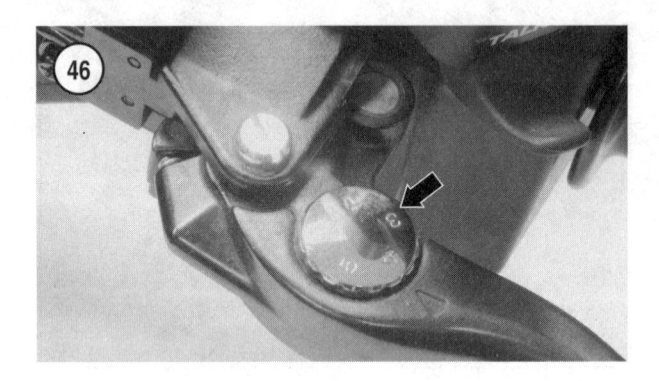

WARNING
Use brake fluid clearly marked DOT 4. Other types may cause clutch failure. Do not intermix different brands or types of brake fluid, as they may not be compatible. Do not intermix silicone-based (DOT 5) brake fluid, as it can cause clutch component damage and lead to clutch system failure.

CAUTION
Handle brake fluid carefully. Do not spill it on painted or plastic surfaces, as it will damage the surface. Wash the area immediately with soap and water and thoroughly rinse it off.

LUBRICATION, MAINTENANCE AND TUNE-UP

entered the clutch system. Bleed the clutch as described in Chapter Six.

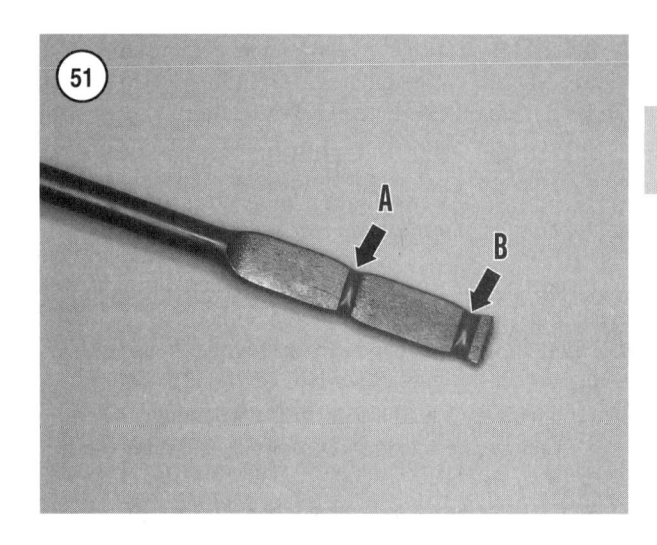

3. Wipe off the master cylinder cover and remove the cover screws. Then remove the cover, set plate and diaphragm.

4. Add fresh DOT 4 brake fluid and fill the reservoir to the level mark inside the reservoir (**Figure 48**).

5. Install the diaphragm, set plate and cover. Install and tighten the cover screws.

6. Start the engine and check the clutch operation.

ENGINE OIL AND FILTER

Oil Level Check

Check the engine oil level with the oil filler cap/dipstick.

1. Place the motorcycle on its centerstand on level ground.

2. Remove the right engine side cover as described in Chapter Seventeen.

3. Start the engine and let it idle for 2-3 minutes.

4. Shut off the engine and let the oil settle for 3 minutes.

CAUTION
Do not check the oil level with the bike on its sidestand; the oil will flow away from the dipstick, and cause a false reading.

5. Remove the oil filler cap/dipstick (**Figure 49**), wipe the gauge and insert the dipstick *without* screwing it in (**Figure 50**).

6. Remove the oil filler cap/dipstick. The oil level should be between the upper (A, **Figure 51**) and lower (B) level marks.

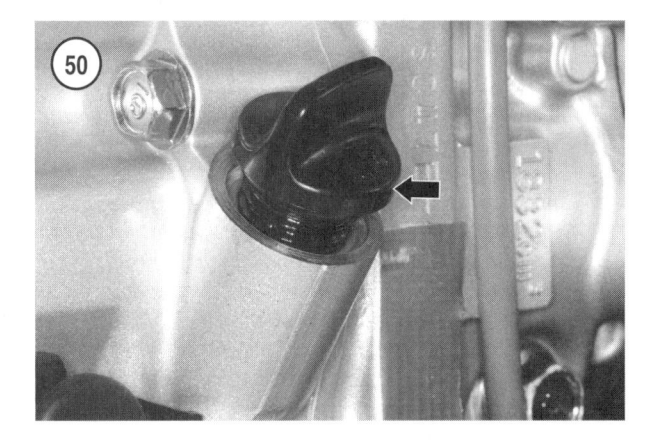

Clutch Fluid Level Check

1. Turn the handlebar so the clutch master cylinder is level.

2. The clutch fluid level must be above the lower level line in the master cylinder window (**Figure 47**). If the clutch fluid level is at or below the lower level line, continue with Step 3.

NOTE
If the reservoir is low, check for fluid leaks. If the reservoir is empty, air has probably

7. If the oil level is near or below the lower level mark (B, **Figure 51**), add the recommended oil (**Table 4**) to correct the level. Add oil slowly to avoid overfilling.

NOTE
*Refer to **Oil and Filter Change** in this section for information on oil selection.*

8. Inspect the O-ring on the oil filler cap/dipstick. Replace it if it is starting to deteriorate or harden.
9. Install the oil filler cap/dipstick (**Figure 49**) and tighten securely.
10. If the oil level is too high, do the following:
    a. Remove the oil filler cap/dipstick and draw out the excess oil using a syringe or suitable pump.
    b. Recheck the oil level and adjust if necessary.
    c. Install the oil filler cap/dipstick and tighten securely.
11. Reinstall the right engine side cover.

Oil and Filter Change

Regular oil and filter changes contribute more to engine longevity than any other maintenance. **Table 1** lists the recommended oil and filter change intervals. These intervals assume that the motorcycle is operated in moderate climates. If the motorcycle is used infrequently, consider a time-based interval for oil changes.

Because the engine oil lubricates the engine, clutch and transmission components, oil requirements for motorcycle engines are more demanding than for automobile engines. Oils specifically designed for motorcycles contain additives to prevent premature viscosity breakdown, protect the engine from oil oxidation resulting from higher engine operating temperatures and provide lubrication qualities designed for engines operating at higher speeds. Consider the following when selecting engine oil:

1. Do not use oil with oil additives or graphite or molybdenum additives. These may adversely affect clutch operation.
2. Do not use vegetable, non-detergent or castor-based racing oils.
3. The Japanese Automobile Standards Organization (JASO) has established an oil classification for motorcycle engines. JASO motorcycle specific oils are identified by the *JASO T 903 Standard*. The JASO label (**Figure 52**) appears on the oil container and identifies the two separate motorcycle oil classifications—MA and MB. JASO classified oil also uses the SAE (Society of Automotive Engineers) viscosity ratings.
4. When selecting an API (American Petroleum Institute) classified oil, select an oil with an SG or higher classification that does not specify it as *ENERGY CONSERVING* on the oil container circular API service label. Use SAE 10W-40 oil.

5. When selecting a JASO classified oil, select an oil with an MA classification that does not specify it as *ENERGY CONSERVING*. Use SAE 10W-40 oil.

NOTE
*There are a number of ways to discard used oil safely. The easiest way is to pour it from the drain pan into a plastic bleach, juice or milk container for disposal. Some service stations and oil retailers accept used oil for recycling. Do not discard oil in household trash or pour it onto the ground. Never add brake fluid, fork oil or any other type of petroleum-based fluid to any engine oil to be recycled. To locate a recycler, contact the American Petroleum Institute (API) at **www.recycleoil.org**.*

NOTE
Warming the engine heats the oil so it flows freely and carries out contamination and sludge.

1. Start the engine and run it until it is at normal operating temperature, then turn it off.

2. Remove the right engine side cover and the front lower fairing as described in Chapter Seventeen.

3. Support the bike on its sidestand when draining the engine oil. This ensures complete draining.

LUBRICATION, MAINTENANCE AND TUNE-UP

WARNING
The engine, exhaust pipes and oil are hot! Work quickly and carefully when removing the oil drain bolt and oil filter.

4. Clean the area around the oil drain bolt and oil filter.
5. Place a clean drip pan under the crankcase and remove the oil drain bolt (A, **Figure 53**) and washer.
6. Remove the oil filler cap/dipstick (**Figure 49**), speed up the oil flow.
7. Allow the oil to drain completely.
8. To replace the oil filter, perform the following:
    a. Install a socket type oil filter wrench squarely onto the oil filter (B, **Figure 53**) and turn the filter counterclockwise until oil begins to run out, then remove the oil filter.
    b. Hold the filter over the drain pan and pour out any remaining oil, then place the old filter in a plastic bag and discard it properly.
    c. Carefully clean the oil filter sealing surface on the crankcase. Do not allow any dirt or other debris to enter the engine.
    d. Lubricate the seal on the new filter with clean engine oil.
    e. Install the new oil filter onto the threaded fitting on the crankcase.
    f. Tighten the filter by hand until it contacts the crankcase. Then tighten it 26 N•m (19 ft.-lb.). If tightening by hand, tighten an additional 3/4 turn.

NOTE
Overtightening the filter will cause it to leak.

9. Replace the drain bolt gasket if leaking or damaged.
10. Install the oil drain bolt (A, **Figure 53**) and gasket and tighten to 34 N•m (25 ft.-lb.).
11. Place the motorcycle on its centerstand on level ground.
12. Insert a funnel into the oil filler hole and fill the engine with the correct weight (**Table 4**) and quantity of oil (**Table 5**).
13. Remove the funnel and screw in the oil filler cap/dipstick and its O-ring (**Figure 49**).

NOTE
*If servicing a rebuilt engine, check the engine oil pressure as described under **Oil Pressure Test** in this section.*

14. Start the engine and let it idle.

CAUTION
The oil pressure indicator should go out within 1-2 seconds. If it stays on, shut off the engine immediately and locate the problem. Do not run the engine with the oil pressure indicator on.

15. Check the oil filter and drain bolt for leaks.
16. Turn the engine off after 2-3 minutes and check the oil level as described in this chapter. Adjust the oil level if necessary.

WARNING
Prolonged contact with oil may cause skin cancer. Wash hands thoroughly with soap and water after contacting engine oil.

17. Install the right engine side cover and front lower fairing.

Oil Pressure Test

Check the engine oil pressure after reassembling the engine or when troubleshooting the lubrication system.

An oil pressure gauge (Honda part No. 07506-3000000, or equivalent) and oil pressure adapter (07510-4220100, or equivalent) are required to test the oil pressure.

1. Remove the left exhaust pipe protector. Refer to *Exhaust System* in Chapter Seventeen.
2. Check that the engine oil level is correct as described in this section. Add oil if necessary.
3. Place the motorcycle on its centerstand on level ground.
4. Remove the rubber boot and disconnect the wire from the oil pressure switch (**Figure 54**). Then loosen and remove the oil pressure switch.
5. Assemble the oil pressure adapter and gauge. Thread the oil pressure gauge adapter into the engine in place of the

oil pressure switch. Make sure the fitting is tight to prevent leaks.

CAUTION
Keep the gauge hose away from the exhaust pipe during this test. If the hose contacts the exhaust pipe, it may melt and spray hot oil.

6. Start the engine and and allow it to reach normal operating temperature.
7. Increase engine speed to 5000 rpm and read the oil pressure on the gauge. The oil pressure should be 530 kPa (77 psi) when the oil temperature is 80° C (176° F).
8. Allow the engine to return to idle, then shut it off and remove the test equipment.
9. If the oil pressure is lower or higher than specified, refer to *Engine Lubrication* in Chapter Two.
10. Install the oil pressure switch as follows:
    a. Clean the oil pressure switch and crankcase threads of all sealant and oil residue.
    b. Apply an RTV sealant to the oil pressure switch threads as shown in **Figure 55**. Do not apply sealant within 3-5 mm (0.1-0.2 in.) from the end of the switch threads.

NOTE
Allow the RTV sealant to set for 10-15 minutes before installing the oil pressure switch.

    c. Install the oil pressure switch and tighten to 12 N•m (106 in.-lb.).
    d. Reconnect the wire onto the switch and cover the switch with its rubber boot.
11. Start the engine observe the oil pressure indicator.

CAUTION
The oil pressure indicator should go out within 1-2 seconds. If it stays on, shut off the engine immediately and locate the problem. Do not run the engine with the oil pressure indicator on.

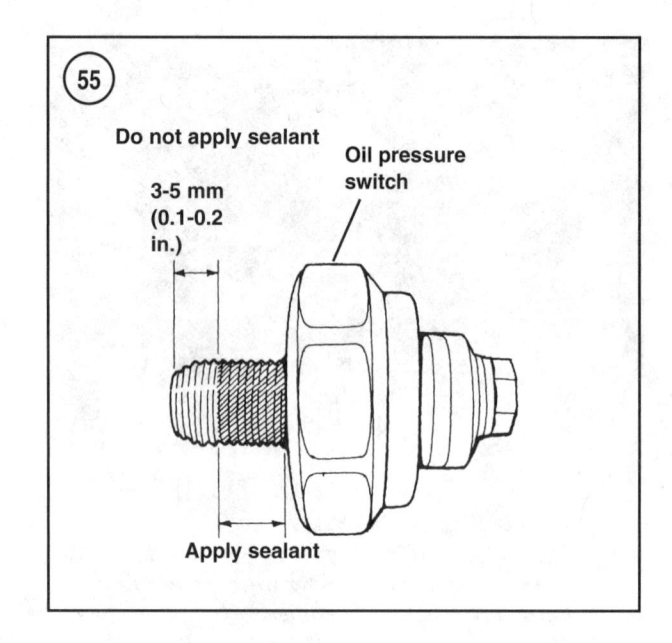

12. Check the engine for oil leaks.

CAUTION
Do not overtighten the switch to correct an oil leak, as this may strip the crankcase threads. If oil leaks from the switch after installing it, remove the switch and reclean the threads. Reseal and reinstall the switch.

13. Install the left exhaust pipe protector as described in Chapter Seventeen.

COOLING SYSTEM

Inspect and service the cooling system at the intervals specified in **Table 1**.

WARNING
When performing any service work on the engine or cooling system, never remove the radiator cap, coolant drain bolt or discon-

LUBRICATION, MAINTENANCE AND TUNE-UP

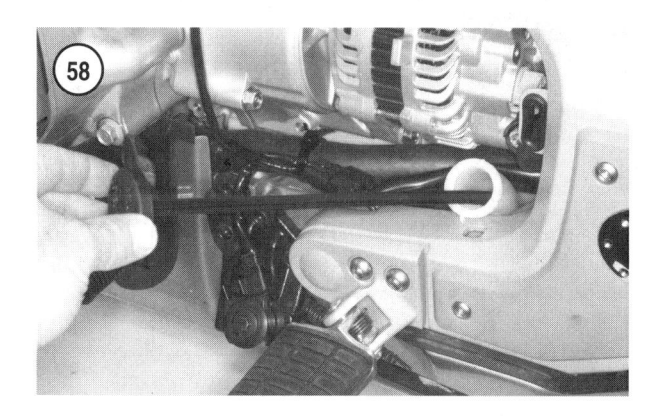

nect any coolant hose while the engine and radiators are hot. Scalding fluid and steam may be blown out under pressure and cause serious injury.

Coolant Type

If adding coolant to the cooling system, use Pro Honda HP Coolant. This is a ready-to-use 50:50 antifreeze/purified, de-ionized water coolant blend. If mixing antifreeze and water, use a 50:50 mixture of distilled water and antifreeze that does not contain silicate inhibitors. Use only soft or distilled water. Never use tap or saltwater, as this will damage engine parts. Distilled (or purified) water can be purchased at supermarkets or drug stores in gallon containers. Never use alcohol-based antifreeze.

CAUTION
Many antifreeze solutions contain silicate inhibitors to protect aluminum parts from corrosion damage. However, these silicate inhibitors can cause premature wear to water pump seals. When selecting an antifreeze, make sure it does not contain silicate inhibitors.

Coolant Test

WARNING
Do not remove the radiator cap when the engine is hot.

1. Remove the right fairing pocket as described in Chapter Seventeen.
2. Remove the radiator cap (**Figure 56**).
3. Test the specific gravity of the coolant with an antifreeze tester to ensure adequate temperature and corrosion protection. A 50:50 mixture is recommended. Never allow the mixture to become less than 40 percent antifreeze. See *Coolant Type* in this section.
4. Reinstall the radiator cap (**Figure 56**).
5. Reinstall the right fairing pocket.

Coolant Level

1. Support the bike on its centerstand.
2. Remove the left engine side cover as described in Chapter Seventeen.
3. Start the engine and allow it to idle until it reaches normal operating temperature. The engine must be running at idle speed when checking the coolant level.
4. Remove the coolant reserve tank cap/dipstick (**Figure 57**). The coolant level should between the UPPER and LOWER level holes on the dipstick (**Figure 58**).
5. If necessary, add coolant as follows:
    a. Turn the engine off.
    b. Add Pro Honda HP Coolant into the reserve tank (not the radiator) to bring the level to the upper mark. See *Coolant Type* in this section.
    c. Inspect the cooling system for leaks as described in Chapter Ten.
6. Reinstall the reserve tank cap/dipstick.
7. Install the left engine side cover.

Coolant Change

Drain and refill the cooling system at the intervals listed in **Table 1**.

It is sometimes necessary to drain the cooling system when servicing the engine. If the coolant is still in good condition, the coolant can be reused if it is not contaminated. Drain the coolant into a *clean* pan and pour the coolant into a container for storage.

WARNING
Waste antifreeze is toxic and may never be discharged into storm sewers, septic systems, waterways or onto the ground. Place used antifreeze in the original container and

dispose of it according to local regulations. Do not store coolant where it is accessible to children or pets.

WARNING
*Do not remove the radiator cap (**Figure 56**) if the engine is hot. The coolant is very hot and is under pressure. Severe scalding will result if hot coolant contacts skin.*

CAUTION
Be careful not to spill antifreeze on painted surfaces as it will damage the surface. Wash immediately with soapy water and rinse thoroughly.

Perform the following procedure when the engine is *cold*.

1. Place the motorcycle on its centerstand on level ground.
2. Remove the left engine side cover as described in Chapter Seventeen.
3. Remove the front lower fairing as described in Chapter Seventeen.
4. Remove the right fairing pocket as described in Chapter Seventeen.
5. Remove the radiator cap (**Figure 56**).
6. Place a drain pan under the coolant drain bolt. Remove the drain bolt (**Figure 59**) and washer and allow the coolant to drain into the pan.
7. Replace the sealing washer if it is leaking or damaged. Reinstall the drain bolt and sealing washer and tighten securely.
8. Remove and drain the coolant reserve tank as described in Chapter Ten. Reinstall the tank.

CAUTION
*Do not use a higher percentage of antifreeze-to-water solution than is recommended under **Coolant Type** in this section. A higher concentration of coolant will actually decrease the performance of the cooling system.*

9. Place a funnel in the radiator filler neck and slowly refill the radiator and engine with a mixture of 50 percent antifreeze and 50 percent distilled water. Add the mixture slowly so it expels as much air as possible from the cooling system. **Table 6** lists engine coolant capacity.

WARNING
Do not run the motorcycle in an enclosed area. The exhaust gasses contain carbon monoxide, a colorless, odorless, poisonous gas. Carbon monoxide levels build quickly in a small enclosed area and can cause unconsciousness and death in a short time.

10. After filling the radiator, bleed the cooling system as follows:
   a. Start the engine and allow it to idle for two to three minutes.
   b. Snap the throttle a few times to bleed air from the cooling system. When the coolant level drops in the radiator, add coolant to bring the level to the bottom of the filler neck.
   c. When the radiator coolant level has stabilized, install the radiator cap (**Figure 56**).
11. Remove the reserve tank cap/dipstick and fill the coolant reserve tank (**Figure 58**) to the UPPER level hole on the dipstick. Install the cap (**Figure 57**).
12. Start the engine and let it run at idle speed until the engine reaches normal operating temperature. Make sure there are no air bubbles in the coolant and the coolant level in the coolant reserve tank stabilizes at the correct level. Add coolant to the coolant reserve tank as necessary.
13. Test ride the bike and readjust the coolant level in the reserve tank as necessary. Check the reserve tank for leaks.

CRANKCASE BREATHER INSPECTION

The engine is equipped with a crankcase emission control system to prevent fumes and gasses from being vented

LUBRICATION, MAINTENANCE AND TUNE-UP

into the atmosphere. However, under various operating conditions, contaminants (water and blow-by gas) that are not burned in the combustion chamber collect in the air box. To remove these contaminants, the air box is equipped with a drain hole and connecting hose. A plug at the bottom of the hose seals the hose. At the intervals specified in **Table 1**, inspect the hose for fluid. If necessary, remove the plug from the end of the hose and drain the contaminants into a container. Check the drain hose more frequently after riding the motorcycle in the rain or after riding long distances under full-throttle. Also check the hose if washing the motorcycle frequently or if the motorcycle is dropped on its side.

1. Remove the left cylinder head side cover as described in Chapter Four.
2. Remove the plug and drain the contaminants into a container (**Figure 60**).
3. Reinstall the plug and secure with the clamp.
4. Install the left cylinder head side cover.

EMISSION CONTROL SYSTEMS

Models are equipped with a secondary air supply and evaporative emission control systems. At the intervals specified in **Table 1**, check all emission control hoses for deterioration, damage or loose connections. Replace any parts or hoses as necessary. Refer to the emission control sections in Chapter Eight for information on inspecting these systems.

FINAL DRIVE OIL

Oil Level Check

1. Place the motorcycle on its centerstand on level ground.
2. Wipe the area around the final drive oil filler plug and unscrew the oil filler plug (A, **Figure 61**).
3. The oil level must be even with the lower edge of the oil filler hole. If the oil level is low, add the recommended type gear oil (**Table 4**) to correct the level.

4. Inspect the oil filler plug O-ring and replace it if it is leaking or damaged. Lubricate a new O-ring with oil.
5. Install and tighten the oil filler plug to 12 N•m (106 in.-lb.).

Oil Change

The recommended oil change interval is listed in **Table 1**.

1. Ride the motorcycle a few miles to warm the oil in the final drive unit.
2. Place the motorcycle on its centerstand on level ground.
3. Place a drain pan under the final drive unit.
4. Remove the oil filler plug (A, **Figure 61**) and the drain plug (B) to drain the oil.
5. With the transmission in neutral, slowly turn the rear wheel to drain oil remaining on the gears.
6. Install a new washer on the drain plug and tighten to 20 N•m (15 ft.-lb.).
7. Add the recommended type gear oil (**Table 4**) to bring the oil level even with the lower edge of the oil filler hole. **Table 7** lists final drive oil capacity.
8. Inspect the oil filler plug O-ring and replace it if it is leaking or damaged. Lubricate a new O-ring with oil.
9. Install and tighten the oil filler plug to 12 N•m (106 in.-lb.).

BATTERY

The original equipment battery is a maintenance-free type. Periodic electrolyte inspections are not required on maintenance-free batteries. Water cannot be added. Refer to Chapter Nine for battery service, testing and replacement procedures.

TIRES AND WHEELS

Tire Pressure

Check and adjust the tire pressure to maintain the tire profile, good traction and handling, and to get the maximum life out of the tire. Check tire pressure when the tires are cold. Never release air pressure from a warm or hot tire to match the recommended tire pressure; doing so causes the tire to be under inflated. Use an accurate tire pressure gauge. **Table 9** lists original equipment tire pressures.

NOTE
After checking and adjusting the air pressure, make sure to reinstall the air valve cap on each tire. The cap prevents dirt and debris

from collecting in the valve stem, which could cause air leakage or an incorrect tire pressure reading.

NOTE
A loss of air pressure may be due to a loose or damaged valve core. Put a few drops of water on the top of the valve core. If the water bubbles, tighten the valve core and recheck. If air is still leaking from the valve after tightening it, replace the valve.

Tire Inspection

Frequently inspect the tires for wear and damage. Inspect the tires for the following:
1. Deep cuts and embedded objects, such as nails and stones. If a nail or other object is in a tire, mark its location with a light crayon before removing it. This helps to locate the hole for repair. Refer to Chapter Eleven for tire changing and repair information.
2. Flat spots.
3. Cracks.
4. Separating plies.
5. Sidewall damage.

WARNING
If a small object has punctured the tire, air leakage may be very slow due to the tendency of tubeless tires to self-seal when punctured. Check the tires carefully.

Tire wear analysis

Analyze abnormal tire wear to determine the cause. Common causes are:
1. Incorrect tire pressure. Check tire pressure and examine the tire tread. Compare the wear in the center of the contact patch with the wear at the edge of the contact patch. Note the following:
   a. If the tire shows excessive wear at the edge of the contact patch, but the wear at the center of the contact patch is normal, the tire has been under inflated. Under inflated tires cause higher tire temperatures, hard or imprecise steering and abnormal wear.
   b. If the tires shows excessive wear in the center of the contact patch, but wear at the edge of the contact patch is normal, the tire has been over inflated. Over inflated tires cause hard riding and abnormal wear.

NOTE
Large amounts of freeway riding causes the tires to exhibit a similar wear pattern as described in substep b.

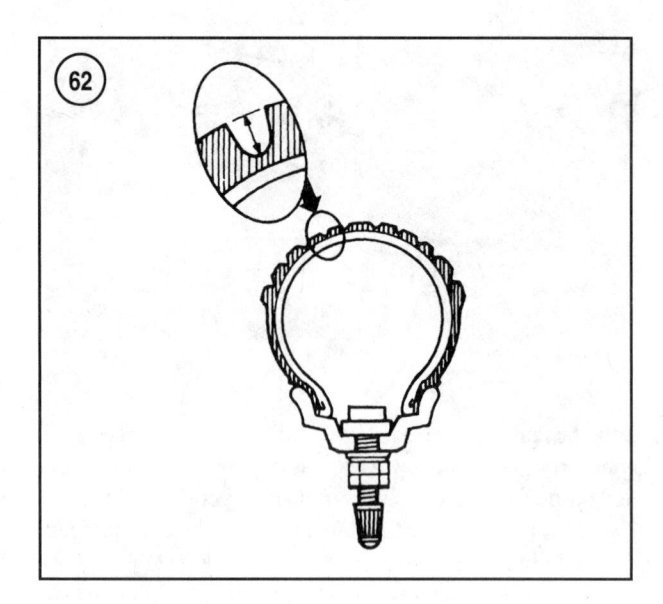

2. Overloading.
3. Incorrect wheel alignment.
4. Incorrect wheel balance. The tire and wheel assembly should be balanced when installing a new tire.
5. Worn or damaged front wheel bearings.

Tread depth

Measure the tread depth (**Figure 62**) in the center of the tire using a small ruler or a tread depth gauge. Honda recommends replacing the original equipment tires before the center tread depth has worn to the following depth:
1. Front: 1.5 mm (0.06 in.).
2. Rear: 2.0 mm (0.08 in.).

The tires are also designed with tread wear indicators that appear when the tires are worn out. When these are visible, the tires are no longer safe and must be replaced.

Wheel Inspection

Frequently inspect the wheels for cracks, warps or dents. A damaged wheel rim-to-tire surface may cause an air leak.

Wheel rim runout is the amount of wobble a wheel shows as it rotates. Check runout with the wheels on the bike. Refer to *Wheel Runout and Balance* in Chapter Eleven for service procedures.

FRONT FORK OIL CHANGE

The front fork legs must be partially disassembled for fork oil replacement and oil level adjustment. Refer to *Front Fork* in Chapter Twelve.

LUBRICATION, MAINTENANCE AND TUNE-UP

BRAKES

Check the brake fluid in each brake master cylinder at the intervals listed in **Table 1**. At the same time, inspect the brake pads for wear. Brake bleeding, servicing the brake components and replacing the brake pads are covered in Chapter Fourteen. Service to ABS components is in Chapter Fifteen. Inspect the brake components more often when towing a trailer or when operating under severe riding conditions.

Brake System Inspection

Check front and rear brake operation as follows:
1. Support the motorcycle on its centerstand.
2. Shift the transmission into neutral.
3. Turn the rear wheel by hand. It should turn easily. Then push the left front brake caliper upward by hand (**Figure 63**) and have an assistant try to turn the rear wheel by hand. It should not turn.
4. Apply the front brake lever and rear brake pedal. Make sure each feels firm. If the lever or pedal feels soft or spongy, air has probably entered the system. Check the brake hoses and bleed the brakes as described in Chapter Fourteen.

Brake Hose Inspection

Inspect the brake hoses between the master cylinder and the brake caliper(s). If there are leaks, tighten the bolt or hose and then bleed the brakes as described in Chapter Fourteen. If this does not stop the leak or if a brake line is obviously damaged, cracked or chafed, replace the brake hose and bleed the system. Refer to Chapter Fourteen for brake system torque specifications.

Brake Fluid Selection

Use DOT 4 brake fluid in the front and rear master cylinder reservoirs.

> *WARNING*
> *Use brake fluid clearly marked DOT 4. Others may cause brake failure. Do not intermix different brands or types of brake fluid as they may not be compatible. Do not intermix silicone based (DOT 5) brake fluid as it can cause brake component damage leading to brake system failure.*

> *CAUTION*
> *Handle brake fluid carefully. Do not spill it on painted or plastic surfaces, as it will damage the surface. Wash the area immediately with soap and water and thoroughly rinse it off.*

Front Brake Fluid Level Check

1. Place the motorcycle on its centerstand. Turn the handlebar and level the front master cylinder.
2. The brake fluid level must be above the lower level line in the master cylinder window (**Figure 64**). If the brake fluid level is at or below the lower level line, continue with Step 3.

> *NOTE*
> *If the reservoir is empty, air has probably entered the brake system. Bleed the brake system as described in Chapter Fourteen.*

3. Wipe off the master cylinder cover and remove the cover screws. Then remove the cover (**Figure 64**), set plate and diaphragm.
4. Add fresh DOT 4 brake fluid to fill the reservoir to the level mark located inside the reservoir.

5. Install the diaphragm, set plate and cover (**Figure 64**). Install and tighten the cover screws.

6. If the brake fluid level was low, check the brake pads for wear as described in this section.

NOTE
A low brake fluid level usually indicates brake pad wear. As the pads wear (become thinner), the brake caliper pistons automatically extend farther out of their bores. As the caliper pistons move outward, the brake fluid level drops in the system. However, if the brake fluid level is low and the brake pads are not worn excessively, check all the brake hoses and lines for leaks.

Rear Brake Fluid Level Check

1. Place the motorcycle on its centerstand on level ground.
2. Remove the right engine side cover as described in Chapter Seventeen.

WARNING
Do not check the rear brake fluid level with the bike resting on its sidestand. A false reading will result.

3. The brake fluid level must be between the upper and lower level marks (**Figure 65**) on the reservoir housing. If the brake fluid level is at or below the lower level line, continue with Step 4.

NOTE
If the reservoir is empty, air has probably entered the brake system. Bleed the brake system as described in Chapter Fourteen.

4. Wipe off the master cylinder cover and unscrew the cover. Then remove the set plate and diaphragm.
5. Add fresh DOT 4 brake fluid to fill the reservoir to the upper level mark on the reservoir.
6. Install the diaphragm and cover. Tighten the cover securely.
7. If the brake fluid level was low, check the brake pads for wear as described in this section.

NOTE
A low brake fluid level usually indicates brake pad wear. As the pads wear (become thinner), the brake caliper pistons automatically extend farther out of their bores. As the caliper pistons move outward, the brake fluid level drops in the system. However, if the brake fluid level is low and the brake pads are not worn excessively, check all the brake hoses and lines for leaks.

8. Reinstall the right engine side cover.

Brake Fluid Change

Every time the reservoir cap is removed, a small amount of dirt and moisture enters the brake fluid. The same thing happens if a leak occurs or when a brake hose is loosened. Dirt can clog the system and cause unnecessary wear. Water in the brake fluid will vaporize at high brake system temperatures, impairing the hydraulic action and reducing the brake's stopping ability. To maintain peak performance, change the brake fluid at the interval specified in **Table 1** or whenever a caliper or master cylinder is overhauled. To change brake fluid, follow the brake bleeding procedure in Chapter Fourteen.

Brake Pad Wear Inspection

Inspect the brake pads for wear at the intervals specified in **Table 1**.

LUBRICATION, MAINTENANCE AND TUNE-UP

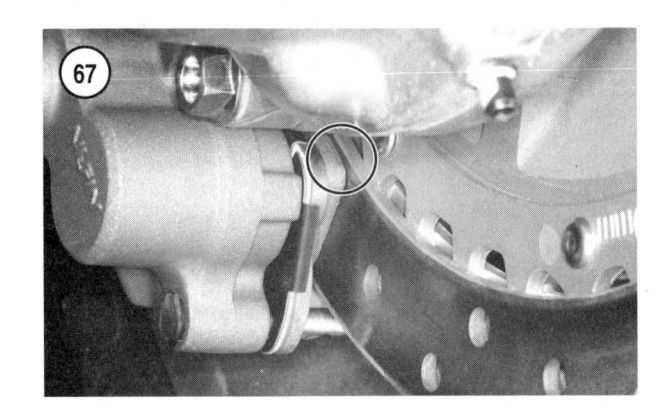

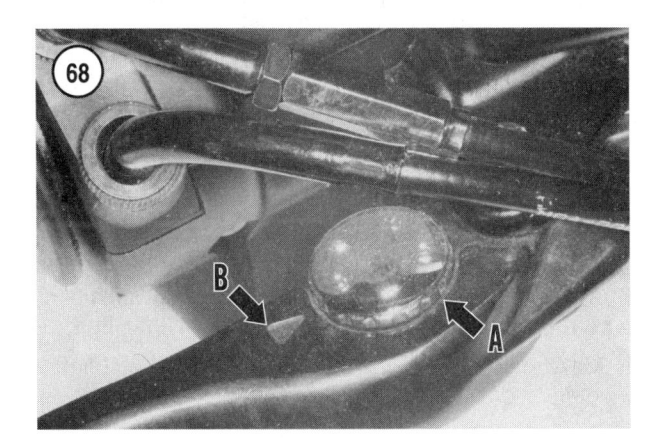

1. Inspect the front (**Figure 66**) and rear (**Figure 67**) brake pads for uneven wear, scoring, oil contamination or other damage. If there is no visible brake pad damage or contamination, perform Step 2.

> *WARNING*
> *A damaged fork oil seal may allow oil to run down the fork tube and contaminate the brake pads. Always check the pads for contamination when a leaking fork seal is detected and replace the seal immediately.*

2. Inspect the brake pad wear limit grooves. See **Figure 66** (front) or **Figure 67** (rear).

3. If either set of pads are worn to the end wear of a limit groove, replace the pads as described in Chapter Fourteen.

> *WARNING*
> *Always replace both pads in each caliper at the same time. On the front brakes, replace the brake pads in both calipers at the same time.*

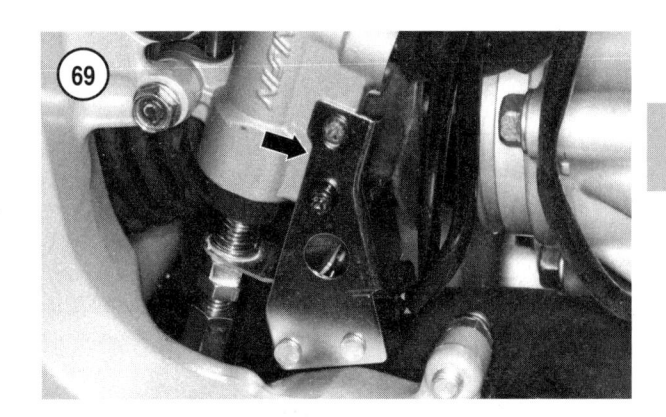

Front Brake Lever Adjustment

The distance between the front brake lever and throttle grip can be adjusted to suit rider preference.

1. Turn the adjuster (A, **Figure 68**) and align one of the index marks on the adjuster with the arrow on the brake lever pivot bolt (B).
2. Support the bike with the front wheel off the ground. Then spin the front wheel by hand and apply the front brake several times. Make sure the front wheel turns without any brake drag and that the front brakes work correctly.

Front Brake Light Switch Adjustment

There is no adjustment for the front brake light switch.

Rear Brake Light Switch Adjustment

There is no adjustment for the rear brake light switch. If the brake light does not come on when operating the rear brake pedal, or the pedal and brake light synchronization is off, check the rear brake light switch (**Figure 69**) for loose mounting screws or other damage. Test the switch as described in Chapter Nine under *Rear Brake Light Switch*.

LIGHTS AND HORN INSPECTION

Check the headlight aim at the intervals specified in **Table 1**. If adjustment is required, check the brake and turn signal light, and horn operation at the intervals specified in **Table 1**. Refer to *Headlight* in Chapter Nine.

With the engine running, check as follows:
1. Pull the front brake lever and check that the brake light comes on.
2. Push the rear brake pedal and check that the brake light comes on.

3. Move the dimmer switch up and down between the high and low positions, and check to see that both headlight elements are working.

4. Move the turn signal switch to the left position and then to the right position and check that all the turn signal lights are working.

5. Operate the horn button and make sure both horns sound loudly.

6. If the horn or any light failed to work properly, refer to Chapter Nine.

SIDESTAND SWITCH AND IGNITION CUT-OFF SYSTEM

Check the sidestand switch (**Figure 70**) and the ignition cut-off system operation at the intervals specified in **Table 1**.

1. Place the motorcycle on its centerstand.
2. Operate the side stand and check its movement and spring tension. Replace the spring if it is weak or damaged.
3. Lubricate the sidestand pivot bolt if necessary.
4. Check the sidestand ignition cut-off system as follows:
    a. Park the bike so both wheels are on the ground.
    b. Sit on the motorcycle and raise the side stand.
    c. Shift the transmission into neutral.
    d. Start the engine and then squeeze the clutch lever and shift the transmission into gear.
    e. Move the sidestand down. When doing so, the engine should stop.
    f. If the engine did not stop as the sidestand was lowered, test the sidestand switch (**Figure 70**) as described in Chapter Nine under *Sidestand Switch*.

STEERING HEAD BEARING INSPECTION

Inspect the steering head bearing adjustment at the intervals specified in **Table 1**.

1. Support the bike on a stand with the front wheel off the ground.
2. Hold onto the handlebars and move them from side to side. Note any binding or roughness.
3. Support the bike so that both wheels are on the ground.
4. Sit on the motorcycle and hold onto the handlebars. Apply the front brake lever and try to push the front fork forward. Try to detect any movement in the steering head area. If so, the bearing adjustment is loose and requires adjustment.
5. If there is any roughness, binding or looseness when performing Step 2 or Step 4, service the steering head bearings as described in Chapter Twelve.

FRONT SUSPENSION INSPECTION

Inspect the front suspension at the intervals specified in **Table 1**.

1. Use a soft, wet cloth to wipe the front fork tubes to remove any dirt and debris. As this debris passes against the fork seals, it will eventually damage the seals and cause an oil leak.
2. Check the front fork for any oil seal leaks or damage.
3. Apply the front brake and pump the fork up and down as vigorously as possible. Check for smooth operation.
4. Make sure the upper and lower fork tube pinch bolts are tight.
5. Check that the handlebar mounting bolts are tight.
6. Make sure the front axle is tight.

WARNING
If any of the previously mentioned fasteners are loose, refer to Chapter Twelve for procedures and torque specifications.

REAR SUSPENSION INSPECTION

Inspect the rear suspension at the intervals specified in **Table 1**.

1. With both wheels on the ground, check the shock absorber by bouncing on the seat several times.
2. Support the bike on its centerstand.
3. With an assistant steadying the bike, push hard on the rear wheel (sideways) to check for side play in the rear swing arm bearings.
4. Check the shock absorber for oil leakage, loose mounting fasteners or other damage.
5. Check for loose or missing suspension fasteners.
6. Make sure the rear axle nut is tight.
7. To adjust the rear shock absorber, refer to Chapter Thirteen.

LUBRICATION, MAINTENANCE AND TUNE-UP

WARNING
If any of the previously mentioned bolts and nuts are loose, refer to Chapter Thirteen for procedures and torque specifications.

FASTENERS

Constant vibration can loosen many fasteners on a motorcycle. Inspect the fasteners at the interval specified in **Table 1**.

1. Check the tightness of all exposed fasteners. Refer to the appropriate chapter for torque specifications.

2. Check that all hose clamps, cable stays and safety clips are properly installed. Replace missing or damaged items.

UNSCHEDULED MAINTENANCE

Front Wheel Bearings

There is no recommended mileage interval for inspecting the front wheel bearings. Check the front wheel bearings whenever the wheel is removed or whenever there is the likelihood of water or other contamination. See Chapter Eleven for service procedures.

Table 1 MAINTENANCE AND LUBRICATION SCHEDULE[1]

At 4000 miles (6400 km)
Clean crankcase breather[2]
Replace engine oil and filter
Inspect brake fluid level[3]
Inspect brake pad wear
Inspect clutch fluid[3]
At 8000 miles (12,800 km)
Inspect fuel hoses
Check throttle operation and lubricate cables
Clean crankcase breather[2]
Inspect coolant level[3]
Inspect cooling system
Inspect secondary air supply system
Inspect final drive oil level
Inspect brake fluid level[3]
Inspect brake system, including brake pad wear
Inspect brake light operation
Inspect clutch fluid level[3]
Inspect clutch system
Inspect reverse operation
Inspect sidestand switch and ignition cut-off system operation
Inspect front and rear suspension
Inspect wheels and tires
Inspect steering head bearings
Inspect headlight operation and aim
Inspect turn signal operation
Inspect horn operation
Inspect fasteners
At 12,000 miles (19,200 km)
Replace air filter element[4]
Clean crankcase breather[2]
Replace engine oil and filter
Inspect evaporative emission control system
Replace brake fluid
Inspect brake pad wear
Replace clutch fluid
At 16,000 miles (25,600 km)
Repeat 8000 miles (12,800 km) service
(continued)

Table 1 MAINTENANCE AND LUBRICATION SCHEDULE[1] (continued)

At 16,000 miles (25,600 km)
Replace spark plugs
At 20,000 miles (25,600 km)
Clean crankcase breather[2]
Replace engine oil and filter
Inspect brake fluid level[3]
Inspect brake pad wear
Inspect clutch fluid level[3]
At 24,000 miles (38,400 km)
Inspect fuel hoses
Check throttle operation and lubricate cables
Replace air filter element[4]
Clean crankcase breather[2]
Replace coolant
Inspect cooling system
Inspect secondary air supply and evaporative emission control systems
Replace final drive oil
Replace brake fluid
Inspect brake system, including brake pad wear
Inspect brake light operation
Replace clutch fluid
Inspect clutch system
Inspect reverse operation
Inspect sidestand switch and ignition cut-off system
Inspect front and rear suspension
Inspect wheels and tires
Inspect steering head bearings
Inspect headlight operation and aim
Inspect turn signal operation
Inspect horn operation
Inspect fasteners
At 32,000 miles (51,200 km)
Check valve clearance[5]

1. Consider this maintenance schedule a guide to general maintenance and lubrication intervals. Harder than normal use, including: exposure to water, high humidity, blowing dirt, towing a trailer or stop-and-go traffic may require more frequent attention to some maintenance items.
2. Increase service intervals when riding at full throttle or in rain.
3. Replace every two years or at the indicated odometer reading, whichever comes first.
4. Service more frequently if riding in wet or dusty conditions.
5. Check more frequently if valve noise increases.

Table 2 TUNE-UP SPECIFICATIONS

Engine compression	1383 kPa (201 psi) at 300 rpm
Engine firing order	1-4-5-2-3-6
Engine idle speed	700 ± 70 rpm
Oil pressure	530 kPa (77 psi) at 5000 rpm
Spark plug type	
NGK	
Standard	BKR6E-11
Cold climate*	BKR5E-11
Extended high speed riding	BKR7E-11
Denso	
Standard	K20PR-U11
Cold climate*	K16PR-U11
Extended high speed riding	K22PR-U11
(continued)	

LUBRICATION, MAINTENANCE AND TUNE-UP

Table 2 TUNE-UP SPECIFICATIONS (continued)

Spark plug gap	1.00-1.10 mm (0.039-0.043 in.)
Throttle lever free play	2-6 mm (0.08-0.24 in.)
Valve clearance	
Intake	0.12-0.18 mm (0.005-0.007 in.)
Exhaust	0.19-0.25 mm (0.008-0.010 in.)

*Cold climate conditions are when the ambient temperature is below 5° C (41° F).

Table 3 SERVICE TORQUE SPECIFICATIONS

	N•m	in.-lb.	ft.-lb.
Clutch reservoir stopper plate screw	2	18	–
Cylinder head cover bolt	12	106	–
Cylinder head side cover bolt	10	88	–
Engine oil drain bolt	34	–	25
Final drive oil drain plug	20	–	15
Final drive oil filler plug	12	106	–
Front brake reservoir cap screw	2	18	–
Oil filter cartridge[1]	26	–	19
Oil pressure switch[2]	12	106	–
Spark plugs	18	–	13
Timing hole cap[3]	18	–	13

1. Lubricate rubber seal and threads with engine oil.
2. Apply RTV sealant to switch threads as described in text.
3. Lubricate threads with grease.

Table 4 RECOMMENDED LUBRICANTS AND FUEL

Brake fluid	DOT 4 brake fluid
Clutch fluid	DOT 4 brake fluid
Control cables	Cable lubricant
Cooling system	Honda HP Coolant or equivalent[1]
Engine oil[2]	
Classification	
JASO T 903 standard rating	MA
API classification	SG or later[3]
Viscosity rating	SAE 10W-40
Final drive unit	Hypoid gear oil, SAE 80
Fork oil	Pro Honda Suspension Fluid SS-8 or equivalent 10 wt fork oil
Fuel	Unleaded gasoline with a pump octane number of 86 or higher[4]

1. Coolant must not contain silicate inhibitors.
2. Do not use oil with molybdenum additives.
3. Use API SG or later classified oils not specified as ENERGY CONSERVING. See text for additional information.
4. Leaded gasoline will damage the catalytic converter.

Table 5 ENGINE OIL CAPACITY

	Liters	U.S. qt.
At oil change only	3.6	3.8
Oil and filter change	3.7	3.9
After engine disassembly	4.6	4.9

Table 6 ENGINE COOLANT CAPACITY

	Liters	U.S. qt.
Radiator and engine	3.53	3.73
Reserve tank	0.65	0.69

Table 7 FINAL DRIVE OIL CAPACITY

	ml	U.S. oz.
At oil change	120	4.1
After final drive disassembly	150	5.1

Table 8 FORK OIL CAPACITY

ml	U.S. oz.
Fork tube runout limit	0.20 mm (0.008 in.)
Fork oil capacity	
Left fork tube	526.5-531.5 ml (17.82-17.98 U.S. oz.)
Right fork tube	482.5-487.5 ml (16.32-16.48 U.S. oz.)
Fork oil level	128 mm (5.0 in.)
Fork oil type	Pro-Honda Suspension Fluid SS-8 or equivalent 10 wt. fork oil
Spring free length	
New	335.3 mm (13.20 in.)
Service limit	328.6 mm (12.94 in.)

Table 9 TIRE INFLATION PRESSURE*

	Front kPa (psi)	Rear kPa (psi)
Rider	250 (36)	280 (41)
Maximum weight capacity	250 (36)	280 (41)

*The tire inflation pressures are for original equipment tires. Aftermarket tires may require different inflation pressure. Refer to the tire manufacturer's specifications.

CHAPTER FOUR

ENGINE TOP END

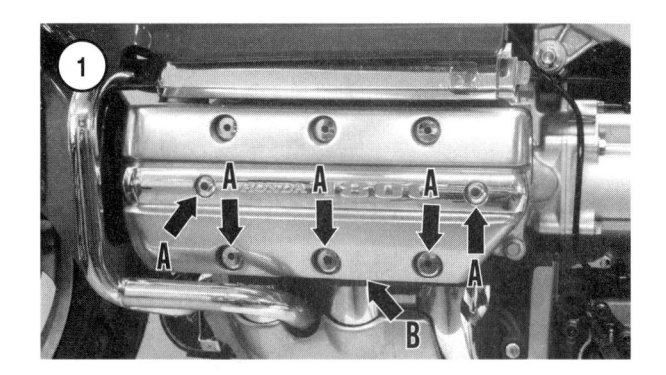

This chapter provides complete service and overhaul procedures, including information for disassembly, removal, inspection, service and reassembly of the camshafts, cylinder heads and valve assemblies. Piston, connecting rod and crankshaft procedures are in Chapter Five. Clutch procedures are in Chapter Six and the transmission procedures are in Chapter Seven.

The engine is a four valve per cylinder, liquid-cooled, flat six-cylinder with overhead camshafts (OHC). The camshafts operate directly on top of the valve lifters and are chain driven by the timing sprocket on the crankshaft. Shims under the valve lifters determine valve clearance.

Before starting any work, review Chapter One.
Engine specifications are in **Tables 1-3** at the end of the chapter.

CYLINDER HEAD COVERS

Each cylinder is equipped with a cylinder head side cover and cylinder head cover.

Removal

1. Support the motorcycle on its centerstand.
2. Use compressed air to remove debris from the area around the cylinder head cover.
3. Remove the rubber plugs, bolts (A, **Figure 1**) and cylinder head side cover (B).
4. Remove the bolt and wire clamp (A, **Figure 2**).
5. Disconnect the spark plug wires from the clip (B, **Figure 2**).

NOTE
*Note the spark plug wire routing (A, **Figure 3**) so the wires can be reinstalled correctly.*

6. Remove the rubber plugs from the exposed cylinder head cover mounting bolts. See **Figure 4** (left side) or **Figure 5** (right side).
7. Remove the bolts, washers and the cylinder head covers (**Figure 4** or **Figure 5**).

Installation

1. Remove all sealer residue from the cylinder head gasket surface.
2. Replace the cylinder head cover mounting bolt washers if damaged.
3. Inspect the gasket (**Figure 6**) around the perimeter of the cylinder head cover. Replace the gasket if it is starting to deteriorate or harden.
4. If necessary, replace the cylinder head cover gasket or reinstall the original gasket as follows:
    a. Remove the old gasket (**Figure 6**) and clean off all gasket sealer residue from the cylinder head cover.
    b. Clean the gasket groove around the perimeter of the cover and around the spark plug holes.

    NOTE
    In substep c, apply gasket sealer onto the side of the new gasket that seats against the cylinder head cover. Do not apply gasket sealer to the side of the gasket that mates to the cylinder head.

    c. Apply Gasgacinch, or a similar adhesive, to the cylinder head cover grooves and upper side of the new gasket, following the sealant manufacturer's instructions.
    d. Install the gasket into the groove in the cylinder head cover (**Figure 6**). Make sure the gasket is seated in the cover groove with no gap.
5. Apply RTV silicone onto the edges of the half-moon cutout in the cylinder head (**Figure 7**).

    NOTE
    Allow the silicone to set for 10-15 minutes before installing the cylinder head cover.

6. Install the cylinder head cover onto the cylinder head. Confirm that the gasket seats squarely into the cylinder head cutout.
7. Install the washers onto the mounting bolts with their UP mark facing up.
8. Install the cylinder head cover bolts (**Figure 4** or **Figure 5**) and washers. Tighten the bolts in two or three steps to 12 N•m (106 in.-lb.) in a crisscross pattern. Install the rubber plugs in the three-exposed mounting bolts.
9. Connect the spark plug wires into the clip (B, **Figure 3**).

ENGINE TOP END

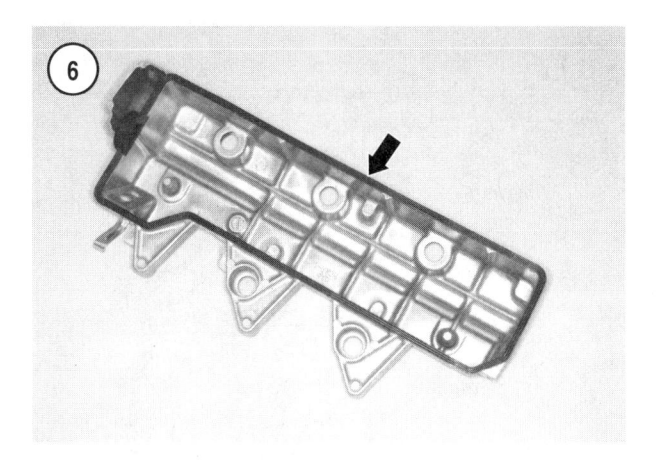

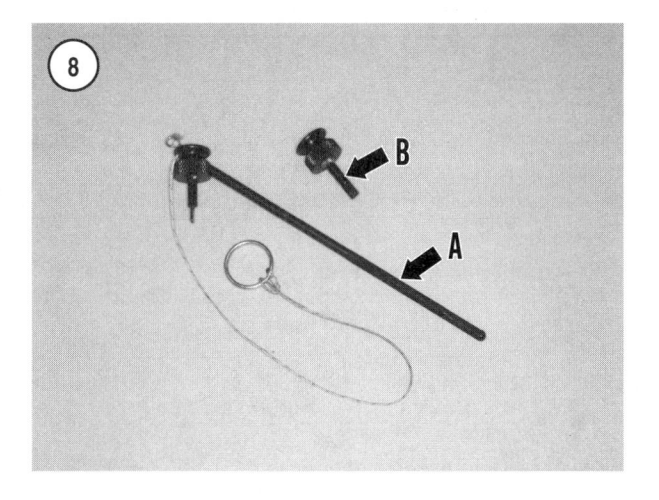

10. Install the wire clamp and secure with the bolt (A, **Figure 2**). Check the spark plug wire routing (A, **Figure 3**).

11. Install the cylinder head side cover (B, **Figure 1**) and tighten the mounting bolts (A) to 10 N•m (88 in.-lb.). Install a rubber plug into each bolt recess.

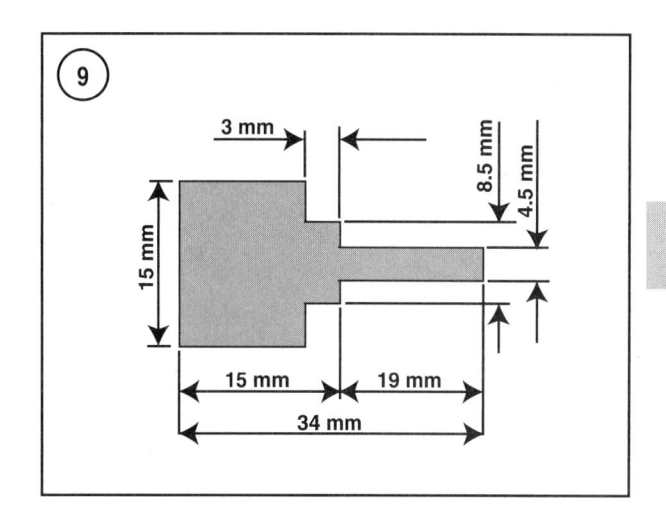

CAMSHAFTS

The camshafts can be serviced with the engine installed in the frame.

Tools

1. Tensioner holder A (Honda part No. 07ZMG-MCAA300 [A, **Figure 8**]). Use this tool when removing the right camshaft with the engine installed in the frame.
2. Tensioner holder B (Honda part No. 07ZMG-MCAA400 [B, **Figure 8**]). Use this tool to adjust the left side cam chain tensioner.
3. If the engine is being serviced out of the frame, the lock plate shown in **Figure 9** can be used instead of tensioner holder A. Fabricate the lock plate from a piece of 1.0 mm thick steel. This tool can also be used in place of tensioner holder B.

Camshaft Removal

This section describes removal of both camshafts. If it is only necessary to remove one camshaft, it is still necessary to remove both cylinder head covers to view the camshaft timing marks.

1. Disconnect the battery negative lead. Refer to *Battery* in Chapter Nine.
2. Remove the center inner fairing and front lower faring. Refer to *Front Lower Fairings* in Chapter Seventeen.
3. Remove the right inner fairing. Refer to *Inner Fairings* in Chapter Seventeen.
4. Disconnect the left horn connectors and remove the left horn.
5. Remove both cylinder head covers. Refer to *Cylinder Head Covers* in this chapter.

6. Remove the right cam chain tensioner bolt (**Figure 10**) and sealing washer. Discard the washer.

7A. If the engine is installed in the frame, install tensioner holder A as follows:

   a. Unscrew the rod from the tool. Wind the string clockwise around the tool and insert the rod through the ring and thread back into the tool (A, **Figure 11**).

   b. Pull the coolant hoses away from the tensioner housing. Install tensioner A into the right tensioner housing so the tabs on the tool blade are just above the slots in the tensioner (B, **Figure 11**). Do not engage the tabs on the tool blade into the slots in the tensioner housing at this time.

   c. Unscrew the rod from the tool and pull the string until the tool stops turning. Push the tool to engage the tabs on the tool blade into the slots in the tensioner housing. This can be confirmed when the tool seats flush against the tensioner housing (C, **Figure 11**).

7B. If the engine is removed from the frame, refer to **Figure 9** and fabricate the lock plate from a piece of 1.0 mm thick steel. Using this tool, rotate the tensioner shaft clockwise to the fully retracted position and lock the right tensioner in this position (**Figure 12**).

8. Remove the left cam chain tensioner sealing bolt (**Figure 13**) and washer. Discard the washer. Use rags to catch any oil that drains from the tensioner housing.

9. Install tensioner B (or the lock plate) into the left tensioner housing so the tabs on the tool blade are just above the slots in the tensioner (B, **Figure 11**). Using this tool, rotate the tensioner shaft clockwise to the fully retracted position and lock the left tensioner in this position.

10. Remove the spark plugs. This makes it easier to rotate the engine. Refer to *Spark Plugs* in Chapter Three.

11. Remove the timing hole cap from the timing cover at the front of the engine.

12. Turn the crankshaft (**Figure 14**) counterclockwise and align the 1.2T | F mark on the pulse generator rotor (A, **Figure 15**) with the index mark on the front crankcase cover (B). Confirm the following:

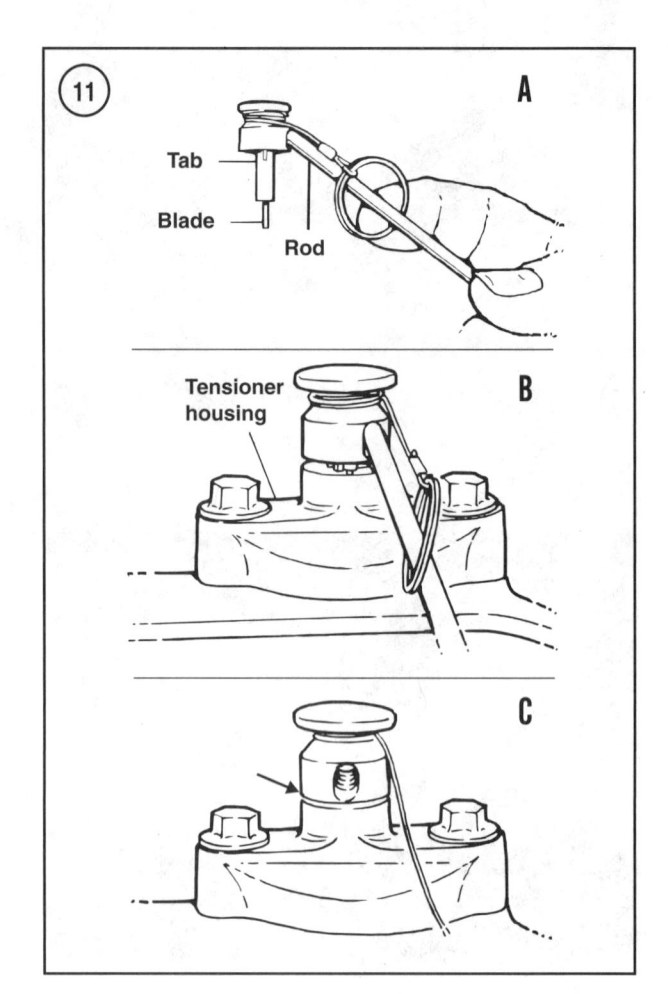

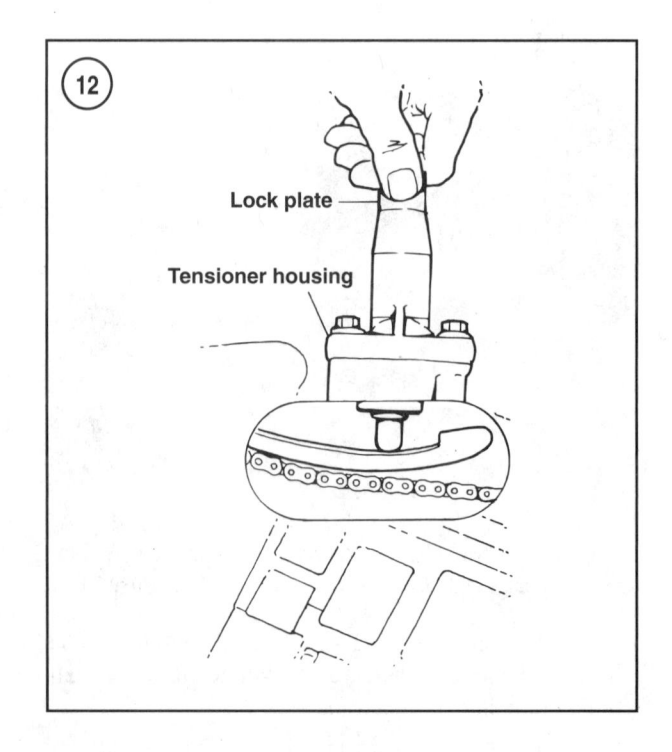

ENGINE TOP END

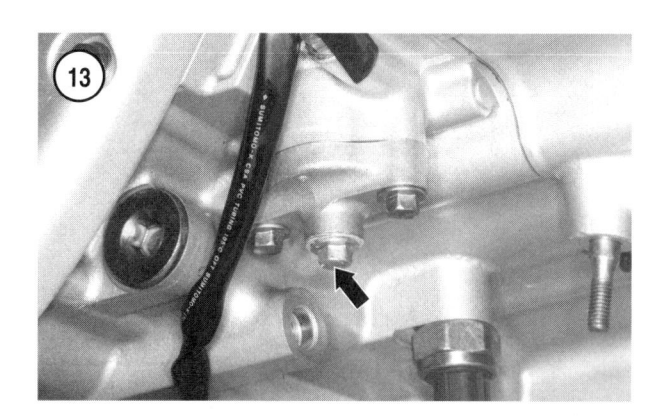

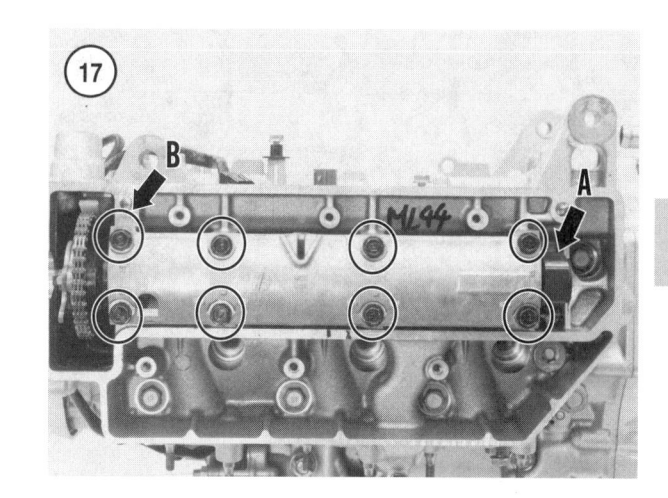

a. The timing marks on the left camshaft sprocket must align with the cylinder head surface (**Figure 16**) and the No. 6 intake camshaft lobe (A, **Figure 17**) must be facing out.

b. If the camshaft lobe is facing inward, rotate the crankshaft counterclockwise 360° (1 full turn) and realign the 1.2T | F mark. The camshaft lobe (A, **Figure 17**) is now facing out.

CAUTION
Due to the valve spring pressure applied against the camshaft holder, failure to loosen the camshaft holder bolts as described in Step 13 and Step 15 may cause the camshaft holder(s) to break. Because the camshaft holders and cylinder heads are matched units, a broken camshaft holder also requires replacing the cylinder head.

13. Remove the left camshaft holder and camshaft as follows:

    a. Loosen the bolts (B, **Figure 17**) gradually, starting with the outside bolts and work inward to prevent damaging the camshaft holder and camshaft.

    b. Tap the camshaft holder and remove it and its dowel pins.

NOTE
The dowel pins are a tight fit in the camshaft holder. Do not remove them unless necessary.

    c. Disengage the camshaft chain from the sprocket and remove the camshaft (**Figure 18**).

CAUTION
When rotating the crankshaft in Step 14, pull the left camshaft chain outward to prevent it from jamming against the crankshaft timing

sprocket. *Doing so could damage the chains and sprocket.*

14. Turn the crankshaft (**Figure 14**) counterclockwise and align the 1.2T │ F mark on the pulse generator rotor (A, **Figure 15**) with the index mark on the front crankcase cover (B). Check that the timing marks on the right camshaft sprocket align with the cylinder head surface (**Figure 19**) and the No. 5 intake camshaft lobe (A, **Figure 20**) is facing out.

15. Remove the right camshaft holder and camshaft as follows:

   a. Loosen the bolts (B, **Figure 20**) gradually, starting with the outside bolts and working inward to prevent damaging the camshaft holder and camshaft.
   b. Tap the camshaft holder and remove it and its dowel pins.

NOTE
The dowel pins are a tight fit in the camshaft holder. Do not remove them unless necessary.

   c. Disengage the camshaft chain from the sprocket and remove the camshaft (**Figure 21**).

16. Before removing the valve lifters and shims, note the following:

   a. Use a divided container to store the valve lifters and shims to install them in their original mounting positions.
   b. The shims may stick to the bottom of the valve lifter. Remove the valve lifters carefully to prevent a shim from falling into the engine.
   c. Remove the valve lifters carefully to avoid damaging the lifter bores in the cylinder head.

17. Remove one of the valve lifters (A, **Figure 22**) and its respective shim (**Figure 23**) and place both of them in the correct location in the holder.

18. Repeat Step 17 for the other valve lifters and shims.

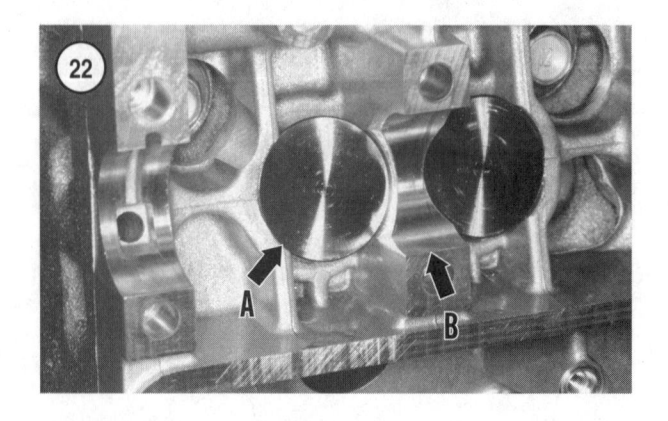

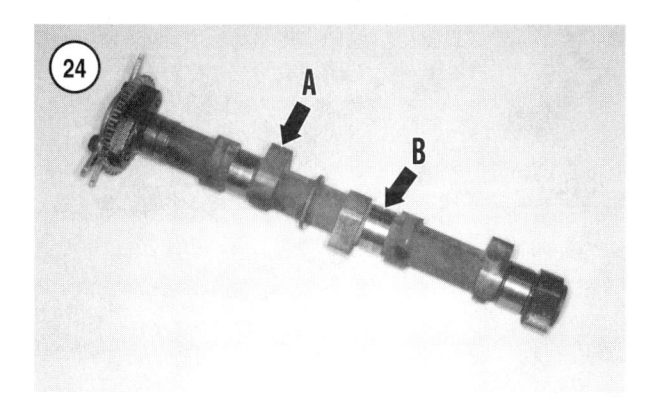

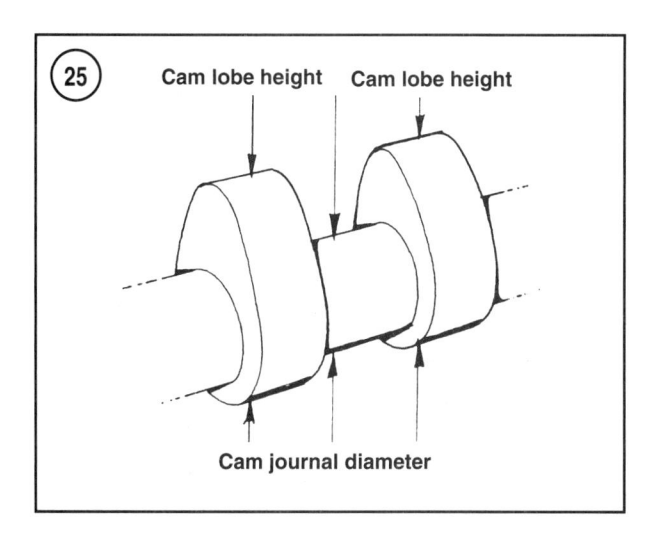

ENGINE TOP END

2. Clean the camshafts in solvent and dry thoroughly. Flush the camshaft oil passages with solvent and compressed air.
3. Check the cam lobes (A, **Figure 24**) for wear. The lobes should not be scored and the edges should be square. Replace the camshaft if the lobes are scored, worn or damaged.
4. Check the camshaft bearing journals (B, **Figure 24**) for wear or scoring. Replace the camshaft if the journals are scored, worn or damaged.
5. If the camshaft lobes or journals are excessively worn or damaged, check the journal surfaces in the cylinder head and in the camshaft holders. Refer to *Camshaft Holder Inspection* in this section.
6. Measure each cam lobe height (**Figure 25**) with a micrometer.
7. Measure each cam journal outside diameter (**Figure 25**) with a micrometer.
8. Support the camshaft journals on a set of V-blocks or crankshaft truing stand and measure runout with a dial indicator. Note the following:
   a. If the runout is out of specification, replace the camshaft and measure the camshaft oil clearance as described in this section.
   b. If the camshaft oil clearance is incorrect with the new camshaft, the original (bent) camshaft damaged the cylinder head and camshaft holder journals.
9. Inspect the camshaft sprockets (**Figure 26**) for broken or chipped teeth. Also, inspect the cam chains and timing sprocket mounted on the crankshaft. See *Cam Chains and Timing Sprocket* in this chapter. To replace the camshaft sprockets, refer to *Camshaft Sprocket Replacement* in this section.

Camshaft Inspection

When measuring the camshafts, compare the actual measurements to the specifications in **Table 2**. Replace worn or damaged parts as described in this section.

1. Before cleaning the camshafts, inspect the oil lubrication holes for contamination. Make sure these holes are clean and open.

Camshaft Holders Inspection

1. Before cleaning the camshaft holders (A, **Figure 27**), inspect the oil lubrication holes for contamination. Small passages and holes in the camshaft holders provide lubrica-

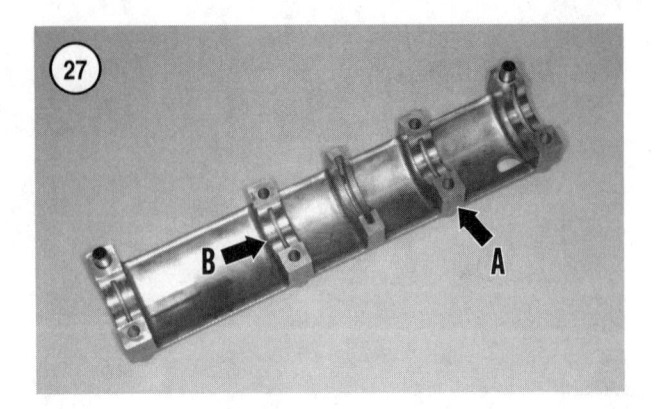

tion for the holder, camshaft and cylinder journals. Make sure these passages and holes are clean and open.

NOTE
Infrequent oil and filter changes may be indicated if the camshaft holder passages are dirty. Contaminated oil passages can cause camshaft and journal failure.

2. Clean and dry all parts.
3. Check the camshaft holders for stress cracks and other damage.
4. Check the camshaft bearing journals (B, **Figure 27**) in the camshaft holders and cylinder head (B, **Figure 22**) for wear and scoring. If there is visual damage, replace the cylinder head and camshaft holder as a set. To determine operational clearance perform the *Camshaft Oil Clearance Measurement* procedure in this section.
5. Check the camshaft holder mounting bolts for damage.

Camshaft Oil Clearance Measurement

This section describes how to measure the clearance between the camshaft and the camshaft holders and cylinder head journal using Plastigage (**Figure 28**). Plastigage is a material that flattens when pressure is applied to it. The marked bands on the envelope are then used to measure the width of the flattened Plastigage. The camshafts and camshaft holder must be installed on the cylinder head when performing this procedure.

Plastigage is available from automotive parts stores in different clearance ranges. When purchasing Plastigage to measure the camshaft oil clearance, select the 0.025-0.076 mm (0.001-0.003 in.) clearance range set.

1. Wipe all oil residue from each cam bearing journal (camshafts, camshaft holders and cylinder head). These surfaces must be clean and dry.
2. The original equipment camshafts are identified by their L (left) and R (right) cast marks (**Figure 29**). 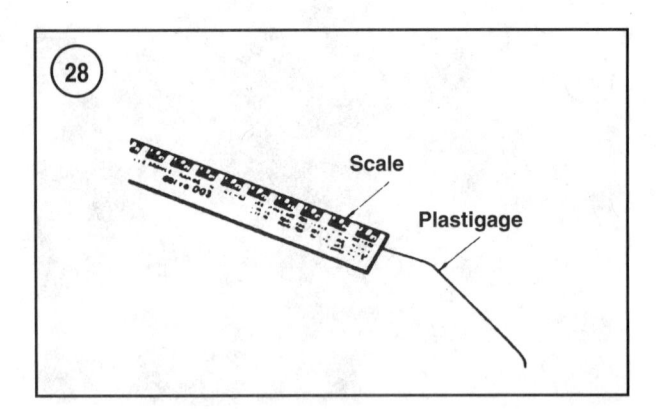 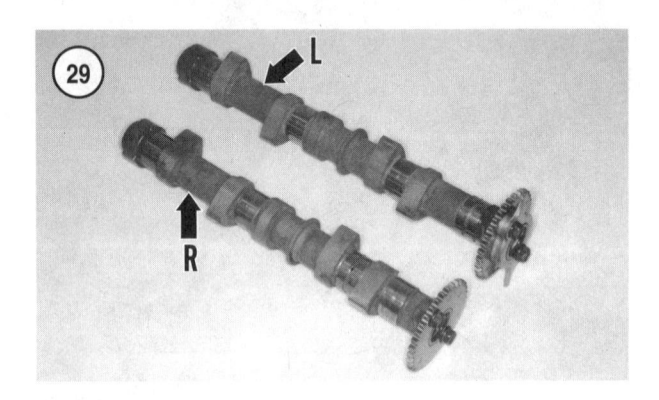
3. Install the camshaft into the cylinder head. Position the cam lobes so the valves will not be pressed open when the cam is installed. Refer to *Camshaft Installation* in this section.
4. Place a strip of Plastigage material on the top of each camshaft bearing journal (**Figure 30**), parallel to the cam.
5. Install and tighten the camshaft holders as described under *Camshaft Installation* in this section.

CAUTION
Do not rotate the camshafts with the Plastigage in place.

CAUTION
Loosen the camshaft holder bolts as described or the camshaft holders may become damaged.

6. Loosen and remove the camshaft holder mounting bolts as described under *Camshaft Removal* in this section.
7. Remove the camshaft holder carefully, making sure the camshafts do not rotate.
8. Measure the widest portion of the flattened Plastigage according to the manufacturer's instructions (**Figure 31**) and compare to the camshaft oil clearance specification in **Table 2**. Note the following:

ENGINE TOP END

30

Plastigage

31

32

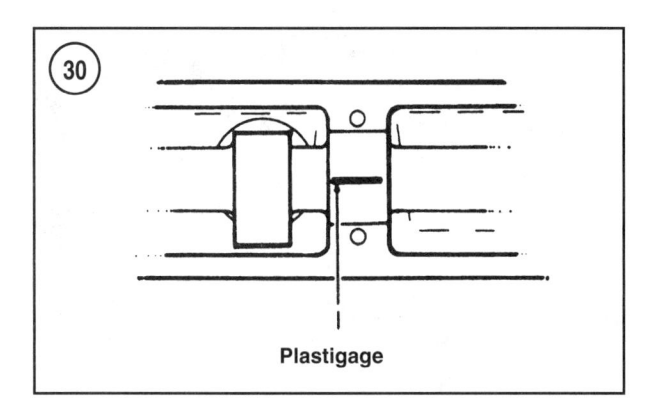

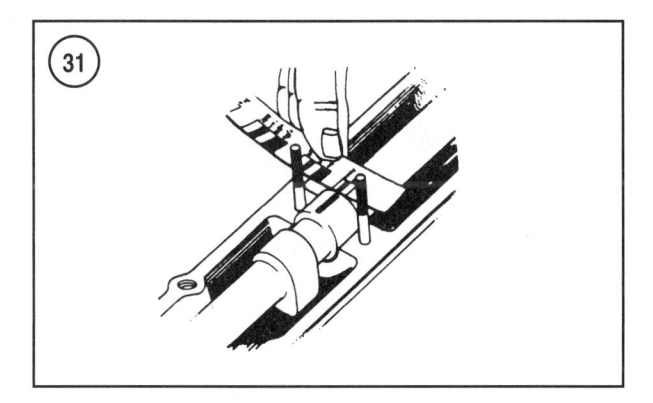

R   L

33

a. If all the measurements are within specification, the cylinder head, camshaft and camshaft holder can be reused.

b. If any measurement exceeds the service limit, replace the camshaft and recheck the oil clearance.

c. If the new measurement exceeds the service limit with the new camshaft, replace the camshaft holder and cylinder head as a set.

9. Remove all Plastigage material from the camshafts, camshaft holders and cylinder head.

Camshaft Sprocket Replacement

Refer to *Camshaft Inspection* in this section to inspect the camshaft sprockets. If the sprocket teeth are worn or damaged, inspect the cam chains and timing sprocket mounted on the crankshaft. Refer to *Cam Chains and Timing Sprocket* in this chapter.

1. Remove the bolts and sprocket. Also, remove the rotor and spacer on the left camshaft.
2. Remove all threadlock residue from the bolt and camshaft threads.
3. Clean and inspect the mounting bolts, rotor and spacer.
4. Align the bolt holes and install the sprockets with their timing marks facing out (**Figure 32**). Install the spacer and rotor on the left camshaft. The original equipment camshafts are identified by their L (left) and R (right) cast marks (**Figure 29**).
5. Apply a medium strength threadlock and tighten the sprocket mounting bolts to 25 N•m (18 ft.-lb.).

Valve Lifter and Shim Inspection

Maintain the correct alignment of the valve lifters and shims when inspecting them. When measuring the valve lifter and bore, compare the measurements to the specifications in **Table 2**.

1. Inspect the valve lifters and shims (**Figure 33**) for wear and damage. Inspect the shims for stress cracks and other damage.
2. Check each valve lifter for any scoring or other damage. The lifter must operate in its cylinder head bore with no binding or chatter. If the side of a lifter is damaged, replace it.
3. Measure the valve lifter outside diameter (**Figure 34**).
4. Check the valve lifter bores in the cylinder head for any scoring or damage. These surfaces must be smooth.

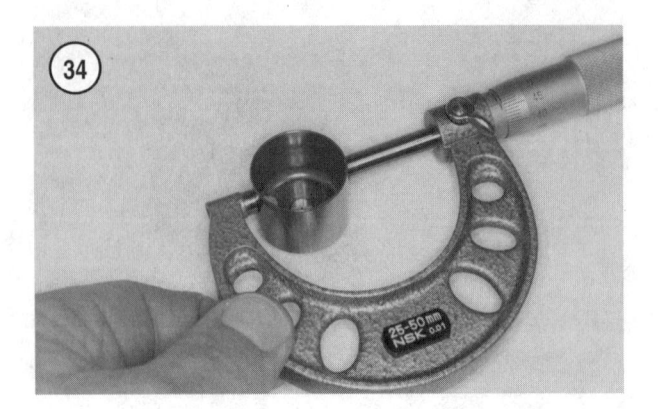

5. Measure the lifter bore inside diameter (**Figure 35**).

Valve Lifter and Shim Installation

Make sure to install the shims and valve lifters in their original operating positions.

NOTE
If the shims and valve lifters were not stored in a marked divided container, the valve clearance will have to be determined after installing the camshafts and camshaft holders.

1. Lubricate the shims with engine oil.
2. Lubricate the outer lifter bore surface with molybdenum oil solution.

NOTE
Molybdenum oil solution is a 50:50 mixture of engine oil and molybdenum disulfide grease.

3. Install a shim into the valve retainer bore (**Figure 23**). Make sure it seats correctly.
4. Install the valve lifter (A, **Figure 22**) over the shim and into the cylinder head bore.
5. Repeat for each shim and valve lifter.

Camshaft Installation

1. If only one camshaft was removed, remove the other cylinder head cover to view the installed camshaft timing marks.
2. If removed, install the valve lifters and shims as described in this section.
3. The original equipment camshafts are identified by their L (left) and R (right) cast marks (**Figure 29**).

NOTE
*In the following steps, rotate the engine with a socket on the pulse generator bolt (**Figure 14**).*

CAUTION
When rotating the crankshaft in the following steps, pull both camshaft chains outward to prevent them from jamming against the crankshaft timing sprocket, which could damage the chains and sprocket.

4. If the engine was rotated after the camshaft(s) were removed, turn the crankshaft (**Figure 14**) counterclockwise and align the 1.2T│F mark on the ignition pulse generator (A, **Figure 36**) with the index mark on the front crankcase cover (B). Note the following:

   a. If only one camshaft was removed, continue with Step 5.
   b. If both camshafts were removed, go to Step 6A to install the left camshaft, then install the right camshaft (Step 6B).

5. Before installing the other camshaft, check that the *installed* camshaft timing marks align with the cylinder head surface and that its M mark is facing out:

ENGINE TOP END

37 RIGHT CAMSHAFT INSTALLATION

Right camshaft — Left camshaft

38 LEFT CAMSHAFT INSTALLATION

Right camshaft — Left camshaft

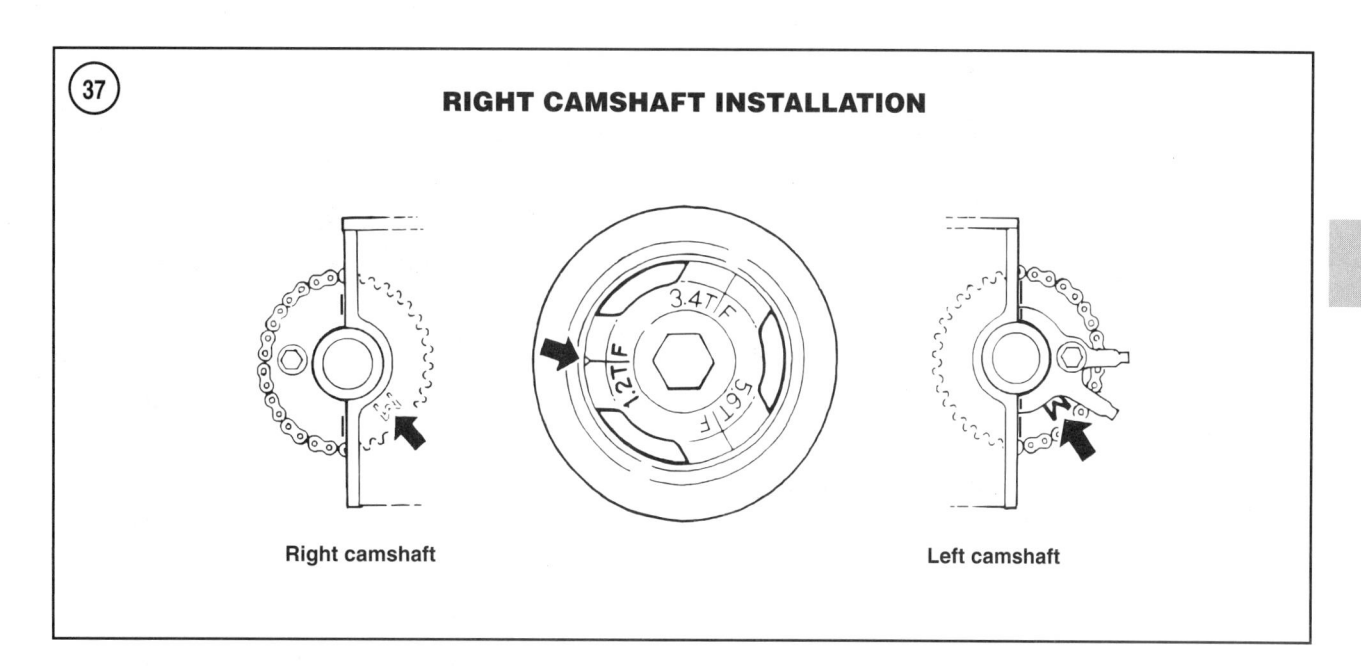

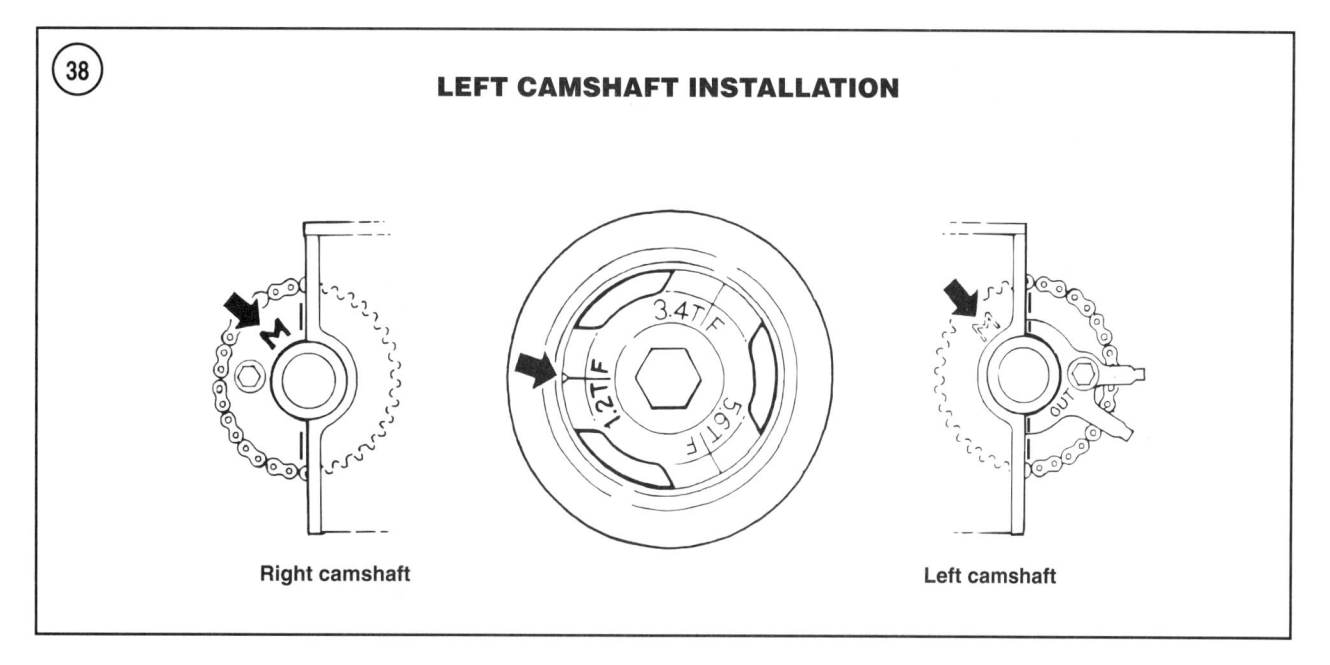

a. If the left camshaft was not removed, its M mark and timing marks should be positioned as shown in **Figure 37**.

b. If the right camshaft was not removed, its M mark and timing marks should be positioned as shown in **Figure 38**.

c. If the M mark for the *installed* camshaft is facing in and cannot be seen, rotate the crankshaft counterclockwise 360° (1 full turn) and realign the 1.2T|F mark. The M mark should be visible and the camshaft's rear lobe should be facing in.

d. The engine is now positioned for installation of the opposite camshaft. Perform either Step 6A (left camshaft) or Step 6B (right camshaft).

6A. Install the left camshaft:

a. Apply molybdenum oil solution to the cylinder head camshaft journals, thrust surfaces and cam lobes.

b. Refer to **Figure 38** and install the left camshaft. Mesh the camshaft sprocket with the cam chain. Align the timing marks on the sprocket with the cylinder surface, making sure the M mark on the sprocket is facing in (**Figure 38**) and the rear cam lobe (A, **Figure 39**) is facing out.

c. Make sure the chain guides are positioned over the sprocket and chain (B, **Figure 39**).

d. Install the camshaft holder as described in Step 7.

e. Install the right camshaft, if removed.

6B. Install the right camshaft:

a. If the left camshaft was installed in Step 6A, rotate the crankshaft counterclockwise 360° (1 full turn) and realign the 1.2T│F mark. The M mark on the left camshaft and timing marks should be positioned as shown in **Figure 37**.

b. Apply molybdenum oil solution to the cylinder head camshaft journals, thrust surfaces and cam lobes.

c. Refer to **Figure 37** and install the right camshaft. Mesh the camshaft sprocket with the cam chain. Align the timing marks on the sprocket with the cylinder surface, making sure the M mark on the sprocket is facing in (**Figure 37**) and the rear cam lobe (A, **Figure 40**) is facing out.

d. Make sure the chain guides are positioned over the sprocket and chain (B, **Figure 40**).

e. Install the camshaft holder as described in Step 7.

7. Install the camshaft holder and tighten the mounting bolts:

NOTE
The camshaft holders are identified by a letter and number. L-1 is the left camshaft holder. R-1 is the right camshaft holder.

a. If removed, install the dowel pins (**Figure 41**) into the camshaft holder.

b. Apply engine oil to the camshaft holder mounting bolt threads and head seating surfaces. Do not install these bolts with dry threads.

c. Install the camshaft holder over the camshaft. Make sure the two dowel pins enter the cylinder head correctly. See **Figure 42** (left side) or **Figure 43** (right side).

d. Install all camshaft holder mounting bolts, but do not tighten them.

CAUTION
The camshaft holder must be tightened so that it moves evenly against the cylinder head surface. Tightening only one side of a holder or overtightening the bolts may break the camshaft holder or camshaft. Make sure to tighten the camshaft holder mounting bolts as described in the following steps.

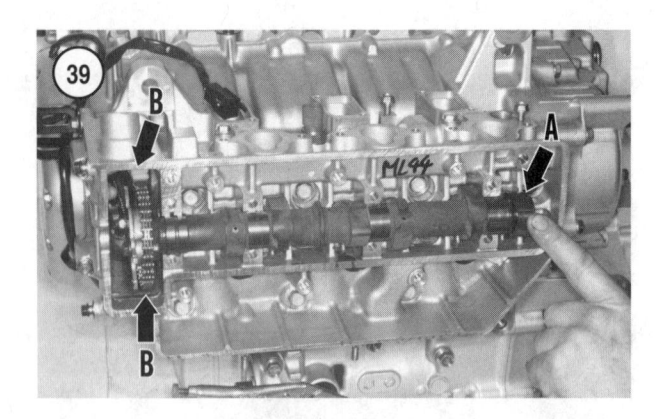

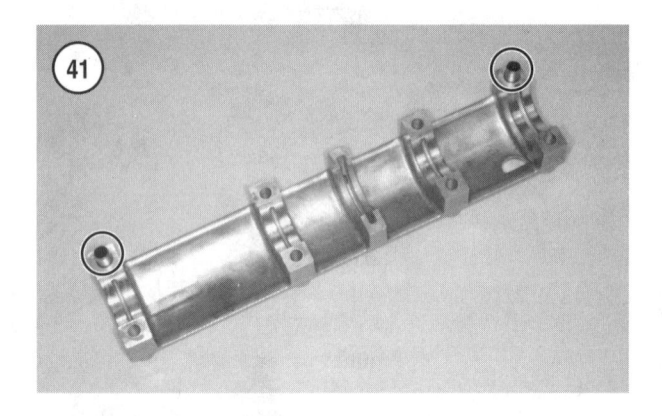

ENGINE TOP END

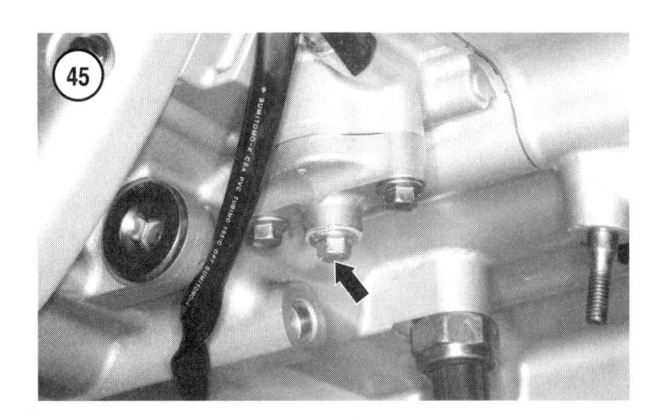

43

44

45

e. Tighten the camshaft holder bolts in the numerical order cast on the camshaft holder, and in several steps to a final torque specification of 12 N•m (106 in.-lb.).

8. Remove the lock plate (or the Honda tensioner holder tools) from each cam chain tensioner to allow the pushrod to extend.

CAUTION
The tensioner should click after removing the lock plate, indicating the pushrod has extended. If not, remove the tensioner and check tensioner operation.

9. Carefully, recheck the camshaft timing. The timing marks on each camshaft sprocket must align with the cylinder head surface when the 1.2T | F mark on the ignition pulse generator (A, **Figure 36**) aligns with the index mark on the front crankcase cover (B).

CAUTION
The timing marks must align correctly at this time; otherwise, camshaft timing will be incorrect. Do not proceed if the camshaft sprocket timing marks are positioned incorrectly.

10. Install the tensioner sealing bolts and new washers (**Figure 44** and **Figure 45**), and tighten to 12 N•m (106 in.-lb.).
11. Lubricate the timing hole cap threads and O-ring with grease and tighten to 18 N•m (13 ft.-lb.).
12. Perform Steps 1-6 under *Camshaft Removal* to complete installation.

CAM CHAINS AND TIMING SPROCKET

The cam chains and timing sprockets (**Figure 46**) can be removed with the engine mounted in the frame.

Removal

1. Drain the engine oil. Refer to *Engine Oil and Filter* in Chapter Three.
2. Remove the camshafts as described in this chapter.
3. Remove the bolts and the front crankcase cover (4, **Figure 47**).
4. If the engine is installed in the frame, remove the ignition pulse generator (7, **Figure 47**) to remove the cover from the engine:
   a. Remove the bolts and wire retainer.
   b. Remove the grommet from the cover groove and remove the pulse generator.
5. Remove the dowel pins and gasket.
6. Loosen the rotor bolt (A, **Figure 48**):
   a. If the engine is installed in the frame, shift the transmission into gear to lock the crankshaft and have an assistant apply the rear brake.
   b. If the engine is removed from the frame, shift the transmission into gear and lock the output shaft with the shaft holder (Honda part No. 07924-PJ40001 [**Figure 49**]).
7. Remove the rotor bolt (A, **Figure 48**), washer and pulse generator rotor (B).
8. Remove the following:
   a. Tensioner pivot bolt and outer collar (A, **Figure 50**).

CHAPTER FOUR

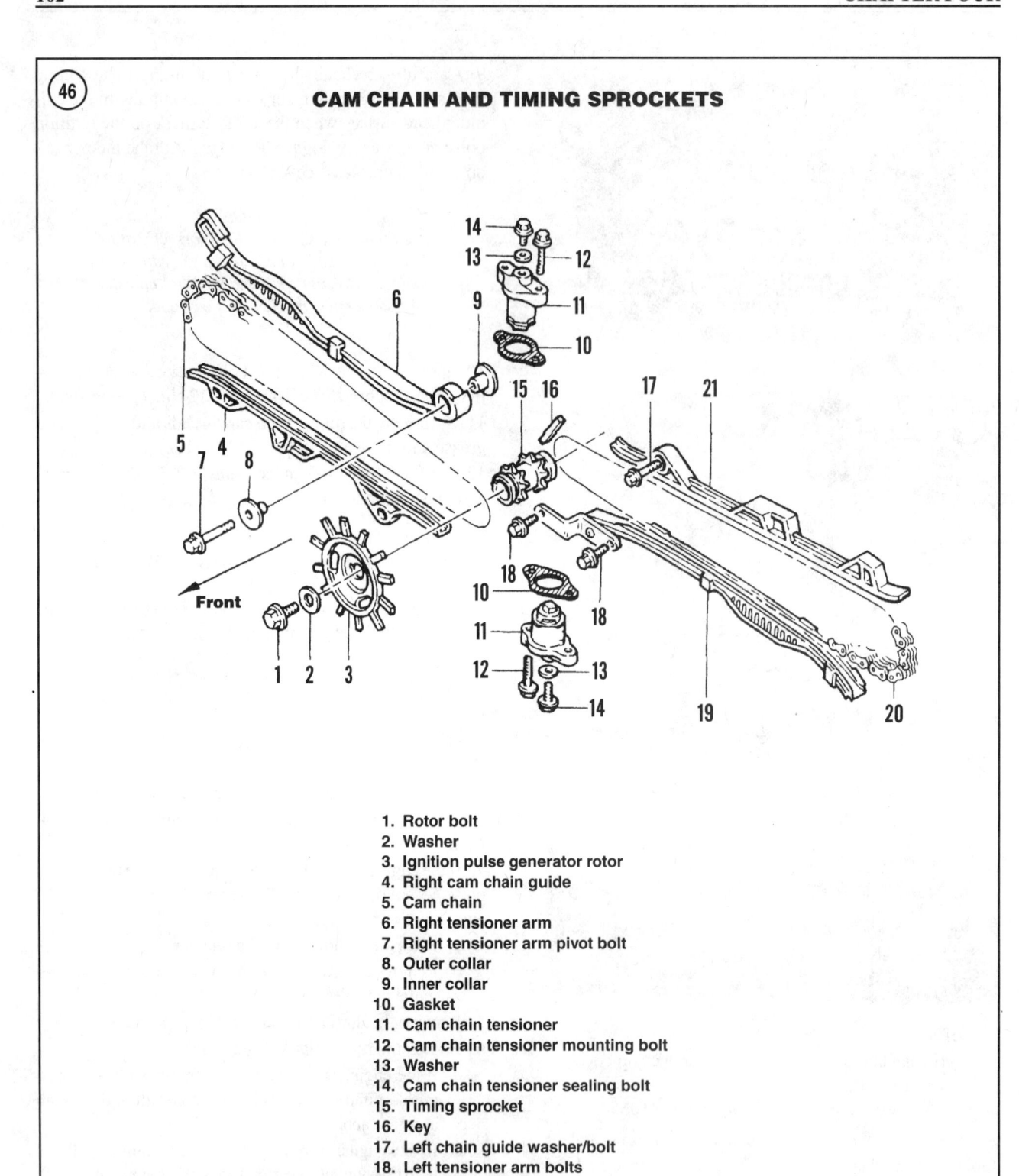

CAM CHAIN AND TIMING SPROCKETS

1. Rotor bolt
2. Washer
3. Ignition pulse generator rotor
4. Right cam chain guide
5. Cam chain
6. Right tensioner arm
7. Right tensioner arm pivot bolt
8. Outer collar
9. Inner collar
10. Gasket
11. Cam chain tensioner
12. Cam chain tensioner mounting bolt
13. Washer
14. Cam chain tensioner sealing bolt
15. Timing sprocket
16. Key
17. Left chain guide washer/bolt
18. Left tensioner arm bolts
19. Left tensioner arm
20. Cam chain
21. Left cam chain guide

ENGINE TOP END

47 FRONT CRANKCASE COVER AND IGNITION PULSE GENERATOR

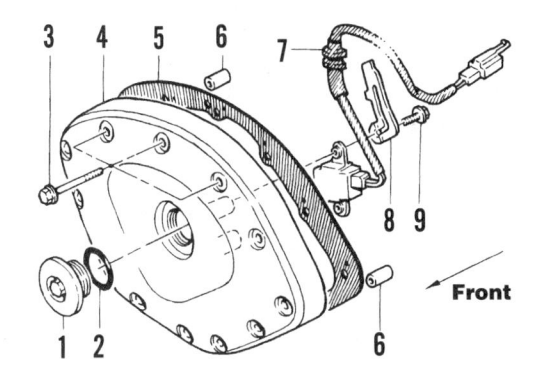

1. Timing hole cap
2. O-ring
3. Bolt
4. Front crankcase cover
5. Gasket
6. Dowel pins
7. Ignition pulse generator
8. Wire retainer
9. Bolt

b. Right tensioner arm (B, **Figure 50**) and inner collar (9, **Figure 46**).
c. Left tensioner arm bolts (C, **Figure 50**).
d. Left tensioner arm (D, **Figure 50**).
e. Right cam chain guide (E, **Figure 50**).

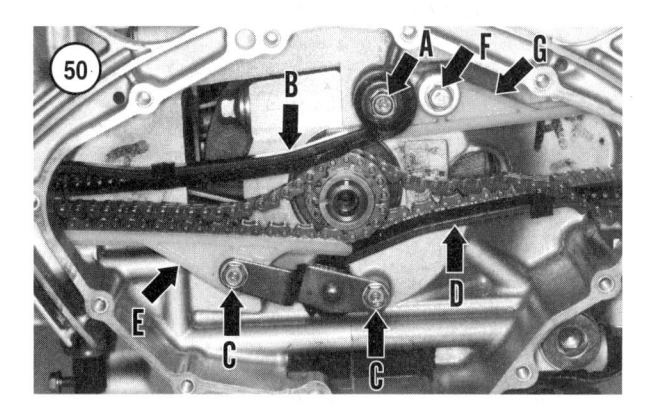

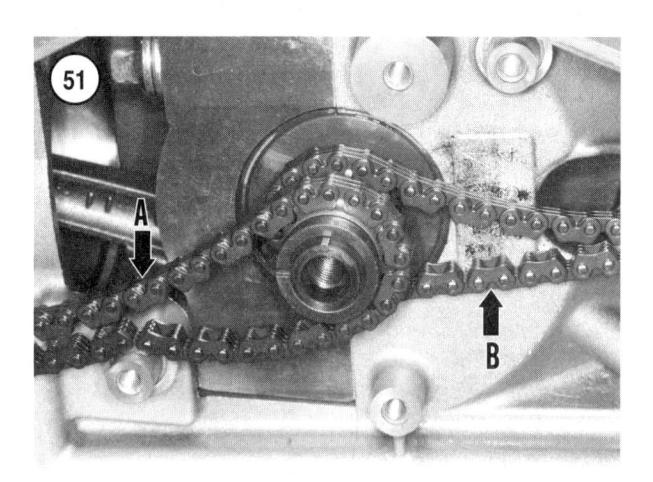

f. Left cam chain guide bolt/washer (F, **Figure 50**) and left cam chain guide (G).

NOTE
The left and right side cam chains are identical. However, identify the chains to install them in their original operating position.

g. Remove the right (A, **Figure 51**) and left (B) cam chains.

h. Note the installed direction and remove the timing sprocket (A, **Figure 52**) and key (B).

Installation

1. Clean and dry all parts.
2. Install the key (B, **Figure 52**) into the crankshaft groove.
3. Install the timing sprocket (A, **Figure 52**) by aligning its groove with the key.
4. Install the left chain (B, **Figure 51**) over the inner sprocket teeth, then install the right chain (A).
5. Install the right cam chain guide (E, **Figure 50**).
6. Install the left cam chain guide (G, **Figure 50**) and the washer/bolt (F). Tighten to 12 N•m (106 in.-lb.).
7. Install the left tensioner arm (D, **Figure 50**) and secure with the two mounting bolts (C). Tighten the bolts securely.
8. Install the inner collar (9, **Figure 46**), right tensioner arm (B, **Figure 50**), outer collar and the bolt (A). Tighten to 12 N•m (106 in.-lb.).
9. Install the ignition pulse generator rotor (B, **Figure 48**) by aligning its groove with the key.
10. Apply engine oil to the rotor bolt threads and seating surface. Refer to Step 6 of *Removal* to prevent the engine from turning over. Install the rotor bolt and washer (A, **Figure 48**) and tighten to 59 N•m (47 ft.-lb.).
11. If the ignition pulse generator was removed, install as follows:
    a. Install the ignition pulse generator into the cover.
    b. Install the grommet into the cover groove. Position the wire retainer over the wire.
    c. Apply a medium strength threadlock onto the bolt threads and tighten to 12 N•m (106 in.-lb.).
12. Clean the front crankcase cover and crankcase gasket surfaces.
13. Install the two dowel pins (B, **Figure 53**) and a new gasket.
14. Apply Yamabond No. 4 (or equivalent semi-drying sealer) to the two areas where the crankcase halves mate together (A, **Figure 53**).
15. Install the front crankcase cover and tighten its mounting bolts to 12 N•m (106 in.-lb.).
16. Refill the engine with oil. Refer to *Engine Oil and Filter* in Chapter Three.
17. Install the camshafts as described in this chapter.

Inspection

1. Clean and dry all parts.
2. Inspect the sliding surface on the tensioner arms and chain guides for wear, chipping and other damage.

ENGINE TOP END

56

Lock plate
Tensioner housing

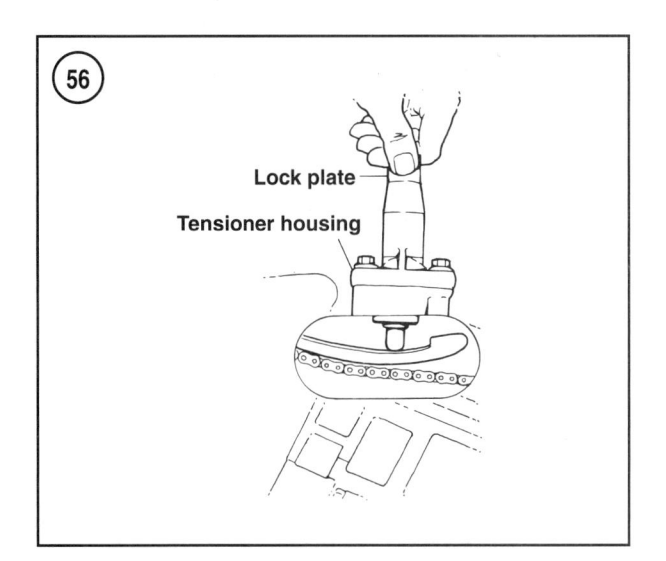

57

3. Inspect the timing sprocket groove and key for damage. Make sure the corners on the key are sharp.
4. Inspect the crankshaft groove for cracks, elongation and other damage.
5. Inspect the timing sprocket teeth (**Figure 54**) for damage. If damaged, check the cam chains and camshaft sprockets for wear.
6. Inspect the chains (**Figure 54**) for wear and damage.
7. Inspect the rotor groove and arms for damage.

CAM CHAIN TENSIONER

The engine is equipped with a cam chain tensioner for each camshaft (**Figure 46**). The left cam chain tensioner is located at the front, bottom side of the left crankcase (**Figure 45**). The right cam chain tensioner is located at the front, top side of the right crankcase (**Figure 55**). The left cam chain tensioner can be replaced with the engine mounted in the frame. Because of a frame member, the right cam chain tensioner cannot be replaced with the engine mounted in the frame.

Removal/Inspection

1. Left cam chain tensioner—Remove the following to access the chain tensioner:
   a. Remove the center inner fairing and front lower fairing. Refer to *Front Lower Fairings* in Chapter Fourteen.
   b. Remove the right inner fairing. Refer to *Inner Fairings* in Chapter Fourteen.
   c. Disconnect the left horn connectors and remove the left horn.
2. Right cam chain tensioner—Remove the engine from the frame. Refer to *Engine* in Chapter Five.

CAUTION
The cam chain tensioner is a non-return type. The internal pushrod will not return to its original position once it has moved out. After the tensioner mounting bolts are loosened, the tensioner pushrod must be reset. If the mounting bolts are loosened, do not simply retighten the mounting bolts. The pushrod has already moved out to an extended position, and it will exert excessive pressure on the chain, which will lead to engine damage.

3. Remove the sealing bolt and washer from the end of the tensioner (**Figure 45**, typical). Discard the washer.

NOTE
*Before removing the cam chain tensioner, the tensioner pushrod must be locked in the retracted position using a stopper plate. Honda offers two separate tensioner holders to do this; however, a lock plate may be fabricated from a piece of thin metal. Refer to **Camshafts** in this chapter for information on these tools.*

4. Using a lock plate or Honda tool, rotate the tensioner pushrod clockwise, then align and insert the tool into the housing grooves to keep the pushrod locked in this position. See **Figure 56**, typical.
5. Remove the mounting bolts, tensioner and gasket from the crankcase. Discard the gasket.
6. The cam chain tensioner is available only as a unit assembly. Do not disassemble it.
7. Inspect the tensioner assembly:
   a. Remove the lock plate or tensioner holder to release the pushrod (**Figure 57**). The pushrod must release forcibly with no binding.
   b. Hold the tensioner housing and try to push the pushrod in by hand. It must not move.

c. Replace the tensioner assembly if there was binding when the pushrod was released, or there was movement when the pushrod was pushed by hand.

Installation

1. Remove all gasket residue from the cam chain tensioner and crankcase surfaces.
2. Using a lock plate or tensioner holder (**Figure 58**), rotate the tensioner pushrod clockwise, then align and insert the tool into the housing grooves to keep the pushrod locked in this position.
3. Install the cam chain tensioner, new gasket and mounting bolts and tighten to 12 N•m (106 in.-lb.).
4. Remove the lock plate from the tensioner to allow the pushrod to extend.

CAUTION
The tensioner should click after removing the lock plate, indicating the pushrod has extended. If not, remove the tensioner and check tensioner operation.

5. Install the sealing bolt and a new washer and tighten to 12 N•m (106 in.-lb.).
6. Reverse Steps 1 and/or Step 2 under *Removal/Inspection* to complete installation.

CYLINDER HEAD

The cylinder heads can be removed with the engine mounted in the frame.

Removal

1. Disconnect the negative battery cable. Refer to *Battery* in Chapter Nine.
2. Remove the engine side covers. Refer to *Engine Side Covers* in Chapter Seventeen.
3. Remove both radiators. Refer to *Radiators* in Chapter Ten.
4. Remove the air filter housing. Refer to *Air Filter Housing* in Chapter Eight.
5. Remove the exhaust system. Refer to *Exhaust System* in Chapter Seventeen.
6. Unbolt and remove the left and right fuel pipe covers (**Figure 59**).
7. Unbolt and remove the left and right radiator mounting brackets.
8. Remove the camshafts as described in this chapter.
9. Unbolt and remove the spark plug wire guide plate (**Figure 60**).

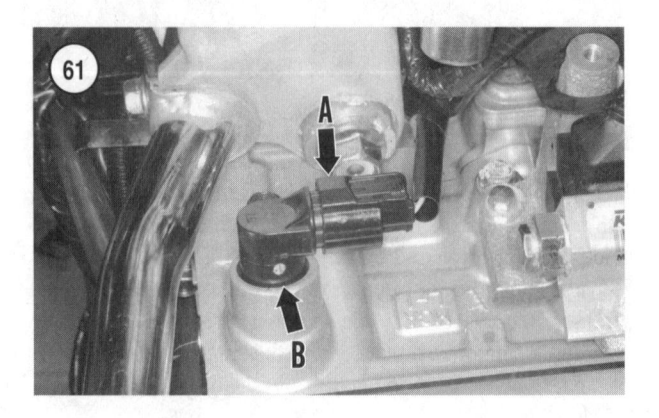

ENGINE TOP END

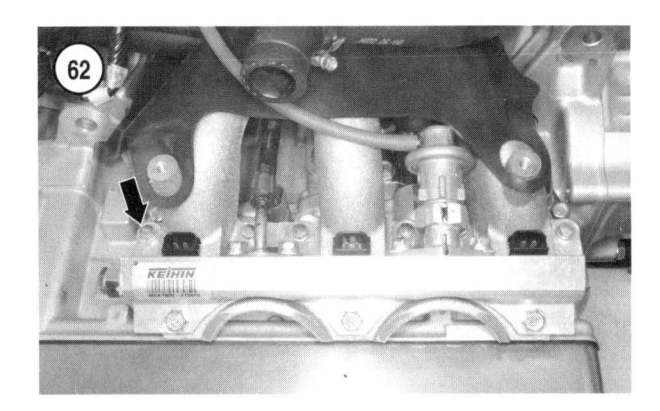

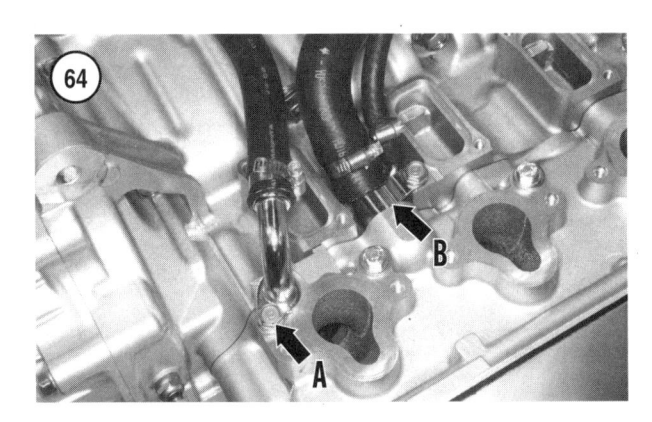

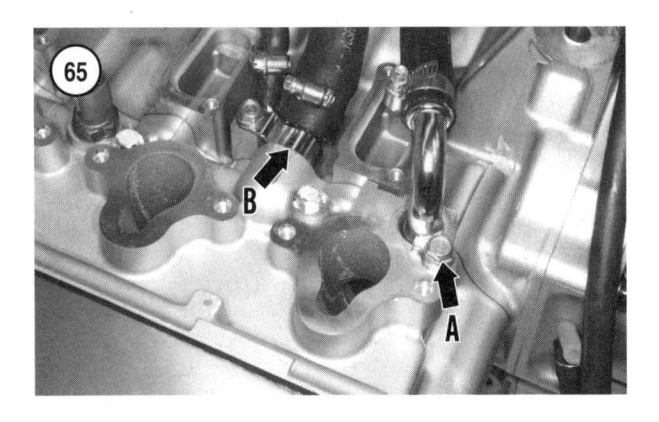

10. Disconnect the spark plugs and reroute the spark plug wires.
11. Perform the following at the left cylinder head:
    a. Disconnect the black 2-pin cam pulse generator connector (A, **Figure 61**).
    b. Remove the bolt and the cam pulse generator (B, **Figure 61**) and its O-ring.
    c. Disconnect the gray 3-pin engine coolant temperature sensor (ECT) connector.

NOTE
The intake manifold must be disconnected from both cylinder heads, even if only one cylinder head is being removed.

12. Remove the bolts securing the intake manifold to the left and right cylinder heads (**Figure 62**, typical). Lift the intake manifold and support it on both sides with wooden blocks. Then remove and discard the intake manifold gaskets.
13. Remove the ECT sensor (**Figure 63**) and its washer from the left cylinder head.
14. Remove the bolt and disconnect the water hose joint at the rear of the cylinder head. See A, **Figure 64** (left) or A, **Figure 65** (right).

NOTE
Steps 15-21 show removal of the left cylinder head. Procedures for the right cylinder head are similar.

CAUTION
Do not remove the cylinder head mounting bolts or remove the cylinder head when the engine is hot. Doing so could cause the cylinder head to warp, requiring its replacement.

15. Remove the three 6–mm cylinder head mounting bolts. See **Figure 66** and **Figure 67**.

CAUTION
Do not let the cylinder head fall after removing the mounting bolts.

16. Loosen the cylinder head bolts (**Figure 68**) in 2-3 steps in a crisscross pattern. Remove the bolts.
17. Remove the cylinder head:
    a. If the engine is installed in the frame, disconnect the water hose joint (B, **Figure 64** or B, **Figure 65**) when removing the cylinder head.
    b. Remove the cylinder head carefully to prevent damaging the intake manifold surfaces.
18. Place the cylinder head on wooden blocks to avoid damaging the gasket surfaces.

19. Remove the cylinder head gasket (A, **Figure 69**) and the two dowel pins (B).

CAUTION
If the dowel pins are tight, do not remove them unless it is necessary. Normally, stuck or rusted dowel pins are damaged during removal.

NOTE
After removing the cylinder head, check the gasket surfaces for signs of leaks. Also, check the head gasket for signs of leakage. A blown gasket could indicate possible cylinder head warp or other damage.

20. If necessary, remove the valve lifters and shims as described in this chapter.
21. Remove the O-ring from the coolant hose joint (if the hose is installed on the engine).

Solvent Test

Before removing the valves from the cylinder head, perform a solvent test to check the valve face-to-valve seat seal.

1. Remove the cylinder head as described in this section.
2. Support the cylinder head with the exhaust port facing up (**Figure 70**). Then pour solvent or kerosene into the ports. Check the combustion chamber for fluid leaking past the exhaust valves. There should be no leaking past the seats in the combustion chamber.
3. Repeat Step 2 for the intake valves.
4. If there is a leak, the combustion chamber will be wet, meaning the valve is not seating correctly. The following conditions cause poor valve seating:
   a. A bent valve stem.
   b. A worn or damaged valve seat.
   c. A worn or damaged valve face.
   d. A crack in the combustion chamber.

Inspection

1. Perform the *Solvent Test* before cleaning or servicing the cylinder head.
2. Remove all traces of gasket residue from the cylinder head (**Figure 71**) and cylinder block gasket surfaces. Do not scratch the gasket surface.
3. Before removing the valves, remove all carbon deposits from the combustion chambers (**Figure 72**) with a wire brush. Take care not to damage the head, valves or spark plug threads.

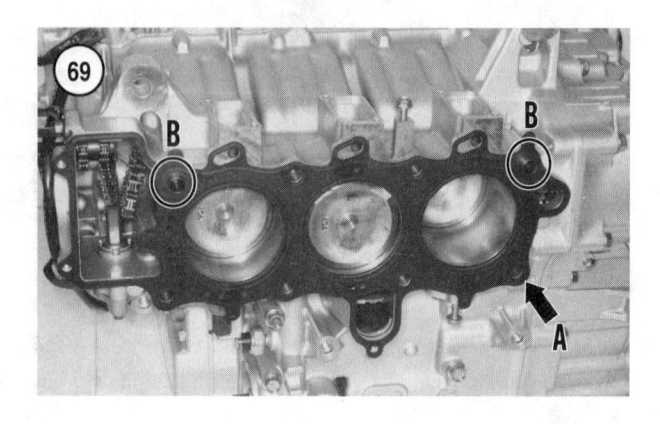

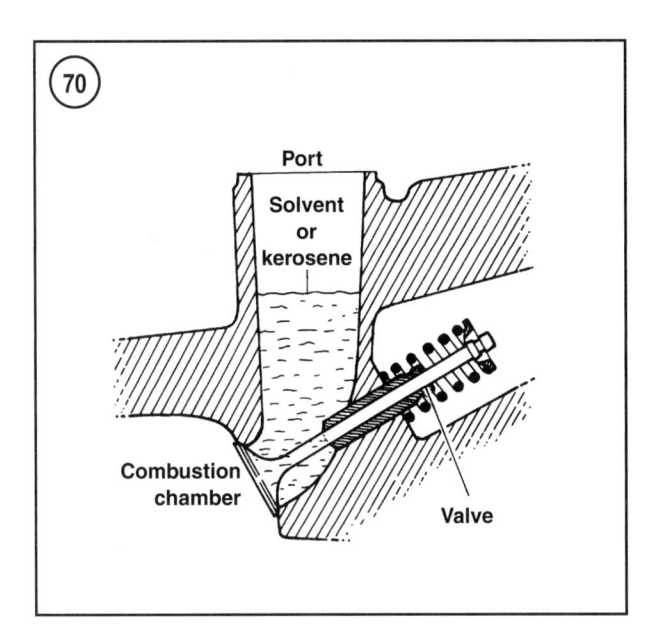

ENGINE TOP END

70

Port — Solvent or kerosene — Combustion chamber — Valve

71

72

73 A, B

CAUTION
If the combustion chambers are cleaned with the valves removed, it will be easy to damage a valve seat. A damaged or even slightly scratched valve seat causes poor valve sealing.

4. Examine the spark plug threads in the cylinder head for damage. If damage is minor or if the threads are contaminated with carbon, use a spark plug thread tap to clean the threads following the manufacturer's instructions. If thread damage is excessive, repair the head by installing a steel thread insert. Purchase thread insert kits at automotive supply stores or have the inserts installed by a Honda dealership or machine shop.

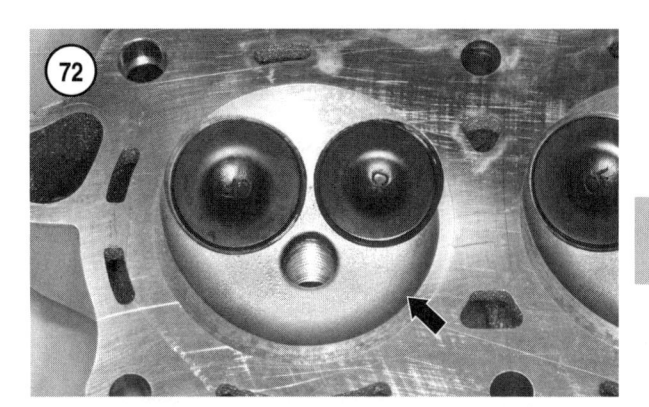

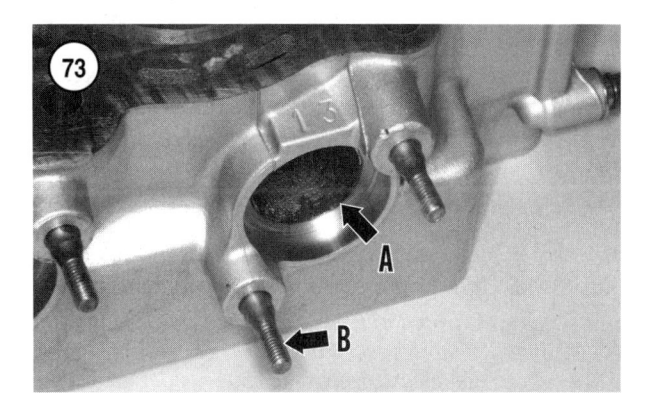

CAUTION
When using a tap to clean spark plug threads, lubricate the tap with an aluminum tap cutting fluid or kerosene.

CAUTION
Aluminum spark plug threads are commonly damaged due to galling, cross-threading and overtightening. To prevent galling, apply an antiseize compound on the plug threads before installation and do not overtighten. Do not lubricate the spark plug threads with engine oil.

5. Clean the entire head in solvent.

NOTE
If the cylinder head was bead-blasted, make sure to clean the head thoroughly with solvent. Residual grit seats in small crevices and other areas and can be hard to remove. Chase each exposed thread with a tap to remove grit from the threads. Residual grit left in the engine will cause premature piston, ring and bearing wear.

6. Check for cracks in the combustion chamber and exhaust ports (A, **Figure 73**). Replace a cracked head.

110 CHAPTER FOUR

74

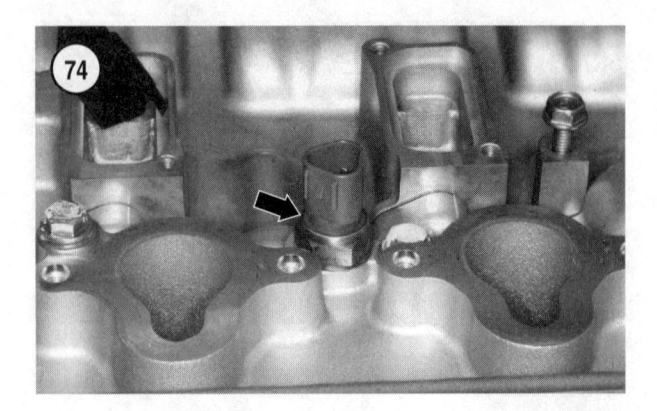

7. Examine the piston crowns. The crowns should show no signs of wear or damage. If the crown appears pecked or spongy-looking, check the spark plugs, valves and combustion chambers for aluminum deposits. If there are deposits, the cylinder(s) is overheating.

CAUTION
Do not clean the piston crowns while the pistons are installed in the cylinders. Carbon scraped from the tops of the pistons will fall between the cylinder wall and piston and onto the piston rings. Because carbon grit is very abrasive, premature cylinder, piston and ring wear will occur. If the piston crowns have heavy deposits of carbon, remove them as described in Chapter Five and clean them. Excessive carbon buildup on the piston crowns reduces piston cooling, raises engine compression and causes overheating.

8. Place a straightedge across the gasket surface at several points. Measure warp by inserting a feeler gauge between the straightedge and cylinder head at each location. Maximum allowable warp is listed in **Table 2**. If the warp exceeds the service limit, resurface or replace the cylinder head. Distortion or nicks in the cylinder head surface could cause an air leak and result in overheating.

9. Check the exhaust pipe studs (B, **Figure 73**) for looseness or thread damage. Slight thread damage can be repaired with a thread file or die. If thread damage is severe, replace the damaged stud(s) as described in Chapter One.

10. Check the valves and valve guides as described under *Valves and Valve Components* in this chapter.

Installation

1. Clean the cylinder head, crankcase and intake manifold gasket surfaces.

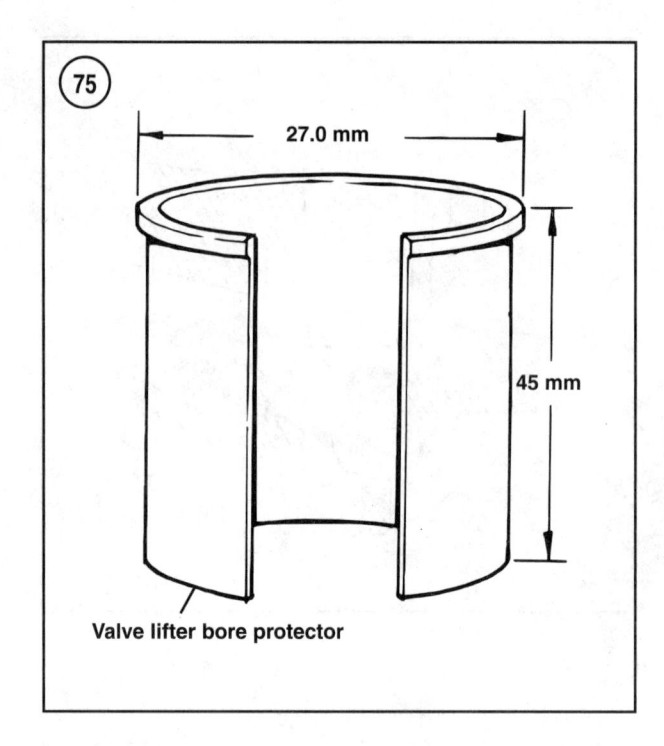

75

27.0 mm

45 mm

Valve lifter bore protector

2. If the intake manifold is installed on the engine, support it with wooden blocks to make room for the cylinder head.
3. If the water pipe is installed on the engine, install a new O-ring onto the pipe. Lubricate the O-ring with coolant.
4. Position the cam chains inside the chain tunnel so they do not interfere with the cylinder head.
5. If removed, install the two dowel pins (B, **Figure 69**).
6. Install a *new* cylinder head gasket (A, **Figure 69**) over the dowel pins and seat against the cylinder block. Make sure all holes align.
7. Clean and dry the cylinder head bolts. Lubricate the 9–mm bolt threads and under the bolt heads with engine oil. Do not lubricate the three 6–mm bolts.
8. One of the 9–mm bolts is shorter than the other seven bolts. Identify this bolt for proper installation.

NOTE
*If the coolant hose (B, **Figure 64** or B, **Figure 65**) is installed on the engine, connect its hose joint into the cylinder head coolant port while installing the cylinder head.*

9. Position the cylinder head against the cylinder block and both dowel pins, seating it against the gasket. Check that the cylinder head is sitting flush against the head gasket.
10. Install the cylinder head 9–mm mounting bolts. Install the shorter bolt (9 × 73 mm) into the hole identified with the arrow in **Figure 68**. Tighten the bolts in several steps to 44 N•m (32 ft.-lb.) in a crisscross pattern.

ENGINE TOP END

76 **VALVE ASSEMBLY**

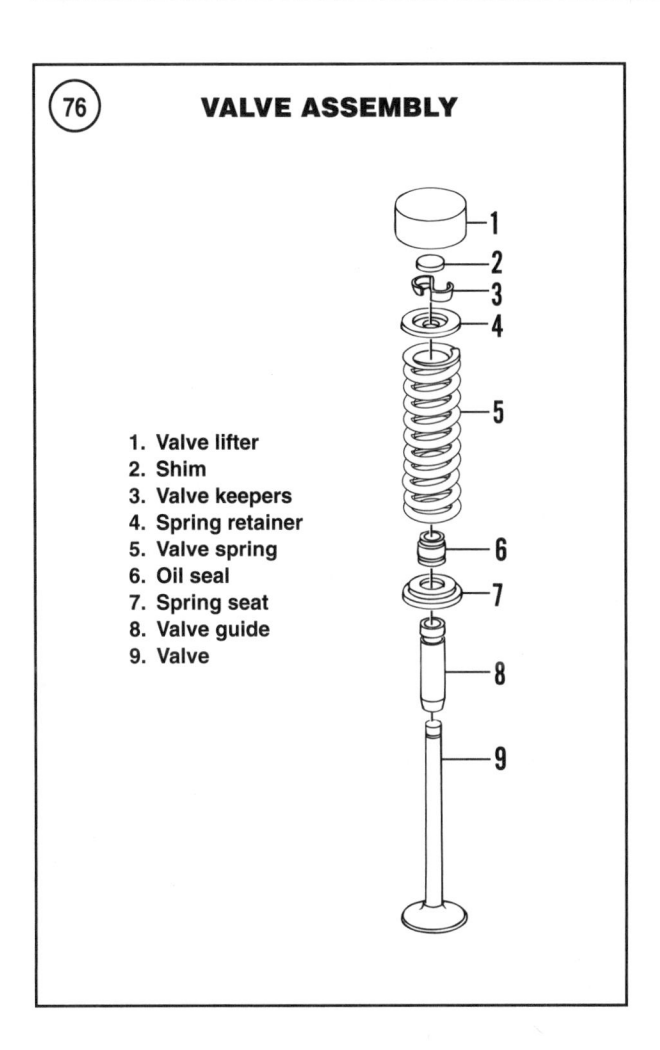

1. Valve lifter
2. Shim
3. Valve keepers
4. Spring retainer
5. Valve spring
6. Oil seal
7. Spring seat
8. Valve guide
9. Valve

77 Protector / Valve lifter bore

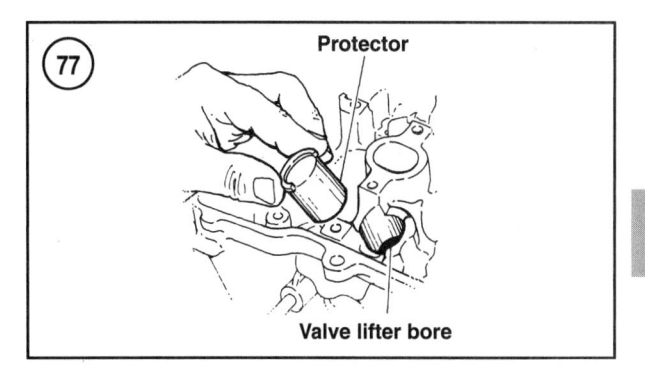

VALVES AND VALVE COMPONENTS

Due to the number of tools and the skills required to perform valve service, it is general practice by those who do their own service to remove the cylinder heads and entrust valve service to a dealership or machine shop.

Tools

To remove and install the valves in this section, the following tools are required:
1. Valve spring compressor.
2. Valve lifter bore protector. This tool is used to protect the valve lifter bore when removing and installing the valves. Make this tool from a 35 mm film container cut to the dimensions shown in **Figure 75**.

Valve Removal

Refer to **Figure 76**.
1. Remove the camshafts, valve lifters and shims as described in this chapter.
2. Remove the cylinder head as described in this chapter.
3. Install the protector into the valve lifter bore (**Figure 77**) of the valve being removed.
4. Install a valve spring compressor (**Figure 78**) squarely over the upper retainer with the other end of the tool placed against the valve head.
5. Tighten the valve spring compressor until the valve keepers separate. Lift the valve keepers out through the valve spring compressor with needlenose pliers or tweezers.
6. Gradually loosen the valve spring compressor and remove it from the head. Remove the upper retainer.
7. Remove the valve spring.

CAUTION
*Remove any burrs from the valve stem grooves (**Figure 79**) before removing the valve; otherwise, the burrs will damage the valve guides.*

11. Install the three 6–mm bolts and tighten securely. See **Figure 66** and **Figure 67**.
12. Reverse Steps 1-14 under *Removal* to complete installation. Note the following:
   a. If the ECT sensor (**Figure 74**) was removed, install with a new gasket and tighten to 25 N•m (18 ft.-lb.).
   b. Install two new intake manifold gaskets. Tighten the bolts in several steps in a crisscross pattern.
   c. Install a new O-ring onto the coolant pipe joints and lubricate with coolant. Tighten the joint bolts securely. See A, **Figure 64** or A, **Figure 65**.
   d. Lubricate a new O-ring with engine oil and install into the groove in the cam pulse generator. Install the cam pulse generator (B, **Figure 61**) and tighten the mounting bolt securely.
   e. Tighten the spark plugs to 18 N•m (13 ft.-lb.).
   f. Check that all coolant hoses and harness connectors are installed correctly.
   g. Start the engine and check for leaks, while bleeding the cooling system as described in Chapter Three.

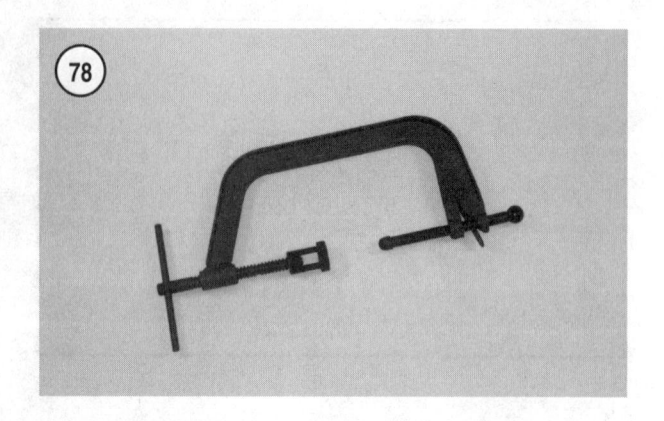

8. Turn the cylinder head over and remove the valve.

NOTE
If a valve is difficult to remove, it may be bent, causing it to stick in its valve guide. Replace the valve and valve guide.

9. Pull the oil seal (**Figure 80**) off the valve guide and discard it.
10. Remove the spring seat.
11. Identify the parts when removing them to reinstall them in their original positions. Refer to **Figure 81** for the intake and exhaust valve parts.
12. Repeat for the remaining intake and exhaust valves.

NOTE
Do not remove the valve guides unless replacing them.

Valve Inspection

When measuring the valve components, compare the actual measurements to the specifications in **Table 2**. Replace parts that are damaged or out of specification as described in this section. Maintain the alignment of the valve components to ensure installation in their original location.

1. Clean the valve components in solvent. Do not damage the valve seating surface.
2. Inspect the valve face (**Figure 82**) for burning, pitting or other signs of wear. Unevenness of the valve face means the valve is not serviceable. If the wear on a valve is too extensive to be corrected by hand-lapping the valve into its seat, replace the valve. The valve face cannot be resurfaced. Replace the valve if defective.
3. Inspect the valve stems for wear and roughness. Check the valve keeper grooves for damage.
4. Measure each valve stem outside diameter with a micrometer (**Figure 83**). Note the following:

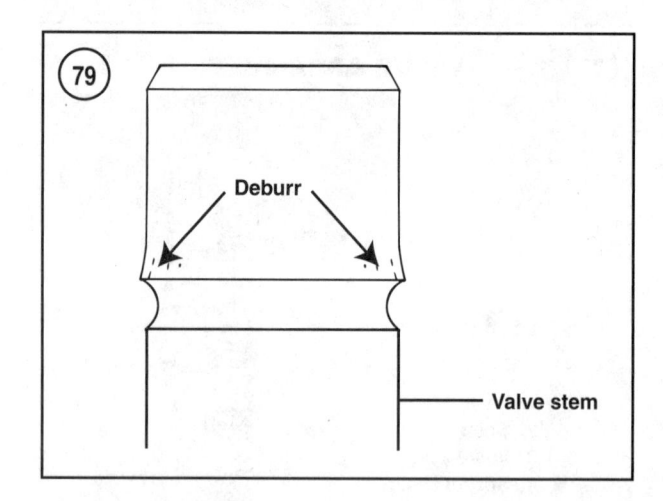

   a. If a valve stem is out of specification, discard the valve.
   b. If a valve stem is within specification, record the measurement so it can be used to determine the valve stem-to-guide clearance in Step 7.

NOTE
Honda recommends reaming the valve guides to remove any carbon buildup before check-

ENGINE TOP END

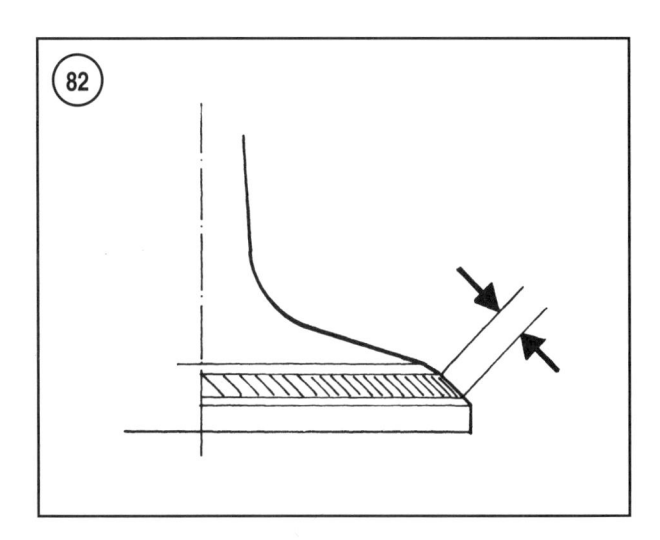

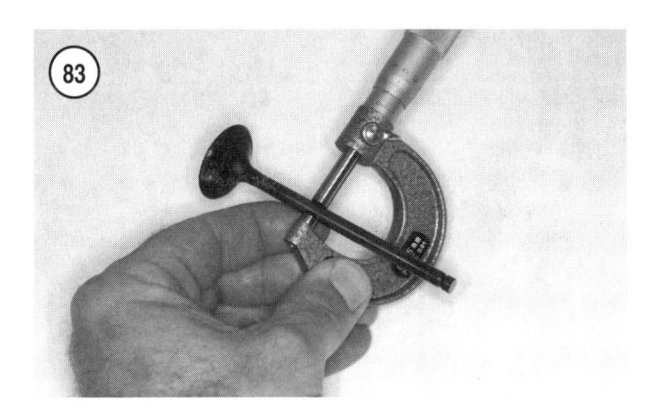

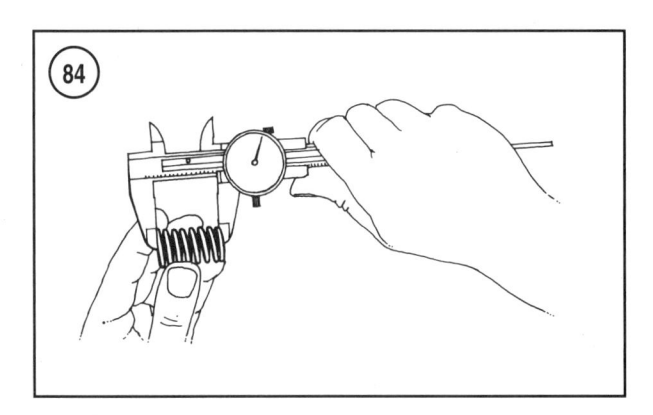

ing and measuring the guides. For the home mechanic it is more practical to remove carbon and varnish from the valve guides with a stiff spiral wire brush. Then clean the valve guides with solvent to wash out all particles. Dry with compressed air.

5. Insert each valve into its respective valve guide and move it up and down by hand. The valve should move smoothly.

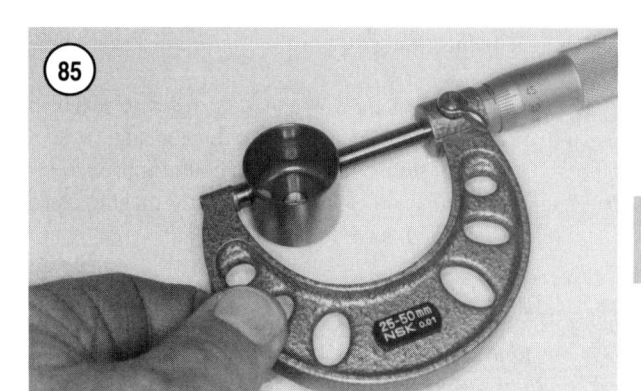

6. Measure each valve guide inside diameter with a small hole gauge and record the measurements. Note the following:

NOTE
Because valve guides wear unevenly (oval shape), measure each guide at different positions. Use the largest bore diameter measurement when determining its size.

   a. If a valve guide is out of specification, replace it as described in this section.
   b. If a valve guide is within specification, record the measurement so it can be used to determine the valve stem-to-guide clearance in Step 7.

7. Subtract the measurement made in Step 4 from the measurement made in Step 6 to determine the valve stem-to-guide clearance. Note the following:

   a. If the clearance is out of specification, determine if a new guide will bring the clearance within specification.
   b. If the clearance will be out of specification with a new guide, replace the valve and guide as a set.

8. Inspect the valve springs as follows:

   a. Inspect each spring for any cracks, distortion or other damage.
   b. Measure the free length of each valve spring with a vernier caliper (**Figure 84**).
   c. Replace defective springs.

9. Check the valve keepers. If they are in good condition, they may be reused; replace in pairs as necessary.

10. Inspect the spring retainer and spring seat for damage.

11. Inspect the valve seats as described under *Valve Seat Inspection* in this section.

12. Measure the valve lifter outside diameter (**Figure 85**) and the lifter bore inside diameter (**Figure 77**) in the cylinder head.

Valve Seat Inspection

The most accurate method for checking the valve seal is to use a marking compound (machinist's dye), available from auto parts and tool stores. Marking compound is used to locate high or irregular spots when checking or making close fits. Follow the manufacturer's directions.

NOTE
Because of the close operating tolerances within the valve assembly, the valve stem and guide must be within tolerance; otherwise the inspection results will be inaccurate.

1. Remove the valves as described in this chapter.
2. Clean the valve seat in the cylinder head and valve mating areas with contact cleaner.
3. Thoroughly clean all carbon deposits from the valve face with solvent and dry thoroughly.
4. Spread a thin layer of marking compound evenly on the valve face.
5. Slowly insert the valve into its guide.
6. Moisten the suction end of a valve lapping tool and attach it to the valve. Insert the valve into the guide.
7. Using the valve lapping tool, tap the valve against the valve seat.
8. Remove the valve and examine the impression left by the marking compound. If the impression on the valve or in the cylinder head is not even and continuous, and the valve seat width (**Figure 86**) is not within the specification in **Table 2**, recondition the valve seat in the cylinder head.
9. Closely examine the valve seat in the cylinder head (**Figure 87**). It should be smooth and even with a polished seating surface.
10. If the valve seat is not in good condition, recondition the valve seat as described in this chapter.
11. Repeat for the other valves.

Valve Guide Replacement

If the valve guides are worn so there is excessive valve stem-to-guide clearance or valve tipping, replace the guides. Special tools and considerable experience are required to properly replace the valve guides in the cylinder head. If these tools are unavailable, remove the cylinder head and have a Honda dealership perform this procedure. When replacing a valve guide, also replace the valve.

Read the entire procedure before attempting valve guide replacement. During some steps it will be necessary to work quickly and have the correct tools on hand.

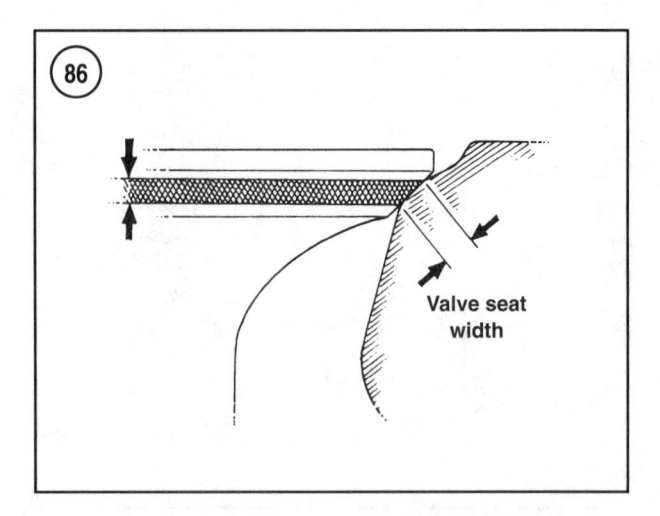

Tools

The following Honda tools (or equivalents) are required to remove and install the valve guides. Confirm part numbers with a Honda dealership before ordering them.
1. Valve guide driver, 5.0 mm (Honda part No. 07942-MA60000 or 07942-8920000).
2. Valve guide reamer, 5.0 mm (Honda part No. 07984-MA60001 or 07984-MA6000D).

Procedure

1. Remove all the valves and valve guide seals from the cylinder head.
2. Place the new valve guides in the freezer for approximately one hour before heating the cylinder head. Chilling them will slightly reduce the outside diameter, while the cylinder will be slightly larger due to heat expansion. This will make valve guide installation much easier. Remove the guides, one at a time, as needed.
3. Remove the ECT sensor from the left cylinder head (**Figure 74**).

ENGINE TOP END

88

Projection height — Valve guide

89

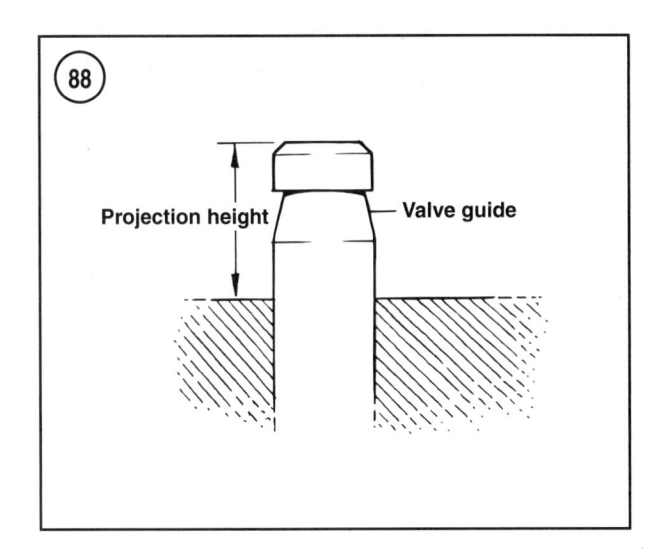

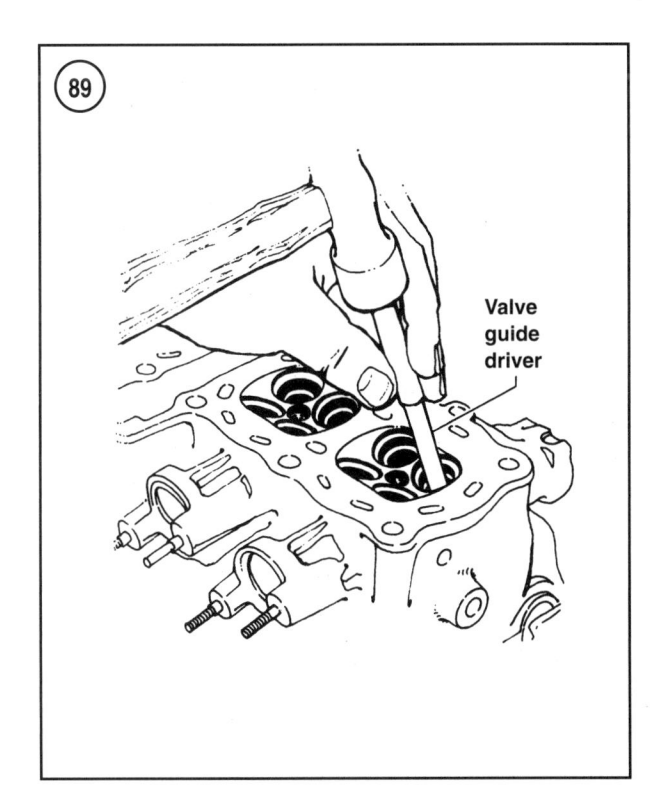

Valve guide driver

NOTE
Flangeless valve guides are used. Step 4 confirms that the guides are installed to the projection height specified in Table 2.

4. Measure the valve guide projection height (**Figure 88**) above the cylinder head surface with a vernier caliper. Record the projection height for each valve guide. Compare to the valve projection height specification in **Table 2**.

5. Place the cylinder head on a hot plate and heat to a temperature of 130-140° C (275-290° F). Do not exceed 150° C (300° F). Monitor the temperature with heat sticks, available at welding supply stores.

CAUTION
Do not heat the cylinder head with a torch; never bring a flame into contact with the cylinder head or valve guide. The direct heat will destroy the case hardening of the valve guide and may warp the cylinder.

WARNING
Wear welding gloves or similar insulated gloves when handling the head. The cylinder head will be very hot.

6. Remove the head from the hot plate. Place the head on wooden blocks with the combustion chambers facing *up*.
7. From the combustion chamber side of the head, drive out the valve guide with the valve guide driver (**Figure 89**). Quickly repeat this step for each guide to be replaced. Reheat the head as required. Discard the valve guides after removing them.

CAUTION
Do not remove the valve guides if the head is not hot enough. Doing so may damage the valve guide bore in the cylinder head.

8. Allow the head to cool.
9. Inspect and clean the valve guide bores. Check for cracks along the bore wall.
10. Reheat the cylinder head as described in Step 5. Then remove it from the hot plate and install it onto the wooden locks with the valve spring side facing *up*.
11. Remove one new valve guide from the freezer.

NOTE
The same Honda valve guide driver tool is used for both removal and installation of the valve guide.

12. Align the valve guide in the bore. Use the valve guide driver and hammer, and drive in the valve guide until the projection height of the valve guide is within the specification in **Table 2** (**Figure 88**).
13. Repeat to install the remaining valve guides.
14. Allow the head to cool to room temperature.
15. Ream each valve guide as follows:
    a. Place the head on wooden blocks with the combustion chambers facing *up*. The guides are reamed from this side.
    b. Coat the valve guide and valve guide reamer with cutting oil.

c. Rotate the reamer *clockwise* into the valve guide (**Figure 90**).

CAUTION
Always rotate the reamer clockwise through the entire length of the guide, both when reaming the guide and when removing the reamer. Rotating the reamer counterclockwise will reverse cut and damage (enlarge) the valve guide bore.

CAUTION
Do not allow the reamer to tilt. Keep the tool square to the hole and apply even pressure and twisting motion during the entire operation.

d. Slowly rotate the reamer through the guide, while periodically adding cutting oil.
e. As the end of the reamer passes through the valve guide, maintain the clockwise motion and work the reamer back out of the guide while continuing to add cutting oil.
f. Clean the reamer of all chips and relubricate with cutting oil before starting on the next guide. Repeat for each guide as required.

16. Thoroughly clean the cylinder head and all valve components in solvent, then clean with detergent and hot water to remove all cutting residue. Rinse in cold water. Dry with compressed air.
17. Measure the valve guide inside diameter with a small bore gauge. The measurement must be within the specification listed in **Table 2**.
18. Apply engine oil to the valve guides to prevent rust.
19. Lubricate a valve stem with engine oil and pass it through each valve guide, verifying that it moves without any roughness or binding.
20. Reface the valve seats as described under *Valve Seat Reconditioning* in this section.
21. Install the ECT sensor (**Figure 74**) onto the left cylinder head. Install with a new gasket and tighten to 25 N•m (18 ft.-lb.).

Valve Seat Reconditioning

Before reconditioning the valve seats, inspect and measure them as described under *Valve Seat Inspection* in this section.

To cut the cylinder head valve seats, use the following valve seat cutters (**Figure 91**):
1. When cutting the 45° seat:
   a. Intake valve 33 mm seat cutter (Honda part No. 07780-0010800 or equivalent).
   b. Exhaust valve 29 mm seat cutter (Honda part No. 07780-0010300 or equivalent).
2. When cutting the 32° seat:
   a. Intake valve 35 mm flat cutter (Honda part No. 07780-0012300 or equivalent).
   b. Exhaust valve 33 mm flat cutter (Honda part No. 07780-0012900 or equivalent).
3. When cutting the 60° seat:
   a. Intake and exhaust valve interior 3.0 mm cutter (Honda part No. 07780-0014000 or equivalent).
   b. 5.0 mm cutter holder (Honda part No. 07781-0010400 or equivalent).
4. Use the 5.0 mm cutter holder (Honda part No. 07781-0010400 or equivalent) to support the cutters when cutting the valve seats.

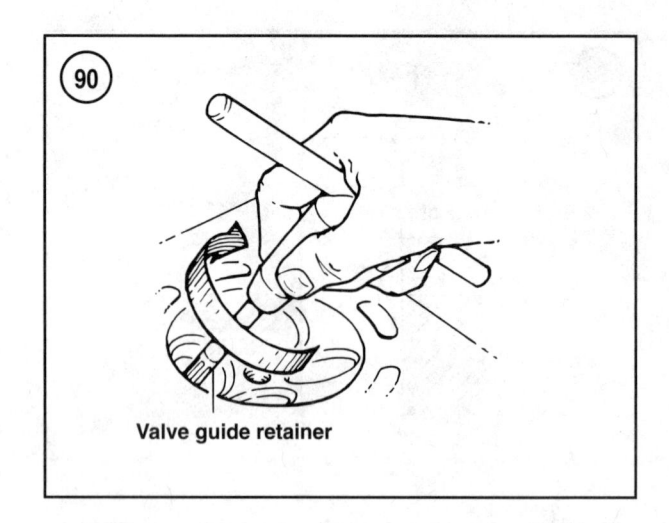

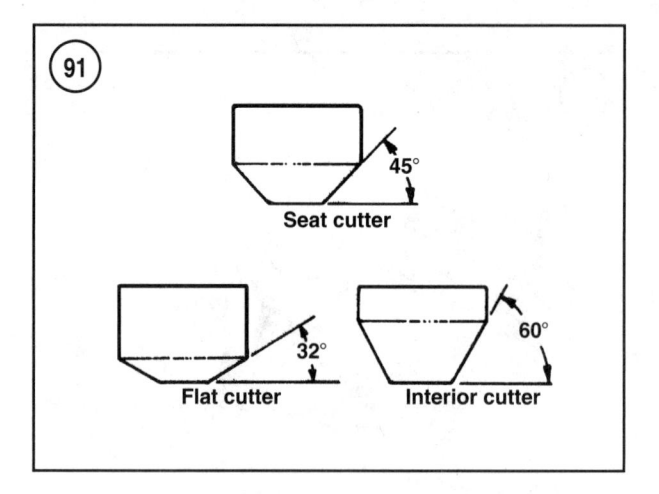

ENGINE TOP END

Figure 92

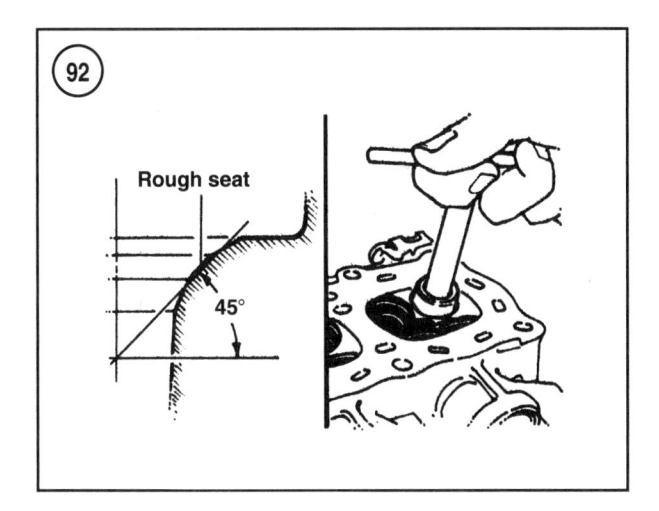

Rough seat — 45°

Figure 93

Old seat width — 32°

Figure 94

Old seat width — 60°

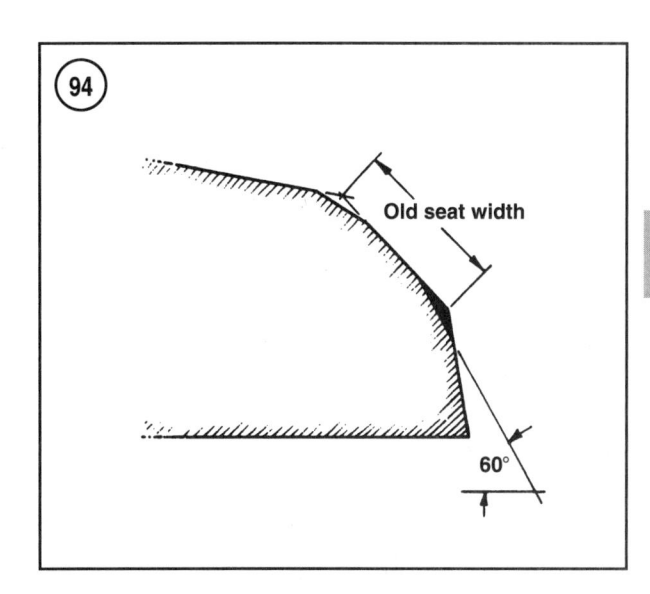

3. Using the 45° cutter, de-scale and clean the valve seat with one or two turns (**Figure 92**).

4. If the seat is still pitted or burned, turn the 45° cutter additional turns until the surface is clean. Avoid removing too much material from the cylinder head.

5. Measure the valve seat (**Figure 86**) with a vernier caliper. Record the measurement to use as a reference point when performing the following:

CAUTION
The 32° cutter removes material quickly. Work carefully and check the progress often.

6. Install the 32° flat cutter onto the solid pilot and lightly cut the seat to remove 1/4 of the existing valve seat (**Figure 93**).

7. Install the 60° interior cutter onto the solid pilot and lightly cut the seat to remove 1/4 of the existing valve seat (**Figure 94**).

8. Measure the valve seat with a vernier caliper (**Figure 86**). Then fit the 45° seat cutter onto the solid pilot and cut the valve seat to the specified width (**Figure 95**) listed in **Table 2**.

9. When the valve seat width is correct, check valve seating as follows:
   a. Clean the valve seat with contact cleaner.
   b. Spread a thin layer of marking compound evenly on the valve face.
   c. Slowly insert the valve into its guide.
   d. Support the valve with two fingers (**Figure 96**) and tap the valve up and down in the cylinder head several times. Do not rotate the valve or a false reading will result.

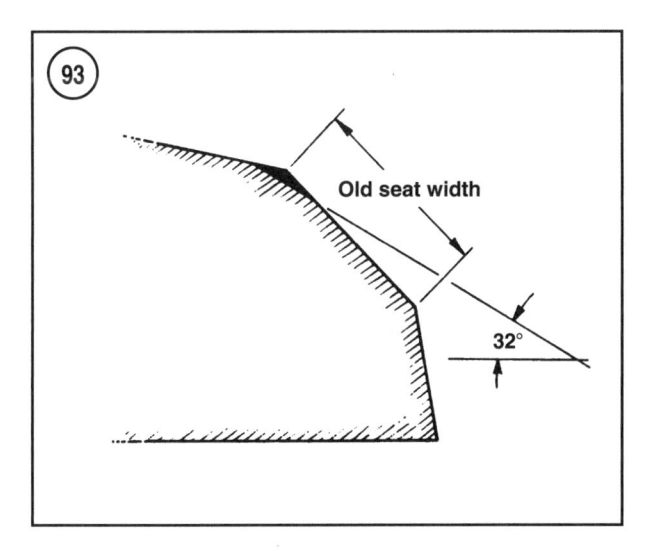

Procedure

NOTE
Follow the manufacturer's instructions when using valve facing equipment.

1. Carefully rotate and insert the solid pilot into the valve guide. Make sure the pilot is correctly seated.

2. Install the 45° seat cutter and cutter holder onto the solid pilot.

CAUTION
Work slowly and make light cuts during reconditioning. Overcutting the valve seats will recede the valves into the cylinder head, reducing the valve adjustment range. If cutting is excessive, the ability to set the valve adjustment may be lost. If this condition happens, replace the cylinder head.

e. Remove the valve and examine the impression left by the marking compound.
f. Measure the valve seat width as shown in **Figure 86**. Refer to **Table 2** for specified valve width.
g. The valve contact should be approximately in the center of the valve seat area.

10. If the contact area is too high on the valve, or if it is too wide, use the 32° flat cutter and remove a portion of the top area of the valve seat material to lower and narrow the contact area on the valve (**Figure 93**).
11. If the contact area is too low on the valve, or too wide, use the 60° interior cutter and remove a portion of the lower area of the valve seat material to raise and narrow the contact area on the valve (**Figure 94**).
12. After obtaining the desired valve seat position and width, use the 45° seat cutter to lightly clean off any burrs that may have been caused by previous cuts
13. When the seat width and contact area is correct, lap the valve as described in this chapter.
14. Repeat Steps 1-13 for all remaining valve seats.
15. Thoroughly clean the cylinder head and all valve components in solvent, then clean with detergent and hot water. Rinse in cold water. Dry with compressed air. Then apply a light coat of clean engine oil to all non-aluminum surfaces to prevent rust.

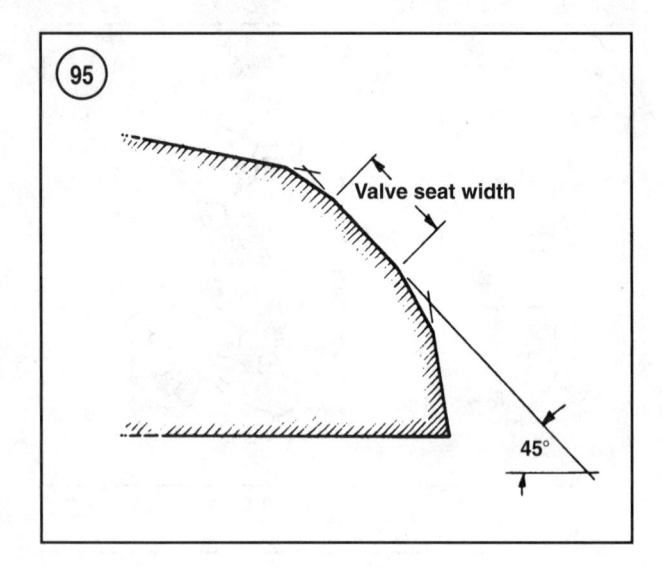

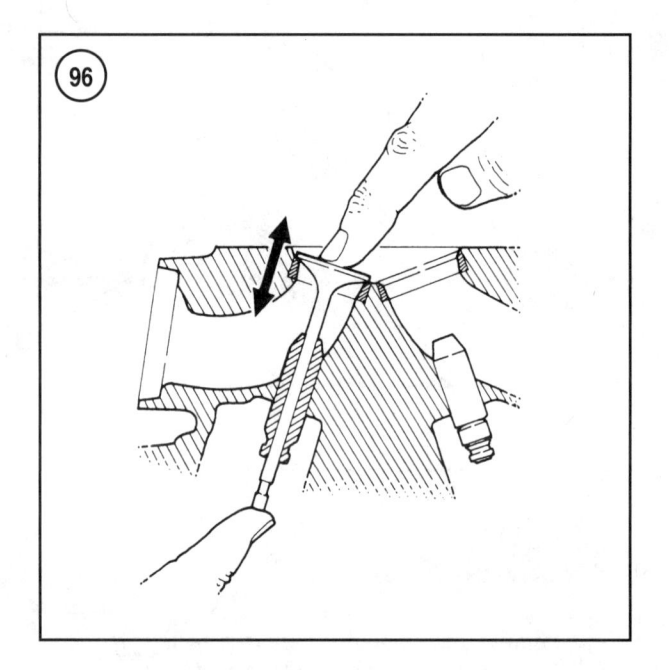

Valve Lapping

Valve lapping can restore the valve seat without reconditioning if the amount of wear or distortion is not too great.

Perform this procedure after determining that the valve seat width and outside diameter are within specifications. Use a valve lapping tool and compound for this procedure.

1. Smear a light coating of fine grade valve lapping compound on the valve face seating surface.
2. Insert the valve into the head.
3. Wet the suction cup of the lapping stick and stick it onto the head of the valve. Spin the tool in both directions, while pressing it against the valve seat, and lap the valve to the seat. Every 5 to 10 seconds, lift and rotate the valve 180° in the valve seat. Continue until the gasket surfaces on the valve and seat are smooth and equal in size.
4. Closely examine the valve seat in the cylinder head (**Figure 87**). It should be smooth and even with a smooth, polished seating ring.
5. Repeat Steps 1-4 for the other valves.
6. Thoroughly clean the cylinder head and all valve components in solvent, then clean with detergent and hot water. Rinse in cold water. Dry with compressed air. Then apply a light coat of clean engine oil to all non-aluminum surfaces to prevent rust.

CAUTION
Any compound left on the valves or in the cylinder head causes excessive wear to the engine components.

7. Install the valve assemblies as described in this chapter.
8. After completing the lapping and reinstalling the valves in the head, perform the *Solvent Test* under *Cylinder Head* in this chapter. There should be no leakage past the seat. If leakage occurs, the combustion chamber will be wet. If fluid leaks past any of the seats, disassemble that valve as-

ENGINE TOP END

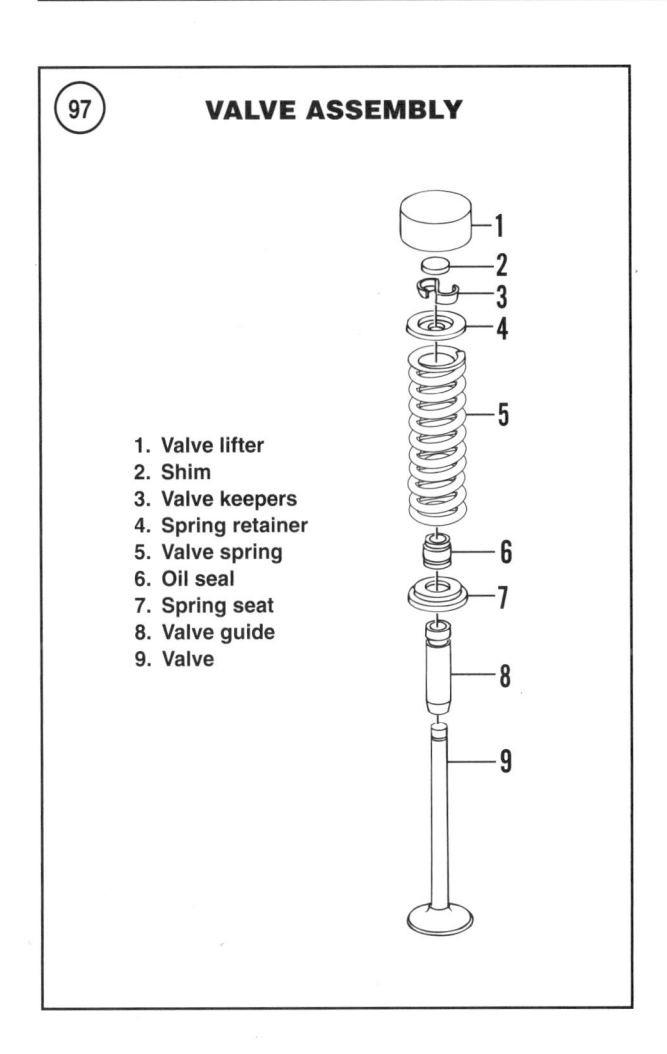

97 VALVE ASSEMBLY

1. Valve lifter
2. Shim
3. Valve keepers
4. Spring retainer
5. Valve spring
6. Oil seal
7. Spring seat
8. Valve guide
9. Valve

sembly and repeat the lapping procedure until there is no leakage.

NOTE
This solvent test does not ensure long-term durability or maximum power. It merely en-sures maximum compression will be available on initial start-up after reassembly.

9. If the cylinder head and valve components are cleaned in detergent and hot water, apply a light coat of engine oil to all bare metal surfaces to prevent rust.

Valve Installation

Following the reference marks made during removal, install the valves in their original locations.

Refer to **Figure 97**.

1. Install the spring seat with its shoulder facing up (**Figure 98**).

NOTE
The oil seals used for the intake and exhaust valves are identical.

2. Lubricate the inside of a *new* oil seal (**Figure 99**) with engine oil. Then push the seal straight down the valve guide until it snaps into the groove in the top of the guide (**Figure 80**). Check that the oil seal is centered and seats squarely on top of the guide. If the seal is cocked to one side, oil will leak past the seal during engine operation.

NOTE
The oil seals must be replaced whenever the valves are removed. Also, if a new seal was installed and then removed, do not reuse it.

3. Install the valve as follows:
   a. Coat a valve stem with molybdenum oil solution.

NOTE
Molybdenum oil solution is a 50:50 mixture of engine oil and molybdenum disulfide grease.

Figure 100

Figure 101

Figure 102

Figure 103

Valve keepers
Valve stem

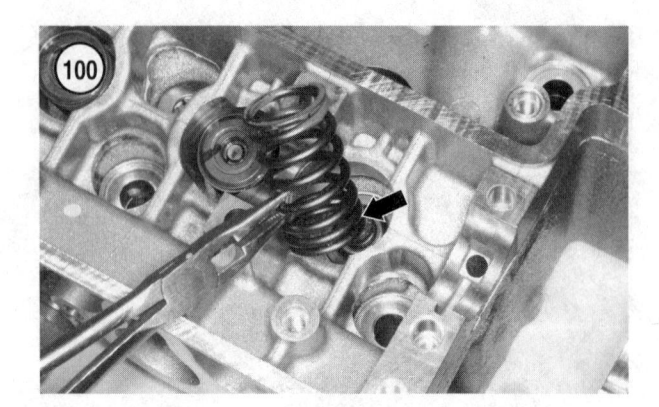

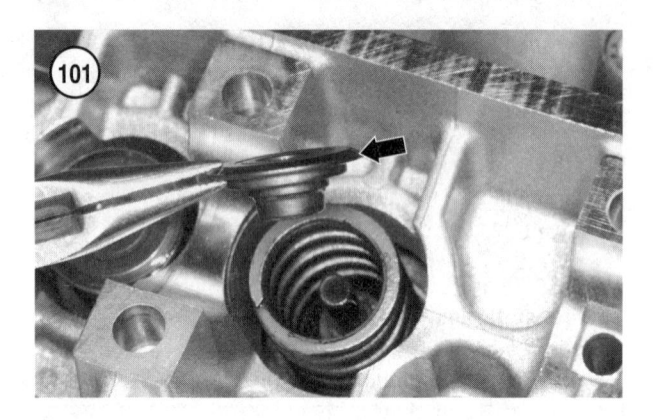

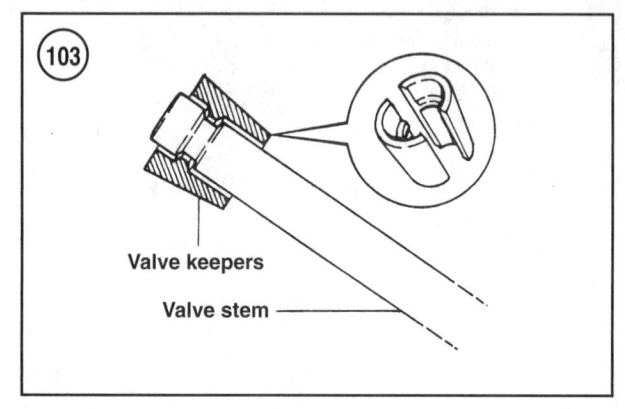

b. Install the valve partway into its guide. Then, slowly turn the valve as it enters the valve stem seal and continue turning it until the valve is installed all the way.

c. Make sure the valve moves up and down smoothly.

4. Install the valve spring with its closer wound coils (**Figure 100**) facing the cylinder head.

5. Install the spring retainer (**Figure 101**) on top of the valve springs.

6. Install the protector (**Figure 77**) into the valve lifter bore. Refer to *Service Tools* at the beginning of this section.

CAUTION
To avoid loss of spring tension, do not compress the springs any more than necessary when installing the valve keepers.

7. Compress the valve spring with a valve spring compressor tool and install the valve keepers (**Figure 102**). Make sure the keepers fit into the rounded groove in the valve stem (**Figure 103**).

8. Gently tap the upper retainer with a plastic hammer to ensure the keepers are properly seated.

9. Repeat Steps 1-8 for the remaining valves.

10. Install the shims and valve lifters as described under *Camshafts* in this chapter.

11. After installing the cylinder head, camshafts and camshaft holders on the engine, check and adjust the valve clearance as described in Chapter Three.

ENGINE TOP END

Table 1 GENERAL ENGINE SPECIFICATIONS

Bore and stroke	74.0 × 71.0 mm (2.91 × 2.80 in.)
Compression ratio	9.8:1
Cylinder alignment	Flat six
Displacement	1832 cc (11.8 cu.-in.)
Engine dry weight	118.3 kg (260.8 lbs.)
Engine firing order	1-4-5-2-3-6
Valve timing	
Intake valve	
Open	-5° BTDC (5° ATDC at 1 mm lift)
Closes	30° ABDC at 1 mm lift
Exhaust valve	
Open	30° BTDC at 1 mm lift
Closes	-5° ATDC (5° BTDC at 1 mm lift)

Table 2 CYLINDER HEAD AND VALVES SERVICE SPECIFICATIONS

	New mm (in.)	Service limit mm (in.)
Cylinder head warp limit	–	0.10 (0.004)
Camshaft		
Journal outside diameter	27.959-27.980 (1.1007-1.1016)	27.96 (1.101)
Journal inside diameter	28.000-28.021 (1.1024-1.1032)	28.05 (1.104)
Lobe height		
Intake	41.610-41.690 (1.6382-1.6413)	41.58 (1.637)
Exhaust	41.680-41.760 (1.6409-1.6441)	41.65 (1.640)
Oil clearance	0.020-0.062 (0.0008-0.0024)	0.10 (0.004)
Runout	–	0.03 (0.001)
Valve		
Valve guide inside diameter		
Intake and exhaust	5.000-5.012 (0.1969-0.1973)	5.04 (0.198)
Valve stem outside diameter		
Intake	4.970-4.995 (0.1957-0.1967)	4.96 (0.195)
Exhaust	4.955-4.980 (0.1951-0.1961)	4.95 (0.195)
Valve stem-to-guide clearance		
Intake	0.0050-0.042 (0.0002-0.0017)	0.075 (0.0030)
Exhaust	0.020-0.057 (0.0008-0.0022)	0.085 (0.0033)
Valve guide projection height above cylinder head		
Intake and exhaust	11.8-12.0 (0.46-0.47)	–
Valve seat width		
Intake and exhaust	0.9-1.1 (0.035-0.043)	1.5 (0.06)
Valve spring free length		
Intake and exhaust	38.20 (1.504)	37.0 (1.46)
Valve lifter		
Valve lifter bore inside diameter	29.010-29.026 (1.1421-1.1428)	29.04 (1.1433)
Valve lifter outside diameter	28.978-28.993 (1.1409-1.1415)	28.97 (1.1404)

Table 3 ENGINE TOP END TORQUE SPECIFICATIONS

	N•m	in.-lb.	ft.-lb.
Cam chain tensioner mounting bolts	12	106	–
Cam chain tensioner sealing bolts	12	106	–
Cam chain tensioner pivot bolt	12	106	–
Camshaft holder bolts	12	106	–
Camshaft sprocket mounting bolts	25	–	18
Cylinder head 9-mm bolts	44	–	32
Cylinder head cover bolts	12	106	–
Cylinder head side cover bolts	10	88	–
Engine coolant temperature (ECT) sensor	25	–	18
Front crankcase cover bolts	12	106	–
Ignition pulse generator bolt	12	106	–
Rotor bolt	59	–	44
Left cam chain guide bolt/washer	12	106	–
Right chain tensioner pivot arm bolt	12	106	–
Spark plugs	18	–	13
Timing hole cap	18	–	13

CHAPTER FIVE

ENGINE LOWER END

This chapter provides service procedures for engine removal and lower end components. These include the rear crankcase cover, primary gears, output shaft, crankcase, connecting rods, pistons, crankshaft and the oil pump/lubrication system. Removal and installation procedures for the transmission and internal shift mechanism assembly are described in Chapter Seven.

Refer to **Table 1-7** at the end of this chapter for engine and torque specifications.

One of the most important aspects of a successful engine overhaul is preparation. Before removing the engine and disassembling the crankcase, degrease the engine and frame. Read the applicable procedures to learn about the special tools and equipment required. A hoist will be required to lift the front of the motorcycle when removing and installing the engine. Identify and store individual assemblies in appropriate storage containers.

References to the left and right sides refer to the position of the parts as viewed by the rider sitting on the seat facing forward, not how the engine may sit on the workbench. Also, review the service methods in Chapter One.

ENGINE SERVICE

1. The following components can be serviced with the engine mounted in the frame:
    a. Throttle body.
    b. Water pump/thermostat housing.
    c. Cylinder head and valves.
    d. Clutch.
    e. Gearshift linkage.

    NOTE
    The engine must be removed to service the gearshift spindle.

    f. Ignition pulse generator.
    g. Starter/reverse motor.
    h. Alternator.
2. The engine must be removed from the frame to service the following components:
    a. Intake manifold.
    b. Oil pump.
    c. Primary gears/output shaft.
    d. Transmission and gearshift spindle.
    e. Reverse shifter/shift drum lock arm.

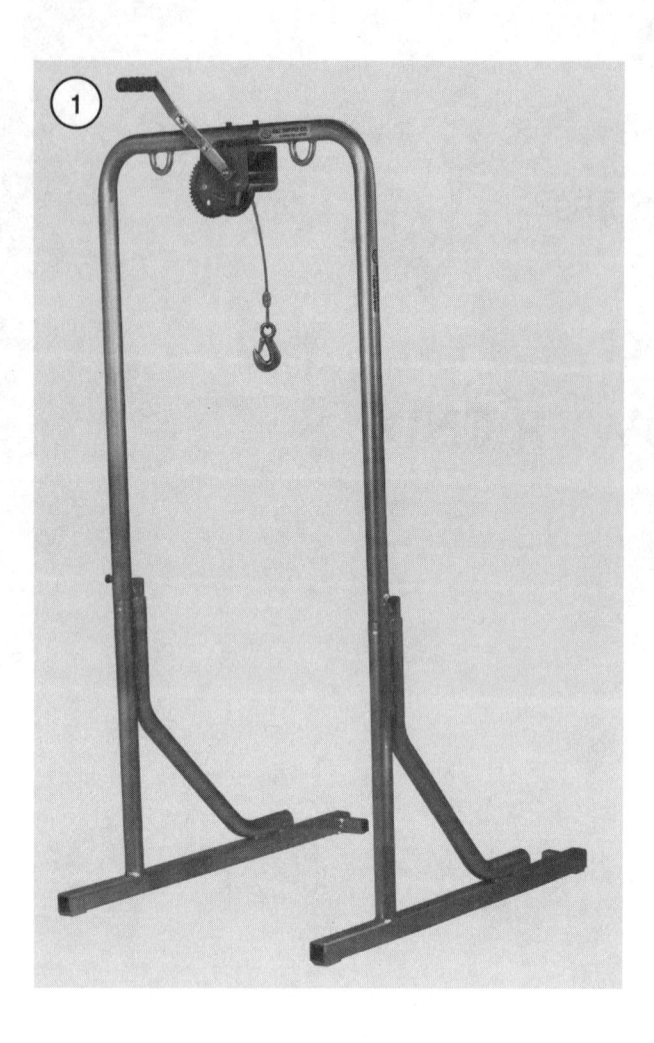

f. Crankcase.
g. Crankshaft.
h. Connecting rods and pistons.

ENGINE

Tools

The following tools are required for engine removal and installation.
1. A hydraulic floor jack to support the engine.
2. A hoist, like the K&L Lift-Gate shown in **Figure 1**, to lift the front of the motorcycle when removing and installing the engine into the frame.
3. The following Honda tools to loosen and tighten some of the engine mounting locknuts:
   a. 20 mm locknut wrench (part No. 07VMA-MBB0101 [A, **Figure 2**]).
   b. Locknut wrench (part No. 07GMA-KT7A200 or 07908-ME90000 [B, **Figure 2**]).

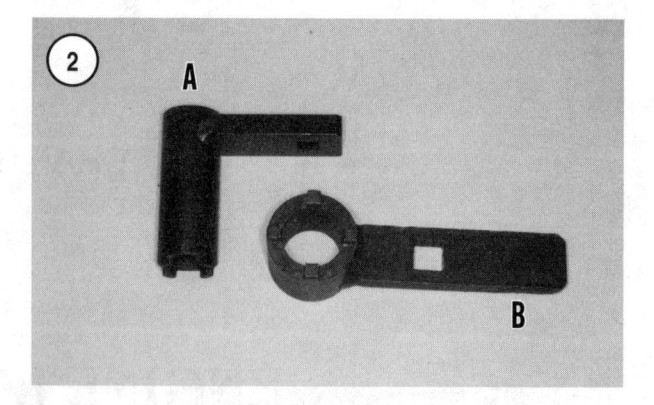

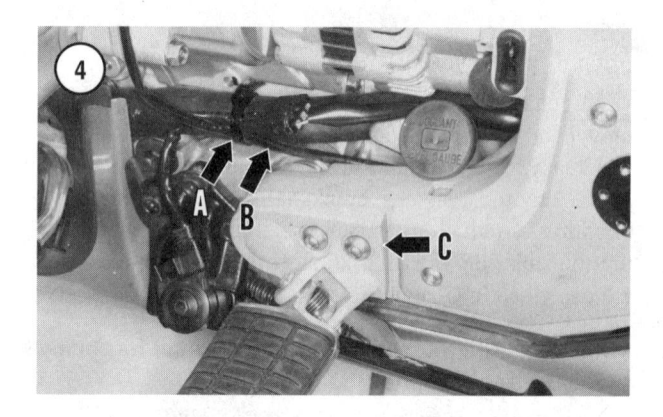

Removal

The engine weighs approximately 118.3 kg (260.8 lbs.) Due to its weight and size, make sure to have a number of assistants available during engine removal and installation.
1. Support the bike on its centerstand.
2. If removing the clutch cover or rear engine cover, drain the engine oil. Refer to *Engine Oil and Filter* in Chapter Three.
3. Remove the following as described in Chapter Seventeen:
   a. Seat.

ENGINE LOWER END

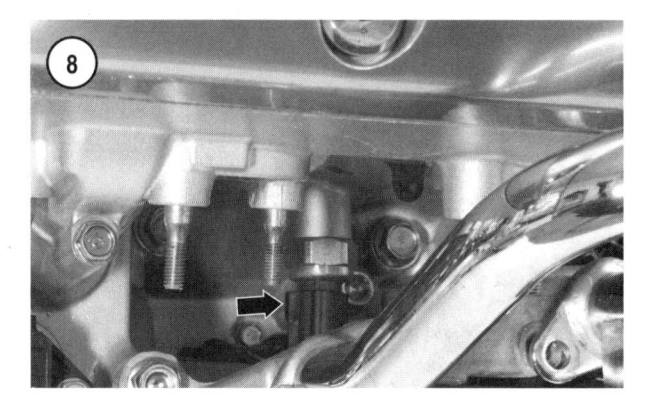

b. Both side covers.
c. Both engine side covers.
d. Front fender B.
e. Center inner fairing.
f. Top shelter.
g. Front fairing.
h. Exhaust system.

4. Disconnect the negative battery cable. Refer to *Battery* in Chapter Nine.
5. Remove the following as described in Chapter Ten:
   a. Radiator reserve tank.
   b. Both radiators.
6. Remove the following as described in Chapter Eight:
   a. Fuel tank.
   b. Air filter housing.
   c. Throttle body.
7. Disconnect the following electrical connectors located inside the connector pouch mounted in front of the throttle body (**Figure 3**):
   a. Ignition pulse generator red 2-pin connector.
   b. Reverse shift actuator red 3-pin and white 2-pin connectors.
   c. Gear position switch black 6-pin connector.
8. Disconnect the following electrical connectors:
   a. Left engine sub-wire harness gray 4-pin and gray 6-pin connectors.
   b. Right engine sub-wire harness gray 4-pin and gray 8-pin connectors.
   c. Speed sensor white 3-pin connector.
   d. Side stand switch 3-pin green or black connector (A, **Figure 4**). Remove the wire band (B, **Figure 4**) securing the harness to the coolant hose.
   e. Starter/reverse motor nut and cable lead.
   f. Alternator nut (A, **Figure 5**), cable lead and 4-pin connector (B).
9. Remove the rear brake reservoir bracket mounting bolt (A, **Figure 6**). Do not disconnect the brake hose.
10. Remove the nut and reverse shift switch connector at the switch (B, **Figure 6**).
11. Disconnect the connector (A, **Figure 7**) at the evaporative emission (EVAP) purge control solenoid valve. Then remove the valve (B, **Figure 7**) from its holder.
12. Disconnect the spark plug wires from the ignition coils (C, **Figure 7**).
13. Disconnect the clamp securing the No. 2 and No. 5 spark plug wires to the bracket, and remove the bracket.
14. Disconnect the knock sensor connectors (**Figure 8**, typical).
15. Disconnect the horn connectors.
16. Remove the EVAP canister:

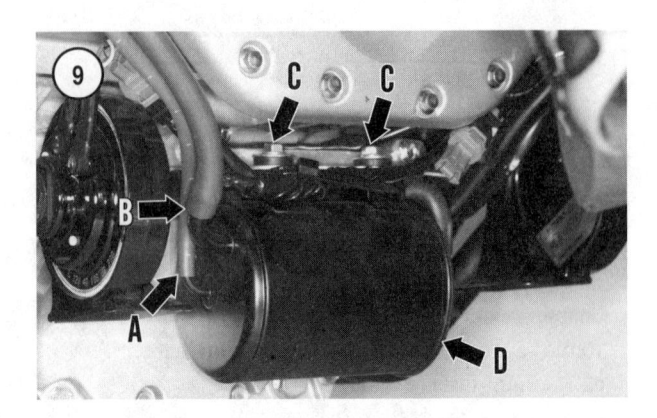

a. Disconnect the No. 1 (A, **Figure 9**) and No. 4 (B) hoses from the EVAP canister. Remove the clamp from the bracket.

b. Remove the bolts (C, **Figure 9**) and lower the canister. Remove the wire clamp from the wire harness and remove the canister (D, **Figure 9**).

17. Remove the coolant drain joint (A, **Figure 10**) and EVAP canister bracket (B) mounting bolts:

18. Disconnect the clutch slave cylinder hose:

a. Drain the clutch fluid. Refer to *Clutch Fluid Draining* in Chapter Six.

b. Remove the banjo bolt (**Figure 11**) and washers. Seal the end of the hose to prevent leaks.

19. Remove the pinch bolt (A, **Figure 12**), pivot bolt (B), washer and the gearshift pedal arm assembly.

20. Unbolt and remove both engine guards (C, **Figure 12**).

21. Unbolt and remove both rider footpegs (C, **Figure 4**).

22. Remove the bolts (**Figure 13**) and disconnect the fuel feed hose joint from the right fuel rail.

23. Place a floor jack underneath the engine. Raise the jack so the pad rests below the bottom of the engine. Place a thick wooden block on the jack to protect the engine.

CAUTION
As the engine mounting bolts are loosened and removed in the following steps, readjust the jack height to remove pressure on the remaining fasteners, frame and engine.

NOTE
*Refer to **Figure 14** to identify the engine mounting fasteners.*

24. Loosen the three left subframe mounting bolts (**Figure 15**).

25. Loosen the three right subframe mounting bolts (**Figure 16**).

26. Remove the left center engine hanger nut and bolt (17, **Figure 14**) and bolt (20).

ENGINE LOWER END

ENGINE MOUNTING FASTENERS

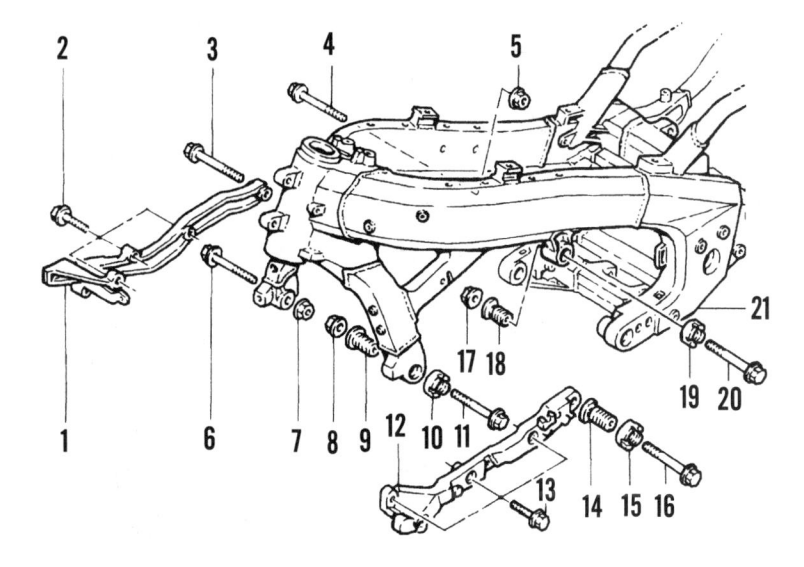

1. Right subrame
2. Right subframe mounting bolts (M10 × 40)
3. Right rear engine hanger bolt (M12 × 84)
4. Right center engine hanger bolt (M10 × 62)
5. Right center engine hanger nut
6. Right front engine hanger bolt (M12 × 72)
7. Right front engine hanger nut
8. Left front engine hanger nut
9. Left front engine hanger adjusting bolt (M22 × 47)
10. Left front engine hanger adjusting bolt locknut
11. Left front engine hanger bolt (M10 × 62)
12. Left subframe
13. Left subframe mounting bolts (M10 × 40)
14. Left rear engine hanger adjusting bolt (M22 × 40)
15. Left rear engine hanger adjusting bolt locknut
16. Left rear engine hanger bolt (M12 × 92)
17. Left center engine hanger nut
18. Left center engine hanger adjusting bolt (M20 × 43.5)
19. Left center engine hanger adjusting bolt locknut
20. Left center engine hanger bolt
21. Frame

27. Remove the left front engine hanger nut (8, **Figure 14**) and bolt (11).

28. Remove the left rear engine hanger bolt (16, **Figure 14**).

29. Loosen the left center engine hanger adjusting bolt locknut (19, **Figure 14**) with the 20 mm locknut wrench. Remove the locknut.

30. Loosen the left center engine hanger adjusting bolt (18, **Figure 14**) with an Allen wrench. Remove the locknut (19, **Figure 14**).

31. Loosen the left front engine hanger adjusting bolt locknut (10, **Figure 14**) using the locknut wrench. Remove the locknut.

32. Loosen the left front engine hanger adjusting bolt (9, **Figure 14**).

33. Loosen the left rear engine hanger adjusting bolt locknut (15, **Figure 14**) using the locknut wrench. Remove the locknut.

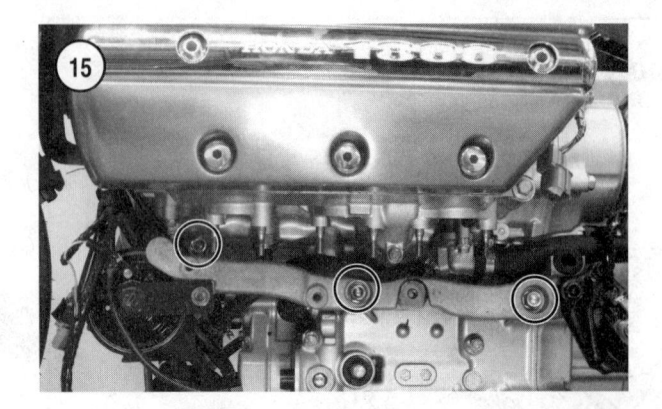

34. Loosen the left rear engine hanger adjusting bolt (14, **Figure 14**).
35. Remove the three left subframe mounting bolts and subframe (**Figure 15**).
36. Remove the right center engine hanger nut (5, **Figure 14**) and bolt (4).
37. Remove the right front engine hanger nut (7, **Figure 14**) and bolt (6).
38. Remove the right rear engine hanger bolt (3, **Figure 14**).
39. Remove the three right subframe mountings bolts and subframe (**Figure 16**).

WARNING
Have an assistant steady the engine on the jack when attaching the engine hoist to the frame.

40. Center a hoist (**Figure 17**) around the front part of the frame. Secure the hoist to the frame steering neck and lift the frame away from the engine.
41. Disconnect the drive shaft boot from the engine.
42. Slide the engine forward to disconnect the output shaft from the drive shaft.
43. Then lower the engine and remove it from the frame (**Figure 18**).

Cleaning and Inspection

1. Remove any corrosion from the engine mount bolts with a wire wheel.
2. Clean and dry the engine mount bolts, nuts, and collars.
3. Clean the engine hanger frame threads.
4. Replace damaged fasteners.
5. Check the coolant hoses for cracks, leaks or other damage. Replace if necessary.
6. Check the wire harness routing in the frame. Check the harness cover and wires for chafing or other damage. Replace harness cable guides and clips as required.
7. Clean the electrical connectors with contact cleaner. 
8. Check the frame mounting bolt holes and threads for damage.
9. Check the drive shaft boot for damage.

Installation

Installation is the reverse of removal. Follow the installation and tightening sequence described below.

1. Check the coolant hoses, spark plug cables and electrical wires and connectors. If necessary, reposition them so they will not interfere or be damaged when installing the engine.

CAUTION
Install and tighten the engine fasteners in the order given; otherwise, the engine may not align properly with the frame. This may cause vibration and handling complaints.

2. Install the engine hanger adjusting bolts into the left side of the frame. Install the adjusting bolts from the inside part of the frame and screw them in fully. See 9, 14 and 18, **Figure 14**.
3. Lubricate the output shaft splines with molybdenum disulfide grease.
4. Install the engine in the frame while engaging the output shaft with the drive shaft.

ENGINE LOWER END

9. Tighten the right center engine hanger bolt (4, **Figure 14**) to 39 N•m (29 ft.-lb.).

10. Tighten the left subframe mounting bolts (13, **Figure 14**) securely.

11. Tighten the left rear engine hanger adjusting bolt (14, **Figure 4**) to 4 N•m (35 in.-lb.) (where it just contacts the engine).

NOTE
*Use the Honda locknut wrenches (**Figure 2**) to tighten the locknuts during the following steps. Because the locknut wrench can extend the length of the torque wrench, the actual torque applied can be greater than the reading set or read on the wrench. The simplest method of using an extension with a torque wrench is to set them at a 90° angle as shown in **Figure 19**. This angle does not increase the effective length of the torque wrench and recalibration is not required. Refer to **Torque Adapters** in Chapter One for more information.*

12. Install the left rear engine hanger adjusting bolt locknut (15, **Figure 15**). Hold the adjusting bolt and tighten the locknut to 54 N•m (40 ft.-lb.) with the locknut wrench (**Figure 19**).

13. Remove the left center engine hanger bolt (20, **Figure 14**). Tighten the left center engine hanger adjusting bolt (18, **Figure 14**) to 4 N•m (35 in.-lb.) (where it just contacts the engine).

14. Install the left center engine hanger adjusting bolt locknut (19, **Figure 14**). Hold the adjusting bolt and tighten the locknut to 54 N•m (40 ft.-lb.) with the 20 mm locknut wrench (**Figure 20**).

15. Remove the left front engine hanger bolt (11, **Figure 14**). Tighten the left front engine hanger adjusting bolt (9, **Figure 14**) to 2 N•m (17 in.-lb.) (where it just contacts the engine).

16. Install the left front engine hanger adjusting bolt locknut (10, **Figure 14**) and tighten to 54 N•m (40 ft.-lb.) with the locknut wrench (**Figure 21**).

17. Install and tighten the left rear engine hanger bolt (16, **Figure 14**) to 64 N•m (47 ft.-lb.).

18. Install and tighten the left front engine hanger bolt (11, **Figure 14**) to 64 N•m (47 ft.-lb.).

19. Install and tighten the left center engine hanger bolt (20, **Figure 14**) to 39 N•m (29 ft.-lb.).

20. Install and tighten the fuel feed hose joint (**Figure 13**):
   a. Lubricate a new O-ring with oil and install it into the fuel feed hose groove.
   b. Install the fuel feed hose joint and tighten the mounting bolts to 10 N•m (88 in.-lb.).

5. Refer to **Figure 14** to install all of the engine mount fasteners and subframes. Do not tighten any of the fasteners until they all have been installed. Then tighten them in the following order, starting with Step 6.

6. Tighten the right subframe mounting bolts (2, **Figure 14**) securely.

7. Tighten the right rear engine hanger bolt (3, **Figure 14**) to 64 N•m (47 ft.-lb.).

8. Tighten the right front engine hanger bolt (6, **Figure 14**) to 64 N•m (47 ft.-lb.).

21. Tighten the rider footpeg (C, **Figure 4**) mounting bolts to 26 N•m (19 ft.-lb.).
22. Install and tighten the gearshift pedal arm assembly (**Figure 12**):
    a. Lubricate the gearshift pedal pivot bolt (A, **Figure 22**) and ball joints (B) with grease.
    b. Align the groove in the gearshift arm with the punch mark on the shift spindle (A, **Figure 12**). Install and tighten the pinch bolt securely.
    c. Install the gearshift pedal and secure with the washer (C, **Figure 22**) and pivot bolt (A). Install the washer between the engine and pedal. Tighten the pivot bolt to 26 N•m (19 ft.-lb.).
23. Secure the clutch hose with the banjo bolt and two new washers (**Figure 11**) and tighten to 34 N•m (25 ft.-lb.).
24. Refill and bleed the clutch system. Refer to *Clutch System Bleeding* in Chapter Six.
25. Fill the engine with the recommended type and quantity of engine oil. Refer to *Engine Oil and Filter* in Chapter Three.
26. Fill the cooling system. Refer to *Cooling System* in Chapter Three.
27. Turn the ignition switch on to pressurize the fuel line. Check for fuel leaks.

NOTE
Do not install the fairing assembly until after starting the engine and checking for leaks.

28. Start the engine and check for oil and coolant leaks. At the same time the engine is running, bleed the cooling system as described in Chapter Three.
29. Shift the transmission into gear and check clutch and transmission operation.
30. Slowly test ride the motorcycle to ensure all systems are operating correctly.

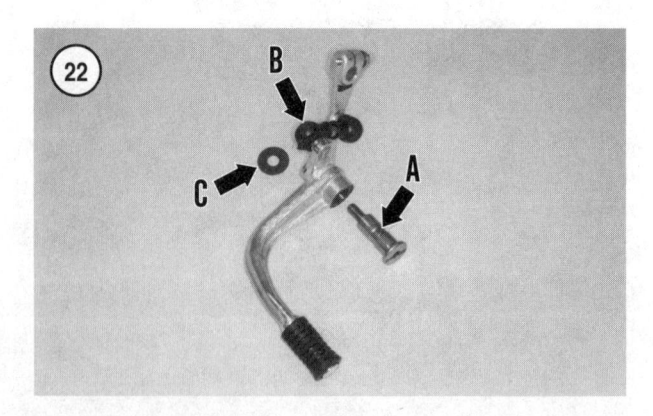

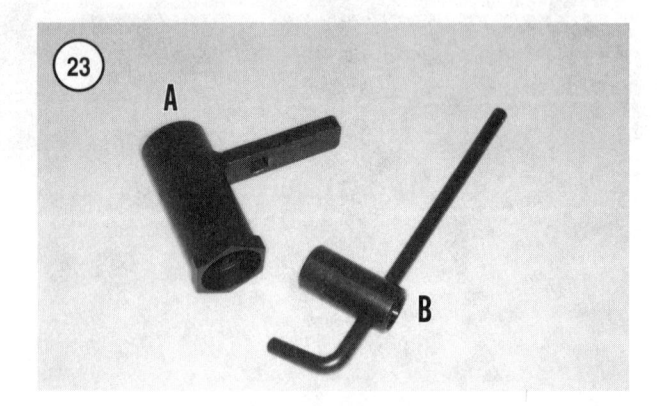

ENGINE LOWER END

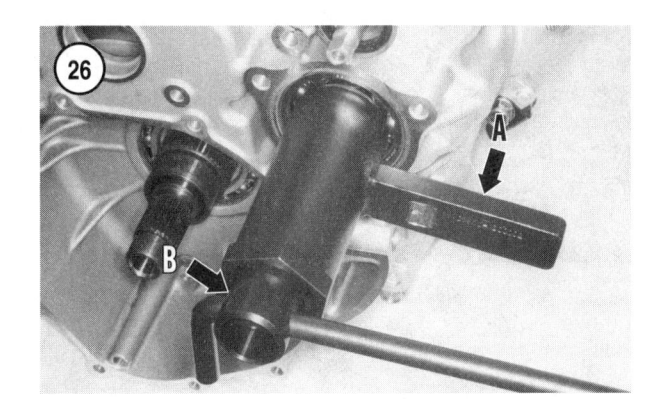

Tools

The following tools are required to loosen and tighten the output shaft locknut:

1. 30 × 64 mm locknut wrench (Honda part No. 07916-MB00002 or 07916-MB00001 [A, **Figure 23**]).
2. Shaft holder (Honda part No. 07924-PJ40001 [US only] or 07PAB-0010200 [B, **Figure 23**]).

Removal

1. Remove the engine from the frame as described in this chapter.
2. Remove the reverse shift arm assembly as described under *Reverse Shift Mechanism* in this chapter.
3. Remove the starter/reverse motor. Refer to *Starter/Reverse Motor Service* in Chapter Nine.
4. Remove the alternator. Refer to *Alternator* in Chapter Nine.
5. Remove the water pump/thermostat. Refer to *Water Pump/Thermostat* in Chapter Ten.
6. Remove the clutch. Refer to *Clutch* in Chapter Six.
7. Remove the bolts and the output shaft bearing holder (**Figure 24**).
8. Remove the output shaft locknut (A, **Figure 25**):
   a. Unstake the dimpled part of the locknut (B, **Figure 25**) with a grinder. Be careful not to damage the output shaft or threads.
   b. Shift the transmission into any gear except neutral.
   c. Install the locknut wrench (A, **Figure 26**) onto the locknut. Mount a breaker bar onto the locknut wrench.
   d. Hold the output shaft with the shaft holder (B, **Figure 26**) and loosen the output shaft locknut. If necessary, place a piece of pipe over the shaft holder rod to help hold it when loosening the locknut.
   e. Remove and discard the locknut (A, **Figure 25**).
9. Remove the rear crankcase cover (**Figure 27**) mounting bolts in several steps and in a crisscross pattern. Remove the cover from the engine.
10. Remove the gasket, two dowel pins (**Figure 28**) and the oil pipe and O-ring (**Figure 29**).
11. Remove the clutch regulator valve (**Figure 30**) and its O-ring.
12. Clean and inspect the rear crankcase cover and components as described in this section.

Installation

1. Make sure all the components installed behind the rear crankcase cover are properly installed.

REAR CRANKCASE COVER

The rear crankcase cover houses the clutch and supports a number of shafts and assemblies. The engine must be removed from the frame to service the rear crankcase cover assembly. The clutch regulator valve is mounted on the back of the rear crankcase cover.

2. Lubricate a new O-ring with engine oil and install it into the groove in the clutch regulator valve. Then install the valve into the cover.
3. Lubricate a new O-ring with engine oil and install it onto the oil pipe. Install the oil pipe and seat the O-ring (**Figure 30**) against the crankcase. If the oil pipe is machined with a shoulder, install the shoulder side into the crankcase.
4. Apply gasket sealant to the two areas where the crankcase halves join together.
5. Install the two dowel pins (**Figure 28**) and a new gasket.
6. Install the rear crankcase cover (**Figure 27**) onto the engine.
7. Install the rear crankcase cover mounting bolts and tighten in several steps to 24 N•m (17 ft.-lb.) in a crisscross pattern.
8. Install a new output shaft locknut as follows:
   a. Clean the locknut and output shaft threads with contact cleaner. These threads must be clean and free of oil.
   b. Thread the locknut onto the output shaft. Install the locknut wrench (A, **Figure 26**) onto the locknut. Mount a torque wrench at a 90° angle on the locknut wrench. This angle does not increase the effective length of the torque wrench and recalibration is not required.

   NOTE
   *Refer to **Torque Adapters** in Chapter One for information on using an extension with a torque wrench.*

   c. Hold the output shaft with the shaft holder (B, **Figure 26**) and tighten the output shaft locknut to 186 N•m (137 ft.-lb.). If necessary, place a piece of pipe over the shaft holder rod to help hold it when tightening the locknut.
   d. Stake the locknut shoulder into the groove in the output shaft (B, **Figure 25**).
9. Replace the output shaft bearing holder O-ring and seal: 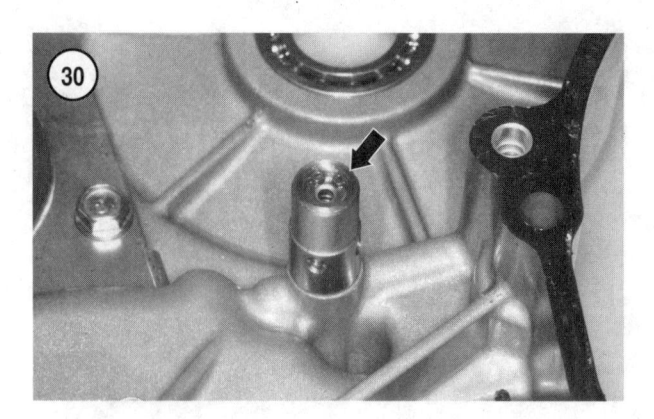 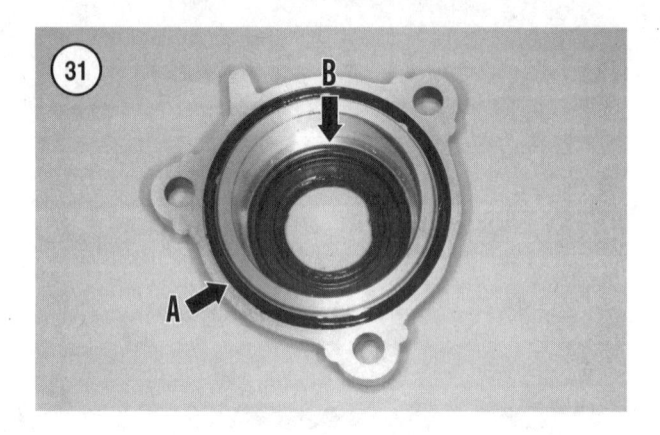
   a. Remove and discard the O-ring (A, **Figure 31**).
   b. Support the holder and drive out the old seal.
   c. Turn the holder over and drive in a new seal (B, **Figure 31**) with its flat side facing out (away from the engine).
   d. Lubricate a new O-ring with grease and install it into the bearing holder groove (A, **Figure 31**).
10. Install the output shaft bearing holder (**Figure 24**):
    a. Lubricate the seal lip with grease.
    b. Install the bearing holder and tighten its mounting bolts to 28 N•m (21 ft.-lb.).
11. Reverse Steps 1-6 under *Removal* in this section.

Inspection

1. Remove the clutch regulator valve (**Figure 30**).
2. Remove and discard the seal (A, **Figure 32**).
3. Clean and dry the rear crankcase cover assembly.
4. Identify the bearings:
   a. Alternator driven gear bearing (A, **Figure 33**).
   b. Output shaft bearing (B, **Figure 33**).
   c. Primary driven gear bearing (C, **Figure 33**).
   d. Reverse shifter shaft needle bearing (B, **Figure 32**).

ENGINE LOWER END

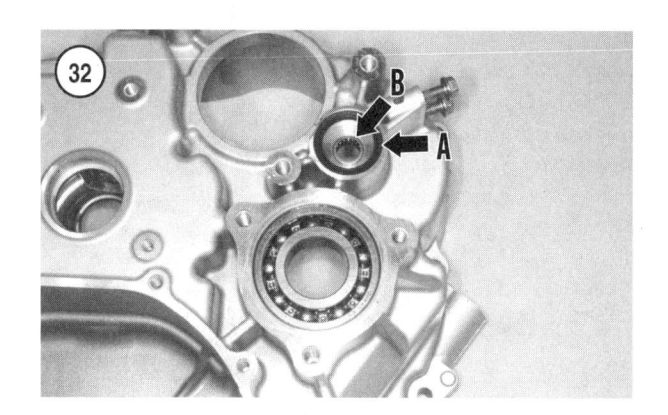

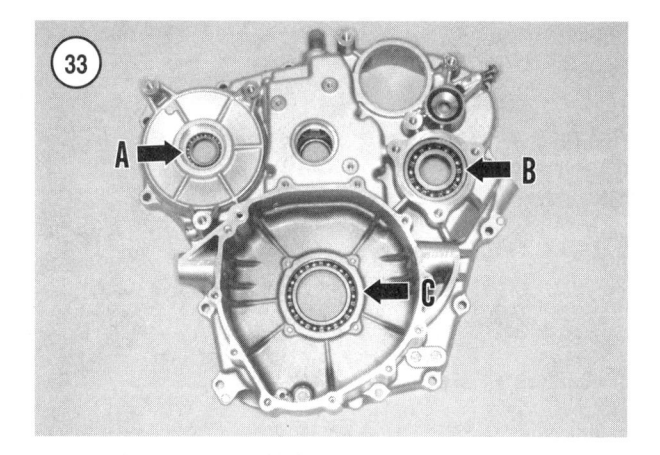

Figure 32

Figure 33

5. Inspect the ball bearings by rotating their inner race slowly by hand. Replace the bearings if there is any roughness or play (axial or radial). Then check that the bearings fit tightly in their mounting bore. If a bearing is loose, its mounting bore may be damaged. Remove the bearing and inspect the mounting bore for damage. A damaged bore will require replacement of the rear crankcase cover. Replace the bearings as described under *Bearing Replacement* in this section.

6. Inspect the needle bearing (B, **Figure 32**) for damage. Replace the needle bearing as described under *Bearing Replacement* in this section.

7. If necessary, remove the bolts, breather plate (**Figure 34**) and gasket. Clean the area under the breather plate. Install the breather plate using a new gasket and tighten the mounting bolts securely.

8. Install a new seal (A, **Figure 32**) by driving it into its bore with its flat side facing out.

9. Install a new O-ring into the clutch regulator valve groove and install the valve into the cover (**Figure 30**).

Figure 34

Clutch Regulator Valve
Inspection

While the clutch regulator valve (**Figure 30**) can be disassembled, replacement parts are not available.

1. Remove the snap ring, spring seal, spring and piston.
2. Clean and dry all parts.
3. Check the spring for fatigue and damaged coils.
4. Check the piston and valve bore for any wear or damage. The sides of the piston and valve bore must be smooth.
5. Replace the clutch regulator valve if there is any damage.
6. Lubricate the piston and bore with engine oil.
7. Install the piston, spring and seal. Compress the spring and secure with the snap ring.

Bearing Replacement

1. Replace damaged bearing(s) as described under *Bearings* in Chapter One and perform the following:
2. Install all of the bearings with their manufacturer's marks facing up.
3. When replacing and installing the reverse shifter shaft needle bearing (B, **Figure 32**), heat the cover to 80° C (176° F).
4. When replacing the primary drive gear bearing (C, **Figure 33**):
    a. Remove the bolts and setting plate (**Figure 35**). Remove all threadlock residue from the bolt and cover threads.
    b. Replace the bearing.
    c. Install the setting plate (**Figure 35**) with its OUT SIDE mark facing out.
    d. Apply a medium strength threadlock onto the bolt threads and tighten to 12 N•m (106 in.-lb.).

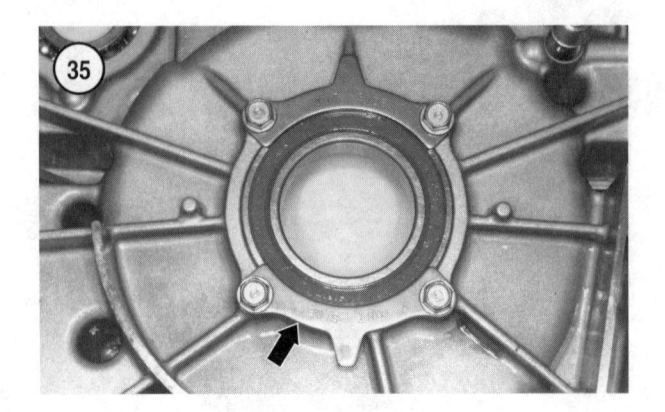

PRIMARY GEARS/OUTPUT SHAFT

Tools

The following tools are required to loosen and tighten the output shaft locknut:

1. Clutch outer holder (Honda part No. 07JMB-MN50100 [**Figure 36**]).
2. Shaft holder (Honda part No. 07JMB-MN50200).

Removal

1. Refer to **Figure 37** to identify the components serviced in this section:
   a. Starter reduction gear and shaft (A).
   b. Alternator driven gear assembly (B).
   c. Starter clutch assembly (C).
   d. Primary driven gear (D).
   e. Primary drive gear (installed behind the primary driven gear).
   f. Alternator drive gear (E).
   g. Output shaft assembly (F).
   h. Driven gear boss and oil pump driven sprocket (installed behind primary driven gear).
   i. Final drive gear (G).
   j. Reverse shifter assembly (H).

ENGINE LOWER END

2. Remove the rear crankcase cover as described in this chapter.
3. Remove the starter reduction gear and shaft (**Figure 38**).
4. Remove the alternator driven gear assembly:
    a. Flat washer (A, **Figure 39**).
    b. Friction spring (B, **Figure 39**).
    c. Insert a screwdriver or punch through one of the holes in the alternator drive gear (A, **Figure 40**).
    d. Align the scissor gear teeth with the tool and remove the alternator driven gear assembly (B, **Figure 40**).
5. Remove the starter clutch:
    a. Temporarily install the clutch outer (A, **Figure 41**) onto the primary driven boss.

NOTE
*The starter clutch mounting bolt (B, **Figure 41**) uses left-hand threads. Turn the bolt **clockwise** to loosen it.*

   b. Hold the clutch outer with the clutch outer holder (C, **Figure 41**) and a breaker bar and loosen the starter clutch mounting bolt (B). Remove the tools.
   c. Remove the bolt (A, **Figure 42**), washer (B) and starter clutch (C).
6. Remove the primary driven gear:
    a. Insert a screwdriver or punch through one of the holes in the primary drive gear (A, **Figure 43**).
    b. Align the scissor gear teeth with the tool and remove the primary driven gear (B, **Figure 43**).
7. Remove the spline washer (A, **Figure 44**) and the primary drive gear (B).
8. Remove the reverse shifter assembly (H, **Figure 37**). See *Reverse Shifter and Shift Drum Lock Arm* in this chapter.
9. Remove the six bolts (A, **Figure 45**) and the alternator drive gear (B).
10. Remove the output shaft assembly (**Figure 46**).
11. Remove the driven gear boss and the oil pump driven sprocket assembly:

a. Temporarily install the clutch outer (A, **Figure 47**) onto the primary driven boss.
b. Hold the clutch outer with the clutch outer holder (B, **Figure 47**) and a breaker bar, and loosen the oil pump driven sprocket bolt (C). Remove the tools.
c. Remove the bolt. Then remove the driven gear boss, chain and oil pump driven gear sprocket as an assembly (**Figure 48**).
d. Remove the bolts and chain cover (**Figure 49**).

12. Remove the final drive gear:
   a. Unstake the dimpled parts of the locknut (A, **Figure 50**) with a grinder. Be careful not to damage the countershaft or threads.
   b. Shift the transmission into any gear except neutral.

NOTE
*The final drive gear locknut (B, Figure 50) uses left-hand threads. Turn the nut **clockwise** to loosen it.*

   c. Hold the mainshaft with the shaft holder (**Figure 51**) and loosen the final drive gear locknut.
   d. Remove and discard the locknut (B, **Figure 50**).
   e. Check for the OUTSIDE stamped on the face of the washer. If there is no mark, identify this side of the washer before removing it.
   f. Remove the washer and final drive gear (C, **Figure 50**).

13. Clean and inspect the individual components as described in this section.

Installation

1. Lubricate the gear teeth and operating surfaces with engine oil.
2. Install the final drive gear:
   a. Install the final drive gear on the countershaft in the direction shown in C, **Figure 50**.
   b. Install the lockwasher with its OUTSIDE mark, or the mark made during removal, facing out.

NOTE
*The final drive gear locknut (B, Figure 50) uses left-hand threads. Turn the nut **counterclockwise** to tighten it.*

   c. Lubricate the threads and seating surface of a new locknut with engine oil and install it.
   d. Hold the mainshaft with the shaft holder (**Figure 51**) and tighten the locknut to 186 N•m (137 ft.-lb.).
   e. Stake the locknut into the countershaft in two places (A, **Figure 50**).

ENGINE LOWER END

3. Install the driven gear boss and oil pump drive sprocket assembly:

   a. Install the chain cover and tighten its mounting bolts securely (**Figure 49**).
   b. Lubricate the needle bearing (**Figure 52**) in the primary driven gear boss with engine oil.
   c. Assemble the driven gear boss, chain and oil pump driven sprocket (**Figure 52**). Position the sprocket with its OUT mark facing out.
   d. Install the assembly as shown in **Figure 48**. 
   e. Apply a medium strength threadlock onto the oil pump driven sprocket bolt threads and install the bolt.
   f. Temporarily install the clutch outer (A, **Figure 47**) onto the primary driven boss.
   g. Hold the clutch outer with the clutch outer holder (B, **Figure 47**) and a breaker bar, and tighten the oil pump driven sprocket bolt (C) to 18 N•m (13 ft.-lb.).

4. Install the output shaft into the crankcase and mesh it with the final drive gear (**Figure 46**).

5. Install the alternator drive gear:

   a. Install the alternator drive onto the crankshaft in the direction shown in B, **Figure 45**.
   b. Lubricate the mounting bolt threads and seating surfaces with engine oil and install hand-tight (A, **Figure 45**).
   c. Secure the crankshaft by holding the rotor bolt (located at the opposite end of the crankshaft) and tighten the alternator drive gear mounting bolts (A, **Figure 45**) to 25 N•m (19 ft.-lb.).

6. Install the reverse shifter assembly (H, **Figure 37**). See *Reverse Shifter and Shift Drum Lock Arm* in this chapter.

7. Install the primary drive gear (B, **Figure 44**) with the thin subgear facing out. Install the spline washer (A, **Figure 44**) and seat against the gear.

8. Install the primary driven gear:

   a. Install the primary driven gear (B, **Figure 43**) with its shoulder side facing out.
   b. Insert a screwdriver or punch through a hole in the primary drive gear (A, **Figure 43**) and wedge it to align the subgear and primary drive gear teeth, then install the primary driven gear by engaging the gears. Remove the tool to release the subgear. The primary driven gear should mesh tightly against the primary drive gear.

9. Install the starter clutch:

   a. Remove the starter driven gear (**Figure 53**) from the starter clutch. Lubricate the needle bearing and sprag clutch surfaces with engine oil.

b. Turn the starter driven gear clockwise to install it into the starter clutch housing.

> **NOTE**
> The starter clutch mounting bolt (A, **Figure 42**) uses left-hand threads. Turn the bolt **counterclockwise** to tighten it.

c. Install the starter clutch assembly (C, **Figure 42**), washer (B) and bolt (A).
d. Temporarily install the clutch outer (A, **Figure 41**) onto the primary driven boss.
e. Hold the clutch outer with the clutch outer holder (C, **Figure 41**) and a breaker bar, and tighten the starter clutch mounting bolt (B, **Figure 41**) to 74 N•m (54 ft.-lb.).

10. Install the alternator driven gear assembly:
    a. Install the alternator driven gear (B, **Figure 40**) with its gear side facing in.
    b. Insert a screwdriver or punch through a hole in the alternator drive gear (A, **Figure 40**) and wedge it to align the subgear and alternator drive gear teeth, then install the alternator driven gear by engaging the gears. Remove the tool to release the subgear. The alternator driven gear should mesh tightly against the alternator drive gear.
    c. Install the friction spring (B, **Figure 39**) and flat washer (A) with its concave side facing in.

11. Lubricate the reduction gear shaft with a molybdenum oil solution. Install the reduction gear and shaft with the larger gear facing out (**Figure 38**).

> **NOTE**
> Molybdenum oil solution is a 50:50 mixture of engine oil and molybdenum disulfide grease.

12. Install the rear crankcase cover as described in this chapter.

Inspection

This section describes inspection procedures for all the parts, except the output shaft. Service the output shaft as described under *Output Shaft* in this chapter.

1. Clean and dry the parts.

> **NOTE**
> Do not disassemble the alternator driven gear assembly *(Figure 54)*.

2. Inspect the parts for excessive wear looseness and damage. Many of the gears are subassemblies consisting of different parts. Inspect all the parts and how the gears are assembled.

3. Inspect the driven gear boss needle bearing (**Figure 55**) for rough movement or visual damage. If necessary, replace the bearing as follows:
    a. Support the driven gear boss in a press with the oil pump drive sprocket side facing down.
    b. Press the needle bearing out of the driven gear boss.
    c. Clean the bore through the driven gear boss.
    d. Support the driven gear boss in a press with the oil pump drive gear side facing up. Center the bearing in the top of the boss with its manufacturer's marks facing up.
    e. Press the new needle bearing into the driven gear boss until the depth from the bearing to the top of the boss (**Figure 55**) is 3.5-4.0 mm (0.14-0.16 in.).

4. Inspect the oil pump drive chain for loose or damaged links.

5. Inspect the splines running through the gears for cracks and other damage.

6. Inspect the fasteners and washers for wear and damage. Discard the locknuts.

ENGINE LOWER END

Figure 56

Figure 57

3. Inspect the bearing (A, **Figure 57**) by rotating its outer race slowly by hand. Replace the bearing if there is any roughness or play (axial or radial). The bearing must fit tightly on the shaft. If loose, check the shaft surface for damage.
4. Inspect the gears for excessive wear and damage.
5. Inspect the damper spring (A, **Figure 56**) and damper lifter (B) for cracks and other damage.
6. Inspect the final driven gear by turning it by hand. Check for any excessive play, which may indicate gear, bushing or shaft wear.

Disassembly

A hydraulic press is required to overhaul the output shaft assembly. Refer to **Figure 58** when performing this procedure.

> *WARNING*
> *The output shaft is under considerable pressure. Do **not** attempt to disassemble the output shaft without a hydraulic press and the correct tools.*

1. Remove the snap ring (A, **Figure 59**) and retainer (B) from the output shaft.
2. Tap the cotters out with a screwdriver (**Figure 60**).
3. Support the final driven gear in a press and press the output shaft out of the bearing (**Figure 61**). Discard the bearing.
4. Disassemble the output shaft in the order shown in **Figure 58**.

Inspection

When measuring the output shaft components, compare the actual measurements to the specifications in **Table 1**. Replace worn or damaged parts as described in this section.
1. Clean and dry all parts.
2. Check the damper lifter (5, **Figure 58**) for damaged splines or worn or damaged engagement lugs.
3. Inspect the damper spring (4, **Figure 58**) for cracks. Measure the damper spring free length.
4. Check the bushing (A, **Figure 62**) surfaces for cracks, pitting and wear. Measure the bushing inside diameter and outside diameter.
5. Inspect the final driven gear (B, **Figure 62**) for excessive wear, burrs, pitting, chipped or missing teeth. Check the engagement slots in the gear for wear or damage. Measure the gear inside diameter.
6. Inspect the reverse driven gear (3, **Figure 58**) as described in Step 5. Also, inspect the damper spring contact area for damage.

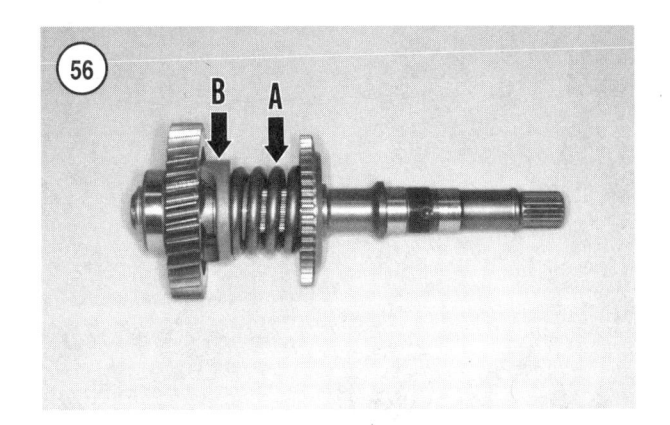

7. Inspect the starter clutch as follows:
   a. Rotate the starter driven gear. The gear should only rotate clockwise (**Figure 53**). If the gear rotates counterclockwise, replace the starter clutch assembly.
   b. Remove the starter driven gear and inspect the needle bearing and gear shoulder for damage.
   c. Lubricate the needle bearing and sprag clutch surfaces with engine oil. Turn the starter driven gear clockwise to install it into the starter clutch housing.

OUTPUT SHAFT

Service the output shaft (**Figure 56**) as described in the following sections. A thorough inspection requires disassembly of the output shaft.

Preliminary Inspection

1. Remove the output shaft as described in this chapter.
2. Clean and dry the output shaft.

CHAPTER FIVE

58 OUTPUT SHAFT

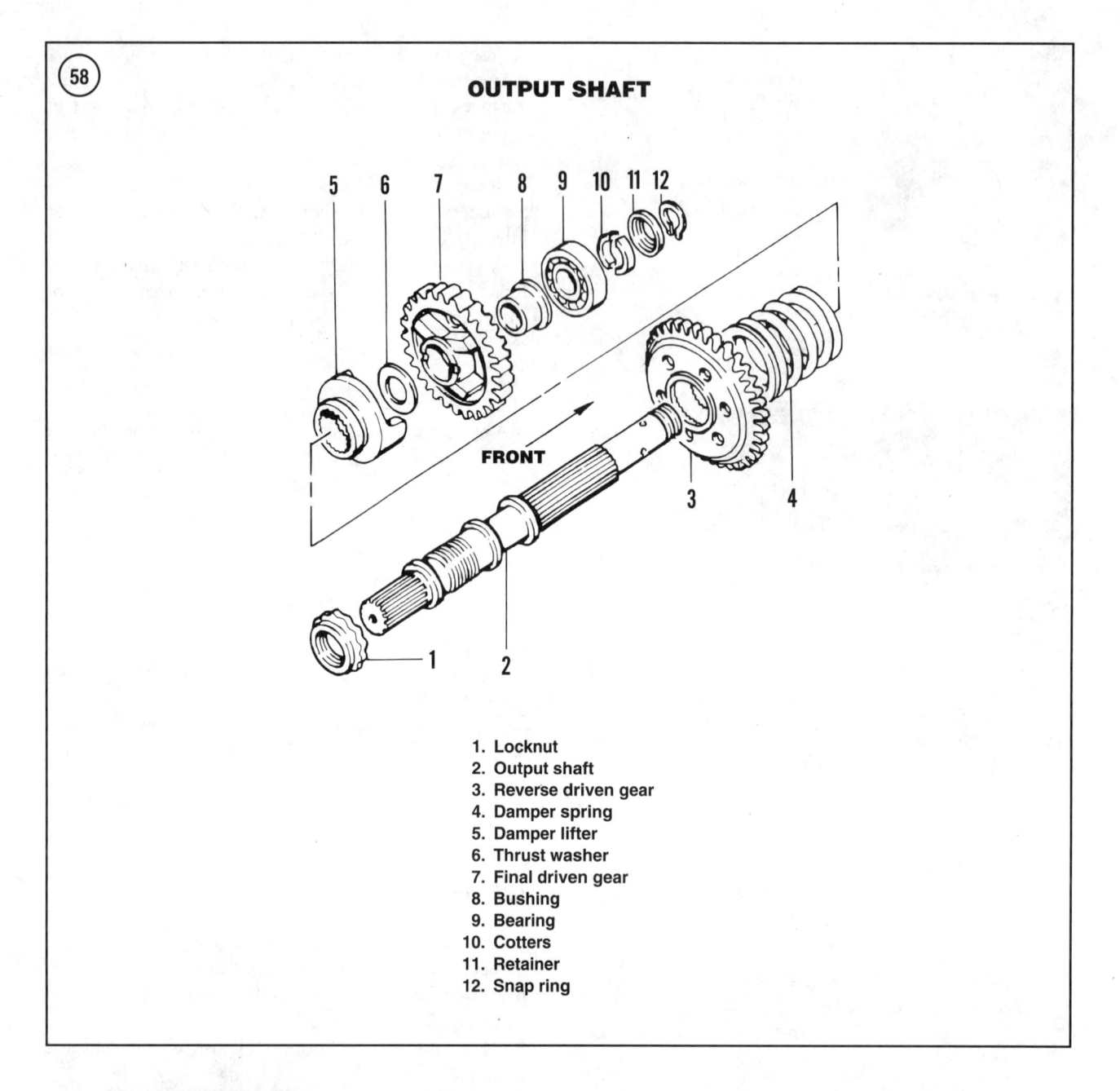

1. Locknut
2. Output shaft
3. Reverse driven gear
4. Damper spring
5. Damper lifter
6. Thrust washer
7. Final driven gear
8. Bushing
9. Bearing
10. Cotters
11. Retainer
12. Snap ring

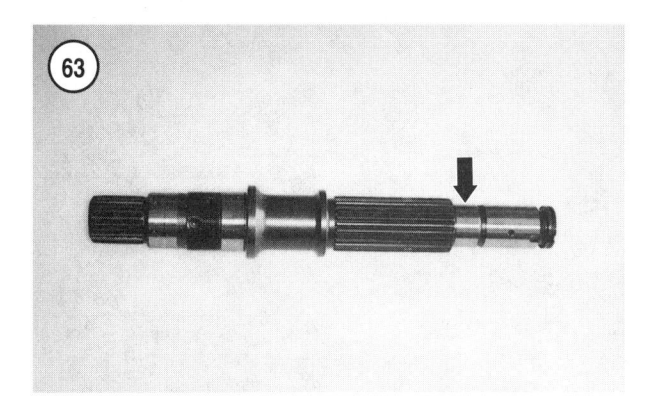

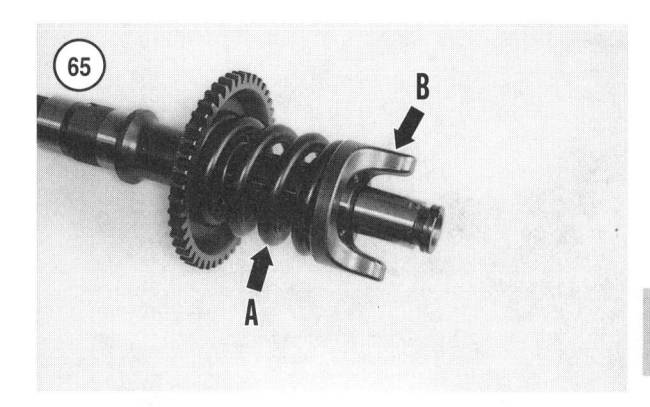

ENGINE LOWER END

7. Inspect the output shaft splines and snap ring grooves for damage. Measure the output shaft diameter where the bushing rides (**Figure 63**).

Assembly

1. Lubricate all bearing surfaces with engine oil.
2. Install the reverse driven gear with its dished side facing toward the spring side of the shaft (**Figure 64**).
3. Install the damper spring (A, **Figure 65**) and seat against the gear.
4. Install the damper lifter with its lugs facing away from the spring (B, **Figure 65**).
5. Install and seat the thrust washer (**Figure 66**) against the damper lifter.
6. Install the final driven gear (**Figure 67**) by aligning its groves with the engagement lugs on the damper lifter. See **Figure 68**.
7. Install the bushing and insert it through the final driven gear (**Figure 69**).
8. Support the output shaft in a press (**Figure 70**).
9. Install a new bearing over the end of the shaft with its closed side facing up (A, **Figure 71**).
10. Place a bearing installer over the shaft and seat it against the inner bearing race (B, **Figure 71**). Press the bearing onto

the shaft until the inner bearing race is approximately even with the lower cotter shoulder (**Figure 72**).

11. Install the two cotters (**Figure 73**) into the cotter groove in the output shaft.

12. Install the retainer (B, **Figure 59**) by aligning its tab with the groove in the shaft. Then turn the retainer and position the tab 180° opposite the shaft groove.

13. Secure the retainer with a new snap ring (B, **Figure 57**).

REVERSE SHIFT ARM

Refer to **Figure 74**.

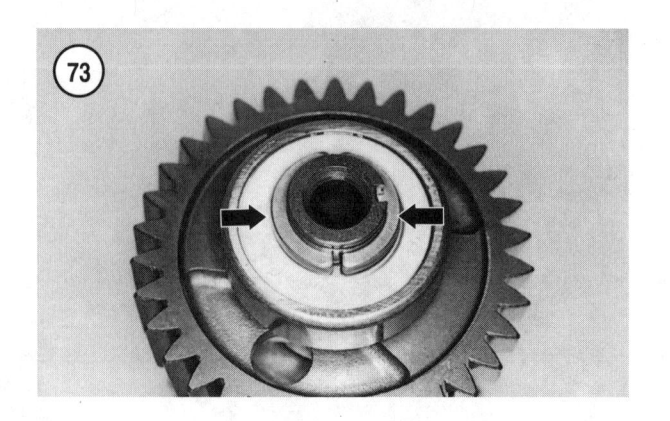

ENGINE LOWER END

REVERSE SHIFT ARM ASSEMBLY

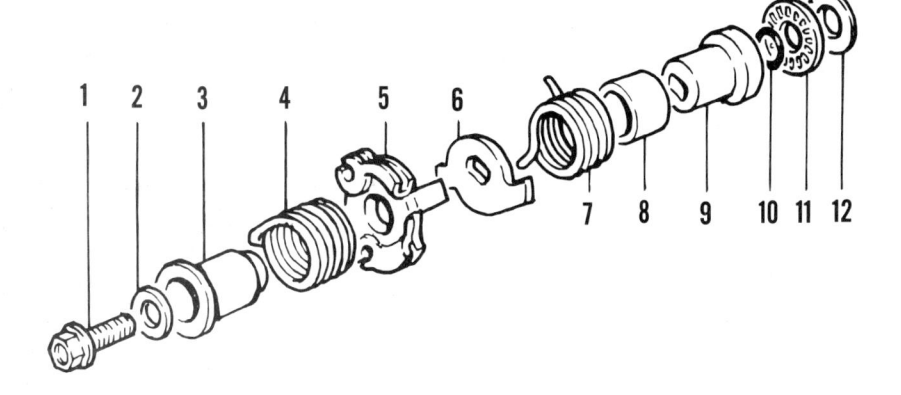

1. Bolt
2. Washer
3. Collar
4. Lost motion spring
5. Reverse shift arm
6. Lost motion plate
7. Shift arm spring
8. Bushing
9. Spring collar
10. O-ring
11. Needle bearing
12. Washer

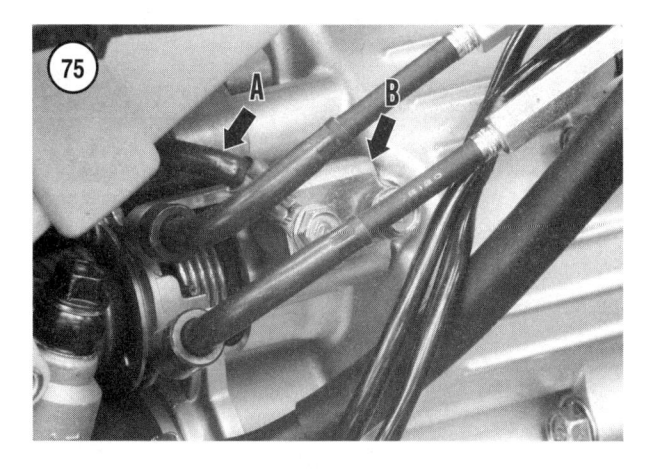

Removal

1. Remove the rear master cylinder. Refer to *Rear Master Cylinder* in Chapter Fourteen.

2. Remove the two bolts, ground cable (A, **Figure 75**) and the reverse shift actuator cable holder (B).

NOTE
The following photographs are shown with the engine removed to better illustrate the steps.

3. Remove the bolt (1, **Figure 74**) and washer (2) and remove the reverse shift arm assembly—collar, lost motion spring and attached cables (**Figure 76**).
4. Disconnect the reverse cables from the reverse shift arm.
5. Refer to **Figure 74** and remove the following:
   a. Lost motion plate (6).
   b. Shift arm spring (7).
   c. Bushing (8).
   d. Spring collar (9) and O-ring (10).
   e. Needle bearing (11).
   f. Washer (12).
6. Remove the oil seal (**Figure 77**).

Inspection

1. Inspect the springs. Replace if broken or weak.
2. Inspect the collar, bushing and spring collar for cracks and scoring.
3. Install a new O-ring (10, **Figure 74**) into the spring collar.
4. Inspect the needle bearing and washer for damage.

Installation

NOTE
The following photographs are shown with the engine removed to better illustrate the steps.

1. Pack the lips of a new seal with grease and install the seal with its flat side facing out (**Figure 77**). Install the seal so it is flush with the bore surface.
2. Install the flat washer (A, **Figure 78**) onto the reverse shifter shaft.
3. Lubricate the needle bearing (B, **Figure 78**) with a molybdenum oil solution and install next to the washer.

NOTE
Molybdenum oil solution is a 50:50 mixture of engine oil and molybdenum disulfide grease.

4. Install the spring collar (A, **Figure 79**) and seat it against the bearing.
5. Lubricate the inner and outer surfaces of the bushing and install over the spring collar (B, **Figure 79**).
6. Install the shift arm spring (A, **Figure 80**) by aligning its inner arm (B) with the bracket on the rear crankcase cover.
7. Install the lost motion plate as shown in **Figure 81**.
8. Insert the collar (3, **Figure 74**) through the lost motion spring. Then install the reverse shift arm onto the end of the collar by inserting the arm on the reverse shift arm between the two spring arms. See **Figure 82**.

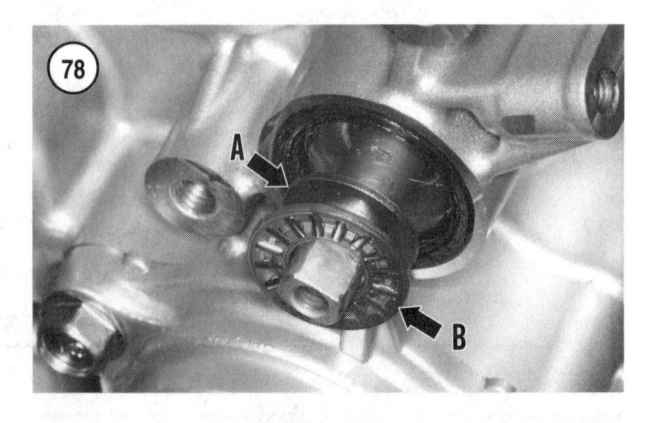

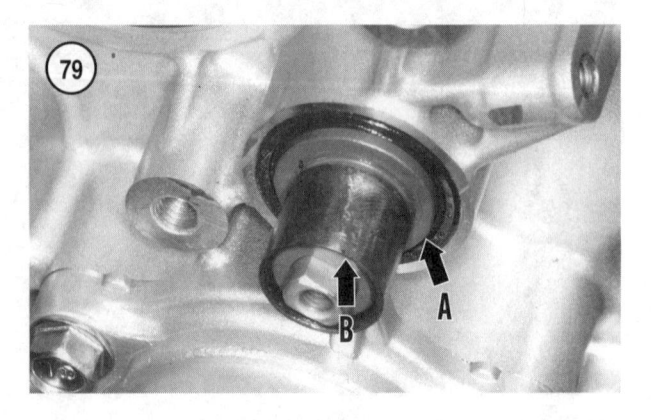

ENGINE LOWER END

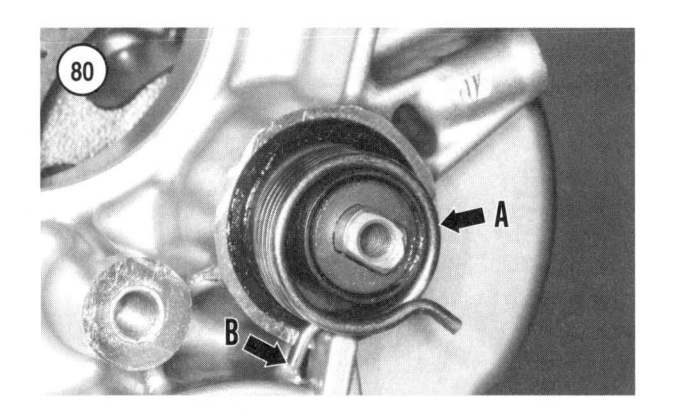

9. Install the reverse shift arm assembly over the shaft as shown in A, **Figure 83**.

10. Apply a medium strength threadlock onto the mounting bolt threads. Install the bolt and washer and tighten to 14 N•m (10 ft.-lb.).

11. Reconnect the reverse cables onto the reverse shift arm, if removed.

12. Apply a medium strength threadlock onto the cable holder bolts. Install the cable holder (B, **Figure 75**) and ground cable (A). Tighten the two mounting bolts to 12 N•m (106 in.-lb.).

13. Adjust the reverse cables. Refer to *Reverse Shift Actuator* in Chapter Nine.

14. Install the rear master cylinder and bleed the brake system as described in Chapter Fourteen.

15. Start the engine and check the reverse operation.

REVERSE SHIFTER AND SHIFT DRUM LOCK ARM

Refer to **Figure 84**.

Removal

1. Remove the rear crankcase cover as described in this chapter.
2. Remove the following:
    a. Starter reduction gear and shaft (**Figure 85**).
    b. Washer (A, **Figure 86**) and reverse idle gear (B).
    c. Reverse shifter shaft assembly (**Figure 87**).
3. Remove the bolts and the starter drive gear holder (**Figure 88**) and both dowel pins (**Figure 89**).
4. Remove the bolt (A, **Figure 90**) and the shift drum lock arm assembly (B).

Reverse Shifter
Disassembly/Inspection/Assembly

Replace worn or damaged parts as described in this section.

1. Remove the pin (**Figure 91**) and the reverse shifter assembly from the shaft (**Figure 92**).
2. Remove the snap ring and disassemble the reverse shifter assembly (**Figure 93**).
3. Clean and dry all parts.
4. Inspect the reverse shifter gear for excessive wear, pitting, chipped or missing gear teeth. Check the pin slot and gear bore for wear and damage.
5. Check the washers for bending, wear or damage.
6. Check the needle bearings for needle or cage damage.

CHAPTER FIVE

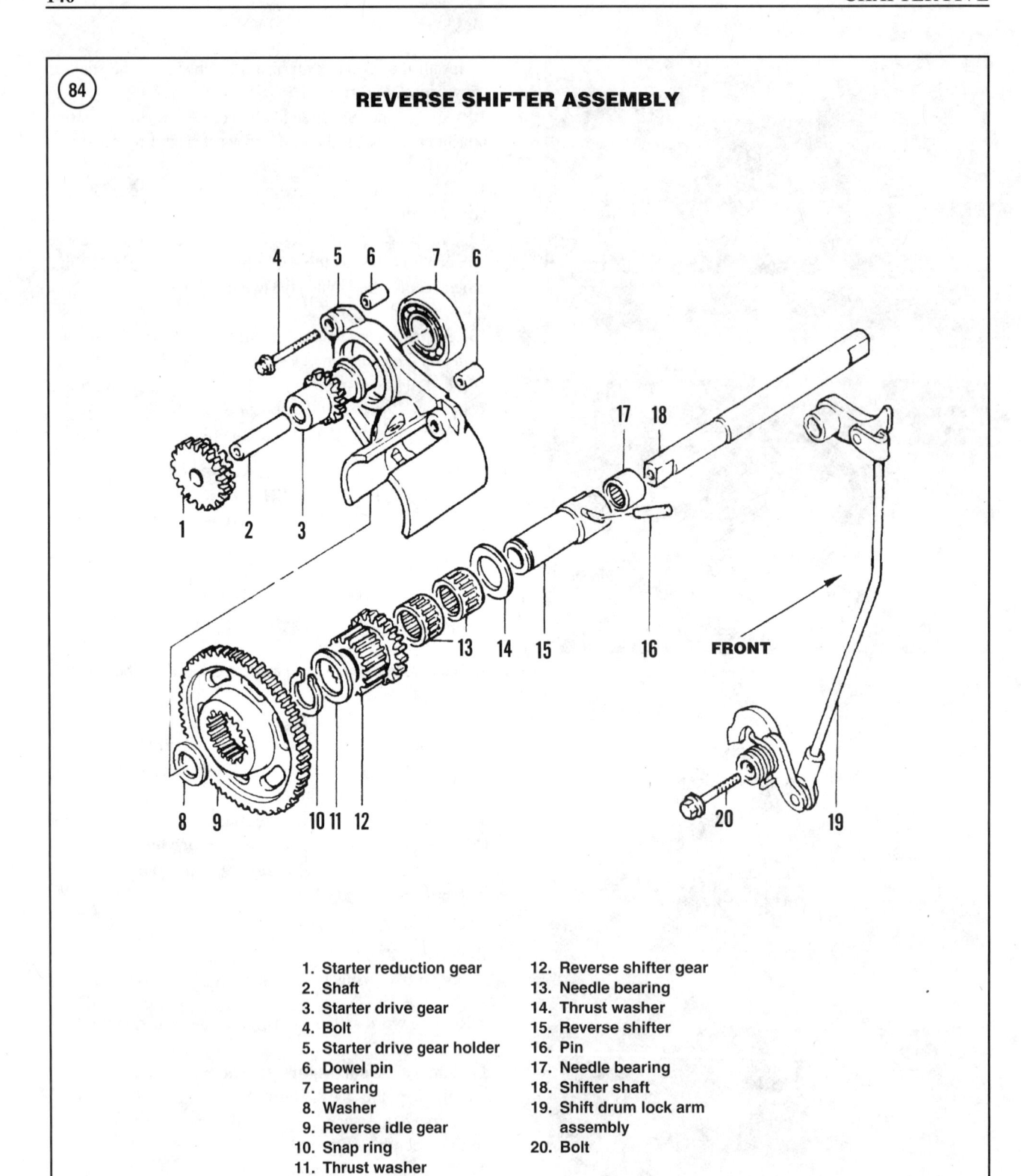

REVERSE SHIFTER ASSEMBLY (84)

1. Starter reduction gear
2. Shaft
3. Starter drive gear
4. Bolt
5. Starter drive gear holder
6. Dowel pin
7. Bearing
8. Washer
9. Reverse idle gear
10. Snap ring
11. Thrust washer
12. Reverse shifter gear
13. Needle bearing
14. Thrust washer
15. Reverse shifter
16. Pin
17. Needle bearing
18. Shifter shaft
19. Shift drum lock arm assembly
20. Bolt

ENGINE LOWER END

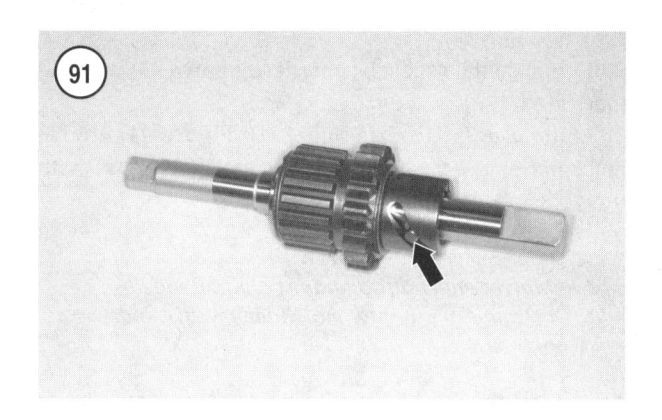

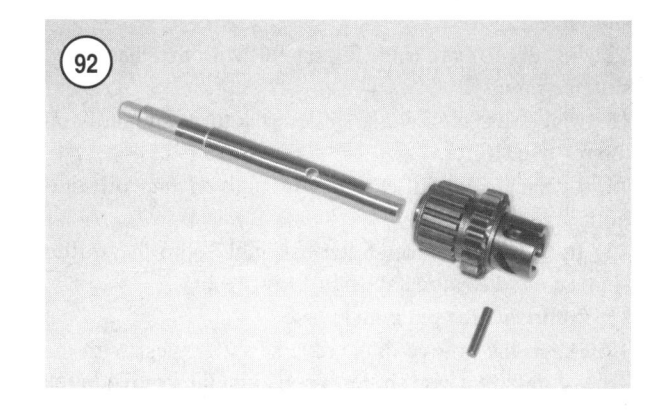

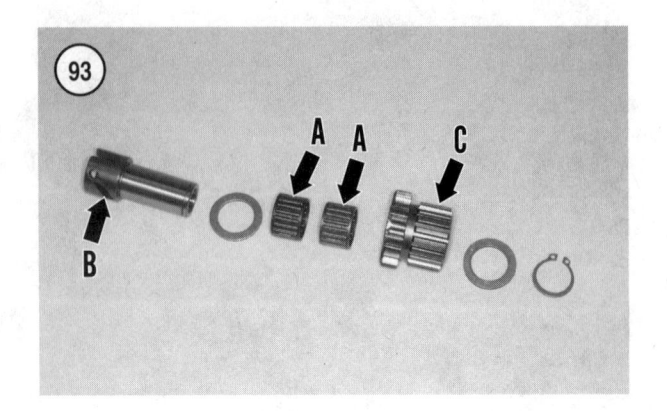

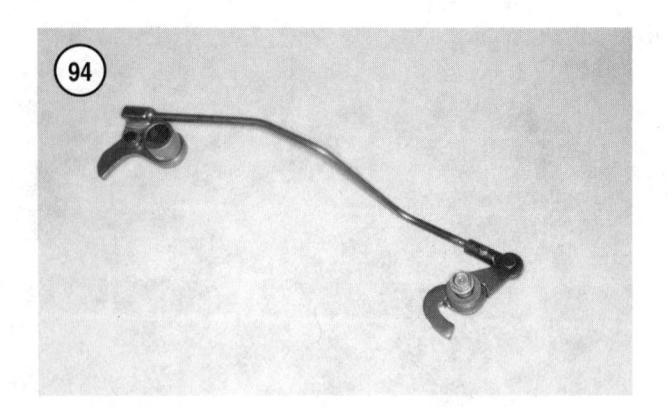

148 CHAPTER FIVE

7. Check the reverse shifter for wear or damage. Check the shaft for pitting, scoring and damage. Check the pin slot and the snap ring groove for wear and damage.

NOTE
Do not disassemble the shift drum lock arm assembly (Figure 94). Replacement parts are not available. If damaged, replace the complete arm.

8. Check the shift drum lock arm (**Figure 94**) for bending or damage. Check the pivot ends for any roughness or damage. Check the spring (A, **Figure 95**) for cracks or damage. Make sure the spring ends lock against the arm as shown.
9. Lubricate the needle bearings (A, **Figure 93**) with engine oil.
10. Lubricate the reverse shifter (B, **Figure 93**) groove and splines, and the reverse shifter gear (C) groove with molybdenum oil solution.

NOTE
Molybdenum oil solution is a 50:50 mixture of engine oil and molybdenum disulfide grease.

11. Install the washer, needle bearings, reverse shifter gear and washer onto the reverse shifter in the order shown in **Figure 93**.
12. Install the snap ring (**Figure 96**) with the chamfered side facing up.
13. Make sure the reverse shifter gear turns smoothly on the reverse shifter.
14. Lubricate the shifter shaft with molybdenum oil solution.
15. Install the reverse shifter assembly onto the shifter shaft in the direction shown in **Figure 92**.
16. Lubricate the pin with grease.
17. Align the groove in the reverse shifter gear with the groove in the reverse shifter. Then align the groove in the reverse shifter assembly with the hole in the shifter shaft and install the pin (**Figure 91**).

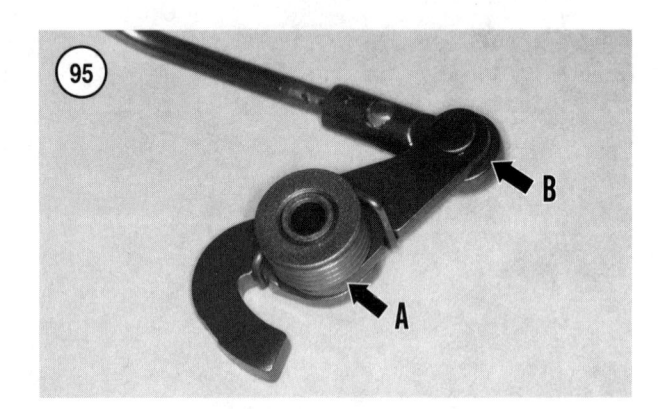

Starter Drive Gear Holder Inspection/Overhaul

1. Check the starter drive gear holder assembly (**Figure 97**):
   a. Check the starter drive gear (A, **Figure 97**) for wear, pitting, chipped or missing gear teeth.

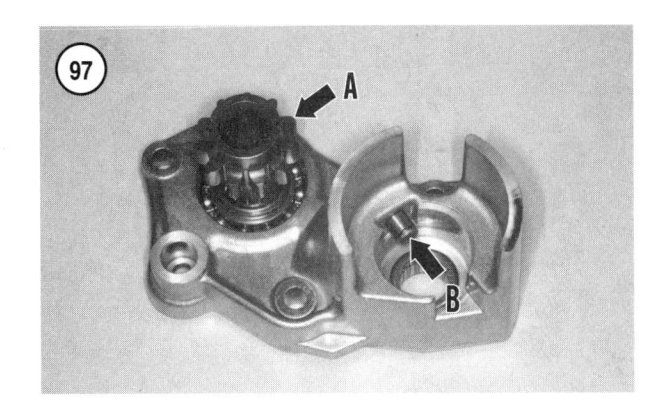

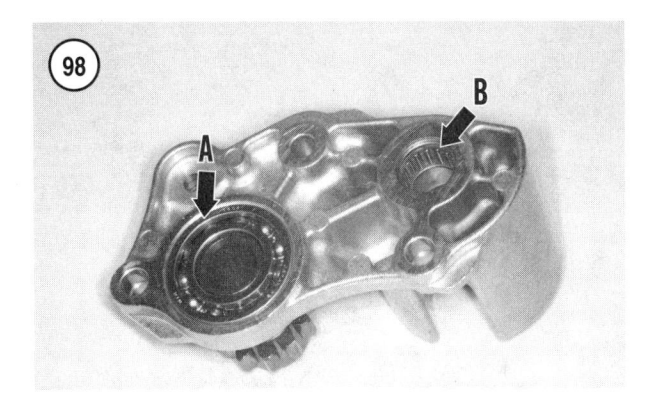

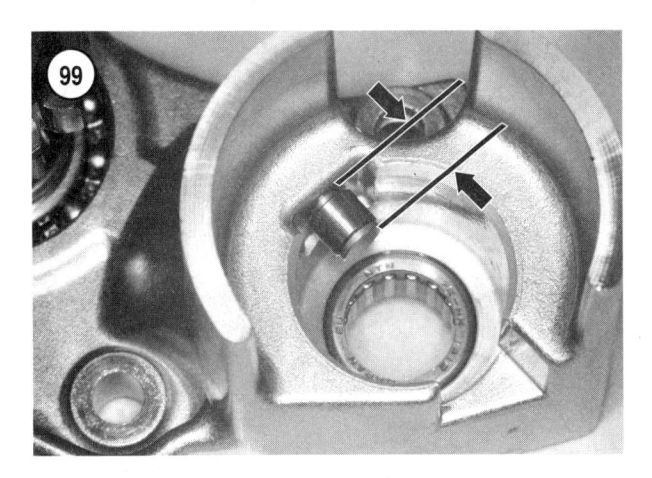

ENGINE LOWER END

d. Replace the starter drive gear and bearing(s), beginning with Step 2.

NOTE
Use a press, suitable adapters and bearing drivers during the following steps.

2. Before removing the bearings, note and record the direction in which the bearing manufacturer's marks face for proper reassembly.
3. Support the holder and press the starter drive gear (A, **Figure 97**) out of its bearing.
4. Press the bearing (A, **Figure 98**) out of the holder.

NOTE
Do not remove the reverse stopper shaft and collar unless the needle bearing will be replaced.

5. Drive the reverse stopper shaft and collar (B, **Figure 97**) out of the holder with a 4-mm pin driver.
6. Support the drive gear holder and press the needle bearing (B, **Figure 98**) out of the holder.
7. Clean and dry the holder assembly.
8. Inspect the bearing mounting bores for scoring, pitting and other damage.
9. Support the holder and press the new bearing (A, **Figure 98**) into the holder.
10. Support the holder and press the new needle bearing (B, **Figure 98**) into the holder.
11. Drive the reverse stopper shaft and collar into the holder so its exposed height from the holder surface (**Figure 99**) is 6.7-7.3 mm (0.26-0.29 in.).
12. Support the bearing inner race and press the starter drive gear (A, **Figure 99**) into the bearing.
13. Check that each bearing race fits tightly in its mounting bore. If a bearing fit is loose, the mounting bore is damaged and the holder must be replaced.

Installation

1. Lubricate the lock arm pivot (B, **Figure 95**) with molybdenum disulfide grease.
2. Install the shift drum lock arm assembly (**Figure 94**):
    a. Apply a medium strength threadlock onto the shift drum lock arm bolt threads.
    b. Install the shift drum lock arm (**Figure 90**) by inserting the upper pivot arm shoulder into the crankcase bore (A, **Figure 100**).
    c. Secure the lower pivot arm with the shift drum lock arm bolt (**Figure 101**) and tighten to 12 N•m (106 in.-lb.).

b. Inspect the ball bearing (A, **Figure 98**) by rotating the starter drive gear by hand. Replace the bearing if there is any roughness or play (axial or radial). Then check that the bearing fits tightly in its mounting bore. A loose bearing may indicate a damaged bearing mounting bore. Remove the bearing and inspect the mounting bore for damage. If the bore is damaged, replace the starter drive gear holder.
c. Inspect the needle bearing (B, **Figure 98**) for damage.

3. Lubricate both starter drive gear holder bearings (A and B, **Figure 98**) with engine oil.
4. Lubricate the reverse stopper shaft and collar (B, **Figure 97**) with molybdenum oil solution.
5. Install the two dowel pins (**Figure 89**).
6. Install the starter drive gear holder (**Figure 88**) and tighten the mounting bolts securely.
7. Install the reverse shifter shaft assembly by aligning the flat on the shaft (**Figure 102**) with the flat in the lock arm hole (B, **Figure 100**).
8. Install the reverse idle gear (B, **Figure 86**) with its OUT mark facing out and then install the washer (A).
9. Lubricate the starter reduction gear shaft with molybdenum oil solution. Install the shaft and starter reduction gear (**Figure 85**).

CRANKCASE

The engine (**Figure 103**) is made up primarily of a cylinder block and two separate cylinder heads. The crankcase is part of the cylinder block that is split vertically on the centerline of the crankshaft. The crankcases are made of die-cast aluminum and matched as a set. The mating of the crankcase halves is a precision fit with no gasket at the joint; only a thin layer of gasket sealer is used to seal the crankcase. Handle the crankcase halves carefully during all service procedures to avoid damaging the bearing and gasket surfaces. There is no separate oil pan as the oil sump is part of the crankcase.

To remove the pistons, the engine must be removed from the frame and the crankcase halves split.

Tools

The following engine tools are required during engine assembly:
1. Piston base A (Honda part No. 07ZMG-MCAA100).
2. Piston base B (Honda part No. 07ZMG-MCAA200).
3. Piston ring compressor (Honda part No. 07JMG-MN5000A). Engine assembly requires three piston ring compressors.
4. Two support blocks: fabricate two wooden support blocks to the dimensions 40 × 40 × 85 mm.

Crankcase Disassembly

1. Remove the engine from the frame as described in this chapter.

ENGINE LOWER END

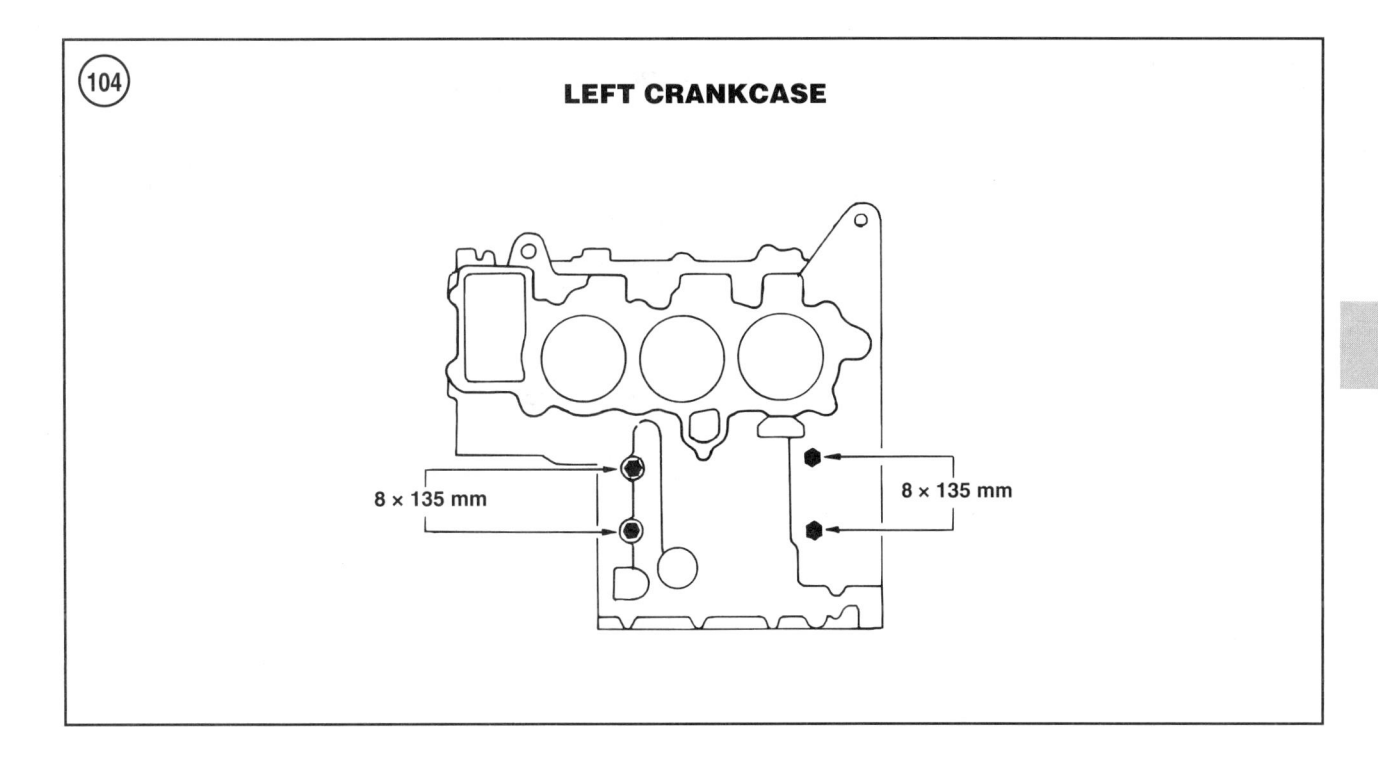

104 **LEFT CRANKCASE**

8 × 135 mm   8 × 135 mm

2. Remove the intake manifold. Refer to *Intake Manifold* in Chapter Eight.

3. Remove the cylinder heads and cam chain. Refer to *Cylinder Heads* in Chapter Four.

4. Remove the gearshift linkage. Refer to *External Shift Linkage* in Chapter Seven.

5. Remove the starter/reverse motor as described in Chapter Nine.

6. Remove the alternator as described in Chapter Nine.

7. Remove the water pump as described in Chapter Ten.

8. Remove the clutch as described in Chapter Six.

9. Remove the rear crankcase cover as described in this chapter.

10. Remove the primary gears and output shaft assembly as described in this chapter.

11. Remove the reverse shift arm assembly as described under *Reverse Shift Arm* in this chapter.

12. Remove the water hose joint from the left crankcase.

13. Remove the four mounting bolts and washers (**Figure 104**) from the left crankcase.

14. Remove the mounting bolts from the right crankcase (**Figure 105**):
   a. Loosen the 6–mm bolts in two steps in a crisscross pattern.
   b. Loosen the 10–mm bolts in two steps in a crisscross pattern.
   c. Remove the bolts and washers.

15. Position the engine with the left crankcase facing down. Place it on a rubber mat or wooden blocks to protect the gasket surfaces.

16. Remove the right crankcase to a point just before the pistons are removed from the right crankcase bores. Have an assistant place shop cloths around the base of each piston to prevent them from falling. Remove the right crankcase from the pistons.

CAUTION
Do not pry the crankcase halves apart. The crankcases are machined as a set. If one is damaged, replace both. If the halves will not separate, check for an unloosened bolt.

17. Remove the dowel pins (**Figure 106**) and oil pipe (**Figure 107**) from the left crankcase.

18. Remove the transmission assembly as described in Chapter Seven.

19. Remove the pistons and connecting rods as described in this chapter.

20. Remove the crankshaft as described in this chapter.

21. Remove the oil pump as described in this chapter.

22. Inspect and clean the crankcase halves as described in this section.

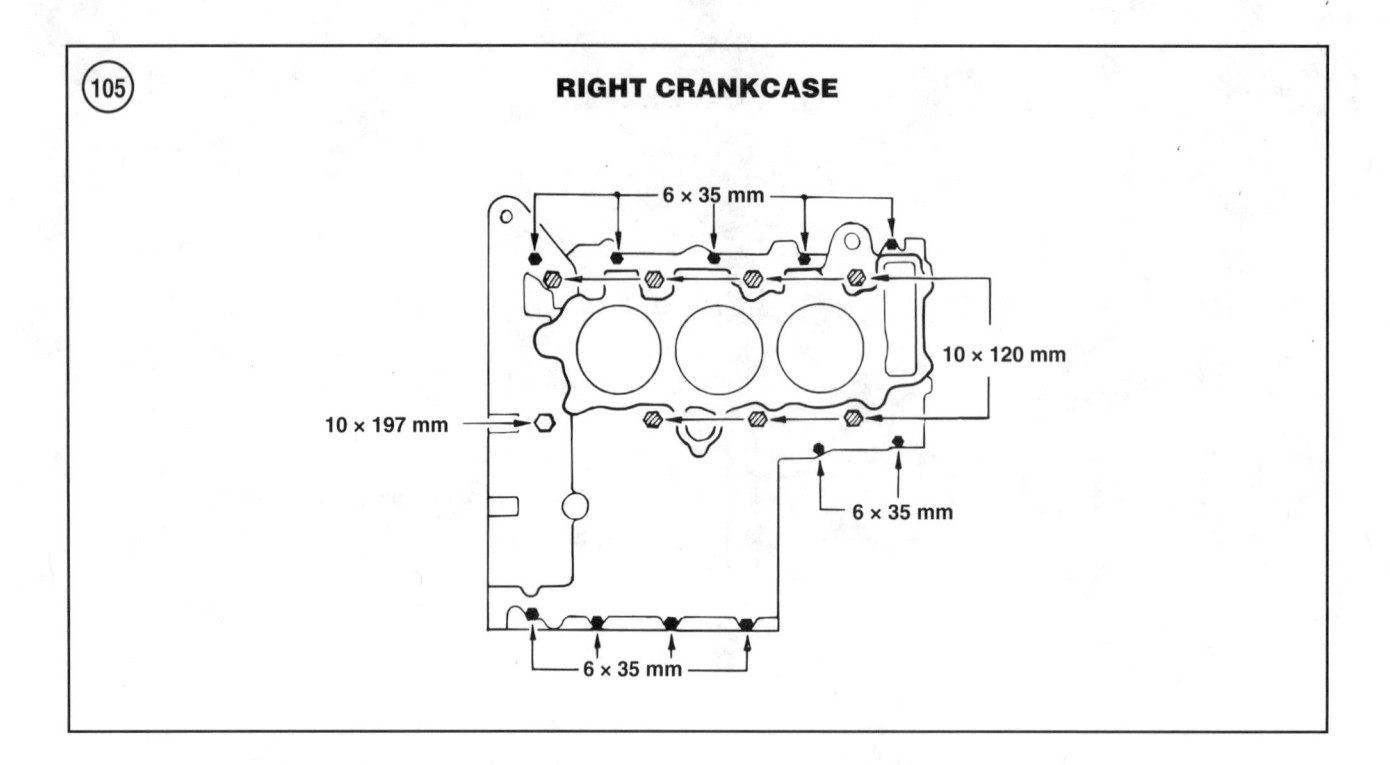

152 CHAPTER FIVE

105 RIGHT CRANKCASE

6 × 35 mm
10 × 120 mm
10 × 197 mm
6 × 35 mm
6 × 35 mm

106

107

Crankcase Cleaning

CAUTION
Place the crankcase halves on wooden blocks or rubber mats and handle them carefully to avoid damaging the machined gasket surfaces.

1. Remove the main journal bearings as described under *Crankshaft* in this chapter.
2. Remove the oil passage plugs (A, **Figure 108**, typical).
3. Remove the oil pressure switch (B, **Figure 108**).

NOTE
The crankcase oil passage plug and oil pressure switch must be removed to ensure thorough cleaning of the crankcase oil passages.

4. Remove the shift shaft oil seal (**Figure 109**).

5. Remove all sealer and gasket residue material from all crankcase gasket surfaces.

6. Inspect the bearings remaining in the crankcase halves. Note the following:

   a. Inspect the ball bearings by rotating their inner race slowly by hand. Replace the bearings if there is any roughness or play (axial or radial). Then check that the bearings fit tightly in their mounting bore. A loose fit bearing may indicate a damaged bearing mounting bore. Remove the bearing and inspect the mounting

ENGINE LOWER END

108

109

110

Do not apply sealer

3-5 mm (0.1-0.2 in.)

Oil pressure switch

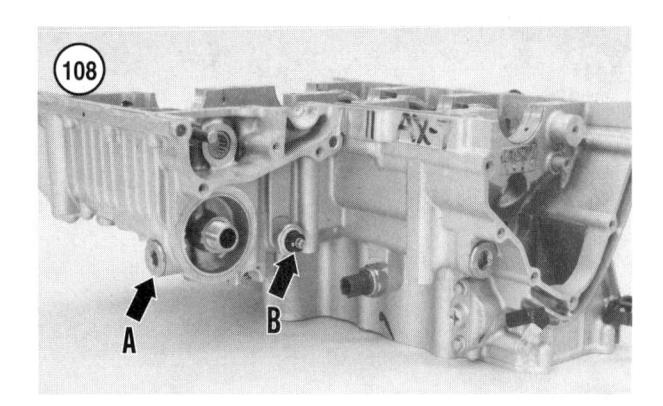

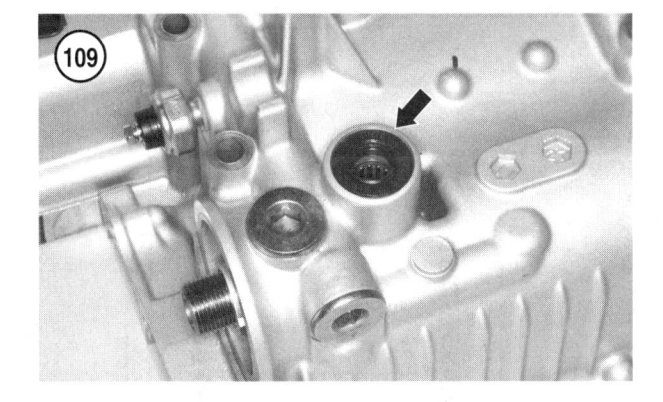

Apply sealer

bore for damage. Replace the crankcase halves if the bore is damaged.

 b. Inspect the needle bearing for cage and needle damage.

 c. Replace the needle bearing as described in Chapter One under *Bearings*.

7. Remove all threadlocking compound from the crankcase threaded holes, where used.

8. Check all bolts and threaded holes for stripping, cross-threading or deposit buildup. Clean out threaded holes with compressed air. Dirt buildup in the bottom of a hole may prevent the bolt from being accurately torqued. Replace damaged bolts and washers.

9. Inspect machined surfaces for burrs, cracks or other damage. Repair minor damage with a fine-cut file or oilstone.

10. Inspect and measure the cylinder bore diameters as described in this section.

11. Thoroughly clean the inside and outside of both crankcase halves and all oil passages with solvent.

12. Clean and dry the oil passages with compressed air. Make sure there is no sealer residue left in any of the oil passages. Use a small flashlight and visually check the oil passages for contamination.

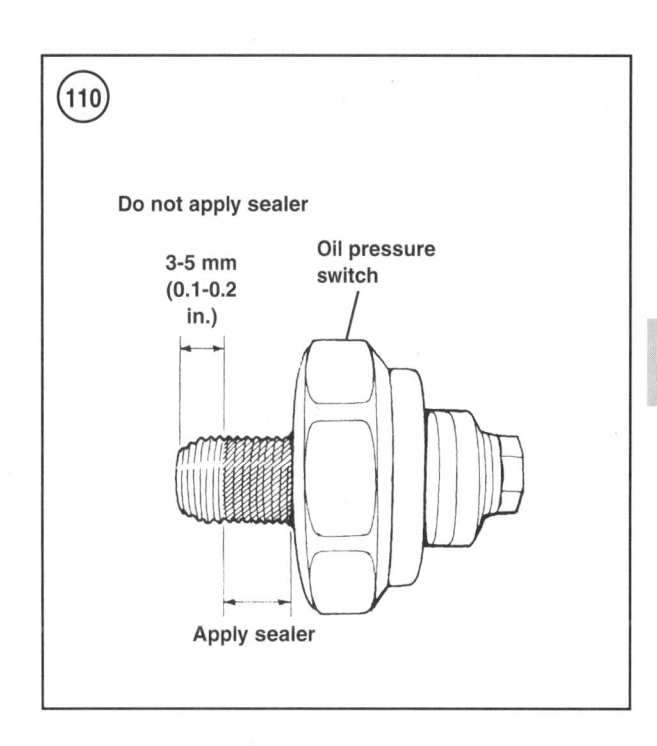

13. Wash each cylinder bore in hot, soapy water and rinse completely. This is the only way to clean the cylinder walls of the fine grit material left from the bore or honing job. After washing the cylinder walls, run a clean white cloth through each cylinder; the cloth should show no traces of grit or other debris. If the cloth is dirty, rewash the cylinder wall. After cleaning the cylinder walls, lubricate each cylinder wall with clean engine oil to prevent the cylinder liners from rusting.

14. Dry the case halves with compressed air. Make sure there is no water or solvent left in the cases as it will contaminate the new engine oil.

15. Oil the cylinder walls to prevent any rust formation.

16. Apply a threadlocking compound onto the threads of the crankcase oil passage plugs (A, **Figure 108**, typical) and tighten them to 44 N•m (32 ft.-lb.).

17. Install the oil pressure switch as follows:

 a. Remove all sealer residue from the switch threads.

 b. Apply ThreeBond 1207B or an equivalent RTV sealer onto the oil pressure switch threads as shown in **Figure 110**.

NOTE
Allow the silicone to set for 10-15 minutes before installing the oil pressure switch.

 c. Install the oil pressure switch (B, **Figure 108**) and tighten to 12 N•m (106 in.-lb.)

18. Using a bearing driver, install a new oil seal until it is flush with the crankcase outer bore surface (**Figure 109**). Pack the seal lips with grease.

Cylinder Inspection

The cylinder bores (**Figure 111**) are an integral part of each crankcase half.

1. Check the cylinder block surface for cracks and damage.
2. Check the cylinder block for warp with a straightedge and flat feeler gauge. Check at different spots across the cylinder block.

NOTE
*If the cylinder block is warped beyond the specification (**Table 2**), refer service to a Honda dealership.*

3. Check the cylinder walls for deep scratches or signs of seizure or other damage.
4. Measure the cylinder bores with a cylinder gauge or inside micrometer at the points shown in **Figure 112**. Measure in two axes—aligned with the piston pin and at 90° to the pin. If the bore diameter, taper or out-of-round for any cylinder exceeds the specifications (**Table 2**), rebore all six cylinders to the next oversize and install new pistons and rings. The following oversize pistons are available:
   a. 0.25 mm (0.010 in.).
   b. 0.50 mm (0.020 in.).
   c. 0.75 mm (0.030 in.).
   d. 1.00 mm (0.040 in.).

NOTE
*Purchase the new pistons before the cylinders are bored so the pistons can be measured; each cylinder must be bored to match one piston only. Piston-to-cylinder clearance is specified in **Table 2**.*

5. Determine piston-to-cylinder clearance as described in *Piston Clearance* in this chapter.

Crankcase Assembly

1. Clean and dry all the crankcase mounting bolts.
2. Install the crankshaft into the left crankcase as described in this chapter.
3. Install the left side cylinder pistons and connecting rods as described in this chapter.
4. Install the right side cylinder pistons and connecting rods as described in this chapter.
5. Install the oil pump as described in this chapter. 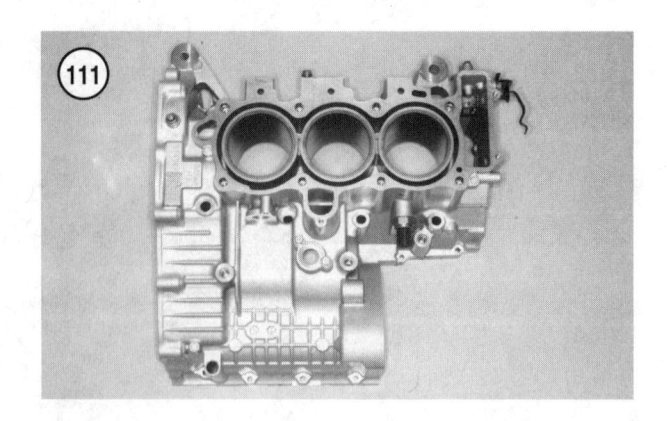 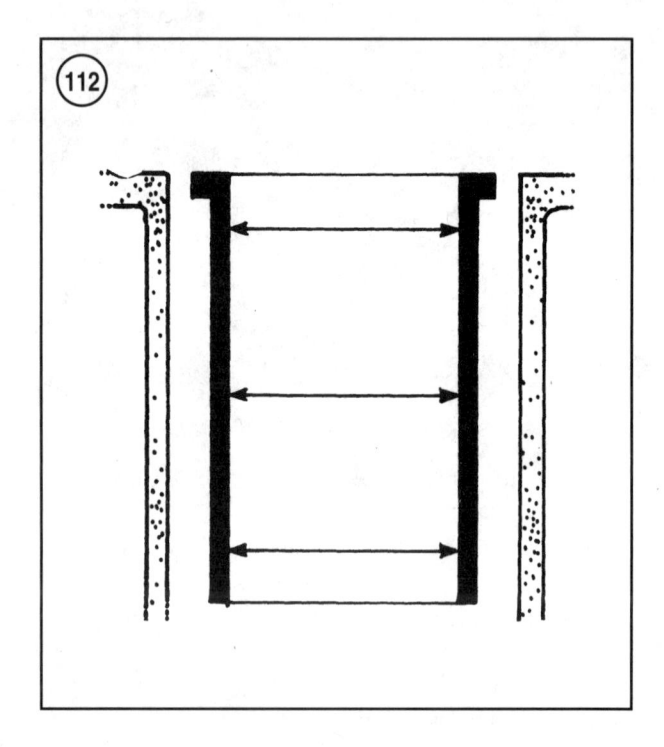
6. Install the transmission assembly as described Chapter Seven.
7. Clean and dry both crankcase gasket surfaces.
8. Install the rotor bolt (**Figure 113**) into the crankshaft. Use this bolt to turn the crankshaft in the following steps.

NOTE
*Refer to **Figure 114** to identify the cylinder number positions.*

9. Rotate the crankshaft to bring the No. 1 piston (**Figure 115**) to TDC (its highest position).
10. Lubricate the piston rings, pistons and cylinder walls with engine oil.
11. Stagger the piston rings as shown in **Figure 116**.

ENGINE LOWER END

113

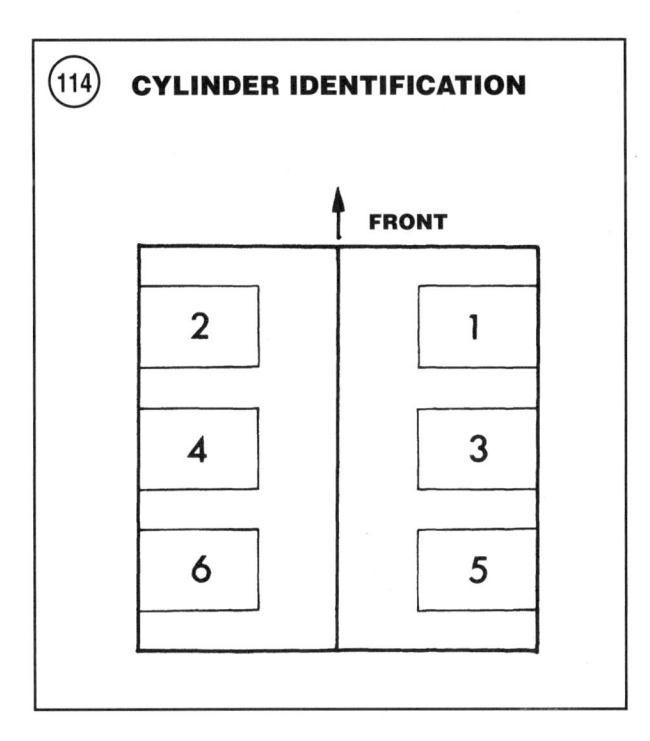

114 CYLINDER IDENTIFICATION

```
         ↑ FRONT
    ┌─────┬─────┐
    │  2  │  1  │
    ├─────┼─────┤
    │  4  │  3  │
    ├─────┼─────┤
    │  6  │  5  │
    └─────┴─────┘
```

12. Support the No. 3 and No. 5 pistons with piston base A (**Figure 117**).

13. Support the No. 1 piston with piston base B (**Figure 117**).

14. Install the piston ring compressors onto the pistons so that the tapered side of each compressor faces up.

15. Align the piston ring compressor cables opposite the transmission side to make it easier to remove them.

16. Stand the two wooden support blocks onto the crankcase at the points shown in **Figure 118**.

17. With the aid of an assistant, align and install the No. 1 cylinder bore over the No. 1 piston while aligning the center shift fork with the mainshaft second/third gear shifter groove (**Figure 119**). As the No. 1 piston and rings enter the cylinder bore, the right crankcase will rest on the support blocks

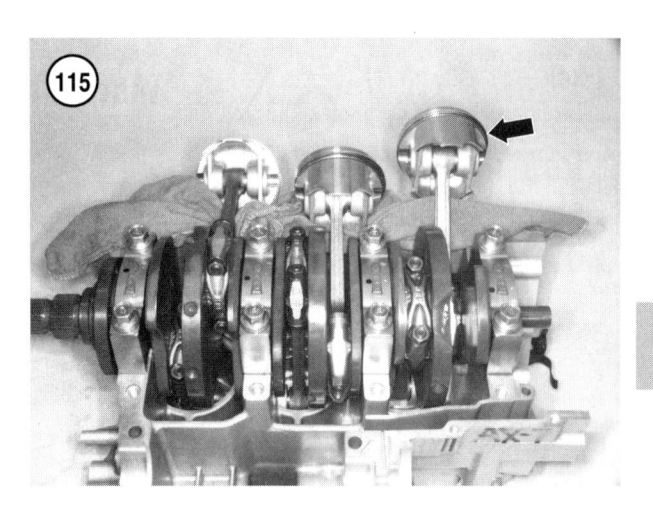

115

18. Remove piston base B from under the No. 1 piston. Then pull both wires on the No. 1 piston ring compressor at the same time to separate the halves from around the piston and remove the compressor halves.

19. Support the right crankcase to prevent it from falling and have an assistant turn the support blocks and rest them on their sides.

20. Align and install the No. 2 and No. 3 cylinder bores over the No. 2 and No. 3 pistons. Maintain pressure to prevent the rings from expanding before they enter the bore. As the piston rings enter the cylinder bores, the right crankcase will rest on the support blocks (**Figure 120**).

21. Remove piston base A (**Figure 120**).

22. Pull both wires on the No. 3 piston ring compressor at the same time to separate the halves from around the piston and remove the compressor halves.

23. Repeat for the No. 5 piston ring compressor.

24. Install the two dowel pins (**Figure 121** and **Figure 122**).

25. Install the oil pipe (**Figure 123**).

NOTE
Use a semi-drying liquid gasket sealer (ThreeBond Silicone Liquid Gasket 1207D or equivalent) to seal the crankcase. When selecting an equivalent, avoid thick and hard-setting materials.

26. Apply a thin coating of gasket sealer to the left crankcase half sealing surfaces indicated in **Figure 124**.

CAUTION
Do not apply sealer to the curved bearing surfaces or oil passage areas, as it will restrict oil flow. Applying sealer to the bearing surfaces changes bearing clearance and causes crankshaft seizure.

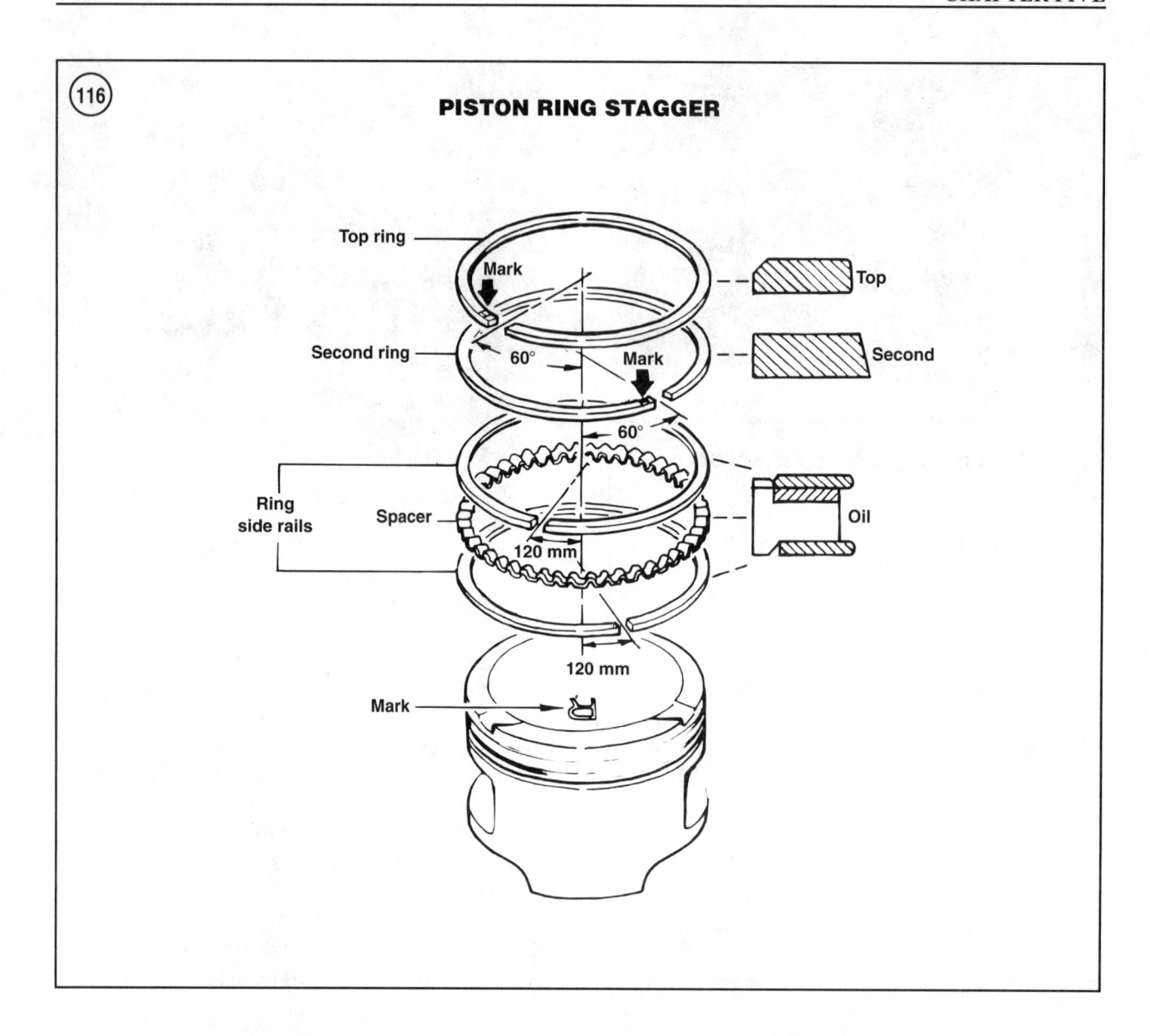

116 PISTON RING STAGGER

- Top ring — Mark
- Second ring — 60° — Mark
- 60°
- Ring side rails — Spacer — 120 mm
- Top
- Second
- Oil
- 120 mm
- Mark

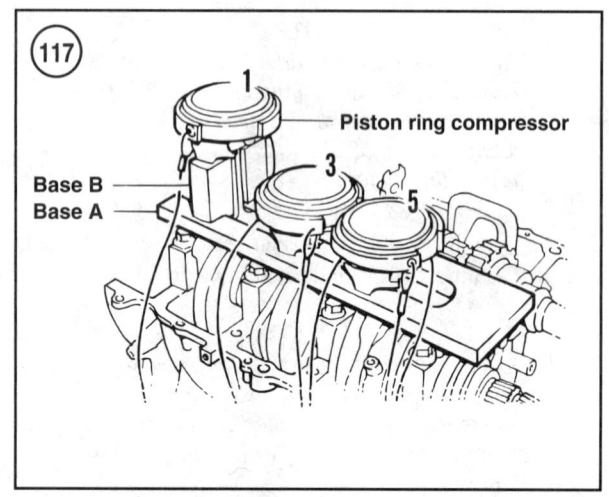

117
- 1 — Piston ring compressor
- Base B
- Base A
- 3
- 5

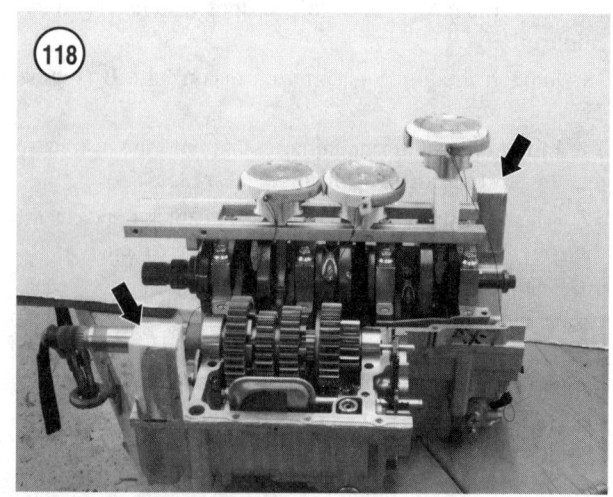

118

ENGINE LOWER END

119

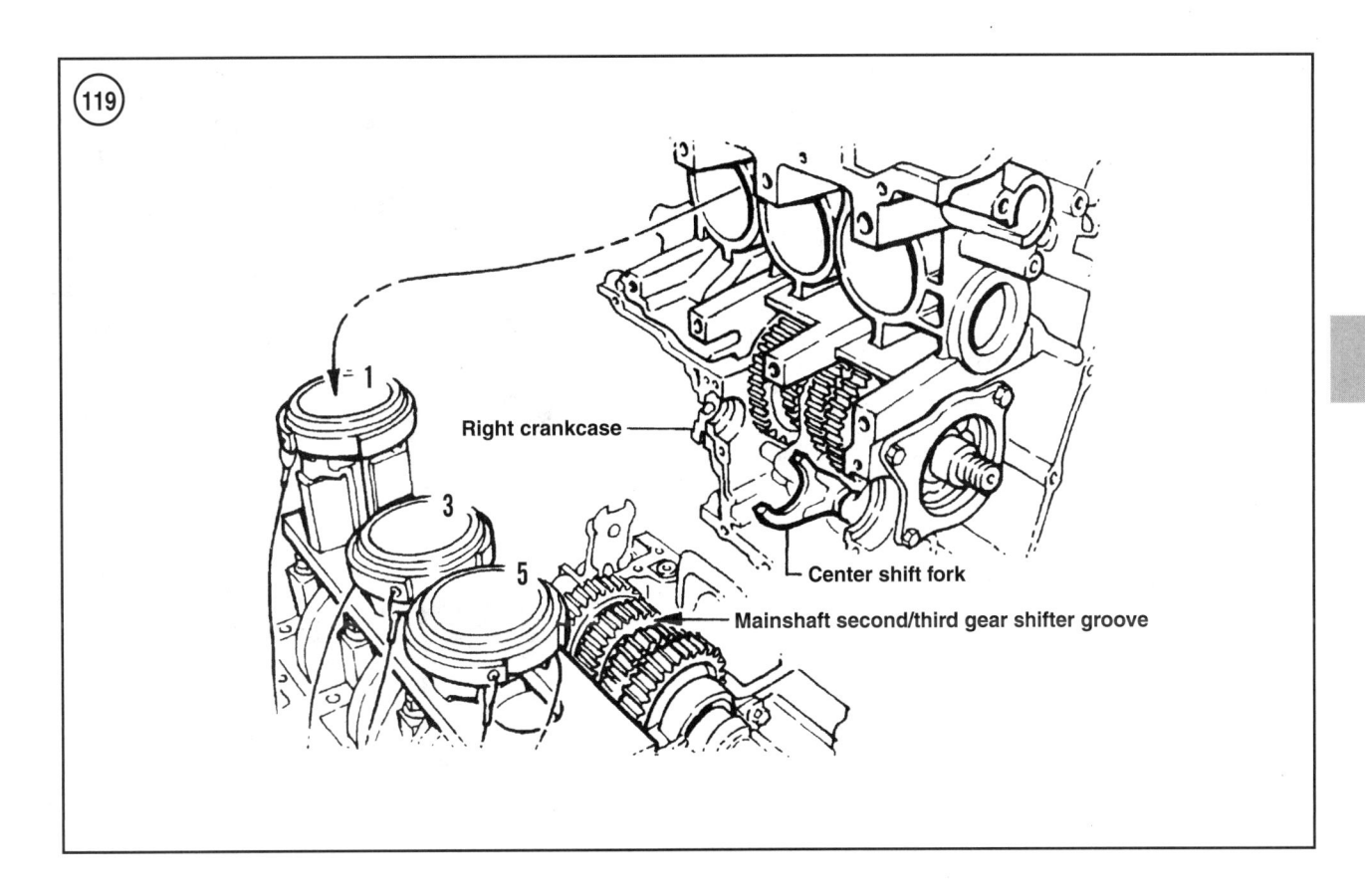

- Right crankcase
- Center shift fork
- Mainshaft second/third gear shifter groove

120

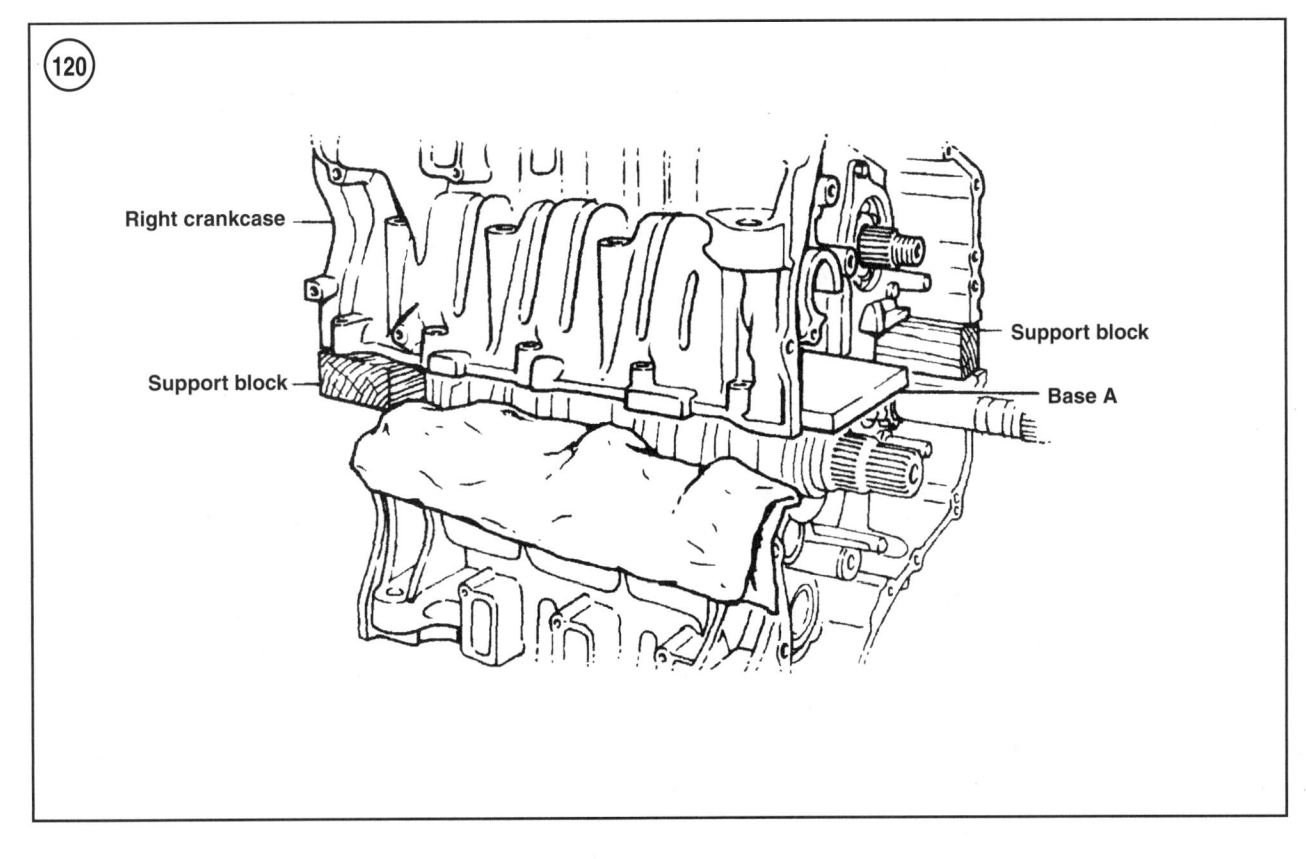

- Right crankcase
- Support block
- Support block
- Base A

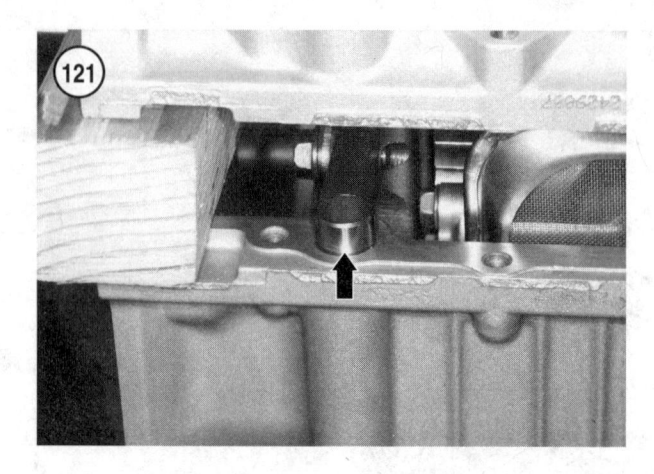

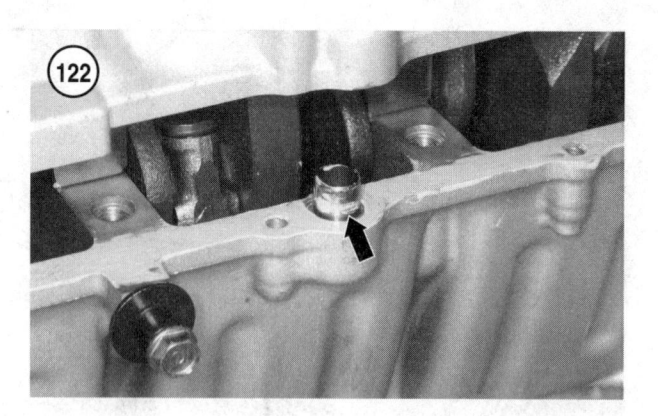

27. Have an assistant hold the right crankcase and remove the two wooden blocks. Then apply sealant to the two areas previously covered by the wooden blocks. Check that the sealant covers the left crankcase sealant surfaces shown in **Figure 124**.
28. Lower the right crankcase to join both halves and tap them together lightly with a plastic mallet. Check the gasket surfaces for any gaps.
29. If the crankcase halves do not fit together completely, note the following:
    a. First check that the transmission shafts are properly installed. Make sure the center shift fork engages the mainshaft second/third gear shifter groove.
    b. If the transmission is not the problem, separate the crankcase halves and investigate the cause of the interference.

CAUTION
Crankcase halves should fit together without force. If the crankcase halves do not fit together completely, do not attempt to pull them together with the crankcase bolts. Do not risk damage by trying to force the case halves together.

NOTE
*Different length crankcase mounting bolts are used. Refer to **Figure 125** and **Figure 126** to identify the crankcase fasteners.*

30. Install and tighten the right crankcase mounting bolts (**Figure 126**):
    a. Install a new washer on each 10–mm crankcase bolt. Apply engine oil to the bolt threads, bolt head sealing surfaces and washers.
    b. Install the 10–mm and 6–mm bolts and hand-tighten.
    c. Tighten the 10–mm bolts in 2-3 steps and in a criss-cross pattern to 34 N•m (25 ft.-lb.).
    d. Tighten the 6 mm bolts to 12 N•m (106 in.-lb.).

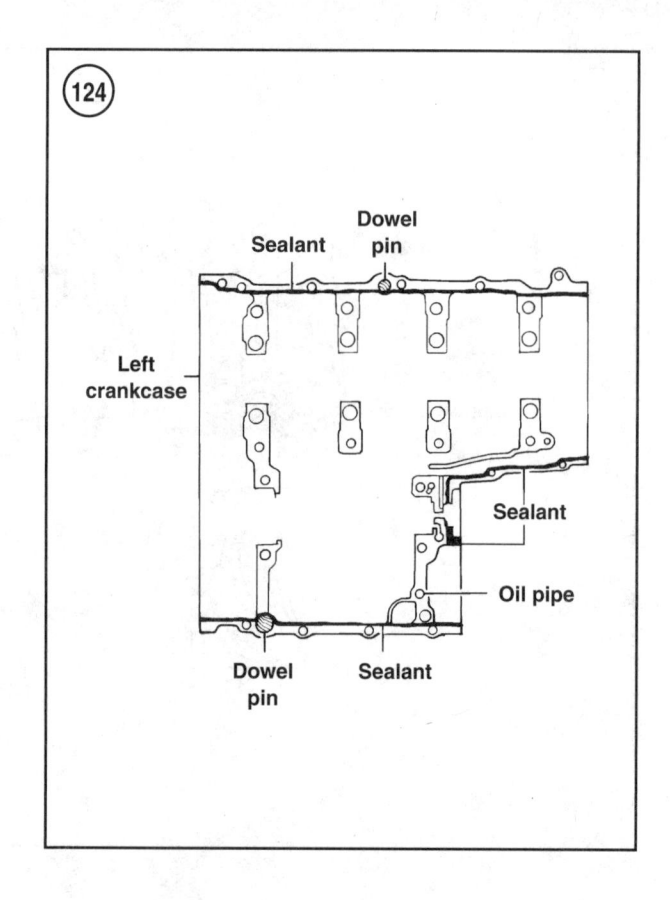

ENGINE LOWER END

125 **LEFT CRANKCASE**

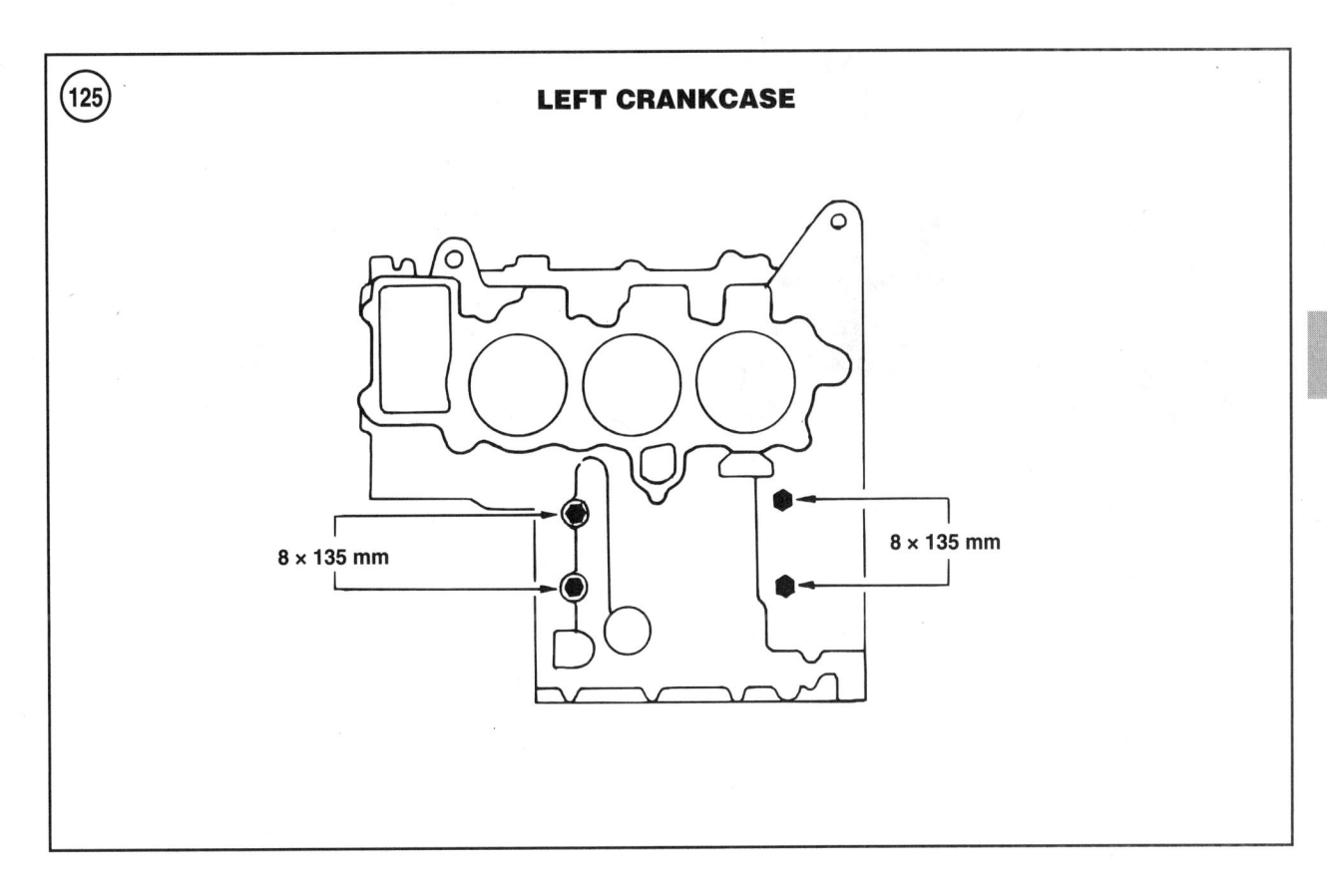

8 × 135 mm

8 × 135 mm

126 **RIGHT CRANKCASE**

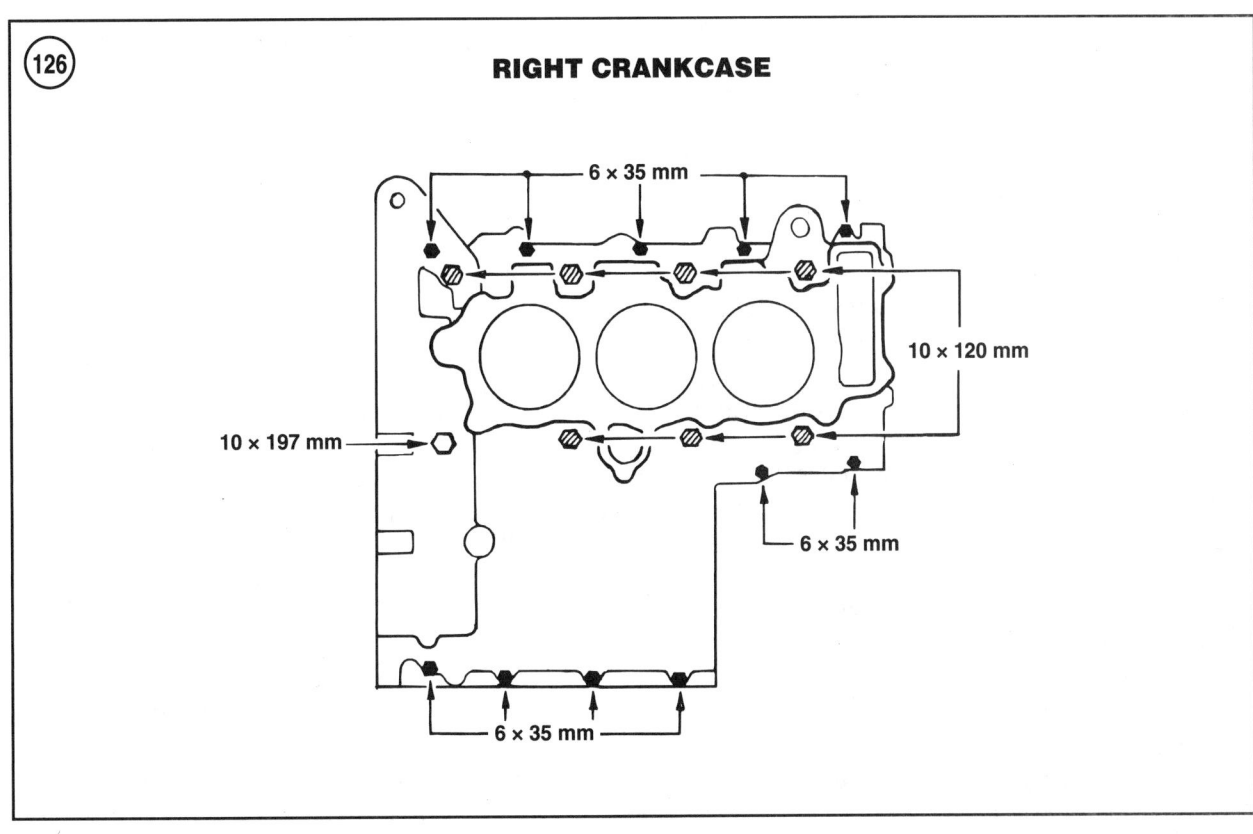

6 × 35 mm

10 × 120 mm

10 × 197 mm

6 × 35 mm

6 × 35 mm

31. Install and tighten the left crankcase mounting bolts:
   a. Install a new washer onto each bolt, and then install and hand-tighten.
   b. Tighten the 8–mm bolts (**Figure 125**) in 2-3 steps and in a crisscross pattern to 25 N•m (19 ft.-lb.).
32. Turn the crankshaft and transmission shafts. There must be no binding or roughness.
33. Reverse Steps 1-12 under *Crankcase Disassembly* to complete assembly and install the engine in the frame.

PISTONS AND CONNECTING RODS

The crankcase must be separated to remove the pistons and connecting rods.

Piston/Connecting Rod Removal

Identify and store each part by cylinder number as it is removed. The pistons, connecting rods and bearing inserts must be installed in their original locations.

1. Mark the cylinder number (**Figure 114**) on the crown of each piston.
2. Insert a flat feeler gauge between a connecting rod and either crankshaft machined web (**Figure 127**). Record the connecting rod side clearance for each connecting rod and compare to the specification listed in **Table 3**. Note the following:
   a. If the clearance is greater than specified, replace the connecting rods and remeasure.
   b. If the clearance is greater than specified with the new connecting rods, replace the crankshaft.
3. Identify the rods, bearing caps and bearings before removing them.
4. On the No. 1, No. 3 and No. 5 cylinders, remove the nuts securing the connecting rod caps and remove the caps, connecting rods and pistons (**Figure 128**).

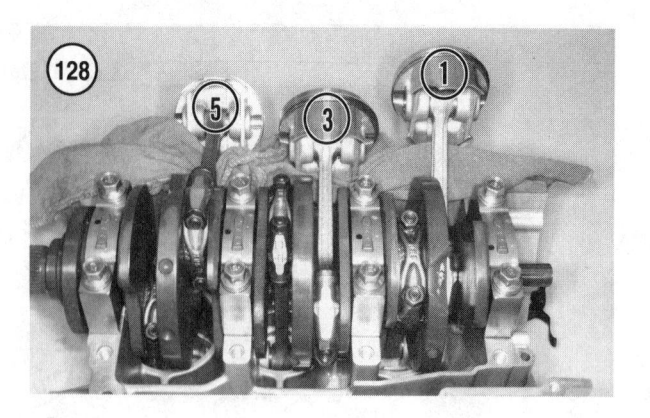

NOTE
The pistons and connecting rods for the No. 2, No. 4 and No. 6 cylinders (left crankcase) are removed through the tops of the cylinder bores (Figure 129).

NOTE
If the connecting rod bearing clearance is to be measured, leave the pistons and rods in the left crankcase. Remove the crankshaft to clean the journals and bearing inserts. Then reinstall the crankshaft and measure the clearance as described in Connecting Rod Bearing Oil Clearance in this section.

CAUTION
Inspect the tops of the cylinder bores on the No. 2, No. 4 and No. 6 cylinders for a ridge where the piston ring reaches the top of its travel (Figure 130). Perform Step 5 only if there is a ridge present.

5. If there is a ridge at the top of any cylinder bore, the ridge must be removed with a ridge reamer. Do one cylinder at a time as follows:
   a. Rotate the crankshaft until the piston to be removed is at the bottom of its travel.

ENGINE LOWER END

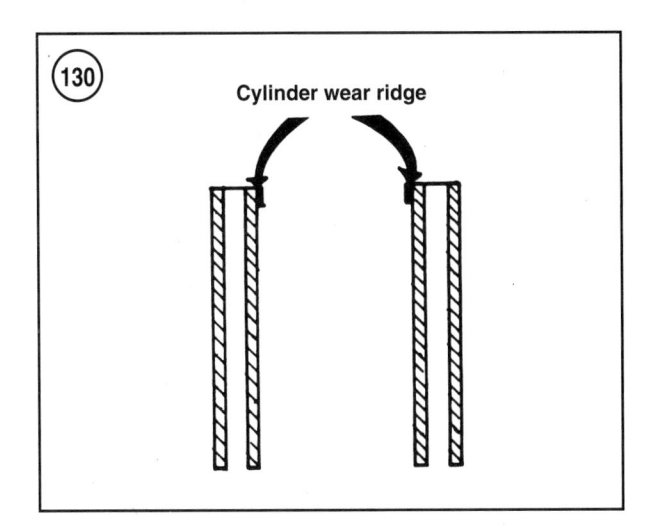

130 Cylinder wear ridge

131

b. Place an oil-soaked shop cloth down into the cylinder and over the piston to collect the cuttings. Remove the ridge and/or deposits from the top of the cylinder bore with a ridge reamer. Follow the manufacturer's instructions.

c. Turn the crankshaft until that piston is at top dead center and remove the rag and the cuttings. Make sure to remove all cuttings as they may scratch the cylinder wall during piston removal.

NOTE
If the ridge is not removed, the piston and rings may be damaged during removal.

6. Remove the nuts securing the connecting rod caps and remove the caps.
7. Mark the piston crowns of the three left-hand pistons with the corresponding cylinder numbers to reinstall the pistons in the correct cylinders. Refer to **Figure 114** for cylinder numbers.
8. Carefully remove the pistons and connecting rods through the tops of the cylinder bores.

9. Reinstall the bearing and cap onto each rod (**Figure 131**).

CAUTION
Keep each bearing insert in its original place in the crankcase, rod or rod cap. If reusing the original bearings, they must be installed exactly as removed to prevent rapid wear or bearing seizure.

Piston Ring Removal

The piston and ring assembly is a three-ring type. The top and second rings are compression rings. The lower ring is an oil control ring assembly consisting of two ring rails and a spacer. See **Figure 132** (left piston) or **Figure 133** (right piston).

When measuring the piston rings, compare the actual measurements to the new and service limit specifications in **Table 2**. Replace parts that are out of specification or show damage as described in this section.

WARNING
The edges of all piston rings are very sharp. Be careful when handling them.

1. Roll the rings around the piston. There must be no binding or roughness.
2. Push the top ring so it is nearly flush with the piston, then measure the side clearance with a flat feeler gauge (**Figure 134**). Repeat for the second ring. If the clearance is greater than specified, replace the rings. If the clearance is still excessive with new rings, replace the piston.

CAUTION
Spreading the piston rings too far may damage them.

NOTE
Store the rings in the order they were removed.

3. Remove the compression rings with a ring expander tool or by spreading the ring ends by hand (**Figure 135**).
4. Remove the oil ring assembly by first removing the upper and then the lower ring rails. Then remove the spacer.

CAUTION
The ring grooves are different widths. When using discarded rings to clean the grooves, grind the ring to fit the groove. Do not remove any metal from the ring groove.

CAUTION
Never use a wire brush to clean the ring grooves.

CHAPTER FIVE

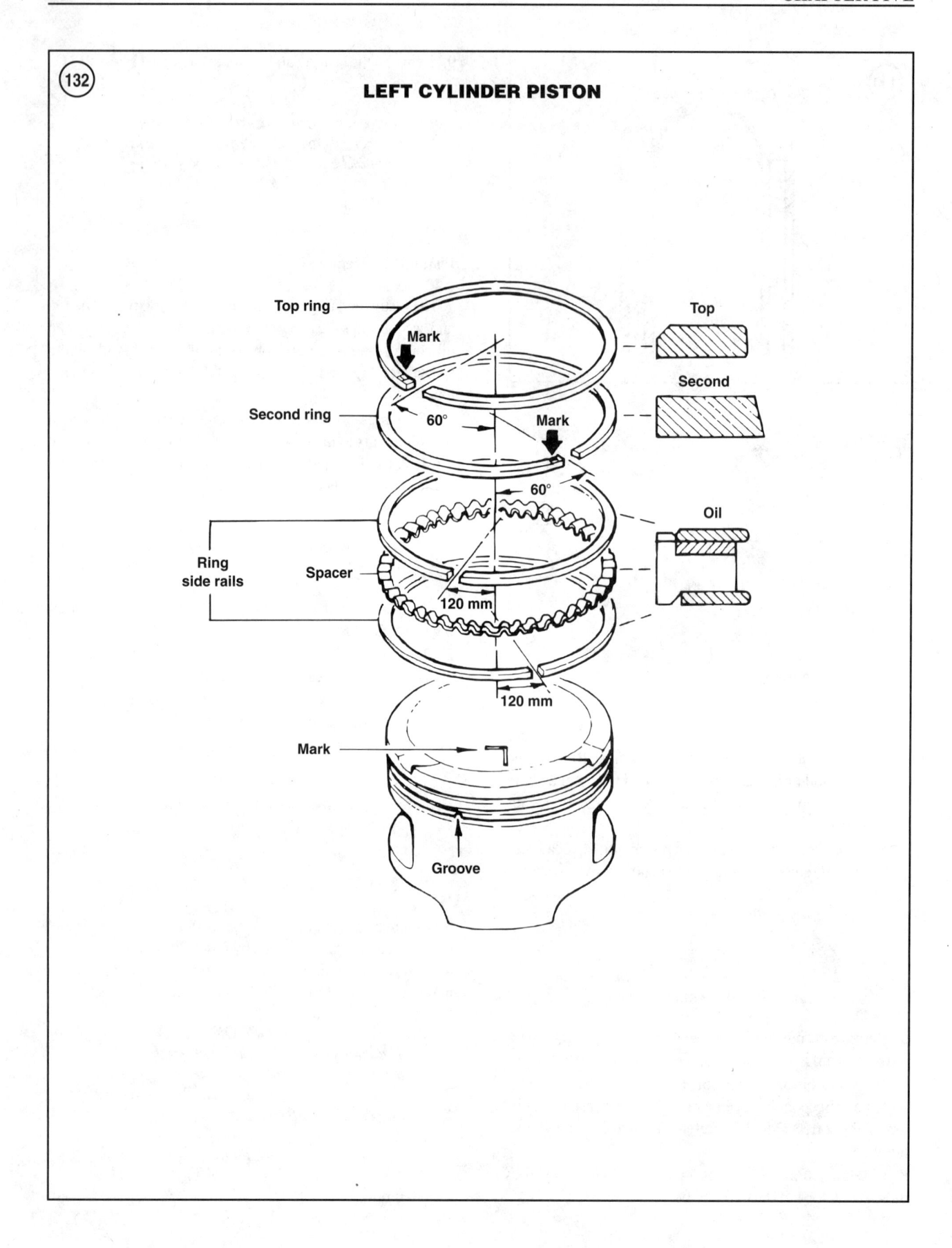

(132)

LEFT CYLINDER PISTON

ENGINE LOWER END 163

133

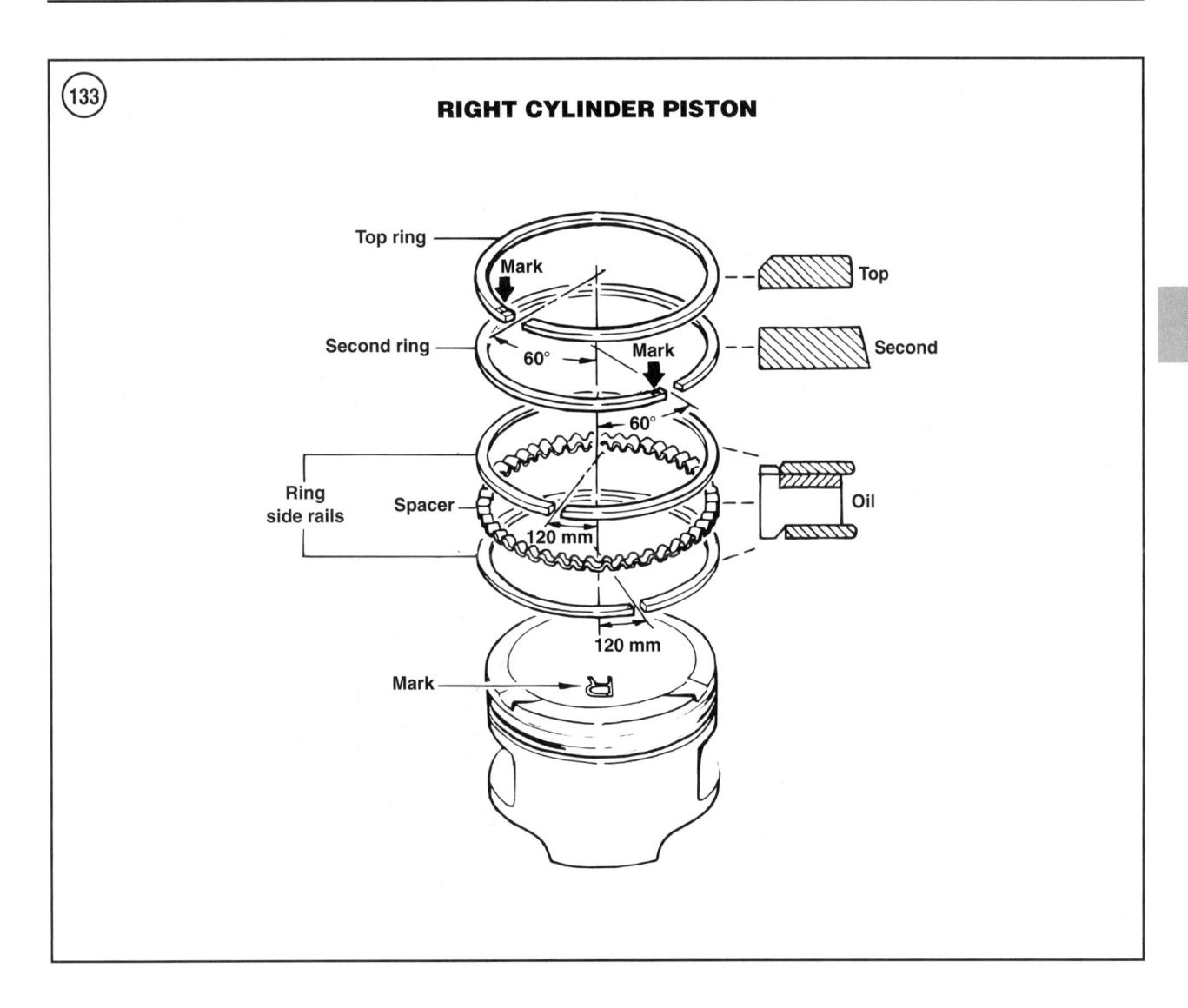

RIGHT CYLINDER PISTON

- Top ring — Mark — Top
- Second ring — 60° — Mark — Second
- Ring side rails — Spacer — 120 mm — Oil
- 120 mm
- Mark

134

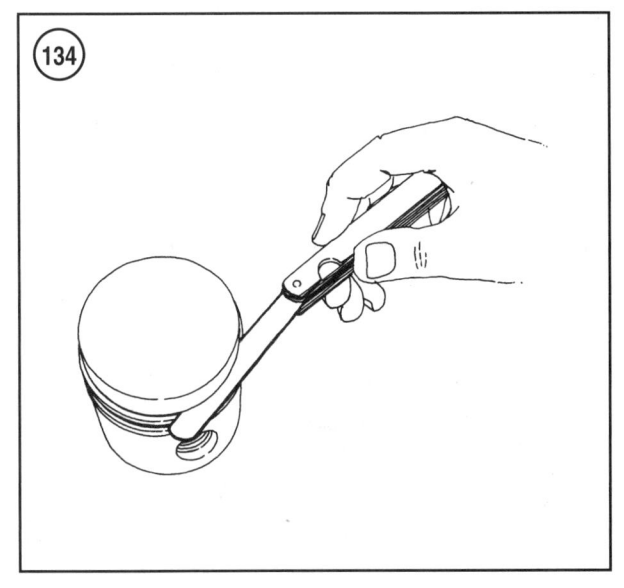

135

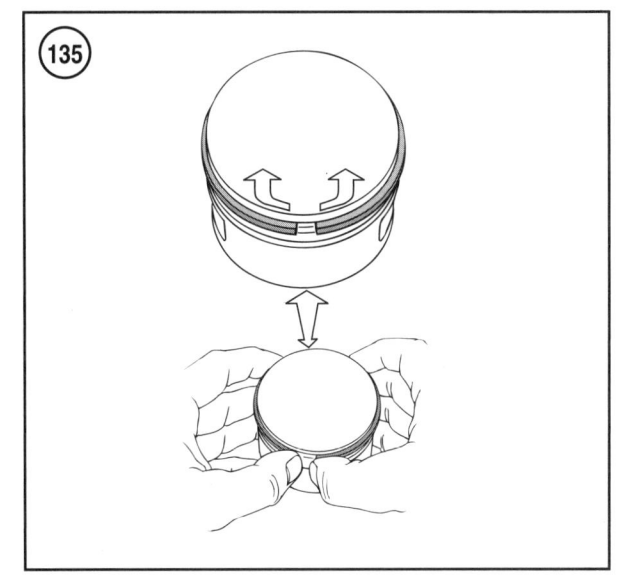

164

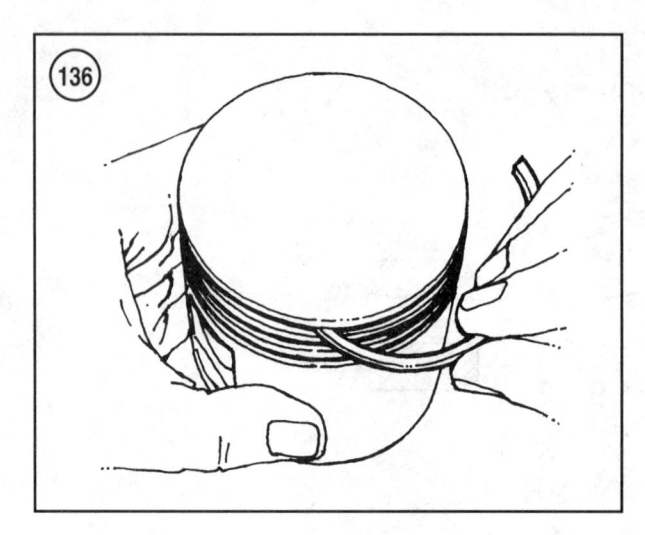

5. Carefully remove all carbon from the ring grooves with a discarded piston ring (**Figure 136**) that has been broken and grounded. Do not remove aluminum material from the ring grooves as this increases ring side clearance.

6. Inspect the grooves carefully for burrs, nicks or broken and cracked lands. Replace the piston if necessary.

7. Roll each ring around its piston groove as shown in **Figure 137**. If there is binding, check for carbon, a damaged ring or a collapsed ring groove.

8. Push a ring into the bottom of the cylinder and square it with the cylinder wall with the piston. Measure the installed ring end gap with a feeler gauge (**Figure 138**) and compare to **Table 2**:
   a. If the gap is too large, measure the cylinder bore inside diameter as described in this chapter. If the cylinder inside diameter is within specifications, replace the rings. If the cylinder inside diameter is too large, rebore the cylinder and install new pistions. Measure the end gap of new rings in the same manner.
   b. If the gap is too small, make sure the correct rings are being installed.

9. Inspect the piston as described in this section.

Piston Ring Installation

1. When installing new rings, deglaze or hone the cylinder. This helps seat the new rings. Refer honing service to a Honda dealership. After honing, measure the ring end gap for each ring.

2. When installing oversized compression rings, check the ring number to ensure that the correct rings are installed. The ring oversize numbers must be the same as the piston oversize numbers.

3. Clean the piston and rings. Dry them with compressed air.

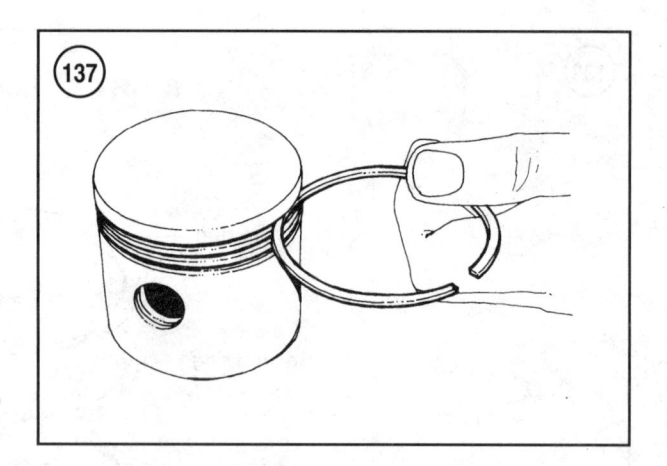

4. Install the piston rings as follows:

WARNING
The piston ring edges are sharp. Be careful when handling them.

a. Refer to **Figure 132** (left) or **Figure 133** (right) to identify the piston rings. The left and right side piston rings are different.

NOTE
When installing aftermarket piston rings, follow the manufacturer's directions.

b. Install the oil control ring assembly into the bottom ring groove. Install the spacer first, and then install each ring rail. Make sure the ends of the spacer butt together (**Figure 139**). They should not overlap. On the left cylinder pistons, align the spacer end stopper with the groove in the piston (**Figure 132**). If reassembling used parts, install the ring rails in their original positions.

ENGINE LOWER END

1. Remove the piston rings as described in this chapter.
2. Carefully clean the carbon from the piston crown (**Figure 140**) with a soft scraper or wire wheel mounted in a drill. Large carbon accumulations reduce piston cooling and cause detonation and piston damage. Renumber the piston as soon as it is cleaned.

CAUTION
Do not remove or damage the carbon ridge around the circumference of the piston above the top ring. If the pistons, rings and cylinders are found to be dimensionally correct and can be reused, removing the carbon ring from the top of the piston or the carbon ridges from the cylinders will promote excessive oil consumption.

CAUTION
Do not use a wire brush on piston skirts or ring lands. The wire brush removes aluminum and increases piston clearance. It also rounds the corners of the ring lands, which causes decreased support for the piston rings.

3. After cleaning the piston, examine the crown. The crown should show no signs of wear or damage. If the crown appears pecked or spongy-looking, check the spark plug, valves and combustion chamber for aluminum deposits. If there are deposits, the engine is overheating.
4. Examine each ring groove for burrs, dented edges and excessive wear. Pay particular attention to the top compression ring groove, as it usually wears more than the other grooves. Because the oil rings are constantly bathed in oil, these rings and grooves wear little compared to compression rings and their grooves. If there is oil ring groove wear or if the oil ring assembly is tight and difficult to remove, the piston skirt may have collapsed due to excessive heat. Replace the piston.
5. Check the oil control holes in the piston for carbon or oil sludge buildup. Clean the holes by hand using a small diameter drill bit.
6. Check the piston skirts for cracks or other damage. If a piston(s) shows signs of partial seizure (bits of aluminum buildup on the piston skirts), replace the pistons and bore the cylinders, if necessary, to reduce the possibility of engine noise and further piston seizure.

NOTE
If the piston skirts are worn or scuffed unevenly from side to side, the connecting rod may be bent or twisted.

7. Measure the piston-to-cylinder clearance as described under *Piston Clearance* in this section.
8. Measure the piston-to-piston pin clearance as described in this section.

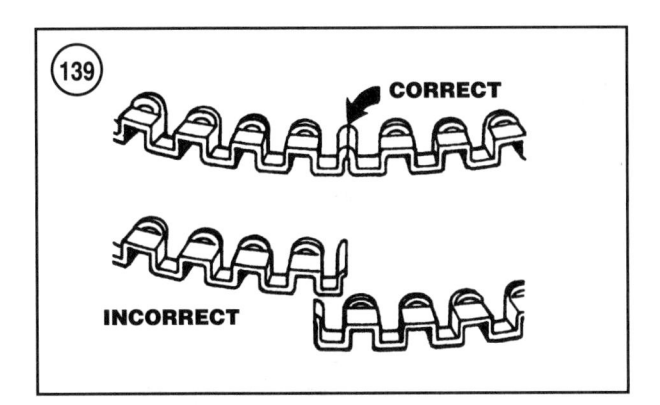

c. Install the compression rings with a ring expander tool or by carefully spreading the ring ends by hand (**Figure 135**).
d. Install the second or middle compression ring with the manufacturer's mark facing up. This ring has a slight taper.
e. Install the top compression ring with the manufacturer's mark facing up. This ring has a slight taper on its top, rear side.

5. Make sure the rings are seated completely in their grooves all the way around the piston and the end gaps are distributed around the piston as shown in **Figure 132** or **Figure 133**.

Piston Inspection

When measuring the piston components, compare the actual measurements to the specifications in **Table 2**. Replace worn or damaged parts as described in this section.

NOTE
The pistons are installed onto the connecting rods (Figure 131) with a press fit. Complete piston inspection requires piston separation. See Piston Removal in this section.

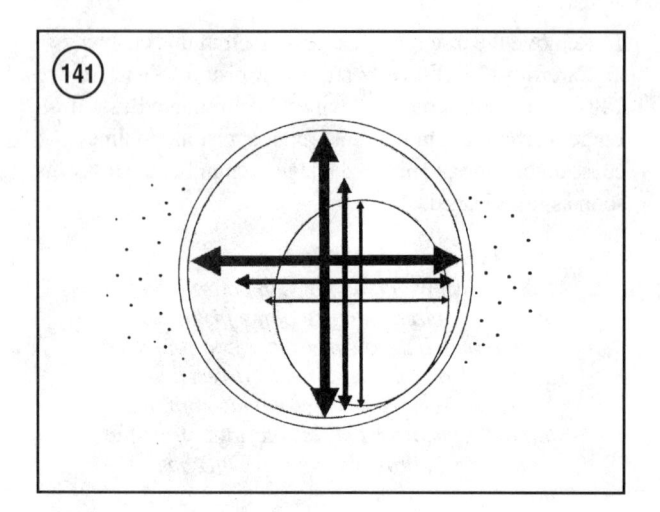

9. If damage or wear indicates piston replacement, select new pistons as described under *Piston Clearance* in this section.

Piston Clearance

1. Make sure the piston and cylinder wall are clean and dry.

2. Measure the cylinder bore with a bore gauge or inside micrometer at three positions shown in **Figure 141**. Measure in line with the piston pin and 90° to the pin. Record the bore inside diameter measurement.

3. Measure the piston outside diameter with a micrometer at a right angle to the piston pin bore. Measure up 10 mm (0.4 in.) from the bottom edge of the piston skirt (**Figure 142**). Record the measurement.

4. Subtract the piston outside diameter from the largest bore diameter; the difference is piston-to-cylinder clearance. If the clearance exceeds the service limit in **Table 2**, determine if the piston, cylinder or both are worn. If necessary, take the cylinder to a Honda dealership that can rebore the cylinder to accept an oversize piston.

5. Install the piston rings as described in this chapter.

Piston Removal

The piston pin is pressed into place. Removing the piston pin from the piston and connecting rod requires using a hydraulic press and a number of Honda special tool. Considerable expense can be saved by removing the piston/connecting rod assemblies and taking them into a dealership for service.

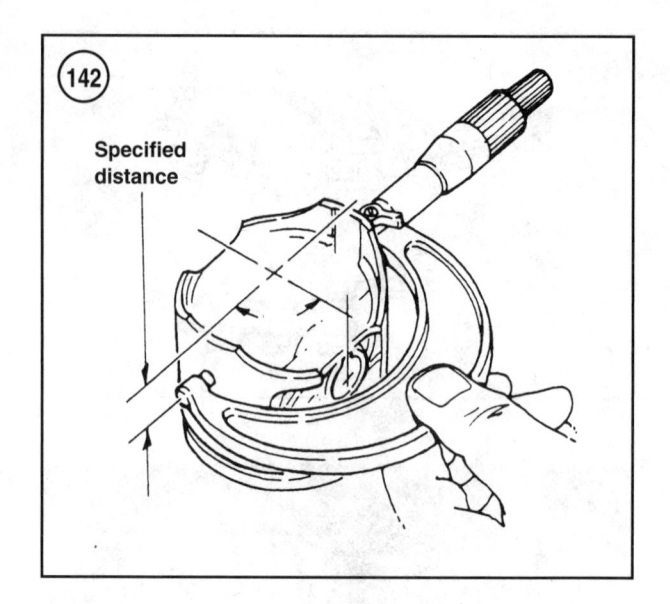

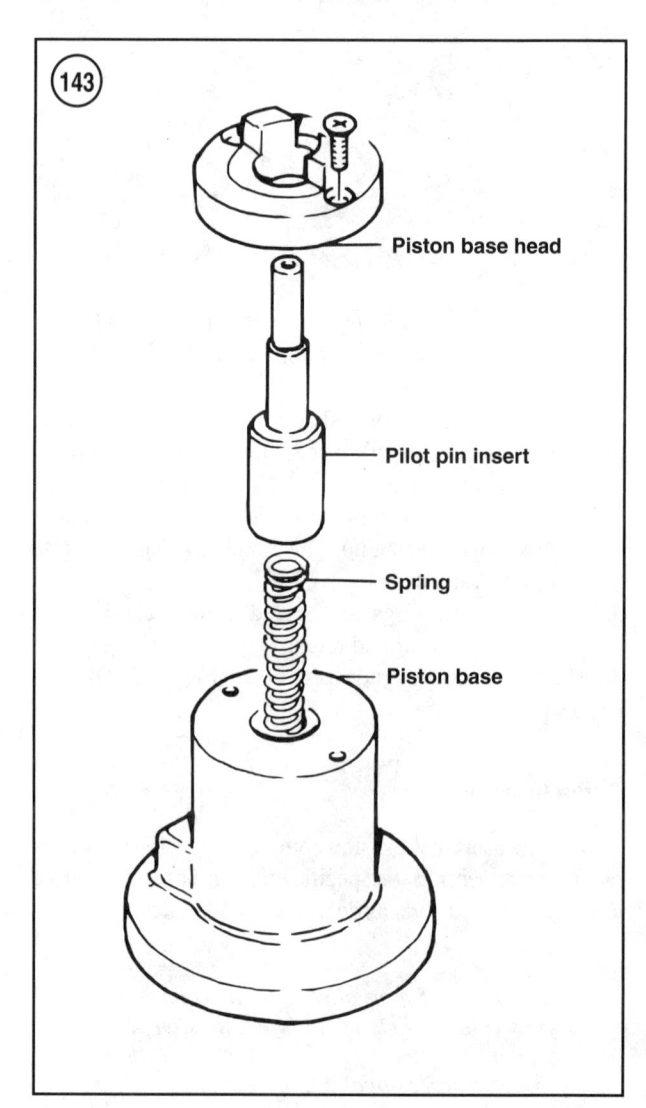

ENGINE LOWER END

144

- Hydraulic press
- Adjustable piston pin driver head
- Adjustable piston pin driver shaft
- 18 mm pivot collar
- Piston base

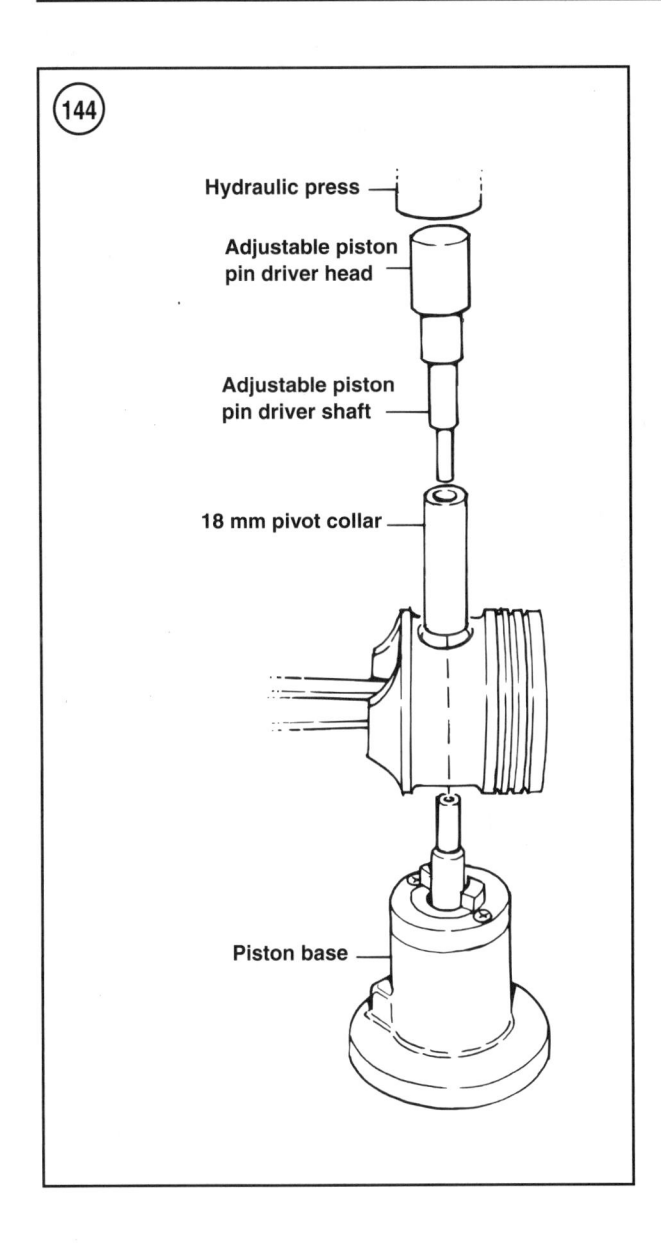

145

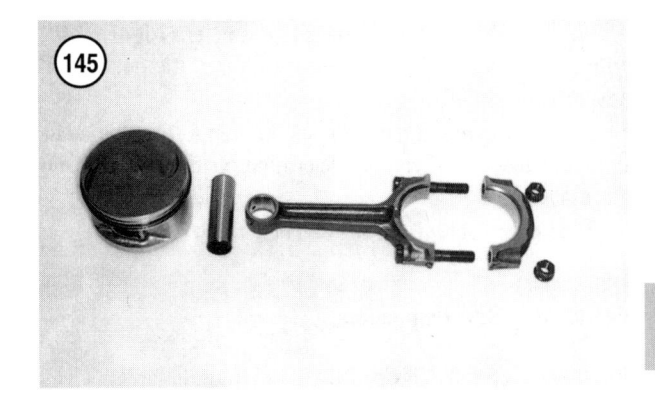

146

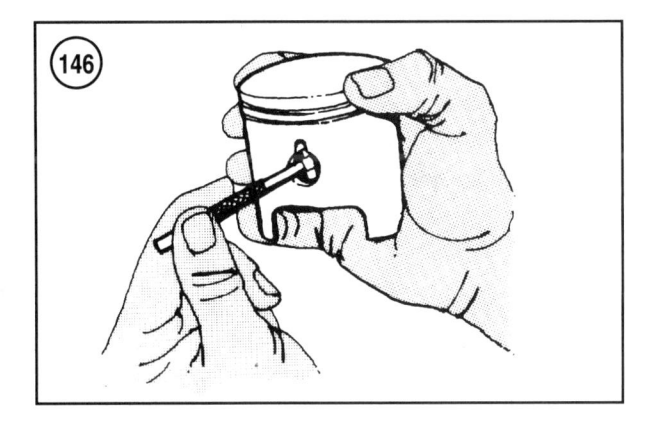

Tools

The following tools are required to separate the pistons from the connecting rods:

1. Piston base (Honda part No. 07973-6570500).
2. Piston base head (Honda part No. 007JGF-001010A).
3. Pilot pin insert (Honda part No. 07973-6570400).
4. Piston base spring (Honda part No. 07973-6570600).
5. 18 mm pivot collar (Honda part No. 07KMF-MT20200).
6. Adjustable piston pin driver head (Honda part No. 07973-6570210).
7. Adjustable piston pin driver shaft (Honda part No. 07973-6570300).
8. Hydraulic press.

Procedure

1. Remove the piston/connecting rod as described in this chapter.
2. Assemble the tools as shown in **Figure 143**.
3. Place the piston on the piston base by aligning the piston boss over the pilot pin.
4. Press the piston pin out with the pivot collar, adjustable piston pin driver head and a hydraulic press (**Figure 144**).
5. Remove the piston and connecting rod and identify them as an assembly (**Figure 145**).
6. Repeat for each piston/connecting rod assembly.

Piston and Piston Pin Clearance

When measuring the piston pins, compare the actual measurements to the specifications in **Table 2**. Replace the piston pins if worn or damaged as described in this section.

1. Clean and dry the piston pins.
2. Inspect the piston pin for chrome flaking or cracks. Replace if necessary.
3. Measure the piston pin hole inside diameter (**Figure 146**) with a telescoping gauge. Replace the piston if the hole is too large.

4. Measure the piston pin outside diameter (**Figure 147**) with a micrometer. Replace the piston pin if the outside diameter is too small.

5. Subtract the measurement made in Step 4 from the measurement made in Step 3 to determine piston-to-piston pin clearance.

6. Repeat for each piston pin and piston.

Connecting Rod Inspection

1. Check each rod (**Figure 145**) for obvious damage such as cracks and burns.
2. Check the piston pin bore for wear or scoring.
3. If the connecting rod straightness is in question, take the connecting rods to a machine shop to check for twisting and bending.
4. Examine the bearing inserts (**Figure 148**) for excessive wear, scoring or burning. They are reusable if in good condition. Make a note of the bearing color (if any) marked on the side of the insert if discarding the bearing.
5. Inspect the connecting rod bearing bolts (**Figure 148**) and check them for cracks or twisting. Replace bolts and nuts in pairs.
6. Check bearing clearance as described in this chapter.

Connecting Rod Bearing Oil Clearance

This section describes how to measure the connecting rod bearing clearance using Plastigage (**Figure 149**) and, if necessary, how to select new bearing inserts. Plastigage is a material that flattens when pressure is applied to it. The marked bands on the envelope are then used to measure the width of the flattened Plastigage to determine bearing clearance.

Plastigage is available in different clearance ranges. Purchase the 0.025-0.076 mm (0.001-0.003 in.) clearance range set.

1. Clean and dry the connecting rod journal.
2. Before measuring the connecting rod bearing oil clearance, measure the taper and out-of-roundness of each crankpin journal with a micrometer (**Figure 150**). Note the following:
    a. To check for taper, measure at several places in a line along the crankpin. Record the measurements for each journal.
    b. To check for out-of-roundness, measure the journal diameter. Record the measurements for each journal.
    c. If the journals are tapered or out-of-round (**Table 3**), the crankshaft may have to be replaced. However, before doing so, consult with a Honda dealership on the appropriate service.

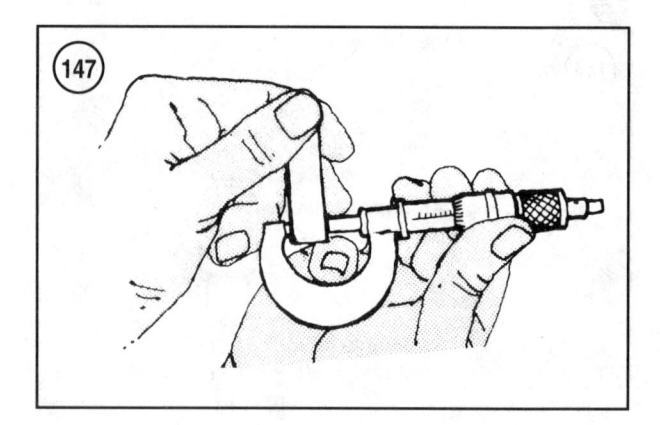

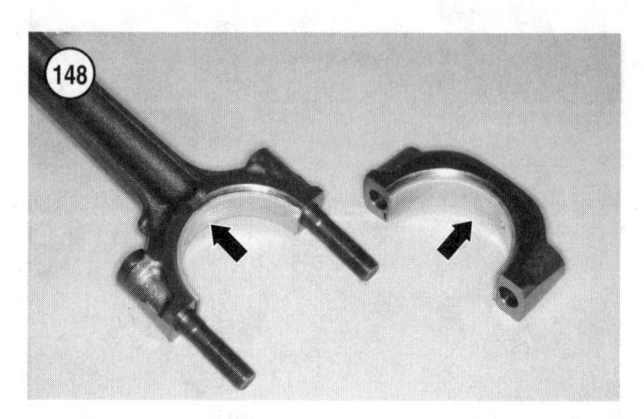

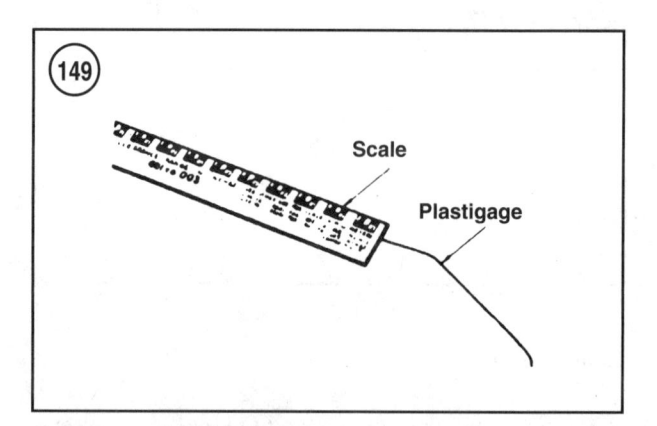

ENGINE LOWER END

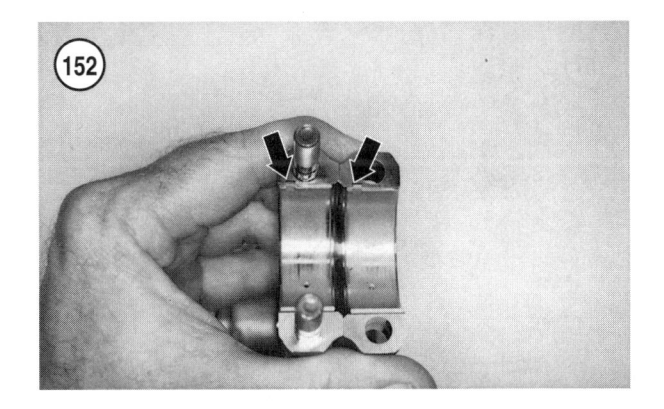

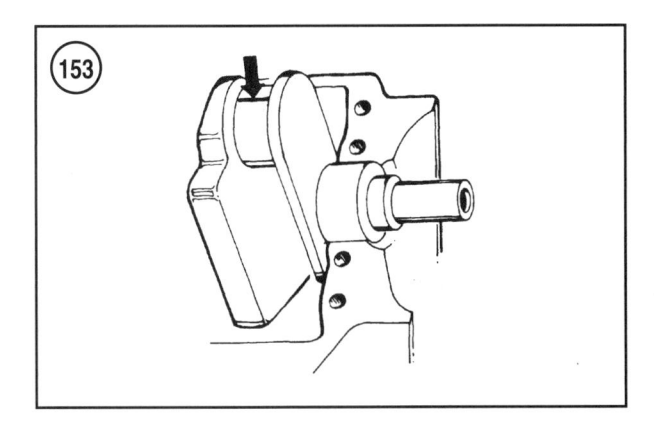

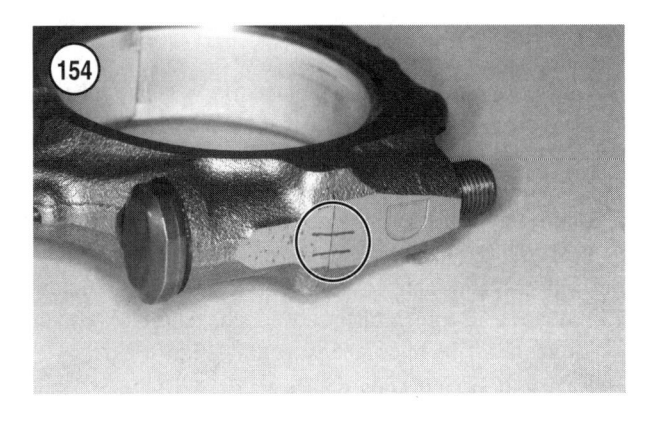

3. Install the crankshaft into the left crankcase (**Figure 151**). Turn the crankshaft so none of the connecting rod journal oil holes face up.

4. Check each rod bearing insert (**Figure 148**) for uneven wear, nicks, seizure and scoring. If the bearings do not show any visible wear, they can be used with the Plastigage to check the bearing clearance. If the journals are in good condition but the bearing inserts are too worn or damaged to be used with the Plastigage, refer to Step 14.

5. Clean the connecting rod and cap bearing surfaces and bearings inserts.

6. Install the rod bearing inserts into the connecting rod and bearing cap. Make sure the anti-rotation tabs on the bearing inserts lock into the rod and cap notches correctly (**Figure 152**).

CAUTION
Install used bearing inserts in their original locations.

7. Install all the connecting rods as described in this section. Be careful not to damage the journal surface with the rod bolts.

8. Place a piece of Plastigage over the rod bearing journal parallel to the crankshaft (**Figure 153**). Do not place the Plastigage material over an oil hole in the crankshaft.

CAUTION
Do not rotate the crankshaft while the Plastigage is in place. This will spread the Plastigage and cause an inaccurate reading.

9. Match the identification code number on the end of the cap with the mark on the rod and install the cap (**Figure 154**).

10. Apply a light coat of oil on the connecting rod bolt threads and install the cap nuts. Tighten the cap nuts in 2-3 stages to 31 N•m (23 ft.-lb.).

11. Loosen the nuts and carefully remove the cap from the connecting rod.

12. Place the envelope scale over the flattened Plastigage (**Figure 155**) to measure it. Compare the different marked bands on the envelope and find one that is closest to the widest width of the flattened Plastigage. The number adjacent to that band is the oil clearance indicated in millimeters or inches. Then measure the Plastigage at both ends of the strip. If the width of the Plastigage varies from one end to the other, the crankpin is tapered. Confirm with a micrometer. See **Table 3** for the correct connecting rod bearing oil clearance. Record the clearance for each journal.

13. Remove the Plastigage strips from the journals with solvent or contact cleaner. Do not scrape the Plastigage off.

170   CHAPTER FIVE

(155)

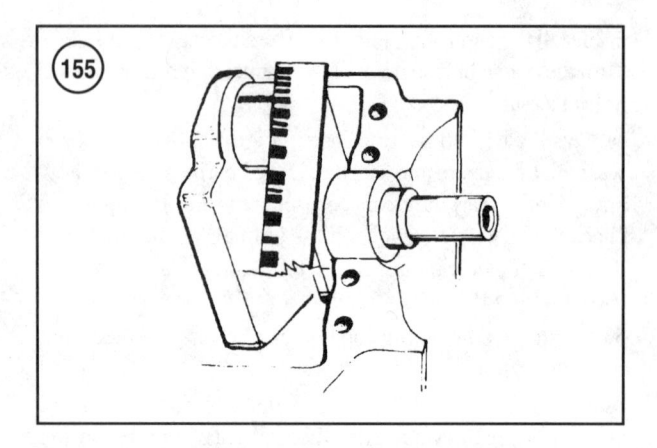

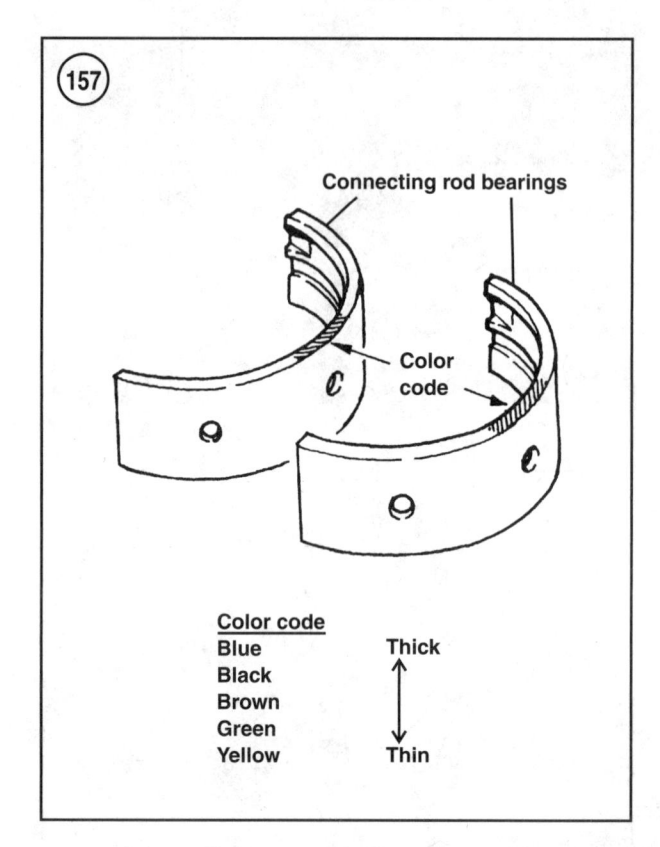

14. If the bearing clearance is greater than specified, use the following steps for new bearing selection.

15. The crankshaft rod journals are marked with letters A, B or C (**Figure 156**) on the counterbalance weights. The front code mark indicates the No. 1 bearing and reads front to rear.

16. The connecting rod and cap are marked with numbers 1, 2 or 3 or letters I, II or III (**Figure 154**).

17. Refer to **Table 4** and select new bearings by cross-referencing the journal outside diameter code (**Figure 156**) in the vertical column to the connecting rod inside diameter code (**Figure 154**) in the horizontal column. Where the two columns intersect, the new bearing color is indicated. The connecting rod bearing inserts are color-coded on the side of the bearing insert. **Figure 157** gives the replacement bearing insert color and thickness.

18. After installing new bearings, recheck clearance by repeating this procedure.

19. Repeat for each cylinder.

Piston Installation

The piston pins are installed with a press. Refer to *Tools* in this section for a description of the tools used during this procedure.

1. If installed, remove the piston rings from the piston as described in this chapter.

2. Thread the piston pin driver shaft onto the piston pin driver head. Set the length to 50 mm (1.969 in.) as shown in **Figure 158**.

3. Install the pivot collar over the pilot pin and into the piston base (**Figure 159**).

4. Set the piston and connecting rod over the pivot collar (**Figure 159**) with the piston crown facing the arrow on the piston base head. Note the following:

    a. Left cylinder pistons (2, 4 and 6): Assemble the piston by positioning the L mark on the piston crown on the *opposite side* of the connecting rod oil hole (**Figure 160**).

    b. Right cylinder pistons (1, 3 and 5): Assembly the piston by positioning the R mark on the piston crown on the *same side* of the connecting rod oil hole (**Figure 160**).

5. Lubricate the piston pin with engine oil.

6. Press the piston pin through the piston and connecting rod with the adjustable piston pin driver assembly (**Figure 159**). Place a 12 mm washer between the adjustable piston pin driver head and hydraulic press arm (**Figure 159**).

7. Repeat for each piston.

ENGINE LOWER END

(158)

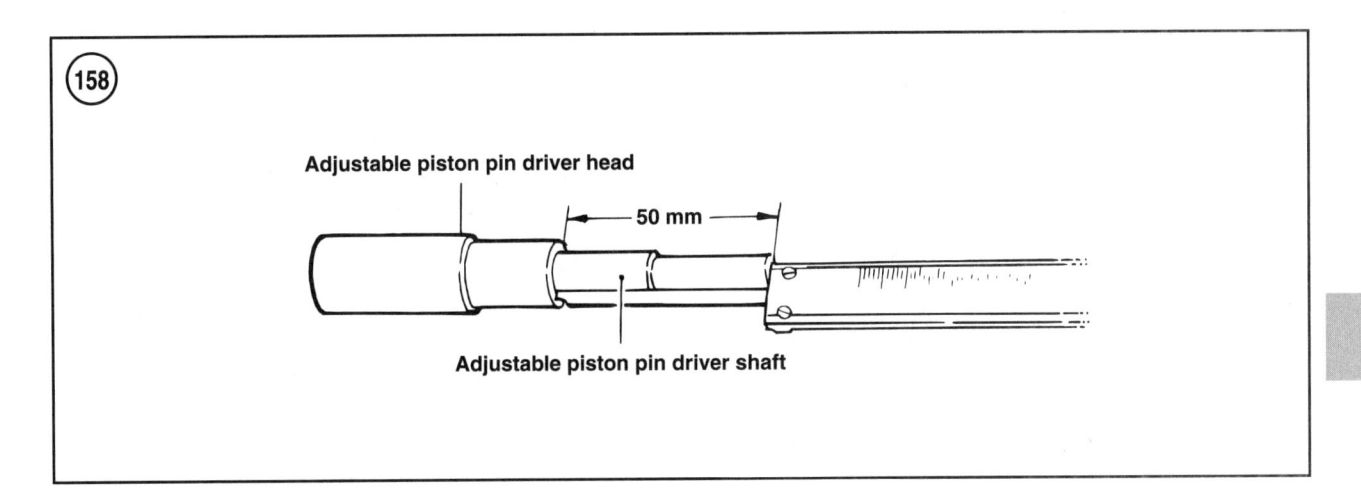

Adjustable piston pin driver head
50 mm
Adjustable piston pin driver shaft

(159)

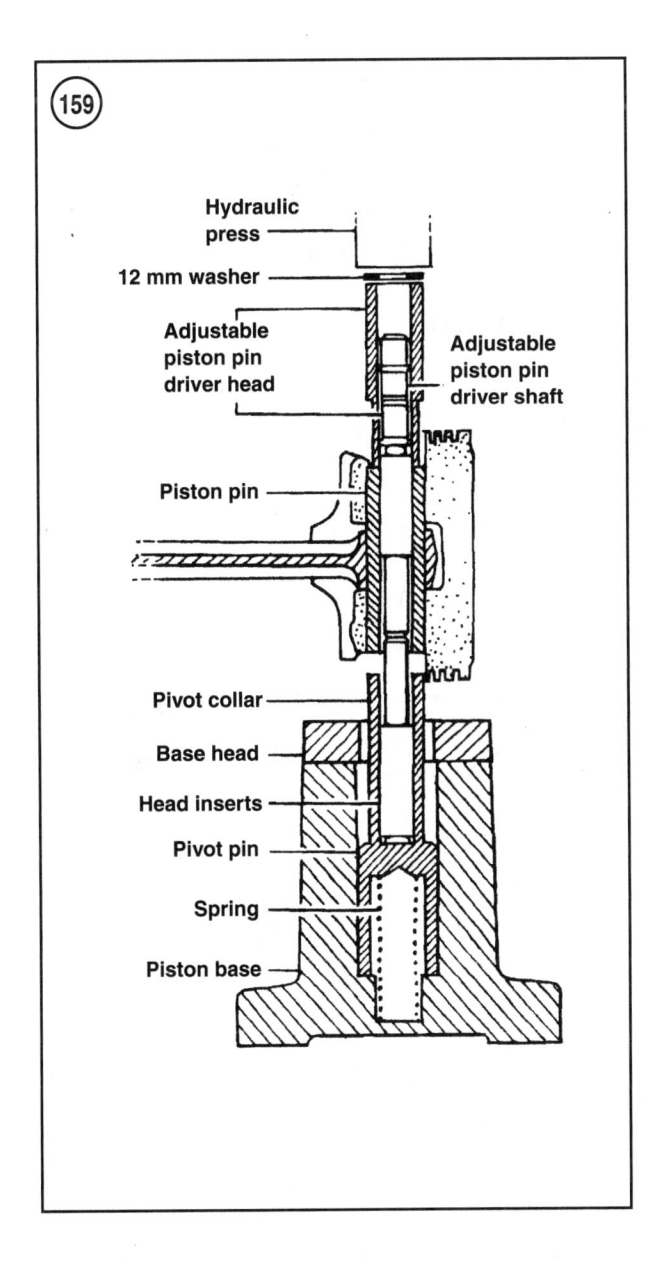

Hydraulic press
12 mm washer
Adjustable piston pin driver head
Adjustable piston pin driver shaft
Piston pin
Pivot collar
Base head
Head inserts
Pivot pin
Spring
Piston base

(160)

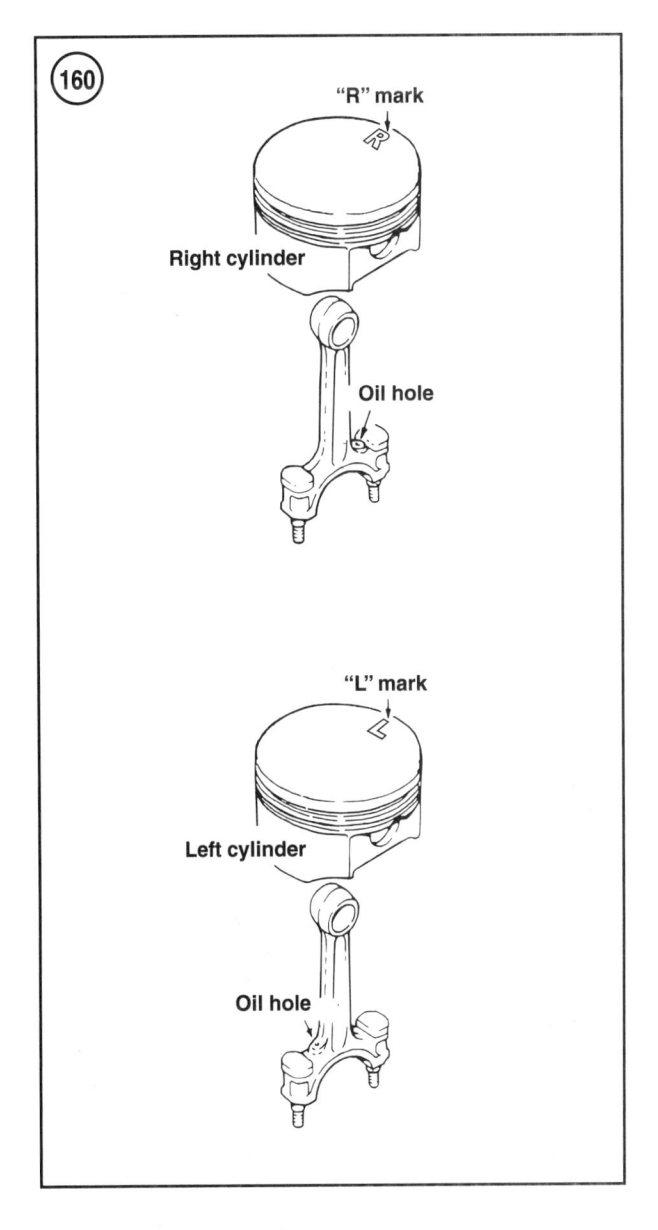

"R" mark
Right cylinder
Oil hole

"L" mark
Left cylinder
Oil hole

8. Install the piston rings as described in this chapter.

Piston and Connecting Rod Installation

1. Install the crankshaft as described in this chapter.
2. Install the piston and piston pins onto their respective connecting rods as described in this chapter.
3. Install the connecting rod bearings as follows:

CAUTION
If reusing the old bearing inserts, install them into their original positions.

   a. Wipe the bearing inserts, and the connecting rod and cap bearing surfaces clean.
   b. Insert the bearing inserts into the rod and cap by aligning the tab on each bearing insert with the notch on the rod and cap (**Figure 154**).
   c. Apply molybdenum oil solution to the bearing insert surfaces.
4. Apply a light coat of engine oil to the cylinder walls and to the piston rings before installation.
5. On the No. 2, No. 4 and No. 6 pistons, perform the following:
   a. Insert short pieces of rubber or vinyl hose over the threads on each connecting rod bolt.
   b. Install a piston ring compressor onto the piston.
   c. Position the piston/rod assembly so the L mark (**Figure 161**) on the piston crown is pointed up toward the top surface of the cylinder block (intake side of engine).
   d. Position the piston into the correct cylinder (**Figure 161**).
   e. Tap the piston into the cylinder (**Figure 162**) while guiding the connecting rod around the crankshaft journal. Remove the piston ring compressor.
   f. Guide the connecting rod onto the crankshaft so the bearing surface does not get damaged by the connecting rod during installation (**Figure 163**).
   g. Remove the hoses (**Figure 163**) from the rod bolts.
   h. Apply molybdenum oil solution to the bearing insert in the rod cap and the crankshaft bearing surface.
   i. Install the connecting rod cap (**Figure 164**) by aligning the identification code number on the end of the cap with the mark on the rod (**Figure 154**).
   j. Lubricate the connecting rod bolt threads and the rod cap nut threads and seating surfaces with engine oil and tighten them evenly in 2-3 steps to 31 N•m (23 ft.-lb.).
6. On the No.1, No. 3 and No. 5 pistons, perform the following:

   a. Position the piston/rod assembly so the R mark (**Figure 165**) on the piston crown is pointed up toward the top surface of the cylinder block (intake side of engine).
   b. Position the piston into the correct cylinder (**Figure 165**) while installing the connecting rod around the crankshaft journal. Guide the connecting rod onto the crankshaft so the bearing surface does not get damaged by the connecting rod during installation.
   c. Apply molybdenum oil solution to the bearing insert in the rod cap and the crankshaft bearing surface.

ENGINE LOWER END

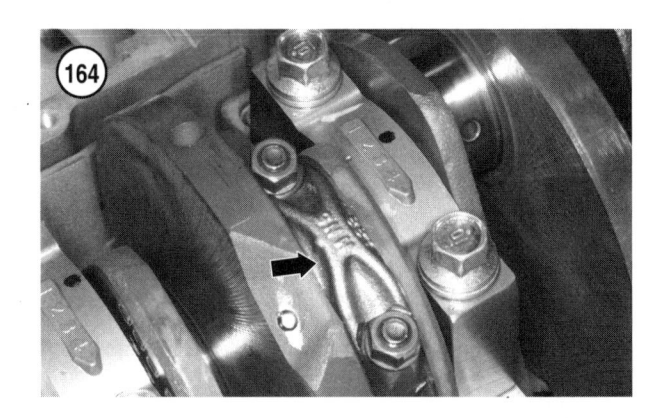

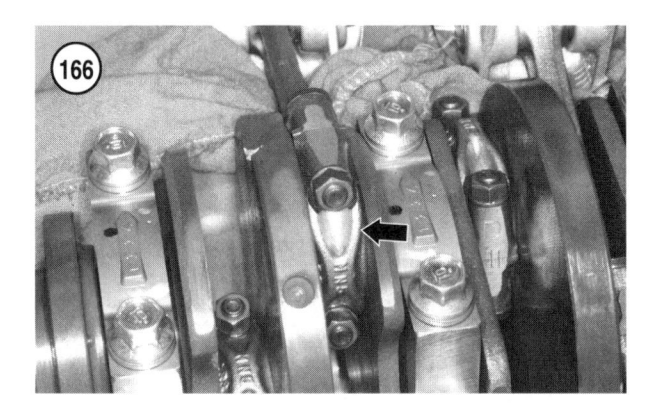

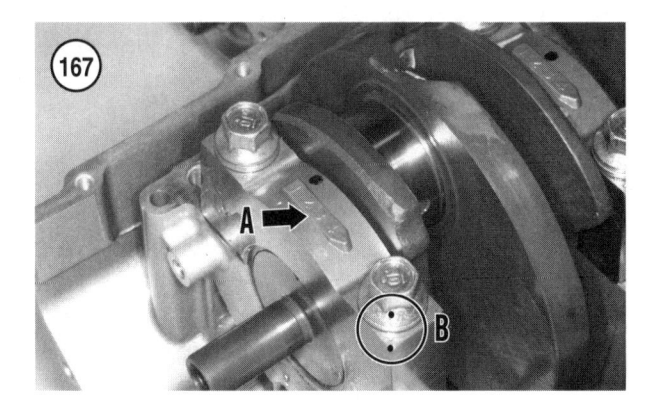

d. Install the connecting rod cap (**Figure 166**) by aligning the identification code number on the end of the cap with the mark on the rod (**Figure 154**).

e. Lubricate the connecting rod bolt threads and the rod cap nut threads and seating surfaces with engine oil and tighten them evenly in 2-3 steps to 31 N•m (23 ft.-lb.).

7. After all rod caps have been installed, have an assistant support the exposed pistons, then rotate the crankshaft several times. Make sure there is no binding.

8. Assemble the crankcase halves as described in this chapter.

CRANKSHAFT

Handle the crankshaft carefully when removing and servicing it in the following sections.

Removal

1. Disassemble the cylinder block as described in this chapter.

2. Remove the No. 1, No. 2 and No. 3 connecting rods from the crankshaft as described under *Piston/Connecting Rod Removal* in this chapter.

3. Remove the rod caps from the No. 2, No. 4 and No. 6 connecting rods from the crankshaft as described under *Piston/Connecting Rod Removal* in this chapter. Cover the rod bolts with plastic hoses (**Figure 163**).

NOTE
*Each main bearing cap (A, **Figure 167**) is marked with the numbers 1, 2, 3 and 4. A dot mark is aligned with one number on each cap, identifying the bearing cap with its position in the engine. In addition, each cap has an arrow that points toward the top of the engine. Check for these marks before removing the main bearing caps.*

4. Make alignments on each bearing cap and bolt (B, **Figure 167**). Keep the bolts with their respective caps. Loosen and remove the main bearing cap bolts and washers.

5. Remove the main bearing caps (**Figure 168**) and their dowel pins. Keep all bearing caps and inserts in order.

6. Remove the thrust bearings (**Figure 169**).

7. Remove the crankshaft (**Figure 170**).

8. Note the bearing inserts (**Figure 171**). Make sure they are completely installed in the crankcase.

Inspection

1. Clean the crankshaft thoroughly with solvent. Clean oil holes with cleaning brushes; flush thoroughly with new solvent and dry with compressed air. Lightly oil all bearing journal surfaces immediately to prevent rust.
2. Carefully inspect each journal (**Figure 172**) for scratches, ridges, scoring and nicks. Very small nicks and scratches may be removed with fine emery cloth. If damage is serious, replace the crankshaft. It cannot be serviced. If the surface on all journals is satisfactory, perform the following.
3. Measure the crankshaft main beaaring journals with a micrometer (**Figure 173**) and check for out-of roundness and taper. See **Table 3** for specifications.
4. Check crankshaft runout as follows:
    a. Support the crankshaft on the main bearing journals identified in A, **Figure 172**.
    b. Set a dial indicator on the main bearing journals identified in B, **Figure 172**.
    c. Turn the crankshaft two revolutions and read the runout on the dial indicator.
    d. Divide the total indicator reading in half to determine the runout and compare to the runout service limit in **Table 3**.
    e. Repeat for the other center main bearing journal.
5. Replace the crankshaft if the specifications in Step 3 and/or Step 4 exceed the service limit. Do not attempt to repair the crankshaft.

Crankshaft Main Bearing Oil Clearance

This section describes how to measure crankshaft main bearing clearance using Plastigage (**Figure 174**). Plastigage is a material that flattens when pressure is applied to it. The marked bands on the envelope are used to measure the width of the flattened Plastigage.

Plastigage is available in different clearance ranges. When purchasing Plastigage, select the 0.025-0.076 mm (0.001-0.003 in.) clearance range set.
1. Clean and dry the main bearing crankpins.
2. Before measuring the main bearing clearance, measure the taper and out-of-roundness of each main bearing journal (**Figure 173**) with a micrometer. Note the following:
    a. To check for taper, measure at both ends of a journal. Do not measure where the radius connects the crank web to the journal. Subtract the smallest diameter from the largest diameter to obtain the maximum amount of taper. Record the measurements for each journal.
    b. To check for out-of-roundness, measure the journal diameter at different places. Record the measure-

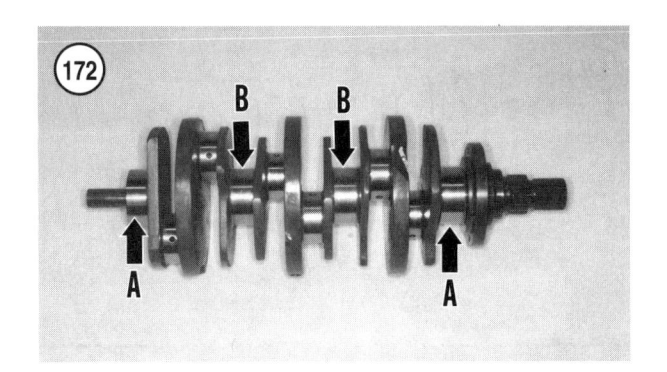

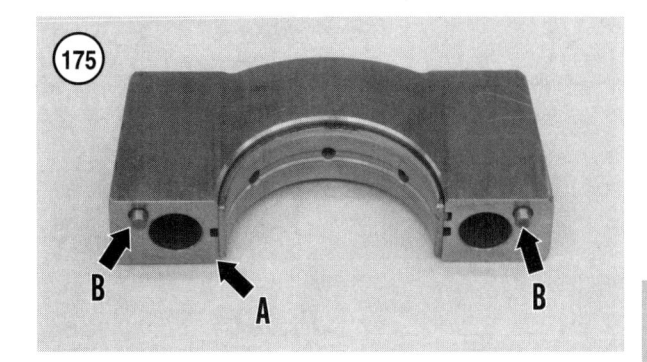

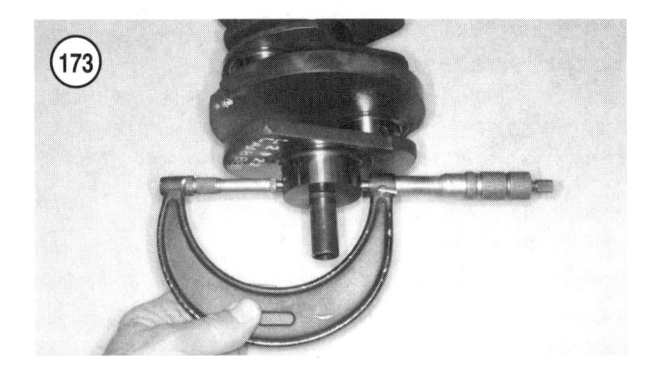

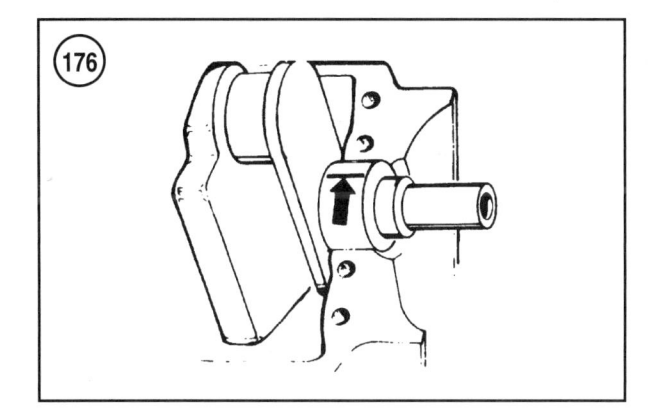

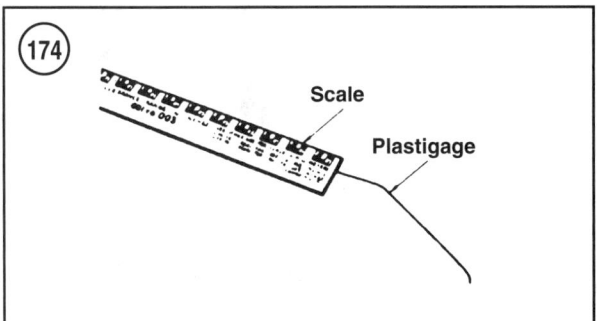

ENGINE LOWER END

172

173

174

Scale
Plastigage

ments. Subtract the smallest diameter from the largest diameter to determine the maximum out-of-round for each.

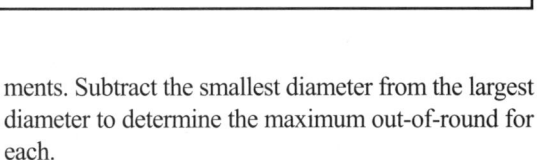

c. If the journal(s) is tapered or out-of-round (**Table 3**), the crankshaft may have to be replaced. However, before doing so, consult with a Honda dealership on the appropriate service.

3. Clean and dry each main bearing insert and its installation surface in the left crankcase and bearing cap.

CAUTION
There must be no oil, dirt, lint or other material on the bearing bores or bearing inserts. If these parts are not clean, an incorrect bearing

175

176

clearance reading may be obtained and cause bearing seizure and engine damage.

4. Check each main bearing insert for uneven wear, nicks, seizure and scoring. If the bearings do not show any visible wear, they can be used with the Plastigage to check the main bearing clearance.

NOTE
The bearing inserts must be installed in their original operating position.

5. Install the existing main bearing inserts into the left crankcase (**Figure 171**) and into the individual bearing caps (A, **Figure 175**). Make sure the anti-rotation tabs on the bearing inserts lock into the case notches correctly.

6. Install the crankshaft (**Figure 170**) so the oil holes in the journals are not facing up.

7. Place a strip of Plastigage material over each main bearing journal parallel to the crankshaft (**Figure 176**). Do not place the Plastigage strip over an oil hole in the crankshaft.

NOTE
Do not rotate the crankshaft while the Plastigage strips are in place.

8. Install the main bearing caps in their correct positions. Refer to *Crankshaft Installation* in this section.

9. Install and tighten the main bearing cap bolts as follows:

   a. Apply engine oil to the bolt threads, sealing surfaces and washers and install them.

   b. Tighten the bolts evenly in 2-3 steps to 20 N·m (14 ft.-lb.).

   c. Tighten each bolt an additional 45 degrees. The bolt and bearing cap alignment marks, made prior to loosening the caps during disassembly, should align (B, **Figure 167**). However, if new bolts were installed, or the original bolts were not marked or were mixed up, tighten the bolts an additional 45 degrees as described in substep d.

   d. Attach a torque angle meter (Lisle part No. 28100, or equivalent) onto a 1/2 in. drive breaker bar (**Figure 177**). Zero the torque angle meter, then tighten each bolt an additional 45°.

10. Remove the main bearing cap bolts and washers.

11. Carefully remove the main bearing caps and measure the width of the flattened Plastigage material following the manufacturer's instructions. Measure both ends of the Plastigage strip (**Figure 178**). A difference of 0.003 mm (0.0001 in.) or more indicates a tapered journal. Confirm with a micrometer. New bearing oil clearance and service limit dimensions are listed in **Table 3**. Remove the Plastigage strip from each bearing journal.

12. If the bearing clearance is greater than specified, use the following steps for new bearing selection.

13. The crankshaft main journals are marked with the code numbers 1, 2 or 3 (**Figure 179**). The numbers identify the main journals, starting with the No. 1 journal and reading front to rear.

14. The left crankcase (next to the front bearing support) is marked with the code letters A, B or C (**Figure 180**). The numbers directly under each letter identify the journal numbers—starting with the No. 1 journal and reading front to rear.

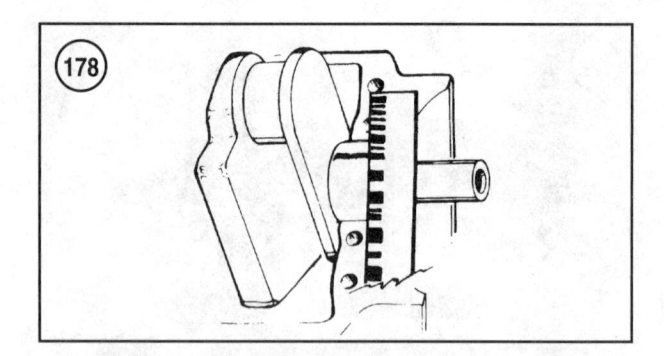

15. Refer to **Table 5** and select new bearings by cross-referencing the main journal outside diameter code number (**Figure 179**) in the vertical column to the crankcase bearing inside diameter code letter (**Figure 180**) in the horizontal column. Where the two columns intersect, the new bearing color is indicated. The main bearing inserts are color-coded on the side of the bearing insert. **Figure 181** gives the replacement bearing insert color and thickness.

16. Repeat for each cylinder.

17. After installing the new bearings, recheck clearance by repeating this procedure.

Crankshaft Installation

1. Install the crankshaft bearing inserts as follows:

ENGINE LOWER END

181

Main bearings
Color code

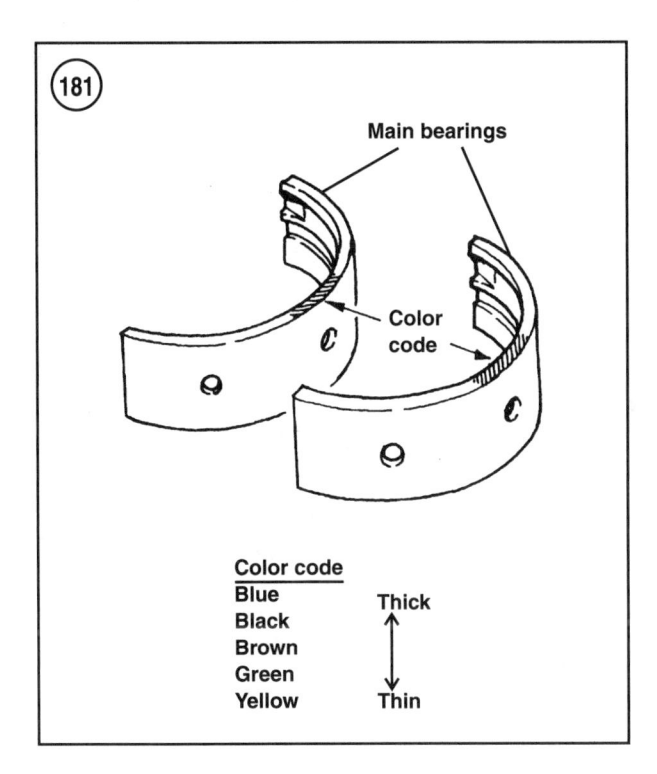

Color code
Blue — Thick
Black ↑
Brown
Green ↓
Yellow — Thin

182

a. Clean and dry each main bearing insert and its installation surface in the left crankcase and bearing cap.

CAUTION
There must be no oil, dirt, lint or other material on the bearing bores or bearing inserts. If these parts are not clean, bearing seizure and engine damage may result.

CAUTION
Install bearing inserts used in their original positions.

b. Install the existing main bearing inserts into the left crankcase (**Figure 171**) and into the individual bearing caps (A, **Figure 175**). Make sure the anti-rotation

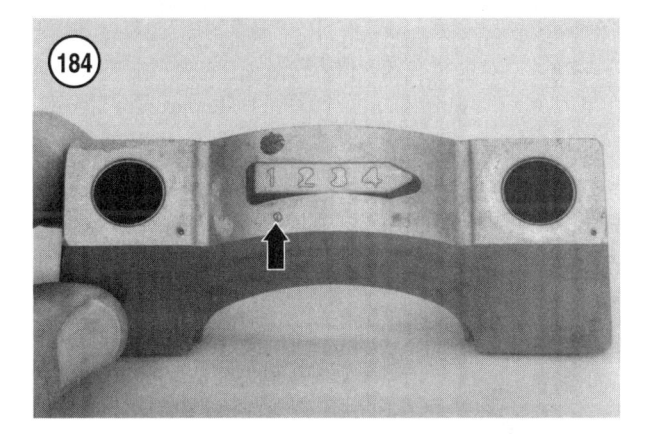

183

184

tabs on the bearing inserts lock into the cap and case notches correctly.

2. Lubricate the main journal and bearing cap inserts with molybdenum oil solution.

3. Install the crankshaft (**Figure 170**).

4. Install the two thrust bearings with their grooved sides facing out (**Figure 182**). Position both thrust bearings so their ends are flush with the crankcase gasket surface (**Figure 183**).

CAUTION
The main bearing caps must be installed in their correct locations or the engine will be damaged by incorrect bearing oil clearance.

5. Install the main bearing caps in their correct positions as follows:

a. Each main bearing cap (**Figure 184**) is marked with the numbers 1, 2, 3 and 4. A dot mark is aligned with one number on each cap, identifying the bearing cap with its position in the engine. The cap with the dot under the number 1 is installed at the front of the engine. Then follow with numbers 2, 3 and 4. See **Figure 185**.

b. The arrows on the bearing caps must point up toward the upper side of the engine.

c. Install two dowel pins into each cap (B, **Figure 175**).
d. Install the main bearing caps (**Figure 186**) in order.
6. Install and tighten the main bearing cap bolts as follows:
   a. Apply engine oil to the bolt threads, sealing surfaces and washers and install them.
   b. Tighten the bolts evenly in 2-3 steps to 20 N·m (14 ft.-lb.).
   c. Tighten each bolt an additional 45 degrees. The bolt and bearing cap alignment marks, made prior to loosening the caps during disassembly, should align (B, **Figure 167**). However, if new bolts were installed, or the original bolts were not marked or were mixed up, tighten the bolts an additional 45 degrees as described in substep d.
   d. Attach a torque angle meter (Lisle part No. 28100, or equivalent) onto a 1/2 in. drive breaker bar (**Figure 177**). Zero the torque angle meter, then tighten each bolt an additional 45°.
7. Reassemble the cylinder block as described in this chapter.

OIL STRAINER AND OIL PRESSURE RELIEF VALVE

The crankcase halves must be disassembled to service the oil strainer, oil pressure relief valve and oil pump.

Oil Strainer
Removal/Installation

1. Separate the crankcase as described in this chapter.
2. Remove the bolt (A, **Figure 187**) and oil strainer (B) from the oil pump.
3. Remove the seal (**Figure 188**).
4. Check the oil strainer screen for clogging or damage. If the screen is clogged, clean it in solvent and thoroughly dry. If the screen cannot be cleaned or if it is damaged, replace the oil strainer assembly.
5. Lubricate a *new* seal and install it into the oil pump (**Figure 188**).
6. Push the oil strainer through the seal and into the oil pump (B, **Figure 187**).
7. Apply a medium strength threadlock onto the oil strainer mounting bolt threads and tighten the bolt (A, **Figure 187**) to 12 N·m (106 in.-lb.).
8. Assemble the crankcase as described in this chapter.

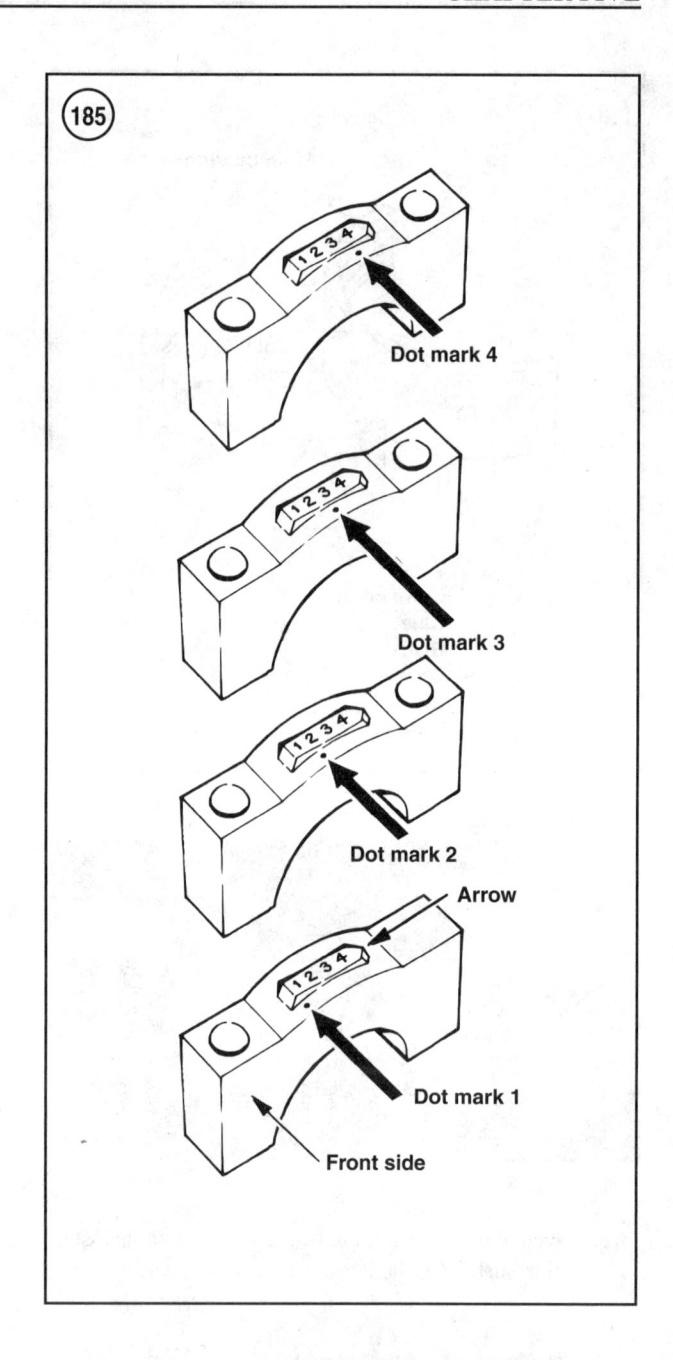

Oil Pressure Relief Valve
Removal/Inspection/Installation

1. Separate the crankcase as described in this chapter.

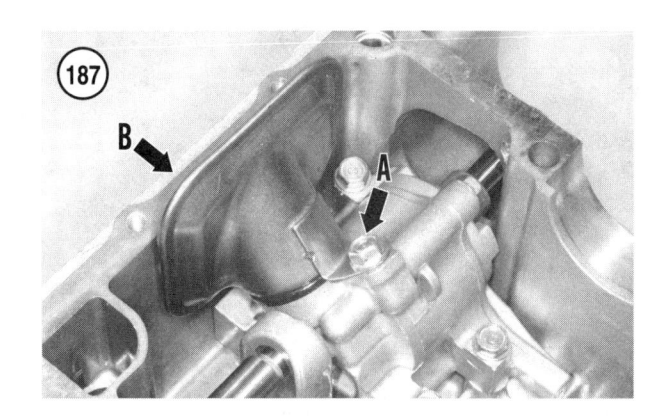

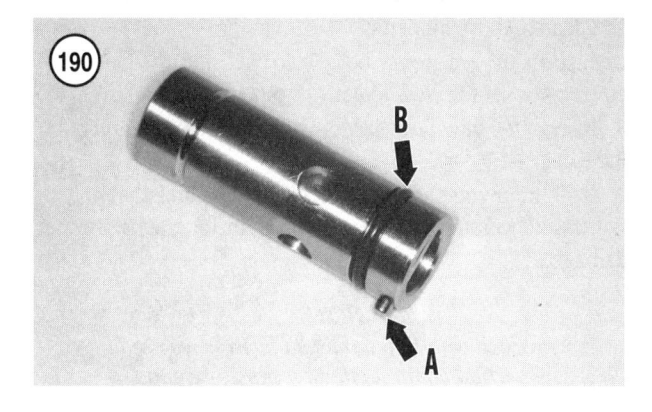

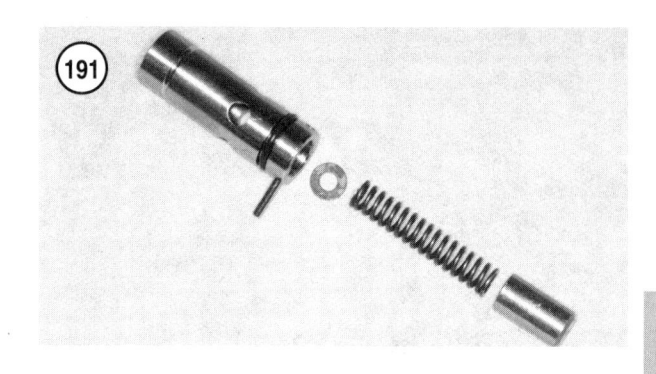

ENGINE LOWER END

179

2. Remove the oil pressure relief valve (**Figure 189**) and O-ring. If the valve is not going to be serviced, store it in a sealed plastic bag until reassembly.

CAUTION
Handle the oil pressure relief valve carefully to prevent dirt from entering the valve and scoring the piston and cylinder.

3. Service the oil pressure relief valve as follows:
    a. Remove the pin (A, **Figure 190**) from the end of the valve. Then remove the piston, spring and washer (**Figure 191**). Do not remove the snap ring from the opposite end of the valve unless necessary.
    b. Clean and dry all parts.
    c. Check the piston and piston pin for scoring, scratches or other damage.
    d. Inspect the spring for cracks, stretched coils or other damage.
    e. If any part shows excessive wear or damage, replace the pressure relief valve assembly. Replacement parts are not available.

CAUTION
The relief valve spring tension helps control oil pressure. A weak or damaged spring can reduce oil pressure and cause engine damage. Likewise, a damaged or stuck piston can reduce oil pressure—the piston will be forced open at a lower oil pressure, reducing the amount of oil the engine receives.

   f. Reverse these steps to assemble the pressure relief valve assembly. Make sure the piston seats against the stop pin. Make sure the snap ring seats in its groove completely.

4. Lubricate a *new* relief valve O-ring (B, **Figure 190**) with engine oil and install it onto the pressure relief valve groove.

5. Push the relief valve into the crankcase until it bottoms out (**Figure 189**).

6. Assemble the crankcase as described in this chapter.

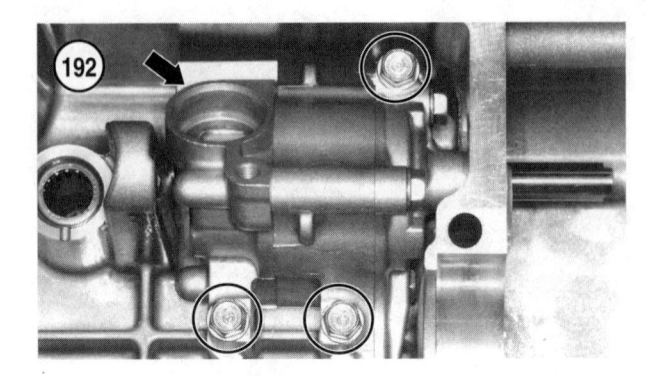

OIL PUMP

The engine cases must be separated to service the oil pump.

Removal/Installation

1. Separate the crankcase as described in this chapter.
2. Remove the transmission as described in Chapter Seven.
3. Remove the bolt (A, **Figure 187**) and oil strainer (B) from the oil pump.
4. Remove the seal (**Figure 188**).
5. Unbolt and remove the oil pump from the crankcase (**Figure 192**).
6. Remove the three dowel pins and two O-rings (**Figure 193**). Discard the O-rings.
7. If necessary, service the oil pump as described in this chapter.
8. Rotate the oil pump drive shaft (A, **Figure 194**) by hand. If there is any binding or roughness, service the oil pump as described in this chapter.
9. If the oil pump is not going to be disassembled, place it in a clean plastic bag until reassembly.
10. Install the three dowel pins (**Figure 193**).
11. Lubricate the two new O-rings with engine oil and install them (**Figure 193**).
12. Install the oil pump over the dowel pins and O-rings. Install and tighten the three oil pump mounting bolts (**Figure 192**) securely.
13. Install the oil strainer as described in this chapter.

Disassembly

Refer to **Figure 195**.

NOTE
When disassembling the oil pump, place the parts on a clean, lint-free cloth or towel.

1. Remove the oil pump as described in this chapter.

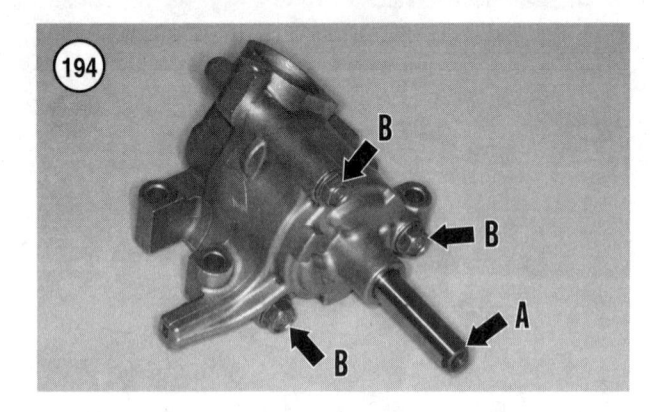

NOTE
Identify the position of the rotor alignment marks when disassembling the oil pump. Install the rotors with the marks facing in the same direction.

2. Remove the mounting bolts (B, **Figure 194**) and separate the oil pump in the order shown in **Figure 195**.

Inspection

When measuring the oil pump components, compare the actual measurements to the specifications in **Table 6**. If any part is damaged or out of specification, replace the damaged part (if available separately) or the oil pump as an assembly.

1. Clean and dry all parts.
2. Inspect the oil pump body/rotor set and the oil pump cover/rotor set for wear, cracks or other damage.
3. Inspect the thrust washer for wear, stress cracks and other damage.
4. Roll the pump shaft on a piece of glass and check it for flatness. Check the pin holes in the shaft for cracks or other damage.

NOTE
Proceed with Step 5 only if the previous visual inspection confirms that all parts are good. If

195

OIL PUMP

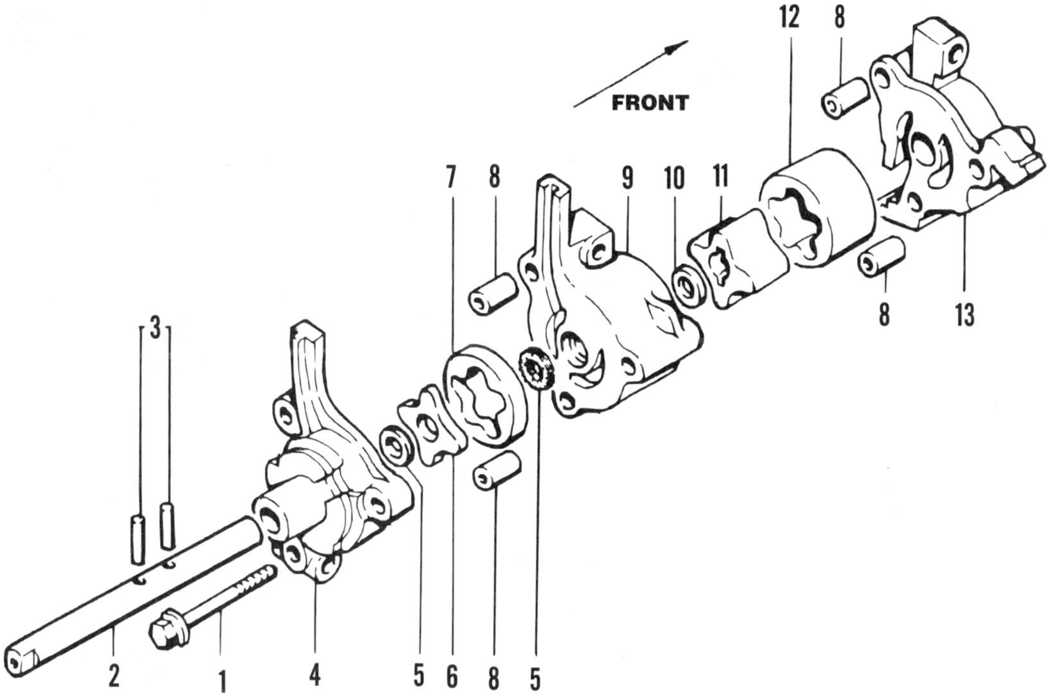

1. Bolt
2. Pump shaft
3. Pins
4. Oil pump cover B
5. Seal
6. Inner rotor B
7. Outer rotor B
8. Dowel pin
9. Oil pump body
10. Thrust washer
11. Inner rotor A
12. Outer rotor A
13. Oil pump cover A

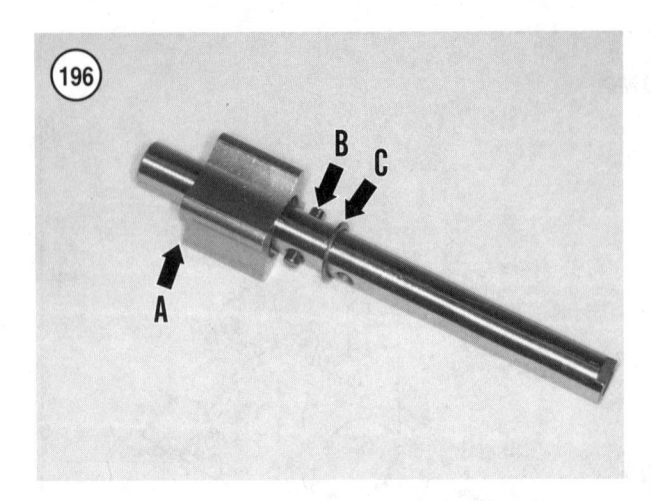

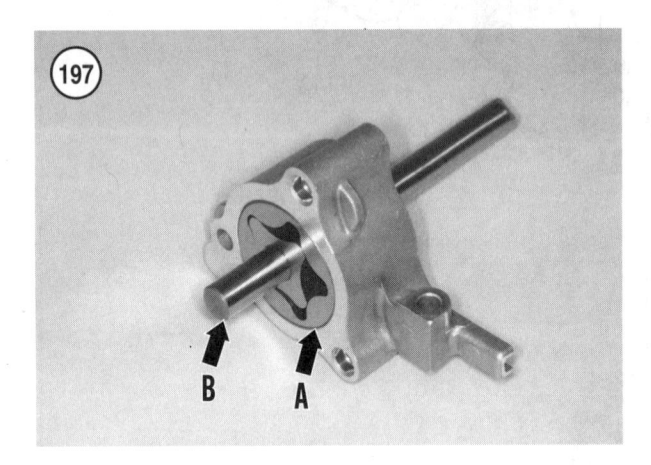

any component is worn or damaged, replace the individual parts if available or replace the oil pump as an assembly.

CAUTION
When assembling the pump halves to measure operating clearances, install the rotors with their alignment marks facing in their original direction.

5. Install inner rotor A (A, **Figure 196**) on the pump shaft. Install the pin (B, **Figure 196**) through the upper hole and install the slot in the rotor over the pin. Install the thrust washer (C, **Figure 196**) and seat against the rotor. Install outer rotor A (A, **Figure 197**) into the pump body and install the shaft and inner rotor A (B, **Figure 197**).
6. Measure the body clearance between the outer rotor and housing with a flat feeler gauge (**Figure 198**).
7. Measure the tip clearance between the inner rotor tip and the outer rotor with a flat feeler gauge (**Figure 199**).
8. Measure the side clearance with a straightedge and flat feeler gauge (**Figure 200**).

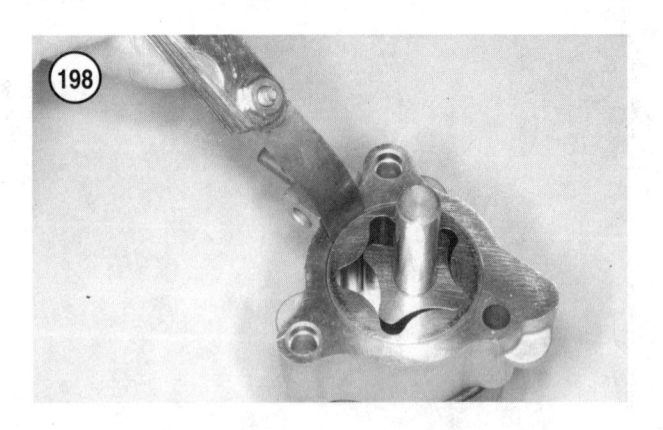

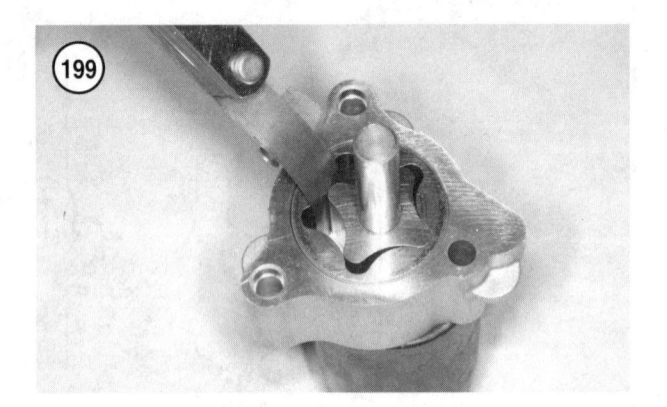

ENGINE LOWER END

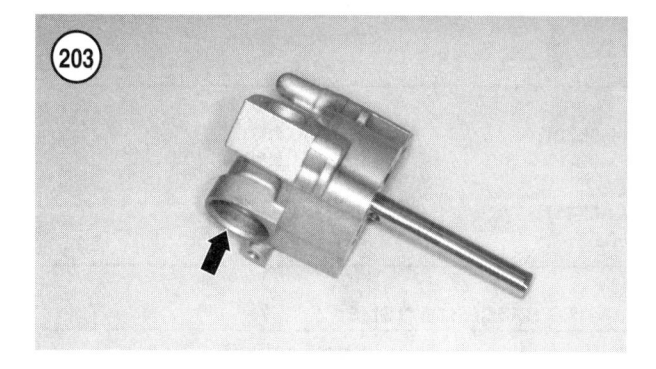

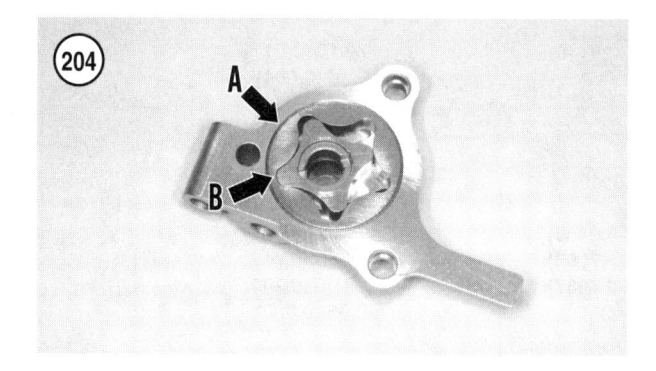

9. Repeat Steps 6-8 for rotor set B.

10. Remove the oil seals from oil pump cover B (A, **Figure 201**) and the pump body (B).

11. Clean the seal bores and check for cracks and other damage.

12. Tap a new seal into oil pump cover B (A, **Figure 201**) and the pump body (B) with the seals closed side facing out. Lubricate the seal lips with grease.

Assembly

1. Lubricate all the components with engine oil.

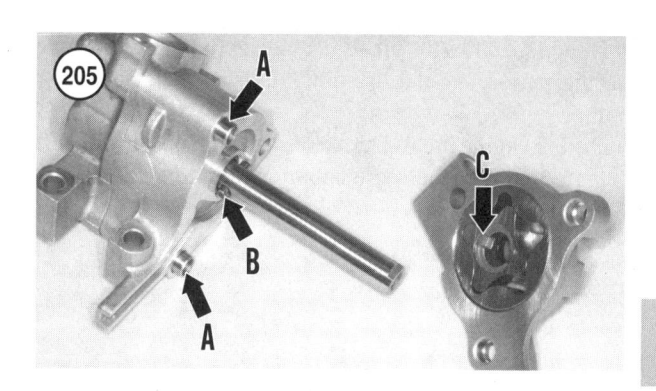

CAUTION
When assembling the oil pump, install the rotors with their alignment marks facing in their original direction.

2. Install inner rotor A (A, **Figure 196**) on the pump shaft. Install the pin (B, **Figure 196**) through the upper hole and install the slot in the rotor over the pin. Install the thrust washer (C, **Figure 196**) and seat against the rotor.
3. Install outer rotor A (A, **Figure 197**) into the pump body and install the shaft and inner rotor A (B).
4. Install the two dowel pins (**Figure 202**) into oil pump cover A.
5. Install oil pump cover A (**Figure 203**) onto the pump body.
6. Install outer rotor B (A, **Figure 204**) into oil pump cover B.
7. Install inner rotor B (B, **Figure 204**) with its slot facing out.
8. Install the two dowel pins (A, **Figure 205**) into the pump body.
9. Install the pin through the shaft hole (B, **Figure 205**).
10. Assemble the oil pump by aligning the slot in inner rotor B (C, **Figure 205**) with the pin (B).
11. Install the oil pump assembly bolts (B, **Figure 194**) and tighten to 13 N•m (115 in.-lb.).
12. Pour new engine oil into the oil strainer opening in the pump body. Rotate the pump shaft several times to make sure the rotors are coated with oil.
13. Store the oil pump in a plastic bag until installation.
14. Install the oil pump as described in this chapter.
15. Check the oil pressure as described in Chapter Three.

BREAK-IN PROCEDURE

If the rings were replaced, new pistons installed, the cylinders rebored or honed or major lower end work performed, break-in the engine just as though it were new. The performance and service life of the engine depends greatly on a careful and sensible break-in.

Honda specifies the break-in range as 0-300 miles. During this period, avoid full-throttle starts and hard acceleration.

During engine break-in, oil consumption may be higher than normal. It is therefore important to frequently check and correct oil level. At no time during the break-in or later should the oil level be allowed to drop below the bottom line on the dipstick; if the oil level is low, the oil will overheat and cause insufficient lubrication and increased wear.

Table 1 OUTPUT SHAFT SPECIFICATIONS

	New mm (in.)	Service limit mm (in.)
Bushing		
Inside diameter	22.026-22.041 (0.8672-0.8678)	22.05 (0.868)
Outside diameter	25.959-25.980 (1.0220-1.0228)	25.95 (1.022)
Damper spring free length	66.0 (2.60)	64.0 (2.52)
Final driven gear inside diameter	26.000-26.013 (1.0236-1.0241)	26.03 (1.025)
Output shaft diameter	22.008-22.021 (0.8665-0.8670)	21.99 (0.866)

Table 2 PISTON, RINGS AND CYLINDER SPECIFICATIONS

	New mm (in.)	Service limit mm (in.)
Cylinder		
Bore inside diameter	74.000-74.015 (2.9134-2.9140)	74.10 (2.917)
Out-of-round	–	0.10 (0.004)
Taper	–	0.10 (0.004)
Warp	–	0.10 (0.004)
Block warp limit	–	0.05 (0.002)
Piston-to-cylinder clearance	0.010-0.045 (0.0004-0.0018)	0.10 (0.004)
Piston and piston pin		
Outside diameter*	73.970-73.990 (2.9122-2.9130)	73.85 (2.907)
Piston pin hole inside diameter	18.010-18.016 (0.7091-0.7093)	18.03 (0.710)
Piston pin outside diameter	17.994-18.000 (0.7084-0.7087)	17.99 (0.708)
Piston-to-piston pin clearance	0.010-0.022 (0.0004-0.009)	0.05 (0.002)
Piston rings		
Piston ring-to-ring groove clearance		
Top	0.025-0.055 (0.0010-0.0022)	0.10 (0.004)
Second	0.015-0.045 (0.0006-0.0018)	0.10 (0.004)
Ring end gap		
Top	0.15-0.30 (0.006-0.012)	0.5 (0.02)
Second	0.30-0.45 (0.012-0.018)	0.6 (0.02)
Oil (side rail)	0.20-0.70 (0.008-0.0028)	0.9 (0.04)

*Refer to text for piston measuring point.

Table 3 CRANKSHAFT SPECIFICATIONS

	New mm (in.)	Service limit mm (in.)
Connecting rod side clearance	0.15-0.30 (0.006-0.012)	0.04 (0.016)
Connecting rod and main journal		
Out-of-round	–	0.005 (0.0002)
Taper	–	0.003 (0.0001)

(continued)

ENGINE LOWER END

Table 3 CRANKSHAFT SPECIFICATIONS (continued)

	New mm (in.)	Service limit mm (in.)
Connecting rod bearing journal oil clearance	0.028-0.046 (0.0011-0.0018)	0.06 (0.002)
Crankshaft runout	–	0.03 (0.001)
Crankshaft main bearing journal oil clearance		
1 and 4	0.012-0.030 (0.0005-0.0012)	0.06 (0.002)
2 and 3	0.020-0.038 (0.0008-0.0015)	0.06 (0.002)

Table 4 CONNECTING ROD BEARING SELECTION

Connecting rod journal outside diameter code	Connecting rod inside diameter code		
	1 (I)	2 (II)	3 (III)
A	Yellow	Green	Brown
B	Green	Brown	Black
C	Brown	Black	Blue

Table 5 CRANKSHAFT MAIN BEARING SELECTION

Crankshaft main bearing journal outside diameter code	Crankcase bearing inside diameter code		
	A	B	C
1	Yellow	Green	Brown
2	Green	Brown	Black
3	Brown	Black	Blue

Table 6 OIL PUMP SPECIFICATIONS

	New mm (in.)	Service limit mm (in.)
Body clearance		
Feed side (rotor set A)	0.15-0.21 (0.006-0.008)	0.35 (0.014)
Scavenge side (rotor set B)	0.15-0.22 (0.006-0.009)	0.35 (0.014)
Side clearance	0.02-0.09 (0.001-0.004)	0.12 (0.005)
Tip clearance	0.15 (0.006)	0.20 (0.008)

Table 7 ENGINE LOWER END TORQUE SPECIFICATIONS

	N•m	in.-lb.	ft.-lb.
Alternator drive gear bolts[1]	25	–	19
Clutch hose banjo bolt	34	–	25
Connecting rod bearing cap nuts[1]	31	–	23
Crankcase oil passage plug	44	–	33
Crankshaft main journal bearing cap nut	Refer to text		
Engine hanger bolts			
Front/rear	64	–	47
Center	39	–	29
Engine oil drain bolt	34	–	25
Final drive gear locknut [1, 2, 3]	186	–	137

(continued)

Table 7 ENGINE LOWER END TORQUE SPECIFICATIONS (continued)

	N•m	in.-lb.	ft.-lb.
Footpeg mounting bolts	26	–	19
Fuel feed hose joint bolt	10	88	–
Gearshift pedal pivot bolt	26	–	19
Left crankcase 8-mm bolts	25	–	19
Left engine hanger adjusting bolt			
Front	2	17	–
Center/rear	4	35	–
Left engine hanger adjusting bolt locknuts	54	–	40
Oil pump assembly bolt	13	115	–
Oil pump driven sprocket bolt[4]	18	–	13
Oil pressure switch[5]	12	106	–
Oil strainer bolt[4]	12	106	–
Output shaft bearing holder mounting bolts	28	–	21
Output shaft locknut[3]	186	–	137
Primary drive gear bearing setting plate bolts[4]	12	106	–
Rear crankcase cover mounting bolts	24	–	17
Reverse shift arm mounting bolt[4]	14	–	10
Reverse shifter actuator cable holder mounting bolts[4]	12	106	–
Right crankcase 10-mm bolts[1]	34	–	25
Right crankcase 6-mm bolts	12	106	–
Shift drum lock arm bolt[4]	12	106	–
Starter clutch mounting bolts[2]	74	–	54

1. Lubricate threads and seating surface with engine oil.
2. Left-hand threads.
3. Stake nut as described in text.
4. Apply medium strength threadlock onto fastener threads.
5. Lubricate threads with RTV as described in text.

CHAPTER SIX

CLUTCH

This chapter provides service procedures for the clutch hydraulic assembly and the clutch. Service specifications are listed in **Table 1** and torque specifications in **Table 2**. Both tables are at the end of the chapter.

BRAKE FLUID

When adding brake fluid, use DOT 4 brake fluid from a sealed container. Purchase brake fluid in small containers and discard any small leftover quantities.

CAUTION
Do not intermix DOT 5 (silicone-based) brake fluid, as it can cause clutch system failure.

Replacement Intervals

Brake fluid is hygroscopic, which means it absorbs water. While the hydraulic system is often thought of as a closed system, moisture is absorbed through its rubber parts (clutch hose, rubber boots and the reservoir diaphragm) and whenever the reservoir cover is removed. Because water is heavier than brake fluid, it collects in the lowest parts of the system and causes rust and corrosion to build inside the master cylinder and slave cylinder assemblies. Contaminated brake fluid reduces the effectiveness of the clutch and causes surface damage to all the parts in the hydraulic system. To maintain clutch performance and prevent moisture damage, change the brake fluid at the intervals specified in Chapter Three.

Brake fluid that exhibits a muddy appearance is severely contaminated and must be flushed from the system. However, because the contamination settles at the bottom of the slave cylinder, and the bleed valve is at the top, the slave cylinder may require removal for overhaul and a thorough cleaning.

PREVENTING BRAKE FLUID DAMAGE

Many of the procedures in this chapter require the handling of brake fluid. Be careful not to spill any fluid, as it stains or damages most surfaces. To prevent brake fluid damage, note the following:

1. Before performing any procedure in which there is the possibility of brake fluid contacting the motorcycle, cover the work area with a large piece of plastic.

2. Before handling brake fluid or working on the clutch system, fill a small container with soap and water and keep it close to the bike. If brake fluid spills, clean the area and rinse thoroughly.

3. To help control the flow of brake fluid when filling the reservoirs, punch a small hole into the seal of a new container next to the pour spout.

CLUTCH FLUID DRAINING

Brake fluid can be drained from the clutch system manually, or with the use of a vacuum pump. Both methods are described in this section.

1. If equipped, remove the EVAP canister and its mounting bracket. Refer to Chapter Eight.
2. Support the bike on its centerstand and turn the handlebar to level the master cylinder reservoir.

WARNING
Brake fluid is a very corrosive chemical. Wear safety glasses and rubber gloves when working on the clutch hydraulic system.

3. Unscrew and remove the reservoir cover (**Figure 1**), diaphragm and float. Use a syringe to siphon the brake fluid from the reservoir and discard it.
4. Remove the dust cap from the bleed valve (**Figure 2**). Remove any dirt from the bleed valve and inside its outlet port.
5A. To drain the clutch system manually:
   a. Use an empty bottle, length of clear hose and wrench for this procedure.
   b. Connect a wrench and hose to the bleed valve (**Figure 3**) onto the slave cylinder. Insert the other end of the hose into a container (**Figure 4**).
   c. Open the bleed valve with a wrench (**Figure 4**) and apply the clutch lever until it moves to the limit of its travel. Close the bleed valve and release the clutch lever. Repeat until brake fluid stops flowing from the hose. Close the bleed valve and disconnect the hose.
5B. To drain the clutch system with a hand-operated vacuum pump (**Figure 5**):
   a. Assemble the vacuum pump following the manufacturer's instructions.
   b. Connect the pump's hose to the bleed valve on the slave cylinder (**Figure 4**).
   c. Operate the vacuum pump to create a vacuum in the hose, then open the bleed valve with a wrench (**Figure 5**) to pull brake fluid from the system. Continue to operate the pump until brake fluid stops flowing from the bleed valve. Close the bleed valve and disconnect the hose. 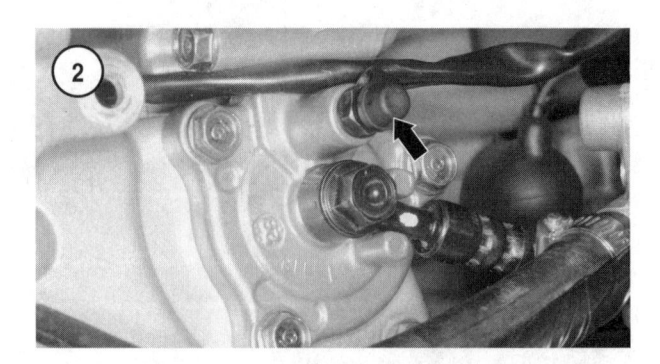 
6. Discard the brake fluid.
7. Reinstall the float, diaphragm and cover.
8. Service the clutch system as described in this chapter.

CLUTCH SYSTEM BLEEDING

Brake fluid is not compressible. As brake fluid is forced out of the master cylinder, it forces the piston in the slave cylinder to move the pushrod and release the clutch. However, air that enters the hydraulic system is compressible. When the clutch lever is applied, air in the system weakens the pressure applied at the slave cylinder. When the clutch lever feels spongy or less responsive, there is probably air in the system. This condition causes increased clutch lever travel and clutch drag, which can cause the gears to grind when the transmission is shifted, especially in the lower gears.

Bleeding the clutch forces brake fluid through the system while allowing the air to escape.

The clutch can be bled manually or with a vacuum pump. Both methods are described in this section.

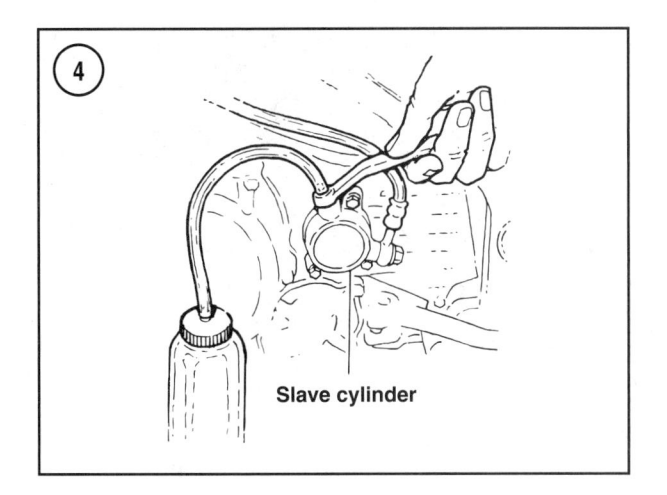

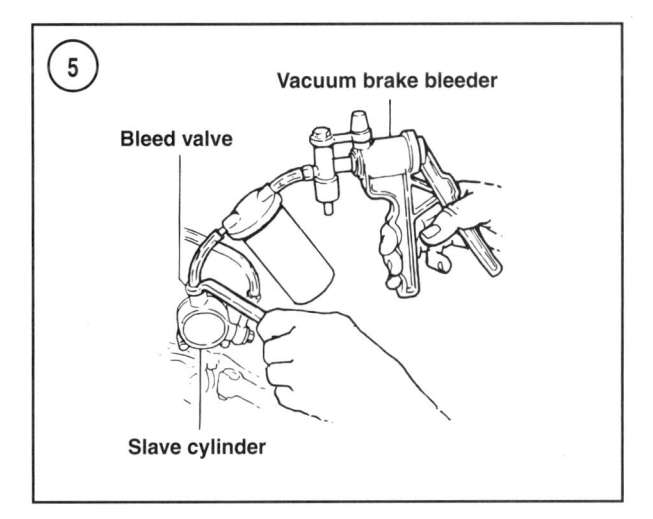

CLUTCH

Slave cylinder

Vacuum brake bleeder
Bleed valve
Slave cylinder

When adding brake fluid during the bleeding process, use DOT 4 brake fluid. Do not reuse brake fluid drained from the system or use DOT 5 (silicone-based) brake fluid. Brake fluid is very harmful to most surfaces, so wipe up any spills immediately with soapy water and rinse completely. Refer to *Brake Fluid* in this chapter for additional information.

When bleeding the clutch, check the fluid level in the master cylinder frequently to make sure it does not run. If air does enter the system, rebleed the system.

Air drawn in around the bleed valve will cause bubbles to form in the vacuum hose. While this is normal, it is misleading because it appears there is air in the system, even when the system has been bled completely. Block off the bleed valve with silicone brake grease.

For additional information on bleeding a hydraulic system, refer to Chapter Fourteen.

Flushing the Clutch System

When flushing the clutch system, do not use any type of fluid (other than DOT 4 brake fluid) as a flushing fluid. Flushing consists of pulling enough new brake fluid through the system until all the old fluid is removed, and the fluid exiting the bleed valve appears clean and air-free. To flush the clutch system, follow one of the bleeding procedures that follow in this section.

Bleeding the Master Cylinder

If the master cylinder was drained of brake fluid or its hose disconnected, bleed the master cylinder before bleeding the clutch system. This procedure removes air pockets between the reservoir and brake hose and assists with system bleeding.

WARNING
Brake fluid is a very corrosive chemical. Wear safety glasses and rubber gloves when bleeding the master cylinder.

1. Cover the area under the master cylinder.

2. Support the bike on its centerstand and turn the handlebar to level the clutch master cylinder reservoir.

3. Remove the reservoir cover (**Figure 1**), diaphragm and float. Fill the reservoir to about 10 mm (3/8 in.) from the top with DOT 4 brake fluid.

4. Have an assistant hold a small catch pan under the master cylinder.

5. Loosen the banjo bolt and brake hose (**Figure 6**) at the master cylinder and allow brake fluid to flow from the port. Then slowly apply the clutch lever until it reaches the end of its travel and hold in this position. Note any aerated fluid exiting from around the hose fitting. Tighten the banjo bolt, then release the clutch lever.

6. Wait a few seconds and repeat Step 5. When bleeding the master cylinder, do not apply the clutch lever when the bolt is tight.
7. Repeat Step 5 and Step 6 until there are no air bubbles exiting the master cylinder.
8. Tighten the banjo bolt to 34 N•m (25 ft.-lb.).
9. Leave the master cylinder in its leveled position, and bleed the clutch system as described under *Manual Bleeding* or *Pressure Bleeding* in this section.

Manual Bleeding

This procedure describes how to bleed the clutch system with an empty bottle, length of clear hose that fits tightly onto the bleed valve, and a wrench (**Figure 4**).

1. Check that all of the clutch system banjo bolts are tight.
2. Remove the EVAP canister and its mounting bracket. Refer to Chapter Eight.

WARNING
Brake fluid is a very corrosive chemical. Wear safety glasses and rubber gloves when bleeding the clutch system.

3. Remove the dust cap from the bleed valve (**Figure 2**). Remove any dirt from the bleed valve and inside its outlet port.
4. Connect a hose to the bleed valve (**Figure 3**) on the slave cylinder. Place the other end of the hose into a clean container. Fill the container with enough new brake fluid to keep the end submerged. Loop the hose higher than the bleed valve to prevent air from being drawn into the slave cylinder during bleeding (**Figure 4**).

CAUTION
Cover all parts that could become contaminated by the accidental spilling of brake fluid. Wash any spilled brake fluid from any surface immediately, as it will damage the finish. Use soapy water and rinse completely.

5. Support the bike on its centerstand and turn the handlebar to level the clutch master cylinder reservoir.
6. Remove the reservoir cover (**Figure 1**), diaphragm and float. Fill the reservoir with DOT 4 brake fluid.
7. If the master cylinder was drained, or its hose disconnected, bleed the master cylinder as described under *Bleeding the Master Cylinder* in this section.
8. With the bleed valve closed, apply the clutch lever until it stops and hold it in this position.

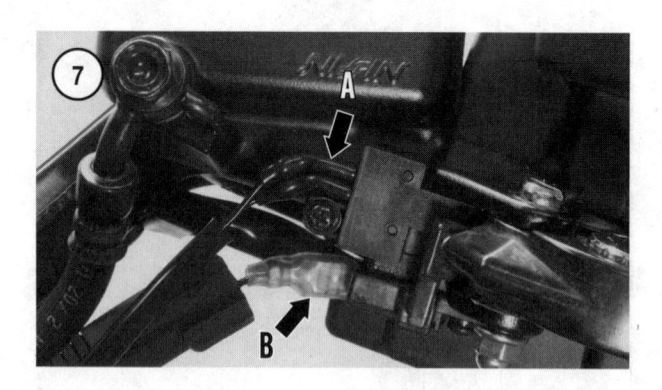

9. Open the bleed valve with the wrench (**Figure 4**) and let the lever move to the limit of its travel, then close the bleed valve. Do not release the clutch lever while the bleed valve is open.
10. Operate the clutch lever a few times. Then release it.

NOTE
Maintain an adequate supply of brake fluid in the reservoir to prevent air from being drawn into the system.

11. Repeat Steps 8-10 until the brake fluid exiting the system is clear and free of air. If the system is difficult to bleed, tap the master cylinder and slave cylinder with a soft mallet to release air bubbles trapped in the system.
12. Test the feel of the clutch lever. It should be firm and offer the same resistance each time it is operated. If the lever feels spongy, air is still trapped in the system and the system must be rebled.
13. When the bleeding procedure is complete, disconnect the hose from the bleed valve. Tighten the bleed valve to 2 N•m (18 in.-lb.).
14. If necessary, add DOT 4 brake fluid to correct the level in the master cylinder reservoir. It must be above the level line. Install the float, diaphragm, and cover. Tighten the cover screws securely.
15. Install the EVAP canister and its mounting bracket.
16. Shift the transmission into neutral and start the engine. Apply the clutch lever and shift the transmission into first gear. If the bike creeps or jumps into gear, air is still trapped in the system. When the bike shifts correctly, test ride the bike slowly at first to make sure the clutch is operating correctly.

Pressure Bleeding

This procedure describes how to bleed the clutch system with a hand-operated vacuum pump (**Figure 5**).

1. Check that all the clutch system banjo bolts are tight.

CLUTCH

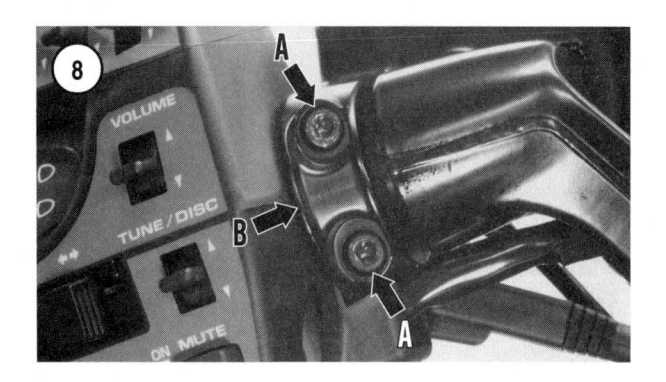

2. Remove the EVAP canister and its mounting bracket. Refer to Chapter Eight.

WARNING
Brake fluid is a very corrosive chemical. Wear safety glasses and rubber gloves when working on the clutch hydraulic system.

3. Remove the dust cap from the bleed valve (**Figure 2**). Remove any dirt from the bleed valve and inside its outlet port.

CAUTION
Cover all parts that could become contaminated by the accidental spilling of brake fluid. Wash any spilled brake fluid from any surface immediately, as it will damage the finish. Use soapy water and rinse completely.

4. Support the bike on its centerstand and turn the handlebar to level the clutch master cylinder reservoir.
5. Assemble the vacuum tool by following the manufacturer's instructions.
6. Connect a wrench and the pump's hose to the bleed valve on the slave cylinder (**Figure 3**). Support the pump with a wire hook to prevent the hose at the slave cylinder from disconnecting.
7. Remove the reservoir cover (**Figure 1**), diaphragm, and float. Fill the reservoir with DOT 4 brake fluid.

NOTE
Maintain an adequate supply of brake fluid in the reservoir to prevent air from being drawn into the system.

8. Operate the vacuum pump to create a vacuum in the hose, then open the bleed valve with a wrench (**Figure 5**) to pull the brake fluid through the system.
9. Close the bleed valve before the brake fluid stops flowing from the system (no more vacuum in line) or before the master cylinder reservoir runs empty.
10. Operate the clutch lever several times, then release it.
11. Repeat Steps 8-10 until the brake fluid exiting the system is clear and air-free. If the system is difficult to bleed, tap the master cylinder and slave cylinder with a soft mallet to release air bubbles trapped in the system.
12. Test the feel of the clutch lever. It should be firm and offer the same resistance each time it is operated. If the lever feels spongy, air is still trapped in the system and the system must be rebled.
13. When the bleeding procedure is complete, disconnect the hose from the bleed valve. Tighten the bleed valve tightness to 2 N•m (18 in.-lb.).
14. If necessary, add DOT 4 brake fluid to correct the level in the master cylinder reservoir. It must be above the level line. Install the float, diaphragm and cover. Tighten the cover screws securely.
15. Shift the transmission into neutral and start the engine. Apply the clutch lever and shift the transmission into first gear. If the bike creeps or jumps into gear, air is still trapped in the system. When the bike shifts correctly, test ride the bike slowly at first to make sure the clutch is operating correctly.

CLUTCH MASTER CYLINDER

Removal

1. Support the bike on its centerstand.
2. Disconnect the clutch switch (A, **Figure 7**) and cruise switch (B) connectors.
3. Cover the top shelter and fairing to prevent damage from brake fluid.

CAUTION
Wash brake fluid off any surface immediately, as it damages the finish. Use soapy water and rinse completely.

4. Remove the master cylinder cover (**Figure 1**), diaphragm and float. Empty the brake fluid reservoir with a syringe. Reinstall the parts.
5. If the slave cylinder is also going to be serviced, drain the clutch system as described in this chapter.
6. Remove the banjo bolt (**Figure 6**) and washers securing the brake hose to the master cylinder. Cover the clutch hose to prevent leakage and contamination.
7. Plug the master cylinder opening.
8. Remove the holder cap, bolts (A, **Figure 8**), holder (B) and master cylinder.
9. If necessary, service the master cylinder as described in this chapter.
10. Clean the handlebar, master cylinder and clamp mating surfaces.

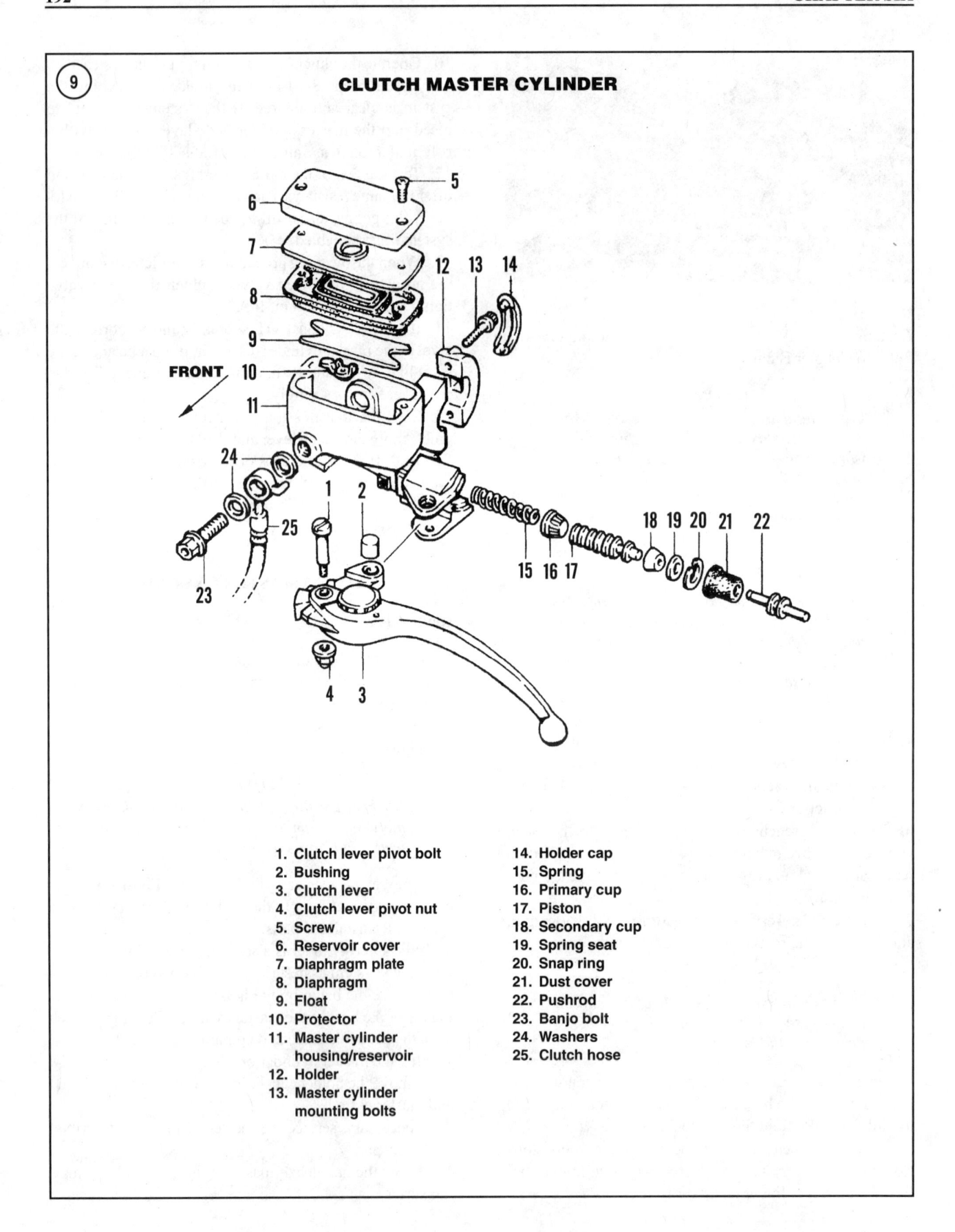

CLUTCH MASTER CYLINDER

1. Clutch lever pivot bolt
2. Bushing
3. Clutch lever
4. Clutch lever pivot nut
5. Screw
6. Reservoir cover
7. Diaphragm plate
8. Diaphragm
9. Float
10. Protector
11. Master cylinder housing/reservoir
12. Holder
13. Master cylinder mounting bolts
14. Holder cap
15. Spring
16. Primary cup
17. Piston
18. Secondary cup
19. Spring seat
20. Snap ring
21. Dust cover
22. Pushrod
23. Banjo bolt
24. Washers
25. Clutch hose

CLUTCH

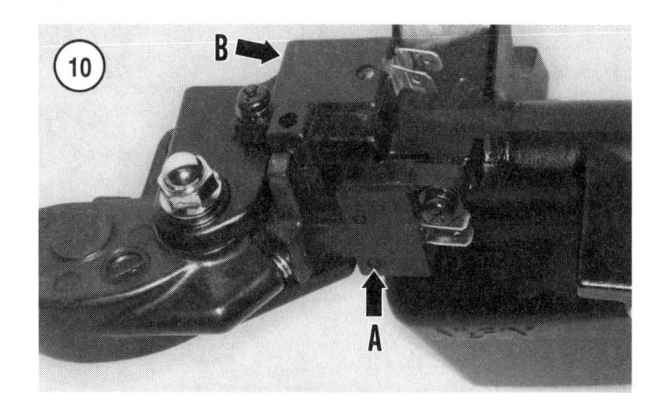

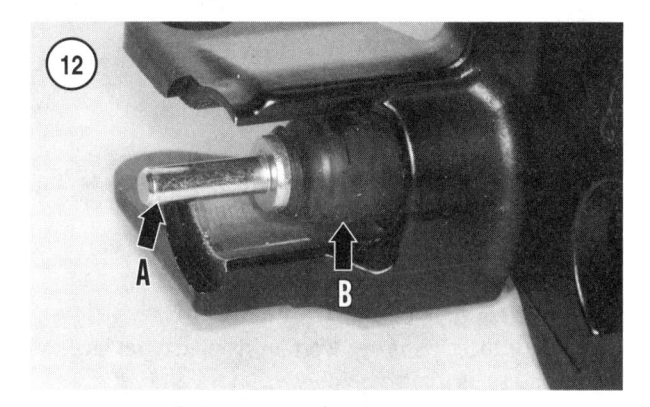

Installation

1. Mount the master cylinder onto the handlebar and align the upper master cylinder and clamp mating surfaces with the punch mark on the handlebar.

2. Install the holder (B, **Figure 8**) and mounting bolts (A). Tighten the upper mounting bolt, then the lower mounting bolt to 12 N•m (106 in.-lb.). Install the holder cap.

3. Bleed the master cylinder as described under *Clutch System Bleeding*. After bleeding the master cylinder, secure the brake hose to the master cylinder with the banjo bolt (**Figure 6**) and two *new* washers. Install a washer on each side of the brake hose. Position the brake hose arm against the master cylinder bracket as shown in **Figure 6** and tighten the banjo bolt to 34 N•m (25 ft.-lb.).

4. Reconnect the clutch switch (A, **Figure 7**) and cruise switch (B) connectors.

5. Check the reservoir and piston bore for leaks.

Disassembly

Refer to **Figure 9**.

1. Remove the master cylinder as described in this section.

2. Remove the following:
   a. Screw and clutch switch (A, **Figure 10**).
   b. Screw and cruise switch (B, **Figure 10**).
   c. Nut, pivot bolt and clutch lever.

3. Remove the master cylinder cover and diaphragm assembly. Drain the reservoir and discard the brake fluid.

4. Remove the protector (**Figure 11**) from the reservoir to uncover the relief and supply ports.

> *CAUTION*
> *Inspect the relief port for any foreign matter or a swollen primary cup. If this port remains closed after the clutch lever is released, pressure will build in the system and cause clutch drag. The relief port is the smaller of the two ports.*

5. Remove the pushrod (A, **Figure 12**) and dust cover (B) from the master cylinder.

> *NOTE*
> *If brake fluid is leaking from the piston bore, the piston cups are worn or damaged. Replace the piston assembly.*

6. Thread a bolt and nut into the brake hose port, and secure the bolt in a vise (**Figure 13**).

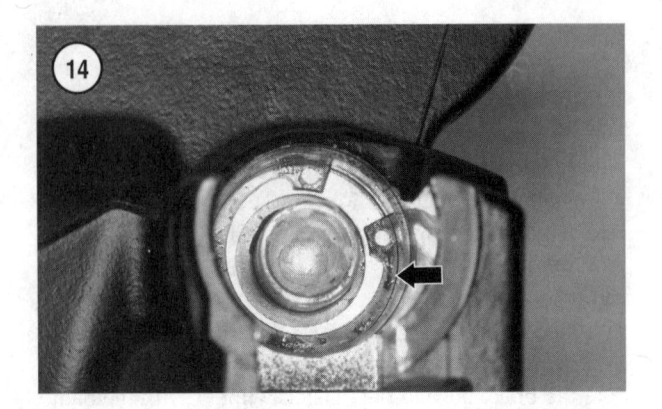

7. Compress the piston and remove the snap ring (**Figure 14**) with snap ring pliers. Allow the piston to spring back and remove the spring seat from the end of the piston.

8. Remove the piston assembly (**Figure 15**) from the master cylinder bore. Do not remove the secondary cup from the piston.

Inspection

When measuring the master cylinder components, compare the actual measurements to the specifications in **Table 1**. Replace worn or damaged parts as described in this section.

1. Clean and dry the master cylinder assembly as follows:
   a. Handle the brake components carefully when servicing them.
   b. Use only DOT 4 brake fluid or isopropyl alcohol to wash rubber parts in the brake system. Never allow any petroleum-based cleaner to contact the rubber parts. These chemicals cause the rubber to swell, requiring their replacement.
   c. Clean the master cylinder snap ring groove carefully. Use a small pick or brush to clean the groove. If a hard varnish residue has built up in the groove, soak the master cylinder in solvent to help soften the residue. Then wash in soapy water and rinse completely.
   d. Blow the master cylinder dry with compressed air.
   e. Place cleaned parts on a clean lint-free cloth until reassembly.

WARNING
Do not get any oil or grease onto any of the components. These chemicals cause the rubber parts in the system to swell, permanently damaging them.

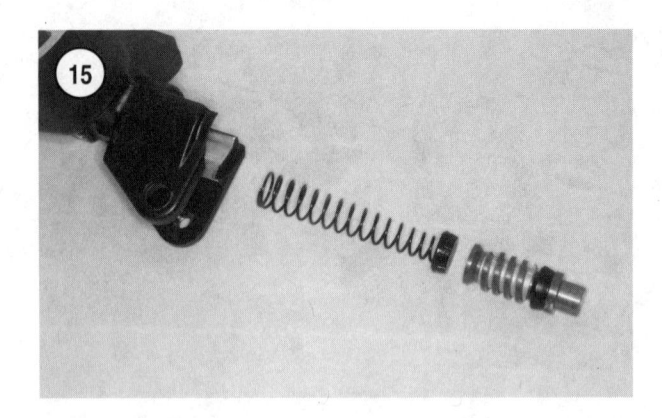

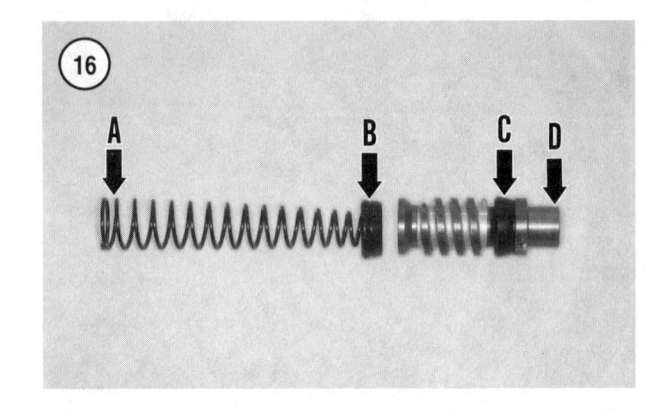

CAUTION
Do not remove the secondary cup from the piston assembly.

2. Check the piston assembly for:
   a. Broken, distorted or collapsed piston return spring (A, **Figure 16**).
   b. Worn, cracked, damaged or swollen primary (B, **Figure 16**) and secondary cups (C).
   c. Worn or pitted piston (D, **Figure 16**).

If any of these parts are worn or damaged, replace the piston and seals as an assembly.

3. Measure the piston outside diameter (**Figure 17**).
4. To assemble a new piston assembly:
   a. The piston, both cups and spring are replaced as an assembly. Lubricate these parts with brake fluid.
   b. Install the primary cup onto the spring (B, **Figure 16**).
   c. Install the secondary cup (C, **Figure 16**) onto the piston. Use the original piston and secondary cup as a reference.
5. Check the pushrod assembly (**Figure 18**) for:
   a. Corroded, weak or damaged spring seat, snap ring or pushrod.
   b. Worn or damaged dust cover.

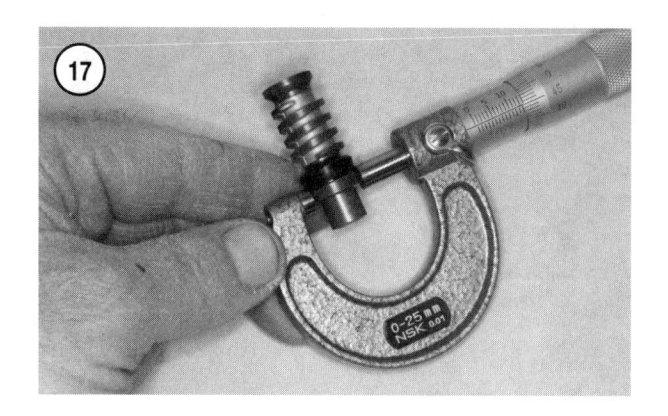

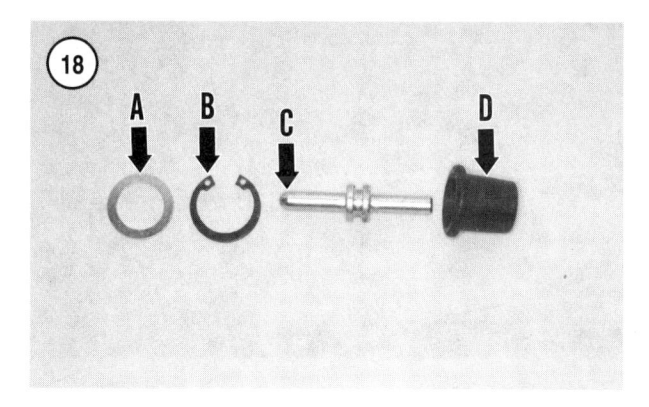

CLUTCH

8. Check for plugged relief and supply ports in the master cylinder. Clean with compressed air.
9. Check the clutch lever assembly for:
   a. Damaged clutch lever.
   b. Excessively worn or damaged pivot bolt.
   c. Damaged adjuster arm assembly.
10. Inspect the diaphragm (8, **Figure 9**) for tears and edge damage. The diaphragm is used to prevent air from entering the reservoir and is folded so that it can move with changes in the brake fluid level. A damaged diaphragm allows moisture to enter the reservoir.
11. Inspect the reservoir cover for damage. Check the small vent notches in the sealing surface for contamination. These must be clear to vent the reservoir to the atmosphere.

Assembly

1. Assemble a new piston assembly as described under *Inspection* in this section.
2. Lubricate the piston assembly and cylinder bore with DOT 4 brake fluid.

> **CAUTION**
> *Do not allow the piston cups to tear or turn inside out when installing them into the master cylinder bore. Both cups are larger than the bore.*

3. Insert the piston assembly—spring end first—into the master cylinder bore (**Figure 15**).
4. Install the spring seat (A, **Figure 18**) over the piston.
5. Compress the piston assembly and install a new snap ring (B, **Figure 18**)—flat side facing out—into the bore groove. See **Figure 14**.

> **CAUTION**
> *The snap ring must seat in the master cylinder groove completely. Push and release the piston a few times to make sure it moves smoothly and that the snap ring does not pop out.*

6. Identify the end of the pushrod with the curved end (C, **Figure 18**). Install the boot over the pushrod (D, **Figure 18**) and seat its outer end into the pushrod groove (**Figure 19**).
7. Lubricate the piston and pushrod ends with silicone brake grease.
8. Install the pushrod/boot assembly (**Figure 12**):
   a. Seat the curved pushrod end into the piston.
   b. Seat the large boot end against the snap ring.
9. Install the clutch lever assembly (**Figure 9**):

   c. Each of these parts can be replaced separately.
6. Check the bore for staining and corrosion. Discard if corroded.

> **NOTE**
> *Most brake fluids are colored to assist in detecting hydraulic leaks. Staining from this coloring should not be confused with corrosion, which can be identified by pits or roughness in the cylinder bore.*

7. Measure the master cylinder bore inside diameter.

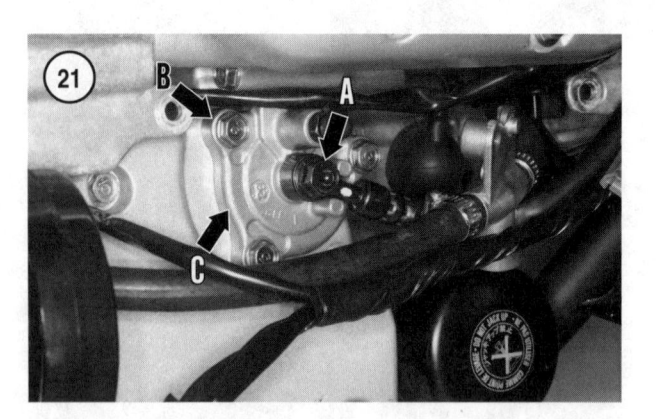

a. Lubricate the pivot bolt shoulder with silicone brake grease. Do not lubricate the bolt threads.
b. Install the bushing (2, **Figure 9**) into the lever.
c. Install the clutch lever by aligning the bushing hole with the pushrod (**Figure 20**).
d. Install and tighten the clutch lever pivot bolt (1, **Figure 9**) to 1 N•m (8.8 in.-lb.). Operate the lever to check for any binding or roughness.
e. Hold the pivot bolt and tighten the clutch lever pivot nut (4, **Figure 9**) to 6 N•m (53 in.-lb.). Operate the lever again.

10. Install the protector into the port channel in the reservoir (**Figure 11**).

NOTE
The protector prevents brake fluid from exiting the reservoir when the clutch lever is being operated with the cover removed (clutch bleeding).

11. Install the clutch switch and screw (A, **Figure 10**).
12. Install the cruise switch and screw (B, **Figure 10**).

CLUTCH SLAVE CYLINDER

The slave cylinder is mounted on the shift linkage cover.

Removal/Installation

1. Remove the EVAP canister and its mounting bracket. Refer to Chapter Eight.
2A. If the clutch hose is not going to be disconnected, squeeze the clutch lever and tie it to the handlebar to prevent the slave cylinder from being forced from the cylinder.
2B. If the clutch hose is going to be disconnected:
   a. Drain the clutch system as described under *Clutch Fluid Draining* in this chapter. 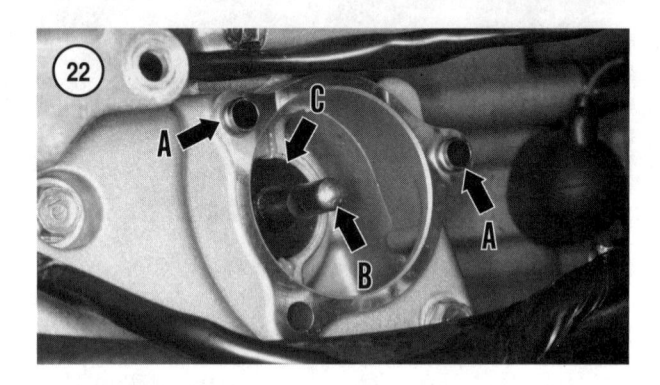
   b. Remove the banjo bolt (A, **Figure 21**) and both washers.
   c. Cover the hose end to prevent leakage.
3. Remove the following:
   a. Mounting bolts (B, **Figure 21**) and slave cylinder (C).
   b. Two dowel pins (A, **Figure 22**).
   c. Clutch pushrod (B, **Figure 22**), if necessary.
4. If the seal (C, **Figure 22**) is leaking, remove it with a screwdriver or seal removal tool.
5. Installation is the reverse of removal. Note the following:
   a. Pack the lips of the new seal with grease and install it with its closed side facing out.
   b. Install the pushrod and lubricate its exposed end with silicone brake grease.
   c. Tighten the slave cylinder mounting bolts (B, **Figure 21**) securely.
   d. Install the hose by aligning it between the two raised tabs on the slave cylinder. Install the banjo bolt and two new washers and tighten to 34 N•m (25 ft.-lb.).
   e. Bleed the clutch as described under *Clutch System Bleeding* in this chapter.
   f. After bleeding the clutch, operate the clutch lever. Make sure it feels firm and offers resistance when

CLUTCH

WARNING
*In the next step, the piston may shoot out of the slave cylinder with considerable force. Keep hands and fingers out of the way while making sure the piston faces down. Apply low air pressure to remove the piston. Do **not** use high air pressure or place the air hose nozzle directly against the hydraulic hose fitting inlet in the slave cylinder body. Hold the air nozzle away from the inlet to allow some of the air to escape during the procedure.*

b. Apply air pressure in short spurts to the banjo bolt hole and remove the piston and piston seal.

Inspection/Reassembly

Replace worn or damaged parts as described in this section.

1. Remove the bleed screw and its cover from the slave cylinder.
2. Remove the spring (A, **Figure 24**).
3. Remove and discard the oil seal (B, **Figure 24**) and piston seal (C).
4. Clean and dry the parts.

NOTE
Most brake fluids are colored to assist in detecting hydraulic leaks. Staining from this coloring should not be confused with corrosion, which can be identified by pits or roughness in the cylinder bore.

5. Check the piston bore for staining and corrosion. Discard if corroded.
6. Check the spring for collapsed, cracked or distorted coils.
7. Inspect the piston for pitting, flat spots or other damage.
8. Lubricate the oil seal and piston seal lips with silicone brake grease.
9. Install the oil seal (B, **Figure 24**) with its groove side facing the piston.
10. Install the piston seal (C, **Figure 24**) with its groove side facing the slave cylinder.
11. Install the small end of the spring (A, **Figure 24**) onto the piston boss.
12. Lubricate the slave cylinder bore and the piston seal with DOT 4 brake fluid.
13. Install the spring and piston into the slave cylinder bore (**Figure 25**). Turn the piston when installing it to avoid damaging the piston seal.

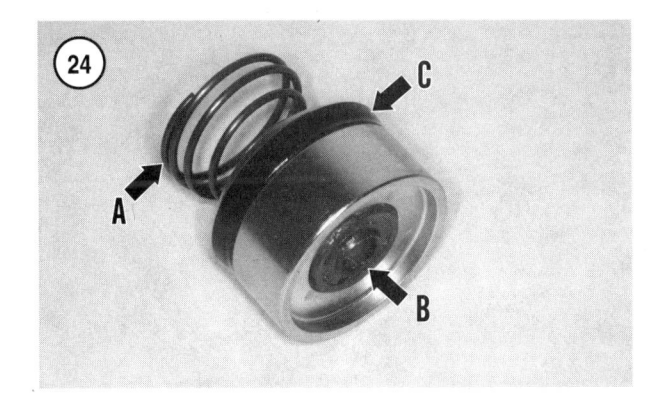

applied. If the clutch lever feels spongy, bleed the clutch system again.

Disassembly

1. Use a rag to turn and remove the piston (**Figure 23**) from the slave cylinder bore. If the piston is tight, perform Step 2.
2. Remove the piston as follows:
   a. Support the slave cylinder on a couple of wooden blocks with the piston facing *down*. Pad the area underneath the piston with shop cloths.

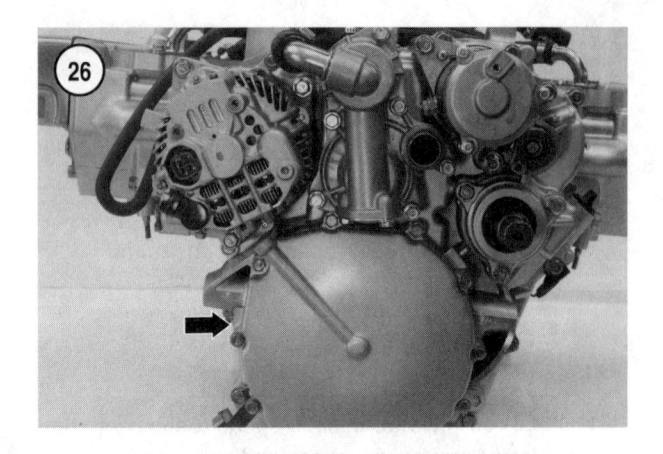

CLUTCH COVER

The clutch cover is mounted at the rear of the engine.

Removal

1. Drain the engine oil. Refer to *Engine Oil and Filter* in Chapter Three.
2. Remove the exhaust system. Refer to *Exhaust System* in Chapter Fourteen.
3. Remove the radiator reserve tank. Refer to *Radiator Reserve Tank* in Chapter Ten.

NOTE
The following photographs are shown with the engine removed to better illustrate the steps.

4. Remove the fuel tank drain tube from the clamp.
5. Remove the bolts, clamp and clutch cover (**Figure 26**).
6. Remove the following:
   a. Gasket.
   b. Two dowel pins (**Figure 27**).
   c. Pipe and O-ring (**Figure 28**).
   d. Scavenge oil strainer (**Figure 29**).
7. Clean the engine and cover gasket surfaces. Flush the cover oil passage with compressed air.
8. Replace the oil seal installed in the cover.

Installation

1. Clean and install the scavenge oil strainer (**Figure 29**).
2. Lubricate a new O-ring with oil and install it (**Figure 28**) and the pipe.
3. Install the two dowel pins (**Figure 27**) and a new gasket.
4. Lubricate the oil seal lip with grease.

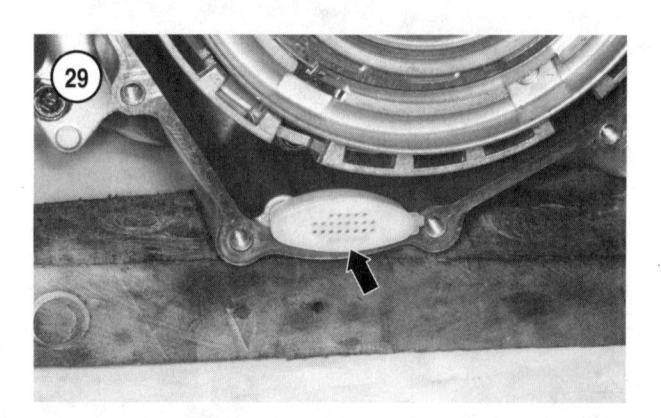

CLUTCH

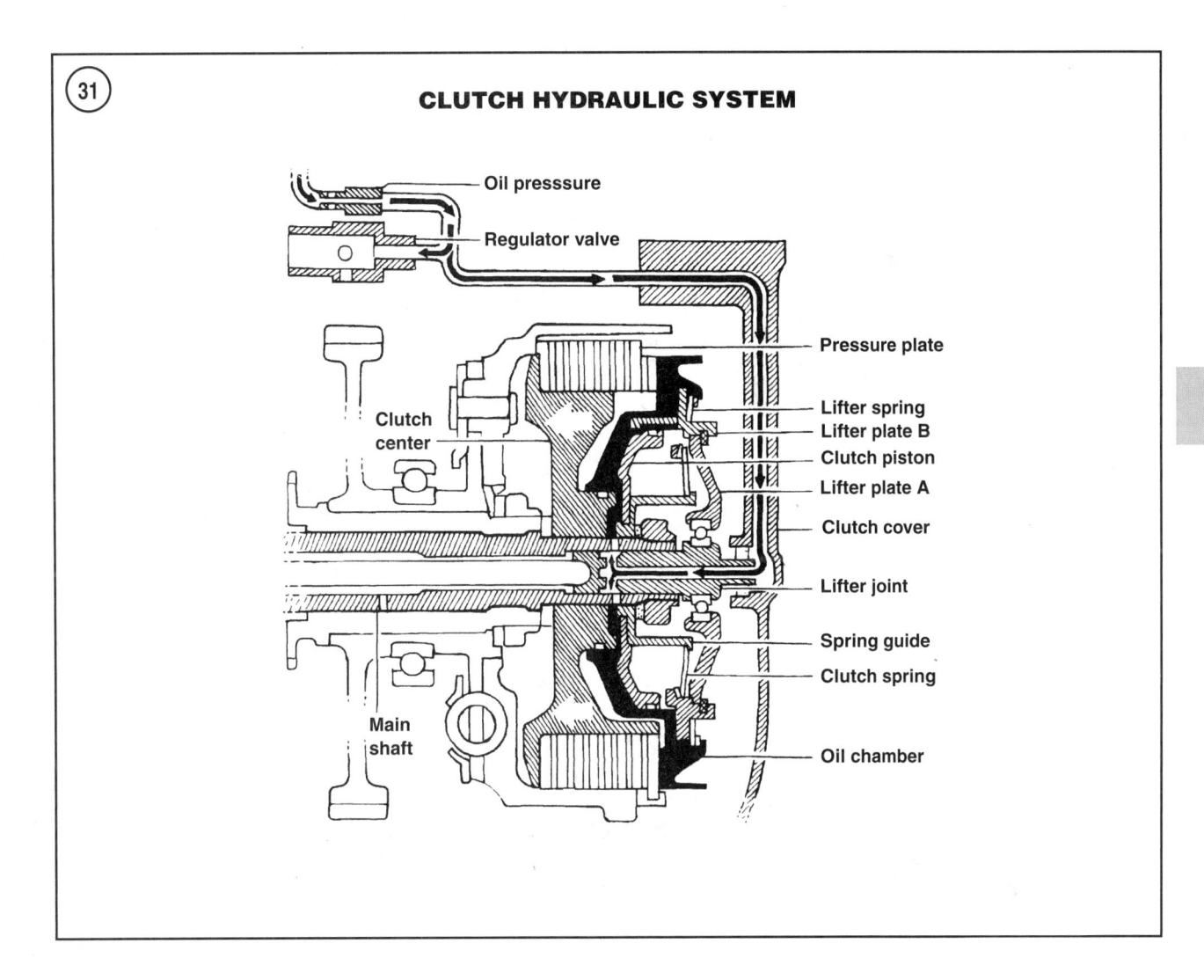

CLUTCH HYDRAULIC SYSTEM

Figure 31: Labeled components include Oil pressure, Regulator valve, Clutch center, Main shaft, Pressure plate, Lifter spring, Lifter plate B, Clutch piston, Lifter plate A, Clutch cover, Lifter joint, Spring guide, Clutch spring, Oil chamber.

5. Install the cover by aligning its seal with the pushrod and dowel pins.
6. Install the mounting bolts and clamp. Tighten the bolts securely in a crisscross pattern.
7. Install the radiator reserve tank. Refer to *Coolant Reserve Tank* in Chapter Ten.
8. Install the exhaust system. Refer to *Exhaust System* in Chapter Fourteen.
9. Refill the engine with the correct type and quantity oil. Refer to *Engine Oil and Filter* in Chapter Three.

Cleaning and Oil Seal Replacement

1. Remove the snap ring (**Figure 30**).
2. Remove the seal with a screwdriver.
3. Clean and dry the cover. Flush the oil passage with compressed air. Make sure there is no solvent remaining in the oil passage.
4. Install the new seal with its flat side facing out.
5. Install a new snap ring with the flat side facing out.
6. Lubricate the seal lip with grease.

CLUTCH

A hydraulically assisted clutch (**Figure 31**) engages and disengages the transmission during shifting. The clutch is a wet, multiplate type with one conically dished clutch spring, eight friction discs and seven clutch plates. The discs and plates are stacked alternately in the clutch outer.

The GL1800 clutch uses both spring pressure and oil pressure to apply force against the pressure plate. This clutch design prevents the need for a stiffer clutch spring and results in a much lighter clutch lever pull for the rider.

When the clutch is engaged, the clutch spring, installed inside lifter plate B, applies force to the pressure plate via lifter plate B. Engine oil, pumped under pressure from the

CLUTCH

32

1. Snap ring
2. Stopper ring
3. Lifter plate A
4. Bearing
5. Snap ring
6. Lifter joint
7. Clutch center locknut
8. Lockwasher
9. Spring guide
10. Stopper ring
11. Lifter spring
12. Clutch spring
13. Spring seat
14. Lifter plate B
15. O-ring
16. Clutch piston
17. Pressure plate
18. Pushrod
19. Friction disc (non-green)
20. Clutch plate
21. Friction disc (green)
22. Friction disc (non-green), larger ID
23. Judder spring
24. Spring seat
25. O-ring
26. Clutch center
27. Wire clip
28. Splined washer
29. Clutch outer locknut
30. Spring washer
31. Clutch outer

CLUTCH

lifter plate B. The outward movement of lifter plate A relieves pressure against the pressure plate, which also allows lifter plate B to move outward. Oil outlet ports in the pressure plate, previously sealed by lifter plate B, are now open and allow oil pressure in the chamber to bleed out. Therefore, as both spring pressure and oil pressure are released from the clutch pack, the clutch plates release and are free to rotate independently. To prevent oil from leaking out of the chamber during clutch disengagement, the lifter spring lightly seats lifter plate B against the pressure plate.

Tools

The following tools are required for clutch service:
1. Shaft holder (Honda part No. 07924-PJ40001). Use this tool to hold the output shaft if the engine is removed from the frame when servicing the clutch center locknut (7, **Figure 32**).
2. Clutch outer holder (Honda part No. 07JBM-MN50100). Use this tool when servicing the clutch outer locknut (29, **Figure 32**).
3. 46 mm locknut wrench (Honda part No. 07JMA-MN50100). Use this tool when servicing the inner locknut (29, **Figure 32**).

Clutch Removal

1. Refer to **Figure 32** and note the following before removing the clutch:
    a. Clutch parts 1-28 can be serviced with the engine installed in the frame.
    b. The clutch outer assembly (29-31) can only be serviced with the engine removed from the frame.
    c. If the clutch outer must be removed, remove the engine from the frame first, then service the clutch assembly. Refer to *Engine Removal* in Chapter Five.
    d. Two new locknuts (7 and 29) are required during installation.
2. Remove the clutch cover as described in this chapter.
3. Remove the stopper ring (A, **Figure 33**) from the groove in lifter plate B.
4. Remove clutch lifter plate A (B, **Figure 33**).
5. Remove the clutch center locknut (**Figure 34**):
    a. Use a chisel or grinder to unstake the locknut.
    b. If the engine is installed in the frame, shift the transmission into any gear and have an assistant apply the rear brake.
    c. If the engine is removed from the frame, install the shaft holder onto the output shaft (**Figure 35**).
    d. Remove the locknut and lockwasher.

oil pump and controlled by a regulator, flows through passages in the crankcase and clutch cover. As the oil passes from the clutch cover through the lifter joint, it enters an oil chamber created by the clutch piston, pressure plate and clutch center. Oil pressure builds inside the chamber and increases the amount of force applied against the clutch spring. Therefore, both spring pressure and oil pressure hold the friction discs and clutch plates together, forcing them to rotate so they do not slip.

When the clutch is disengaged, the lifter joint is forced outward by the pushrod. The outer end of the lifter joint pushes against lifter plate A, which is assembled inside

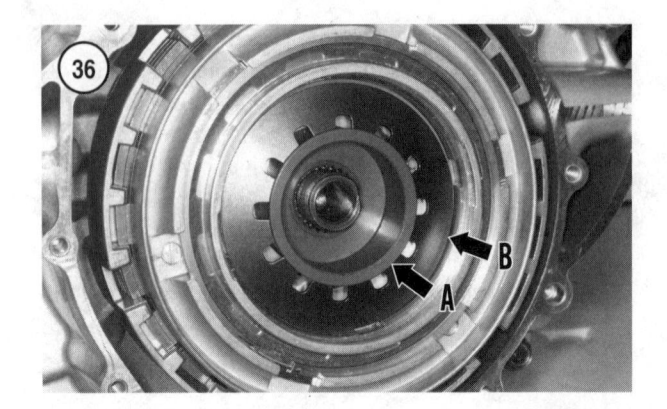

6. Remove the following:
   a. Spring guide (A, **Figure 36**).
   d. Clutch spring (B, **Figure 36**).
   e. Spring seat (13, **Figure 32**).
7. Remove the clutch center assembly (**Figure 37**) from the clutch outer. This includes parts 10, 11 and 14-26 in **Figure 32**.
8. To disassemble the clutch center assembly, refer to *Clutch Center Disassembly* in this section.
9. Remove the splined washer (A, **Figure 38**).
10. Inspect the clutch outer splines (B, **Figure 38**) for grooves, notches or other damage caused by the friction discs. Damage to the slots can prevent the friction discs from separating when the clutch is released, causing clutch drag.
11. To remove the clutch outer:
   a. Remove the engine from the frame. Refer to *Engine Removal* in Chapter Five.
   b. Use a chisel or grinder to unstake the clutch outer locknut (**Figure 39**).
   c. Install the clutch outer holder and secure it with a breaker bar (**Figure 40**).
   d. Remove the locknut with the 46 mm locknut wrench (**Figure 41**).
   e. Remove the spring washer (**Figure 42**) and the clutch outer (**Figure 43**). 
12. Clean and inspect the parts as described under *Inspection* in this section.

Clutch Installation

1. Install the clutch outer:
   a. Clean the primary driven gear boss threads of all oil and threadlock residue. These threads must be clean and dry.

CLUTCH

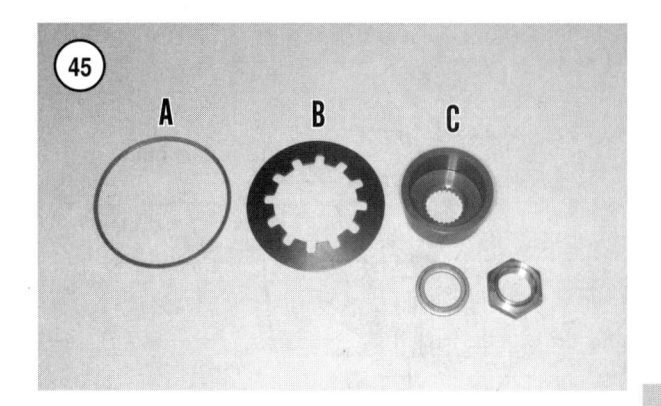

b. Apply a molybdenum oil solution to the to the friction spring on the backside of the clutch outer.

NOTE
Molybdenum oil solution is a 50:50 mixture of engine oil and molybdenum disulfide grease.

c. Install the clutch outer (**Figure 43**).
d. Install the spring washer (**Figure 42**) with its OUTSIDE mark facing out.
e. Apply a medium strength threadlock onto the clutch outer locknut threads and install it.
f. Install the clutch outer holder (**Figure 40**) and tighten the clutch outer locknut to 186 N•m (137 ft.-lb.) using the 46 mm locknut wrench (**Figure 41**).
g. Stake the locknut into the driven gear boss groove (**Figure 39**).

2. Install the splined washer (A, **Figure 38**).
3. Assemble the clutch center assembly as described under *Clutch Center Assembly* in this section.
4. Install the clutch center (**Figure 37**) by aligning the tabs on the *outer* friction disc with the clutch outer slots as shown in **Figure 44**.
5. Install the spring seat (A, **Figure 45**) with its concaved side facing out and seat it against the back of lifter plate B. See **Figure 46**.
6. Install the clutch spring (B, **Figure 45**) with its concaved side facing in and seat it against the spring seat. See **Figure 46** and B, **Figure 36**.
7. Install the spring guide (C, **Figure 45**) and seat it against the clutch spring (A, **Figure 36**).
8. Install the lockwasher (8, **Figure 32**) and seat it against the spring guide.
9. Install and tighten the new clutch center locknut:
   a. Apply engine oil to the new clutch center locknut threads and its seating surface, and install it against the lockwasher.

46

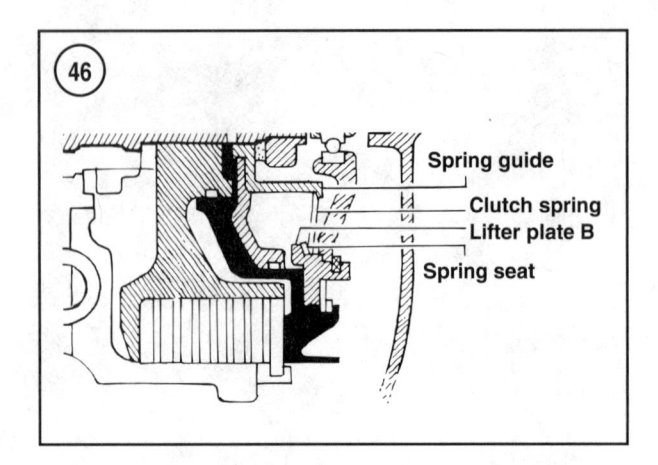

Spring guide
Clutch spring
Lifter plate B
Spring seat

47

48

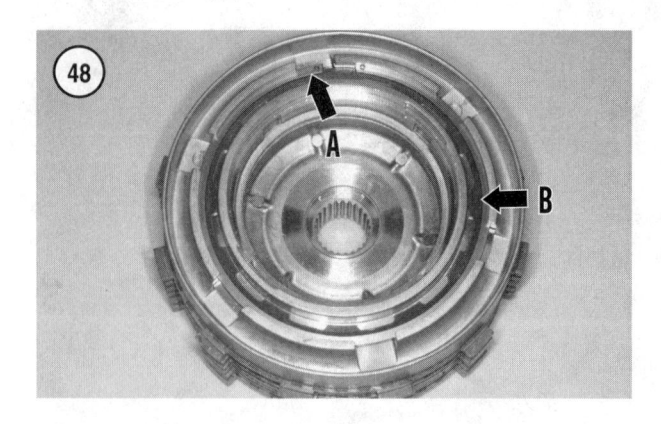

49

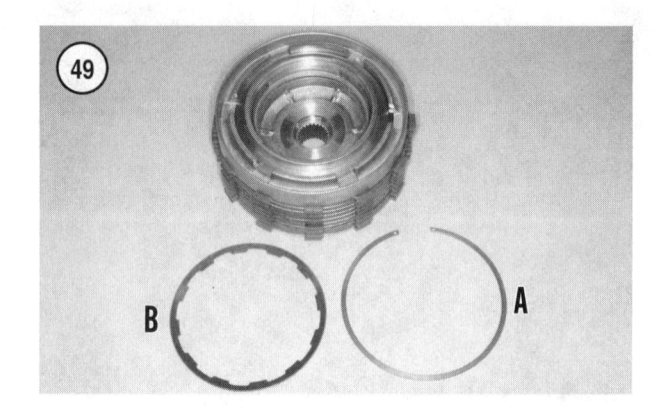

50

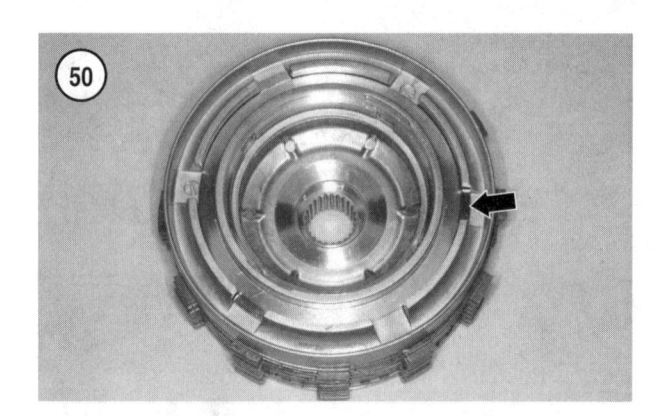

b. If the engine is in the frame, shift the transmission into any gear and have an assistant apply the rear brake.
c. If the engine is removed from the frame, install the shaft holder onto the output shaft (**Figure 35**).
d. Tighten the clutch center locknut to 127 N•m (94 ft.-lb.).
e. Stake the locknut into the mainshaft groove (**Figure 34**).

10. Apply engine oil onto the lifter joint sliding surface (**Figure 47**).

11. Install lifter plate A (B, **Figure 33**) and secure it with the stopper ring (A, **Figure 33**). Make sure the stopper ring seats in the groove completely.

12. Install the clutch cover as described in this chapter.

13. Install the engine into the frame, if removed. Refer to *Engine Installation* in Chapter Five.

Clutch Center Disassembly

Refer to **Figure 32**.
1. Remove the clutch center assembly as described under *Clutch Removal* in this section.

2. Remove the stopper ring (A, **Figure 48**) with a screwdriver. See A, **Figure 49**.

3. Remove the lifter spring (B, **Figure 48**). See B, **Figure 49**.

4. Remove lifter plate B (**Figure 50**).

5. Remove the clutch piston (**Figure 51**) and its O-ring from the pressure plate.

NOTE
If the clutch piston is stuck, remove the pressure plate first, then the clutch piston.

CLUTCH

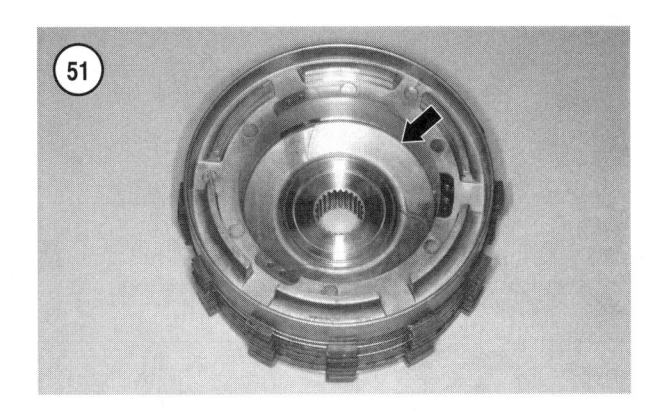

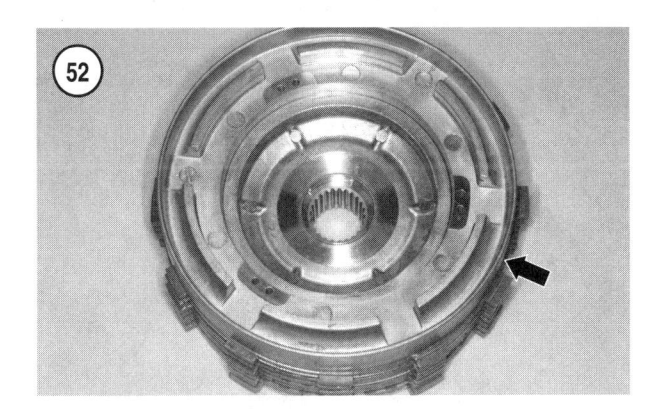

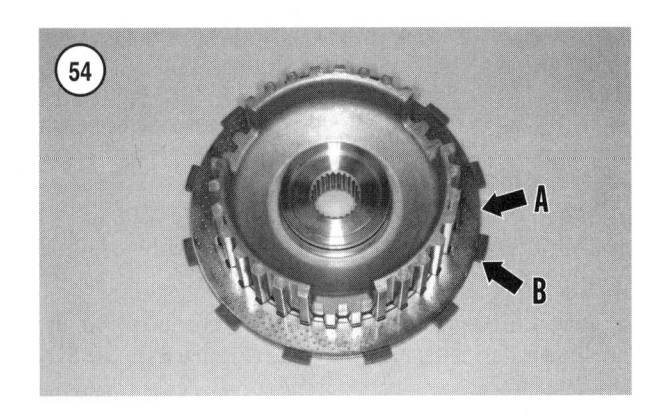

6. Remove the pressure plate (**Figure 52**) from the clutch center.
7. Identify the clutch plates as follows:
    a. Three different types of friction discs are used. Refer to **Figure 32** to identify the discs and their alignment.
    b. The last clutch plate and friction disc (22, **Figure 32**) are secured to the clutch center with the wire clip (27, **Figure 32**).
    c. All the clutch plates (20, **Figure 32**) are identical.
    d. Compare the friction discs and clutch plates with **Figure 32** during removal to identify them.
8. Remove the seven friction discs (19 and 21, **Figure 32**) and the six clutch plates (20, **Figure 32**) from the clutch center. See **Figure 53**.
9. Remove the judder spring set, clutch plate and friction disc as follows (**Figure 54**):
    a. The wire clip secures the assembly to the clutch center and must be removed first.
    b. Compress the clutch plate and push the wire clip ends (**Figure 55**) through the clutch center. When both ends are free, start at one end and remove the wire clip from the grooves in the clutch center. Compress the clutch plate as required to free the wire clip.
    c. Remove the clutch plate (A, **Figure 54**), friction disc (B), judder spring and spring seat (**Figure 56**).
10. Clean and inspect the parts as described under *Clutch Inspection* in this section.

Clutch Center Assembly

Refer to **Figure 32**.

1. Lubricate the friction discs and clutch plates with engine oil. Soak new friction discs in engine oil.

2. Lubricate a new O-ring with oil and install it into the clutch center groove (**Figure 57**).

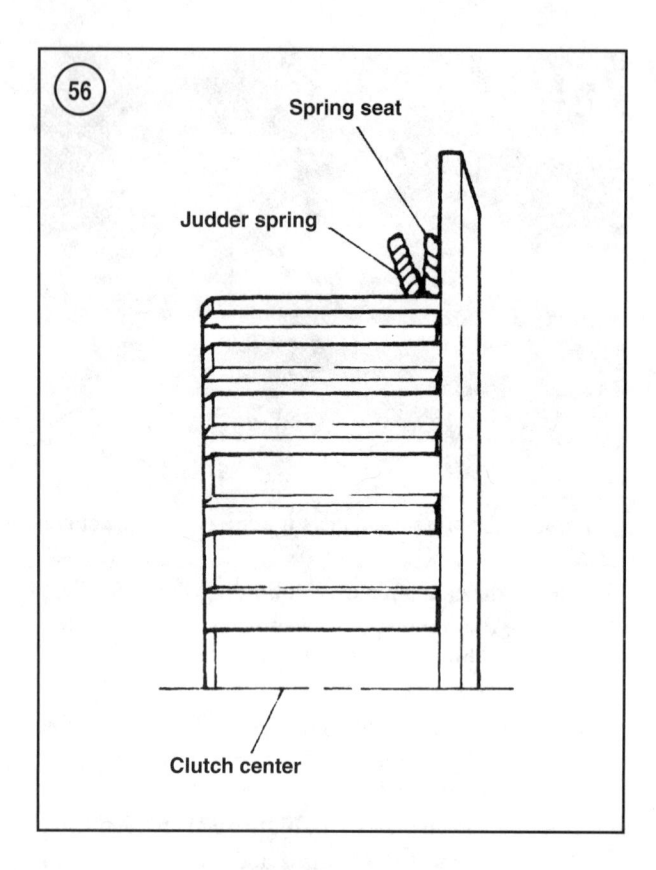

3. Install the friction disc (22, **Figure 32**) clutch plate and judder spring set:
   a. Install the spring seat and judder spring with its concaved side facing out (**Figure 56**).
   b. Install the friction disc (B, **Figure 54**).
   c. Install the clutch plate (A, **Figure 54**).
   d. Install a new wire clip (27, **Figure 32**) by inserting one end of the wire into the hole in the clutch center (**Figure 58**). Then compress the clutch plate and feed the wire clip into the grooves in the clutch center. When the wire clip is seated firmly in all the clutch center grooves, install its remaining end into the hole (**Figure 55**).

4. Install the clutch plates as follows:
   a. Refer to Step 7 under *Clutch Center Disassembly* to identify the friction discs.
   b. Alternately install the friction discs and clutch plates in the order shown in **Figure 32**.

5. Install the pressure plate by aligning its three ramps with the grooves in the clutch center (**Figure 59**). See **Figure 52**.

6. Lubricate a new O-ring with engine oil and install it into the clutch piston groove (**Figure 60**).

7. Install the clutch piston into the pressure plate until it seats fully (**Figure 51**). Apply pressure evenly to ensure proper seating. Do not pinch the O-ring.

8. Install lifter plate B by aligning its three tabs with the three oil holes in the pressure plate (**Figure 61**). See **Figure 50**.

CLUTCH 207

Clutch Inspection

When measuring the clutch components, compare the actual measurements to the specifications in **Table 1**. Replace worn or damaged parts as described in this section. When troubleshooting a dragging or slipping clutch, inspect the clutch parts before cleaning them. Refer to *Clutch* in Chapter Two to isolate certain clutch problems with causes. Always determine the cause of the failure before installing new parts. Then clean and dry all parts as required.

Lifter joint, lifter bearing and lifter plate A

1. Inspect the lifter joint (**Figure 47**) and replace it if either shaft end is grooved or damaged.

2. Turn the lifter joint and make sure it turns smoothly and without any excessive play. Then pull the lifter joint to check the bearing fit in its bore. If the bearing is loose, the bore is probably damaged.

3. Inspect lifter plate A for cracks and other damage.

4. If necessary, replace the lifter joint and/or lifter bearing as follows:

   a. Remove the outer snap ring (A, **Figure 63**) and the lifter joint (B).
   b. To replace the bearing, remove the inner snap ring and press out the bearing. Inspect the bearing bore for damage. Press in the new bearing with its manufacturer's marks facing out. Make sure the bearing is a tight fit in its bore. If not, replace lifter plate A.
   c. Install the inner snap ring fully into the groove.
   d. Install the lifter joint (B, **Figure 63**) through the bearing and secure it with the snap ring (A).

5. Clean the lifter joint bore with compressed air.

6. Lubricate the lifter joint and bearing.

9. Install the lifter spring (B, **Figure 49**) so its concaved side faces out. See B, **Figure 48**.

NOTE
Make sure the stopper ring groove is free of debris.

10. Install the stopper ring (A, **Figure 49**) into the ring groove in the pressure plate. See A, **Figure 48**. Make sure the ring seats fully in the groove (**Figure 62**).

11. Install the clutch center assembly as described under *Clutch Installation* in this section.

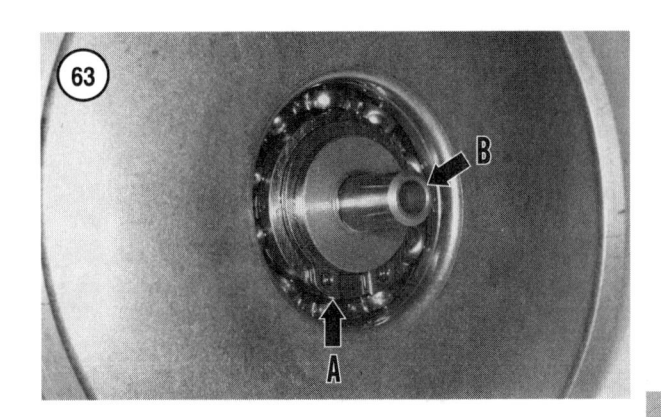

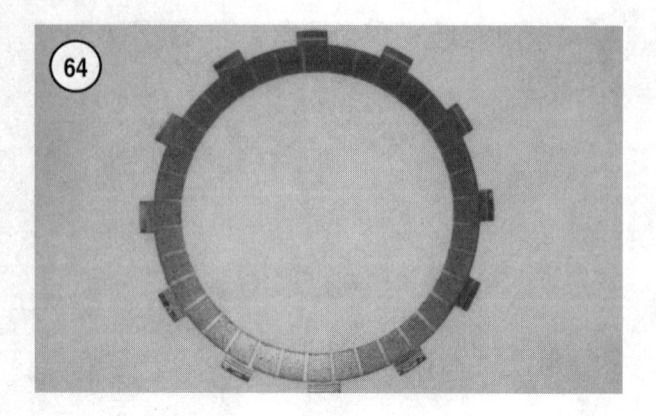

Clutch center

1. Inspect the friction discs (**Figure 64**) as follows:

   NOTE
   If any friction disc is damaged or out of specification, replace all the friction discs as a set.

   a. Inspect the friction material for excessive or uneven wear, cracks and other damage. Check the disc tangs for surface damage. The sides of the disc tangs where they contact the clutch outer slots must be smooth; otherwise, the discs cannot engage and disengage correctly.

   NOTE
   If the disc tangs are damaged, inspect the clutch outer slots carefully as described later in this section.

   b. Measure the thickness of each friction disc with a vernier caliper (**Figure 65**). Measure at several places around the disc.

2. Inspect the clutch plates (**Figure 66**) as follows:

   a. Inspect the clutch plates for damage and color change. Overheated clutch plates will have a blue discoloration.

   b. Check the clutch plates for an oil glaze buildup. Remove by lightly sanding both sides of each plate with 400 grit sandpaper placed on a surface plate or piece of glass.

   c. Place each clutch plate on a surface plate or piece of glass and check for warp with a feeler gauge (**Figure 67**). If any are warped beyond the service limit, replace the entire set.

   d. The clutch plate inner teeth mesh with the clutch center splines. Check the clutch plate teeth for any roughness or damage. The teeth contact surfaces

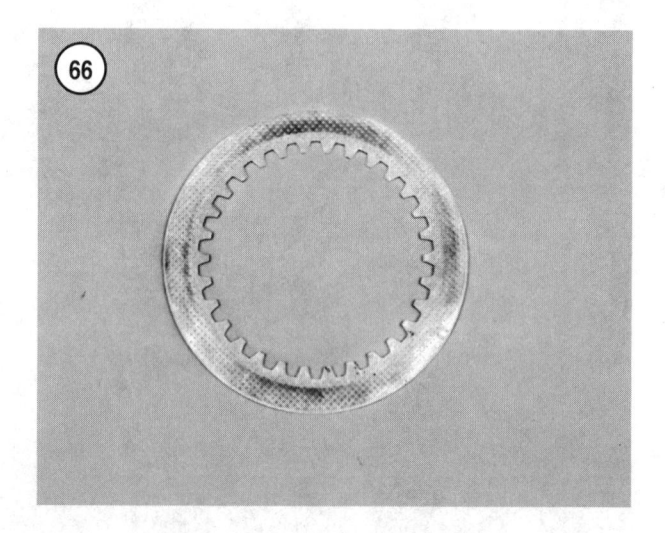

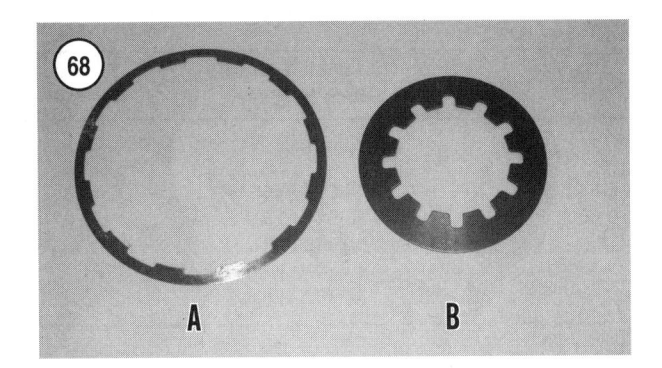

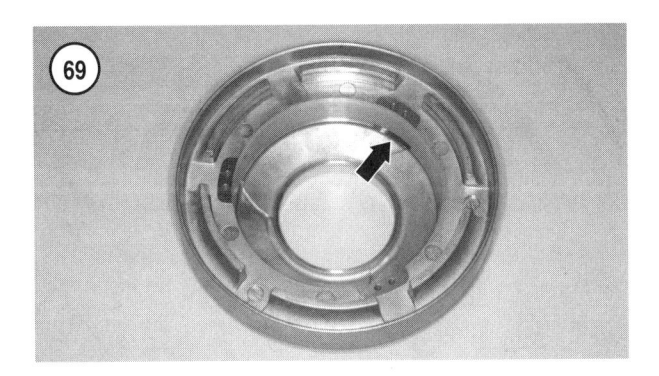

CLUTCH

3. Inspect the clutch center:
    a. The clutch plate teeth slide in the clutch center splines. Inspect the splines for rough spots, grooves or other damage. Repair minor damage with a file or oil stone. If the damage is excessive, replace the clutch center.
    b. Inspect the wire clip groove and splines for damage.
    c. Replace the O-ring.
4. Check the lifter spring (A, **Figure 68**) and clutch spring (B) for cracks, distortion or other damage. Measure the height of each spring and replace if less than the service limit. A weak clutch spring (B, **Figure 68**) causes the clutch to slip.
5. Inspect the pressure plate splines and ramps for damage. Clean the oil passages (**Figure 69**) with solvent and compressed air.
6. Inspect lifter plate B for damage. Clean the oil passages with solvent and compressed air.
7. Inspect the clutch piston (**Figure 60**) for cracks and other damage. Replace the O-ring.
8. Inspect the spring seat (24, **Figure 32**) and judder spring (23) for distortion and damage.

Clutch outer

1. The friction disc tangs slide in the clutch outer grooves. Inspect the grooves for cracks or galling. Repair minor damage with a file. If the damage is excessive, replace the clutch outer.
2. Check the splines for damage.

must be smooth; otherwise, the plates cannot engage and disengage correctly.

NOTE
If the clutch plate teeth are damaged, inspect the clutch center splines carefully.

Table 1 and Table 2 are on the following page.

Table 1 CLUTCH SPECIFICATIONS

	New mm (in.)	Service limit mm (in.)
Clutch lifter spring free height	2.9 (0.11)	2.5 (0.10)
Clutch master cylinder		
Cylinder inside diameter	14.000-14.043 (0.5512-0.5529)	14.055 (0.5533)
Piston outside diameter	13.957-13.984 (0.5495-0.5506)	13.945 (0.5490)
Clutch plate warp	–	0.30 (0.012)
Clutch spring free height	4.8 (0.19)	4.6 (0.18)
Friction disc thickness	3.72-3.88 (0.146-0.153)	3.5 (0.14)

Table 2 CLUTCH TORQUE SPECIFICATIONS

	N•m	in.-lb.	ft.-lb.
Clutch center locknut[1]	127	–	94
Clutch cruise switch screw	1	8.8	–
Clutch hose banjo bolt	34	–	25
Clutch lever pivot bolt	1	8.8	–
Clutch lever pivot nut	6	53	–
Clutch master cylinder holder bolts	12	106	–
Clutch outer locknut[2]	186	–	137
Clutch switch screw	1	8.8	–
Slave cylinder bleed valve	2	18	–

1. Lubricate locknut threads and seating surface with engine oil. Stake nut.
2. Apply a medium strength threadlock onto fastener threads.

CHAPTER SEVEN

SHIFT MECHANISM AND TRANSMISSION

This chapter covers service for the external shift mechanism, transmission and internal shift mechanism. The external shift mechanism can be serviced with the engine assembled and installed in the frame. Transmission and internal shift mechanism service requires crankcase disassembly (Chapter Five).

Transmission specifications are listed in. **Tables 1-5** at the end of the chapter.

EXTERNAL SHIFT LINKAGE

The external shift linkage assembly consists of the stopper arm, shift drum cam and parts of the gearshift arm assembly. This assembly is mounted at the front of the engine, behind the slave cylinder, and can be serviced with the engine in the frame.

Troubleshooting

If the bike is experiencing shifting problems, refer to *Gearshift Linkage* in Chapter Two before removing the shift mechanism in this section.

Shift Linkage Cover
Removal

The shift linkage cover is mounted on the front of the engine, behind the slave cylinder. The gear position switch is mounted inside the shift linkage cover.

1. Support the motorcycle on its centerstand and shift the transmission into neutral.
2. Drain the engine oil. Refer to *Engine Oil and Filter* in Chapter Three.
3. Remove the EVAP canister and its mounting bracket. Refer to Chapter Eight.
4. Remove the slave cylinder without disconnecting the clutch hose:
   a. Place a block between the clutch lever and handlebar to prevent the slave cylinder from being forced from the cylinder.
   b. Remove the mounting bolts (A, **Figure 1**) and slave cylinder (B).
   c. Remove the two dowel pins (A, **Figure 2**).
   d. Remove the clutch pushrod (B, **Figure 2**), if necessary.

NOTE
The following photographs are shown with the engine removed to better illustrate the procedure.

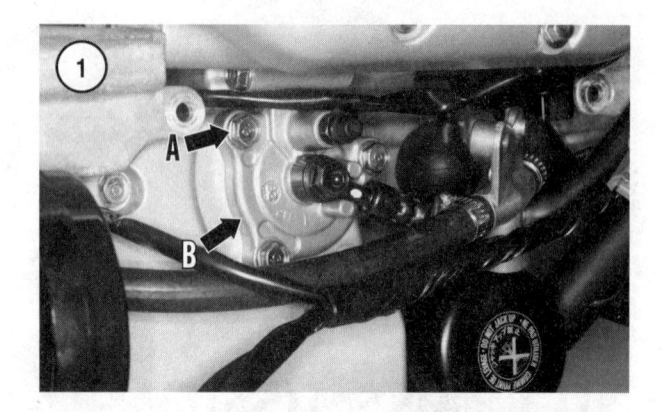

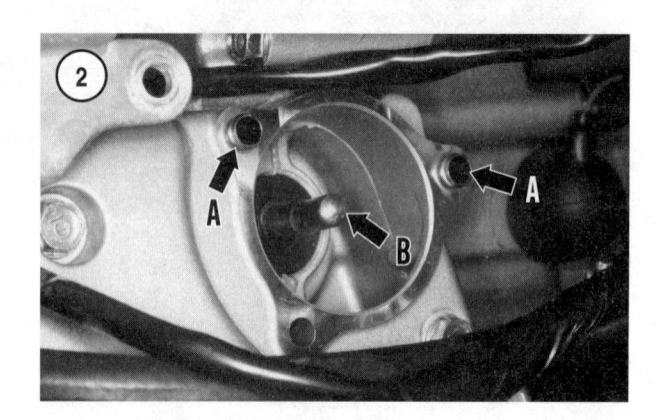

5. Remove the bolts, gearshift linkage cover (**Figure 3**), dowel pins (A, **Figure 4**) and gasket.

NOTE
The gear position switch black 6-pin connector is mounted next to the throttle body. The air filter housing must be removed to disconnect this connector. If it is necessary to service the shift linkage cover, unbolt and remove the switch (A, Figure 5) from inside the cover.

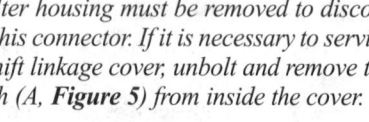

Shift Linkage Cover
Installation

1. Remove all gasket residue from the shift linkage cover and crankcase mating surfaces.
2. Apply Yamabond No. 4 (or equivalent semi-drying sealer) to the three areas where the crankcase halves mate together.
3. Install the two dowel pins (A, **Figure 4**) and a new gasket.
4. If removed, install the gear position switch (A, **Figure 5**) and tighten its mounting bolts.
5. Install the wire grommet into the cover groove (B, **Figure 5**).

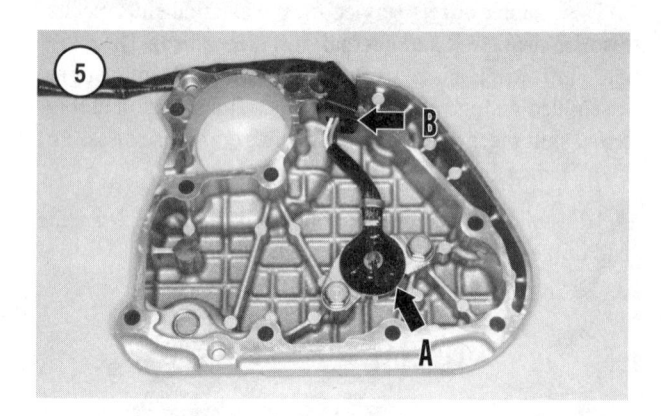

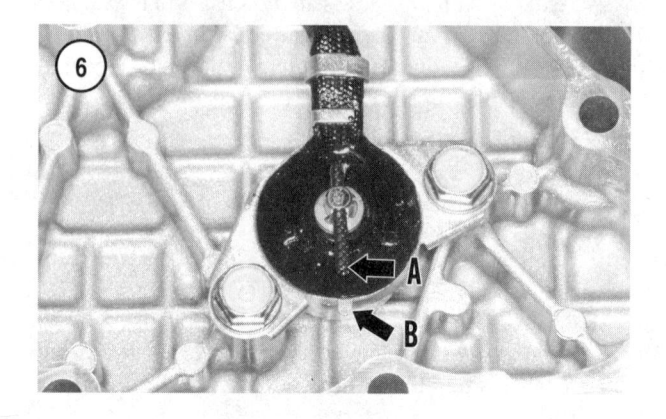

SHIFT MECHANISM AND TRANSMISSION

EXTERNAL SHIFT LINKAGE

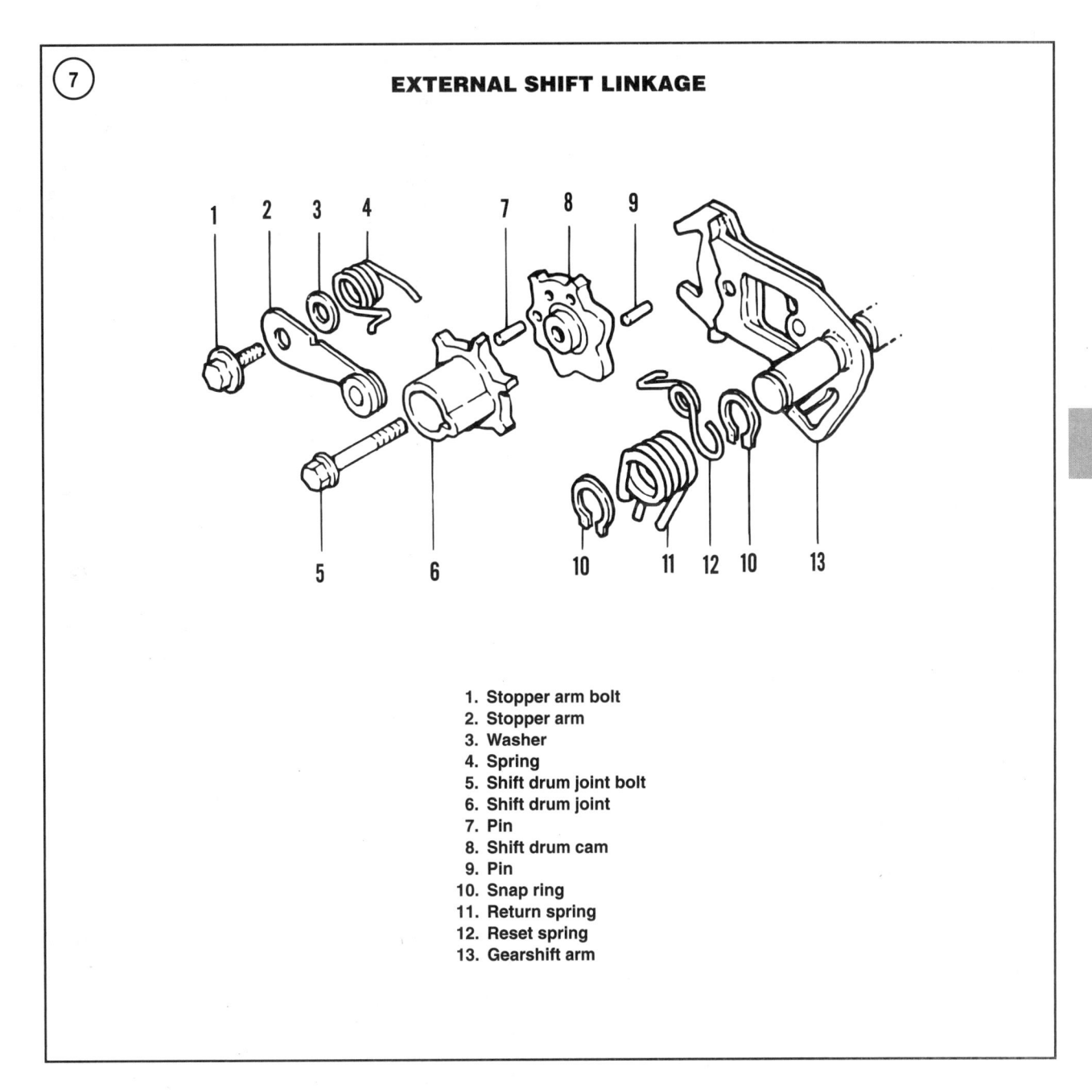

1. Stopper arm bolt
2. Stopper arm
3. Washer
4. Spring
5. Shift drum joint bolt
6. Shift drum joint
7. Pin
8. Shift drum cam
9. Pin
10. Snap ring
11. Return spring
12. Reset spring
13. Gearshift arm

6. Shift the transmission into neutral. This positions the pin groove (B, **Figure 4**) in the shift drum joint face down.

7. Align the long end of the switch pin (A, **Figure 6**) with the tab on the switch (B).

8. Install the cover. Make sure the long end of the switch pin (A, **Figure 6**) fits into the pin groove in the shift drum joint (B, **Figure 4**).

9. Install and tighten the shift linkage cover mounting bolts.

10. Reverse Steps 1-4 of *Shift Linkage Cover Removal*.

11. Check the cover for oil leaks.

12. Check clutch operation.

Shift Linkage Removal

Refer to **Figure 7**.

1. Remove the shift linkage cover as described in this section.

2. Remove the stopper arm bolt (A, **Figure 8**) and the shift drum joint (B).

3. Remove the four dowel pins (A, **Figure 9**) and the shift drum cam (B).
4. Remove the pin (**Figure 10**) from the shift drum.
5. Remove the bolt (A, **Figure 11**) and the stopper arm assembly (B).

> *NOTE*
> *The engine must be removed and the crankcase separated to remove the gearshift arm (C, Figure 4).*

6. Remove the snap ring (A, **Figure 12**) and the return spring (B), if necessary.

Shift Linkage
Inspection

Replace worn or damaged parts as described in this section.
1. Clean and dry all parts. Remove all threadlock residue from the shift drum joint bolt threads.
2. Inspect the stopper arm assembly (**Figure 13**) for:
   a. Damaged pivot bolt shoulder. Remove any burrs or rough spots from the bolt shoulder.
   b. Damaged stopper arm. Check the stopper arm for a damaged pivot hole or a stuck or binding roller.
   c. Bent or damaged washer. A damaged washer can prevent the stopper arm from moving correctly.
   d. Weak or damaged spring.
3. Check the shift drum joint (A, **Figure 14**) for:
   a. Damaged pin holes.
   b. Damaged outer pin groove.
4. Check the shift drum cam (B, **Figure 14**) for:
   a. Damaged pin holes.
   b. Worn or damaged cam ramps.
5. Inspect the return spring for cracks and other damage.

Shift Linkage
Installation

Refer to **Figure 7**.
1. If removed, install the return spring. Spread the spring arms around the spindle arm and locating pin (B, **Figure 12**). Secure the spring with a *new* snap ring (A, **Figure 12**).
2. Install the stopper arm assembly (**Figure 13**) as follows:
   a. Assemble the stopper arm as shown in **Figure 13**. Install the washer (3, **Figure 7**) on the backside of the stopper arm.
   b. Install the stopper arm assembly (B, **Figure 11**) and tighten the bolt (A, **Figure 11**) to 12 N•m (106 in.-lb.).

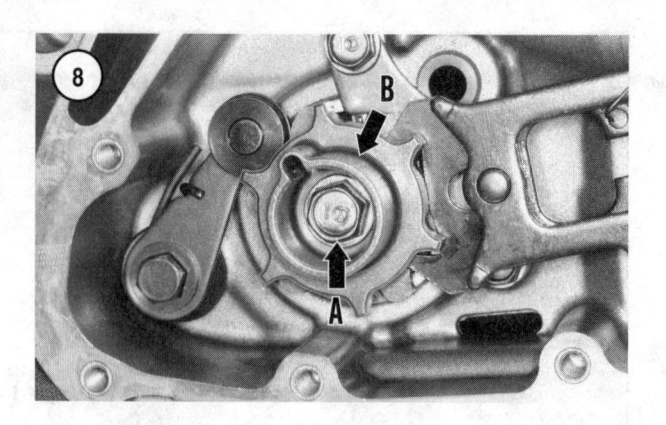

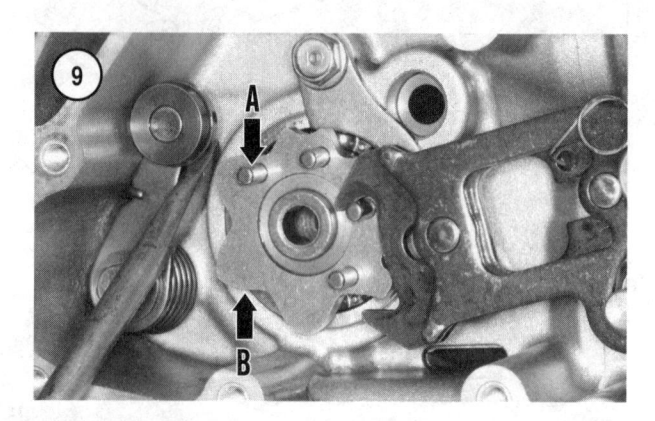

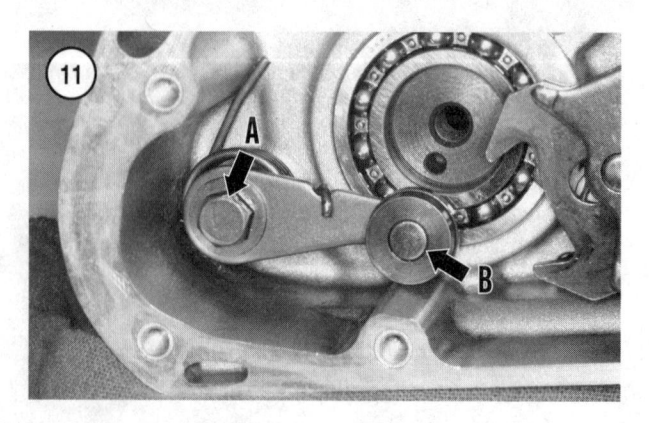

SHIFT MECHANISM AND TRANSMISSION

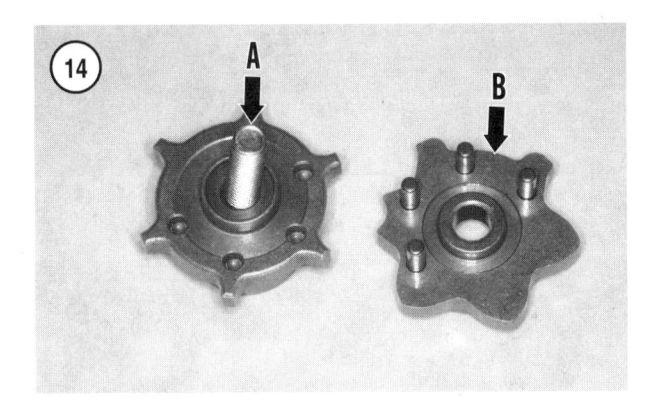

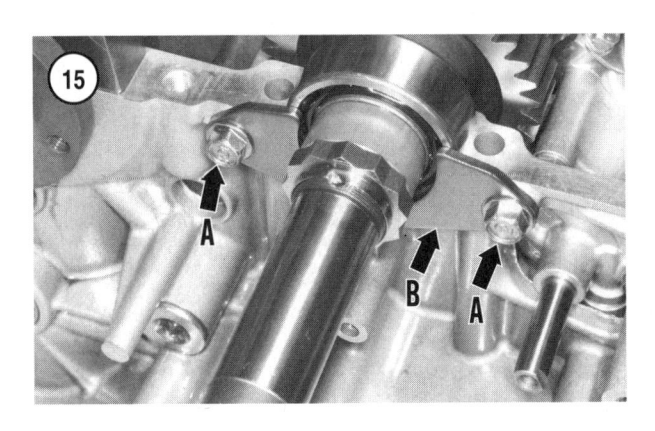

c. Pivot the stopper arm up with a screwdriver (**Figure 10**). If the stopper arm does not move, it is not centered correctly on its bolt shoulder. Loosen the bolt and reposition the stopper arm. Retighten the stopper arm bolt (substep b) and recheck the stopper arm's movement.

3. Install the pin (**Figure 10**) into the shift drum.

4. Align the groove in the shift drum cam with the pin (**Figure 10**) and install the shift drum cam (B, **Figure 9**).

5. Install the four pins (A, **Figure 9**) into the shift drum cam.

6. Install the shift drum joint (B, **Figure 8**) by aligning its pin holes with the pins.

7. Apply a medium strength threadlock onto the shift drum joint bolt (A, **Figure 8**) and tighten to 27 N•m (20 ft.-lb.).

8. Shift the transmission into neutral.

9. Install the shift linkage cover as described in this section.

TRANSMISSION ASSEMBLY

The mainshaft and gearshift spindle assembly are mounted in the left crankcase. The countershaft and shift fork/shift drum assembly are mounted in the right crankcase. After separating the crankcase assembly, remove the transmission assembly in the following order:

1. Mainshaft.
2. Gearshift spindle.
3. Shift forks and shift drum.
4. Countershaft.

MAINSHAFT

The mainshaft is installed in the left crankcase.

Removal

1. Separate the crankcase. Refer to *Crankcase* in Chapter Five.

2. Remove the bolts (A, **Figure 15**) and the mainshaft bearing plate (B).

3. Remove the clutch pushrod (A, **Figure 16**) and seal (B).

4. Remove the mainshaft (**Figure 17**) from the left crankcase.

5. Remove the dowel pin (**Figure 18**).

6. Service the mainshaft as described in this section.

Installation

1. Install the pistons and crankshaft before installing the mainshaft. Refer to *Pistons and Connecting Rods* and *Crankshaft* in Chapter Five.
2. Install the pin (**Figure 18**).
3. Install the mainshaft by aligning the hole in the outer race (**Figure 19**) with the pin (**Figure 18**). Make sure the outer race seats flush in the crankcase.
4. Install a new seal (B, **Figure 16**) with its closed side facing out. Install the pushrod (A, **Figure 16**) through the seal and into the mainshaft.
5. Install the mainshaft bearing plate as follows:
   a. Apply a medium strength threadlock onto the bearing plate bolt threads.
   b. Install the bearing plate (B, **Figure 15**) with its OUTSIDE mark facing out and tighten the mounting bolts (A) to 26 N•m (19 ft.-lb.).
6. Assemble the crankcase. Refer to *Crankcase* in Chapter Five.

Disassembly

Refer to **Figure 20**.
1. Clean the assembled mainshaft in solvent and dry with compressed air.
2. Disassemble the mainshaft in the following order:
   a. Outer race and needle bearing.
   b. Thrust washer.
   c. First gear.
   d. Lockwasher and spline washer.
   e. Fourth gear and bushing.
   f. Thrust washer.
   g. Second/third combination gear.
   h. Snap ring and spline washer.
   i. Fifth gear and bushing.

NOTE
Do not remove the bearing (A, Figure 21) except to replace it.

Inspection

When measuring the mainshaft components, compare the measurements to the specifications in **Table 2**. Replace worn or damaged parts as described in this section.

Maintain the alignment of the mainshaft components when cleaning and inspecting the parts in this section.
1. Clean and dry the mainshaft and bearing assembly (**Figure 21**). Flush the oil control holes.
2. Inspect the mainshaft (**Figure 21**) for:
   a. Worn or damages splines.

SHIFT MECHANISM AND TRANSMISSION

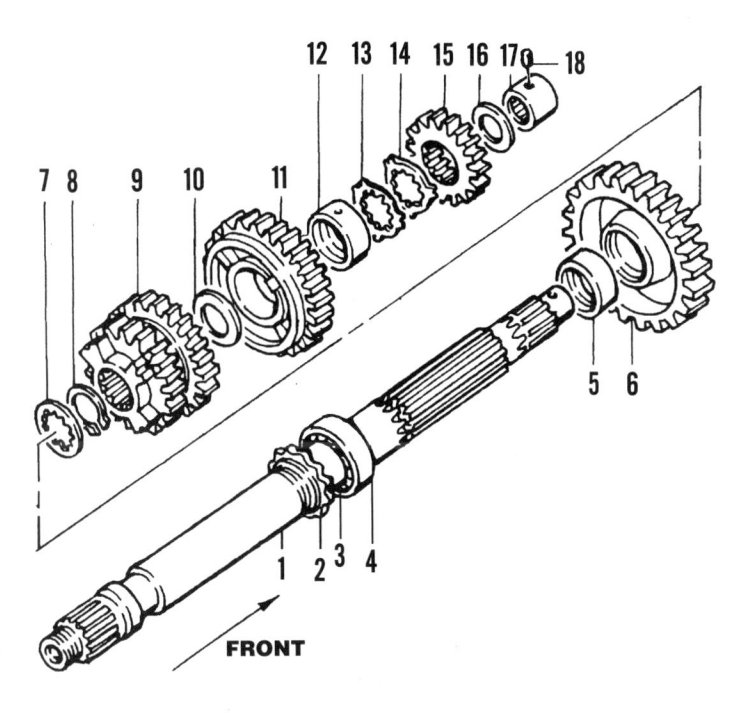

⑳ MAINSHAFT

1. Mainshaft
2. Locknut
3. Collar
4. Bearing
5. Fifth gear bushing
6. Fifth gear
7. Spline washer
8. Snap ring
9. Second/third combination gear
10. Thrust washer
11. Fourth gear
12. Fourth gear bushing
13. Spline washer
14. Lockwasher
15. First gear
16. Thrust washer
17. Needle bearing
18. Pin

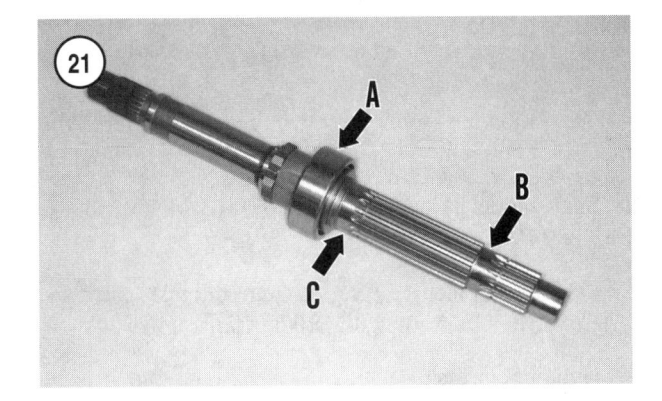

b. Excessively worn or damaged bearing surfaces.

c. Cracked or rounded-off snap ring grooves.

3. Hold the mainshaft and turn the bearing outer race (A, **Figure 21**) by hand. If the bearing is loose, turns roughly or is damaged, replace it as described under *Bearing Replacement* in this section.

NOTE
If the mainshaft bearing is a loose fit on the shaft, check the bearing's mounting position on the shaft carefully for any cracks, exces-

22

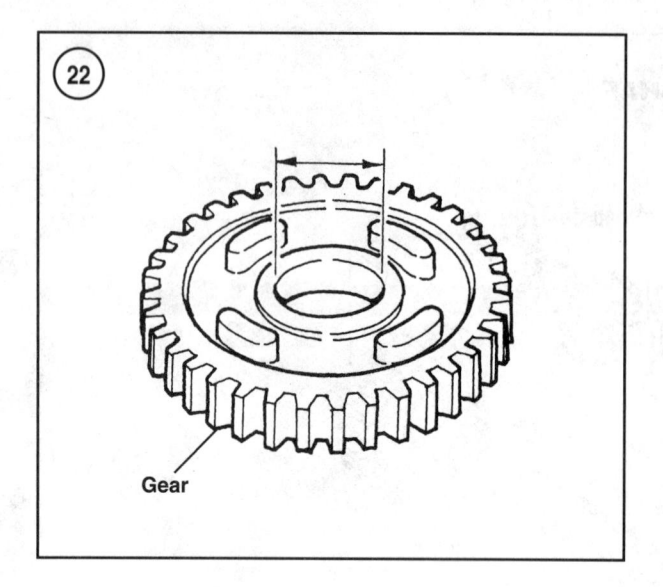

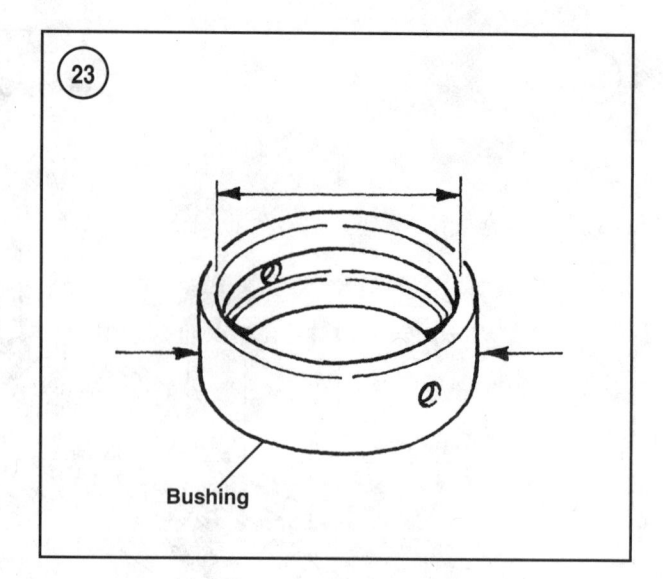

Gear

23

Bushing

sive wear or other damage. The mainshaft may require replacement at the same time.

4. Measure the mainshaft outside diameter at its fourth (B, **Figure 21**) and fifth (C) gear operating positions.

5. Check each mainshaft gear for:
   a. Missing, broken or chipped teeth.
   b. Worn, damaged, or rounded-off gear dogs.
   c. Worn or damaged splines.
   d. Cracked or scored gear bore.

6. Check the mainshaft bushings for:
   a. Excessively worn or damaged bearing surface.
   b. Worn or damaged splines.
   c. Cracked or scored bore.

7. Measure the fourth and fifth gear inside diameters (**Figure 22**).

8. Measure the fourth and fifth gear bushing inside and outside diameters (**Figure 23**).

9. Using the measurements recorded in Step 8 and Step 9, determine the bushing-to-shaft and gear-to-bushing clearances. If necessary, replace worn parts to obtain clearances within the service limit specifications.

10. Make sure each gear slides or turns on the mainshaft without any binding or roughness.

11. Check the washers for burn marks, excessive wear or other damage.
    a. The thrust washers should be smooth and show no signs of wear.
    b. The teeth on the spline washers should be uniform and have a positive fit on the mainshaft.

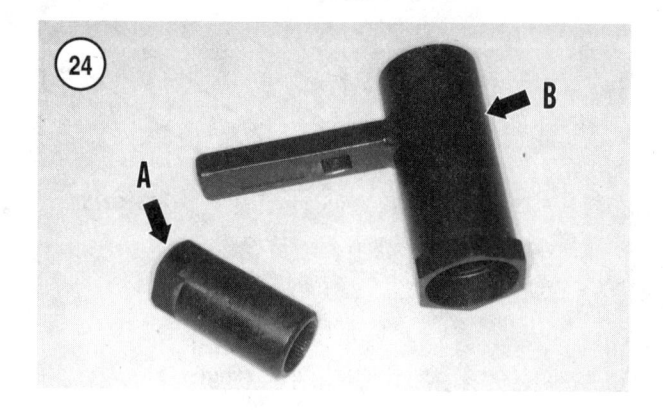

24

A   B

Bearing Replacement

A new mainshaft locknut and bearing are required. Check with a Honda dealership on the availability of both parts before removing the old bearing.

Tools

The following tools are required to replace the mainshaft bearing:

1. Shaft holder (Honda part No. 07JMB-MN50200 [A, **Figure 24**]).

2. Locknut wrench, 30 × 64 mm (Honda part No. 07916-MB00002 or 07916-MB00001[B, **Figure 24**]).

3. Hydraulic press.

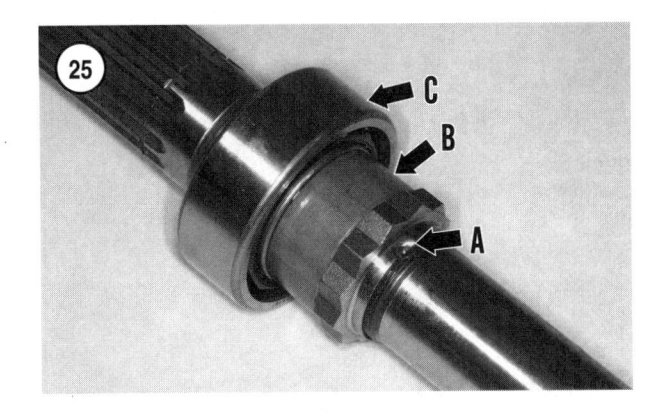

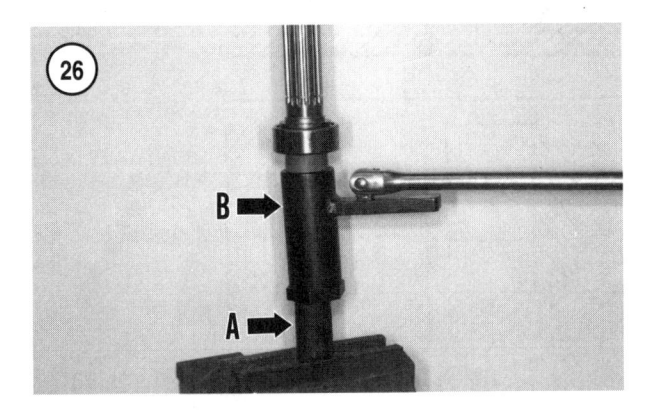

SHIFT MECHANISM AND TRANSMISSION

5. Remove the locknut (A, **Figure 25**) and collar (B). Discard the locknut.

6. Support the bearing in a press. Press the mainshaft out of the bearing. Discard the bearing.

7. Clean and dry the mainshaft. Check the bearing mounting area for cracks and other damage.

8. Support the mainshaft in the press. Press the *new* bearing on the mainshaft until it bottoms. Apply pressure against the bearing's inner race. Install the bearing with its manufacturer's marks facing in the direction recorded in Step 1.

9. Install the collar (B, **Figure 25**).

10. Lubricate the threads of a *new* locknut with engine oil. Thread the locknut onto the mainshaft. Secure the mainshaft with the shaft holder mounted in a vise.

11. Tighten the mainshaft locknut as follows:
    a. Depending on what angle the locknut wrench (B, **Figure 26**) is mounted on the torque wrench, the actual torque applied to the locknut may change. If necessary, refer to *Torque Adapters* in Chapter One to recalculate the torque setting.
    b. Tighten the mainshaft locknut to 186 N•m (137 ft.-lb.).

12. Stake the locknut (**Figure 27**) into the groove in the mainshaft.

13. Hold the mainshaft and turn the outer bearing race by hand. Make sure the bearing was not damaged during installation.

Assembly

Throughout this procedure, the orientation of many parts (**Figure 20**) is made in relationship with the mainshaft bearing (A, **Figure 21**).

1. Before mainshaft assembly, note the following:
    a. Install a *new* snap ring.
    b. Install the snap ring and washers with their chamfered edge facing *away* from the thrust load. See **Figure 28**.
    c. Align the snap ring gap with the transmission shaft groove as shown in **Figure 29**.
    d. Install the snap ring with a pair of snap ring pliers while holding the back of the snap ring with a pair of pliers (**Figure 30**). Then slide the snap ring down the shaft and seat it into its correct groove. Check the snap ring to make sure it seats in its groove completely.

2. Lubricate all sliding surfaces with engine oil.

3. Install the fifth gear bushing and fifth gear (**Figure 31**). The gear dogs must face away from the bearing (**Figure 32**).

Procedure

1. Record the direction of the manufacturer's marks on the bearing face so the bearing can be installed facing in the same direction.

2. Unstake the locknut (A, **Figure 25**) with a grinder.

3. Secure the mainshaft with the shaft holder (A, **Figure 26**) mounted in a vise.

4. Loosen the locknut with the locknut wrench (B, **Figure 26**).

MAINSHAFT ASSEMBLY

Figure 28 — Mainshaft Assembly: First gear, Fourth gear, Second/third combination gear, Fifth gear.

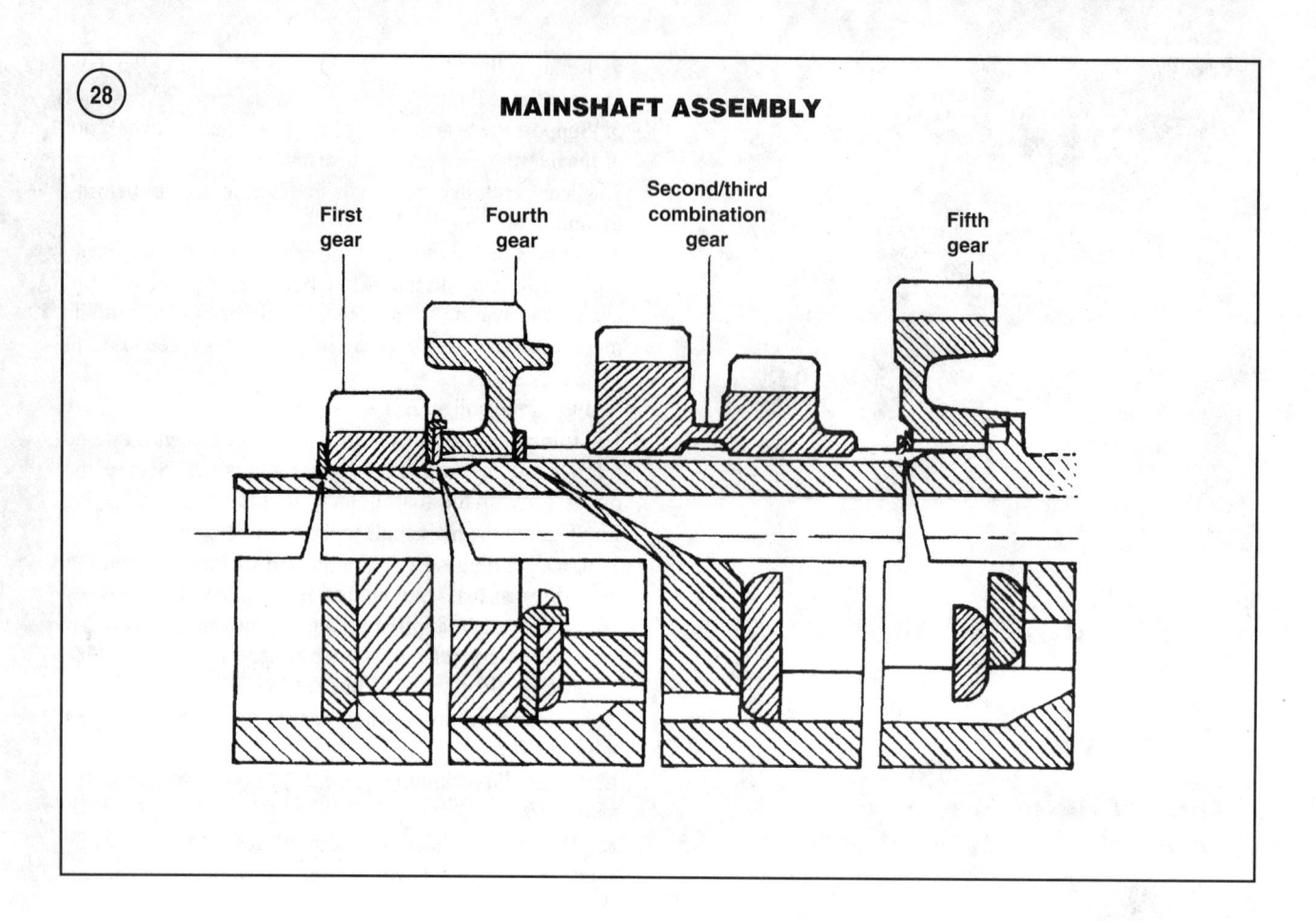

Figure 29 — Snap ring, Shaft groove.

Figure 30

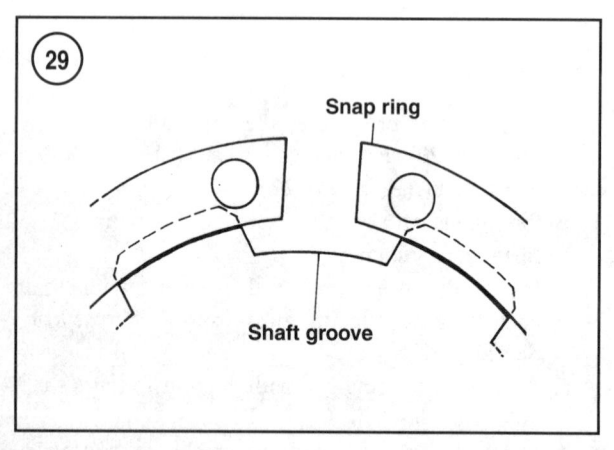

4. Install the spline washer (A, **Figure 33**) and a *new* snap ring (B). The flat side of both parts must face away from the bearing. Seat the snap ring in the groove next to fifth gear (**Figure 34**).

5. Install the second/third combination gear with the smaller gear (**Figure 35**) facing toward the bearing.

6. Install the thrust washer (A, **Figure 36**). The flat side must face toward the bearing.

7. Install the fourth gear bushing (B, **Figure 36**) and fourth gear (C). The shoulder side of the gear must face away from the bearing (**Figure 37**).

8. Install the spline washer and lockwasher set:
   a. Install the spline washer (A, **Figure 38**) with its flat side facing away from the bearing. Then turn the washer so its splines align with the raised splines on

SHIFT MECHANISM AND TRANSMISSION

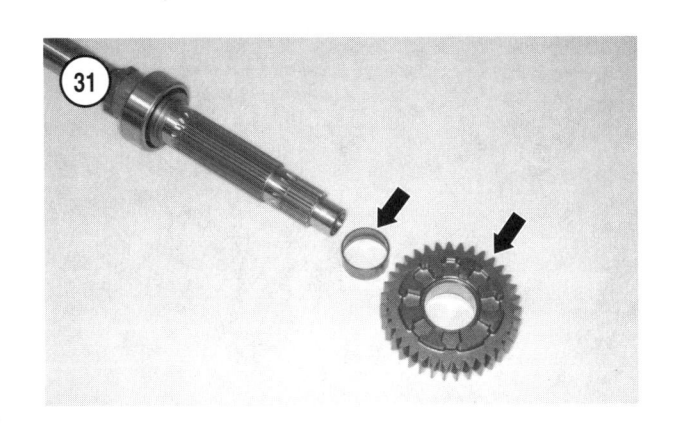

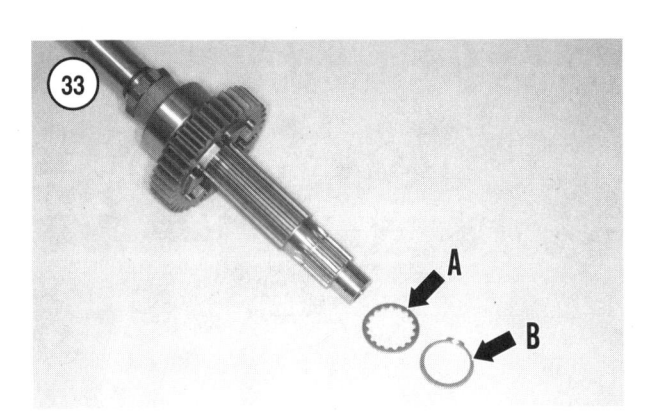

221

the mainshaft (A, **Figure 39**). When this alignment is made, the spline washer cannot be removed from the mainshaft.

b. Install the lockwasher (B, **Figure 38**) with its bent arms (B, **Figure 39**) facing in and insert them into the spline washer notches. See **Figure 40**.

NOTE
If the lockwasher arms do not align, the spline washer (A, Figure 39) is not properly installed. Reinstall the spline washer as described in substep a.

9. Install first gear (**Figure 41**) and seat it against the lockwasher.
10. Check the direction of the gears on the shaft (**Figure 42**).
11. Install the thrust washer (A, **Figure 43**). The flat side of the washer must face away from the bearing.
12. Install the needle bearing (B, **Figure 43**) and its outer race (C).

GEARSHIFT SPINDLE ASSEMBLY

The gearshift spindle assembly (**Figure 44**) is mounted in the left crankcase.

Removal

1. Separate the crankcase. Refer to *Crankcase* in Chapter Five.
2. Straighten the lockwasher tab. Remove the bolt (A, **Figure 45**) and lockwasher.
3. Hold the joint arm (B, **Figure 45**) and remove the gearshift arm (C) and thrust washer (D). Remove the joint arm (B, **Figure 45**).
4. Remove the gearshift spindle (**Figure 46**).
5. Inspect and service the gearshift spindle assembly, needle bearings and seal as described in this section.

SHIFT MECHANISM AND TRANSMISSION

GEARSHIFT SPINDLE ASSEMBLY

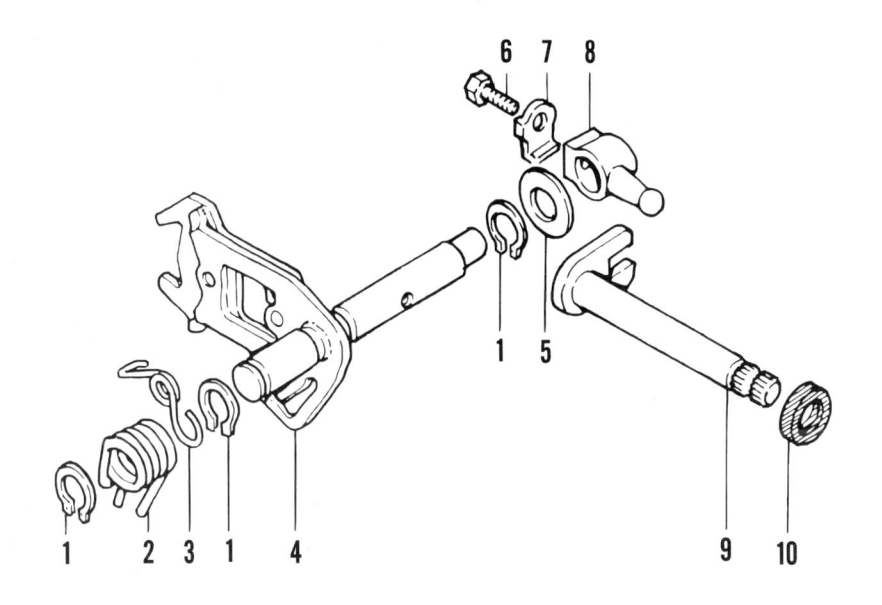

1. Snap ring
2. Return spring
3. Reset spring
4. Gearshift arm
5. Thrust washer
6. Joint arm bolt
7. Lockwasher
8. Joint arm
9. Gearshift spindle
10. Seal

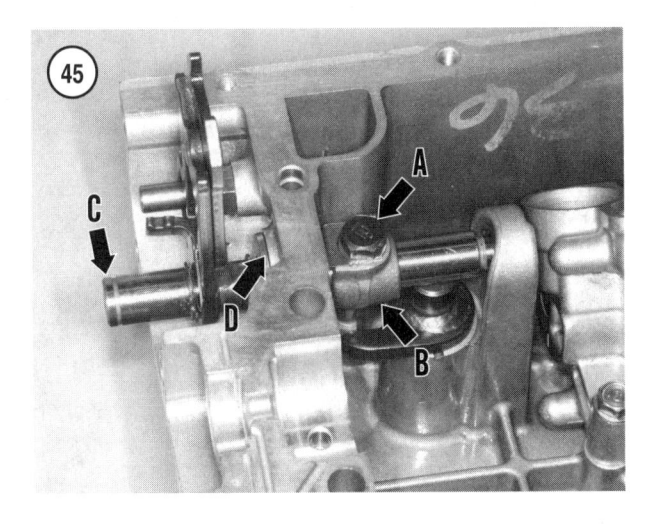

Installation

1. Lubricate the needle bearings (installed in the crankcase) with engine oil.
2. Pack the oil seal lips (**Figure 47**) with grease.
3. Install the gear shift spindle through the needle bearing and oil seal and position it as shown in **Figure 46**. Install the spindle slowly to avoid tearing the seal lips.
4. Lubricate the gearshift arm and washer with engine oil.
5. Engage the joint arm with the gearshift spindle in the direction shown in **Figure 48**.
6. Install the gearshift arm (with washer attached) through the crankcase and joint arm (**Figure 45**).
7. Install a new lockwasher onto the joint arm bolt.
8. Align the joint arm and gearshift arm holes. Install the joint arm bolt and lockwasher (A, **Figure 45**) and tighten to 25 N•m (18 ft.-lb.). Bend the lockwasher tabs against the bolt head.
9. Assemble the crankcase. Refer to *Crankcase* in Chapter Five.

Inspection

1. Clean and dry all parts.
2. Remove the seal (**Figure 47**) with a screwdriver.
3. Inspect the gearshift arm (A, **Figure 49**) and gearshift spindle (B) shafts for bending or other damage.
4. Inspect the joint arm ball (C, **Figure 49**) for wear and damage. Check the engagement area on the gearshift spindle (D, **Figure 49**) for the same conditions.
5. Check the gearshift arm (A, **Figure 49**) for:
   a. Worn or damaged shifting arms.
   b. Damaged snap ring grooves.
   c. Worn or damaged springs.
6. Inspect the needle bearings (**Figure 50**, typical) for damage.

NOTE
Confirm the availability of the gearshift spindle needle bearings with a Honda dealership before removing the original bearings.

7. Using a bearing driver, install a new seal until it is flush with the crankcase outer bore surface (**Figure 47**). Pack the seal lips with grease.

SHIFT FORKS AND SHIFT DRUM

The shift forks and shift drum assembly (**Figure 51**) are installed in the right crankcase.

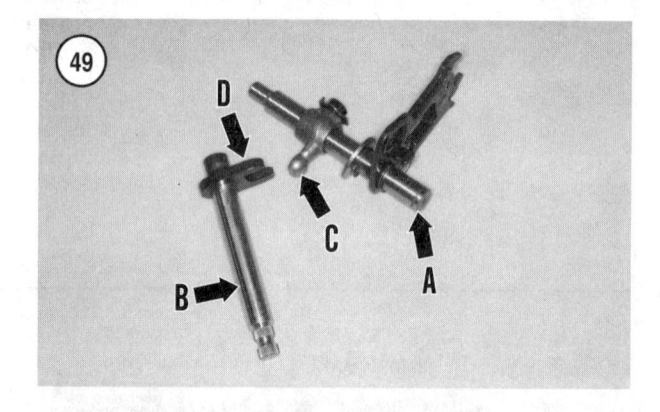

SHIFT MECHANISM AND TRANSMISSION

51 SHIFT FORKS AND SHIFT DRUM

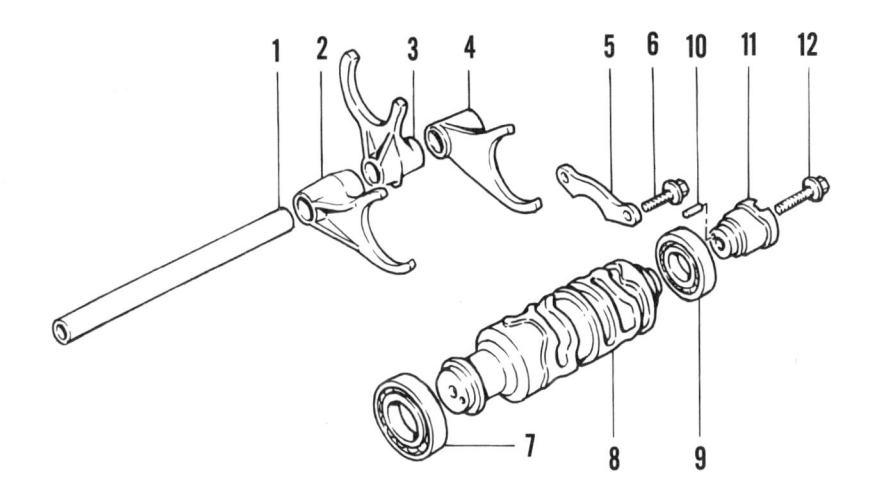

1. Shift fork shaft
2. Front shift fork (F)
3. Center shift fork (C)
4. Rear shift fork (R)
5. Shift drum bearing plate
6. Bolt
7. Bearing
8. Shift drum
9. Bearing
10. Pin
11. Reverse lock cam
12. Bolt

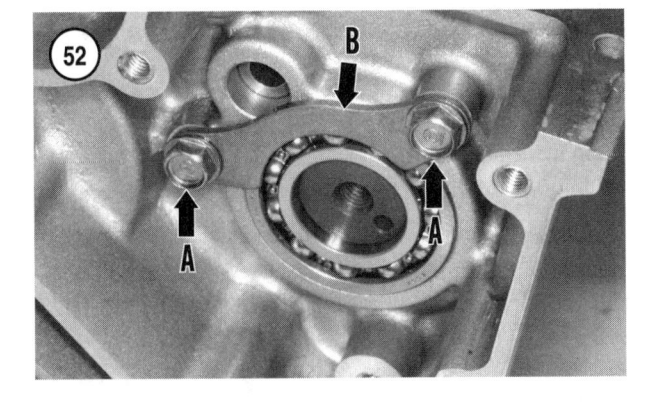

Removal

1. Separate the crankcase. Refer to *Crankcase* in Chapter Five.
2. Remove the bolts (A, **Figure 52**) and the shift drum bearing plate (B).
3. Remove the shift fork shaft (A, **Figure 53**) and the three shift forks (B).
4. Remove the shift drum bearing (**Figure 54**) and shift drum (**Figure 55**).
5. Inspect the shift fork shaft, shift forks and shift drum as described in this section.

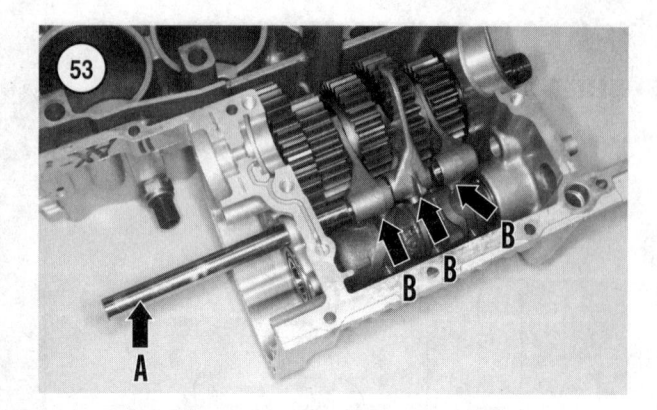

Installation

1. Clean and dry all parts.
2. Lubricate all sliding surfaces with engine oil.
3. Install the shift drum (**Figure 55**) and its bearing (**Figure 54**). Seat the bearing fully in the bearing bore.
4. Identify each shift fork with its letter mark (**Figure 56**):
   a. F: Front shift fork.
   b. C: Center shift fork.
   c. R: Rear shift fork.
5. Install the shift forks:
   a. Install the shift forks with their identification marks (**Figure 56**) facing toward the front side of the engine.
   b. Install the pin on each shift fork into the corresponding shift drum groove.
   c. Install the F and R shift forks into the countershaft gear shifter grooves. The C shift fork operates in the mainshaft second/third combination gear shifter groove. Refer to **Figure 57**.
   d. Install the shift fork shaft (A, **Figure 53**) through each shift fork. Make sure the pin on each shift fork rides in its corresponding shift drum groove.
6. Apply a medium strength threadlock onto the shift drum bearing plate bolts. Install the plate (B, **Figure 52**) and tighten the mounting bolts (A) to 12 N•m (106 in.-lb.).

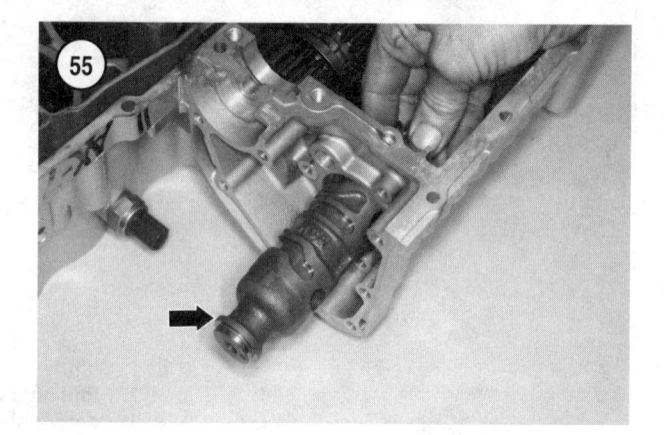

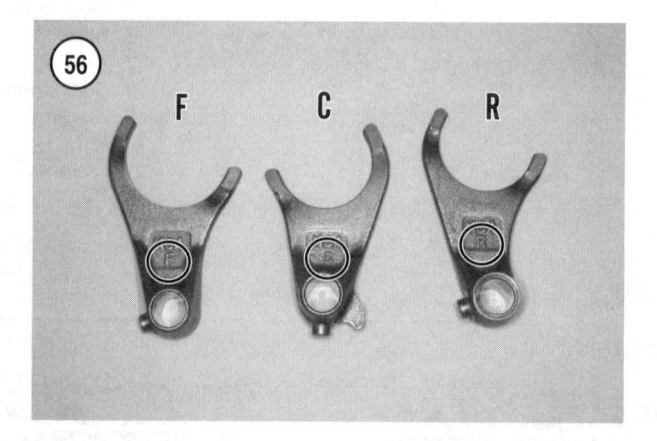

Shift Forks and Shaft Inspection

When measuring the shift forks and shaft, compare the actual measurements to the specifications in **Table 3**. Replace worn or damaged parts as described in this section.

1. Clean and dry the shift forks and shaft.
2. Inspect each shift fork (**Figure 56**) for wear or damage. Examine the shift forks at the points where they contact the slider gear (A, **Figure 58**). These surfaces must be smooth with no signs of wear, bending, cracks, heat discoloration or other damage.

3. Check each shift fork pin (B, **Figure 58**) for cracks, excessive wear or other damage. If there is damage or wear, check the corresponding shift drum groove for damage.

4. Measure the thickness of each shift fork claw (A, **Figure 58**).

SHIFT MECHANISM AND TRANSMISSION

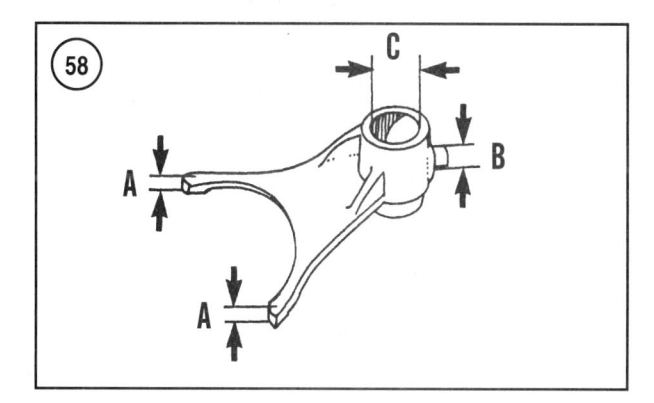

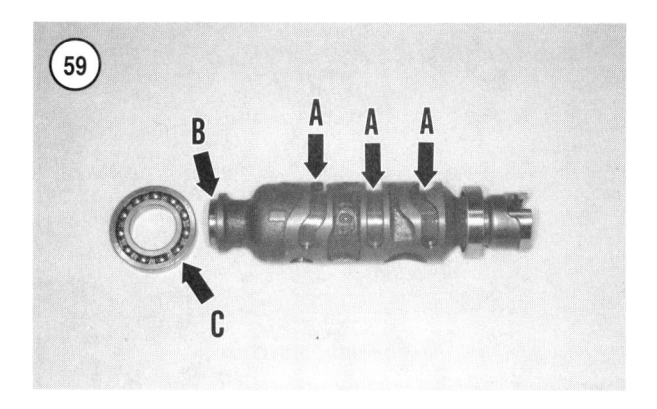

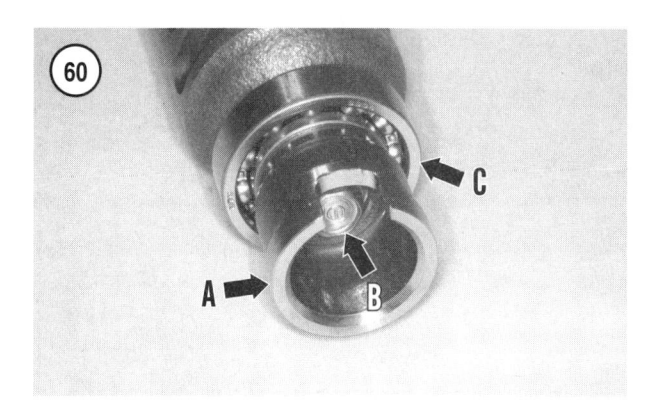

5. Measure the inside diameter of each shift fork (C, **Figure 58**).
6. Check the shift fork shaft for bending or other damage. Roll the shift fork shaft on a surface plate or piece of glass and check for any clicking or other conditions that indicate a bent shaft. Slide each shift fork on the shaft. If there is binding with all three shift forks, check for a bent shaft. If only one shift fork binds, check for a damaged bore.
7. Measure the shift fork shaft outside diameter at each of the three shift fork operating positions.

Shift Drum Inspection

Replace the shift drum, bearings or reverse lock cam if they show damage as described in this section.
1. Clean and dry the shift drum.
2. Check the shift drum for:
   a. Excessively worn or damaged cam grooves (A, **Figure 59**).
   b. Excessively worn or damaged bearing surfaces (B, **Figure 59**).
3. Spin the shift drum bearings (C, **Figure 59**) by hand. If a bearing turns roughly or if there is any catching or other damage, replace it.
4. Inspect the reverse lock cam (A, **Figure 60**) for damage.
5. To replace the reverse lock cam and/or the rear bearing:
   a. Remove the reverse lock cam bolt (B, **Figure 60**), cam (A), bearing (C) and pin.
   b. Install the bearing onto the shift drum shoulder.
   c. Install the pin.
   d. Install the reverse lock cam by aligning its hole with the pin.
   e. Apply a medium strength threadlock onto the reverse lock cam bolt (B, **Figure 60**) and tighten to 12 N•m (106 in.-lb.).

COUNTERSHAFT

The countershaft is installed in the right crankcase.

Removal

The countershaft is partially disassembled during removal.
   Refer to **Figure 61**.
1. Separate the crankcase. Refer to *Crankcase* in Chapter Five.
2. Remove the shift forks and shift drum. Refer to *Shift Forks and Shift Drum* in this chapter.
3. Remove the bolts (A, **Figure 62**) and the countershaft bearing plate (B).

CHAPTER SEVEN

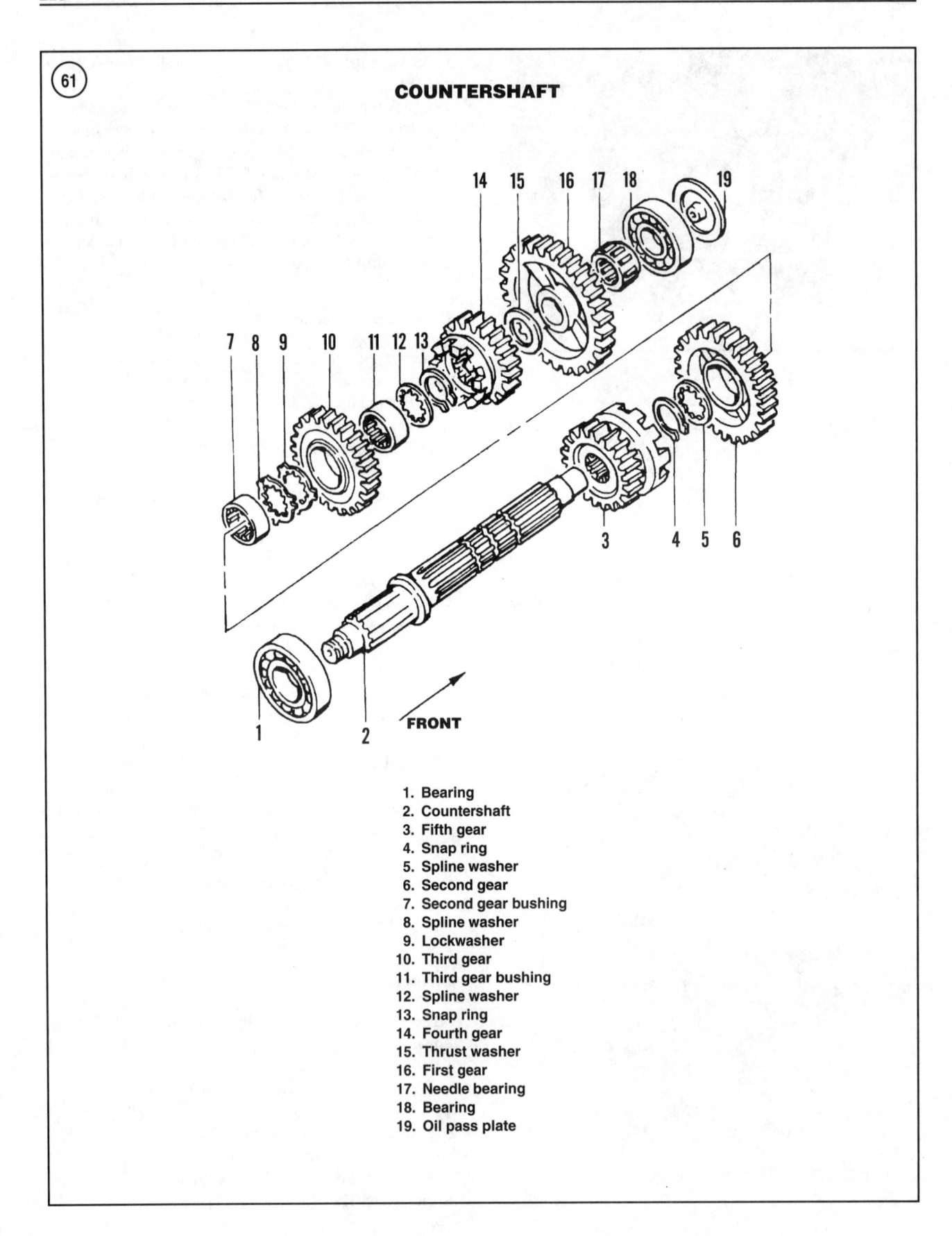

COUNTERSHAFT

1. Bearing
2. Countershaft
3. Fifth gear
4. Snap ring
5. Spline washer
6. Second gear
7. Second gear bushing
8. Spline washer
9. Lockwasher
10. Third gear
11. Third gear bushing
12. Spline washer
13. Snap ring
14. Fourth gear
15. Thrust washer
16. First gear
17. Needle bearing
18. Bearing
19. Oil pass plate

SHIFT MECHANISM AND TRANSMISSION

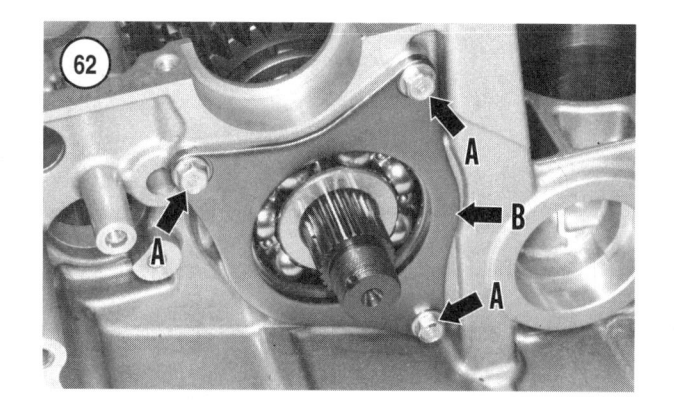

4. Slide the countershaft toward the rear and remove the rear side bearing (**Figure 63**).

5. Remove first gear (**Figure 64**), the needle bearing (**Figure 65**) and thrust washer (A, **Figure 66**). Do not remove fourth gear (B, **Figure 66**).

6. Lift and remove the countershaft from the crankcase (**Figure 67**).

7. Remove the front side bearing (18, **Figure 61**) and the oil pass plate (19) from the crankcase bore.

8. Service the countershaft as described in this section.

Installation

1. Install the oil pass plate (19, **Figure 61**) and the front side bearing (18) into the crankcase bore.

2. Install the countershaft (without first gear) into the crankcase (**Figure 67**).

3. Install the thrust washer (A, **Figure 66**). The flat side must face toward fourth gear.

4. Lubricate the needle bearing with oil and install it (**Figure 65**).

5. Install first gear (**Figure 64**) so its flat side faces away from fourth gear.

6. Install the rear side bearing onto the countershaft and seat it into the crankcase bore (**Figure 63**).

7. Check the direction of the gears on the shaft (**Figure 68**).
8. Install the countershaft bearing plate:
   a. Apply a medium strength threadlock onto the bearing plate bolt threads.
   b. Install the bearing plate (B, **Figure 62**) and tighten the mounting bolts (A) to 12 N•m (106 in.-lb.).
9. Install the shift drum and shift forks. Refer to *Shift Forks and Shift Drum* in this chapter.
10. Assemble the crankcase. Refer to *Crankcase* in Chapter Five.

Disassembly

Refer to **Figure 61**.
1. Clean the assembled countershaft in solvent and dry with compressed air.
2. Disassemble the countershaft in the following order:

NOTE
First gear, the needle bearing and the thrust washer were removed during countershaft removal.

   a. Fourth gear.
   b. Snap ring and spline washer.
   c. Third gear and third gear bushing.
   d. Lockwasher and spline washer.
   e. Second gear and second gear bushing.
   f. Spline washer and snap ring.
   g. Fifth gear.

Inspection

When measuring the countershaft components, compare the actual measurements to the specifications in **Table 4**. Replace worn or damaged parts as described in this section.

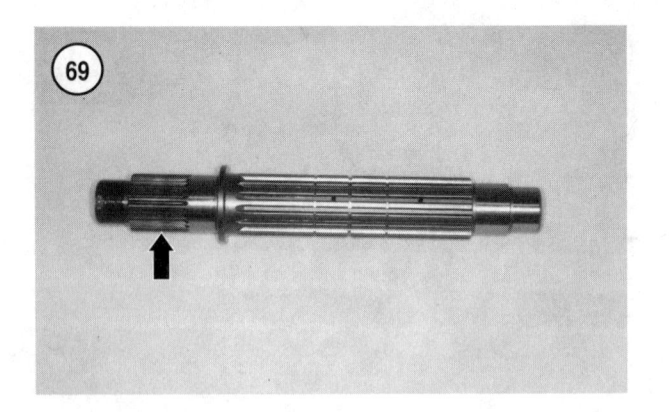

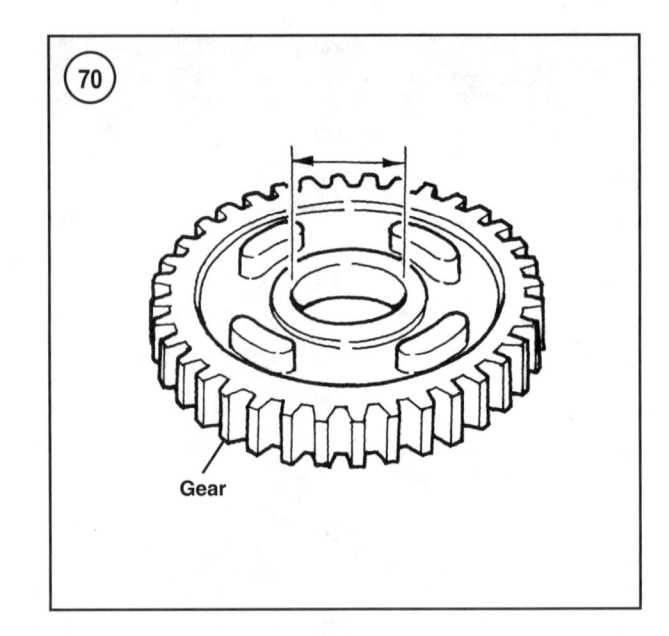

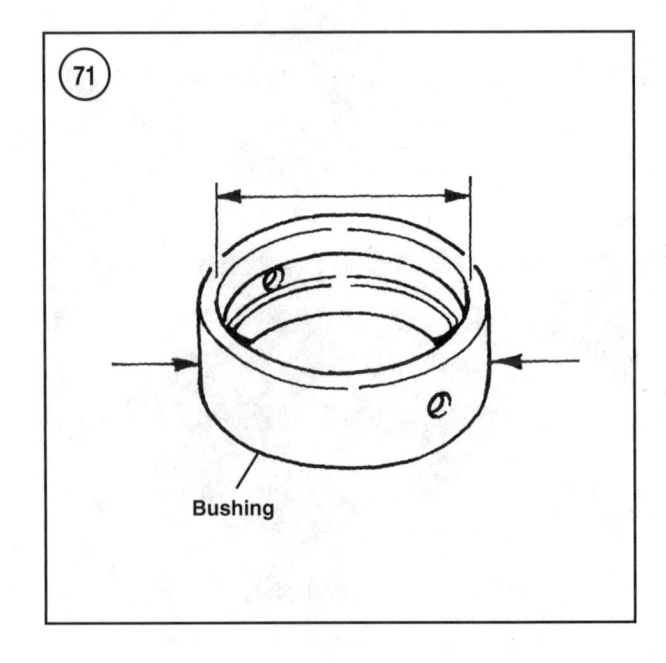

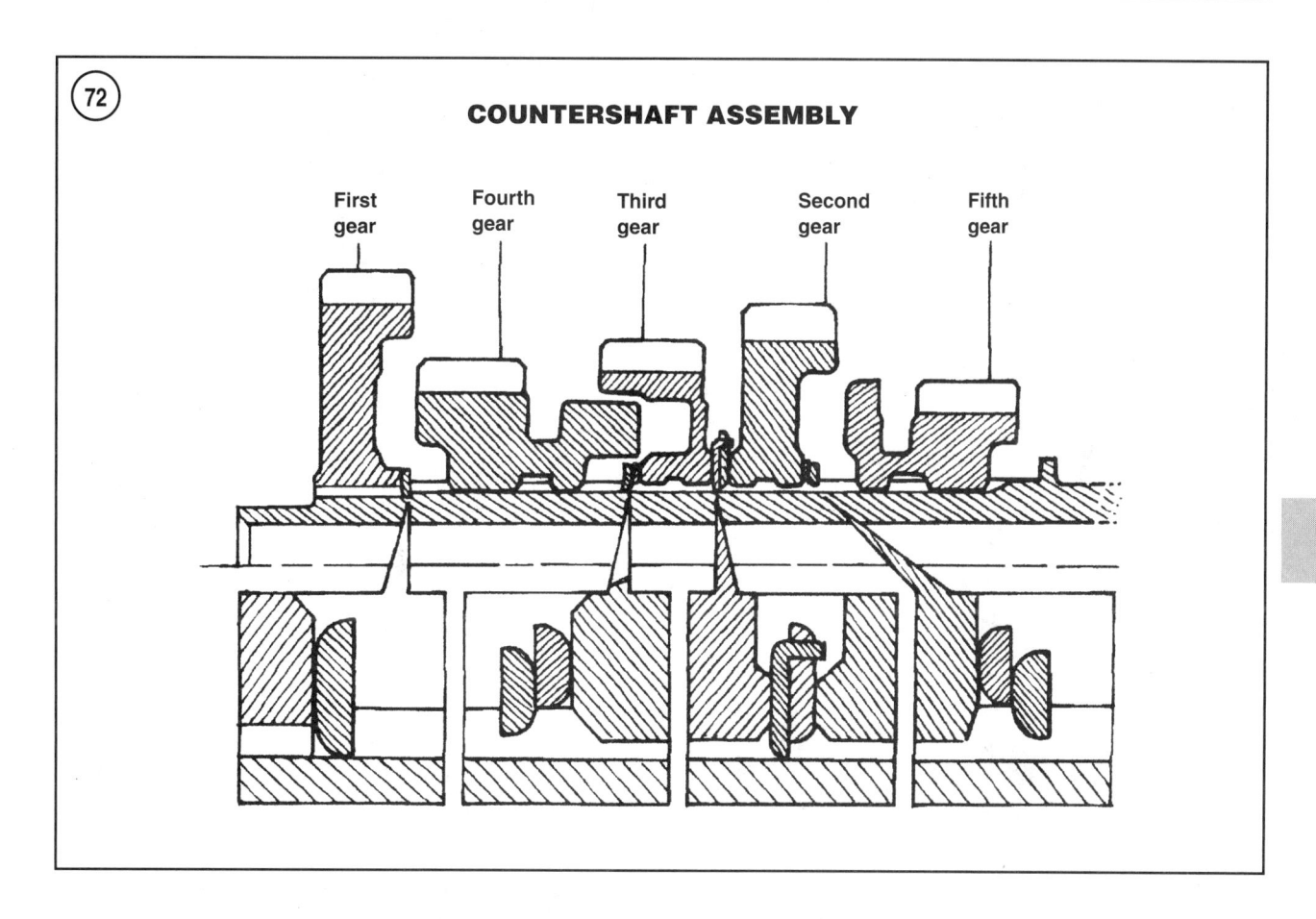

SHIFT MECHANISM AND TRANSMISSION

72

COUNTERSHAFT ASSEMBLY

First gear — Fourth gear — Third gear — Second gear — Fifth gear

Maintain the alignment of the countershaft components when cleaning and inspecting the parts in this section.

1. Clean and dry the countershaft. Flush the oil control holes.
2. Inspect the countershaft (**Figure 69**) for:
    a. Worn or damaged splines.
    b. Excessively worn or damaged bearing surfaces.
    c. Cracked or rounded-off snap ring grooves.
3. Check each countershaft gear for:
    a. Missing, broken or chipped teeth.
    b. Worn, damaged, or rounded-off gear lugs.
    c. Worn or damaged splines.
    d. Cracked or scored gear bore.
4. Check the countershaft bushings for:
    a. Excessively worn or damaged bearing surface.
    b. Worn or damaged splines.
    c. Cracked or scored bore.
5. Measure the second and third gear inside diameters (**Figure 70**).
6. Measure the second and third gear bushing outside diameters (**Figure 71**).
7. Using the measurements recorded in Step 5 and Step 6, determine the gear-to-bushing clearances. If necessary, replace worn parts to obtain the clearances within the service limit specifications.
8. Make sure each gear slides or turns on the countershaft without any binding or roughness.
9. Check the washers for burn marks, excessive wear or other damage.
    a. The thrust washer should be smooth and show no signs of wear.
    b. The teeth on the spline washers should be uniform and have a positive fit on the countershaft.

Assembly

Throughout this procedure, the orientation of many parts is made in relationship to the splined end of the countershaft (**Figure 69**).

Refer to **Figure 61**.

1. Prior to countershaft assembly, note the following:
    a. Install *new* snap rings.
    b. Install the snap ring and thrust washers with their chamfered edge facing *away* from the thrust load. See **Figure 72**.

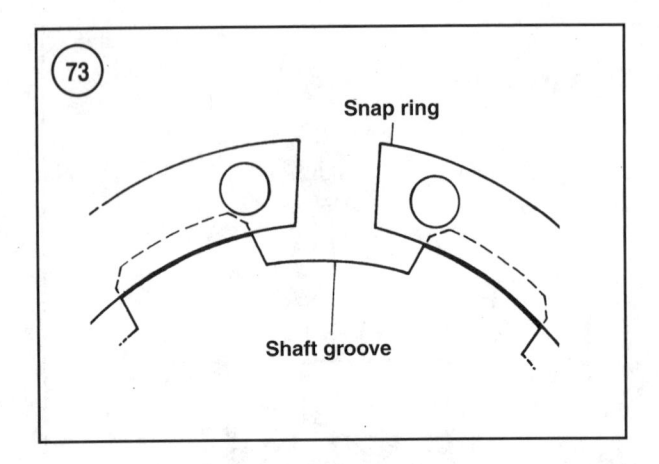

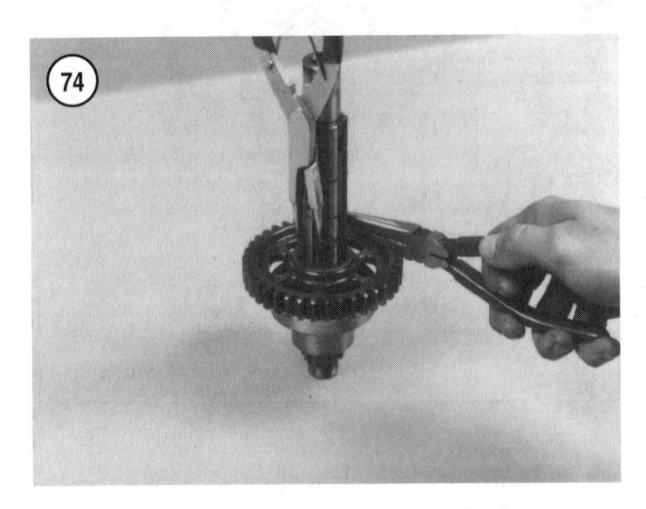

c. Align the snap ring gap with the transmission shaft groove as shown in **Figure 73**.
d. Install snap rings with a pair of snap ring pliers while holding the back of the snap ring with a pair of pliers (**Figure 74**). Then slide the snap ring down the shaft and seat it into its correct groove. Check the snap ring to make sure it seats in its groove completely.

2. Lubricate all sliding surfaces with engine oil.
3. Install fifth gear (**Figure 75**). The gear dogs must face away from the splined shaft end.
4. Install a *new* snap ring (A, **Figure 76**). The flat side of the snap ring must face toward the splined shaft end. The snap ring must seat in the groove in the shaft (**Figure 77**).
5. Install the spline washer (B, **Figure 76**). The flat side of the washer must face toward the splined shaft end.
6. Install the second gear bushing (A, **Figure 78**) and seat it against the washer.
7. Install second gear (B, **Figure 78**). The flat side of the gear must face away from the splined shaft end.
8. Install the spline washer and lockwasher set:

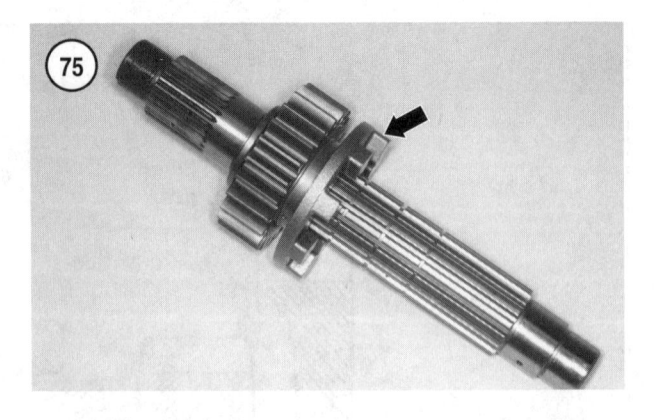

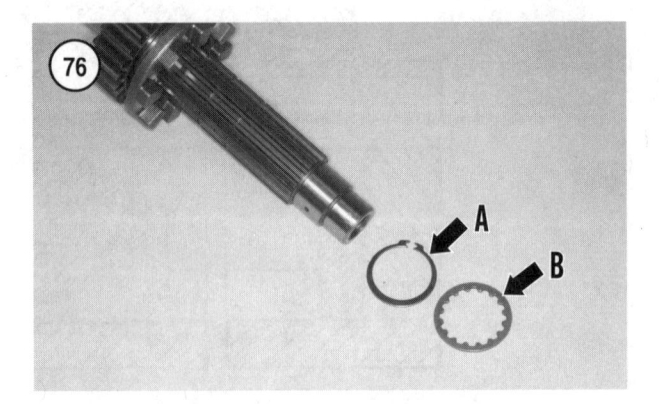

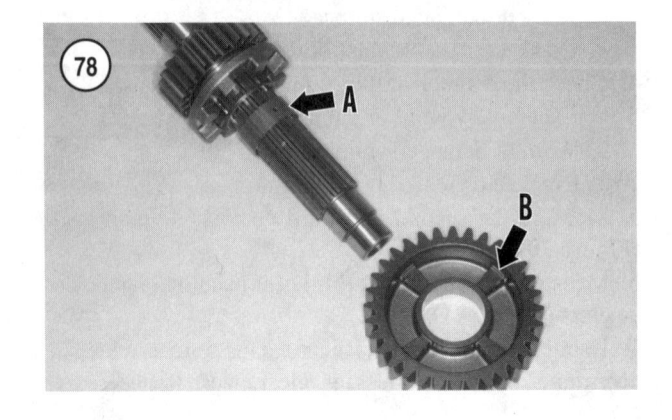

SHIFT MECHANISM AND TRANSMISSION

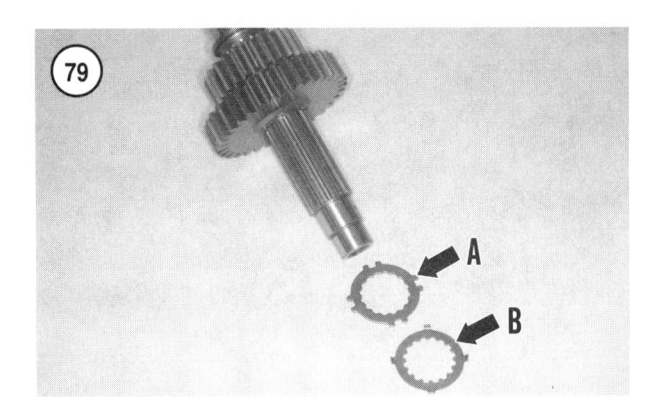

a. Install the spline washer (A, **Figure 79**) with its flat side facing away from the splined shaft end. Turn the washer so its splines align with the raised splines on the countershaft (A, **Figure 80**). When this alignment is made, the spline washer cannot be removed from the countershaft.

b. Install the lockwasher (B, **Figure 79**) with its bent arms (B, **Figure 80**) facing in and insert them into the spline washer notches. See **Figure 81**.

NOTE
*If the lockwasher arms do not align, the spline washer (A, **Figure 80**) is not properly installed. Reinstall the spline washer as described in substep a.*

9. Install the third gear bushing (A, **Figure 82**) and seat it against the lockwasher.

10. Install third gear (B, **Figure 82**). The gear dogs must face away from the splined shaft end.

NOTE
*The gear dog side of third gear is deeper than the opposite side of the gear. See **Figure 83** and **Figure 72**.*

11. Install the spline washer (A, **Figure 84**). The flat side of the spline washer must face away from the splined shaft end.

12. Install a *new* snap ring (B, **Figure 84**). The flat side of the snap ring must face away from the splined shaft end. The snap ring must seat in the groove in the shaft (**Figure 85**).

13. Install fourth gear (**Figure 86**) with its shifter groove facing toward the splined end of the shaft.

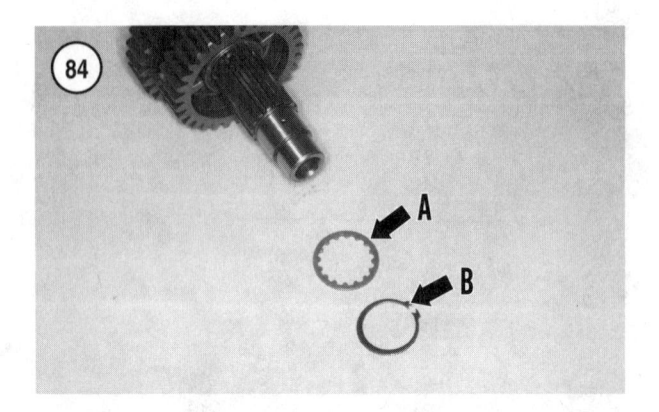

NOTE
*Do not install the washer, needle bearing or first gear (**Figure 87**). These will be installed during countershaft installation.*

14. Check the direction of the gears on the shaft (**Figure 88**).

TRANSMISSION SHIFTING CHECK

Check the shifting operation with the engine mounted in the frame or with it sitting on the workbench. Always check the shifting after servicing any of the shift mechanism or transmission components or after assembling the crankcase halves.

1. Install the external shift linkage as described under *External Shift Linkage* in this chapter.
2. Turn the mainshaft while moving the shift drum by hand and align the raised shift drum cam detent with the stopper lever roller as shown in **Figure 89**. The raised shift drum cam detent position is its neutral position. The lower detent positions on the shift drum cam represent different gear positions. When the shift drum and stopper lever are aligned as shown in **Figure 89**, the transmission is

SHIFT MECHANISM AND TRANSMISSION

in neutral. In this position, the mainshaft and countershaft assemblies can be turned separately, indicating they are not engaged or meshed together.

3. Check each gear position by turning the mainshaft and shift drum by hand. Each time the stopper lever roller seats into a different shift drum cam detent ramp, the transmission has changed gear. When the transmission is in gear, the countershaft and mainshaft are engaged and will turn together.

4. If the transmission does not shift properly into each gear, refer to *Gearshift Linkage Troubleshooting* and *Transmission* in Chapter Two.

Table 1 TRANSMISSION SPECIFICATIONS

Transmission type	Constant mesh, 5-speed with reverse
Primary reduction	1.591 (78/49)
Secondary reduction (output drive)	1.028 (36/35)
Gear ratios	
First gear	2.375 (38/16)
Second gear	1.454 (32/22)
Third gear	1.068 (31/29)
Fourth gear	0.843 (27/32)
Fifth gear	0.686 (24/35)

Table 2 MAINSHAFT SPECIFICATIONS

	New mm (in.)	Service limit mm (in.)
Bushing inside diameter		
Fourth gear	28.007-28.028 (1.1026-1.1035)	28.04 (1.104)
Fifth gear	32.007-32.028 (1.2601-1.2609)	32.04 (1.261)
Bushing outside diameter		
Fourth gear	30.950-30.975 (1.2186-1.2195)	30.93 (1.218)
Fifth gear	34.950-34.975 (1.3760-1.3770)	34.93 (1.375)
Bushing-to-shaft clearance	0.007-0.041 (0.0003-0.0016)	0.08 (0.003)
Gear inside diameter		
Fourth gear	31.000-31.025 (1.2205-1.2215)	31.04 (1.222)
Fifth gear	35.000-35.025 (1.3780-1.3789)	35.04 (1.380)
Gear-to-bushing clearance	0.025-0.075 (0.0010-0.0030)	0.10 (0.004)
Mainshaft outside diameter		
Fourth gear position	27.987-28.000 (1.1018-1.1024)	27.96 (1.101)
Fifth gear position	31.987-32.000 (1.2593-1.2598)	31.96 (1.258)

Table 3 SHIFT FORK AND SHIFT SHAFT SPECIFICATIONS

	New mm (in.)	Service limit mm (in.)
Shift fork claw thickness	5.93-6.00 (0.233-0.236)	5.6 (0.22)
Shift fork inside diameter	14.000-14.018 (0.5512-0.5519)	14.04 (0.553)
Shift fork shaft outside diameter	13.966-13.984 (0.5498-0.5506)	13.90 (0.547)

Table 4 COUNTERSHAFT SPECIFICATIONS

	New mm (in.)	Service limit mm (in.)
Bushing outside diameter		
Second and third gears	32.950-32.975 (1.2972-1.2982)	32.93 (1.296)
Gear inside diameter		
Second and third gears	33.000-33.025 (1.2992-1.3002)	33.04 (1.301)
Gear-to-bushing clearance	0.025-0.075 (0.0010-0.0030)	0.10 (0.004)

Table 5 SHIFT MECHANISM AND TRANSMISSION TORQUE SPECIFICATIONS

	N•m	in.-lb.	ft.-lb.
Countershaft bearing plate bolts[1]	12	106	–
Joint arm bolt	25	–	18
Mainshaft locknut[2]	186	–	137
Mainshaft plate bolts[1]	26	–	19
Return spring locating pin	25	–	18
Reverse lock cam bolt[1]	12	106	–
Shift drum joint bolts[1]	27	–	20
Shift drum bearing plate bolts[1]	12	106	–
Stopper arm bolt	12	106	–

1. Apply a medium strength threadlock onto fastener threads.
2. Lubricate threads and seating surface with engine oil.

CHAPTER EIGHT

FUEL INJECTION AND EMISSION CONTROL SYSTEMS

The chapter describes service procedures for the fuel injection and emission control systems.

Table 1 lists fuel system component abbreviations. **Table 2** lists fuel system test specifications.

Tables 1-4 are at the end of the chapter.

FUEL INJECTION SYSTEM PRECAUTIONS

Servicing the electronic fuel injection system (**Figure 1**) requires special precautions to prevent damage to the throttle body and to the engine control module (ECM). Common fuel system and electrical service procedures acceptable to other motorcycles with different fuel and electrical systems may damage parts of these systems.

1. The fuel system is pressurized. Wear eye protection when working on the fuel system, especially when depressurizing the system. Refer to *Depressurizing the Fuel System* in this chapter.
2. Do not clean the throttle bore inside diameter surfaces with any type of commercial carburetor cleaner, as it will damage the molybdenum surface coating.
3. Except for servicing the fuel injectors and starter valve assembly, the throttle body should not be disassembled. Certain bolts on the throttle body are pre-set by the manufacturer and must not be loosened or removed. Refer to *Throttle Body* in this chapter.
4. Turn the ignition switch off before disconnecting any connectors in the fuel or engine management systems. ECM damage may occur if electrical components are disconnected/connected with the ignition switch on.

FUEL AND VACUUM HOSE IDENTIFICATION

The fuel injection system uses a number of fuel and vacuum hoses. To allow easier reassembly, use a vacuum hose identifier kit to identify the hoses and connection points before disconnecting them. The one shown (Lisle part No. 74600 [**Figure 2**]) consists of 48 color-coded hose fittings in 1/8 to 1/2 in. sizes. Automotive and aftermarket parts suppliers carry this kit, or similar equivalents.

CHAPTER EIGHT

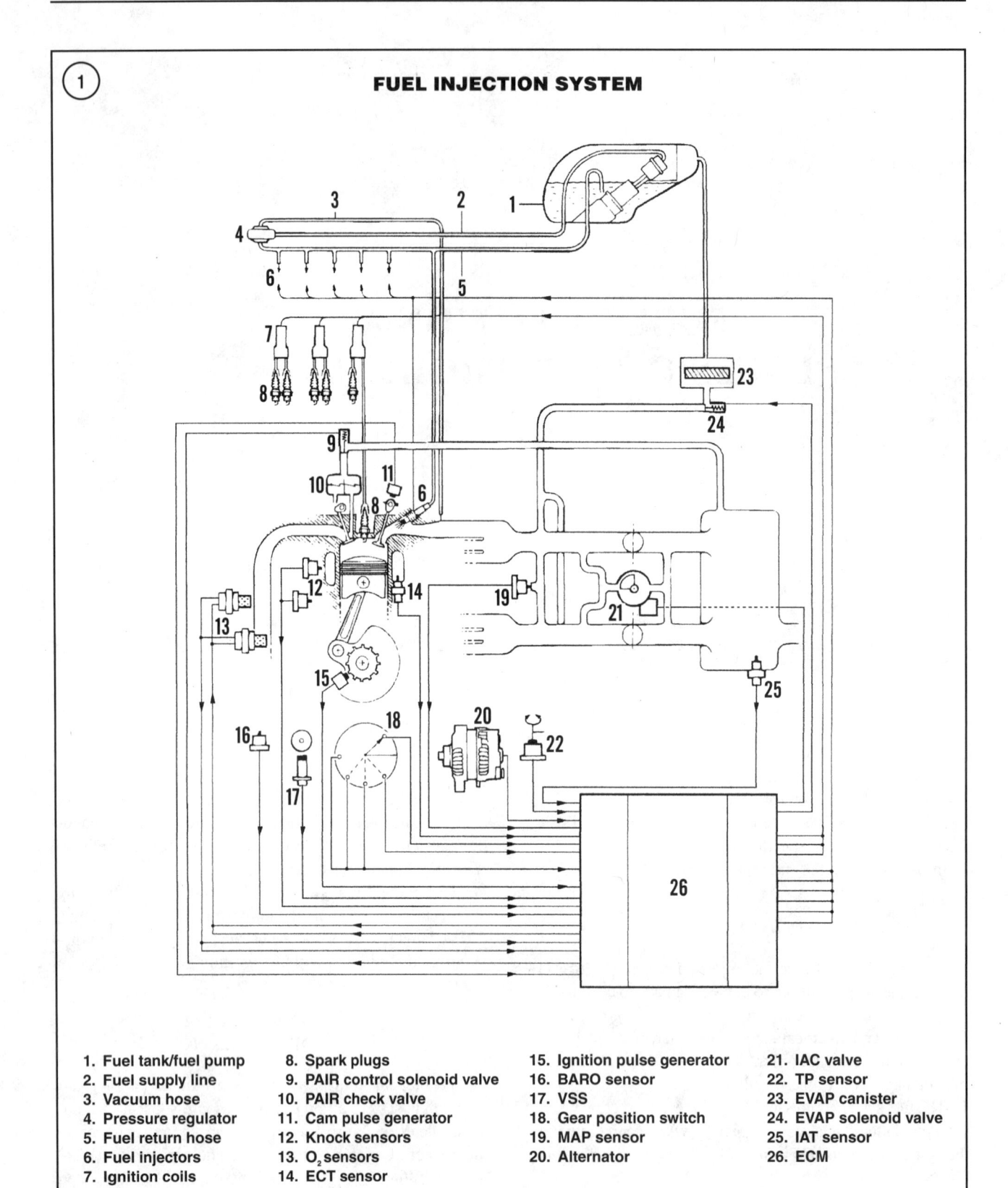

①

FUEL INJECTION SYSTEM

1. Fuel tank/fuel pump
2. Fuel supply line
3. Vacuum hose
4. Pressure regulator
5. Fuel return hose
6. Fuel injectors
7. Ignition coils
8. Spark plugs
9. PAIR control solenoid valve
10. PAIR check valve
11. Cam pulse generator
12. Knock sensors
13. O_2 sensors
14. ECT sensor
15. Ignition pulse generator
16. BARO sensor
17. VSS
18. Gear position switch
19. MAP sensor
20. Alternator
21. IAC valve
22. TP sensor
23. EVAP canister
24. EVAP solenoid valve
25. IAT sensor
26. ECM

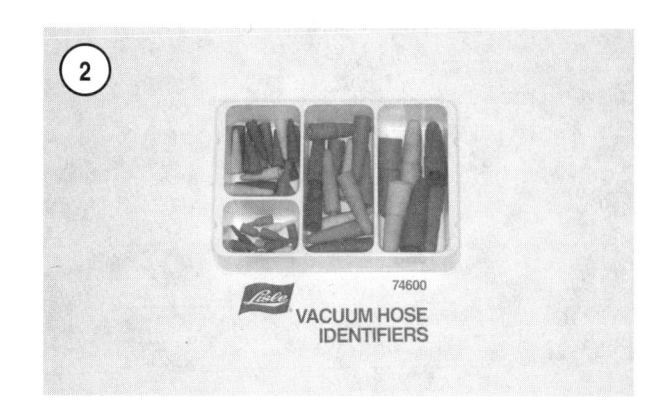

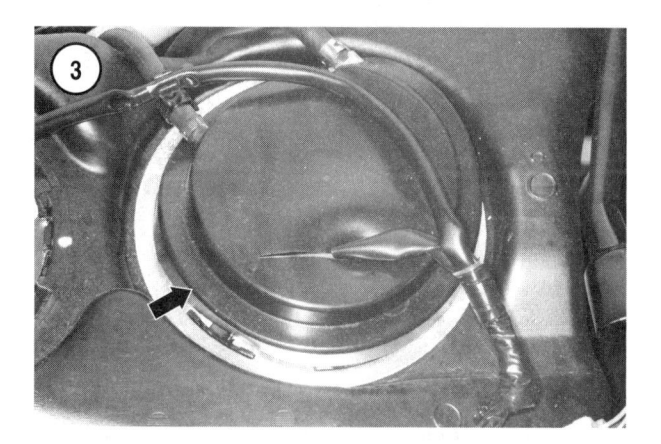

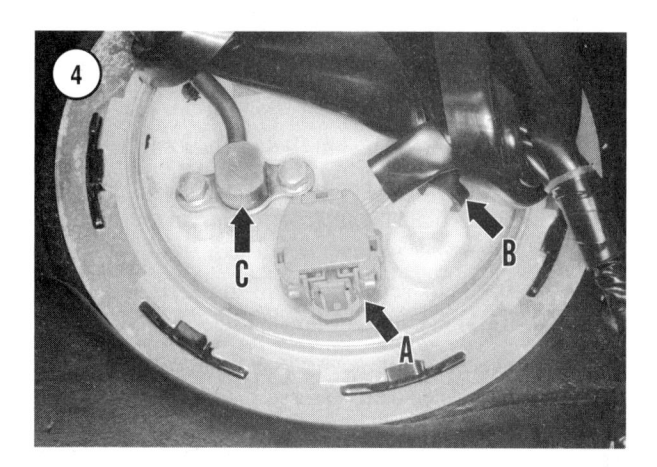

FUEL INJECTION AND EMISSION CONTROL SYSTEMS

WARNING
Always wear eye protection when working on the fuel system.

WARNING
The ignition switch must be turned off when depressurizing the fuel system.

1. To depressurize at the fuel pump:
   a. Remove the seat. Refer to *Seat* in Chapter Seventeen.
   b. Remove the cover (**Figure 3**) from the top of the fuel pump.

   WARNING
   Fuel vapors will be present when depressurizing the fuel system, and some fuel may spill. Because gasoline is extremely flammable, perform this procedure away from all open flames, including appliance pilot lights and sparks. Do not smoke or allow someone who is smoking in the work area as an explosion and fire may occur. Always work in a well-ventilated area. Wipe up spills immediately.

   c. Disconnect the 5-pin connector (A, **Figure 4**) at the fuel pump.
   d. Disconnect the fuel return hose (B, **Figure 4**).
   e. Remove the two bolts securing the fuel supply hose (C, **Figure 4**) to the fuel pump.
   f. Place a rag over the fuel supply hose fitting, then pull the fitting up to disconnect it. Soak up the spilled fuel with the rag.
   g. Check both hose ends for dirt and damage.
   h. Lubricate a new O-ring with engine oil and install it onto the fuel supply hose joint. Install the hose joint (C, **Figure 4**) and tighten the bolts to 10 N•m (88 in.-lb.).
   i. Install the fuel return hose (B, **Figure 4**) and reconnect the 5-pin connector.
   j. Turn the ignition switch on and check for fuel leaks at the pump.
   k. Reinstall the cover at the fuel pump.
   l. Install the seat.

2. To depressurize at the left fuel rail:
   a. Remove the left fuel injector cover.
   b. Clean off the sealing bolt (A, **Figure 5**) and the immediate area around the fuel rail to prevent dirt from entering the system.
   c. Place several rags over and around the sealing bolt to catch any spilled fuel in the following steps.
   d. Slowly loosen and remove the 6–mm sealing bolt (A, **Figure 5**) and washer to depressurize the fuel system.

DEPRESSURIZING THE FUEL SYSTEM

The fuel system is under pressure at all times, even when the engine is not operating. The fuel system is not equipped with a port for relieving the fuel pressure, so whenever a fuel line or fitting is loosened or removed, gasoline will spray out unless the system is depressurized first. Before disconnecting any fuel line or fitting, perform the following.

Do not remove the banjo bolt (B, **Figure 5**) and washer.

e. Install the 6–mm sealing bolt (A, **Figure 5**) and a new sealing washer and tighten to 12 N•m (106 in.-lb.).

f. Install the left fuel injector cover.

3. Properly dispose of any fuel-soaked rags.

FUEL PRESSURE TEST

Fuel pressure is calculated so the correct amount of fuel is discharged from the injectors at all times. Because fuel pressure at the injectors must be kept constant, a rise or drop in fuel pressure can greatly affect engine operation. When the injectors do not discharge enough fuel, a lean air/fuel mixture results. When the injectors discharge too much fuel, the air/fuel ratio is rich.

The ECM is not programmed to display a *specific* diagnostic trouble code (DTC) if the fuel delivery system is operating incorrectly. There are no sensors to monitor the fuel pressure or fuel flow. Check the fuel pressure when troubleshooting symptoms may be caused by improper fuel discharge or flow through the injectors. Likewise, check the fuel pressure whenever troubleshooting a drivability or fuel mileage problem.

Use a fuel pressure gauge (Honda part No. 07406-0040002) or equivalent, and a new 6–mm sealing washer (Honda part No. 90430-PD6-003) to check the fuel pressure.

WARNING
Fuel vapors are present when checking the fuel pressure, and some fuel may spill. Because gasoline is extremely flammable, perform this procedure away from all open flames, including appliance pilot lights and sparks. Do not smoke or allow someone who is smoking in the work area, as an explosion and fire may occur. Always work in a well-ventilated area. Wipe up spills immediately.

1. Remove the front fairing. Refer to *Front Fairing* in Chapter Seventeen.

2. Unbolt and remove the left fuel injector cover.

3. Clean off the 6-mm sealing bolt (A, **Figure 5**) and the immediate area around the fuel rail to prevent dirt from entering the system.

4. Place several cloths over and around the sealing bolt to catch and spilled fuel in the following steps.

5. Slowly loosen and remove the 6–mm sealing bolt (A, **Figure 5**) and washer to depressurize the fuel system. Do not remove the banjo bolt (B, **Figure 5**) and washer from the fuel rail.

6. Install the fuel pressure gauge and sealing washer into the banjo bolt in place of the sealing bolt (**Figure 6**). Tighten securely.

7. Pinch the pressure regulator vacuum hose with a hose pincher (**Figure 7**).

NOTE
*To make a hose pincher, slip a length of fuel hose over the jaws of a pair of long jaw locking pliers (**Figure 8**).*

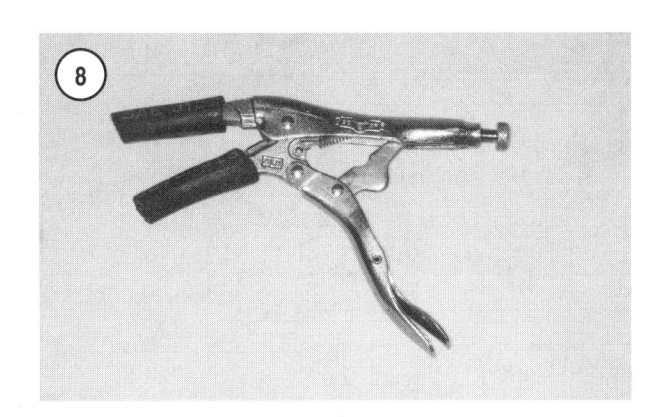

FUEL INJECTION AND EMISSION CONTROL SYSTEMS

shooting idle and off-idle drivability problems.

10. Turn the engine off.
11. If the fuel pressure reading is lower than specified in Step 8, check for the following:
    a. Leaking fuel line.
    b. Clogged fuel filter.
    c. Damaged fuel pump.
    d. Damaged pressure regulator.
12. If the fuel pressure reading is higher than specified in Step 8, check for the following:
    a. Plugged fuel return line.
    b. Damaged fuel pump.
    c. Damaged pressure regulator.
13. Remove the fuel pressure gauge and sealing washer.
14. Install the 6–mm sealing bolt (A, **Figure 5**) and a new sealing washer and tighten to 12 N•m (106 in.-lb.).
15. Remove the hose pincher (**Figure 7**).
16. Turn the ignition switch on and check for fuel leaks at the sealing bolt.
17. Reverse Step 1 and Step 2 to complete installation.

FUEL FLOW TEST

WARNING
Fuel vapors are present when checking the fuel flow, and some fuel may spill. Because gasoline is extremely flammable, perform this procedure away from all open flames, including appliance pilot lights and sparks. Do not smoke or allow someone who is smoking in the work area as an explosion and fire may occur. Always work in a well-ventilated area. Wipe up spills immediately.

1. Remove the seat. Refer to *Seat* in Chapter Seventeen.
2. Remove the screws and relay box cover (**Figure 9**).
3. Remove the fuel pump relay (**Figure 10**).

NOTE
The pressure regulator modifies fuel pressure according to intake manifold vacuum. At wide open throttle, vacuum supplied at the regulator is very low and requires a higher fuel pressure to keep the regulator valve open. Pinching the hose as described in Step 7 simulates this condition.

8. Start the engine and read the fuel pressure gauge with the engine running at idle speed. The standard high fuel pressure reading is 343 kPa (50 psi).
9. Release the hose pincher and continue to allow the engine to run at idle speed. This fuel pressure reading should drop approximately 10 psi from the reading obtained in Step 8.

NOTE
The low pressure measurement specified in Step 9 is not listed by the manufacturer. This step is included to duplicate a condition where high intake manifold vacuum is supplied at the pressure regulator valve. In this situation, a higher vacuum at the regulator moves the diaphragm in the regulator, thus reducing the amount of fuel pressure required at the regulator valve. This step is critical when trouble-

CHAPTER EIGHT

11

Relay box

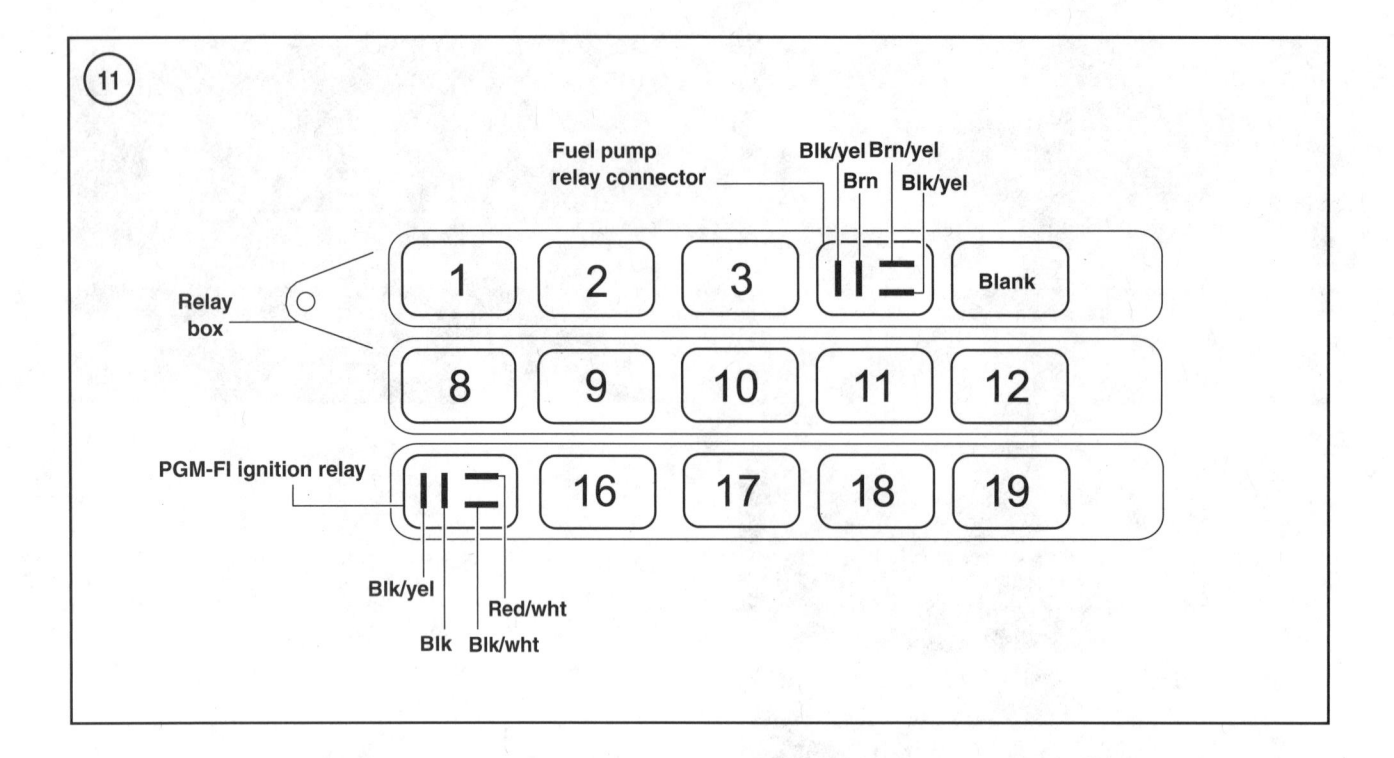

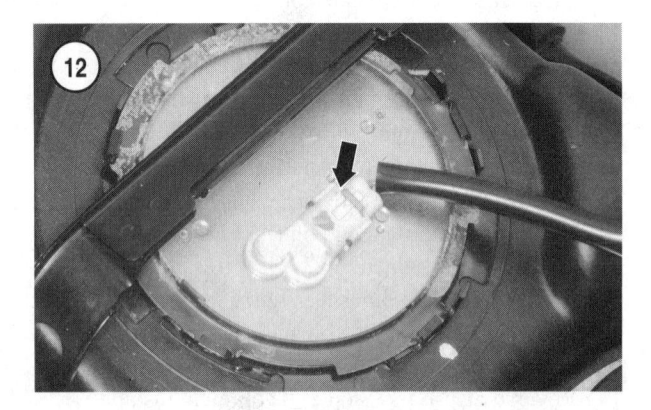

Fuel pump relay connector — 1, 2, 3
Blk/yel, Brn, Brn/yel, Blk/yel — (between 3 and Blank)
Blank

8, 9, 10, 11, 12

PGM-FI ignition relay — (before 16), 16, 17, 18, 19
Blk/yel, Blk, Blk/wht, Red/wht

12

13

4. Connect a jumper wire between the black/yellow and brown relay box terminals (**Figure 11**).

CAUTION
Fuel will spill from the fuel return hose when it is disconnected. Catch the fuel in a plastic jar and dispose of it properly.

5. Disconnect the fuel return hose at the fuel tank (B, **Figure 4**). Plug the fuel tank hose joint.

6. Place the loose end of the fuel return hose into a large plastic graduated beaker.

7. Turn the ignition switch on for ten seconds, then shut it off.

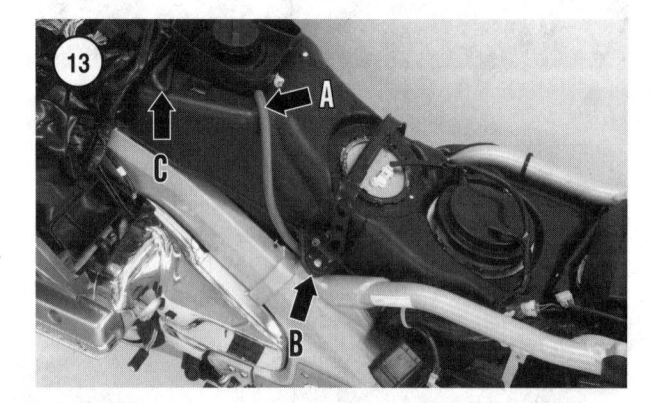

8. The specified fuel flow for 10 seconds, with a fully charged battery, is 133 ml (4.5 U.S. oz.). Pour the measured fuel into an approved gasoline storage container.

9. If the fuel flow is less than specified, check for the following:
   a. Plugged fuel hose or fuel return hose.
   b. Clogged fuel filter.

NOTE
The engine is not equipped with a separate fuel filter. An integral, non-replaceable fuel filter is mounted in the fuel pump.

   c. Damaged fuel pump.
   d. Damaged pressure regulator.

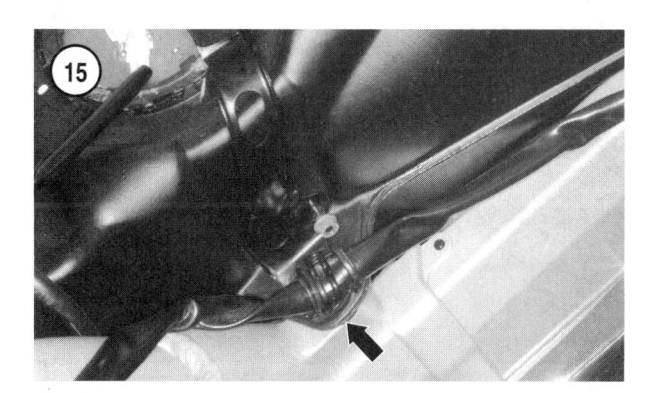

FUEL INJECTION AND EMISSION CONTROL SYSTEMS

10. Reconnect the fuel return hose and secure it with its hose clamp.
11. Reverse Steps 1-4 to complete installation.

FUEL TANK

Removal/Installation

WARNING
Fuel vapors will be present when removing the fuel tank, and some fuel may spill. Because gasoline is extremely flammable, perform this procedure away from all open flames, including appliance pilot lights and sparks. Do not smoke or allow someone who is smoking in the work area as an explosion and fire may occur. Always work in a well-ventilated area. Wipe up spills immediately.

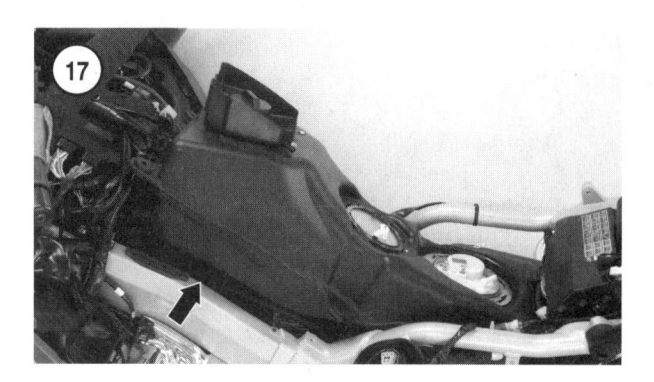

WARNING
Gasoline is extremely flammable and must not be stored in an open container. Store gasoline in a sealed gasoline storage container, away from heat, sparks or flames.

1. Remove the top shelter (Chapter Seventeen).
2. Remove the battery, case and battery tray (Chapter Nine).
3. Pull the cover (**Figure 3**) off the fuel pump and perform the following:
   a. Disconnect the gray 5-pin connector (A, **Figure 4**).
   b. Spread the hose clamp and disconnect the fuel return hose (B, **Figure 4**) from the nozzle on the fuel pump.
   c. Remove the two bolts and remove the fuel feed hose (C, **Figure 4**) and its O-ring.
4. Disconnect the fuel level sender white 2-pin connector (**Figure 12**).
5. Disconnect the fuel tank tray drain hose (A, **Figure 13**).
6. Disconnect the fuel tank breather hose (**Figure 14**).
7. Remove the mounting bolts and remove the seat bracket (B, **Figure 13**). Note the clamp and wire harness installed on the right bolt (**Figure 15**).
8. Remove the two front mounting bolts. Note the wire harness routing around the right front mounting bolt (**Figure 16**).

CAUTION
The fuel tank fits tightly in the frame. Check for any wiring harnesses or cable guides and push them away from the tank during removal. Also, do not allow the fuel tank to slide too far down the frame or it may damage the suspension level actuator sensor and require replacement of the rear shock absorber assembly.

9. Tilt the fuel tank to the right and remove it from the left side of the frame (**Figure 17**).

10. Position the following over the frame rails before installing the fuel tank into the frame (**Figure 18**):
    a. Breather hose.
    b. Drain hose.
    c. Fuel pump and fuel sender connectors.
    d. Fuel hoses.
    e. Make sure the wiring harness is positioned along the outside of the right frame rail.
11. Tilt the fuel tank to the right and install it from the left side of the frame. Install the right rear corner of the fuel tank first, then work the fuel tank into position.

CAUTION
The left front edge of the fuel tank may catch on the ECM holder and wiring harnesses when pushing the fuel tank down in the front. Observe this area and redirect the fuel tank around the ECM holder and wiring harnesses to avoid pinching a wire.

NOTE
Check the alignment of the hoses and wiring harnesses before installing the fuel tank mounting bolts. Make sure each has sufficient length to reach its mounting position on the fuel tank.

12. Install the front mounting bolts and washers. Just start the bolts. Do not tighten them.
13. Check the alignment of the fuel tank and rear frame mounting holes. Reposition the tank, then install the mounting bracket—nut studs facing forward—and the mounting bolts (B, **Figure 13**). Install the wire harness clamp on the right rear bolt (**Figure 15**).
14. Tighten the fuel tank mounting bolts securely.
15. Route the breather hose (C, **Figure 13**) across the front of the drain pan and connect it to the nozzle (**Figure 14**).
16. Reconnect the drain hose to the nozzle on the drain pan (A, **Figure 13**).
17. Lubricate a new O-ring with engine oil and install it over the fuel supply hose joint (**Figure 19**).
18. Install the fuel supply hose and fuel return hose through the rubber cap.
19. Reconnect the fuel return hose (B, **Figure 4**) and secure it with the hose clamp.
20. Install the fuel supply hose joint (C, **Figure 4**) onto the fuel pump and secure it with the two mounting bolts. Tighten the bolts evenly to seat the fuel supply hose joint, then tighten to 10 N•m (88 in.-lb.).
21. Reconnect the gray 5-pin fuel pump harness connector (A, **Figure 4**).
22. Seat the rubber cap (**Figure 3**) over the fuel pump.
23. Reconnect the white 2-pin fuel level sensor connector (**Figure 12**).

FUEL INJECTION AND EMISSION CONTROL SYSTEMS

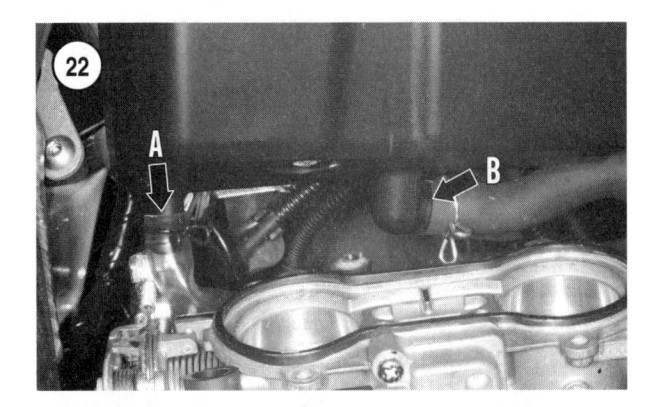

24. Reinstall the battery tray, case and battery. Refer to *Battery* in Chapter Nine.
25. Turn the ignition switch on and check the fuel tank hoses for leaks before installing the seat.
26. Install the top shelter. Refer to *Top Shelter* in Chapter Seventeen.

Inspection

1. Inspect all the hoses for cracks, deterioration and other damage. Replace damaged hoses with hoses of the same type and size materials. The hoses must be flexible and strong enough to withstand engine heat and vibration.
2. Replace weak or damaged hose clamps.
3. Inspect the rubber grommets for deterioration or other damage. Replace if necessary.

AIR FILTER HOUSING

Removal/Installation

1. Remove the fuel tank as described in this chapter.
2. Remove the air filter. Refer to *Air Filter* in Chapter Three.

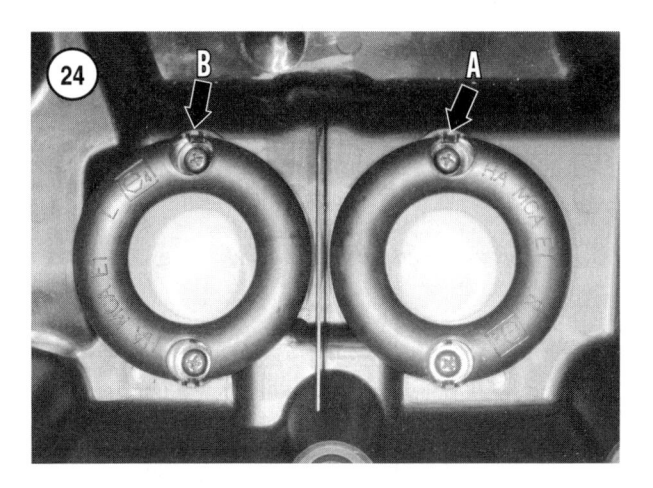

3. Disconnect the two crankcase breather hoses (**Figure 20**) from the rear of the air box.
4. Remove the screws (**Figure 21**) securing the air funnels and air box to the throttle body.

NOTE
The screws and air funnels cannot be removed until the air box is removed from the engine.

5. Lift the air box and disconnect the air box drain hose (A, **Figure 22**) and the PAIR supply hose (B) from the bottom of the air box.

NOTE
There is not a lot of room to work with when disconnecting the hoses. If it is necessary to provide slack in the air box drain hose, remove the center inner fairing and the front inner fairing (Chapter Seventeen). Then remove the clamp and plug from the drain hose and slide the hose toward the air box.

6. Remove the air box.
7. Locate the rubber gasket (**Figure 23**) installed between the air box and throttle body.
8. To service the air funnels (**Figure 24**):
    a. Remove the screws and turn the air funnels to remove them. Turn the left air funnel (A, **Figure 24**) counterclockwise. Turn the right air funnel (B, **Figure 24**) clockwise.
    b. Reverse to install, noting the L (left side) and R (right side) marks. Do not switch them.
9. Remove all sealer residue from the air box and throttle body mating surfaces.
10. Apply Gasgacinch to the side of the gasket that will seat against the air box. Follow the manufacturer's directions for application and drying time. Then install the gas-

ket into the groove in the air box (**Figure 23**). Seat the gasket squarely into the groove.

CAUTION
The gasket must seat squarely into the air box groove or the engine will leak air. If the gasket is rolled or pinched at any point, remove and reinstall it correctly.

11. Clean the air box with compressed air.
12. Check the engine cavity around the throttle body for hose and wiring harness routing that could interfere with air box installation.
13. Reconnect the air box drain hose (A, **Figure 22**) and the PAIR supply hose (B) to the bottom of the air box. Secure each hose with its clamp.

NOTE
After reconnecting the hoses, make sure the rubber gasket in the bottom of the air box was not dislodged or damaged.

14. Lower the air box, then align the two rear mounting screws with the threaded holes in the throttle body, and install the air box onto the throttle body. Start the four mounting screws (**Figure 21**) and tighten them evenly to pull the air box squarely against the throttle body. Make sure each screw is tight.
15. Check the air box and the throttle body valves for any debris, then install the air filter and cover (Chapter Three).
16. Reconnect the two crankcase breather hoses at the rear of the air box (**Figure 20**). Secure each hose with a clamp.
17. Install the air filter cover and the remaining parts (Chapter Three).
18. Install the fuel tank as described in this chapter.

FUEL INJECTORS

Testing

The injectors (**Figure 25**) can be tested without removing them from the fuel rails (**Figure 26**).

1. Remove the injector covers (**Figure 27**).
2. Start the engine and allow it to idle.
3. Use a mechanic's stethoscope and listen for operating sounds at each fuel injector.
4. If there is a difference in operating sound at one or more injectors, remove and inspect the fuel injector(s) as described in this section.

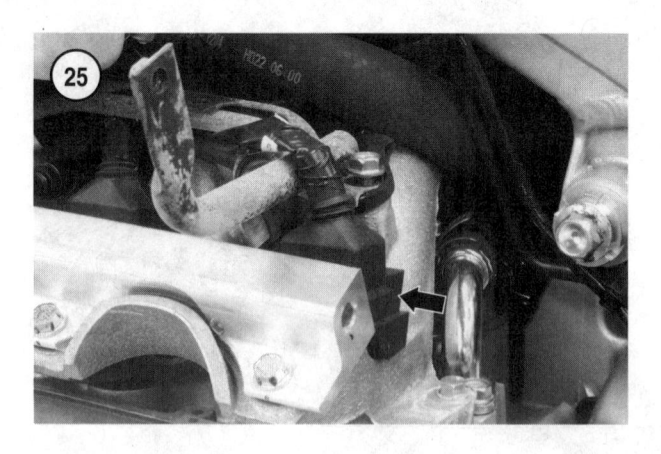

Removal

Refer to **Figure 26** for the fuel hose routing.
1. Remove the radiators. Refer to *Radiators* in Chapter Ten.
2. Remove the injector covers (**Figure 27**).
3. Depressurize the fuel system at the left fuel rail as described in this chapter.
4. Remove the bolts and the radiator stay (**Figure 28**).
5. Slide the cover off each fuel injector connector (A, **Figure 29**).
6. Disconnect the connector (B, **Figure 29**) at each fuel injector.
7A. Left fuel rail—Remove the mounting bolts and the fuel pipe (A, **Figure 30**) and pressure regulator stay (B).
7B. Right fuel rail—Remove the mounting bolts and both fuel pipes (**Figure 31**).
8. Remove and discard the O-rings installed on the fuel pipes (2, **Figure 26**) and pressure regulator stay (18).
9. Clean the fuel rails with compressed air to prevent dirt from entering the intake manifold.
10. Remove the three mounting bolts and remove the fuel rail (C, **Figure 30**) and injectors. See **Figure 32**.
11. Cover the intake manifold injector bore openings to prevent dirt from entering the engine.
12. Remove the clip (A, **Figure 33**) from the injector.
13. Remove the injector (B, **Figure 33**) and O-rings from the fuel rail.
14. Repeat for each injector.

WARNING
*The injector O-rings (**Figure 34**) are a different size and color. Use the installed position of the old O-rings to identify the new O-rings.*

Inspection

1. Visually inspect the fuel injectors for damage.

FUEL INJECTION AND EMISSION CONTROL SYSTEMS

㉖ **FUEL HOSE, PRESSURE REGULATOR AND FUEL RAIL ASSEMBLY**

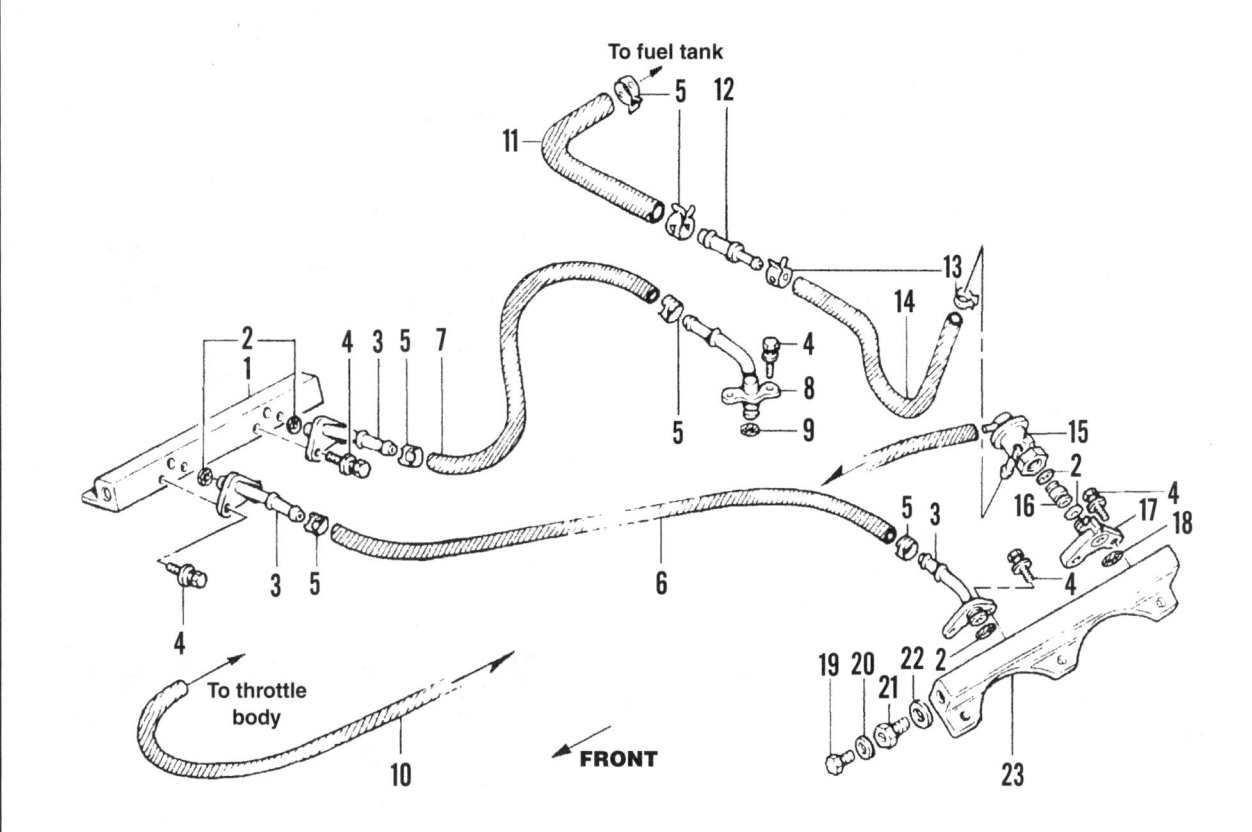

1. Right fuel rail
2. O-ring
3. Fuel pipe
4. Bolt
5. Hose clamp
6. Fuel supply hose
7. Fuel supply hose
8. Fuel pipe
9. O-ring
10. Vacuum hose
11. Fuel return hose
12. Fuel joint
13. Hose clamp
14. Fuel return hose
15. Pressure regulator
16. Joint pipe
17. Pressure regulator stay
18. O-ring
19. 6-mm sealing bolt
20. Sealing washer
21. Banjo bolt
22. Sealing washer
23. Left fuel rail

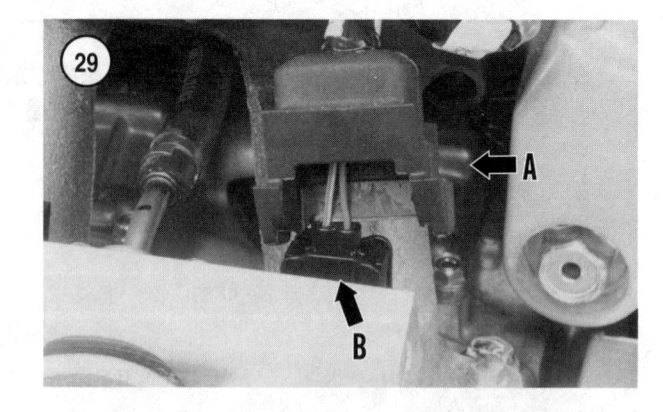

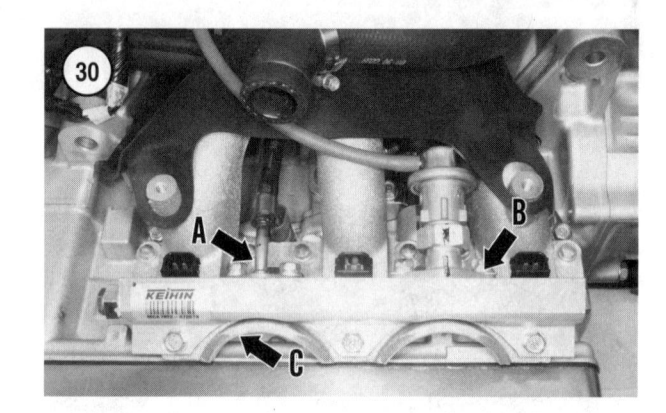

2. Inspect the spray nozzle for clogging or other damage.
3. Use an ohmmeter set at R × 1 and measure resistance across the two fuel injector terminals (**Figure 35**). The specification is 11.1-12.3 ohms.
4. Repeat for each fuel injector.

Installation

1. Clean the fuel rails in solvent and dry with compressed air.
2. Lubricate the new O-rings (**Figure 34**) with engine oil and install them into the injector grooves. Match the color marks on the new O-rings with the old O-rings previously removed.
3. Install the injector into the fuel rail by aligning the tab on the injector with the hole in the fuel rail (**Figure 36**). Secure the injector with the clip (A, **Figure 33**).
4. Repeat for each fuel injector.
5. Install the fuel rail by inserting the fuel injectors into the intake manifold.
6. Install the fuel rail mounting bolts and tighten to 10 N•m (88 in.-lb.).
7. Lubricate the new fuel rail O-rings (2, **Figure 26**) with engine oil and install them into the fuel rail grooves.

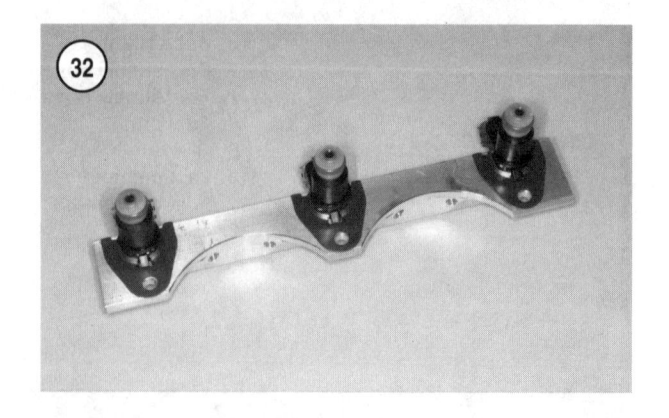

FUEL INJECTION AND EMISSION CONTROL SYSTEMS

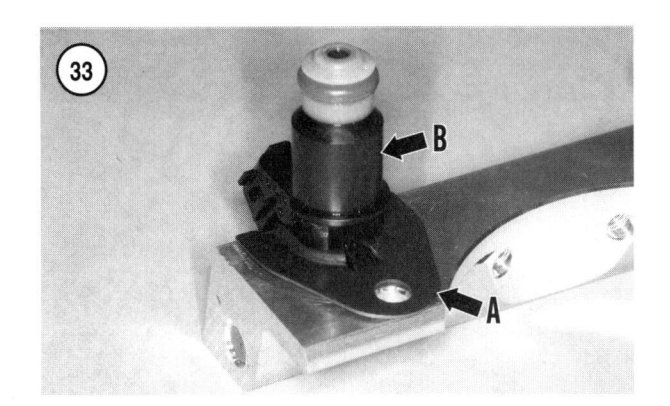

8. Lubricate a new pressure regulator stay O-ring (18, **Figure 26**) with engine oil and install into the stay groove.

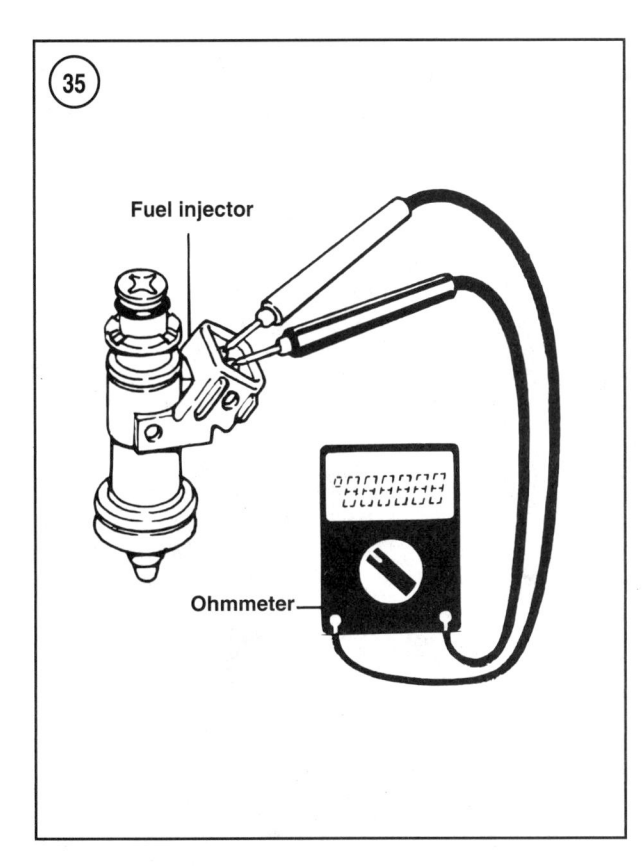

9A. Left fuel rail—Install the fuel pipe (A, **Figure 30**) and pressure regulator stay (B) mounting bolts and tighten to 10 N•m (88 in.-lb.).

9B. Right fuel rail—Install the fuel pipe (**Figure 31**) mounting bolts and tighten to 10 N•m (88 in.-lb.).

10. Reconnect the connector (B, **Figure 29**) at each fuel injector.
11. Slide the cover over each fuel injector connector (A, **Figure 29**).
12. Install the radiator stay (**Figure 28**) and tighten its mounting bolts securely.
13. Install the injector covers (**Figure 27**).
14. Install the radiators. Refer to *Radiators* in Chapter Ten.

FUEL PRESSURE REGULATOR

The fuel pressure regulator is mounted on the left fuel rail (**Figure 26**).

Replacement

1. Remove the left side radiator. Refer to *Radiators* in Chapter Ten.
2. Unbolt and remove the left injector cover.
3. Unbolt and remove the left radiator stay (**Figure 28**).
4. Relieve the fuel pressure at the left fuel rail as described in this chapter. Install the 6–mm sealing bolt (A, **Figure 5**) and a new sealing washer and tighten to 12 N•m (106 in.-lb.).
5. Clean the pressure regulator with compressed air to prevent dirt from entering the system.
6. Hold the pressure regulator (A, **Figure 37**) and loosen the nut (B).
7. Remove the pressure regulator and joint pipe (16, **Figure 26**).

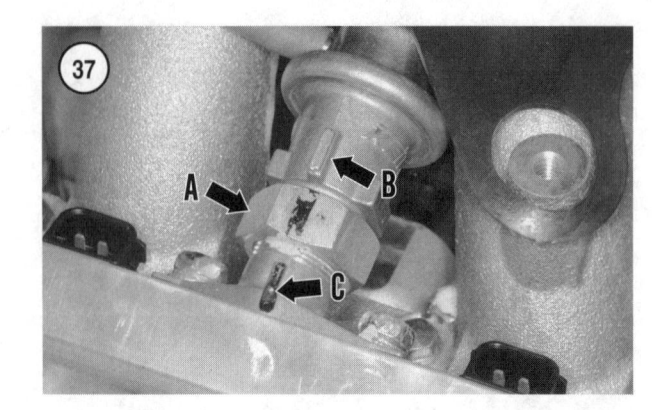

8. Disconnect the vacuum hose (10, **Figure 26**) and fuel return hose (14) from the pressure regulator.

9. Remove and discard the O-ring installed on the joint pipe (16, **Figure 26**).

10. Lubricate two new O-rings with engine oil and install them onto the joint pipe grooves.

11. Install the joint pipe into the pressure regulator stay.

12. Reconnect the fuel return hose onto the pressure regulator and secure with the clamp. Make sure the connection is tight.

13. Reconnect the vacuum hose onto the pressure regulator.

14. Install the pressure regulator by aligning its index mark (A, **Figure 37**) with the index mark on the stay (C). Then hold the regulator in place and tighten the nut (B, **Figure 37**) to 27 N•m (20 ft.-lb.).

15. Turn the ignition switch on and check for fuel leaks at the sealing bolt and fuel return hose.

16. Reverse Steps 1-3 to complete installation.

THROTTLE BODY

Removal

1. Remove the radiators. Refer to *Radiators* in Chapter Ten.

FUEL INJECTION AND EMISSION CONTROL SYSTEMS

43

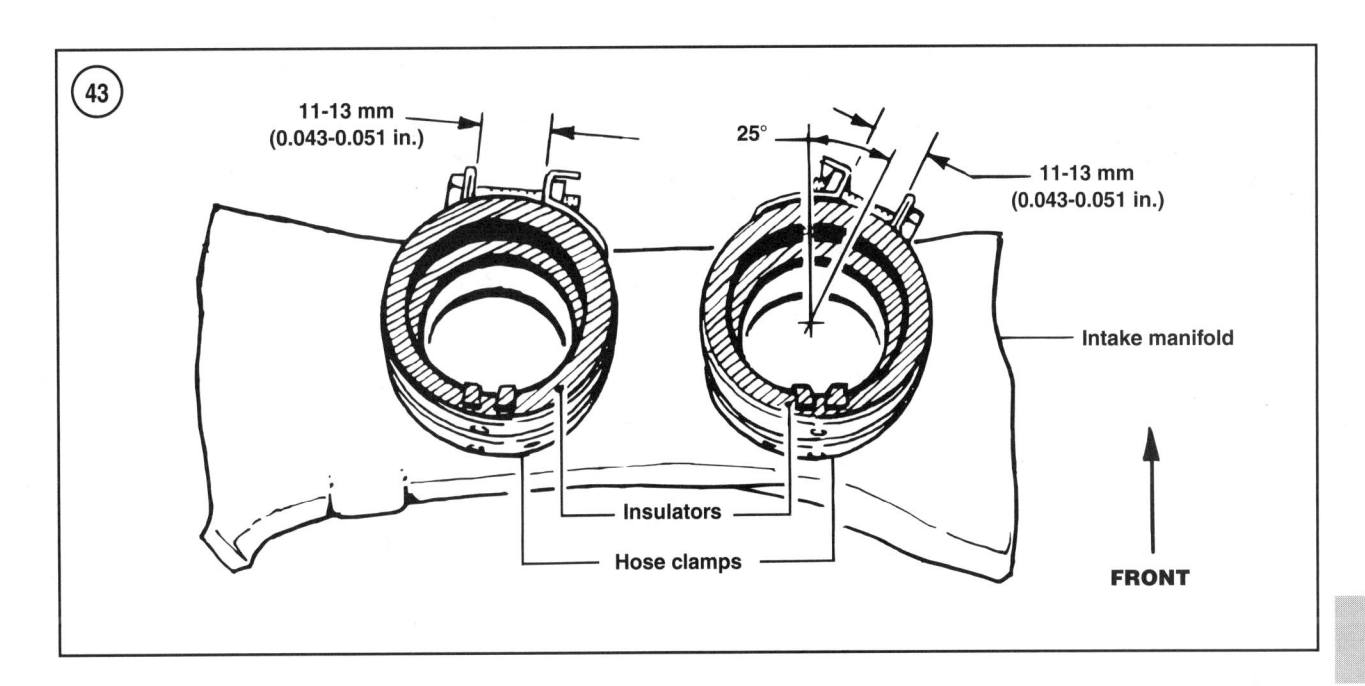

11-13 mm (0.043-0.051 in.)
25°
11-13 mm (0.043-0.051 in.)
Intake manifold
Insulators
Hose clamps
FRONT

44

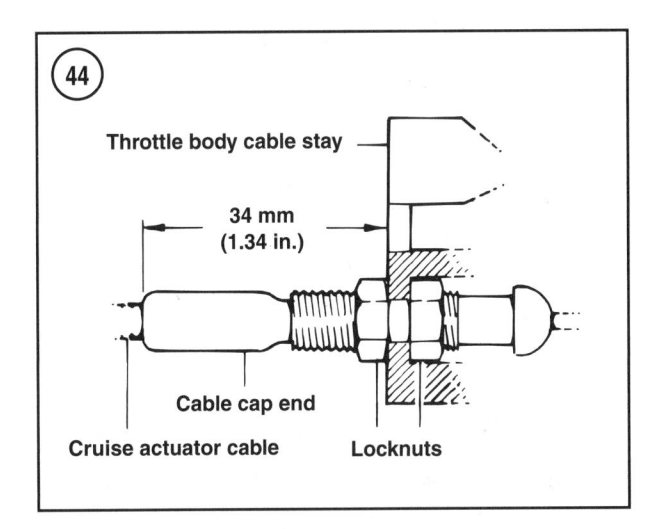

Throttle body cable stay
34 mm (1.34 in.)
Cable cap end
Cruise actuator cable
Locknuts

2. Remove the air filter housing as described in this chapter.
3. Loosen the throttle body insulator hose clamp screws.
4. Pull on the throttle body (**Figure 38**) to release it from the insulators.
5. Disconnect the TP sensor connector (**Figure 39**).
6. Disconnect the IAC valve connector (**Figure 40**).

NOTE
Identify the cables with masking tape and a marking pen before disconnecting them.

7. Disconnect the cruise actuator cable (A, **Figure 41**).
8. Disconnect the throttle cables (B, **Figure 41**).
9. Disconnect the MAP sensor connector (A, **Figure 42**).
10. Disconnect the No. 5 hoses (B, **Figure 42**).
11. Disconnect the vacuum hose (C, **Figure 42**).
12. Remove the throttle body.
13. Plug the intake manifold openings to prevent dirt from falling into the engine.

Installation

1. Install the insulators onto the intake manifold and position the hose clamps as shown in **Figure 43**.
2. Connect the throttle cable ends onto the throttle drum and tighten the cable securely (B, **Figure 41**).
3. Connect the cruise actuator cable onto the actuator drum and against the throttle body cable stay (A, **Figure 41**). Position the cable so the distance from the throttle body cable stay to the end of the cable cap is 34 mm (1.34 in.) (**Figure 44**). Tighten the locknuts and recheck the adjustment.
4. Reconnect the vacuum hose (C, **Figure 42**).
5. Reconnect the No. 5 hoses (B, **Figure 42**).
6. Reconnect the MAP sensor connector (A, **Figure 42**).
7. Reconnect the IAC valve connector (**Figure 40**).
8. Reconnect the TP sensor connector (**Figure 39**).
9. Align the throttle body spigots with the insulators and push the throttle body down firmly. Tighten the hose clamp screws until the distance between the hose tabs is 11-13 mm (0.043-0.051 in.). See **Figure 43**.
10. Check the throttle cable adjustment. Refer to *Throttle Cable Operation and Adjustment* in Chapter Three.
11. Check the cruise actuator cable adjustment. Refer to *Cruise Actuator* in Chapter Sixteen.
12. Reverse Step 1 and Step 2 to complete installation.

CHAPTER EIGHT

Disassembly/Reassembly

1. Only certain items can be replaced on the throttle body. Before servicing the throttle body assembly, note the following:
    a. The throttle body is pre-set by the manufacturer. Do not disassemble or adjust the throttle body in any way other than as described in this section.
    b. The throttle body is equipped with three sensors: TP sensor, IAC valve and MAP sensor. The TP sensor is not removable.
    c. Do not loosen or tighten the white painted bolts and screws on the throttle body. Doing so may cause throttle and idle valve synchronization failure. See A, **Figure 45** and **Figure 46**.
    d. Do not adjust the throttle stop screw (B, **Figure 45**).
    e. Do not adjust the air screws (**Figure 47**).
    f. Tighten all bolts and screws to the specifications listed in the text.
2. To replace the MAP sensor (A, **Figure 48**):
    a. Remove the two screws (B, **Figure 48**), MAP sensor and O-ring.
    b. Install a new O-ring into the throttle body groove.
    c. Install the MAP sensor and tighten the screws to 2 N•m (18 in.-lb.).
3. To replace the IAC valve (A, **Figure 49**):
    a. Remove the two screws (B, **Figure 49**), IAC valve (A) and O-ring.
    b. Install a new O-ring into the throttle body groove.
    c. Install the IAC valve and tighten the screws to 4 N•m (35 in.-lb.).
4. To replace the cable stay (A, **Figure 50**), remove the two screws (B). Install the cable stay and tighten the screws to 4 N•m (35 in.-lb.).

FUEL PUMP

The fuel pump is mounted inside the fuel tank.

Operational Test

1. Remove the seat. Refer to *Seat* in Chapter Seventeen.
2. Turn the ignition switch on and listen for fuel pump operation. If the fuel pump does not operate, turn the ignition switch off and continue with Step 3.
3. Remove the cover (**Figure 51**) from the top of the fuel pump.
4. Disconnect the 5-pin connector (A, **Figure 52**) at the fuel pump.
5. Connect a voltmeter between the brown (+) and green (–) terminals on the wiring harness side of the 5-pin connector (**Figure 53**). Turn the ignition switch on while reading the

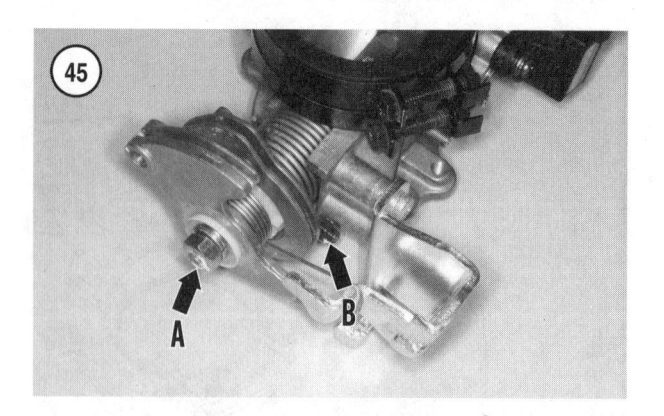

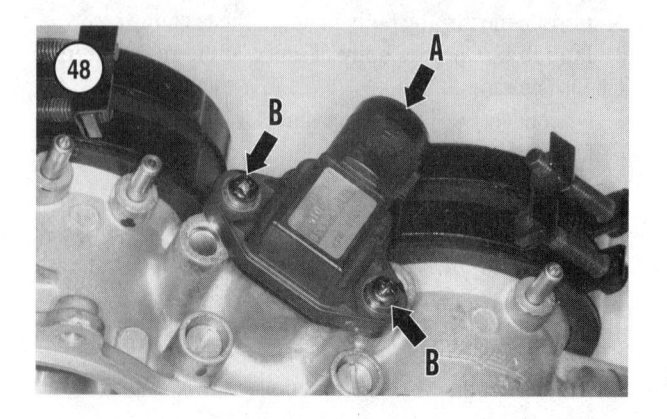

FUEL INJECTION AND EMISSION CONTROL SYSTEMS

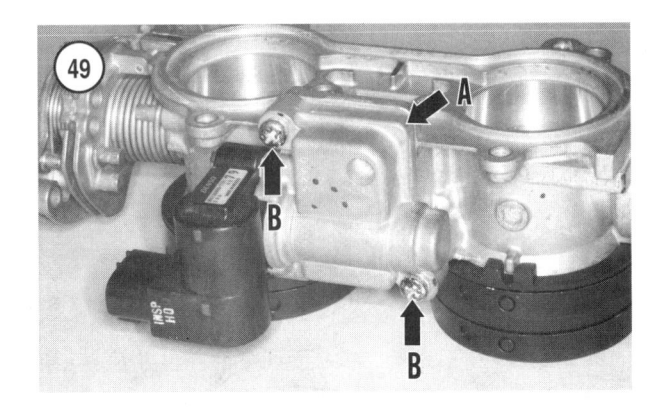

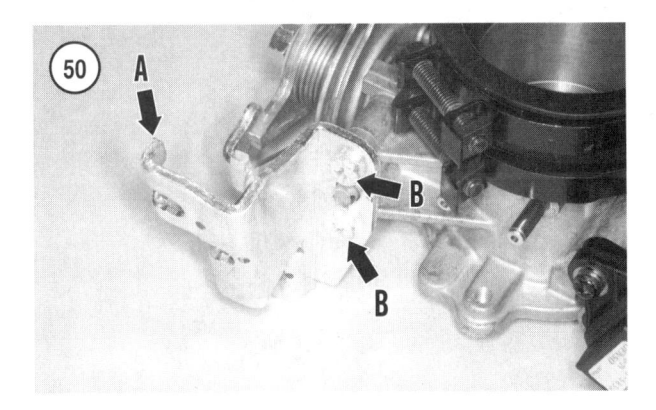

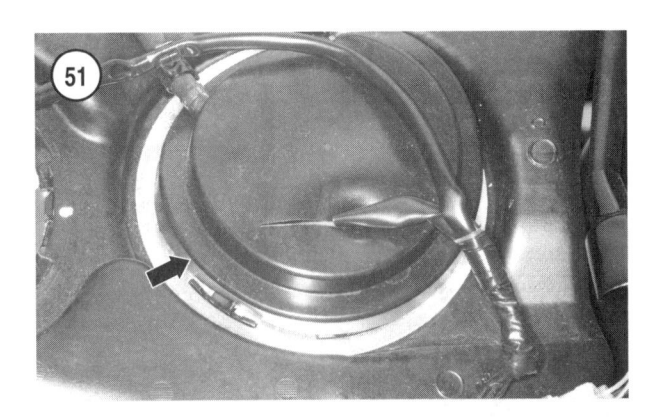

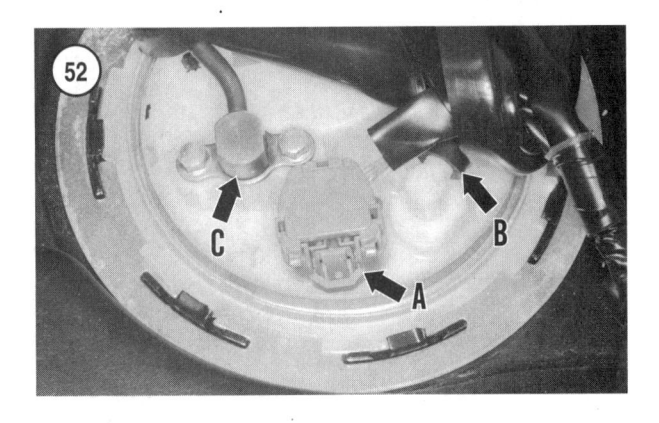

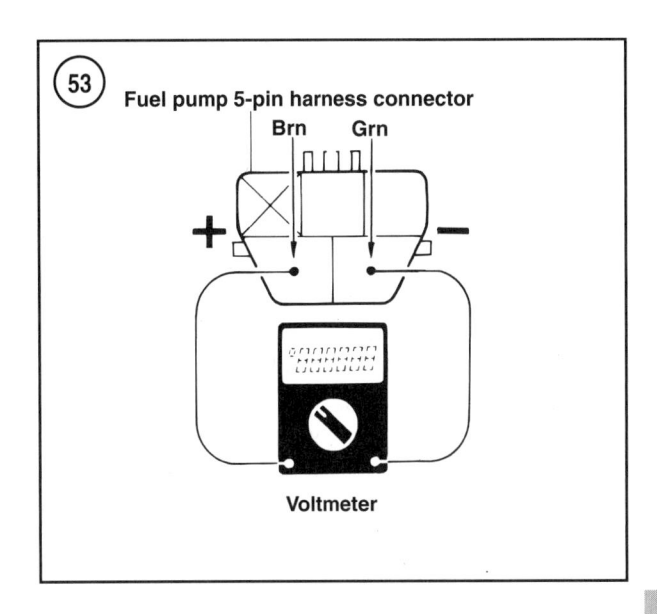

voltmeter. It should read 13.0-13.2 volts (battery voltage) for a few seconds. Turn the ignition switch off. Note the following:
   a. If there is battery voltage, replace the fuel pump.
   b. If there is no battery voltage, go to Step 6.
6. If there is no battery voltage, check the following:
   a. Check for an open circuit in the green wire between the 5-pin fuel pump connector and ground.
   b. Check for an open circuit in the brown wire between the 5-pin fuel pump connector and the fuel pump relay.
   c. If the green and brown wires test correctly, test the fuel pump relay circuit as described under *Fuel Pump Relay* in this chapter.
7. Reverse Steps 1-4 to complete installation.

Removal/Installation

The fuel pump can be replaced with the fuel tank mounted on the motorcycle.

A fuel sender/pump wrench (Honda part No. 07ZMA-MCAA201 or 07ZMA-MCAA200) is required to remove and install the fuel pump retainer plate.

Refer to **Figure 54**.

> *WARNING*
> *Fuel vapors are present when depressurizing the fuel system and when removing and installing the fuel pump. Also some fuel may spill when doing these procedures. Because gasoline is extremely flammable, perform this procedure away from all open flames, including appliance pilot lights and sparks. Do not smoke or allow someone who is smoking in*

the work area as an explosion and fire may occur. Always work in a well-ventilated area. Wipe up spills immediately.

1. Remove the seat. Refer to *Seat* in Chapter Seventeen.
2. Remove the cover (**Figure 51**) from the top of the fuel pump.
3. Disconnect the 5-pin connector (A, **Figure 52**) at the fuel pump.
4. Disconnect the fuel return hose (B, **Figure 52**).
5. Remove the two bolts securing the fuel supply hose (C, **Figure 52**) to the fuel pump.
6. Place a rag over the fuel supply hose fitting, then pull the fitting up to disconnect it. Soak up the spilled fuel with the rag.
7. Using the fuel sender/pump wrench, turn the fuel pump retainer plate (2, **Figure 54**) counterclockwise to loosen and remove it.
8. Remove the spacer ring (3, **Figure 54**) and fuel pump (4) from the fuel tank.
9. Remove the base gasket (5, **Figure 54**).
10. Install a new base gasket (5, **Figure 54**).
11. Install the fuel pump into the fuel tank by aligning the lugs on the bottom of the pump with the grooves in the fuel tank.
12. Install the spacer ring and a new retainer plate.
13. Turn the retainer plate with the fuel sender/pump wrench until it stops.
14. Check both hose ends for dirt and damage.
15. Lubricate a new O-ring with engine oil and install it onto the fuel supply hose joint (**Figure 55**). Install the hose joint (C, **Figure 52**) and tighten the bolts to 10 N•m (88 in.-lb.).
16. Install the fuel return hose (B, **Figure 52**) and reconnect the 5-pin connector (A).
17. Turn the ignition switch on and check for fuel leaks at the fuel pump.
18. Install the seat.

FUEL PUMP RELAY

A single 4-terminal electrical relay is used to control the fuel pump system. The relay is in the main relay box (**Figure 56**) underneath the seat.

The relays are easily unplugged or plugged into the relay terminal strip. Because only 4-terminal and 5-terminal relays are used, there is no danger of installing a relay in the wrong position. The 4-terminal relays are identical and the 5-terminal relays are identical (**Figure 56**). Refer to *Relay Box* in Chapter Nine for more information on identifying and testing the relays.

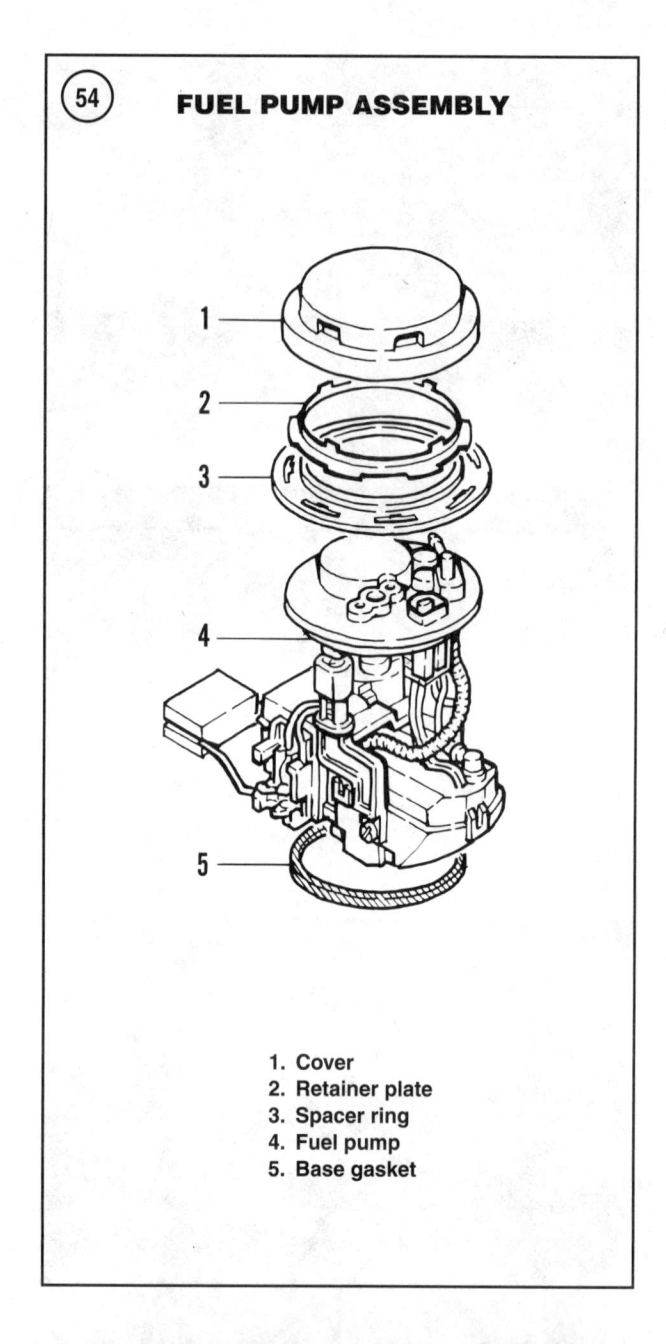

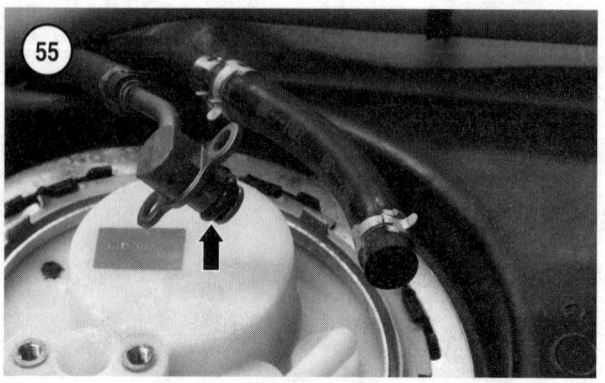

54 FUEL PUMP ASSEMBLY

1. Cover
2. Retainer plate
3. Spacer ring
4. Fuel pump
5. Base gasket

FUEL INJECTION AND EMISSION CONTROL SYSTEMS

56

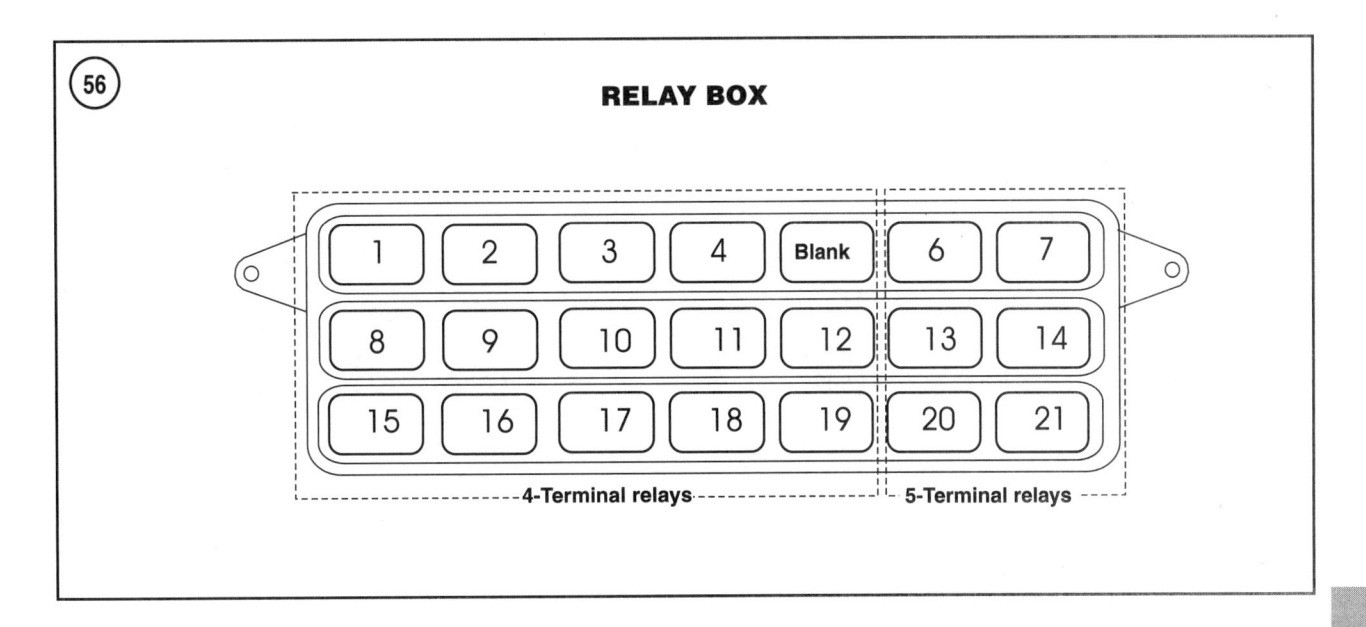

RELAY BOX

```
  1   2   3   4  Blank  6   7
  8   9  10  11  12  13  14
 15  16  17  18  19  20  21
```
·······4-Terminal relays·······| ···5-Terminal relays···

57

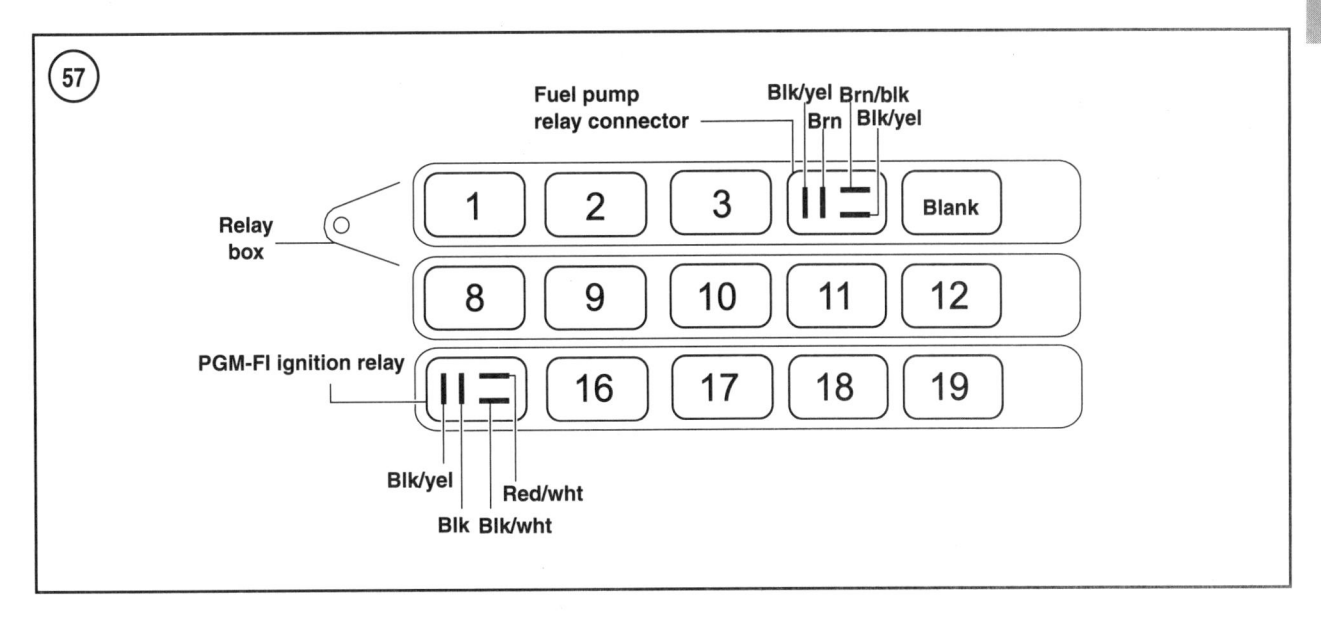

Fuel pump relay connector — Blk/yel, Brn, Brn/blk, Blk/yel

Relay box

```
  1   2   3   [ ][=]  Blank
  8   9  10  11  12
 [ ][=] 16  17  18  19
```

PGM-FI ignition relay — Blk/yel, Blk, Red/wht, Blk/wht

Relay and Circuit Testing

Refer to the wiring diagrams at the end of the manual to identify the fuel pump and relay circuits referred to in this procedure.

1. Remove the seat. Refer to *Seat* in Chapter Seventeen.

2. Remove the relay box cover screws and cover.

3. Replace the fuel pump relay (4, **Figure 56**) with a known good 4-terminal relay. Turn the ignition switch on or to ACC. The fuel pump should operate for a few seconds. Note the following:

   a. If the fuel pump operates, the original fuel pump relay is faulty. Install a new 4-terminal relay and repeat the test.
   b. If the fuel pump did not operate, turn the ignition switch off and continue with Step 4.

NOTE
*The meter test connections called out in Steps 4-8 are made at the relay connector terminals inside the relay box (**Figure 57**).*

4. Turn the ignition switch off and remove the fuel pump relay (4, **Figure 56**). Turn the ignition switch on and measure voltage between the fuel pump relay black/yellow terminal (+)

(**Figure 57**) and ground (–). Make this test for both fuel pump relay black/yellow terminals:
   a. No battery voltage: Refer to Step 5.
   b. Battery voltage: Refer to Step 6.
5. Turn the ignition switch off and remove the PGM-FI ignition relay (15, **Figure 56**). Check the continuity of the black/yellow wire between the fuel pump relay and the PGM-FI ignition relay (**Figure 56**). Make the test for both fuel pump relay black/yellow terminals:
   a. Continuity recorded at both tests: Check the PGM-FI ignition relay circuits. Refer to *PGM-FI Ignition System Relay* in Chapter Nine.
   b. No Continuity: Check for an open circuit in the black/yellow wire.
6. Turn the ignition switch off. Connect a voltmeter between the two black/yellow wire fuel pump relay terminals (**Figure 57**). Turn the ignition switch on. There should be battery voltage for a few seconds. Turn the ignition switch off and note the following:
   a. No battery voltage: Refer to Step 7.
   b. Battery voltage for a few seconds: The relay wire circuit is good. Check for contaminated or damaged terminals in the 5-pin connector (A, **Figure 52**).
7. Remove the top shelter. Refer to *Top Shelter* in Chapter Seventeen.
8. Disconnect the black 22-pin ECM connector (A, **Figure 58**). Check for contaminated or damaged terminals in the 22-pin harness connector. Check the continuity of the brown/black wire between the fuel pump relay and the black 22-pin ECM harness connector:
   a. No continuity: Check for an open circuit in the brown/black wire.
   b. Continuity: The ECM is faulty. Install a new ECM and check the fuel pump operation.
9. Reinstall the relay(s).
10. Install the relay box cover and secure it with the mounting screws.
11. Install the seat. Refer to *Seat* in Chapter Seventeen.

BARO AND MAP SENSORS

Output Voltage Testing

1. Remove the top shelter. Refer to *Top Shelter* in Chapter Seventeen.
2. Turn the ignition switch off.
3. Disconnect the black (A, **Figure 58**) and gray (B) 22-pin ECM harness connectors.
4A. Connect the Honda test harness (breakout box) to the black and gray ECM harness connectors as described under *ECM Test Harness* in this chapter.

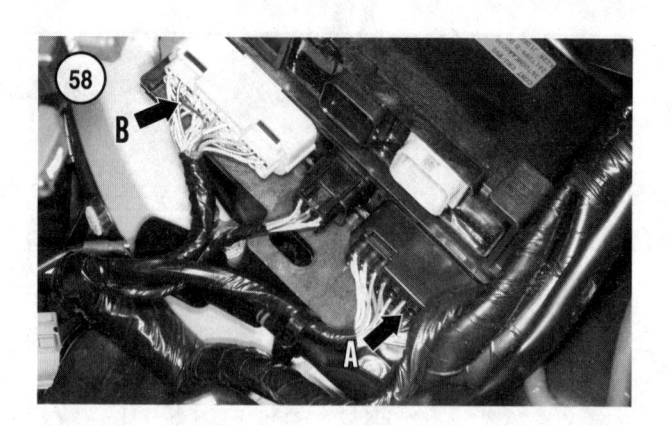

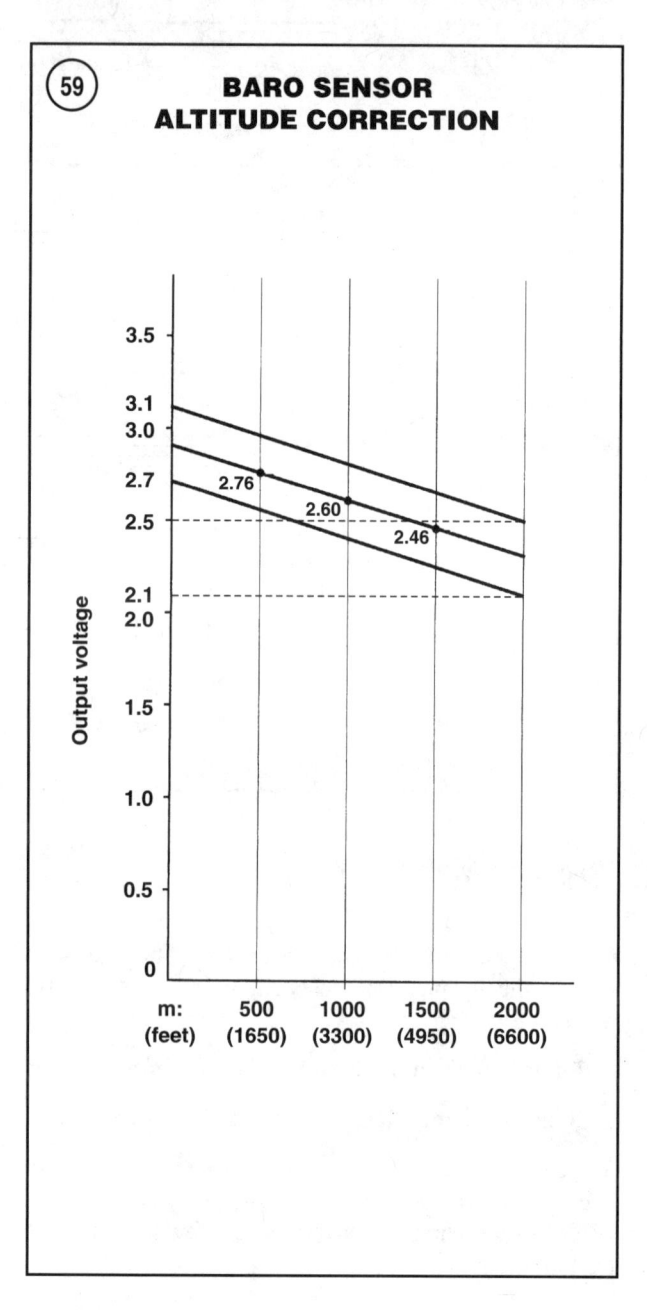

FUEL INJECTION AND EMISSION CONTROL SYSTEMS

6. Reverse Steps 1-3 to complete installation.

BARO Sensor
Replacement

1. Remove the top shelter. Refer to *Top Shelter* in Chapter Seventeen.
2. Disconnect the BARO sensor 3-pin connector (A, **Figure 60**).
3. Remove the screw and BARO sensor (B, **Figure 60**).
4. Installation is the reverse of removal.

MAP Sensor
Replacement

The MAP sensor (A, **Figure 61**) is mounted on the front of the throttle body. Replacement requires removing the throttle body. Refer to *Throttle Body* in this chapter.

IDLE AIR CONTROL (IAC) VALVE

Replacement

The IAC valve (B, **Figure 61**) is mounted on the rear of the throttle body. Replacement requires removing the throttle body. Refer to *Throttle Body* in this chapter.

THROTTLE POSITION (TP) SENSOR

Replacement

The TP sensor (C, **Figure 61**) is mounted on the right side of the throttle body. Because the TP sensor is not available separately, replace the throttle body as an assembly. Refer to *Throttle Body* in this chapter.

CAM PULSE GENERATOR

Replacement

1. Remove the left side injector cover.
2. Disconnect the black 2-pin cam pulse generator harness connector (A, **Figure 62**).
3. Remove the bolt and the cam pulse generator (B, **Figure 62**).
4. Lubricate a new O-ring with engine oil and install it into the groove in the new cam pulse generator.
5. Installation is the reverse of removal.

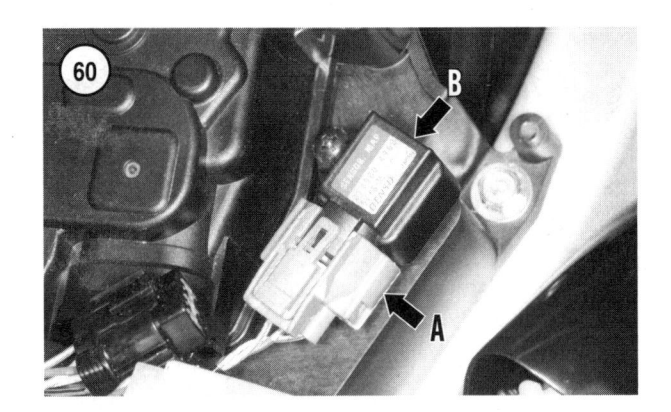

4B. If the Honda test harness is unavailable, refer to *ECM Test Harness* in this chapter to identify the ECM pin terminals called out in the following steps.

5. Perform the voltage tests as follows:

NOTE
*The voltage specifications listed in this test are calculated for sea level conditions. Because output voltage readings change with altitude, use the correction chart in **Figure 59** when performing the tests at altitudes other than sea level.*

a. Turn the ignition switch on and the engine stop switch to RUN.
b. Measure voltage between the BARO sensor terminals B15 (+) and C4 (–). The standard voltage reading is 2.7-3.1 volts.
c. Measure voltage between the MAP sensor terminals B4 (+) and C4 (–). The standard voltage reading is 2.7-3.1 volts.
d. If the voltage readings are incorrect, replace either the BARO or MAP sensor as described in this section.

KNOCK SENSORS

A separate knock sensor is mounted in each cylinder head. See **Figure 63** (left side) and **Figure 64** (right side).

Replacement

1. Remove the front exhaust pipe protector. Refer to *Exhaust System* in Chapter Seventeen.
2. Remove the front lower fairing. Refer to *Front Lower Fairings* in Chapter Seventeen.
3. Disconnect the 1-pin connector at the knock sensor (**Figure 63** or **Figure 64**).
4. Loosen and remove the knock sensor.
5. Installation is the reverse of removal. Fit a length of hose onto the sensor's connector end to help guide the sensor into position and to start threading it into the cylinder head. Tighten the knock sensor to 31 N•m (23 ft.-lb.).

OXYGEN (O_2) SENSORS

Each exhaust pipe assembly is equipped with an O_2 sensor (**Figure 65**).

Replacement

1. Remove the top shelter. Refer to *Top Shelter* in Chapter Seventeen.
2. Trace the wire harness from the O_2 sensor to its 4-pin connector and disconnect the connector. The left side O_2 sensor 4-pin connector is black and the right side connector is white.
3. Remove the O_2 sensor (**Figure 66**) from the exhaust pipe. If reusing the O_2 sensor, wrap it in a thick, clean cloth to prevent contamination.

CAUTION
O_2 sensors can be damaged by contact with silicone, oil and many types of other chemicals.

4. Installation is the reverse of removal.
5. Tighten the O_2 sensor to 25 N•m (18 ft.-lb.).

Inspection

When replacing an O_2 sensor, examine the color and condition of the sensor tip:
1. Black sooty deposits indicate the engine was operating with a rich air/fuel mixture.
2. Dark brown deposits indicate excessive oil consumption.
3. White deposits indicate a possible head gasket leak and antifreeze entering the combustion chamber.

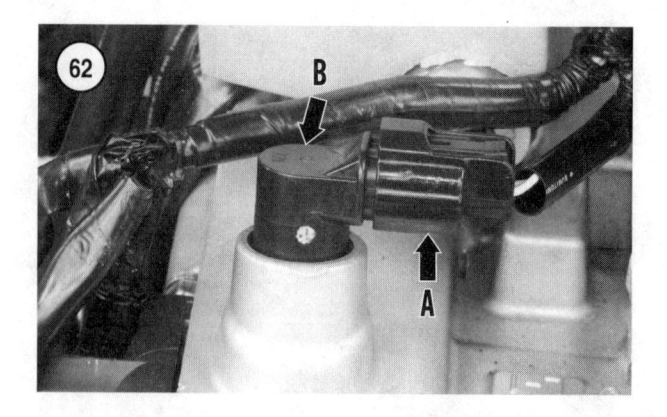

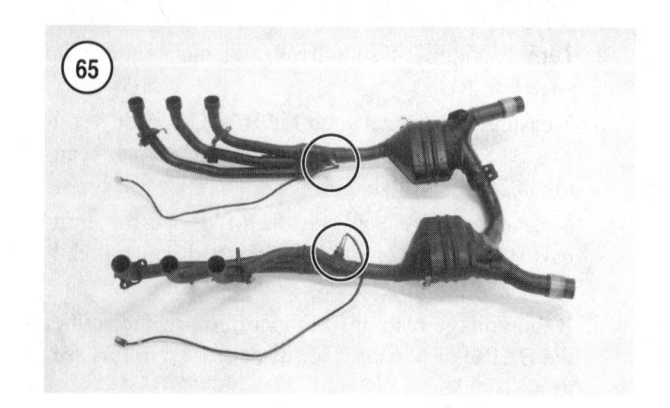

FUEL INJECTION AND EMISSION CONTROL SYSTEMS

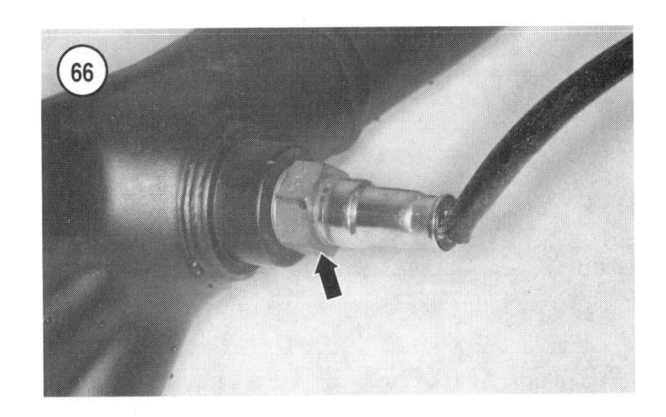

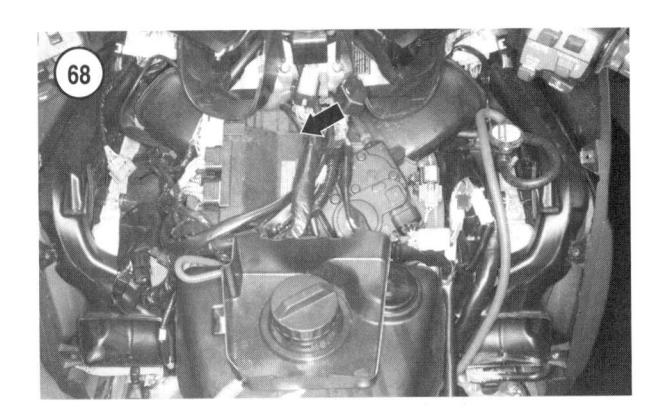

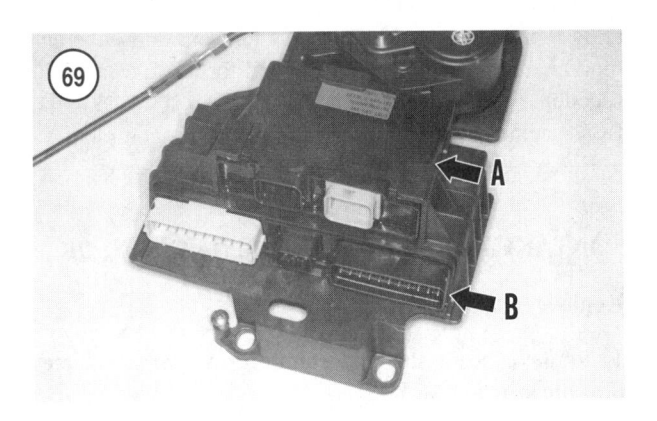

VEHICLE SPEED SENSOR (VSS)

Replacement

1. Remove the top shelter. Refer to *Top Shelter* in Chapter Seventeen.
2. Remove the right front exhaust pipe protector. Refer to *Exhaust System* in Chapter Seventeen.
3. Disconnect the VSS white 3-pin connector from the connector pouch near the right air vent.

NOTE
***Figure 67** shows the vehicle speed sensor with the exhaust pipe removed for clarity.*

4. Unbolt and remove the VSS (**Figure 67**) and its O-ring.
5. Installation is the reverse of removal, plus the following:
    a. Lubricate a new O-ring with engine oil and install it on the vehicle speed sensor.
    b. Tighten the vehicle speed sensor mounting bolts (**Figure 67**) securely.

ENGINE CONTROL MODULE (ECM)

The ECM (**Figure 68**) is mounted in a control unit holder on top of the air filter cover. The cruise/reverse control module (A, **Figure 69**) is mounted over the ECM (B).

Power and Ground Circuit Inspection

1. Remove the top shelter. Refer to *Top Shelter* in Chapter Seventeen.
2. Turn the ignition switch OFF.
3. Disconnect the black (A, **Figure 58**) and gray (B) 22-pin ECM harness connectors.
4A. Connect the Honda test harness to the black and gray ECM harness connectors as described under *ECM Test Harness* in this chapter.
4B. If the Honda test harness is unavailable, refer to *ECM Test Harness* in this chapter to identify the ECM pin terminals called out in the following steps.
5. Perform the power input tests as follows:
    a. Turn the ignition switch on and the engine stop switch to RUN.
    b. Measure voltage between test pin box No. 1 terminal and ground. There should be battery voltage.
    c. If there is no voltage, check for an open circuit in the black/yellow wire between the ECM and PGM-FI ignition relay.
    d. If the black/yellow wire is tests correctly, check the PMG-FI ignition relay circuits as described in Chapter Nine.
6. Perform the ground circuit tests as follows:

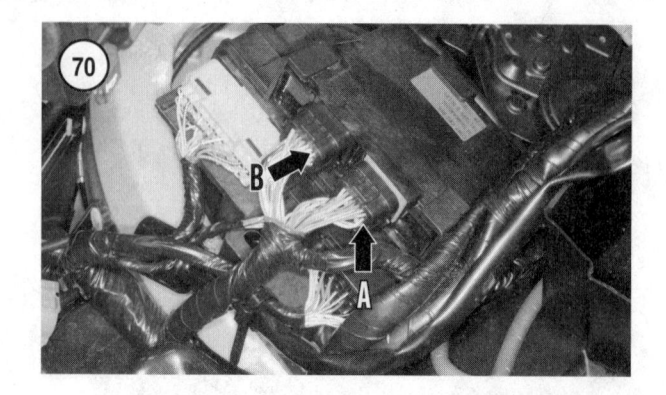

a. Check for continuity between the test pin box No. 9 terminal and ground.
b. Check for continuity between the test pin box No. 19 terminal and ground.
c. Check for continuity between the test pin box No. 20 terminal and ground.
d. Continuity should be recorded during each test.
e. If there is no continuity, check the green wires for an open circuit or loose or damaged connections.

7. Reverse Steps 1-4 to complete installation.

Removal/Installation

1. Remove the top shelter. Refer to *Top Shelter* in Chapter Seventeen.
2. Turn the ignition switch off.
3. Disconnect the black 26-pin (A, **Figure 70**) and gray 26-pin (B) harness connectors from the cruise/reverse control module.
4. Disconnect the following harness connectors at the ECM:
    a. Gray 22-pin (A, **Figure 71**).
    b. Black 6-pin (B, **Figure 71**).
    c. Black 22-pin (C, **Figure 71**).
5. Carefully lift the holder clamps and remove the cruise/reverse control module (A, **Figure 69**).
6. Carefully lift the holder clamps and remove the ECM (B, **Figure 69**). See **Figure 72**.

CAUTION
The pins in the control modules are now exposed. Handle the assembly carefully to avoid damaging them.

7. Installation is the reverse of these steps. If a new ECM was installed, perform the *ECM Initialization* procedure in this section.

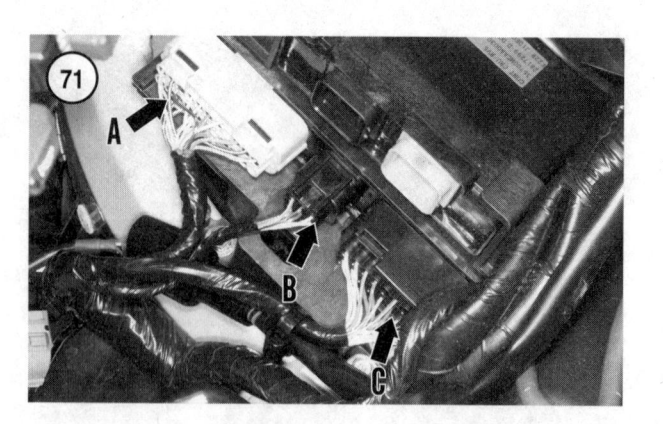

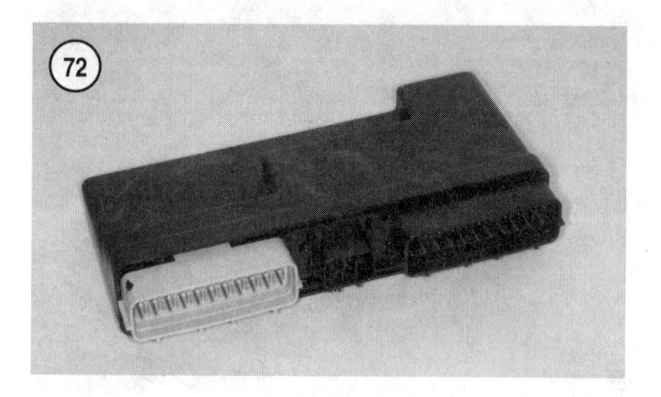

ECM Initialization

This procedure must be performed after installing a new ECM to enter idle information into the ECM memory. Otherwise, the engine will be unable to maintain the correct engine idle speed.

1. Make sure the new ECM is installed and the engine is ready to start.

NOTE
Do not operate the throttle anytime during Step 2.

2. Start the engine and allow it to warm up to normal operating temperature. When the fans turn on, allow the engine to continue to run at idle speed for an additional 90 seconds. At this time, the idle information is stored in the ECM memory and the ECM initialization is complete.
3. Either ride the motorcycle or turn the engine off.

INTAKE AIR TEMPERATURE (IAT) SENSOR

Replacement

1. Remove the top shelter. Refer to *Top Shelter* in Chapter Seventeen.

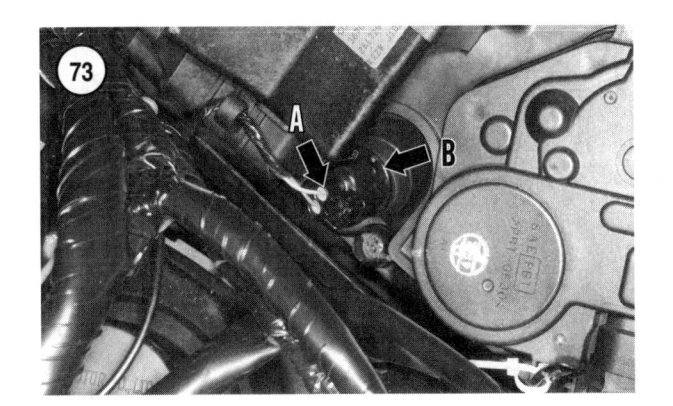

FUEL INJECTION AND EMISSION CONTROL SYSTEMS

2A. Disconnect the 2-pin connector (A, **Figure 73**) and remove the IAT sensor (B) from the air filter cover. See **Figure 74**.

2B. If the control unit holder interferes with removal of the IAT sensor, remove the holder as described in Chapter Three under *Air Filter*, then remove the IAT sensor.

3. Installation is the reverse of removal.

ENGINE IDLE SPEED ADJUSTMENT

The engine control module (ECM) determines the engine idle speed based on inputs from the IAT, ECT, TP and MAP sensors. The idle air control valve (IAC), mounted on the throttle body, bypasses the closed throttle valves through a channel in the throttle body to supply the intake manifold with the correct amount of air to set the idle speed.

Idle Speed Inspection

WARNING
Do not start and run the motorcycle in an enclosed area. The exhaust gasses contain carbon monoxide, a colorless, tasteless, poisonous gas. Carbon monoxide levels build quickly in an enclosed area and can cause unconsciousness and death in a short time.

1. Place the motorcycle on its centerstand and shift the transmission into neutral.
2. Start the engine and allow it to warm to normal operating temperature.

NOTE
Before checking the engine idle speed, make sure the MIL does not indicate a diagnostic trouble code (DTC). If a DTC is observed, record the code and test the fuel injection system as described in this chapter.

3. Check the engine idle speed. The correct idle speed is 700 ± 70 rpm. If engine idle speed is below 630 rpm, perform the *Low Engine Idle Speed Adjustment*. If idle speed is above 770 rpm, perform the *High Engine Idle Speed Adjustment*.

Low Engine Idle Speed Adjustment

1. Check the charging system output (Chapter Nine). If the charging output is correct, continue with Step 2.
2. Remove the top shelter. Refer to *Top Shelter* in Chapter Seventeen.
3. Remove the left engine side cover. Refer to *Engine Side Covers* in Chapter Seventeen.
4. Turn the ignition switch off and disconnect the following connectors:
   a. ECM black 22-pin connector (C, **Figure 71**).
   b. Alternator 4-pin connector (**Figure 75**).
5. Check continuity between the black 22-pin ECM connector (C, **Figure 71**) C8 terminal and the alternator connector (**Figure 75**) black wire terminal. Refer to *ECM Test Harness* in this chapter to identify the C8 terminal in the black ECM 22-pin harness side connector.
   a. Continuity: Refer to Step 6.

b. No continuity: Repair the open circuit in the black wire between the ECM and alternator connectors.

6. Remove the air filter housing as described in this chapter.
7. Reinstall the fuel tank.
8. Reconnect the connectors disconnected in Steps 4 and 6.
9. Start the engine and allow it to warm to normal operating temperature.
10. Disconnect the idle air control valve 3-pin connector (A, **Figure 76**). Then tape over the two throttle body air passages (B, **Figure 76**). The engine idle speed should decrease:

   a. Idle speed did not decrease: Replace the IAC valve as described in this chapter.

   b. Idle speed decreased: Turn both air screws (C, **Figure 76**) counterclockwise (out) in equal amounts to raise the engine idle speed to 500-600 rpm.

NOTE
__Figure 47__ shows the air screws with the throttle body removed from the engine.

   c. If the engine idle speed cannot be set within the range listed in substep b, refer to Step 11.

11. Remove the throttle body as described in this chapter. Then remove the air screws and clean the air screw passages. Reinstall the throttle body and readjust the air screws as described in Step 9 and Step 10.
12. Remove the tape placed over the throttle body air passages.
13. Reverse Steps 1-7 to complete installation.

High Engine Idle Speed Adjustment

1. Check the charging system output as described in Chapter Nine. If the charging output is correct, continue with Step 2.
2. Turn the engine off and allow it to cool until the temperature gauge needle registers below the C level line.
3. Start the engine and allow it to idle until the radiator cooling fans come on, then stop the engine and restart it. Allow the engine to idle for 10-20 seconds, then recheck the idle speed and note the following:

   a. If the idle speed is 630-770 rpm, the system is working correctly.

   b. If the idle speed is above 770 rpm, refer to Step 4.

4. Remove the air filter housing as described in this chapter.
5. Perform the following:

   a. Check that the No. 5 hoses and vacuum hose are connected to the throttle body. Refer to *Throttle Body* in this chapter.

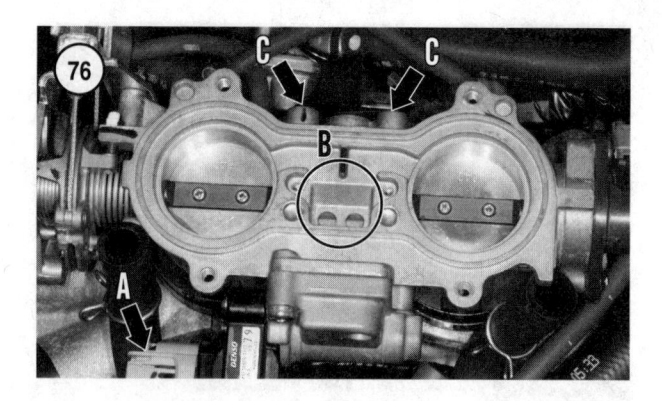

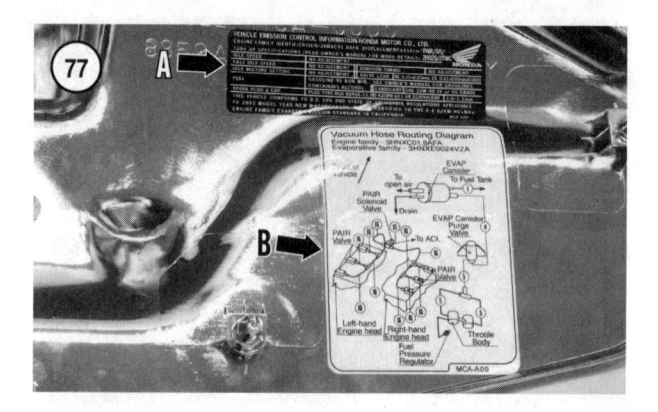

   b. Check that the vacuum hose is connected to the pressure regulator. Refer to *Pressure Regulator* in this chapter.

   c. If all the hoses are in good condition and properly connected, refer to Step 6.

6. Reinstall the fuel tank.
7. Reconnect the ECM, IAT sensor and BARO sensor connectors disconnected in Step 4.
8. Start the engine and allow it to warm to normal operating temperature.
9. Disconnect the idle air control valve 3-pin connector (A, **Figure 76**). Then tape over the two throttle body air passages (B, **Figure 76**). The engine idle speed should be 500-600 rpm:

FUEL INJECTION AND EMISSION CONTROL SYSTEMS

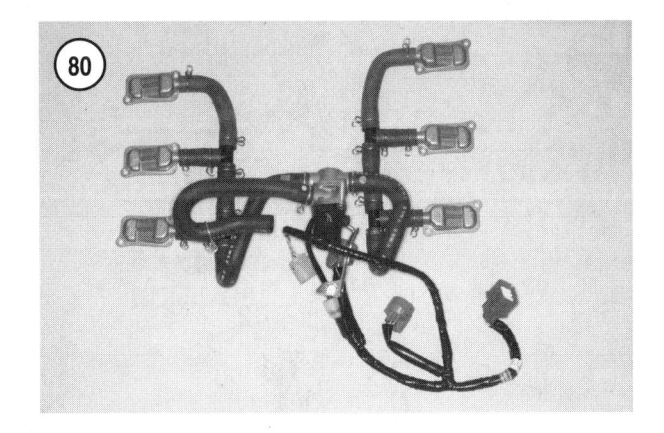

a. If the engine idle speed is higher than 600 rpm, refer to Step 10.
b. If the engine idle speed is correct, refer to Step 11.

10. Turn both air screws (C, **Figure 76**) clockwise (in) in equal amounts to lower the engine idle speed to 500-600 rpm. Note the following:

NOTE
Figure 47 shows the air screws with the throttle body removed from the engine.

a. If idle speed adjusts to 500-600 rpm, refer to Step 11.
b. If the idle speed does not lower to 500-600 rpm and the air screws are turned all the way in, install a new throttle body assembly. Refer to *Throttle Body* in this chapter.

11. Turn the engine off. Remove the tape placed over the throttle body air passages.

12. Start the engine and check the idle speed. It should be between 1500 to 1600 rpm. Note the following:
    a. If the idle speed is correct, go to Step 13.
    b. If the idle speed is incorrect, replace the IAC valve as described in this chapter.

13. Turn the engine off and reconnect the IAC valve 3-pin connector (A, **Figure 76**).

14. Start the engine and check the idle speed. It should be between 630-770 rpm. Note the following:
    a. If the idle speed is correct, the system is working correctly.
    b. If the idle speed is incorrect, replace the IAC valve as described in this chapter.

15. Reinstall the air filter housing and the remaining parts removed to complete assembly.

EMISSION CONTROL SYSTEM LABELS

The Vehicle Emission Control Information label attached to the inside of the right engine side cover (A, **Figure 77**) lists tune-up information. The vacuum hose routing diagram (B, **Figure 77**) is useful when identifying hoses used in the emission control system.

SECONDARY AIR SUPPLY SYSTEM

The secondary air supply system lowers emissions output by introducing fresh air into the exhaust port. This function is performed by the Pulse Secondary Air Injection (PAIR) control valve. The introduction of air raises the exhaust temperature, which consumes some of the unburned fuel in the exhaust and changes most of the hydrocarbons and carbon monoxide into carbon dioxide and water vapor. The PAIR control valve is part of the fuel injection system and controlled by the ECM. Reed valves installed in the crankcase prevent a reverse flow of air through the system.

No adjustments are required for the PAIR system. However, inspect the system as described in this section at the intervals listed in Chapter Three.

Inspection

1. Start the engine and warm it to normal operating temperature. Turn the engine off.
2. Remove the air filter element. Refer to *Air Filter* in Chapter Three.
3. Check the secondary air intake port (**Figure 78**). The port should be clean. If there are carbon particles in and around the port, check the pulse PAIR check valves as described in this section.
4. Remove the air filter housing as described in this chapter.
5. Install the fuel tank.
6. Reconnect the connectors disconnected in Step 2.
7. Start the engine. Open the throttle slightly and check that air is being drawn through the No. 15 air supply hose (A, **Figure 79**). If there is no air suction through the hose, check the air supply hoses (**Figure 80**) for contamination.

If the hoses are clear, test the PAIR control solenoid valve as described in this section.

NOTE
Figure 80 shows the PAIR control solenoid valve and air supply hoses removed from the engine for clarity.

8. Reverse these steps to install the removed parts and complete installation.

**PAIR Control Solenoid Valve
Inspection/Replacement**

1. Remove the air filter housing as described in this chapter.
2. Disconnect the 2-pin connector and air supply hoses from the PAIR control solenoid valve (B, **Figure 79**) and remove the valve.

NOTE
Figure 81 shows the PAIR control solenoid valve removed from the engine for clarity.

3. Inspect the operation of the PAIR control solenoid valve (**Figure 82**) as follows:
   a. Wipe off and then blow through the inlet port.
   b. Air should flow from the two outlet ports. Replace the valve if air does not flow from the outlet ports.
   c. Connect a 12 volt battery to the solenoid valve connector black/yellow terminal (+) and to the orange/green terminal (−).
   d. Blow through the inlet port once again.
   e. Air should not flow from the outlet ports with battery voltage applied to the valve. Replace the valve if air flows from the outlet ports.
   f. Disconnect the battery leads from the valve.
4. Reverse Step 1 and Step 2 to complete installation.

**PAIR Check Valve
Removal/Inspection/Installation**

1. Remove the intake manifold as described in this chapter.
2. Remove the bolts, check valve covers (**Figure 83**) and hoses from the crankcase. **Figure 80** shows the check valve covers with the hoses attached as an assembly.
3. Remove the PAIR check valve (**Figure 84**) from each port.
4. Inspect the PAIR check valves (**Figure 85**) and replace if any of the following conditions are present:

NOTE
The reed valves must not be serviced or disassembled. Do not try to bend the stopper. If any part of the valve is defective, replace the valve as an assembly.

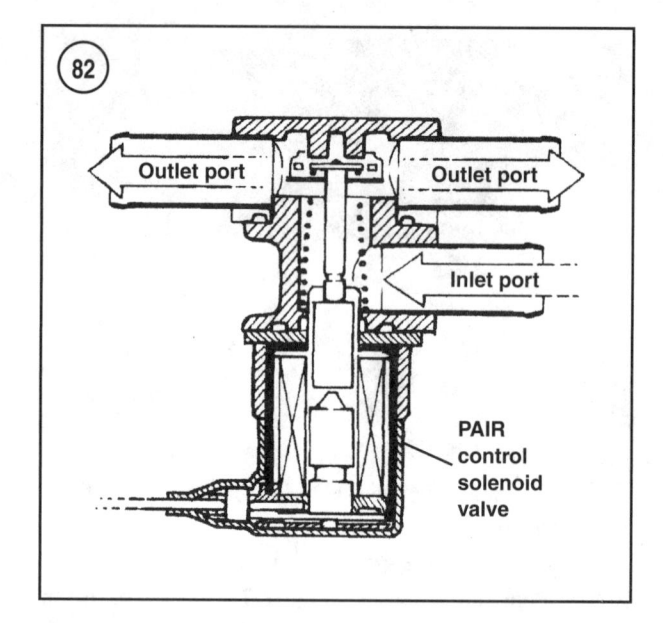

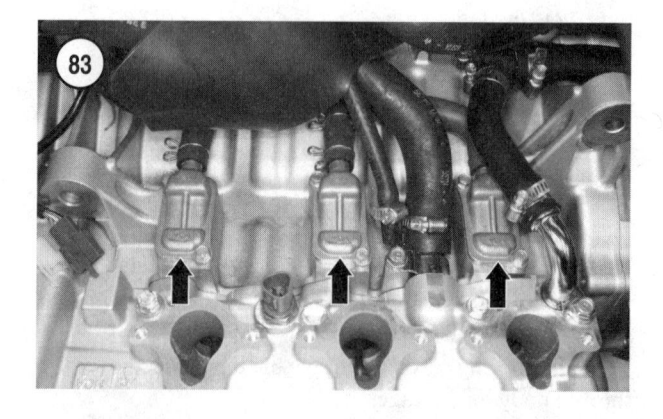

   a. Damaged rubber seat.
   b. Damaged reed valve.
   c. Clearance between the reed valve and its seat.
   d. Loose or damaged stopper.
5. Install the PAIR check valve into the crankcase as shown in **Figure 84**.

FUEL INJECTION AND EMISSION CONTROL SYSTEMS

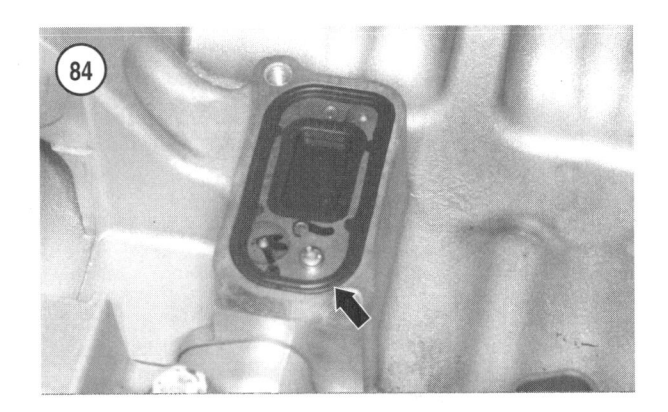

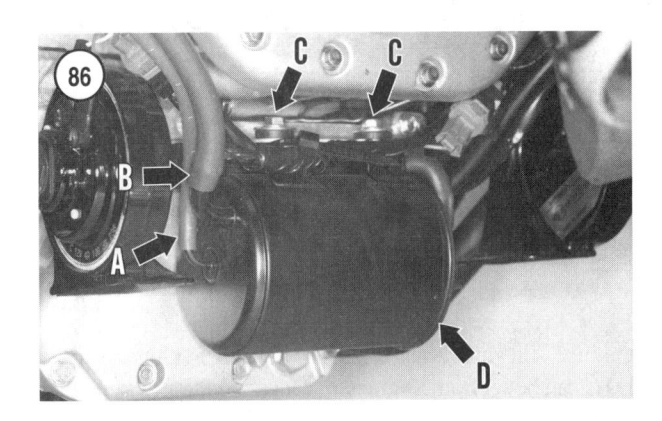

6. Install the PAIR check valve covers and hoses (**Figure 83**). Tighten the mounting bolts to 5 N•m (44 in.-lb.).

7. Install the intake manifold as described in this chapter.

EVAPORATIVE EMISSION CONTROL SYSTEM

An evaporative emission control system (EVAP) is installed on all models.

A vacuum hose routing diagram label is mounted on the inside of the right engine side cover (B, **Figure 77**). Fuel vapors from the fuel tank are routed into a EVAP canister, where it is stored when the engine is not running. When the engine is running, these vapors are drawn into the EVAP purge control solenoid valve and through the throttle body and into the engine to be burned. Make sure all hose clamps are tight. Check all hoses for deterioration and replace as necessary.

Inspect the EVAP control system at the intervals specified in Chapter Three.

EVAP Canister
Removal/Installation

1. Remove the front lower fairing. Refer to *Front Lower Fairings* in Chapter Seventeen.

2. Disconnect the No. 1 (A, **Figure 86**) and No. 4 (B) hoses from the EVAP canister. Remove the clamp from the bracket.

3. Remove the bolts (C, **Figure 86**) and lower the canister. Remove the wire clamp from the wire harness (**Figure 87**) and remove the canister (D, **Figure 86**).

4. Inspect the canister for damage.

5. Inspect the hoses for cuts, damage or soft spots. Replace any damaged hoses.

6. Installation is the reverse of removal.

EVAP Purge Control Solenoid Valve
Inspection/Replacement

1. Remove the center inner fairing. Refer to *Front Lower Fairings* in Chapter Seventeen.

2. Disconnect the 2-pin connector (A, **Figure 88**), No. 4 tube (B) and No. 5 tube (located behind the valve) at the EVAP purge control solenoid valve (C).

3. Remove the valve from its mounting bracket.

4. Inspect the operation of the EVAP purge control solenoid valve (**Figure 89**) as follows:

   a. Wipe off and then blow through the inlet port.
   b. Air should not flow through the outlet port. Replace the valve if air flows through the outlet port.

c. Connect a 12 volt battery to the solenoid valve connector black/yellow terminal (+) and to the yellow/black terminal (−).
d. Blow through the inlet port once again.
e. Air should flow from the outlet port with battery voltage applied to the valve. Replace the valve if air did not flow from the outlet port.
f. Disconnect the battery leads from the valve.
5. Reverse Steps 1-3 to complete installation.

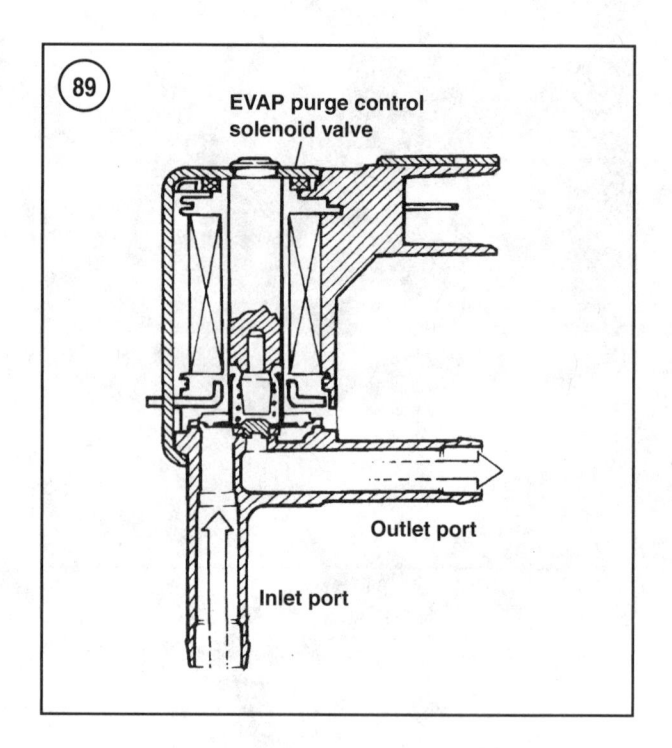

INTAKE MANIFOLD

Removal/Installation

1. Remove the engine from the frame as described in Chapter Five.
2. Make an identification mark on the intake manifold so it can be installed facing in its original direction.
3. Remove the bolts and the intake manifold (**Figure 90**) from the cylinder heads. Remove the intake manifold gaskets.
4. Inspect the intake manifold (**Figure 91**) for cracks and other damage.
5. Install the intake manifold, using a new gasket on each side. Tighten the intake manifold mounting bolts securely.
6. Install the engine into the frame as described in Chapter Five.

MALFUNCTION INDICATOR LAMP (MIL)

The fuel injection system is equipped with self-diagnostic capability. Under normal operating conditions, the malfunction indicator lamp (MIL) (**Figure 92**) lights for a few seconds and then goes off when the ignition switch is turned on and the engine stop switch is in the RUN position. This check informs the rider that the MIL indicator is operational. If the MIL indicator stays on or comes on at any time when the engine is running, a fault has occurred in the PGM-FI system. The engine control module (ECM) stores a diagnostic trouble code (DTC) in the system's memory. Under most conditions, the bike still runs when the MIL indicator is illuminated. A DTC can be set from a variety of causes: ignition system, fuel system, secondary air injection system, or a combination of these systems and conditions. Start fuel injection troubleshooting by systematically retrieving the DTCs and following the information listed in the specified flow chart.

Diagnostic Trouble Code (DTC)

Diagnostic trouble codes (DTC) are retrieved from the ECM by triggering a series of timed flashes across the MIL (**Figure 92**). The number of flashes displayed by the MIL indicator indicates the stored DTC (**Table 4**).

Note the following before retrieving DTC(s):

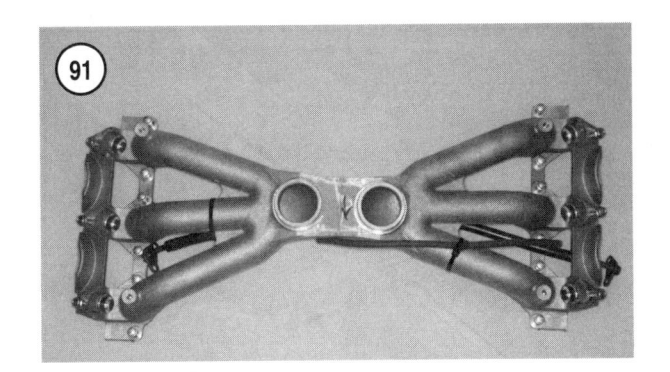

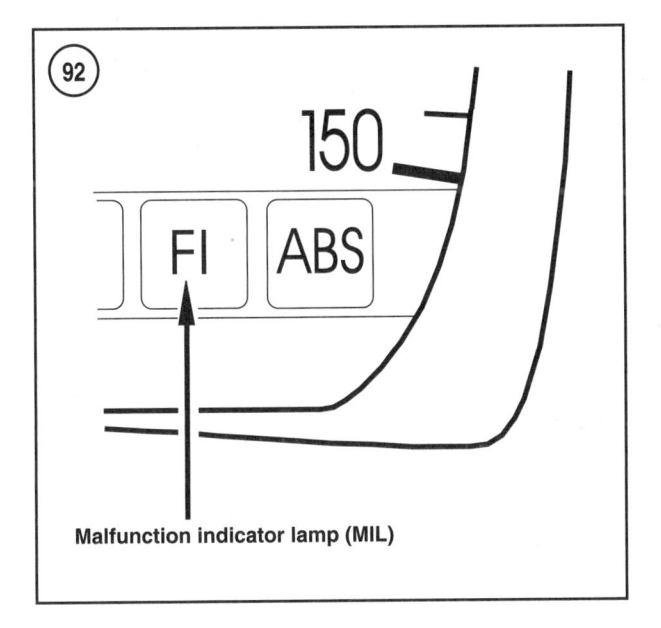

FUEL INJECTION AND EMISSION CONTROL SYSTEMS

91

92

Malfunction indicator lamp (MIL)

93

1. The DTC set in the ECU memory will remain in memory until the problem is repaired and the DTC is erased. Turning the ignition switch off during or after retrieving the DTC does not erase it. However, to view the DTC again after turning the ignition switch off and then on, refer to *Reading DTC(s)* in this section.
2. The ECU can store more than one DTC. The system displays the DTC, starting with the lower number first. For example, DTC 9 would be displayed before DTC 23.
3. Always write down the DTC number(s).
4. After troubleshooting and servicing the fuel injection system, erase the DTC to ensure the problem has been successfully repaired.

Self-Diagnostic Check (MIL Lamp Check)

This is a standard check to make before starting the engine.

1. Support the motorcycle on its sidestand. 
2. Turn the ignition switch on and the engine stop switch to the RUN position.
3. The MIL should light for a few seconds and then turn off. Note the following:
   a. If the MIL did not turn on, there is a problem with the MIL lighting circuit. Refer to *MIL Lamp Circuit Test* in this section.
   b. If the MIL turned off after a few seconds, there are no codes stored in the ECM.

   NOTE
   If the engine does not start, operate the starter for at least 10 seconds to trigger codes that cause a no-start condition. If the MIL now blinks, a code has been set.

   c. If the MIL stayed on, a DTC has been set in the ECM. Perform the *Reading DTC(s)* procedure in this section.

Reading DTC(s)

1. Turn the ignition switch off.
2. Remove the seat. Refer to *Seat* in Chapter Seventeen.
3A. On 2001-2003 models, locate the black 3-pin service check connector in the pouch on the left side of the relay box. Connect the two terminals with a jumper wire (**Figure 93**).
3B. On 2004-2005 models, locate the black 4-pin data link connector in the pouch on the left side of the relay box. Remove the dummy plug from the connector and short the green and brown wire terminals with the SCS service connector (Honda part No. 070PZ-ZY30100) (**Figure 94**).
4. Turn the ignition switch on and the engine stop switch to RUN and observe the MIL as the diagnostic system immediately enters its start mode:
   a. If the MIL comes on and stays on, there are no codes stored in the ECM. Disconnect the jumper wire (2001-2003) or disconnect the data link connector

and reinstall the dummy plug (2004-2005). Reinstall the seat.

NOTE
Figure 95 shows a graphic representation of observing the MIL indicator when there are no codes stored in memory.

NOTE
If the engine does not start, the engine must be turned over to generate a DTC from memory. Turn the ignition switch on and the engine stop switch to RUN. Operate the starter for more than 10 seconds. Check the MIL to see if it blinks. If so, continue with substep b.

b. If the MIL starts blinking, DTC(s) are stored in the ECM. Continue with Step 5.

5. Determine the diagnostic failure code(s) as follows:

  a. The system indicates codes with a series of long (1.3 second) and short (0.5 second) flashes. Each long flash equals 10. Each short flash equals 1. For example, a 1.3 second illumination (1.3 second × 10 = 10) followed by three 0.5 second illuminations (0.5 second × 3 = 3) indicate a failure code of 13 (10 + 3).

NOTE
Figure 96 shows a graphic representation of observing the MIL indicator when reading DTC 13.

  b. When more than one failure code occurs, the system will display the codes from lowest to highest. A short gap separates individual codes.

NOTE
Figure 97 shows a graphic representation of retrieving two codes: code 7 and code 23.

6. After retrieving and writing down the DTC, refer to the diagnostic trouble code chart in **Figure 98**. The chart lists the DTC number, the problem area(s) and specifies a flow chart to isolate the problem. Refer to *DTC Flow Charts* in this section for additional information.

7. After repairing the problem, erase the DTC(s) as described under *Erasing DTC(s)* in this section.

Erasing DTC(s)

After reading a DTC and repairing the system, erase the code(s) from memory as follows. If the motorcycle was repaired at a Honda dealership, the code(s) will have been erased at the dealership.

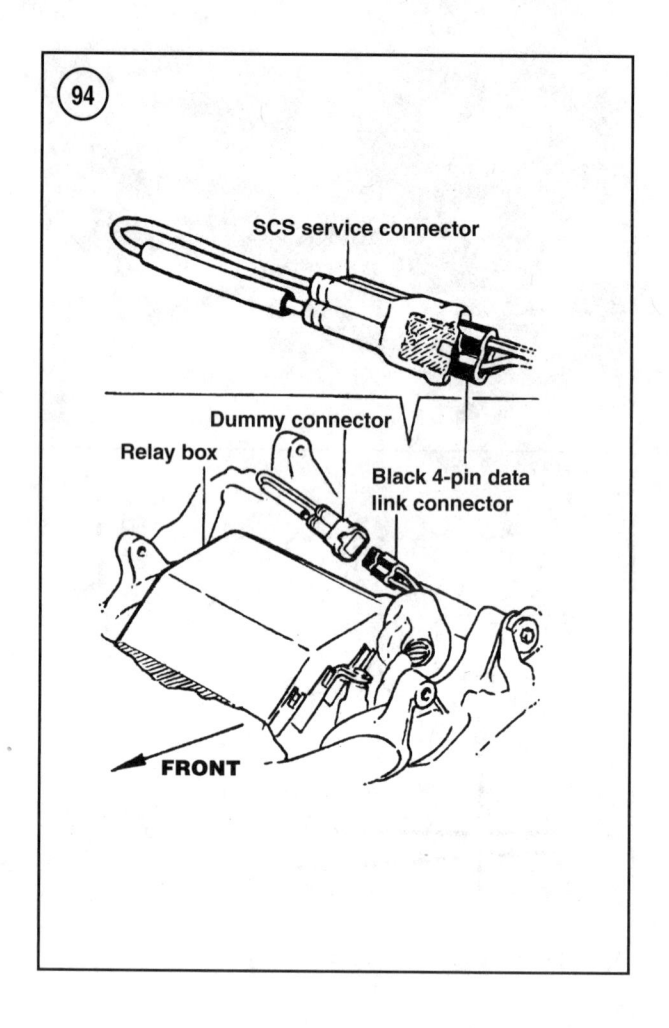

NOTE
Read the following steps through once, then perform the procedure.

1. Turn the ignition switch off.
2. Remove the seat. Refer to *Seat* in Chapter Seventeen.
3A. Locate the black 3-pin service check connector in the pouch on the left side of the relay box. Connect the two terminals with a jumper wire (**Figure 93**).
3B. On 2004-2005 models, locate the black 4-pin data link connector in the pouch on the left side of the relay box. Remove the dummy plug from the connector and short the green and brown wire terminals with the SCS service connector (Honda part No. 070PZ-ZY30100).
4. Turn the ignition switch on and disconnect the jumper wire (**Figure 94**) or the SCS service connector (**Figure 95**).
5. The MIL will come on for approximately 5 seconds. While the light is on, short the connector terminals again with the jumper wire or install the SCS service connector to erase the memory. The MIL will go off, then start blink-

FUEL INJECTION AND EMISSION CONTROL SYSTEMS

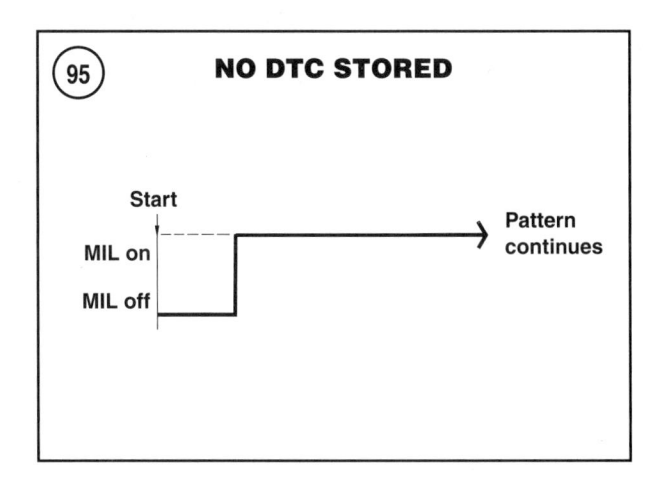

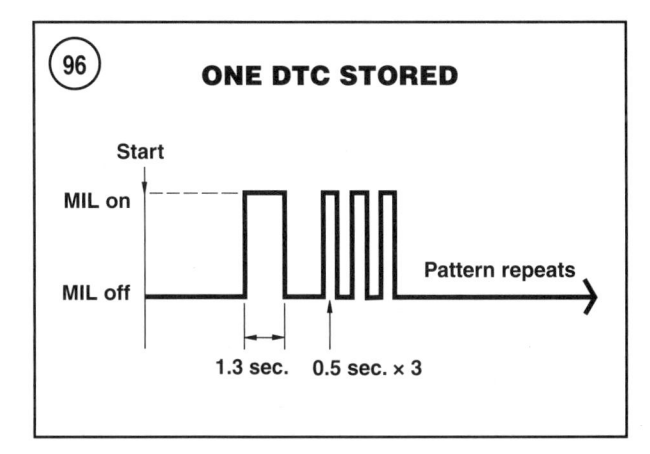

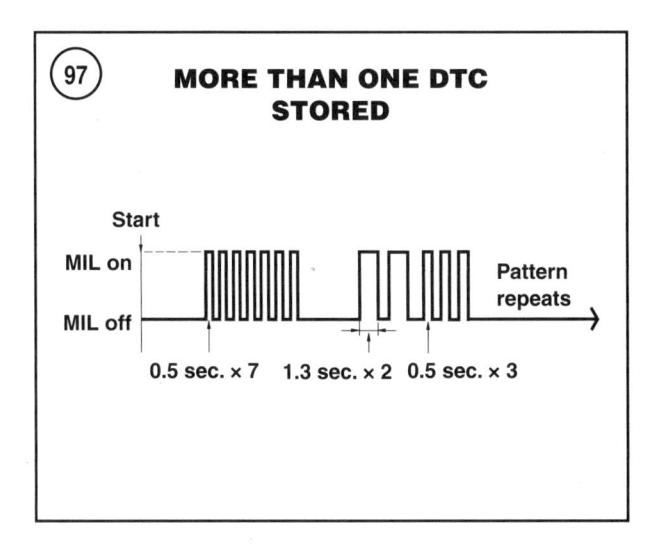

95 NO DTC STORED

96 ONE DTC STORED
1.3 sec.   0.5 sec. × 3

97 MORE THAN ONE DTC STORED
0.5 sec. × 7   1.3 sec. × 2   0.5 sec. × 3

ing. Count the number of blinks. If the MIL blinks 33 times, the code was not erased from memory. Either the erase procedure was not correctly performed, or there is still a code set in memory.

NOTE
In Step 5, the connector must be jumped within the 5-second time period when the MIL is illuminated. If not, the MIL will not blink.

6. Turn the ignition switch off and remove the jumper wire or disconnect the SCS service connector. On 2004-2005 models, install the dummy plug.
7. Repeat the *Self-Diagnostic Check* in this section to confirm that the code(s) was successfully cleared from memory.

DTC Flow Charts

To troubleshoot the fuel and ignition systems, perform the *Reading DTC(s)* procedure described in this section.

If any codes were stored, refer to **Figure 98** and match the *DTC* column with the *DTC flowchart* column. Then, refer to the indicated flowchart(s) (**Figures 99-121**) and perform the relevant procedures.

If the MIL lamp does not come on or stays on, refer to **Table 4** and *MIL Lamp Circuit Test* in this chapter.

If component replacement is called for in a flowchart, refer to *Electrical Component Replacement* in Chapter Nine.

When using the flow charts, note the following:
1. Refer to the wiring diagrams at the end of this manual to identify the connectors. Refer to the appropriate section in this chapter and in Chapter Nine to locate the individual components. Refer to **Table 1** for component abbreviations.
2. Turn the ignition switch off before disconnecting and reconnecting connectors in the fuel injection and ignition systems. The term *Ignition Switch ON* in the flow charts means to turn the switch on, but not to start the engine. The chart will instruct when to start the engine.
3. Cold start means the engine is cold (20° C [68° F]) and has not been started for several hours. When the engine is started, the coolant temperature needle should align with the C (cold) indicator mark on the coolant temperature gauge.
4. Hot start means the engine is hot. The engine is considered hot when the fans come on while the engine is idling in neutral.
5. Do not disconnect or reconnect electrical connectors when the ignition switch is on. A voltage spike may damage the ECM.
6. Refer to *ECM Test Harness* in this section for information identifying the ECM harness connector terminals.
7. Make sure the electrical connectors are secure and free of corrosion.
8. Check the wiring harness for damage.

DIAGNOSTIC TROUBLE CODES

DTC	Problem	DTC flowchart*
1	Loose MAP sensor connector, open or short circuit in MAP sensor wire or faulty MAP sensor	99
7	Loose ECT sensor connector, open or short circuit in ECT sensor wire or faulty ECT sensor	100
8	Loose TP sensor connector, open or short circuit in TP sensor wire or faulty TP sensor	101
9	Loose IAT sensor connector, open or short circuit in IAT sensor wire or faulty IAT sensor	102
10	Loose BARO sensor connector, open or short circuit in BARO sensor wire or faulty BARO sensor	103
11	Loose VSS connector, open or short circuit in VSS wire or faulty VSS	104
12	Faulty No. 1 fuel injector	105
13	Faulty No. 2 fuel injector	106
14	Faulty No. 3 fuel injector	107
15	Faulty No. 4 fuel injector	108
16	Faulty No. 5 fuel injector	109
17	Faulty No. 6 fuel injector	110
18	Loose cam pulse generator connector, open or short circuit in cam pulse generator wire or faulty cam pulse generator	111
19	Loose ignition pulse generator connector, open or short circuit in ignition pulse generator wire or faulty ignition pulse generator	112
21	Faulty right oxygen sensor or circuit	113
22	Faulty left oxygen sensor or circuit	114
23	Faulty right oxygen sensor heater circuit or sensor	115
24	Faulty left oxygen sensor heater circuit or sensor	116
25	Faulty right knock sensor circuit or sensor	117
26	Faulty left knock sensor circuit or sensor	118
29	Faulty IAC valve	119
33	Faulty E-2 prom in ECM	120
41	Faulty gear position switch, clutch switch, sidestand switch or gear position switch circuit	121

*Refer to *Electrical Component Replacement* in Chapter Nine if component replacement is indicated in a flowchart.

FUEL INJECTION AND EMISSION CONTROL SYSTEMS

(99)

DTC1: LOOSE MAP SENSOR CONNECTOR, OPEN OR SHORT CIRCUIT IN MAP SENSOR WIRE OR FAULTY MAP SENSOR

The manifold absolute pressure (MAP) sensor signals the ECM on engine load. The ECM uses this information to adjust the air/fuel ratio and ignition timing. Any condition that affects MAP sensor input can cause a drivability problem.

1. Typical faults are:
   a. Incorrect air/fuel ratio.
   b. Increased fuel consumption.
   c. Engine starts but stalls.
2. Before performing this test, note the following:
   a. The MAP sensor is mounted on the throttle body.
   b. Make sure to check the wiring connectors, terminals and wires. Refer to the wiring diagrams at the end of the manual.
   c. Refer to *ECM Test Harness* in this chapter when connecting the breakout box to the ECM connectors.

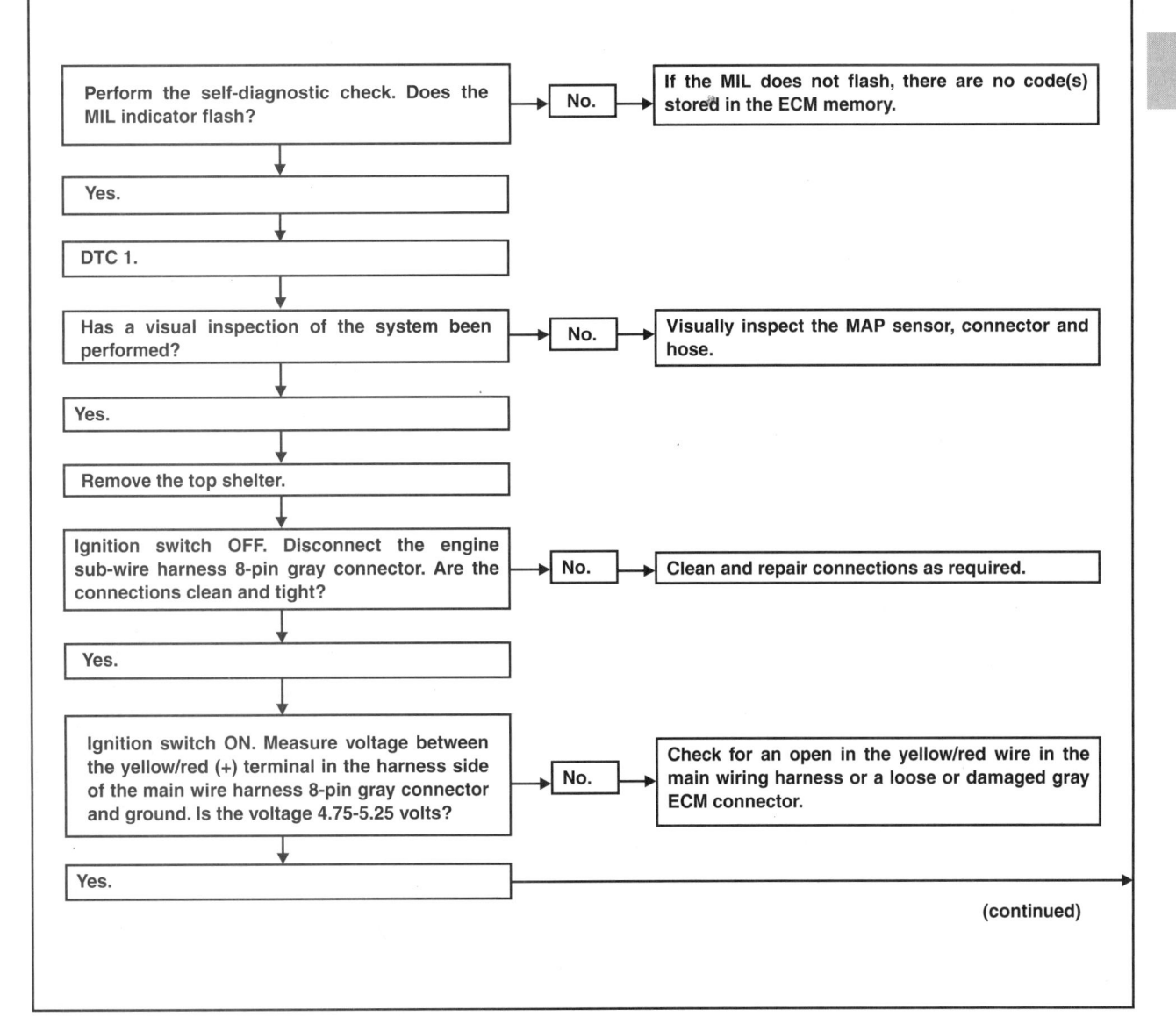

Perform the self-diagnostic check. Does the MIL indicator flash? → **No.** → If the MIL does not flash, there are no code(s) stored in the ECM memory.

↓ Yes.

DTC 1.

↓

Has a visual inspection of the system been performed? → **No.** → Visually inspect the MAP sensor, connector and hose.

↓ Yes.

Remove the top shelter.

↓

Ignition switch OFF. Disconnect the engine sub-wire harness 8-pin gray connector. Are the connections clean and tight? → **No.** → Clean and repair connections as required.

↓ Yes.

Ignition switch ON. Measure voltage between the yellow/red (+) terminal in the harness side of the main wire harness 8-pin gray connector and ground. Is the voltage 4.75-5.25 volts? → **No.** → Check for an open in the yellow/red wire in the main wiring harness or a loose or damaged gray ECM connector.

↓ Yes.

(continued)

(99) (continued)

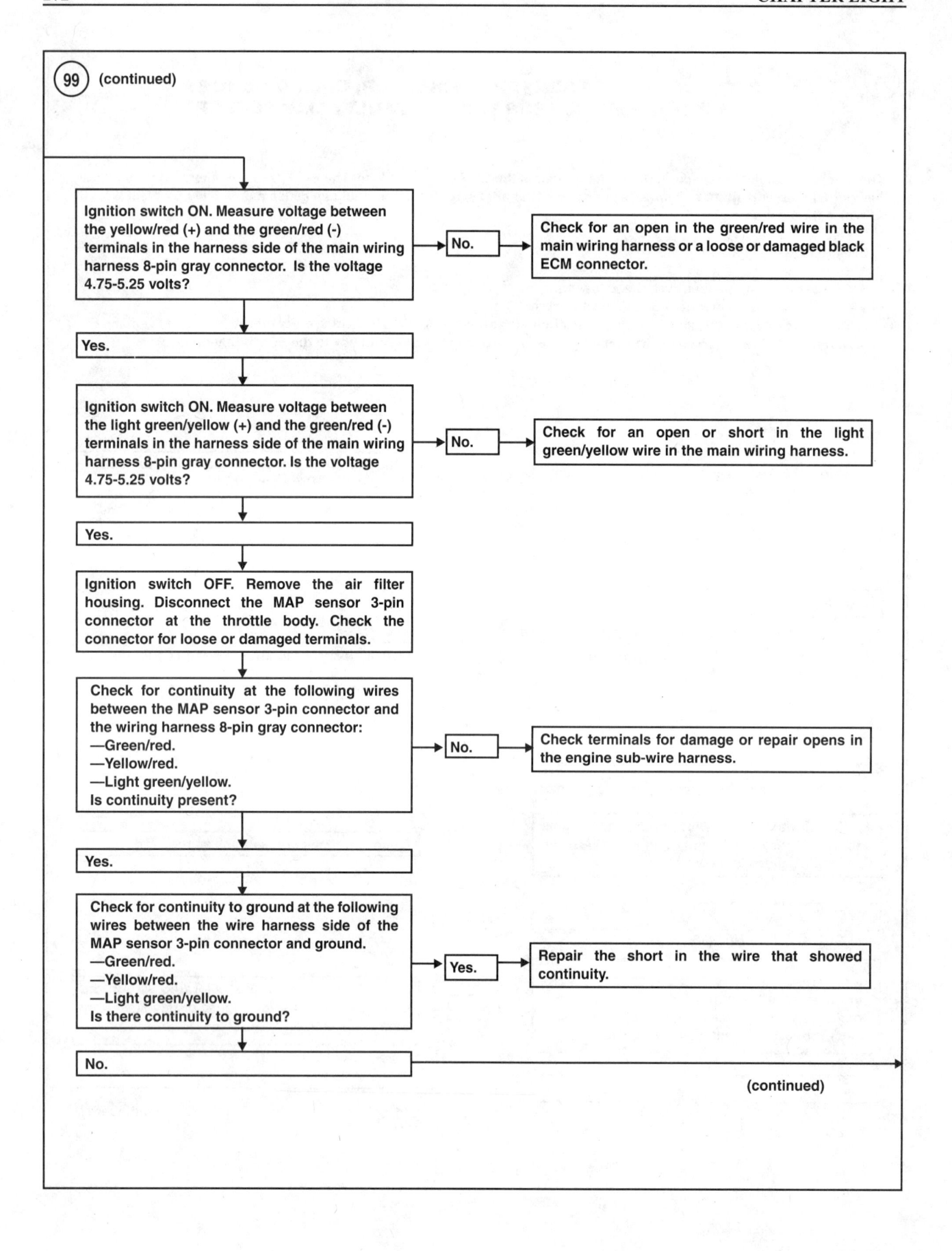

```
↓
[Ignition switch ON. Measure voltage between the yellow/red (+) and the green/red (-) terminals in the harness side of the main wiring harness 8-pin gray connector. Is the voltage 4.75-5.25 volts?] --No.--> [Check for an open in the green/red wire in the main wiring harness or a loose or damaged black ECM connector.]
↓ Yes.
[Ignition switch ON. Measure voltage between the light green/yellow (+) and the green/red (-) terminals in the harness side of the main wiring harness 8-pin gray connector. Is the voltage 4.75-5.25 volts?] --No.--> [Check for an open or short in the light green/yellow wire in the main wiring harness.]
↓ Yes.
[Ignition switch OFF. Remove the air filter housing. Disconnect the MAP sensor 3-pin connector at the throttle body. Check the connector for loose or damaged terminals.]
↓
[Check for continuity at the following wires between the MAP sensor 3-pin connector and the wiring harness 8-pin gray connector:
—Green/red.
—Yellow/red.
—Light green/yellow.
Is continuity present?] --No.--> [Check terminals for damage or repair opens in the engine sub-wire harness.]
↓ Yes.
[Check for continuity to ground at the following wires between the wire harness side of the MAP sensor 3-pin connector and ground.
—Green/red.
—Yellow/red.
—Light green/yellow.
Is there continuity to ground?] --Yes.--> [Repair the short in the wire that showed continuity.]
↓ No.
(continued)
```

FUEL INJECTION AND EMISSION CONTROL SYSTEMS

99 (continued)

↓

Perform the following:
1. Reconnect the engine sub-wire harness 8-pin connector.
2. Reconnect the MAP sensor 3-pin connector.
4. Install the air filter housing.
5. Disconnect the ECM connectors.
6. Connect the breakout box to the ECM connectors.

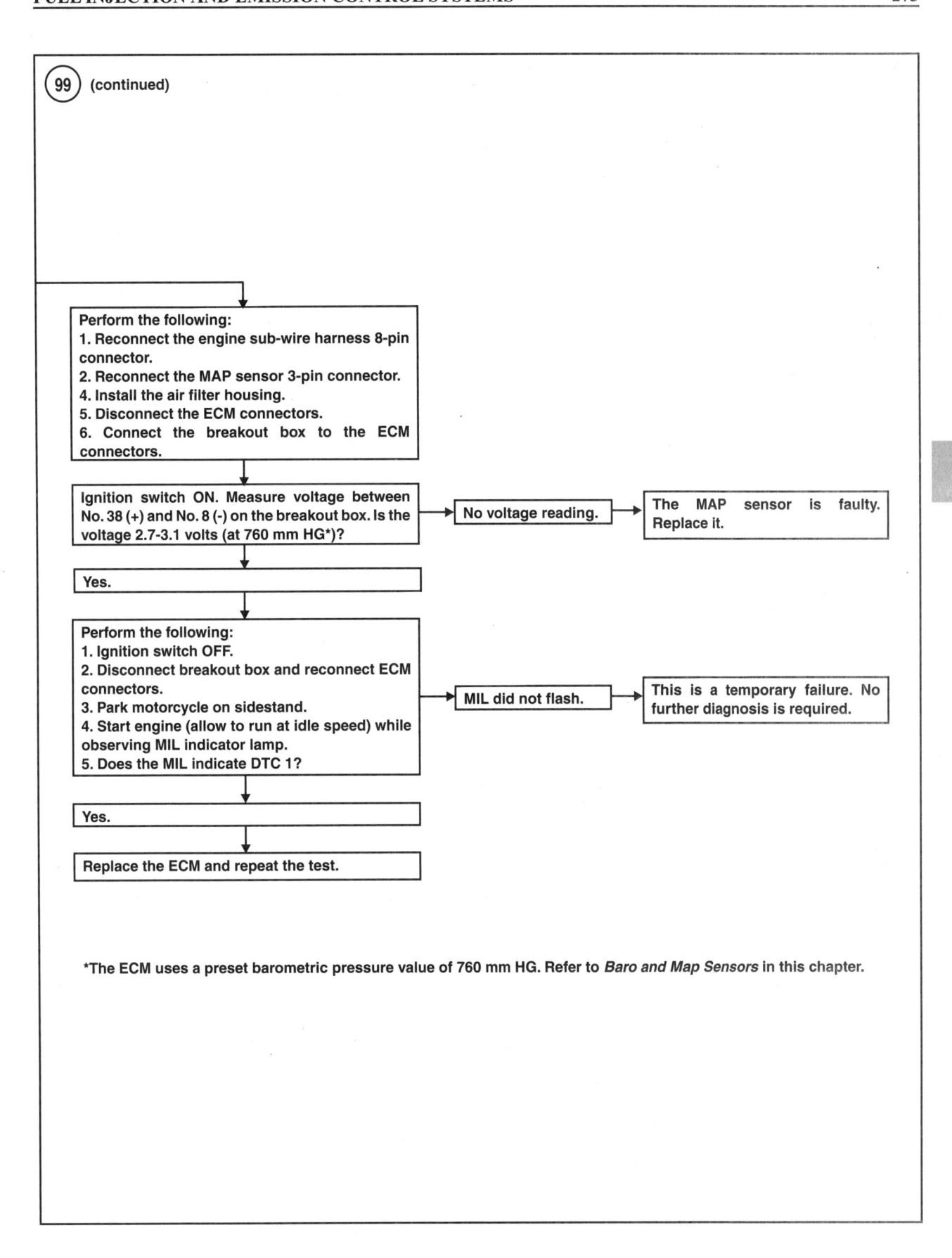

↓

Ignition switch ON. Measure voltage between No. 38 (+) and No. 8 (-) on the breakout box. Is the voltage 2.7-3.1 volts (at 760 mm HG*)? → No voltage reading. → The MAP sensor is faulty. Replace it.

↓ Yes.

Perform the following:
1. Ignition switch OFF.
2. Disconnect breakout box and reconnect ECM connectors.
3. Park motorcycle on sidestand.
4. Start engine (allow to run at idle speed) while observing MIL indicator lamp.
5. Does the MIL indicate DTC 1? → MIL did not flash. → This is a temporary failure. No further diagnosis is required.

↓ Yes.

Replace the ECM and repeat the test.

*The ECM uses a preset barometric pressure value of 760 mm HG. Refer to *Baro and Map Sensors* in this chapter.

DTC 7: LOOSE ECT SENSOR CONNECTOR, OPEN OR SHORT CIRCUIT IN ECT SENSOR WIRE OR FAULTY ECT SENSOR

The engine coolant temperature (ECT) sensor sends a voltage signal to the ECM for engine temperature. At low coolant temperatures, the voltage signal is high. At high coolant temperatures, the signal is low.

1. A faulty ECT or ECT circuit can cause the following drivability problems:
   a. Hard engine starting (especially when the engine is cold).
   b. Engine stalling or stumbling.
   c. Poor fuel economy.
   d. Incorrect air/fuel ratio (too rich or too lean).
   e. Incorrect operation of the emission control system.
2. Before performing this test, note the following:
   a. The ECT sensor depends on a properly operating cooling system.
   b. Make sure the coolant level is correct.
   c. Check the cooling fans for proper operation.
   d. Test the radiator cap and cooling system as described in Chapter Ten.
   e. Make sure to check the wiring connectors, terminals and wires (ECT sensor 3-pin and engine sub-wire harness 6-pin gray connectors). The ECT sensor is mounted at the top of the left side of the engine, under the intake manifold. Refer to the wiring diagram at the end of the manual.
3. Refer to Chapter Nine for ECT sensor testing.

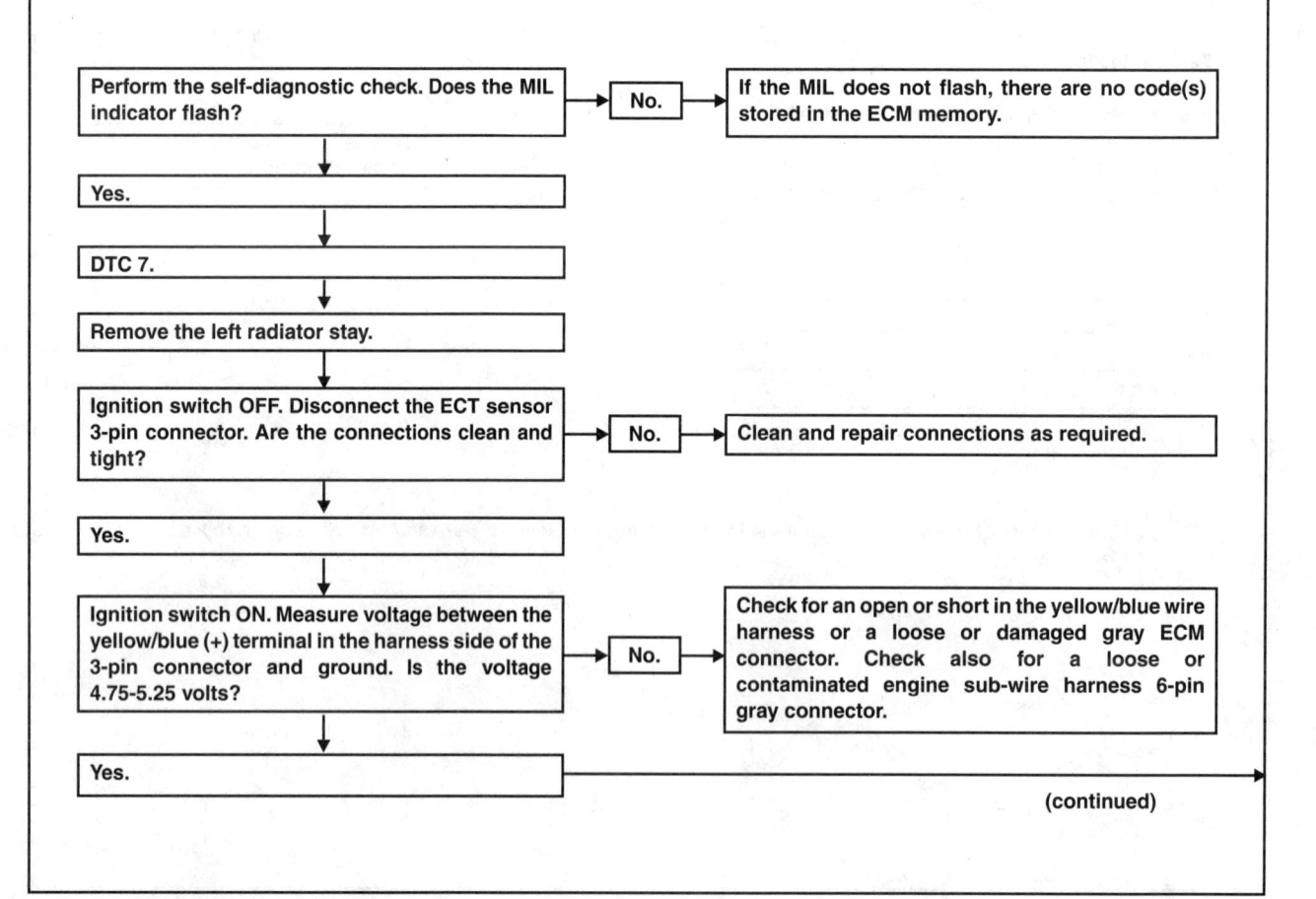

Perform the self-diagnostic check. Does the MIL indicator flash? → **No.** → If the MIL does not flash, there are no code(s) stored in the ECM memory.

↓ Yes.

DTC 7.

↓

Remove the left radiator stay.

↓

Ignition switch OFF. Disconnect the ECT sensor 3-pin connector. Are the connections clean and tight? → **No.** → Clean and repair connections as required.

↓ Yes.

Ignition switch ON. Measure voltage between the yellow/blue (+) terminal in the harness side of the 3-pin connector and ground. Is the voltage 4.75-5.25 volts? → **No.** → Check for an open or short in the yellow/blue wire harness or a loose or damaged gray ECM connector. Check also for a loose or contaminated engine sub-wire harness 6-pin gray connector.

↓ Yes.

(continued)

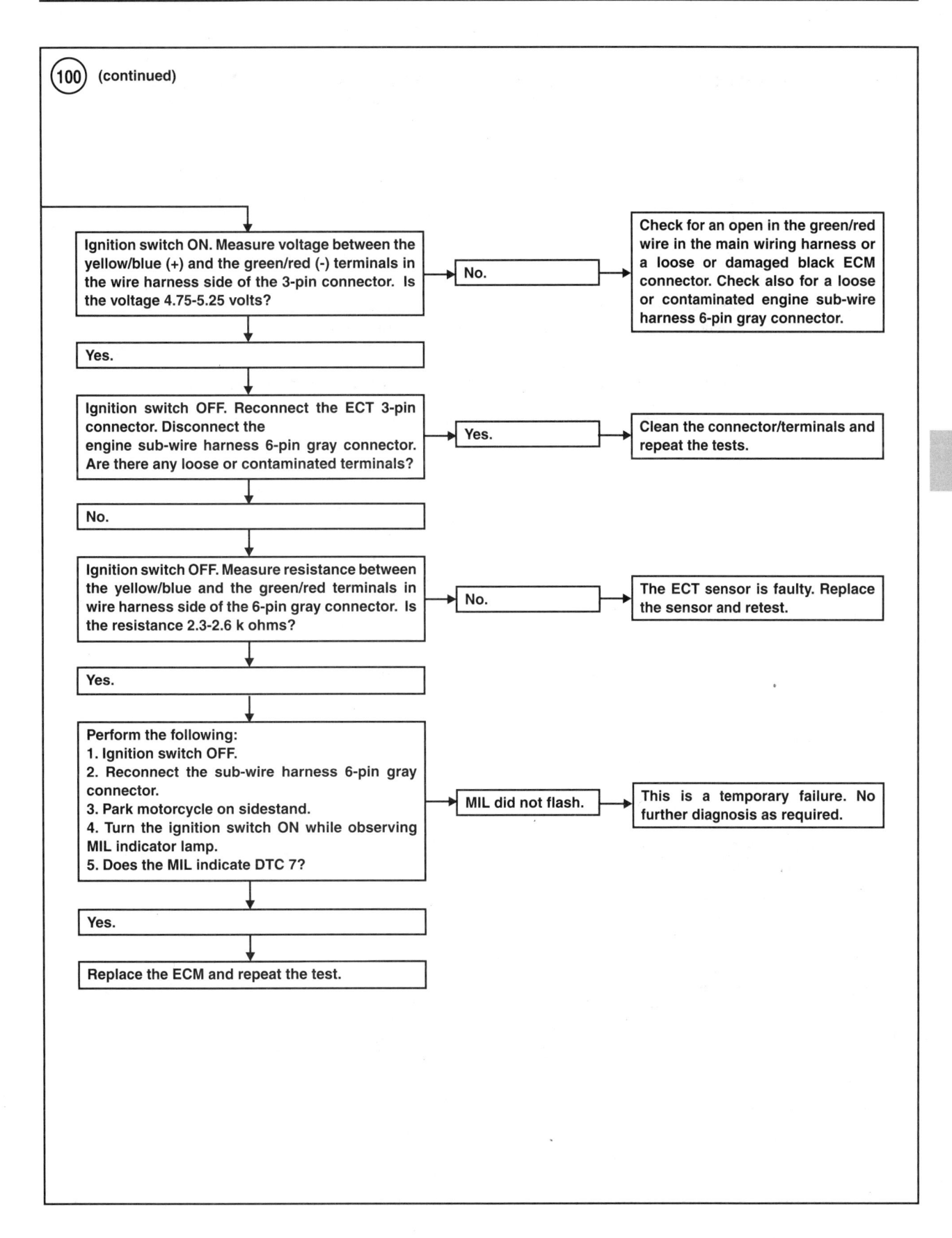

FUEL INJECTION AND EMISSION CONTROL SYSTEMS

(100) (continued)

Ignition switch ON. Measure voltage between the yellow/blue (+) and the green/red (-) terminals in the wire harness side of the 3-pin connector. Is the voltage 4.75-5.25 volts?

→ **No.** → Check for an open in the green/red wire in the main wiring harness or a loose or damaged black ECM connector. Check also for a loose or contaminated engine sub-wire harness 6-pin gray connector.

↓ **Yes.**

Ignition switch OFF. Reconnect the ECT 3-pin connector. Disconnect the engine sub-wire harness 6-pin gray connector. Are there any loose or contaminated terminals?

→ **Yes.** → Clean the connector/terminals and repeat the tests.

↓ **No.**

Ignition switch OFF. Measure resistance between the yellow/blue and the green/red terminals in wire harness side of the 6-pin gray connector. Is the resistance 2.3-2.6 k ohms?

→ **No.** → The ECT sensor is faulty. Replace the sensor and retest.

↓ **Yes.**

Perform the following:
1. Ignition switch OFF.
2. Reconnect the sub-wire harness 6-pin gray connector.
3. Park motorcycle on sidestand.
4. Turn the ignition switch ON while observing MIL indicator lamp.
5. Does the MIL indicate DTC 7?

→ **MIL did not flash.** → This is a temporary failure. No further diagnosis as required.

↓ **Yes.**

Replace the ECM and repeat the test.

(101) DTC 8: LOOSE TP SENSOR CONNECTOR, OPEN OR SHORT CIRCUIT IN TP SENSOR WIRE OR FAULTY TP SENSOR

The throttle position (TP) sensor sends a voltage signal to the ECM on the rate of throttle opening and throttle position. The ECM uses these signals, along with others, to adjust the air/fuel mixture and ignition timing.

1. A faulty TP sensor, TP sensor circuit or incorrect TP sensor mounting position can cause the following drivability problems:
   a. Poor engine acceleration.
   b. Engine hesitation.
   c. Incorrect engine idle speed.
2. Before performing this test, note the following:
   a. The TP sensor is mounted on the throttle body and rotates with the throttle shaft.
   b. The TP sensor mounting position is fixed and must not be changed. If the TP sensor is faulty, the throttle body must be replaced as an assembly. The TP sensor is not available separately.
   c. Make sure to check the wiring connectors, terminals and wires (3-pin and engine sub-wire harness 8-pin gray connectors). Refer to the wiring diagrams at the end of the manual.
   d. Refer to *ECM Test Harness* in this chapter when connecting the breakout box to the ECM connectors.

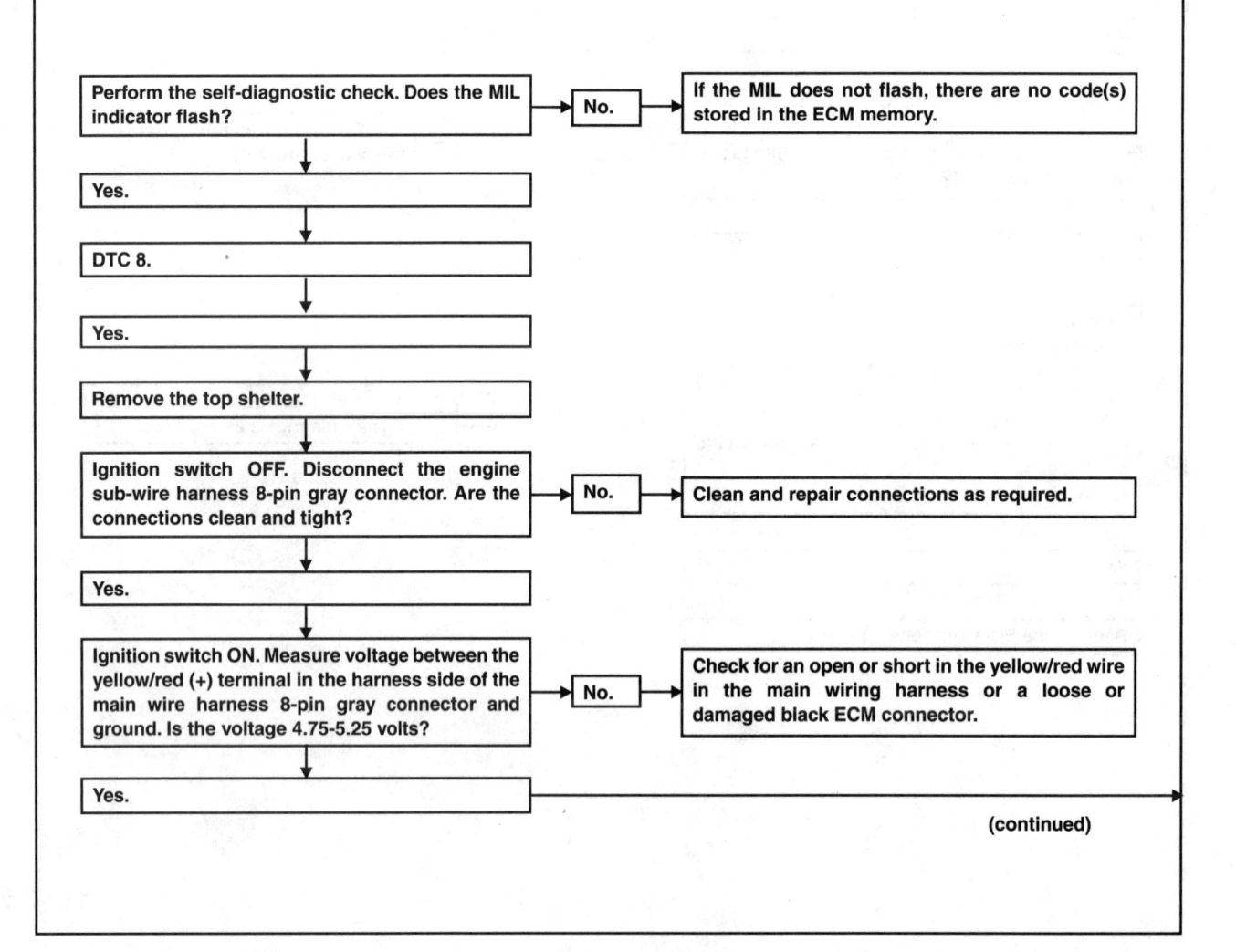

Perform the self-diagnostic check. Does the MIL indicator flash? → **No.** → If the MIL does not flash, there are no code(s) stored in the ECM memory.

↓ Yes.

DTC 8.

↓ Yes.

Remove the top shelter.

↓

Ignition switch OFF. Disconnect the engine sub-wire harness 8-pin gray connector. Are the connections clean and tight? → **No.** → Clean and repair connections as required.

↓ Yes.

Ignition switch ON. Measure voltage between the yellow/red (+) terminal in the harness side of the main wire harness 8-pin gray connector and ground. Is the voltage 4.75-5.25 volts? → **No.** → Check for an open or short in the yellow/red wire in the main wiring harness or a loose or damaged black ECM connector.

↓ Yes.

(continued)

FUEL INJECTION AND EMISSION CONTROL SYSTEMS

(101) (continued)

↓

Ignition switch ON. Measure voltage between the yellow/red (+) terminal and the green/red wire (-) in the harness side of the main wire harness 8-pin gray connector and ground. Is the voltage 4.75-5.25 volts? → **No.** → Check for an open in the green/red wire in the main wiring harness or a loose or damaged black ECM connector.

↓ **Yes.**

Ignition switch OFF. Disconnect the gray ECM 22-pin connector. Check for continuity between the light green wire terminal in the wire harness side of the 8-pin gray connector and ground. Is continuity present? → **Yes.** → There should be no continuity. Check for a short in the light green wire in the main wiring harness.

↓ **No.**

Remove the air filter housing. Disconnect the TP sensor 3-pin connector and check it for contamination and loose terminals.

↓

Check for continuity at the following wires between the TP sensor 3-pin connector (sensor side) and the wiring harness 8-pin gray connector:
—Green/red.
—Yellow/red.
—Light green.
→ **No.** → Check terminals for damage or repair opens in the engine sub-wire harness.

↓ **Yes.**

Check for continuity to ground of the following wires between the wire harness side of the TP sensor 3-pin connector and ground.
—Green/red.
—Yellow/red.
—Light green.
Is there continuity to ground? → **Yes.** → There should be no continuity. Repair the short in the wire that showed continuity.

↓ **No.**

Perform the following:
1. Reconnect the engine sub-wire harness 8-pin gray connector.
2. Reconnect the TP sensor 3-pin connector.
4. Install the air filter housing.
5. Connect the breakout box to the ECM connectors.

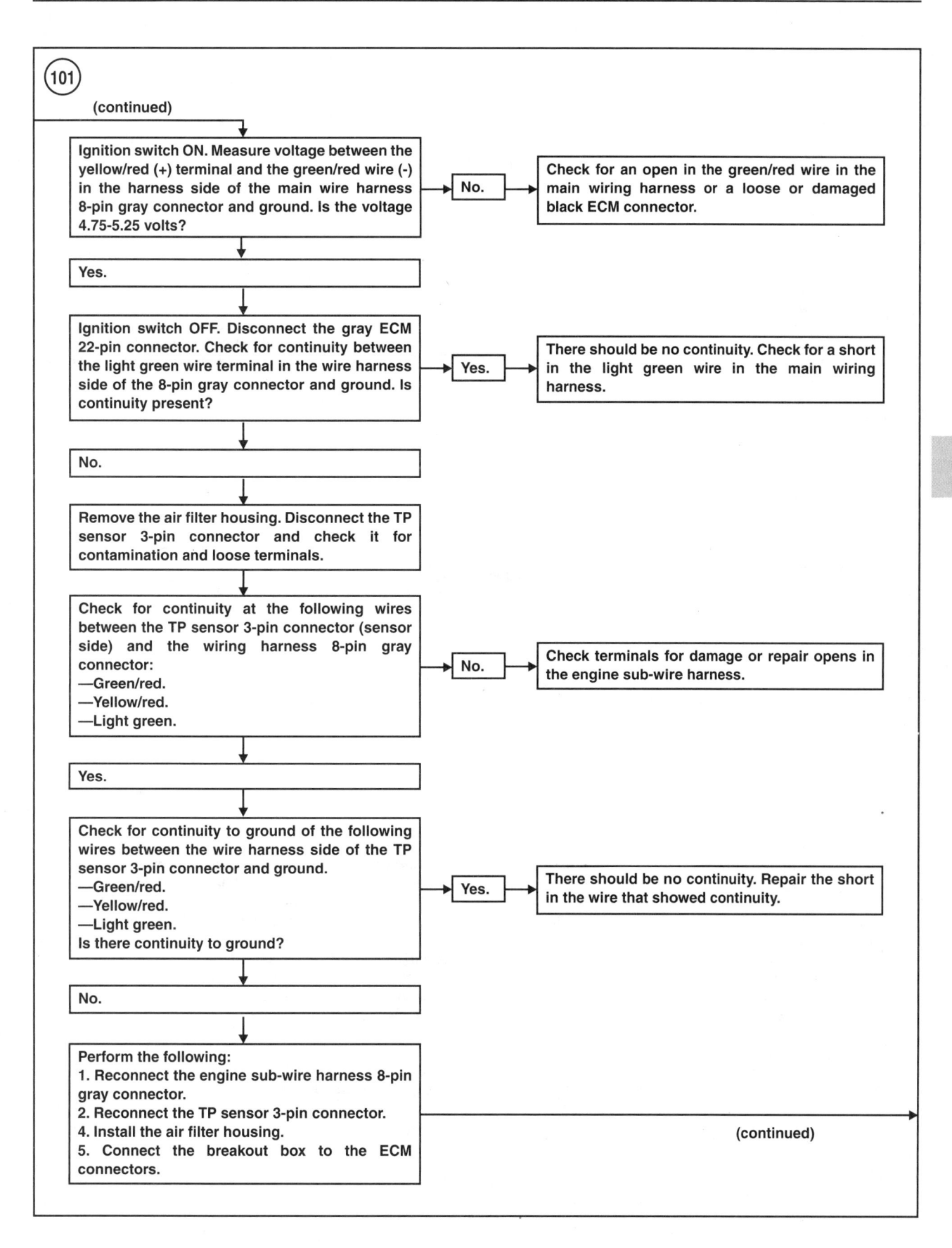

→ **(continued)**

101 (continued)

↓

Ignition switch ON. Measure voltage between pins No. 51 (+) and No. 8 (-) on the breakout box while *slowly* opening the throttle from its fully closed to wide open position[1]. Is the voltage 0.4-0.6 volts with the throttle full closed and 4.2-4.8 volts with the throttle fully open?[2] → No voltage reading. → The TP sensor is faulty. Replace the throttle body and retest.

↓ Yes.

Perform the following:
1. Ignition switch OFF.
2. Disconnect breakout box and reconnect ECM connectors.
3. Park motorcycle on sidestand.
4. Turn the ignition switch ON while observing MIL indicator lamp.
5. Does the MIL indicate DTC 8? → MIL did not flash. → This is a temporary failure. No further diagnosis is required.

↓ Yes.

Replace the ECM and repeat the test.

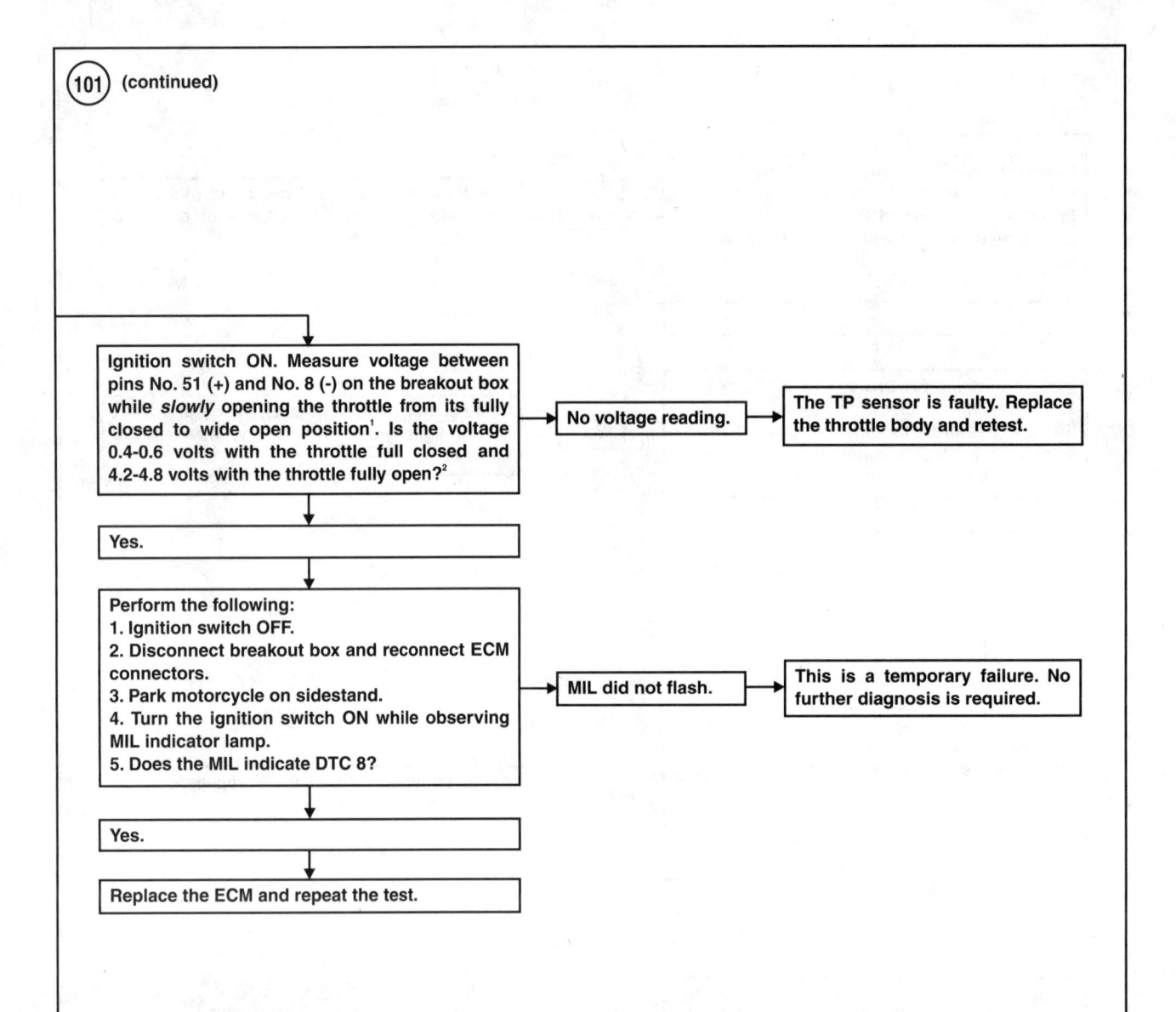

1. The voltmeter readings should increase evenly. If the readings become erratic, indicate a lower reading (after a higher reading is noted), or open circuit readings, the TP sensor may be damaged. Repeat the test while closing the throttle. The voltmeter readings should decrease evenly. Repeat the test while an assistant moves and twists the TP sensor connector and wiring harness. If the meter readings do not increase/decrease correctly, check the wiring harness and connector for damage.

2. The throttle position voltage readings specified are based on a voltage reading of 5.0 volts. If the voltage readings recorded at the beginning of this procedure were not 5.0 volts, convert the 0.4-0.6 (throttle fully closed) and 4.2-4.8 (throttle fully open) voltage readings. For example, if the voltage recorded at the beginning of the procedure was 4.0 volts, determine the new throttle fully closed voltage readings for 0.4 and 0.6 volts as follows:

$0.4 \times 4.0/5.0 = 0.32$ volts
$0.6 \times 4.0/5.0 = 0.48$ volts

In this example, the correct throttle fully closed reading is 0.32-0.48 volts.
Repeat to recalculate the 4.2-4.8 voltage readings for the throttle fully open position.

FUEL INJECTION AND EMISSION CONTROL SYSTEMS

(102) DTC 9: LOOSE IAT SENSOR CONNECTOR, OPEN OR SHORT CIRCUIT IN IAT SENSOR WIRE OR FAULTY IAT SENSOR

The intake air temperature (IAT) sensor measures the temperature of the air as it enters the engine. The ECM monitors input from the IAT to adjust the air/fuel mixture. A fault in the IAT circuit can cause hard starting when the engine is cold.

1. Before making this test, note the following:
   a. The IAT sensor is mounted on the air filter cover.
   b. Check the wiring connectors, terminals and wires (2-pin connector). Refer to wiring diagrams at the end of the manual.

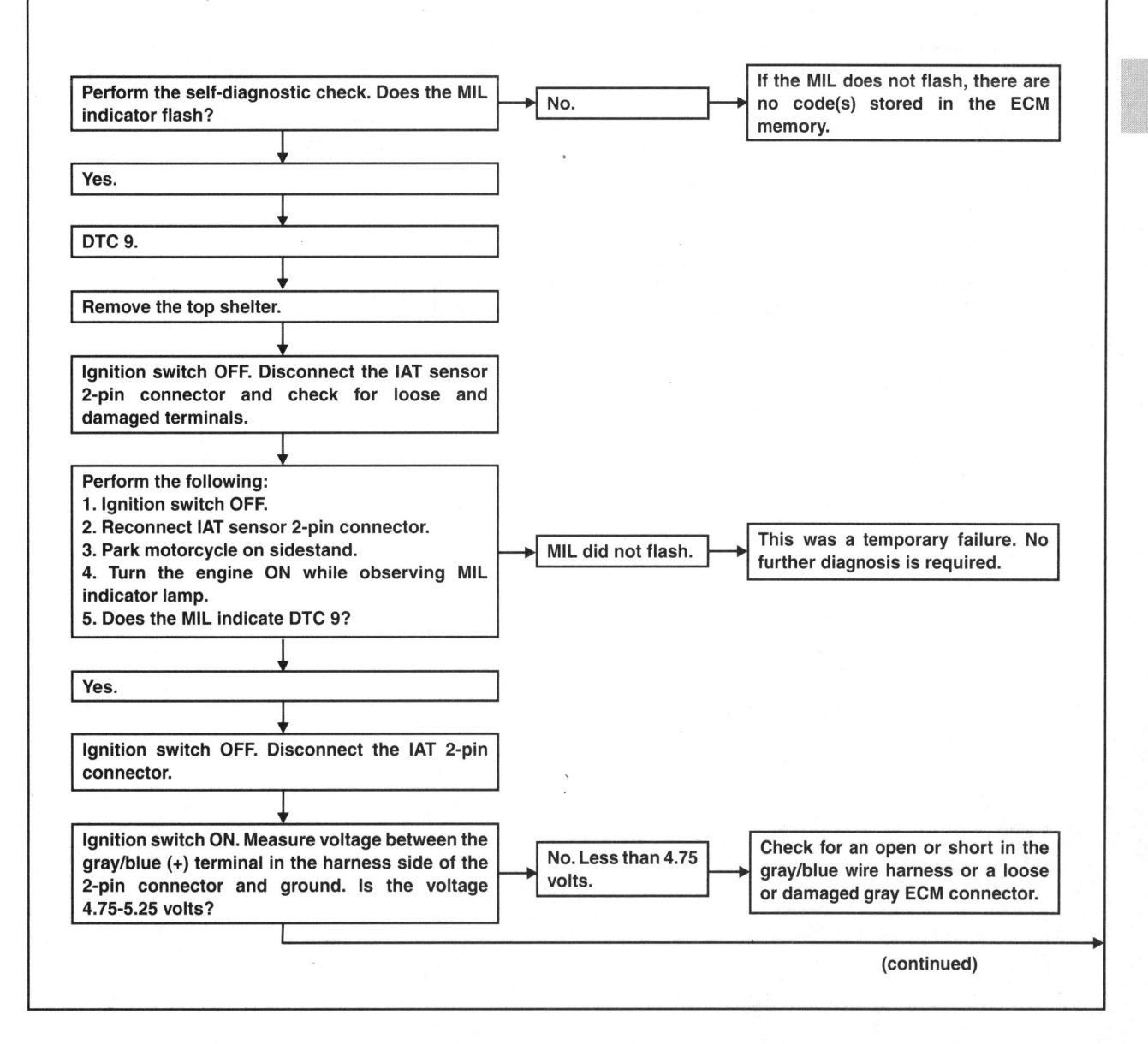

Perform the self-diagnostic check. Does the MIL indicator flash?
→ No. → If the MIL does not flash, there are no code(s) stored in the ECM memory.

Yes.
↓
DTC 9.
↓
Remove the top shelter.
↓
Ignition switch OFF. Disconnect the IAT sensor 2-pin connector and check for loose and damaged terminals.
↓
Perform the following:
1. Ignition switch OFF.
2. Reconnect IAT sensor 2-pin connector.
3. Park motorcycle on sidestand.
4. Turn the engine ON while observing MIL indicator lamp.
5. Does the MIL indicate DTC 9?
→ MIL did not flash. → This was a temporary failure. No further diagnosis is required.

Yes.
↓
Ignition switch OFF. Disconnect the IAT 2-pin connector.
↓
Ignition switch ON. Measure voltage between the gray/blue (+) terminal in the harness side of the 2-pin connector and ground. Is the voltage 4.75-5.25 volts?
→ No. Less than 4.75 volts. → Check for an open or short in the gray/blue wire harness or a loose or damaged gray ECM connector.

(continued)

102 (continued)

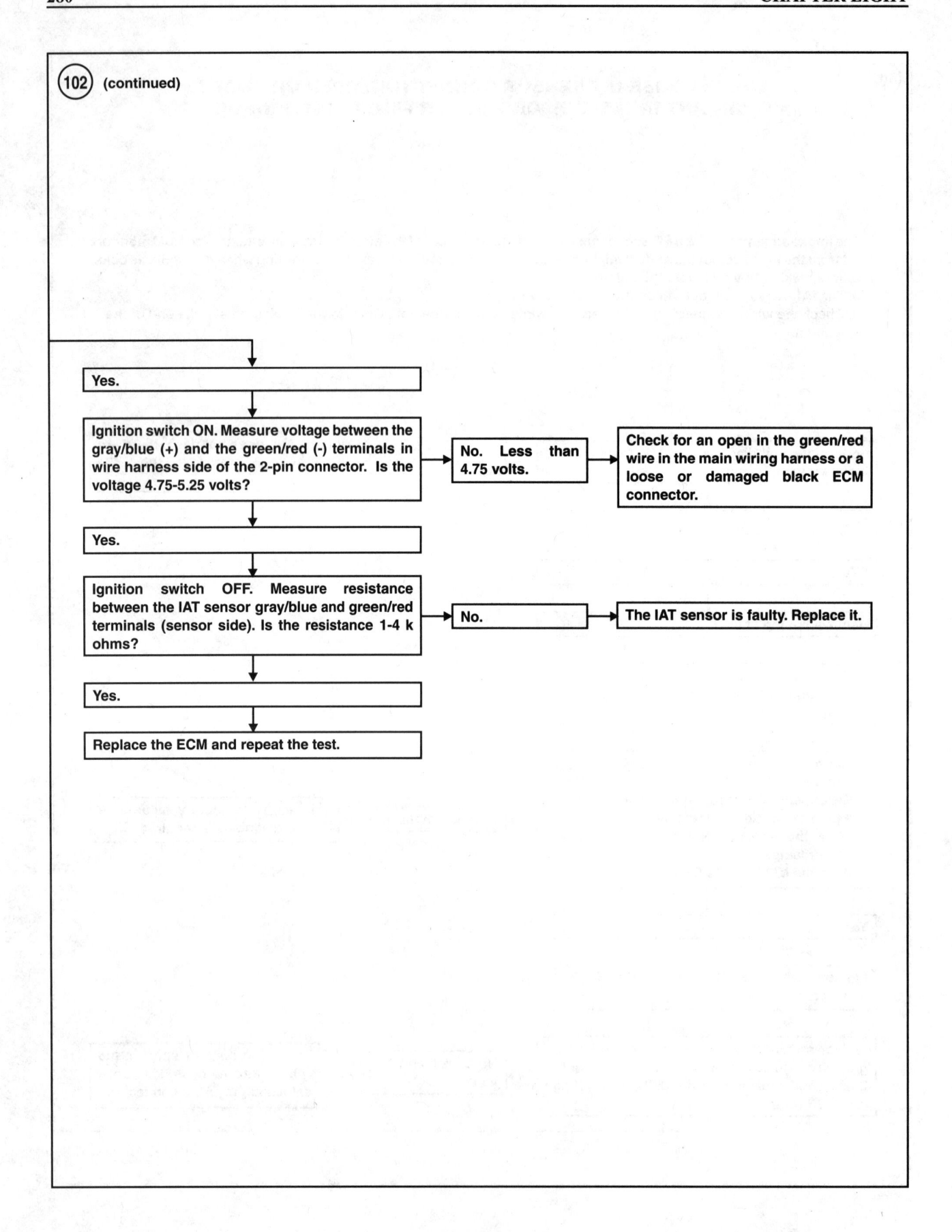

↓

Yes.

↓

Ignition switch ON. Measure voltage between the gray/blue (+) and the green/red (-) terminals in wire harness side of the 2-pin connector. Is the voltage 4.75-5.25 volts? → **No. Less than 4.75 volts.** → Check for an open in the green/red wire in the main wiring harness or a loose or damaged black ECM connector.

↓

Yes.

↓

Ignition switch OFF. Measure resistance between the IAT sensor gray/blue and green/red terminals (sensor side). Is the resistance 1-4 k ohms? → **No.** → The IAT sensor is faulty. Replace it.

↓

Yes.

↓

Replace the ECM and repeat the test.

FUEL INJECTION AND EMISSION CONTROL SYSTEMS

103

DTC 10: LOOSE BARO SENSOR CONNECTOR, OPEN OR SHORT CIRCUIT IN BARO SENSOR WIRE OR FAULTY BARO

The barometric pressure (BARO) sensor senses barometric pressure changes due to changing altitude and weather conditions. A signal is sent to the ECM, which uses the information to adjust the air/fuel ratio and ignition timing. A fault in the BARO circuit may go unnoticed at low altitudes, but can cause the engine to idle roughly at higher altitudes.

1. Before making this test, note the following:
   a. The BARO sensor is mounted on the air filter cover.
   b. Check the wiring connectors, terminals and wires (3-pin connector). Refer to the wiring diagrams at the end of the manual.
   c. Refer to *ECM Test Harness* in this chapter when connecting the breakout box to the ECM connectors.

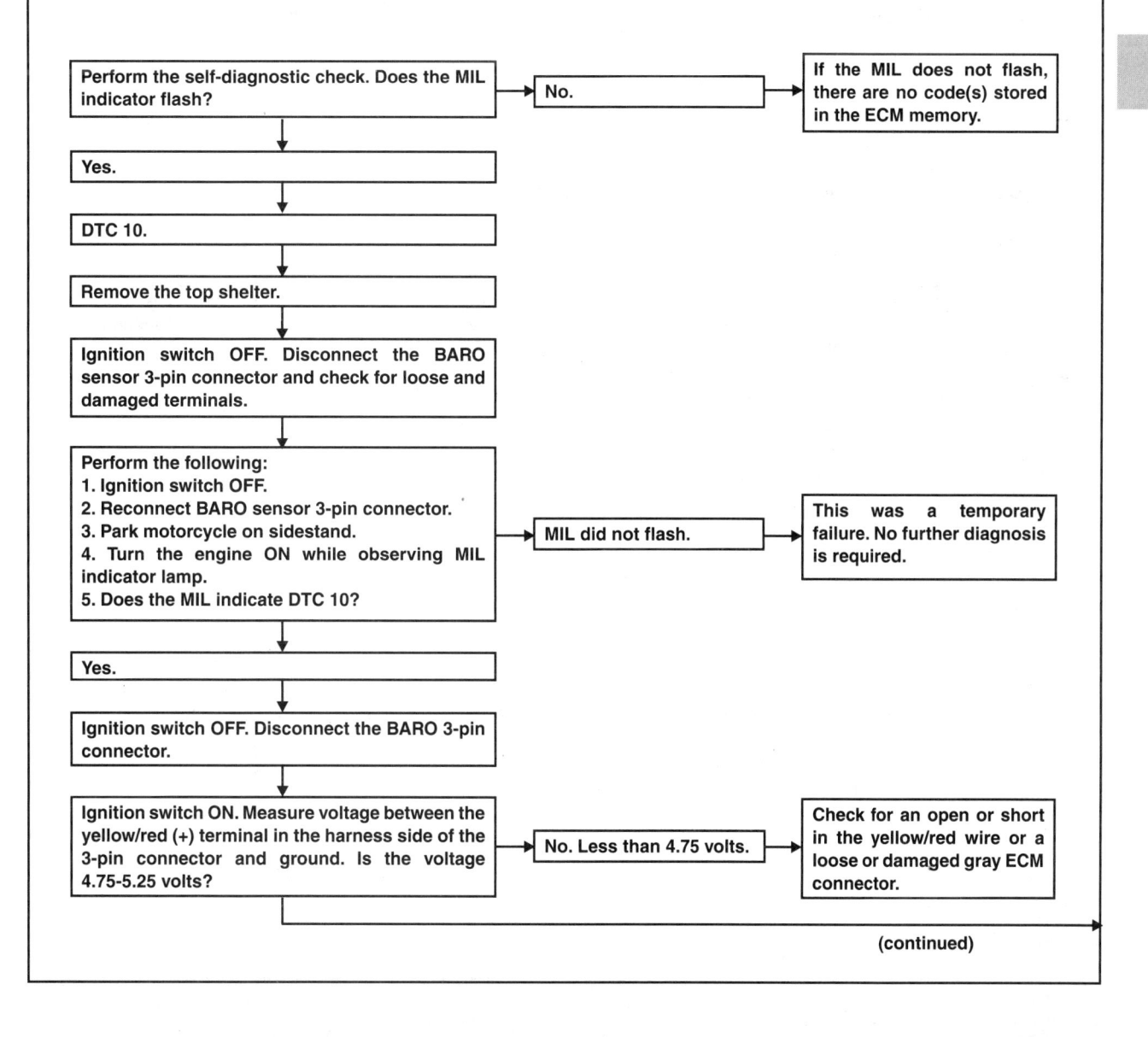

Perform the self-diagnostic check. Does the MIL indicator flash? → **No.** → If the MIL does not flash, there are no code(s) stored in the ECM memory.

↓ Yes.

DTC 10.

↓

Remove the top shelter.

↓

Ignition switch OFF. Disconnect the BARO sensor 3-pin connector and check for loose and damaged terminals.

↓

Perform the following:
1. Ignition switch OFF.
2. Reconnect BARO sensor 3-pin connector.
3. Park motorcycle on sidestand.
4. Turn the engine ON while observing MIL indicator lamp.
5. Does the MIL indicate DTC 10?

→ **MIL did not flash.** → This was a temporary failure. No further diagnosis is required.

↓ Yes.

Ignition switch OFF. Disconnect the BARO 3-pin connector.

↓

Ignition switch ON. Measure voltage between the yellow/red (+) terminal in the harness side of the 3-pin connector and ground. Is the voltage 4.75-5.25 volts? → **No. Less than 4.75 volts.** → Check for an open or short in the yellow/red wire or a loose or damaged gray ECM connector.

↓

(continued)

103 (continued)

```
                        ↓
              ┌──────────────────┐
              │ Yes.             │
              └──────────────────┘
                        ↓
┌──────────────────────────────────────┐                              ┌──────────────────────────┐
│ Ignition switch ON. Measure voltage  │                              │ Check for an open in the │
│ between the yellow/red (+) and the   │  → No. Less than 4.75 volts. →│ green/red wire in the main│
│ green/red (-) terminals in wire      │                              │ wiring harness or a loose │
│ harness side of the 3-pin connector. │                              │ or damaged black ECM     │
│ Is the voltage 4.75-5.25 volts?      │                              │ connector.               │
└──────────────────────────────────────┘                              └──────────────────────────┘
                        ↓
              ┌──────────────────┐
              │ Yes.             │
              └──────────────────┘
                        ↓
┌──────────────────────────────────────┐                              ┌──────────────────────────┐
│ Ignition switch ON. Measure voltage  │                              │ Check for an open in the │
│ between the light green/black (+)    │  → No. Less than 4.75 volts. →│ light green/black wire in │
│ and the green/red (-) terminals in   │                              │ the main wiring harness. │
│ wire harness side of the 3-pin       │                              │                          │
│ connector. Is the voltage            │                              │                          │
│ 4.75-5.25 volts?                     │                              │                          │
└──────────────────────────────────────┘                              └──────────────────────────┘
                        ↓
              ┌──────────────────┐
              │ Yes.             │
              └──────────────────┘
                        ↓
┌──────────────────────────────────────┐
│ Perform the following:               │
│ 1. Ignition switch OFF.              │
│ 2. Reconnect the BARO sensor 3-pin   │
│    connector.                        │
│ 3. Connect the breakout box to the   │
│    ECM connectors.                   │
└──────────────────────────────────────┘
                        ↓
┌──────────────────────────────────────┐                              ┌──────────────────────────┐
│ Ignition switch ON. Measure voltage  │                              │ The BARO sensor is       │
│ between pins No. 49 (+) and 8 (-) on │  → No voltage reading.       →│ faulty. Replace the BARO │
│ the breakout box. Is the voltage     │                              │ sensor and retest.       │
│ 2.7-3.1 volts (at 760 mm HG*)?       │                              │                          │
└──────────────────────────────────────┘                              └──────────────────────────┘
                        ↓
              ┌──────────────────┐
              │ Yes.             │
              └──────────────────┘
                        ↓
┌──────────────────────────────────────┐
│ Replace the ECM and repeat the test. │
└──────────────────────────────────────┘
```

*The ECM uses a preset barometric pressure value of 760 mm HG. Refer to *BARO and MAP Sensors* in this chapter for additional information.

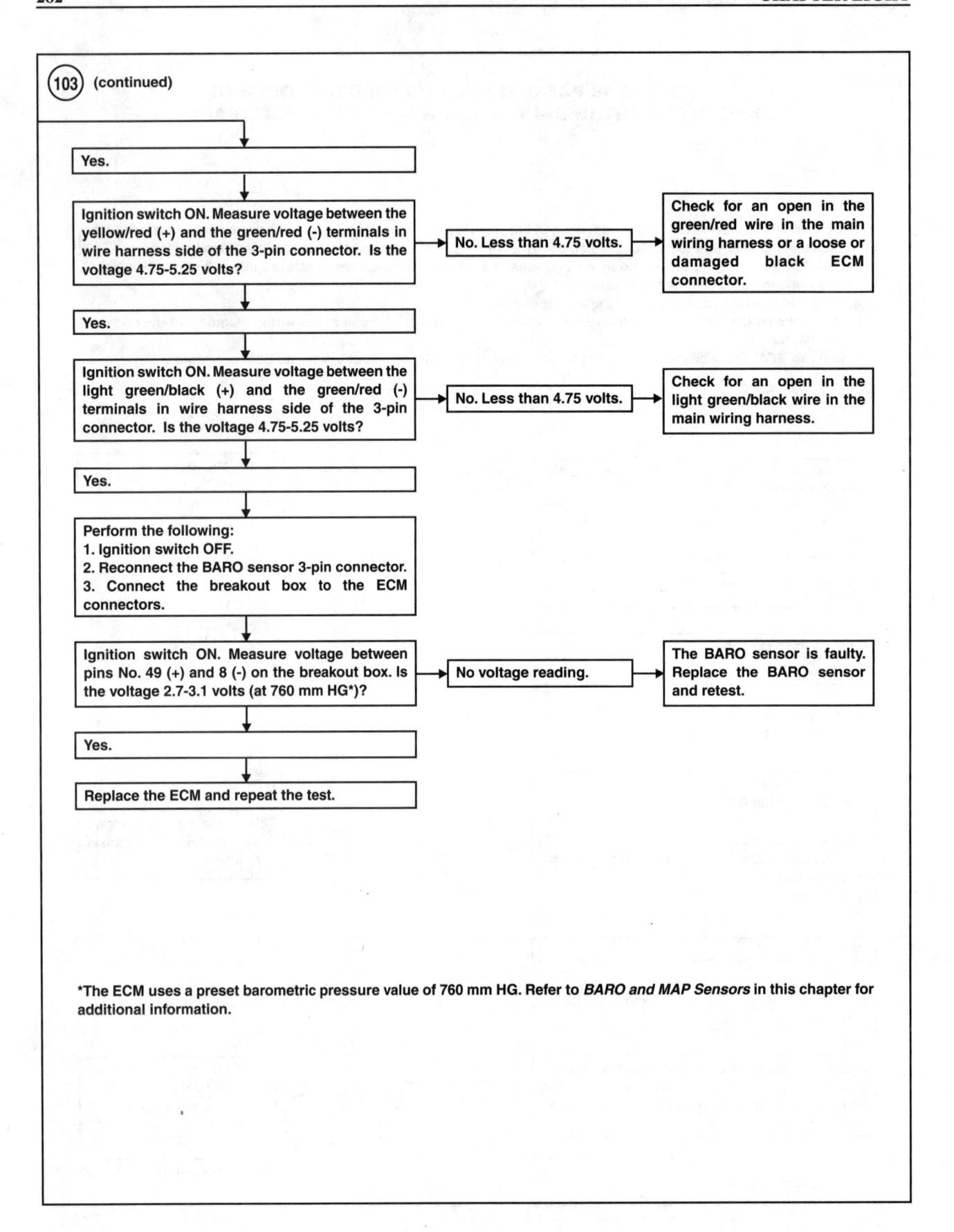

FUEL INJECTION AND EMISSION CONTROL SYSTEMS

(104)

DTC 11: LOOSE VSS CONNECTOR, OPEN OR SHORT CIRCUIT IN VSS WIRE OR FAULTY VSS

The vehicle speed sensor (VSS) is a magnetic sensor that signals the ECM on the speed of the motorcycle. A faulty VSS may affect speedometer or cruise control operation. Chipped or broken teeth on the transmission mainshaft can affect the VSS.

1. Before making this test, test, note the following:
   a. The VSS is mounted on the right rear side of the engine, underneath the cylinder head.
   b. Check the wiring connectors, terminals and wires. Refer to wiring diagrams at the end of the manual.
   c. Refer to the *ECM Test Harness* in this chapter when connecting the breakout box to the ECM connectors.

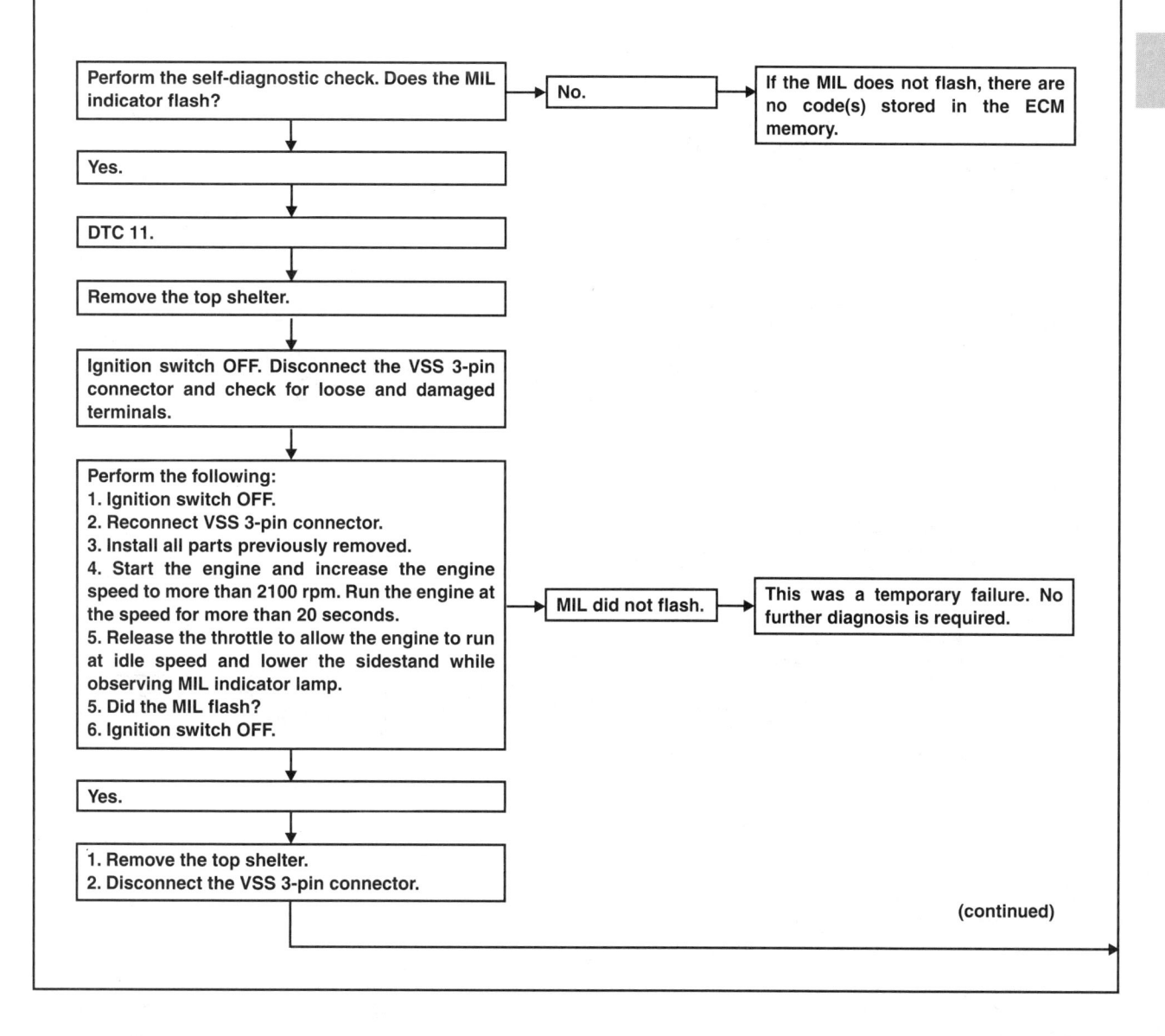

Perform the self-diagnostic check. Does the MIL indicator flash? → **No.** → If the MIL does not flash, there are no code(s) stored in the ECM memory.

↓

Yes.

↓

DTC 11.

↓

Remove the top shelter.

↓

Ignition switch OFF. Disconnect the VSS 3-pin connector and check for loose and damaged terminals.

↓

Perform the following:
1. Ignition switch OFF.
2. Reconnect VSS 3-pin connector.
3. Install all parts previously removed.
4. Start the engine and increase the engine speed to more than 2100 rpm. Run the engine at the speed for more than 20 seconds.
5. Release the throttle to allow the engine to run at idle speed and lower the sidestand while observing MIL indicator lamp.
5. Did the MIL flash?
6. Ignition switch OFF.

→ MIL did not flash. → This was a temporary failure. No further diagnosis is required.

↓

Yes.

↓

1. Remove the top shelter.
2. Disconnect the VSS 3-pin connector.

(continued)

(104) (continued)

↓

Ignition switch ON. Measure voltage between the brown/white (+) terminal in the harness side of the 3-pin connector and ground. Does the voltmeter read battery voltage (13.0-13.2 volts)? → No voltage reading. → Check for an open or short in the brown/white wire.

↓ Yes.

Ignition switch ON. Measure voltage between the brown/white (+) and the green (-) terminals in wire harness side of the 3-pin connector. Does the voltmeter read battery voltage (13.0-13.2 volts)? → No voltage reading. → Check for an open in the green wire.

↓ Yes.

Ignition switch OFF. Disconnect the ECM 22-pin gray connector. Check for continuity between the white/black terminal in the wire harness side of the VSS 3-pin connector and ground. Is continuity present? → Yes. → There should be no continuity. Check for a short in the white/black wire.

↓ No.

Ignition switch OFF. Check for continuity between the white/black terminals in the wire harness side of the VSS 3-pin connector and the wire harness side of the ECM 22-pin gray connector. Is continuity present? → No. → Check for an open in the white/black wire.

↓ Yes.

Reconnect the VSS 3-pin connector. Connect the breakout box to the ECM connectors.

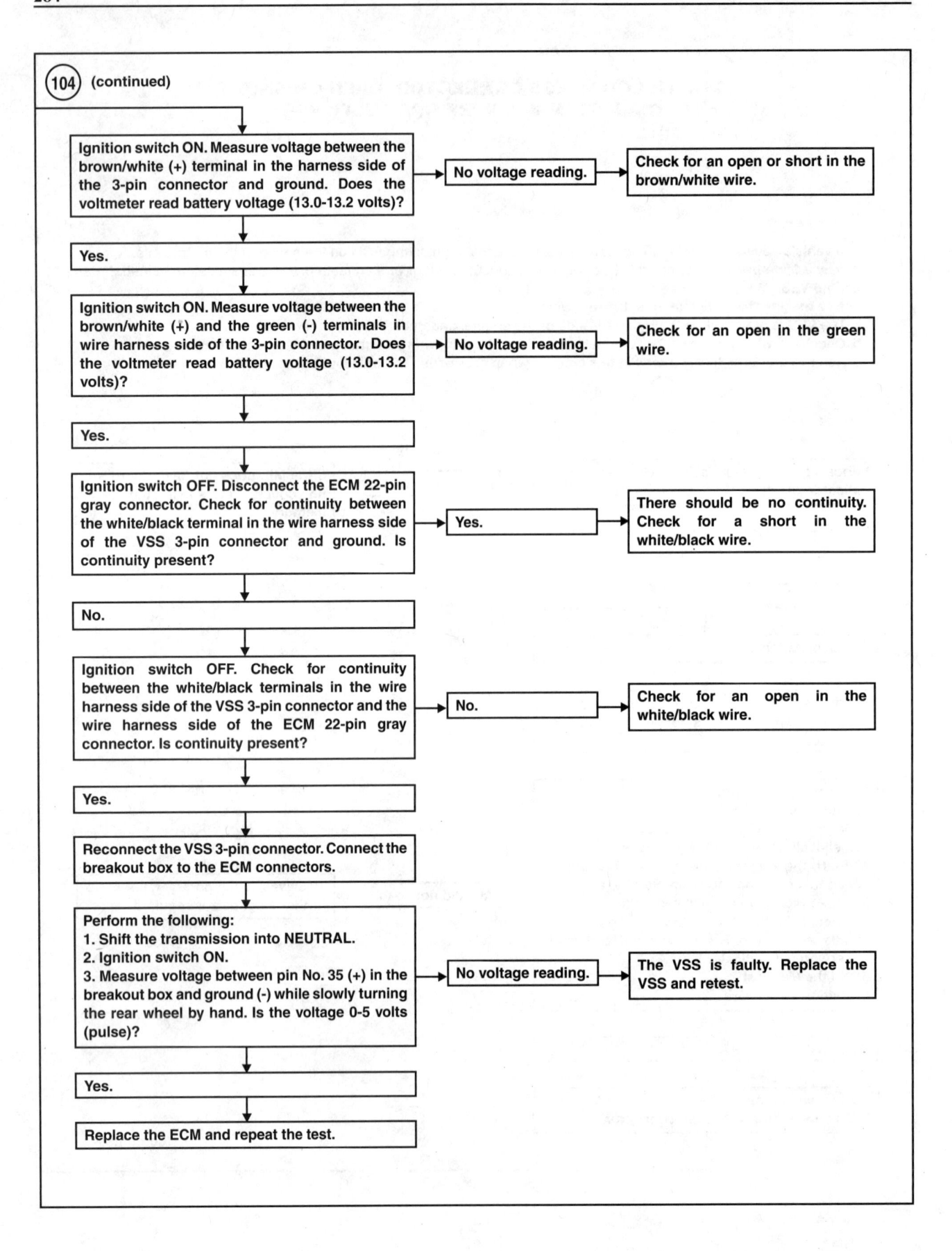

↓

Perform the following:
1. Shift the transmission into NEUTRAL.
2. Ignition switch ON.
3. Measure voltage between pin No. 35 (+) in the breakout box and ground (-) while slowly turning the rear wheel by hand. Is the voltage 0-5 volts (pulse)? → No voltage reading. → The VSS is faulty. Replace the VSS and retest.

↓ Yes.

Replace the ECM and repeat the test.

FUEL INJECTION AND EMISSION CONTROL SYSTEMS

(105)   **DTC 12: FAULTY NO. 1 FUEL INJECTOR**

1. Before making this test, note the following:
   a. The No. 1 fuel injector is mounted on the right, front side of the engine.
   b. Check the wiring connectors, terminals and wires (2-pin connector). Refer to the wiring diagrams at the end of this manual.
   c. To check for an intermittent problem, carefully move the wiring harness and connector(s) with the meter attached to the circuit.
   d. Refer to *Fuel Injectors* in this chapter.

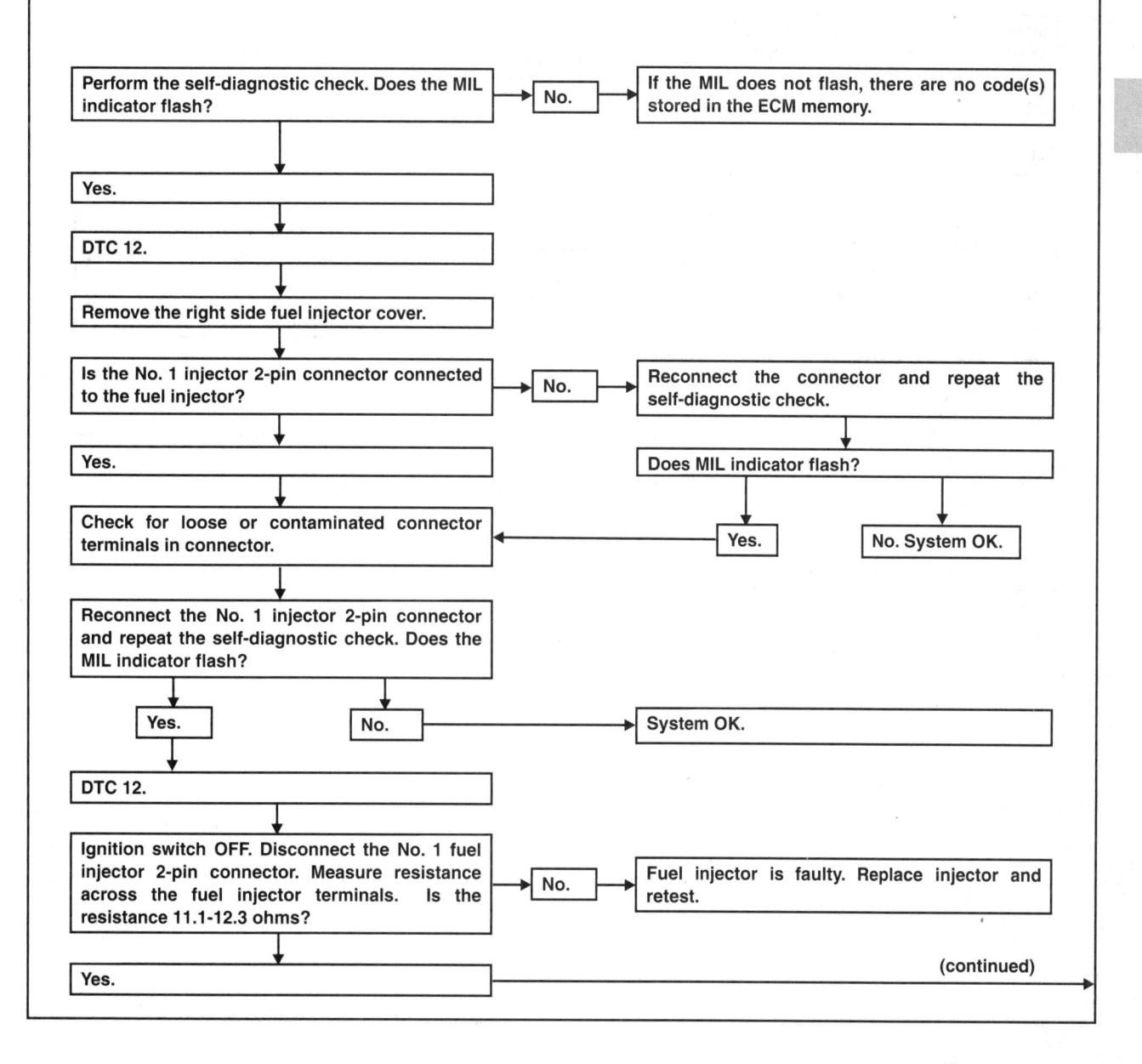

Perform the self-diagnostic check. Does the MIL indicator flash? → **No.** → If the MIL does not flash, there are no code(s) stored in the ECM memory.

↓ Yes.

DTC 12.

↓

Remove the right side fuel injector cover.

↓

Is the No. 1 injector 2-pin connector connected to the fuel injector? → **No.** → Reconnect the connector and repeat the self-diagnostic check.
↓ Yes.   ↓ Does MIL indicator flash?
                                              Yes. / No. System OK.

Check for loose or contaminated connector terminals in connector. ←

↓

Reconnect the No. 1 injector 2-pin connector and repeat the self-diagnostic check. Does the MIL indicator flash?

↓ Yes. / No. → System OK.

DTC 12.

↓

Ignition switch OFF. Disconnect the No. 1 fuel injector 2-pin connector. Measure resistance across the fuel injector terminals. Is the resistance 11.1–12.3 ohms? → **No.** → Fuel injector is faulty. Replace injector and retest.

↓ Yes.   *(continued)*

(105) (continued)

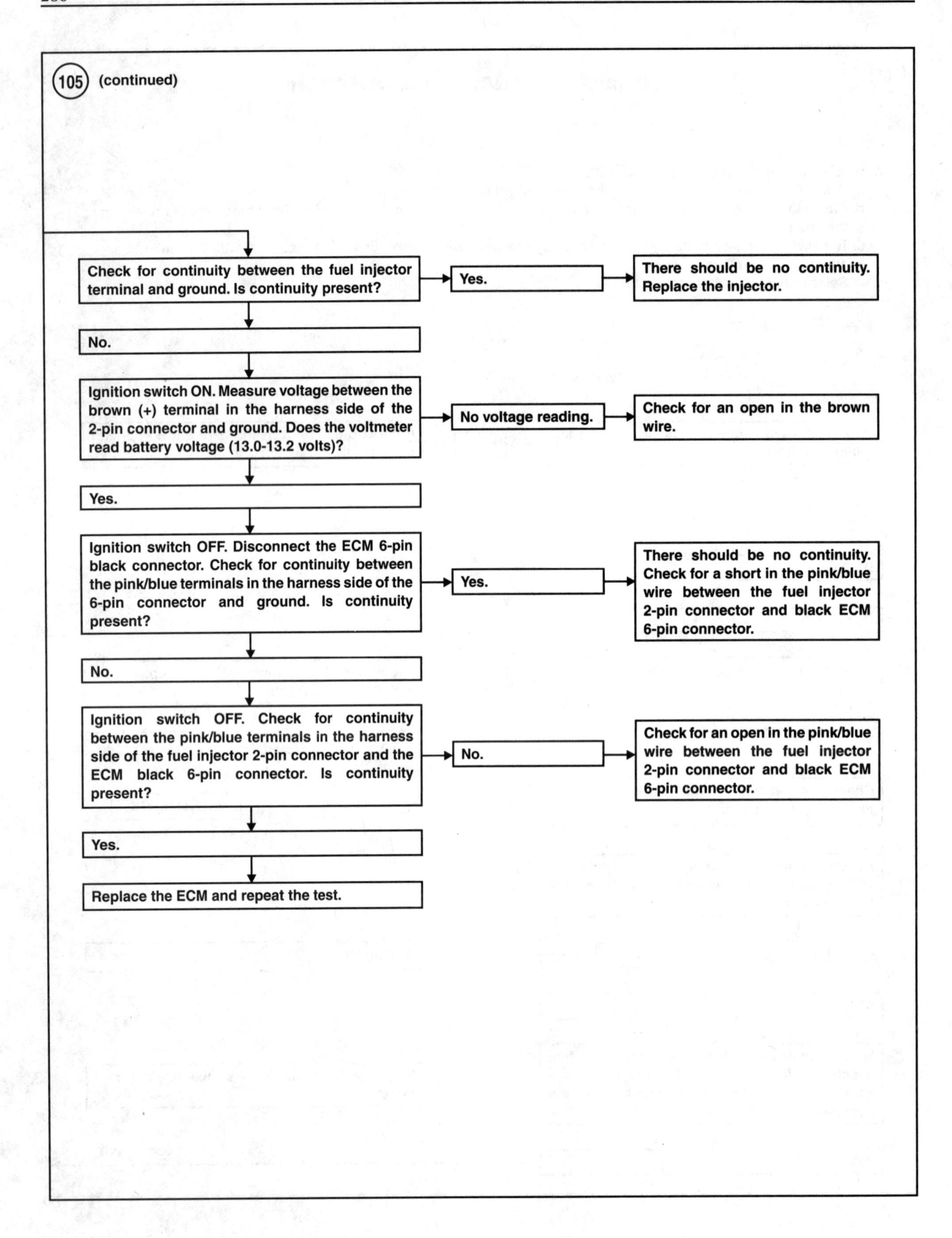

Check for continuity between the fuel injector terminal and ground. Is continuity present? → **Yes.** → There should be no continuity. Replace the injector.

↓ No.

Ignition switch ON. Measure voltage between the brown (+) terminal in the harness side of the 2-pin connector and ground. Does the voltmeter read battery voltage (13.0-13.2 volts)? → **No voltage reading.** → Check for an open in the brown wire.

↓ Yes.

Ignition switch OFF. Disconnect the ECM 6-pin black connector. Check for continuity between the pink/blue terminals in the harness side of the 6-pin connector and ground. Is continuity present? → **Yes.** → There should be no continuity. Check for a short in the pink/blue wire between the fuel injector 2-pin connector and black ECM 6-pin connector.

↓ No.

Ignition switch OFF. Check for continuity between the pink/blue terminals in the harness side of the fuel injector 2-pin connector and the ECM black 6-pin connector. Is continuity present? → **No.** → Check for an open in the pink/blue wire between the fuel injector 2-pin connector and black ECM 6-pin connector.

↓ Yes.

Replace the ECM and repeat the test.

FUEL INJECTION AND EMISSION CONTROL SYSTEMS

(106) **DTC 13: FAULTY NO. 2 FUEL INJECTOR**

1. Before making this test, note the following:
   a. The No. 2 fuel injector is mounted on the left, front side of the engine.
   b. Check the wiring connectors, terminals and wires (2-pin connector). Refer to the wiring diagrams at the end of this manual.
   c. To check for an intermittent problem, carefully move the wiring harness and connector(s) with the meter attached to the circuit.
   d. Refer to *Fuel Injectors* in this chapter.

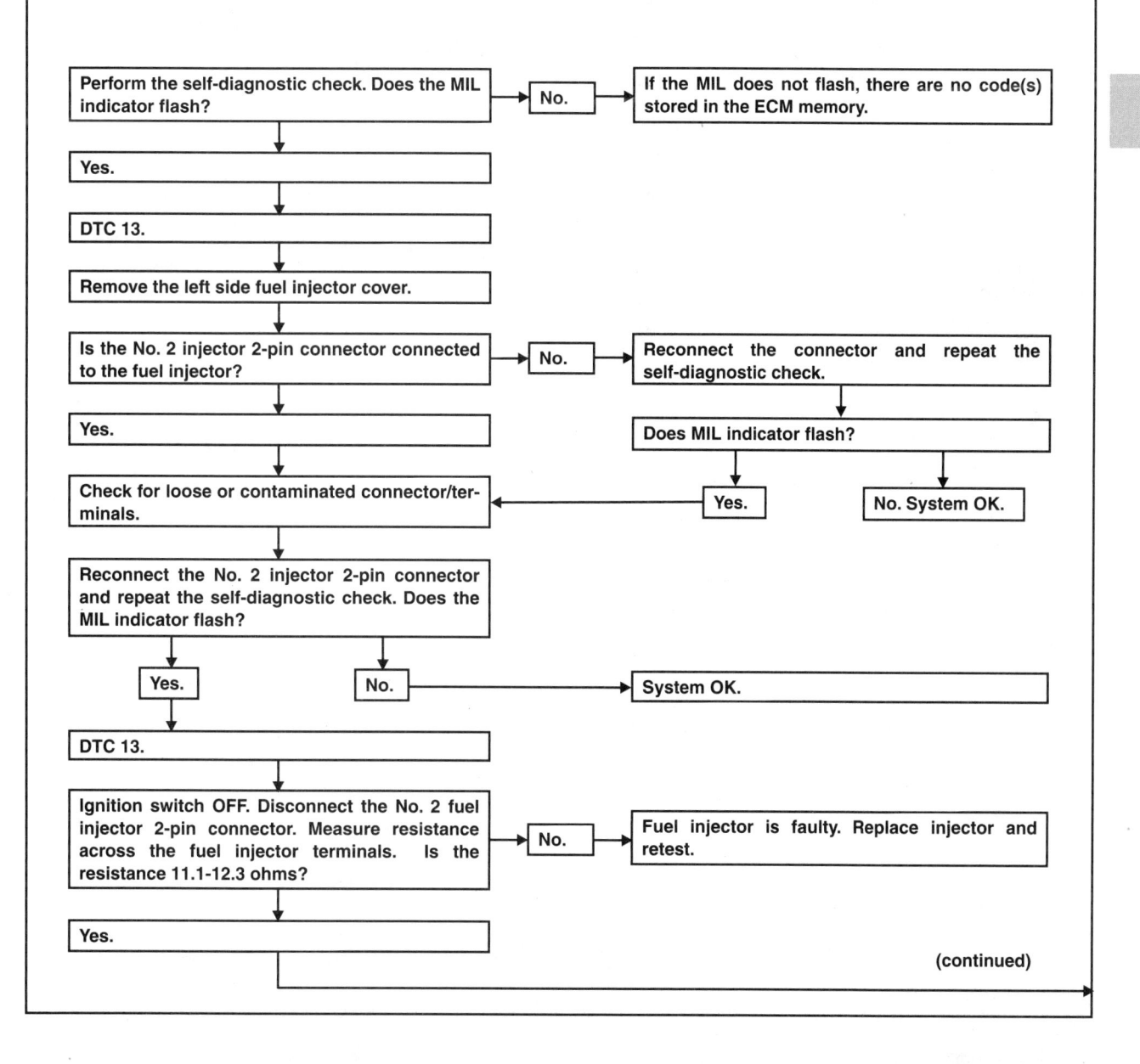

- Perform the self-diagnostic check. Does the MIL indicator flash?
  - **No.** → If the MIL does not flash, there are no code(s) stored in the ECM memory.
  - **Yes.**
- DTC 13.
- Remove the left side fuel injector cover.
- Is the No. 2 injector 2-pin connector connected to the fuel injector?
  - **No.** → Reconnect the connector and repeat the self-diagnostic check. Does MIL indicator flash?
    - **Yes.** → Check for loose or contaminated connector/terminals.
    - **No. System OK.**
  - **Yes.**
- Check for loose or contaminated connector/terminals.
- Reconnect the No. 2 injector 2-pin connector and repeat the self-diagnostic check. Does the MIL indicator flash?
  - **Yes.**
  - **No.** → System OK.
- DTC 13.
- Ignition switch OFF. Disconnect the No. 2 fuel injector 2-pin connector. Measure resistance across the fuel injector terminals. Is the resistance 11.1-12.3 ohms?
  - **No.** → Fuel injector is faulty. Replace injector and retest.
  - **Yes.**

(continued)

106 (continued)

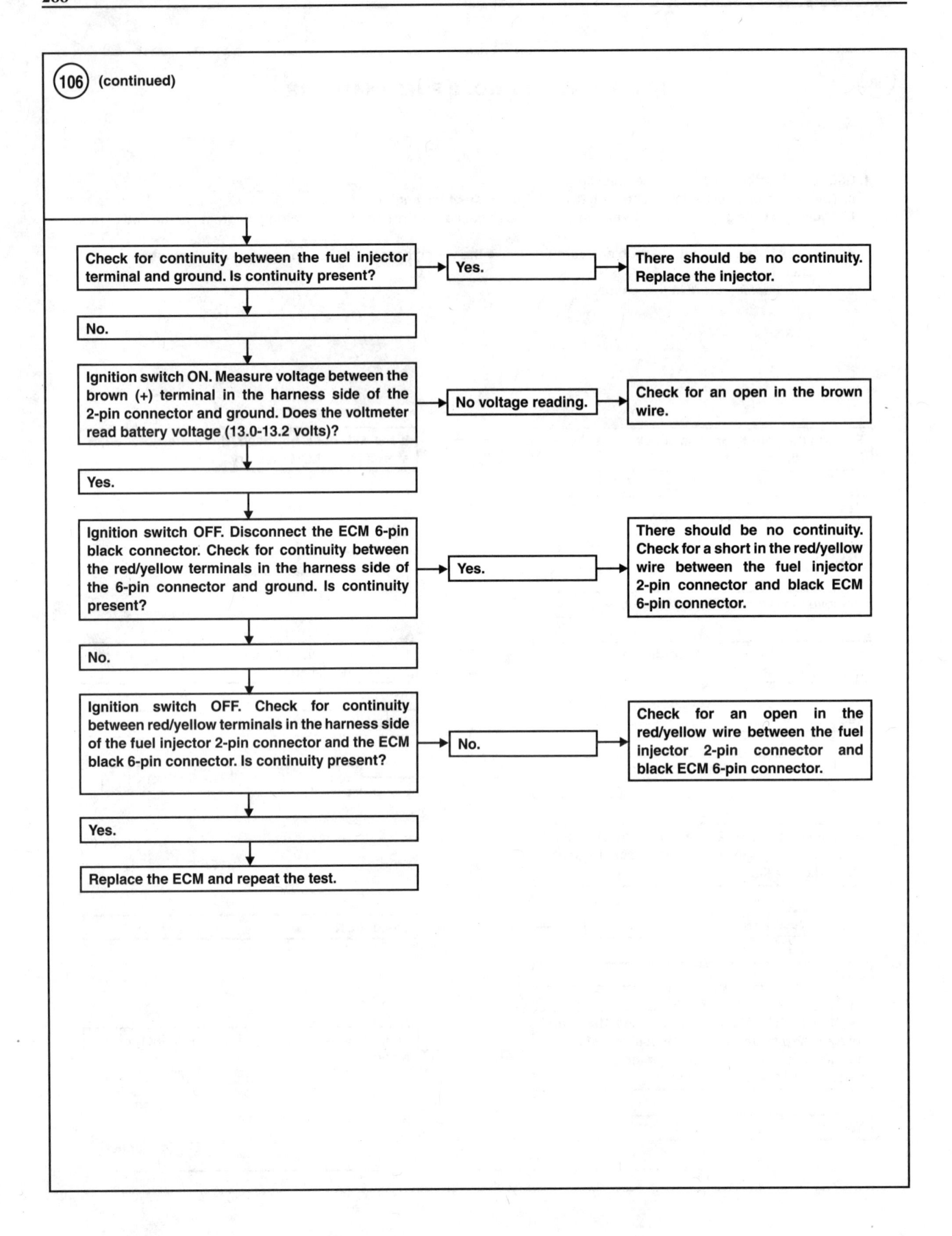

```
┌─────────────────────────────────┐       ┌──────────────────────────────────┐
│ Check for continuity between    │ Yes.  │ There should be no continuity.   │
│ the fuel injector terminal and  │──────▶│ Replace the injector.            │
│ ground. Is continuity present?  │       │                                  │
└─────────────────────────────────┘       └──────────────────────────────────┘
              │
              ▼
        ┌──────┐
        │ No.  │
        └──────┘
              │
              ▼
┌─────────────────────────────────┐       ┌──────────────────────────────────┐
│ Ignition switch ON. Measure     │ No    │ Check for an open in the brown   │
│ voltage between the brown (+)   │voltage│ wire.                            │
│ terminal in the harness side of │reading│                                  │
│ the 2-pin connector and ground. │──────▶│                                  │
│ Does the voltmeter read battery │       │                                  │
│ voltage (13.0-13.2 volts)?      │       │                                  │
└─────────────────────────────────┘       └──────────────────────────────────┘
              │
              ▼
        ┌──────┐
        │ Yes. │
        └──────┘
              │
              ▼
┌─────────────────────────────────┐       ┌──────────────────────────────────┐
│ Ignition switch OFF. Disconnect │       │ There should be no continuity.   │
│ the ECM 6-pin black connector.  │ Yes.  │ Check for a short in the         │
│ Check for continuity between    │──────▶│ red/yellow wire between the fuel │
│ the red/yellow terminals in the │       │ injector 2-pin connector and     │
│ harness side of the 6-pin       │       │ black ECM 6-pin connector.       │
│ connector and ground. Is        │       │                                  │
│ continuity present?             │       │                                  │
└─────────────────────────────────┘       └──────────────────────────────────┘
              │
              ▼
        ┌──────┐
        │ No.  │
        └──────┘
              │
              ▼
┌─────────────────────────────────┐       ┌──────────────────────────────────┐
│ Ignition switch OFF. Check for  │       │ Check for an open in the         │
│ continuity between red/yellow   │ No.   │ red/yellow wire between the fuel │
│ terminals in the harness side   │──────▶│ injector 2-pin connector and     │
│ of the fuel injector 2-pin      │       │ black ECM 6-pin connector.       │
│ connector and the ECM black     │       │                                  │
│ 6-pin connector. Is continuity  │       │                                  │
│ present?                        │       │                                  │
└─────────────────────────────────┘       └──────────────────────────────────┘
              │
              ▼
        ┌──────┐
        │ Yes. │
        └──────┘
              │
              ▼
┌─────────────────────────────────┐
│ Replace the ECM and repeat the  │
│ test.                           │
└─────────────────────────────────┘
```

FUEL INJECTION AND EMISSION CONTROL SYSTEMS

(107)

DTC 14: FAULTY NO. 3 FUEL INJECTOR

1. Before making this test, note the following:
   a. The No. 3 fuel injector is mounted on the right side of the engine.
   b. Check the wiring connectors, terminals and wires (2-pin connector). Refer to the wiring diagrams at the end of the manual.
   c. To check for an intermittent problem, carefully move the wiring harness and connector(s) with the meter attached to the circuit.
   d. Refer to *Fuel Injectors* in this chapter.

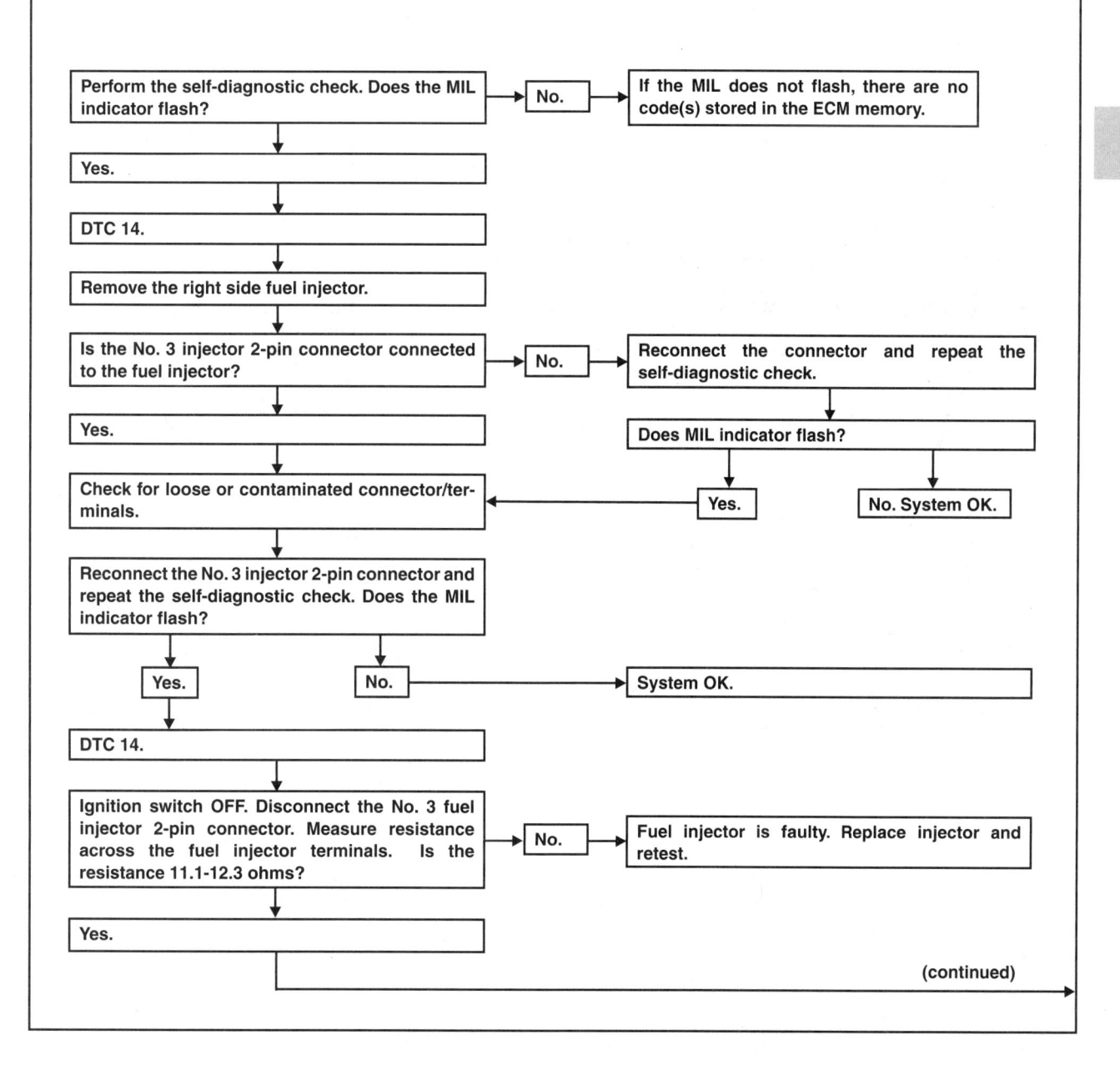

Perform the self-diagnostic check. Does the MIL indicator flash? → No. → If the MIL does not flash, there are no code(s) stored in the ECM memory.

↓ Yes.

DTC 14.

↓

Remove the right side fuel injector.

↓

Is the No. 3 injector 2-pin connector connected to the fuel injector? → No. → Reconnect the connector and repeat the self-diagnostic check.
↓ Yes. ↓ Does MIL indicator flash?
   Yes. → (back to check) | No. System OK.

Check for loose or contaminated connector/terminals.

↓

Reconnect the No. 3 injector 2-pin connector and repeat the self-diagnostic check. Does the MIL indicator flash?

   Yes. ↓   No. → System OK.

DTC 14.

↓

Ignition switch OFF. Disconnect the No. 3 fuel injector 2-pin connector. Measure resistance across the fuel injector terminals. Is the resistance 11.1-12.3 ohms? → No. → Fuel injector is faulty. Replace injector and retest.

↓ Yes.

(continued)

(107) (continued)

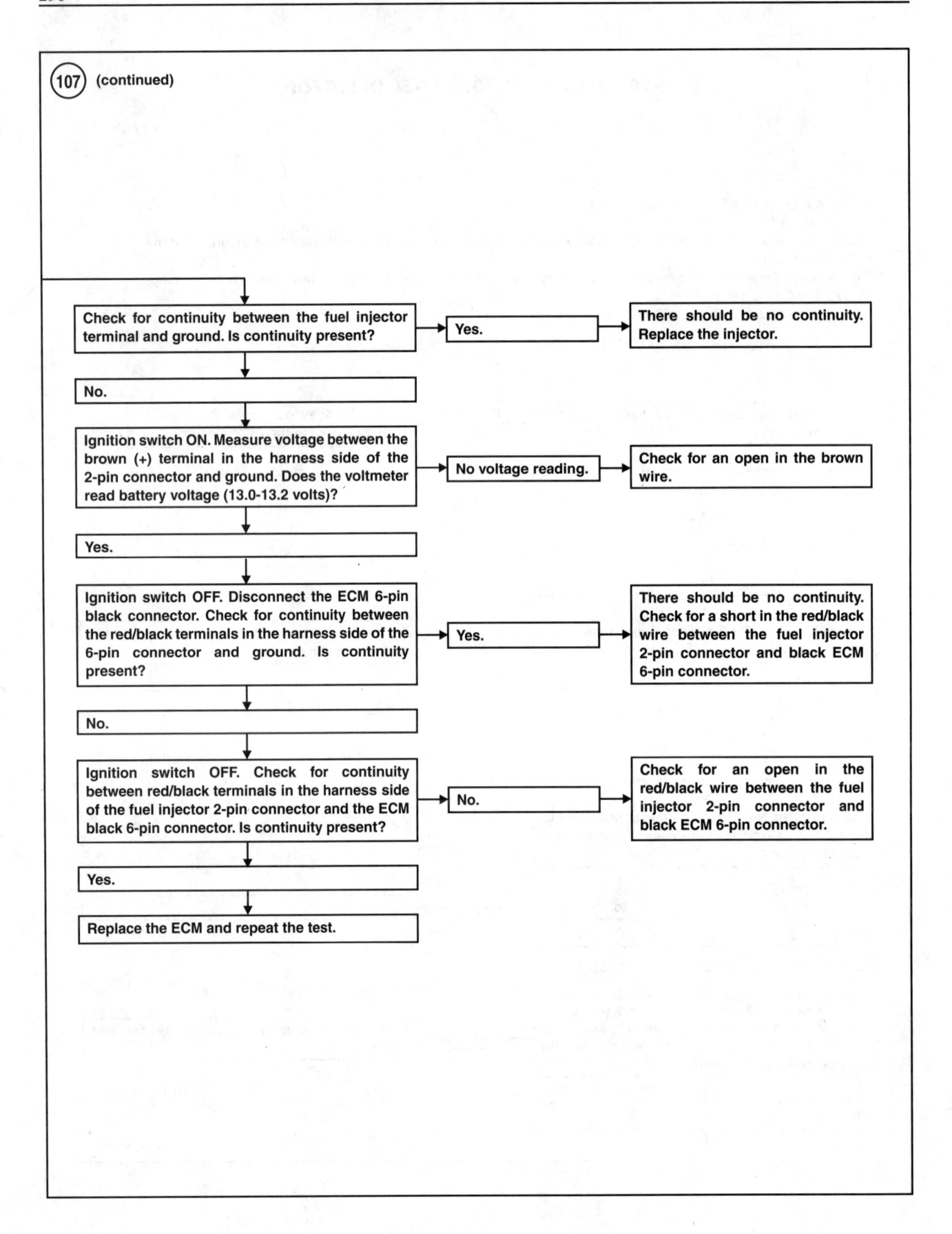

Check for continuity between the fuel injector terminal and ground. Is continuity present? → **Yes.** → There should be no continuity. Replace the injector.

↓ **No.**

Ignition switch ON. Measure voltage between the brown (+) terminal in the harness side of the 2-pin connector and ground. Does the voltmeter read battery voltage (13.0-13.2 volts)? → **No voltage reading.** → Check for an open in the brown wire.

↓ **Yes.**

Ignition switch OFF. Disconnect the ECM 6-pin black connector. Check for continuity between the red/black terminals in the harness side of the 6-pin connector and ground. Is continuity present? → **Yes.** → There should be no continuity. Check for a short in the red/black wire between the fuel injector 2-pin connector and black ECM 6-pin connector.

↓ **No.**

Ignition switch OFF. Check for continuity between red/black terminals in the harness side of the fuel injector 2-pin connector and the ECM black 6-pin connector. Is continuity present? → **No.** → Check for an open in the red/black wire between the fuel injector 2-pin connector and black ECM 6-pin connector.

↓ **Yes.**

Replace the ECM and repeat the test.

FUEL INJECTION AND EMISSION CONTROL SYSTEMS

(108)   **DTC 15: FAULTY NO. 4 FUEL INJECTOR**

1. Before making this test, note the following:
   a. The No. 4 fuel injector is mounted on the left side of the engine.
   b. Check the wiring connectors, terminals and wires (2-pin connector). Refer to the wiring diagrams at the end of this manual.
   c. To check for an intermittent problem, carefully move the wiring harness and connector(s) with the meter attached to the circuit.
   d. Refer to *Fuel Injectors* in this chapter.

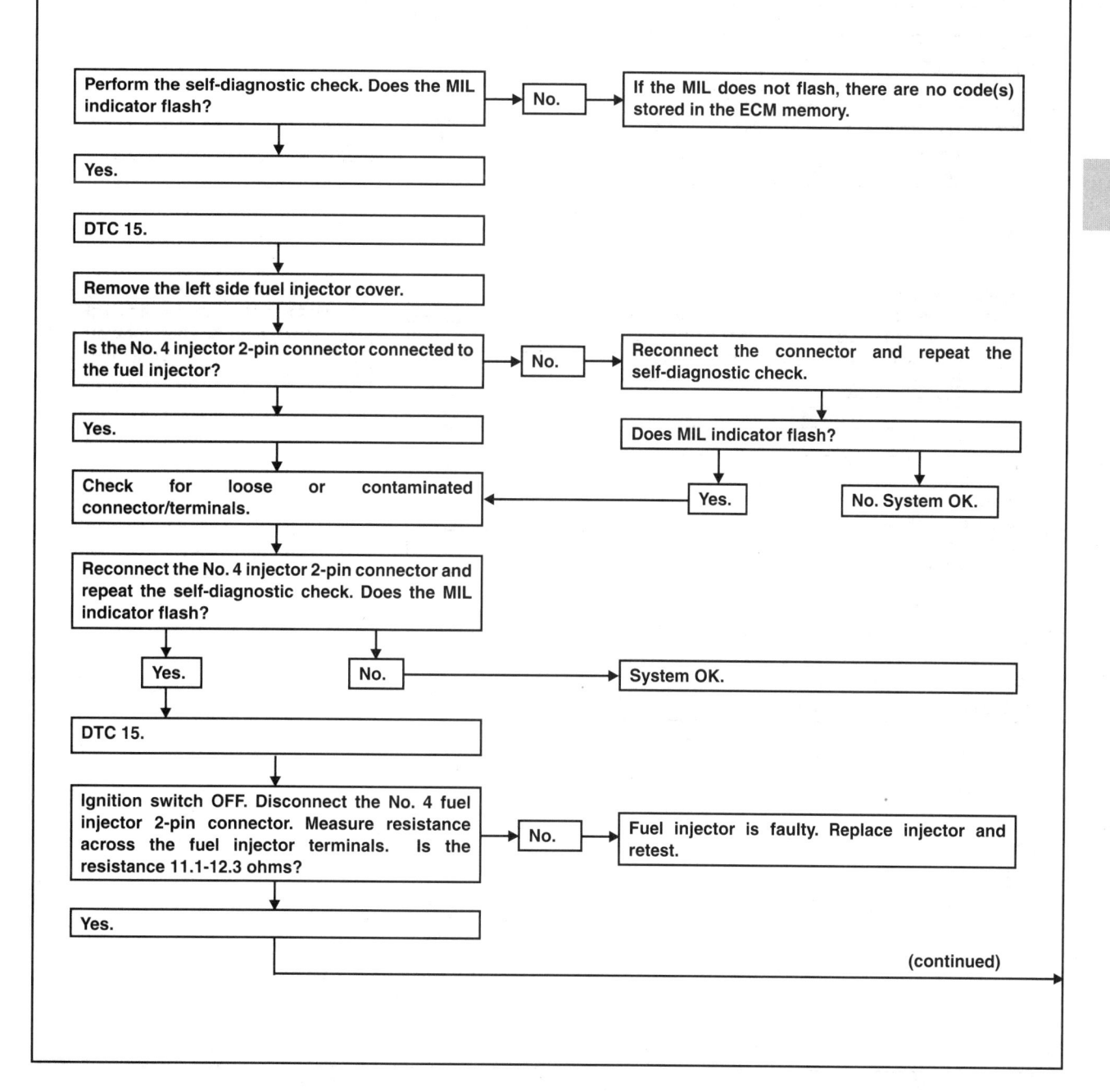

Perform the self-diagnostic check. Does the MIL indicator flash? → **No.** → If the MIL does not flash, there are no code(s) stored in the ECM memory.

↓ Yes.

DTC 15.

↓

Remove the left side fuel injector cover.

↓

Is the No. 4 injector 2-pin connector connected to the fuel injector? → **No.** → Reconnect the connector and repeat the self-diagnostic check.

↓ Yes.   ↓

                                     Does MIL indicator flash?

Check for loose or contaminated connector/terminals. ←——— Yes.   No. System OK.

↓

Reconnect the No. 4 injector 2-pin connector and repeat the self-diagnostic check. Does the MIL indicator flash?

↓ Yes.   No. → System OK.

DTC 15.

↓

Ignition switch OFF. Disconnect the No. 4 fuel injector 2-pin connector. Measure resistance across the fuel injector terminals. Is the resistance 11.1-12.3 ohms? → **No.** → Fuel injector is faulty. Replace injector and retest.

↓ Yes.

(continued)

291

(108) (continued)

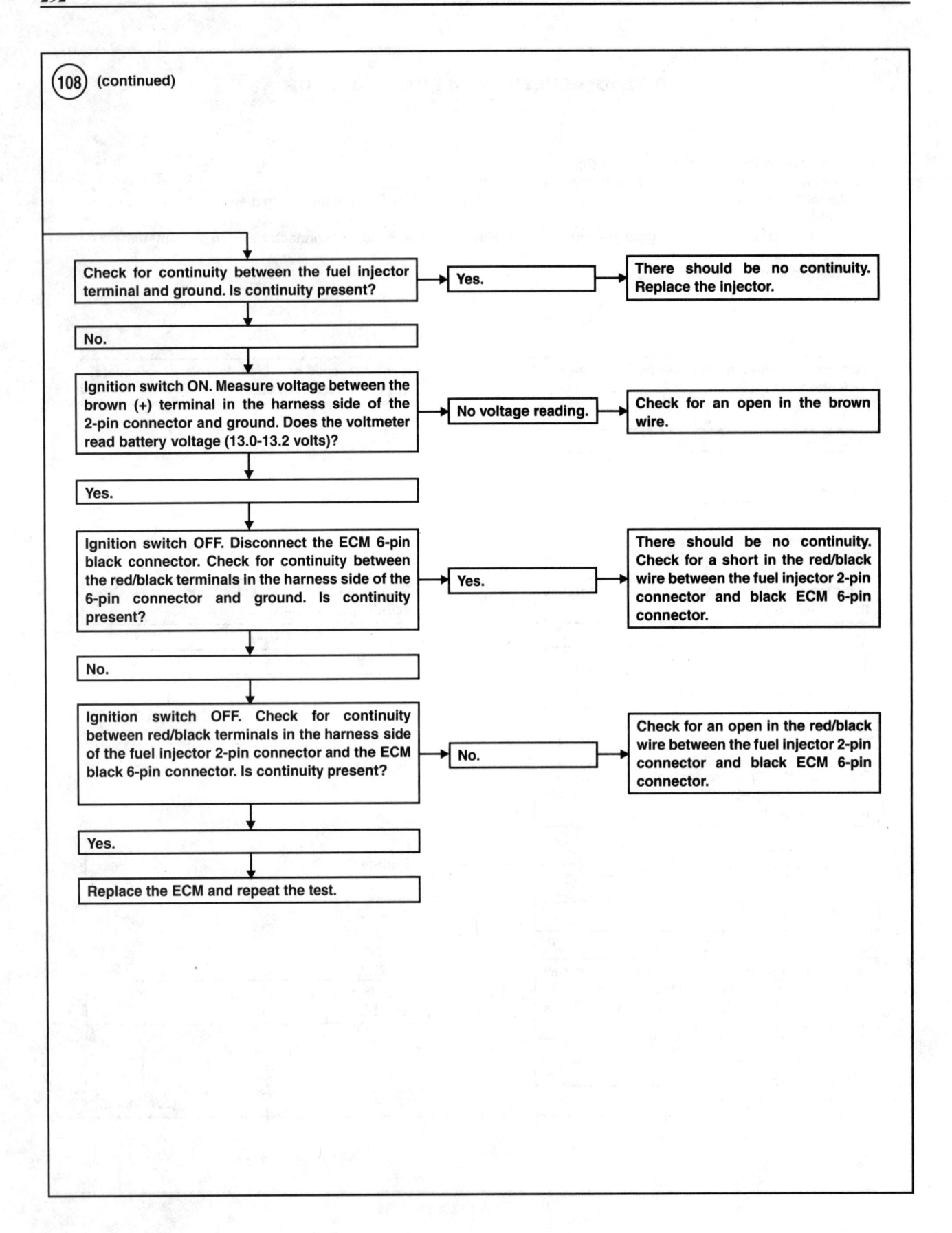

↓

Check for continuity between the fuel injector terminal and ground. Is continuity present? — **Yes.** → There should be no continuity. Replace the injector.

↓ **No.**

Ignition switch ON. Measure voltage between the brown (+) terminal in the harness side of the 2-pin connector and ground. Does the voltmeter read battery voltage (13.0-13.2 volts)? — **No voltage reading.** → Check for an open in the brown wire.

↓ **Yes.**

Ignition switch OFF. Disconnect the ECM 6-pin black connector. Check for continuity between the red/black terminals in the harness side of the 6-pin connector and ground. Is continuity present? — **Yes.** → There should be no continuity. Check for a short in the red/black wire between the fuel injector 2-pin connector and black ECM 6-pin connector.

↓ **No.**

Ignition switch OFF. Check for continuity between red/black terminals in the harness side of the fuel injector 2-pin connector and the ECM black 6-pin connector. Is continuity present? — **No.** → Check for an open in the red/black wire between the fuel injector 2-pin connector and black ECM 6-pin connector.

↓ **Yes.**

Replace the ECM and repeat the test.

FUEL INJECTION AND EMISSION CONTROL SYSTEMS

⑩⁹ DTC 16: FAULTY NO. 5 FUEL INJECTOR

1. Before making this test, note the following:
   a. The No. 5 fuel injector is mounted on the right, front side of the engine.
   b. Check the wiring connectors, terminals and wires (2-pin connector). Refer to the wiring diagrams at the end of this manual.
   c. To check for an intermittent problem, carefully move the wiring harness and connector(s) with the meter attached to the circuit.
   d. Refer to *Fuel Injectors* in this chapter.

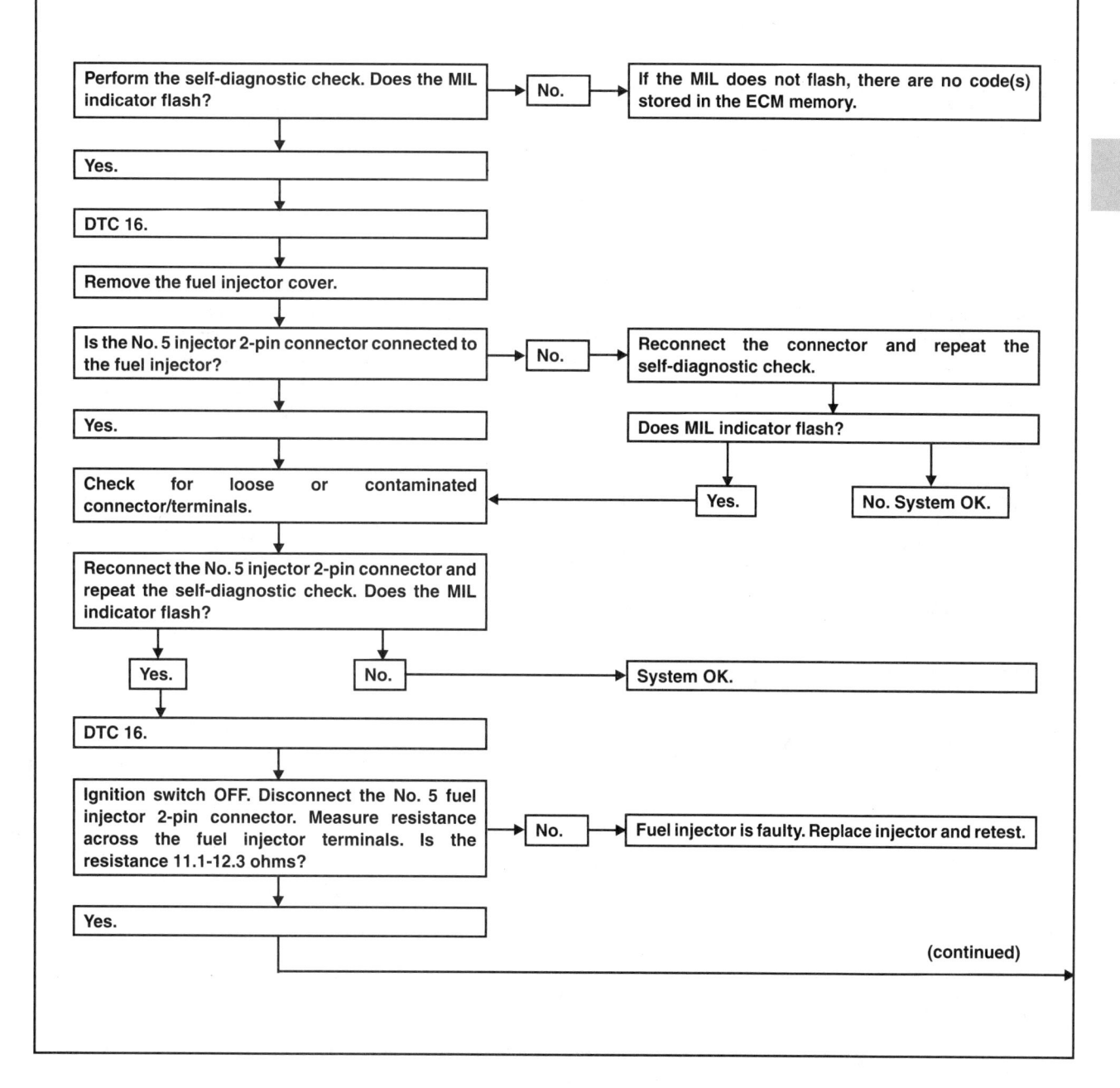

Perform the self-diagnostic check. Does the MIL indicator flash? → No. → If the MIL does not flash, there are no code(s) stored in the ECM memory.

↓ Yes.

DTC 16.

↓

Remove the fuel injector cover.

↓

Is the No. 5 injector 2-pin connector connected to the fuel injector? → No. → Reconnect the connector and repeat the self-diagnostic check.

↓ Yes.   ↓ Does MIL indicator flash?

                                                       Yes.   No. System OK.

Check for loose or contaminated connector/terminals. ←

↓

Reconnect the No. 5 injector 2-pin connector and repeat the self-diagnostic check. Does the MIL indicator flash?

↓ Yes.   No. → System OK.

DTC 16.

↓

Ignition switch OFF. Disconnect the No. 5 fuel injector 2-pin connector. Measure resistance across the fuel injector terminals. Is the resistance 11.1-12.3 ohms? → No. → Fuel injector is faulty. Replace injector and retest.

↓ Yes.

(continued)

(109) (continued)

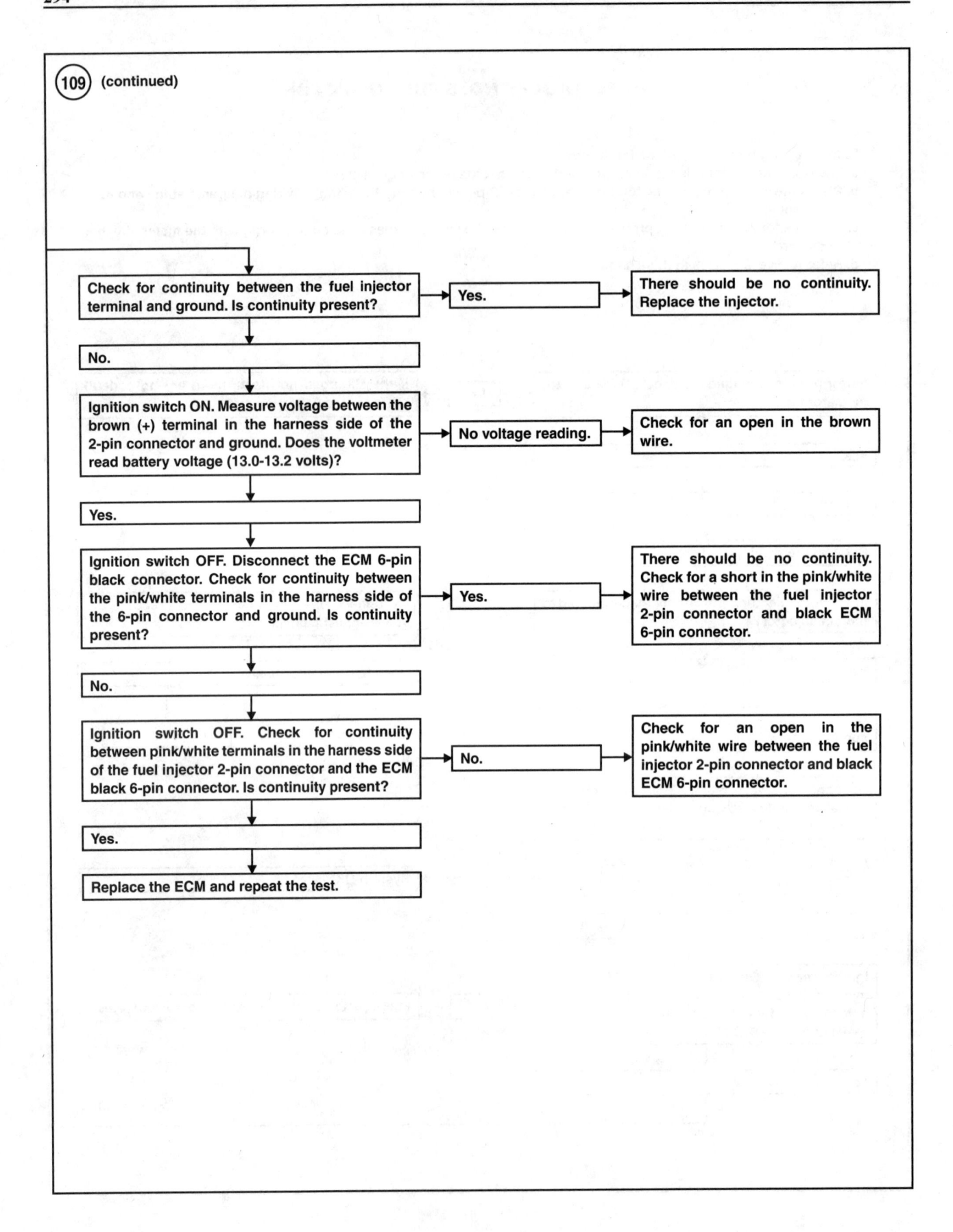

Check for continuity between the fuel injector terminal and ground. Is continuity present? → **Yes.** → There should be no continuity. Replace the injector.

↓ **No.**

Ignition switch ON. Measure voltage between the brown (+) terminal in the harness side of the 2-pin connector and ground. Does the voltmeter read battery voltage (13.0-13.2 volts)? → **No voltage reading.** → Check for an open in the brown wire.

↓ **Yes.**

Ignition switch OFF. Disconnect the ECM 6-pin black connector. Check for continuity between the pink/white terminals in the harness side of the 6-pin connector and ground. Is continuity present? → **Yes.** → There should be no continuity. Check for a short in the pink/white wire between the fuel injector 2-pin connector and black ECM 6-pin connector.

↓ **No.**

Ignition switch OFF. Check for continuity between pink/white terminals in the harness side of the fuel injector 2-pin connector and the ECM black 6-pin connector. Is continuity present? → **No.** → Check for an open in the pink/white wire between the fuel injector 2-pin connector and black ECM 6-pin connector.

↓ **Yes.**

Replace the ECM and repeat the test.

FUEL INJECTION AND EMISSION CONTROL SYSTEMS

(110)

DTC 17: FAULTY NO. 6 FUEL INJECTOR

1. Before making this test, note the following:
   a. The No. 6 fuel injector is mounted on the right, front side of the engine.
   b. Check the wiring connectors, terminals and wires (2-pin connector). Refer to the wiring diagrams at the end of this manual.
   c. To check for an intermittent problem, carefully move the wiring harness and connector(s) with the meter attached to the circuit.
   d. Refer to *Fuel Injectors* in this chapter.

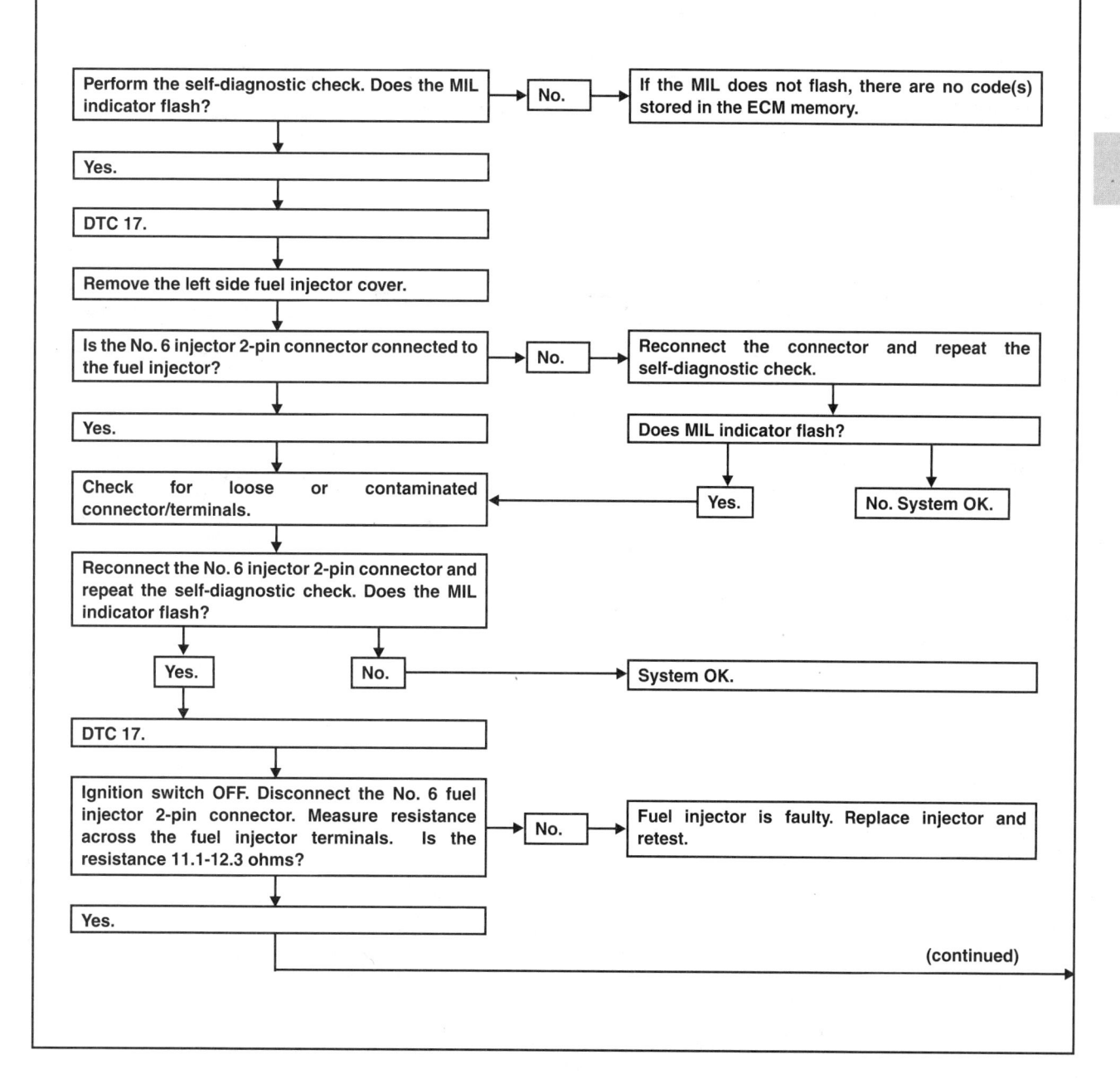

```
Perform the self-diagnostic check. Does the MIL indicator flash?
  → No. → If the MIL does not flash, there are no code(s) stored in the ECM memory.
  ↓ Yes.

DTC 17.
  ↓

Remove the left side fuel injector cover.
  ↓

Is the No. 6 injector 2-pin connector connected to the fuel injector?
  → No. → Reconnect the connector and repeat the self-diagnostic check.
                ↓
           Does MIL indicator flash?
              ↓ Yes.        ↓ No. System OK.
  ↓ Yes.

Check for loose or contaminated connector/terminals. ←
  ↓

Reconnect the No. 6 injector 2-pin connector and repeat the self-diagnostic check. Does the MIL indicator flash?
  ↓ Yes.    → No. → System OK.

DTC 17.
  ↓

Ignition switch OFF. Disconnect the No. 6 fuel injector 2-pin connector. Measure resistance across the fuel injector terminals. Is the resistance 11.1-12.3 ohms?
  → No. → Fuel injector is faulty. Replace injector and retest.
  ↓ Yes.

(continued)
```

(110) (continued)

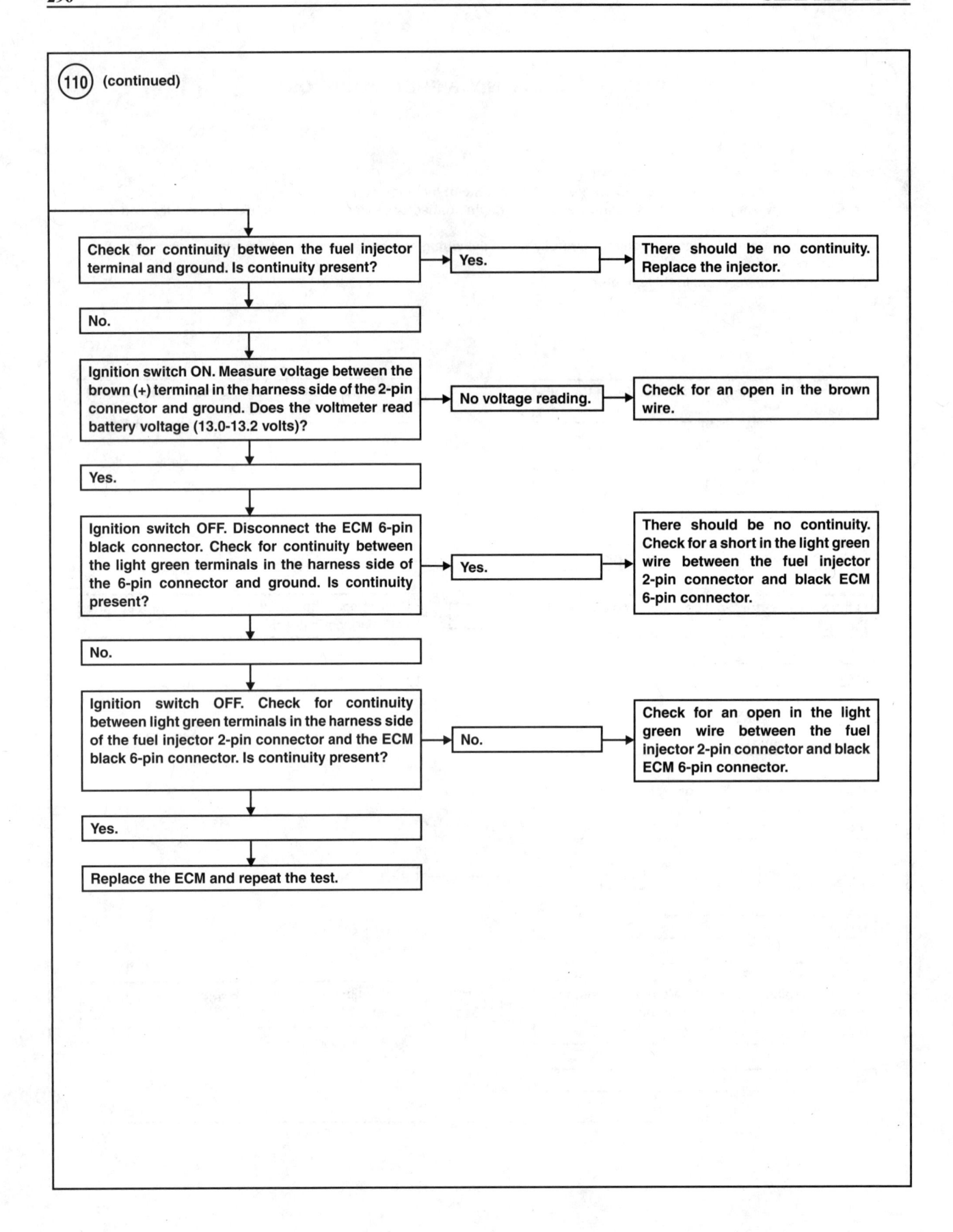

- Check for continuity between the fuel injector terminal and ground. Is continuity present?
  - Yes. → There should be no continuity. Replace the injector.
  - No.
- Ignition switch ON. Measure voltage between the brown (+) terminal in the harness side of the 2-pin connector and ground. Does the voltmeter read battery voltage (13.0-13.2 volts)?
  - No voltage reading. → Check for an open in the brown wire.
  - Yes.
- Ignition switch OFF. Disconnect the ECM 6-pin black connector. Check for continuity between the light green terminals in the harness side of the 6-pin connector and ground. Is continuity present?
  - Yes. → There should be no continuity. Check for a short in the light green wire between the fuel injector 2-pin connector and black ECM 6-pin connector.
  - No.
- Ignition switch OFF. Check for continuity between light green terminals in the harness side of the fuel injector 2-pin connector and the ECM black 6-pin connector. Is continuity present?
  - No. → Check for an open in the light green wire between the fuel injector 2-pin connector and black ECM 6-pin connector.
  - Yes.
- Replace the ECM and repeat the test.

FUEL INJECTION AND EMISSION CONTROL SYSTEMS

(111)

DTC 18: LOOSE CAM PULSE GENERATOR CONNECTOR, OPEN OR SHORT CIRCUIT IN CAM PULSE GENERTOR WIRE OR FAULTY CAM PULSE GENERATOR

1. Before making this test, note the following:
   a. The cam pulse generator is mounted on the left side of the engine. Refer to *Cam Pulse Generator* in this chapter.
   b. Refer to *ECM Test Harness* in this chapter when connecting the breakout box to the ECM connectors.
   c. Refer to *Peak Voltage Tests and Equipment* in Chapter Nine for information on using peak voltage testers.
   d. Check the wiring connectors, terminals and wires (2-pin connector). Refer to the wiring diagrams at the end of this manual.

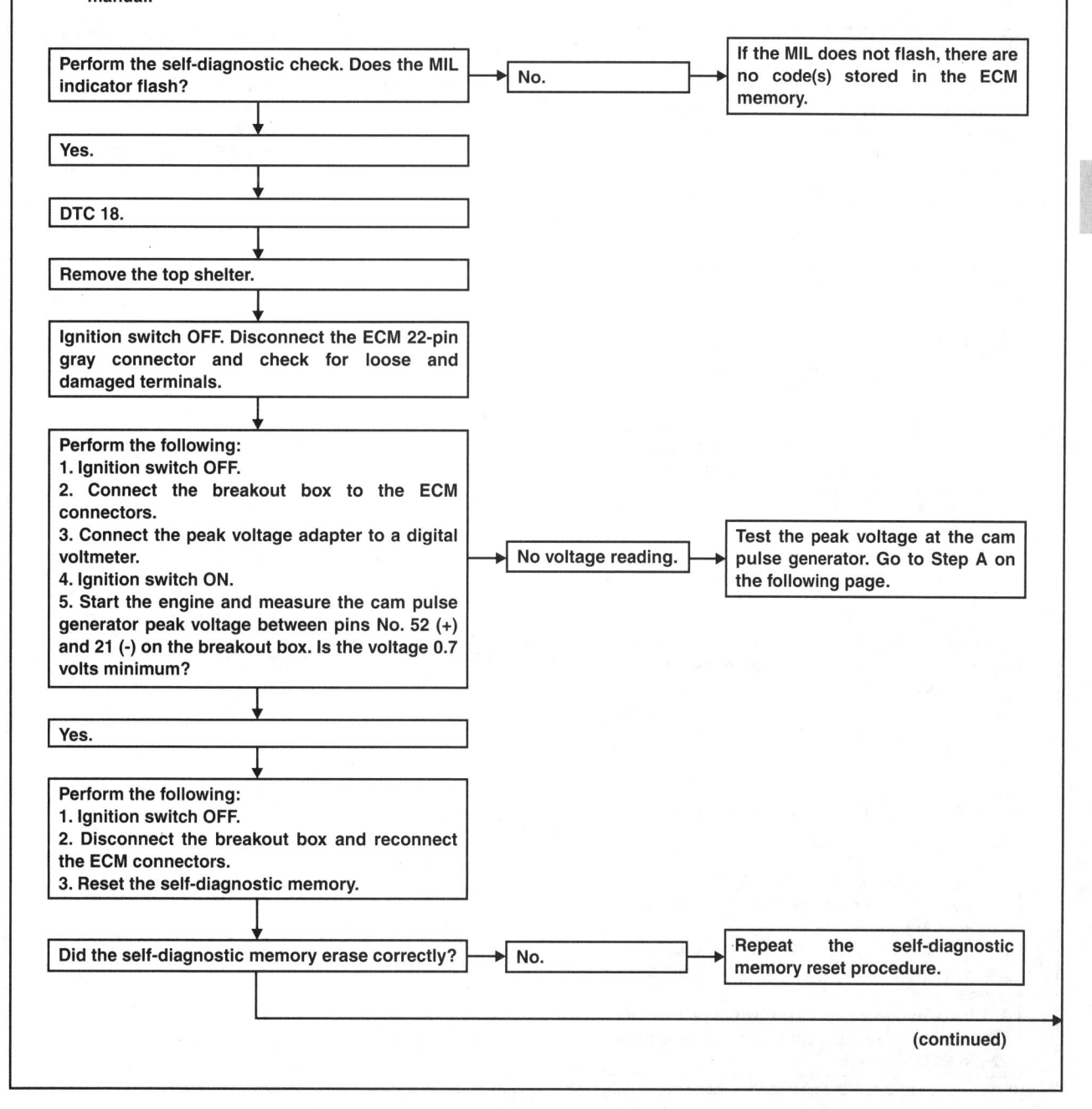

```
Perform the self-diagnostic check. Does the MIL
indicator flash?  ──No.──► If the MIL does not flash, there are
                           no code(s) stored in the ECM
                           memory.
        │
       Yes.
        ▼
    DTC 18.
        ▼
Remove the top shelter.
        ▼
Ignition switch OFF. Disconnect the ECM 22-pin
gray connector and check for loose and
damaged terminals.
        ▼
Perform the following:
1. Ignition switch OFF.
2. Connect the breakout box to the ECM
connectors.
3. Connect the peak voltage adapter to a digital
voltmeter.                                  ──No voltage reading.──► Test the peak voltage at the cam
4. Ignition switch ON.                                                pulse generator. Go to Step A on
5. Start the engine and measure the cam pulse                         the following page.
generator peak voltage between pins No. 52 (+)
and 21 (-) on the breakout box. Is the voltage 0.7
volts minimum?
        │
       Yes.
        ▼
Perform the following:
1. Ignition switch OFF.
2. Disconnect the breakout box and reconnect
the ECM connectors.
3. Reset the self-diagnostic memory.
        ▼
Did the self-diagnostic memory erase correctly? ──No.──► Repeat the self-diagnostic
                                                         memory reset procedure.
```

(continued)

(111) (continued)

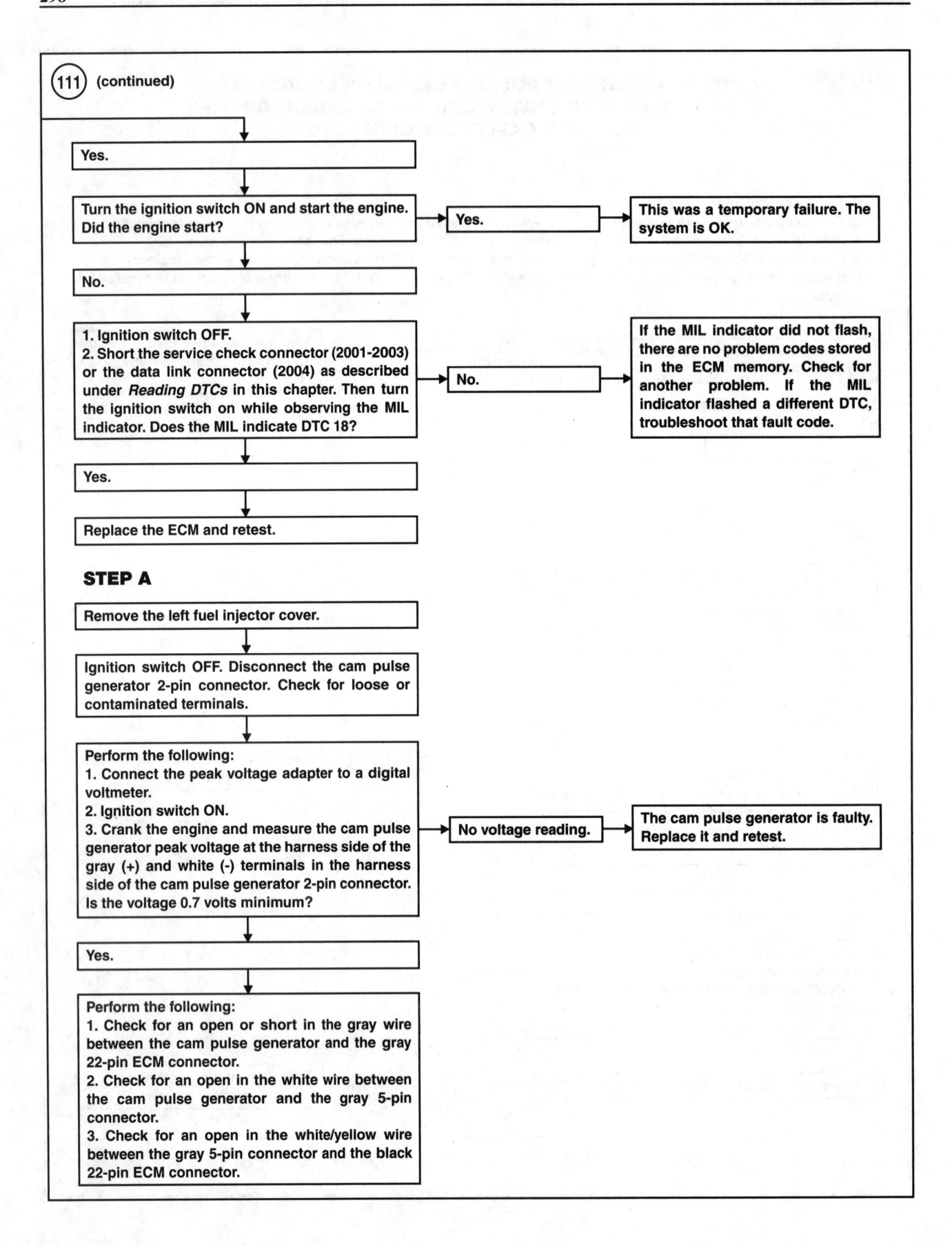

↓

Yes.

↓

Turn the ignition switch ON and start the engine. Did the engine start? → Yes. → This was a temporary failure. The system is OK.

↓ No.

1. Ignition switch OFF.
2. Short the service check connector (2001-2003) or the data link connector (2004) as described under *Reading DTCs* in this chapter. Then turn the ignition switch on while observing the MIL indicator. Does the MIL indicate DTC 18? → No. → If the MIL indicator did not flash, there are no problem codes stored in the ECM memory. Check for another problem. If the MIL indicator flashed a different DTC, troubleshoot that fault code.

↓ Yes.

Replace the ECM and retest.

STEP A

Remove the left fuel injector cover.

↓

Ignition switch OFF. Disconnect the cam pulse generator 2-pin connector. Check for loose or contaminated terminals.

↓

Perform the following:
1. Connect the peak voltage adapter to a digital voltmeter.
2. Ignition switch ON.
3. Crank the engine and measure the cam pulse generator peak voltage at the harness side of the gray (+) and white (-) terminals in the harness side of the cam pulse generator 2-pin connector. Is the voltage 0.7 volts minimum? → No voltage reading. → The cam pulse generator is faulty. Replace it and retest.

↓ Yes.

Perform the following:
1. Check for an open or short in the gray wire between the cam pulse generator and the gray 22-pin ECM connector.
2. Check for an open in the white wire between the cam pulse generator and the gray 5-pin connector.
3. Check for an open in the white/yellow wire between the gray 5-pin connector and the black 22-pin ECM connector.

FUEL INJECTION AND EMISSION CONTROL SYSTEMS

(112)

DTC 19: LOOSE IGNITION PULSE GENERATOR CONNECTOR, OPEN OR SHORT CIRCUIT IN IGNITION PULSE GENERTOR WIRE OR FAULTY IGNITION PULSE GENERATOR

1. Before making this test, note the following:
    a. The ignition pulse generator is mounted on the inside of the front crankcase cover. Refer to *Ignition Pulse Generator* in Chapter Nine.
    b. Refer to *ECM Test Harness* in this chapter when connecting the breakout box to the ECM connectors.
    c. Refer to *Peak Voltage Tests and Equipment* in Chapter Nine for information on using peak voltage testers.
    d. Check the wiring connectors, terminals and wires. Refer to the wiring diagrams at the end of this manual.

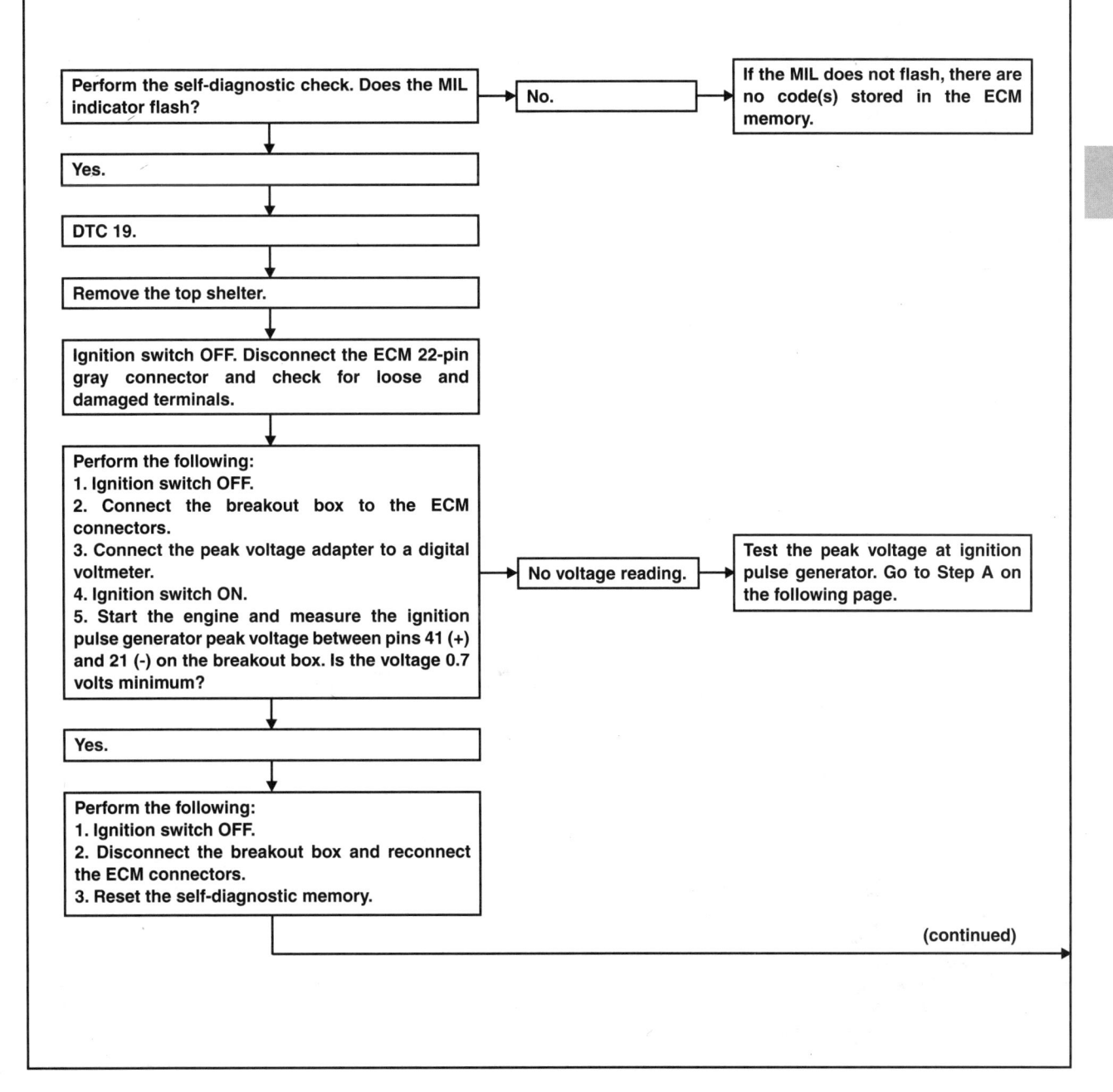

Perform the self-diagnostic check. Does the MIL indicator flash?
→ No. → If the MIL does not flash, there are no code(s) stored in the ECM memory.

↓ Yes.

DTC 19.

↓

Remove the top shelter.

↓

Ignition switch OFF. Disconnect the ECM 22-pin gray connector and check for loose and damaged terminals.

↓

Perform the following:
1. Ignition switch OFF.
2. Connect the breakout box to the ECM connectors.
3. Connect the peak voltage adapter to a digital voltmeter.
4. Ignition switch ON.
5. Start the engine and measure the ignition pulse generator peak voltage between pins 41 (+) and 21 (-) on the breakout box. Is the voltage 0.7 volts minimum?
→ No voltage reading. → Test the peak voltage at ignition pulse generator. Go to Step A on the following page.

↓ Yes.

Perform the following:
1. Ignition switch OFF.
2. Disconnect the breakout box and reconnect the ECM connectors.
3. Reset the self-diagnostic memory.

(continued)

(112) (continued)

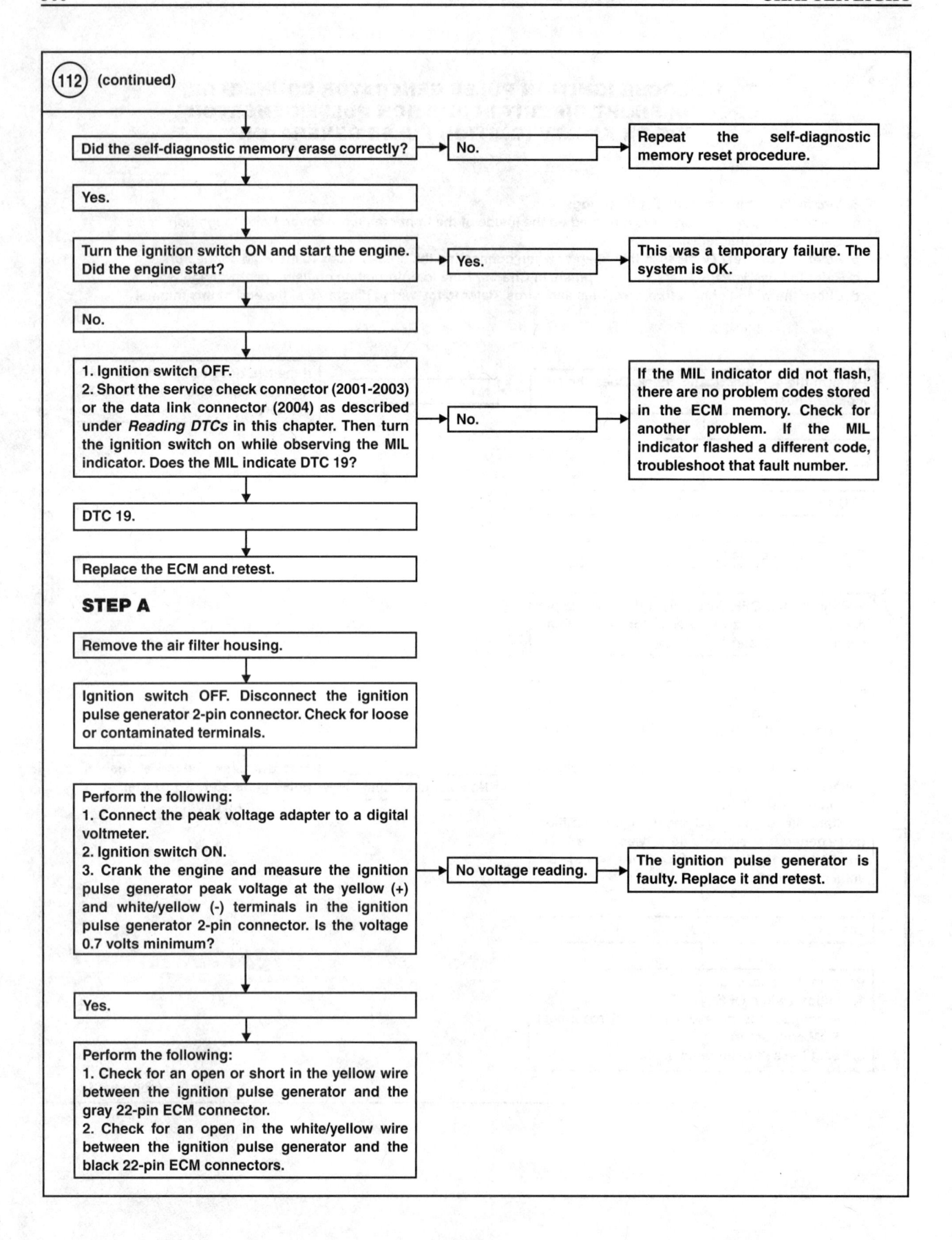

Did the self-diagnostic memory erase correctly? → **No.** → Repeat the self-diagnostic memory reset procedure.

↓ Yes.

Turn the ignition switch ON and start the engine. Did the engine start? → **Yes.** → This was a temporary failure. The system is OK.

↓ No.

1. Ignition switch OFF.
2. Short the service check connector (2001-2003) or the data link connector (2004) as described under *Reading DTCs* in this chapter. Then turn the ignition switch on while observing the MIL indicator. Does the MIL indicate DTC 19? → **No.** → If the MIL indicator did not flash, there are no problem codes stored in the ECM memory. Check for another problem. If the MIL indicator flashed a different code, troubleshoot that fault number.

↓ DTC 19.

Replace the ECM and retest.

STEP A

Remove the air filter housing.

↓

Ignition switch OFF. Disconnect the ignition pulse generator 2-pin connector. Check for loose or contaminated terminals.

↓

Perform the following:
1. Connect the peak voltage adapter to a digital voltmeter.
2. Ignition switch ON.
3. Crank the engine and measure the ignition pulse generator peak voltage at the yellow (+) and white/yellow (-) terminals in the ignition pulse generator 2-pin connector. Is the voltage 0.7 volts minimum? → **No voltage reading.** → The ignition pulse generator is faulty. Replace it and retest.

↓ Yes.

Perform the following:
1. Check for an open or short in the yellow wire between the ignition pulse generator and the gray 22-pin ECM connector.
2. Check for an open in the white/yellow wire between the ignition pulse generator and the black 22-pin ECM connectors.

FUEL INJECTION AND EMISSION CONTROL SYSTEMS

DTC 21: FAULTY RIGHT OXYGEN SENSOR OR CIRCUIT

The oxygen sensors send a voltage signal to the ECM on the amount of oxygen in the exhaust gas. The ECM uses this signal to adjust the air/fuel mixture.

1. Note the following when testing the oxygen sensor:
   a. The right hand oxygen sensor is mounted in the right exhaust pipe.
   b. The right side oxygen sensor 4-pin connector is mounted inside the connector pouch located on the right side of the air filter cover.
   c. Never apply voltage to the oxygen sensor. Oxygen sensors generate their own voltage and to do so externally will damage the sensor.
   d. Use a digital voltmeter with an input impedance of at least 10 Megaohms.
   e. When the engine is first started, the ECM detects a cold start condition and ignores the oxygen sensor readings. Bring the engine to normal operating temperature before testing the oxygen sensor.
2. After the engine has warmed to normal operating temperature, the oxygen sensor voltage readings should fluctuate in response to changes of the oxygen content in the exhaust gasses. If the voltage does not respond or change, the oxygen sensor may be defective.
3. A problem in the fuel system can cause false oxygen sensor voltage readings. Note the following:
   a. Consistently low voltage readings (taken at closed and wide-open throttle settings) may be caused by a lean fuel mixture. A rich fuel mixture can cause high voltage readings.
   b. Contamination on the oxygen sensor prevents its sensor element from correctly detecting the amount of oxygen in the exhaust gases. This will result with a high voltage reading (during testing), while causing the sensor to send a false (rich) signal to the ECM. Refer to *Oxygen Sensor* in this chapter for information on how to visually inspect and interpret the oxygen sensor color and operating conditions.
4. A problem in the ignition system can cause false oxygen sensor test readings. For example, a fouled spark plug, or any condition that causes a misfire, will result in unburned air/fuel in the exhaust.
5. Check the wiring connectors, terminals and wires (4-pin connector).
6. Refer to *ECM Test Harness* in this chapter when connecting the breakout box to the ECM connectors.

Refer to wiring diagrams at the end of the manual.

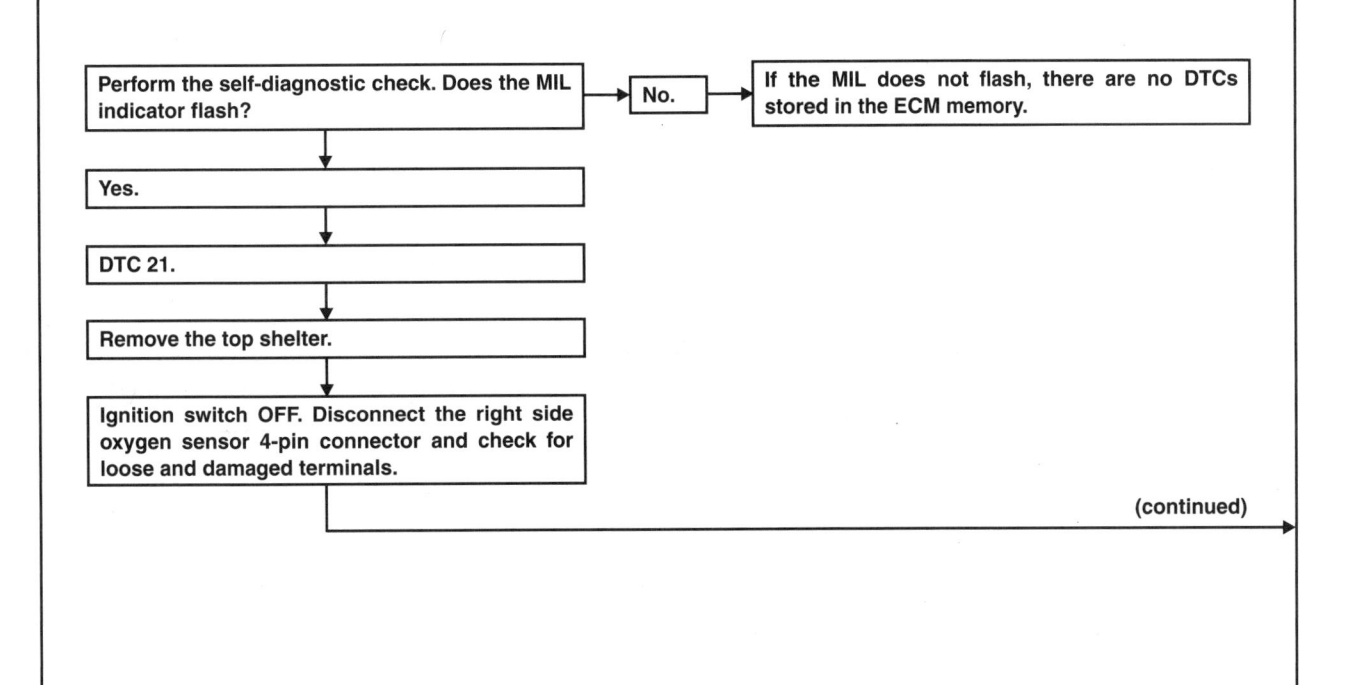

Perform the self-diagnostic check. Does the MIL indicator flash? → No. → If the MIL does not flash, there are no DTCs stored in the ECM memory.

↓ Yes.

DTC 21.

↓

Remove the top shelter.

↓

Ignition switch OFF. Disconnect the right side oxygen sensor 4-pin connector and check for loose and damaged terminals.

(continued)

(113) (continued)

Perform the following:
1. Ignition switch OFF.
2. Reconnect the right side oxygen sensor 4-pin connector.
3. Park the motorcycle on its sidestand.
4. Start the engine and allow the engine to idle until the radiator fan turns on.
5. Snap the throttle quickly to increase engine speed to 5000 rpm, then close the throttle and allow the engine to return to idle while observing the MIL indicator lamp.
6. Did the MIL indicate DTC 21?

→ MIL did not flash. → This was a temporary failure. No further diagnosis is required.

Yes.
↓

Ignition switch OFF. Disconnect the right side oxygen sensor 4-pin connector and the ECM 22-pin gray connector. Check for continuity between the black/red terminal in the wire harness side of the oxygen sensor 4-pin connector and ground. Is continuity present?

→ Yes. → There should be no continuity. Check for a short in the black/red wire between the oxygen sensor connector and the ECM connector.

Yes.
↓

Ignition switch OFF. Connect the breakout box to the ECM connectors. Check for continuity between the black/red terminal in the harness side of the oxygen sensor 4-pin connector and pin 39 in the breakout box. Is continuity present?

→ No. → Check for an open in the black/red wire between the oxygen sensor connector and the ECM connector.

Yes.
↓

Ignition switch OFF. Check for continuity between the green/red terminals in the harness side of the oxygen sensor 4-pin connector and pin 8 in the breakout box. Is continuity present?

→ No. → Check for and open in the green/red wire between the oxygen sensor connector and the ECM connector.

Yes.
↓

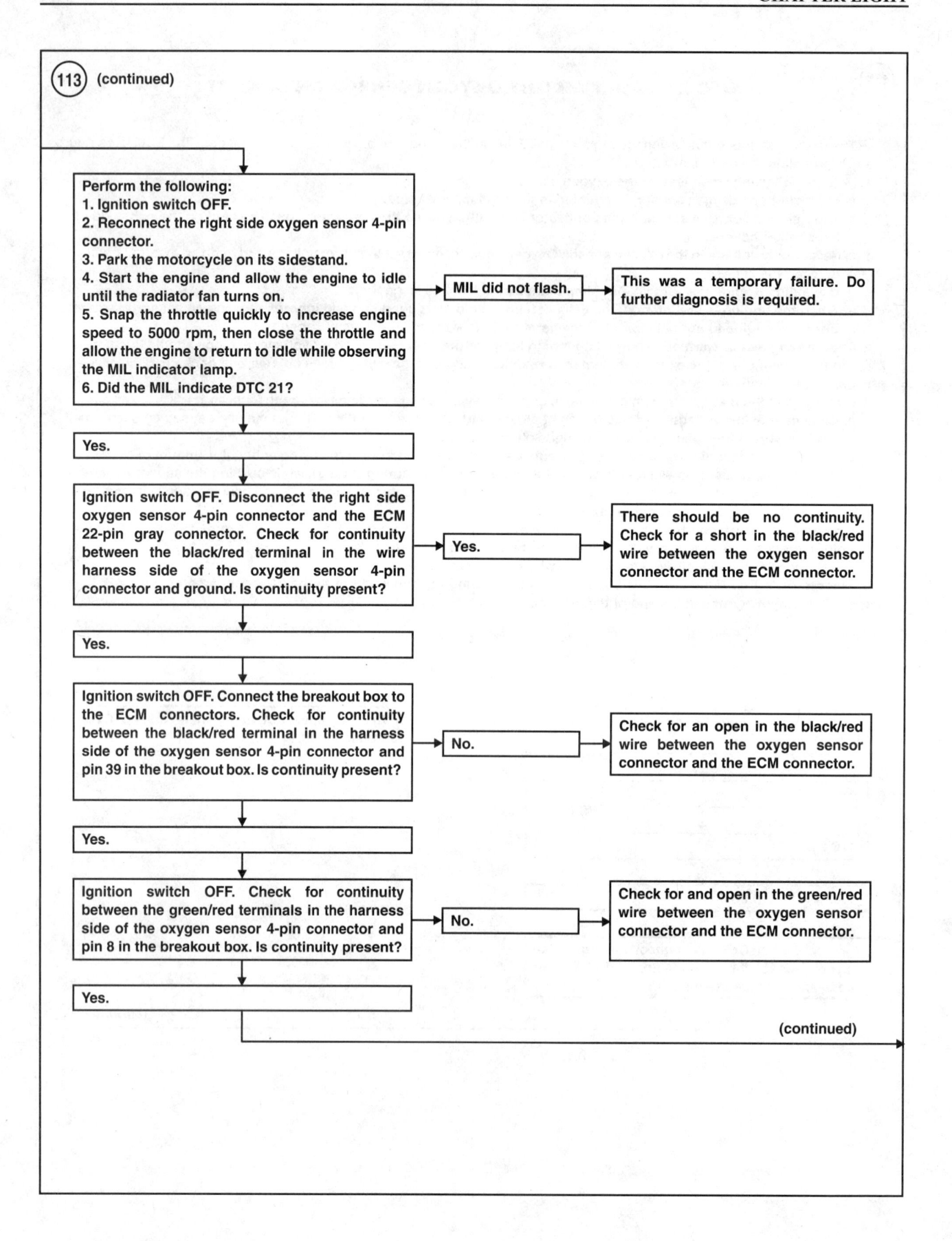

(continued)

FUEL INJECTION AND EMISSION CONTROL SYSTEMS

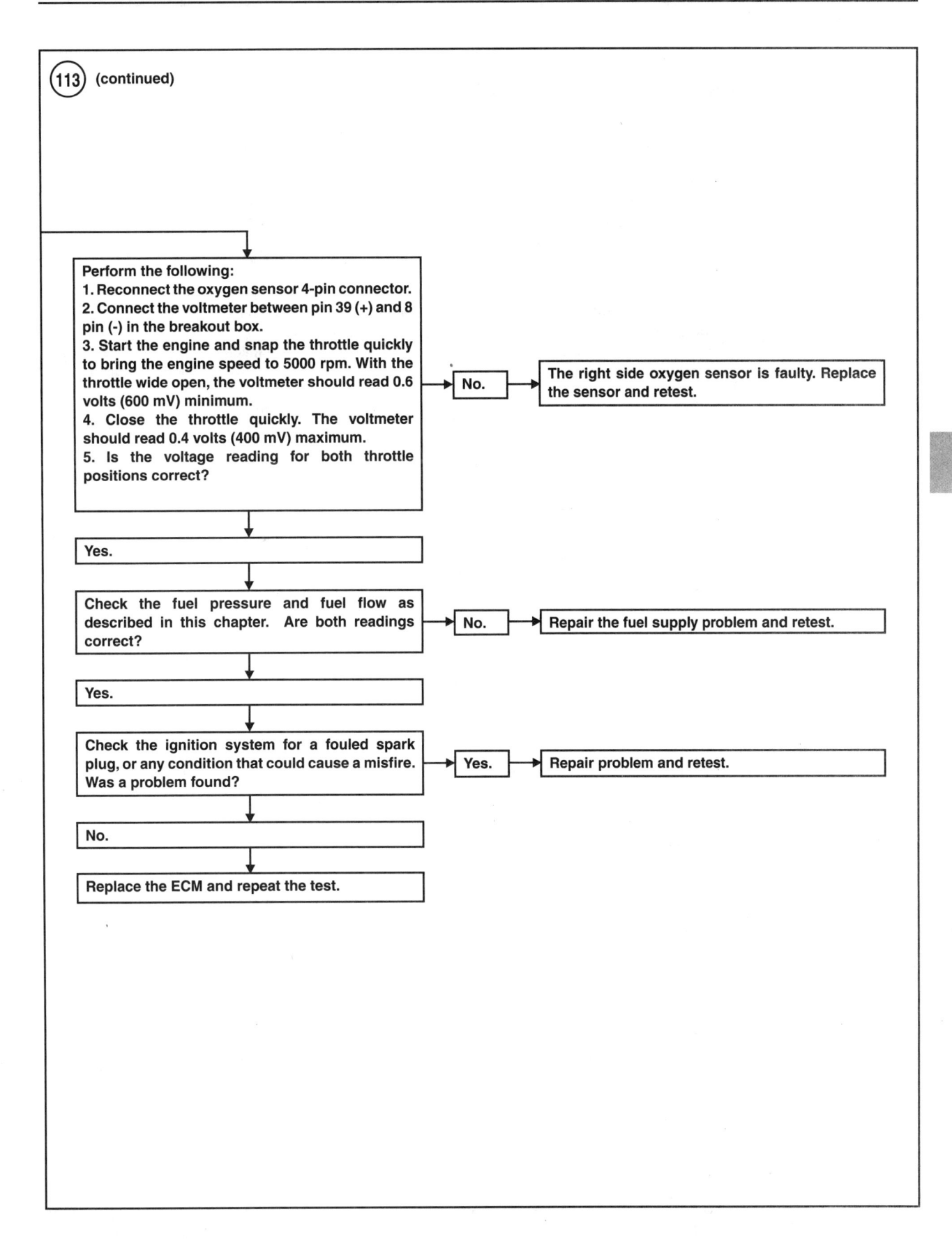

113 (continued)

↓

Perform the following:
1. Reconnect the oxygen sensor 4-pin connector.
2. Connect the voltmeter between pin 39 (+) and 8 pin (-) in the breakout box.
3. Start the engine and snap the throttle quickly to bring the engine speed to 5000 rpm. With the throttle wide open, the voltmeter should read 0.6 volts (600 mV) minimum.
4. Close the throttle quickly. The voltmeter should read 0.4 volts (400 mV) maximum.
5. Is the voltage reading for both throttle positions correct?

→ **No.** → The right side oxygen sensor is faulty. Replace the sensor and retest.

↓ Yes.

Check the fuel pressure and fuel flow as described in this chapter. Are both readings correct?

→ **No.** → Repair the fuel supply problem and retest.

↓ Yes.

Check the ignition system for a fouled spark plug, or any condition that could cause a misfire. Was a problem found?

→ **Yes.** → Repair problem and retest.

↓ No.

Replace the ECM and repeat the test.

DTC 22: FAULTY LEFT OXYGEN SENSOR OR CIRCUIT

The oxygen sensors send a voltage signal to the ECM on the amount of oxygen in the exhaust gas. The ECM uses this signal to adjust the air/fuel mixture.

1. Note the following when testing the oxygen sensor:
   a. The left hand oxygen sensor is mounted in the left exhaust pipe.
   b. The left side oxygen sensor 4-pin connector is mounted inside the connector cover located on the left side of the air filter cover.
   c. Never apply voltage to the oxygen sensor. Oxygen sensors generate their own voltage and to do so externally will damage the sensor.
   d. Use a digital voltmeter, with an input impedance of at least 10 Megaohms.
   e. When the engine is first started, the ECM detects a cold start condition and ignores the oxygen sensor readings. Bring the engine to normal operating temperature before testing the oxygen sensor.
2. After the engine has warmed to normal operating temperature, the oxygen sensor voltage readings should fluctuate in response to changes of the oxygen content in the exhaust gases. If the voltage does not respond or change, the oxygen sensor may be defective.
3. A problem in the fuel system can cause false oxygen sensor voltage readings. Note the following:
   a. Consistently low voltage readings (taken at closed and wide-open throttle settings) may be caused by a lean fuel mixture. A rich fuel mixture can cause high voltage readings.
   b. Contamination on the oxygen sensor prevents its sensor element from correctly detecting the amount of oxygen in the exhaust gases. This will result with a high voltage reading (during testing), while causing the sensor to send a false (rich) signal to the ECM. Refer to *Oxygen Sensor* in this chapter for information on how to visually inspect and interpret the oxygen sensor color and operating conditions.
4. A problem in the ignition system can cause false oxygen sensor test readings. For example, a fouled spark plug, or any condition that causes a misfire, will result in unburned air/fuel in the exhaust.
5. Check the wiring connectors, terminals and wires (4-pin connector).
6. Refer to *ECM Test Harness* in this chapter when connecting the breakout box to the ECM connectors.

Refer to the wiring diagrams at the end of the manual.

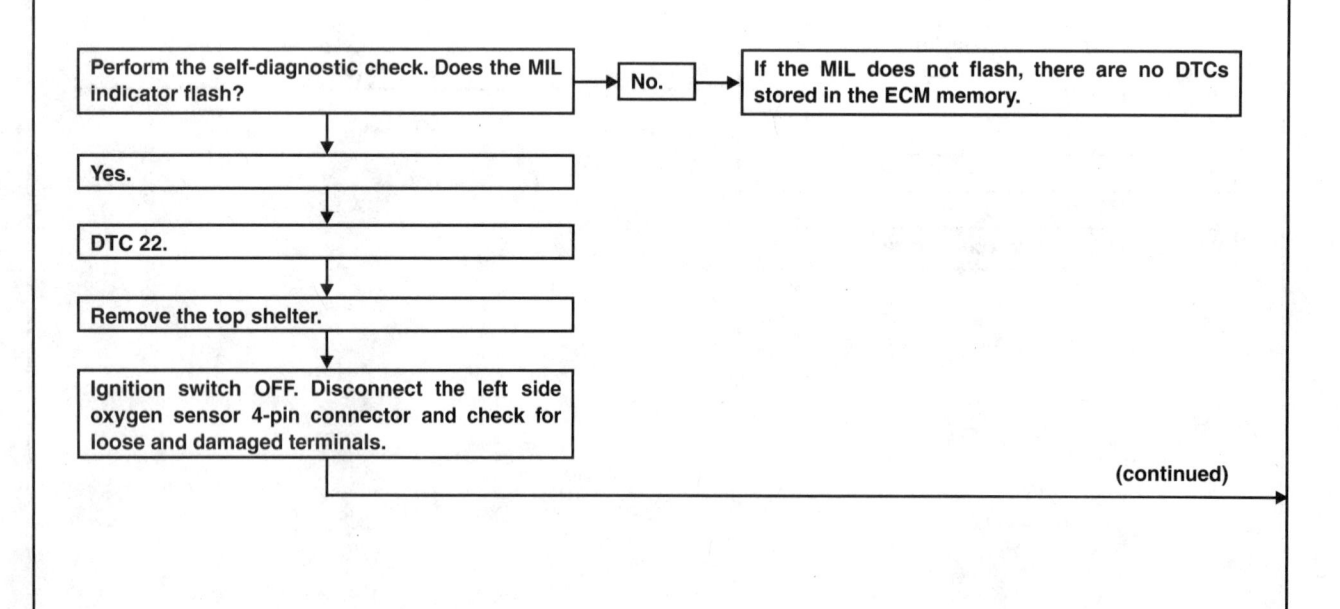

Perform the self-diagnostic check. Does the MIL indicator flash? → No. → If the MIL does not flash, there are no DTCs stored in the ECM memory.

↓

Yes.

↓

DTC 22.

↓

Remove the top shelter.

↓

Ignition switch OFF. Disconnect the left side oxygen sensor 4-pin connector and check for loose and damaged terminals.

(continued)

FUEL INJECTION AND EMISSION CONTROL SYSTEMS

(114) (continued)

↓

Perform the following:
1. Ignition switch OFF.
2. Reconnect the left side oxygen sensor 4-pin connector.
3. Park the motorcycle on its sidestand.
4. Start the engine and allow the engine to idle until the radiator fan turns on.
5. Snap the throttle quickly to increase engine speed to 5000 rpm, then close the throttle and allow the engine to return to idle while observing the MIL indicator lamp.
6. Did the MIL indicate DTC 22?

→ MIL did not flash. → This was a temporary failure. Do further diagnosis is required.

↓ Yes.

Ignition switch OFF. Disconnect the left side oxygen sensor 4-pin connector and the ECM 22-pin gray connector. Check for continuity between the black/orange terminal in the wire harness side of the oxygen sensor 4-pin connector and ground. Is continuity present?

→ Yes. → There should be no continuity. Check for a short in the black/orange wire between the oxygen sensor connector and the ECM connector.

↓ Yes.

Ignition switch OFF. Connect the breakout box to the ECM connectors. Check for continuity between the black/orange terminals in the harness side of the oxygen sensor 4-pin connector and pin 50 in the breakout box. Is continuity present?

→ No. → Check for an open in the black/orange wire between the oxygen sensor connector and the ECM connector.

↓ Yes.

Ignition switch OFF. Check for continuity between the green/red terminals in the harness side of the oxygen sensor 4-pin connector and pin 8 in the breakout box. Is continuity present?

→ No. → Check for an open in the green/red wire between the oxygen sensor connector and the ECM connector.

↓ Yes.

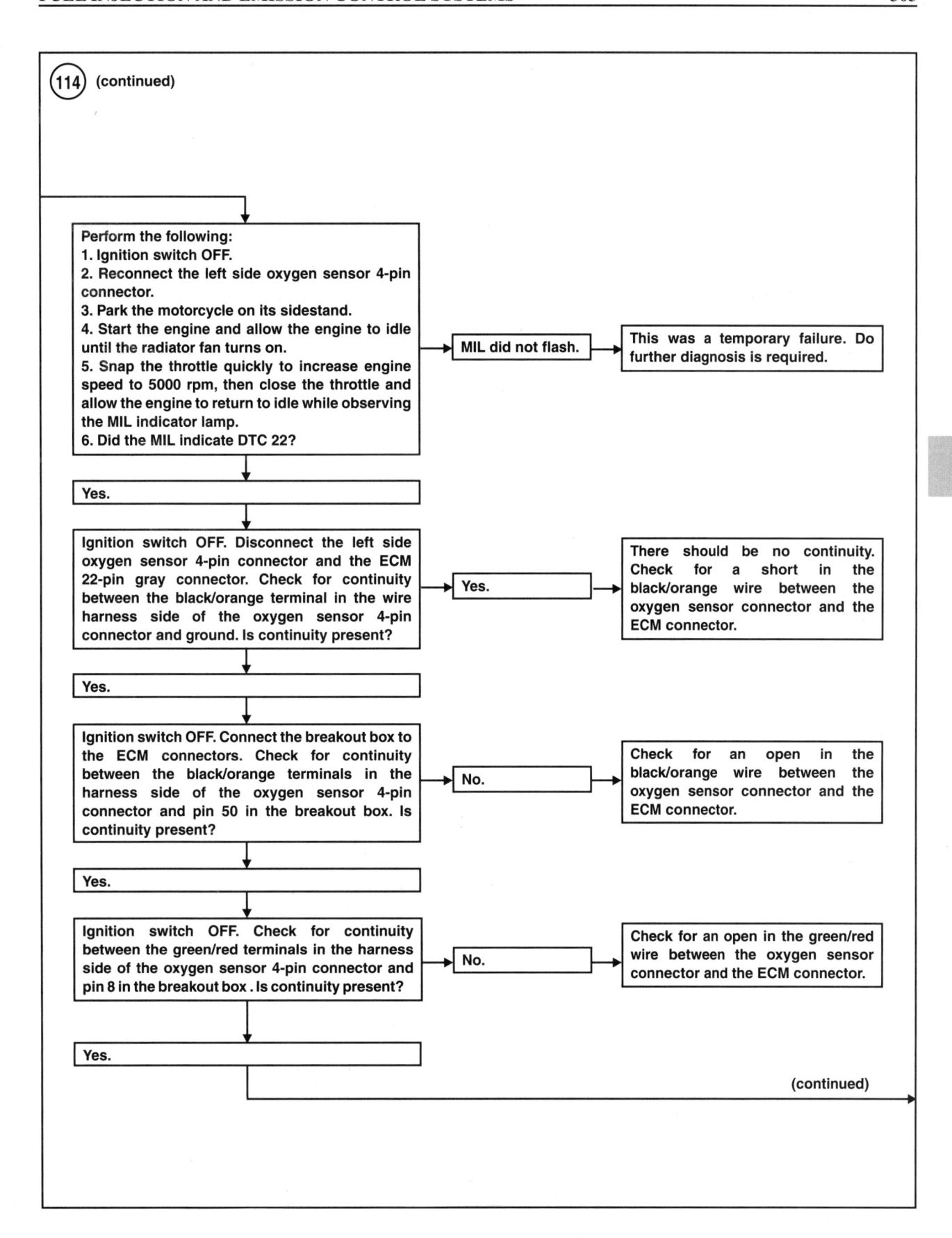

(continued)

(114) (continued)

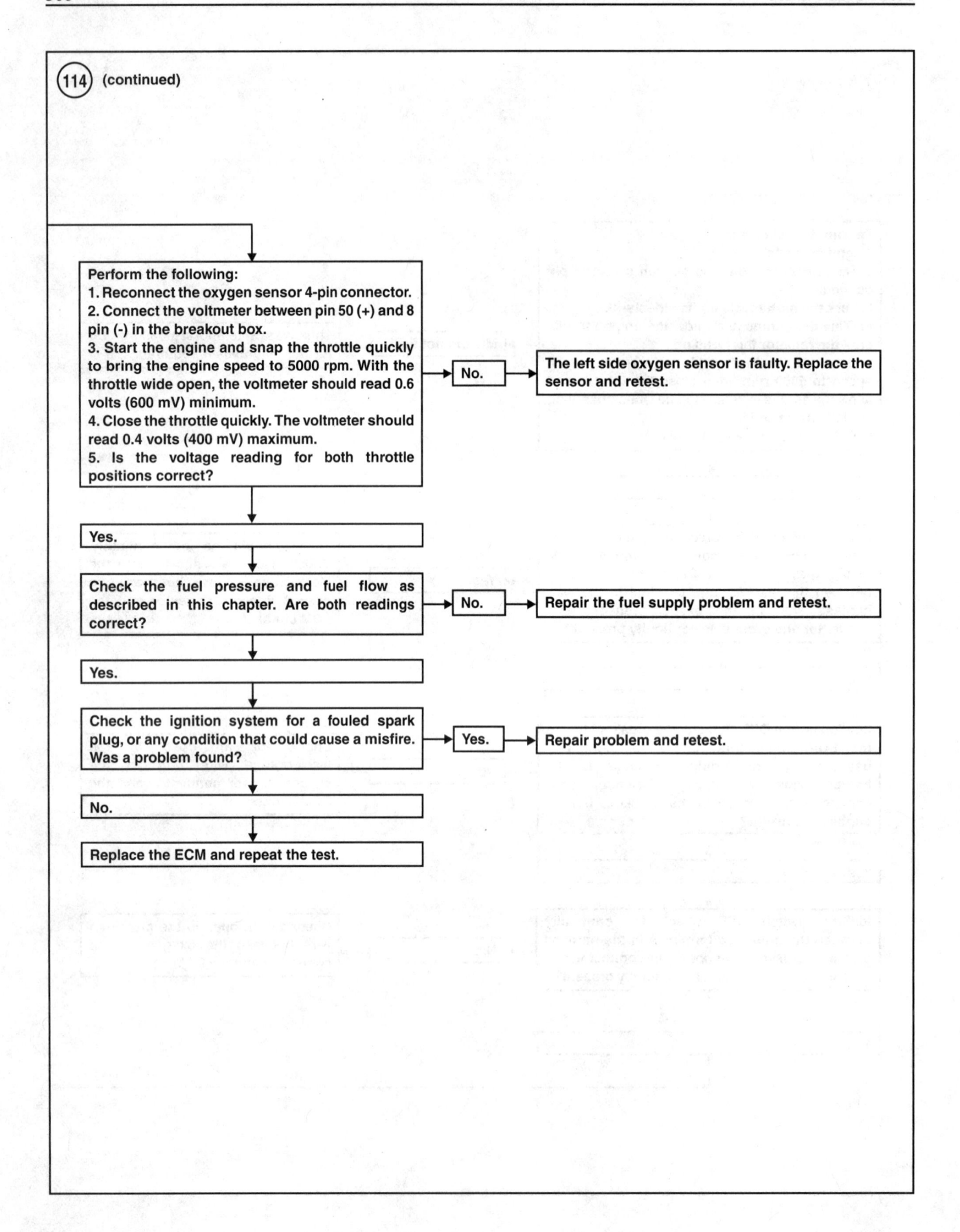

Perform the following:
1. Reconnect the oxygen sensor 4-pin connector.
2. Connect the voltmeter between pin 50 (+) and 8 pin (-) in the breakout box.
3. Start the engine and snap the throttle quickly to bring the engine speed to 5000 rpm. With the throttle wide open, the voltmeter should read 0.6 volts (600 mV) minimum.
4. Close the throttle quickly. The voltmeter should read 0.4 volts (400 mV) maximum.
5. Is the voltage reading for both throttle positions correct?

→ No. → The left side oxygen sensor is faulty. Replace the sensor and retest.

Yes.

Check the fuel pressure and fuel flow as described in this chapter. Are both readings correct?

→ No. → Repair the fuel supply problem and retest.

Yes.

Check the ignition system for a fouled spark plug, or any condition that could cause a misfire. Was a problem found?

→ Yes. → Repair problem and retest.

No.

Replace the ECM and repeat the test.

FUEL INJECTION AND EMISSION CONTROL SYSTEMS

(115) DTC 23: FAULTY RIGHT OXYGEN SENSOR HEATER CIRCUIT OR SENSOR

The oxygen sensors are equipped with an electric heating element to reduce sensor warm-up time and maintain higher sensor temperatures when the engine is idling. Keeping the sensor hot also burns off deposits collected on the sensor housing, which helps to prevent incorrect sensor readings. A faulty heating element increases sensor warm-up time and causes the ECM to stay in open loop (warm-up) longer, thus supplying a richer air/fuel mixture to the engine.

1. Before making this test, note the following:
    a. The right hand oxygen sensor is mounted in the right exhaust pipe.
    b. The right side oxygen sensor 4-pin connector is mounted inside the connector cover located on the right side of the air filter cover.
    c. An ohmmeter can be used to check the resistance of the heating element. Do *not* check the resistance of the sensor element. Refer to the wiring diagrams at the end of the manual to verify connector and wire colors.
    d. Check the wiring connectors, terminals and wires.
    e. Refer to *ECM Test Harness* in this chapter when connecting the breakout box to the ECM connectors.

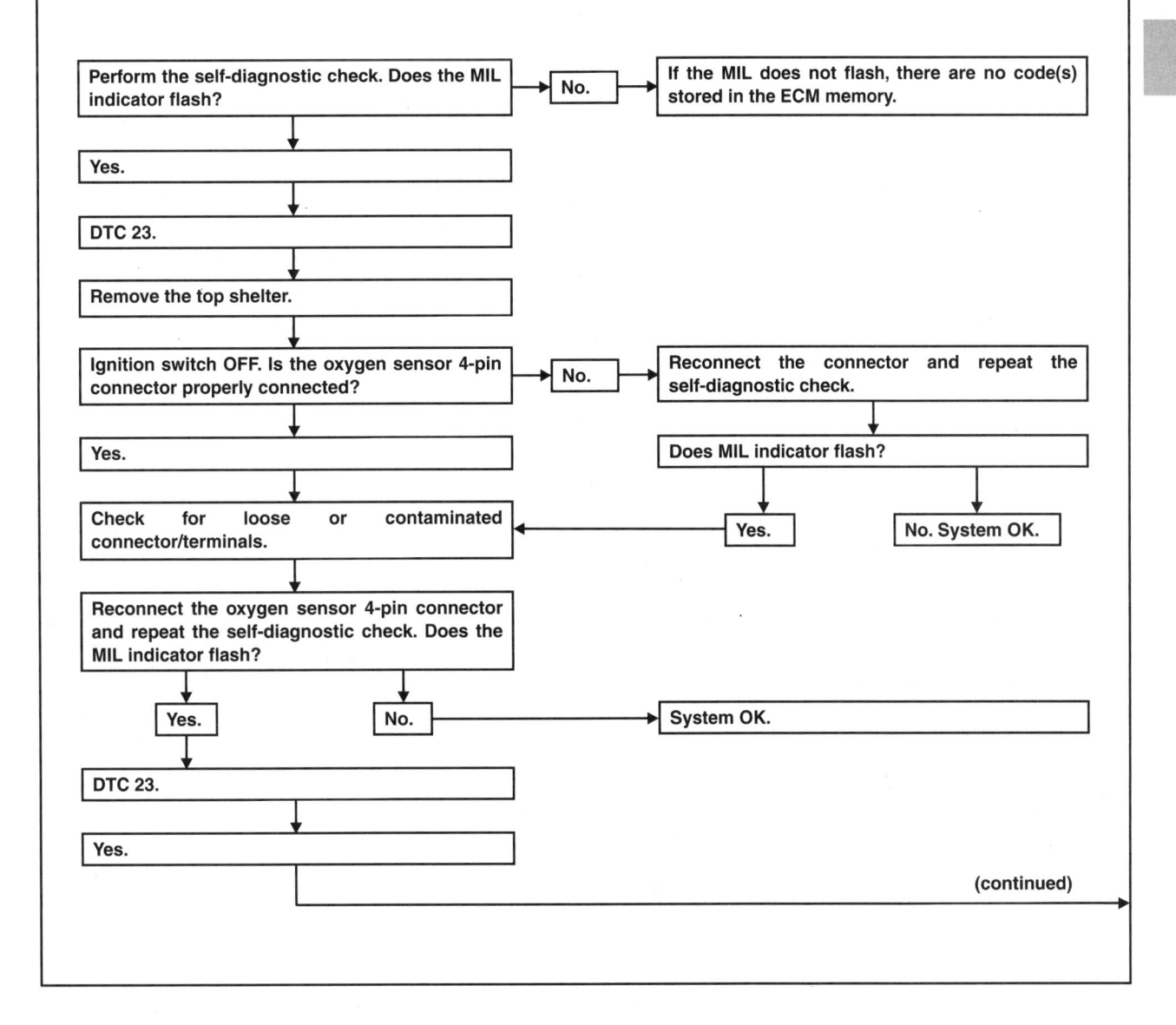

Perform the self-diagnostic check. Does the MIL indicator flash? → **No.** → If the MIL does not flash, there are no code(s) stored in the ECM memory.

↓ Yes.

DTC 23.

↓

Remove the top shelter.

↓

Ignition switch OFF. Is the oxygen sensor 4-pin connector properly connected? → **No.** → Reconnect the connector and repeat the self-diagnostic check.

↓ Yes. → Does MIL indicator flash? → Yes. / No. System OK.

Check for loose or contaminated connector/terminals. ← Yes.

↓

Reconnect the oxygen sensor 4-pin connector and repeat the self-diagnostic check. Does the MIL indicator flash?

↓ Yes.   ↓ No. → System OK.

DTC 23.

↓

Yes.

(continued)

115 (continued)

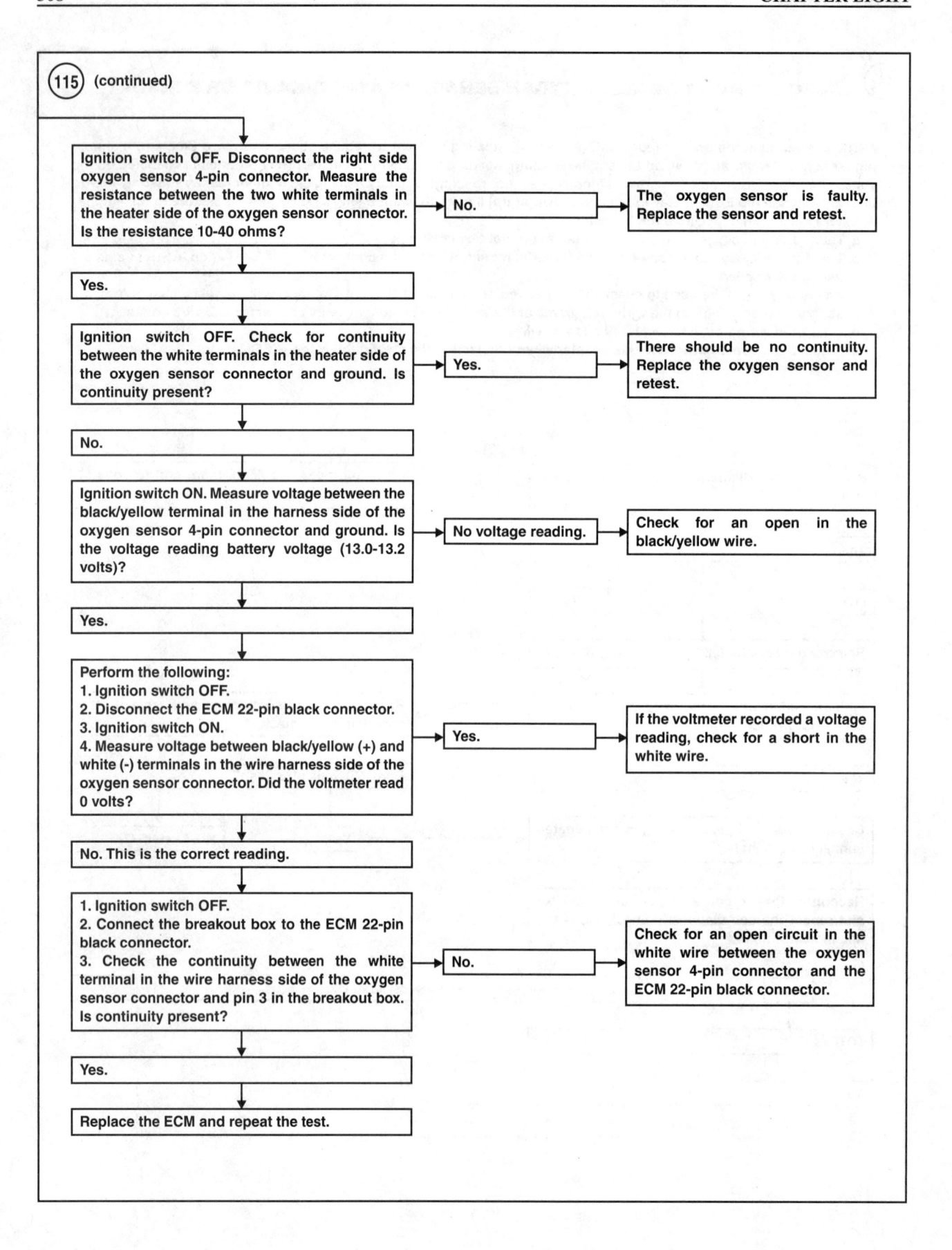

Ignition switch OFF. Disconnect the right side oxygen sensor 4-pin connector. Measure the resistance between the two white terminals in the heater side of the oxygen sensor connector. Is the resistance 10-40 ohms?

→ **No.** → The oxygen sensor is faulty. Replace the sensor and retest.

↓ Yes.

Ignition switch OFF. Check for continuity between the white terminals in the heater side of the oxygen sensor connector and ground. Is continuity present?

→ **Yes.** → There should be no continuity. Replace the oxygen sensor and retest.

↓ No.

Ignition switch ON. Measure voltage between the black/yellow terminal in the harness side of the oxygen sensor 4-pin connector and ground. Is the voltage reading battery voltage (13.0-13.2 volts)?

→ **No voltage reading.** → Check for an open in the black/yellow wire.

↓ Yes.

Perform the following:
1. Ignition switch OFF.
2. Disconnect the ECM 22-pin black connector.
3. Ignition switch ON.
4. Measure voltage between black/yellow (+) and white (-) terminals in the wire harness side of the oxygen sensor connector. Did the voltmeter read 0 volts?

→ **Yes.** → If the voltmeter recorded a voltage reading, check for a short in the white wire.

↓ No. This is the correct reading.

1. Ignition switch OFF.
2. Connect the breakout box to the ECM 22-pin black connector.
3. Check the continuity between the white terminal in the wire harness side of the oxygen sensor connector and pin 3 in the breakout box. Is continuity present?

→ **No.** → Check for an open circuit in the white wire between the oxygen sensor 4-pin connector and the ECM 22-pin black connector.

↓ Yes.

Replace the ECM and repeat the test.

FUEL INJECTION AND EMISSION CONTROL SYSTEMS

⑪⑯ DTC 24: FAULTY LEFT OXYGEN SENSOR HEATER CIRCUIT OR SENSOR

The oxygen sensors are equipped with an electric heating element to reduce sensor warm-up time and maintain higher sensor temperatures when the engine is idling. Keeping the sensor hot also burns off deposits collected on the sensor housing, which helps to prevent incorrect sensor readings. A faulty heating element increases sensor warm-up time and causes the ECM to stay in open loop (warm-up) longer, thus supplying a richer air/fuel mixture to the engine. This condition increases fuel consumption.

1. Before making this test, note the following:
   a. The left hand oxygen sensor is mounted in the left exhaust pipe.
   b. The left side oxygen sensor 4-pin connector is mounted inside the connector cover located on the left side of the air filter cover.
   c. An ohmmeter can be used to check the resistance of the heating element. Do *not* check the resistance of the sensor element. Refer to wiring diagrams at the end of the manual to verify connector and wire colors.
   d. Check the wiring connectors, terminals and wires.
   e. Refer to *ECM Test Harness* in this chapter when connecting the breakout box to the ECM connectors.

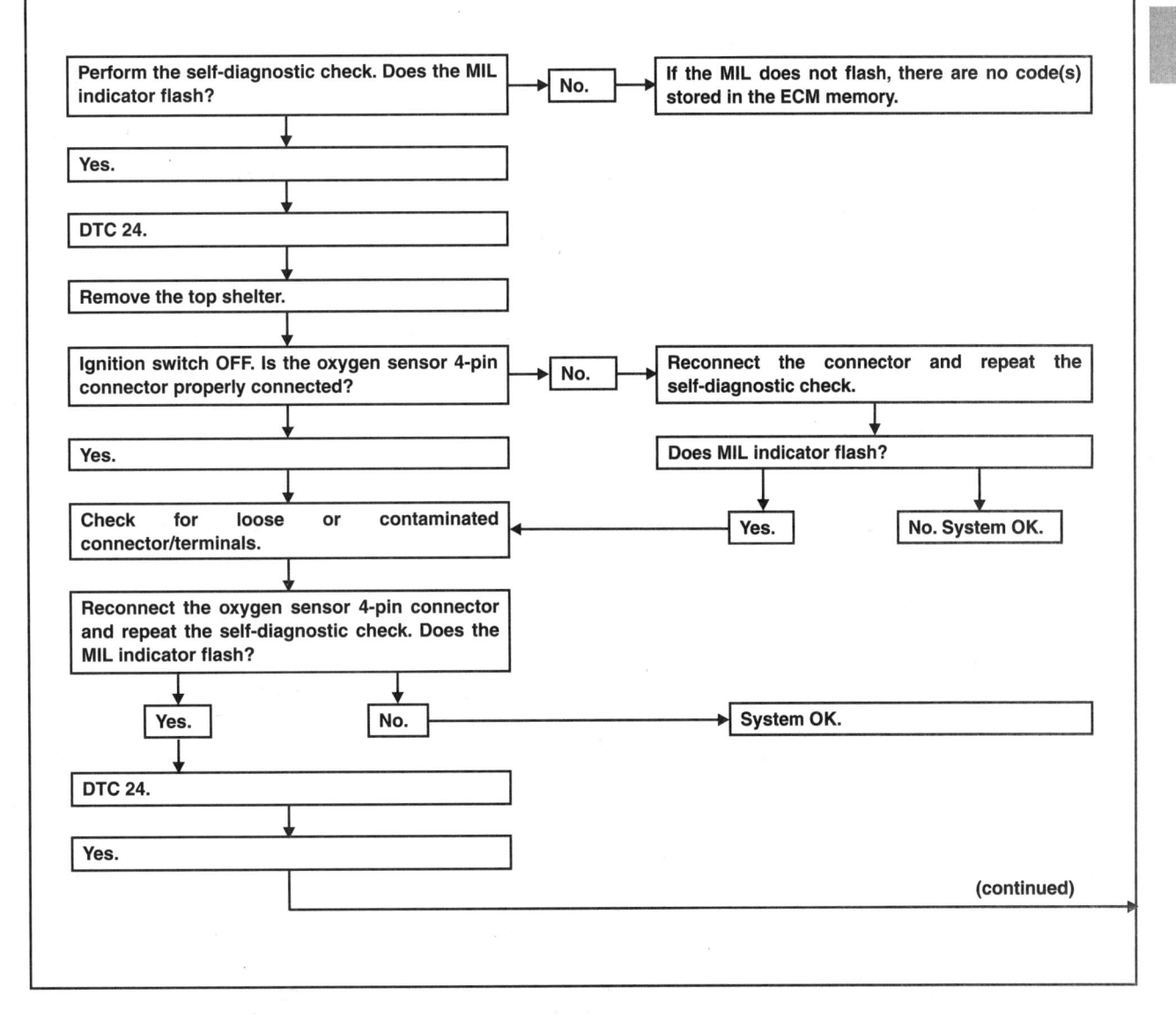

```
Perform the self-diagnostic check. Does the MIL indicator flash?
   │
   ├── No. ──▶ If the MIL does not flash, there are no code(s) stored in the ECM memory.
   │
   ▼ Yes.
DTC 24.
   │
   ▼
Remove the top shelter.
   │
   ▼
Ignition switch OFF. Is the oxygen sensor 4-pin connector properly connected?
   │
   ├── No. ──▶ Reconnect the connector and repeat the self-diagnostic check.
   │              │
   │              ▼
   │           Does MIL indicator flash?
   │              │
   │              ├── Yes. ──▶ (to Check for loose...)
   │              └── No. System OK.
   │
   ▼ Yes.
Check for loose or contaminated connector/terminals.
   │
   ▼
Reconnect the oxygen sensor 4-pin connector and repeat the self-diagnostic check. Does the MIL indicator flash?
   │
   ├── No. ──▶ System OK.
   │
   ▼ Yes.
DTC 24.
   │
   ▼ Yes.
(continued)
```

116 (continued)

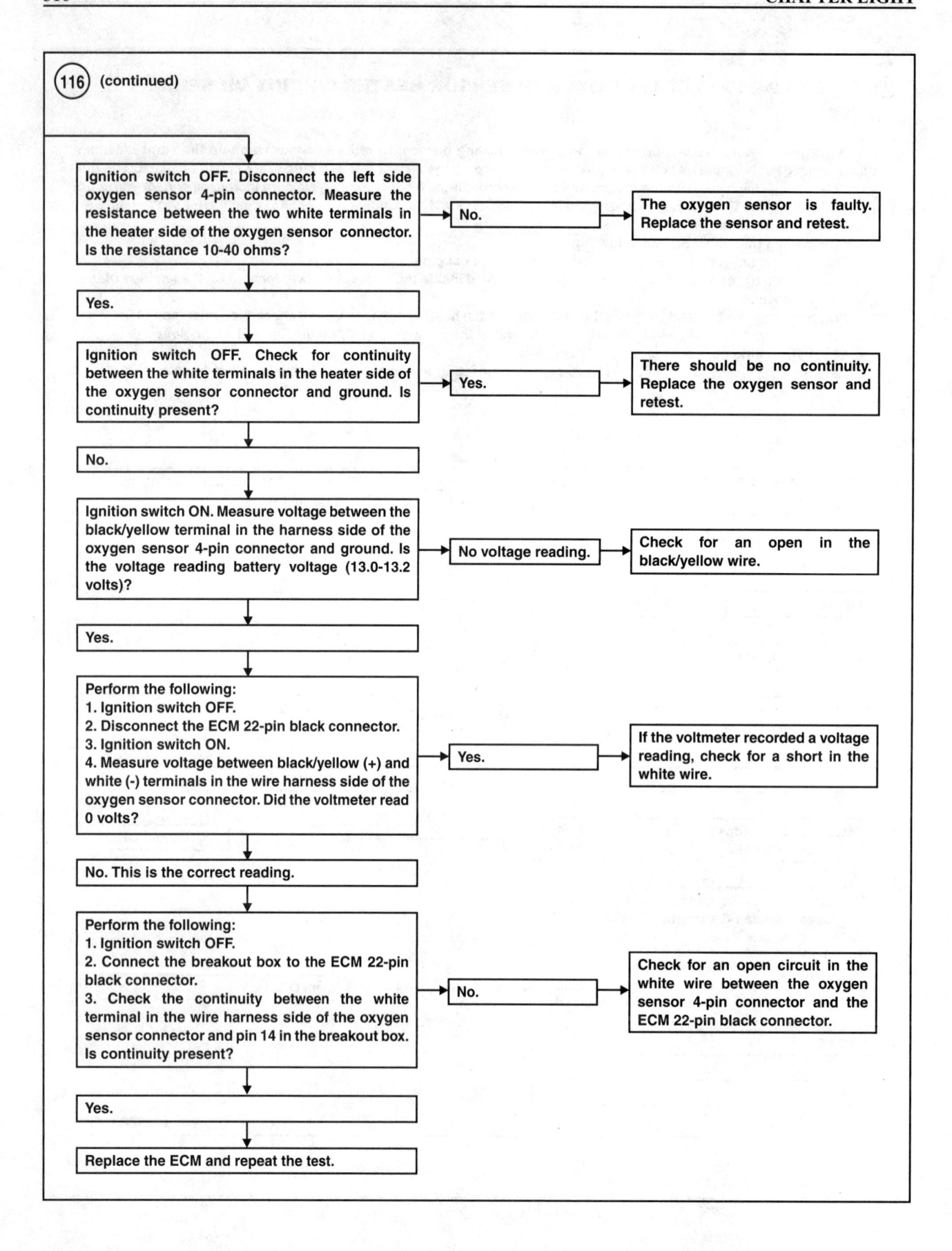

↓

Ignition switch OFF. Disconnect the left side oxygen sensor 4-pin connector. Measure the resistance between the two white terminals in the heater side of the oxygen sensor connector. Is the resistance 10-40 ohms? → **No.** → The oxygen sensor is faulty. Replace the sensor and retest.

↓ Yes.

Ignition switch OFF. Check for continuity between the white terminals in the heater side of the oxygen sensor connector and ground. Is continuity present? → **Yes.** → There should be no continuity. Replace the oxygen sensor and retest.

↓ No.

Ignition switch ON. Measure voltage between the black/yellow terminal in the harness side of the oxygen sensor 4-pin connector and ground. Is the voltage reading battery voltage (13.0-13.2 volts)? → **No voltage reading.** → Check for an open in the black/yellow wire.

↓ Yes.

Perform the following:
1. Ignition switch OFF.
2. Disconnect the ECM 22-pin black connector.
3. Ignition switch ON.
4. Measure voltage between black/yellow (+) and white (-) terminals in the wire harness side of the oxygen sensor connector. Did the voltmeter read 0 volts? → **Yes.** → If the voltmeter recorded a voltage reading, check for a short in the white wire.

↓ No. This is the correct reading.

Perform the following:
1. Ignition switch OFF.
2. Connect the breakout box to the ECM 22-pin black connector.
3. Check the continuity between the white terminal in the wire harness side of the oxygen sensor connector and pin 14 in the breakout box. Is continuity present? → **No.** → Check for an open circuit in the white wire between the oxygen sensor 4-pin connector and the ECM 22-pin black connector.

↓ Yes.

Replace the ECM and repeat the test.

FUEL INJECTION AND EMISSION CONTROL SYSTEMS

(117)

DTC 25: FAULTY RIGHT KNOCK SENSOR CIRCUIT OR SENSOR

The knock sensor signals the ECM when the engine detonation is detected. The ECM then retards the ignition timing to prevent detonation. The knock sensor only affects the ignition timing. When the knock sensor is working correctly, the engine is capable of operating with the maximum amount of timing advance for the prevailing engine and weather conditions. This results in optimum engine performance and fuel mileage. Failure in the knock sensor system can cause drivability problems. For example, if the knock sensor excessively retards the ignition timing, engine performance will be reduced, thus increasing fuel consumption. If the knock sensor does not correctly sense detonation, engine damage can occur. Remember also that engine detonation can result from engine causes not related to the knock sensor.

1. Before making this test, note the following:
    a. The right side knock sensor is mounted in the right side of the engine block, next to the horn.
    b. Contamination or poor terminal contact in the knock sensor and ECM connectors can cause a voltage drop that will reduce the signal sent to the ECM. A faulty (reduced) signal prevents the ECM from controlling detonation.
2. To check the knock sensor operation with a timing light, perform the following:
    a. Remove the timing cover and connect a timing light to the engine as described under *Ignition Timing* in Chapter Three.
    b. Remove the necessary body components to access the area around the knock sensor. Refer to *Knock Sensor* in this chapter. Do not disconnect the knock sensor connector or remove the knock sensor.
    c. Start the engine, allow to idle, and check the ignition timing. Use a hammer and *lightly* tap the engine while observing the timing marks. The timing should retard momentarily, and then advance. If there is no change, check the connectors and wiring for loose connections, contamination or damage.
    d. Refer to the wiring diagram at the end of the manuual. Perform the procedures in the troubleshooting chart to isolate the problem.

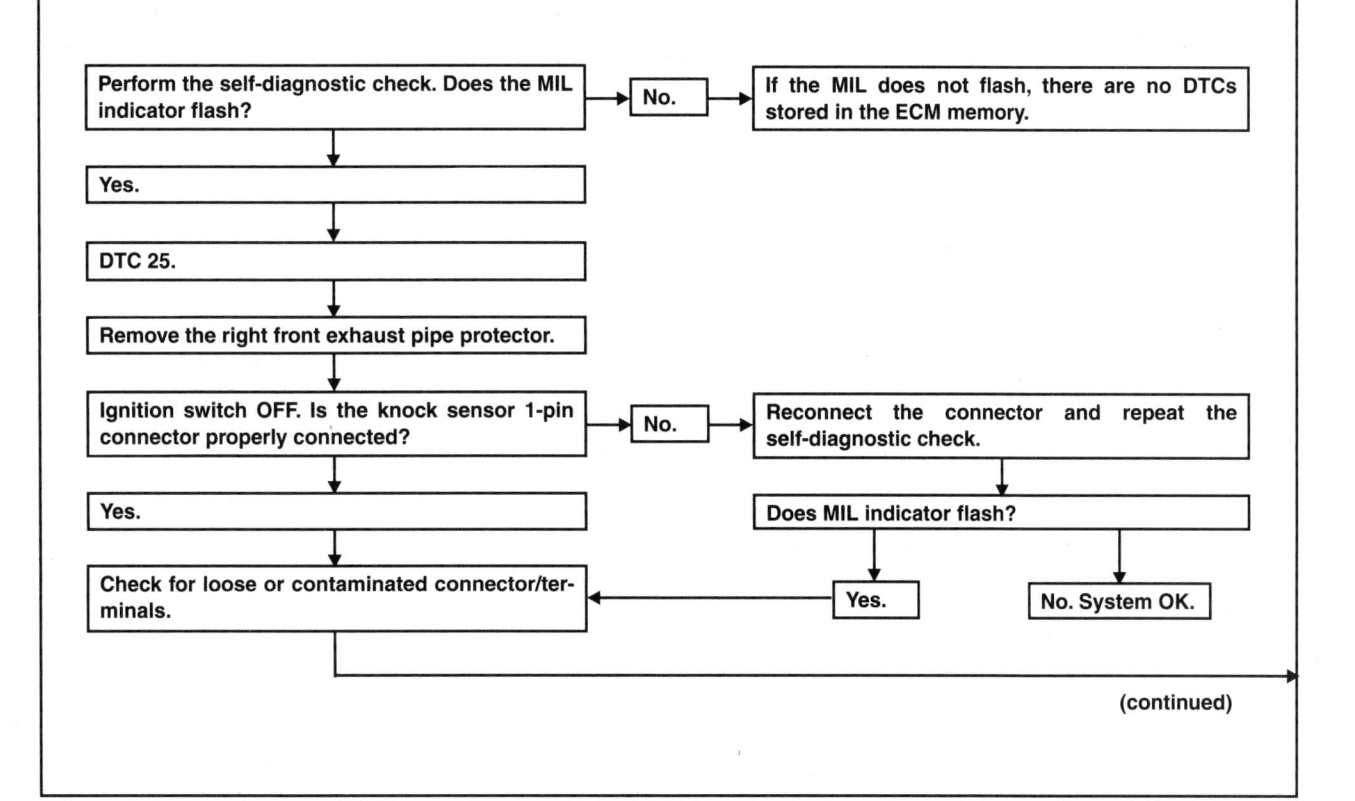

Perform the self-diagnostic check. Does the MIL indicator flash? → **No.** → If the MIL does not flash, there are no DTCs stored in the ECM memory.

↓ Yes.

DTC 25.

↓

Remove the right front exhaust pipe protector.

↓

Ignition switch OFF. Is the knock sensor 1-pin connector properly connected? → **No.** → Reconnect the connector and repeat the self-diagnostic check.

↓ Yes.

Does MIL indicator flash?
- Yes. → Check for loose or contaminated connector/terminals.
- No. System OK.

(continued)

117. (continued)

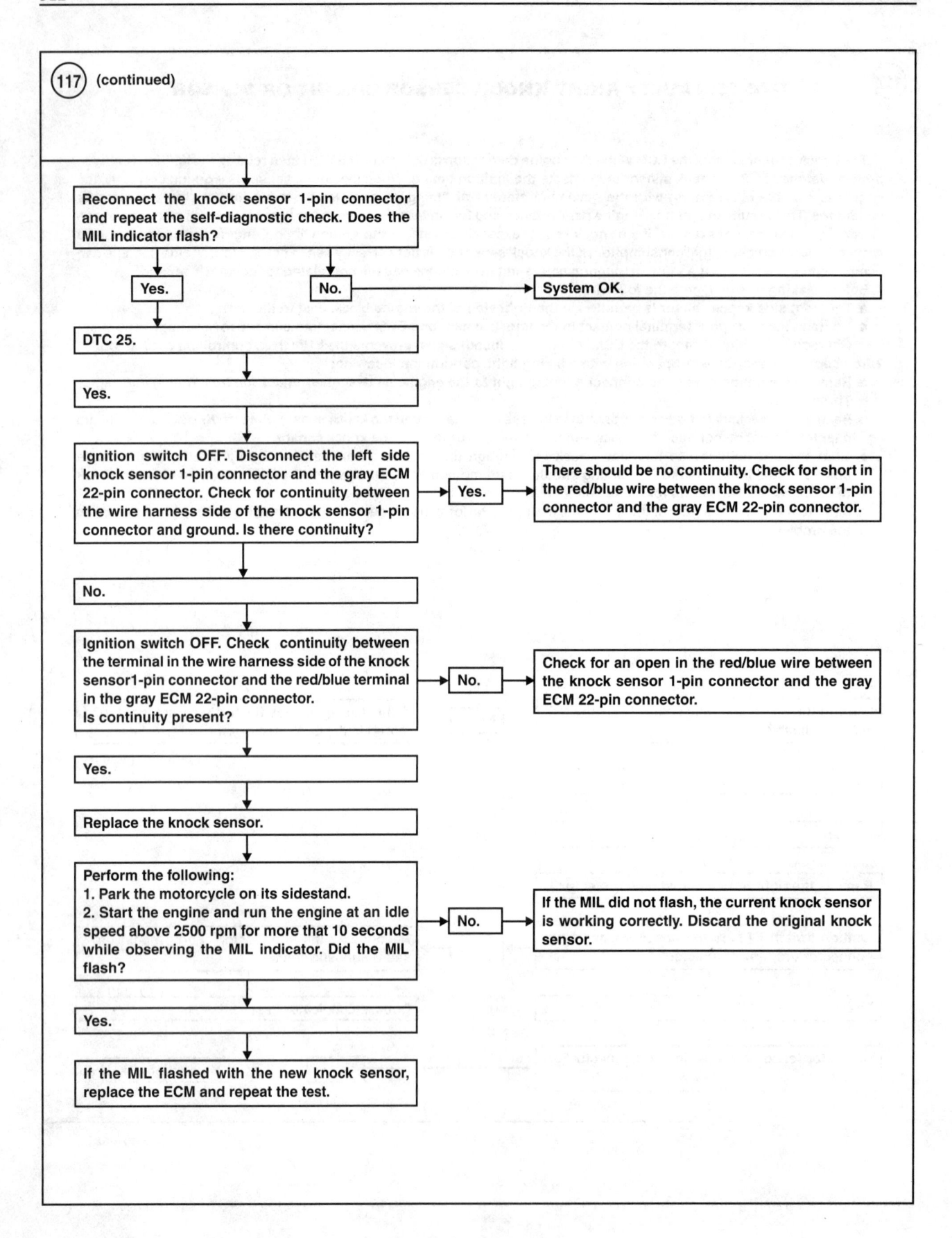

```
             │
             ▼
  ┌──────────────────────────────┐
  │ Reconnect the knock sensor    │
  │ 1-pin connector and repeat    │
  │ the self-diagnostic check.    │
  │ Does the MIL indicator flash? │
  └──────────────────────────────┘
        │                │
     ┌──┴──┐          ┌──┴──┐
     │Yes. │          │ No. │──────────────────────────►  System OK.
     └──┬──┘          └─────┘
        │
  ┌─────┴─────┐
  │ DTC 25.   │
  └─────┬─────┘
     ┌──┴──┐
     │Yes. │
     └──┬──┘
        ▼
  ┌──────────────────────────────────────┐
  │ Ignition switch OFF. Disconnect the  │      ┌─────┐    ┌─────────────────────────────────┐
  │ left side knock sensor 1-pin         │      │Yes. │───►│ There should be no continuity.  │
  │ connector and the gray ECM 22-pin    │─────►└─────┘    │ Check for short in the red/blue │
  │ connector. Check for continuity      │                 │ wire between the knock sensor   │
  │ between the wire harness side of the │                 │ 1-pin connector and the gray    │
  │ knock sensor 1-pin connector and     │                 │ ECM 22-pin connector.           │
  │ ground. Is there continuity?         │                 └─────────────────────────────────┘
  └──────────────────────────────────────┘
        │
     ┌──┴──┐
     │ No. │
     └──┬──┘
        ▼
  ┌──────────────────────────────────────┐
  │ Ignition switch OFF. Check           │      ┌─────┐    ┌─────────────────────────────────┐
  │ continuity between the terminal in   │      │ No. │───►│ Check for an open in the red/   │
  │ the wire harness side of the knock   │─────►└─────┘    │ blue wire between the knock     │
  │ sensor 1-pin connector and the red/  │                 │ sensor 1-pin connector and the  │
  │ blue terminal in the gray ECM 22-pin │                 │ gray ECM 22-pin connector.      │
  │ connector.                           │                 └─────────────────────────────────┘
  │ Is continuity present?               │
  └──────────────────────────────────────┘
        │
     ┌──┴──┐
     │Yes. │
     └──┬──┘
        ▼
  ┌──────────────────────────────┐
  │ Replace the knock sensor.    │
  └──────────────┬───────────────┘
                 ▼
  ┌──────────────────────────────────────┐
  │ Perform the following:               │      ┌─────┐    ┌─────────────────────────────────┐
  │ 1. Park the motorcycle on its        │      │ No. │───►│ If the MIL did not flash, the   │
  │    sidestand.                        │─────►└─────┘    │ current knock sensor is working │
  │ 2. Start the engine and run the      │                 │ correctly. Discard the original │
  │    engine at an idle speed above     │                 │ knock sensor.                   │
  │    2500 rpm for more that 10 seconds │                 └─────────────────────────────────┘
  │    while observing the MIL indicator.│
  │    Did the MIL flash?                │
  └──────────────────────────────────────┘
        │
     ┌──┴──┐
     │Yes. │
     └──┬──┘
        ▼
  ┌──────────────────────────────────────┐
  │ If the MIL flashed with the new      │
  │ knock sensor, replace the ECM and    │
  │ repeat the test.                     │
  └──────────────────────────────────────┘
```

FUEL INJECTION AND EMISSION CONTROL SYSTEMS

(118) DTC 26: FAULTY LEFT KNOCK SENSOR CIRCUIT OR SENSOR

The knock sensor signals the ECM when the engine detonation is detected. The ECM then retards the ignition timing to prevent detonation. The knock sensor only affects the ignition timing. When the knock sensor is working correctly, the engine is capable of operating with the maximum amount of timing advance for the prevailing engine and weather conditions. This results in optimum engine performance and fuel mileage. Failure in the knock sensor system can cause drivability problems. For example, if the knock sensor excessively retards the ignition timing, engine performance will be reduced, thus increasing fuel consumption. If the knock sensor does not correctly sense detonation, engine damage can occur. Remember also that engine detonation can result from engine causes not related to the knock sensor.

1. Before making this test, note the following:
   a. The right side knock sensor is mounted in the right side of the engine block, next to the horn.
   b. Contamination or poor terminal contact in the knock sensor and ECM connectors can cause a voltage drop that will reduce the signal sent to the ECM. A faulty (reduced) signal prevents the ECM from controlling detonation.
2. To check the knock sensor operation with a timing light, perform the following:
   a. Remove the timing cover and connect a timing light to the engine as described under *Ignition Timing* in Chapter Three.
   b. Remove the necessary body components to access the area around the knock sensor. Refer to *Knock Sensor* in this chapter. Do not disconnect the knock sensor connector or remove the knock sensor.
   c. Start the engine, allow to idle, and check the ignition timing. Use a hammer and *lightly* tap the engine while observing the timing marks. The timing should retard momentarily, and then advance. If there is no change, check the connectors and wiring for loose connections, contamination or damage.
   d. Refer to the wiring diagram at the end of the manuual. Perform the procedures in the troubleshooting chart to isolate the problem.

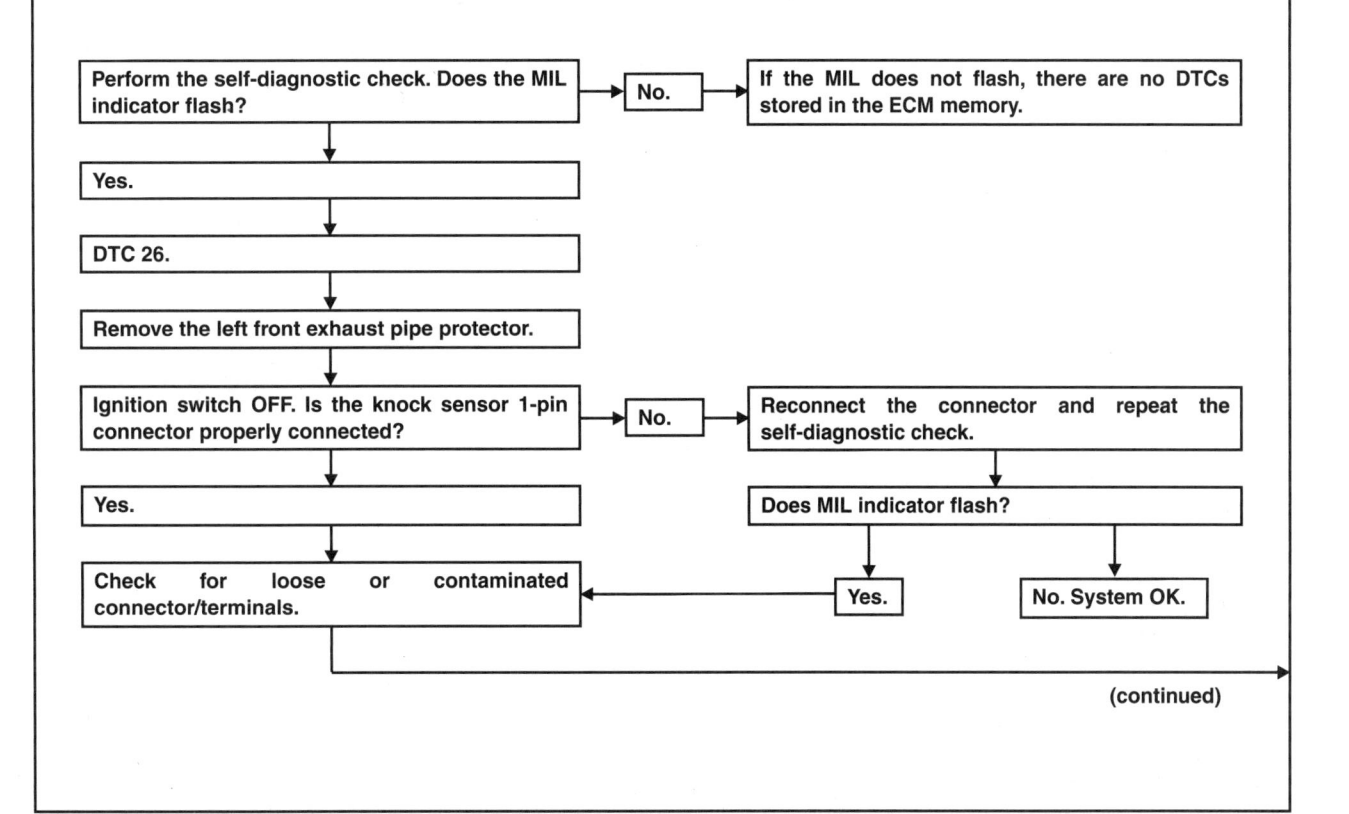

Perform the self-diagnostic check. Does the MIL indicator flash? → No. → If the MIL does not flash, there are no DTCs stored in the ECM memory.

↓ Yes.

DTC 26.

↓

Remove the left front exhaust pipe protector.

↓

Ignition switch OFF. Is the knock sensor 1-pin connector properly connected? → No. → Reconnect the connector and repeat the self-diagnostic check.

↓ Yes.   Does MIL indicator flash?

Check for loose or contaminated connector/terminals. ← Yes.   No. System OK.

(continued)

118. (continued)

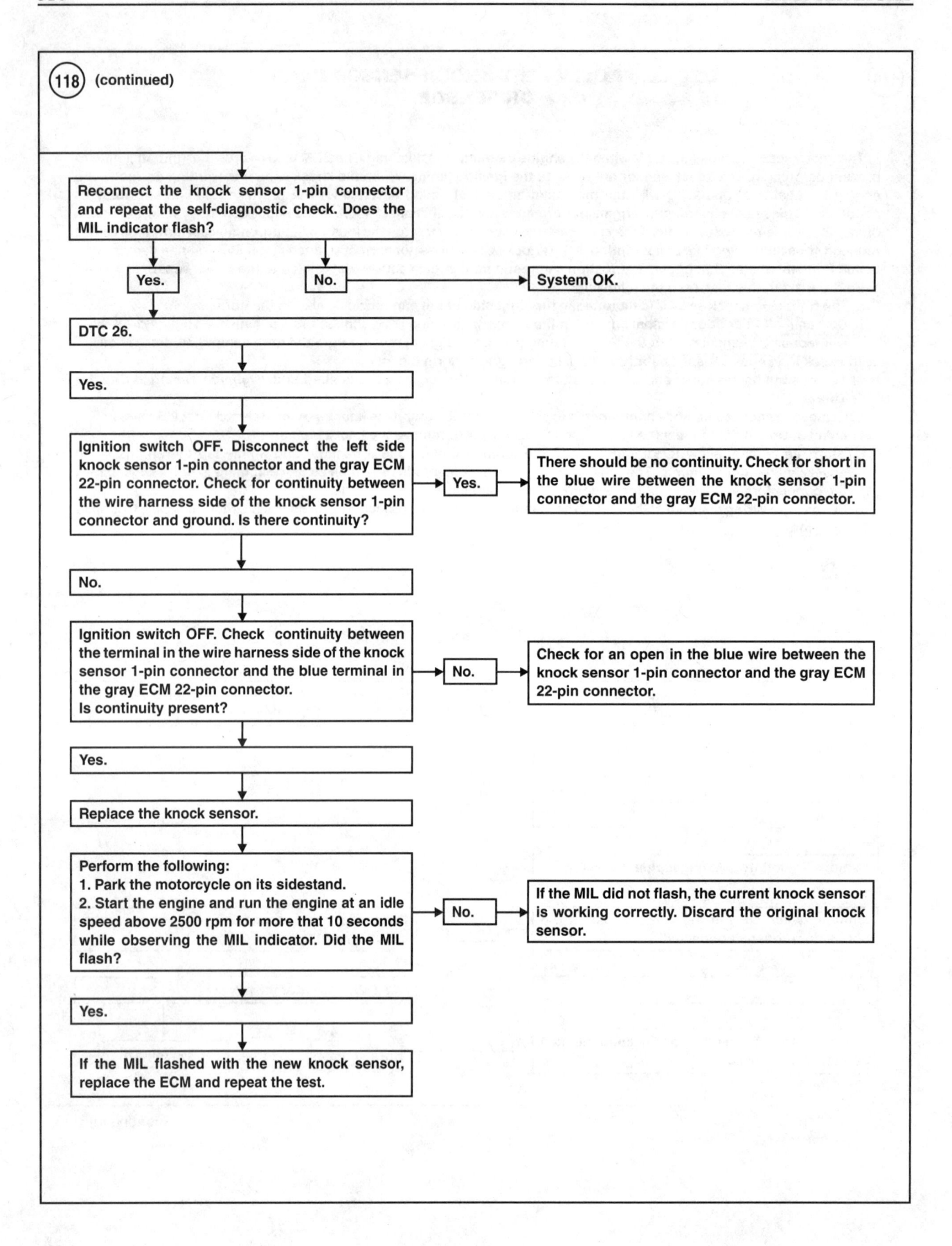

Reconnect the knock sensor 1-pin connector and repeat the self-diagnostic check. Does the MIL indicator flash?

- No. → System OK.
- Yes. → DTC 26.

Yes. ↓

Ignition switch OFF. Disconnect the left side knock sensor 1-pin connector and the gray ECM 22-pin connector. Check for continuity between the wire harness side of the knock sensor 1-pin connector and ground. Is there continuity?

- Yes. → There should be no continuity. Check for short in the blue wire between the knock sensor 1-pin connector and the gray ECM 22-pin connector.

No. ↓

Ignition switch OFF. Check continuity between the terminal in the wire harness side of the knock sensor 1-pin connector and the blue terminal in the gray ECM 22-pin connector.
Is continuity present?

- No. → Check for an open in the blue wire between the knock sensor 1-pin connector and the gray ECM 22-pin connector.

Yes. ↓

Replace the knock sensor.

↓

Perform the following:
1. Park the motorcycle on its sidestand.
2. Start the engine and run the engine at an idle speed above 2500 rpm for more that 10 seconds while observing the MIL indicator. Did the MIL flash?

- No. → If the MIL did not flash, the current knock sensor is working correctly. Discard the original knock sensor.

Yes. ↓

If the MIL flashed with the new knock sensor, replace the ECM and repeat the test.

FUEL INJECTION AND EMISSION CONTROL SYSTEMS

⑲ DTC 29: FAULTY IAC VALVE

The idle air control (IAC) valve, mounted on the rear side of the throttle body, controls engine idle speed by increasing or decreasing the amount of air that bypasses the throttle valves when they are fully closed. The IAC valve receives a signal from the ECM that controls its on/off voltage. When starting a cold engine, the IAC valve also works as a fast idle valve. A faulty IAC valve or circuit can cause hard starting (both cold and hot starts), rough idle and stalling.

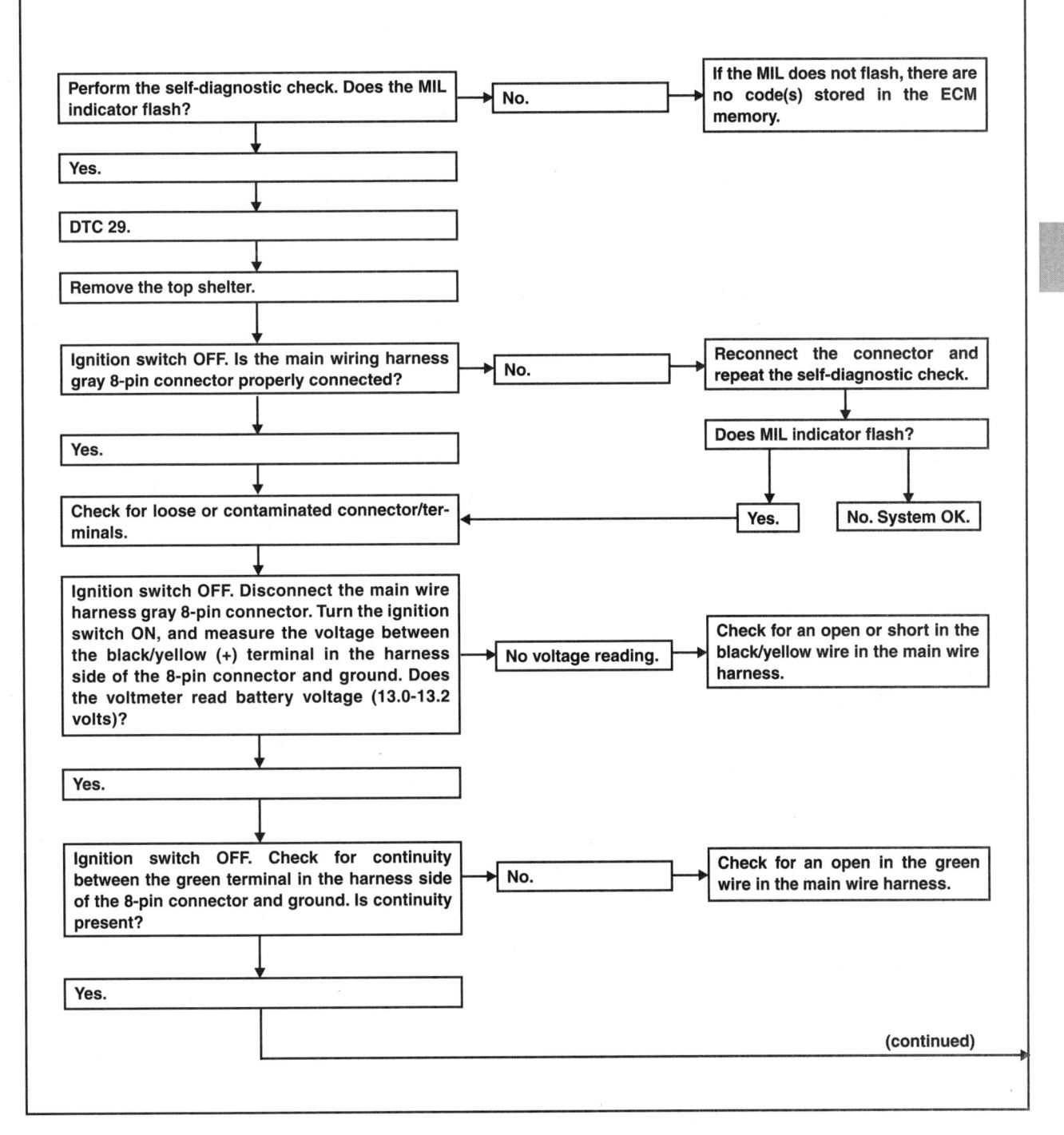

```
┌─────────────────────────────────────────┐         ┌─────────────────────────────────────┐
│ Perform the self-diagnostic check. Does │  No.    │ If the MIL does not flash, there are│
│ the MIL indicator flash?                │────────▶│ no code(s) stored in the ECM memory.│
└─────────────────────────────────────────┘         └─────────────────────────────────────┘
          │ Yes.
          ▼
┌──────────────┐
│ DTC 29.      │
└──────────────┘
          │
          ▼
┌─────────────────────────┐
│ Remove the top shelter. │
└─────────────────────────┘
          │
          ▼
┌─────────────────────────────────────────┐         ┌─────────────────────────────────────┐
│ Ignition switch OFF. Is the main wiring │  No.    │ Reconnect the connector and repeat  │
│ harness gray 8-pin connector properly   │────────▶│ the self-diagnostic check.          │
│ connected?                              │         └─────────────────────────────────────┘
└─────────────────────────────────────────┘                         │
          │ Yes.                                                    ▼
          │                                         ┌─────────────────────────────────────┐
          │                                         │ Does MIL indicator flash?           │
          │                                         └─────────────────────────────────────┘
          │                                             │ Yes.          │ No. System OK.
          ▼                                             ▼
┌─────────────────────────────────────────┐
│ Check for loose or contaminated         │
│ connector/terminals.                    │
└─────────────────────────────────────────┘
          │
          ▼
┌─────────────────────────────────────────┐         ┌─────────────────────────────────────┐
│ Ignition switch OFF. Disconnect the     │         │ Check for an open or short in the   │
│ main wire harness gray 8-pin connector. │ No      │ black/yellow wire in the main wire  │
│ Turn the ignition switch ON, and        │ voltage │ harness.                            │
│ measure the voltage between the         │ reading.│                                     │
│ black/yellow (+) terminal in the        │────────▶│                                     │
│ harness side of the 8-pin connector and │         │                                     │
│ ground. Does the voltmeter read battery │         │                                     │
│ voltage (13.0-13.2 volts)?              │         │                                     │
└─────────────────────────────────────────┘         └─────────────────────────────────────┘
          │ Yes.
          ▼
┌─────────────────────────────────────────┐         ┌─────────────────────────────────────┐
│ Ignition switch OFF. Check for          │  No.    │ Check for an open in the green wire │
│ continuity between the green terminal   │────────▶│ in the main wire harness.           │
│ in the harness side of the 8-pin        │         └─────────────────────────────────────┘
│ connector and ground. Is continuity     │
│ present?                                │
└─────────────────────────────────────────┘
          │ Yes.
          ▼
                                                              (continued)
```

119 (continued)

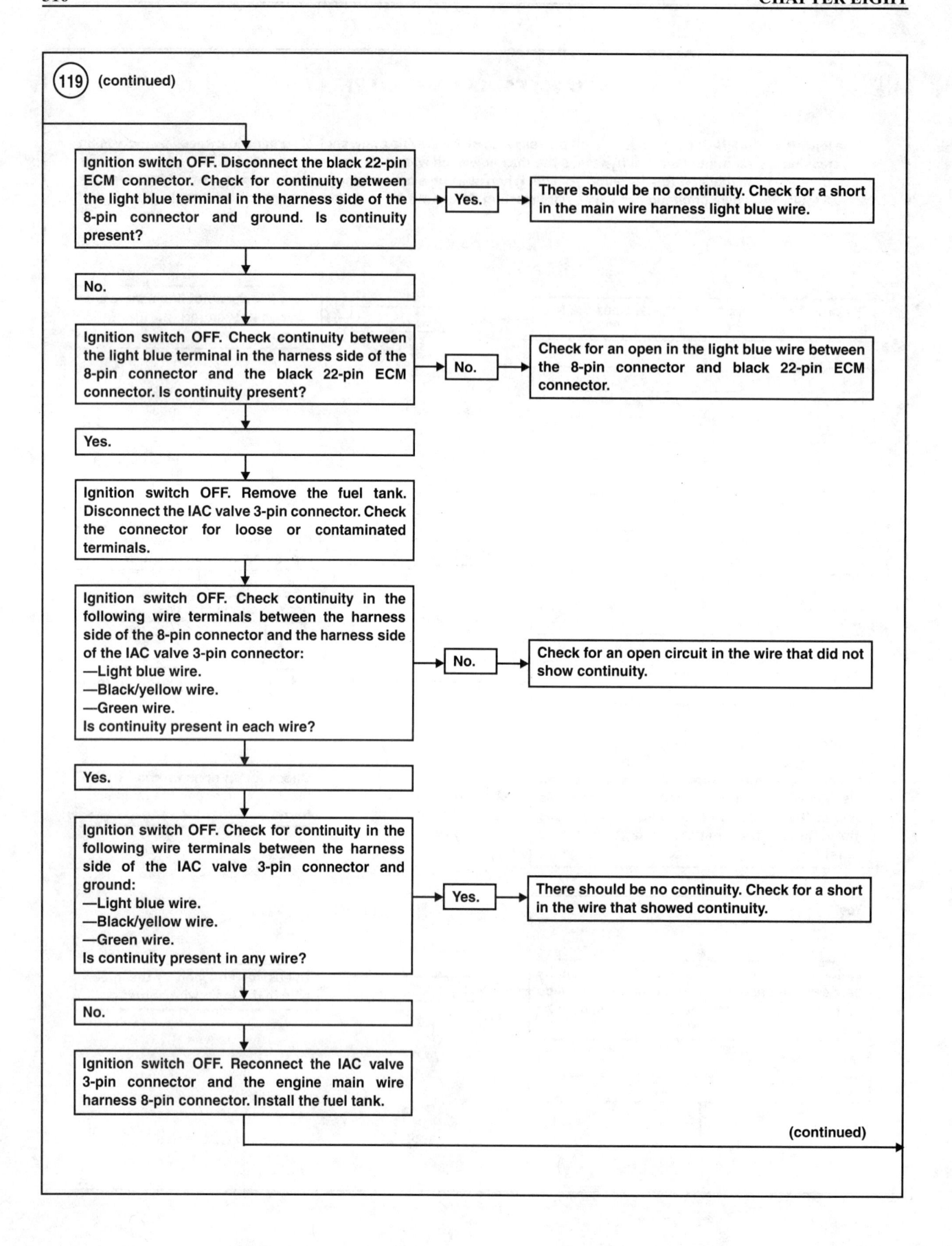

Ignition switch OFF. Disconnect the black 22-pin ECM connector. Check for continuity between the light blue terminal in the harness side of the 8-pin connector and ground. Is continuity present?

→ **Yes.** → There should be no continuity. Check for a short in the main wire harness light blue wire.

↓ **No.**

Ignition switch OFF. Check continuity between the light blue terminal in the harness side of the 8-pin connector and the black 22-pin ECM connector. Is continuity present?

→ **No.** → Check for an open in the light blue wire between the 8-pin connector and black 22-pin ECM connector.

↓ **Yes.**

Ignition switch OFF. Remove the fuel tank. Disconnect the IAC valve 3-pin connector. Check the connector for loose or contaminated terminals.

↓

Ignition switch OFF. Check continuity in the following wire terminals between the harness side of the 8-pin connector and the harness side of the IAC valve 3-pin connector:
—Light blue wire.
—Black/yellow wire.
—Green wire.
Is continuity present in each wire?

→ **No.** → Check for an open circuit in the wire that did not show continuity.

↓ **Yes.**

Ignition switch OFF. Check for continuity in the following wire terminals between the harness side of the IAC valve 3-pin connector and ground:
—Light blue wire.
—Black/yellow wire.
—Green wire.
Is continuity present in any wire?

→ **Yes.** → There should be no continuity. Check for a short in the wire that showed continuity.

↓ **No.**

Ignition switch OFF. Reconnect the IAC valve 3-pin connector and the engine main wire harness 8-pin connector. Install the fuel tank.

(continued)

FUEL INJECTION AND EMISSION CONTROL SYSTEMS

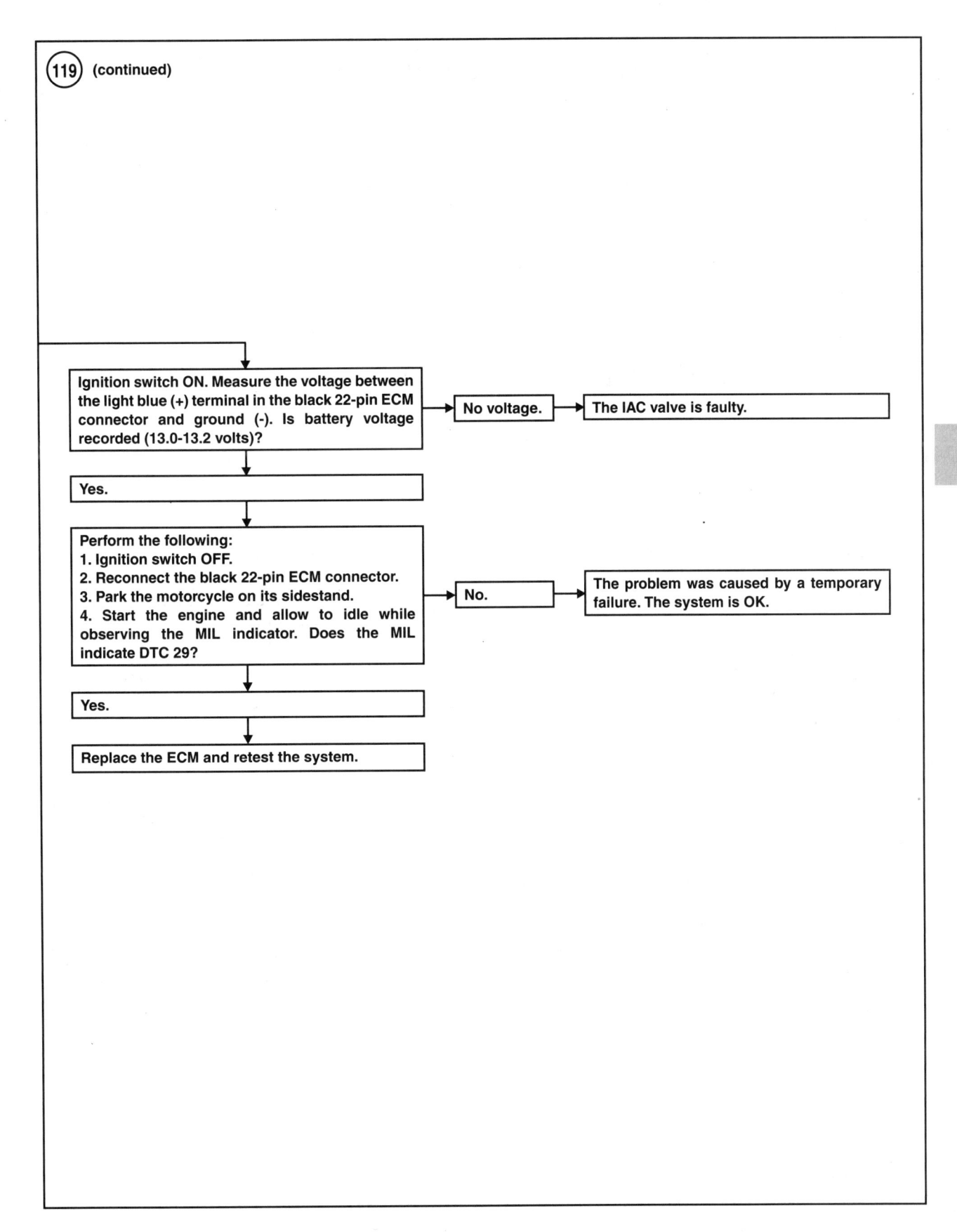

(119) (continued)

↓

Ignition switch ON. Measure the voltage between the light blue (+) terminal in the black 22-pin ECM connector and ground (-). Is battery voltage recorded (13.0-13.2 volts)? → **No voltage.** → The IAC valve is faulty.

↓ Yes.

Perform the following:
1. Ignition switch OFF.
2. Reconnect the black 22-pin ECM connector.
3. Park the motorcycle on its sidestand.
4. Start the engine and allow to idle while observing the MIL indicator. Does the MIL indicate DTC 29? → **No.** → The problem was caused by a temporary failure. The system is OK.

↓ Yes.

Replace the ECM and retest the system.

DTC 33: FAULTY E2-PROM IN ECM

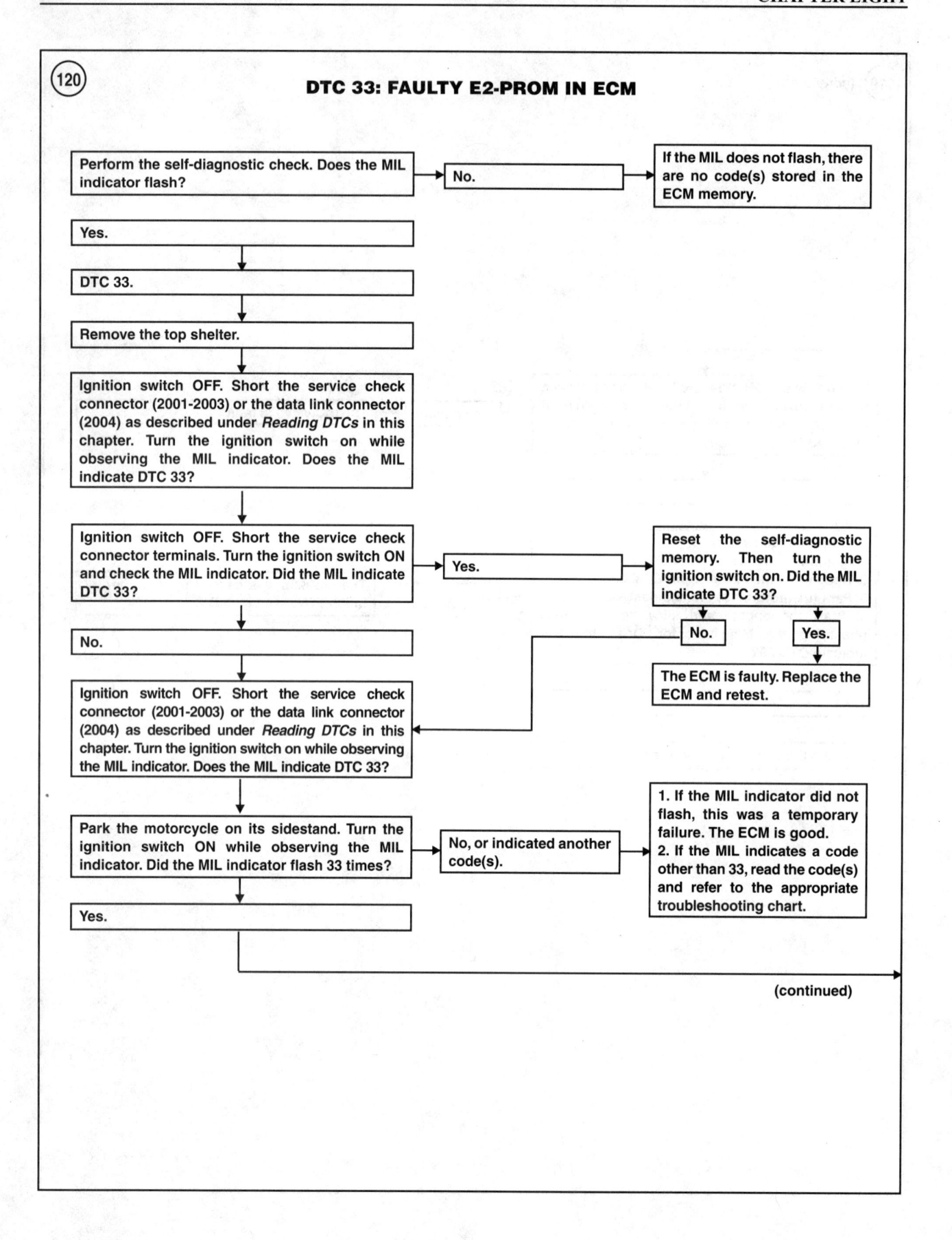

Perform the self-diagnostic check. Does the MIL indicator flash?
→ **No.** → If the MIL does not flash, there are no code(s) stored in the ECM memory.

↓ **Yes.**

DTC 33.

↓

Remove the top shelter.

↓

Ignition switch OFF. Short the service check connector (2001-2003) or the data link connector (2004) as described under *Reading DTCs* in this chapter. Turn the ignition switch on while observing the MIL indicator. Does the MIL indicate DTC 33?

↓

Ignition switch OFF. Short the service check connector terminals. Turn the ignition switch ON and check the MIL indicator. Did the MIL indicate DTC 33?
→ **Yes.** → **Reset the self-diagnostic memory. Then turn the ignition switch on. Did the MIL indicate DTC 33?**
- **No.** → (returns to flow)
- **Yes.** → **The ECM is faulty. Replace the ECM and retest.**

↓ **No.**

Ignition switch OFF. Short the service check connector (2001-2003) or the data link connector (2004) as described under *Reading DTCs* in this chapter. Turn the ignition switch on while observing the MIL indicator. Does the MIL indicate DTC 33?

↓

Park the motorcycle on its sidestand. Turn the ignition switch ON while observing the MIL indicator. Did the MIL indicator flash 33 times?
→ **No, or indicated another code(s).** →
1. If the MIL indicator did not flash, this was a temporary failure. The ECM is good.
2. If the MIL indicates a code other than 33, read the code(s) and refer to the appropriate troubleshooting chart.

↓ **Yes.**

(continued)

FUEL INJECTION AND EMISSION CONTROL SYSTEMS

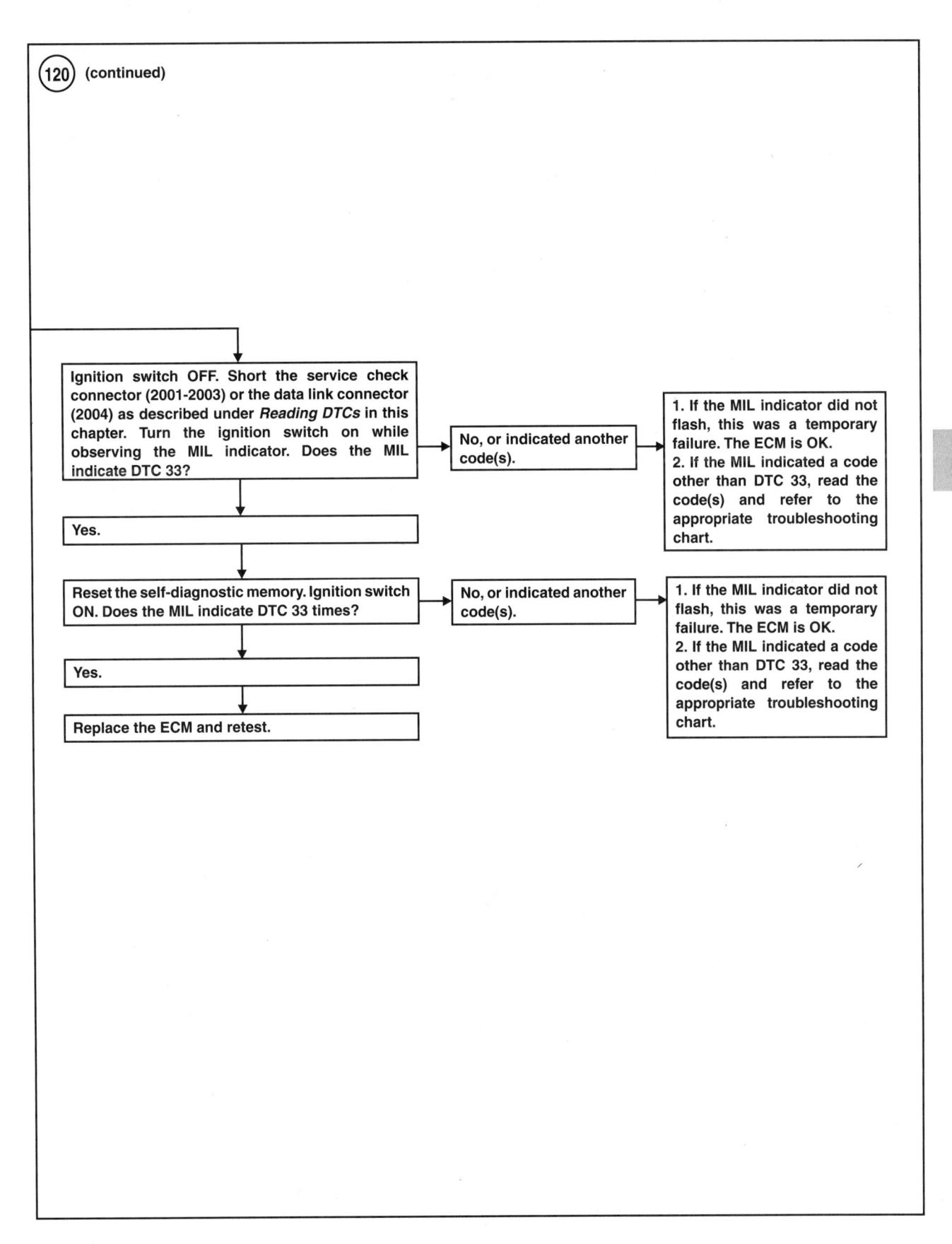

120 (continued)

↓

Ignition switch OFF. Short the service check connector (2001-2003) or the data link connector (2004) as described under *Reading DTCs* in this chapter. Turn the ignition switch on while observing the MIL indicator. Does the MIL indicate DTC 33? → No, or indicated another code(s). → 1. If the MIL indicator did not flash, this was a temporary failure. The ECM is OK.
2. If the MIL indicated a code other than DTC 33, read the code(s) and refer to the appropriate troubleshooting chart.

↓ Yes.

Reset the self-diagnostic memory. Ignition switch ON. Does the MIL indicate DTC 33 times? → No, or indicated another code(s). → 1. If the MIL indicator did not flash, this was a temporary failure. The ECM is OK.
2. If the MIL indicated a code other than DTC 33, read the code(s) and refer to the appropriate troubleshooting chart.

↓ Yes.

Replace the ECM and retest.

121

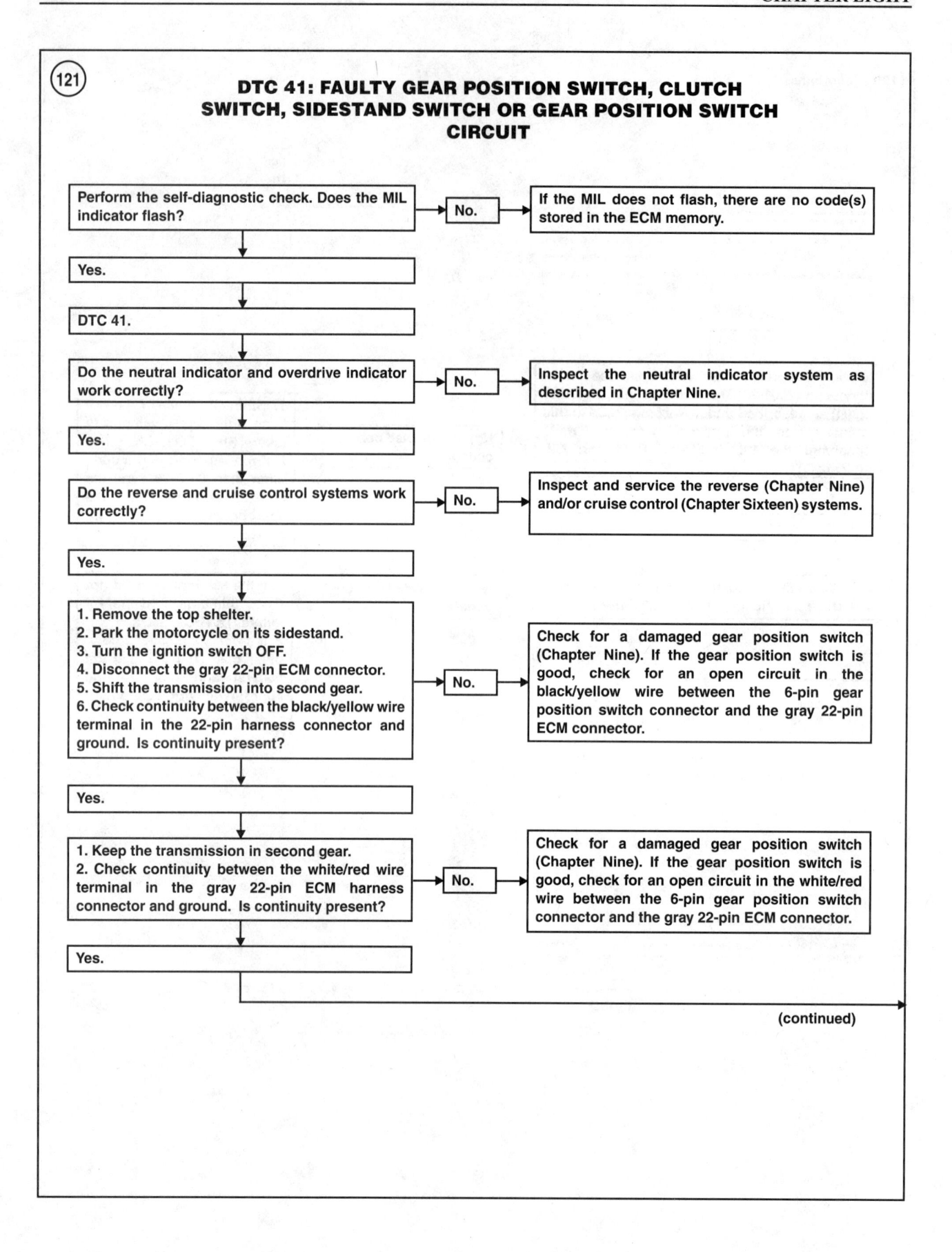

DTC 41: FAULTY GEAR POSITION SWITCH, CLUTCH SWITCH, SIDESTAND SWITCH OR GEAR POSITION SWITCH CIRCUIT

Perform the self-diagnostic check. Does the MIL indicator flash? → **No.** → If the MIL does not flash, there are no code(s) stored in the ECM memory.

↓ Yes.

DTC 41.

↓

Do the neutral indicator and overdrive indicator work correctly? → **No.** → Inspect the neutral indicator system as described in Chapter Nine.

↓ Yes.

Do the reverse and cruise control systems work correctly? → **No.** → Inspect and service the reverse (Chapter Nine) and/or cruise control (Chapter Sixteen) systems.

↓ Yes.

1. Remove the top shelter.
2. Park the motorcycle on its sidestand.
3. Turn the ignition switch OFF.
4. Disconnect the gray 22-pin ECM connector.
5. Shift the transmission into second gear.
6. Check continuity between the black/yellow wire terminal in the 22-pin harness connector and ground. Is continuity present?

→ **No.** → Check for a damaged gear position switch (Chapter Nine). If the gear position switch is good, check for an open circuit in the black/yellow wire between the 6-pin gear position switch connector and the gray 22-pin ECM connector.

↓ Yes.

1. Keep the transmission in second gear.
2. Check continuity between the white/red wire terminal in the gray 22-pin ECM harness connector and ground. Is continuity present?

→ **No.** → Check for a damaged gear position switch (Chapter Nine). If the gear position switch is good, check for an open circuit in the white/red wire between the 6-pin gear position switch connector and the gray 22-pin ECM connector.

↓ Yes.

(continued)

FUEL INJECTION AND EMISSION CONTROL SYSTEMS

121 (continued)

↓

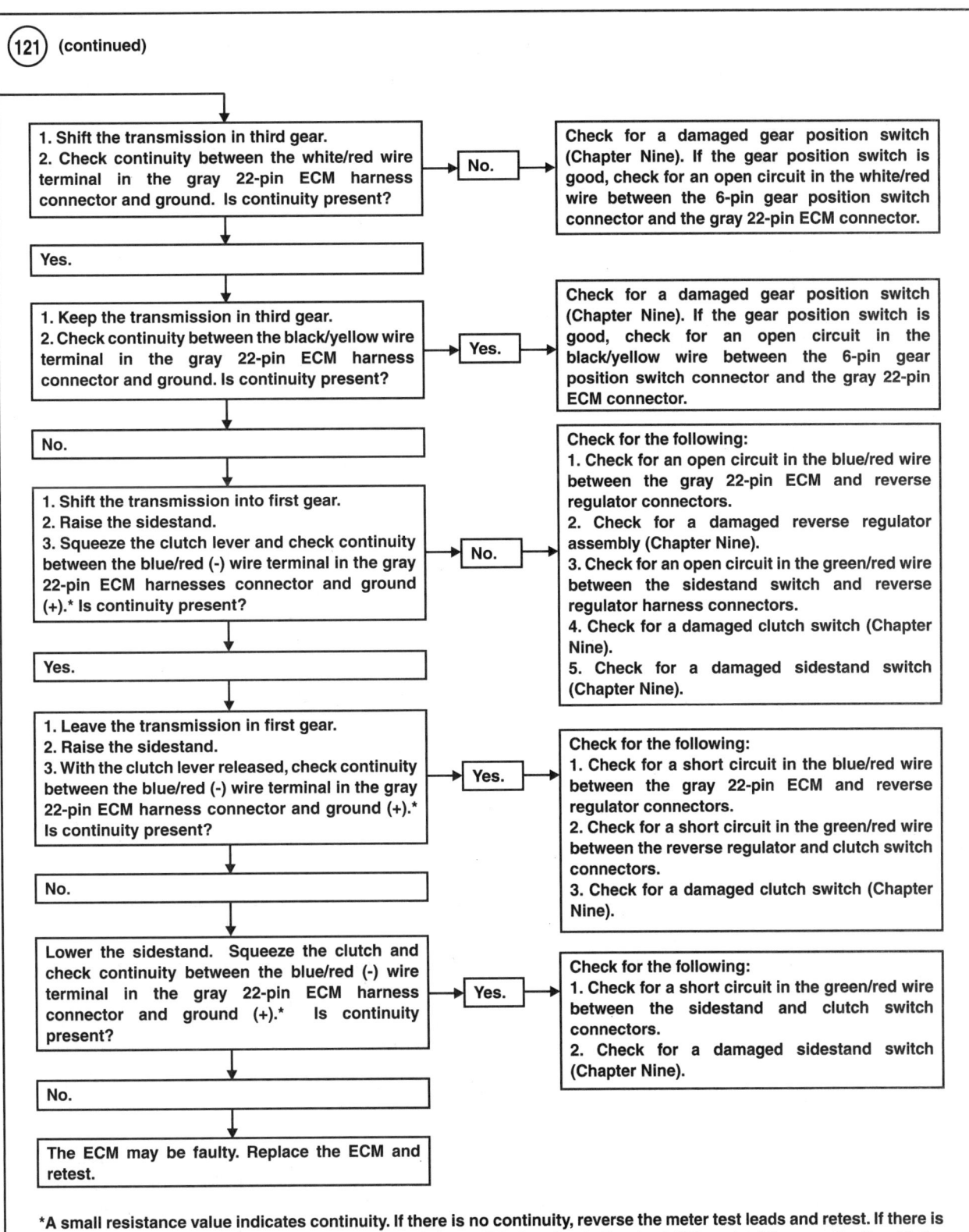

1. Shift the transmission in third gear.
2. Check continuity between the white/red wire terminal in the gray 22-pin ECM harness connector and ground. Is continuity present?

→ **No.** → Check for a damaged gear position switch (Chapter Nine). If the gear position switch is good, check for an open circuit in the white/red wire between the 6-pin gear position switch connector and the gray 22-pin ECM connector.

↓ Yes.

1. Keep the transmission in third gear.
2. Check continuity between the black/yellow wire terminal in the gray 22-pin ECM harness connector and ground. Is continuity present?

→ **Yes.** → Check for a damaged gear position switch (Chapter Nine). If the gear position switch is good, check for an open circuit in the black/yellow wire between the 6-pin gear position switch connector and the gray 22-pin ECM connector.

↓ No.

1. Shift the transmission into first gear.
2. Raise the sidestand.
3. Squeeze the clutch lever and check continuity between the blue/red (-) wire terminal in the gray 22-pin ECM harnesses connector and ground (+).* Is continuity present?

→ **No.** → Check for the following:
1. Check for an open circuit in the blue/red wire between the gray 22-pin ECM and reverse regulator connectors.
2. Check for a damaged reverse regulator assembly (Chapter Nine).
3. Check for an open circuit in the green/red wire between the sidestand switch and reverse regulator harness connectors.
4. Check for a damaged clutch switch (Chapter Nine).
5. Check for a damaged sidestand switch (Chapter Nine).

↓ Yes.

1. Leave the transmission in first gear.
2. Raise the sidestand.
3. With the clutch lever released, check continuity between the blue/red (-) wire terminal in the gray 22-pin ECM harness connector and ground (+).* Is continuity present?

→ **Yes.** → Check for the following:
1. Check for a short circuit in the blue/red wire between the gray 22-pin ECM and reverse regulator connectors.
2. Check for a short circuit in the green/red wire between the reverse regulator and clutch switch connectors.
3. Check for a damaged clutch switch (Chapter Nine).

↓ No.

Lower the sidestand. Squeeze the clutch and check continuity between the blue/red (-) wire terminal in the gray 22-pin ECM harness connector and ground (+).* Is continuity present?

→ **Yes.** → Check for the following:
1. Check for a short circuit in the green/red wire between the sidestand and clutch switch connectors.
2. Check for a damaged sidestand switch (Chapter Nine).

↓ No.

The ECM may be faulty. Replace the ECM and retest.

*A small resistance value indicates continuity. If there is no continuity, reverse the meter test leads and retest. If there is continuity with the test leads reversed, continue with the YES direction in the chart. If there is still no continuity with the test leads reversed, continue with the NO direction in the chart.

122

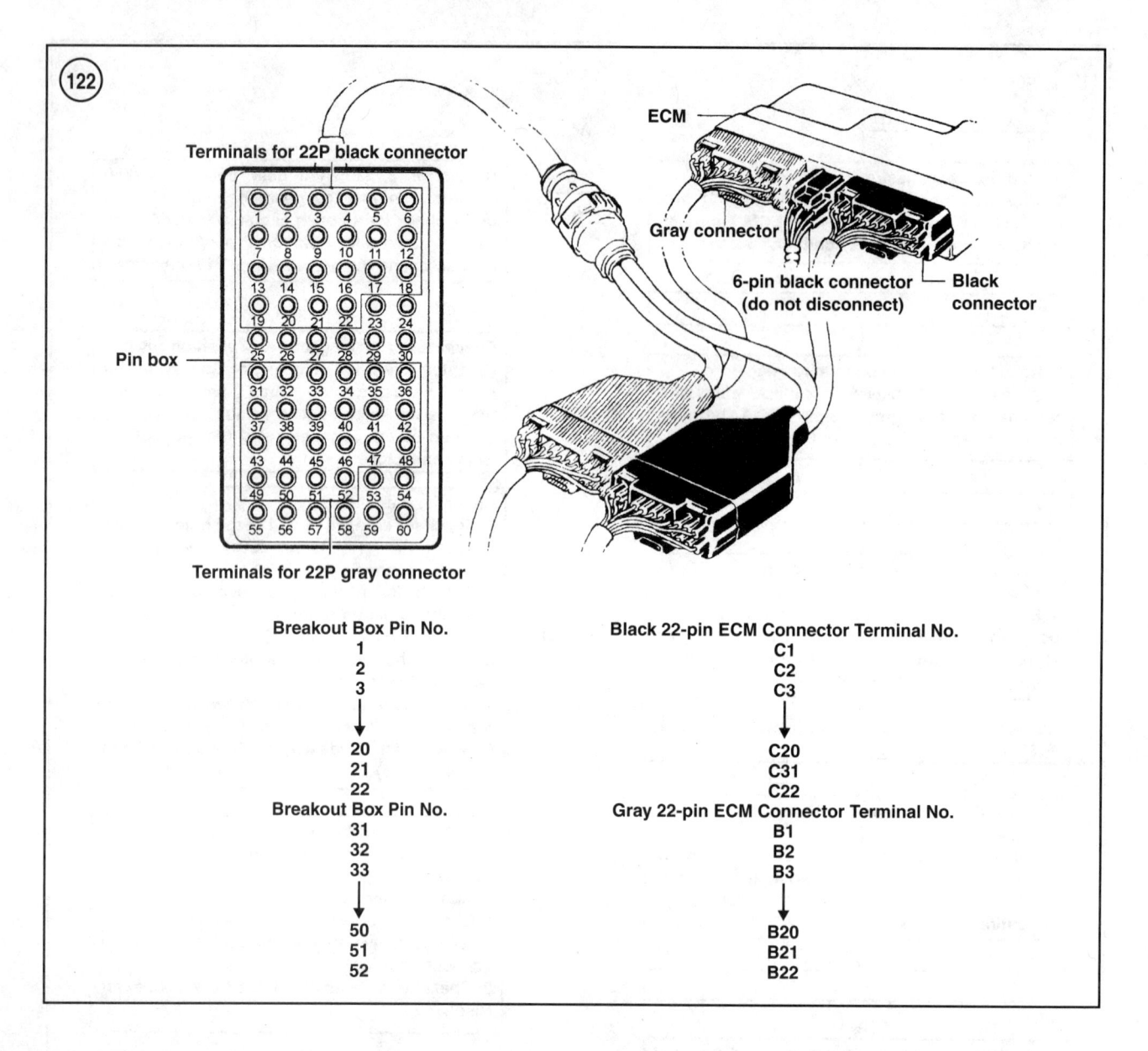

Terminals for 22P black connector

Pin box

Terminals for 22P gray connector

ECM

Gray connector

6-pin black connector (do not disconnect)

Black connector

Breakout Box Pin No.	Black 22-pin ECM Connector Terminal No.
1	C1
2	C2
3	C3
↓	↓
20	C20
21	C31
22	C22

Breakout Box Pin No.	Gray 22-pin ECM Connector Terminal No.
31	B1
32	B2
33	B3
↓	↓
50	B20
51	B21
52	B22

9. If the affected system has a vacuum function, make sure the hose(s) are attached securely and in good condition (no leaks).

10. If a DTC refers to a particular component, refer to the tests in this chapter, or Chapter Nine, to determine if further testing is possible.

11. If the above tests fail to locate the problem, refer troubleshooting and repair to a Honda dealership.

12. Do not perform tests with a battery charger attached to the battery or circuit. Test and charge the battery as described in Chapter Nine.

13. Before replacing the ECM, take the motorcycle to a Honda dealership for further testing. Have the dealership confirm that the ECM is faulty before purchasing a replacement. Refer to *Electrical Component Replacement* in Chapter Nine.

14. After troubleshooting, clear any DTC(s) and perform self-diagnosis to ensure that the problem has been corrected.

ECM Test Harness

Testing the fuel injection system consists of making resistance and voltage tests at the ECM wiring harness terminals. To effectively do this, use the test harness (part No. 07WMZ-MBGA000 [**Figure 122**]). The test harness (sometimes referred to as a breakout box) plugs into the ECM terminals and then connects to the gray and black 22-pin ECM harness connectors. The tests can then be made at a harness

FUEL INJECTION AND EMISSION CONTROL SYSTEMS

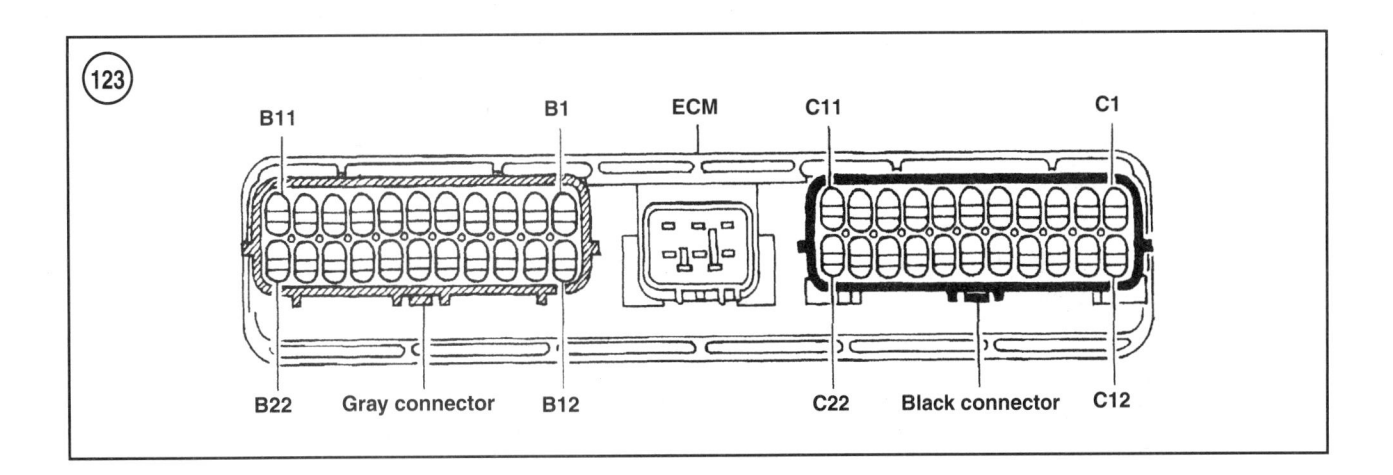

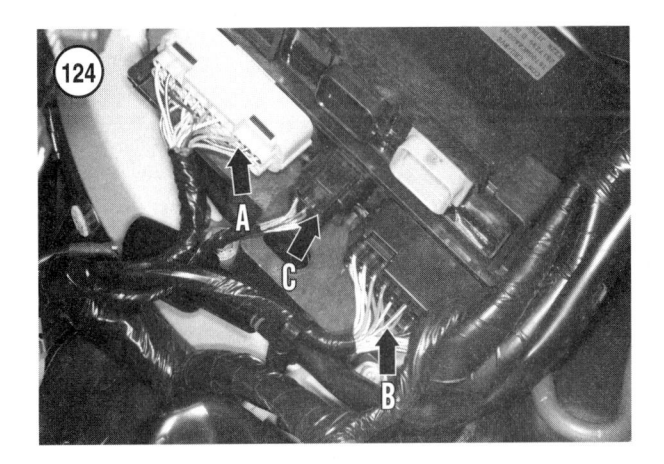

123 B11 — B1 — ECM — C11 — C1
B22 — Gray connector — B12 — C22 — Black connector — C12

124 A, C, B

pin box without the possibility of damaging the terminals at the connectors. Refer to **Figure 123** to identify the pin numbers at the ECM with the connectors unplugged.

1. Note the following to identify the pin box (**Figure 122**) and ECM (**Figure 123**) terminals:
   a. Terminals No. 1 through No. 22 in the pin box are used to test the terminals in the black 22-pin ECM harness connector (C1-C22, **Figure 123**).
   b. Terminals No. 31 through No. 52 in the pin box (**Figure 122**) are used to test the terminals in the gray 22-pin ECM harness connector (B1-B22, **Figure 123**).
   c. If the test harness is not available, disconnect the gray and black 22-pin ECM harness connectors and make the checks at the back of the connectors. **Figure 122** shows the pin box and its identification numbers. Use the chart in **Figure 122** to convert the pin box numbers (called out in the flow charts) to the ECM pin numbers identified in **Figure 123**. For example, if pin box number 21 is called out in a test procedure, it cross-references to the 22-pin black connector C21 terminal (**Figure 123**). Only those checks made with the engine not running can be made. The test harness is required to make running checks.

2. Always turn the ignition switch off before disconnecting or reconnecting the connectors at the ECM.

3. Remove the top shelter. Refer to *Top Shelter* in Chapter Seventeen.

4. Disconnect the gray (A, **Figure 124**) and black (B) 22-pin ECM connectors. Do not disconnect the black 6-pin connector (C, **Figure 124**).

5. Connect the test harness to the gray and black 22-pin ECM connectors as shown in **Figure 122**.

6. Perform the tests as described in the flow chart or procedure.

7. Reverse these steps to remove the test harness and reconnect the ECM connectors.

MIL Lamp Circuit Test

If the MIL does not light when the ignition switch is turned on and with the engine stop switch set to RUN, refer to **Table 4** and perform the following:

1. Support the motorcycle on its sidestand and turn the ignition switch on. Check the operation of the sidestand and oil pressure indicators.
   a. If they do not operate correctly, check the combination meter power circuit. Refer to *Combination Meter* in Chapter Nine.
   b. If they operate correctly, refer to Step 2.

2. Remove the top shelter. Refer to *Top Shelter* in Chapter Seventeen.

3. Turn the ignition switch off.

4. Disconnect the black 22-pin ECM harness connector (B, **Figure 124**).

5. Connect a jumper wire between the black 22-pin ECM harness connector white/blue wire and ground. Turn the ignition on. The MIL should light. Note the following:

a. If the MIL came on, the ECM is faulty. Replace the ECM and retest.
b. If the MIL did not come on, check the white/blue wire between the ECM and combination meter for a short circuit. If the white/blue wire is good, replace the combination meter. Refer to *Combination Meter* in Chapter Nine.

6. Reverse Step 2 and Step 4 to complete installation

Table 1 FUEL INJECTION SYSTEM COMPONENT ABBREVIATIONS

Barometric pressure sensor	BARO sensor
Diagnostic trouble code	DTC
Engine control module	ECM
Engine coolant temperature sensor	ECT sensor
Idle air control valve	IAC valve
Intake air temperature sensor	IAT sensor
Malfunction indicator lamp	MIL
Manifold absolute pressure sensor	MAP sensor
Oxygen sensor	O_2 sensor
PAIR	Pulsed secondary air injection system
Throttle position sensor	TP sensor
Vehicle speed sensor	VSS

Table 2 FUEL INJECTION SYSTEM TEST SPECIFICATIONS[1]

Item	Specification
Cam pulse generator peak voltage	0.7 volts minimum
ECT sensor resistance	2.3-2.6 k ohms
Engine idle speed	700 +/-70 rpm
Fuel injector resistance	11.1-12.3 ohms
Fuel pressure	See text
Fuel pump flow[2]	133 ml (4.5 U.S. Oz.)/10 seconds
Ignition pulse generator peak voltage	0.7 volts minimum
IAT sensor resistance	2.2-2.7 k ohms (1-4 k ohms when testing at harness)
Manifold absolute pressure at idle	400-450 mm Hg (15.7-17.7 in./HG)
Throttle body identification number	GQ61A
Throttle grip free play	2-6 mm (0.08-0.24 in.)
TP sensor resistance	4-6 k ohms

1. Resistance and voltage readings at 20° C (68° F).
2. Battery must by fully charged.

Table 3 FUEL INJECTION SYSTEM TORQUE SPECIFICATIONS

	N•m	in.-lb.	ft.-lb.
Cable stay mounting bolts at throttle body	4	35	–
Fuel supply hose joint bolts	10	88	–
Fuel pipe mounting bolts	10	88	–
Fuel pipe banjo bolt	33	–	24
Fuel rail mounting bolts	10	88	–
Fuel rail sealing bolt	12	106	–
IAC valve	4	35	–
Knock sensor	31	–	23
Map sensor	2	18	–
Oxygen sensor	25	–	18
PAIR check valve cover bolts	5	44	–
Pressure regulator stay bolt	10	88	–
Pressure regulator	27	–	20

FUEL INJECTION AND EMISSION CONTROL SYSTEMS

Table 4 DIAGNOSTIC TROUBLE CODES (DTC)

DTC	Symptoms	Cause(s)
MIL does not come on	Engine does not start	Open circuit in the ECM power input and ground lines. Damaged fuel injection relay or circuit wires. Damaged bank angle sensor or circuit wires. Damaged engine stop switch. Blown fuel injection/ignition fuse. Blown stop switch fuse. Damaged engine stop switch or circuit wires. Damaged ECM.
MIL does not come on	Engine operates normally	Open circuit in MIL wire. Damaged combination meter. Damaged ECM.
MIL stays on	Engine operates normally	Short circuit in service check connector (2001-2003) or data link connector (2004-2005). Short circuit in MIL wire. Damaged ECM.
1	Increased fuel consumption Engine stalls	Loose or poor contacts on MAP sensor. Open or short circuit in MAP sensor wire. Damaged MAP sensor.
7	Hard start at low temperature Engine stalls or stumbles	Loose or poor contacts on ETC sensor. Open or short circuit in ECT sensor wire. Damaged ETC sensor.
8	Poor engine response when operating the throttle quickly	Loose or poor contacts on TP sensor. Open or short circuit in TP sensor wire. Damaged TP sensor.
9	Hard start at low temperature	Loose or poor contacts on IAT sensor wire. Open or short circuit in IAT sensor wire. Faulty IAT sensor.
10	Engine operates normally at low altitude, but idles roughly at high altitude	Loose or poor contacts on BARO sensor. Open or short circuit in BARO sensor wire. Damaged BARO sensor.
11	Engine operates normally Erratic speedometer and/or cruise control operation	Loose or poor contacts on VSS. Open or short circuit in VSS wire. Damaged VSS.
12	Engine does not start	Loose or poor contacts on No. 1 fuel injector. Open or short circuit in No. 1 fuel injector wire. Damaged No. 1 fuel injector.
13	Engine does not start	Loose or poor contacts on No. 2 fuel injector. Open or short circuit in No. 2 fuel injector wire. Damaged No. 2 fuel injector.
14	Engine does not start	Loose or poor contacts on No. 3 fuel injector. Open or short circuit in No. 3 fuel injector wire. Damaged No. 3 fuel injector.
15	Engine does not start	Loose or poor contacts on No. 4 fuel injector. Open or short circuit in No. 4 fuel injector wire. Damaged No. 4 fuel injector.

(continued)

Table 4 DIAGNOSTIC TROUBLE CODES (DTC) (continued)

DTC	Symptoms	Cause(s)
16	Engine does not start	Loose or poor contacts on No. 5 fuel injector. Open or short circuit in No. 5 fuel injector wire. Damaged No. 5 fuel injector.
17	Engine does not start	Loose or poor contacts on No. 6 fuel injector. Open or short circuit in No. 6 fuel injector wire. Damaged No. 6 fuel injector.
18	Engine does not start	Loose or poor contacts on cam pulse generator wire. Open or short circuit in cam pulse generator wire. Damaged cam pulse generator.
19	Engine does not start	Loose or poor contacts on ignition pulse generator wire. Open or short circuit in ignition pulse generator wire. Damaged ignition pulse generator.
21	Engine operates normally	Open or short circuit in right oxygen sensor wire. Damaged right oxygen sensor.
22	Engine operates normally	Open or short circuit in left oxygen sensor wire. Damaged left oxygen sensor.
23	Engine operates normally	Open or short circuit in right oxygen sensor heater wire. Damaged right oxygen sensor.
24	Engine operates normally	Open or short circuit in left oxygen sensor heater wire. Damaged left oxygen sensor.
25	Engine operates normally	Loose or poor contacts on right knock sensor connector. Open or short circuit in right knock sensor wire. Damaged right knock sensor.
26	Engine operates normally	Loose or poor contacts on left knock sensor connector. Open or short circuit in left knock sensor wire. Damgaed left knock sensor.
29	Engine is hard to start, stalls, idles roughly	Loose or poor contacts on IAC valve connector. Open or short circuit in IAC valve wire. Damaged IAC valve.
33	Engine operates normally; ECM does not hold diagnostic data	Damaged E–2 prom in ECM
41	Engine operates normally	Loose or poor contacts in gear position switch and circuits. Open or short circuit in gear position switch wire(s). Damaged gear position switch. Damaged sidestand switch. Damaged clutch switch.

CHAPTER NINE

ELECTRICAL SYSTEM

This chapter contains service and test procedures for the electrical system components. Specification **Tables 1-9** are at the end of this chapter.

Fuel system components are covered in Chapter Eight.

ELECTRICAL COMPONENT REPLACEMENT

Most motorcycle dealerships and parts suppliers will not accept the return of any electrical part. If you cannot determine the *exact* cause of any electrical system malfunction, have a Honda dealership retest that specific system to verify your test results. If you purchase a new electrical component(s), install it, and then find that the system still does not work properly, you will probably be unable to return the unit for a refund.

Consider any test results carefully before replacing a component that tests only *slightly* out of specification, especially resistance. A number of variables can affect test results dramatically. These include: the testing meter's internal circuitry, ambient temperature and conditions under which the machine has been operated. All instructions and specifications have been checked for accuracy; however, successful test results depend to a great degree upon individual accuracy.

ELECTRICAL CONNECTORS

The GL1800 is equipped with numerous electrical components, connectors and wiring harnesses. Corrosion-causing moisture can enter these electrical connectors and cause poor electrical connections leading to component failure. Troubleshooting an electrical circuit with one or more corroded electrical connectors can be time-consuming and frustrating.

When reconnecting electrical connectors, pack them with a dielectric grease compound. Dielectric grease is specially formulated for sealing and waterproofing electrical connections without interfering with current flow. Use only this compound or an equivalent designed for this specific purpose. Do not use a substitute that may interfere with the current flow within the electrical connector. Do not use silicone sealant.

BATTERY

A sealed, maintenance-free battery is installed on all models. The battery electrolyte level cannot be serviced. When replacing the battery, use a sealed type; do not install a non-sealed battery. Never remove the sealing caps from the top of the battery. The battery does not require periodic electrolyte inspection or refilling. See **Table 1** for battery specifications.

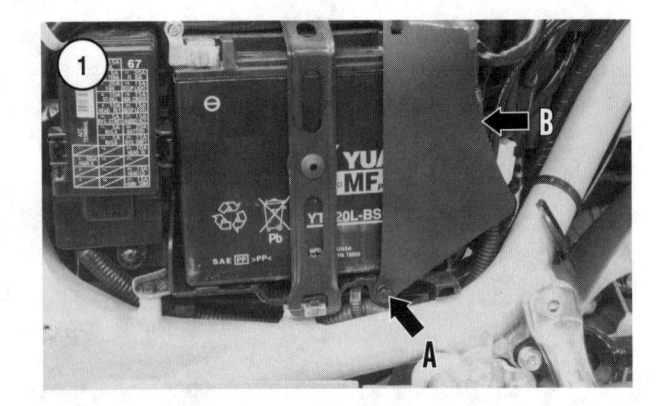

To prevent accidental shorts that could blow a fuse when working on the electrical system, always disconnect the negative battery cable from the battery.

WARNING
Even though the battery is a sealed type, protect your eyes, skin and clothing; electrolyte is corrosive and can cause severe burns and permanent injury. The battery case may be cracked and leaking electrolyte. If electrolyte gets into your eyes, flush your eyes thoroughly with clean, running water and get immediate medical attention. Always wear safety goggles when servicing the battery.

WARNING
While batteries are being charged, highly explosive hydrogen gas forms in each cell. Some of this gas escapes through the vent openings and may form an explosive atmosphere in and around the battery. This condition can persist for several hours. Sparks, an open flame or a lighted cigarette can ignite the gas, causing an internal battery explosion and possible serious injury.

NOTE
When replacing the battery, make sure to turn in the old battery for recycling. The lead plates and the plastic case can be recycled. Most motorcycle dealerships accept old batteries in trade when purchasing a new one. Never place an old battery in household trash; it is illegal, in most states, to place any acid or lead (heavy metal) contents in landfills.

Safety Precautions

Take the following precautions to prevent an explosion:

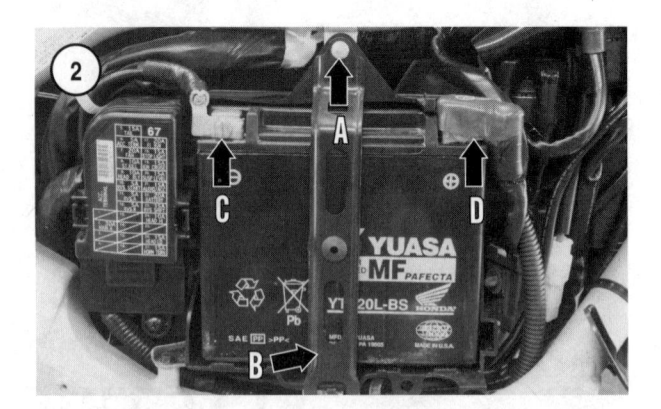

1. Do not smoke or permit any open flame near any battery being charged or which has been recently charged.
2. Do not disconnect live circuits at the battery. A spark usually occurs when a live circuit is broken.
3. Take care when connecting or disconnecting a battery charger. Make sure the power switch is off before making or breaking connections. Poor connections are a common cause of electrical arcs, which cause explosions.
4. Keep children and pets away from charging equipment and batteries.

Battery Removal/Installation

1. Read *Safety Precautions* in this section.
2. Record the frequencies of the radio preset buttons.
3. Turn the ignition switch off.
4. Remove the left side cover as described in Chapter Seventeen under *Side Covers*.
5. Remove the battery trim clip (A, **Figure 1**) and battery cover (B).
6. Remove the holder bolt (A, **Figure 2**) and lower the holder (B).
7. Disconnect the negative battery cable (C, **Figure 2**), then the positive cable (D), from the battery terminals.
8. Remove the battery.

ELECTRICAL SYSTEM

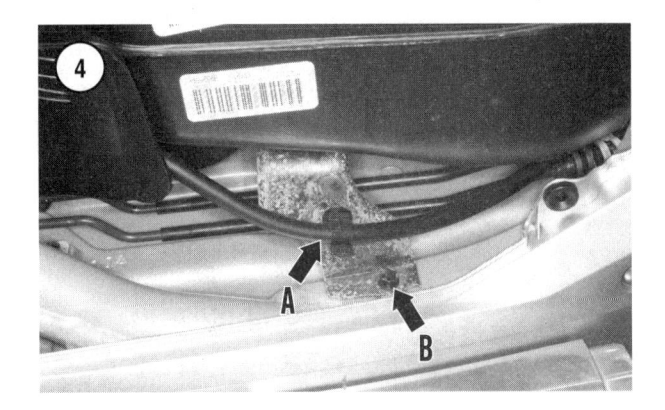

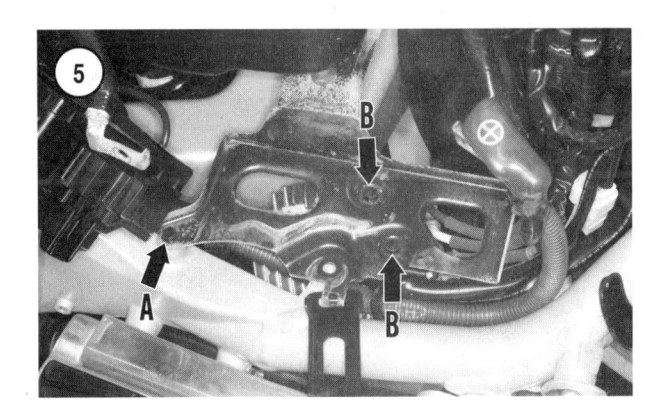

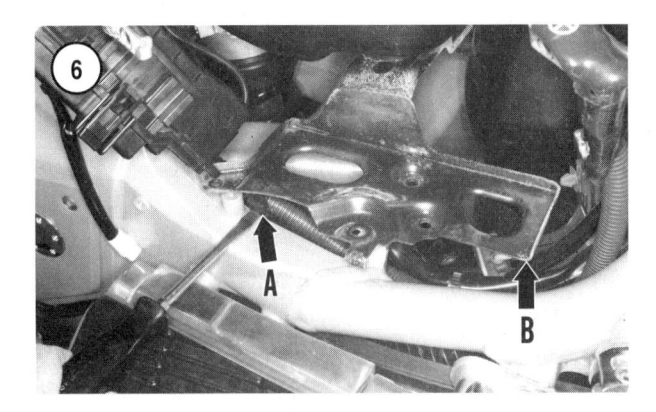

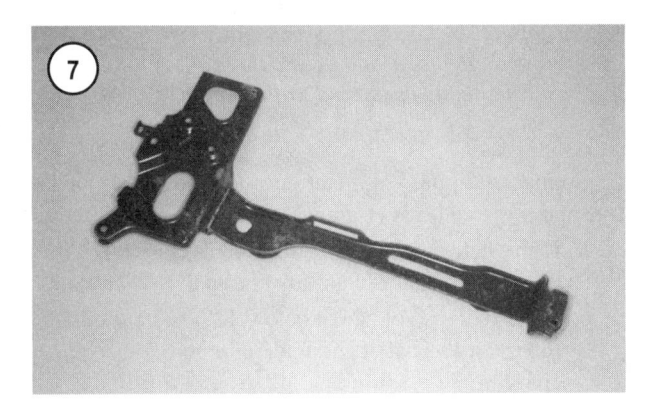

9. After servicing or replacing the battery, install it by reversing these removal steps, plus the following:
   a. Always connect the positive battery cable first (D, **Figure 2**), then the negative cable (C).

   CAUTION
   Make sure the battery cables are connected to their proper terminals. Connecting the battery backward reverses the polarity and damages components in the electrical system. When installing a replacement battery, confirm that the negative (–) terminal is mounted on the left side of the battery.

   b. Coat the battery leads with dielectric grease or petroleum jelly.
10. Reset the clock and radio.
11. Install the left side cover.

Battery Box and Tray Removal/Installation

1. Remove the battery as described in this section.
2. Remove the right side cover. Refer to *Side Covers* in Chapter Seventeen.
3. Remove the battery box:
   a. Remove the lower battery box mounting bolt (A, **Figure 3**) and collar. Remove the battery box (B).
   b. Remove the screw (C, **Figure 3**) securing the fuse box to the battery box.
4. At the right side of the battery tray:
   a. Remove the shock hose from the plastic clamp (A, **Figure 4**).
   b. Remove the mounting bolt (B, **Figure 4**).
5. At the left side of the battery tray:
   a. Remove the battery tray mounting bolt (A, **Figure 5**).
   b. Remove the two reverse resistor mounting bolts (B, **Figure 5**). Do not disconnect the cable or connector from the reverse resistor.
   c. Open the plastic clamp (A, **Figure 6**) and remove the wire harness from the clamp. The clamp stays mounted on the battery tray.

   CAUTION
   The battery tray is a tight fit. Remove it carefully to avoid scratching the swing arm or damaging the brake hoses mounted on the top of the swing arm.

6. Push the battery tray forward to disconnect it from the tab on rear fender B, and remove the tray from the left side of the frame (B, **Figure 6**). See **Figure 7**.

7. Install the battery tray from the left side. Work it into place and install its slot into the tab on rear fender B.
8. At the right side of the battery tray:
   a. Install the mounting bolt (B, **Figure 4**).
   b. Install the shock hose (A, **Figure 4**) into the plastic clamp.
9. At the left side of the battery tray:
   a. Close the clamp (A, **Figure 6**) around the fuse box wiring harness.
   b. Install the two reverse resistor mounting bolts (B, **Figure 5**) and secure the resistor mounting bracket beneath the battery tray.
   c. Check the routing of the reverse resistor wiring harness and other wire harnesses behind the battery tray.
   d. Install the left side mounting bolt (A, **Figure 5**) and tighten securely.
10. To install the battery box:
    a. Working from the rear of the battery box, slide the fuse box onto the battery box rails, then slide the battery box into position.
    b. Install the lower battery mounting bolt (A, **Figure 3**), collar and tighten securely.
    c. Secure the fuse box to the battery box with the screw (C, **Figure 3**).
11. Install the battery as described in this section.
12. Tighten the left battery tray mounting bolt.
13. Install the right side cover.

Battery Cleaning/Inspection

The battery electrolyte level cannot be serviced. *Never* remove the sealing bar cap (**Figure 8**) from the top of the battery. The battery does not require periodic electrolyte inspection or refilling.

1. Read *Safety Precautions* in this section.
2. Remove the battery from the motorcycle as described in this section. Do not clean the battery while it is mounted in the motorcycle.
3. Clean the battery case with a solution of warm water and baking soda. Rinse thoroughly with clean water.
4. Inspect the physical condition of the battery. Look for bulges or cracks in the case, leaking electrolyte or corrosion buildup.
5. Check the battery terminal bolts, spacers and nuts for corrosion and damage. Clean parts with a solution of baking soda and water and rinse thoroughly. Replace any that are damaged.
6. Check the battery cable clamps for corrosion and damage. If corrosion is minor, clean the battery cable clamps with a stiff brush. Replace excessively worn or damaged cables.

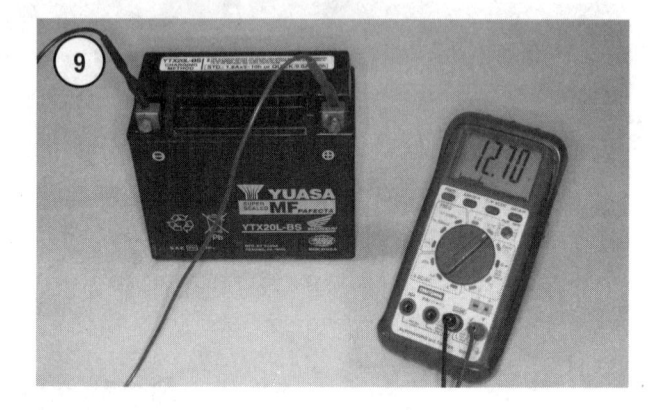

Voltage Test

The maintenance-free battery can be tested while mounted in the motorcycle. Use a digital voltmeter for this procedure. See **Table 2** for battery voltage readings for maintenance-free batteries.

1. Read *Safety Precautions* in this section.

> *NOTE*
> *To prevent false test readings, do not test the battery if the battery terminals are corroded. Remove and clean the battery and terminals as described in this chapter, then reinstall it.*

2. Connect a digital voltmeter between the battery negative and positive leads (**Figure 9**). Note the following:
   a. If the battery voltage is 13.0-13.2 volts (at 20° C [68° F]), the battery is fully charged. See **Table 2**.
   b. If the battery voltage is below 12.3 volts (at 20° C [68° F]), the battery is undercharged and requires charging. See **Table 2**.

ELECTRICAL SYSTEM

4. Perform the *Voltage Test* in this section. Charge the battery if the reading is below 12.8 volts.

5. Connect the load tester to the battery terminals—red lead to the positive terminal, and the black lead to the negative terminal.

6. Determine the correct amount of load to apply to the battery. Refer to the manufacturer's instructions, plus the following:

   a. All GL1800 models use a 12 volt/18 amp/hour battery (**Table 1**).
   b. The amount of load to use is determined by the original capacity of the battery. This is measured in cold cranking amperes (CCA) or in the amp/hour rating. The correct load to apply is one/half the CCA rating or three times the amp/hour rating (180 amps × 3 = 54 amps).

7. Test the battery and determine the results following the manufacturer's instructions.

8. If the battery fails the load test, recharge the battery and retest. If the battery fails the load test again, replace the battery.

Charging

Refer to *New Battery Setup* in this section if the battery is new.

To recharge a maintenance-free battery, use a digital voltmeter and a charger (**Figure 11**) with an adjustable or automatically variable amperage output. If this equipment is not available, have the battery charged by a shop with the proper equipment. Excessive voltage and amperage from an unregulated charger can damage the battery and shorten service life.

The battery should only self-discharge approximately one percent of its given capacity each day. If a battery not in use, without any loads connected, loses its charge within one week after charging, the battery is defective.

If the motorcycle is not used for long periods of time, an automatic battery charger with variable voltage and amperage outputs is recommended for optimum battery service life.

WARNING
During the charging process, highly explosive hydrogen gas is released from the battery. Charge the battery only in a well-ventilated area away from any open flames (including pilot lights on home gas appliances). Do not allow any smoking in the area. Never check the charge of the battery by connecting screwdriver blades or other metal objects between the terminals;

3. If the battery is undercharged, recharge it as described in this section. Then test the charging system as described under *Charging System* in this chapter.

Load Test

A battery load test (or capacity test) is the most accurate way to determine battery condition. This is an independent test where a battery load tester checks a battery's performance under full current load. A battery load tester places an electrical load on the battery to determine if it can provide current while maintaining the minimum required voltage. **Figure 10** shows the TestMate battery load tester available from Motion Pro.

A battery load test procedure will vary depending on the type of tester. A battery must be at least 75 percent charged before making a load test. Use the following guidelines to supplement the manufacturer's instructions:

1. Remove the battery as described in this chapter.

2. Inspect the battery case for any leaks or damage. Do not test the battery if the case is damaged.

3. Clean the battery terminals.

the resulting spark can ignite the hydrogen gas.

CAUTION
Always remove the battery from the motorcycle before connecting the battery charger. Never recharge a battery in the frame; corrosive gasses emitted during the charging process will damage surfaces.

1. Remove the battery as described in this chapter.
2. Connect the positive (+) charger lead to the positive battery terminal and the negative (–) charger lead to the negative battery terminal.
3. Set the charger at 12 volts and switch it on. Charge the battery at a slow charge rate of 1/10 its given capacity. To determine the charge rate, divide the battery amp hour capacity by 10. For the GL1800, 20 amp hour divided by 10 = 2.0 amp current.

CAUTION
When using an adjustable battery charger, follow the manufacturer's instructions. Do not use a larger output battery charger or increase the charge rate on an adjustable battery charger to reduce charging time. Doing so can cause permanent battery damage. For the GL1800, purchase a charger with an amp current rating closest to 2.0 amps.

4. After the battery has been charged for 4-5 hours, turn the charger off, disconnect the leads and allow the battery to set for a minimum of 30 minutes. Then check the battery with a digital voltmeter. A fully charged battery will read 13.0 volts or higher 30 minutes after taken off the charger. If the battery reading is less than 13.0 volts, charge it again.

Storage

When the motorcycle is ridden infrequently or put in storage for an extended amount of time, the battery must be periodically charged to ensure it will be capable of working correctly when returned to service. Use an automatic battery charger with variable voltage and amperage outputs.

1. Remove the battery as described in this chapter.
2. Clean the battery and terminals with a solution of baking soda and water.
3. Inspect the battery case for any cracks, leaks or bulging. Replace the battery if the case is leaking or damaged.
4. Clean the battery box in the motorcycle.

5A. For temporary storage and infrequent motorcycle use, install the battery into the motorcycle but do not connect the battery terminals. Check the battery every two weeks and charge as necessary.
5B. For extended storage (longer than one month), remove the battery and charge to 100%. Then store the battery in a cool, dry place. Continue to charge the battery once a month when stored in temperatures below 16° C (60° F) and every two weeks when stored in temperatures above 16° C (60° F).

New Battery Setup

If a new battery is not filled and charged, the battery will require activation (initialization). Follow the battery manufacturer's instructions while observing the following guidelines when activating a new battery:

WARNING
Wear safety goggles when servicing and handling the battery in this section.

1. Use the electrolyte that comes with the battery. Do not use electrolyte from a common container.
2. Fill the battery using all the electrolyte included with the battery kit.
3. Allow the battery to sit for one hour. This allows the plates to absorb the electrolyte for optimum performance.
4. Loosely install the sealing cap over the battery filling holes.
5. Charge the battery following the manufacturer's instructions.

CAUTION
A new battery must be fully charged before installation. Failure to do so reduces the life of the battery. Using a new battery without an initial charge causes permanent battery damage. That is, the battery will never be able to hold more than an 80% charge. Charging a new battery after using it will not bring its charge to 100%. When purchasing a new battery from a dealership or parts store, verify its charge status. If necessary, have them perform the activation or initial charge before accepting the battery.

6. Press the sealing cap firmly to seal each of the battery fill holes. Make sure the cap seats flush into the battery.

ELECTRICAL SYSTEM

CAUTION
Never remove the sealing cap or add electrolyte to the battery.

CHARGING SYSTEM

The charging system supplies power to operate electrical system components and keeps the battery charged. The charging system consists of the battery, alternator and an IC regulator mounted inside the alternator. The alternator is mounted onto and driven by the engine. Alternator speed is determined by gears connected to the crankshaft. A 100-amp main fuse protects the circuit.

When the engine starts, an alternating current (AC) voltage is created in the alternator field windings. This AC voltage is rectified to direct current (DC) voltage by the diodes within the alternator. The IC regulator regulates voltage output.

Troubleshooting

Because the GL1800 charging system is not equipped with an indicator system (light or gauge), slow cranking or short bulb life may be the first indicator of a charging system problem.
1. A fully charged battery is required to accurately test the charging system. Perform the battery load test as described under *Battery* in this chapter. If the battery is damaged or weak, the charging system may not be at fault.
2. Check for a blown ignition cruise fuse (No. 19) as described under *Fuses* in this chapter.
3. Refer to the indicated troubleshooting charts in **Figure 12** and **Figure 13**. Use the wiring diagrams at the end of this manual to identify and locate the appropriate connectors.
4. After making a repair, repeat the *Charging Voltage Test* in this section to confirm the charging system is working correctly.

Current Drain Test
(Milliampere Test)

The GL1800 is equipped with a number of electrical components that drain (draw) current when the ignition system is off. These include the digital clock, trip meter and odometer. These types of current draws are often referred to as key-off or parasitic loads. Key-off loads are normal for large touring motorcycles with many electrical accessories. For this reason, Honda lists an acceptable amount of current drain for the GL1800. If the bike is ridden regularly, and the key-off loads do not exceed the specified amount, they will not drain the battery. If the bike sits for longer than two weeks, key-off loads can weaken and drain the battery. Adding accessories to the electrical system also increases the current drain, depending on how long the motorcycle is not used. A faulty component or circuit also increases current drain on the battery. For these reasons, perform this test before troubleshooting the charging system to determine if the current drain is normal or excessive—if a battery discharges, the charging system may not be at fault.

NOTE
When installing electrical accessories, do not wire them into a circuit where they stay on all the time. Refer to the manufacturer's instructions.

Perform this test before performing the charging voltage test.
1. Turn the ignition switch off.
2. Remove the left side cover. Refer to *Side Covers* in Chapter Seventeen.
3. Disconnect the negative battery cable (C, **Figure 2**).

CAUTION
Before connecting the ammeter into the circuit, set the meter to its highest amperage scale. This step prevents a large current flow from damaging the meter or blowing the meter's fuse.

4. Connect an ammeter between the negative battery cable and the negative battery terminal (**Figure 14**). Do not turn the ignition switch on once this connection is made.
5. Switch the ammeter to its lowest scale and note the reading. The maximum current drain is 5 mA or less. A current drain that exceeds 5 mA must be found and repaired.
6. If the current drain rate is excessive, consider the following probable causes:
   a. Faulty alternator. Disconnect the connectors at the alternator. If the current reading drops to acceptable limits, test the rectifier for a damaged diode(s). Refer to *Alternator* in this chapter.
   b. Damaged battery.
   c. Short circuit in the system.
   d. Loose, dirty or faulty electrical connectors.
   e. Aftermarket electrical accessories added to the electrical system.
7. To find the short circuit that is causing a current drain, refer to the wiring diagrams at the end of the manual. Then continue to measure the current drain while disconnecting different connectors in the electrical system one by one.

⑫ CHARGING SYSTEM: WEAK OR DAMGAED BATTERY

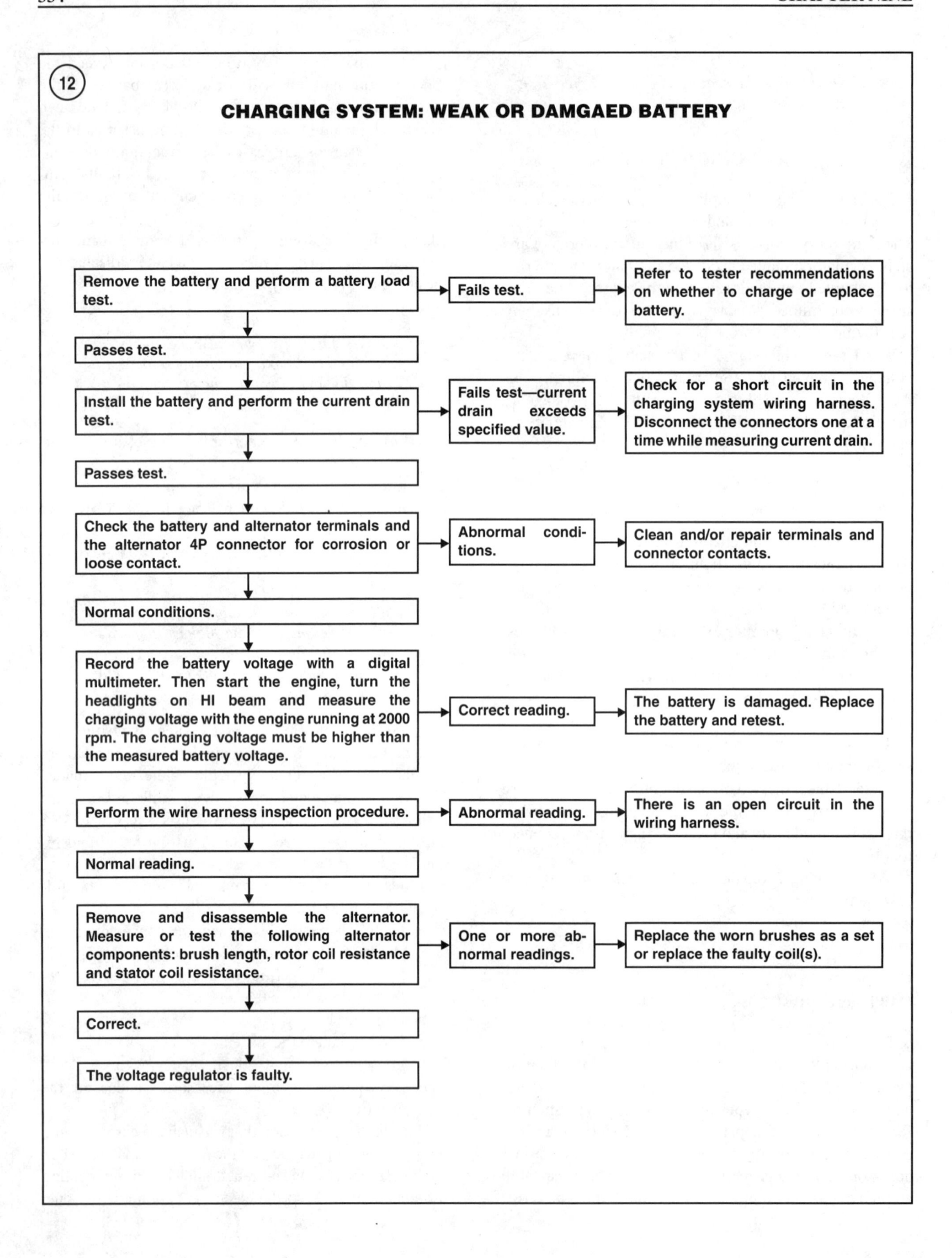

Remove the battery and perform a battery load test. → Fails test. → Refer to tester recommendations on whether to charge or replace battery.

↓ Passes test.

Install the battery and perform the current drain test. → Fails test—current drain exceeds specified value. → Check for a short circuit in the charging system wiring harness. Disconnect the connectors one at a time while measuring current drain.

↓ Passes test.

Check the battery and alternator terminals and the alternator 4P connector for corrosion or loose contact. → Abnormal conditions. → Clean and/or repair terminals and connector contacts.

↓ Normal conditions.

Record the battery voltage with a digital multimeter. Then start the engine, turn the headlights on HI beam and measure the charging voltage with the engine running at 2000 rpm. The charging voltage must be higher than the measured battery voltage. → Correct reading. → The battery is damaged. Replace the battery and retest.

↓

Perform the wire harness inspection procedure. → Abnormal reading. → There is an open circuit in the wiring harness.

↓ Normal reading.

Remove and disassemble the alternator. Measure or test the following alternator components: brush length, rotor coil resistance and stator coil resistance. → One or more abnormal readings. → Replace the worn brushes as a set or replace the faulty coil(s).

↓ Correct.

The voltage regulator is faulty.

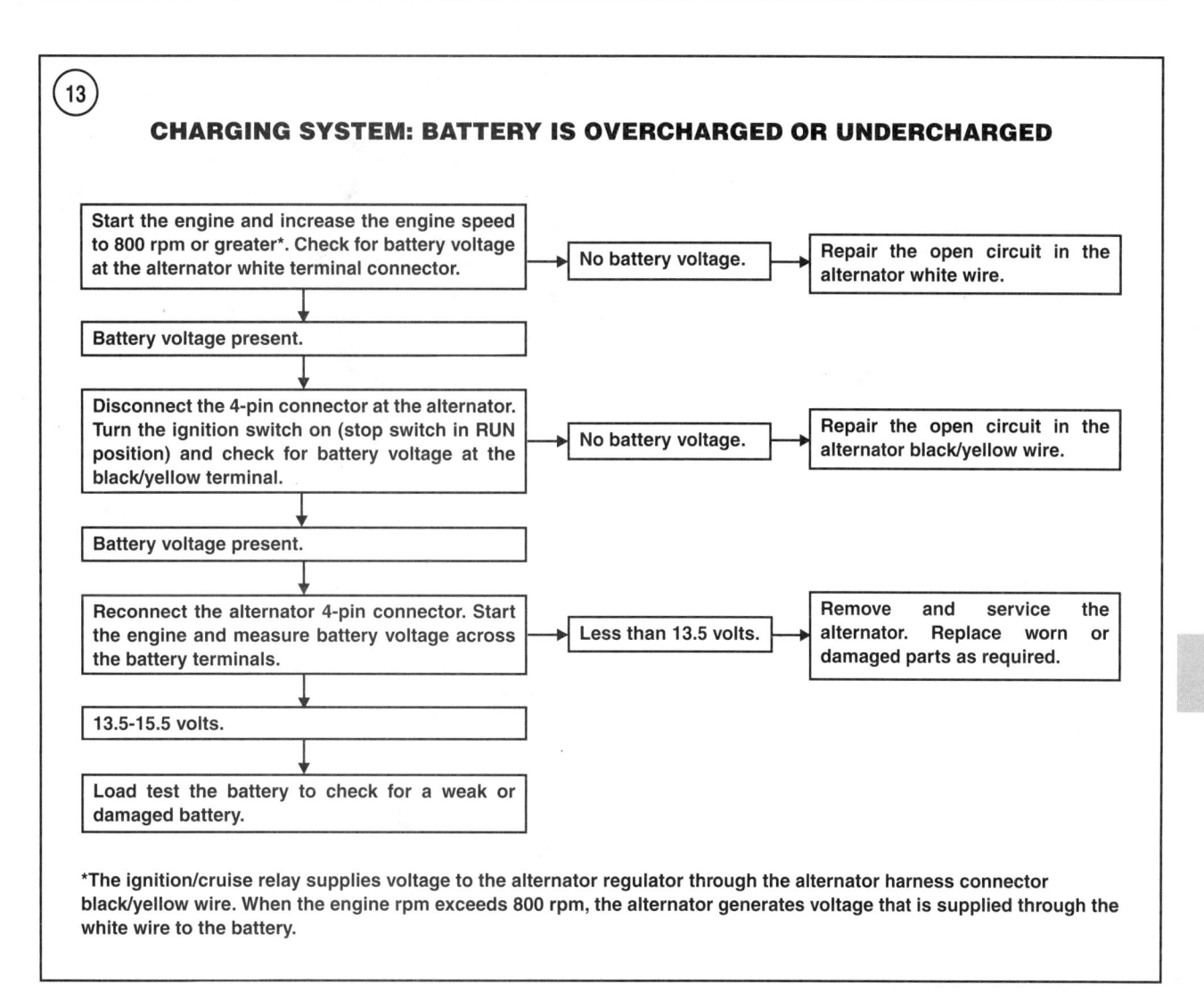

ELECTRICAL SYSTEM

⑬ CHARGING SYSTEM: BATTERY IS OVERCHARGED OR UNDERCHARGED

Start the engine and increase the engine speed to 800 rpm or greater*. Check for battery voltage at the alternator white terminal connector.
→ No battery voltage. → Repair the open circuit in the alternator white wire.

↓ Battery voltage present.

Disconnect the 4-pin connector at the alternator. Turn the ignition switch on (stop switch in RUN position) and check for battery voltage at the black/yellow terminal.
→ No battery voltage. → Repair the open circuit in the alternator black/yellow wire.

↓ Battery voltage present.

Reconnect the alternator 4-pin connector. Start the engine and measure battery voltage across the battery terminals.
→ Less than 13.5 volts. → Remove and service the alternator. Replace worn or damaged parts as required.

↓ 13.5-15.5 volts.

Load test the battery to check for a weak or damaged battery.

*The ignition/cruise relay supplies voltage to the alternator regulator through the alternator harness connector black/yellow wire. When the engine rpm exceeds 800 rpm, the alternator generates voltage that is supplied through the white wire to the battery.

---

⑭ Ammeter / Negative battery cable

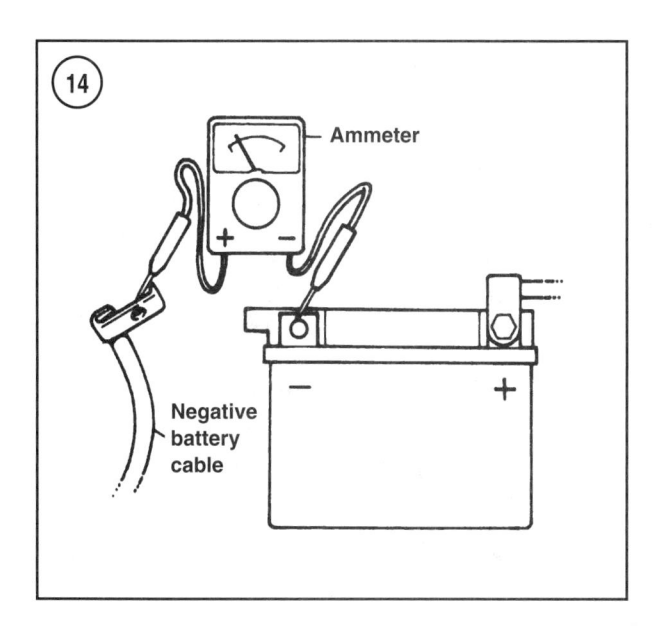

When the current drain rate returns to an acceptable level, the circuit is indicated. Test the circuit further to find the problem.

8. Disconnect the ammeter.
9. Reconnect the negative battery cable (C, **Figure 2**).
10. Install the left side cover.

Charging Voltage Test

This procedure tests charging system operation. It does not measure maximum charging system output. **Table 3** lists charging system test specifications.

To obtain accurate test results, the battery must be fully charged (13.0-13.2 volts).

1. Remove the left side cover as described under *Side Covers* in Chapter Seventeen.

2. Remove the battery trim clip (A, **Figure 1**) and battery cover (B).

3. Start and run the engine until it reaches normal operating temperature, then turn the engine off.

4. Connect a digital voltmeter to the battery terminals as shown in **Figure 15**. To prevent a short, make sure the voltmeter leads attach firmly to the battery terminals. Record the minimum or measured battery voltage reading on the voltmeter.

CAUTION
Do not disconnect either battery cable when making this test. Doing so may damage the voltmeter or electrical accessories.

5. Start the engine and allow it to idle until it reaches normal operating temperature. Turn the headlight to HI beam. Gradually increase engine speed to 2000 rpm and read the voltage indicated on the voltmeter. The voltmeter should show a reading greater than the measured battery voltage recorded in Step 4 and less than 15.5 volts.

NOTE
*If the battery is often discharged, but the charging voltage tested normal during Step 5, the battery may be damaged. Perform a battery load-test as described under **Battery** in this chapter.*

6. If the voltage reading is incorrect, perform the *Wiring Harness Test* in this section, while noting the following:
   a. If the charging voltage is too low, check for an open or short circuit in the charging system wiring harness, an open or short in the alternator, high resistance in the alternator-to-battery cable or a damaged regulator/rectifier.
   b. If the charging voltage is too high, check for a poor regulator/rectifier ground, damaged regulator/rectifier or a damaged battery.

Wiring Harness Test

This procedure tests the integrity of the wires and connectors attached to the alternator. Refer to **Figure 13**.

1. Disconnect the negative battery cable at the battery as described under *Battery* in this chapter.
2. Slide the rubber cover off the alternator terminal nut. Then remove the nut and disconnect the alternator cable (A, **Figure 16**) at the alternator.
3. Disconnect the 4-pin connector (B, **Figure 16**) at the alternator.
4. Reconnect the negative battery cable at the battery.
5. Check the battery charge circuit as follows:
   a. Connect a voltmeter between the disconnected alternator cable terminal (+) and a good engine ground (−).
   b. The voltmeter should read battery voltage at all times (ignition switch on or off).
   c. If the voltage is less than specified, check the alternator cable for damage.
   d. Disconnect the voltmeter leads.
6. Check the battery voltage circuit as follows:
   a. Connect a voltmeter between the alternator 4-pin connector black/yellow terminal (+) and ground (−).
   b. The voltmeter should read battery voltage with the ignition switch turned on and the engine stop switch in the RUN position.
   c. Turn the ignition switch off.
   d. If the voltage is less than specified, check the black/yellow wire and connector for damage.
   e. Disconnect the voltmeter leads.
7. Disconnect the negative battery cable at the battery.
8. Reconnect the 4-pin connector (B, **Figure 16**) at the alternator.

ELECTRICAL SYSTEM

9. Reconnect the alternator cable (A, **Figure 16**) at the alternator. Tighten the nut securely and cover the terminal with the rubber cap.
10. Reconnect the negative battery cable at the battery.

ALTERNATOR

Removal/Installation

1. Perform the tests under *Charging System* before removing the alternator.
2. Disconnect the negative battery cable at the battery as described under *Battery* in this chapter.
3. Remove the fuel tank as described in Chapter Eight.
4. Slide the rubber cover off the alternator terminal nut. Then remove the nut and disconnect the alternator cable (A, **Figure 16**) at the alternator.
5. Disconnect the 4-pin connector (B, **Figure 16**).
6. Remove the alternator mounting bolts. **Figure 17** shows two of the three mounting bolts.
7. Pull the alternator back to disconnect it from the engine and remove it.
8. Service the alternator as described in this section.
9. Reverse these steps to install the alternator, plus the following:
    a. Replace the alternator O-ring (**Figure 18**) if damaged.
    b. Lubricate the O-ring with engine oil.
    c. Tighten the alternator mounting bolts to 29 N•m (21 ft.-lb.).
    d. Perform the *Charging Voltage Test* in this chapter.

Disassembly

Refer to **Figure 19**.
1. Remove the alternator as described in this chapter.
2. Draw an alignment mark across the front housing, stator coil and rear housing.
3. Loosen, but do not remove, the four rotor mounting screws (A, **Figure 20**).
4. Remove the four through bolts (B, **Figure 20**).

NOTE
Record the position and number of shims, if used, so they can be installed in their original mounting positions.

5. Remove the four rotor mounting screws (A, **Figure 20**) and separate the front housing (C) from the rotor.
6. Remove the nut (A, **Figure 21**) and terminal set (B).
7. Thread a knock puller into the threads in the front end of the rotor (A, **Figure 22**). Support the rear housing and operate the puller to remove the rotor assembly (B, **Figure 22**).
8. Remove the three screws (**Figure 23**) and separate the rear housing from the stator coil and regulator/rectifier assembly.
9. The stator coil must be desoldered from the regulator/rectifier for testing and replacement. If necessary, refer to *Stator Coil Testing* in this section.

Inspection

CAUTION
Do not clean the rotor, rotor bearings, regulator/rectifier, slip rings or brush assemblies in solvent. Wipe these components with a clean cloth.

1. Inspect the brushes as described in this section.
2. Inspect the rotor coil as follows:
    a. Inspect the slip rings (A, **Figure 24**) for abnormal wear, roughness or discoloration; these conditions cannot be repaired and requires rotor coil replacement.
    b. Measure the slip ring outside diameter. Replace the rotor if the outside diameter is less than the service limit specified in **Table 3**.
3. Use an ohmmeter to test the rotor coil as follows:

CHAPTER NINE

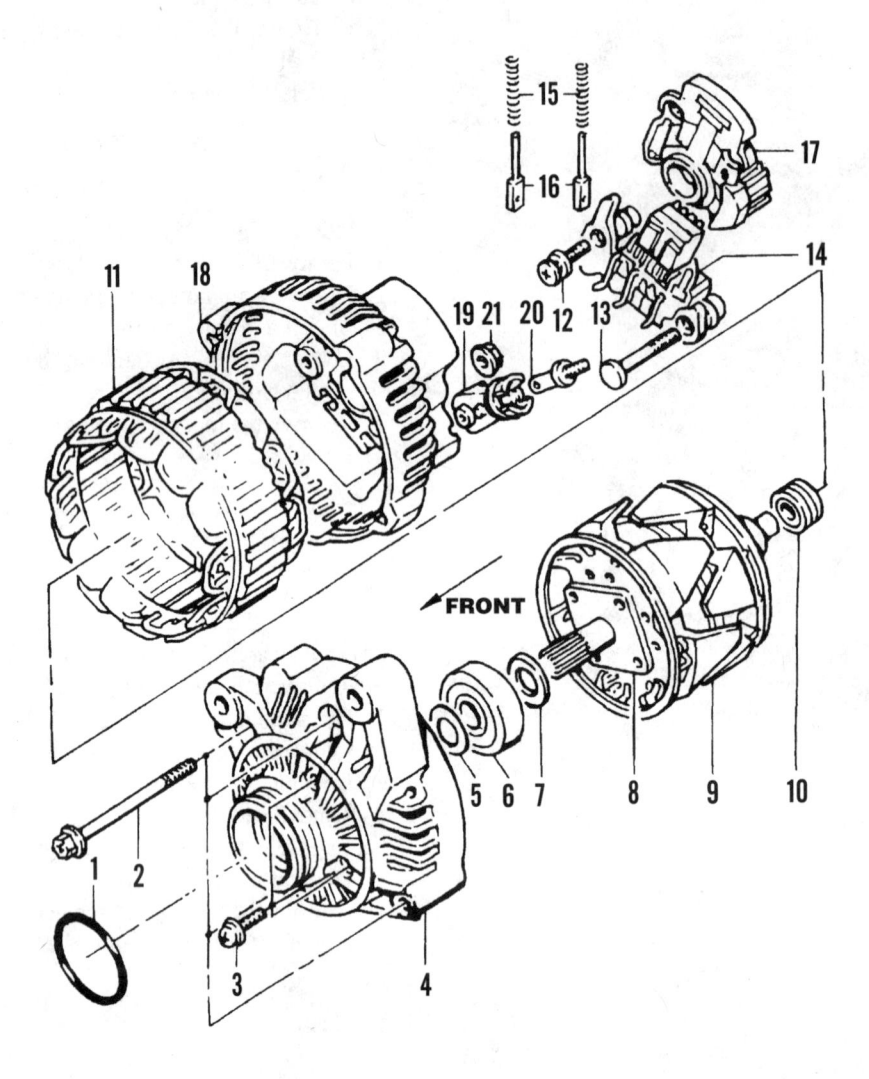

(19) ALTERNATOR

1. O-ring
2. Bolt
3. Rotor mounting screw
4. Front housing
5. Shim
6. Bearing
7. Shim
8. Retainer
9. Rotor
10. Bearing
11. Stator coil
12. Screw
13. Screw
14. Rectifier
15. Spring
16. Brush
17. Regulator
18. Rear housing
19. Holder
20. Terminal
21. Nut

ELECTRICAL SYSTEM

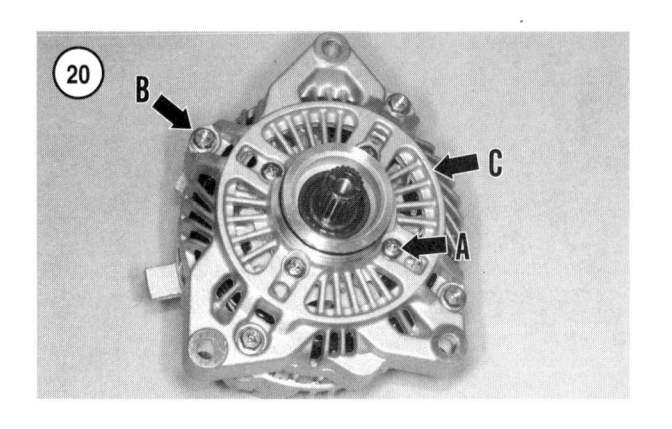

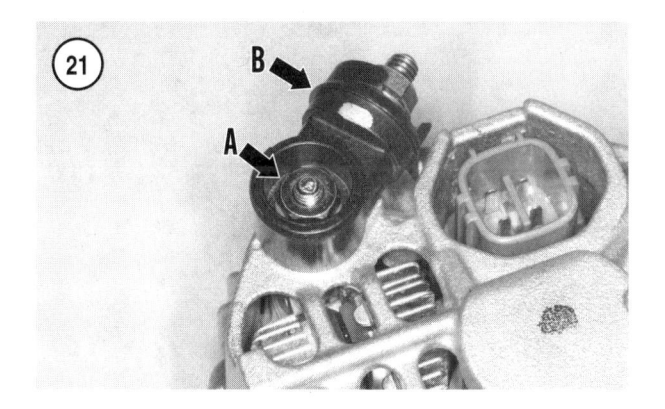

a. Check for continuity between the slip ring and rotor shaft (**Figure 25**); there should be no continuity. If there is continuity, the rotor winding is grounded to the rotor shaft.

b. Measure the rotor coil resistance between the slip rings (**Figure 26**). The standard reading is 2.5-2.9 ohms. A higher reading indicates a broken wire. A low reading indicates a shortened winding. Replace the rotor coil if the reading is incorrect.

4. Wipe the rotor bearings (B, **Figure 24** and **Figure 27**) with a clean, dry cloth, then turn them by hand. Each bear-

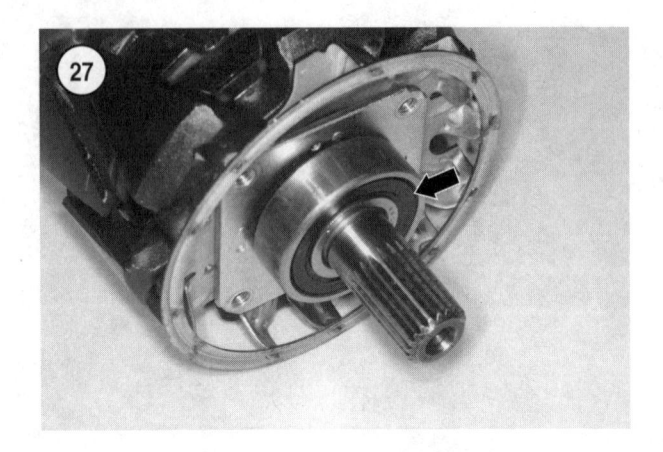

ing should turn smoothly and quietly. If the bearings turn roughly, they may be damaged or the sealed lubricant has started to dry. If necessary, replace both bearings as described under *Rotor Bearing Replacement* in this section.

5. Inspect the front and rear housings (**Figure 28**) for damage.
6. Inspect the seal (**Figure 29**) in the front housing for damage. Make sure a replacement seal is available before removing the seal. If a replacement seal is available, install the seal with its open side (**Figure 29**) facing out. Lubricate the seal lip with grease.
7. Inspect the rotor (**Figure 30**) for damage.

Rotor Bearing Replacement

Replace the front and rear bearings using a bearing puller and bearing driver as follows:

1. Disassemble the alternator as described in this section.
2. Before removing the bearings, make sure they are a tight fit on the shaft. If a bearing can be moved on the shaft, the shaft is probably worn.
3. Remove the front and rear bearings with a bearing puller. See B, **Figure 24** and **Figure 27**.
4. Examine the bearing mounting areas on the shaft. Check for galling, cracks and wear.
5. Press the new front (B, **Figure 24**) and rear (**Figure 27**) bearings onto the shaft.

Brush Inspection and Replacement

1. Inspect the brushes for contamination, damage and wear as follows:
   a. Visually inspect the brushes (A, **Figure 31**). There must be no oil or grease. Spray contaminated brushes with an electrical contact cleaner. Allow to dry.

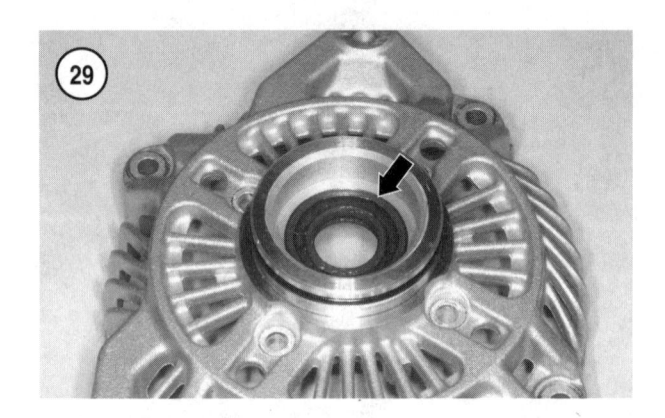

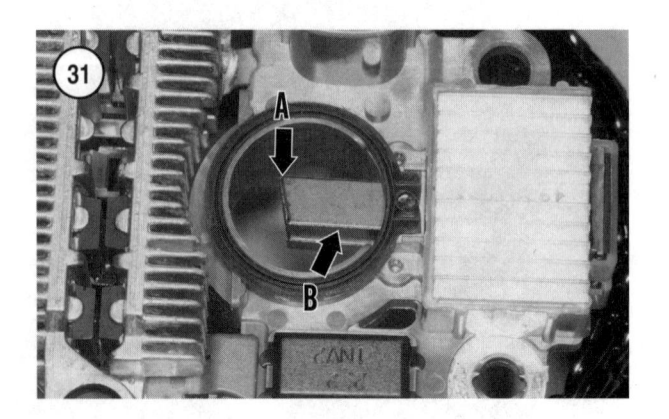

ELECTRICAL SYSTEM

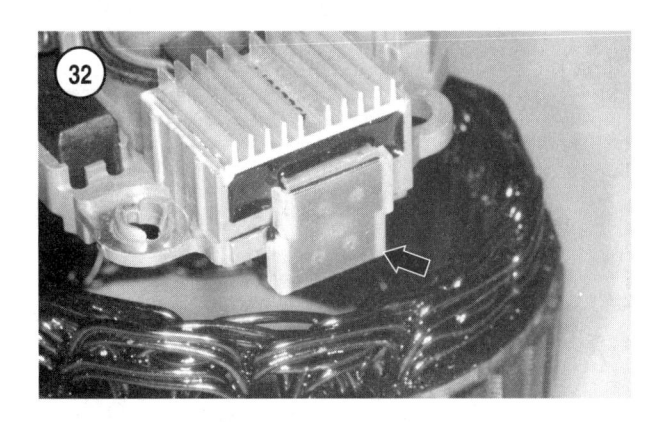

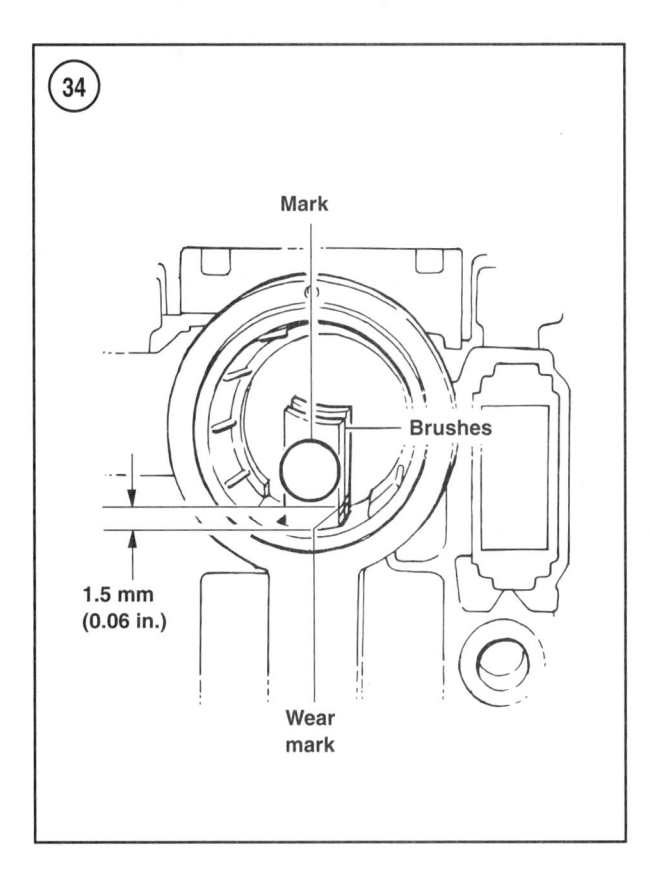

b. Check the brushes for any cracks, chips and other damage.

c. Replace the brushes if their exposed length is equal with or close to the wear lines scribed on the brushes (B, **Figure 31**).

d. If necessary, replace both brushes as a set as described in this procedure.

2. Remove the cap (**Figure 32**) and insulator (**Figure 33**) from around the brush holder.

3. Desolder the brushes and pull the brushes out of the brush holder.

4. Install the new brushes into the brush holder. The marked side of each brush must face toward the front housing as shown in **Figure 34**.

5. Position the brushes so that their wear lines are 1.5 mm (0.06 in.) above the bottom of the brush holder as shown in **Figure 34**.

NOTE
To avoid causing heat damage to the regulator/rectifier, use a 32 watt soldering iron set to a low temperature (180-200° C [356-392° F]).

CAUTION
When soldering the brushes, do not use an excessive amount of solder. Also, make sure the solder does not enter the brush holder or the brushes will not work correctly.

6. Align the solder end of the brushes with the brush holder surface.

7. Solder the brushes in place. Cut off any surplus wire, making sure to remove the cut ends from the alternator.

Stator Coil Testing

Desolder and remove the stator coil for testing.

CAUTION
To prevent excessive heat from damaging the diodes in the regulator/rectifier, work quickly when heating the terminals with a soldering gun. Hold the stator coil wires with needle nose pliers (between the soldered joint and regulator/rectifier) and use them as a heat sink.

1. Melt the three stator coil soldering joints (**Figure 35**) at the regulator/rectifier, and remove the stator coil assembly.

NOTE
Allow the stator coil to cool (20° C [68° F]) before testing.

2. Measure the resistance between the desoldered stator coil wires. Check all three wires. Compare resistance with specifications in **Table 3**. If the ohmmeter reads infinite resistance, the windings are shorted and the coil must be replaced.

3. To check the stator windings for grounds, connect the ohmmeter between one of the stator coil wires and to the silver part on the stator core. There should be no continuity. If there is continuity, the circuit is grounded and the stator coil must be replaced.

Regulator/Rectifier Testing

1. The stator coil must be desoldered from the regulator/rectifier before the diodes can be checked. Refer to *Stator Coil Testing* in this section.

2. Test the diodes in the rectifier (**Figure 36**) by performing the following:
   a. Check for continuity between the B and each P terminal. Reverse the connection and check the continuity in the opposite direction. There should be continuity in one direction, but no continuity in the other direction.
   b. Check for continuity between ground (E) and each P terminal. Reverse the connection and check the continuity in the opposite direction. There should be continuity in one direction, but no continuity in the other direction.
   c. All diodes should show continuity in one direction only. If any diode is faulty (same reading in both directions), replace the regulator/rectifier assembly (**Figure 37**).

NOTE
A shorted diode will test with low readings in both directions. A grounded diode will test with high readings in both directions.

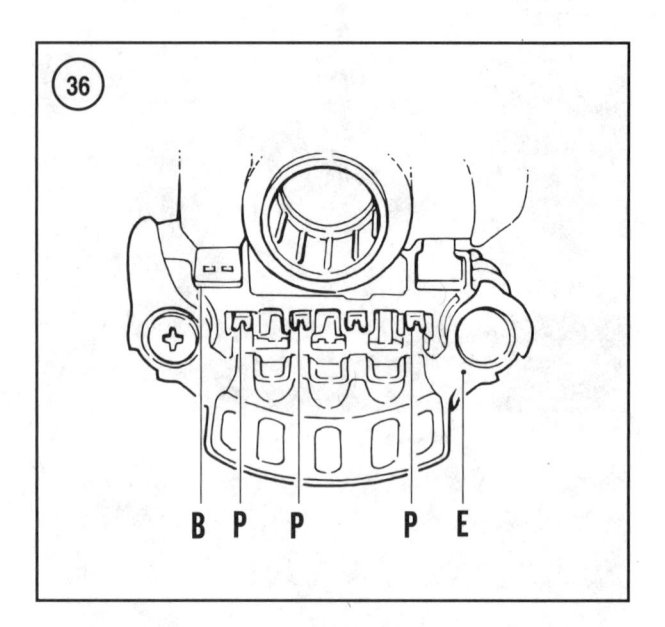

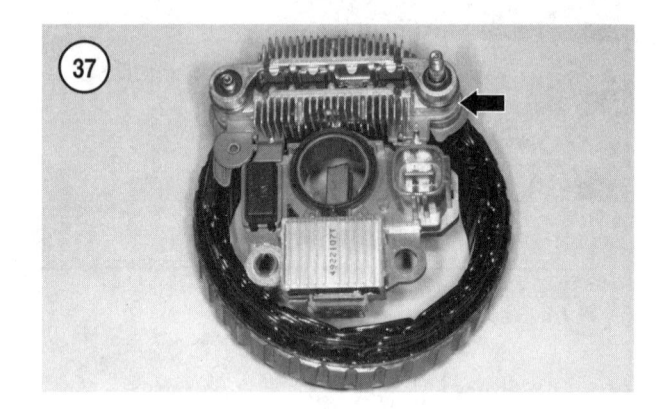

3. Solder the regulator/rectifier to the stator coil as described under *Reassembly* in this section.

Reassembly

NOTE
If shims were used, install them in their original mounting positions recorded during disassembly.

1. Use a high-temperature (300° C [572° F]) high lead solder and solder the stator and regulator/rectifier wires (**Figure 35**) onto the diode terminals.

2. Compress the brushes into their holders and hold them in place with a small rod (**Figure 38**).

3. Install the rear housing (A, **Figure 39**) over the stator coil and rod (B) installed in Step 2.

4. Install the three screws (**Figure 23**) and tighten securely.

ELECTRICAL SYSTEM

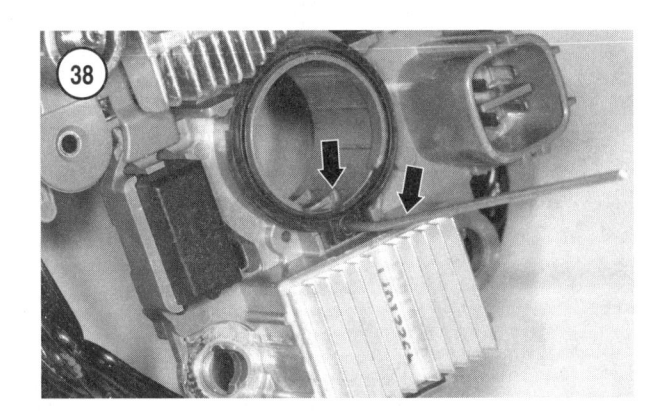

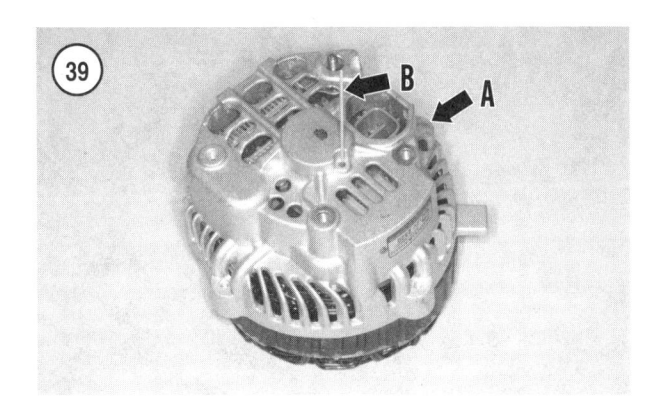

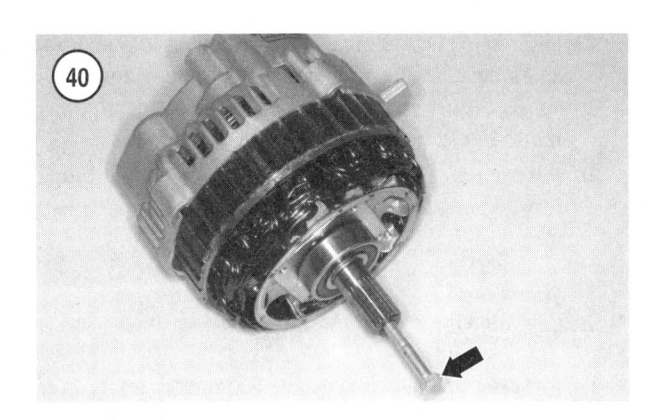

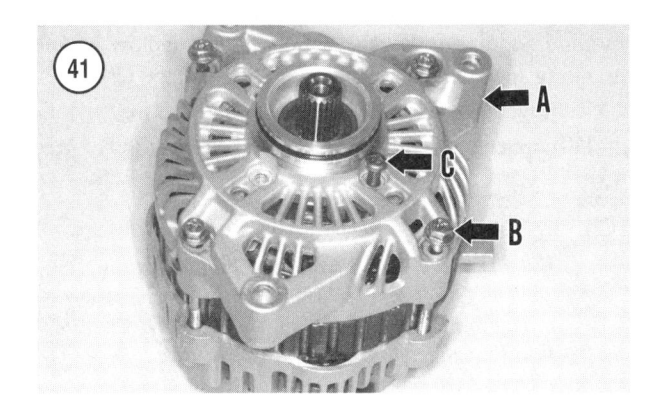

5. Install the terminal (B, **Figure 21**) and tighten the nut (A) to 8 N•m (71 in.-lb.). Install the cap.

6. Install the rotor in the direction shown in **Figure 22**. Thread a bolt into the rotor shaft (**Figure 40**) and tap the rear bearing into the rear housing. Remove the bolt.

7. Lubricate the front housing seal lip (**Figure 29**) with grease.

8. Align and install the front housing (A, **Figure 41**) over the rotor shaft and front bearing. If necessary, tap the front housing to seat the front bearing in the housing bore.

9. Remove the rod (B, **Figure 39**) to release the brushes.

10. Check that the bolt holes in the rear housing align with the gaps in the stator core.

11. Install the four through bolts (B, **Figure 41**), but do not tighten them.

12. Insert a long 6 × 1.00 mm bolt through the front housing and thread it into the retainer (C, **Figure 41**).

NOTE
Figure 42 shows the retainer with the front housing removed for clarity.

13. Pull the bolt to lift and align the retainer with the front housing, and install two of the rotor mounting screws. Remove the long bolt and install the remaining two screws. Tighten the screws (A, **Figure 20**) securely.

14. Tighten the alternator through bolts (B, **Figure 20**) securely.

IGNITION SYSTEM TROUBLESHOOTING

Peak Voltage Tests and Equipment

WARNING
High voltage is present during ignition system operation. Do not touch ignition com-

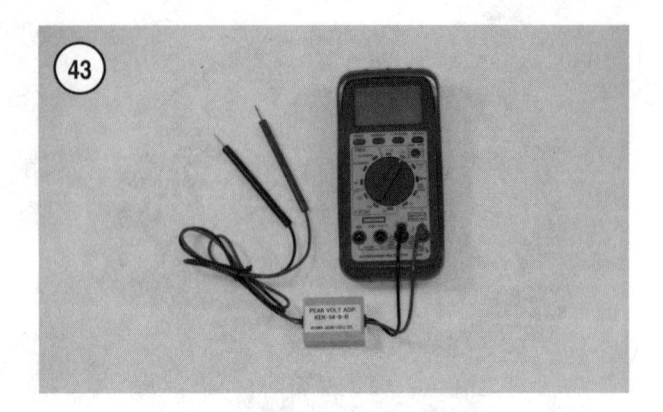

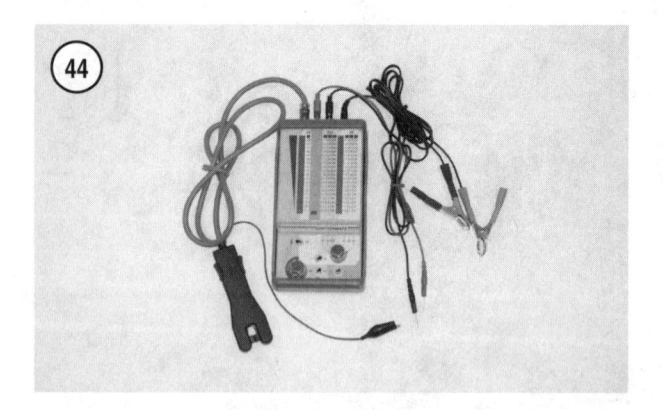

ponents, wires or test leads while cranking or running the engine.

Peak voltage tests check the voltage output of the ignition coil and pulse generator at normal cranking speed. These tests make it possible to accurately test the voltage output under operating conditions.

The peak voltage specifications (**Table 4**) are minimum values. If the measured voltage meets or exceeds the specification, the test results are satisfactory. In some cases, the voltage may greatly exceed the minimum specification.

In order to check peak voltage, use a peak voltage tester. One of the following testers, or an equivalent, can be used to perform peak voltage tests described in this section. Refer to the manufacturer's instructions when using these tools.

1. Peak voltage adapter (Honda part No. 07HGJ-0020100). Use this tool in combination with a digital multimeter with a minimum impedance of 10M ohms/DCV. A meter with a lower impedance will not display accurate measurements. See **Figure 43**.
2. Ignition Mate (Motion Pro part No. 08-0193). See **Figure 44**.

Preliminary Checks

Before testing the ignition system, make the following checks:
1. Make sure the battery is fully charged and in good condition. A weak battery causes a slow engine cranking speed.
2. Perform the *Spark Test* in Chapter Two. If there is a crisp, blue spark, the ignition system is working correctly. Test each spark plug:
   a. If there is no spark at all six spark plugs, check for a disconnected connector.
   b. If the spark test shows there is no spark at one coil group (No. 1 and No. 2, No. 3 and No. 4 or No. 5 and No. 6), switch the ignition coils and repeat the spark test. If the inoperative cylinders now have spark, the original ignition coil is defective. Replace it and retest.
   c. Also check for fouled or damaged spark plugs, loose spark plug caps or water in the spark plugs caps.

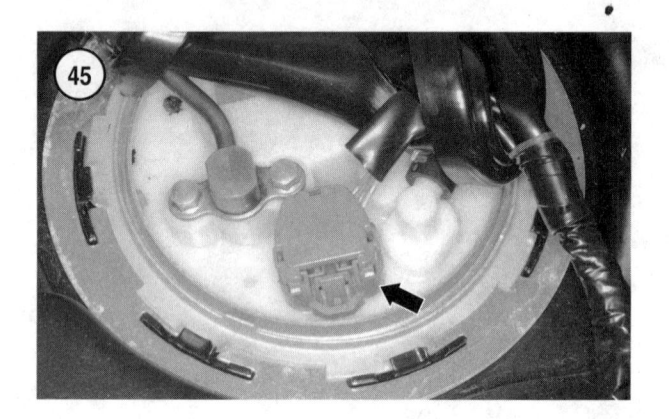

Ignition Coil Signal Peak Voltage Test

This test requires a peak voltage tester as described under *Peak Voltage Tests and Equipment* in this section.

1. Check the battery to make sure it is fully charged and in good condition. A weak battery causes a slow engine cranking speed and an inaccurate peak voltage test.
2. Remove the seat. Refer to *Seat* in Chapter Seventeen.
3. Disconnect the fuel pump 5-pin connector (**Figure 45**).

WARNING
Perform Step 3 to disable the fuel injection system. Otherwise, fuel will be injected into the cylinders when the engine is turned over, flooding the cylinders and creating explosive fuel vapors.

ELECTRICAL SYSTEM

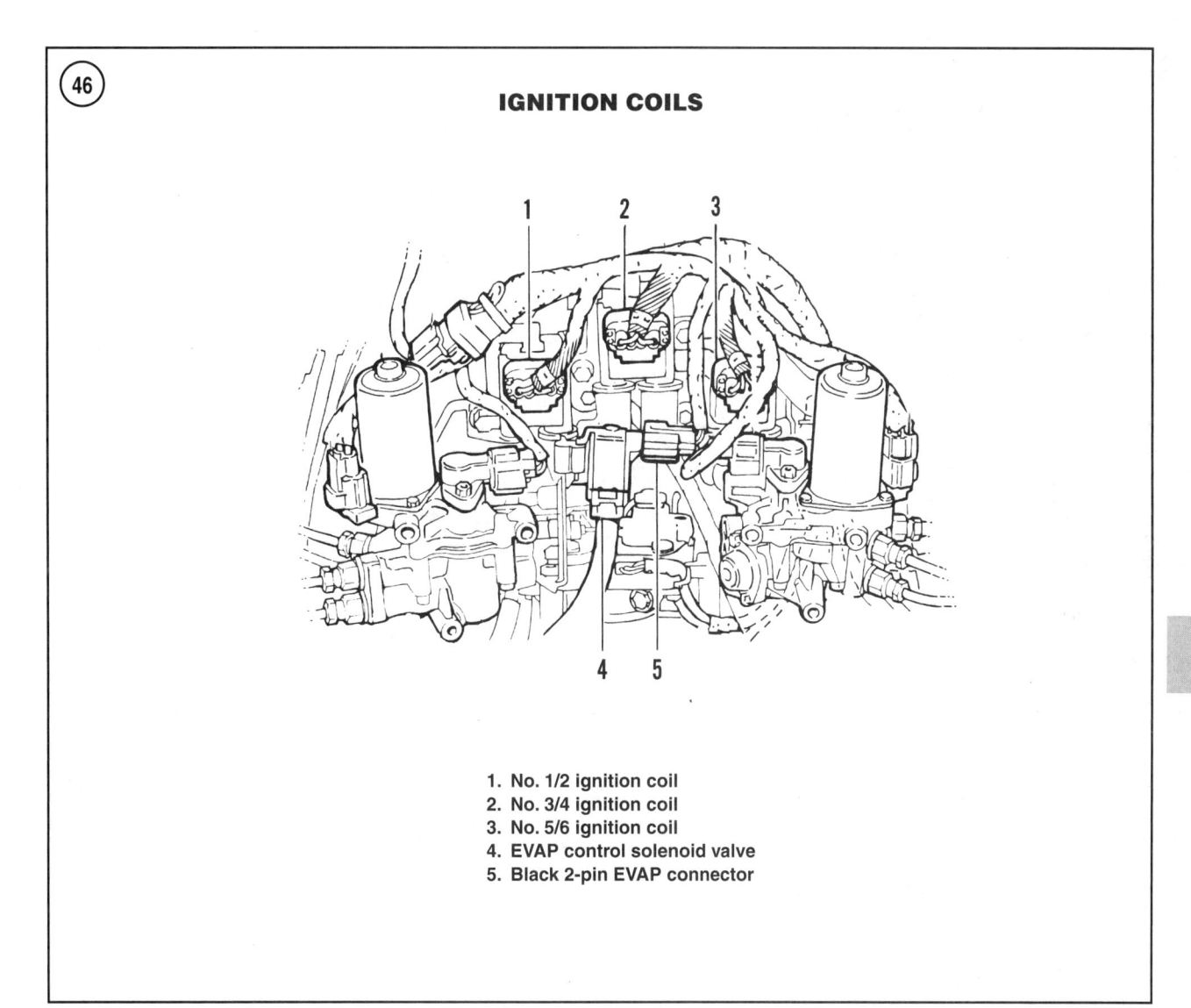

46 **IGNITION COILS**

1. No. 1/2 ignition coil
2. No. 3/4 ignition coil
3. No. 5/6 ignition coil
4. EVAP control solenoid valve
5. Black 2-pin EVAP connector

4. Check engine compression as described in Chapter Three. If the compression is low in one or more cylinders, the following test results will be inaccurate.

5. Check all the ignition component electrical connectors and wiring harnesses. Make sure the connectors are clean and properly connected.

6. Check that each spark plug cap is tightened securely.

7. Remove the center inner fairing. Refer to *Front Lower Fairings* in Chapter Seventeen.

8. If using the Honda peak voltage adapter, connect it to the multimeter as shown in **Figure 43**.

NOTE
*If using the Ignition Mate (**Figure 44**) tester or a similar peak voltage tester, follow the manufacturer's instructions for connecting the tester to the ignition coil.*

9. Disconnect the No. 1 and No. 2 ignition coil 4-pin connector (**Figure 46**). Connect the positive test lead to the yellow/white terminal and the negative test lead to ground.

10. Shift the transmission into neutral.

11. Turn the ignition switch on and the engine stop switch to RUN.

WARNING
High voltage is present during ignition system operation. Do not touch spark plugs, ignition components, connectors or test leads while cranking the engine.

47

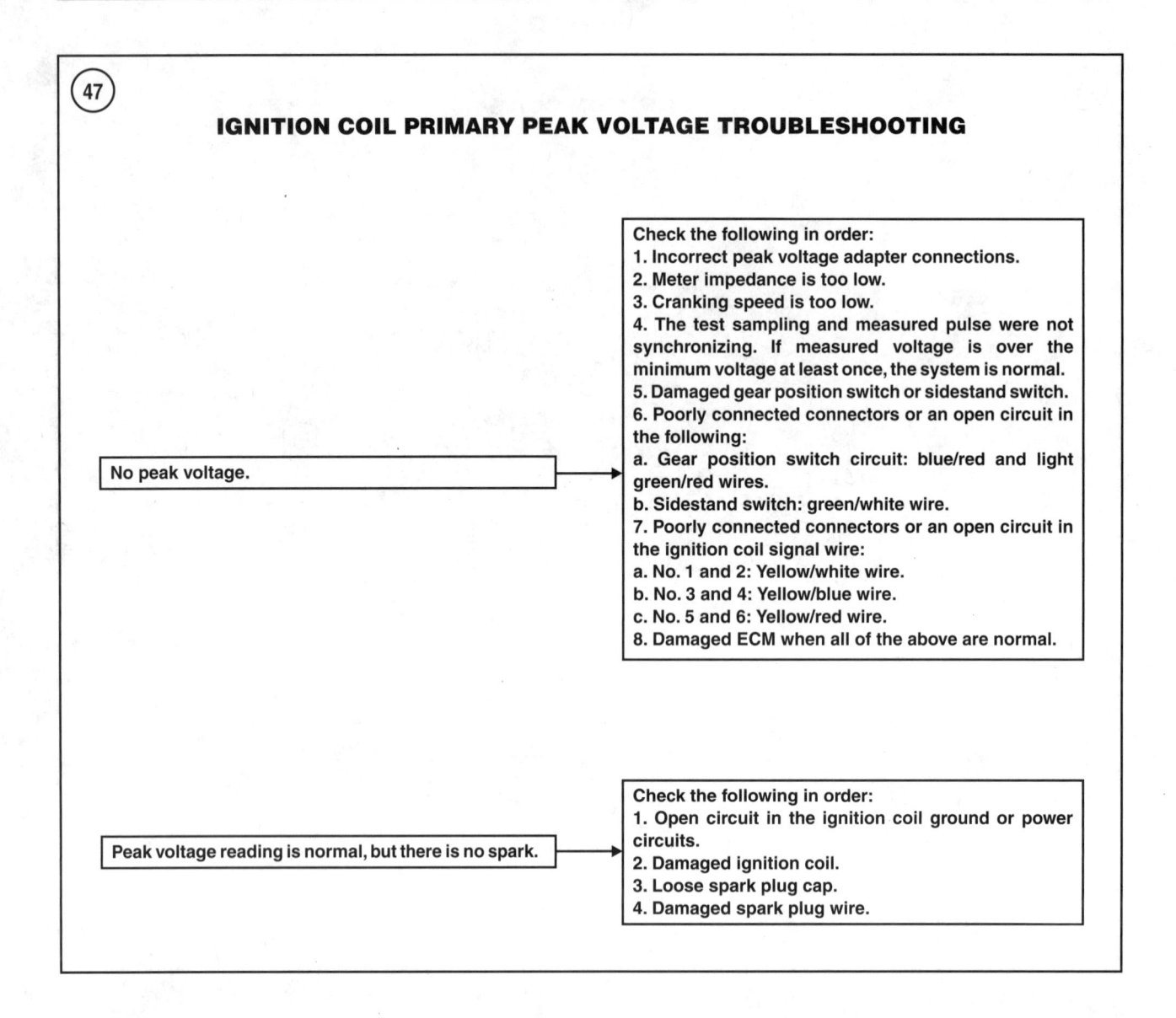

IGNITION COIL PRIMARY PEAK VOLTAGE TROUBLESHOOTING

No peak voltage. →

Check the following in order:
1. Incorrect peak voltage adapter connections.
2. Meter impedance is too low.
3. Cranking speed is too low.
4. The test sampling and measured pulse were not synchronizing. If measured voltage is over the minimum voltage at least once, the system is normal.
5. Damaged gear position switch or sidestand switch.
6. Poorly connected connectors or an open circuit in the following:
a. Gear position switch circuit: blue/red and light green/red wires.
b. Sidestand switch: green/white wire.
7. Poorly connected connectors or an open circuit in the ignition coil signal wire:
a. No. 1 and 2: Yellow/white wire.
b. No. 3 and 4: Yellow/blue wire.
c. No. 5 and 6: Yellow/red wire.
8. Damaged ECM when all of the above are normal.

Peak voltage reading is normal, but there is no spark. →

Check the following in order:
1. Open circuit in the ignition coil ground or power circuits.
2. Damaged ignition coil.
3. Loose spark plug cap.
4. Damaged spark plug wire.

---

12. Press the starter button while reading the meter. The meter should indicate a minimum peak voltage reading of 2.5-5.0 volts DC. If the reading is less than this measurement, refer to **Figure 47**.

NOTE
*All peak voltage specifications are **minimum voltages**. If the measured voltage meets or exceeds the specification, the test results are satisfactory. On some components, the voltage may greatly exceed the minimum specification.*

13. Reconnect the No. 1 and No. 2 ignition coil 4-pin connector.
14. Disconnect the No. 3 and No. 4 ignition coil 4-pin connector (**Figure 46**). Connect the positive test lead to the yellow/blue terminal and the negative test lead to ground. Repeat the test in Steps 10-12.
15. Reconnect the No. 3 and No. 4 ignition coil 4-pin connector.
16. Disconnect the No. 5 and No. 6 ignition coil 4-pin connector (**Figure 46**). Connect the positive test lead to the yellow/red terminal and the negative test lead to ground. Repeat the test in Steps 10-12.
17. Reconnect the No. 5 and No. 6 ignition coil 4-pin connector.
18. Reverse Steps 2-8 to complete installation.

Ignition Pulse Generator Peak Voltage Test

This test requires a peak voltage tester as described under *Peak Voltage Tests and Equipment* in this section.

ELECTRICAL SYSTEM

48

1. Check the battery to make sure it is fully charged and in good condition. A weak battery causes a slow engine cranking speed and an inaccurate peak voltage test.
2. Remove the seat. Refer to *Seat* in Chapter Seventeen.
3. Disconnect the fuel pump 5-pin connector (**Figure 45**).

WARNING
Step 3 must be performed to disable the fuel injection system. Otherwise, fuel will be injected into the cylinders when the engine is turned over, flooding the cylinders and creating explosive fuel vapors.

4. Check engine compression as described in Chapter Three. If the compression is low in one or more cylinders, the following test results will be inaccurate.
5. Check all the ignition component electrical connectors and wiring harnesses. Make sure the connectors are clean and properly connected.
6A. Connect the test harness to the black and gray ECM harness connectors as described in Chapter Eight under *ECM Test Harness*.
6B. If the Honda test harness is unavailable, refer to Chapter Eight under *ECM Test Harness* to identify the ECM pin terminals called out in the following steps.
7. If using the Honda peak voltage adapter, connect it to the multimeter as shown in **Figure 43**.

NOTE
*If using the Ignition Mate (**Figure 44**) tester or a similar peak voltage tester, follow the manufacturer's instructions for connecting the tester to the ignition coil.*

8. Connect the test leads between pins No. 41 (+) and No. 21 (-) on the breakout box.
9. Shift the transmission into neutral.
10. Turn the ignition switch on and the engine stop switch to RUN.

WARNING
High voltage is present during ignition system operation. Do not touch spark plugs, ignition components, connectors or test leads while cranking the engine.

11. Press the starter button while reading the meter. The meter should indicate a minimum peak voltage reading of 0.7 volts DC. If the reading is less than this measurement, continue with Step 12.

NOTE
*All peak voltage specifications are **minimum voltages**. If the measured voltage meets or exceeds the specification, the test results are satisfactory. On some components, the voltage may greatly exceed the minimum specification.*

12. Measure the peak voltage at the ignition pulse generator connector:
   a. Turn the ignition switch off.
   b. Remove the air filter housing. Refer to *Air Filter Hosing* in Chapter Eight.
   c. Disconnect the ignition pulse generator 2-pin red connector (**Figure 48**).
   d. Connect the positive test lead to the yellow (+) connector terminal and the negative test lead to the white/yellow (–) connector terminal.

NOTE
Connect the test leads to the ignition pulse generator side connector terminals, not to the wire harness connector terminals.

   e. Turn the ignition switch on and the engine stop switch to RUN.
   f. Press the starter button while reading the meter. The meter should indicate a minimum peak voltage reading of 0.7 volts DC. If the reading is now correct, check the yellow wire between the ignition pulse generator and the gray 22-pin ECM connector for an open or short circuit. Then check the white/yellow wire between the ignition pulse generator and the black 22-pin ECM connector. At the same time, check the connectors for loose terminals or contamination. If the reading is still incorrect, refer to **Figure 49**.

13. Install the previously removed parts to complete assembly.

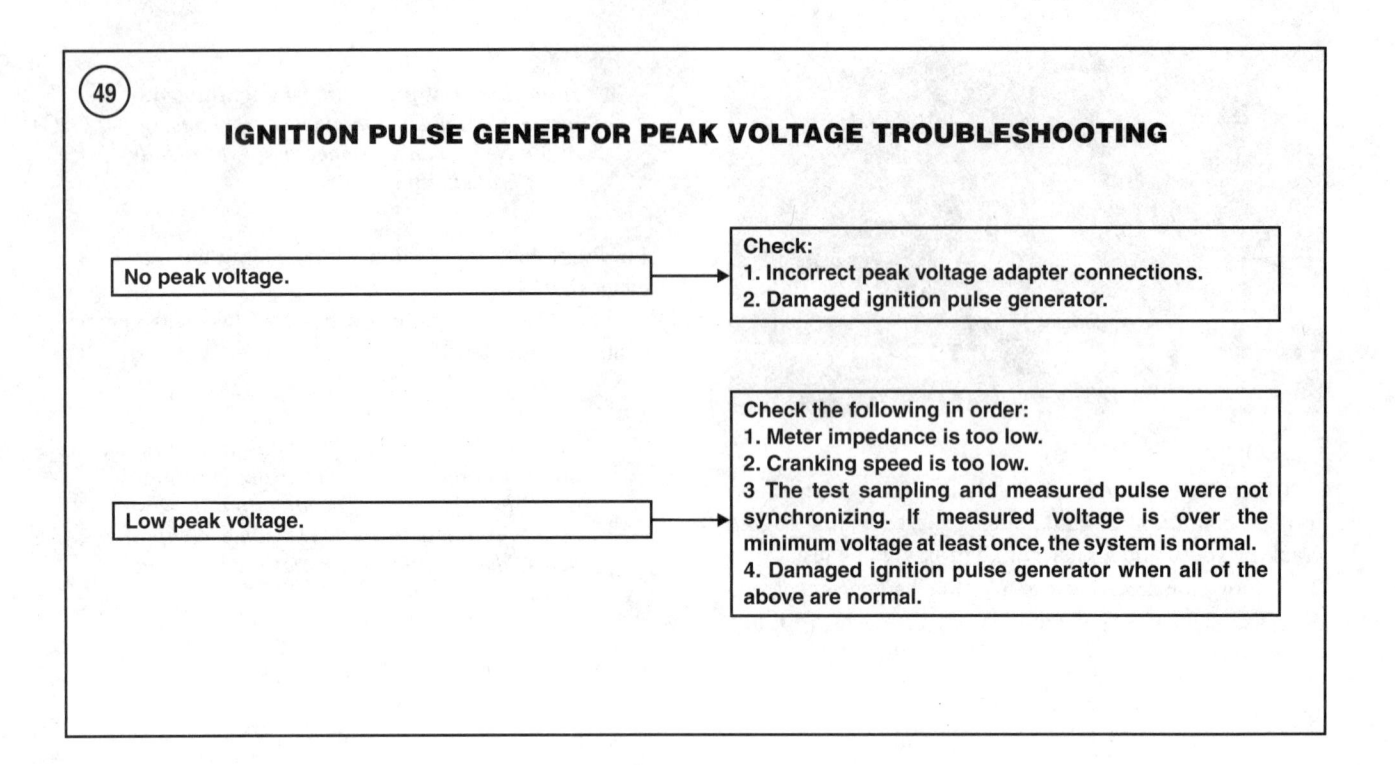

⑭

IGNITION PULSE GENERTOR PEAK VOLTAGE TROUBLESHOOTING

No peak voltage. → Check:
1. Incorrect peak voltage adapter connections.
2. Damaged ignition pulse generator.

Low peak voltage. → Check the following in order:
1. Meter impedance is too low.
2. Cranking speed is too low.
3 The test sampling and measured pulse were not synchronizing. If measured voltage is over the minimum voltage at least once, the system is normal.
4. Damaged ignition pulse generator when all of the above are normal.

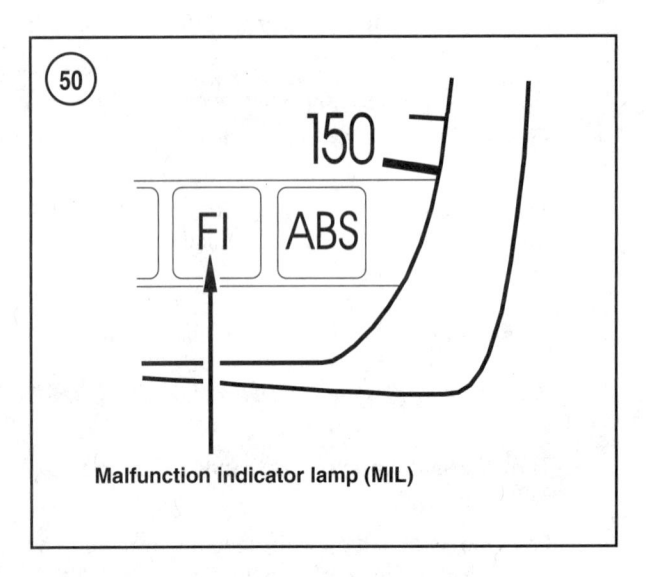

⑮

Malfunction indicator lamp (MIL)

IGNITION COIL

Power/Ground Circuit Test

If the peak voltage readings are correct, but there is no spark at one group of spark plugs (1 and 2, 3 and 4, or 5 and 6), perform the following:
1. Remove the center inner fairing. Refer to *Front Lower Fairings* in Chapter Seventeen.
2. Turn the ignition switch off.
3. Disconnect the 4-pin connector at the affected ignition coil (**Figure 46**).
4. Turn the ignition switch on and observe the MIL light (**Figure 50**):
    a. If the MIL comes on for a few seconds and then turns off, go to Step 5.
    b. If the MIL stays on, check the PGM-FI ignition relay circuits as described in this chapter.
5. Turn the ignition switch on and measure for battery voltage between the 4-pin ignition coil harness connector black/yellow terminal (+) and ground (−).
    a. Yes. The input circuit is working correctly. Go to Step 6.
    b. No. Check the black/yellow wire between the ignition coil and PGM-FI ignition relay for an open circuit.
6. Turn the ignition switch off and check for continuity between the 4-pin ignition coil harness connector green terminal and ground.
    a. Yes. The ground circuit is working correctly.
    b. No. Check the green wire between the ignition coil and the engine ground terminal for an open circuit.
7. Reverse Steps 1-3 to complete installation.

Ignition Coil Replacement

Refer to **Figure 51**.

ELECTRICAL SYSTEM

4. Disconnect the 4-pin connectors from each ignition coil.
5. Disconnect the spark plug wires from the ignition coils.
6. Unbolt and remove the ignition coil from its mounting bracket.
7. Installation is the reverse of removal.

Spark Plug Secondary Wire Replacement

Refer to **Figure 46** and **Figure 51**.

1. Remove the center inner fairing. Refer to *Front Lower Fairings* in Chapter Seventeen.
2. Turn the ignition switch off.
3. Remove the EVAP purge control solenoid valve from its mounting stay.
4. Remove the No. 3 and No. 4 ignition coil.
5. Remove the wire clamp from the mounting bracket and remove the bracket from the No. 2 and No. 5 spark plugs.
6. Remove the cylinder head side covers. Refer to *Cylinder Head Covers* in Chapter Four.

NOTE
Sketch the spark plug routing on a piece of paper before disconnecting the spark plug caps and removing the spark plug wires.

7. Disconnect the spark plug caps from the spark plugs.
8. Remove the front lower inner cover, wire clamps and spark plug wires.
9. Installation is the reverse of removal.

IGNITION PULSE GENERATOR

Removal/Installation

Refer to **Figure 52**.

1. Remove the front lower fairing. Refer to *Front Lower Fairings* in Chapter Seventeen.
2. Remove the air filter housing. Refer to *Air Filter Housing* in Chapter Eight.
3. Disconnect the ignition pulse generator 2-pin red connector (**Figure 48**).
4. Drain the engine oil. Refer to *Engine Oil and Filter* in Chapter Three.
5. Remove the bolts and the front crankcase cover (**Figure 52**).
6. Remove the bolts and wire retainer.
7. Remove the grommet from the cover groove and remove the pulse generator.
8. Remove the dowel pins and gasket.

FRONT CRANKCASE COVER AND IGNITION PULSE GENERATOR

1. Timing hole cap
2. O-ring
3. Bolt
4. Front crankcase cover
5. Gasket
6. Dowel pins
7. Ignition pulse generator
8. Wire retainer
9. Bolt

---

1. Remove the center inner fairing. Refer to *Front Lower Fairings* in Chapter Seventeen.
2. Turn the ignition switch off.
3. Remove the EVAP purge control solenoid valve from its mounting stay (**Figure 46**).

9. Clean the cover and crankcase gasket surfaces of all sealer and gasket residue.
10. Install the ignition pulse generator into the cover.
11. Install the grommet into the cover groove. Position the wire retainer over the wire.
12. Apply a medium strength threadlock onto the bolt threads and tighten to 12 N•m (106 in.-lb.).
13. Apply Yamabond No. 4 (or equivalent semi-drying sealer) to the two areas where the crankcase halves mate together (A, **Figure 53**).
14. Install the two dowel pins (B, **Figure 53**) and a new gasket.
15. Install the front crankcase cover and tighten its mounting bolts to 12 N•m (106 in.-lb.).
16. Refill the engine with oil. Refer to *Engine Oil and Filter* in Chapter Three.
17. Reverse Steps 1-3 to complete installation.

IGNITION PULSE GENERATOR ROTOR

Removal/Installation

Remove and install the ignition pulse generator rotor as described in Chapter Four under *Cam Chains and Timing Sprocket*. Disregard the information for camshaft removal.

PGM-FI IGNITION RELAY

A single 4-terminal electrical relay is used to control the PGM-FI ignition relay. The relay is located in the main relay box underneath the seat (**Figure 54**). Refer to *Relay Box* in this chapter for more information on identifying and testing the relays.

Relay and Circuit Testing

If the MIL (**Figure 50**) stays on and there is no input voltage at the ignition coil, test the ignition relay circuit as follows. Refer to the wiring diagrams at the end of the manuual to identify the ignition system relay circuit referred to in this procedure.

NOTE
*Refer to **Ignition Coil** in this chapter for information on checking ignition coil circuit input voltage.*

1. Remove the seat. Refer to *Seat* in Chapter Seventeen.
2. Remove the relay box cover screws and cover.
3. Replace the PGM-FI ignition relay (15, **Figure 54**) with a known good 4-terminal relay. Turn the ignition switch on and check the MIL (**Figure 50**):
   a. If the MIL comes on for a few seconds and then turns off, the original relay was faulty. Install a new 4-terminal relay and repeat the test.
   b. If the MIL came on and stayed on, turn the ignition switch off and continue with Step 4.

NOTE
*The meter test connections called out in Steps 4-7 are made at the PGM-FI ignition relay connector terminals inside the relay box (**Figure 55**).*

4. Turn the ignition switch off and remove the PGM-FI ignition relay (15, **Figure 54**). Turn the ignition switch on and check for battery voltage between the relay black terminal (+) and ground (–).
   a. No: Go to Step 5.
   b. Yes: Go to Step 6.
5. Turn the ignition switch off. Remove the right side cover. Refer to *Side Covers* in Chapter Seventeen. Remove the fuse box cover and check the No. 17 fuse. See *Fuses* in this chapter.
   a. If the fuse is good, check the black wire installed between the PGM-FI ignition relay and the fuse box for an open circuit.
   b. If the fuse is blown, install a new fuse and the PGM-FI ignition relay. Turn the ignition switch on. If the fuse blows, check the black wire and other related wires leading from the relay for a short circuit.
6. Turn the ignition switch on and check for battery voltage between the PGM-FI ignition relay black/white terminal (+) and ground (–).
   a. Yes: Go to Step 7.
   b. No: Check the black/white wire connection at the engine stop switch.

ELECTRICAL SYSTEM

54 RELAY BOX

```
[ 1 ][ 2 ][ 3 ][ 4 ][Blank][ 6 ][ 7 ]
[ 8 ][ 9 ][10 ][11 ][ 12 ][13 ][14 ]
[15 ][16 ][17 ][18 ][ 19 ][20 ][21 ]
```
------- 4-Terminal relays ------- | -- 5-Terminal relays --

55

Fuel pump relay connector — Blk/yel, Brn, Brn/blk, Blk/yel

Relay box

```
[ 1 ][ 2 ][ 3 ][||=][Blank]
[ 8 ][ 9 ][10 ][11 ][ 12 ]
[||=][16 ][17 ][18 ][ 19 ]
```

PGM-FI ignition relay

Blk/yel, Blk, Red/wht, Blk/wht

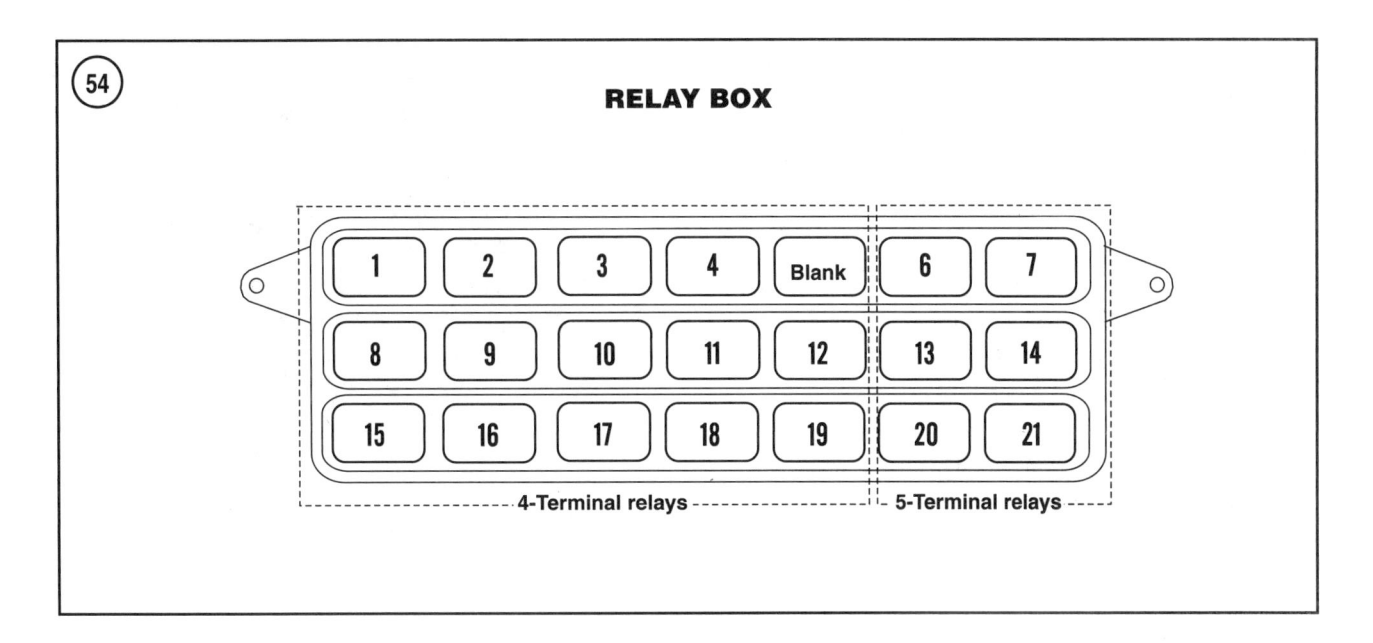

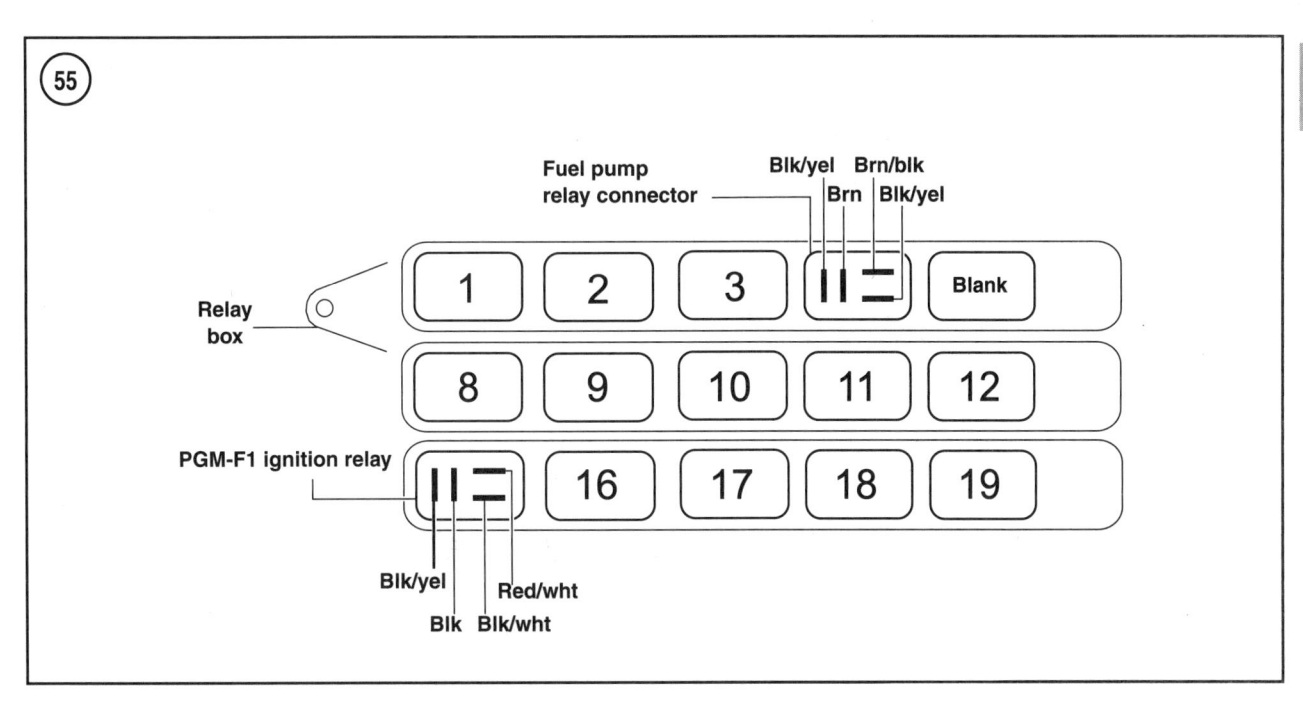

7. Turn the ignition switch off and check for continuity between the red/white wire terminal and ground.
   a. Yes. The circuit is working correctly. Check the circuit's related connectors for loose pins or contamination.
   b. No. Test the bank angle sensor as described in the following section.
8. Install the PGM-FI ignition relay (15, **Figure 54**).
9. Install the relay box cover and secure with the mounting screws.
10. Install the seat. Refer to *Seat* in Chapter Seventeen.

BANK ANGLE SENSOR

Testing/Replacement

1. Remove the front fairing. Refer to *Front Fairing and Fairing Molding* in Chapter Seventeen.
2. Locate the green 3-pin bank angle sensor connector (**Figure 56**) disconnected during fairing removal.

3. Turn the ignition switch off and check for continuity between the green wire terminal in the green 3-pin bank angle sensor harness side terminal (**Figure 56**) and ground.
   a. Yes: Go to Step 4.
   b. No: Check for an open circuit in the green wire between the bank angle sensor and its ground connection.
4. Turn the ignition switch on and check for voltage between the green 3-pin bank angle sensor red/white (+) harness side terminal and ground (–).
   a. Yes: Go to Step 5.
   b. No. Check for an open circuit in the red/white wire.
5. Turn the ignition switch on and check for battery voltage between the green 3-pin bank angle sensor white (+) harness side terminal and ground (–).
   a. Yes: Go to Step 6.
   b. No. Check for an open circuit in the white wire.
6. Remove the mounting screws (A, **Figure 57**) and the bank angle sensor (B) from the right headlight on the front fairing.
7. Connect the green 3-pin connector to the bank angle sensor.
8. Connect the positive voltmeter lead to the red/white terminal (+) and the negative lead to the green terminal (–).
9. Hold the bank angle sensor in its normal operating position (horizontal as shown in B, **Figure 57**) and turn the ignition switch on. The voltmeter should read 0-1 volt. Then turn the bank angle sensor 43° to either the left or right side. At this angle, the voltmeter should read battery voltage. If the bank angle sensor failed to operate as described, replace the bank angle sensor.

NOTE
When repeating this test, the ignition switch must be turned off, then back to on.

10. Install the bank angle sensor as shown in B, **Figure 57** and tighten the mounting screws to 2 N•m (18 in.-lb.).
11. Install the front fairing (Chapter Seventeen).

ENGINE CONTROL MODULE

Refer to Chapter Eight.

IGNITION TIMING

Refer to Chapter Three.

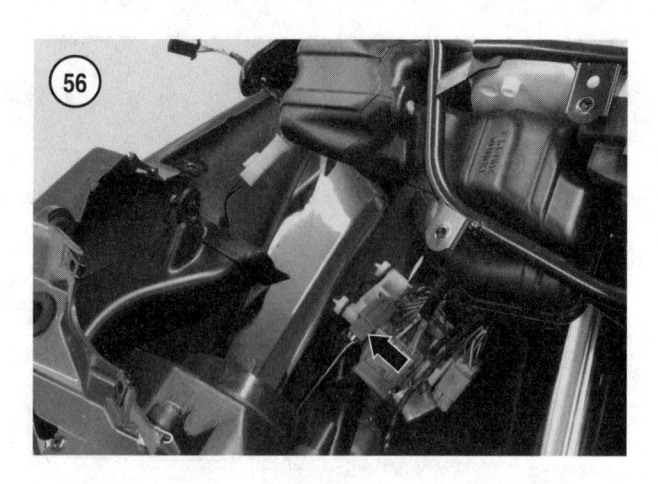

STARTER/REVERSE SYSTEM TROUBLESHOOTING

Starter Operates Slowly

1. Test the battery as described in this chapter.
2. Check for the following:
   a. Check for loose or corroded battery terminals.
   b. Check for loose or corroded battery ground cable.
   c. Loose starter cable.
3. The starter may be faulty. Remove, disassemble and bench test the starter as described in this chapter.

Starter Relay Switch Clicks But Engine Does Not Turn

Check for the following:
1. Crankshaft cannot turn because of mechanical failure.
2. Damaged starter idle gear.
3. Damaged starter drive or driven gear.

Starter Operates But Engine Does Not Turn Over

Check for a damaged starter clutch (Chapter Five).

Starter/Reverse Motor Does Not Operate

1. Check the following:
   a. Test the battery as described in this chapter.
   b. Make sure the reverse shift switch is off when attempting to operate the starter for engine starting.
   c. Check for a blown No. 8, No. 10 or No. 12 fuse as described in this chapter.
2. Check for loose or contaminated connections at the battery.

ELECTRICAL SYSTEM

57

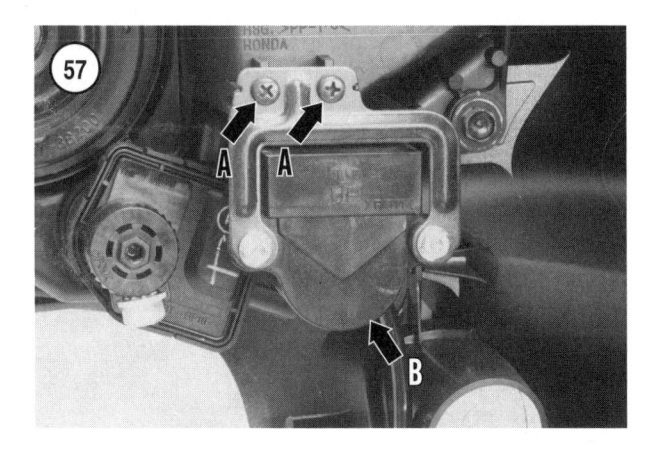

3. Check for a loose or contaminated cable terminal at starter relay A.
4. Check for a loose or contaminated cable terminal at starter relay B.
5. Check for an open circuit in the starter relay switch cable between starter relay A and B.
6. Turn the ignition switch on and the engine stop switch to RUN. Push the starter/reverse switch. Listen for a click at starter relay A and starter relay B.
   a. No. Test starter relay A and/or starter relay B circuits as described in this chapter.
   b. Yes. Go to Step 7.
7. Remove the fuel tank. Refer to *Fuel Tank* in Chapter Eight. Check for loose or contaminated connections at the starter. Then check the starter cable for an open circuit between the starter and starter relay B.
8. Momentarily connect a thick cable from the positive battery cable directly to the starter terminal. The starter should operate.
   a. No: Starter is faulty.
   b. Yes: Starter relay B is faulty.

Reverse Shift Actuator Will Not Turn to the Reverse Position When the Reverse Shift Switch is Set to On

1. Check for the following:
   a. Blown No. 23 fuse.
   b. The neutral indicator light comes on when the ignition switch is on and the transmission is in neutral.
   c. The motorcycle is at a full stop.
   d. The reverse shift actuator cables are properly adjusted. See *Reverse Shift Actuator Cable Adjustment* in this chapter.
2. Shift the transmission into neutral. Turn the ignition switch on. Set the reverse shift switch to on. The reverse actuator should operate.
   a. No. Go to Step 3.
   b. Yes. Perform *Test A* in this section.
3. The reverse indicator light should blink when performing Step 2.
   a. Yes. Go to Step 4.
   b. No. Perform *Test B* in this section.
4. Check the reverse shift relays system as described under *Reverse Shift Relays* in this chapter. The relays and system should operate.
   a. Yes. Go to Step 5.
   b. No. Replace the faulty relay(s) or repair the damaged wire as described under *Reverse Shift Relays*.
5. Perform test No. 40 under *Cruise/Reverse Control Module, Main Wiring Harness Check* in Chapter Sixteen.
   a. If the test is correct, go to Step 6.
   b. If not, check for an open circuit in the blue/black wire between the cruise/reverse control module and reverse shift relay 1. Refer to the wiring diagrams at the end of this manual.
6. Perform test No. 34 under *Cruise/Reverse Control Module, Main Wiring Harness Check* in Chapter Sixteen.
   a. If the test is correct, go to Step 7.
   b. If not, check for an open circuit in the blue/white wire between the cruise/reverse control module and reverse shift relay 2. Then check for an open circuit in the red wire between reverse shift relay 1 and reverse shift relay 2. Refer to the wiring diagrams at the end of this manual.
7. Perform test No. 27 under *Cruise/Reverse Control Module, Main Wiring Harness Check* in Chapter Sixteen.
   a. If the test is correct, go to Step 8.
   b. If not, check for an open circuit in the yellow/white wire between the cruise/reverse control module and reverse shift relay 3. Refer to the wiring diagrams at the end of this manual.
8. Perform test No. 40 under *Cruise/Reverse Control Module, Main Wiring Harness Check* in Chapter Sixteen.
   a. If the test is correct, go to *Test C* in this section.
   b. If not, check for an open circuit in the pink wire between the cruise/reverse control module and reverse shift relay 3. Refer to the wiring diagrams at the end of this manual.

Test A

1. Confirm that the reverse indicator light is on.
   a. Yes. This was probably an intermittent failure. The system is normal.
   b. No. The indicator stays off. Perform the *Reverse Shift Actuator Operates, But the Reverse Indicator Stays Off* in this section.
   c. No. The indicator blinks. Go to Step 2.

2. Perform test No. 25 under *Cruise/Reverse Control Module, Main Wiring Harness Check* in Chapter Sixteen.
    a. If the test is correct, go to Step 3.
    b. If not, check for an open circuit in the white/blue wire between the cruise/reverse control module and cruise switch. Refer to the wiring diagrams at the end of this manual.
3. Perform test No. 17 under *Cruise/Reverse Control Module, Main Wiring Harness Check* in Chapter Sixteen.
    a. If the test was incorrect, check for an open circuit in the green/orange wire between the cruise/reverse control module and reverse switch. Refer to the wiring diagrams at the end of this manual.
    b. If the test was correct, replace the cruise/reverse control module. Refer to *Cruise/Reverse Control Module* in Chapter Sixteen.

Test B

1. Perform test No. 47 under *Cruise/Reverse Control Module, Main Wiring Harness Check* in Chapter Sixteen.
    a. If the test was correct, go to Step 2.
    b. If not, check for an open circuit in the light/green wire between the cruise/reverse control module and the gear position switch. Refer to the wiring diagrams at the end of this manual.
2. Perform test No. 23 under *Cruise/Reverse Control Module, Main Wiring Harness Check* in Chapter Sixteen.
    a. If the test was correct, go to Step 3.
    b. If not, check for an open circuit in the yellow/white wire between the cruise/reverse control module and the right handlebar switch. Refer to the wiring diagrams at the end of this manual.
3. Perform test No. 20 and No. 52 under *Cruise/Reverse Control Module, Main Wiring Harness Check* in Chapter Sixteen.
    a. If the test was incorrect, check for an open circuit in the green wire between the cruise/reverse control module and the ground terminal. Refer to the wiring diagrams at the end of this manual.
    a. If the test was correct, replace the cruise/reverse control module. Refer to *Cruise/Reverse Control Module* in Chapter Sixteen.

Test C

1. Perform test No. 8 under *Cruise/Reverse Control Module, Main Wiring Harness Check* in Chapter Sixteen.
    a. If the test was correct, go to Step 2.
    b. If not, check for an open circuit in the green wire between the cruise/reverse control module and the ground terminal. Refer to the wiring diagrams at the end of this manual.
2. Check the reverse shift actuator motor as described under *Reverse Shift Actuator* in this chapter.
    a. If the test was incorrect, replace the reverse shift actuator motor.
    b. If the test was correct, replace the cruise/reverse control module. Refer to *Cruise/Reverse Control Module* in Chapter Sixteen.

Reverse Shift Actuator Operates, But the Reverse Indicator Stays Off

1. Check the speed limiter fuse.
    a. If the fuse is blown, replace it.
    b. If the fuse is good, go to Step 2.
2. Perform test No. 24 under *Cruise/Reverse Control Module, Main Wiring Harness Check* in Chapter Sixteen.
    a. If the test was correct, go to Step 3.
    b. If not, check for an open circuit in the white/red wire between the cruise/reverse control module and the combination meter. Check the connectors for contamination or loose terminals. If the wiring harness and connectors are good, the combination meter is damaged. See *Combination Meter* in this chapter. Refer to the wiring diagrams at the end of this manual.
3. Perform test No. 19 under *Cruise/Reverse Control Module, Main Wiring Harness Check* in Chapter Sixteen.
    a. If the test was incorrect, check for an open circuit in the yellow wire between the cruise/reverse control module and the speed limiter fuse at the fuse box. Refer to the wiring diagrams at the end of this manual.
    b. If the test was correct, replace the cruise/reverse control module. Refer to *Cruise/Reverse Control Module* in Chapter Sixteen.

Reverse Shift Actuator Will Not Rotate to Neutral When the Reverse Switch is Turned Off

Refer to *Reverse Shift Actuator* in this chapter to access the reverse shift actuator when checking it in its neutral and reverse positions.

1. Turn the ignition switch off and push the reverse shift switch to its off position.
2. Turn the ignition switch on. Check that the neutral indicator is on and the reverse shift actuator is in its neutral position.
    a. If both work correctly, the system is working correctly.

ELECTRICAL SYSTEM

b. If not, go to Step 3.

3. Turn the ignition switch off. Remove the center inner fairing. Refer to *Front Lower Fairings* in Chapter Seventeen.

4. Turn the reverse shift actuator pulley clockwise to its reverse position with a 22 mm wrench.
   a. If the pulley turns, go to Step 5.
   b. If not, check the reverse actuator as described in this chapter. If the reverse actuator is working, check the reverse shift mechanism as described in Chapter Five under *Reverse Shift Mechanism*.

5. Turn the ignition switch on. Push the reverse shift switch to its on position.
   a. If the reverse indicator light comes on, go to Step 6.
   b. If not, perform the *Reverse Shift Actuator Operates, But the Reverse Indicator Stays Off* procedure in this section.

6. Perform test No. 27 under *Cruise/Reverse Control Module, Main Wiring Harness Check* in Chapter Sixteen.
   a. If the test is correct, go to Step 7.
   b. If not, check for an open circuit in the yellow/white wire between the cruise/reverse control module and the reverse shift relay 3. If good, check for an open circuit in the red wire between the reverse shift relays 1 and 3. Refer to the wiring diagrams at the end of this manual.

7. Perform test No. 34 under *Cruise/Reverse Control Module, Main Wiring Harness Check* in Chapter Sixteen.
   a. If the test was incorrect, check for an open circuit in the blue/white wire between the cruise/reverse control module and the reverse shift relay 2. Refer to the wiring diagrams at the end of this manual.
   b. If the test was correct, replace the cruise/reverse control module. Refer to *Cruise/Reverse Control Module* in Chapter Sixteen.

Starter Operates for Engine Starting But Will Not Turn for Reverse Operation

1. Confirm the following:
   a. The engine is running.
   b. The sidestand is raised up all the way.
   c. The transmission is in neutral.
   d. The reverse shift switch is pushed to on and the reverse indicator light is on.

NOTE
The reverse system will turn off (and reverse indicator will turn off) when the speed limiter is activated during reverse operation, followed by the electrical motor brake operation or when the motor is overloaded (over 3 seconds). To operate in reverse again, push the reverse shift switch to off, then push it to its on position.

2. Perform test No. 48 under *Cruise/Reverse Control Module, Main Wiring Harness Check* in Chapter Sixteen.
   a. If the test is correct, go to Step 3.
   b. If not, check for an open circuit in the blue/red and blue wires between the cruise control module and the oil pressure switch. If good, check for a damaged oil pressure switch diode as described in this chapter. Refer to the wiring diagrams at the end of this manual.

3. Perform the *Operational Check* under *Starter Relay B* in this chapter.
   a. If continuity is present, go to Step 4.
   b. If not, check for an open circuit in the red wire between the reverse resistor and starter relay B. If this circuit is good, check for an open circuit in the black wire between the reverse resistor and starter relay B. If both tests and all related connectors are good, the reverse resistor is damaged. Replace as described in this chapter.

4. Perform the *Regulated Voltage Check* under *Reverse Regulator* in this chapter.
   a. If the test results are correct, this was a temporary failure. The system is normal.
   b. If not, go to Step 5.

5. Perform test No. 18 and No. 43 under *Cruise/Reverse Control Module, Main Wiring Harness Check* in Chapter Sixteen.
   a. If both tests are correct, go to Step 6.
   b. If not, check for an open circuit in the yellow/red wire between the cruise/reverse control module and the starter reverse switch. If good, check for an open circuit in the yellow/red wire between the cruise/reverse control module and the reverse relay. Refer to the wiring diagrams at the end of this manual.

6. Perform the *Reverse Regulator Circuit Check* under *Reverse Regulator* in this chapter.
   a. If the tests are correct, go to Step 7.
   b. If not, check for an open circuit in the light blue wire between the cruise/reverse control module and the reverse regulator assembly.
   c. If not, check for an open circuit in the green wire between the reverse regulator assembly and the engine ground terminal.
   d. If not, check for an open circuit in the white/blue wire between the reverse regulator assembly and the reverse position switch in the reverse shift actuator.

e. If not, check for an open circuit in the brown/red wire between starter relay A and the reverse regulator assembly.

7. Replace the reverse regulator and perform the *Regulated Voltage Check* (Step 4) again.
   a. If the results are correct, the original reverse regulator was damaged.
   b. If not, check the cruise/reverse control module connectors for contamination or a damaged terminal. If good, the cruise/reverse control module is damaged. Replace the cruise/reverse control module. Refer to *Cruise/Reverse Control Module* in Chapter Sixteen.

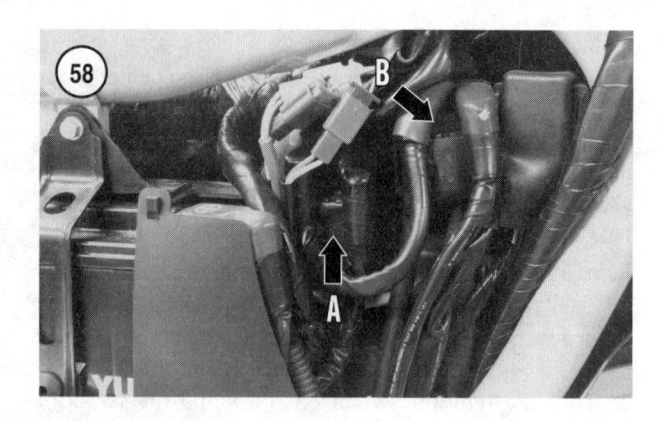

Motorcycle Starts Running In Reverse But Then Stops

1. Check the No. 11 fuse.
   a. If the fuse is blown, replace it.
   b. If good, go to Step 2.
2. Perform test No. 12 under *Cruise/Reverse Control Module, Main Wiring Harness Check* in Chapter Sixteen.
   a. If the test is correct, go to Step 3.
   b. If not, check for an open circuit in the pink wire between the reverse fuse A and the starter relay B in the fuse box. If good, check for an open circuit in the pink/white wire between the cruise/reverse control module and reverse fuse A in the fuse box. Refer to the wiring diagrams at the end of this manual.
3. Check the cruise/reverse control module connectors for contamination or a damaged terminal. If good, the cruise/reverse control module is damaged. Replace the cruise/reverse control module. Refer to *Cruise/Reverse Control Module* in Chapter Sixteen.

Reverse Speed is Too Slow (Below 0.6 mph); No Uphill Power

1. Test the battery as described in this chapter. If good, go to Step 2.
2. Perform the *System Test* under *Power Control Relay* in this chapter.
   a. If the results are correct, go to Step 3.
   b. If not, check for damaged power control relay circuits or a damaged power control relay.
3. Check the reverse resistor as described under *Reverse Resistor* in this chapter.
   a. If the results are correct, the cruise/reverse control module is damaged. Replace the cruise/reverse control module. Refer to *Cruise/Reverse Control Module* in Chapter Sixteen.
   b. If not, the reverse resistor is faulty. Replace the reverse resistor.

Speed Limiter Will Not Work When the Reverse Running Speed is Above 1.6 mph

1. Perform the *System Test* under *Speed Limiter Relay* in this chapter.
   a. If the results are correct, go to Step 2.
   b. If not, check for damaged speed limiter relay circuits or a damaged speed limiter.
2. Check continuity in the yellow wire between the speed limiter fuse terminal (in fuse box) and the speed limiter relay connector.
   a. If continuity is present, go to Step 3.
   b. If not, check for an open circuit in the yellow wire between the fuse box and the speed limiter relay.
3. Check the reverse resistor as described under *Reverse Resistor* in this chapter.
   a. If the results are correct, the cruise/reverse control module is damaged. Replace the cruise/reverse control module. Refer to *Cruise/Reverse Control Module* in Chapter Sixteen.
   b. If not, the reverse resistor is faulty. Replace the reverse resistor.

STARTER RELAY A

System Check

1. Remove the left saddlebag. Refer to *Saddlebags* in Chapter Seventeen.
2. Push the reverse shift switch to the off position.
3. Shift transmission into neutral.
4. Turn the ignition switch on and the engine stop switch to RUN.

ELECTRICAL SYSTEM

59 STARTER RELAY TESTING

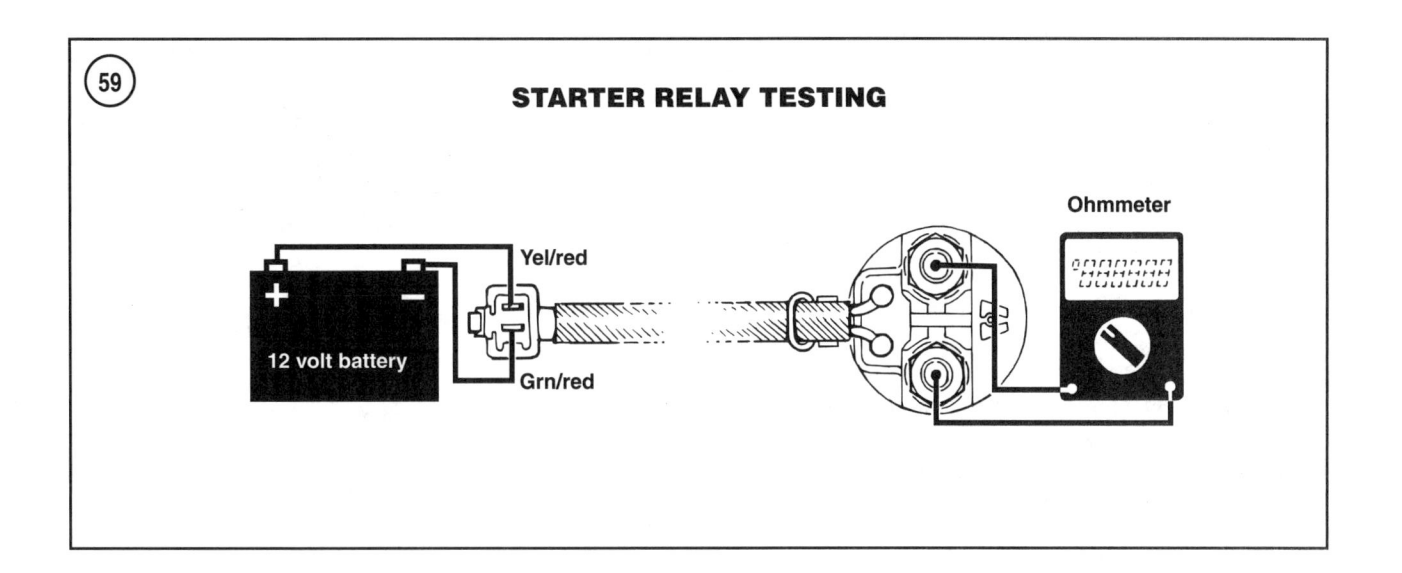

5. Push the starter/reverse switch. Starter relay A (A, **Figure 58**) should click. If the relay did not click, continue with Step 6.

6. Disconnect the red 2-pin starter relay A connector.

7. Relay coil ground circuit test: Turn the ignition switch on. Check for continuity between the brown/red terminal in the harness side of the red 2-pin connector and ground. There should be continuity.

8. Relay coil input voltage circuit test: Connect a voltmeter between the yellow/red (+) terminal in the harness side of the red 2-pin connector and ground (–). Turn the ignition switch on and push the starter/reverse switch. Battery voltage should be present.

Operational Check

1. Remove the left saddlebag. Refer to *Saddlebags* in Chapter Seventeen.

2. Disconnect the two large cable leads at the starter relay A (A, **Figure 58**).

3. Disconnect the red 2-pin starter relay A connector.

4. Connect an ohmmeter across the two large relay terminals. The ohmmeter should show no continuity

5. Connect a fully charged 12-volt battery to the starter relay switch red 2-pin connector—positive battery terminal to the yellow/red wire (+) and the negative battery terminal to the green/red (–) wire terminal (**Figure 59**). The ohmmeter should show continuity.

6. Replace starter relay A if it failed this test.

Replacement

1. Remove the left saddlebag. Refer to *Saddlebags* in Chapter Seventeen.

2. Disconnect the red 2-pin starter relay A connector.

3. Disconnect the two large cable leads at starter relay A (A, **Figure 58**) and remove the relay.

4. Installation is the reverse of removal.

STARTER RELAY B

System Check

1. Remove the left saddlebag. Refer to *Saddlebags* in Chapter Seventeen.

2. Push the reverse shift switch to its off position.

3. Shift transmission into neutral.

4. Turn the ignition switch on and the engine stop switch to RUN.

5. Push the starter/reverse switch. Starter relay B (B, **Figure 58**) should click. If the relay did not click, continue with Step 6.

6. Disconnect the white 2-pin starter relay B connector.

7. Relay coil ground circuit test: Turn the ignition switch on. Check for continuity between the green/red terminal in the harness side of the red 2-pin connector and ground. There should be continuity.

8. Relay coil input voltage circuit test: Connect a voltmeter between the yellow/red (+) terminal in the harness side of the red 2-pin connector and ground (–). Turn the ignition switch on and push the starter/reverse switch. Battery voltage should be present.

Operational Check

1. Remove the left saddlebag. Refer to *Saddlebags* in Chapter Seventeen.
2. Disconnect the two large cable leads at (B, **Figure 58**).
3. Disconnect the white 2-pin starter relay B connector.
4. Connect an ohmmeter (R × 1) across the two large relay terminals. The ohmmeter should show no continuity.
5. Connect a fully charged 12-volt battery to the starter relay switch white 2-pin connector—positive battery terminal to the yellow/red wire (+) and the negative battery terminal to the green/red (−) wire terminal (**Figure 59**). The ohmmeter should show continuity.
6. Replace starter relay B if it failed this test.

Replacement

1. Remove the left saddlebag. Refer to *Saddlebags* in Chapter Seventeen.
2. Disconnect the red 2-pin starter relay A connector.
3. Disconnect the two large cable leads at starter relay B (B, **Figure 58**) and remove the relay.
4. Installation is the reverse of removal.

REVERSE SWITCH

Removal/Installation

1. Remove the right engine side cover. Refer to *Engine Side Covers* in Chapter Seventeen.
2. Remove the cover and disconnect the wire at the reverse switch (**Figure 60**).
3. Remove the reverse switch and its washer.
4. Installation is the reverse of removal. Install the reverse switch with a new washer and tighten to 12 N•m (106 in.-lb.).

Testing

1. Check that the switch shaft moves smoothly. If it sticks, replace the switch.
2. Connect an ohmmeter across the switch (**Figure 61**). The ohmmeter should show continuity.
3. Replace the reverse switch if it failed this test.

REVERSE RELAY

The reverse relay is located in the main relay box underneath the seat (18, **Figure 54**). Refer to *Relay Box* in this chapter for more information on identifying and testing the relays.

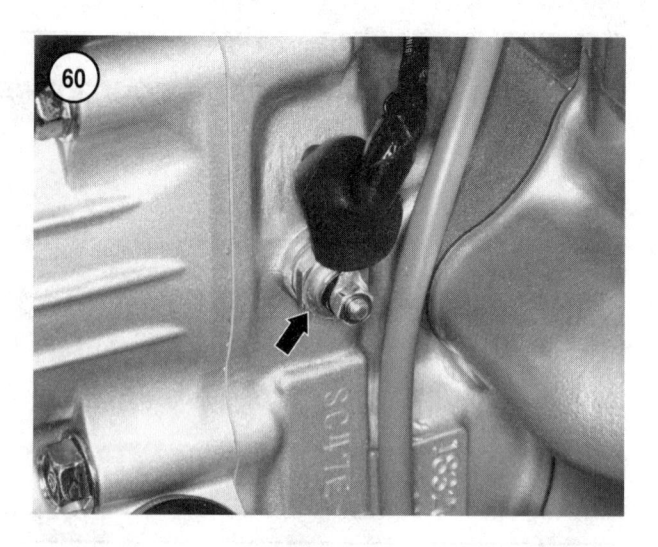

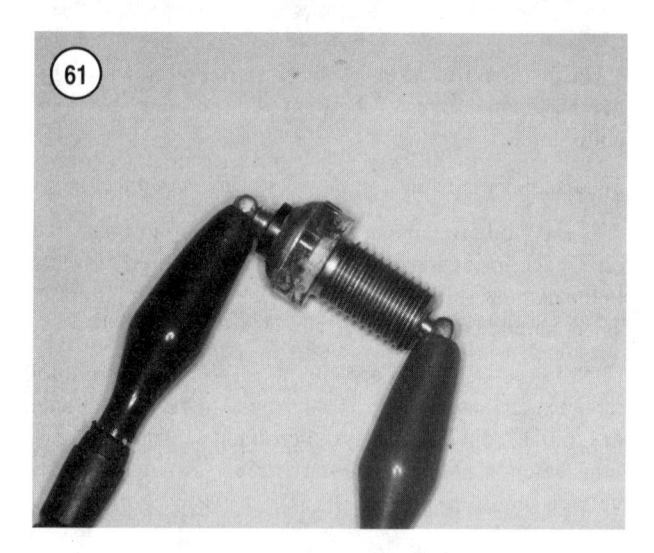

Relay and Circuit Testing

Refer to wiring diagrams at the end of the manual to identify the reverse relay circuit.

1. Remove the seat. Refer to *Seat* in Chapter Seventeen.
2. Remove the relay box cover screws and cover.
3. Replace the reverse relay (18, **Figure 54**) with a known good 4-terminal relay. Turn the ignition switch on and push the starter/reverse switch.
   a. If the starter relay clicks, the original reverse relay was faulty. Install a new 4-terminal relay and repeat the test.
   b. If not, go to Step 4.

NOTE
The meter test connections called out in Steps 4-6 are made at the reverse relay con-

ELECTRICAL SYSTEM

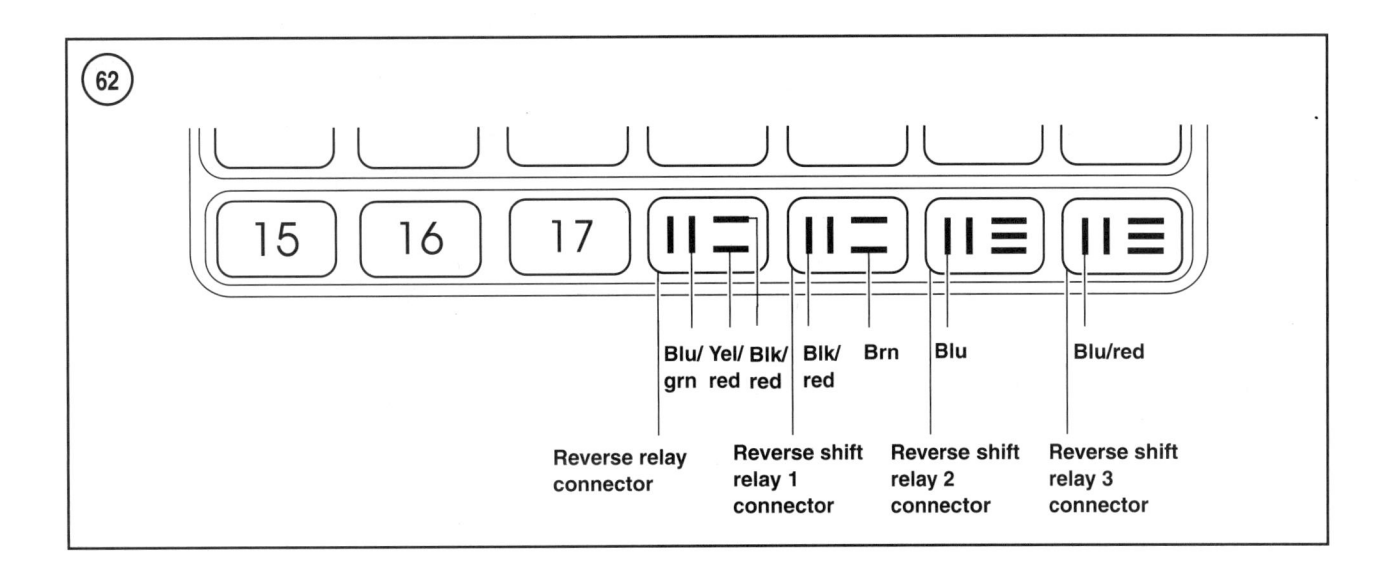

62

```
┌─────────────────────────────────────────────────────────┐
│  [15]  [16]  [17]  [||=] [||=] [||=] [||=]              │
│                     │││   │ │   │  │   │                │
│               Blu/ Yel/ Blk/ Blk/ Brn Blu  Blu/red      │
│               grn red  red  red                         │
│                                                          │
│               Reverse  Reverse  Reverse  Reverse         │
│               relay    shift    shift    shift           │
│               connector relay 1 relay 2  relay 3         │
│                        connector connector connector    │
└─────────────────────────────────────────────────────────┘
```

*nector terminals inside the relay box (**Figure 62**).*

4. Turn the ignition switch off and remove the reverse relay (18, **Figure 54**). Turn the ignition switch on and check for battery voltage between the reverse relay yellow/red terminal (+) and ground (−).

 a. Yes. Go to Step 5.

 b. No. Check for an open circuit in the yellow/red wires between the reverse shift switch and the reverse relay. If the wires and connectors are good, diode D7 in the reverse regulator may be faulty. Test the diode as described under *Reverse Regulator Assembly* in this chapter.

5. Turn the ignition switch off and check for continuity between the black/red wire terminal and ground.

 a. Yes. Go to Step 6.

 b. No. Check for an open circuit in the black/red wire between the reverse regulator assembly and reverse relay.

 c. No. Check for a damaged D4 or D5 diode in the reverse regulator assembly. Test the diodes as described under *Reverse Regulator Assembly* in this chapter.

 d. No. Check for an open circuit in the green/orange wire between the reverse switch and reverse regulator assembly.

 e. If the checks in substeps b-d are good, the reverse switch is damaged. Replace the reverse switch as described in this chapter.

6. Turn the ignition switch off and check for continuity between the blue/green wire terminal and ground.

 a. Yes. The system is normal. Check all the related connectors for contamination or damaged pin terminals.

 b. No. Check for an open circuit in the blue/green wire between the reverse regulator assembly and reverse relay.

 c. No. Check for a damaged D2, D3 or D11 diode in the reverse regulator assembly. Test the diodes as described under *Reverse Regulator Assembly* in this chapter.

 d. No. Check for an open circuit in the green/red wire between the clutch switch and reverse regulator assembly.

 e. Check for an open circuit in the green/white wire between the sidestand switch and clutch switch.

 f. If the checks in substeps b-e are good, either the clutch switch or sidestand switch are faulty.

7. Install the reverse relay (18, **Figure 54**).

8. Install the relay box cover and secure with the mounting screws.

9. Install the seat. Refer to *Seat* in Chapter Seventeen.

REVERSE SHIFT RELAYS

Three reverse shift relays are used. These relays are located in the main relay box (**Figure 54**) underneath the seat:

1. Reverse shift relay 1 (19).
2. Reverse shift relay 2 (20).
3. Reverse shift relay 3 (21).

The relays are easily unplugged or plugged into the relay terminal strip. Because only 4-terminal and 5-terminal relays are used, there is no danger of installing a relay in the wrong position. The 4-terminal relays are identical, and the 5-terminal relays are identical. Refer to *Relay Box* in this chapter for more information on identifying and testing the relays.

Relay and Circuit Testing

Refer to the wiring diagrams at the end of the manual to identify the reverse shift relay circuits.
1. Before testing the reverse shift relays, confirm that the neutral indicator lights when the ignition switch is turned on and the transmission is in neutral. If so, go to Step 2.
2. Turn the ignition switch off.
3. Remove the seat. Refer to *Seat* in Chapter Seventeen.
4. Remove the relay box cover screws and cover .
5. Turn the ignition switch off. Replace reverse shift relay 1 (19, **Figure 54**) with a known good 4-terminal relay. Turn the ignition switch on or to ACC and shift the transmission into neutral. Note the following:
    a. If the reverse shift actuator operates, the original reverse shift relay 1 was faulty. Install a new 4-terminal relay and repeat the test.
    b. If the reverse actuator does not operate, turn the ignition switch off and go to Step 6.
6. Push the reverse shift switch off and turn the ignition switch off. Replace reverse shift relay 2 (20, **Figure 54**) with a known good 5-terminal relay. Turn the ignition switch on and push the reverse shift switch to ON. Note the following:
    a. If the reverse shift actuator operates, the original reverse shift relay 3 was faulty. Install a new 5-terminal relay and repeat the test.
    b. If the reverse actuator does not operate, turn the ignition switch off and go to Step 7.
7. Push the reverse shift switch off and turn the ignition switch off. Replace reverse shift relay 3 (21, **Figure 54**) with a known good 5-terminal relay. Turn the ignition switch on and push the reverse shift switch on. Note the following:
    a. If the reverse shift actuator operates, the original reverse shift relay 3 was faulty. Install a new 5-terminal relay and repeat the test.
    b. If the reverse actuator does not operate, turn the ignition switch off and go to Step 8.

NOTE
*The meter test connections called out in Step 8 and Step 9 are made at the relay connector terminals inside the relay box (**Figure 62**).*

8. Turn the ignition switch off and remove reverse shift relay 1 (19, **Figure 54**). Measure voltage between the red/black terminal (+) and ground (–):
    a. Battery voltage: Go to Step 9.
    b. No battery voltage: Check for an open in the red/black wire between the No. 23 fuse and the relay box.
9. Turn the ignition switch on. Measure voltage between the brown terminal (+) and ground (–):
    a. Battery voltage: Go to Step 10.
    b. No battery voltage: Check for an open in the brown wire between the No. 10 fuse and the relay box.
10. Turn the ignition switch off. Remove reverse shift relay 2 and reverse shift relay 3. Check continuity in the red wires between relays 1 and 2, and then between 1 and 3.
    a. If continuity is present during both tests, perform *Reverse Shift Actuator Will Not Turn to the Reverse Position When the Reverse Shift Switch is Set to ON* under *Starter/Reverse System Troubleshooting* in this chapter.
    b. If not, check for an open in the red wire between the affected relay(s).
11. Turn the ignition switch off and reinstall the relays.
12. Install the relay box cover and secure with the mounting screws.
13. Install the seat. Refer to *Seat* in Chapter Seventeen.

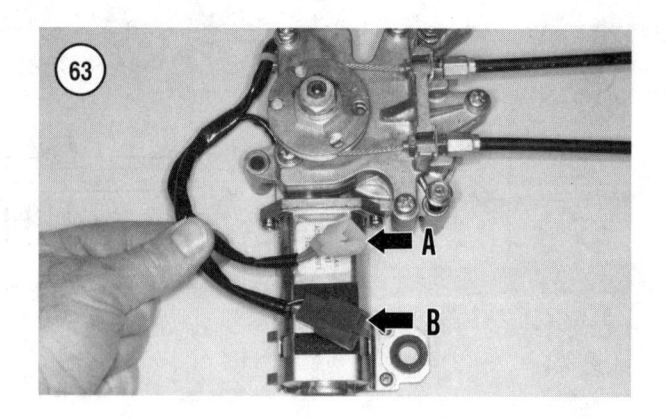

REVERSE SHIFT ACTUTOR

Motor Inspection

1. Turn the ignition switch off.
2. Remove the seat. Refer to *Seat* in Chapter Seventeen.
3. Remove the relay box cover screws and cover .
4. Remove reverse shift relays 2 and 3.
5. Shift the transmission into neutral.

NOTE
*The test connections called out in Step 6 are made at the relay connector terminals inside the relay box (**Figure 62**).*

ELECTRICAL SYSTEM

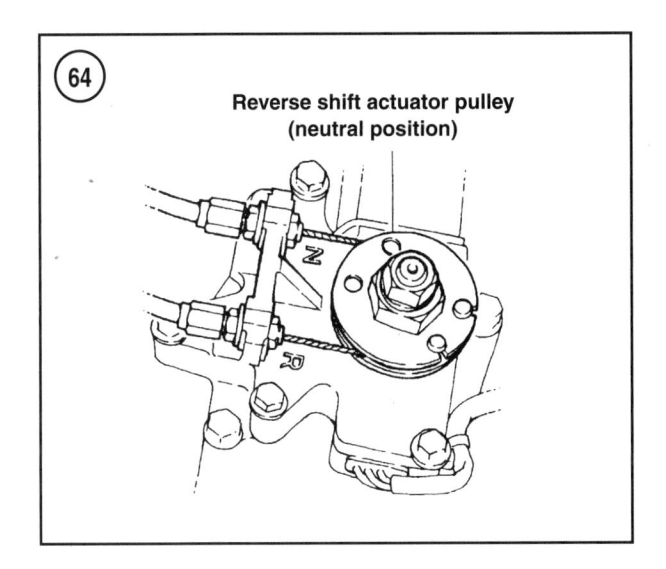

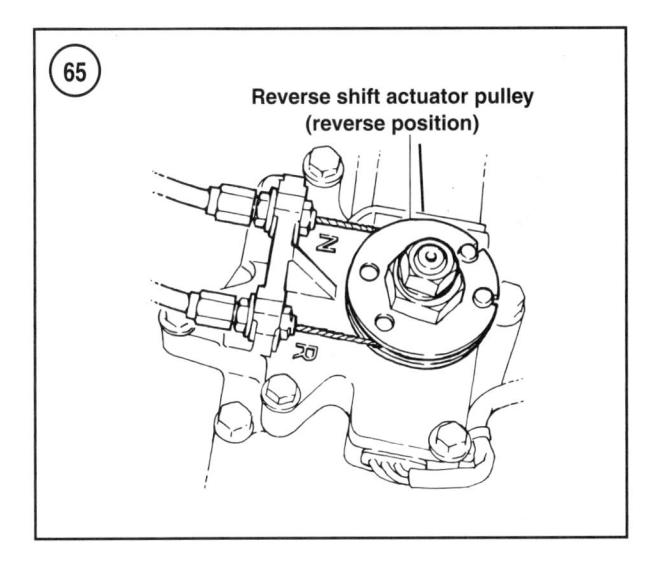

64 Reverse shift actuator pulley (neutral position)

65 Reverse shift actuator pulley (reverse position)

6. Connect a fully charged 12-volt battery to the reverse shift relay 2 and reverse shift relay 3 connector terminals. Connect the positive battery terminal to the shift relay 2 blue terminal (+) and the negative battery terminal to the shift relay 3 blue/red connector.

   a. If the reverse shift activator does not operate, go to Step 7.
   b. If the reverse shift actuator operates, check for an open circuit in the blue wire between the reverse shift actuator and reverse shift relay 2. If good, check for an open circuit between the blue/red wire between the reverse shift actuator and reverse shift relay 3.

7. Remove the air filter housing. Refer to *Air Filter Housing* in Chapter Eight.

8. Disconnect the reverse shift actuator 2-pin white connector (A, **Figure 63**).

9. Disconnect the cables from the reverse shift actuator as described under *Removal/Installation* in this section.

10. Apply battery voltage to the relay terminals as described in Step 6.

    a. If the reverse shift actuator does not operate, replace the reverse shift actuator.
    b. If the reverse shift activator does operate, check the reverse shift mechanism as described in Chapter Five under *Reverse Shift Mechanism*.

11. Reverse the disassembly steps to complete installation.

Reverse Position Switch Check

The reverse position switch is an integral part of the reverse shift actuator.

1. Remove the air filter housing. Refer to *Air Filter Housing* in Chapter Eight.

2. Disconnect the white 2-pin (A, **Figure 63**) and red 3-pin (B) reverse shift actuator connectors.

3. Input voltage check: Turn the ignition switch on and measure voltage in the white 3-pin harness connector between the brown terminal (+) and ground (−). If there is no battery voltage, check for an open circuit in the brown wire between the No. 10 fuse and the reverse shift actuator.

4. Operation check: Perform the following:

   a. Check that the reverse shift actuator is in neutral (**Figure 64**).
   b. Check for continuity between the black/white and brown terminals in the red 3-pin connector. There should be continuity.
   c. Connect a fully charged 12-volt battery to the actuator side of the white 2-pin connector. Momentarily connect the positive battery terminal to the blue (+) terminal and the negative battery terminal to the blue/red terminal. This will turn the actuator to its reverse position (**Figure 65**).
   d. Check for continuity between the white/blue and brown terminals in the red 3-pin connector. There should be continuity.

5. Replace the reverse shift actuator if it failed any part of this test.

6. Reverse the disassembly steps to complete installation.

66

Neutral cable
Clearance 0 mm (0 in.)
Clearance 0.3-0.8 mm (0.01-0.03 in.)
Reverse cable

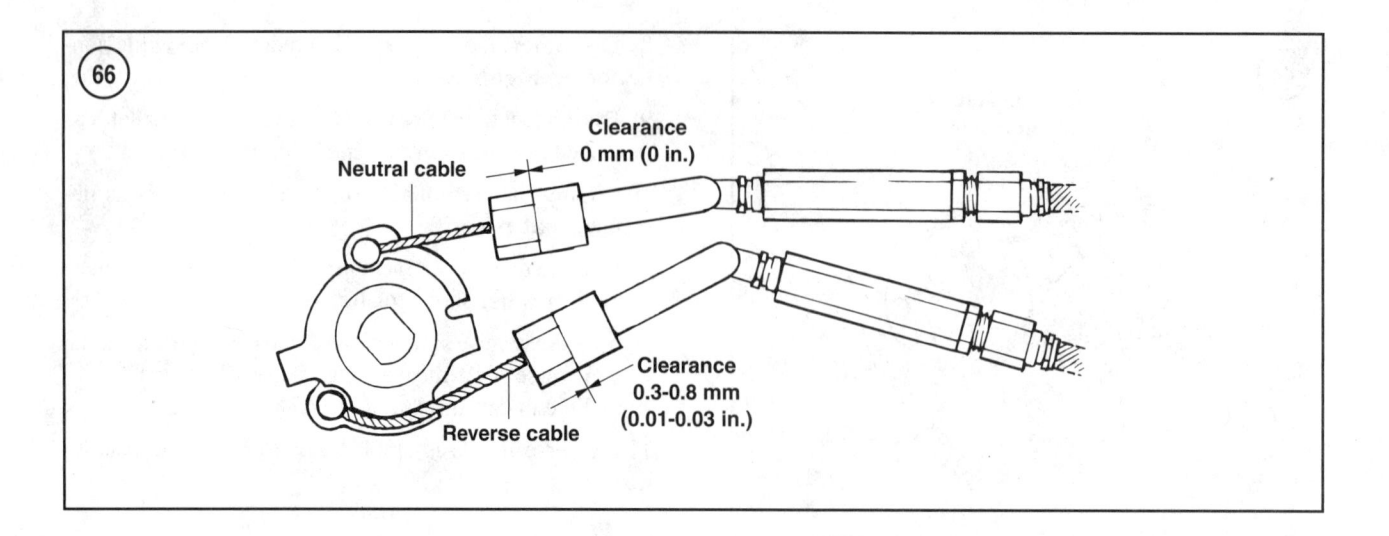

Reverse Shift Actuator Removal/Installation

1. Remove the right engine side cover. Refer to *Engine Side Covers* in Chapter Seventeen.
2. Remove the center inner fairing. Refer to *Front Lower Fairings* in Chapter Seventeen.
3. On GL1800A models, remove the ABS modulators. Refer to *ABS Modulators* in Chapter Fifteen.
4. Loosen the reverse shift actuator cable locknuts and adjusters (**Figure 66**).
5. Loosen the cable locknuts and adjusters and disconnect the neutral (A, **Figure 67**) and reverse (B) cables from the reverse shift actuator.
6. Remove the three mounting bolts (C, **Figure 67**) and the reverse shift actuator (D).
7. Installation is the reverse of removal. Note the following:
   a. The reverse shift actuator housing is marked with the letters N (neutral) and R (reverse) to identify the cable positions. Reconnect the neutral (A, **Figure 67**) and reverse (B) cables.
   b. Adjust the actuator cables as described in the following procedure.

Reverse Shift Actuator Cable Adjustment

1. Remove the right engine side cover. Refer to *Engine Side Covers* in Chapter Seventeen.
2. Remove the center inner fairing. Refer to *Front Lower Fairings* in Chapter Seventeen.
3. Check that the reverse shift actuator is in neutral (**Figure 64**).
4. Check that the neutral cable clearance (**Figure 66**) between the cable holder and cable is 0 mm.

67

5. Check that the reverse cable clearance (**Figure 66**) between the cable holder and cable is 0.3-0.8 mm (0.01-0.03 in.).
6. To adjust the cables, loosen the cable locknuts and turn the cable adjusters as required. Recheck each cable adjustment.
7. Shift the transmission into neutral.
8. Turn the ignition switch on. Then push the reverse shift switch on to turn the reverse actuator pulley to its reverse position as shown in **Figure 65**.
9. With the ignition switch still on, check that there is clearance between the lost motion plate and reverse shift arm as shown in **Figure 68**. If there is no clearance, recheck the cable adjustment.
10. Push the reverse shift switch off. The reverse actuator pulley should return to its neutral position (**Figure 64**).
11. Turn the ignition switch off.
12. Reverse Step 1 and Step 2 to complete installation.

ELECTRICAL SYSTEM

68

Clearance
Reverse shift arm
Lost motion plate

69

70

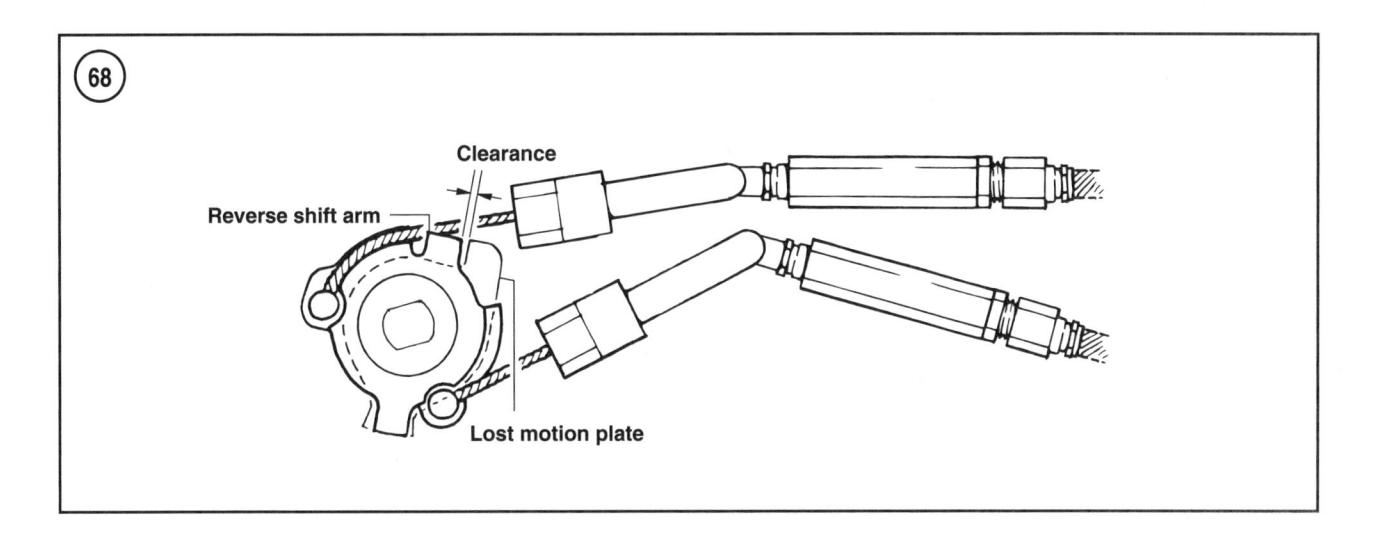

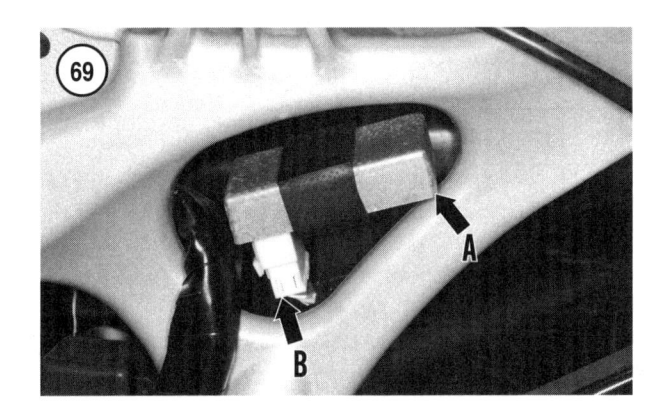

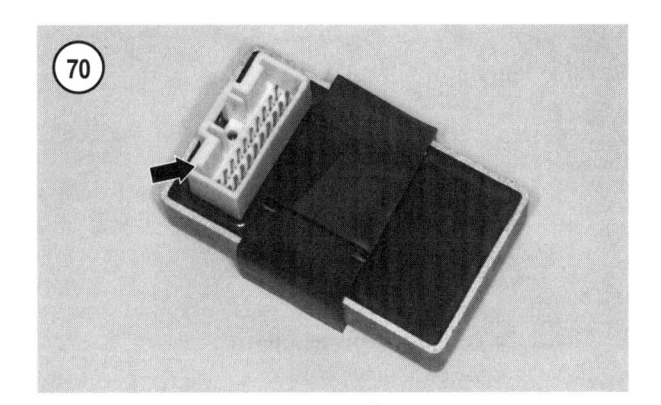

REVERSE REGULATOR

Removal/Installation

1. Remove the left saddlebag. Refer to *Saddelbags* in Chapter Seventeen.

2. Remove the reverse regulator (A, **Figure 69**) from its mounting position.

3. Disconnect the white 20-pin connector (B, **Figure 69**).

4. Installation is the reverse of removal.

Diode Inspection

The reverse regulator (**Figure 70**) consists of the following eleven separate diodes:

1. D1: Ground of starter relay B.
2. D2: Ground of reverse relay (neutral position side). Gear position (neutral) signal of ECM.
3. D3: Ground of reverse relay (clutch switch and sidestand switch).
4. D4: Ground of reverse relay coil (reverse switch side).
5. D5: Ground of reverse relay coil (neutral position side).
6. D6: Neutral indicator ground.
7. D7: Reverse relay coil power source.
8. D8: Power source of headlight and headlight relay in reverse running.
9. D9: Gear position (overdrive) signal of ECM.
10. D10: Gear position (fourth) signal of ECM.
11. D11: Reverse relay ground.
12. Test the diodes in the reverse rectifier by performing the following. Make all the tests at the reverse regulator pins (**Figure 71**).

   a. Remove the reverse diode as described in this section.
   b. Switch an ohmmeter to the R × 1 scale.
   c. Check for continuity between the D1 and D11 terminals. Reverse the connection and check the continuity in the opposite direction. All diodes should show continuity in one direction only. If any diode is faulty (same reading in both directions), replace the reverse regulator/rectifier assembly (**Figure 70**).

(71) REVERSE REGULATOR DIODE CHECK

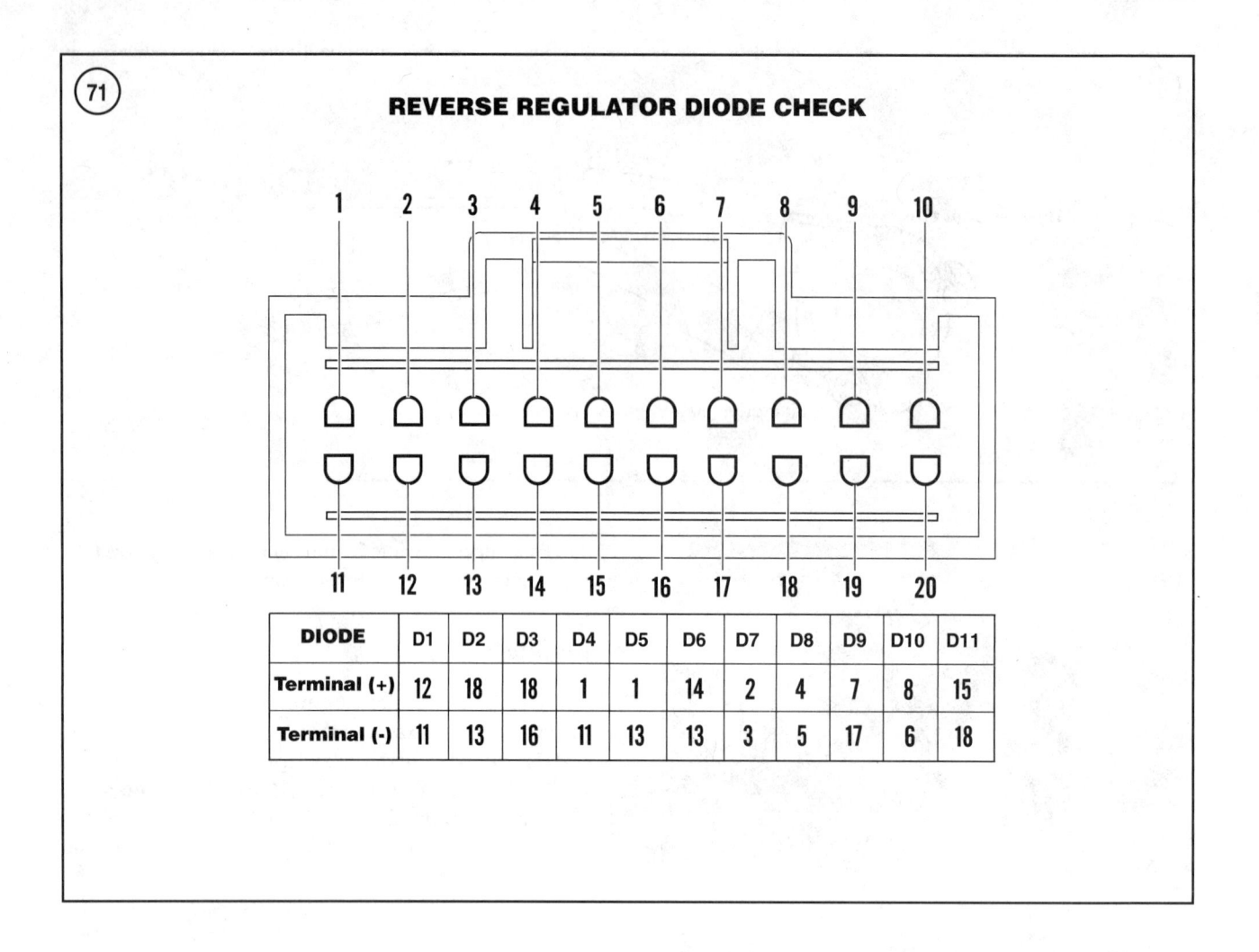

DIODE	D1	D2	D3	D4	D5	D6	D7	D8	D9	D10	D11
Terminal (+)	12	18	18	1	1	14	2	4	7	8	15
Terminal (−)	11	13	16	11	13	13	3	5	17	6	18

Regulated Voltage Check

1. Remove the left saddlebag. Refer to *Saddlebags* in Chapter Seventeen.
2. Support the motorcycle on its centerstand. The rear wheel must be off the ground during this test.
3. Shift the transmission into neutral and raise the sidestand.
4. Start the engine. Push the reverse shift switch on and check that the reverse indicator comes on.
5. Measure voltage at the red 2-pin starter relay A connector (A, **Figure 58**) as follows:
    a. Connect the positive test lead to the brown/red terminal (+) and the negative test lead (−) to ground.
    b. Push the starter/reverse switch while reading the voltmeter. The voltage should read approximately 0 volts for the first 0.3 seconds, then read a minimum of 4.0 volts. The rear wheel will turn in reverse if the reverse system is working properly.
6. Push the reverse shift switch off and turn the ignition switch off.

Reverse Regulator Circuit Check

1. Remove the reverse regulator assembly as described in this section.

NOTE
Make all the tests (Steps 2-5) on the wiring harness side connectors.

2. Check the ground circuit as follows:
    a. Switch the ohmmeter to R × 1.
    b. Connect the ohmmeter between the green wire and engine ground.
    c. The ohmmeter must read continuity.
    d. If there is no continuity or high resistance, check the green wire for damage.
3. Check the reverse position switch circuit as follows:
    a. Reconnect the white 20-pin reverse regulator connector (B, **Figure 69**).
    b. Shift the transmission into neutral and turn the ignition switch on.

ELECTRICAL SYSTEM

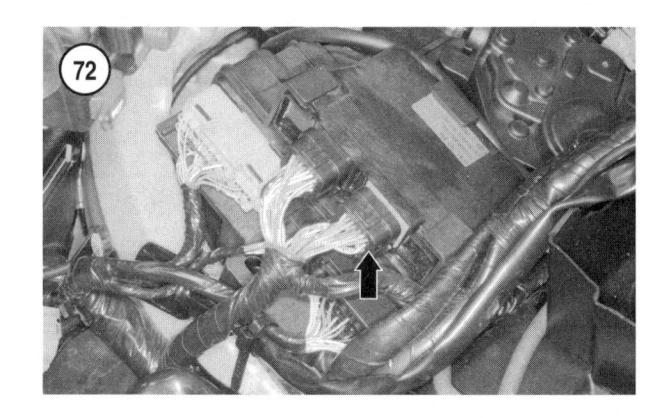

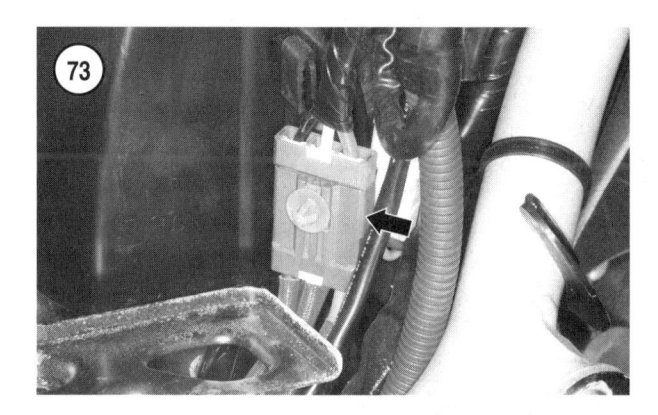

c. Push the reverse shift switch on. Check that the reverse indicator comes on.
d. Turn the ignition switch off and disconnect the white 20-pin reverse regulator connector (B, **Figure 69**).
e. Measure voltage between the white/blue terminal (+) and ground (–).
f. The voltmeter must show battery voltage.
g. Turn the reverse shift actuator to its neutral position.

4. Check the starter relay A circuit as follows:
   a. Turn the ignition switch off.
   b. Disconnect the red 2-pin starter relay A connector.
   c. Check for continuity in the brown/red terminals between the red 2-pin connector and the 20-pin reverse regulator connector.
   d. The ohmmeter must read continuity.

5. Check the control module circuit as follows:
   a. Remove the top shelter. Refer to *Top Shelter* in Chapter Seventeen.
   b. Turn the ignition switch off.
   c. Disconnect the gray 26-pin cruise/reverse control module connector (**Figure 72**).
   d. Check for continuity between the light blue wire between the 20-pin reverse regulator connector and the 26-pin cruise/reverse control module connector.
   e. The ohmmeter must read continuity.
   f. If there is no continuity or high resistance, check the light blue wire for damage.

6. Install the reverse regulator assembly as described in this section.
7. Reconnect the connectors.
8. Install the top shelter.

REVERSE RESISTOR ASSEMBLY

Removal/Installation

1. Remove the battery and battery case. See *Battery* in this chapter.
2. Disconnect the reverse resistor 3-pin connector (**Figure 73**).
3. Remove the two bolts (**Figure 74**) securing the reverse resistor to the battery tray.
4. Remove the nut and cable (A, **Figure 75**) from the reverse resistor and remove the reverse resistor.
5. Installation is the reverse of removal.

Resistance Test

1. Switch an ohmmeter to R × 1.

2. Measure resistance between the cable terminal (A, **Figure 76**) and each 3-pin connector terminal (B) as follows:

NOTE
Perform each test at an ambient temperature of 20° C (68° F). Do not test if the reverse resistor is hot.

  a. Black connector terminal: 0.20-0.25 ohms.
  b. Red connector terminal: 0.20-0.25 ohms.
  c. White connector terminal: 0.15-0.20 ohms.
3. Replace the reverse resistor if it failed any test in Step 2.

POWER CONTROL RELAYS

The power control relays are mounted beside the battery and underneath starter relay A and B (**Figure 77**).

System Test

1. Remove the left saddlebag. Refer to *Saddlebags* in Chapter Seventeen.
2. Remove the top shelter. Refer to *Top Shelter* in Chapter Seventeen.
3. Shift the transmission into neutral.
4. Turn the ignition switch on. Push the reverse shift switch on. Check that the reverse indicator came on.
5. Turn the ignition switch off.
6. Disconnect the gray 26-pin cruise/reverse control module connector (**Figure 72**) and continue with the following tests:

Power control relay 1 test

1. Connect a jumper wire between the gray 26-pin harness connector orange terminal and ground. Turn the ignition switch on. The power control relay 1 should click.
  a. No. Go to Step 2.
  b. Yes. Go to Step 3.
2. Turn the ignition switch off. Replace power control relay 1 with power control relay 2 and repeat Step 1.
  a. The power control relay should click. If so, replace power control relay 1 with a new one and reinstall power control relay 2. Go to Step 3.
  b. If not, check for an open circuit in the white/blue wire between power control relay 1 and the reverse shift actuator (reverse position switch). If good, check for an open circuit in the orange wire between the power control relay 1 and the cruise/reverse control module.

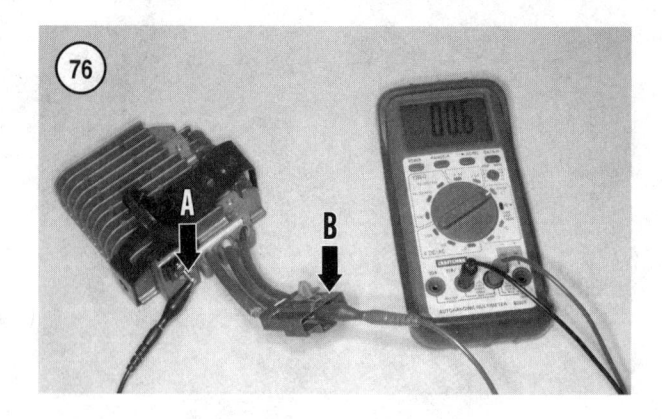

3. Turn the ignition switch off and disconnect the reverse resistor connector (**Figure 73**).
4. With the jumper wire connected as described in Step 1, check for continuity between the black and red wire terminals in the wire harness side of the reverse resistor connector. There should be continuity. If there is no continuity, check for an open circuit in the black and red wires between the reverse resistor and the power control relay 1.
5. Remove the jumper wire.

Power control relay 2 test

1. Connect a jumper wire between the gray 26-pin harness connector white terminal and ground. Turn the ignition switch on. The power control relay 2 should click.
  a. Yes. Go to Step 3.
  b. No. Go to Step 2.
2. Turn the ignition switch off. Replace power control relay 2 with power control relay 1 and repeat Step 1. The power control relay should click.
  a. No. Check for an open circuit in the white/blue wire between power control relay 2 and the reverse shift actuator (reverse position switch). If good, check for an open circuit in the white wire between the power control relay 2 and the cruise/reverse control module.
  b. Yes. Replace power control relay 1 with a new one and reinstall power control relay 2. Go to Step 3.
3. Turn the ignition switch off and disconnect the reverse resistor connector (**Figure 73**).
4. With the jumper wire connected as described in Step 1, check for continuity between the black and red wire terminals in the wire harness side of the reverse resistor connector. There should be continuity. If there is no continuity, check for an open circuit in the black and red wires between the reverse resistor and the power control relay 2.
5. Remove the jumper wire.

ELECTRICAL SYSTEM

77
Front →
Power control relay 1
Power control relay 2
Speed limiter relay
1-pin white connector
(To starter relay switch)

78
Power control relay and speed limiter relay
Ohmmeter
12 volt battery

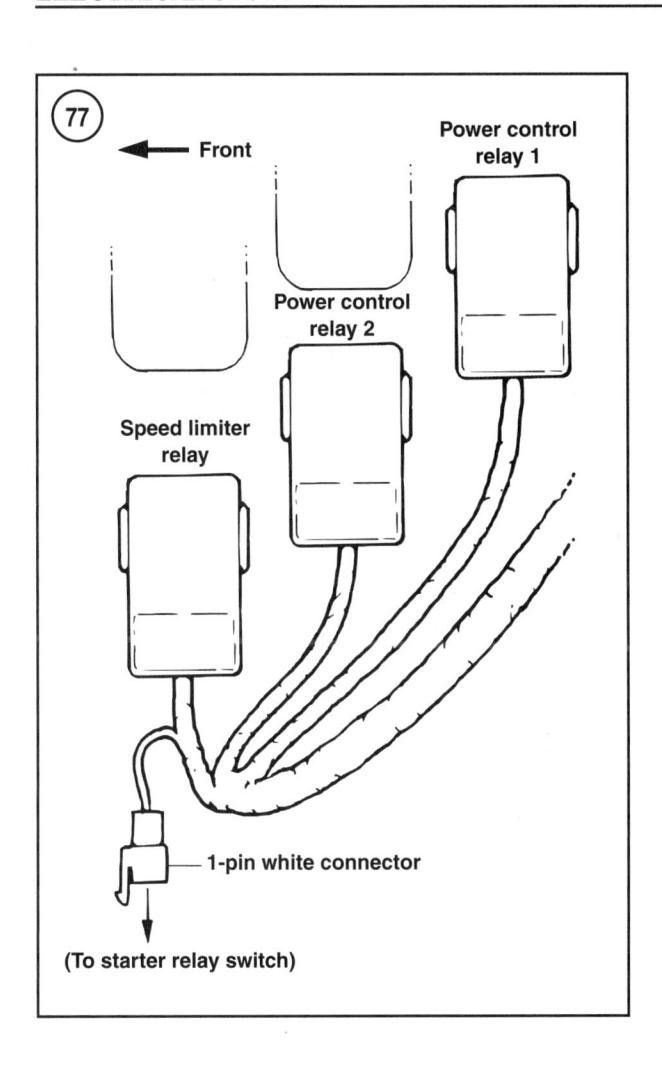

Operational Check

1. Remove the power control relay (**Figure 77**).

2. Connect an ohmmeter and a fully charged 12-volt battery to the relay terminals as shown in **Figure 78**.

3. There should be continuity with the battery connected and no continuity with the battery disconnected.

4. Replace the power control relay if it failed either part of this test.

SPEED LIMITER RELAY

The speed limiter relay is mounted beside the battery and underneath starter relay switches A and B (**Figure 77**).

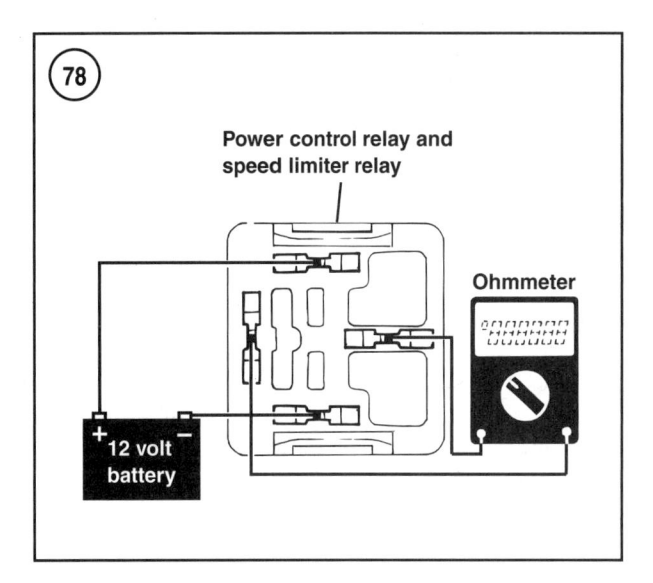

System Test

1. Remove the left saddlebag. Refer to *Saddlebags* in Chapter Seventeen.

2. Remove the top shelter. Refer to *Top Shelter* in Chapter Seventeen.

3. Shift the transmission into neutral.

4. Turn the ignition switch on. Push the reverse shift switch on. Check that the reverse indicator came on.

5. Turn the ignition switch off.

6. Disconnect the gray 26-pin cruise/reverse control module connector (**Figure 72**) and continue with the following test:

Speed limiter relay

1. Connect a jumper wire between the gray 26-pin harness connector gray terminal and ground. Turn the ignition switch on. The speed limiter relay should click.
   a. Yes. Go to Step 3.
   b. No. Go to Step 2.

2. Turn the ignition switch off. Replace the speed limiter relay with a new relay and repeat Step 1. The speed limiter relay should click.
   a. Yes. Replace the original relay with a new one. Go to Step 3.
   b. No. Check for an open circuit in the white/blue wire between the speed limiter relay and the reverse shift actuator (reverse position switch). If good, check for an open circuit in the gray wire between the speed limiter relay and the cruise/reverse control module.

3. Turn the ignition switch off and disconnect the reverse resistor connector (**Figure 73**).
4. Check for continuity in the white wire between the speed limiter relay and reverse resistor connectors. There should be continuity. If there is no continuity, check for an open circuit in the white wire.
5. Remove the jumper wire.

Operation Check

1. Remove the speed limiter relay (**Figure 77**).
2. Connect an ohmmeter and a fully charged 12-volt battery to the relay terminals as shown in **Figure 78**.
3. There should be continuity with the battery connected and no continuity with the battery disconnected.
4. Replace the speed limiter relay if it failed either part of this test.

STARTER/REVERSE MOTOR SERVICE

CAUTION
Do not operate the starter for more than five seconds at a time. Wait approximately 10 seconds between starting attempts.

Starter Removal/Installation

1. Disconnect the negative battery cable at the battery. See *Battery* in this chapter.
2. Remove the fuel tank. Refer to *Fuel Tank* in Chapter Eight.
3. Remove the reverse shift arm. Refer to *Reverse Shift Mechanism* in Chapter Five.
4. Disconnect the starter cable at the starter.
5. Remove the three starter mounting bolts and remove the starter (**Figure 79**).
6. Install by reversing these removal steps, plus the following:
   a. Lubricate the starter O-ring (**Figure 80**) with engine oil.
   b. Remove all corrosion from the starter cable.
   c. Tighten the starter mounting bolts to 29 N•m (21 ft.-lb.).

Starter Disassembly

Refer to **Figure 81**.
1. Scribe alignment marks on the starter housing, reverse reduction gear case and rear cover (**Figure 82**) for reassembly.
2. Remove the starter case through bolts (**Figure 83**).
3. Remove the reduction gear case and reduction gear assembly (**Figure 84**).

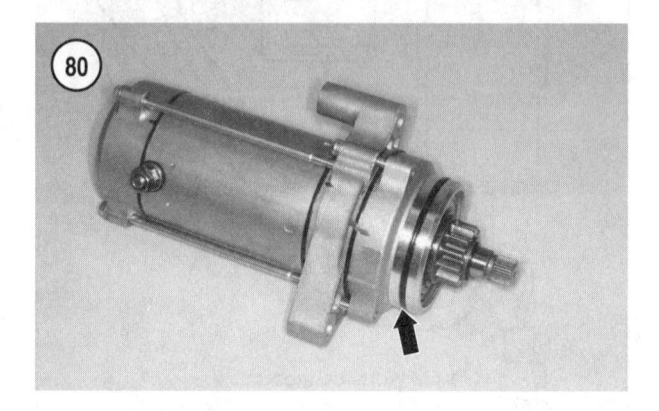

NOTE
*The number of shims (2, **Figure 81**) varies.*

4. Slide the rear cover (**Figure 85**) off the armature shaft and remove the shim(s). See **Figure 86**.
5. Remove the small washer (14, **Figure 81**), large washer and gear holder (A, **Figure 87**) and armature (B).

CAUTION
Do not immerse the armature coil or case in solvent, as the solvent may damage the insulation. Wipe the windings with a cloth lightly moistened in solvent and dry with compressed air.

6. Clean all grease, dirt and carbon from the components.
7. Inspect the starter as described in this chapter.

Brush Inspection

1. Measure the length of each brush (**Figure 88**). If any brush is too short (**Table 5**), replace the brushes as a set.

ELECTRICAL SYSTEM

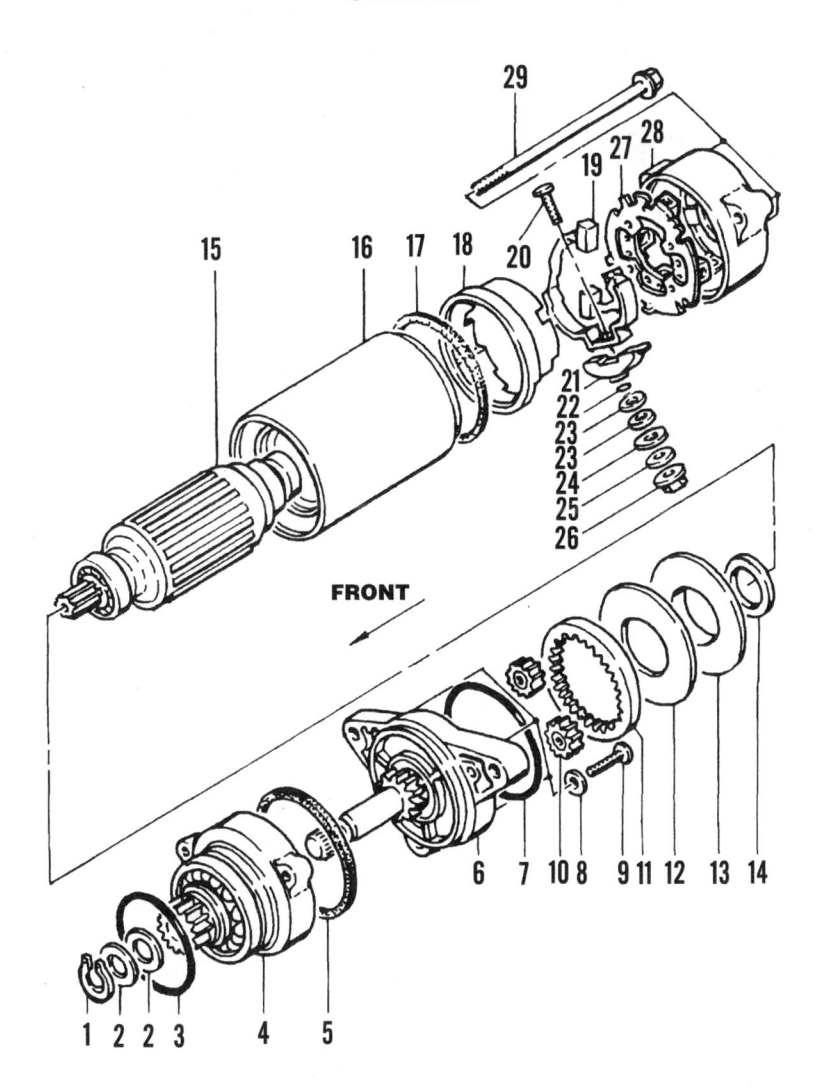

81 STARTER

1. Snap ring
2. Shim(s)
3. O-ring
4. Reverse reduction gearcase
5. O-ring
6. Starter reduction gearcase
7. O-ring
8. Washer
9. Screw
10. Reduction gear
11. Reduction gear
12. Washer
13. Gear holder
14. Washer
15. Armature
16. Housing
17. O-ring
18. Cover plate
19. Brush set
20. Cable bolt
21. Insulator
22. O-ring
23. Insulator washers
24. Insulator washer
25. Steel washer
26. Nut
27. Brush plate
28. Rear cover
29. Throughbolt

2. Inspect the brush springs for fatigue, cracks or other damage. Replace the brush plate if the springs are excessively worn or damaged.

3. Use an ohmmeter to make the following tests:
   a. Check for continuity between the cable terminal and the insulated brush wire (**Figure 89**); there should be continuity.
   b. Check for continuity between the housing and insulated brush; there should be no continuity

4. If necessary, replace the brush set and brush plate (**Figure 90**) as described in the following procedure.

Brush and Brush Plate Replacement

The brush set (19, **Figure 81**) and brush plate (27) can be replaced separately. The brush springs are mounted on the brush plate and cannot be replaced separately.

CAUTION
Before removing the nuts and washers in Step 1, record their description and order. Reinstall them in the same order to insulate this set of brushes from the case.

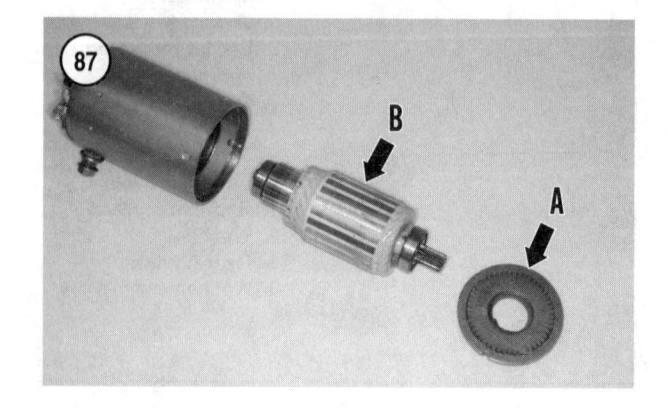

ELECTRICAL SYSTEM 371

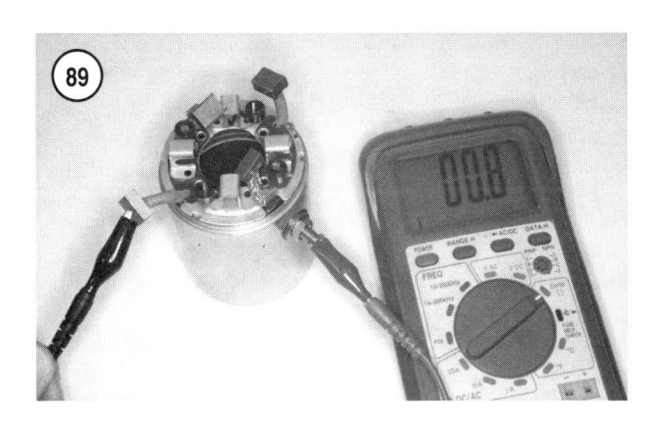

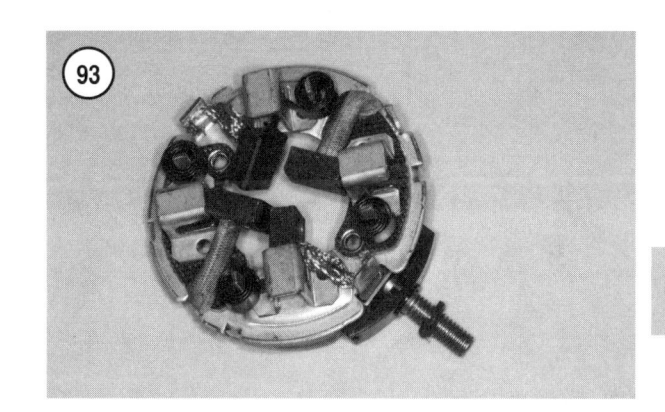

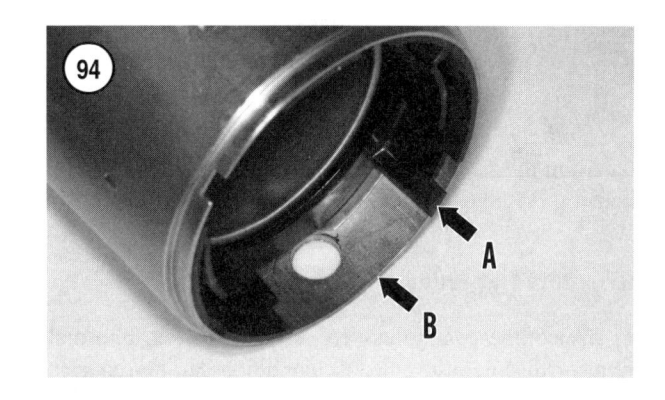

1. Remove the nut (**Figure 91**) and washers securing the brush set and terminal set to the housing (**Figure 92**).

2. Assemble the new parts onto the brush plate as shown in **Figure 93**.

3. Center the cover plate into the housing as shown in A, **Figure 94**.

4. Refer to **Figure 92** and install the cover plate into the housing. Center the insulator (21, **Figure 81**) into the gap shown in B, **Figure 94**.

5. Refer to **Figure 92** and install the washer set (**Figure 81**) as follows:
    a. O-ring (22).

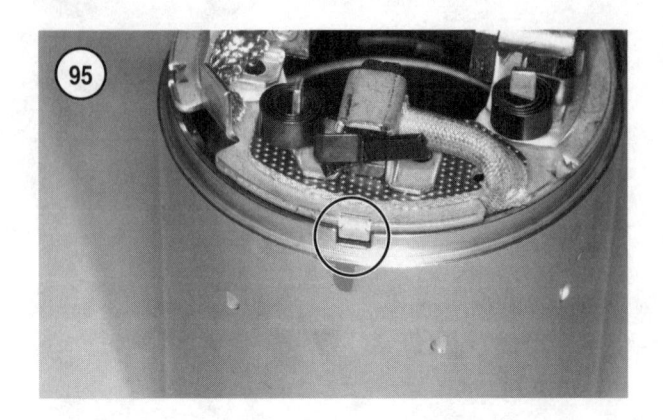

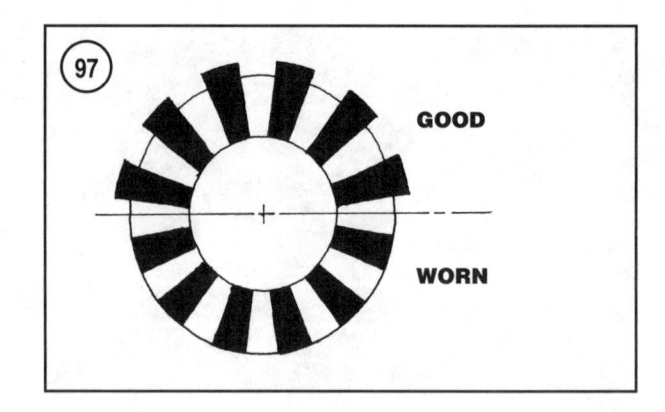

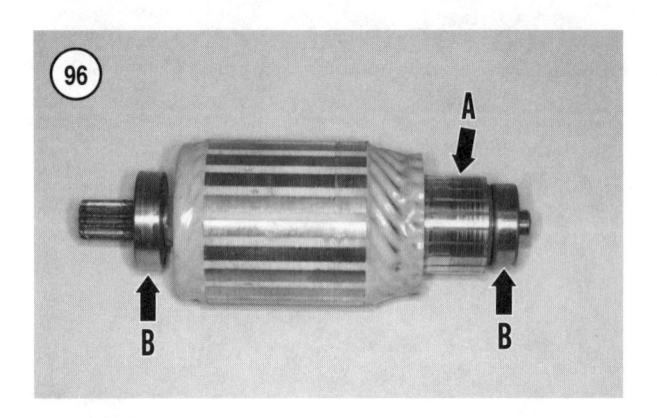

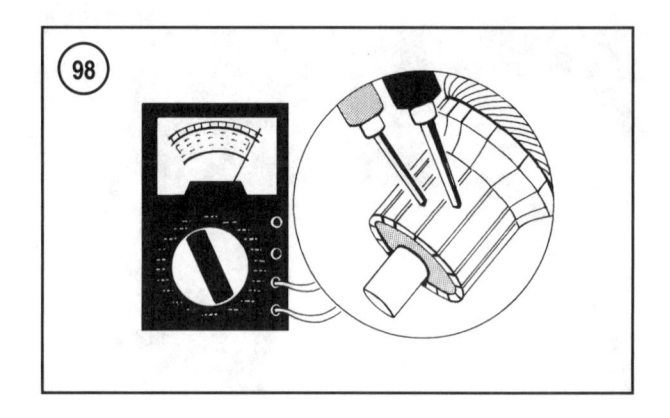

b. Two small insulated washers (23).
c. Large insulator washer (24).
d. Steel washer (25) and nut (26). Tighten the nut securely.

6. Align the tab on the brush plate with the notch in the housing (**Figure 95**).

Armature Inspection

1. Inspect the commutator (A, **Figure 96**) for abnormal wear or discoloration; neither condition can be repaired and requires armature replacement.
2. The mica in a good commutator is below the surface of the copper bars. On a worn commutator the mica and copper bars may be worn to the same level (**Figure 97**). If necessary, undercut the mica between each pair of bars.
3. Inspect the bearings (B, **Figure 96**) for roughness or damage. The bearings are not available separately.
4. Use an ohmmeter to make the following tests:
   a. Check for continuity between the commutator bars (**Figure 98**); there should be continuity between pairs of bars.
   b. Check for continuity between the commutator bars and the shaft (**Figure 99**); there should be no continuity.

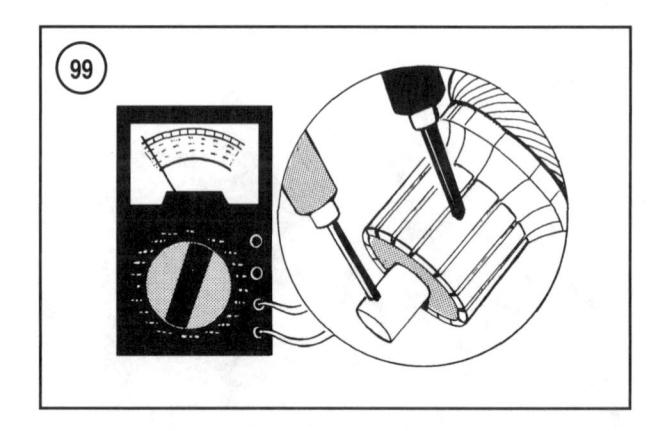

5. If the commutator fails either test in Step 4, or if the armature bearings are damaged, replace the starter assembly. The armature and bearings are not available separately.

Reduction Gear Case
Disassembly/Inspection/Reassembly

The reduction gear case (**Figure 100**) can be disassembled for inspection and lubrication only. Replacement parts are not available.

ELECTRICAL SYSTEM

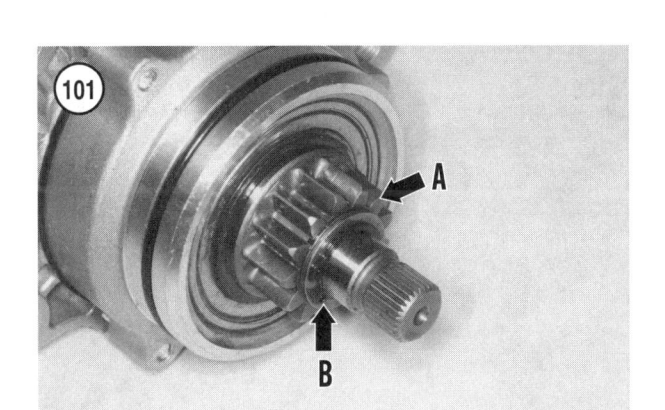

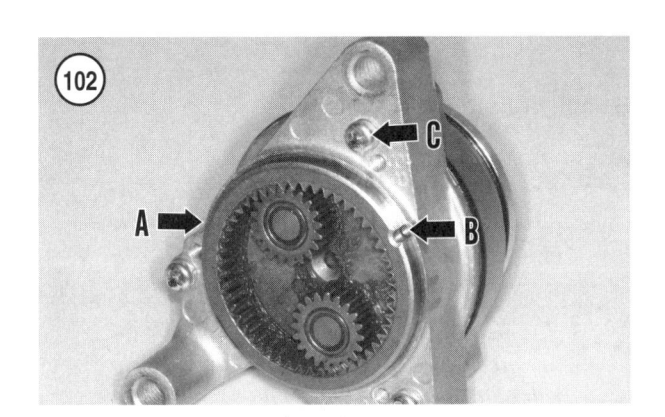

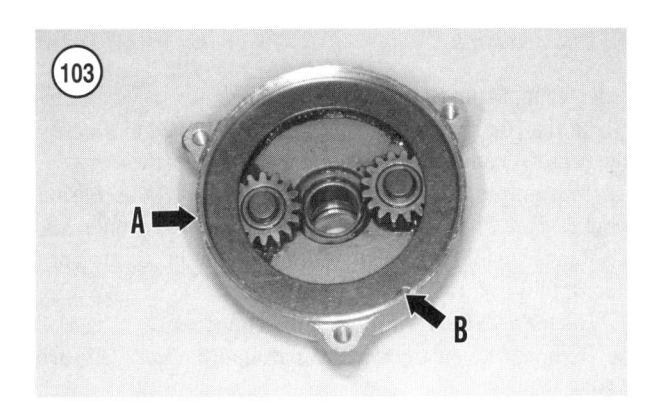

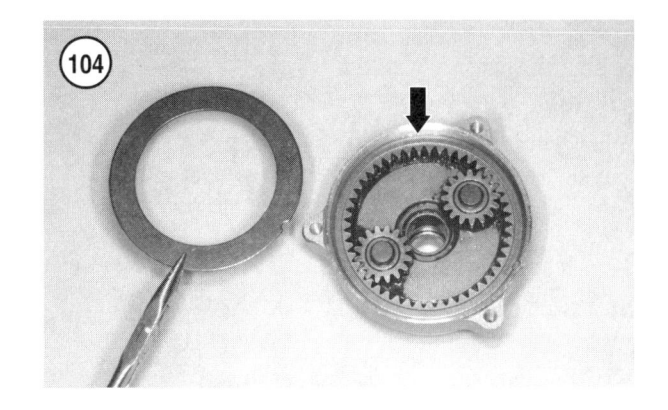

1. Inspect the reverse drive gear (A, **Figure 101**) for damage.

2. Remove the snap ring (B, **Figure 101**) and washers from the starter shaft.

3. Remove the starter reduction gears (A, **Figure 102**) and pin (B).

4. Remove the screws (C, **Figure 102**) and separate the starter reduction gearcase from the reverse gearcase. Remove the O-ring (5, **Figure 81**).

5. Remove the washer (A, **Figure 103**) and inspect the reverse gears (**Figure 104**) for damage.

6. Remove the shaft from the starter reduction gearcase.

7. Inspect the shaft and starter reduction gears (**Figure 105**) for damage.

8. Replace the reduction gear case as an assembly if there is any damage.

9. Replace the O-ring (5, **Figure 81**) if damaged.

10. Reverse these steps to assemble the reduction gear case. Note the following:

    a. Lubricate the O-ring installed inside the starter reduction gearcase and all the reduction gears with molybdenum disulfide grease.

    b. Install the washer by aligning its groove with the pin (B, **Figure 103**).

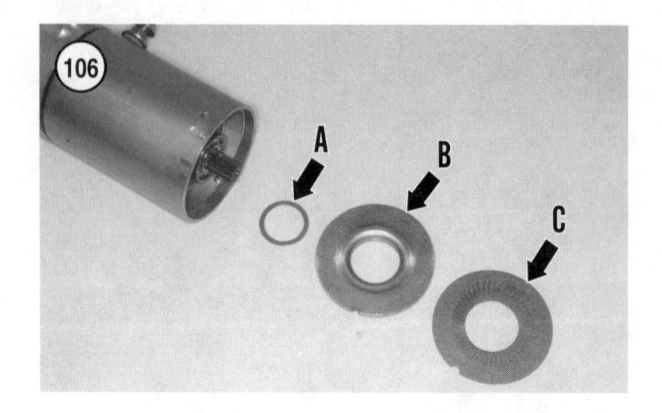

c. Assemble the reduction gearcase halves by aligning the groove with the pin.

d. Tighten the gearcase screw (C, **Figure 102**) securely.

e. Align the outer starter reduction gear with the pin (B, **Figure 102**).

Starter Assembly

1. Install the O-ring (17, **Figure 81**) into the housing groove.

2. Set the brushes onto the brush holder.

3. Install the brush plate into the housing.

4. Align the tab on the brush plate with the notch in the housing (**Figure 95**).

5. Push the brushes back and install the armature (B, **Figure 87**). Release the brushes so they seat against the commutator.

6. Install the washer (**Figure 86**) into the end cover bore.

7. Install the end cover (**Figure 85**) by aligning its slot with the tab on the brush plate.

8. Install the washer (A, **Figure 106**), gear holder (B) and washer (C).

9. Install the reduction gearcase assembly (**Figure 84**) by aligning the pin in the housing with the groove in the outer reduction gear.

10. Check the alignment marks made before disassembly (**Figure 82**).

11. Install and tighten the three through bolts (**Figure 83**) securely.

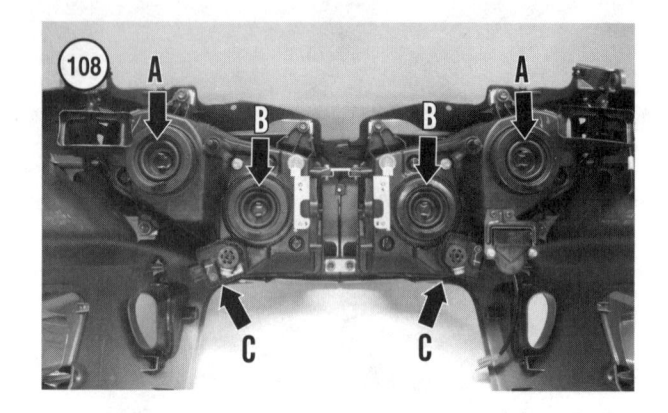

HEADLIGHT

Headlight Bulb Replacement

WARNING
If the headlight just burned out or it was just turned off it will be hot! Do not touch the bulb until it cools off.

CAUTION
*All models use quartz-halogen bulbs (**Figure 107**). Traces of oil on this type of bulb will reduce the life of the bulb. Do not touch the bulb glass. Clean any oil or other chemicals from the bulb with an alcohol-moistened cloth.*

Refer to **Table 6** for bulb specifications.

Figure 108 shows the front fairing removed to identify the high (A) and low (B) beam headlight positions.

1. To remove a high beam bulb, remove the fairing pocket. Refer to *Fairing Pockets* in Chapter Seventeen.

2. Reach inside the fairing and locate the headlight connector (**Figure 109**, typical).

3. Disconnect the headlight bulb connector.

4. Remove the dust cover from around the bulb (**Figure 110**).

ELECTRICAL SYSTEM

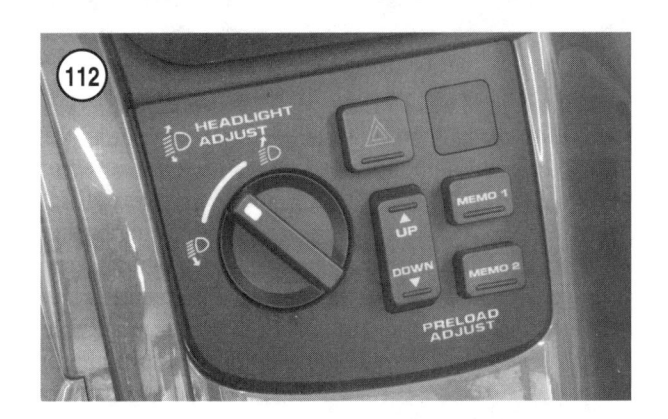

5. Push the bulb retainer (**Figure 111**) down and remove the bulb.
6. Check the connector for dirty or loose fitting terminals.
7. Align the tabs on the new bulb with the notches in the bulb holder and install the bulb.
8. Install the dust cover (**Figure 110**) so its TOP mark is at the top of the housing. Make sure the dust cover fits snugly around the bulb.
9. Plug the connector into the back of the bulb.
10. Install the fairing pocket, if removed.
11. Start the engine and check the headlight operation. If necessary, perform the *Headlight Adjustment* in this section.

Headlight Adjustment

Proper headlight beam adjustment is critical to both rider and to on-coming drivers. Always check the headlight beam adjustment when changing the weight load on the motorcycle.

Use the headlight beam adjustment knob (**Figure 112**) on the left side of the top shelter.
1. Start the engine.
2. To raise the headlight beam, turn the adjustment knob clockwise.
3. To lower the headlight beam, turn the adjustment knob counterclockwise.

Headlight Housing
Removal/Installation

1. Remove the front fairing. Refer to *Front Fairing and Fairing Molding* in Chapter Seventeen.
2. Remove the screws and the bank angle sensor (A, **Figure 113**) from the right headlight housing.
3. Remove the two bolts, nuts and joint plate beside the headlight housing being removed. See A, **Figure 114** (left) or B, **Figure 113** (right).

4. Remove the four bolts and the headlight housing. See B, **Figure 114** (left) or C, **Figure 113** (right).
5. Installation is the reverse of removal.
6. Check the headlight adjustment.

HEADLIGHT ADJUSTERS

Figure 108 shows the front fairing removed to identify the headlight adjusters (C).

System Check

If the headlight adjuster does not work, perform the following:
1. Turn the ignition switch on.
    a. If the low beams are on, go to Step 2.
    b. If not, check the No. 9 fuse. Replace the fuse if blown.
2. Turn the ignition switch off. Remove the inner fairings. See Chapter Seventeen under *Inner Fairings*. Disconnect the gray 3-pin connector from each headlight adjuster. See **Figure 115** (left) and **Figure 116** (right). Check each connector for contamination or loose terminals.
3. Switch an ohmmeter to R × 1 and check for continuity between the gray 3-pin harness connector green terminal and ground.
    a. If continuity is present, go to Step 4.
    b. If not, repair the open in the green wire.
4. Start the engine and run at idle speed. Check for battery voltage between the gray 3-pin harness connector brown/red terminal (+) and ground (−).
    a. Yes. Go to Step 5.
    b. No. Check the brown/red wire between the headlight adjuster and its relay for an open circuit. If good, perform the tests under *Headlight Adjuster Relay* in this chapter.
5. Connect a voltmeter between the gray 3-pin harness connector light green wire terminal (+) and ground (−). With the engine still running at idle speed, turn the headlight adjust knob (**Figure 112**). The voltage should vary from 1.2 to 10.8 volts.
    a. Yes: Replace the headlight adjuster.
    b. No: Check the light green wire between the headlight adjuster and headlight adjusting switch for an open or short circuit. If good, perform the tests under *Headlight Adjusting Switch* in this chapter.

Headlight Adjuster Replacement

1. Turn the ignition switch off.

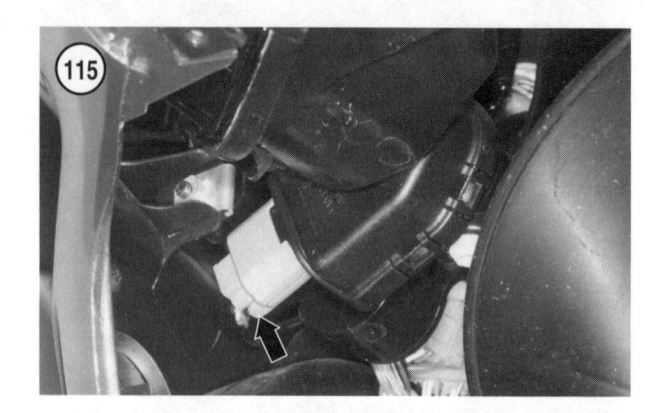

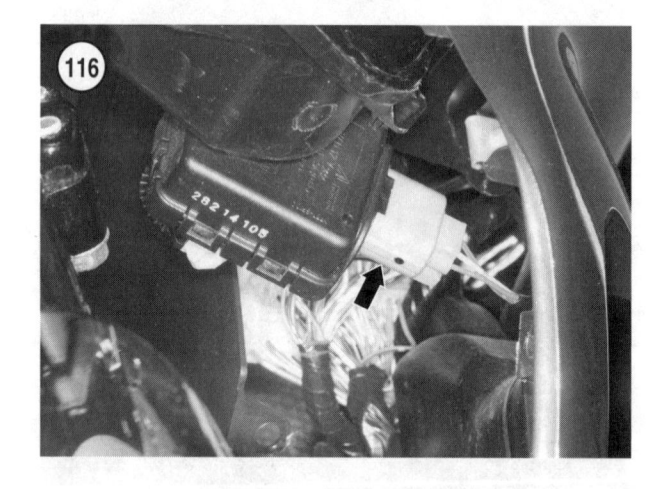

2. Remove the inner fairings. Refer to *Inner Fairings* in Chapter Seventeen. Disconnect the gray 3-pin connector from each headlight adjuster. See **Figure 115** (left) and **Figure 116** (right).

3A. Left headlight adjuster: Turn the headlight adjuster (**Figure 117**) counterclockwise; see A, **Figure 118**. Then turn the adjusting bolt (B, **Figure 18**) counterclockwise and remove the headlight adjuster from the headlight housing.

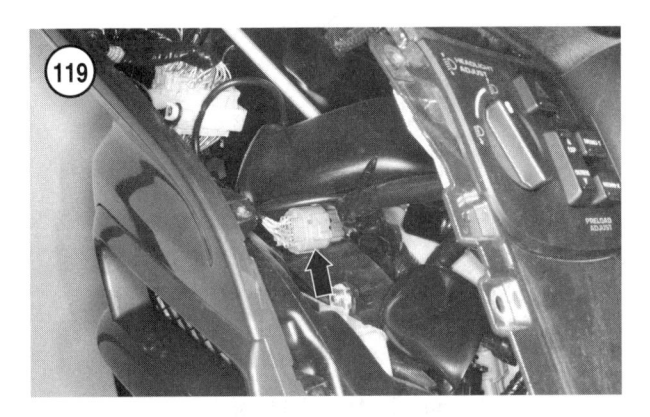

ELECTRICAL SYSTEM

System Check

1. Turn the ignition switch off.
2. Remove the left fairing pocket. Refer to *Fairing Pockets* in Chapter Seventeen.
3. Disconnect the gray 14-pin left panel switch connector. **Figure 119** shows the connector with the top shelter partially removed for clarity.
4. Switch an ohmmeter to R × 1000 and measure resistance between the blue/green and yellow/red terminals in the gray 14-pin left panel switch harness connector (headlight adjusting switch). The resistance should be 4.0-6.0 K ohms.

   a. Yes. Go to Step 4.
   b. No. The headlight adjusting switch is faulty. Replace the panel switch assembly as described in this chapter.

5. Switch an ohmmeter to R × 1. Connect the test leads between the yellow/blue and blue/green terminals on the switch side of the gray 14-pin connector. Operate the headlight adjust switch (**Figure 112**) while reading the ohmmeter. The resistance should increase when the adjuster is turned from its low to high position. The resistance should decrease when the adjuster is turned from its high to low position.

   a. Yes. Go to Step 6.
   b. No. Replace the panel switch assembly as described in this chapter.

6. Check for continuity between the gray 14-pin harness connector green terminal and ground.

   a. Yes. Go to Step 7.
   b. No. Check the green wire between the 14-pin harness connector and ground for an open circuit.

7. Start the engine and run at idle speed. Check for battery voltage between the gray 14-pin harness connector brown/red terminal (+) and ground (−).

   a. Yes. The system is normal.
   b. No. Check the brown/red wire between the headlight adjuster relay and panel switch connector for an open circuit. If good, check the headlight adjuster relay system as described in the following procedure.

3B. Right headlight adjuster: Turn the headlight adjuster clockwise. Then turn the adjusting bolt counterclockwise and remove the headlight adjuster from the headlight housing.

4. Installation is the reverse of removal. Check the headlight beam adjustment as described in this chapter.

HEADLIGHT ADJUSTING SWITCH

The headlight adjusting switch is mounted in the panel switch assembly (**Figure 112**).

HEADLIGHT ADJUSTER RELAY

The headlight adjuster relay (7, **Figure 120**) is located in the main relay box underneath the seat. Refer to *Relay Box* in this chapter for more information on identifying and testing the relays.

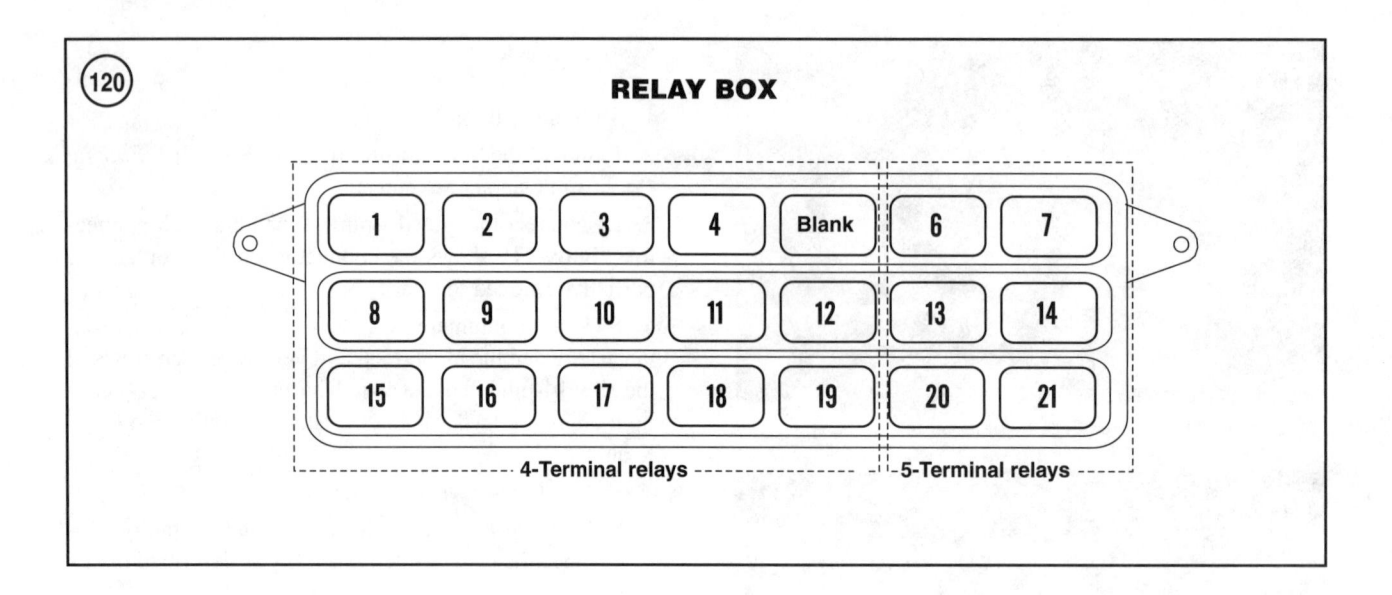

120 RELAY BOX

4-Terminal relays — *5-Terminal relays*

(Positions 1–14 are 4-terminal relays; positions 6, 7, 13, 14, 20, 21 area shown with 5-terminal relays. Position 5 is Blank.)

Relay and Circuit Testing

Refer to the wiring diagrams at the end of the manual to identify the headlight adjuster relay circuit referred to in this procedure.

1. Turn the ignition switch on. The oil indicator should come on.
   a. Yes. Go to Step 2.
   b. No. Check the oil pressure switch as described in Chapter Three under *Engine Oil and Filter*.
2. Remove the seat. Refer to *Seat* in Chapter Seventeen.
3. Remove the screws and relay box cover.
4. Replace the headlight adjuster relay (7, **Figure 120**) with a known good 5-terminal relay. Start the engine and allow it to idle. Operate the headlight beam adjust knob (**Figure 112**). The headlight adjusters should work.
   a. Yes. The original headlight adjuster relay was faulty. Install a new 5-terminal relay and repeat the test.
   b. No. Go to Step 5.

NOTE
The meter test connections called out in Steps 4-6 are made at the relay connector terminals inside the relay box (Figure 121).

5. Turn the ignition switch off and remove the headlight adjuster relay (7, **Figure 120**).
6. Check for continuity between the blue/white wire terminal (**Figure 121**) and ground. Start the engine and allow it to idle. There should be no continuity.
   a. No continuity: Go to Step 7.
   b. Continuity: Check for an open circuit in the blue/white wire between the oil pressure switch diode (D17) and the headlight adjuster relay. If good, replace the oil pressure switch diode as described in this section.

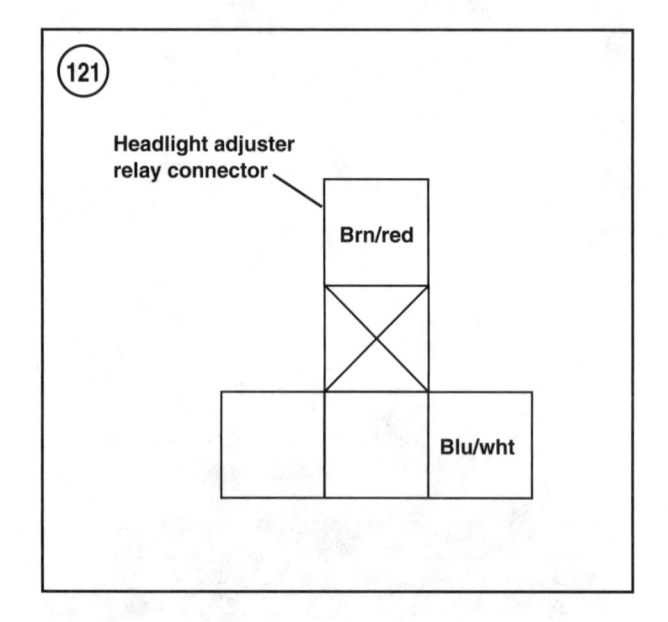

121 Headlight adjuster relay connector — Brn/red, Blu/wht

7. Turn the engine off. Then turn the ignition switch on but do not start the engine. Check for battery voltage between the headlight adjuster relay brown/red terminal (+) (**Figure 121**) and ground (−).
   a. If battery voltage is present, the system is normal. Check the connectors for contamination or loose terminals.
   b. If not, check for an open circuit in the brown/red wire between the No. 9 fuse and the headlight adjuster relay.

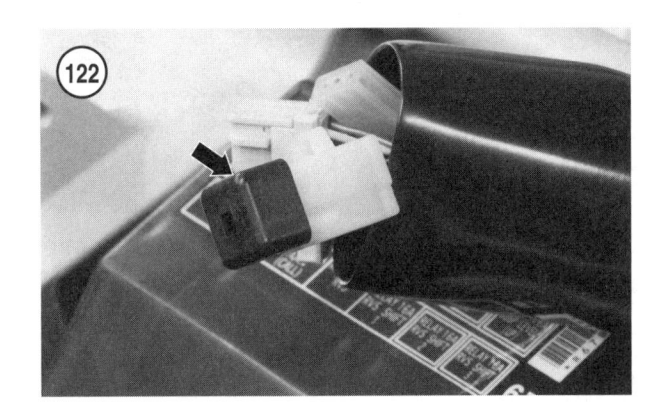

ELECTRICAL SYSTEM

2. Disconnect the oil pressure switch diode (**Figure 122**) from its connector.
3. Set an ohmmeter to the R × 1 scale.
4. Check continuity between the positive and negative diode terminals (**Figure 123**). At each pair of opposite terminals (positive and negative), check the continuity in one direction, reverse the test leads, and check the continuity in the opposite direction. Each pair should have continuity in one direction and no continuity when the test leads are reversed.
5. Replace the diode if it fails this test.

FRONT TURN SIGNAL LIGHT

Bulb Replacement

1. Remove the screw from the bottom of the rearview mirror housing and remove the mirror.
2. Turn and remove the bulb holder (**Figure 124**) and replace the blown bulb.
3. Install the bulb holder by aligning its arrow mark with the arrow mark on the housing. Turn the holder to lock.
4. Install the mirror and secure with the screw.

TRUNK BRAKE/TAILLIGHT

Bulb Replacement

1. Open the trunk.
2. Remove the three Acorn nuts (**Figure 125**) securing the lens housing to the trunk and remove the lens housing from the outside of the trunk (**Figure 126**). Do not remove the Phillips screws.
3. Disconnect the two connectors (**Figure 126**) and remove the housing.
4. Turn the bulb holder and remove it from the housing. Turn the bulb holder so the arrow mark on the holder turns toward the round mark on the housing.

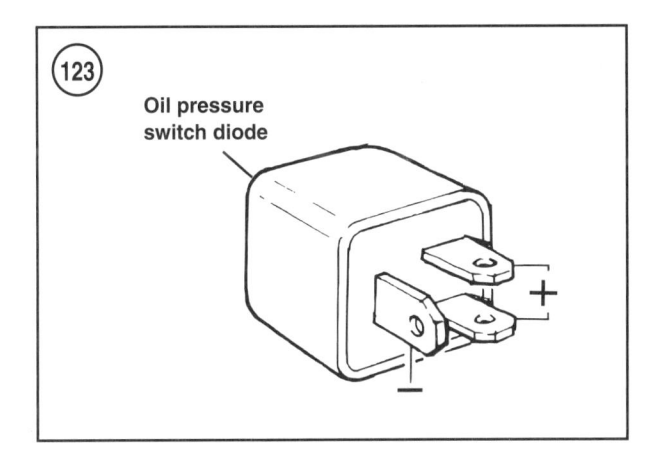

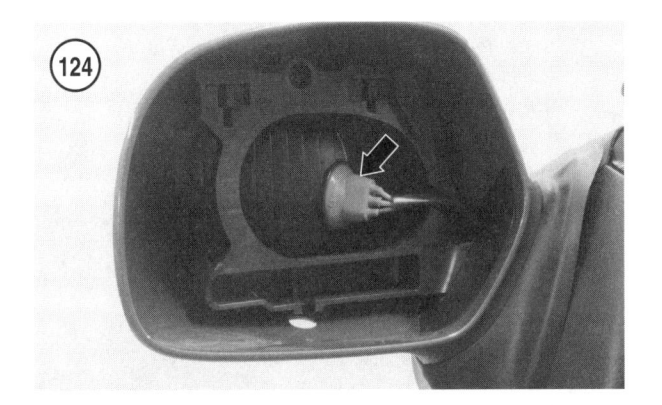

8. Install the headlight adjuster relay (**Figure 120**).
9. Install the relay box cover and secure with the mounting screws.
10. Install the seat. See Chapter Seventeen under *Seat*.

Oil Pressure Switch Diode
Testing/Replacement

1. Remove the seat. Refer to *Seat* in Chapter Seventeen.

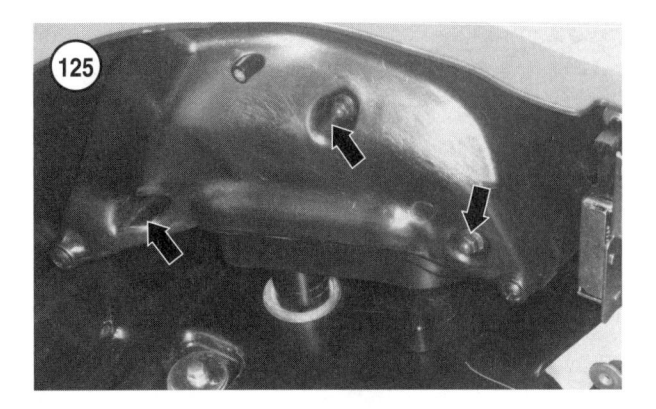

5. Pull the bulb straight out of the holder to remove. Insert a new bulb.

6. Align the arrow mark on the bulb holder with the round mark on the housing and install the holder into the housing. Turn the holder and align the arrow mark on the holder with the arrow mark on the housing.

7. Turn the ignition switch on and check the taillight and brake light operation.

LICENSE PLATE LIGHT

Bulb Replacement

The license plate lens is mounted through an opening in the bottom of the trunk lower cover.

1. If necessary, remove the license plate and frame. Do not remove the license plate holder from rear fender A.
2. Remove the screws and lens cover (**Figure 127**).
3. Pull the bulb straight out of its socket. Do not turn or twist the bulb.
4. Install the bulb by reversing these steps. The two lens screws are self-tapping. Do not overtighten them as this may strip the mating threads in the lens holder.
5. Turn the ignition switch on and check the license plate light operation.

SADDLEBAG COMBINATION LIGHT

Brake/Taillight and Turn Signal Bulbs

The lens housing must be removed to replace these bulbs.

1. Open the saddlebag.
2. Remove the three Acorn nuts (**Figure 128**) inside and toward the rear of the saddlebag, and pull the lens housing away from the saddlebag (**Figure 129**).

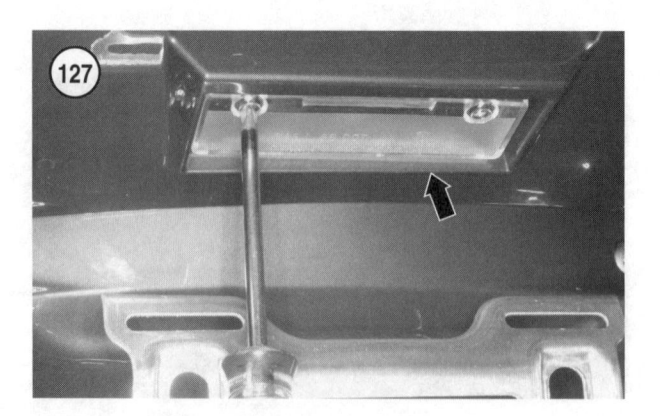

CAUTION
The threaded studs mounted on the lens housing can scratch the saddlebag. Place a towel between the saddlebag and lens housing before replacing the blown bulb.

NOTE
If disconnecting both connectors, note the TURN decal mounted on the turn signal wire harness. These bulb holders can be installed incorrectly. Identify the connectors before disconnecting them. If necessary, use the wiring diagram to identify the wires and connector.

3. Turn the bulb holder (A or B, **Figure 130**) so the arrow mark on the holder aligns with the round mark on housing and remove from housing. Both connectors can be removed without having to disconnect the 2-pin connector.

4. Pull the bulb straight out of the holder to remove. Insert a new bulb.

5. Align the arrow mark on the bulb holder with the round mark on the housing and install the holder into the housing. Turn the holder and align the arrow mark on the holder with the arrow mark on the housing.

ELECTRICAL SYSTEM

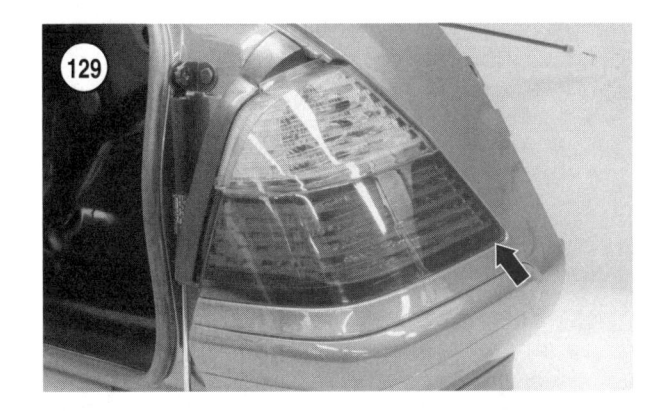

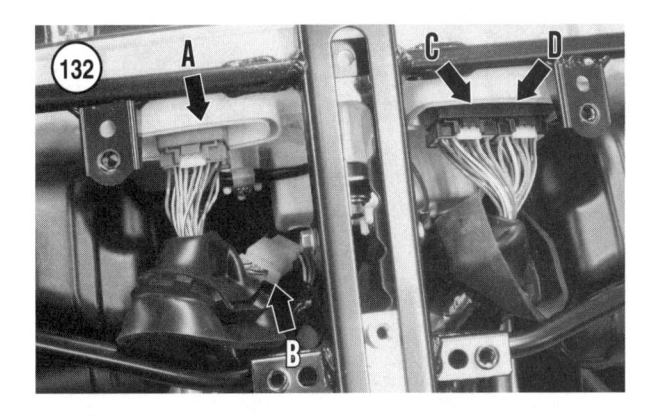

6. Turn the ignition switch on and check the taillight and brake light operation.
7. Position the lens housing flush against the saddlebag and secure with the three Acorn nuts.

COMBINATION METER

Removal/Installation

1. Disconnect the negative battery cable as described in this chapter.
2. Remove the front fairing. Refer to *Front Fairing and Fairing Molding* in Chapter Seventeen.
3. Remove the three nuts securing the combination meter (**Figure 131**) to the faring bracket.
4. Slide the boots away from the two combination meter connectors.
5. Disconnect the following connectors at the combination meter **Figure 132**:
   a. Blue 20-pin connector (A).
   b. White 3-pin connector (B).
   c. Black 20-pin connector (C).
   d. Black 16-pin connector (D).
6. Remove the combination meter.
7. Installation is the reverse of removal. Note the following:
   a. Seat the connector boots fully into the combination meter housing grooves.
   b. Start the engine and check all the illumination lights and gauges for proper operation.

Power/Ground Circuit Test

1. Remove the combination meter.
2. Check the combination meter electrical connectors for dirty or loose fitting terminals.

> *NOTE*
> *Perform the following tests on the combination meter wiring harness side connector terminals, not on the meter side.*

3. Power source circuit: Turn the ignition switch on and check for battery voltage between the blue 20-pin connector brown/white terminal (+) and ground (−).
   a. Yes. The power circuit is good.
   b. No. Check the brown/white wire between the combination meter and taillight relay for an open circuit. If good, test the 4-terminal taillight relay as described under *Relay Box* in this chapter.
4. Accessory power circuit: Turn the ignition switch on and check for battery voltage between the blue 20-pin

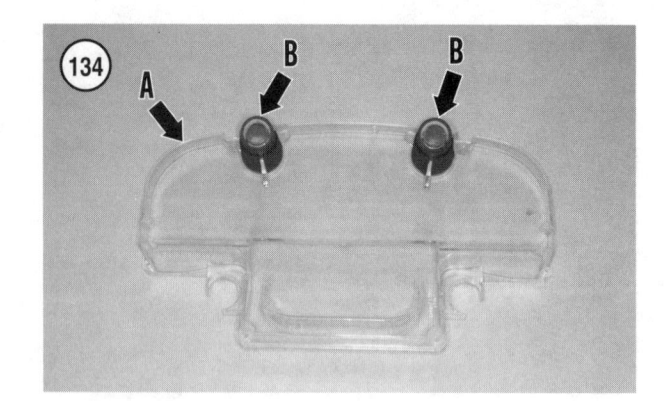

connector light green/black terminal (+) and ground (−). Then check with the ignition switch in its ACC position.
   a. Yes. The accessory power circuit is good.
   b. No. Check the light green/black wire between the combination meter and accessory relay for an open circuit. If good, test the 4-terminal accessory relay as described under *Relay Box* in this chapter.
5. Back-up power circuit: Turn the ignition switch off and check for battery voltage between the blue 20-pin connector red/yellow terminal (+) and ground (−).
   a. Yes. The back-up power circuit is good.
   b. No. Check for a blown No. 22 fuse. If good, check the red/yellow wire between the combination meter and No. 22 fuse for an open circuit.
6. Ground circuit: Turn the ignition switch off. Check for continuity between the black 20-pin connector green terminal and ground.
   a. Yes. The ground circuit is good.
   b. No. Check the green wire between the combination meter and ground terminal for an open circuit.
7. Sensor ground circuit: Turn the ignition switch off. Check for continuity between the blue 20-pin connector green terminal and ground.
   a. Yes. The sensor ground circuit is good.
   b. No. Check the green wire between the combination meter and ground terminal for an open circuit.

Combination Meter Disassembly/Reassembly

This section describes disassembly of the combination meter assembly (**Figure 133**) and replacement of the gauges and circuit board.
1. Remove the screws and meter lens (A, **Figure 134**).

NOTE
The turn signal lenses (B, Figure 134) are an integral part of the meter lens assembly. Do not remove them.

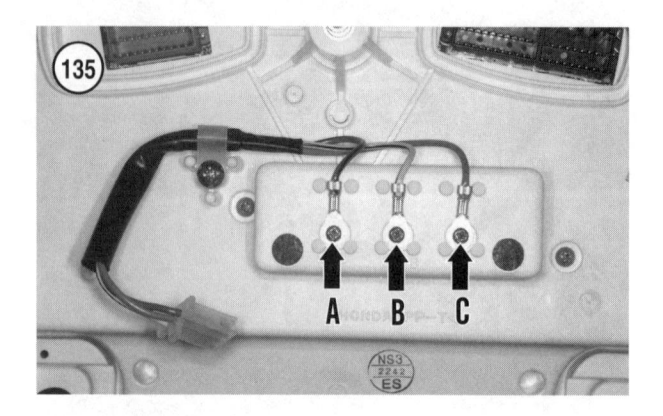

ELECTRICAL SYSTEM

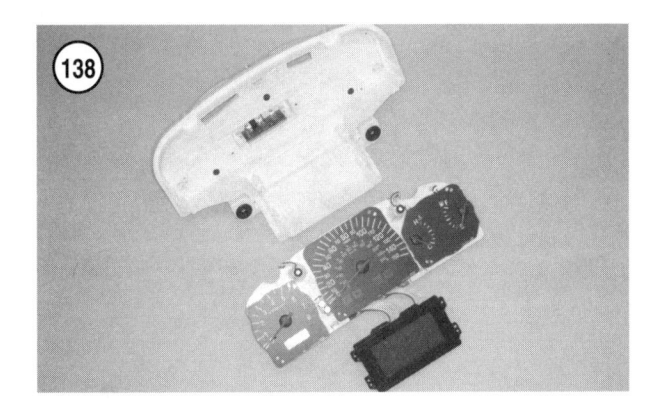

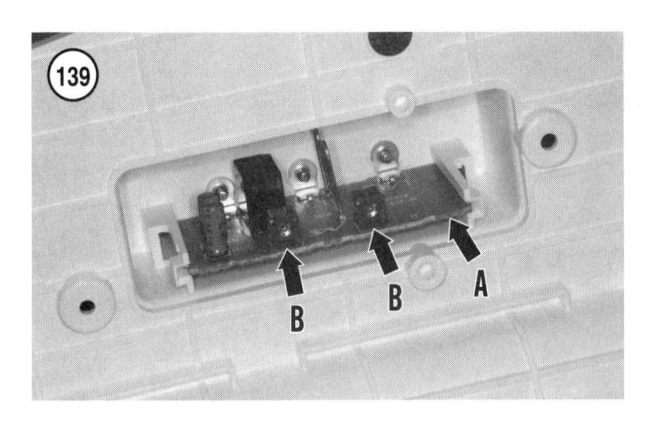

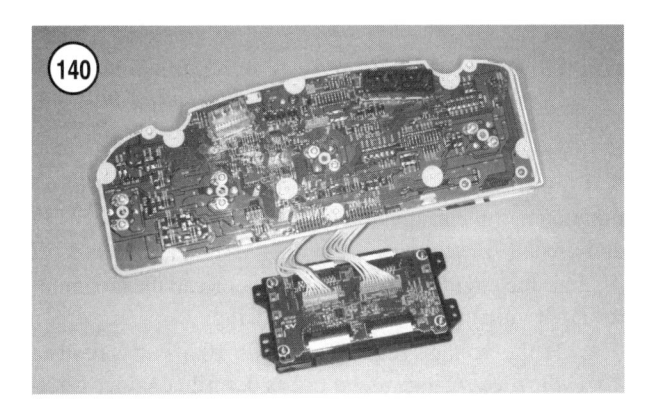

2. If the turn signal relay circuit board is to be replaced, remove the three terminal screws (**Figure 135**) from the backside of the lower case.

3. Turn the combination meter over and remove the screws securing the meter visor to the lower case and remove the meter visor (**Figure 136**).

4. Turn the combination meter over.

5. Remove the screws securing the LCD (**Figure 137**) to the lower case.

6. Lift the meter/gauge and LCD assembly from the lower case and remove them (**Figure 138**).

7. To replace the turn signal relay circuit board (A, **Figure 139**):
   a. Slide the board out of the lower case.
   b. Install the new board into the lower case grooves with the terminal eyelets (B, **Figure 139**) facing toward the rear of the lower case.

8. To replace the LCD:
   a. Position the meter/gauge and LCD assembly with the circuit board side facing up (**Figure 140**).
   b. Disconnect the two connectors (**Figure 141**) at the LCD.
   c. Connect the two connectors at the new LCD unit.

9. To replace the speedometer or tachometer:
   a. Position the meter/gauge and LCD assembly with the circuit board side facing up (**Figure 140**).
   b. Remove the screws securing the speedometer or tachometer to the circuit board.
   c. Disconnect and remove the unit.
   d. Reverse to install.

10. Place the meter/gauge and LCD assembly into the lower case (**Figure 137**). Make sure the alignment pins in the meter/gauge assembly fit into the holes in the lower case.

11. Align the two holes in the LCD frame with the two pins in the lower case. Then install and tighten the four LCD screws.

12. If the turn signal relay circuit board was replaced, reconnect the wires and tighten the three terminal screws (**Figure 135**) at the backside of the lower case. Identify the wire colors as follows:
   a. Blue/black (A).
   b. Light green/black (B).
   c. Gray (C).

13. Install the meter visor to hold the meter/gauge assembly in place. Hold the meter visor (to prevent it from falling off) and turn the assembly over so the backside faces up.

14. Install the meter/gauge screws and tighten securely.

15. Turn the combination meter over and install the meter lens and its mounting screws.

SPEEDOMETER

Circuit Test

If the speedometer does not operate, perform the following procedures:

1. Turn the ignition switch on. The fuel gauge, tachometer and coolant temperature gauge should operate correctly.
   a. Yes. Go to Step 2.
   b. No. Perform the power source circuit and sensor ground circuit tests under *Combination Meter, Power/Ground Circuit Test* in this chapter.
2. Perform the *Self-Diagnostic Check* under *Malfunction Indicator Lamp (MIL)* in Chapter Eight. The MIL should flash.
   a. No. Go to Step 3.
   b. Yes—DTC 11. Refer to *DTC Fuel Injection Flow Charts* in Chapter Eight and perform the steps under flow chart *DTC 11: Loose VSS Connector, Open or Short Circuit in VSS Wire or Faulty VSS*.
3. Remove the combination meter as described in this chapter.
4. Support the motorcycle on its centerstand and shift the transmission into neutral.
5. Turn the ignition switch on and measure voltage between the black 20-pin white/black (+) and green (–) connector terminals. Turn the rear wheel slowly while reading the voltmeter. The voltmeter should alternately read between 0-5 volts.
   a. Yes. Replace the meter/gauge assembly as described under *Combination Meter* in this chapter.
   b. No. Check for an open circuit in the white/black wire between the combination meter and speed sensor.
6. Install all parts previously removed.

TACHOMETER

Circuit Test

If the tachometer does not operate, perform the following procedures:

1. Turn the ignition switch on. The fuel gauge, speedometer and coolant temperature gauge should operate correctly.
   a. Yes. Go to Step 2.
   b. No. Perform the power source circuit and sensor ground circuit tests under *Combination Meter, Power/Ground Circuit Test* in this chapter.
2. Remove the combination meter as described in this chapter.

NOTE
*Refer to **Peak Voltage Tests and Equipment** under **Ignition System Troubleshooting** in this chapter when performing Step 3.*

3. Connect a peak voltage tester to the black 20-pin combination meter wiring harness connector yellow/green (+) and green (–) terminals.
4. Start the engine and allow it to idle. Read the tachometer input voltage and note the following:
   a. If the voltage reading exceeds 10.5 volts, replace the meter/gauge assembly as described under *Combination Meter* in this chapter.
   b. If there is a voltage reading but it is less than 10.5 volts, replace the ECM as described in Chapter Eight.
   c. If there is no voltage reading, go to Step 5.
5. Turn the ignition switch OFF and disconnect the gray 22-pin ECM connector. Refer to *Engine Control Module (ECM)* in Chapter Eight.
6. Check for continuity in the yellow/green wire between the gray 22-pin ECM and the black 20-pin combination meter harness connectors.
   a. Yes. Go to Step 7.
   b. No. Check the yellow/green wires for an open circuit or connector damage.

ELECTRICAL SYSTEM

ECT sensor terminal (144)

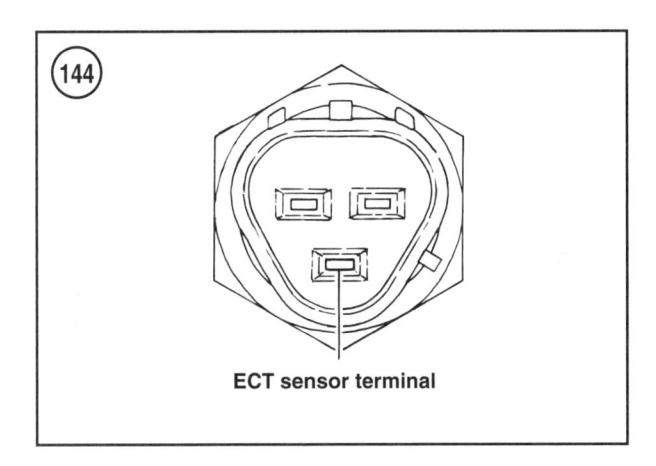

7. Check for continuity in the yellow/green wire between the gray 22-pin ECM harness connector and ground. Then check for continuity between yellow/green connector terminal in the black 20-pin combination meter harness connector and ground. There should be no continuity.
   a. If continuity is present during either test, repair the short in the yellow/green wire(s).
   b. If not, the ECM is faulty. Replace the ECM and retest.
8. Install by reversing these steps.

COOLANT TEMPERATURE GAUGE

Circuit Test

If the coolant temperature gauge does not operate, perform the following procedures:
1. Turn the ignition switch on. The fuel gauge, speedometer and tachometer should operate correctly.
   a. Yes. Go to Step 2.
   b. No. Perform the power source circuit and sensor ground circuit tests under *Combination Meter, Power/Ground Circuit Test* in this chapter.
2. Remove the left radiator stay. Refer to *Radiators* in Chapter Ten.
3. Turn the ignition switch off.
4. Disconnect the 3-pin ECT sensor connector (**Figure 142**).
5. Connect a jumper wire between the green/black terminal in the harness side of the 3-pin connector and ground. Turn the ignition switch on. The needle on the coolant temperature gauge should move to H.
   a. Yes. Test the ECT sensor as described in this chapter.
   b. No. Go to Step 6.
6. Remove the combination meter as described in this chapter.

7. Check for continuity in the green/black wire between the 3-pin ECT and blue 20-pin combination meter harness side connectors.
   a. Yes. The circuit is normal.
   b. No. Replace the meter/gauge assembly as described under *Combination Meter* in this chapter.
8. Install all parts previously removed.

ECT SENSOR

Inspection/Replacement

The engine must be cold for this test, preferably not operated for at least 12 hours.
1. Remove both radiators. Refer to *Radiators* in Chapter Ten.
2. Remove the intake manifold mounting bolts. Then raise and support the manifold to access the ECT sensor.
3. Loosen and remove the ECT (**Figure 143**) from the left cylinder head.
4. Place the ECT in a pan filled with a 50:50 mixture of coolant (water/antifreeze). Support the ECT so its sensor tip, not its threads, are not covered by the coolant. Make sure the bottom part of the ECT is at least 40 mm (1.57 in.) away from the bottom of the pan.
5. Place a shop thermometer in the pan. Use a thermometer that is rated higher than the test temperature (**Table 7**).
6. Heat the coolant and check resistance between the ECT terminal identified in **Figure 144** and the threads (ground) on the ECT. Maintain the coolant at the temperatures specified in **Table 7** for three minutes before determining the actual resistance readings.
7. If the resistance value is out of specification by more than 10%, replace the ECT.
8. Install the ECT into the cylinder head (**Figure 143**) using a *new* gasket and tighten to 25 N•m (18 ft.-lb.).
9. Reverse Step 1 and Step 2 to complete installation.
10. Fill and bleed the cooling system.

FUEL GAUGE (LOW FUEL INDICATOR)

Circuit Test

If the low fuel indicator does not work, perform the following.

WARNING
Fuel vapors are present when testing and replacing the fuel level sensor. Also, when working with the fuel level sensor, some fuel may spill. Because gasoline is extremely

flammable, perform this procedure away from all open flames, including appliance pilot lights and sparks. Do not smoke or allow someone who is smoking in the work area, as an explosion and fire may occur. Always work in a well-ventilated area. Wipe up spills immediately.

1. Turn the ignition switch on. The fuel gauge, speedometer, tachometer and coolant temperature gauge should operate correctly.
   a. Yes. Go to Step 2.
   b. No. Perform the power source circuit and sensor ground circuit tests under *Combination Meter, Power/Ground Circuit Test* in this chapter.
2. Remove the seat. Refer to *Seat* in Chapter Seventeen.
3. Turn the ignition switch off and disconnect the white 2-pin fuel level sensor (**Figure 145**) connector.
4. Turn the ignition switch on. The fuel gauge should move to E and the fuel level indicator should come on.
   a. Yes. Test the fuel level sensor as described in the following procedure.
   b. No. Go to Step 5.
5. Turn the ignition switch off. Connect a jumper wire between the gray/black terminal in the harness side of the white 2-pin connector and ground. Turn the ignition switch on. The gauge needle should move to F, but the fuel level indicator should not light.
   a. Yes. Test the fuel level sensor as described in the following section.
   b. No. If only one of the tests was correct (either the gauge or indicator), replace the meter/gauge assembly as described under *Combination Meter* in this chapter.
   c. No. Both the indicator and gauge did not work correctly. Go to Step 6.
6. Remove the combination meter as described in this chapter.
7. Check for continuity in the gray/black wire between the white 2-pin fuel level sensor harness connector and the blue 20-pin combination meter harness connector.
   a. Yes. Go to Step 8.
   b. No. Check the gray/black wires for an open circuit or connector damage.
8. Check for continuity between the gray/black wire in the white 2-pin fuel level sensor harness connector and ground. Then check for continuity between the gray/black wire in the blue 20-pin combination meter harness connector and ground. There should be no continuity.
   a. If continuity is present in either test, repair the short in the gray/black wire(s).
   b. If not, replace the meter/gauge assembly as described under *Combination Meter* in this chapter.

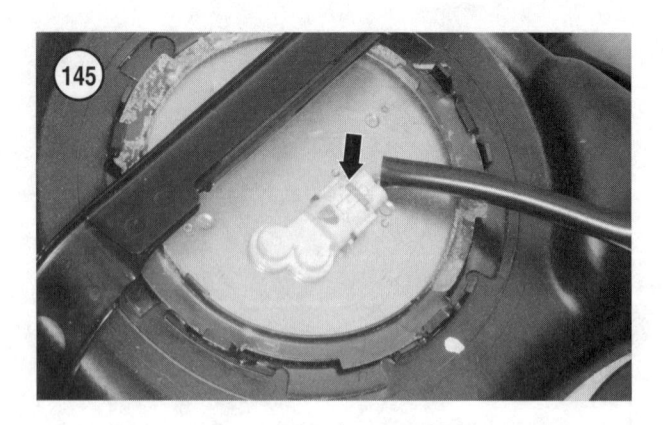

9. Install by reversing these steps.

FUEL LEVEL SENSOR

WARNING
Fuel vapors are present when testing and replacing the fuel level sensor in this section. Also, when working with the fuel level sensor, some fuel may spill. Because gasoline is extremely flammable, perform this procedure away from all open flames, including appliance pilot lights and sparks. Do not smoke or allow someone who is smoking in the work area, as an explosion and fire may occur. Always work in a well-ventilated area. Wipe up spills immediately.

Fuel Level Sensor Test/Replacement

The fuel level sensor can be replaced with the fuel tank mounted on the motorcycle.

The fuel sender/pump wrench (Honda part No. 07ZMA-MCAA201 or 07ZMA-MCAA200) is required to remove and install the fuel level sensor retainer plate.

1. Remove the seat. Refer to *Seat* in Chapter Seventeen.
2. Turn the ignition switch off and disconnect the white 2-pin fuel level sensor (**Figure 145**).
3. Remove the mounting bolts and remove the seat bracket installed over the fuel tank. Note the clamp and wire harness installed on the right mounting bolt; see Chapter Eight under *Fuel Tank*.
4. Using the fuel sender/pump wrench, turn the retainer plate (**Figure 146**) counterclockwise to loosen and remove it.
5. Remove the fuel level sensor and base gasket (**Figure 146**) from the fuel tank.
6. Switch an ohmmeter to R × 1. Measure resistance across the two fuel level sensor terminals by moving the

ELECTRICAL SYSTEM

146 FUEL LEVEL SENSOR

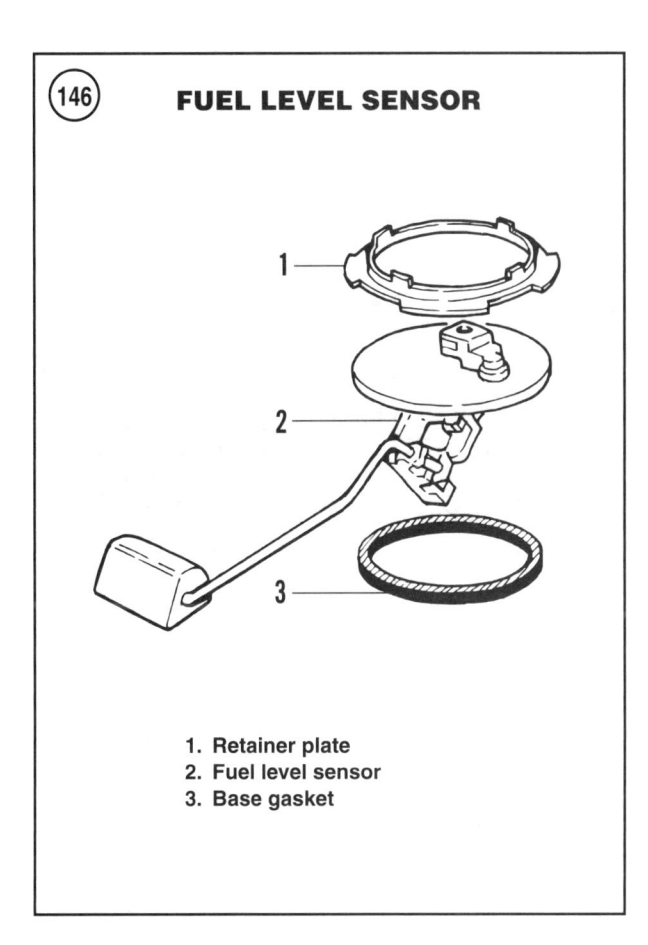

1. Retainer plate
2. Fuel level sensor
3. Base gasket

float arm from its empty to full position; compare the resistance to the test specifications in **Table 7**. If either resistance reading is out of specification, replace the fuel level sensor.

7. Install a new base gasket (3, **Figure 146**).

8. Install the fuel level sensor into the fuel tank by aligning the lugs on the sensor base with the grooves in the fuel tank.

9. Install the retainer plate and tighten it with the fuel sender/pump wrench until it stops.

10. Reverse Steps 1-3 to complete installation.

Fuel Pump Side Fuel Level Sensor Test

1. Remove the fuel pump. Refer to *Fuel Pump* in Chapter Eight.

2. Switch an ohmmeter to R × 1. Measure resistance across the two fuel pump terminals by moving the float arm from its empty to full position and compare to the test specifications in **Table 7**. If either resistance reading is out of specification, replace the fuel pump.

3. Reverse Step 1.

OIL PRESSURE INDICATOR AND OIL PRESSURE SWITCH

Circuit Test 1

Perform this test if the oil pressure indicator does not come on when the ignition switch is turned on.

1. Remove the left front exhaust pipe protector. Refer to *Exhaust System* in Chapter Seventeen.

2. Disconnect the wire at the oil pressure switch (**Figure 147**).

3. Ground the oil pressure switch wire with a jumper wire.

4. Turn the ignition switch on. The oil pressure indicator should light.
   a. Yes. Replace the oil pressure switch and retest.
   b. No. Check the blue/red wire between the oil pressure switch and combination meter for an open circuit.

5. Reverse Step 1 and Step 2 to complete installation.

Circuit Test 2

Perform this test if the oil pressure indicator stays on when the engine is running.

1. Remove the left front exhaust pipe protector. Refer to *Exhaust System* in Chapter Seventeen.

2. Disconnect the wire at the oil pressure switch (**Figure 147**).

3. Switch an ohmmeter to R × 1. Check for continuity between the oil pressure switch wire and ground. There should be no continuity.
   a. No continuity present, check the oil pressure. Refer to *Engine Oil and Filter* in Chapter Three. If the oil pressure is normal, replace the oil pressure switch.

b. If continuity is present, check the blue/red wire between the oil pressure switch and combination meter for a short circuit.

Oil Pressure Switch Replacement

1. Remove the left front exhaust pipe protector. Refer to *Exhaust System* in Chapter Seventeen.
2. Disconnect the wire at the oil pressure switch (**Figure 147**).
3. Loosen and remove the oil pressure switch.
4. Clean the oil pressure switch and crankcase threads of all sealer and oil residue.
5. Apply an RTV sealer to the oil pressure switch threads as shown in **Figure 148**. Do not apply sealer within 3-5 mm (0.1-0.2 in.) from the end of the switch threads.

CAUTION
Allow the RTV sealer to set for 10-15 minutes before installing the oil pressure switch.

6. Install the oil pressure switch and tighten to 12 N•m (106 in.-lb.).
7. Reconnect the wire onto the switch and cover the switch with its rubber boot.
8. Follow the sealer manufacturer's recommendations for drying time, then start the engine and check for leaks.

CAUTION
The oil pressure indicator should go out within 1-2 seconds. If it stays on, shut off the engine immediately and locate the problem. Do not run the engine with the oil pressure indicator on.

CAUTION
Do not overtighten the switch to correct an oil leak, as this may strip the crankcase threads. If the switch is leaking oil after installing it, remove the switch and reclean the threads. Reseal and reinstall the switch.

9. Install the left front exhaust pipe protector (Chapter Seventeen).

NEUTRAL INDICATOR

Circuit Test

1. Shift the transmission into neutral and turn the ignition switch on. The neutral indicator should come on.
    a. Yes. The system is normal.
    b. No. Go to Step 2.
2. Remove the combination meter as described in this chapter.
3. Turn the ignition switch on and check for battery voltage between the black 20-pin combination meter black/white (+) terminal and ground (−).
    a. Yes. Go to Step 4.
    b. No. Check the black/red and black/white wire between the combination meter and reverse position switch for an open circuit. This wire changes color at a connector. Then check the black/brown wire between the reverse position switch and reverse shift switch as described in this chapter. If both circuits are good, test the reverse position switch and reverse shift switch as described in this chapter.
4. Check for continuity between the light green wire in the 20-pin combination meter harness connector and ground.
    a. Yes. Replace the meter/gauge assembly as described under *Combination Meter* in this chapter.
    b. No. Check the light green/red and light green wires between the gear position switch and combination meter for an open circuit. If good, check the gear position switch as described in this chapter. If good, check for a faulty D6 diode as described under *Reverse Regulator* in this chapter.
5. Install all previously removed parts.

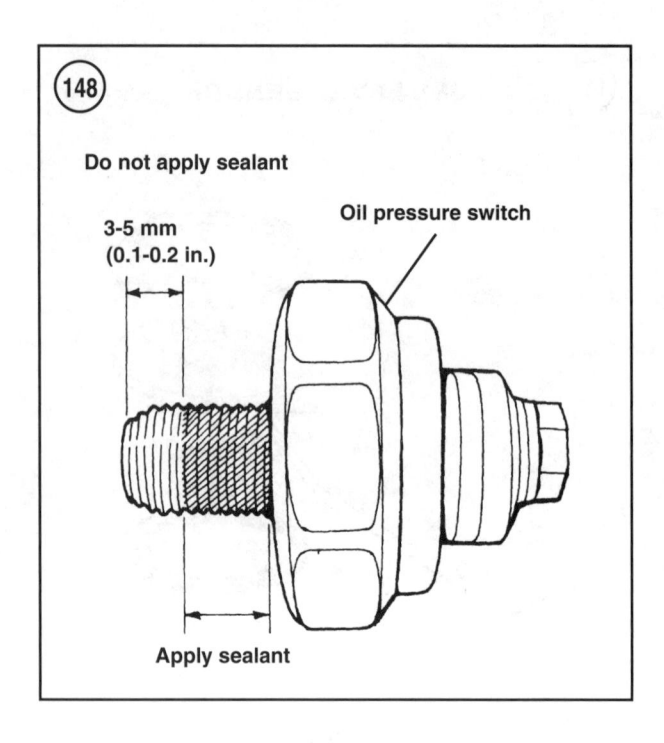

ELECTRICAL SYSTEM

149

TURN SIGNAL RELAY

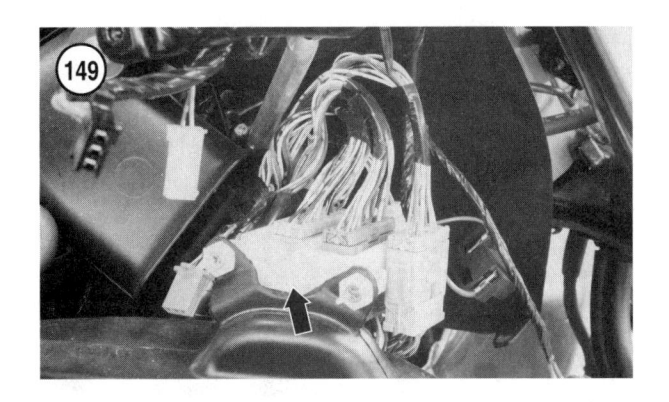

The turn signal relay is a small circuit board (A, **Figure 139**) mounted inside the combination meter. Refer to *Combination Meter Disassembly/Reassembly* in this chapter to replace the relay.

NOTE
When troubleshooting a faulty turn signal, also check fuses no. 21 and no. 27.

Power Circuit Test

1. Remove the front fairing. Refer to *Front Fairing and Fairing Molding*) in Chapter Seventeen.
2. Turn the ignition switch to ON or ACC and measure voltage between the white 3-pin combination meter (B, **Figure 132**) light green/black (+) connector terminal and ground (–). Battery voltage should be present.
3. Reverse Step 1.

Function Test

1. Remove the front fairing. Refer to *Front Fairing and Fairing Molding* in Chapter Seventeen.
2. Turn the ignition switch OFF. Make sure the white 3-pin combination meter connector (B, **Figure 132**) is plugged into the main wiring harness. Ground the blue/green wire terminal in the 3-pin connector (B, **Figure 132**) with a jumper wire.
3. Turn the ignition switch to ON or ACC. The turn signal lights should blink.
4. Reverse Step 1.

TURN SIGNAL CANCEL UNIT

Power/Ground Circuit Test

1. Remove the front fairing. Refer to *Front Fairing and Fairing Molding* in Chapter Seventeen.

2. Disconnect the gray 6-pin main wiring harness connector from the right side connector holder (**Figure 149**).
3. Check the connectors for dirty or loose fitting terminals.

NOTE
Perform the following tests on the gray 6-pin wiring harness side connector terminals, not on the combination meter side.

4. Ground circuit: Turn the ignition switch off. Check for continuity between the green terminal and ground.
   a. Yes. The ground circuit is good.
   b. No. Check the green wire between the 6-pin connector and ground terminal for an open circuit.
5. Turn signal on circuit: Turn the ignition switch off. Check for continuity between the pink terminal and ground when operating the turn signal switch.
   a. If continuity is present when operating the switch to either the left or right side, the circuit is normal.
   b. If not, check the pink wire between the 6-pin connector and the left handlebar switch for an open circuit.
6. Turn signal off circuit: Turn the ignition switch off. Check for continuity between the light green/white terminal and ground when the turn signal switch is pushed.
   a. Yes. The circuit is normal.
   b. No. Check the light green/white wire between the 6-pin connector and the left handlebar switch for an open circuit.
7. Power source circuit: Turn the ignition switch on and check for battery voltage between the 6-pin connector white/green terminal (+) and ground (–).
   a. Yes. The power circuit is good.
   b. No. Check the white/green wire between the 6-pin connector and the horn turn relay. If good, install a new 4-terminal horn turn relay as described under *Relay Box* in this chapter.

NOTE
Perform the tests in Steps 4-7 and repair any damage before testing the turn signal cancel circuit in Step 8.

8. Turn signal cancel circuit: Reconnect the 6-pin main wiring harness connector (**Figure 149**). Turn the ignition switch on and check for battery voltage between the 6-pin connector blue/black terminal (+) and ground (–). There should be no battery voltage when moving the turn signal switch button from side-to-side. There should be battery voltage when the turn signal switch button is pushed.
   a. If the test results are correct, replace the turn signal cancel unit as described in this section.
   b. If not, check for an open or short circuit in the blue/green and blue/black wires between the com-

9

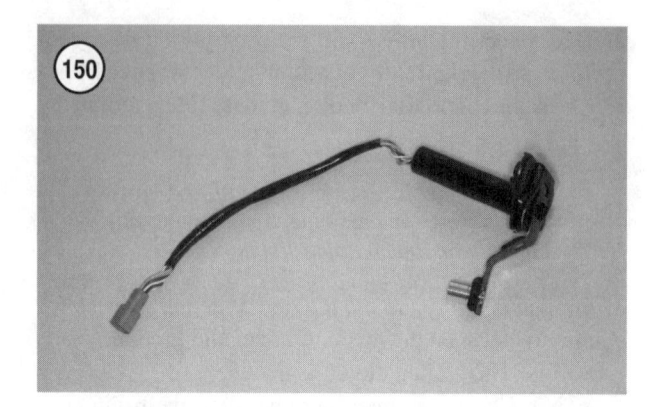

bination meter white 3-pin connector and the 6-pin wiring harness connector. If the wires/connectors are good, check for a faulty turn signal diode as described under *Position Light Relay* in this chapter.

9. Speed pulse circuit: Reconnect the 6-pin main wiring harness connector (**Figure 149**). Support the motorcycle on its centerstand and shift the transmission into neutral. Turn the ignition switch on and measure voltage between the 6-pin connector white/black terminal (+) and ground (−). Turn the rear wheel slowly while reading the voltmeter. The voltmeter should alternately read between 0-5 volts.

   a. Yes. The circuit is normal.

   b. No. Check for an open or short circuit in the white/black wire between the black 20-pin combination meter connector and the 6-pin main wiring harness connector.

10. Install previously removed parts.

Turn Signal Cancel Unit Test

The turn signal cancel unit (**Figure 150**) is installed inside the steering stem.

1. Check the angle sensor plate (A, **Figure 151**) installed on the bottom of the turn signal cancel unit for damage.

2. Remove the screw (B, **Figure 151**) securing the turn signal cancel unit to the frame. Check that the angle sensor plate (A, **Figure 151**) rotates smoothly.

Turn Signal Cancel Unit Replacement

1. Remove the meter panel. Refer to *Meter Panel* in Chapter Seventeen.

2. Remove the screws and the handlebar cover (**Figure 152**).

3. Remove the cover and disconnect the green 7-pin turn signal cancel unit connector (**Figure 153**).

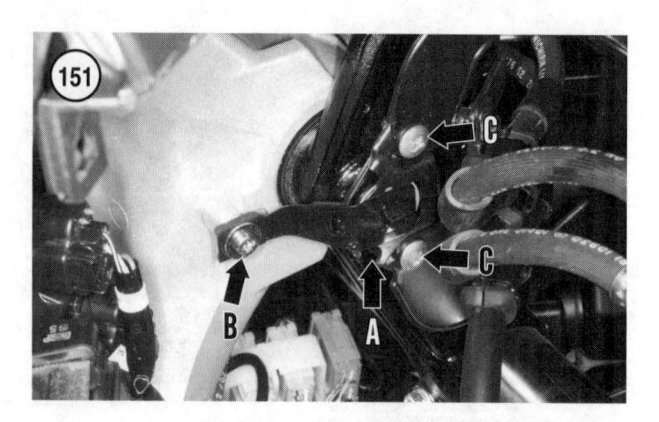

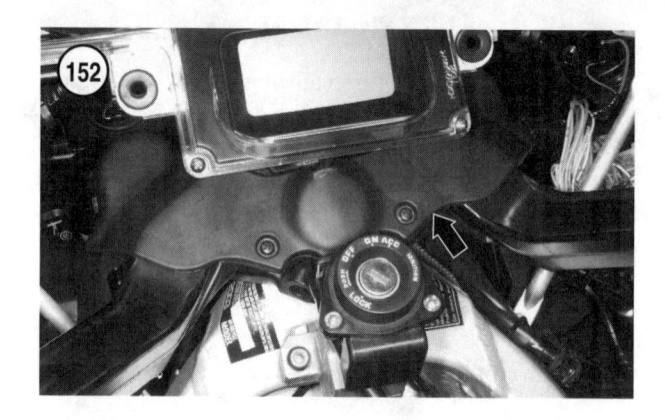

4. Remove the three screws (B and C, **Figure 151**) and remove the turn signal cancel unit (**Figure 150**) from the steering stem.

5. Installation is the reverse of removal. Turn the ignition switch on and check the turn signal operation.

POSITION LIGHT RELAY

Circuit Test

1. Remove the front fairing. Refer to *Front Fairing and Fairing Molding* in Chapter Seventeen.

2. Remove the position light relay (**Figure 154**) from the wiring harness.

NOTE
Make all the following tests at the position light relay 5-pin harness side connector.

3. Turn the ignition switch on and check for battery voltage between the white/green terminal (+) and ground (−).

   a. Yes. Go to Step 4.

   b. No. Check the white/green wire between the 6-pin connector and the horn turn relay. If good, install a new 4-terminal horn turn relay as described under *Relay Box* in this chapter.

ELECTRICAL SYSTEM

4. Turn the ignition switch on and measure voltage between the light blue/white terminal (+) and ground (−). There should be a pulsed voltage reading (voltage and no voltage) when the turn signal switch is pushed to the right side.
   a. If the test is correct, go to Step 5.
   b. If not, check the light blue/white wire for an open circuit. If good, test the turn signal switch as described in this chapter.
5. Turn the ignition switch on and measure voltage between the orange/white terminal (+) and ground (−). There should be a pulsed voltage reading (voltage and no voltage) when the turn signal switch is pushed to the left side.
   a. If the test is correct, go to Step 6.
   b. If not, check the orange wire for an open circuit. If good, test the turn signal switch as described in this chapter.

NOTE
The turn signal diode is mounted underneath the position light relay.

6. Turn the ignition switch off and disconnect the turn signal diode. Check for continuity in the blue/white wire between the position light relay 3-pin connector and the position light relay 5-pin connector. There should be continuity.

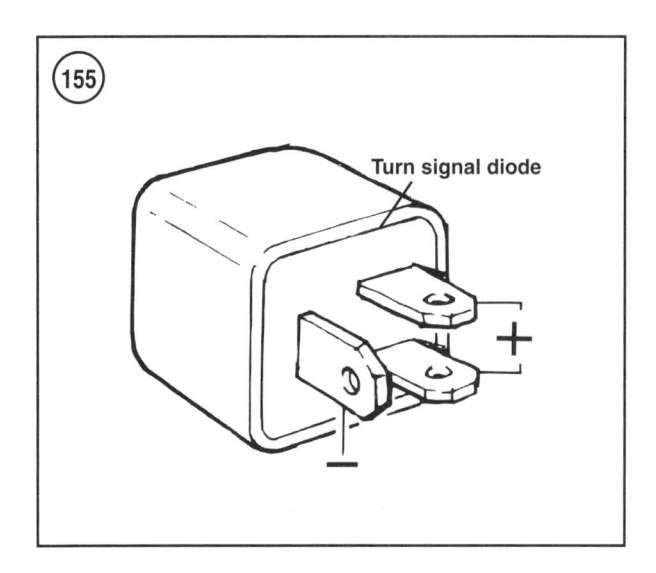

7. Install parts previously removed.

Turn Signal Diode Check

1. Remove the front fairing. Refer to *Front Fairing and Fairing Molding* in Chapter Seventeen.

NOTE
The turn signal diode is mounted underneath the position light relay.

2. Turn the ignition switch off and disconnect the turn signal diode.
3. Set an ohmmeter to the R × 1 scale.
4. Check continuity between the positive and negative diode terminals (**Figure 155**). At each pair of opposite terminals (+ and -), check the continuity in one direction, reverse the test leads, and check the continuity in the opposite direction. Each pair should have continuity in one direction and no continuity when the test leads are reversed.
5. Replace the diode if it fails this test.

OPEN AIR TEMPERATURE SENSOR

System Check

1. Remove the top shelter. Refer to *Top Shelter* in Chapter Seventeen.
2. Disconnect the white 2-pin open air temperature sensor connector (A, **Figure 156**).
3. Turn on the ignition switch and measure voltage between the yellow/blue (+) and green/black (−) terminals in the wire harness side of the 2-pin connector. The voltage should be approximately 5 volts.

a. Yes. Go to Step 4.
b. No. Check for an open circuit in the yellow/blue and green/black wires between the white 2-pin open air temperature sensor connector and the blue 20-pin combination meter connector.

4. Turn off the ignition switch. Switch an ohmmeter to R × 1000 and measure resistance between the two open air temperature sensor connector terminals on the sensor side. Compare the resistance readings at the different temperatures specified in **Table 7**.
5. Reverse Step 1 and Step 2 to complete installation.

Replacement

1. Remove the front fairing. Refer to *Front Fairing and Fairing Moldings* in Chapter Seventeen.
2. Remove the right upper air duct (B, **Figure 156**).
3. Remove the screw and open air temperature sensor (**Figure 157**).
4. Installation is the reverse of removal.

MULTI-DISPLAY CONTROL SWITCH

Removal/Installation

1. Remove the meter panel. Refer to *Meter Panel* in Chapter Seventeen.
2. Remove the two screws and the multi-display control switch (A, **Figure 158**).
3. Installation is the reverse of removal.

Continuity Test

1. Remove the meter panel. Refer to *Meter Panel* in Chapter Seventeen.
2. Disconnect the switch connector (A, **Figure 158**) Switch an ohmmeter to R × 1.
3. Mode switch: Connect the ohmmeter between the black/yellow and green/yellow terminals. There should be continuity when the MODE button (A, **Figure 159**) is pushed.
4. TRIP switch: Connect the ohmmeter between the brown/yellow and green/yellow terminals. There should be continuity when the TRIP button (B, **Figure 159**) is pushed.
5. DISP switch: Connect the ohmmeter between the blue/yellow and green/yellow terminals. There should be continuity when the DISP button (C, **Figure 159**) is pushed.
6. Replace the multi-display control switch if it fails any test.
7. Installation is the reverse of removal.

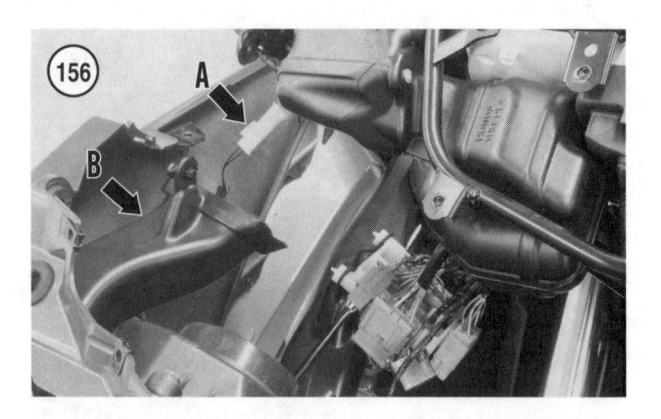

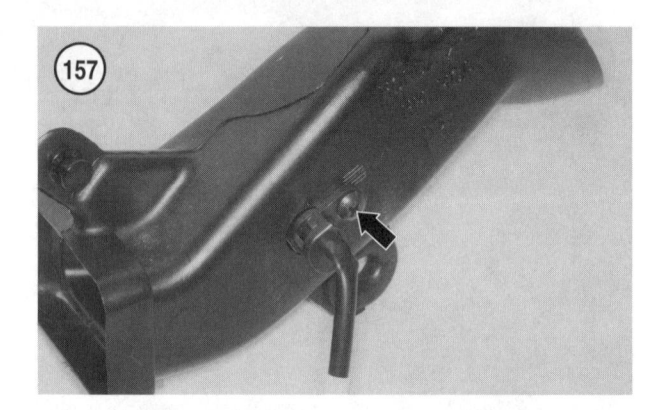

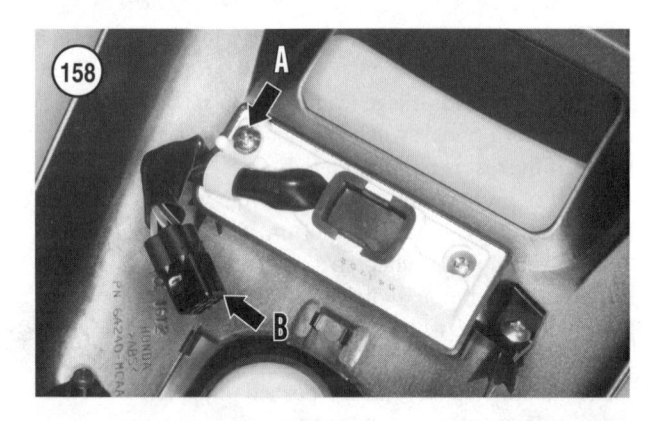

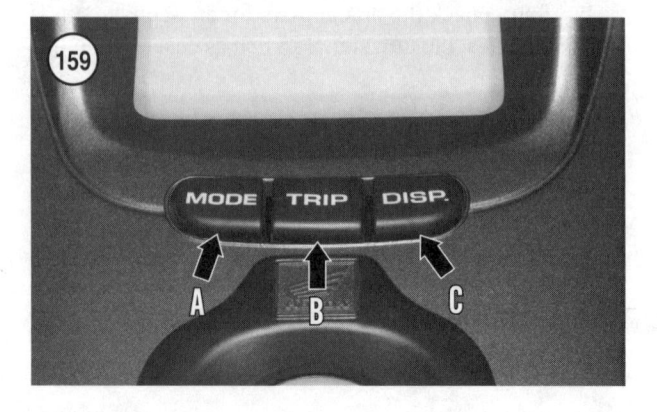

ELECTRICAL SYSTEM

6. Release the three switch tabs and remove the left panel switch assembly (**Figure 161**).
7. Installation is the reverse of removal.

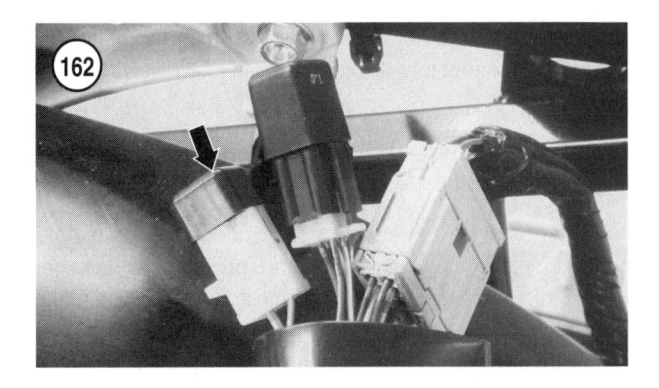

HAZARD SWITCH

Continuity Test

1. Remove the left fairing pocket. Refer to *Fairing Pockets* in Chapter Seventeen.
2. Turn off the ignition switch.
3. Disconnect the gray 14-pin left panel switch connector (**Figure 160**).
4. Switch an ohmmeter to R × 1.
5. Connect the ohmmeter between the pink/white and gray terminals (**Figure 160**). There should be continuity when the hazard button is pushed and no continuity when the button is released.
6. Connect the ohmmeter between the green and blue/black terminals (**Figure 160**). There should be continuity when the hazard button is pushed and no continuity when the button is released.
7. If the switch failed either continuity test, replace the left panel switch assembly as described in this chapter.
8. Installation is the reverse of these steps.

LEFT PANEL SWITCH

Removal/Installation

1. Remove the left fairing molding. Refer to *Front Fairing and Fairing Molding* in Chapter Seventeen.
2. Remove the left fairing pocket. Refer to *Fairing Pockets* in Chapter Seventeen.
3. Disconnect the ignition switch.
4. Disconnect the 14-pin gray left panel switch connector (**Figure 160**).
5. Remove the screw from the upper left corner of the panel switch.

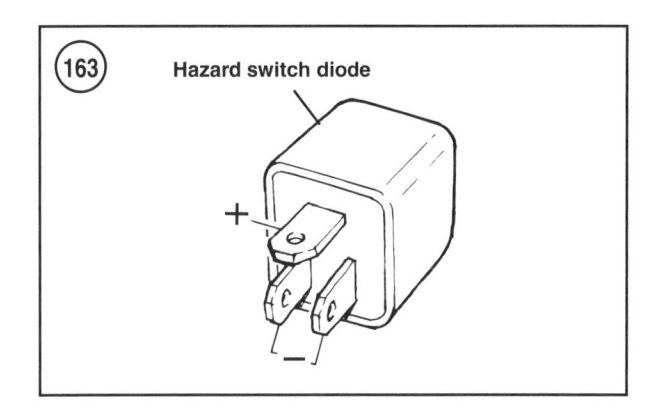

HAZARD SWITCH DIODE

Testing/Replacement

1. Remove the seat. Refer to *Seat* in Chapter Seventeen.
2. Turn off the ignition switch and disconnect the turn signal diode (**Figure 162**).
3. Set an ohmmeter to the R × 1 scale.
4. Check continuity between the positive and negative diode terminals (**Figure 163**). At each pair of opposite ter-

164 HORN SWITCH

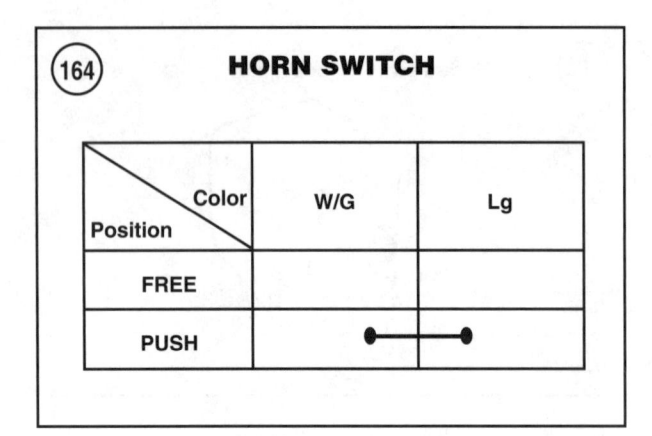

Position \ Color	W/G	Lg
FREE		
PUSH	•——————•	

minals (+ and -), check the continuity in one direction, reverse the test leads, and check the continuity in the opposite direction. Each pair should have continuity in one direction and no continuity when the test leads are reversed.

5. Replace the diode if it fails this test.

HANDLEBAR SWITCHES

Continuity Test

Test the switches for continuity using an ohmmeter or a self-powered test light. Operate the switch in each of its operating positions and compare the results with the switch continuity diagram included with the wiring diagrams at the end of the manual. For example, **Figure 164** shows the continuity diagram for the horn switch. When the horn button is pressed, there should be continuity between the white/green and light green terminals. The line joining the two terminals shows continuity (**Figure 164**). An ohmmeter connected between these two terminals should show continuity or a test light should illuminate. When the horn button is free, there should be no continuity between the same terminals.

NOTE
*Refer to **Audio Switches** in this chapter to test the audio switches mounted in the left-hand handlebar switch housing.*

1. Refer to the appropriate switch procedure in this chapter to access the switch connectors.

2. Check the fuse as described under *Fuse* in this chapter.

3. Check the battery as described under *Battery* in this chapter. Charge the battery to the correct state of charge, if required.

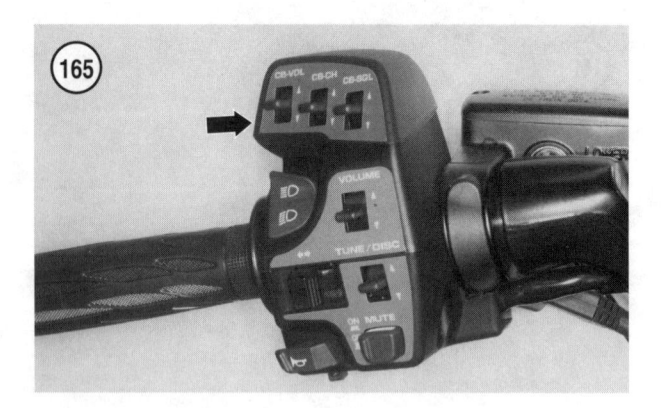

4. Disconnect the negative battery cable at the battery if the switch connectors are not disconnected from the circuit.

5. When separating two connectors, pull on the connector housings and not the wires.

6. After locating a defective circuit, check the connectors to make sure they are clean and properly connected. Check all wires going into a connector housing to make sure each wire is positioned properly and that the wire end is not loose.

7. Before disconnecting two connectors, check them for any locking tabs or arms that must be pushed or opened. If two connectors are difficult to separate, do not force them because damage may occur.

8. When reconnecting electrical connector halves, push them together until they click or snap into place.

9. If the switch is operating erratically, chances are the contacts are oily, dirty or corroded. Disassemble the switch housing as described in this section to access the switch contacts. Clean the contacts as required.

10. If a switch or button does not perform properly, replace the switch as described in its appropriate section.

ELECTRICAL SYSTEM

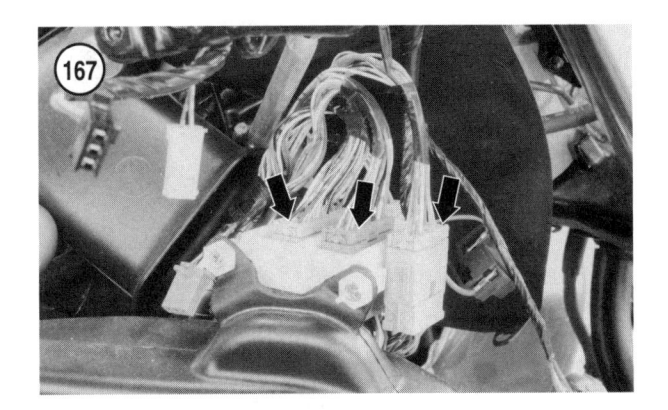

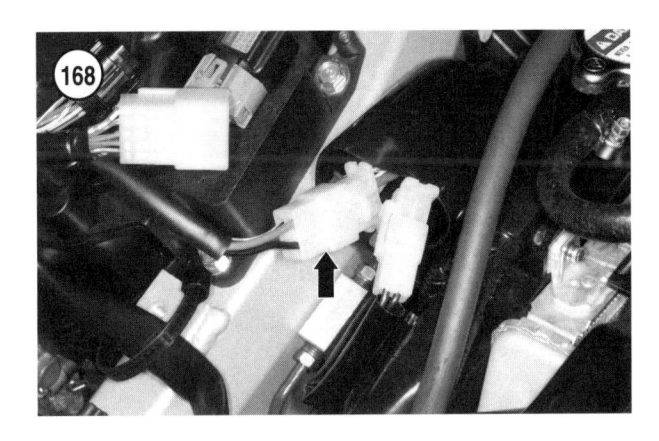

Replacement

1. Remove the front fairing. Refer to *Front Fairing and Fairing Molding* in Chapter Seventeen.

2. Remove the left (**Figure 165**) and right (**Figure 166**) side handlebar switches from the handlebars as described in Chapter Twelve under *Handlebars*.

3A. On 2001-2003 models, disconnect the handlebar switches (**Figure 167**, typical):
   a. Green 12-pin left handlebar switch connector.
   b. Blue 14-pin left handlebar switch connector.
   c. Green 18-pin right handlebar switch connector.

3B. On 2004 models, disconnect the handlebar switches (**Figure 167**, typical):
   a. Blue 12-pin left handlebar switch connector.
   b. Gray 14-pin left handlebar switch connector.
   c. Blue 18-pin right handlebar switch connector.

4. Installation is the reverse of removal. Operate each switch to make sure the assembly is correct.

WARNING
Do not ride the motorcycle until each switch function is working properly.

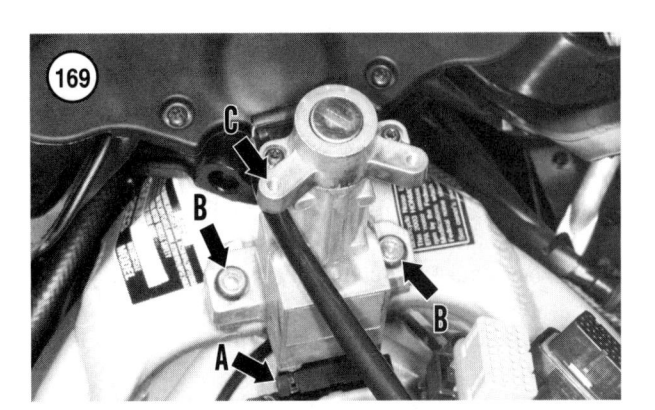

IGNITION SWITCH

Testing

1. Remove the top shelter. Refer to *Top Shelter* in Chapter Seventeen.

2. Disconnect the white 4-pin ignition switch connector (**Figure 168**).

3. Test the ignition switch as described under *Handlebar Switches, Continuity Test* in this chapter.

4. Reverse Step 1 and Step 2.

Replacement

1. Remove the top shelter. Refer to *Top Shelter* in Chapter Seventeen.

2. Disconnect the white 4-pin ignition switch connector (**Figure 168**).

3. Disconnect the connector (A, **Figure 169**) from the bracket on the ignition switch.

4. Remove the two bolts (B, **Figure 169**) and the ignition switch (C).

5. Installation is the reverse of these steps. Tighten the ignition switch mounting bolts to 25 N•m (18 ft.-lb.).

6. Check the ignition switch for proper operation.

CRUISE CONTROL SWITCHES

Refer to Chapter Sixteen to test and replace the cruise control switches.

CLUTCH SWITCH

Testing/Replacement

The clutch switch is mounted on the clutch lever housing.

1. Disconnect the two electrical connectors at the clutch switch (A, **Figure 170**).
2. Switch an ohmmeter to the R × 1 scale, and then connect the ohmmeter leads across the two clutch switch terminals.
3. Read the ohmmeter scale while pulling in and releasing the clutch lever. Note the following:
    a. There must be continuity with the clutch lever pulled in and no continuity with the lever released.
    b. Replace the clutch switch if it fails to operate as described.
4. Remove the clutch switch mounting screw (B, **Figure 170**) and clutch switch.
5. Install a new clutch switch by reversing these removal steps.

FRONT BRAKE LIGHT SWITCH

Testing/Replacement

The front brake light switch and front cruise cancel switch are enclosed in the same switch housing mounted on the bottom of the front master cylinder. The two lower (larger) terminals (A, **Figure 171**) plug into the cruise cancel switch. The two upper (smaller) terminals (B, **Figure 171**) plug into the brake light switch. Service the cruise part of the switch as described in Chapter Sixteen.
1. Disconnect the connectors from the upper switch terminals (**Figure 172**).
2. Check for continuity between the switch terminals. There should be continuity when applying the brake lever and no continuity when releasing the brake lever. Replace the switch if faulty.
3. If necessary, remove the screw and replace the switch (**Figure 173**).
4. Installation is the reverse of removal.
5. Make sure all connectors are plugged tightly into the switch.
6. Reposition the switch cover.
7. Turn the ignition switch on and operate the front brake lever to check the rear brake light.

WARNING
Do not ride the motorcycle until the rear brake light works correctly.

REAR BRAKE LIGHT SWITCH

Testing/Replacement

The rear brake light switch and rear cruise cancel switch are enclosed in the same switch housing (**Figure**

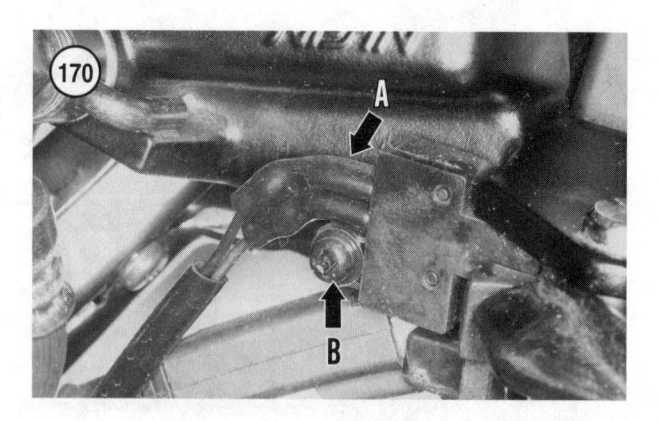

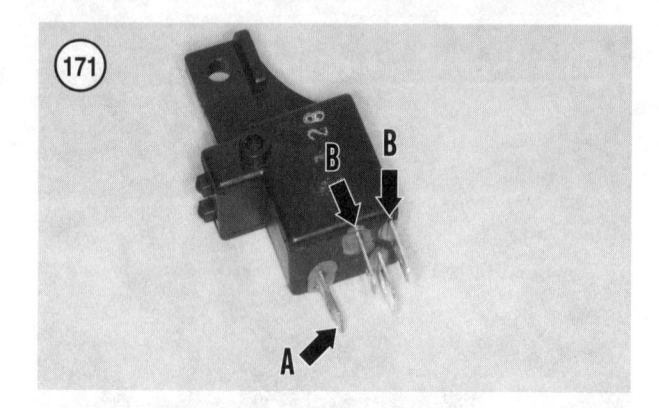

ELECTRICAL SYSTEM

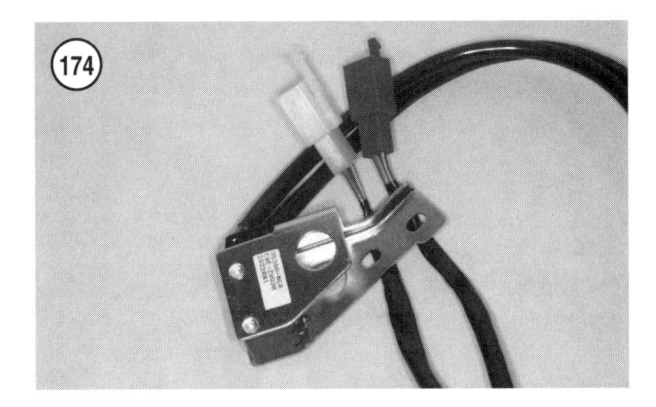

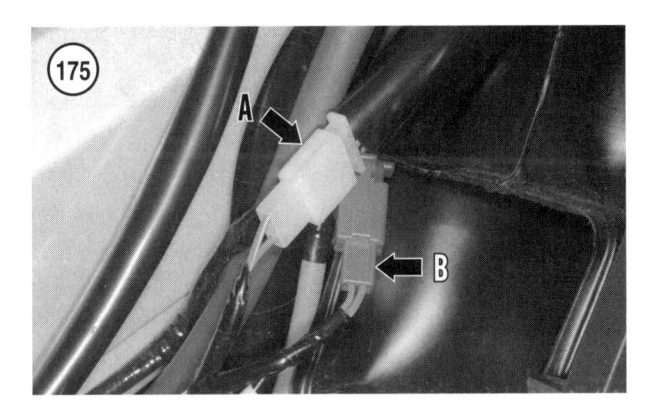

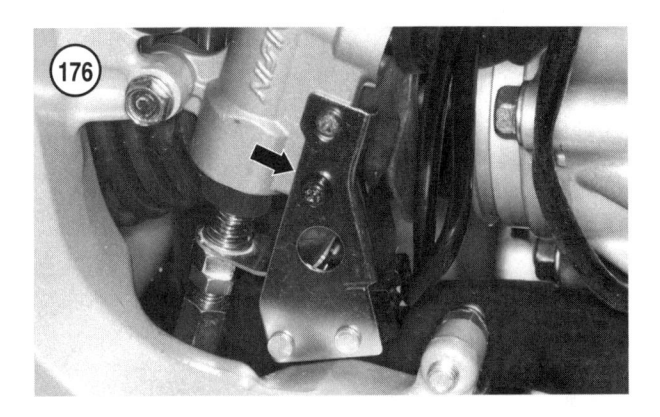

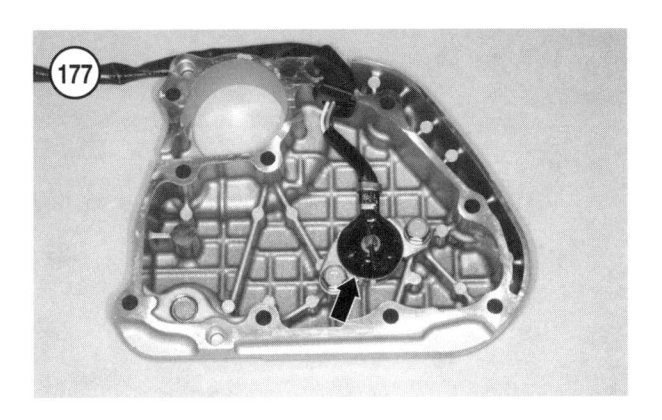

174). Service the cruise part of the switch as described in Chapter Sixteen.

1. Remove the top shelter. Refer to *Top Shelter* in Chapter Seventeen.
2. Disconnect the rear brake light switch white 2-pin connector (A, **Figure 175**). The red connector (B, **Figure 175**) is used for the cruise part of the switch.
3. Check for continuity between the switch terminals. There should be no continuity when releasing the brake pedal and continuity when applying the brake pedal. Replace the switch if faulty.
4. To replace the switch:
    a. Remove the right engine side cover. Refer to *Engine Side Covers* in Chapter Seventeen.
    b. Disconnect the switch connectors (A and B, **Figure 175**).
    c. Remove the screws and switch (**Figure 176**).
    d. Use a medium strength threadlock on the switch screw threads.
    e. Install the switch (**Figure 176**) onto the master cylinder and secure with the mounting screws. Do not tighten.
    f. Turn on the ignition switch. Adjust the switch so that the brake light comes on when the brake pedal moves the push rod 0.7-1.7 mm (1/32-1/16 in.)
    g. Tighten the switch screws (**Figure 176**) to 2 N•m (17 in.-lb.).
    h. Recheck the brake light adjustment and turn the ignition switch off.
5. Installation is the reverse of removal.

WARNING
Do not ride the motorcycle until the rear brake light works correctly.

GEAR POSITION SWITCH

The gear position switch (**Figure 177**) is mounted inside the shift linkage cover.

Testing/Replacement

1. Support the motorcycle on its centerstand.
2. Remove the air filter housing. Refer to *Air Filter Housing* in Chapter Eight.
3. Disconnect the black 6-pin gear position switch connector located inside the connector pouch identified in **Figure 178**.
4. Switch an ohmmeter to R × 1. Turn the rear wheel by hand and shift the transmission into gear to check each position for continuity. Continuity should be recorded during each test:

a. Neutral: Light green/red to ground.
b. First gear: No test required.
c. Second gear: Black/yellow to ground.
d. Third gear: White/red to ground.
e. Fourth gear: Red/white to ground.
f. Fifth gear (overdrive): Green/orange to ground.

5. If any test was incorrect, replace the gear position switch as described in Chapter Seven under *External Shift Linkage, Shift Linkage Cover*.
6. Installation is the reverse of these steps.

SIDESTAND SWITCH

Testing

1. Remove the left engine side cover. Refer to *Engine Side Covers* in Chapter Seventeen.
2. Disconnect the black 3-pin sidestand switch connector (A, **Figure 179**).
3. Test the sidestand switch as described under *Handlebar Switches, Continuity Test* in this chapter.
4. Reverse Step 1 and Step 2.

Replacement

1. Support the bike on its centerstand.
2. Remove the left engine side cover. Refer to *Engine Side Covers* in Chapter Seventeen.
3. Disconnect the black 3-pin sidestand switch connector (A, **Figure 179**).
4. Remove the bolt and sidestand switch (B, **Figure 179**).
5. Clean the switch mounting area on the sidestand.
6. Install the sidestand switch by aligning the switch pin with the hole in the sidestand and the notch in the switch with the pin in the mounting bracket. See **Figure 180**.
7. Install a *new* sidestand switch mounting bolt and tighten to 10 N•m (88 in.-lb.).
8. Reverse Steps 1-3 to complete installation. Test the sidestand switch before operating the motorcycle.

HORNS

Removal/Installation

1. Remove the front lower fairing. Refer to *Front Lower Fairings* in Chapter Seventeen.
2. Disconnect the electrical connectors from the horn (**Figure 181**, typical).
3. Remove the bolt and the horn assembly.
4. Install by reversing these removal steps. Make sure the electrical connections are secure and corrosion-free.

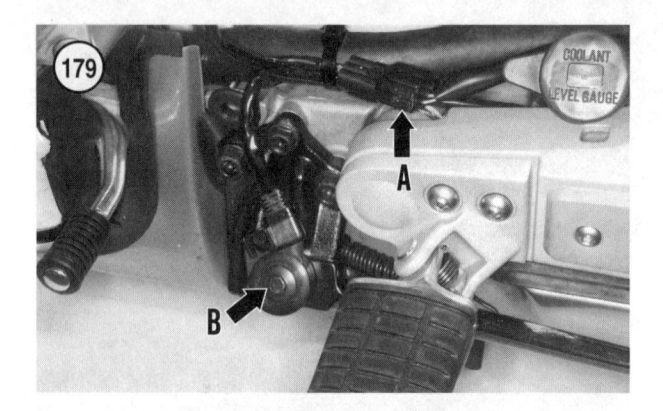

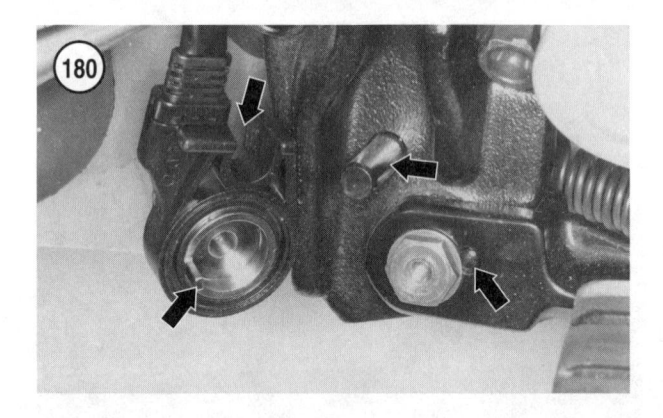

ELECTRICAL SYSTEM

(182) RELAY BOX

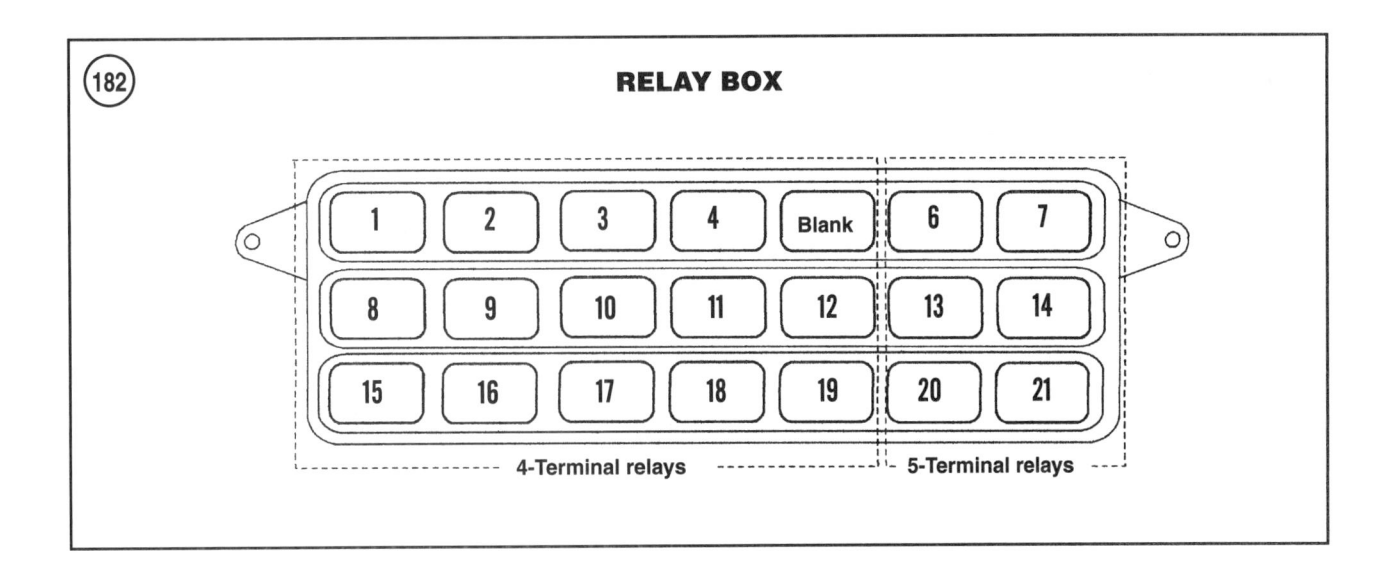

(Diagram showing relay box with positions 1-21, with position 5 labeled "Blank". Positions 1-4, 8-11, 15-18 are 4-Terminal relays; positions 6-7, 12-14, 19-21 are 5-Terminal relays.)

Horn Testing

1. Disconnect the electrical connectors from the horn.
2. Connect a 12-volt battery across the horn terminals. The horn must sound loudly.
3. Replace the horn if it did not sound loudly in Step 2.
4. Repeat for the other horn.
5. Check the horn operation. If the horn does not work properly, test the horn as described in this section.

WARNING
Do not ride the bike until both horns work properly.

RELAY BOX

Procedures described throughout this and other chapters test individual relay circuits. This section describes procedures how to test individual 4-terminal and 5-terminal relays. All the relays are installed in the relay box underneath the seat (**Figure 182**).

The relays are easily unplugged or plugged into the relay terminal strip. Because only 4-terminal and 5-terminal relays are used, there is no danger of installing a relay in the wrong position. The 4-terminal relays are identical, and the 5-terminal relays are identical.

NOTE
A questionable relay can be quickly checked by exchanging it with a known good relay. To do this, first identify the relay as a 4-terminal or 5-terminal type. Unplug the relay and plug in a known good relay.

1. Turn off the ignition switch.
2. Remove the seat. Refer to *Seat* in Chapter Seventeen.
3. Remove the screws and relay box cover (**Figure 183**).
4. Unplug the questionable relay (**Figure 182** and **Figure 184**).
5. Switch an ohmmeter to R × 1.

185

4-terminal relay

No. 1
No. 2
No. 3 No. 4

186

5-terminal relay

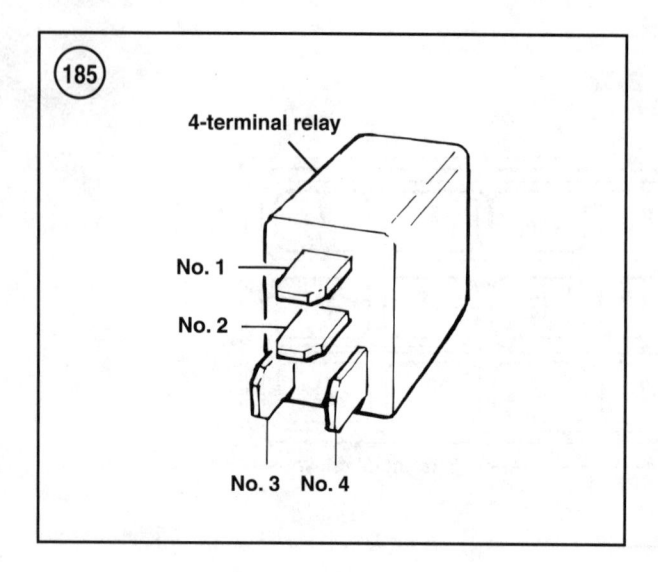

No. 1
No. 2
No. 3 No. 5
No. 4

6. 4-terminal relay test (**Figure 185**): Check for continuity across terminals 1 and 2; there should be no continuity. Connect a 12-volt battery across terminals 3 (+) and 4 (−) and check for continuity across terminals 1 and 2; there should be continuity.

7. 5-terminal relay test (**Figure 186**): Perform the following:
   a. Check for continuity across terminals 1 and 2; there should be no continuity.
   b. Check for continuity across terminals 1 and 4; there should be continuity.
   c. Connect a 12-volt battery across terminals 3 (+) and 5 (−) and check for continuity across terminals 1 and 2; there should be continuity.

8. Replace any relay that failed to test as described in Step 6 or Step 7.

9. Installation is the reverse of removal.

FUSES

Whenever a fuse blows, determine the cause before replacing the fuse. Usually, the trouble is a short circuit in the wiring. Worn-through insulation or a short to ground from a disconnected wire may cause this condition.

CAUTION
If replacing a fuse, make sure the ignition switch is turned off. This step lessens the chance of a short circuit.

CAUTION
Never substitute any metal object for a fuse. Never use a higher amperage fuse than specified. An overload could cause a fire and the complete loss of the motorcycle.

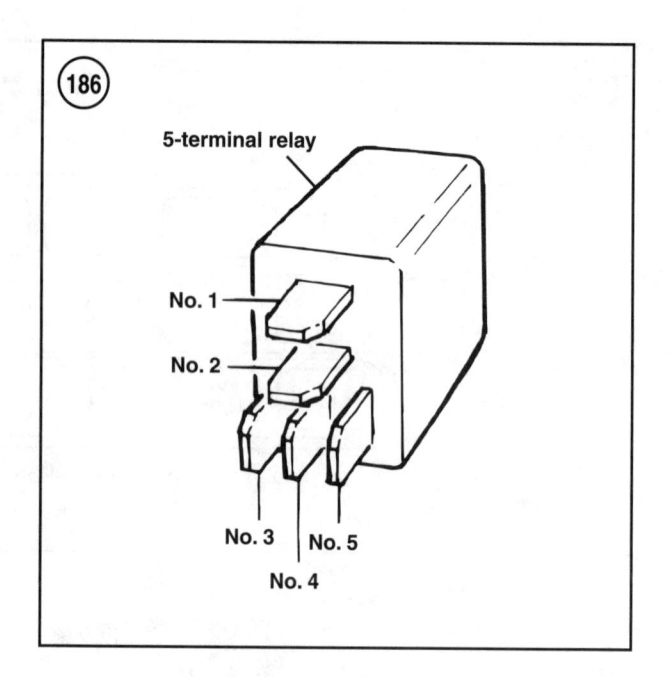

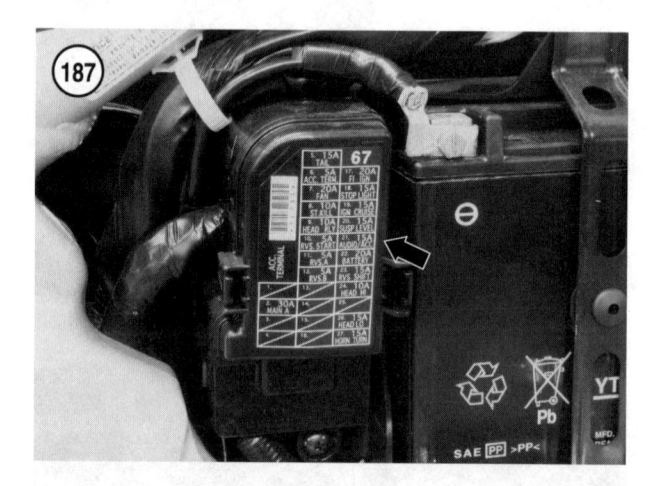

187

All of the fuses are mounted inside the fuse box located behind the left side cover. To identify an individual fuse and its amperage, refer to the printed information on the fuse box cover (**Figure 187**). Refer to **Figure 188** and **Table 8**.

1. Turn off the ignition switch.

2. Remove the left side cover. Refer to *Side Covers* in Chapter Seventeen.

3. Remove the fuse box cover (**Figure 187**).

4. Remove and inspect the fuse. Replace the fuse if it has blown (**Figure 189**).

5. Install and secure the fuse box cover.

6. Install the left side cover.

ELECTRICAL SYSTEM

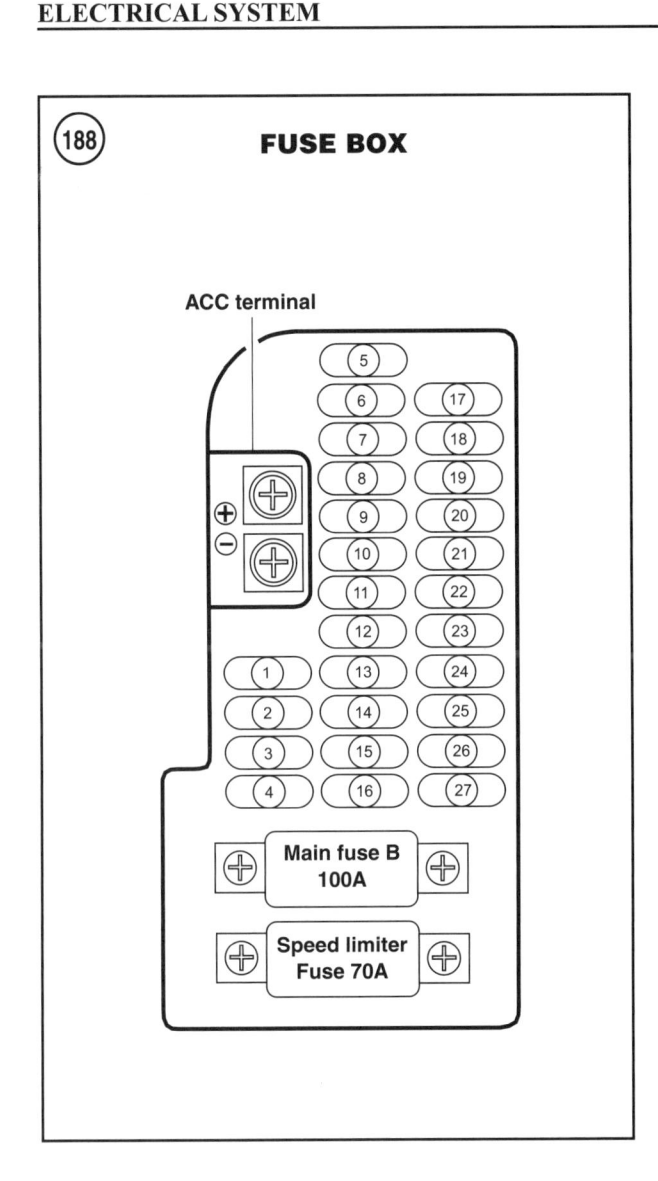

188 FUSE BOX

ACC terminal

Main fuse B 100A

Speed limiter Fuse 70A

189 Blown fuse

AUDIO SYSTEM

Troubleshooting

1. Make sure the battery is fully changed. Check that the battery connections are clean and free of corrosion. Check the battery condition as described in this chapter.

2. Tune the radio to a station with a strong signal.

3. Check that the No. 21 and No. 22 fuses are good.

4. Check that the multi-display works correctly.

5. Check the wiring harnesses for proper routing and condition. Make sure they are properly secured.

6. All electrical connectors must be clean and properly connected.

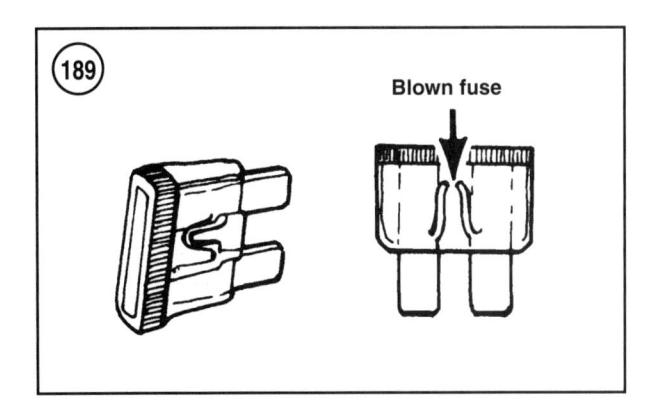

Audio System Flow Charts

To troubleshoot the audio system, refer to **Figure 190** and find the symptom or condition and turn to the indicated diagnostic flow chart (**Figures 191-195**). Perform the relevant procedures until the problem is resolved

When using the flow charts, note the following:
1. Refer to the wiring diagrams at the end of this manual to identify the connectors.
2. Unless otherwise specified, perform the test procedures with the ignition switch off.
3. Do not disconnect or reconnect electrical connectors when the ignition switch is on.
4. Do not perform tests with a battery charger attached to the battery or circuit. Test and charge the battery as described in this chapter.

AUDIO UNIT

Power/Ground Circuit Test

1. Remove the top shelter. Refer to *Top Shelter* in Chapter Seventeen.
2. Check the connectors for dirty or loose fitting terminals.

NOTE
Perform the following tests on the black 34-pin wiring harness side connector terminals (A, Figure 196), not on the audio unit side.

3. Ground circuit: Turn the ignition switch off. Check for continuity between the green terminal and ground.
   a. Yes. The ground circuit is good.
   b. No. Check the green wire between the connector and ground terminal for an open circuit.
4. Back-up circuit: Turn the ignition switch on and check for battery voltage between the red/yellow terminal (+) and ground (−).
   a. Yes. The back-up circuit is good.

⑲⓪ AUDIO TROUBLESHOOTING CHART

Symptom or condition	Figure number
No sound from one or both speakers	191
Weak or noisy reception	192
Radio works, but audio switch does not	193
Radio works, but audio display does not	194
Auto volume control (AVC) does not work	195

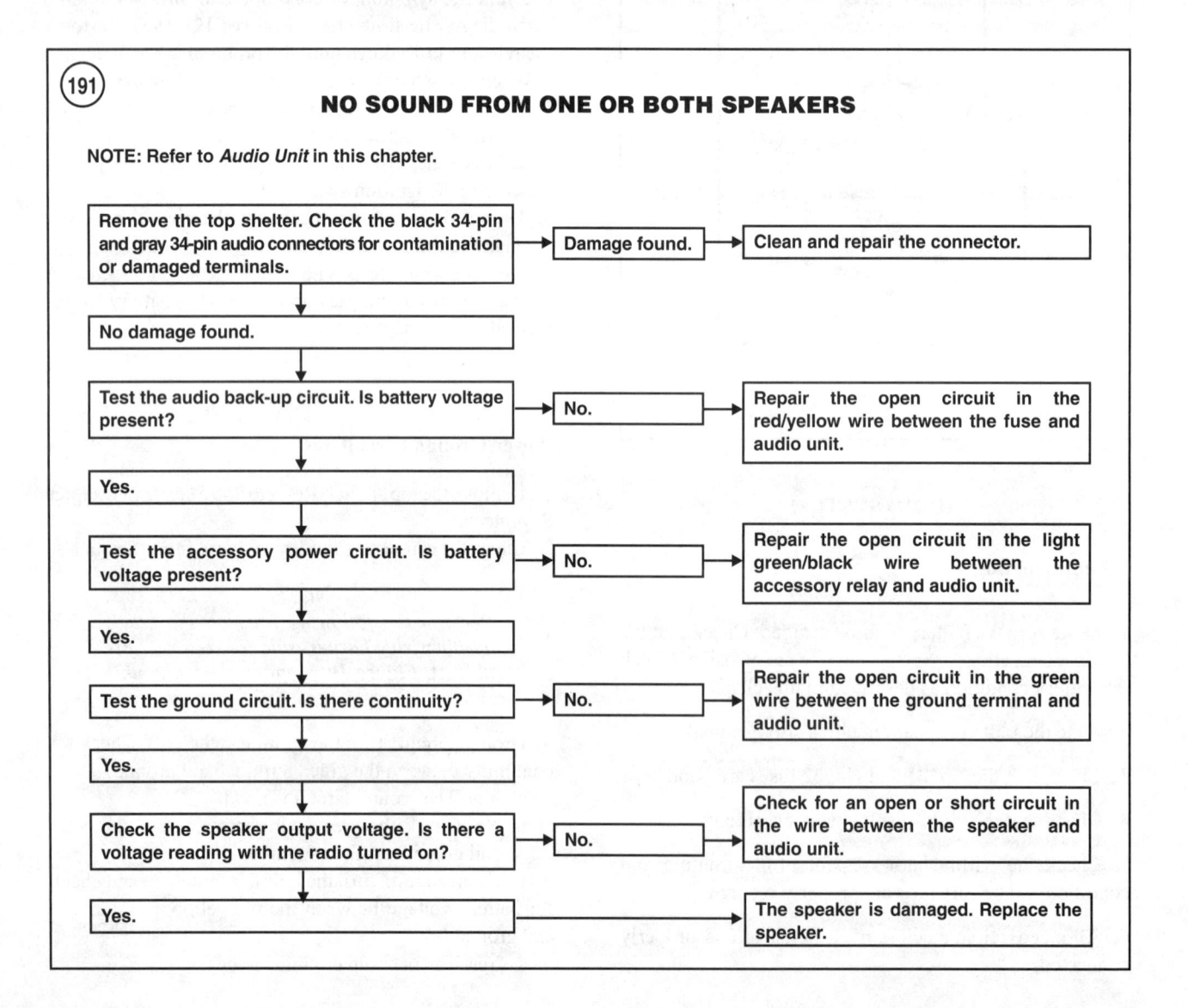

⑲① NO SOUND FROM ONE OR BOTH SPEAKERS

NOTE: Refer to *Audio Unit* in this chapter.

Remove the top shelter. Check the black 34-pin and gray 34-pin audio connectors for contamination or damaged terminals. → Damage found. → Clean and repair the connector.

↓ No damage found.

Test the audio back-up circuit. Is battery voltage present? → No. → Repair the open circuit in the red/yellow wire between the fuse and audio unit.

↓ Yes.

Test the accessory power circuit. Is battery voltage present? → No. → Repair the open circuit in the light green/black wire between the accessory relay and audio unit.

↓ Yes.

Test the ground circuit. Is there continuity? → No. → Repair the open circuit in the green wire between the ground terminal and audio unit.

↓ Yes.

Check the speaker output voltage. Is there a voltage reading with the radio turned on? → No. → Check for an open or short circuit in the wire between the speaker and audio unit.

↓ Yes. → The speaker is damaged. Replace the speaker.

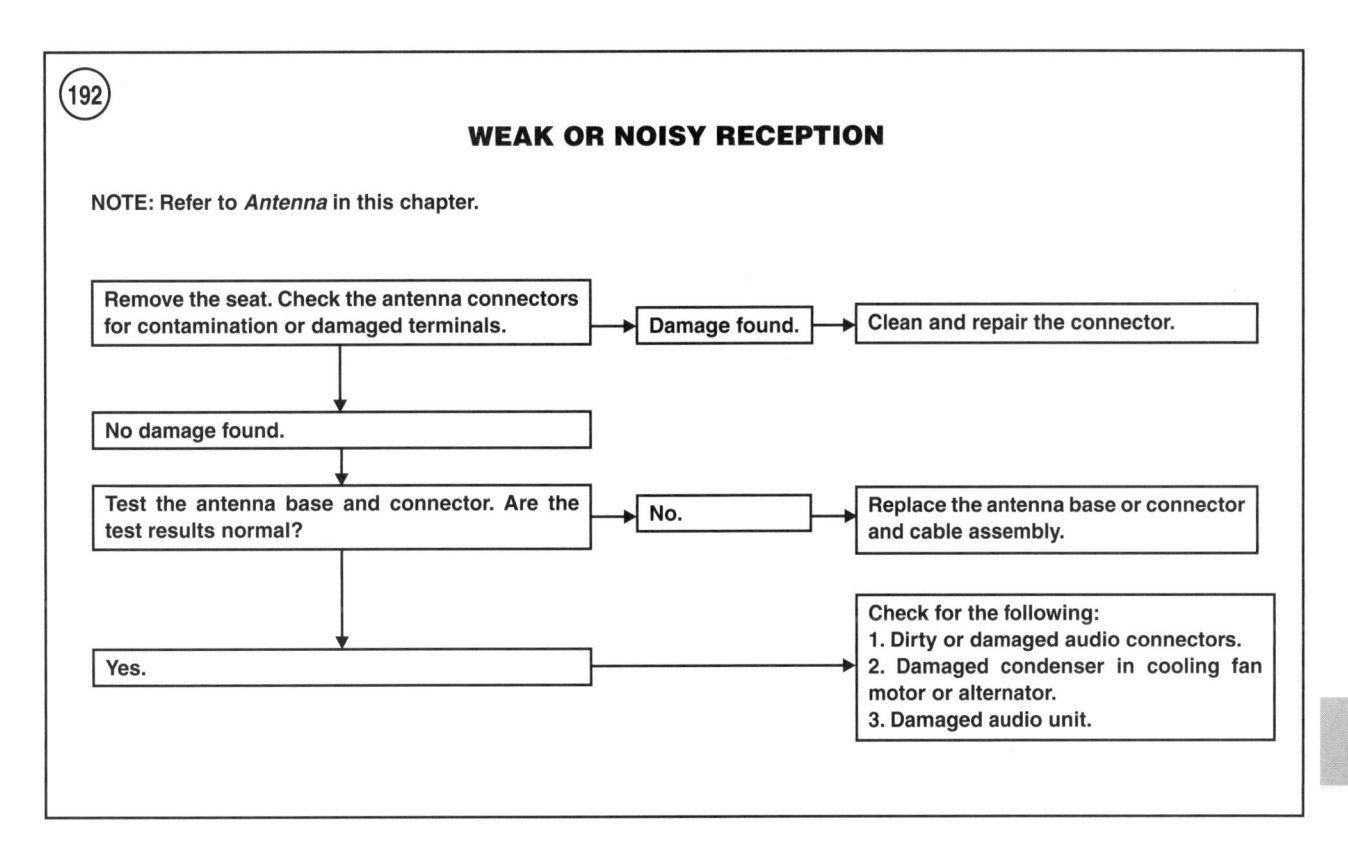

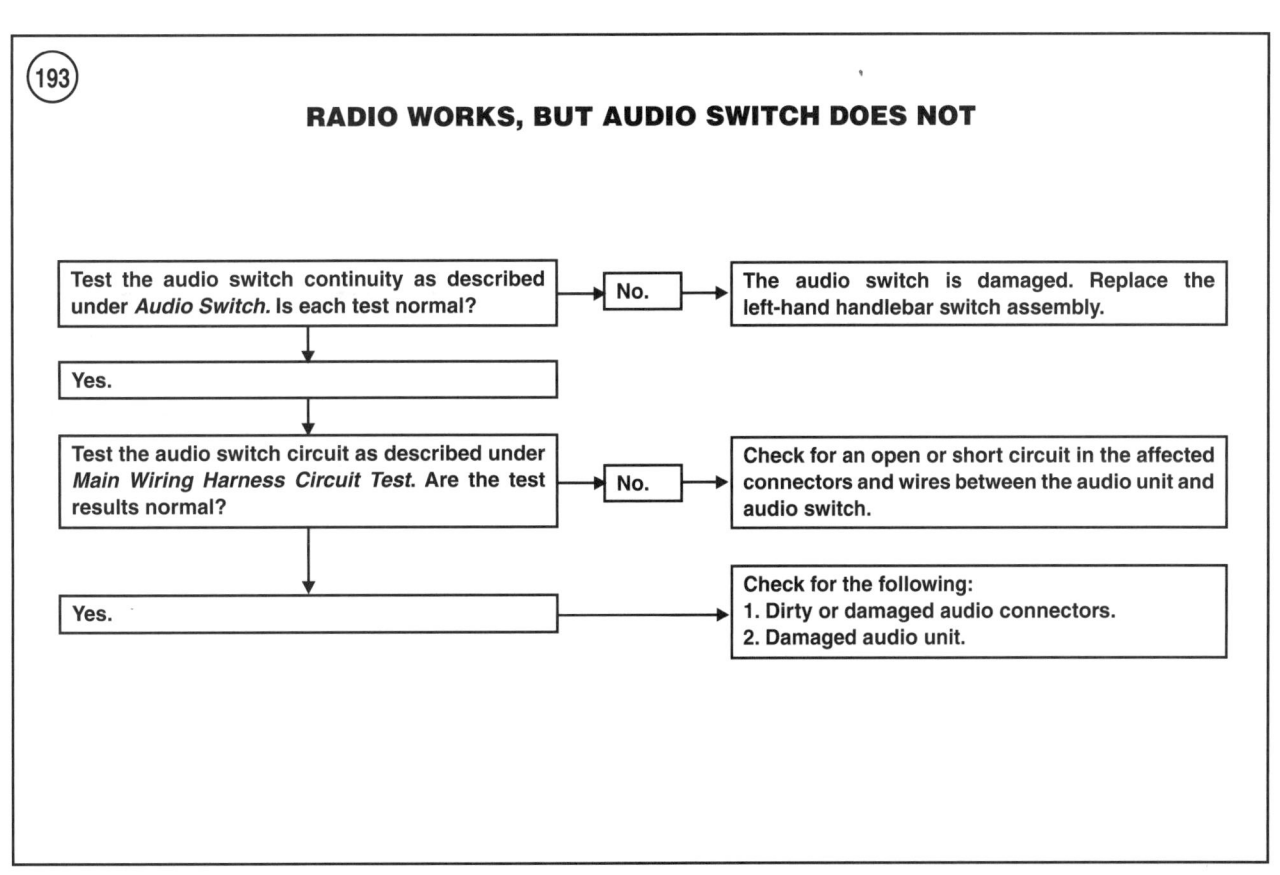

ELECTRICAL SYSTEM

(192) WEAK OR NOISY RECEPTION

NOTE: Refer to *Antenna* in this chapter.

- Remove the seat. Check the antenna connectors for contamination or damaged terminals. → Damage found. → Clean and repair the connector.
- No damage found.
- Test the antenna base and connector. Are the test results normal? → No. → Replace the antenna base or connector and cable assembly.
- Yes. → Check for the following:
  1. Dirty or damaged audio connectors.
  2. Damaged condenser in cooling fan motor or alternator.
  3. Damaged audio unit.

(193) RADIO WORKS, BUT AUDIO SWITCH DOES NOT

- Test the audio switch continuity as described under *Audio Switch*. Is each test normal? → No. → The audio switch is damaged. Replace the left-hand handlebar switch assembly.
- Yes.
- Test the audio switch circuit as described under *Main Wiring Harness Circuit Test*. Are the test results normal? → No. → Check for an open or short circuit in the affected connectors and wires between the audio unit and audio switch.
- Yes. → Check for the following:
  1. Dirty or damaged audio connectors.
  2. Damaged audio unit.

194

RADIO WORKS, BUT AUDIO DISPLAY DOES NOT

Test the audio display circuit as described under *Main Wiring Harness Circuit Test*. Are the test results normal? → No. → Check for an open or short circuit in the affected connectors and wires between the audio unit and combination meter.

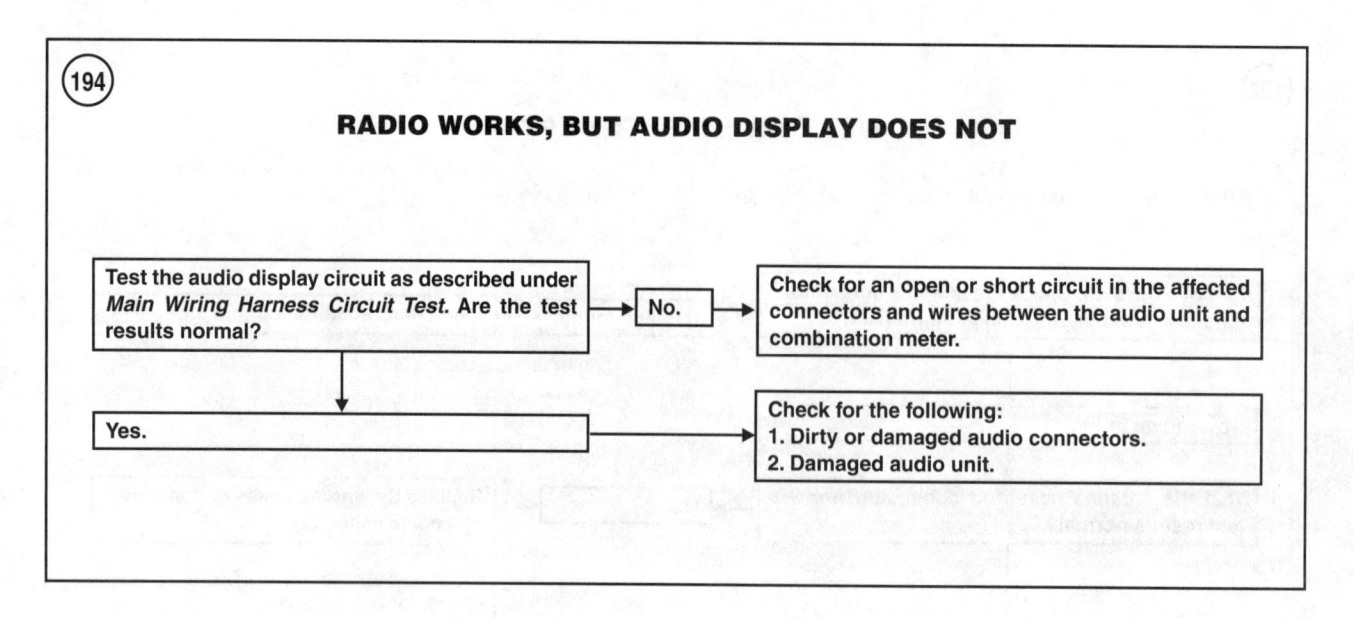

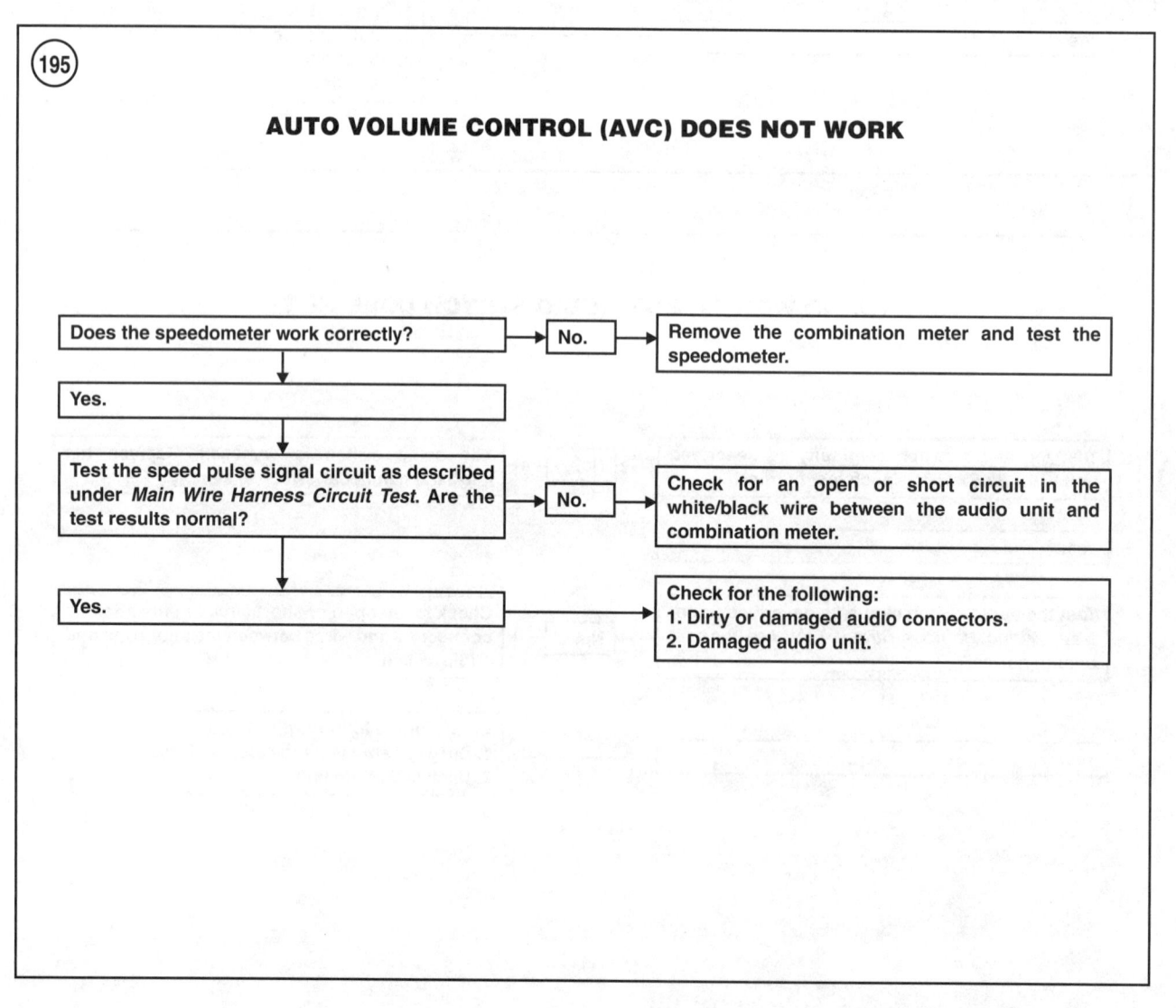

Yes. → Check for the following:
1. Dirty or damaged audio connectors.
2. Damaged audio unit.

195

AUTO VOLUME CONTROL (AVC) DOES NOT WORK

Does the speedometer work correctly? → No. → Remove the combination meter and test the speedometer.

Yes.

Test the speed pulse signal circuit as described under *Main Wire Harness Circuit Test*. Are the test results normal? → No. → Check for an open or short circuit in the white/black wire between the audio unit and combination meter.

Yes. → Check for the following:
1. Dirty or damaged audio connectors.
2. Damaged audio unit.

ELECTRICAL SYSTEM

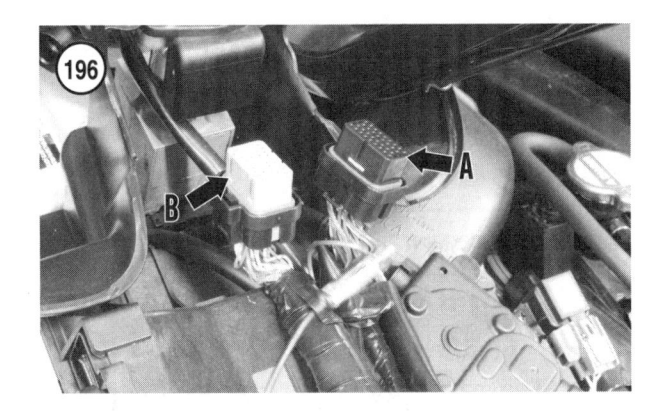

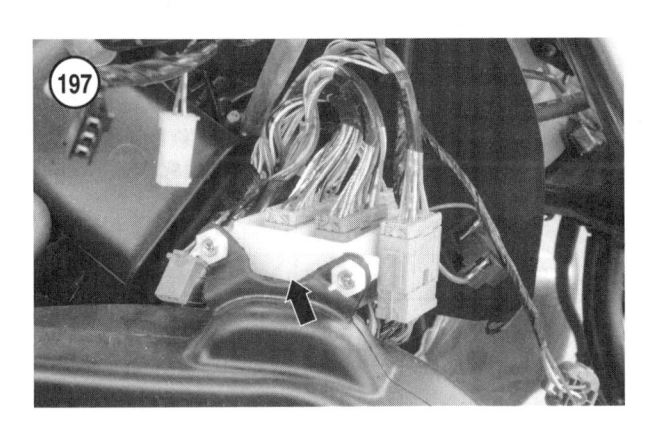

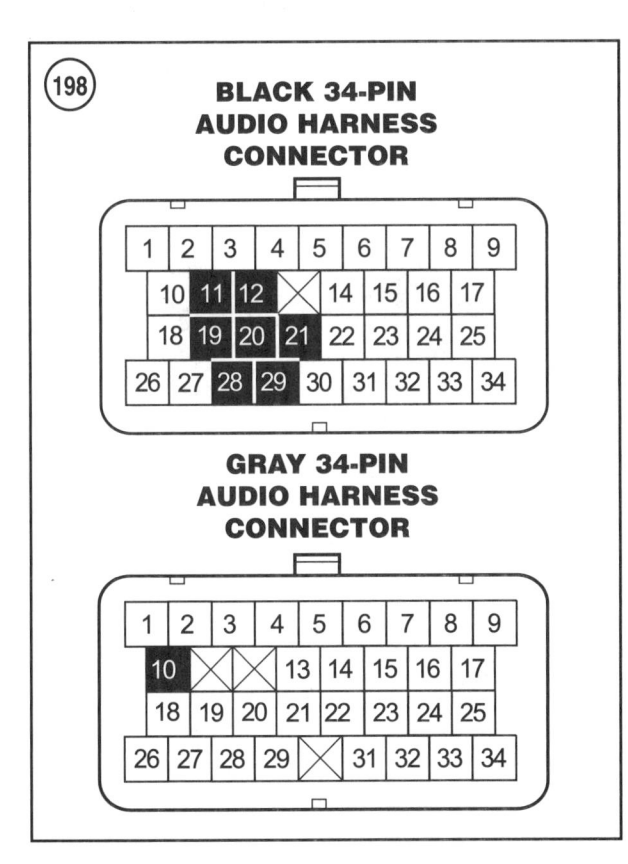

b. No. Check the red/yellow wire between the audio unit and fuse for an open circuit.

5. Accessory circuit: Turn the ignition switch on and measure for battery voltage between the light green/black terminal (+) and ground (–). Repeat the test with the ignition switch turned to ACC.

   a. If battery voltage is present during both tests, the accessory circuit is good.

   b. If not, check the light green/black wire between the audio unit and relay for an open circuit.

6. Install previously removed parts.

Main Wire Harness Circuit Test

1. Remove the top shelter. Refer to *Top Shelter* in Chapter Seventeen.

2. Remove the front fairing. Refer to *Front Fairing and Fairing Molding* in Chapter Seventeen.

3A. On 2001-2003 models, disconnect the left handlebar switch green 12-pin and blue 14-pin (**Figure 197**) connectors.

3B. On 2004 models, disconnect the left handlebar switch blue 12-pin and gray 14-pin connectors (**Figure 197**, typical).

NOTE
Steps 4-7 test the audio circuit.

4. Compare the black 34-pin and gray 34-pin audio connectors in **Figure 198** with the connectors in **Figure 196**. Then compare the 12-pin and 14-pin connectors in **Figure 199** with the connectors in **Figure 197**. Then check continuity between the 34-pin audio connectors in **Figure 198** and the handlebar audio switch connectors in **Figure 199** that have the same number (same color). There should be continuity between the terminals with the same numbers.

5. Check for continuity between the green terminal in the 12-pin connector (**Figure 199**) and ground. There should be continuity.

NOTE
Steps 6-8 test the audio display circuit.

6. Remove the combination meter as described in this chapter.

7. Check for continuity between the black 34-pin audio connector (A, **Figure 196**) and the black 16-pin combination meter connector (D, **Figure 132**) at the terminals indicated in **Figure 200**. There should be continuity between each same wire color.

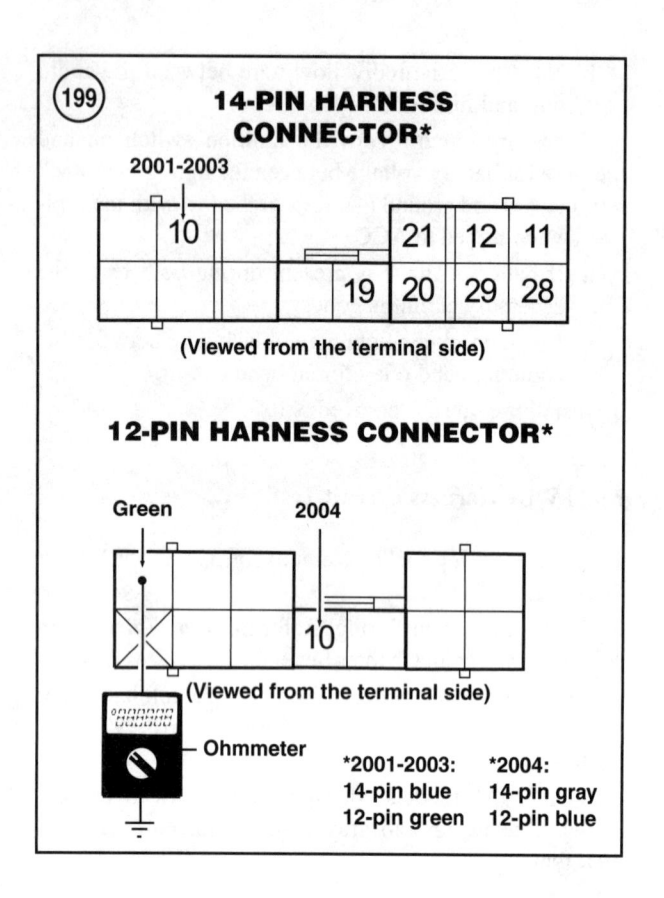

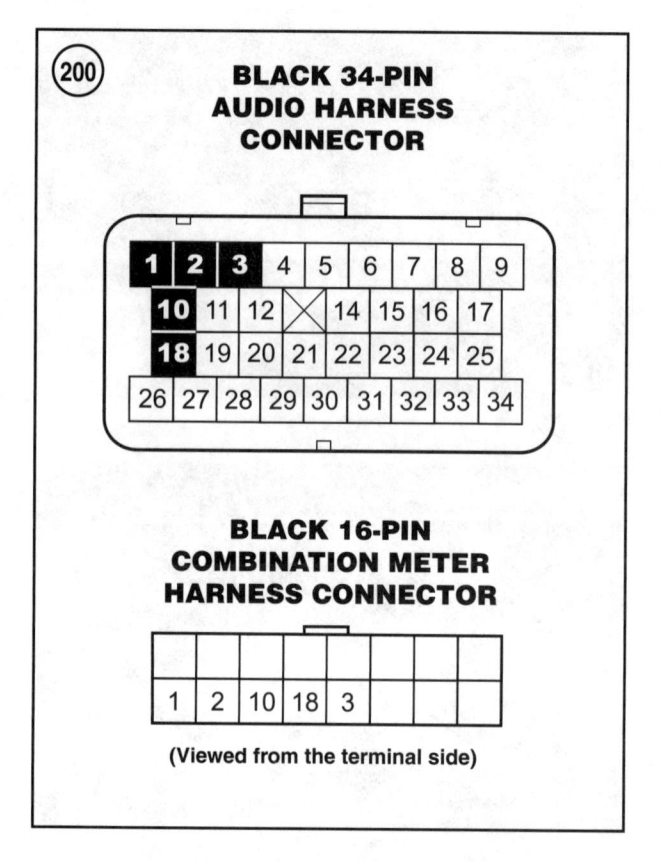

199

14-PIN HARNESS CONNECTOR*

2001-2003

	10		21	12	11
			19	20	29	28

(Viewed from the terminal side)

12-PIN HARNESS CONNECTOR*

Green   2004

Ohmmeter

*2001-2003:   *2004:
14-pin blue   14-pin gray
12-pin green 12-pin blue

200

BLACK 34-PIN AUDIO HARNESS CONNECTOR

1	2	3	4	5	6	7	8	9
10	11	12	×	14	15	16	17
18	19	20	21	22	23	24	25
26	27	28	29	30	31	32	33	34

BLACK 16-PIN COMBINATION METER HARNESS CONNECTOR

| 1 | 2 | 10 | 18 | 3 | | | |

(Viewed from the terminal side)

8. Check for a short circuit between the connector terminals identified in **Figure 200** and ground. There should be no continuity.

NOTE
Step 9 tests the speed pulse signal circuit.

9. Support the motorcycle on its centerstand. Turn the ignition switch on and measure voltage between the gray 34-pin audio harness connector No. 24 terminal (+) (**Figure 198**) and ground. Turn the rear wheel slowly while reading the voltmeter. The voltmeter should alternately read between 0-5 volts.

NOTE
Step 10 tests the starter/reverse switch circuit.

10. Turn the ignition switch on and the ignition switch to RUN. Push the starter/reverse switch and measure voltage between the gray 34-pin audio harness connector No. 34 terminal (+) (**Figure 198**) and ground. The voltmeter should read battery voltage.

11. Installation is the reverse of these steps.

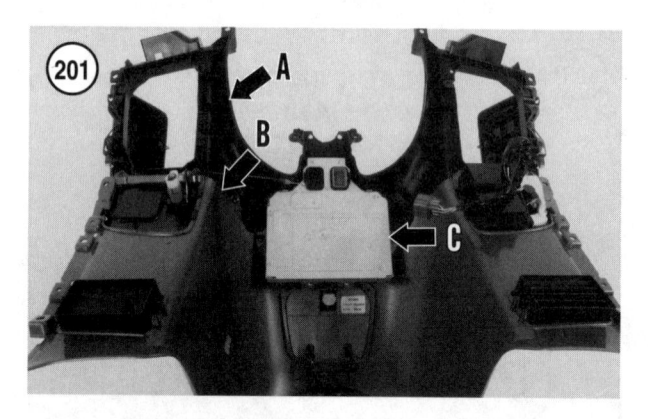

201

Speaker Output Voltage Test

1. Remove the speakers as described in this chapter.
2. Turn a multimeter to its AC scale.

NOTE
In Step 3 and Step 4, connect the meter test leads to the speaker wire harness connector terminals.

3. Left speaker: Connect the positive test lead to the blue/green (+) terminal and the negative test lead to the gray/black (–) terminal.

ELECTRICAL SYSTEM

202
14-PIN LEFT HANDLEBAR SWITCH CONNECTOR (2001-2003 BLUE/2004 GRAY) 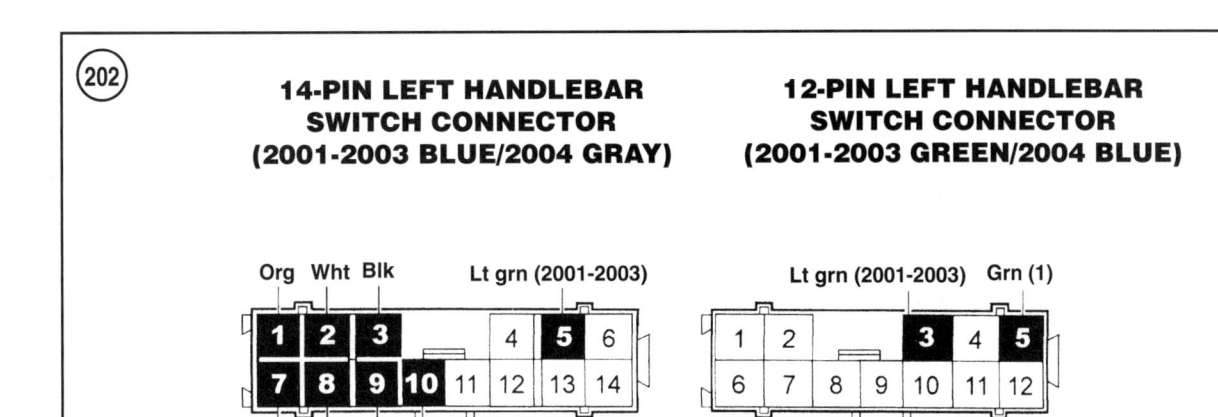

Org Wht Blk — Lt grn (2001-2003)
```
| 1 | 2 | 3 |   | 4 | 5 | 6 |
| 7 | 8 | 9 |10 |11 |12 |13 |14|
```
Red Yel Brn Grn (2)

12-PIN LEFT HANDLEBAR SWITCH CONNECTOR (2001-2003 GREEN/2004 BLUE)

Lt grn (2001-2003)   Grn (1)
```
| 1 | 2 |   | 3 | 4 | 5 |
| 6 | 7 | 8 | 9 |10 |11 |12|
```

(Viewed from the switch side)

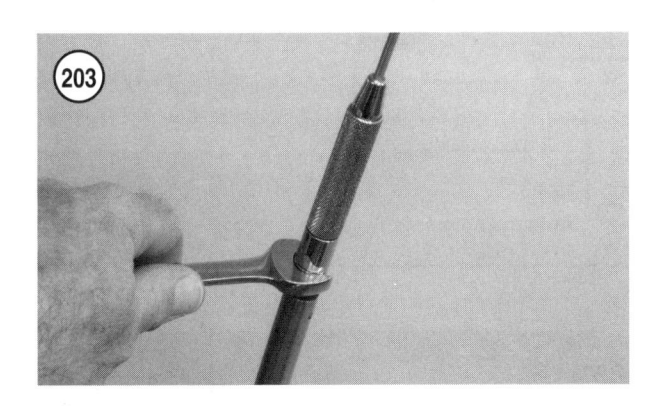

4. Right speaker: Connect the positive test lead to the red/green (+) terminal and the negative test lead to the brown/black (–) terminal.

5. Turn the ignition switch and the radio on.

6. The voltage should increase or decrease when turning the radio volume control knob.

Audio Unit Removal/Installation

1. Remove the top shelter. Refer to *Top Shelter* in Chapter Seventeen.

2. Remove the screws and separate the front (A, **Figure 201**) and rear (B) top shelter halves.

3. Remove the two screws and the audio unit (C, **Figure 201**).

4. Installation is the reverse of these steps.

AUDIO SWITCH

Testing

The audio switch is part of the left-hand handlebar switch assembly.

1. Remove the front fairing. Refer to *Front Fairing and Fairing Molding* in Chapter Seventeen.

2. Disconnect the left handlebar switch 12-pin and 14-pin (**Figure 197**) connectors.

3. Test the audio switches as described under *Handlebar Switches, Continuity Test* in this chapter. When doing so, refer to **Figure 202** to identify the 12-pin and 14-pin left handlebar switch connector terminals (**Figure 197**) with the wire colors listed on the individual audio switch continuity diagrams at the end of this manual.

4. Installation is the reverse of removal. Operate each switch to make sure the assembly is correct.

Replacement

Refer to *Handlebar Switches* in this chapter to replace the left-hand handlebar switch.

ANTENNA

Removal/Installation

1. Remove the seat. Refer to *Seat* in Chapter Seventeen.

2. Loosen the nut and remove the antenna (**Figure 203**) from the base.

3. Open the trunk lid.

4. Remove the screws and the trunk side pocket (A, **Figure 204**).

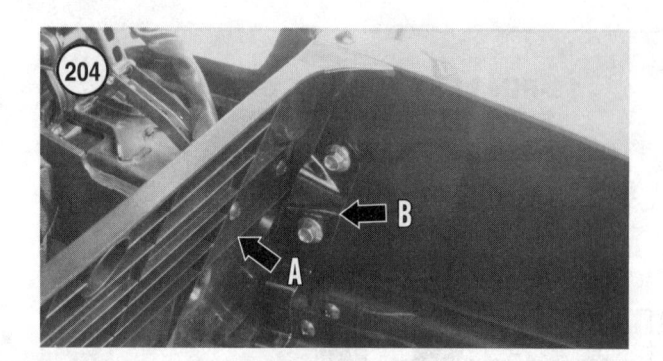

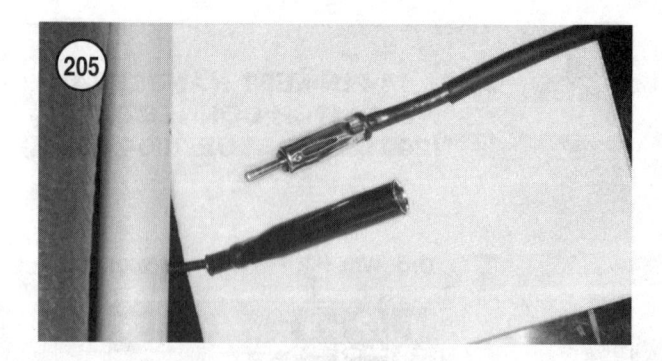

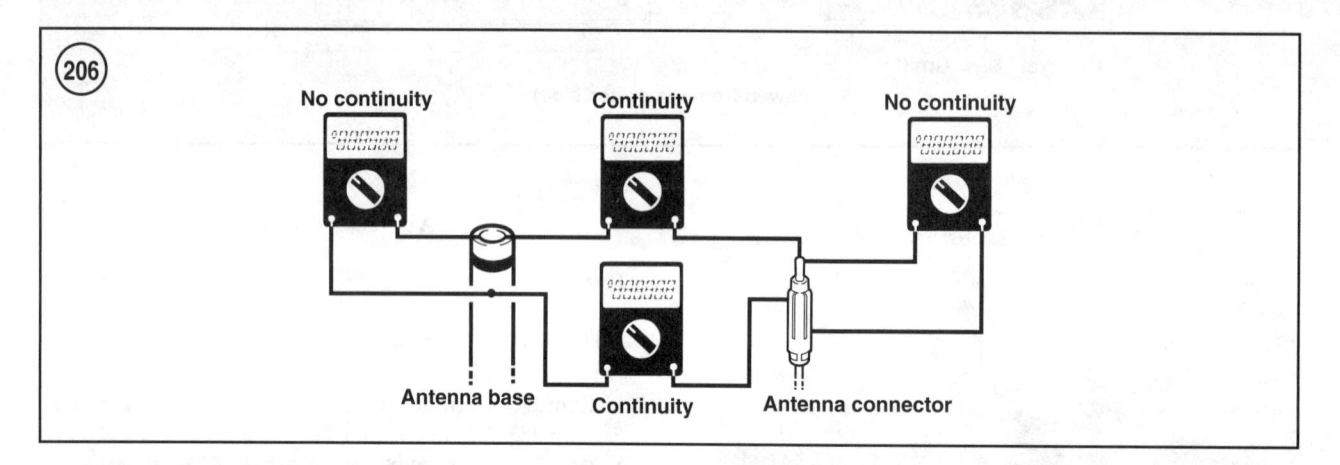

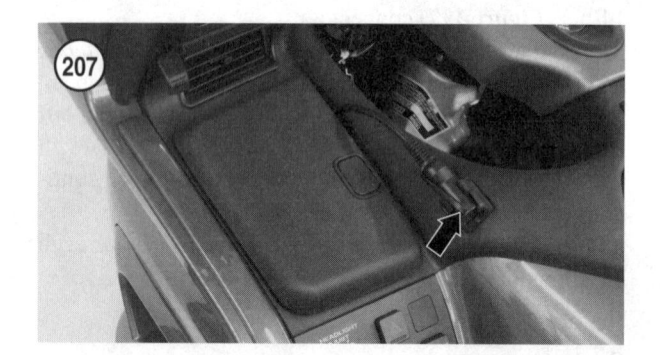

5. Disconnect the antenna connector (**Figure 205**) mounted under the frame cross-member and remove the antenna wire.
6. Remove the bolts, antenna stay (B, **Figure 204**) and antenna base.
7. Installation is the reverse of removal. Tighten the antenna base bolts to 14 N•m (10 ft.-lb.).

Inspection

1. Remove the seat. Refer to *Seat* in Chapter Seventeen.
2. Open the trunk lid.

3. Disconnect the antenna connector (**Figure 205**) mounted under the frame cross-member.
4. Switch an ohmmeter to R × 1.
5. Test the antenna and antenna connector as shown in **Figure 206**.

HEADSET JUNCTION CABLE

Testing/Replacement

1. To test the front headset (**Figure 207**), remove the left fairing pocket. Refer to *Fairing Pockets* in Chapter Seventeen.

ELECTRICAL SYSTEM

209

Headset junction cable

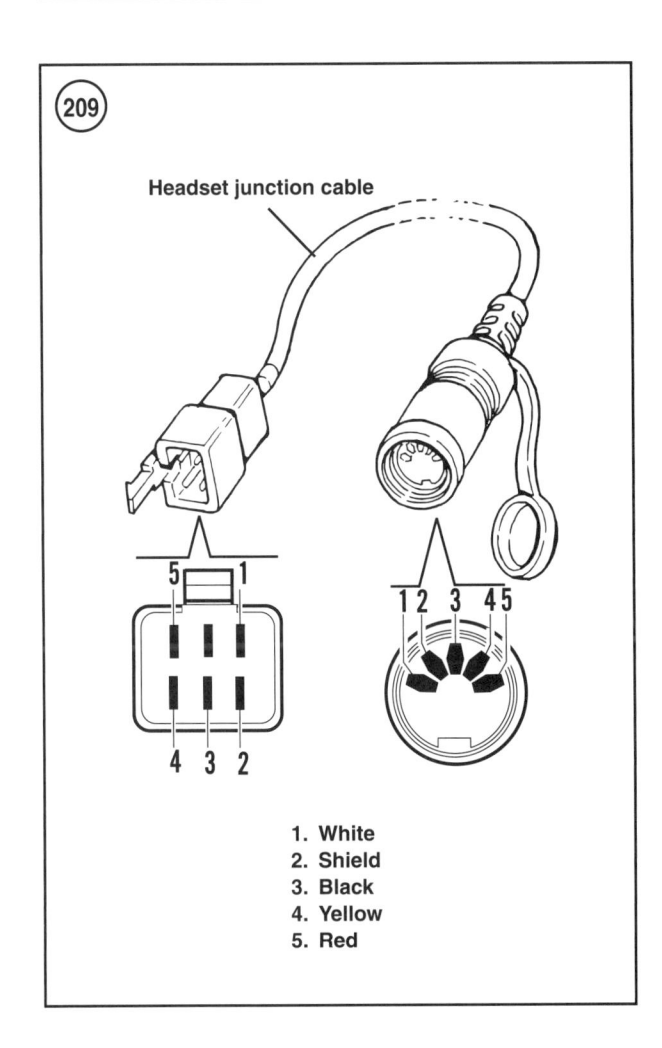

1. White
2. Shield
3. Black
4. Yellow
5. Red

210

2. To test the rear headset (**Figure 208**), remove the seat. See Chapter Seventeen under *Seat*.

3. Disconnect the red 6-pin headset connector.

4. Check for continuity between the connector terminals and junction connector terminals identified in **Figure 209**. Continuity should be recorded at each test. Replace the headset junction cable if any test is incorrect.

5. Installation is the reverse of these steps.

211

212

SPEAKERS

Removal/Installation

1. Remove the meter panel. Refer to *Meter Panel* in Chapter Seventeen.

2. Remove the four screws (**Figure 210**).

3. Disconnect the connectors (**Figure 211**) and remove the speaker.

4. Installation is the reverse of these steps. The speaker terminals are different widths.

Testing

Switch an analogue ohmmeter to R × 1 and connect its test leads to the speaker terminals (**Figure 212**). The speaker should click when the ohmmeter leads are connected. If not, replace the speaker.

WIRING DIAGRAMS

Wiring diagrams for all models are located at the end of the manual.

Table 1 BATTERY SPECIFICATIONS

Type	Maintenance-free (sealed)
Capacity	12 volts, 18 amp hour
Current drain (draw)	5 mA maximum
Voltage (at 20° C [68° F])	
Fully charged	13.0-13.2 volts
Needs charging	Below 12.3 volts
Charging current	
Normal	1.8 amps × 5-10 hours
Quick	9.0 amps × 1 hour

Table 2 MAINTENANCE-FREE BATTERY VOLTAGE READINGS

State of charge	Voltage reading
100%	13.0-13.2
75%	12.8
50%	12.5
25%	12.2
0%	12.0 volts or less

Table 3 ALTERNATOR AND CHARGING SYSTEM SPECIFICATIONS*

Alternator	
Type	Triple phase output alternator
Charging system output	1100 watts maximum
Stator coil resistance	0.07-0.09 ohms
Rotor coil resistance	2.5-2.9 ohms
Rotor coil slip ring outside diameter	
New	22.7 mm (0.89 in.)
Service limit	21.2 mm (0.83 in.)

*Test at 20° C (68° F). Do not test when the engine or component is hot.

Table 4 IGNITION SYSTEM SPECIFICATIONS

Ignition coil peak voltage	2.5-5.0 volts
Ignition pulse generator peak voltage	0.7 volts minimum

Table 5 STARTING SYSTEM SPECIFICATIONS*

Reverse resistor resistance	
Black and red terminals	0.20-0.25 ohms
White terminal	0.15-0.20 ohms
Starter/reverse motor brush length	
New	12.5 mm (0.49 in.)
Service limit	6.0 mm (0.24 in.)

*Test at an 20° C (68° F). Do not test when the engine or component is hot.

Table 6 BULB SPECIFICATIONS

Item	Wattage × quantity
Brake/taillight	21/5W × 6
Front turn signal/running light	21/5W × 2
Headlight HI beam	55W × 2
Headlight LO beam	55W × 2
Left handlebar switch light	1.4W × 7
Left panel switch light	1.4W × 4
License light	3CP (5W × 1)
Rear turn signal light	21W × 2
Right handlebar switch light	1.4W × 4

ELECTRICAL SYSTEM

Table 7 SENSOR AND SWITCH TEST SPECIFICATIONS

Item	Test readings
ECT sensor	
At 80° C (176° F)	47-57 ohms
At 120° C (248° F)	14-18 ohms
Fuel level sensor[1]	
Float at empty position	61.4-63.4 ohms
Float at full position	1.5-2.5 ohms
Fuel pump side fuel level sensor[1]	
Float at empty position	28.6-30.6 ohms
Float at full position	1.5-2.5 ohms
Open air temperature sensor[2]	
-5° C (23° F)	21k ohms
0° C (32° F)	16 k ohms
10° C (50° F)	10 k ohms
20° C (68° F)	6 k ohms
30° C (86° F)	4 k ohms
40° C (104° F)	2.5 k ohms

1. Test at an ambient temperature of 20° C (68° F). Do not test when the engine or component is hot.
2. Readings are approximate.

Table 8 FUSE SPECIFICATIONS

Fuse No.	Circuit/Fuse	Amperage
1	Not used	–
2	Main fuse A	30
3	ABS motor fuse (front)	30
4	ABS motor fuse (rear)	30
5	Taillight fuse	15
6	Accessory terminal	–
7	Fan fuse	20
8	Stop switch fuse	10
9	Headlight relay fuse	10
10	Reverse start fuse	5
11	Reverse fuse A	–
12	Reverse fuse B	5
13	ABS main fuse	5
14	Not used	–
15	Not used	–
16	Not used	–
17	PGM-FI ignition fuse	20
18	Stop light fuse	15
19	Ignition cruise fuse	15
20	Suspension level fuse	15
21	Audio/ACC fuse	15
22	Battery fuse	20
23	Reverse shift fuse	15
24	Headlight HI fuse	10
25	Not used	–
26	Headlight LO fuse	15
27	Horn/turn fuse	15

Table 9 ELECTRICAL SYSTEM TORQUE SPECIFICATIONS

	N•m	in.-lb.	ft.-lb.
Alternator mounting bolts	29	–	21
Alternator terminal nut	8	71	–
Antenna base bolts	14	–	10

(continued)

Table 9 ELECTRICAL SYSTEM TORQUE SPECIFICATIONS (continued)

	N•m	in.-lb.	ft.-lb.
Bank angle sensor mounting screws	2	18	–
ECT sensor	25	–	18
Front crankcase cover	12	106	–
Ignition pulse generator[1]	12	106	–
Ignition switch	25	–	18
Oil pressure switch[2]	12	106	–
Rear brake light/cruise cancel switch holder[1]	2	18	–
Reverse switch	12	106	–
Sidestand switch mounting bolt	10	88	–
Starter mounting bolts	29	–	21

1. Apply a medium strength threadlock onto fastener threads.
2. Apply RTV sealant to switch threads as described in text.

CHAPTER TEN

COOLING SYSTEM

This chapter describes the repair and replacement of cooling system components. **Table 1** and **Table 2** at the end of the chapter lists cooling system specifications. Refer to Chapter Three for routine cooling system maintenance.

The pressurized cooling system consists of the radiator cap, radiators, water pump, thermostat, electric cooling fans and coolant reservoir tank. The water pump and thermostat are combined into the same housing.

The water pump requires no routine maintenance and replacement parts are unavailable. Replace the complete water pump/thermostat housing if it is defective.

Drain and flush the cooling system at the interval listed in Chapter Three. Refill with a mixture of ethylene glycol antifreeze (formulated for aluminum engines) and distilled water. Do not reuse the old coolant, as it deteriorates with use. Do not operate the cooling system with only distilled water (even in climates where antifreeze protection is not required); doing so will promote internal engine corrosion. Refer to *Coolant Change* in Chapter Three.

WARNING
*Never remove the radiator cap (**Figure 1**), the coolant drain plug or disconnect any coolant hose while the engine and radiator are hot. Scalding fluid and steam may blow out under pressure and cause serious injury. The cooling system must be cool before removing any system component.*

WARNING
Antifreeze (coolant) is a toxic waste material. Drain into a suitable container and dispose of it according to local toxic waste regulations. Do not store coolant where it is accessible to children or pets.

CHAPTER TEN

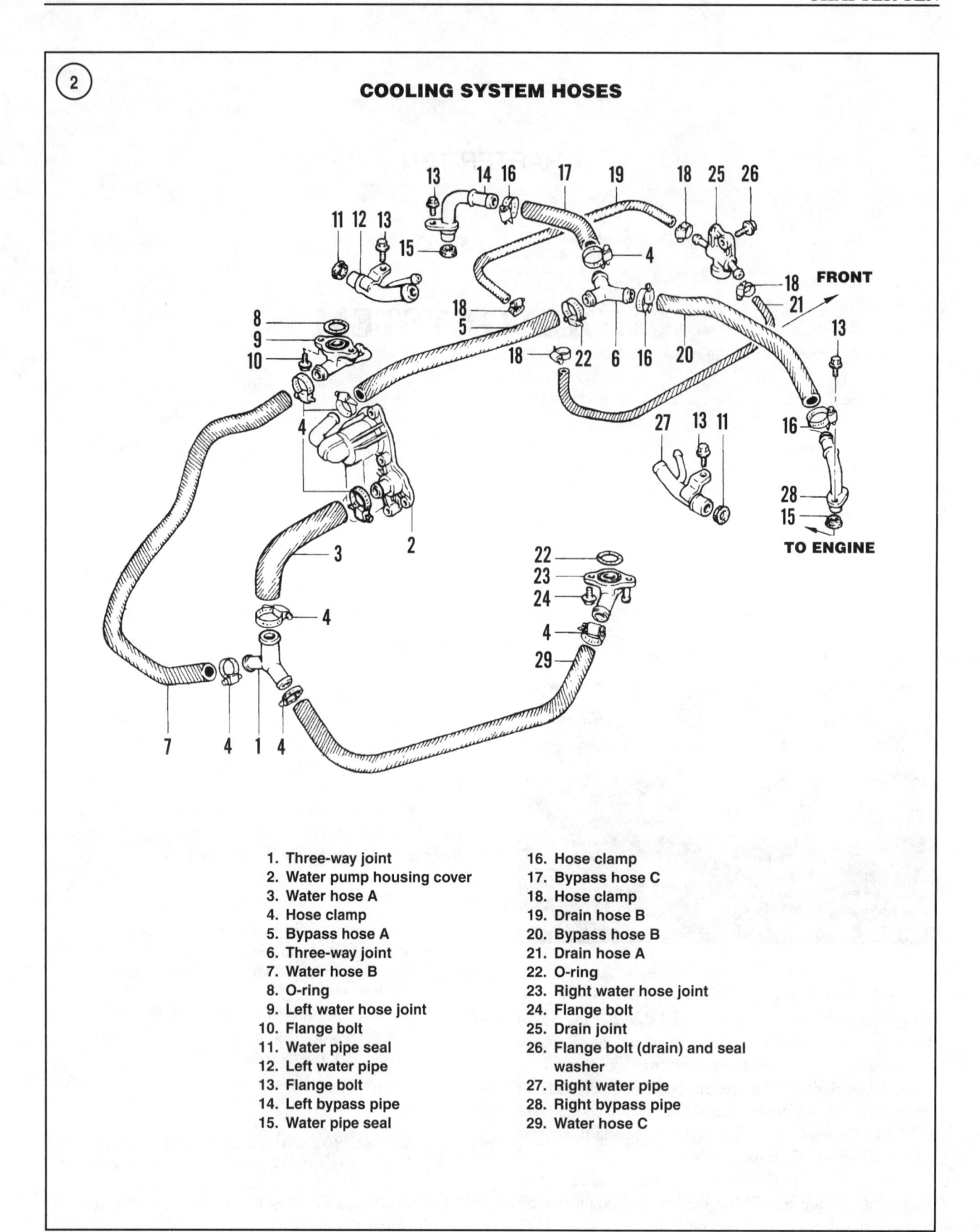

② **COOLING SYSTEM HOSES**

1. Three-way joint
2. Water pump housing cover
3. Water hose A
4. Hose clamp
5. Bypass hose A
6. Three-way joint
7. Water hose B
8. O-ring
9. Left water hose joint
10. Flange bolt
11. Water pipe seal
12. Left water pipe
13. Flange bolt
14. Left bypass pipe
15. Water pipe seal
16. Hose clamp
17. Bypass hose C
18. Hose clamp
19. Drain hose B
20. Bypass hose B
21. Drain hose A
22. O-ring
23. Right water hose joint
24. Flange bolt
25. Drain joint
26. Flange bolt (drain) and seal washer
27. Right water pipe
28. Right bypass pipe
29. Water hose C

COOLING SYSTEM

TEMPERATURE WARNING SYSTEM

A dial type coolant temperature gauge is located in the combination meter. A needle moves across the dial to show the engine coolant temperature. The engine is warm enough to ride when the needle rises above the C (cold) mark. Under normal conditions, the needle operates midway between the C (cold) and H (hot) marks. Conditions that can cause the needle to rise higher include riding in high ambient temperatures, continuous stop-and-go traffic and climbing in foothill and mountain areas. When the needle nears the H (hot) mark, park in a safe spot and turn the engine off to let it cool. Determine the cause of the overheating before operating the motorcycle. Refer to *Engine* in Chapter Two for additional information.

COOLING SYSTEM INSPECTION

1. If steam is observed at the muffler after the engine has sufficiently warmed up, a head gasket might be damaged. If enough coolant leaks into a cylinder(s), the cylinder could hydrolock and prevent the engine from being cranked. Coolant may also be present in the engine oil. If the oil is foamy or milky-looking, there is coolant in the oil. If so, correct the problem before returning the motorcycle to service.
2. Refer to *Cooling System* in Chapter Three and check the coolant level.
3. Remove the fairing components and fuel tank to access the hoses.
4. Check the radiator(s) for clogged or damaged fins.
5. Check the radiator for loose or missing mounting bolts.
6. Check all coolant hoses for cracks or damage. With the engine cold, squeeze the hoses by hand. If a hose collapses easily, it is damaged and must be replaced. Make sure the hose clamps are tight, but not so tight that they cut the hoses. Refer to *Cooling System Hoses* in this chapter.
7. Make sure the overflow tube is connected to the radiator (next to the radiator cap) and is not clogged or damaged.
8. Because of the extensive bodywork on the GL1800, the engine and hoses are hidden. The first sign of a leak may be coolant puddles underneath the motorcycle or steam coming from the engine when it is running. If there is coolant loss, check the cooling system hoses, water pump and water jackets (cylinder and cylinder head) carefully.
9. To check the cooling system for leaks, pressure test it as described in this chapter.

COOLING SYSTEM HOSES

After removing any cooling system component, inspect the adjoining hose(s) to determine if replacement is necessary. Hoses deteriorate with age and should be inspected carefully for conditions that may cause them to fail. Loss of coolant causes the engine to overheat, and spray from a leaking hose can injure the rider. A collapsed hose prevents coolant circulation and causes overheating. Observe the following when servicing hoses:

1. Refer to **Figure 2** for a diagram of the engine cooling hoses. For a diagram of the hoses attached to the radiator and coolant reserve tank, refer to *Radiators* in this chapter.
2. Make sure the cooling system is cool before removing any coolant hose or component.
3. Use original equipment replacement hoses; they are formed to a specific shape and dimension for a correct fit.
4. Loosen the hose clamps on the hose that is to be replaced. Slide the clamps back off the component fittings.

CAUTION
Do not use excessive pressure when attempting to remove a stubborn hose from a radiator. Also, use caution when loosening hoses with hose pliers. The aluminum radiator hose joints are easily damaged.

5. Twist the hose to release it from the joint. If the hose has been on for some time, it may be difficult to break loose. If so, insert a small screwdriver between the hose and joint and spray WD-40 or a similar lubricant into the opening and carefully twist the hose to break it loose.

NOTE
Remove all lubricant residue from the hose and hose fitting before reinstalling the hose.

6. Examine the fittings for cracks or other damage. Repair or replace as necessary. If the fitting is good, use a wire brush and clean off any hose residue that may have transferred to the fitting. Wipe clean with a cloth.
7. Inspect the hose clamps for rust and corrosion and replace if necessary.
8. If a hose is difficult to install on the joint, soak the end in hot water to make it more pliable. Do not use any lubricant when installing hoses.
9. With the hose correctly installed, position the clamp approximately one-half inch from the end of the hose and tighten the clamp. Position the clamps so they can be checked for tightness when other components are installed near them.

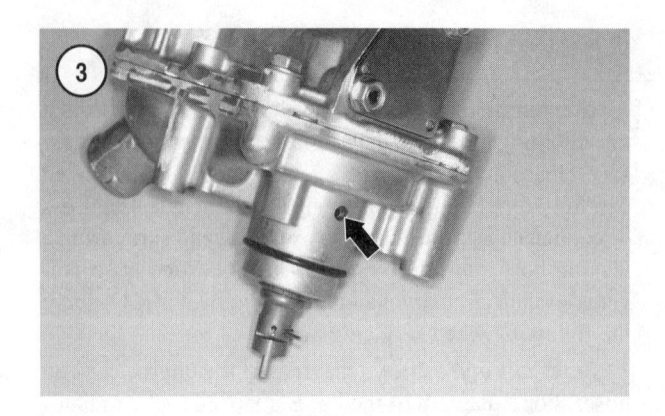

WATER PUMP MECHANICAL SEAL INSPECTION

The water pump is equipped with an inspection hole (**Figure 3**). When coolant leaks from the hole, the mechanical seal in the water pump is damaged. To view the inspection hole:

1. Remove the coolant reserve tank as described in this chapter.
2. Check for signs of coolant or coolant stains on the clutch cover (**Figure 4**). The inspection hole cannot be viewed directly when the water pump/thermostat housing is installed on the engine. **Figure 4** shows the water pump/thermostat housing and clutch cover with the engine removed from the frame. If there is leakage, replace the water pump/thermostat housing as described in this chapter.
3. Reinstall the coolant reserve tank as described in this chapter.
4. Clean up any spilled coolant so it does not contact the rear tire.

PRESSURE TEST

Perform the following test when troubleshooting the cooling system. Perform the test when the engine is cold. A hand pump tester is used to pressurize the system.

WARNING
Never remove the radiator cap (Figure 1), the coolant drain plugs or disconnect any coolant hose while the engine and radiator are hot. Scalding fluid and steam may blow out under pressure and cause serious injury.

1. Remove the right fairing pocket. Refer to *Fairing Pockets* in Chapter Seventeen.
2. With the engine cold, remove the radiator cap (**Figure 1**).
3. Add coolant to the radiator to bring the level up to the filler neck.
4. Check the rubber washers on the radiator cap (**Figure 5**). Replace the cap if the washers show signs of deterioration, cracking or other damage. If the radiator cap is good, perform Step 5.

CAUTION
Do not exceed 20 psi (137 kPa) or the cooling system components may become damaged.

5. Lubricate the rubber washer on the bottom of the radiator cap with coolant and install it onto a cooling system pressure tester (**Figure 6**, typical). Apply 16-20 psi (108-137 kPa) and check for a pressure drop. Replace the cap if it cannot hold this pressure.
6. Mount the pressure tester onto the filler neck (**Figure 7**, typical) and pressure test the cooling system to 16-20 psi (108-137 kPa). If the system cannot hold this pressure, check for coolant leaks at the following components:
   a. Filler neck.
   b. Coolant hoses. Either from loose hose connectors or damaged hoses.

COOLING SYSTEM

6

Cooling system tester — Radiator cap

7

Cooling system tester

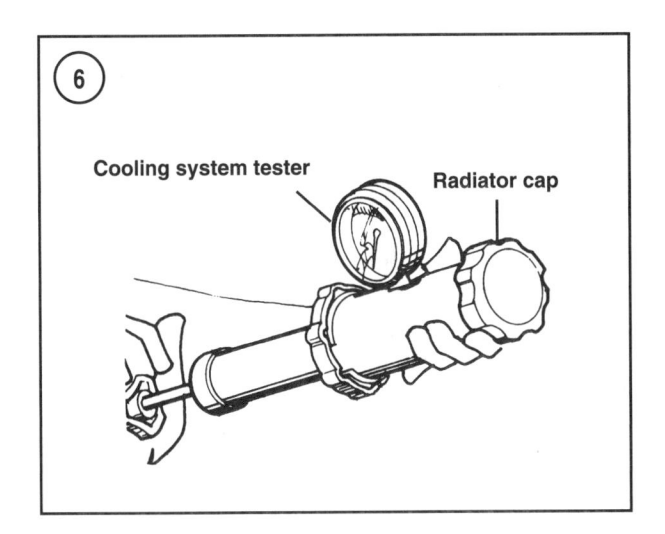

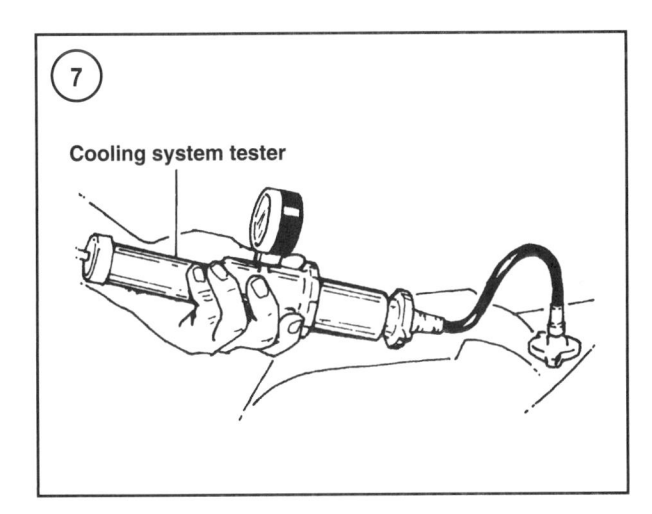

c. Damaged or deteriorated O-rings installed in coolant hose connectors. See **Figure 2**.
d. Damaged water pump mechanical seal.
e. Water pump.
f. Loose coolant drain bolt.
g. Warped cylinder head or cylinder mating surfaces.

NOTE
If the test pressure drops rapidly, but there are no visible coolant leaks, coolant may be leaking into one of the cylinder heads. Perform a compression test as described in Chapter Three.

7. Check all cooling system hoses for damage or deterioration. Replace any questionable hose. Make sure all hose clamps are tight.
8. Remove the tester and install the radiator cap.
9. Reverse Step 1.

COOLANT RESERVE TANK

The coolant reserve tank (**Figure 8**) is mounted behind the engine.

Removal/Installation

1. Park the motorcycle on its centerstand.
2. Remove the left engine side cover. Refer to *Engine Side Covers* in Chapter Seventeen.
3. Note the hose routing around the coolant reserve tank before removing the tank (**Figure 9**).
4. Remove the bolt and collar and lower the tank below the frame.
5. Disconnect the siphon hose and overflow hose from the tank.
6. Drain the tank of all coolant, then flush and clean the tank with water.
7. Inspect the coolant reserve tank and replace it if it is leaking or damaged.
8. Installation is the reverse of removal. Note the following:
   a. Hook the tank groove against the frame (**Figure 10**).
   b. Install the collar and tighten the mounting bolt securely.
   c. Make sure the reserve tank hose is not pushing against the drive shaft boot. If so, the boot may contact the drive shaft and melt.
   d. Refill the coolant reserve tank with a 50:50 mixture of antifreeze and distilled water as described in Chapter Three.
   e. Check the hoses for leaks.

RADIATORS

Right Radiator Removal/Installation

The radiator and fan are removed as an assembly.
Refer to **Figure 8**.

1. Remove the top shelter. Refer to *Top Shelter* in Chapter Seventeen.
2. Remove the front fairing. Refer to *Front Fairing* in Chapter Seventeen.
3. Drain the engine coolant. Refer to *Coolant Change* in Chapter Three.
4. Disconnect the white 2-pin fan motor connector (**Figure 11**).
5. Remove the screws securing the connector holder to the air duct (**Figure 12**).
6. Remove the trim clips and air duct (A, **Figure 13**).
7. Release the hooks and remove the radiator grille (B, **Figure 13**).

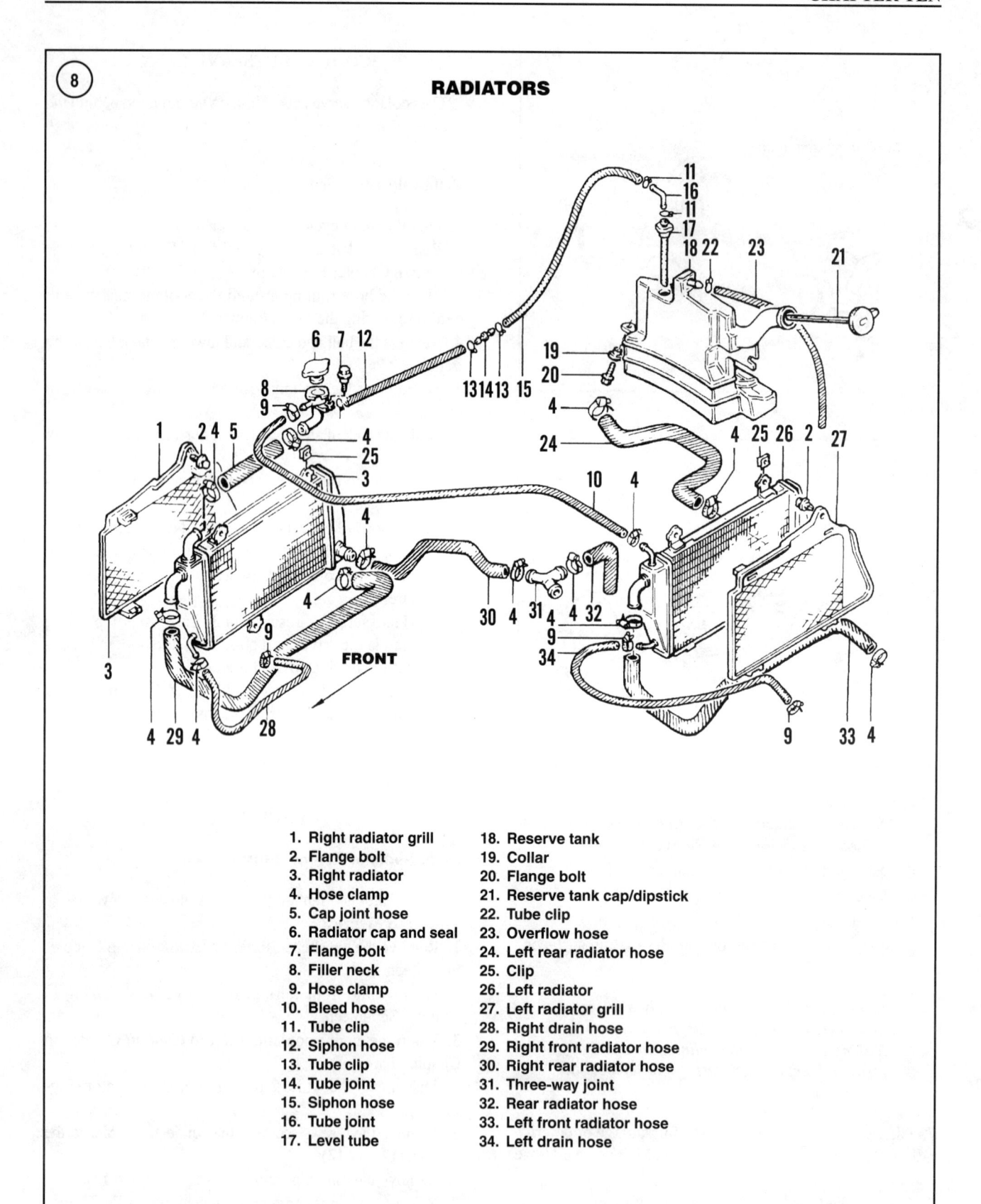

CHAPTER TEN

RADIATORS

1. Right radiator grill
2. Flange bolt
3. Right radiator
4. Hose clamp
5. Cap joint hose
6. Radiator cap and seal
7. Flange bolt
8. Filler neck
9. Hose clamp
10. Bleed hose
11. Tube clip
12. Siphon hose
13. Tube clip
14. Tube joint
15. Siphon hose
16. Tube joint
17. Level tube
18. Reserve tank
19. Collar
20. Flange bolt
21. Reserve tank cap/dipstick
22. Tube clip
23. Overflow hose
24. Left rear radiator hose
25. Clip
26. Left radiator
27. Left radiator grill
28. Right drain hose
29. Right front radiator hose
30. Right rear radiator hose
31. Three-way joint
32. Rear radiator hose
33. Left front radiator hose
34. Left drain hose

COOLING SYSTEM

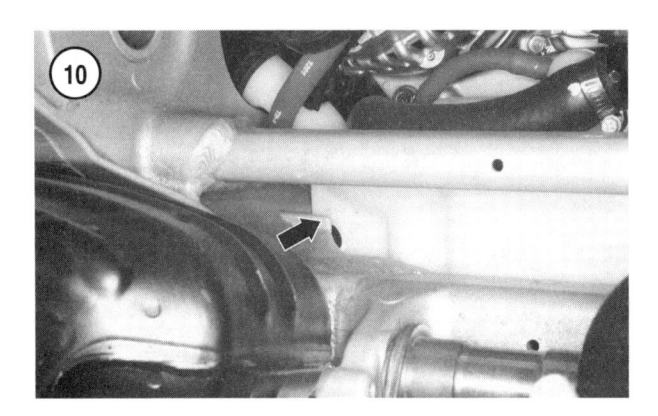

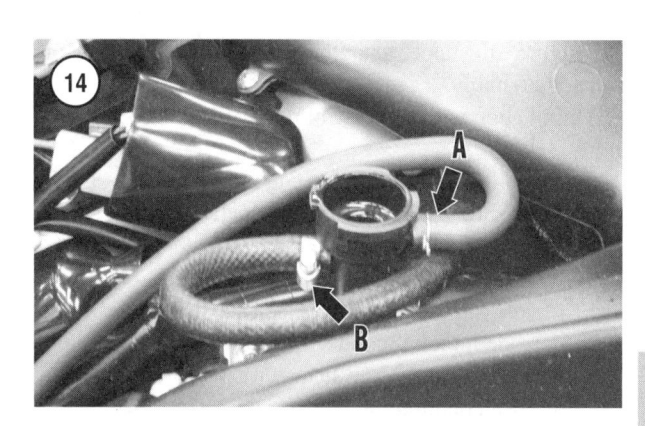

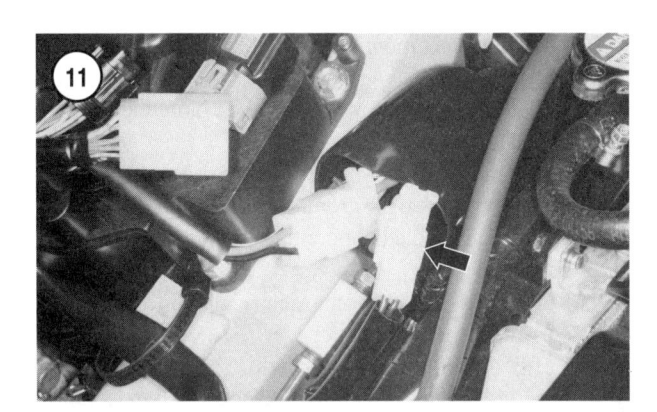

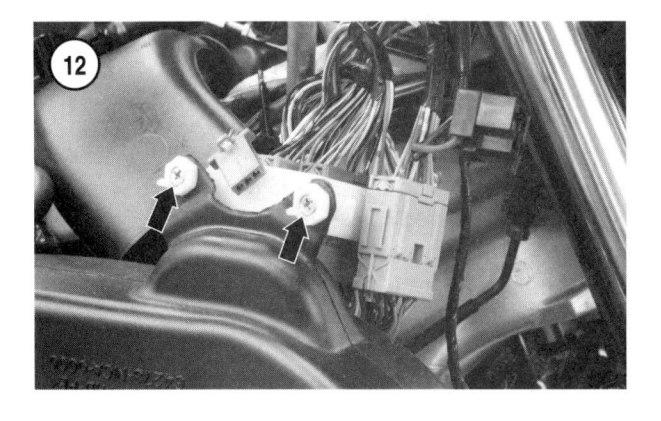

8. Disconnect the siphon hose (A, **Figure 14**) and bleed hose (B) from the filler neck.
9. Remove the filler neck mounting bolt.

NOTE
The radiator can be removed with the upper radiator hose and filler neck attached.

10. Disconnect the lower radiator hose at the rear of the radiator.
11. Disconnect the upper radiator hose (A, **Figure 15**) and drain hose (B) from the front of the radiator.

12. Remove the upper mounting bolts and remove the right radiator assembly.
13. Installation is the reverse of removal. Note the following:
    a. Install the radiator by inserting the boss on the bottom of the shroud into the grommet on the mounting stay.
    b. Tighten the radiator mounting bolts to 14 N•m (10 ft.-lb.).
    c. Make sure the fan motor electrical connector is not corroded, then reconnect.
    d. Refill the cooling system with the recommended type and quantity of coolant as described in Chapter Three.
    e. Check the hoses for leaks.

Left Radiator
Removal/Installation

The radiator and fan are removed as an assembly. Refer to **Figure 8**.
1. Remove the top shelter. Refer to *Top Shelter* in Chapter Seventeen.
2. Remove the front fairing. Refer to *Front Fairing* in Chapter Seventeen.
3. Drain the engine coolant. Refer to *Coolant Change* in Chapter Three.
4. Disconnect the black 2-pin fan motor connector (**Figure 16**).
5. Remove the screws securing the connector holder to the air duct.
6. Remove the trim clips and air duct.
7. Release the hooks and remove the radiator grille from the front of the radiator.
8. Disconnect the hoses at the radiator:
    a. Bleed hose (A, **Figure 17**).
    b. Upper radiator hose (B, **Figure 17**).
    c. Drain hose (C, **Figure 17**).
    d. Lower radiator hose (**Figure 18**).

9. Remove the upper mounting bolts and remove the left radiator assembly.
10. Installation is the reverse of removal. Note the following:
    a. Install the radiator by inserting the boss on the bottom of the shroud into the grommet on the mounting stay.
    b. Tighten the radiator mounting bolts to 14 N•m (10 ft.-lb.).
    c. Make sure the fan motor electrical connector is not corroded, then reconnect.
    d. Refill the cooling system with the recommended type and quantity of coolant as described in Chapter Three.
    e. Check the hoses for leaks.

Inspection

1. Inspect the radiator cap seals (**Figure 5**) for deterioration or damage. Check the spring for damage. Pressure test the radiator cap as described under *Pressure Test* in this chapter. Replace the radiator cap if necessary.
2. Flush the exterior of the radiators (**Figure 19**) with a water hose on low pressure. Spray both the front and the back to remove dirt and debris. Carefully use a whiskbroom or stiff paintbrush to remove any stubborn dirt.

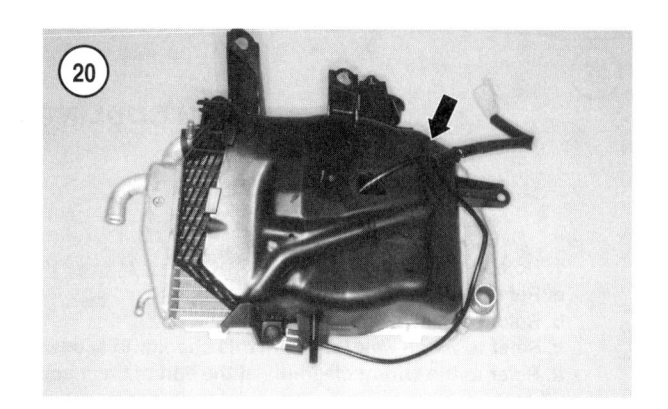

COOLING SYSTEM

COOLING FAN TROUBLESHOOTING CHART

Problem/symptom	Diagnostic flowchart
Both cooling fans do not run	Figure 22
Cooling fans do not shut off	Figure 23
Only one cooling fan runs	Figure 24

CAUTION
Do not press too hard or the cooling fins and tubes may become damaged and leak.

3. Carefully straighten out any bent cooling fins with a broad-tipped screwdriver.

4. Check for cracks or leakage (usually a moss-green colored residue) at the filler neck, the inlet and outlet hose joints and the upper and lower tank seams.

COOLING FANS

An electric cooling fan is mounted on the backside of each radiator (**Figure 20**). The cooling fans are controlled by the fan control relay, the engine coolant temperature sensor (ECT) and the engine control module (ECM). The ECT, mounted in the left cylinder head water jacket, converts coolant temperature into voltage and signals the ECM on temperature changes.

Troubleshooting

Figure 21 lists particular symptom-based problems for the fan control circuit. Find the complaint that best describes the actual condition, then turn to the indicated diagnostic flowchart (**Figures 22-24**) and perform the relevant procedures until the problem is resolved.

Refer to the wiring diagrams at the end of the manual to identify connectors and components in the system.

Before troubleshooting the fan control circuit, check the following:

1. The battery must be fully charged. Refer to *Battery* in Chapter Nine.
2. Check for a blown No. 7 fuse. Refer to *Fuses* in Chapter Nine.

Removal/Installation

1. Remove the radiator(s) as described in this chapter.

㉒ BOTH COOLING FANS DO NOT RUN

1. Note the following:
a. Refer to *Fuses* in Chapter Nine.
b. Refer to *Seat* in Chapter Seventeen.
c. Refer to *Fan Relay Testing* in this chapter to access the fan relay and identify the relay terminals.
d. Refer to the wiring diagrams at the end of the manual to identify the fan motor control connectors, fuse and relay.

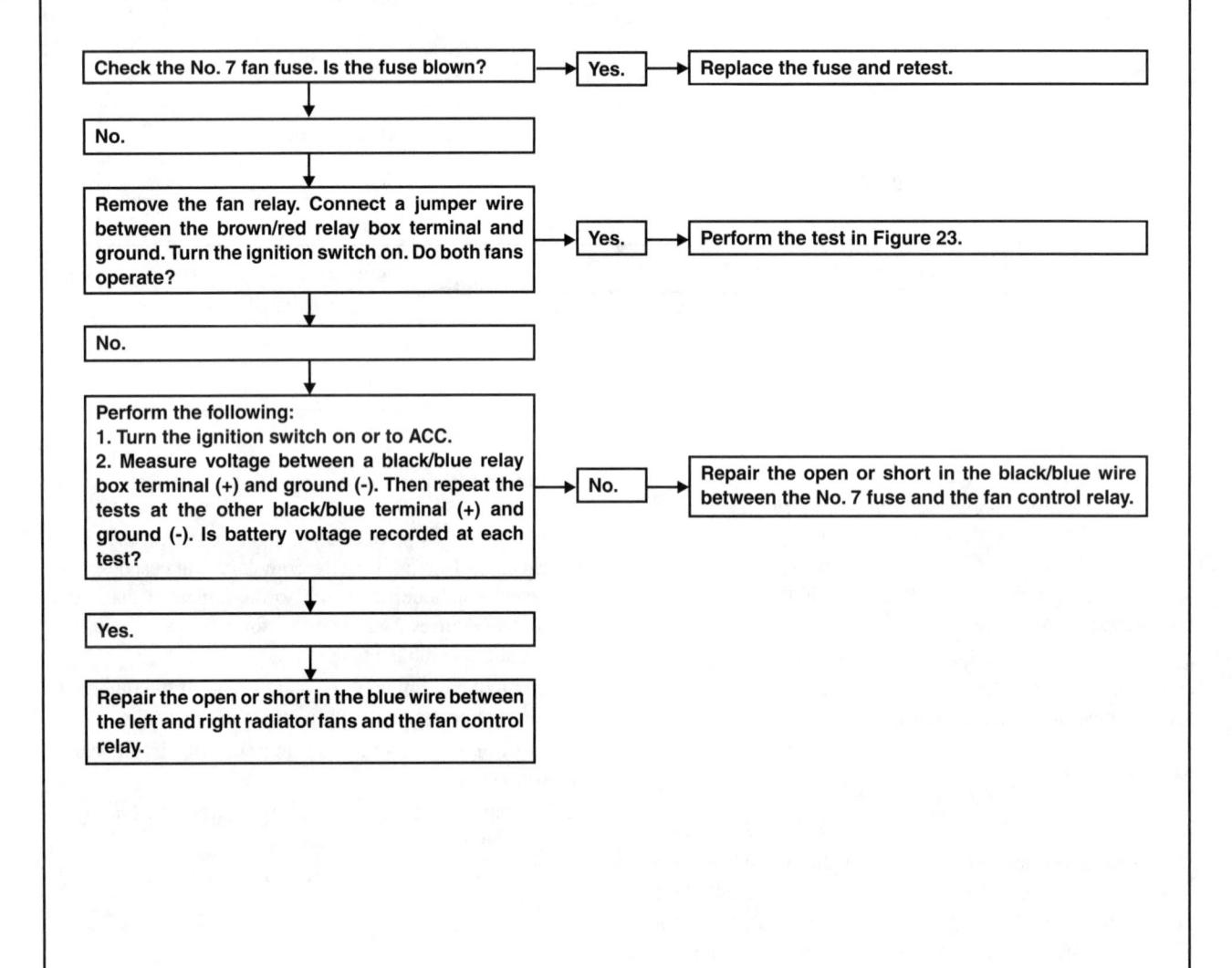

Check the No. 7 fan fuse. Is the fuse blown? → Yes. → Replace the fuse and retest.

↓ No.

Remove the fan relay. Connect a jumper wire between the brown/red relay box terminal and ground. Turn the ignition switch on. Do both fans operate? → Yes. → Perform the test in Figure 23.

↓ No.

Perform the following:
1. Turn the ignition switch on or to ACC.
2. Measure voltage between a black/blue relay box terminal (+) and ground (-). Then repeat the tests at the other black/blue terminal (+) and ground (-). Is battery voltage recorded at each test? → No. → Repair the open or short in the black/blue wire between the No. 7 fuse and the fan control relay.

↓ Yes.

Repair the open or short in the blue wire between the left and right radiator fans and the fan control relay.

COOLING SYSTEM

(23)

COOLING FANS DO NOT SHUT OFF

1. Note the following:
a. Refer to *ECT Sensor* in Chapter Nine.
b. Use a non-contact infrared thermometer or a contact pyrometer when measuring the coolant temperature at the ECT. If these tools are not available, remove and bench test the ECT as described in Chapter Nine. Test results depend on accurate coolant temperature measurement.
c. Refer to *Top Shelter* in Chapter Seventeen.
d. Refer to *Engine Control Module* in Chapter Nine.
e. Refer to the wiring diagrams at the end of the manual to identify the fan motor control circuit.

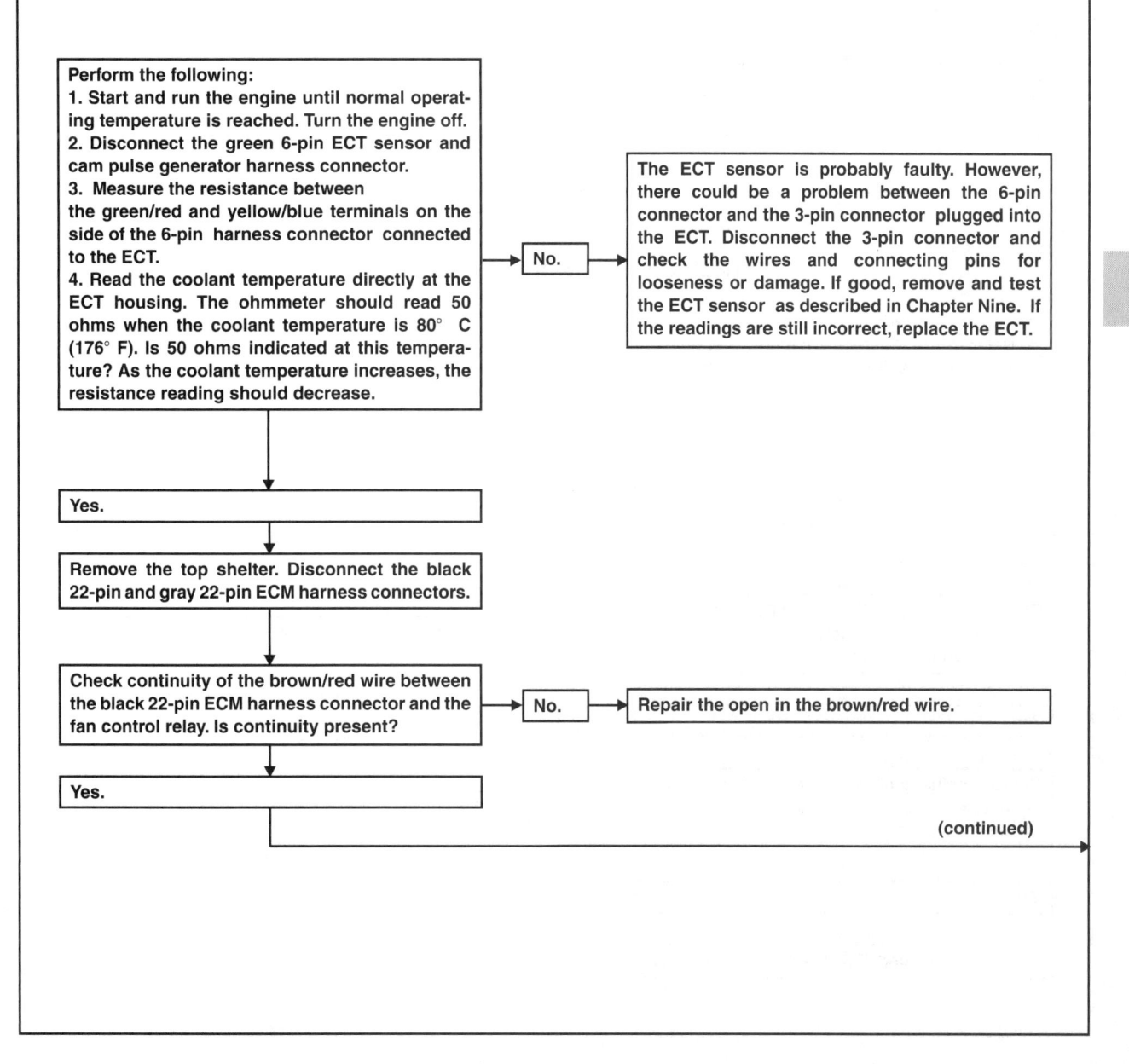

Perform the following:
1. Start and run the engine until normal operating temperature is reached. Turn the engine off.
2. Disconnect the green 6-pin ECT sensor and cam pulse generator harness connector.
3. Measure the resistance between the green/red and yellow/blue terminals on the side of the 6-pin harness connector connected to the ECT.
4. Read the coolant temperature directly at the ECT housing. The ohmmeter should read 50 ohms when the coolant temperature is 80° C (176° F). Is 50 ohms indicated at this temperature? As the coolant temperature increases, the resistance reading should decrease.

→ **No.** → The ECT sensor is probably faulty. However, there could be a problem between the 6-pin connector and the 3-pin connector plugged into the ECT. Disconnect the 3-pin connector and check the wires and connecting pins for looseness or damage. If good, remove and test the ECT sensor as described in Chapter Nine. If the readings are still incorrect, replace the ECT.

↓ **Yes.**

Remove the top shelter. Disconnect the black 22-pin and gray 22-pin ECM harness connectors.

↓

Check continuity of the brown/red wire between the black 22-pin ECM harness connector and the fan control relay. Is continuity present? → **No.** → Repair the open in the brown/red wire.

↓ **Yes.**

(continued)

(23) (continued)

Check continuity of the green/red wire between the black 22-pin ECM harness connector and the 3-pin ECT harness connector. Is continuity present? → **No.** → Repair the open in the green/red wire.

Yes.

Check continuity of the yellow/blue wire between the gray 22-pin ECM harness connector and the 3-pin ECT harness connector. Is continuity present? → **No.** → Repair the open in the yellow/blue wire.

Yes.

Replace the ECM.

---

(24) ONLY ONE COOLING FAN RUNS

1. Note the following:
   a. Refer to *Top Shelter* in Chapter Seventeen to remove and install the top shelter.
   b. Refer to the wiring diagrams at the end of the manual to identify the fan motor control circuit.

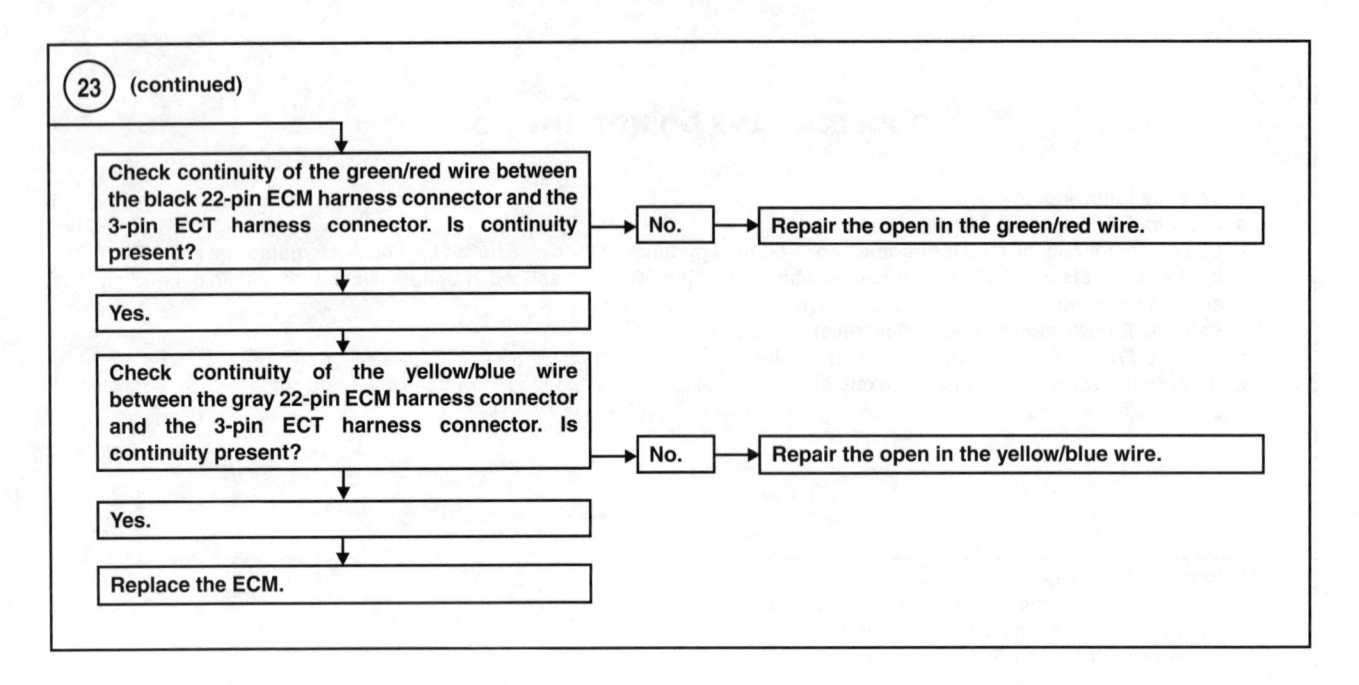

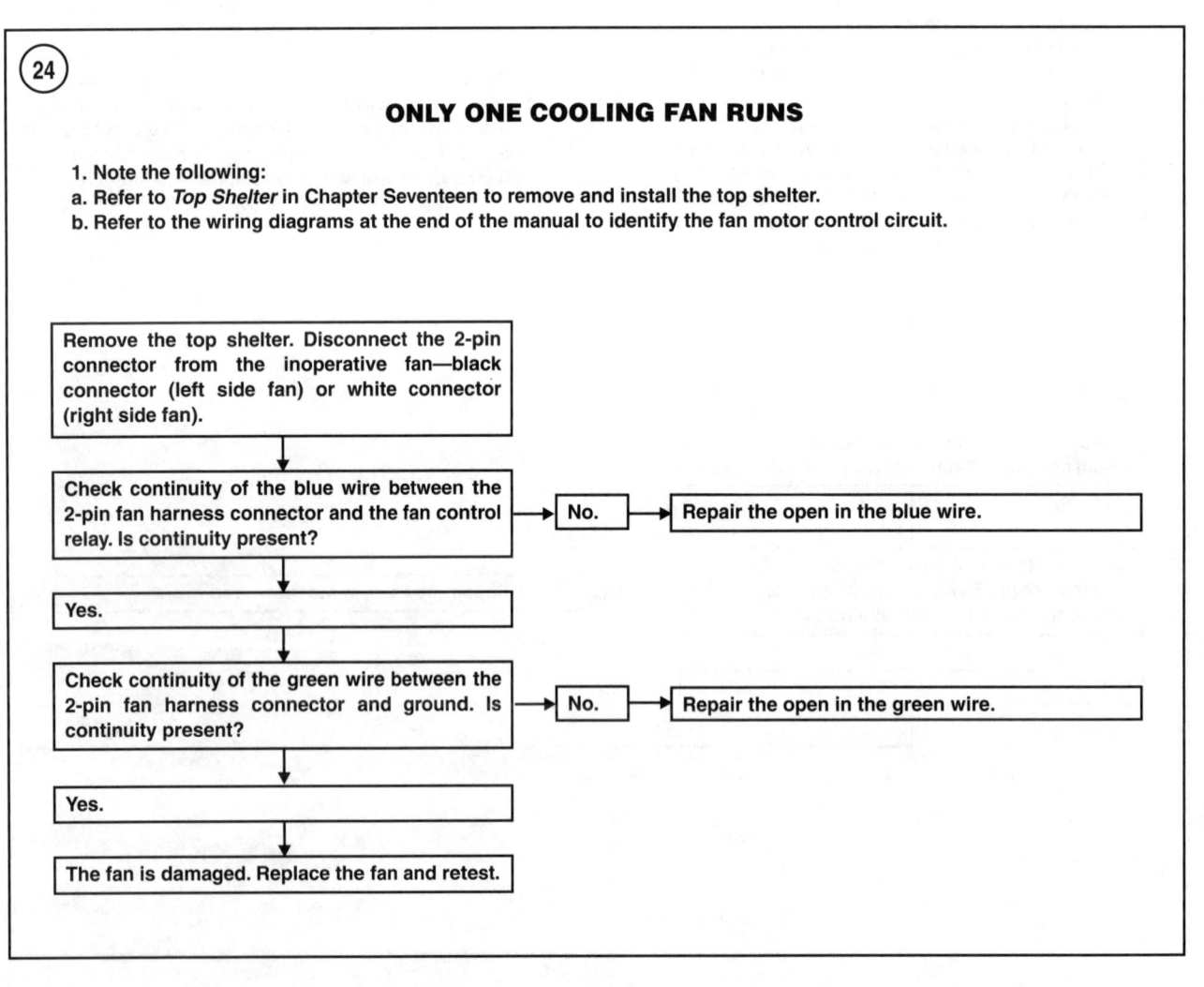

Remove the top shelter. Disconnect the 2-pin connector from the inoperative fan—black connector (left side fan) or white connector (right side fan).

Check continuity of the blue wire between the 2-pin fan harness connector and the fan control relay. Is continuity present? → **No.** → Repair the open in the blue wire.

Yes.

Check continuity of the green wire between the 2-pin fan harness connector and ground. Is continuity present? → **No.** → Repair the open in the green wire.

Yes.

The fan is damaged. Replace the fan and retest.

COOLING SYSTEM

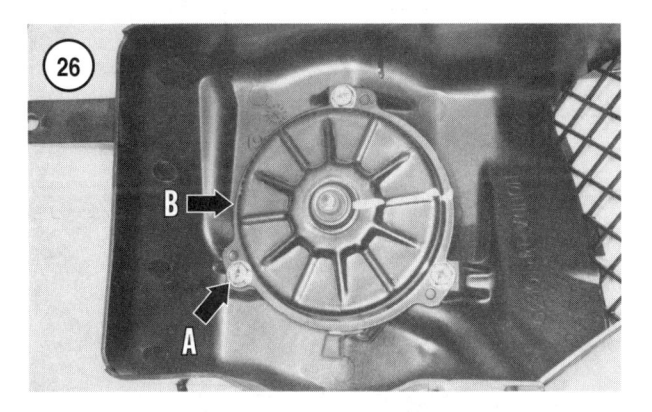

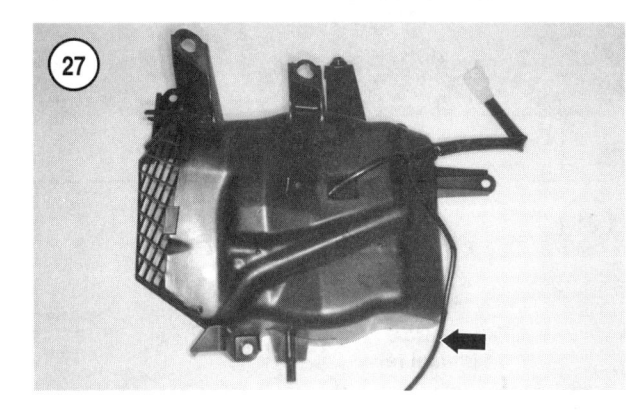

2. Remove the mounting bolts and the shroud/fan motor assembly (**Figure 20**) from the radiator.
3. Remove the nut (A, **Figure 25**) and fan blade (B) from the fan motor.
4. Remove the bolts (A, **Figure 26**) and fan motor (B) from the shroud.
5. Inspect the shroud and replace if damaged.
6. Route the fan motor connector and condenser wire through the shroud (**Figure 27**).
7. Align the groove in the fan blade with the fan motor shaft and install the fan blade (B, **Figure 25**). Apply a medium strength threadlock onto the cooling fan shaft threads and tighten the nut (A, **Figure 25**) securely.
8. Install the shroud/fan motor onto the radiator and secure with the three mounting bolts. Mount the condenser at the lower mounting bolt position (**Figure 28**). Tighten the bolts securely.
9. Install the radiator(s) as described in this chapter.

FAN RELAY TESTING

A single relay controls the radiator fan system. This relay is located in the relay box (3, **Figure 29**) underneath the seat. The fan relay is a 4-terminal relay. Refer to *Relay Box* in Chapter Nine for more information on identifying and testing the relays.

NOTE
*The fan relay can be quickly checked by exchanging it with a known good relay. To do this, first identify a different 4-terminal relay in the relay box (**Figure 29**) and unplug it. Unplug the fan relay and plug in the good relay.*

Refer to the wiring diagrams at the end of the manual to identify the fan motor and fuse circuits referred to in this procedure.

Perform this test if both fans do not operate.

1. Before testing the circuit, start the engine and check the malfunction indicator lamp (MIL). Note the following:
   a. If the MIL stays on, read the code as described in Chapter Eight under *Malfunction Indicator Lamp (MIL)*. If diagnostic trouble code (DTC) 7 is stored, refer to the *DTC Fuel Injection Flow Charts* in Chapter Eight and perform the appropriate test procedure.
   b. If the MIL light did not stay on and both cooling fans do not operate, continue with Step 2.
   c. Turn the engine off.
2. Turn the ignition switch off.
3. Remove the seat. Refer to *Seat* in Chapter Seventeen.

29 RELAY BOX

1	2	3	4	Blank	6	7
8	9	10	11	12	13	14
15	16	17	18	19	20	21

4-Terminal relays — 5-Terminal relays

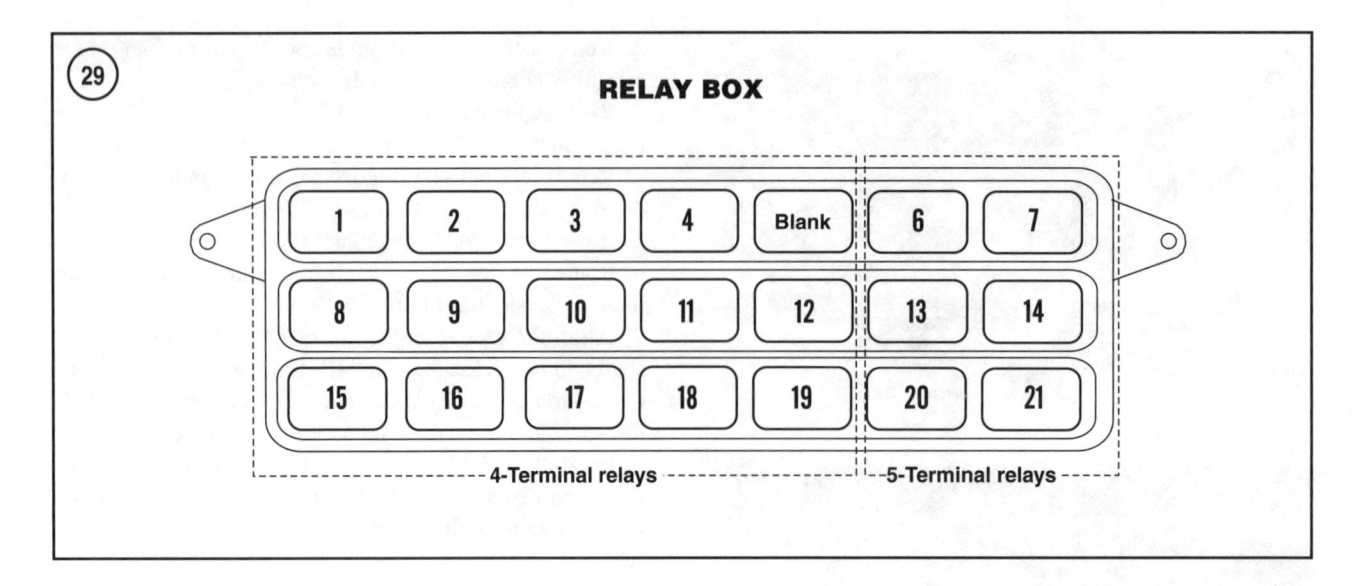

4. Remove the relay box cover screws and cover (**Figure 30**).

5. Remove the fan control relay (3, **Figure 29**).

NOTE
*The meter test connections called out in Steps 6-8 are made at the fan relay connector terminals inside the relay box (**Figure 31**).*

6. Turn the ignition switch on and check for battery voltage between the fan control relay black/blue terminal (+) and ground (−). There should be battery voltage:

   a. Battery voltage: Go to Step 7.
   b. No battery voltage: Check for a blown No. 7 fuse. If the fuse is good, check for an open in the black/blue wire between the fan control relay and the No. 7 fuse.

7. Turn the ignition switch off. Connect a jumper wire between the blue and black/blue fan control relay terminals. Turn the ignition switch on. Both fans should operate. Note the following:

   a. Yes: Go to Step 8.
   b. No: Check for an open circuit in the blue and green wires between the fan control relay and both fans.
   c. Only one fan operates: Check for an open circuit in the blue and green wires between the fan control relay and the inoperative fan. If the circuit is good, the fan is probably faulty. Make a jumper harness and switch the fan 2-pin connectors to retest the fans. If the inoperative fan still does not work, the fan is faulty and must be replaced.

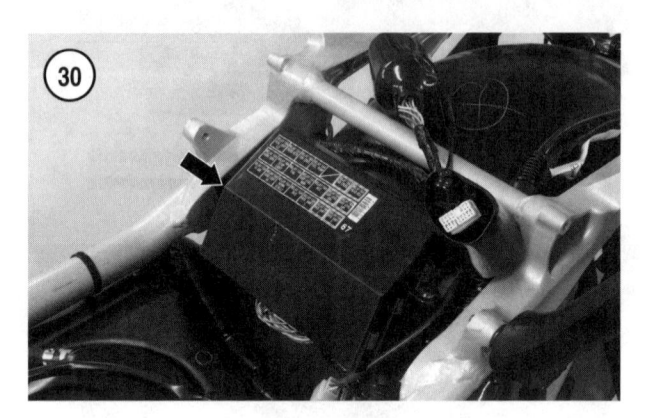

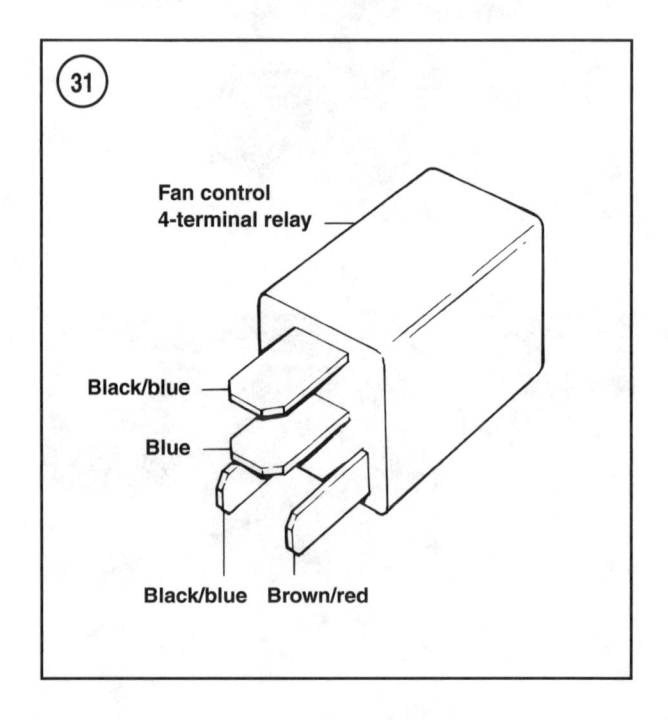

30

31

Fan control 4-terminal relay

Black/blue
Blue
Black/blue   Brown/red

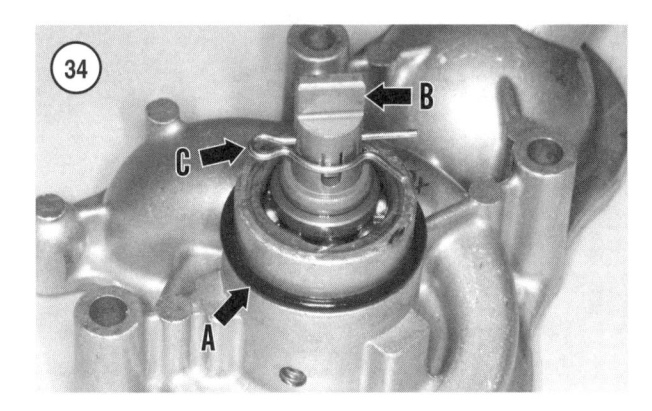

COOLING SYSTEM

10. Install the relay box cover (**Figure 30**) and secure with the mounting screws.
11. Install the seat. Refer to *Seat* in Chapter Seventeen.

WATER PUMP/THERMOSTAT

The water pump/thermostat housing (**Figure 32**) is mounted on the backside of the engine. The housing can be serviced with the engine mounted in the frame.

The water pump requires no routing maintenance and replacement parts (impeller and shaft assembly) are unavailable. The thermostat and housing O-rings can be replaced separately.

Removal/Installation

1. Drain the cooling system as described in Chapter Three under *Coolant Change*.

NOTE
If the engine is removed from the frame, the water pump/thermostat housing can be removed with the alternator and starter/reverse motor installed on the engine.

2. Remove the alternator. Refer to *Alternator* in Chapter Nine.
3. Remove the starter/reverse motor. Refer to *Starter/Reverse Motor* in Chapter Nine.
4. Label and disconnect the hoses at the water pump/thermostat housing (**Figure 32**).

NOTE
The following photographs are shown with the engine removed to better illustrate the steps.

5. Remove the bolts identified in A, **Figure 33** to remove the water pump/thermostat housing.
6. Turn the pump shaft by hand. If there is any roughness or binding, replace the water pump/thermostat as a unit.
7. If necessary, service the thermostat as described in this section.
8. Installation is the reverse of removal. Note the following:
   a. Lubricate a new O-ring (A, **Figure 34**) with engine oil and install it into the housing groove.
   b. Inspect the joint (B, **Figure 34**) for excessive wear or damage. If the edges are not sharp, replace the joint.
   c. Install a new lock pin (C, **Figure 34**), making sure it secures the joint (B) tightly to the pump shaft.
   d. Install the water pump by aligning the flat on the shaft end (B, **Figure 34**) with the slot in the starter clutch bolt.

8. Replace the fan control relay (3, **Figure 29**) with a known good 4-terminal relay. Start the engine and allow it to warm up. Note the following:

   a. If both cooling fans operate, replace the fan control relay.

   b. If the cooling fans do not operate, check for an open circuit in the brown/red wire between the fan control relay and the black 22-pin ECM connector.

9. Turn the ignition switch off and reinstall the fan control relay (3, **Figure 29**).

35 WATER PUMP/THERMOSTAT

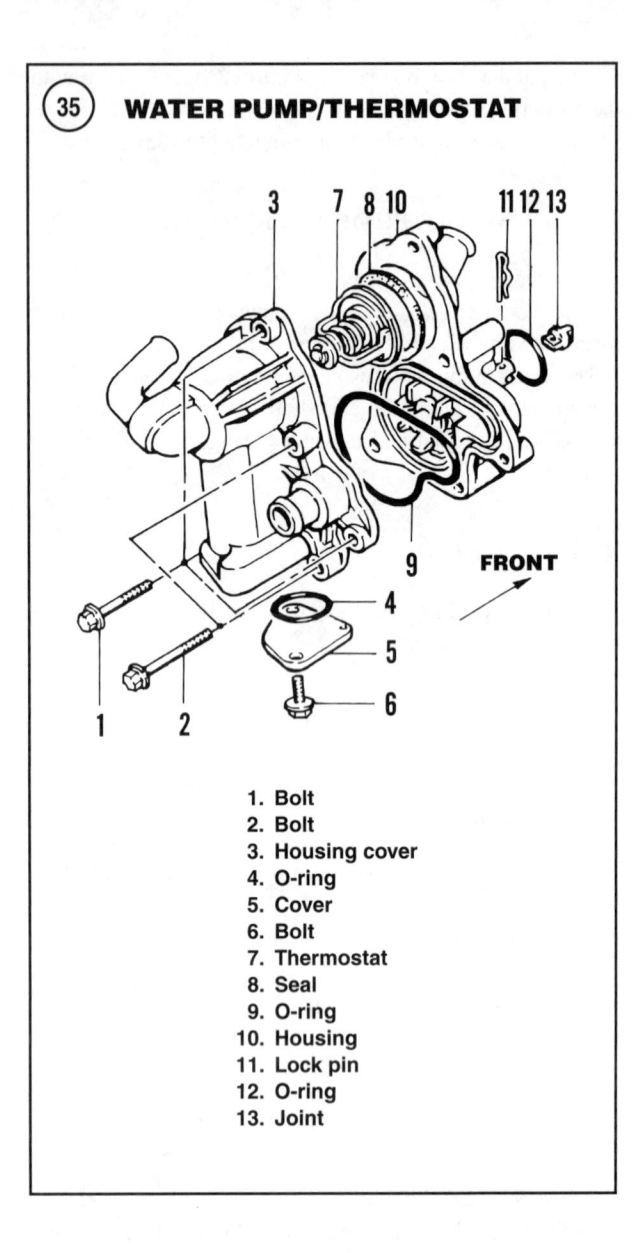

1. Bolt
2. Bolt
3. Housing cover
4. O-ring
5. Cover
6. Bolt
7. Thermostat
8. Seal
9. O-ring
10. Housing
11. Lock pin
12. O-ring
13. Joint

e. Tighten the three mounting bolts (A, **Figure 33**) to 13 N•m (115 in.-lb.).
f. Refill the cooling system with the recommended type and quantity of coolant as described under *Coolant Change* in Chapter Three.

Disassembly/Inspection/Reassembly

Refer to **Figure 35**.

The housing cover (3, **Figure 35**), thermostat (7), seal (8) and O-rings are available separately. The housing (10, **Figure 35**), containing the water pump impeller and shaft, is not available separately. If any part of the water pump is defective, replace the water pump/thermostat as a unit.

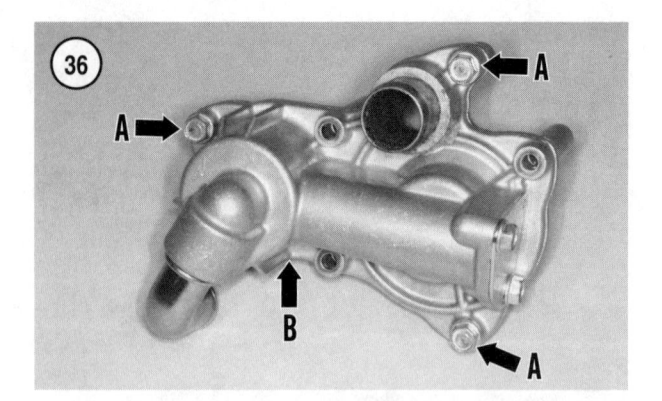

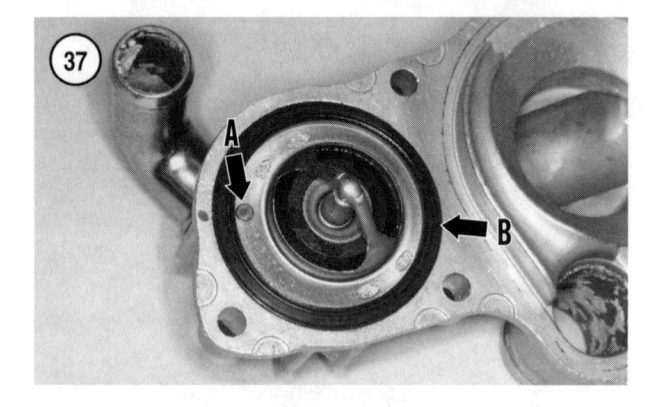

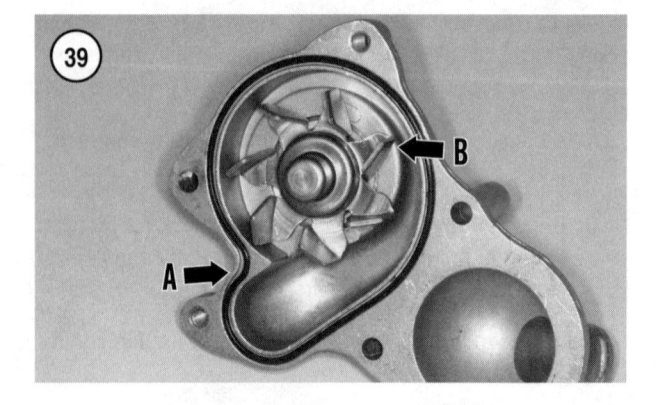

COOLING SYSTEM

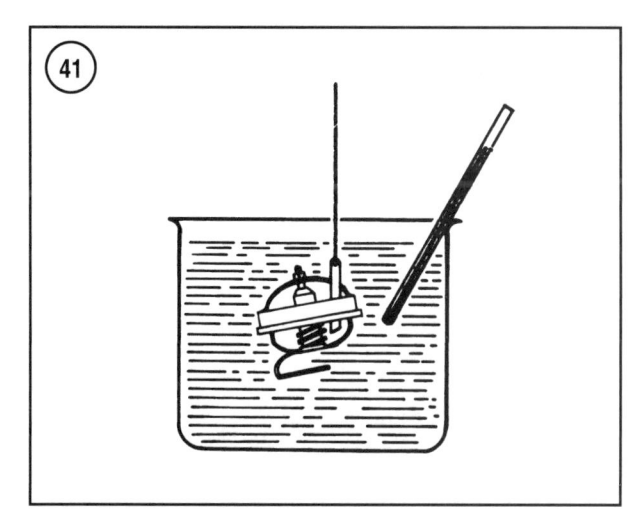

6. Check the impeller blades (B, **Figure 39**) for corrosion or damage. If corrosion is minor, clean the blades. If corrosion is excessive, or if the blades are cracked or broken, replace the water pump/thermostat housing.
7. Turn the impeller shaft by hand. If the bearing turns roughly or is damaged, replace the water pump/thermostat housing.
8. Install a new O-ring into the pump groove (A, **Figure 39**).
9. Install a new seal onto the thermostat flange (**Figure 38**).
10. Install the thermostat with its vent hole (A, **Figure 37**) facing the top of the housing and its support arm (A, **Figure 40**) aligned with the housing groove (B).
11. Install the housing cover (B, **Figure 36**) and tighten the mounting bolts (A) to 13 N•m (115 in.-lb.).
12. Install the water pump/thermostat housing as described in this section.

Thermostat
Inspection

Test the thermostat in boiling water to ensure proper operation.

NOTE
Do not allow the thermometer or thermostat to touch the sides or bottom of the pan, or a false reading will result.

1. Suspend the thermostat in a pan of water (**Figure 41**).
2. Place a thermometer in the pan of water. Use a thermometer rated higher than the test temperature.
3. Gradually heat and stir the water until it reaches 76-80° C (169-176° F). The thermostat should start to open at this temperature.
4. Continue to heat the water. At 90° C (194° F), the valve lift should measure a minimum of 8 mm (0.3 in.). If the valve lift is less than the minimum, the thermostat is defective.

NOTE
After the specified temperature is reached, it may take 3-5 minutes for the valve to open completely.

5. Replace the thermostat if it remains open at room temperature or stays closed after the specified temperature has been reached during the test.

1. Remove the water pump/thermostat housing as described in this section.
2. Remove the bolts (A, **Figure 36**) and the housing cover (B).
3. Note the position of the vent hole in the thermostat flange (A, **Figure 37**), then remove the thermostat and seal (B).
4. Inspect the thermostat:
   a. Remove the seal (**Figure 38**) and discard it.
   b. Inspect the thermostat for rust, scale contamination and visible damage.
   c. With the thermostat at room temperature, check to see if light passes between the valve and seat. If so, replace the thermostat.
   d. If the thermostat appears good, test it as described in this section.
5. Remove and discard the O-ring (A, **Figure 39**).

Table 1 and Table 2 are on the following page.

Table 1 COOLING SYSTEM SPECIFICATIONS

Coolant	
Standard concentration	50:50 mixture of coolant and purified water
Type	Honda HP coolant or an equivalent*
Coolant capacity	
Radiator and engine	3.53L (3.73 U.S. qt.)
Reserve tank	0.65L (0.69 U.S. qt.)
Thermostat	
Begins to open	76-80° C (169-176° F)
Fully open	90° C (194° F)
Valve lift (minimum)	8 mm (0.3 in.)

*Use a high quality ethylene coolant that does not contain silicate inhibitors.

Table 2 COOLING SYSTEM TORQUE SPECIFICATIONS

	N•m	in.-lb.	ft.-lb.
Radiator mounting bolts	14	–	10
Water pump/thermostat housing bolts	13	115	–
Water pump cover plate bolt	13	115	–

CHAPTER ELEVEN

WHEELS AND TIRES

1

This chapter describes service procedures for the wheels, front wheel bearings and tires.

Tire and wheel specifications are listed in **Table 1**. **Tables 1-4** are at the end of the chapter.

BIKE LIFT

Many procedures in this chapter require lifting either the front or rear wheel off the ground. While the centerstand can be used to lift the rear wheel, a jack is required to lift the front wheel. The K&L MC450 Center Jack is a jack (**Figure 1**) that works well. When using the MC450 center jack, place the motorcycle on its centerstand. Place a block of wood across the jack and position the jack underneath the front part of the engine. Operate the jack and lift the front part of the motorcycle until the front wheel just clears the ground. The K&L Center Jack can be ordered through most motorcycle dealerships.

FRONT WHEEL

Removal

CAUTION
Use care when removing, handling and installing the wheel. The brake discs can easily be damaged by side impacts due to their thin design. A disc that is not true will cause brake pulsation. Protect the discs if transporting the wheel for tire service.

1. Support the motorcycle with the front wheel off the ground. Refer to *Bike Lift* in this chapter.
2. Remove front fender A and the fender covers. Refer to *Front Fender Assembly* in Chapter Seventeen.
3. Remove the left brake caliper by doing the following:
   a. Cover the area between the brake caliper and rim with a rag.

432

②

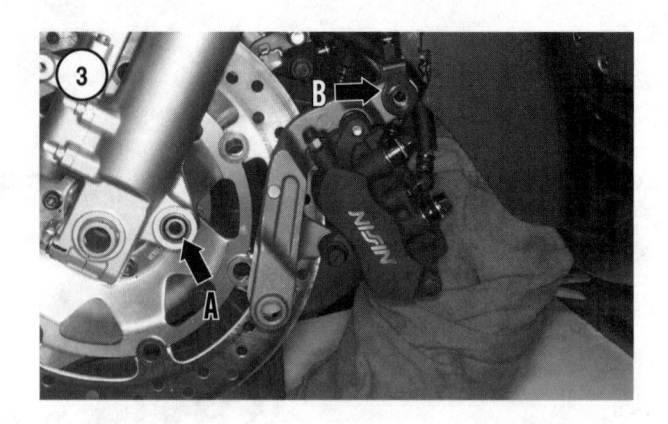

③

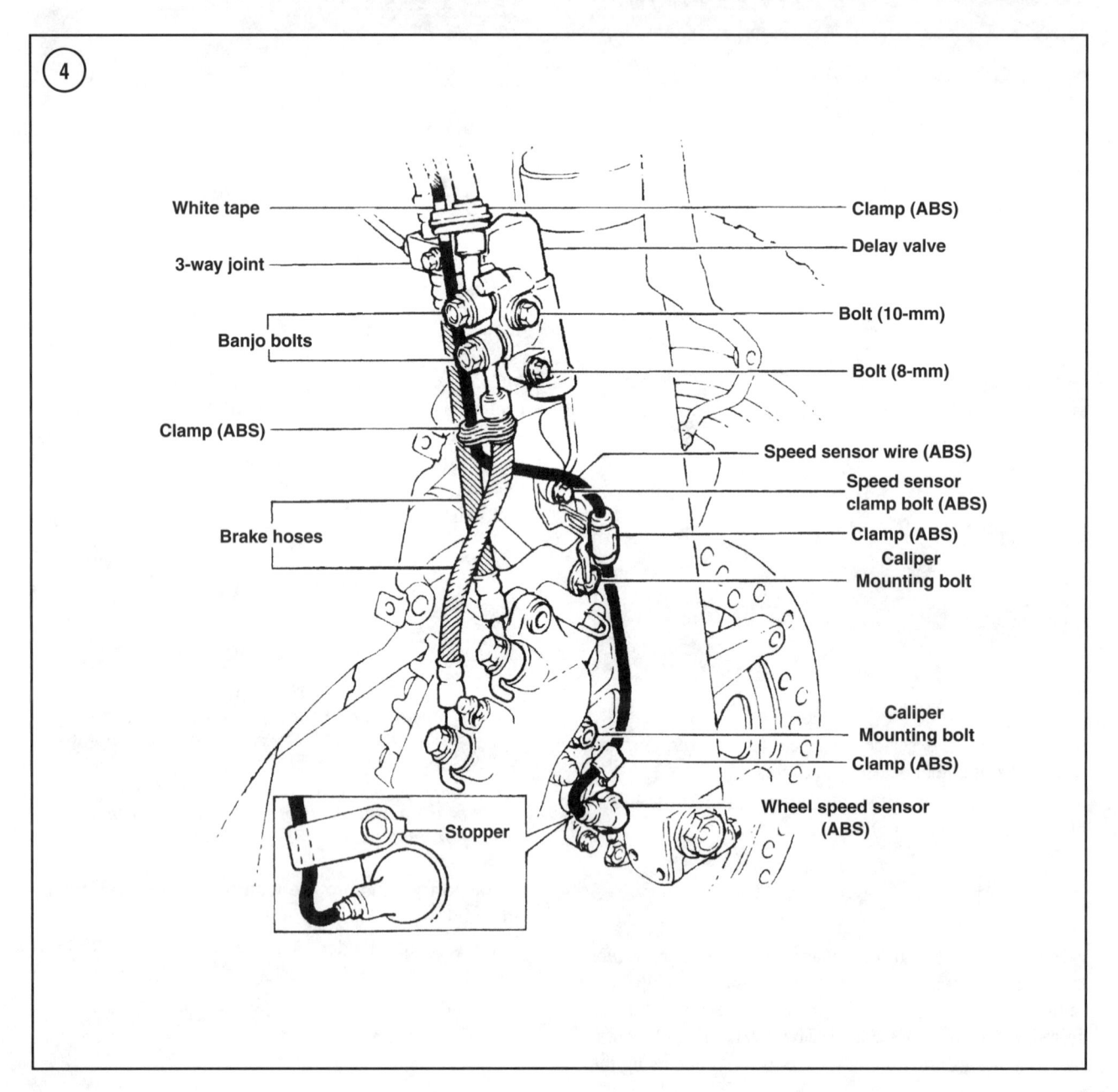

④

- White tape
- 3-way joint
- Banjo bolts
- Clamp (ABS)
- Brake hoses
- Stopper
- Clamp (ABS)
- Delay valve
- Bolt (10-mm)
- Bolt (8-mm)
- Speed sensor wire (ABS)
- Speed sensor clamp bolt (ABS)
- Clamp (ABS)
- Caliper Mounting bolt
- Caliper Mounting bolt
- Clamp (ABS)
- Wheel speed sensor (ABS)

WHEELS AND TIRES

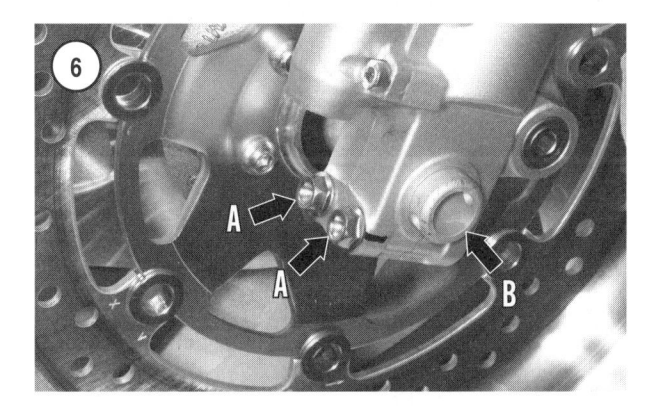

b. Remove the secondary master cylinder joint bolt (A, **Figure 2**) and the caliper pivot bolt (B), then remove the brake caliper.
c. Pull the brake caliper back until it contacts the rim, then twist it outward and remove it from the brake disc (**Figure 3**).
d. Secure a rag to the brake caliper with a rubber band and support the caliper with a length of stiff wire.

4. On GL1800A models, remove the speed sensor bolt and clamp (**Figure 4**) mounted above the right brake caliper.
5. Remove the right brake caliper by doing the following:
   a. Remove the mounting bolts and brake caliper (**Figure 4**).
   b. Secure a rag to the brake caliper with a rubber band and support the caliper with a length of stiff wire.
6. Loosen the right side axle pinch bolts (A, **Figure 5**).
7. Remove the front axle bolt (B, **Figure 5**).
8. Loosen the left side axle pinch bolts (A, **Figure 6**).
9. Remove the front axle (B, **Figure 6**) and pull the wheel forward.

NOTE
Before removing the front wheel, note the direction of the rim and tire rotation arrows. The wheel must be reinstalled so the arrows point in the direction of forward rotation.

NOTE
The left and right side wheel collars are different. Identify the collars before removing them.

10. Remove the left side collar (**Figure 7**).
11. Remove the right side collar (**Figure 8**).

CAUTION
Do not set the wheel down on the disc surface, as it may become damaged.

NOTE
Insert a spacer in the calipers to hold the brake pads in place. Then, if the brake lever or pedal is inadvertently squeezed, the pistons will not be forced out of the calipers.

Inspection

Replace worn or damaged parts as described in this section.

1. Clean the axle and collars in solvent to remove all grease and dirt.
2. Remove any corrosion on the front axle and collars with a piece of fine emery cloth.

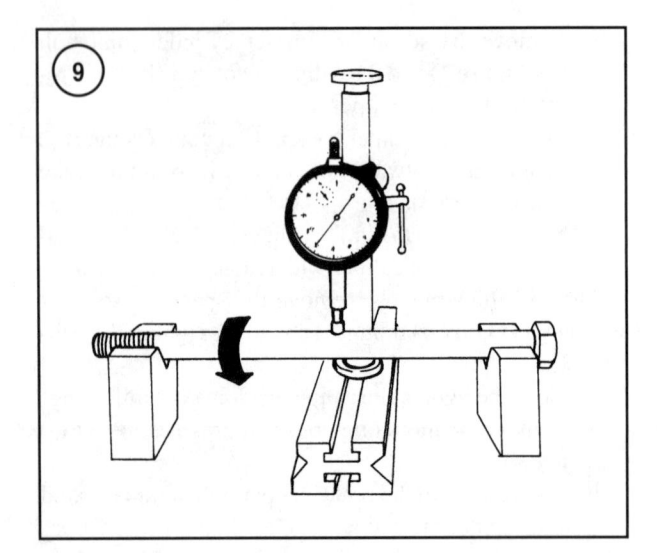

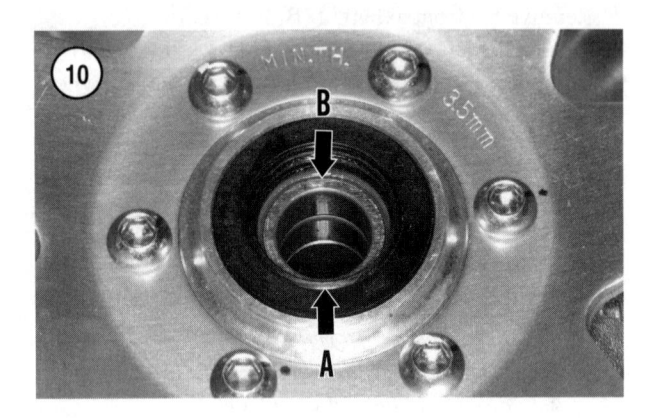

3. Check the axle surface for any cracks or other damage. Check the axle operating areas for any nicks or grooves that can cut and damage the seals.

4. Check the axle bolt and axle threads for damage. Replace the axle and axle bolt if their corners are damaged.

5. Check the axle runout with a set of V-blocks and dial indicator (**Figure 9**). Runout is 1/2 the total indicator reading. Replace the axle if its runout exceeds the service limit in **Table 2**. Do not attempt to straighten it.

6. Clean the seals with a rag. Then inspect the seals (A, **Figure 10**) for excessive wear, hardness, cracks or other damage. If necessary, replace the seals as described under *Front Hub* in this chapter.

7. Turn each bearing inner race (**Figure 11**) by hand. The bearing must turn smoothly. Some axial play is normal, but radial play must be negligible. If one bearing is damaged, replace both bearings as a set. Refer to *Front Hub* in this chapter.

8. Check the brake disc bolts for tightness. Correct tightening torque is 20 N•m (15 ft.-lb.).

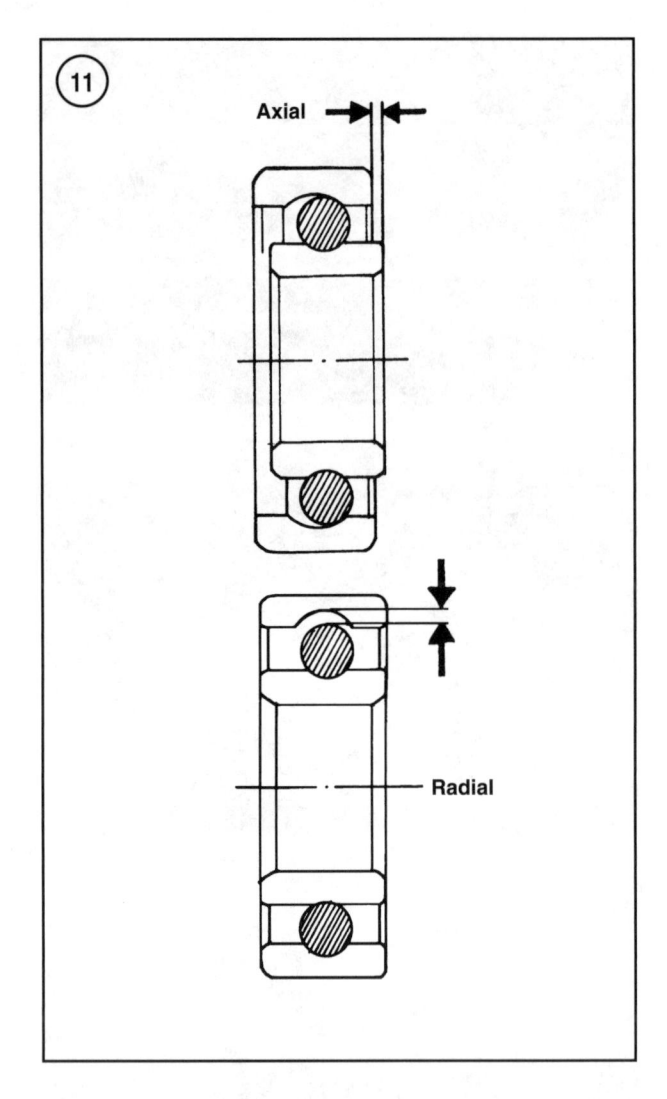

9. On GL1800A models, check the front pulser ring mounting bolts for tightness. The correct torque specification is 8 N•m (71 in.-lb.).

10. Check wheel runout as described in this chapter.

Installation

1. Make sure the axle bearing surfaces on the fork sliders and axle are free from burrs and nicks.

2. Lightly coat the axle with bearing grease and set aside until installation.

3. Lubricate the seal lips (A, **Figure 10**) with waterproof grease.

4. Install the left collar (**Figure 7**).

5. Install the right collar (**Figure 8**).

6. Install the wheel between the sliders with the wheel rim arrow (**Figure 12**) pointing forward (normal rotation).

WHEELS AND TIRES

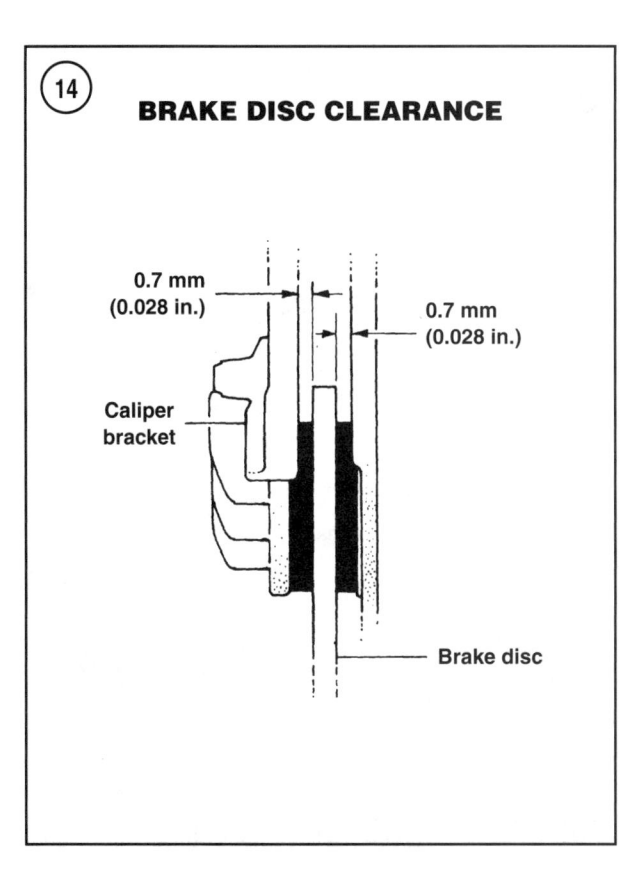

7. Install the axle (B, **Figure 6**) from the left side. Check that the index line on the axle aligns with the slider (**Figure 13**).
8. Install the axle bolt (B, **Figure 5**). Then hold the left side of the axle and tighten the axle bolt to 59 N•m (44 ft.-lb.).
9. Tighten the right side axle pinch bolts (A, **Figure 5**) to 22 N•m (16 ft.-lb.).
10. Install the right side brake caliper as follows:
    a. Remove the spacer block from between the brake pads in the caliper.
    b. Install the right side brake caliper and secure with *new* mounting bolts (**Figure 4**).
    c. Tighten the right side brake caliper mounting bolts to 31 N•m (23 ft.-lb.).
11. On GL1800A models, reinstall the speed sensor wire clamp and tighten the bolt securely (**Figure 4**).
12. Install the left side brake caliper as follows:
    a. Lubricate the left caliper pivot bearings and collar with grease. Install the collar (A, **Figure 3**) into the bearings.
    b. Lubricate the upper pivot collar with grease and install it in the caliper bracket (B, **Figure 3**).
    c. Remove the spacer block from between the brake pads in the caliper.
    d. Install the caliper assembly over the brake disc.
    e. Install a *new* secondary master cylinder joint bolt (A, **Figure 2**) and a *new* left front brake caliper pivot bolt (B).
    f. Tighten the left brake caliper secondary master cylinder joint bolt (A, **Figure 2**) to 25 N•m (18 ft.-lb.).
    g. Tighten the left front brake caliper pivot bolt (B, **Figure 2**) to 31 N•m (23 ft.-lb.).
13. Lower the front wheel and remove the jack. Mount the motorcycle and lower the centerstand. Apply the front brake and pump the fork several times to seat the axle.
14. Tighten the left side axle pinch bolts (A, **Figure 6**) to 22 N•m (16 ft.-lb.).

WARNING
Step 15 determines if there is adequate brake disc to caliper clearance. Failure to provide adequate clearance may cause brake disc damage and reduced braking efficiency. Both conditions may cause brake failure.

15. Check the brake disc clearance as follows:
    a. Insert a 0.7 mm (0.03 in.) feeler gauge between the side of each brake disc and caliper bracket as shown in **Figure 14**.
    b. The feeler gauge should pass through the gap easily. If the clearance is too small, loosen the axle pinch bolts for the side being checked and pull the slider

along the axle by hand to obtain the correct clearance. When the clearance is correct, tighten the axle pinch bolts to 22 N•m (16 ft.-lb.).
c. Turn the front wheel and apply the front and rear brake several times, then recheck the clearance.

WARNING
If the correct clearance cannot be obtained, check the brake disc for loose mounting bolts or excessive runout (Chapter Fourteen). Do not ride the motorcycle until the brake disc clearance is correct and both front brake calipers work correctly.

16. On GL1800A models, check the front wheel speed sensor air gap. Refer to *Wheel Speed Sensor* in Chapter Fifteen.
17. Install front fender A and the front fender side covers. Refer to *Front Fender Assembly* in Chapter Seventeen.
18. Apply the front and rear brake several times to make sure the proper brake pressure is felt at the brake level and pedal.

REAR WHEEL

Removal/Installation

1. Support the bike on its centerstand.
2. Remove rear fender A. Refer to *Rear Fender A* in Chapter Seventeen.
3. Remove the two bolts and the fender brace (**Figure 15**).
4. Park the motorcycle on its sidestand so both wheels are on the ground.
5. Apply the rear brake and loosen the rear wheel nuts (**Figure 16**).
6. Support the motorcycle on its centerstand.
7. Remove the rear wheel nuts and rear wheel.
8. Reverse these steps to install the rear wheel, plus the following.
    a. Perform the *Inspection* procedure to clean and inspect the wheel.
    b. Install the rear wheel nuts (**Figure 17**) and tighten hand-tight.
    c. Tighten the rear wheel nuts (**Figure 16**) in a crisscross pattern to 108 N•m (80 ft.-lb.).
    d. On GL1800A models, check the rear wheel speed sensor air gap if the rear brake disc was removed. Refer to *Wheel Speed Sensor* in Chapter Fifteen.

Inspection

Replace worn or damaged parts as described in this section.

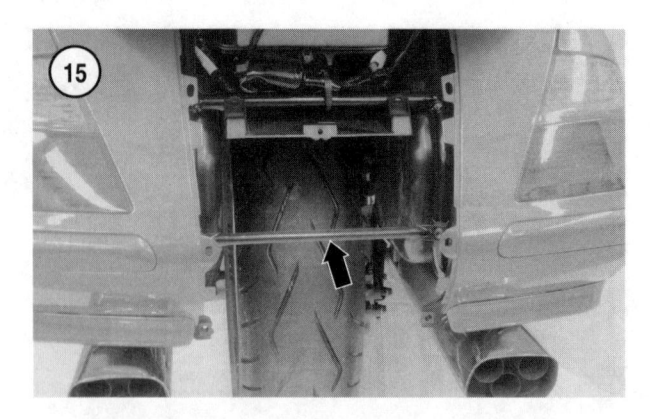

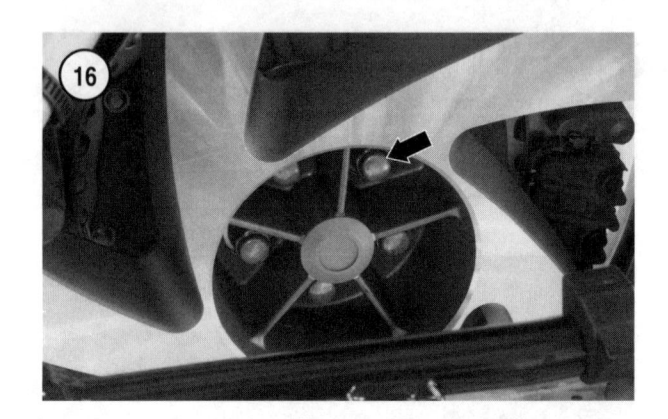

WHEELS AND TIRES

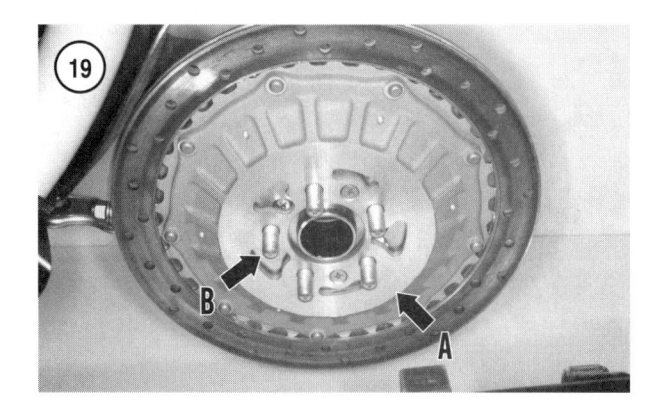

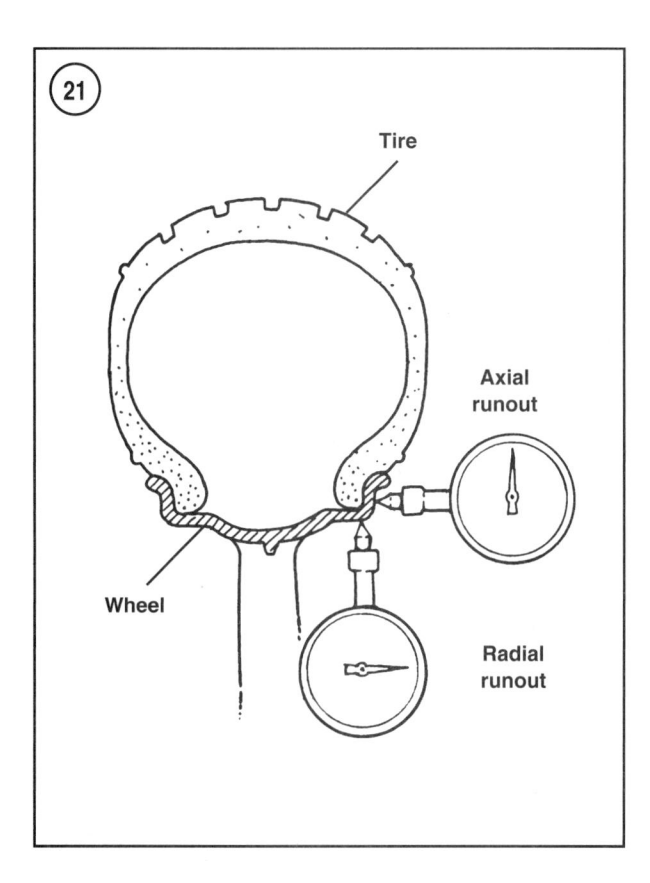

1. Clean the hub (**Figure 18**) and brake disc (A, **Figure 19**) mating surfaces.
2. Clean the studs (B, **Figure 19**) and rear wheel nuts. Inspect the threads for corrosion and damage.
3. Check the wheel bolt holes for cracks, hole elongation or other damage. Check both sides of the wheel.
4. Check wheel runout as described in this chapter.

WHEEL RUNOUT AND BALANCE

Proper wheel inspection includes visual inspection, checking runout and wheel balance. Checking runout and wheel balance requires a truing or wheel balancing stand. If these tools are not available, refer the service to a Honda dealership.

Replace the wheel if it is dented or damaged in any way. While a new wheel is not cheap, its cost does not compare to the injury that could occur if a damaged wheel becomes unstable or fails while riding. If there is any doubt as to wheel condition, take it to a Honda dealership and have them inspect the wheel. It is also a good idea to have a specialist inspect any used wheels that may be considered for purchase.

Rim Runout Check

1. Clean the wheel rim to remove all road grit and other debris. Any material left on the rim affects its runout. This includes any surface roughness caused by peeled or uneven paint and corrosion.
2. Inspect the entire wheel for dents, bending or cracks. Check the rim and rim sealing surface for scratches that could cause the tire to leak air.
3. Remove the axle cap from the rear wheel (**Figure 20**).

NOTE
The runout check can be performed with the tire mounted on the rim.

4. Mount the wheel on a truing stand. See **Figure 21** for the dial indicator inspection points.
5. Spin the wheel slowly by hand and measure the radial (up and down) runout with a dial indicator as shown in **Figure 21**. If the runout exceeds 2.0 mm (0.08 in.), go to Step 7.
6. Spin the wheel slowly by hand and measure the axial (side to side) runout with a dial indicator as shown in **Figure 21**. If the runout exceeds 2.0 mm (0.08 in.), go to Step 7.
7A. Front Wheel: If the runout is excessive, remove the wheel from the truing stand and turn each bearing inner race (B, **Figure 10**) by hand. If necessary, remove the seal to check the bearings closely. Each bearing must turn

smoothly and be a tight fit in its mounting bore. Some axial play is normal, but radial play must be negligible (**Figure 11**). Then check the bearing for visual damage. If a bearing turns roughly, replace both bearings as a set. If a bearing is loose in its mounting bore, the hub is probably damaged. Remove the bearings and check the mounting bore for any cracks, gouges or other damage. Refer to *Front Hub* in this chapter. If the front wheel bearings and hub are in good condition but the runout is out of specification, replace the damaged wheel.

7B. Rear Wheel: If the runout is excessive, the wheel (**Figure 22**) is probably damaged. Refer further inspection to a Honda dealership.

Wheel Balance

A wheel that is not balanced is unsafe because it seriously affects the steering and handling of the motorcycle. Depending on the degree of unbalance and the speed of the motorcycle, anything from a mild vibration to a violent shimmy may occur, which may cause loss of control. An imbalanced wheel also causes abnormal tire wear.

Motorcycle wheels can be checked for balance either statically (single plane balance) or dynamically (dual plane balance). This section describes how to static balance the wheels using a wheel balancing stand. To obtain a higher degree of accuracy, take both wheels to a dealership and have them balance the wheels with a two plane computer dynamic wheel balancer. This machine spins the wheel to accurately detect any imbalance.

Balance weights are used to balance the wheel and are attached to the rim. Weight kits are available from motorcycle dealerships. When static balancing, purchase the crimp type that can be attached to the raised ridge along the center of the rim (**Figure 23**). When a wheel is dynamically balanced, different size weights may be placed on both sides of the wheel.

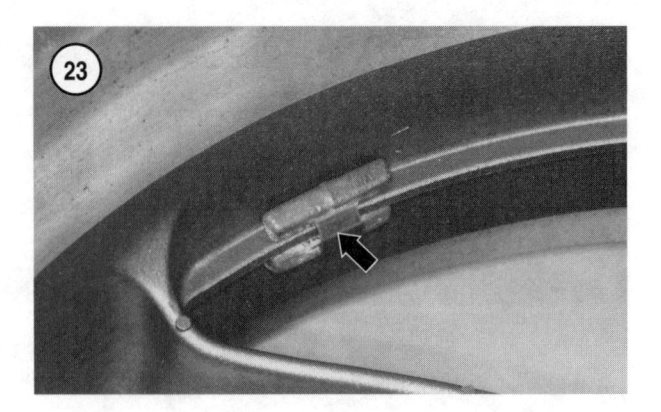

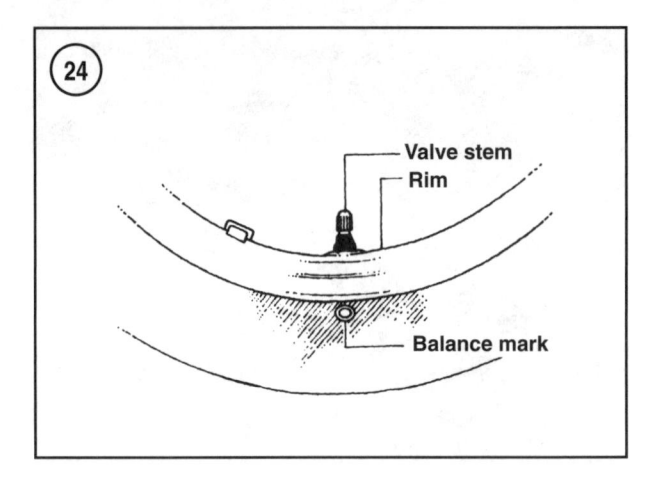

The wheel must be able to rotate freely when checking wheel balance. Because excessively worn or damaged wheel bearings affect the accuracy of this procedure, check the front wheel bearings as described under *Front Hub* in this chapter. Check the rear wheel hub (**Figure 18**) for cracks and other damage. Also, confirm that the tire balance mark, a paint mark on the tire, is aligned with the valve stem (**Figure 24**).

NOTE
Leave the brake discs and pulser ring (GL1800A) mounted on the front wheel when checking and adjusting wheel balance.

1. Remove the wheel as described in this chapter.

2. Front wheel: Clean the seals and inspect the wheel bearings as described under *Front Wheel Inspection* in this chapter.

3. Clean the tire and rim. Remove any stones or pebbles stuck in the tire tread.

4. Mount the wheel on an inspection stand (**Figure 25**).

WHEELS AND TIRES

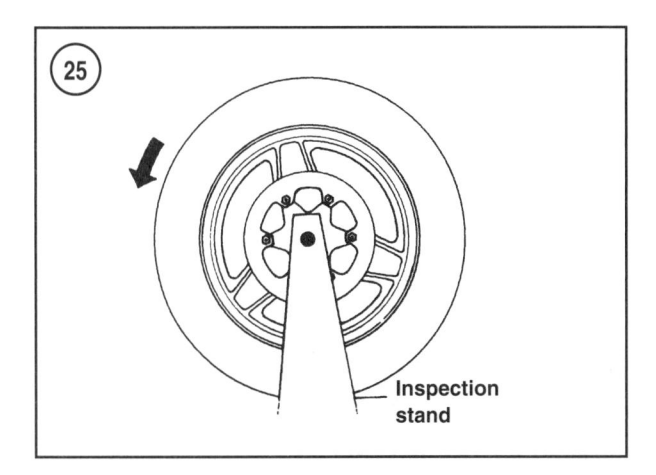

Figure 25 — Inspection stand

Figure 26

NOTE
*Due to the design of the rear wheel, the wheel must be balanced, either statically or dynamically, on a wheel balance machine. These machines require a special adapter (**Figure 26**, typical).*

NOTE
To check the original balance of the wheel, leave the weights in place on the rim.

5. Spin the wheel by hand and let it coast to a stop. Mark the tire at its bottom point with chalk.

6. Spin the wheel several more times. If the same spot on the tire stops at the bottom each time, the wheel is out of balance. This is the heaviest part of the tire. When an unbalanced wheel is spun, it always comes to rest with the heaviest part at the bottom.

7. Attach a test weight to the wheel at the point opposite the heaviest spot and spin the wheel again.

8. Experiment with different weights until the wheel, after spinning, comes to rest at a different position each time. When a wheel is correctly balanced, the weight of the tire and wheel assembly is distributed equally around the wheel.

9. Remove the test weight and install the correct size weight to the rim. Make sure it is crimped tightly to the rim's surface (**Figure 23**).

10. Record the weight, number and position of the weights on the wheel. Then, if the bike experiences a handling or vibration problem later, first check for any missing weights.

11. Install the wheel as described in this chapter.

FRONT HUB

The front hub contains the seals, wheel bearings and distance collar (**Figure 27**).

Pre-Inspection

Inspect each wheel bearing as follows:

1. Support the bike with the front wheel off the ground. Make sure the axle is tightened securely.
    a. Hold the wheel along its sides (180° apart) and try to rock it back and forth. If there is any noticeable play at the axle, the wheel bearings are worn or damaged and require replacement. Then have an assistant apply the front brake. While the brake is applied, rock the wheel again. On excessively worn bearings, play is detected at the bearings even though the wheel is locked in position.
    b. Push both front calipers in by hand to move the brake pads away from the brake disc. This makes it easier to spin the wheel when performing substep c.
    c. Spin the wheel and listen for excessive wheel bearing noise. A grinding or catching noise indicates worn bearings.
    d. Apply the front brake several times to reposition the front brake pads in the calipers.
    e. To check any questionable bearing, continue with Step 2.

CAUTION
Do not remove the wheel bearings for inspection purposes, as they may become damaged during removal. Remove the wheel bearings only if they require replacement.

2. Remove the front wheel as described in this chapter.

CAUTION
When handling the wheel assembly in the following steps, do not lay the wheel down where it is supported by the brake disc as

FRONT WHEEL

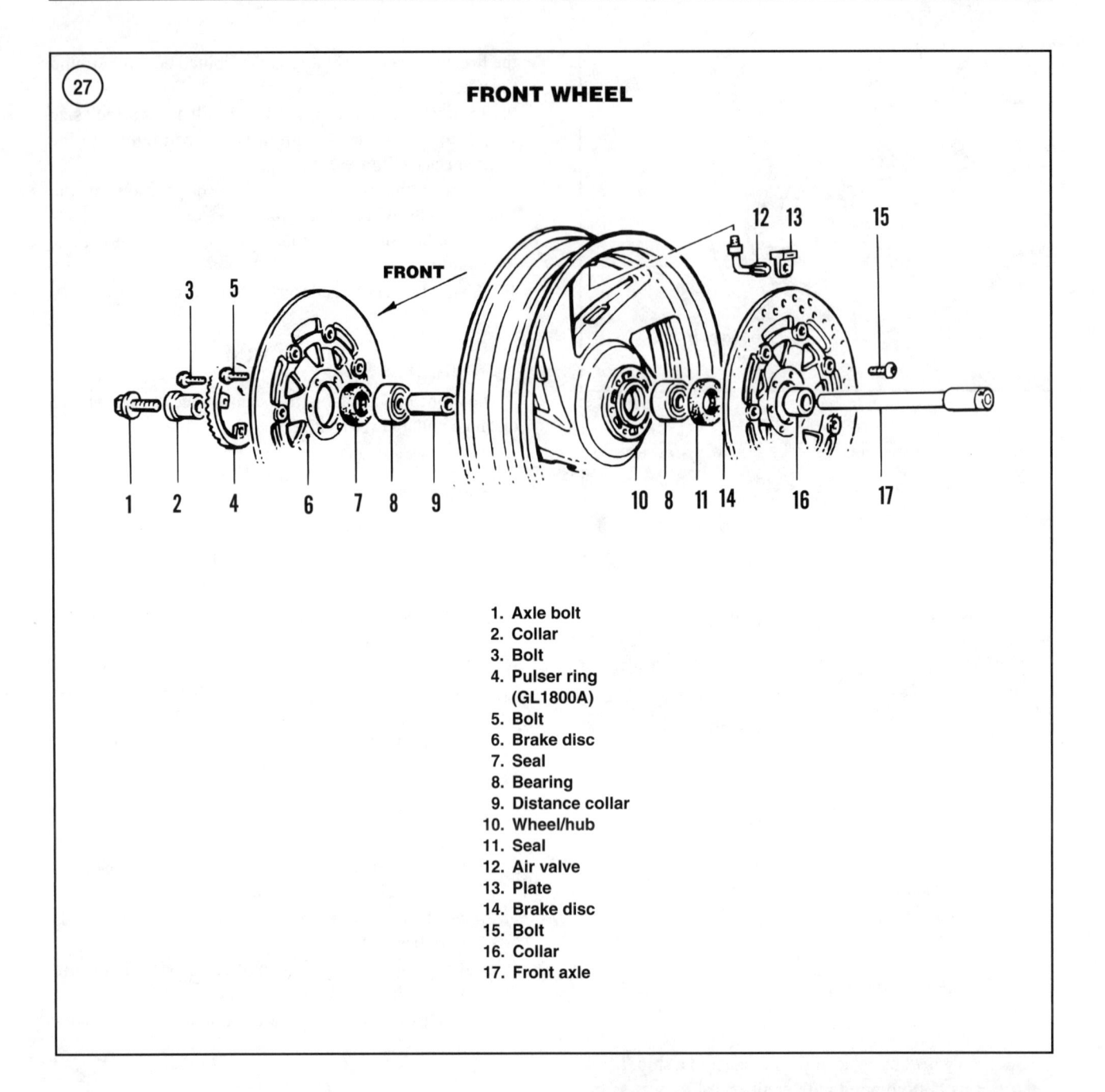

1. Axle bolt
2. Collar
3. Bolt
4. Pulser ring (GL1800A)
5. Bolt
6. Brake disc
7. Seal
8. Bearing
9. Distance collar
10. Wheel/hub
11. Seal
12. Air valve
13. Plate
14. Brake disc
15. Bolt
16. Collar
17. Front axle

this could damage the disc. Support the wheel on two wooden blocks.

3. Pry the seals out of the hub (**Figure 28**). Support the tool with a rag to avoid damaging the hub or brake disc.

4. Remove any burrs created during seal removal with emery cloth. Do not enlarge the mounting bore.

NOTE
Before removing the wheel bearings, check the tightness of the bearings in the hub by

WHEELS AND TIRES

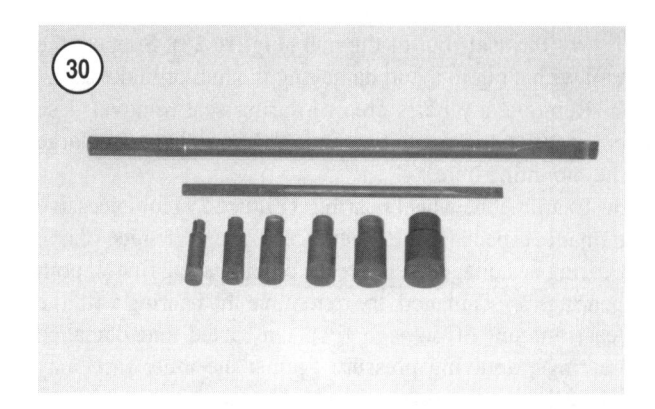

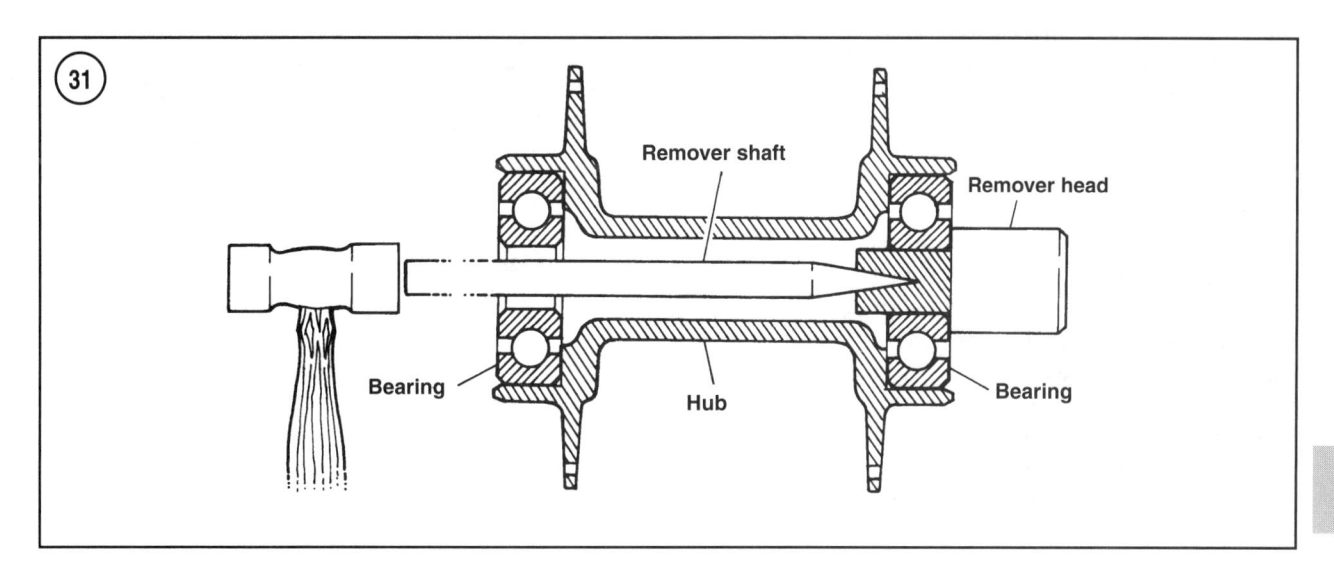

pulling the bearing up and then from side to side. The outer bearing race should be a tight fit in the hub with no movement. If the outer bearing race is loose and wobbles, the bearing bore in the hub may be cracked or damaged. Remove the bearings as described in this procedure and check the hub bore carefully. If any cracks or damage are found, replace the hub. It cannot be repaired.

5. Turn each bearing inner race (A, **Figure 29**) by hand. The bearing must turn smoothly with no roughness, catching, binding or excessive noise. Some axial play is normal, but radial play must be negligible (**Figure 11**).

6. Check the bearing's outer seal (B, **Figure 29**) for buckling or other damage that would allow dirt to enter the bearing.

7. If one bearing is damaged, replace both bearings as a set.

Disassembly

This section describes front wheel bearing removal using an expanding collet type bearing removal tool. To remove a bearing where the inner race assembly has fallen out, refer to *Removing Damaged Bearings* in this section.

The bearing removal tool shown in this section is manufactured by Motion Pro (**Figure 30**) and can be ordered through most motorcycle dealerships. The Motion Pro tool can be purchased as a set, or the individual pieces can be purchased separately. To remove the front wheel bearings, use the 20 mm remover head and the larger driver rod. Similar tools are also available from Honda and Kowa Seiki. **Figure 31** shows how the tools are assembled to remove a typical wheel bearing.

CAUTION
When handling the wheel assembly in the following steps, do not lay the wheel down where it is supported by the brake disc. Support the wheel on two wooden blocks.

1. Pry the seals out of the hub (**Figure 28**). Support the tool with a rag to avoid damaging the hub or brake disc.
2. Remove any burrs created during seal removal. Use emery cloth to smooth the mounting bore. Do not enlarge the mounting bore.
3. Examine the wheel bearings (**Figure 29**) for excessive damage, especially the inner race. If the inner race of one bearing is damaged, remove the other bearing first. If both bearings are damaged, try to remove the bearing with the least amount of damage first. On rusted and damaged bearings, applying pressure against the inner race may cause the race to pop out, leaving the outer race in the hub.

WARNING
Wear safety glasses when removing the bearings in the following steps.

4. Select the 20 mm remover head and insert it into one of the bearings (**Figure 32**).
5. From the opposite side of the hub, insert the remover shaft into the slot in the backside of the remover head (**Figure 33**). Then position the hub with the remover head tool resting against a solid surface and strike the remover shaft to force it into the slit in the remover head. This expands the remover head tool against the inner bearing race to lock it in place. See **Figure 31**.
6. Position the wheel and strike the end of the remover shaft with a hammer to drive the bearing from the hub (**Figure 34**). Remove the bearing and tool. Release the remover head from the bearing.
7. Remove the distance collar from the hub.
8. Remove the opposite bearing the same way.
9. Clean and dry the hub and distance collar.
10. Discard both bearings.

Inspection

1. Check the hub mounting bore for cracks or other damage. If one bearing is loose, the mounting bore is damaged. Replace the wheel.
2. Inspect the distance collar for cracks, corrosion or other damage. Then check the distance collar ends. If the ends appear compressed or damaged, replace the distance collar. Do not try to repair the distance collar by cutting or grinding its end surfaces, as this shortens the distance collar.

CAUTION
The distance collar operates against the wheel bearing inner races to prevent them from moving inward when the axle is tightened. If the ends of the distance collar are damaged, shortened, or if it is not installed

in the hub, the inner bearing races move inward and bind as the axle is tightened, causing bearing damage and seizure.

Assembly

1. Before installing the new bearings and seals, note the following:
    a. Install both bearings with their closed side facing out. If a bearing is sealed on both sides, install the bearing with its manufacturer's marks facing out.
    b. Install both seals with their closed side facing out.
    c. When grease is specified in the following steps, use a water-resistant bearing grease.
2. Remove any dirt or debris from the hub before installing the bearings.
3. Pack the open side of each bearing with grease.

NOTE
When installing the bearings, install the right side bearing first, then the left side bearing.

4. Place the right side bearing squarely against the bore opening with its closed side facing out. Select a driver (**Figure 35**) with an outside diameter slightly smaller than

WHEELS AND TIRES

Removing Damaged Bearings

When worn and rusted wheel bearings are used too long, the inner race can break apart and fall out of the bearing, leaving the outer race pressed in the hub. Because the outer race seats against a shoulder inside the hub, its removal is difficult because only a small part of the race is accessible above the hub's shoulder. This presents a small and difficult target to drive against. To remove a bearing's outer race under these conditions, first heat the hub evenly with a propane torch. Then drive out the outer race with a drift and hammer. Grind a clearance tip on the end of the drift, if necessary, to avoid damaging the hub's mounting bore. Check this before heating the hub. When removing the race, apply force at opposite points around the race to prevent it from rocking and binding in the mounting bore once it starts to move. After removing the race, inspect the hub mounting bore carefully for cracks or other damage.

TIRE SAFETY

Tire wear and performance is greatly affected by tire pressure. Have a good tire gauge on hand and make a habit of frequent pressure checks. Refer to **Table 1** for original equipment tire specifications. If using another tire brand, follow their recommendation.

Follow a sensible break-in period when running on new tires. New tires will exhibit significantly less adhesion ability until scrubbed in. Do not subject a new tire to hard corning, hard acceleration or hard braking for the first 100 miles (160 km).

TIRE CHANGING

Tire changing is an important part of the motorcycle's safety and operation. Due to the size of the tire and tight bead/rim seal, changing tires can be extremely difficult. Incorrect installation due to a lack of patience, skill or equipment can damage both the tire and/or the wheel. Many experienced owners who do most or all of their own maintenance and service choose to have their tires changed by a professional technician with specialized tire changing equipment designed for alloy wheels.

The following procedure is provided for those who choose to do the work.

Removal

WARNING
The wheels can be damaged easily during tire removal. Work carefully to avoid damaging the tire beads, inner liner of the tire or

the bearing's outside diameter. Then drive the bearing into the bore until it bottoms out. See **Figure 29**, typical.

5. Turn the wheel over and install the distance collar. Center it against the bearing inner race.

6. Place the left side bearing squarely against the bore opening with its closed side facing out. Using the same driver, drive the bearing partway into the bearing bore. Then stop and make sure the distance collar is centered in the hub. If not, install the axle through the hub to align the distance collar with the bearing. Then remove the axle and continue installing the bearing until it bottoms in the hub.

7. Insert the axle through the hub and turn it by hand. Check for any roughness or binding, which indicates bearing damage.

8. Pack the lip of each seal with grease.

9. Place a seal squarely against one of the bore openings with its closed side facing out. Then drive the seal in the bore until it is flush with the outside of the hub's mounting bore.

10. Repeat Step 9 for the other seal.

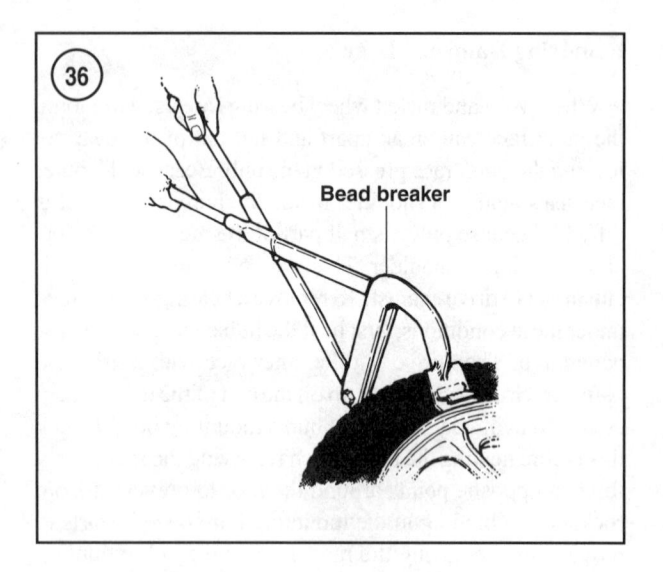

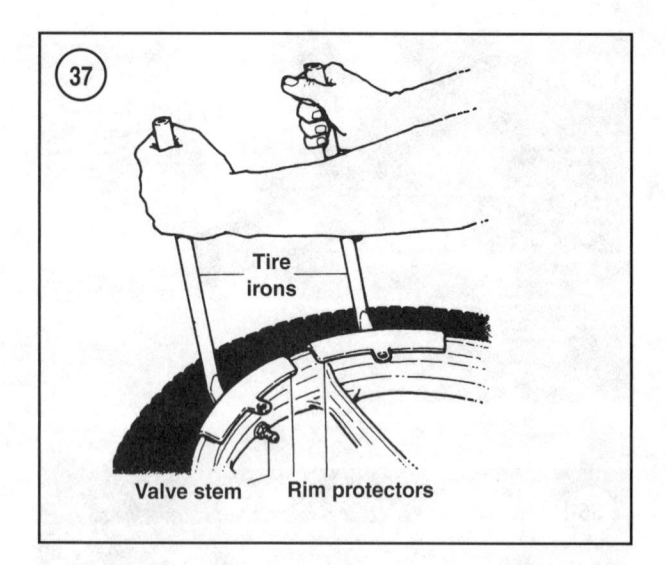

the wheel rim flange (sealing surfaces). As described in the text, insert rim protectors between the tire irons and rim to protect the rim from damage.

WARNING
The original equipment cast wheels are designed to use tubeless tires. Do not install a tube inside of a tubeless tire as excessive heat may build up in the tire and cause the tube to burst.

NOTE
Tires are harder to replace when the rubber is hard and cold. If the weather is hot, place the wheels and new tires in the sun or in a closed automobile. The heat helps soften the rubber, easing removal and installation. If the weather is cold, place the tires and wheels inside a warm building.

NOTE
It is easier to replace tires when the wheel is mounted on some type of raised platform. A popular item used by many home mechanics is a metal drum. Before placing the wheel on a drum, cover the drum edge with a length of garden or heater hose, split lengthwise and secured in place with plastic ties. When changing a tire at ground level, support the wheel on two wooden blocks to prevent the brake disc (front wheel) from contacting the floor.

1. If the tire is going to be reused, mark the valve stem location on the tire (**Figure 24**) so the tire can be installed in the same position for easier balancing.

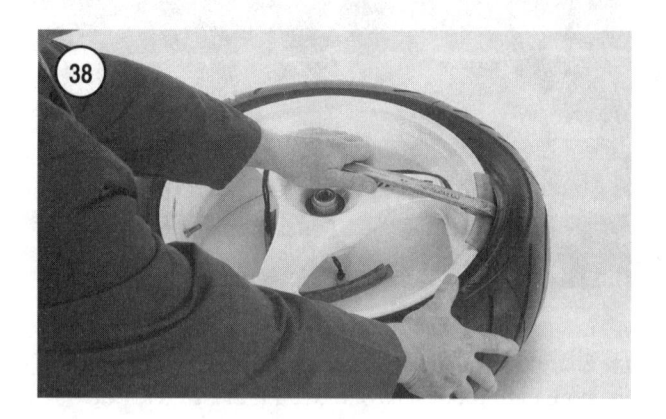

2. Remove the valve core to deflate the tire.

WARNING
The inner rim and tire bead areas are sealing surfaces on a tubeless tire. Do not scratch the inside of the rim or damage the tire bead. Do not force the bead off the rim with any type of leverage, such as a long tire iron. It is very easy to damage the tire bead surface on the tire and rim. It is also possible to crack or break the alloy wheel. Removing tubeless tires from their rims can be difficult because of the exceptionally tight bead and rim seal. If unable to break the tire bead with a bead breaker, take the wheel to a motorcycle dealership and have them change the tire.

3. Use a bead breaker and break the bead all the way around the tire (**Figure 36**). Do not try to force the bead with tire irons. Make sure that both beads are clear of the rim beads.

WHEELS AND TIRES

4. Lubricate the tire beads with soapy water on the side to be removed first.

CAUTION
*Always use rim protectors (**Figure 37**) between the tire irons and the rim to protect the rim from damage.*

5. Insert the tire iron under the bead next to the valve (**Figure 38**). Force the bead on the opposite side of the tire into the center of the rim and pry the bead over the rim with the tire iron.

6. Insert a second tire iron next to the first to hold the bead over the rim. Then work around the tire with the first tool prying the bead over the rim (**Figure 39**). Work slowly by taking small bites with the tire irons. Taking large bites or using excessive force can damage the tire bead or rim.

NOTE
If the tire is tight and hard to pry over the rim, use a third tire iron and a rim protector. Use one hand and arm to hold the first two tire irons, then use the other hand to operate the third tire iron when prying the tire over the rim.

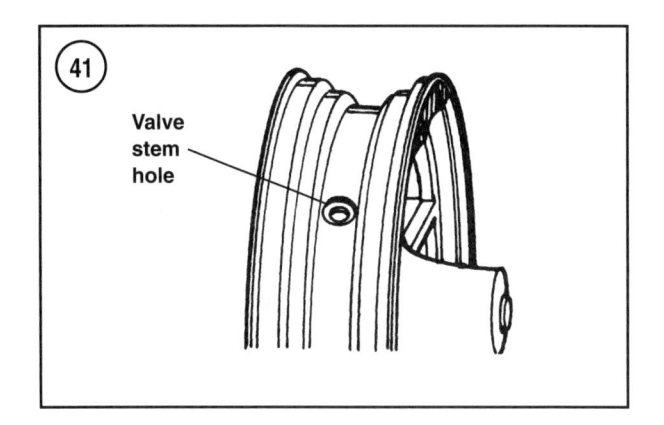

7. Turn the wheel over. Insert a tire iron between the second bead and the same side of the rim that the first bead was pried over (**Figure 40**). Force the bead on the opposite side from the tool into the center of the rim. Pry the second bead off the rim, working around the wheel with the two rim protectors and tire irons.

8. Remove the valve stem and discard it. Remove all rubber residue from the valve stem hole (**Figure 41**, typical) and inspect the hole for cracks and other damage.

9. Remove old balance weights from the rim surface.

Inspection

1. Carefully clean the rim bead with a brush—do not use excessive force or damage the rim sealing surface. Then inspect the sealing surface for any cracks, corrosion or other damage.

WARNING
If there is any doubt as to wheel condition, take it to a Honda dealership for a thorough inspection.

WARNING
Carefully consider whether a tire should be replaced. If there is any doubt about the condition of the existing tire, replace it with a new one. Do not take a chance on a tire failure at any speed.

2. If any one of the following is observed, replace the tire:
   a. A puncture or split with a total length of diameter exceeding 6 mm (0.24 in.).
   b. A scratch or split on the sidewall.
   c. Any type of ply separation.
   d. Tread separation or excessive/abnormal wear pattern.

e. Tread depth of less than the minimum value specified in **Table 1** for original equipment tires. Aftermarket tire tread depth minimum may vary.
f. Scratches on either sealing bead.
g. The cord is cut in any place.
h. Flat spots in the tread from skidding.
i. Any abnormality in the inner liner.

Installation

1. Install a new air valve.
2A. If installing the original tire, carefully inspect the tire for any damage.
2B. If installing a new tire, remove all stickers from the tire tread.
3. Lubricate both beads of the tire with soapy water.
4. Make sure the correct tire, either front or rear, is installed on the correct wheel and the direction arrow on the tire faces in the same direction as the wheel direction arrow (**Figure 42**).
5. If remounting the old tire, align the mark made in Step 1 of *Removal* with the valve stem. If installing a new tire, align the colored spot near the bead (indicating the lightest point of the tire) with the valve stem (**Figure 24**).
6. Place the backside of the tire into the center of the rim. The lower bead should go into the center of the rim and the upper bead outside. Use both hands to push the backside of the tire into the rim (**Figure 43**) as far as possible. Use tire irons when it becomes difficult to install the tire by hand.
7. Press the upper bead into the rim opposite the valve. Pry the bead into the rim on both sides of the initial point with a tire tool, working around the rim to the valve stem (**Figure 44**). If the tire wants to pull up on one side, either use another tire iron or a knee to hold the tire in place. The last few inches are usually the toughest to install. If possible, continue to push the tire into the rim by hand. Relubricate the bead if necessary. If the tire bead wants to pull out from under the rim, use both knees to hold the tire in place. If necessary, use a tire iron for the last few inches (**Figure 45**).
8. Check the bead on both sides of the tire for an even fit around the rim. Align the paint spot (**Figure 24**) near the bead indicating the lightest point of the tire with the valve stem.
9. Lubricate both sides of the tire with soapy water.

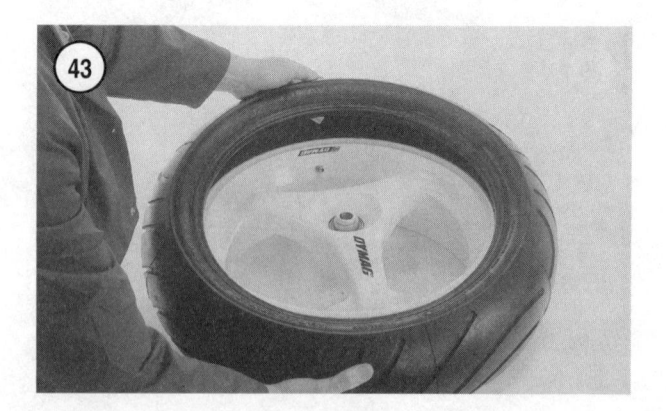

WARNING
Always wear eye protection when seating the tire beads onto the rim. Never exceed 56 psi (386 kPa) inflation pressure as the tire could burst causing severe injury. Never stand directly over the tire while inflating it.

10. Inflate the tire until the beads seat into place. A loud pop should be heard as each bead seats against its side of the rim.
11. After inflating the tire, check to see that the beads are fully seated and that the tire rim lines are the same distance from the rim all the way around the tire (**Figure 46**). If one or both beads do not seat, deflate the tire, relubricate the rim and beads with soapy water and reinflate the tire.

WHEELS AND TIRES

12. Inflate the tire to the required pressure listed in **Table 3**. Screw on the valve stem cap.
13. Balance the wheel as described in this chapter.

WARNING
*After installing new tires, follow the tire manufacturer's instructions for breaking-in (scuffing) the tire. Refer to **Tire Safety** in this chapter.*

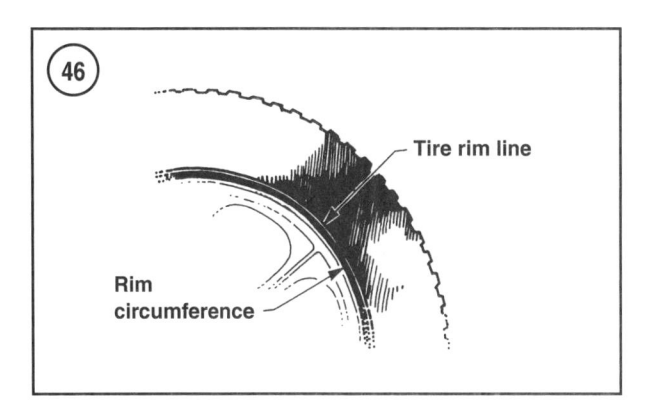

TIRE REPAIRS

Only use tire plugs as an *emergency* repair. Refer to the manufacturer's instructions to install and note the vehicle's weight and speed restrictions. After performing an emergency tire repair with a plug, consider the repair temporary and replace the tire at the earliest opportunity.

Refer all tire repairs to a Honda dealership or other qualified motorcycle technician.

Table 1 TIRE AND WHEEL SPECIFICATIONS

Tire size	
Front	130/70 R18 63H
Rear	180/60 R16 74H
Tire brands	
Bridgestone	
Front	G707 Radial
Rear	G704 Radial
Dunlop	
Front	D250F
Rear	D250
Minimum tire tread depth	1.5 mm (0.06 in.)

Table 2 WHEEL AND AXLE SERVICE SPECIFICATIONS

	Service limit mm (in.)
Axle runout	0.20 (0.008)
Wheel runout	
Axial (side-to-side)	2.0 (0.08)
Radial (up-and-down)	2.0 (0.08)

Table 3 TIRE INFLATION PRESSURE*

	Front kPa (psi)	Rear kPa (psi)
Up to 90 kg (200 lbs. load)	250 (36)	250 (36)
Maximum weight capacity	250 (36)	250 (36)

*Tire inflation pressures are for original equipment tires. Aftermarket tires may require different inflation pressure.

Table 4 WHEEL TORQUE SPECIFICATIONS

	N•m	in.-lb.	ft.-lb.
Front axle bolt	59	–	44
Front axle pinch bolts	22	–	16
Front brake disc bolts	20	–	15
Front pulser ring bolt (ABS)*	8	71	–
Left front brake caliper pivot bolt*	31	–	23
Left front brake caliper-to-secondary master cylinder joint bolt*	25	–	18
Rear wheel nuts	108	–	80
Right front brake caliper mounting bolt*	31	–	23

*ALOC fastener. Install new fastener during assembly.

CHAPTER TWELVE

FRONT SUSPENSION AND STEERING

This chapter describes procedures for the repair and maintenance of the handlebar, front fork and steering components. Refer to Chapter Ten for front wheel and tire service.

Front suspension and steering specifications are listed in **Tables 1-3** at the end of the chapter.

WARNING
Replace all fasteners used on the front suspension and steering components with parts of the same type. Do not use a replacement part of lesser quality or substitute design; this may affect the performance of the system or cause failure of the part, which leads to loss of control of the bike. Use the torque specifications listed in Table 3 during installation to ensure proper retention of these parts.

NOTE
*The ALOC fasteners identified in **Table 3** use a pre-applied threadlock. Honda specifies replacing ALOC fasteners during installation. If a replacement ALOC fastener is unavailable, remove all threadlock from the original fastener and apply a medium strength threadlock to the threads during installation.*

HANDLEBARS

These models are equipped with separate handlebar assemblies that bolt directly to the top bridge. Covers installed on the inside part of each handlebar protect and guide cables and wiring harnesses.

NOTE
Before removing the handlebars, make a drawing of the cables and wiring harnesses from the handlebars and through the frame. This information proves helpful when reinstalling the handlebars and connecting the cables.

Left Handlebar Removal

This section explains how to remove all the components from the left handlebar and then the handlebar. If it is only necessary to remove the handlebar without removing the clutch master cylinder or left grip assembly, skip those steps which are not required.

1. If necessary, remove the outer weight and handlebar grip from the handlebar as described in this chapter.

2. Remove the meter panel. Refer to *Meter Panel* in Chapter Seventeen.
3. Unscrew and remove the handlebar center cover (**Figure 1**).
4. Disconnect the clutch switch connectors (A, **Figure 2**).
5. Disconnect the clutch cruise switch connectors (B, **Figure 2**).
6. Remove the clutch master cylinder as follows:
   a. Top off the brake fluid (DOT 4) in the master cylinder to help prevent air from entering the system when removing and repositioning the master in this step.
   b. Remove the plastic cap from the mounting bolts.

   NOTE
   Do not disconnect the clutch hose when removing the master cylinder.

   c. Loosen and remove the master cylinder mounting bolts (A, **Figure 3**), holder (B) and master cylinder.

   CAUTION
   Support the master cylinder in an upright position during this procedure to prevent air from entering the system.

   NOTE
   Place a thick towel across the motorcycle for placing the handlebar after removing its mounting bolts.

7. Remove the handlebar mounting bolts (**Figure 4**) and handlebar. Lay the handlebar across the towel.

   NOTE
   Reposition the master cylinder to keep it upright.

8. Remove the screws and handlebar cover (**Figure 5**).

   NOTE
   Record the clutch hose and wiring harness routing through the handlebar.

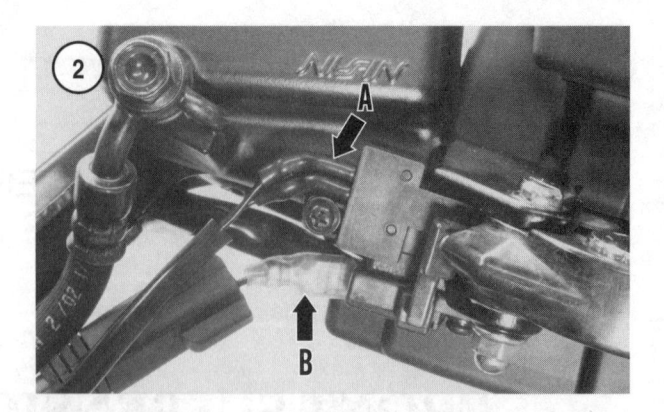

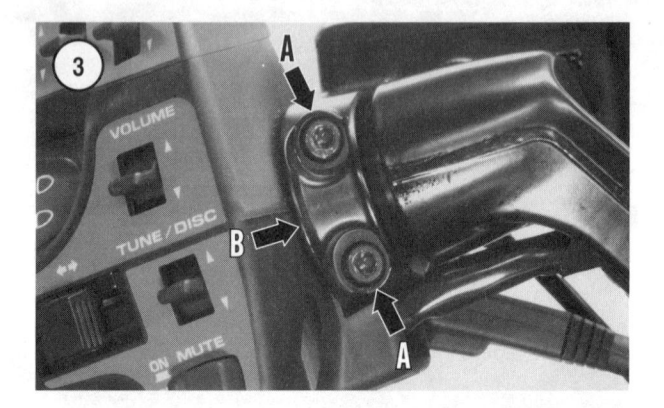

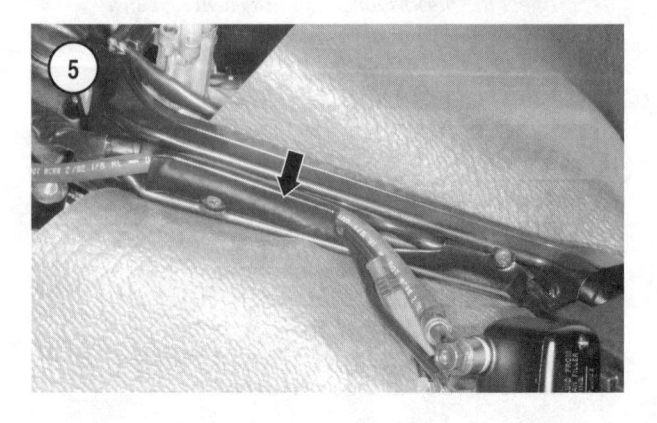

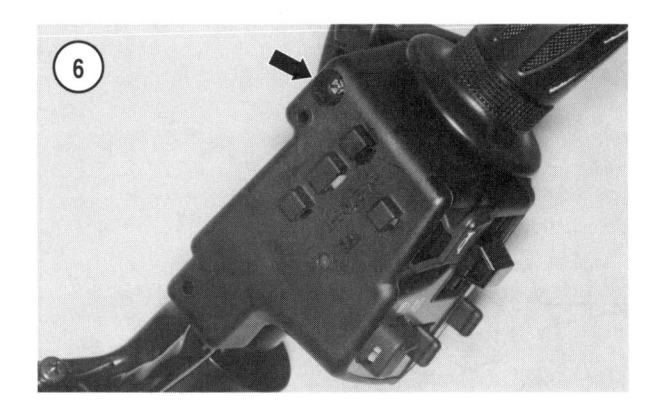

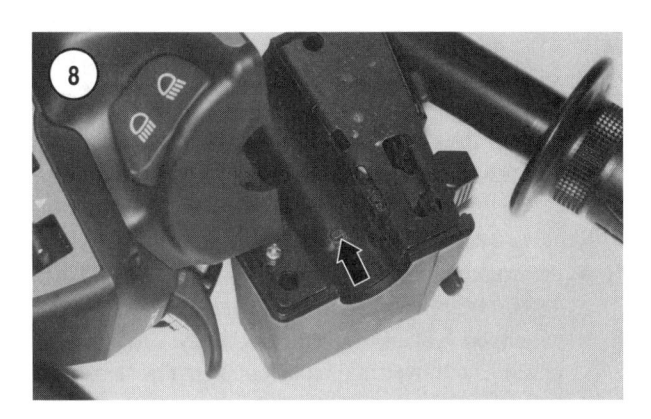

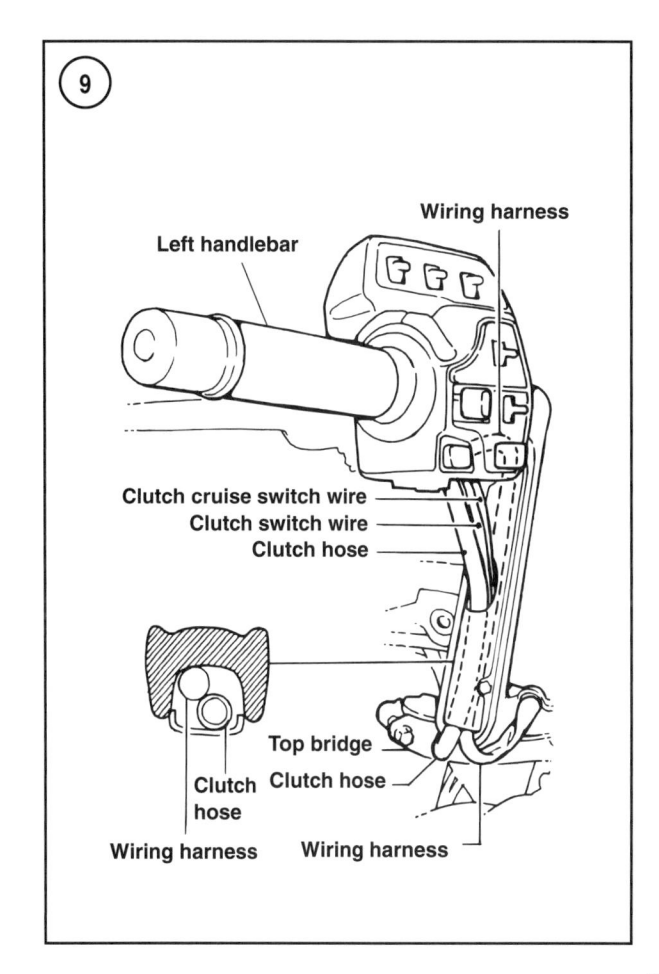

FRONT SUSPENSION AND STEERING

9. Remove the left handlebar switch:
   a. Remove the screws (**Figure 6**) from the lower switch housing half. Lift the upper half from the handlebar.
   b. Remove the screw (A, **Figure 7**), setting plate (B) and left handlebar switch.
10. Remove the left handlebar.

Left Handlebar Installation

1. Install the left handlebar switch:
   a. Align the pin in the lower switch half (**Figure 8**) with the hole in the handlebar and install the lower switch half.
   b. Hook the setting plate (B, **Figure 7**) into the switch and secure with the screw (A).
   c. Install the upper switch half and secure with the lower mounting screws (**Figure 6**). Tighten the screws securely.
2. Route the clutch hose and wiring harness through the handlebar. Refer to **Figure 9** for 2001-2003 models. 2004 model is similar. Install the handlebar cover (**Figure 5**) and mounting screws.
3. Position the handlebar onto the top bridge and center around the clutch hose and left handlebar switch wire harness as shown in **Figure 10**. Install the handlebar mounting bolts (**Figure 4**) and tighten to 26 N•m (19 ft.-lb.).
4. Install the clutch master cylinder:
   a. Align the end of the master cylinder with the punch mark on the handlebar. Install the holder (B, **Figure 3**) and the two mounting bolts (A). Tighten the upper mounting bolt first, then the lower mounting bolt to 12 N•m (106 in.-lb.).

HOSE AND WIRING HARNESS ROUTING

⑩

- Clutch hose
- Wire band (white tape position)
- Turn signal cancel unit wire
- Left handlebar switch wire
- Right handlebar switch wire
- Brake hose
- Left handlebar
- Right handlebar
- Top bridge
- Turn signal cancel unit
- Left handlebar switch wire
- Right handlebar switch wire

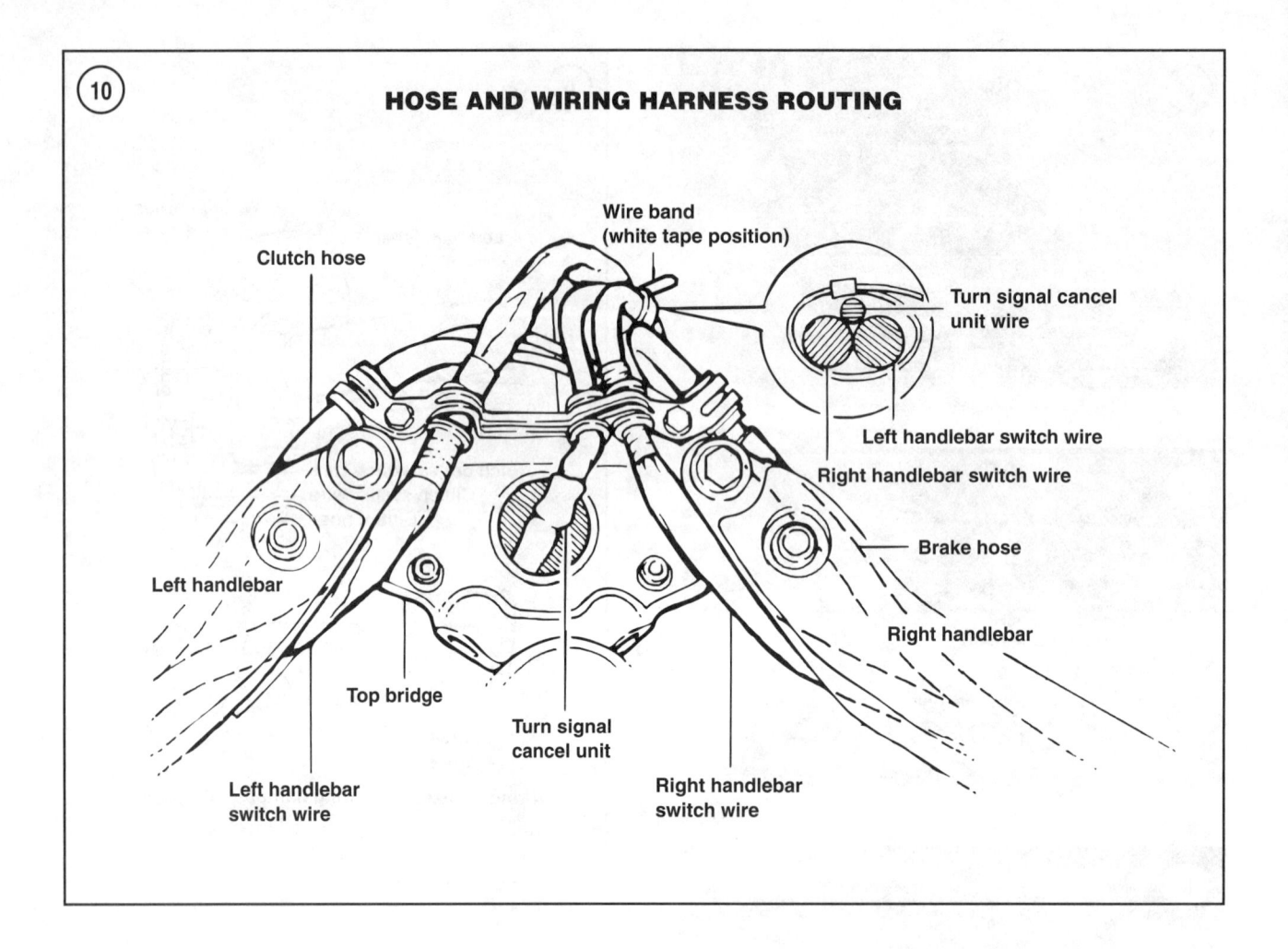

---

b. Install the plastic cap into the mounting bolts.
5. Connect the clutch cruise switch connectors (B, **Figure 2**).
6. Connect the clutch switch connectors (A, **Figure 2**).
7. Check all the left handlebar switches for proper operation.
8. Install the handlebar center cover (**Figure 1**).
9. Install the meter panel. Refer to *Meter Panel* in Chapter Seventeen.

Right Handlebar Removal

This section explains how to remove all of the components from the right handlebar and then the handlebar. If it is only necessary to remove the handlebar without removing the brake master cylinder or throttle housing, skip those steps that are not required.

1. If necessary, remove the handlebar grip from the throttle housing as described in this chapter.
2. Remove the meter panel. See Chapter Seventeen, *Meter Panel*.
3. Unscrew and remove the handlebar center cover (**Figure 1**).
4. Remove the screw and the handlebar weight.
5. At the front brake master cylinder:
   a. Top off the brake fluid (DOT 4) in the master cylinder to help prevent air from entering the system when the master is removed and repositioned in this step.
   b. Remove the plastic cap from the mounting bolts.
   c. Loosen, but do not remove, the master cylinder mounting bolts (**Figure 11**).

NOTE
Do not disconnect the brake hose when removing the master cylinder.

6. Remove the throttle cables from the handlebar clamp (**Figure 12**).

NOTE
Place a thick towel across the motorcycle for placing the handlebar after removing its mounting bolts.

FRONT SUSPENSION AND STEERING

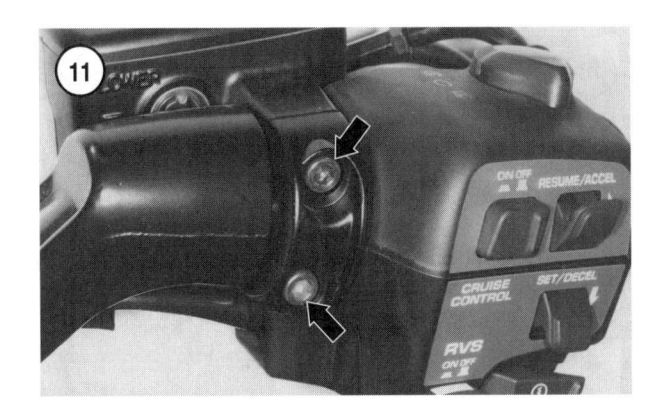

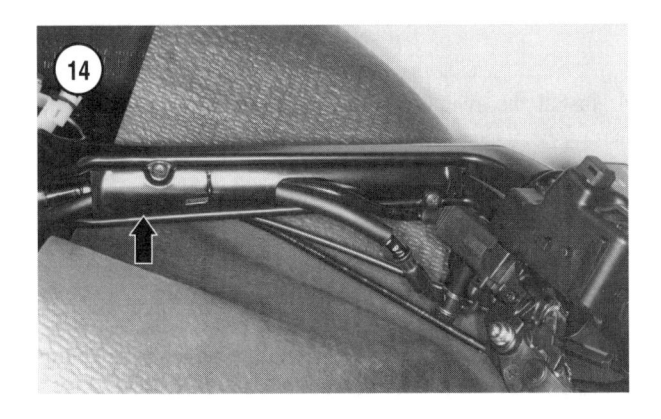

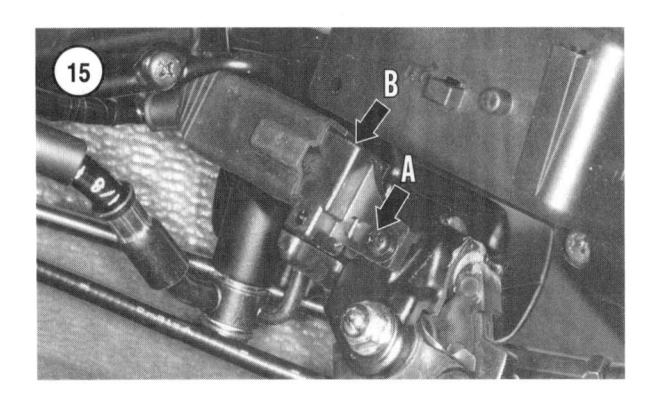

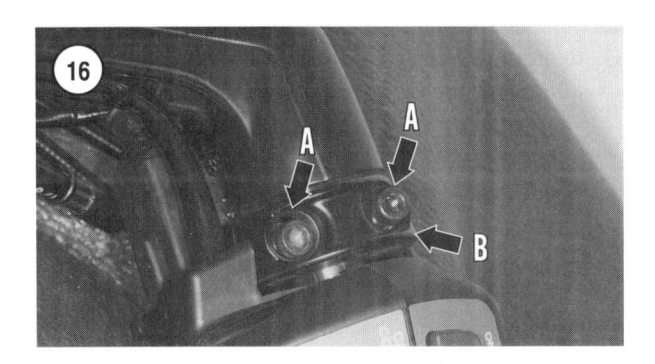

7. Remove the handlebar mounting bolts (**Figure 13**) and handlebar. Lay the handlebar across the towel.
8. Remove the screws and handlebar cover (**Figure 14**).

NOTE
Record the brake hose and wiring harness routing through the handlebar.

NOTE
The front brake light switch and front cruise cancel switch are enclosed in the switch housing mounted on the bottom of the front master cylinder. The two upper (smaller) terminals plug into the brake light switch. The two lower (larger) terminals plug into the cruise cancel switch. While the connectors can be disconnected with the switch mounted on the master cylinder handlebar, it is easier to remove the switch and leave the connectors attached to the switch.

9. Remove the screw (A, **Figure 15**) and switch housing (B).
10. Remove the master cylinder bolts (A, **Figure 16**), holder (B) and master cylinder.

NOTE
Reposition the master cylinder to keep it upright.

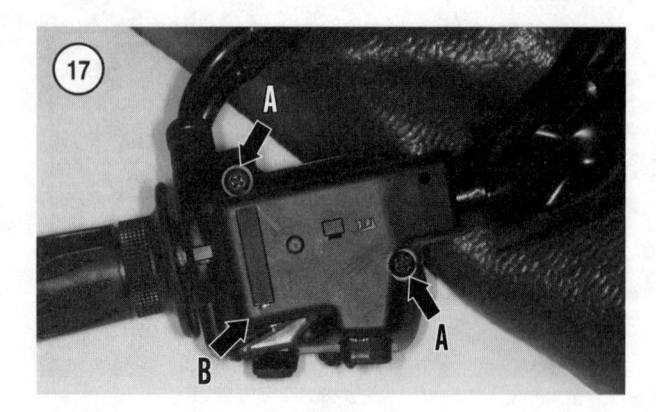

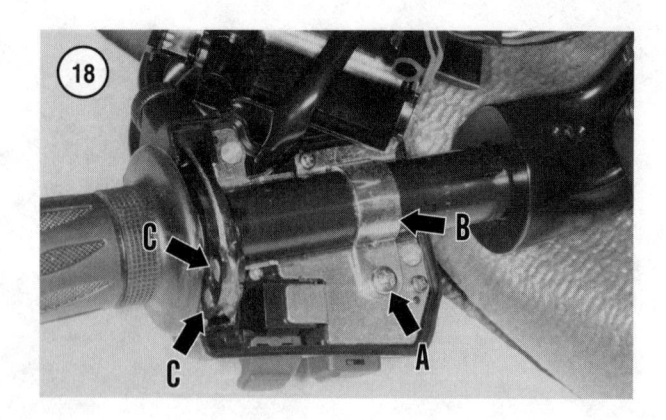

11. Remove the right handlebar switch/throttle housing:
   a. Remove the screws (A, **Figure 17**) from the lower switch housing half and lift it away from the handlebar.
   b. Remove the screw (A, **Figure 18**) and setting plate (B).
   c. Disconnect the throttle cables (C, **Figure 18**) from the throttle grip and remove the grip from the handlebar.
12. Remove the right handlebar.
13. If necessary, replace the right grip as described in this section.

Right Handlebar Installation

1. Assemble and install the right handlebar switch/throttle housing:
   a. Clean the right handlebar/throttle grip area of all old grease.
   b. Lubricate the throttle grip flange and sliding surfaces with grease.
   c. Lubricate the cable ends with grease and install them onto the throttle grip.
   d. Align the pin in the upper switch half with the hole in the handlebar and install the upper switch half. Make sure the switch half seats over the throttle grip.
   e. Hook the setting plate (B, **Figure 18**) into the switch and secure with the screw (A).
   f. Install the lower switch half (B, **Figure 17**) and secure with the lower mounting screws (A). Tighten the front screw first, then the rear screw.
2. Route the brake hose and wiring harness through the handlebar. Refer to **Figure 19** for 2001-2003 models. 2004 model is similar.
3. Align the end of the master cylinder with the punch mark on the handlebar. Install the holder (B, **Figure 16**) and the two mounting bolts (A). Tighten the bolts finger-tight. 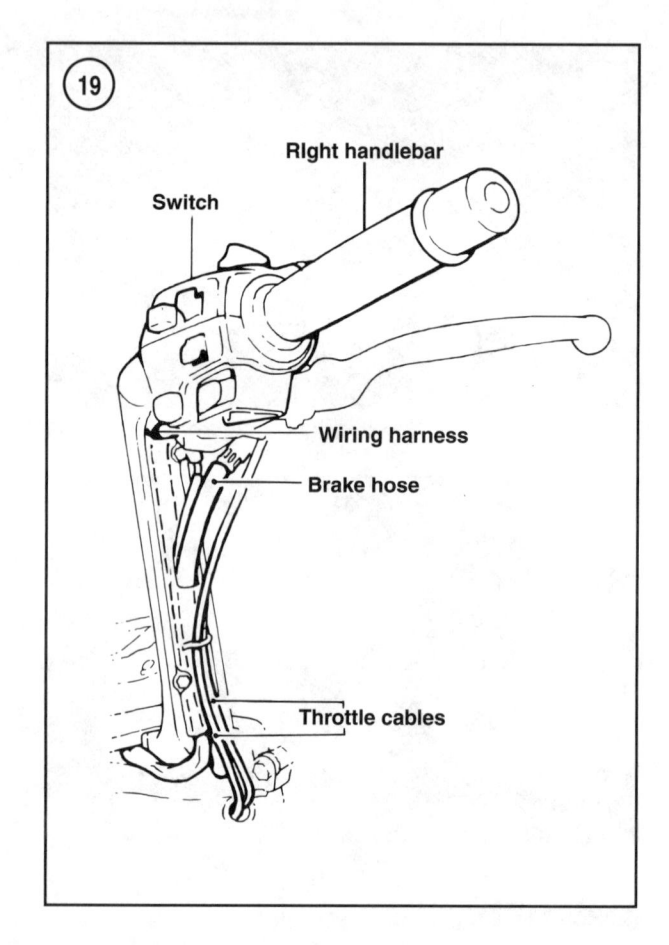
4. Install the switch (B, **Figure 15**) and secure with the screw (A).
5. Recheck the cable and wiring harness routing, then install the handlebar cover (**Figure 14**) and secure with the mounting screws.
6. Position the handlebar onto the top bridge and center around the brake hose and right handlebar switch wire harness as shown in **Figure 10**. Install the handlebar mounting bolts (**Figure 13**) and tighten to 26 N•m (19 ft.-lb.).

FRONT SUSPENSION AND STEERING

HANDLEBAR WEIGHT ASSEMBLY

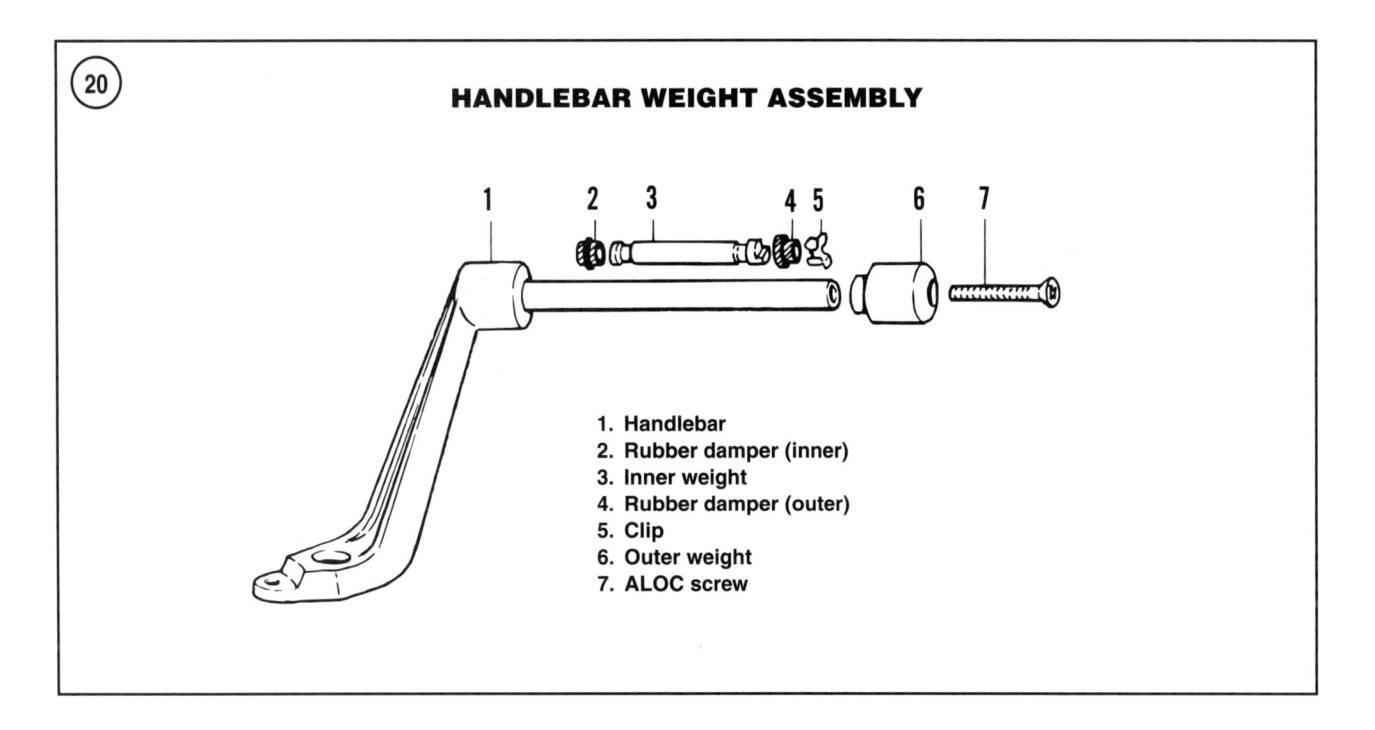

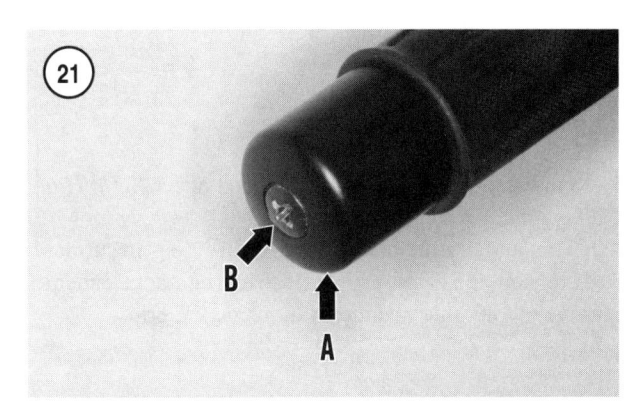

1. Handlebar
2. Rubber damper (inner)
3. Inner weight
4. Rubber damper (outer)
5. Clip
6. Outer weight
7. ALOC screw

7. Secure the throttle cables into the handlebar clamp (**Figure 12**).
8. Check that the end of the master cylinder aligns with the punch mark on the handlebar. Tighten the upper mounting bolt first, then the lower mounting bolt to 12 N•m (106 in.-lb.). Install the plastic cap into the mounting bolts.
9. Check all the right handlebar switches for proper operation.
10. Install the outer handlebar weight as described in this section.
11. Install the handlebar center cover (**Figure 1**).
12. Install the meter panel. Refer to *Meter Panel* in Chapter Seventeen.
13. Open and release the throttle grip. Make sure it opens and closes (snaps back) without any binding or roughness. Then support the bike on its centerstand and turn the handlebar from side to side, checking throttle operation at both steering lock positions.

> *WARNING*
> *An improperly installed throttle grip assembly may cause the throttle to stick open. Failure to properly assemble and adjust the throttle cables and throttle grip could cause a loss of steering control. Do not start or ride the motorcycle until the throttle grip is correctly installed and snaps back when released.*

Handlebar Weights

Each handlebar is equipped with an inner and outer handlebar weight assembly (**Figure 20**). The handlebar weights can be serviced with the handlebars mounted on the bike.

1. Outer weight assembly:
    a. Hold the outer weight (A, **Figure 21**) and remove the mounting screw (B) and weight.
    b. Install the outer weight against the inner weight—align the grooves and bosses (A, **Figure 22**) on both parts. Install a new ALOC screw and tighten to 10 N•m (88 in.-lb.). Make sure there is a clearance gap between the hand grip and handlebar weight.
2. Inner weight assembly:
    a. Remove the outer weight.

b. Remove the clip (B, **Figure 22**) from the end of the handlebar.
c. Insert the screw into the weight and pull the weight assembly from the handlebar.

> **NOTE**
> *The inner and outer rubber dampers are different. Identify these parts before removing them.*

d. Service the weight assembly as required. Replace the rubber dampers if excessively worn or damaged. The inner and outer rubber dampers are different. Install as shown in **Figure 20**.
e. Install the weight assembly into the handlebar with the threaded end of the weight facing out.
f. Install the clip (B, **Figure 22**) into the end of the handlebar to secure the weight assembly.
g. Install the outer weight (Step 1).

Handlebar Grips Inspection

The handlebar grips must be secured tightly to the left handlebar and to the throttle grip (right side). Replace cut or damaged grips as water may enter between the grip and its mounting surface and cause the grip to slip. This could cause steering control loss. Replace the handlebar grips as described in the following procedure.

Handlebar Grip Replacement

1. Remove the outer handlebar weights as described in this section.
2. To remove the hand grips, do the following:

> **NOTE**
> *If reusing the hand grips, remove them carefully to avoid puncturing or tearing them.*

a. Carefully insert a thin flat-blade screwdriver between the handlebar or throttle and grip.
b. Spray electrical contact cleaner into the space created by the screwdriver. Then remove the screwdriver and quickly twist the grip back and forth to break the cement bond between the grip and handlebar or grip and throttle.
c. Slide the grip off the handlebar or throttle.

> **NOTE**
> *Steps 3-11 describe how to cement the hand grips onto the handlebar and throttle grip. When using a grip cement (Honda Grip Cement or ThreeBond 1501C Griplock), follow the manufacturer's instructions carefully.*

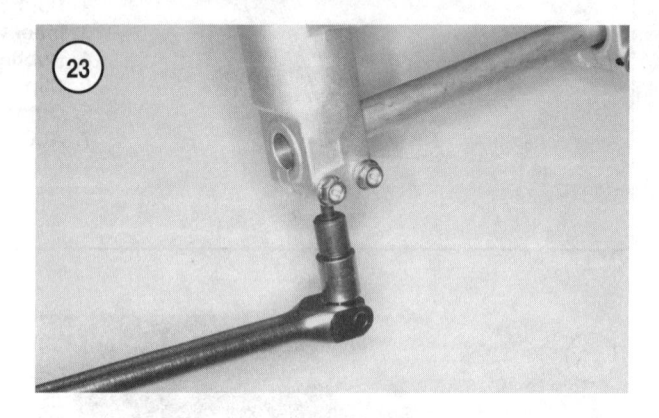

3. Remove the throttle grip as described under *Right Handlebar Removal* in this section.
4. Clean and dry the handlebar and throttle grip surface.
5. If reusing the grips, use an electrical contact cleaner to remove all old glue residue from inside the grips.

> **NOTE**
> *If the original hand grips are torn or damaged, install new grips.*

6. Apply the grip cement to the left handlebar, the outer surface of the throttle and inside the hand grips.

> **NOTE**
> *The left and right side hand grips are different. Install the grip with the larger inside diameter over the throttle (right side). Also, if the grip surfaces are not symmetrical, install the grips with their raised or directional hand pads positioned correctly.*

7. Install the hand grip over the handlebar (left side) or throttle (right side). Make sure there is clearance between the end of the grip and the switch housing. Remove all excess grip cement from the end of each grip.
8. Follow the grip cement manufacturer's instructions regarding drying time before operating the motorcycle.

FRONT SUSPENSION AND STEERING

WARNING
Loose or damaged hand grips can slide off and cause steering control loss. Make sure the hand grips are correctly installed and cemented in place before operating the motorcycle.

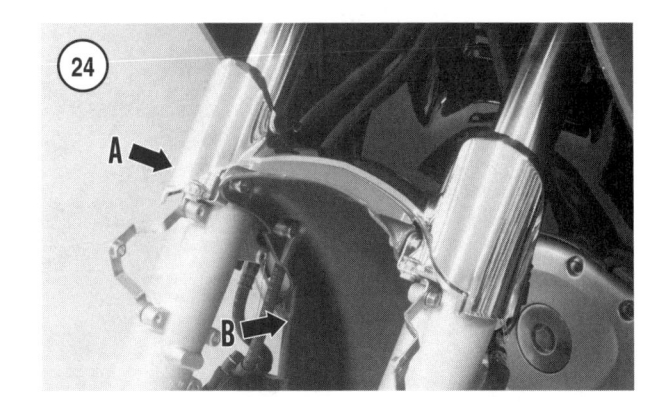

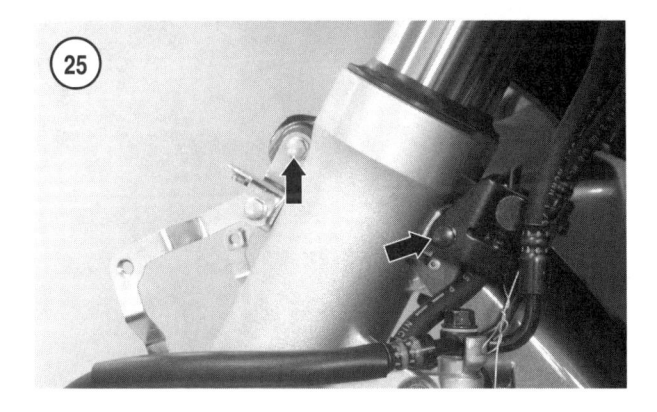

9. When the grip cement is dry, install the throttle grip as described under *Right Handlebar Installation* in this section.

10. Install the outer handlebar weights as described in this section.

11. Open the throttle grip and then release it. Make sure it opens and closes (snaps back) without any binding or roughness.

WARNING
An improperly installed throttle grip assembly may cause the throttle to stick open. Failure to properly assemble and adjust the throttle cables and throttle grip could cause steering control loss. Do not start or ride the motorcycle until the throttle grip is correctly installed and snaps back when released.

FRONT FORK

The following sections describe complete service and adjustment of the front fork. To prevent damaging the fork when servicing it, note the following:

1. To avoid rounding off the shoulders on the fork caps, use a six-point socket when loosening and tightening the fork caps.
2. Do not overtighten the fork tube pinch bolts, as this can damage the fork bridge threads and fork tubes. Always use the listed torque specifications.
3. The fork sliders are easily scratched. Handle them carefully during all service procedures.
4. Use a metal plate (as described in the text) when holding the fork tubes in a vise.

Front Fork Removal

NOTE
A number of different fasteners secure the front fender assembly, front brake calipers, delay valve and secondary master cylinder to the fork tubes. Store and identify the fasteners in a divided container.

1. Remove the front wheel. Refer to *Front Wheel* in Chapter Eleven.
2. If the fork is going to be overhauled:
   a. Insert the front axle through both fork tubes. Position the axle so the fork Allen bolt is accessible.
   b. Loosen, but do not remove, the fork Allen bolt in the bottom of each fork tube (**Figure 23**). Remove the axle.
3. To remove the front fender top cover (A, **Figure 24**):
   a. Remove the trim clip in top, middle of cover.
   b. Remove the two mounting bolts.
4. Remove front fender B (B, **Figure 24**):
   a. Remove the two mounting bolts from left fork slider and brake hose pipe bracket (**Figure 25**).
   b. Remove the two mounting bolts from right fork slider (A, **Figure 26**).
   c. Slide the fender down behind fork tubes and remove.

NOTE
*Do not remove the four lower anti-dive housing mounting bolts or housing (**Figure 27**) from the fork tube. There are no replacement parts or seals for the housing.*

NOTE
Do not remove the brake hose banjo bolts when removing the anti-dive plunger case or the brake components in the following steps.

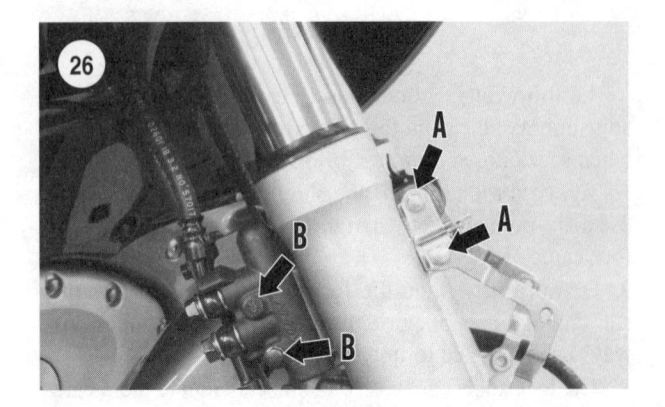

5. Support the front brake assembly with a piece of stiff wire.
6. Remove the delay valve mounting bolt (B, **Figure 26**) at the right fork tube.
7. Remove the following from the left fork tube:
   a. Anti-dive plunger case bolts (A, **Figure 28**) and case (B).
   b. Brake hose joint mounting bolt (A, **Figure 29**).
   c. Secondary master cylinder mounting bolts (B, **Figure 29**).
8. Pull the front brake assembly away from the fork tubes. Adjust the support wire installed in Step 5 to remove any weight from the brake hoses.
9. Loosen the top fork tube pinch bolt (A, **Figure 30**).

NOTE
Look up through the front fairing and note the routing of any hoses or cables around the fork tubes.

10. If the fork tube is going to be serviced, loosen the fork cap (B, **Figure 30**). Do not remove it.
11. While supporting the fork tube, loosen the lower fork tube pinch bolts (**Figure 31**) and remove the fork tube.

NOTE
Rust and corrosion built up around the fork tube and steering stem clamp surfaces can lock the fork tube in place. Support the fork tube and spray the top area of each clamp with a penetrating oil.

12. Remove the other fork tube.

Front Fork Installation

1. Make sure all fork tube pinch bolts and fork bridge threads are clean.

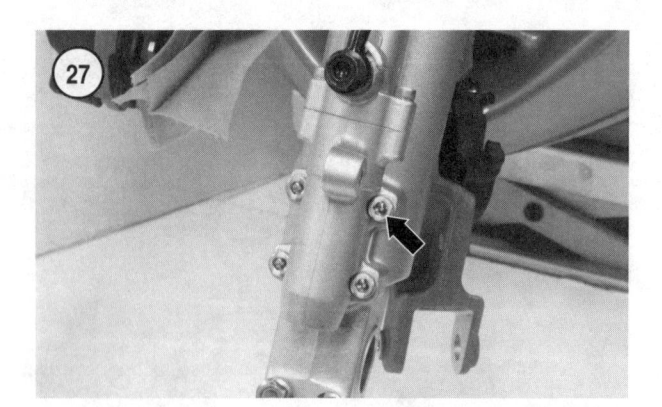

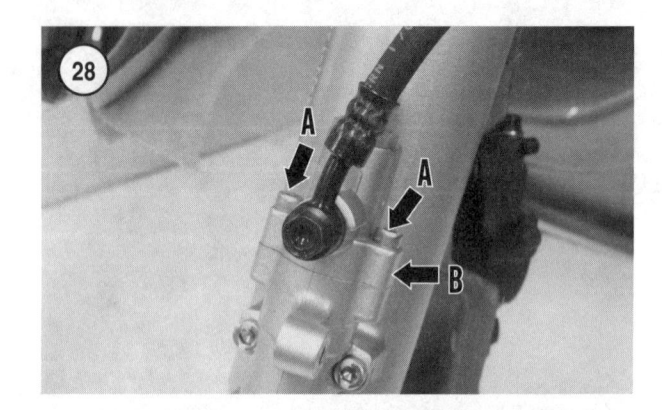

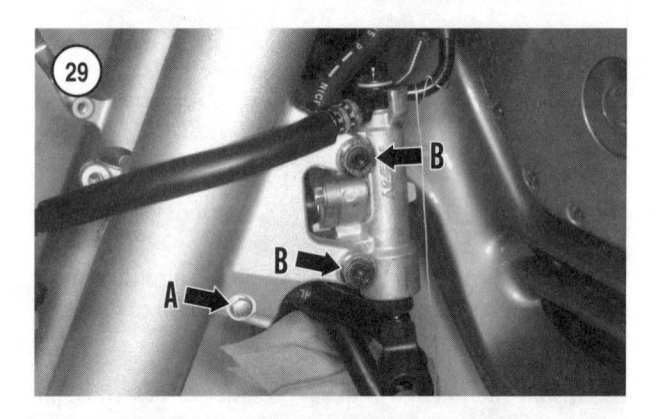

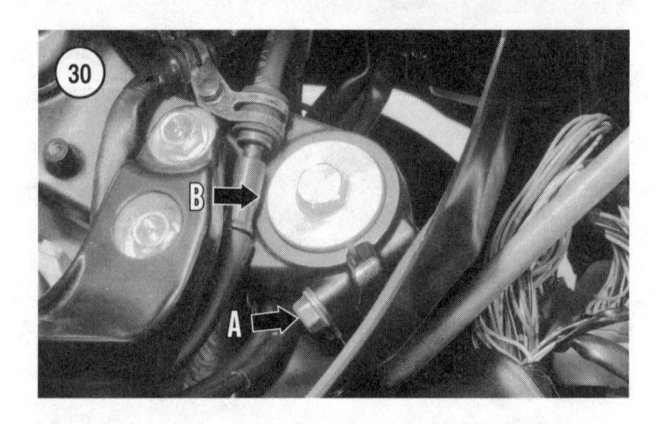

FRONT SUSPENSION AND STEERING

to the notes made before removing the fork tubes.

3. Tighten the lower fork tube pinch bolts (**Figure 31**) to 29 N•m (22 ft.-lb.).
4. If the fork cap (B, **Figure 30**) was loosened, tighten it to 23 N•m (17 ft.-lb.).

NOTE
*Align the throttle cable guide as shown in **Figure 33**.*

5. Tighten the upper fork tube pinch bolt (A, **Figure 30**) to 26 N•m (19 ft.-lb.).
6. Repeat for the other fork tube.

NOTE
Because the secondary master cylinder and anti-dive plunger case use ALOC bolts (threads pre-coated with a threadlock), Honda specifies installing new ALOC bolts during installation. If reusing the original fasteners, clean the threads and apply a medium strength threadlock.

7. Install the brake components onto the left fork slider as follows:
    a. Mount the secondary master cylinder and tighten its bolts (B, **Figure 29**) to 31 N•m (23 ft.-lb.).
    b. Install the brake hose joint and tighten the mounting bolt (A, **Figure 29**) to 12 N•m (106 in.-lb.).
8. Install the plunger case:
    a. Clean the anti-dive plunger case and housing mating surfaces (**Figure 34**) with compressed air. Make sure all debris is removed from these surfaces and from the housing bore on the slider.
    b. Lubricate the anti-dive plunger tip (**Figure 34**) with silicone brake grease.
    c. Mount the anti-dive plunger case (B, **Figure 28**) using new mounting bolts (A) and tighten to 4 N•m (35 in.-lb.).

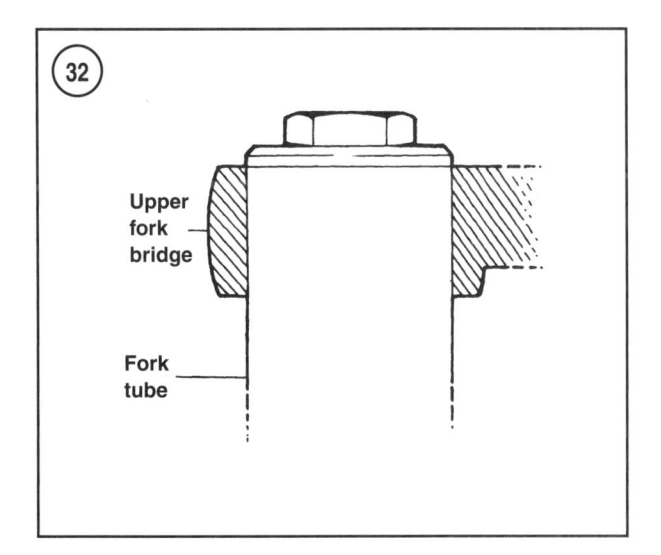

2. Slide the fork tube through the lower and upper fork bridges. Position the fork so the top of the fork tube is flush with the top bridge surface (**Figure 32**).

NOTE
Make sure any cables or wiring harnesses are routed correctly around the fork tubes. Refer

CHAPTER TWELVE

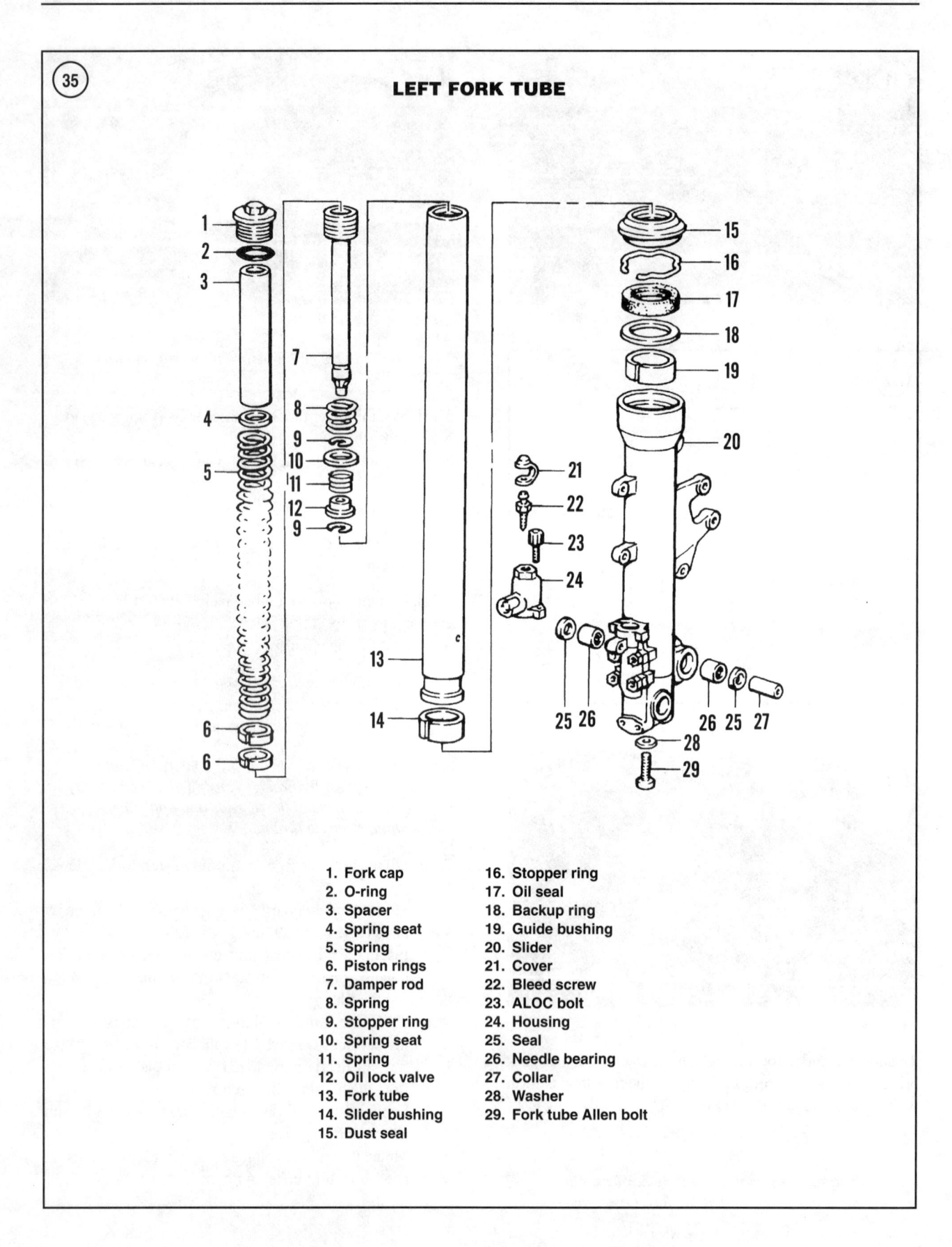

(35) LEFT FORK TUBE

1. Fork cap
2. O-ring
3. Spacer
4. Spring seat
5. Spring
6. Piston rings
7. Damper rod
8. Spring
9. Stopper ring
10. Spring seat
11. Spring
12. Oil lock valve
13. Fork tube
14. Slider bushing
15. Dust seal
16. Stopper ring
17. Oil seal
18. Backup ring
19. Guide bushing
20. Slider
21. Cover
22. Bleed screw
23. ALOC bolt
24. Housing
25. Seal
26. Needle bearing
27. Collar
28. Washer
29. Fork tube Allen bolt

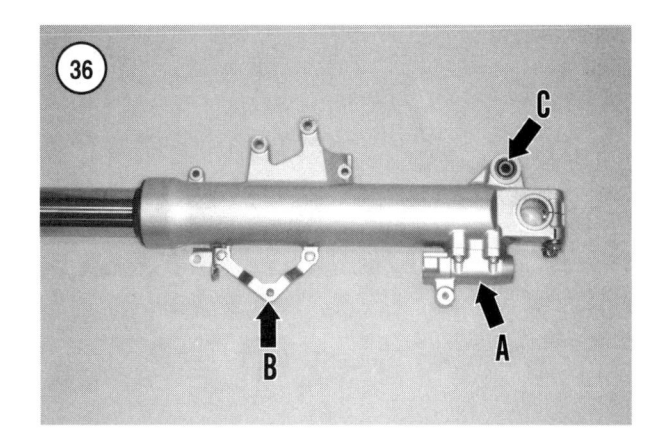

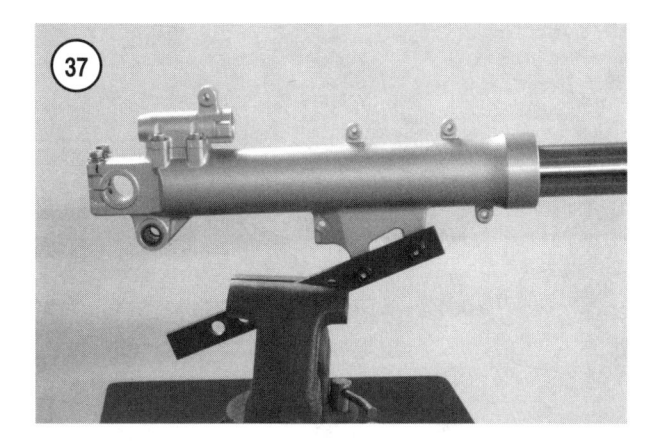

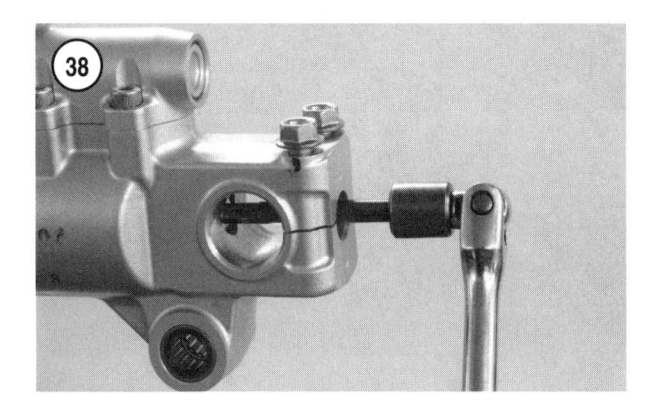

FRONT SUSPENSION AND STEERING

12. Install the fender top cover (A, **Figure 24**) and secure with the two mounting bolts and trim clip.
13. Install the front wheel. Refer to *Front Wheel* in Chapter Eleven.

> **WARNING**
> *After installing the front wheel, operate the front brake lever to reposition the caliper pistons. If the brake lever feels spongy, bleed the brakes as described in Chapter Fourteen.*

Left Fork Tube Disassembly

This section describes complete disassembly of the left front fork (**Figure 35**). If only changing the fork oil and/or setting the oil level, begin at Step 1 and follow the required steps listed in the text.

> **NOTE**
> *Do not remove the anti-dive plunger case (A, **Figure 36**) in the following steps. Replacement parts are not available for the case.*

1. Remove the front fender mounting plate (B, **Figure 36**).
2. Remove the collar (C, **Figure 36**) from the needle bearings.
3. Bolt a flat metal plate onto the slider and clamp the metal plate (**Figure 37**) in a vise to support the fork tube during disassembly and reassembly.

> **NOTE**
> *If only changing the fork oil and/or setting the oil level, disregard the steps pertaining to loosening the fork tube Allen bolt in Step 4.*

> **NOTE**
> *Disregard Step 4 if the fork tube Allen bolt was loosened during fork removal.*

4. Loosen, but do not remove, the fork tube Allen bolt (**Figure 38**).
5. Loosen and remove the fork cap (**Figure 39**) and its O-ring.
6. Remove the spacer (A, **Figure 40**), spring seat (B) and fork spring (C).
7. Turn the fork tube over a drain pan and pour out the fork oil by operating the fork tube several times.

> **NOTE**
> *If only changing the fork oil and/or setting the oil level, go to the **Fork Oil Adjustment** procedure at the end of this section. If disassembling the fork, continue with Step 8.*

9. Mount the delay valve and secure with the lower mounting bolt (B, **Figure 26**). Do not tighten the bolt.
10. Install rear fender B (B, **Figure 24**) and its mounting bolts. Tighten the bolts securely.
11. Tighten the delay valve mounting bolts (B, **Figure 26**) to 12 N•m (106 in.-lb.).

> **NOTE**
> *Remove the wire used to support the brake assembly.*

8. Remove the Allen bolt (29, **Figure 35**), previously loosened, and washer (28) from the base of the slider. Discard the washer.

CAUTION
Do not use excessive force when removing the dust seal in Step 9.

9. Carefully pry the seal out of the slider with a suitable tool. Move the tool around the seal in small increments. It is easy to scratch and damage the slider at the top of the dust seal bore. When selecting a starting point, choose the side facing in (toward the wheel).

10. Pry the dust seal (**Figure 41**) out of the slider and remove.

11. Slip the tip of a small screwdriver behind the stopper ring and carefully pry the ring out of the slider groove (**Figure 42**).

NOTE
On this type of fork, a pressed-in bushing in the slider and a bushing on the fork tube keep the slider and fork tube from separating. To remove the fork tube from the slider, use these parts as a slide hammer as described in Step 12 and Step 13.

12. Hold the slider in one hand and the fork tube in the other hand (**Figure 43**). Pull the fork tube out until resistance is felt. The slider and guide bushings are now contacting each other.

13. Hold the fork tube and pull hard on the slider, using quick in-and-out strokes (**Figure 43**). Doing so withdraws the oil seal, backup ring and guide bushing from the slider. See **Figure 44**.

14. Remove the slider and pour any remaining oil in the oil pan.

15. Disassemble the oil lock valve and remove the damper rod as follows:

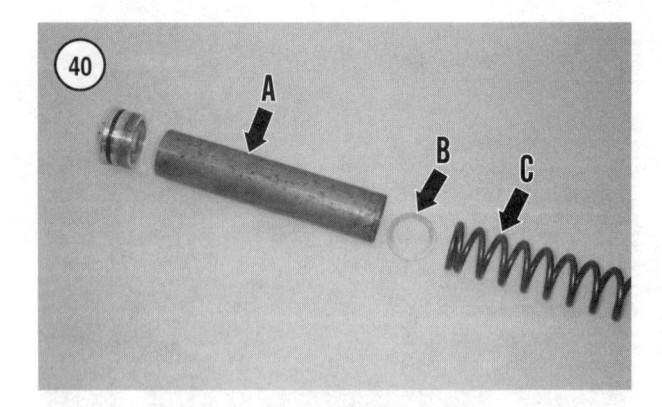

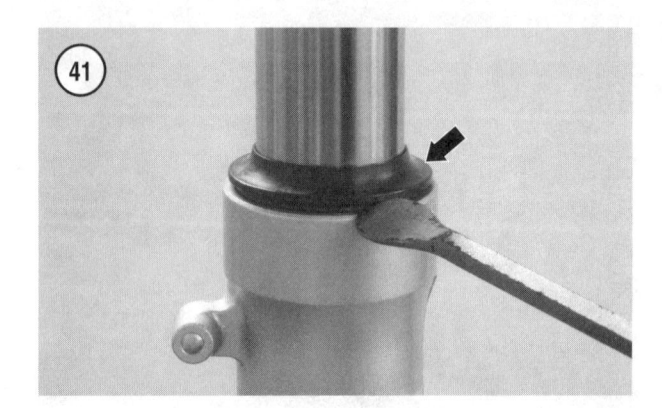

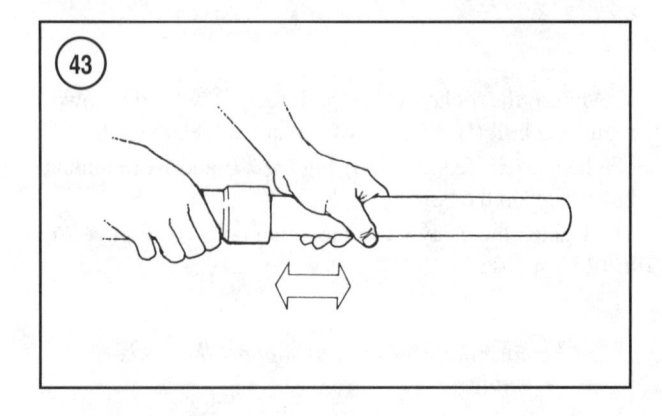

FRONT SUSPENSION AND STEERING

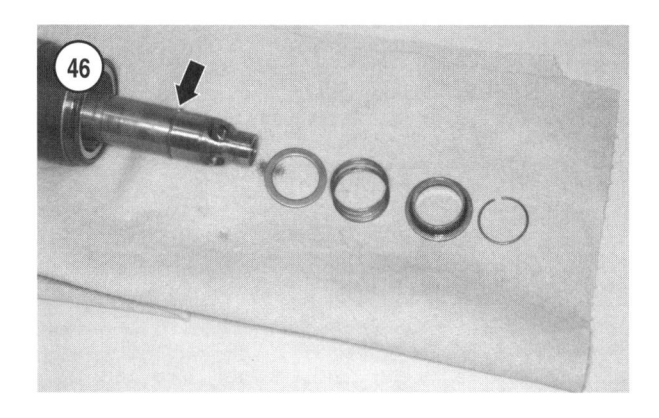

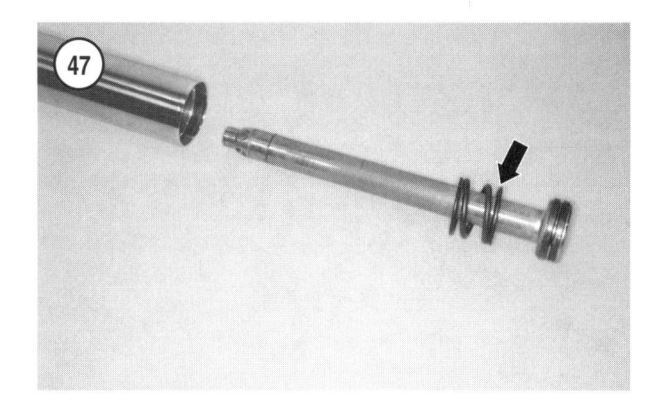

a. Remove the stopper ring (**Figure 45**) from the groove in the end of the damper rod.

b. Remove the oil lock valve, spring and spring seat. Do not remove the upper stopper ring (**Figure 46**) unless necessary.

c. Turn the fork tube over and remove the damper rod and its spring (**Figure 47**).

16. Slide the oil seal, backup ring and guide bushing (**Figure 44**) off the fork tube.

17. Spread the groove in the slider bushing (**Figure 48**) and slide it off the fork tube.

18. Clean and inspect the components as described in this section.

19. To service the needle bearing assembly (26, **Figure 35**), refer to *Brake Caliper Pivot Bearing Replacement* in this section.

Left Fork Tube Reassembly

Refer to **Figure 35**.

1. Before assembling the parts, make sure there is no solvent left in the slider or on any part.

2. Coat all parts, except the Allen bolt and damper rod threads, with new fork oil before installation.

3. Install the spring (**Figure 47**) onto the damper rod. Then slide the damper rod assembly through the fork tube.

4. Assemble the oil lock assembly as follows:

   a. Install a new stopper ring (if removed) into the upper damper rod groove (A, **Figure 49**).

   b. Install the spring seat (B, **Figure 49**) and spring (C).

   c. Install the oil lock valve (D, **Figure 49**) with its shoulder facing the spring. See **Figure 50**.

   d. Install a new stopper ring (E, **Figure 49**) into the lower damper rod groove. See **Figure 51**.

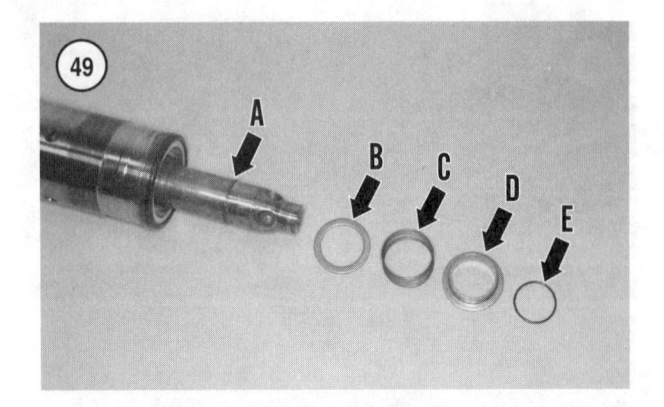

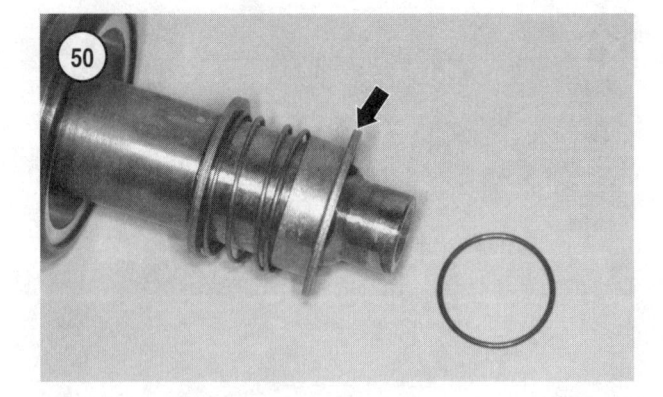

5. Install the slider bushing (**Figure 52**) onto the fork tube, if removed.
6. Install the fork tube/damper rod assembly (**Figure 53**) into the slider.
7. Mount the slider in a vise (**Figure 37**).
8. Install a new washer onto the Allen bolt.
9. Install a medium strength threadlock onto the fork tube Allen bolt threads. Then thread the Allen bolt (**Figure 54**) into the bottom of the fork damper and tighten to 20 N•m (15 ft.-lb.).

NOTE
*If the fork damper turns with the Allen bolt, temporarily install the fork spring, spring seat, spacer, and fork cap in the order shown in **Figure 35**. Installing these parts applies pressure against the fork damper and prevents it from turning. Remove these parts after tightening the Allen bolt.*

10. Secure the slider in the vise so the fork tube faces straight up.
11. Install the guide bushing and backup ring as follows:
    a. Slide the guide bushing (A, **Figure 55**) and backup ring (B) down the fork tube. Install the backup ring with its chamfered side facing down and set it on top of the bushing.
    b. Use a fork seal driver (**Figure 56**) to drive the bushing into the fork slider until it bottoms out in the recess in the slider. The knocking sound made by the driver changes when the bushing bottoms out.

NOTE
*Motion Pro fork seal drivers (**Figure 56**), or equivalent, can be purchased from aftermarket suppliers. To select a driver, first measure the outside diameter of one fork tube.*

NOTE
To avoid damaging the fork seal and dust seal when installing them over the top of the fork

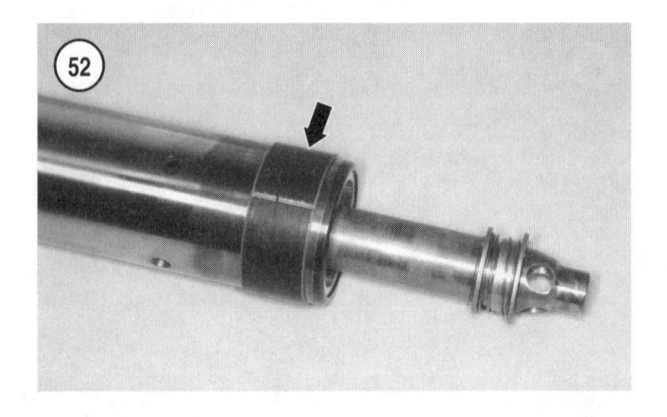

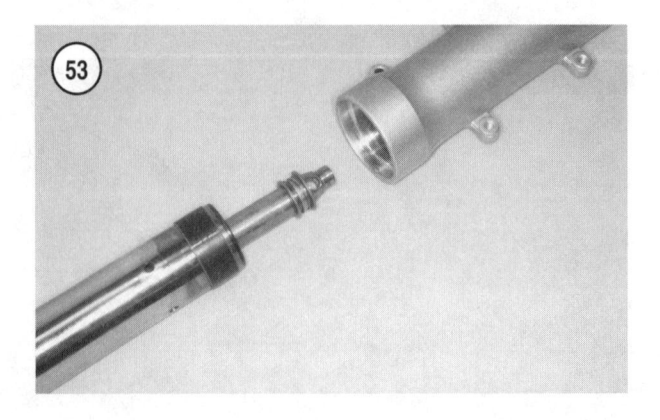

FRONT SUSPENSION AND STEERING

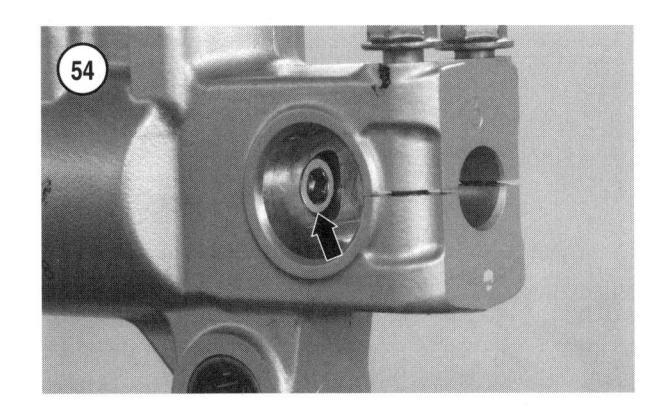

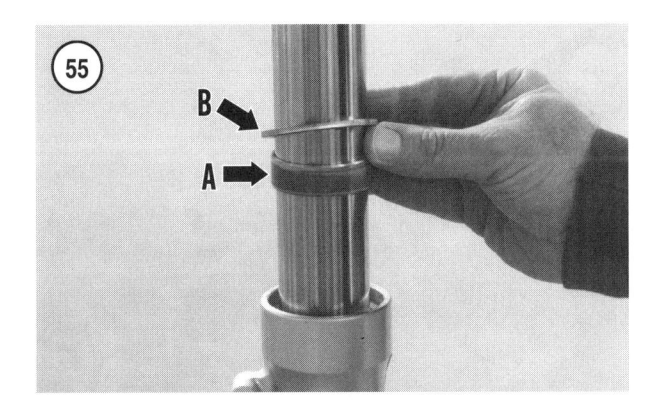

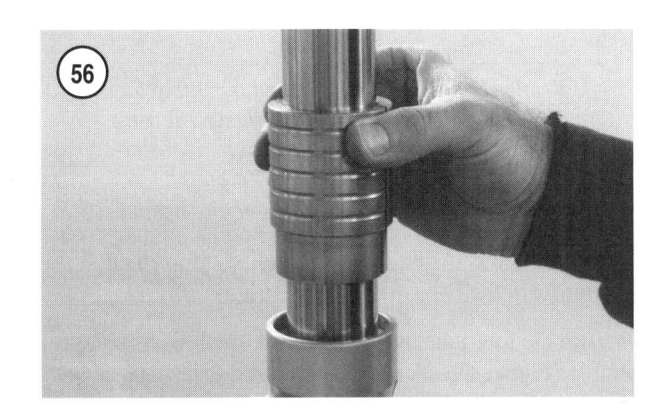

tube, first place a plastic bag over the fork tube and coat it thoroughly with fork oil.

12. Install a new fork seal as follows:
    a. Lubricate the seal lips with fork oil.
    b. Install the seal (**Figure 57**) over the fork tube with its manufacturer's name and size code facing up. Slide it down the fork tube and center it into the top of the slider until its outer surface is flush with the slider's outer bore surface.
    c. Drive the oil seal into the slider with the same tool (**Figure 56**) used in Step 11.
    d. Continue to install the seal until the groove in the slider can be seen above the top surface of the seal.
13. Slide the stopper ring (**Figure 42**) over the fork tube and install it into the groove in the slider. Make sure the stopper ring is completely seated in the slider groove.

NOTE
If the stopper ring cannot seat completely into the slider groove, the seal is not installed far enough into the slider.

14. Slide the dust seal down the fork tube and seat it into the slider (**Figure 58**).
15. Fill the fork with oil and set the oil level as described under *Fork Oil Adjustment* in this section.
16. Install the fork spring with the closely spaced coil side (**Figure 59**) facing down.
17. Install the spring seat (A, **Figure 60**) and spacer (B).
18. Install a new O-ring onto the fork cap, if needed.
19. Lubricate the fork cap O-ring with fork oil. Install the fork cap (**Figure 39**) and tighten hand-tight.

NOTE
The fork cap will be tightened completely after the fork tube is installed onto the motorcycle.

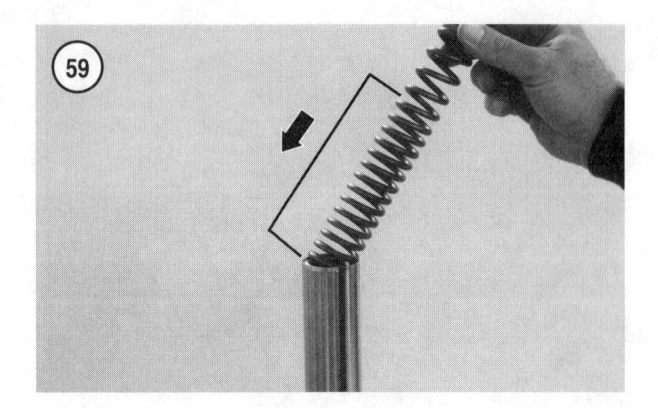

20. Install the front fender mounting plate (B, **Figure 36**).
21. Lubricate the needle bearings and seals (A, **Figure 61**) with grease and install the collar (B).

Right Fork Tube
Disassembly

This section describes complete disassembly of the right front fork (**Figure 62**). If only changing the fork oil and/or setting the oil level, begin at Step 1 and follow the required steps listed in the text.

1. Remove the front fender mounting plate (**Figure 63**).
2. Bolt a flat metal plate onto the slider and clamp the metal plate (**Figure 37**, typical) in a vise to support the fork tube during disassembly and reassembly.

NOTE
If only changing the fork oil and/or setting the oil level, disregard the steps pertaining to loosening the fork tube Allen bolt in Step 3.

NOTE
Disregard Step 3 if the fork tube Allen bolt was loosened during fork removal.

3. Loosen, but do not remove, the fork tube Allen bolt (**Figure 38**).
4. Hold the fork tube and loosen the fork cap (**Figure 64**). Pull the fork cap out of the fork tube along with the piston rod on the fork damper.
5. Hold the locknut with a wrench (**Figure 65**), then turn the fork cap counterclockwise to loosen and remove it from the piston rod. See **Figure 66**.
6. Remove the spacer, spring seat and fork spring (**Figure 67**).
7. Turn the fork tube over a drain pan and pour out the fork oil by operating the fork tube several times. Then pump the piston rod to remove oil from the fork damper.

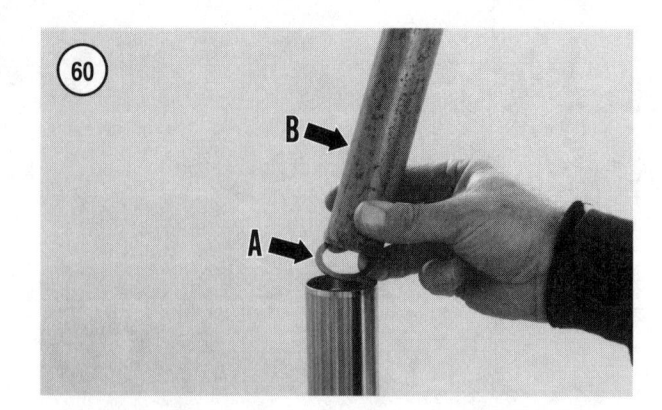

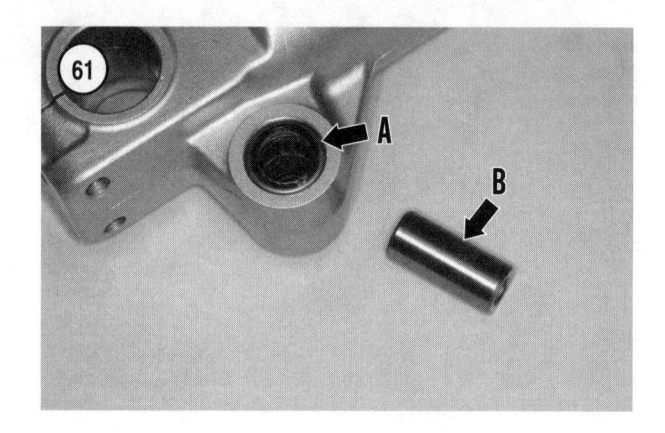

NOTE
*If only changing the fork oil and/or setting the oil level, go to the **Fork Oil Adjustment** procedure at the end of this section. If disassembling the fork, continue with Step 8.*

8. Remove the Allen bolt, previously loosened, and washer from the base of the slider (**Figure 68**).
9. Turn the fork tube over and remove the fork damper assembly (**Figure 69**). Pump the piston rod to remove any remaining oil in the damper.

CAUTION
Do not use excessive force when removing the dust seal in Step 10. Carefully pry the seal out of the slider with a suitable tool. Move the tool around the seal in small increments. It is easy to scratch and damage the slider at the top of the dust seal bore. When selecting a starting point, choose the side facing the wheel.

10. Pry the dust seal (**Figure 70**) out of the slider and remove it.

FRONT SUSPENSTION AND STEERING

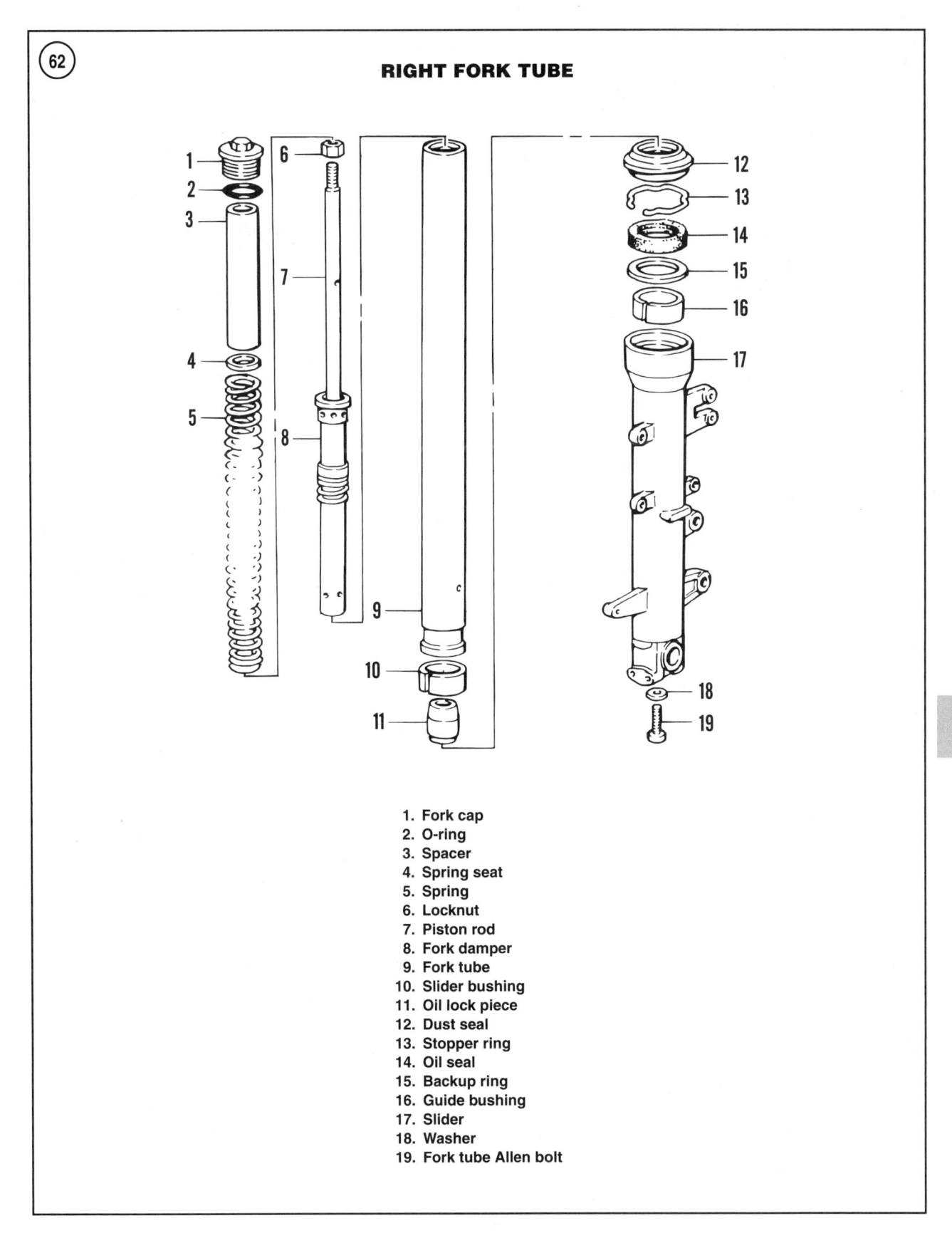

62

RIGHT FORK TUBE

1. Fork cap
2. O-ring
3. Spacer
4. Spring seat
5. Spring
6. Locknut
7. Piston rod
8. Fork damper
9. Fork tube
10. Slider bushing
11. Oil lock piece
12. Dust seal
13. Stopper ring
14. Oil seal
15. Backup ring
16. Guide bushing
17. Slider
18. Washer
19. Fork tube Allen bolt

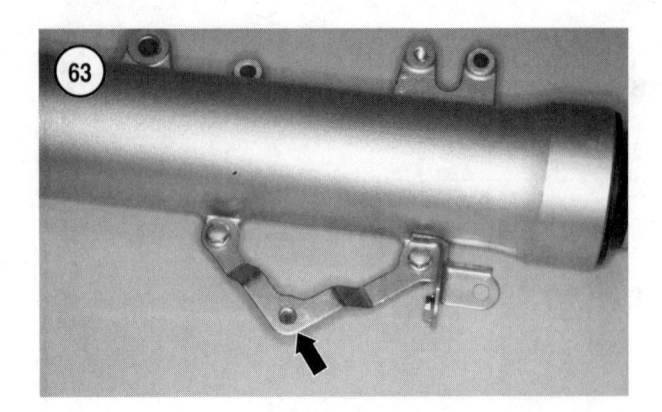

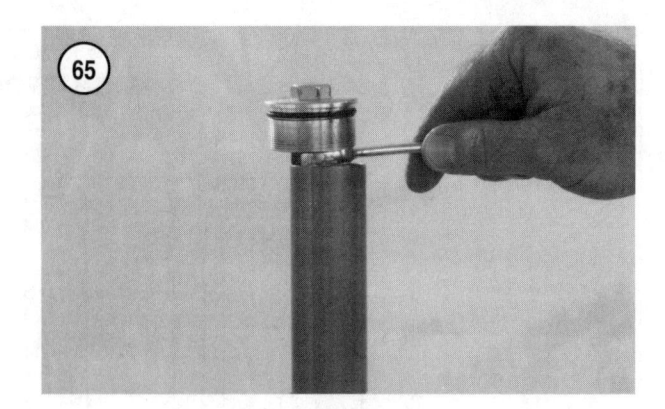

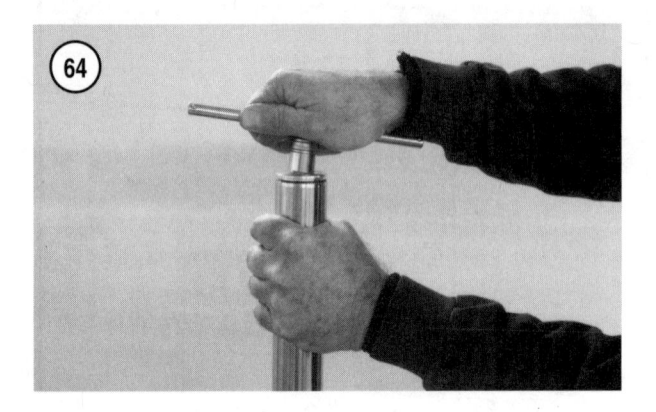

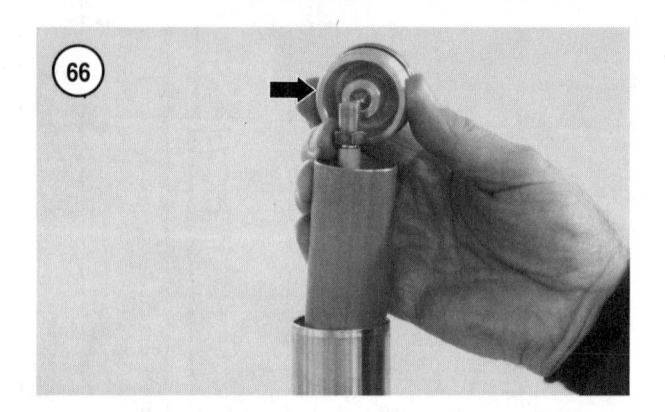

11. Slip the tip of a small screwdriver behind the stopper ring and carefully pry the ring out of the slider groove (**Figure 71**).

NOTE
On this type of fork, a pressed-in bushing in the slider and a bushing on the fork tube keeps the slider and fork tube from separating. To remove the fork tube from the slider, use these parts as a slide hammer as described in Step 12 and Step 13.

12. Hold the slider in one hand and the fork tube in the other hand (**Figure 72**). Pull the fork tube out until resistance is felt. The slider and guide bushings are now contacting each other.

13. Hold the fork tube and pull hard on the slider, using quick in-and-out strokes (**Figure 72**). Doing so withdraws the oil seal, backup ring and guide bushing (A, **Figure 73**) from the slider. The oil lock piece (B, **Figure 73**) is also accessible and can be removed.

14. Remove the slider and pour any remaining oil into an oil pan.

15. Slide the oil seal, backup ring and guide bushing (A, **Figure 73**) off the fork tube.

FRONT SUSPENSION AND STEERING

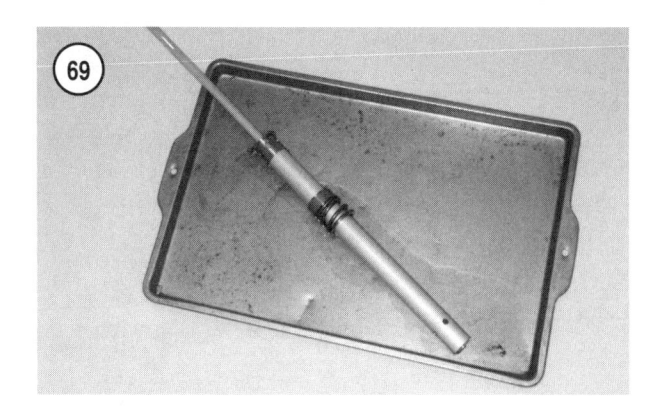

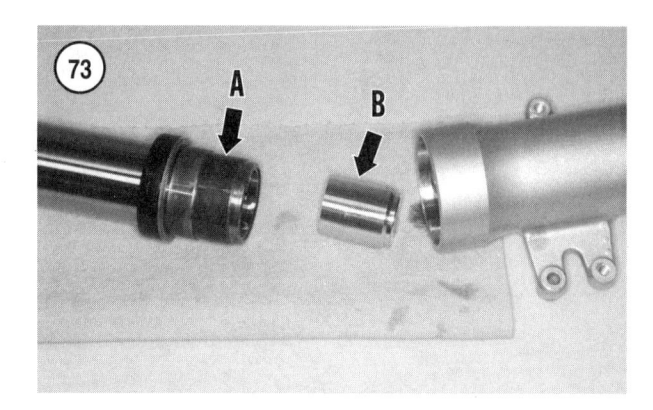

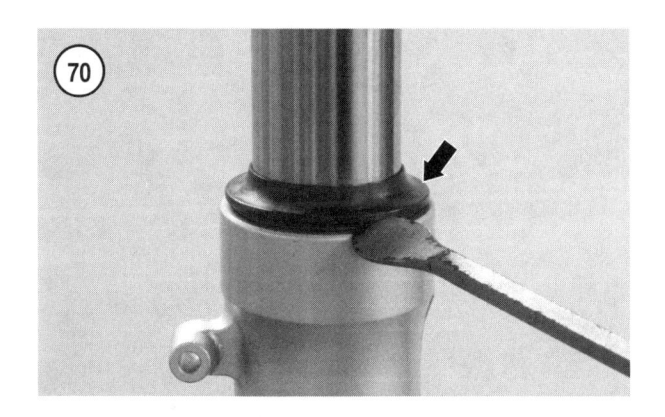

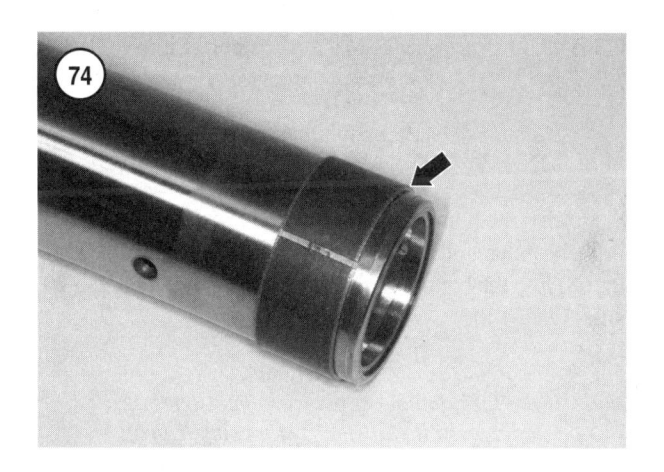

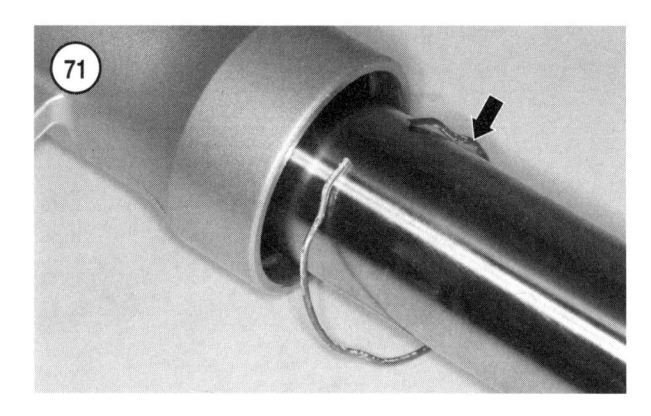

16. Spread the groove in the slider bushing (**Figure 74**) and slide it off the fork tube.
17. Inspect the components as described in this chapter.

Right Fork Tube Assembly

Refer to **Figure 62**.

1. Before assembling the parts, make sure there is no solvent left in the slider or on any part.
2. Coat all parts, except the Allen bolt and fork damper threads, with new fork oil before installation.
3. Install the slider bushing (**Figure 74**) onto the fork tube, if removed.
4. Install the fork damper assembly (**Figure 75**) through the fork tube.
5. Spray a cloth with contact cleaner and use it to clean the fork damper threads of all oil residue (**Figure 76**).
6. Install the oil lock piece (**Figure 77**) onto the end of the fork damper. Then slide the fork tube assembly into the slider. Make sure the oil lock piece seats into the bottom of the slider, and the fork damper seats into the oil lock piece.
7. Mount the slider in a vise (**Figure 37**, typical).

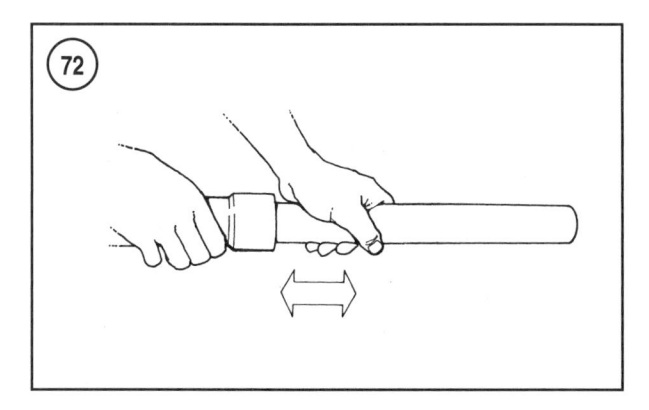

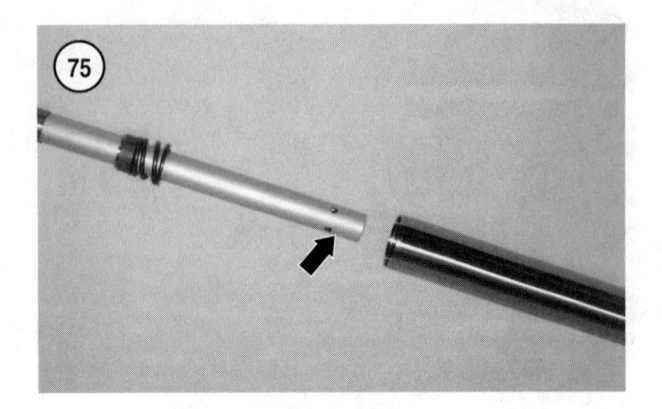

8. Install a new washer onto the Allen bolt (**Figure 68**).

NOTE
Do not use an excessive amount of threadlock in Step 9.

9. Install a medium strength threadlock compound onto the fork tube Allen bolt threads. Then thread the Allen bolt (**Figure 68**) into the bottom of the fork damper and tighten to 20 N•m (15 ft.-lb.).

NOTE
*If the fork damper turns with the Allen bolt, temporarily install the fork spring, spring seat, spacer and fork cap in the order shown in **Figure 62**. During this step, it is unnecessary to thread the fork cap onto the piston rod. Instead, push the piston rod down and bottom it out of the way. Installing these parts applies pressure against the fork damper and prevents it from turning. Remove these parts after tightening the Allen bolt.*

10. Secure the slider in the vise so the fork tube faces straight up.
11. Install the fork slider bushing and backup ring as follows:
  a. Slide the guide bushing (A, **Figure 78**) and backup ring (B) down the fork tube. Install the backup ring with its chamfered side facing down and set it on top of the bushing.
  b. Use a fork seal driver (**Figure 79**) to drive the bushing into the fork slider until it bottoms out in the recess in the slider. The knocking sound made by the driver changes when the bushing bottoms out.

NOTE
*Motion Pro fork seal drivers (**Figure 79**), or equivalent, can be purchased from aftermarket suppliers. To select a driver, first measure the outside diameter of one fork tube.*

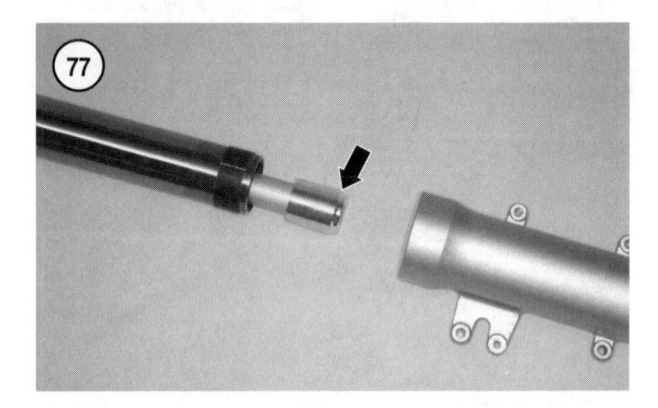

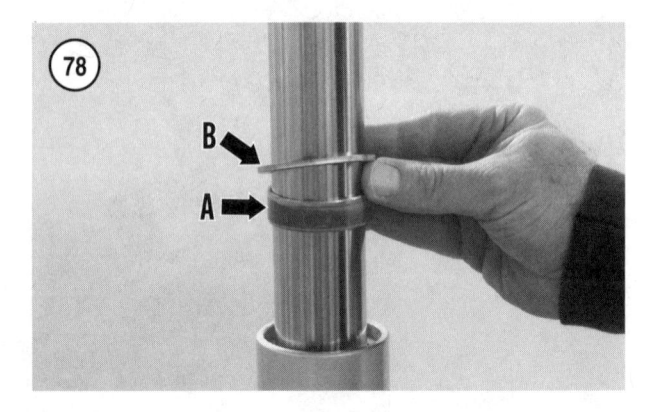

FRONT SUSPENSTION AND STEERING

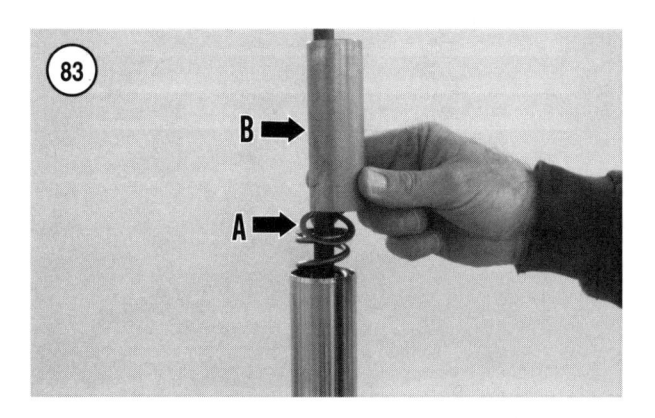

CAUTION
To avoid damaging the fork seal and dust seal when installing them over the top of the fork tube, first place a plastic bag over the fork tube and coat it thoroughly with fork oil.

12. Install a new fork seal as follows:
    a. Lubricate the seal lips with fork oil.
    b. Install the seal (**Figure 80**) over the fork tube with its manufacturer's name and size code facing up. Slide it down the fork tube and center it into the top of the slider until its outer surface is flush with the slider's outer bore surface.
    c. Drive the oil seal into the slider with the same tool (**Figure 79**) used in Step 11.
    d. Continue to install the seal until the groove in the slider can be seen above the top surface of the seal.
13. Slide the stopper ring (**Figure 71**) over the fork tube and install it into the groove in the slider. Make sure the stopper ring is completely seated in the slider groove.

NOTE
If the stopper ring cannot seat completely into the slider groove, the seal is not installed far enough into the slider.

14. Slide the dust seal down the fork tube and seat it into the slider (**Figure 81**).
15. Fill the fork with oil and set the oil level as described under *Fork Oil Adjustment* in this section.

NOTE
If a damper rod bleeding tool was used to bleed the fork, leave it threaded onto the end of the piston rod. Otherwise, tie a length of stiff wire onto the piston rod (around the locknut) to help retrieve the piston rod in the following steps.

16. Install the fork spring with the closely spaced coil side (**Figure 82**) facing down.
17. Install the spring seat (A, **Figure 83**) and spacer (B).
18. Install a new O-ring onto the fork cap, if needed.
19. Lubricate the fork cap O-ring with fork oil.
20. Install and tighten the fork cap as follows:
    a. Raise the piston rod and remove the bleed tool or piece of wire from the piston rod.
    b. Thread the locknut (A, **Figure 84**) all the way down the piston rod.
    c. Thread the fork cap (B, **Figure 84**) onto the piston rod until it bottoms out against the locknut.
    d. Hold the locknut with a wrench (A, **Figure 85**) and tighten the fork cap (B) to 20 N•m (15 ft.-lb.).

21. Thread the fork cap into the fork tube and tighten hand-tight.

NOTE
The fork cap will be tightened completely after the fork tube is installed onto the motorcycle.

22. Install the front fender mounting plate.

Fork Inspection

When measuring the fork components, compare the actual measurements to the specifications in **Table 2**. Replace worn or damaged parts as described in this section.

1. Thoroughly clean all parts, except the fork damper used on the right fork tube (**Figure 86**), in solvent and dry them. Remove all threadlocking compound from the fork damper and Allen bolt threads.

NOTE
Cleaning the fork damper in solvent allows it to absorb some of the solvent, which is difficult to remove. Solvent left in the fork damper will contaminate the fork oil. Instead, wipe the fork damper off with a clean cloth and set aside for inspection and reassembly.

2. Check the fork tube for excessive wear or scratches. Check the chrome for flaking or other damage that could damage the oil seal.
3. Check the fork tube for straightness. Place the fork tube on V-blocks and measure runout with a dial indicator. If the runout is excessive, replace the fork tube.
4. Check the slider for dents or exterior damage. Check the stopper ring groove for cracks or damage. Check the oil seal mounting bore for dents or other damage.

CAUTION
*Do not disassemble or modify the fork damper. Do not remove the piston ring (A, **Figure 87**) or rebound spring (B) from the damper housing. Doing so may damage the fork damper assembly.*

5. Check the right fork damper assembly (**Figure 86**) as follows:
   a. Check the fork damper housing and piston rod for straightness.
   b. Hold the fork damper and operate the piston rod by hand. Make sure there is no binding or roughness, which indicates a bent piston rod or damaged fork damper.

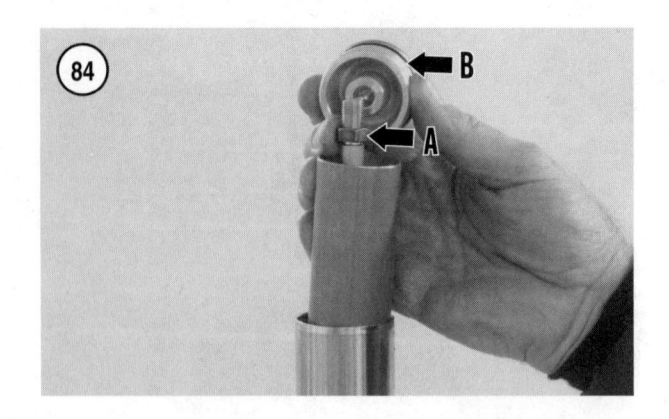

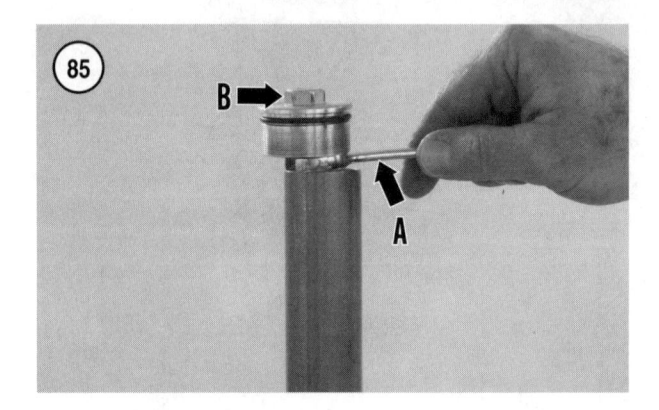

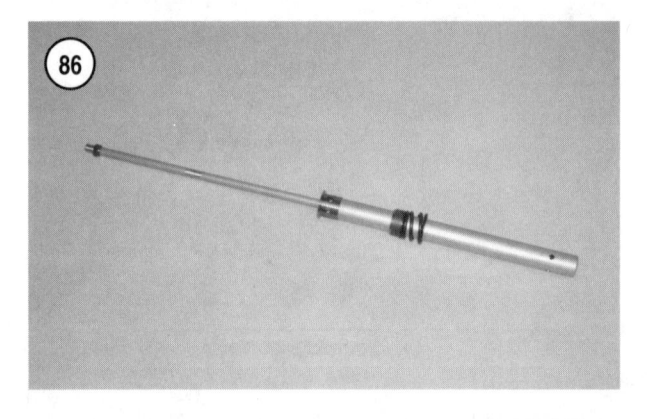

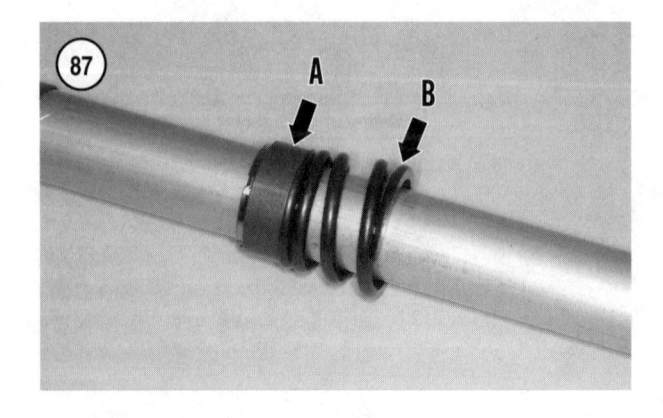

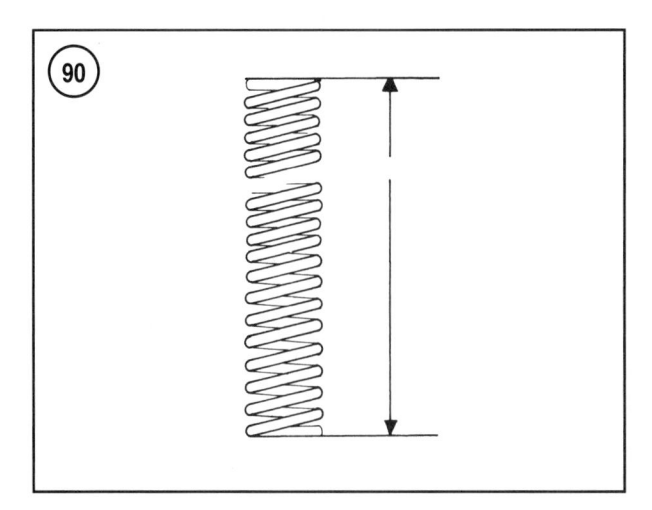

FRONT SUSPENSTION AND STEERING

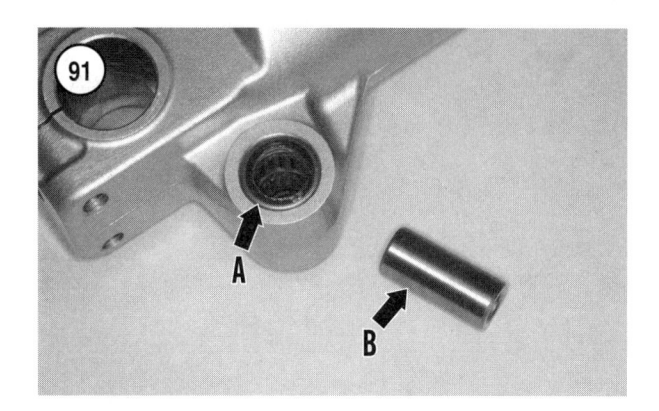

b. Replace the piston rings (**Figure 89**) in the top of the damper if they are damaged or are excessively worn. Replace both piston rings as a set.
c. Check the stop ring grooves in the bottom of the damper rod for cracks or other damage.
d. Inspect the oil lock assembly (**Figure 49**) for worn or damaged parts.

NOTE
The left and right fork tubes use the same fork spring.

7. Measure the free length of the fork spring (**Figure 90**) with a tape measure. Replace the spring if it is worn to the service limit. Replace the left and right side springs if they are unequal in length.
8. Inspect the slider and guide bushings for scoring, excessive wear or damage. Check for discoloration and material coating damage. If the coating is worn off so that the copper base material is showing on approximately 3/4 of the total surface, the bushing is excessively worn. Replace both bushings as a set.
9. Check the backup ring for cracks or distortion.
10. Replace O-rings that are excessively worn, cracked or swollen.
11. Inspect the brake caliper pivot bearings and collar (left fork tube) as follows:
   a. Remove the collar (A, **Figure 91**) and seals (B). Discard the seals.
   b. Clean and dry the collar and bearings.
   c. Inspect the collar for flat spots, cracks and other damage.
   d. Inspect the bearings (**Figure 92**) for visual wear and other damage. Install the collar into the bearings. If the collar turns roughly or is hard to install, replace the bearings and collar as described in the following section.
   e. Install the seals (**Figure 93**) with their flat side facing out. Lubricate the bearings with grease.

c. Check the piston (A, **Figure 87**) and rebound spring (B) for any cracks or other damage. If these parts are damaged, replace the fork rod as an assembly.

6. Inspect the left damper rod assembly (**Figure 88**) as follows:

   a. Inspect the damper rod and spring for cracks and other damage.

Brake Caliper Pivot Bearing Replacement (Left Fork Tube)

1. Remove the left fork tube as described in this chapter.
2. Remove the pivot collar (A, **Figure 91**) and seals (B).
3. Support the fork tube in a press (**Figure 94**) and press out both pivot bearings at the same time.
4. Press in the new bearings until the outer bearing cage on each bearing is 3.5 mm (0.14 in.) below the bore surface.
5. Lubricate the bearing needles with grease.
6. Lubricate the seal lips with grease.
7. Install the seals (**Figure 93**) into the mounting bore with their closed side facing out.
8. Install the pivot collar (B, **Figure 91**).

Fork Oil Adjustment

This section describes steps on filling the fork with oil and setting the oil level. Refer to **Table 2** for the recommended type, quantity and level of fork oil.

1. Remove the fork spring and drain the fork tube as described under *Left Fork Tube Disassembly* or *Right Fork Tube Disassembly* in this chapter.
2. On the right fork tube:
   a. Thread a fork bleeding tool onto the piston rod (**Figure 95**).
   b. If a bleeding tool is unavailable, hook a piece of stiff wire on the piston rod and under the locknut.
   c. Push the piston rod down and bottom out.

NOTE
If the right fork tube was not disassembled, turn the fork tube over and operate the bleeding tool or wire (installed in Step 2) to operate the piston rod and drain the damper housing. Continue until the piston rod moves freely, indicating the damper is free of oil.

3. Push the fork tube down and bottom it out against the slider. Support the slider so it cannot tip over.
4. Slowly pour the fork oil (**Table 2**) into the fork.

NOTE
As oil replaces air during the bleeding procedure (Steps 5-8), the oil level in the fork drops. Continue to add oil to maintain a high oil level in the fork. When bleeding the fork tube, do not be concerned with maintaining or achieving the proper oil capacity. Setting the oil level (Step 9) determines the actual amount of oil used in each fork tube.

5. Right fork tube: Slowly pump the piston rod (**Figure 95**). At first, the piston rod will be easy to move. As oil enters the damper, operating the piston rod will become more difficult. Continue until the tension is constant for each stroke.
6. Hold the slider with one hand and slowly extend the fork tube. Repeat this until the fork tube moves smoothly with the same amount of tension through the compression and rebound travel strokes. Then stop with the fork tube bottomed out.

FRONT SUSPENSTION AND STEERING

95

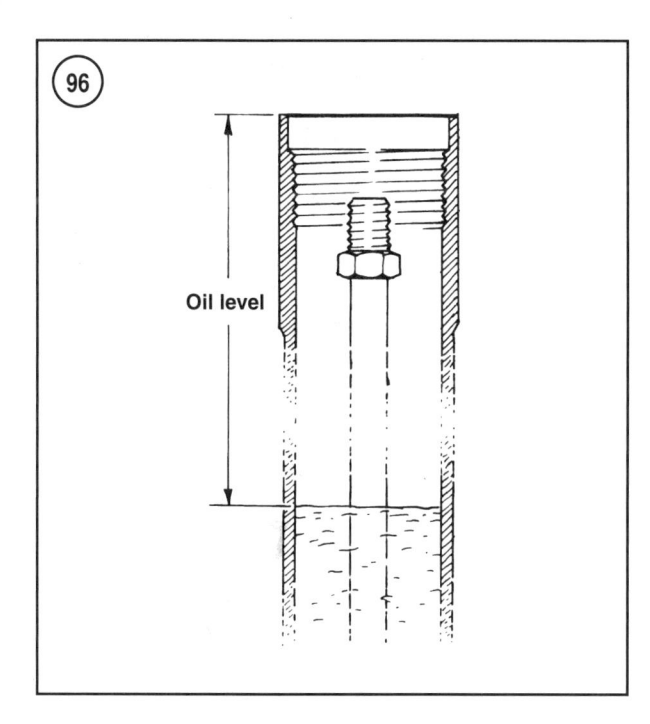

96

Oil level

7. Right fork tube: Bottom the fork tube into the slider. Pump the piston rod approximately 8-10 times, or until the piston rod moves smoothly with noticeable and consistent tension through the compression and rebound travel strokes. Stop with the piston rod at the bottom of its stroke.

8. Set the fork tube aside for approximately five minutes to allow any suspended air bubbles in the oil to surface.

9. Set the oil level (**Figure 96**) as follows:

   a. Make sure the fork tube is bottomed against the slider and placed in a vertical position. On the right fork tube, push the piston rod all the way down.

   b. Use an oil level gauge (**Figure 97**) and set the oil level to the specification.

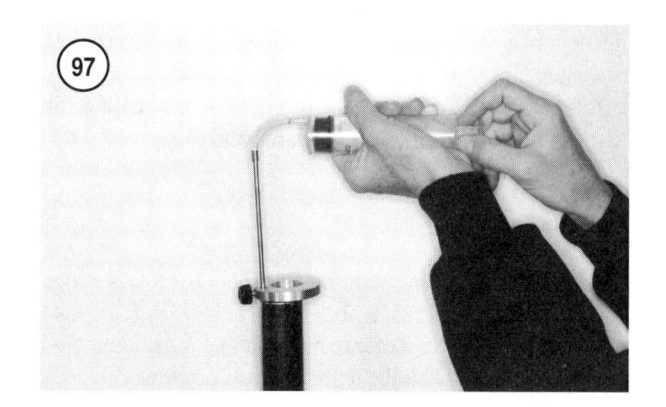

97

NOTE
If no oil is drawn out when setting the oil level, there is not enough oil in the fork tube. Add more oil and reset the level.

   c. Remove the oil level gauge.

10. Leave the bleeding tool or piece of wire on the right fork tube piston rod. These will be used during fork assembly.

11. Complete fork assembly as described under *Left Fork Tube Assembly* or *Right Fork Tube Assembly* in this chapter.

STEERING HEAD AND STEM

The steering head (**Figure 98**) uses retainer-type steel bearings. Each bearing consists of three pieces: upper race, lower race and bearing. The bearings can be lifted out of their operating positions after removing the steering stem. Do not remove the lower inner race (pressed onto the steering stem) or the bearing races (pressed into the frame) unless they are to be replaced.

Regular maintenance consists of steering inspection, adjustment and bearing lubrication. When the steering cannot be adjusted correctly, the bearings may require replacement. However, to determine bearing condition, the steering assembly must be removed and inspected. Inspect the steering adjustment and the bearings at the interval listed in the maintenance schedule (Chapter Three).

Servicing and adjusting the steering stem and bearings is a labor intensive procedure. However, it is critical to ensure correct and satisfactory steering and handling of the motorcycle. This section describes complete service and adjustment procedures for the steering head assembly.

Tools

Use a steering stem socket (Honda part No. 07916-3710100) or equivalent and a spring scale to adjust the steering stem/bearing. These tools are shown in the appropriate procedure.

Refer to *Steering Head Bearing Races* in this chapter for bearing race replacement procedures and tools.

Troubleshooting

Before removing the steering assembly to troubleshoot a steering complaint, refer to *Front Steering and Suspension* in Chapter Two. Refer to the area that most identifies the problem and check the items listed as possible causes.

Removal

Refer to **Figure 98**.
1. Remove the following as described in Chapter Seventeen:
   a. Top shelter.
   b. Meter panel.
   c. Windshield.
   d. Left and right inner fairings.
   e. Front fairing.
2. Remove the combination meter. Refer to *Combination Meter* in Chapter Nine.
3. Remove the top shelter left and right inner covers (**Figure 99**).
4. Secure the motorcycle with the front tire off the ground. Refer to *Bike Stands* in Chapter Ten.
5. Remove the screws and the handlebar center cover (**Figure 100**).
6. Lift the cover (A, **Figure 101**) from the steering nut and disconnect the turn signal cancel connector (**Figure 102**).
7. Remove the two bolts (B, **Figure 101**) and the brake hose and wire harness guide bracket (C) from the upper bridge.

NOTE
To avoid having to remove the master cylinders, cables and wiring harnesses from the handlebars, support each handlebar with some type of vertical stand (camera tripod) like that shown in Figure 103. By using a stand for each handlebar, the handlebars can be moved forward so that their hoses and wiring harnesses do not interfere with the movement of the steering stem assembly. However, because of their routing, the throttle housing and throttle cables must be removed from the right handlebar.

8. Remove the left handlebar mounting bolts (D, **Figure 101**) and position the handlebar so the clutch master cylinder (**Figure 103**) is positioned upright.

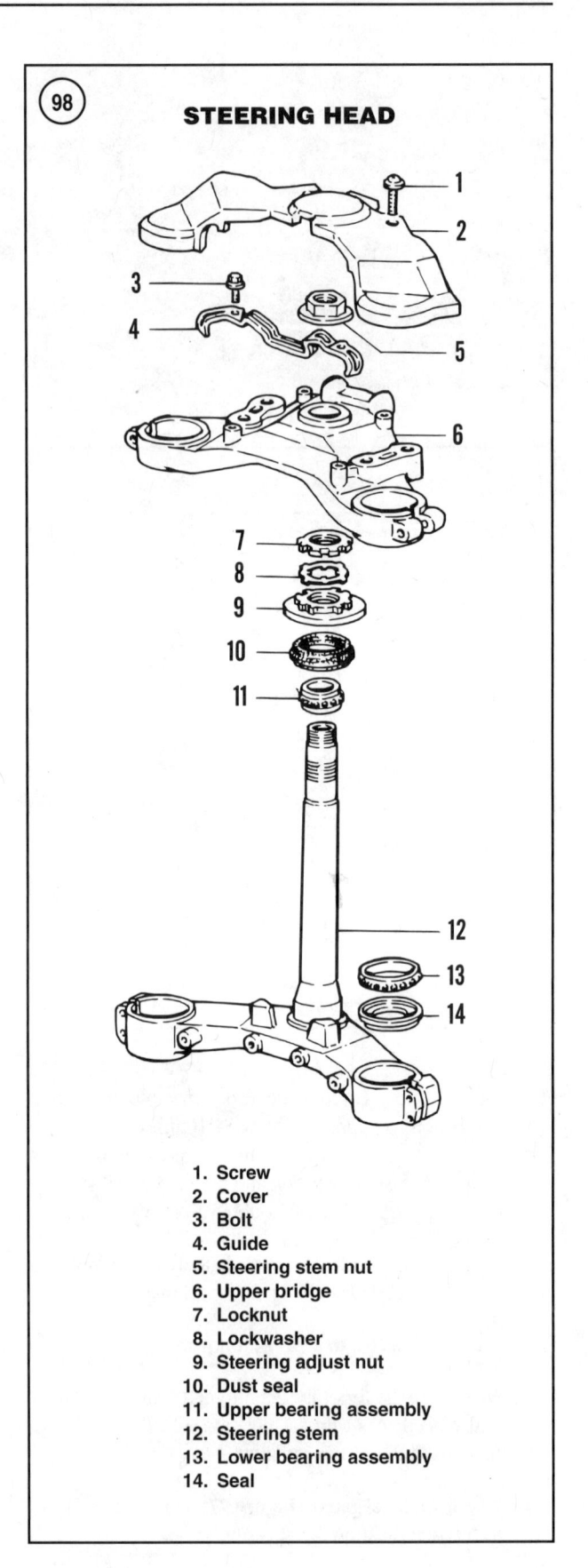

98

STEERING HEAD

1. Screw
2. Cover
3. Bolt
4. Guide
5. Steering stem nut
6. Upper bridge
7. Locknut
8. Lockwasher
9. Steering adjust nut
10. Dust seal
11. Upper bearing assembly
12. Steering stem
13. Lower bearing assembly
14. Seal

FRONT SUSPENSTION AND STEERING

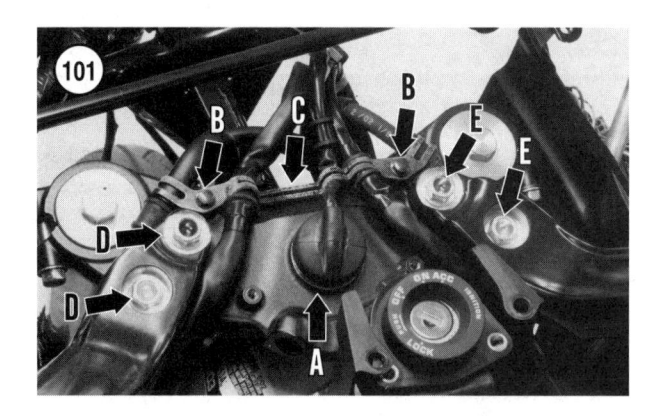

9. Remove the right handlebar and reposition the throttle cables as follows:

NOTE
The throttle cables are routed underneath the steering stem while the brake hoses are routed over the steering stem. Thus, the throttle assembly must be removed from the handlebar before the throttle cables can be rerouted to prevent their interference with the steering stem.

NOTE
*Refer to **Handlebar** in this chapter for photographs and information on removing the right handlebar.*

a. Remove the screw and handlebar weight.
b. Remove the handlebar mounting bolts (E, **Figure 101**).
c. Turn the handlebar over and remove the cover screws and cover.
d. Remove the throttle housing screws.
e. Remove the screw securing the switch housing plate to the housing.
f. Slide the throttle assembly (with the throttle cables attached) off the handlebar.
g. Support the handlebar (**Figure 104**) on a stand to pull the brake hose and wiring harnesses away from the steering stem.
h. Locate the throttle cable guide mounted onto the upper bridge/fork tube pinch bolt. Secure the throttle cables with a zip tie (**Figure 105**), then remove them from the guide.

10. Remove the screws and the air deflector (A, **Figure 106**).

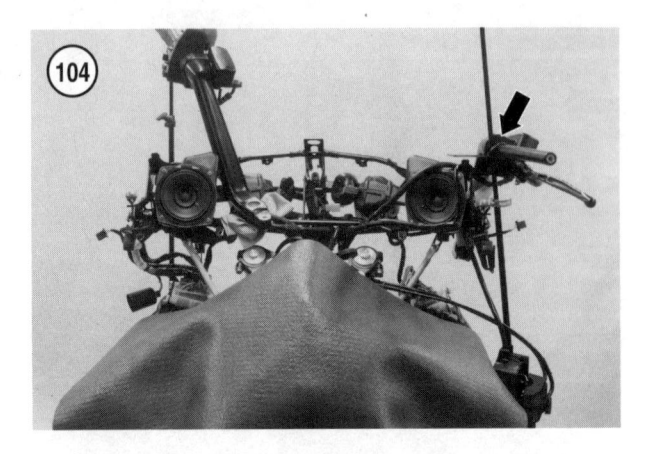

11. Remove the brake hose bracket mounting bolts (B, **Figure 106**) from the lower bridge. Make sure the hoses do not contact the steering bracket.

12. Remove the screw (**Figure 107**) securing the turn signal cancel assembly to the frame.

NOTE
At this point in the procedure, there should be no cables, hoses or wiring harnesses interfering with the movement of the steering stem. Check by turning the steering stem. If so, reposition the item so the steering stem can move with no interference.

13. Before loosening the steering stem nut, check the steering adjustment as described under *Steering Bearing Preload* in this chapter.

14. Loosen, but do not remove, the steering stem nut (A, **Figure 108**).

15. Remove the front wheel. Refer to *Front Wheel* in Chapter Eleven.

16. Remove the front forks. Refer to *Front Fork* in this chapter.

17. Tie the front brake hoses/caliper assembly to the upper fairing bracket to pull the hoses away from the steering stem.

18. Remove the steering stem nut (A, **Figure 108**) and upper bridge (B).

19. Pry the lockwasher tabs (A, **Figure 109**) away from the locknut grooves. Then remove the locknut (B, **Figure 109**) and lockwasher. Install a new lockwasher during reassembly.

NOTE
Before loosening the steering adjust nut, turn the steering stem from lock-to-lock to check the steering adjustment.

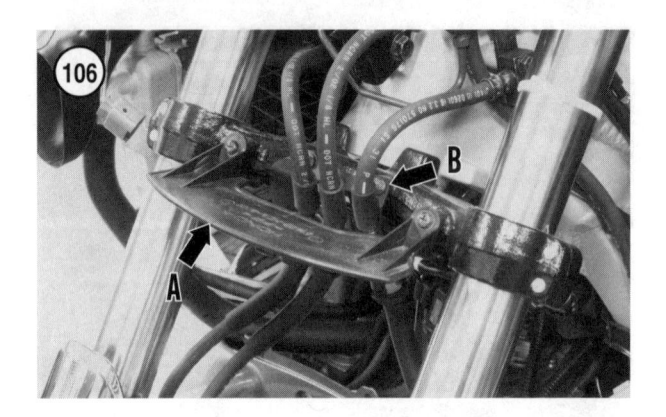

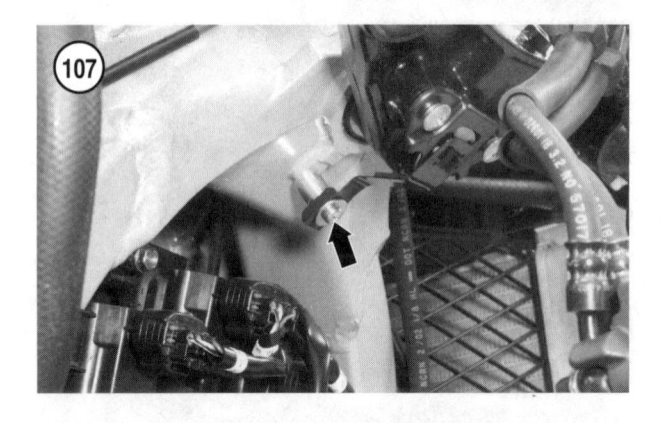

FRONT SUSPENSTION AND STEERING

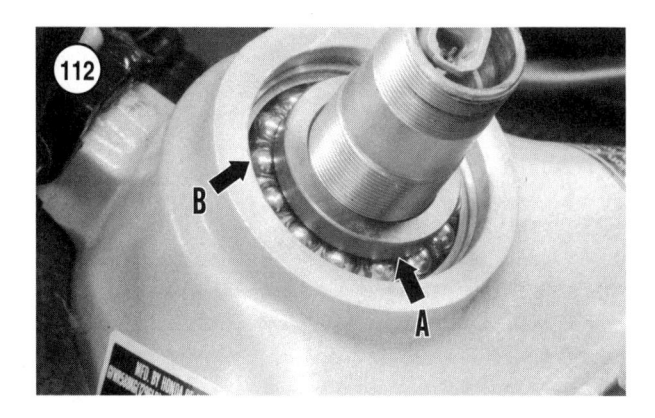

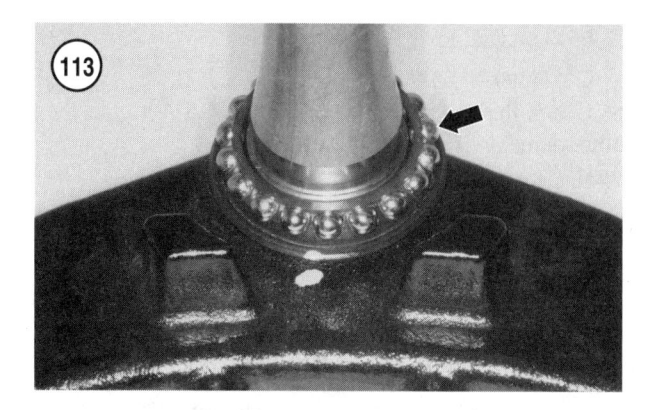

20. Loosen the steering adjust nut with the steering stem socket (A, **Figure 110**).
21. Support the steering stem and remove the following:
    a. Steering adjust nut (B, **Figure 110**).
    b. Dust seal (**Figure 111**).
    c. Bearing race (A, **Figure 112**).
    d. Lower the steering stem and remove it and the lower bearing assembly (**Figure 113**) from the frame.
    e. Upper bearing (B, **Figure 112**).

NOTE
The upper outer race, lower outer race and lower inner race are installed with a press fit. Only remove these parts when replacing the bearing assembly.

Inspection

Replace parts that show excessive wear or damage as described in this section.

WARNING
Improperly repairing damaged frame and steering components can cause steering control loss. If there is apparent frame, steering stem or fork bridge damage, consult with a Honda dealership or qualified frame shop for professional inspection and possible repair.

1. Clean and dry all parts. If necessary, remove the two screws and the turn signal cancel unit (**Figure 114**) from the steering stem.
2. Check the frame for cracks and fractures.
3. Inspect the steering stem nut, locknut, and steering adjust nut for excessive wear or damage.
4. Inspect the upper dust seal for tearing, deterioration or other damage.
5. Check the steering stem (A, **Figure 115**) for:
    a. Cracked or bent stem.

b. Damaged lower bridge.
c. Damaged threads.

6. Check the upper bridge (B, **Figure 115**) for cracks or other damage. Replace if necessary.

7. Inspect the bearing assemblies as follows:
   a. Inspect the bearing races (A, **Figure 116**, typical) for excessive wear, pitting, cracks or other damage. To replace the lower inner bearing race or the outer bearing races, refer to *Steering Head Bearing Races* in this chapter.
   b. Inspect the upper (B, **Figure 116**) and lower (C, **Figure 116**) bearings for dents, pitting, excessive wear, corrosion, retainer damage or discoloration.
   c. Replace the upper and lower bearing assemblies at the same time.

NOTE
Each bearing assembly consists of the bearing and an inner and outer race. Always replace the bearings in upper and lower sets.

8. When reusing bearings, clean them thoroughly with a bearing degreaser and dry thoroughly. Repack each bearing with grease.

Installation/Adjustment

1. If removed, install the turn signal cancel unit (**Figure 114**) and secure with the two mounting screws.

2. Make sure the upper and lower outer bearing races are properly seated in the steering head. Then lubricate each bearing race with grease.

3. Lubricate the lower outer bearing race and dust seal lip with grease.

4. Thoroughly lubricate each bearing (B and C, **Figure 116**) with bearing grease.

5. Install the lower steering bearing (**Figure 113**) onto the lower inner bearing race.

6. Lubricate the upper dust seal lip with grease.

7. Lubricate the upper inner race with grease.

8. Install the steering stem into the steering head and hold in place. Make sure the lower bearing is centered inside the lower outer race.

9. Install the upper bearing and seat it into its outer race (**Figure 117**).

10. Install the upper inner race (A, **Figure 112**) and seat it into the bearing.

11. Install the upper dust seal (**Figure 111**) and seat it over the bearing assembly.

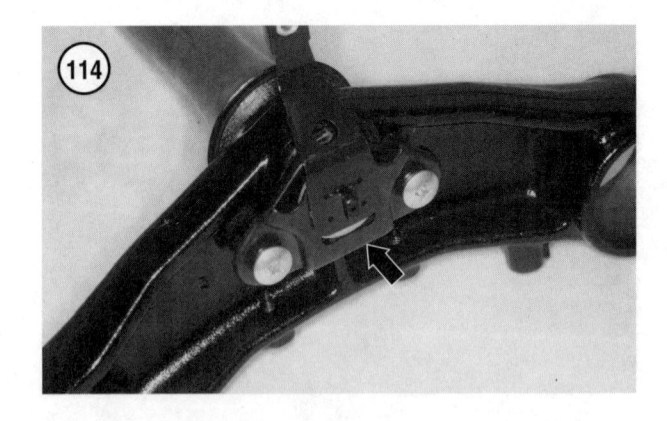

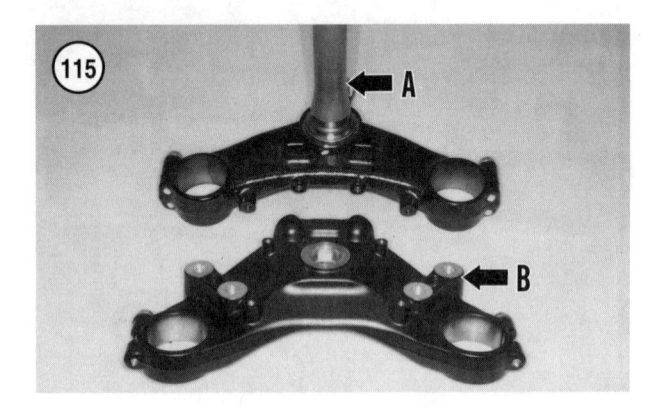

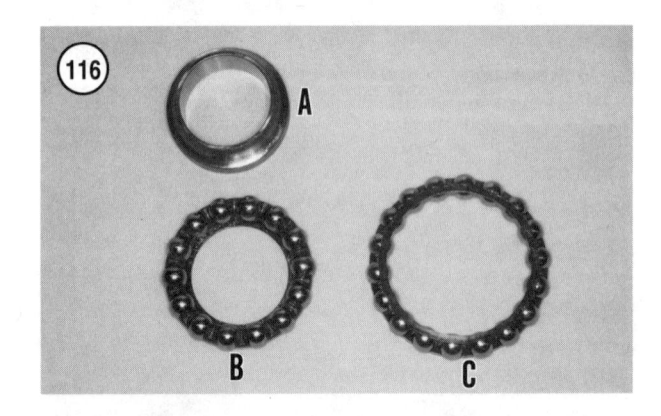

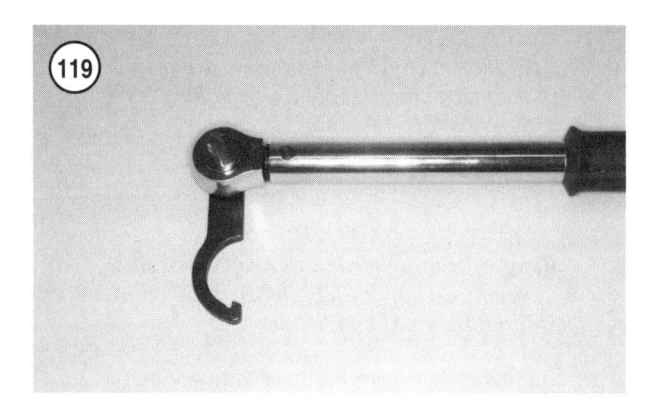

FRONT SUSPENSTION AND STEERING

NOTE
The steering stem, steering stem nut and steering adjust nut threads must be clean to obtain accurate fastener torque and bearing preload adjustment. Remove all dirt, grease or other residue.

12. Lubricate the steering adjust nut threads with oil and thread it onto the steering stem (**Figure 118**). Tighten finger-tight.
13. Tighten the steering adjust nut as follows:
    a. Use the steering stem socket (A, **Figure 110**) to seat the bearings in the following steps. See *Tools* in this section.

NOTE
*If the Honda tool is not available, use a spanner wrench and torque wrench (**Figure 119**) to seat the bearings. Refer to **Torque Wrench Adapters** in Chapter One for information on using these tools.*

   b. Tighten the steering adjust nut (**Figure 118**) to 27 N•m (20 ft.-lb.).
   c. Loosen the steering adjust nut and retighten to 27 N•m (20 ft.-lb.).
   d. Turn the steering stem from lock-to-lock four to ten times to seat the bearings. The steering stem must pivot smoothly. Retighten the steering adjust nut to 27 N•m (20 ft.-lb.).
   e. Repeat substep d.

NOTE
If the steering stem does not pivot smoothly, one or both bearing assemblies may be damaged. Remove the steering stem and inspect the bearings.

NOTE
Do not continue with Step 14 until the steering stem turns correctly. If there is any excessive play or roughness, recheck the steering adjustment.

14. Align the tabs of a *new* lockwasher with the grooves in the steering adjust nut and install the lockwasher (**Figure 120**). The outer two tabs will be bent into the locknut grooves.

CAUTION
Never reinstall a used lockwasher, as the tabs may break off, making the lockwasher ineffective.

15. Install and tighten the locknut (B, **Figure 109**) as follows:
    a. Install the locknut (B, **Figure 109**) and tighten finger-tight.
    b. Hold the steering adjust nut (to keep it from turning) and tighten the locknut approximately 1/4 turn (90°) to align its grooves with the outer lockwasher tabs.
    c. Bend the outer lockwasher tabs (A, **Figure 109**) up into the locknut grooves.
16. Install the upper bridge (B, **Figure 108**) and the steering stem nut (A). Tighten the nut only finger-tight at this time.

17. Install both fork tubes as described in this chapter and tighten the upper and lower pinch bolts hand-tight.
18. Tighten the steering stem nut (A, **Figure 108**) to 103 N•m (76 ft.-lb.).
19. Turn the steering stem from lock-to-lock. Make sure it moves smoothly. There must be no play or binding. Note the following:
    a. If the steering stem turns correctly, continue with Step 20.
    b. If the steering stem is too loose or tight, remove the steering stem nut, fork tubes and upper bridge. Then readjust the steering adjust nut. Repeat until the steering play feels correct. Damaged bearings and races can also cause tightness.

CAUTION
If the steering adjustment is too loose, the steering becomes unstable and causes front wheel wobble. If the steering adjustment is too tight, the bearings eventually score or notch the races. The steering then becomes sluggish as the damaged bearings and races operate against each other. Both conditions hamper steering performance.

NOTE
Arriving at the proper steering adjustment usually comes down to the feel of the steering stem as it is moved from side-to-side. The number of attempts required to arrive at the correct steering adjustment (feel) can vary considerably.

20. Install the front wheel and the brake calipers as described Chapter Eleven.
21. Perform the *Steering Bearing Preload* procedure in this chapter.

NOTE
The steering bearing preload check measures the amount of weight required to move the steering stem. This check confirms whether the steering adjustment is correct.

22. Install and tighten the turn signal cancel screw (**Figure 107**).
23. Install the brake hose bracket bolts (B, **Figure 106**) and tighten to 12 N•m (106 in.-lb.).
24. Install the air deflector (A, **Figure 106**) around the front brake hoses and secure it with the two screws mounted on the lower bridge. Then install and tighten the upper two mounting screws.
25. Install the left side handlebar and tighten its mounting bolts (D, **Figure 101**) to 26 N•m (19 ft.-lb.).

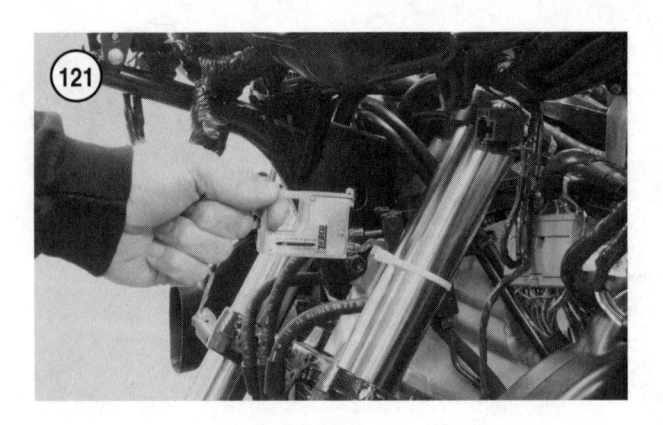

NOTE
*Refer to **Figure 10** for the correct alignment of the hydraulic hoses and wiring harnesses around both handlebars.*

26. Install the right side handlebar:

NOTE
*Refer to **Handlebar** in this chapter for photographs and additional information on installing the right handlebar.*

   a. Cut the cable tie originally used to secure the throttle cables (if used), then install the throttle cables into the guide on the right fork tube (**Figure 105**).
   b. Remove the handlebar from the method used to support it.
   c. Attach the throttle cables to the throttle if previously disconnected.
   d. Slide the throttle over the right handlebar.
   e. Align the pin in the upper switch with the hole in the handlebar.
   f. Install the lower switch housing clamp and secure it with the screw. Tighten the screw securely.
   g. Install the upper switch housing and secure it with the two mounting screws. Tighten securely.
   h. Install the handlebar and secure it with its two mounting bolts. Check the brake hose, wiring harness and throttle cable routing when positioning the handlebar. Secure the throttle cables into the handlebar clamp.
   i. Tighten the right side handlebar mounting bolts (E, **Figure 101**) to 26 N•m (19 ft.-lb.).
27. Install the brake hose and wire harness guide bracket (C, **Figure 101**) and secure them with the two mounting bolts (B). Make sure the hoses and wiring harnesses align and fit through the raised notches in the bracket.
28. Reconnect the turn signal connector (**Figure 102**). Then position the cover (A, **Figure 101**) over the steering stem nut.

FRONT SUSPENSTION AND STEERING

122

- Front tube
- Spring scale
- Forward

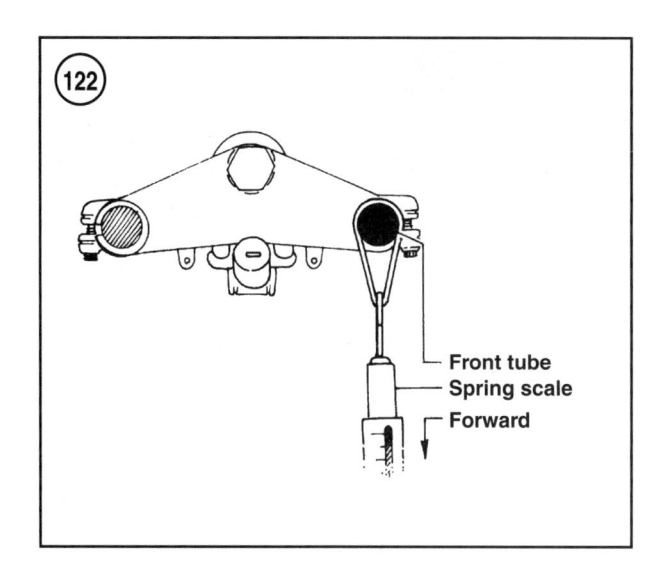

29. Turn the handlebar from side-to-side. Check the cables, hoses and wiring harnesses for proper routing. If there is any binding, check and correct the alignment as required.
30. Install the handlebar center cover (**Figure 100**) and tighten the mounting screws securely.
31. Install the top shelter left and right inner covers (**Figure 99**).
32. Install the combination meter. Refer to *Combination meter* in Chapter Nine.
33. Install the following as described in Chapter Seventeen:
    a. Front fairing.
    b. Left and right inner fairings.
    c. Windshield.
    d. Meter panel.
    e. Top shelter.
34. Check that the front and rear brakes work properly.

WARNING
Do not ride the motorcycle until the horn, cables and brakes all work properly.

STEERING BEARING PRELOAD

Proper steering bearing preload is important because it controls bearing play and steering control. If the preload is too loose, excessive play in the steering causes the wheel to wobble. This is noticed as a slight to severe side-to-side movement or oscillation of the handlebars. A wobble may occur at all vehicle speeds, or start and then stop at certain speeds. In all respects, it is a frustrating condition and can be difficult to troubleshoot. If the preload is too tight, the bearings and races will suffer unnecessary wear and cause stiff and uneven steering, requiring the rider to make a greater steering effort when turning the handlebars. Dry or damaged bearings can cause similar conditions.

Check the steering head for looseness at the intervals specified in Chapter Three or whenever the following symptoms or conditions exist:
1. The handlebars vibrate more than normal.
2. The front fork makes a clicking or clunking noise when the front brake is applied.
3. The steering feels tight or slow.
4. The motorcycle does not steer straight on level road surfaces.

Inspection

When installing the steering stem assembly, the steering bearings are preloaded (bearing placed under pressure) by carefully tightening the steering adjust nut and the steering stem nut. To check bearing preload, Honda specifies using a spring scale attached to one of the fork tubes. This method measures the amount of weight required to move the steering stem with the front end assembled and the front wheel off the ground. When measuring bearing preload with a spring scale, the steering stem must be free to turn without interference from cables, hoses or wiring harnesses. On the GL1800, this requires prior removal of the front fairing and handlebars.

Use a spring scale for this procedure.
1. Perform Steps 1-12 under *Steering Head and Stem Removal* in this chapter.

NOTE
Perform this procedure with the front fork and wheel mounted on the motorcycle.

2. Support the motorcycle with the front wheel off the ground.
3. Turn the steering stem from side-to-side. There should be no interference from a cable or hose when rotating the steering stem. If there still is interference, reposition or remove the affecting part as required.
4. Attach a plastic tie onto one of the fork tubes (between the fork bridges). Then attach a spring scale onto the plastic tie (**Figure 121**).
5. Center the wheel. Position the spring scale at a 90° angle with the steering stem (**Figure 122**). Pull the spring scale and note the reading on the scale when the steering stem begins to turn. This reading is steering stem bearing preload. See **Table 1** for the correct steering preload reading.
6. If the preload reading is incorrect, adjust the steering assembly as described under *Steering Head and Stem* in this

chapter. Perform the adjustment with the front fork and front wheel mounted on the motorcycle.

7. When the bearing preload is correct, perform Steps 22-34 under *Steering Head and Stem, Installation/Adjustment* in this chapter.

WARNING
Do not ride the motorcycle until the horn, cables and brakes all work properly.

STEERING HEAD BEARING RACES

Due to the aluminum frame design, a number of special tools are required to remove and install the steering head bearing races. See **Figure 123** (upper) and **Figure 124** (lower). Due to the expense of these tools, consider having a Honda dealership perform the procedure. Replacing the bearing races without the proper tools may damage the frame.

Do not replace the bearing races unless they are worn or damaged. They are easily damaged during removal.

Tools

A hydraulic press and the following Honda tools are required to replace the steering head races.
1. Main bearing driver attachment: 07946-ME90200.
2. Fork seal driver weight: 07947-KA50100.
3. Oil seal driver: 07965-MA60000.
4. Installer shaft: 07VMF-KZ30200.
5. Installer attachment A: 07VMF-MAT0100.
6. Installer attachment B: 07VMF-MCAA100.
7. Remover attachment A: 07VMF-MAT0300.
8. Remover attachment B: 07VMF-MAT0400.

Outer Bearing Race Replacement

Do not remove the upper (**Figure 123**) or lower (**Figure 124**) outer bearing races unless they are going to be replaced. The bearing/race assembly must be replaced in sets.

CAUTION
If there is any binding when removing or installing the bearing races, stop and release all tension from the bearing race. Check the tool alignment to make sure the bearing race is moving evenly in its mounting bore. Otherwise, the bearing race may gouge the frame mounting bore and cause permanent damage.

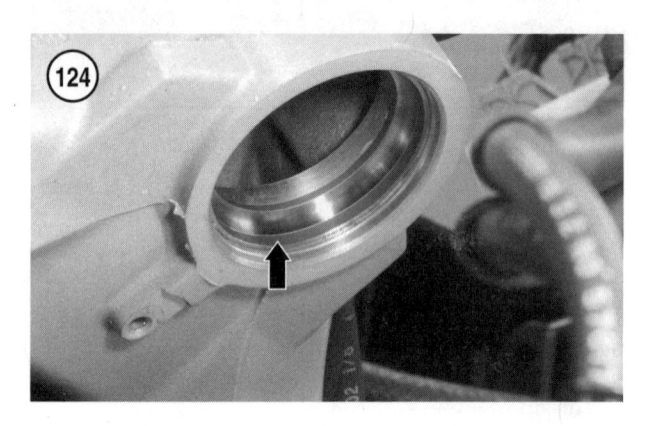

1. To remove the upper outer race (**Figure 123**), do the following:
   a. Assemble the tools onto the steering head as shown in **Figure 125**. Make sure to align remover attachment A with the groove in the steering head.
   b. Hold the end of the installer shaft with a wrench and turn the upper nut slowly to remove the upper outer race.
   c. Disassemble the tools from the steering head and bearing race.
2. To remove the lower outer race (**Figure 124**), do the following:
   a. Assemble the tools onto the steering head as shown in **Figure 126**. Make sure to align remover attachment B with the groove in the steering head.
   b. Hold the end of the installer shaft with a wrench and turn the lower nut slowly to remove the lower outer race.
   c. Disassemble the tools from the steering head and bearing race.
3. Clean each bearing race bore, then check for cracks or other damage.
4. To install the new upper outer race, do the following:
   a. Place the new race squarely into the bore opening with its tapered side facing out. Then assemble the

FRONT SUSPENSTION AND STEERING

125 UPPER BEARING OUTER RACE REMOVAL

- Upper nut
- Main bearing driver attachment
- Fork seal driver weight
- Upper bearing outer race
- Remover attachment A
- Lower nut
- Frame
- Installer shaft
- Installer attachment B

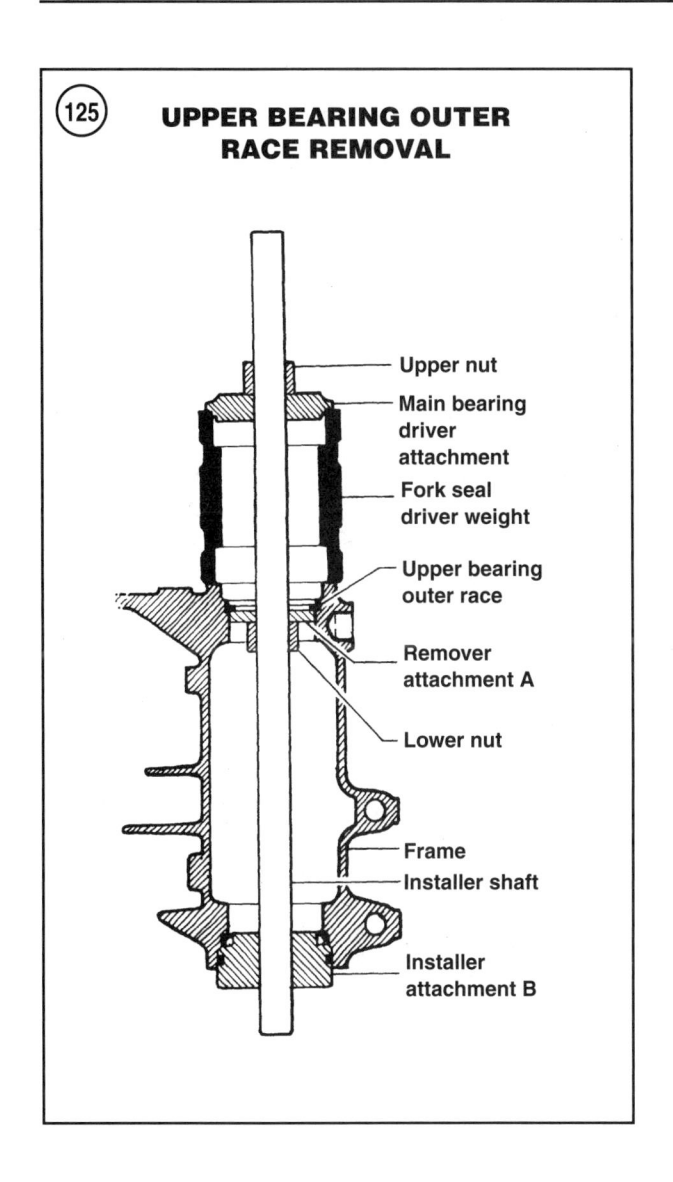

126 LOWER BEARING OUTER RACE REMOVAL

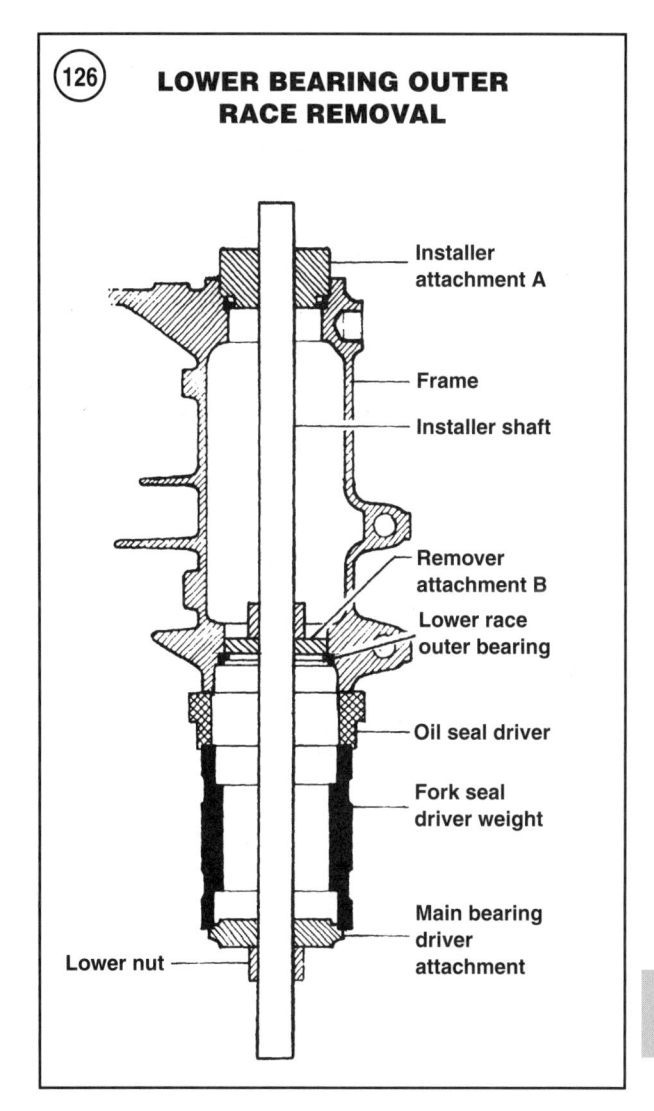

- Installer attachment A
- Frame
- Installer shaft
- Remover attachment B
- Lower race outer bearing
- Oil seal driver
- Fork seal driver weight
- Main bearing driver attachment
- Lower nut

tools through the bearing race and steering head as shown in **Figure 127**. Recheck the bearing race and tool alignment before applying pressure to the race.

  b. Hold the end of the installer shaft with a wrench and slowly turn the lower nut to align the groove in installer attachment A with the upper mounting bore in the steering head. This ensures that the tool and bearing race are square with the steering head mounting bore.
  c. Then hold the end of the installer shaft and turn the lower nut to press the bearing race into the steering head. Continue until the bearing race bottoms in the mounting bore.
  d. Carefully remove the tools from the steering head.
5. To install the new lower outer race, do the following:
  a. Place the new race squarely into the bore opening with its tapered side facing out. Then assemble the

tools through the bearing race and steering head as shown in **Figure 128**. Recheck the bearing race and tool alignment before applying pressure to the race.

  b. Hold the end of the installer shaft with a wrench and slowly turn the upper nut to align the groove in the installer attachment B with the lower mounting bore in the steering head. This ensures that the tool and bearing race are square with the steering head mounting bore.
  c. Then hold the end of the installer shaft and turn the upper nut to press the bearing race into the steering head. Continue until the bearing race bottoms in the mounting bore.
  d. Carefully remove the tools from the steering head.
6. Lubricate the upper (**Figure 123**) and lower (**Figure 124**) bearing races with grease.

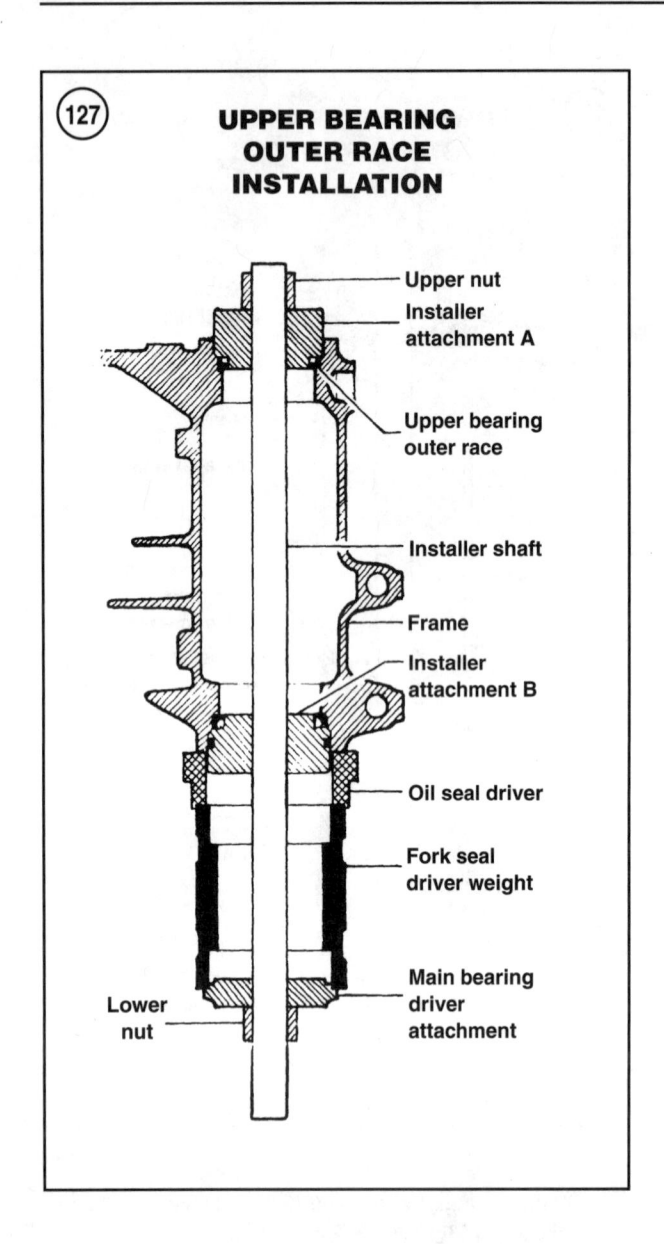

Figure 127 UPPER BEARING OUTER RACE INSTALLATION — Upper nut, Installer attachment A, Upper bearing outer race, Installer shaft, Frame, Installer attachment B, Oil seal driver, Fork seal driver weight, Main bearing driver attachment, Lower nut

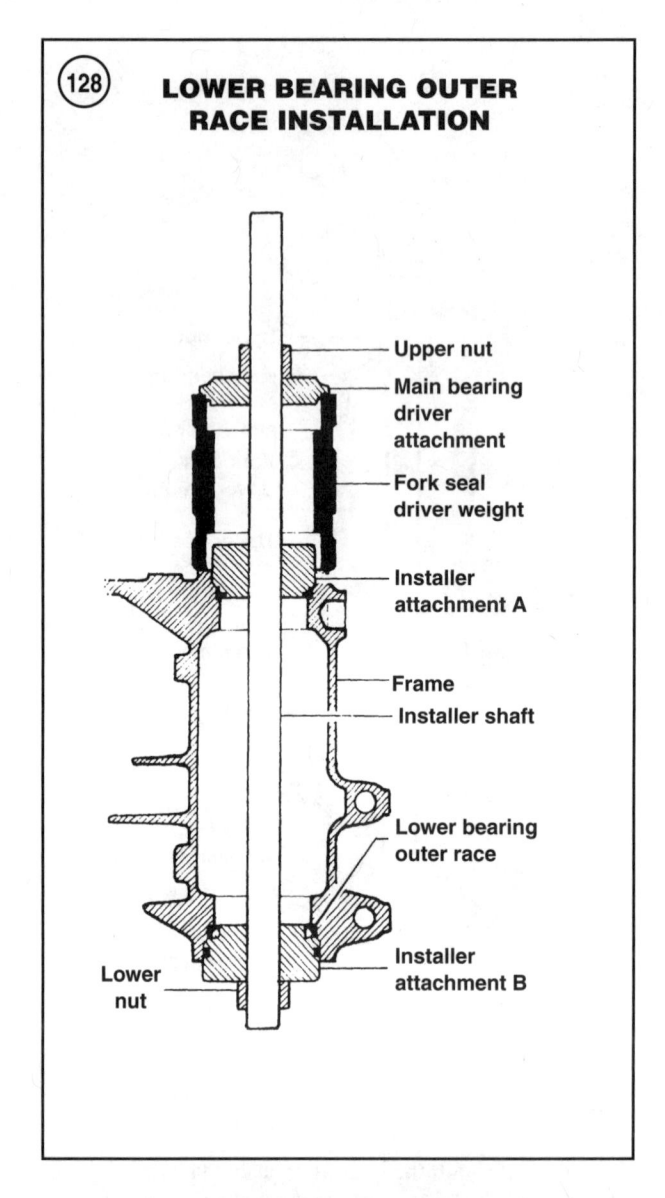

Figure 128 LOWER BEARING OUTER RACE INSTALLATION — Upper nut, Main bearing driver attachment, Fork seal driver weight, Installer attachment A, Frame, Installer shaft, Lower bearing outer race, Installer attachment B, Lower nut

Lower Inner Bearing Race Replacement

The lower inner race (A, **Figure 129**) is a press fit on the steering stem. Replace the lower dust seal (B, **Figure 129**) when replacing the lower inner race.

1. Thread the steering stem nut onto the steering stem (**Figure 130**).

NOTE
Installing the steering stem nut as described in Step 1 helps prevent damaging the steering stem threads when removing the lower inner bearing race.

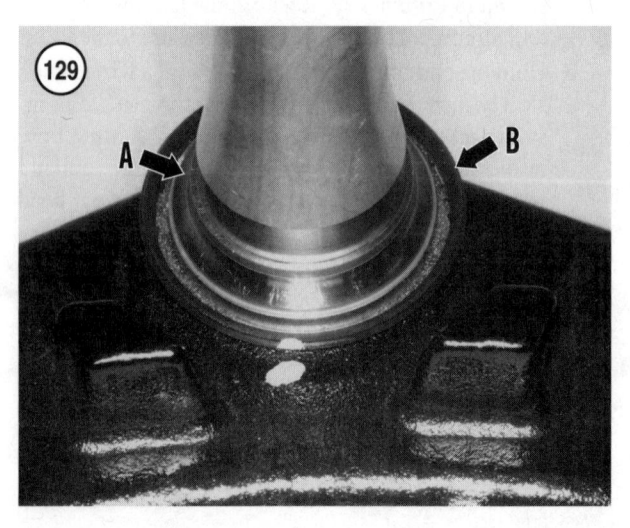

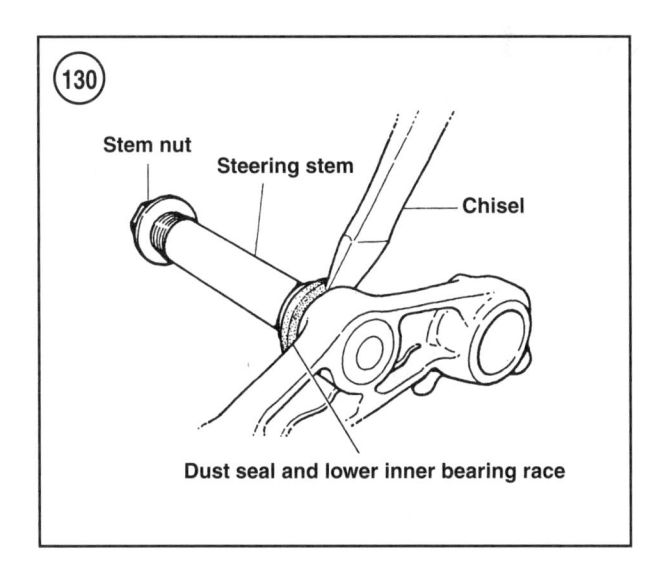

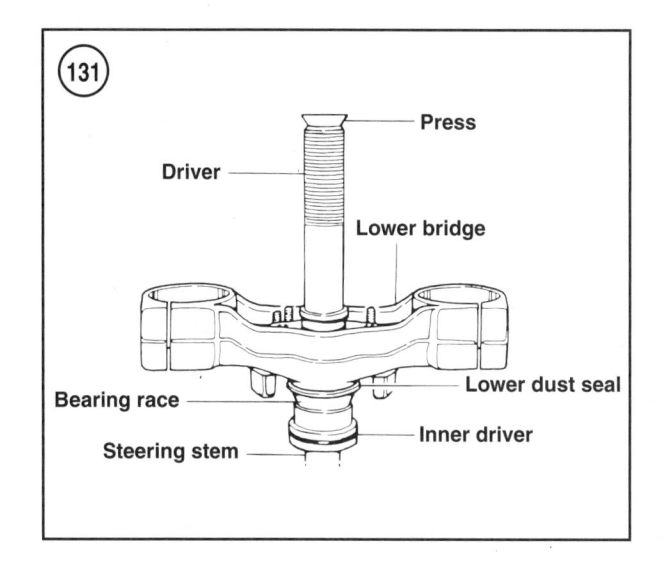

FRONT SUSPENSTION AND STEERING

130
Stem nut, Steering stem, Chisel, Dust seal and lower inner bearing race

131
Press, Driver, Lower bridge, Lower dust seal, Bearing race, Inner driver, Steering stem

WARNING
Striking a chisel with a hammer can cause flying chips. Wear safety glasses in Step 2 to prevent eye injury.

2. Remove the lower inner bearing race and dust seal with a chisel as shown in **Figure 130**. To prevent damaging the steering stem, remove the bearing race evenly by applying pressure against the bearing race a little at a time while working at different points around the bearing.
3. Discard the lower inner bearing race and dust seal.
4. Clean the steering stem with solvent and dry thoroughly.
5. Inspect the steering stem race surface for cracks or other damage. Replace the steering stem if necessary.
6. Install a new lower dust seal over the steering stem.
7. Slide the new lower inner bearing race—bearing surface facing up—onto the steering stem until it stops.
8. Install the steering stem upside down in a press while supporting the lower dust seal and lower inner race as shown in **Figure 131**. Do not allow the bearing driver to contact the bearing race surface. Support the bottom of the steering stem with a bearing driver centered under the press ram.
9. Press the lower inner race onto the steering stem until it bottoms.
10. Remove the steering stem from the press.
11. Lubricate the bearing race with grease.

Table 1 STEERING AND FRONT SUSPENSION SPECIFICATIONS

Front axle travel	122 mm (4.8 in.)
Front axle runout limit	0.20 mm (0.008 in.)
Steering	
Caster angle	29° 15 minutes
Trail length	109 mm (4.3 in.)
Steering stem bearing preload	8.8-13.7 N (0.9-1.4 kg/2.0-3.1 lbs.)

Table 2 FRONT FORK SERVICE SPECIFICATIONS

Fork tube runout limit	0.20 mm (0.008 in.)
Fork oil capacity	
Left fork tube	526.5-531.5 ml (17.82-17.98 U.S. oz.)
Right fork tube	482.5-487.5 ml (16.32-16.48 U.S. oz.)
(continued)	

Table 2 FRONT FORK SERVICE SPECIFICATIONS (continued)

Fork oil level	128 mm (5.0 in.)
Fork oil type	Pro-Honda Suspension Fluid SS-8 or equivalent 10 wt. fork oil
Spring free length	
New	335.3 mm (13.20 in.)
Service limit	328.6 mm (12.94 in.)

Table 3 FRONT SUSPENSION AND STEERING TORQUE SPECIFICATIONS

	N•m	in.-lb.	ft.-lb.
Anti-dive plunger case bolts[1]	4	35	–
Brake hose bracket bolts	12	106	–
Brake hose joint	12	106	–
Clutch master cylinder mounting bolts	12	106	–
Delay valve mounting bolts	12	106	–
Fork cap	23	–	17
Fork tube Allen bolt[2]	20	–	15
Fork cap/locknut (right fork)	20	–	15
Fork tube pinch bolts			
Upper	26	–	19
Lower	29	–	22
Front brake master cylinder holder bolts	12	106	–
Handlebar mounting bolts	26	–	19
Handlebar weight mounting screw[1]	10	88	–
Secondary master cylinder mounting bolts	31	–	23
Steering adjust nut[3]	27	–	20
Steering adjust locknut	see text		
Steering stem nut	103	–	76

1. ALOC fastener. Install a new fastener during installation.
2. Apply a medium strength threadlock onto fastener threads.
3. Lubricate threads and seating surface with engine oil.

CHAPTER THIRTEEN

REAR SUSPENSION AND FINAL DRIVE

This chapter describes repair and replacement procedures for the rear suspension and final drive components. Refer to Chapter Eleven for rear wheel and tire service.

Tables 1-3 are at the end of this chapter.

WARNING
Replace all fasteners used on the rear suspension and final drive components with parts of the same type. Do not use a replacement part of lesser quality or substitute design; this may affect the performance of the system or result in failure of the part and lead to loss of control of the bike. Use the torque specifications listed during installation to ensure proper retention of these parts.

NOTE
The threads on the ALOC fasteners are coated with a threadlock. Honda specifies replacing ALOC fasteners during installation. If a replacement ALOC fastener is unavailable, remove all threadlock from the original fastener and apply a medium strength threadlock to the threads during installation.

NOTE
Many of the nuts used on the rear suspension are U-nuts that use a thin metal insert to prevent the nut from backing off the fastener if it loosens. Replace U-nuts when they are worn out.

SHOCK ABSORBER

A single shock absorber with an attached spring preload actuator is used on all models. An oil hose connects the actuator assembly to the shock absorber (**Figure 1**).

CAUTION
The shock absorber and actuator assembly is not serviceable. There are no replacement parts for the shock absorber or actuator assembly. Do not loosen the oil hose bolts in an attempt to separate the parts.

Spring Preload

Spring preload can be adjusted by operating the rear spring preload adjustment switch mounted on the left side of the top shelter (**Figure 2**). The preload adjustment changes can be read directly on the multi-display as the switch is operated.

The spring preload system can be adjusted to 26 different positions (0-25). The standard position is 0.

Operational requirements

The following conditions must be met before the spring preload system can be operated:
1. The ignition switch must be on or in the ACC position.
2. The motorcycle must be stopped with the transmission in neutral.
3. The reverse system must be turned off.

Adjustment

1. Park the motorcycle on its centerstand.

> *NOTE*
> *Make sure all electrical accessories (including the audio system) are turned off when adjusting the spring preload. Otherwise, the battery can be quickly discharged.*

2. Turn the ignition switch on or to the ACC position.
3. Press and hold the UP or DOWN side of the manual height switch (**Figure 2**). The electric motor will operate as the different spring preload settings are flashed on the multi-display (**Figure 3**, typical). Continue until the desired preload setting is reached (0-25). Note the following:
   a. When the spring preload adjustment system is working correctly, the different adjustment positions are displayed on the multi-display and the electric motor will operate.
   b. If the multi-display reads SUS ADJ ERROR as the preload adjust switch is being operated, a problem has occurred and the spring preload adjust system will not operate. Refer to *Troubleshooting* in this section.

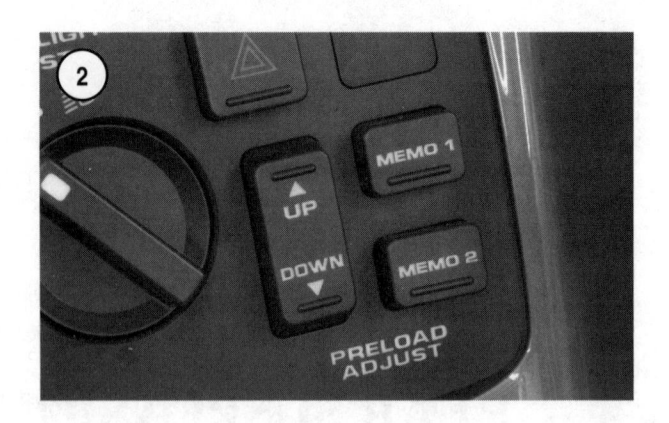

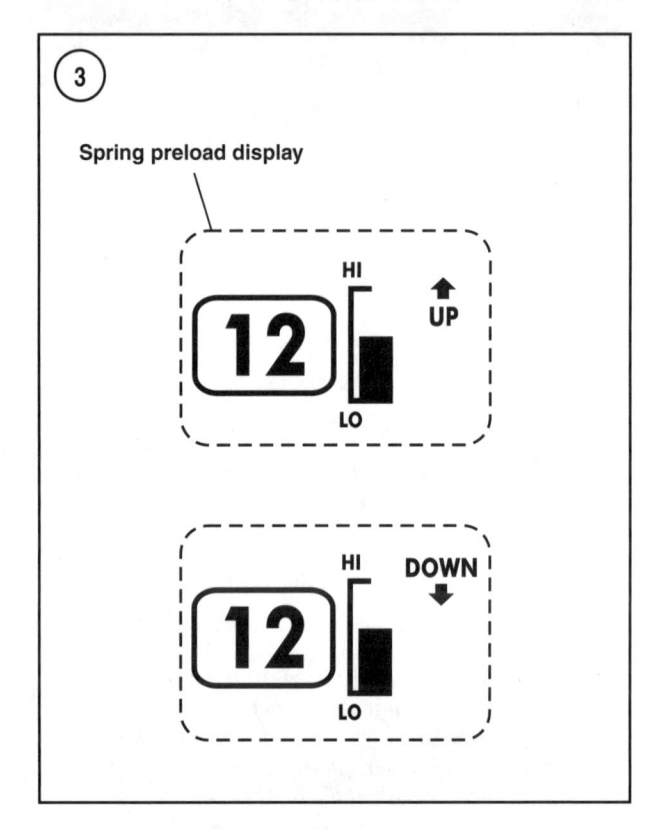

> *NOTE*
> *After making significant spring preload changes to compensate for load changes or trailer use, check the headlight adjustment. Refer to **Headlight Adjustment** in Chapter Nine.*

Storing and retrieving preload settings

Two spring preload adjustment settings can be stored into memory by operating the MEMO1 and MEMO2 buttons on the panel switch (**Figure 2**). Perform the following:
1. Park the motorcycle on its centerstand.

REAR SUSPENSION AND FINAL DRIVE

④ SPRING PRELOAD ADJUSTMENT TROUBLESHOOTING

Problem/Symptom	Diagnostic flowchart
Suspension level actuator does not operate	Figure 5
Suspension level cannot be memorized with either memory switch	Figure 6
Suspension level does not change, even though actuator operates and multi-display shows that the pre-load position is changing	Figure 7

NOTE
Make sure all electrical accessories (including the audio system) are turned off when adjusting the spring preload. Otherwise, the battery can be quickly discharged.

2. Turn the ignition switch on or to the ACC position.
3. Adjust the spring preload to the desired position.
4. Press and hold the MEMO1 or MEMO2 button. STORE MEMO 1 or STORE MEMO 2 will appear beside the numbered setting and blink on the multi-display. When the blinking stops, the position is stored in the ECU memory and the multi-display will read MEMO 1 or MEMO 2, then turn off.
5. To retrieve a setting from memory, push and release the MEMO1 or MEMO2 button. The CALL MEMO 1 or CALL MEMO 2 displays and blinks. The motor will operate as the preload setting is changed to the memory setting. When the memory setting is reached, the motor will stop and the display will read MEMO 1, then turn off.

NOTE
A dead battery or disconnecting the battery terminals erases the memory readings.

Troubleshooting

When there is a problem in the spring preload adjust system, the multi-display reads SUS ADJ ERROR as the preload adjust switch is being operated. Mechanical problems, such as a leaking adjust hydraulic system, can occur and may not trigger the error reading on the multi-display.

Figure 4 lists particular symptom-based problems. Find the complaint that most matches the condition, then turn to the indicated diagnostic flowcharts (**Figures 5-7**) and perform the relevant procedures until the problem is resolved.

Refer to the wiring diagrams at the end of this manual to identify connectors and components in the system.
Before troubleshooting the spring preload adjustment system, check the following:
1. The battery must be fully charged. Refer to *Battery* in Chapter Nine.
2. Check for a blown No. 20 fuse. Refer to *Fuses* in Chapter Nine.
3. Check that the audio system operates correctly.
4. The neutral indicator must come on when the ignition switch is on with the transmission in neutral.
5. Make sure the reverse shift switch is pushed to the off position and the reverse indicator is not on.

Removal

The shock absorber and actuator assembly (**Figure 1**) are removed as an assembly.

CAUTION
The shock absorber and actuator assembly are not serviceable. There are no replacement parts for the shock absorber or actuator assembly. Do not loosen the oil hose bolts in an attempt to separate the parts.

1. Support the bike on its centerstand.
2. Remove the right saddlebag. Refer to *Saddlebags* in Chapter Seventeen.
3. Remove the fuel tank. Refer to *Fuel Tank* in Chapter Eight.
4. Remove the rear wheel. Refer to *Rear Wheel* in Chapter Eleven.

Procedure continued on page 499.

⑤ SUSPENSION LEVEL ACTUATOR DOES NOT OPERATE

1. Note the following:
   a. The panel switch assembly is mounted on the left side of the top shelter.
   b. The manual height switch, part of the panel switch assembly, is marked with UP and DOWN buttons.
   c. Refer to *Fairing Pockets* in Chapter Seventeen to remove and install the left fairing pocket.
   d. Refer to *Saddlebags* in Chapter Seventeen to remove the right side saddlebag.
   d. The gray 14-pin panel switch connector is located inside the left fairing pocket opening in the top shelter.
   e. Refer to *Panel Switch* in Chapter Nine to replace the panel switch.
   f. Refer to *Combination Meter* in Chapter Nine to access the combination meter electrical connectors.
   g. Refer to *Suspension Level Relays* in this chapter to identify the main, UP and DOWN relays.
   h. A separate 12-volt battery and test leads will be required to check the actuator motor operation.
   i. Refer to Figure 8 to identify and disconnect the gray 3-pin angle sensor connector and gray 2-pin control motor connector.

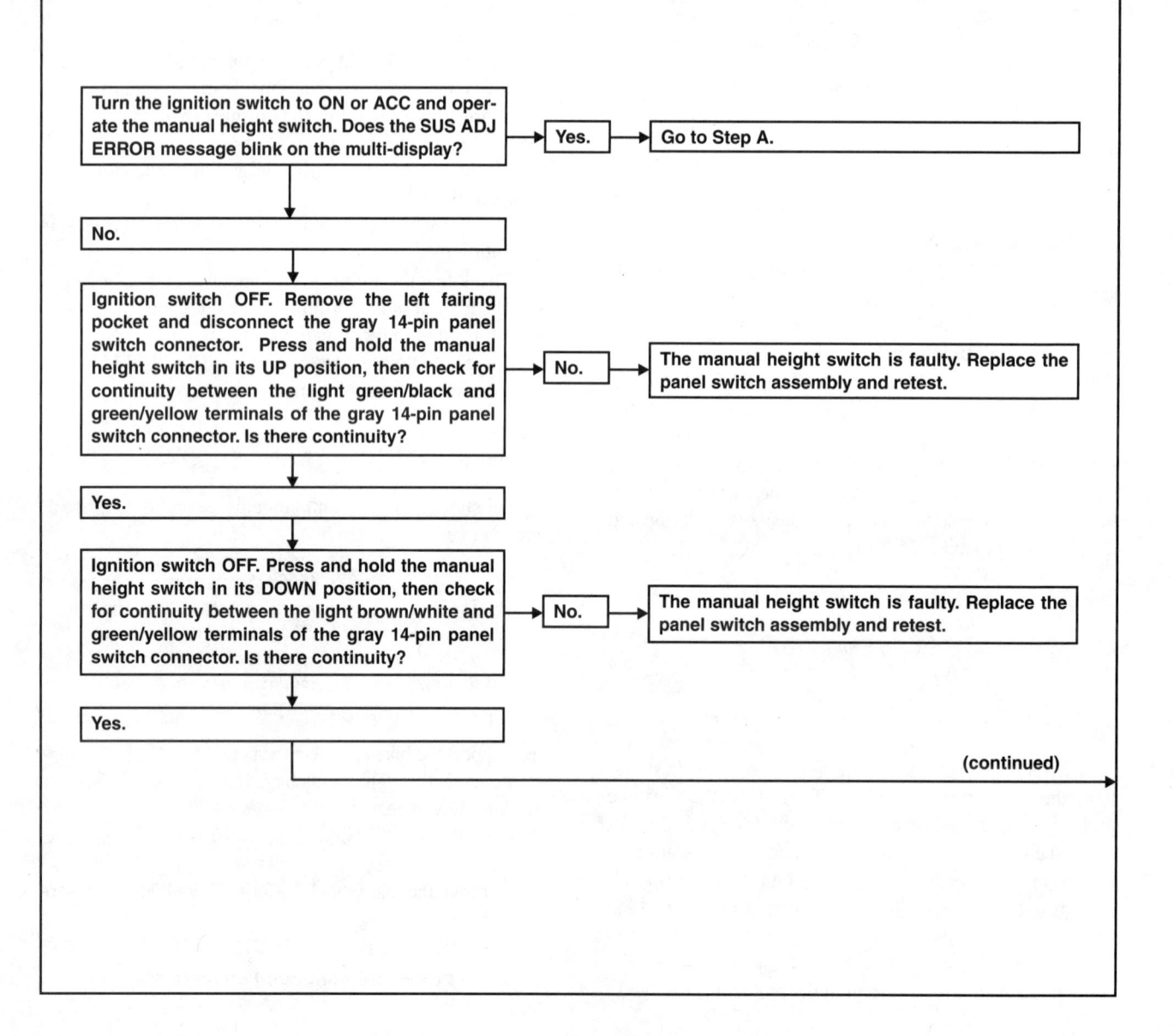

Turn the ignition switch to ON or ACC and operate the manual height switch. Does the SUS ADJ ERROR message blink on the multi-display? → Yes. → Go to Step A.

↓ No.

Ignition switch OFF. Remove the left fairing pocket and disconnect the gray 14-pin panel switch connector. Press and hold the manual height switch in its UP position, then check for continuity between the light green/black and green/yellow terminals of the gray 14-pin panel switch connector. Is there continuity? → No. → The manual height switch is faulty. Replace the panel switch assembly and retest.

↓ Yes.

Ignition switch OFF. Press and hold the manual height switch in its DOWN position, then check for continuity between the light brown/white and green/yellow terminals of the gray 14-pin panel switch connector. Is there continuity? → No. → The manual height switch is faulty. Replace the panel switch assembly and retest.

↓ Yes.

(continued)

REAR SUSPENSION AND FINAL DRIVE

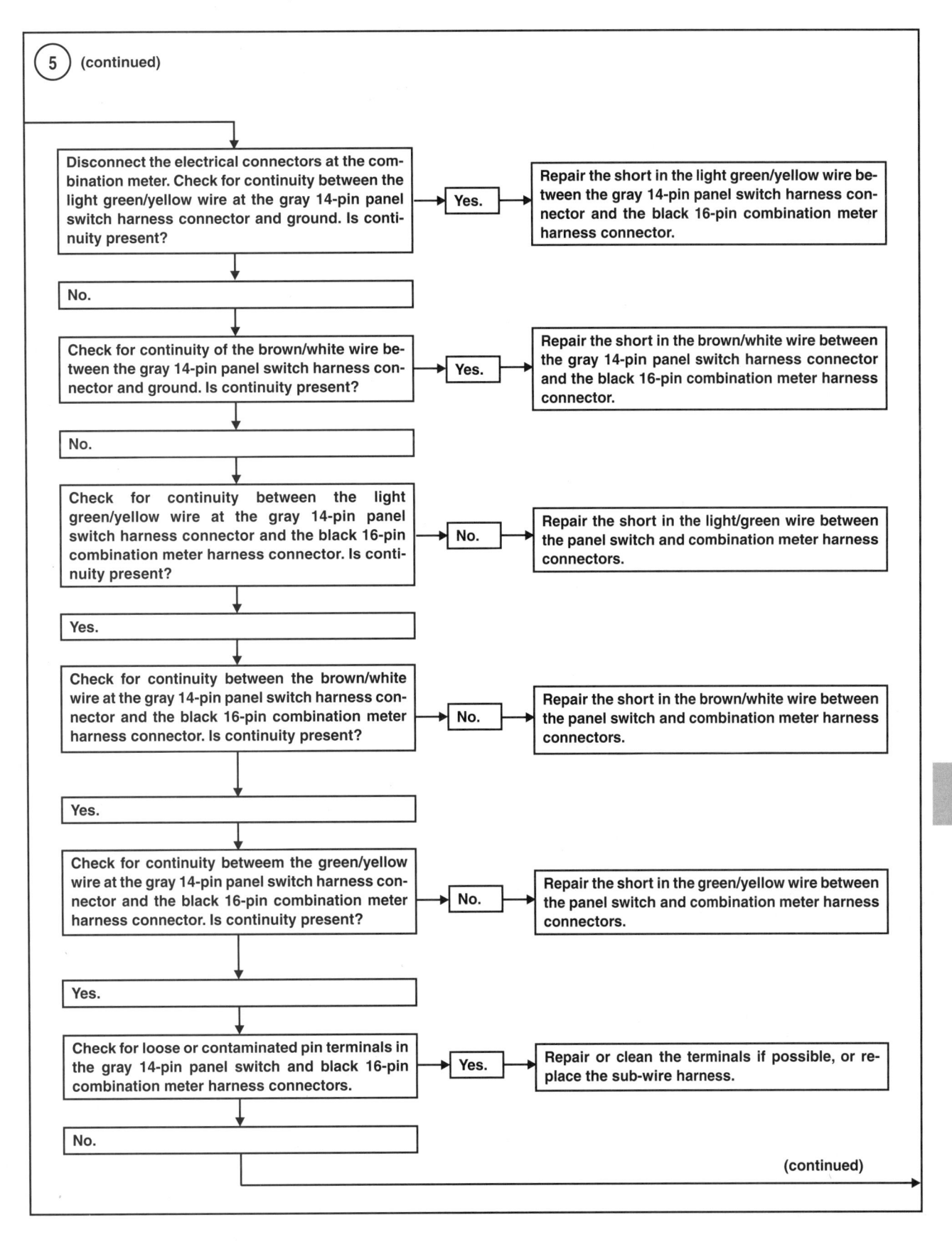

⑤ (continued)

↓

Disconnect the electrical connectors at the combination meter. Check for continuity between the light green/yellow wire at the gray 14-pin panel switch harness connector and ground. Is continuity present? → **Yes.** → Repair the short in the light green/yellow wire between the gray 14-pin panel switch harness connector and the black 16-pin combination meter harness connector.

↓ **No.**

Check for continuity of the brown/white wire between the gray 14-pin panel switch harness connector and ground. Is continuity present? → **Yes.** → Repair the short in the brown/white wire between the gray 14-pin panel switch harness connector and the black 16-pin combination meter harness connector.

↓ **No.**

Check for continuity between the light green/yellow wire at the gray 14-pin panel switch harness connector and the black 16-pin combination meter harness connector. Is continuity present? → **No.** → Repair the short in the light/green wire between the panel switch and combination meter harness connectors.

↓ **Yes.**

Check for continuity between the brown/white wire at the gray 14-pin panel switch harness connector and the black 16-pin combination meter harness connector. Is continuity present? → **No.** → Repair the short in the brown/white wire between the panel switch and combination meter harness connectors.

↓ **Yes.**

Check for continuity beteewm the green/yellow wire at the gray 14-pin panel switch harness connector and the black 16-pin combination meter harness connector. Is continuity present? → **No.** → Repair the short in the green/yellow wire between the panel switch and combination meter harness connectors.

↓ **Yes.**

Check for loose or contaminated pin terminals in the gray 14-pin panel switch and black 16-pin combination meter harness connectors. → **Yes.** → Repair or clean the terminals if possible, or replace the sub-wire harness.

↓ **No.**

(continued)

13

⑤ (continued)

```
↓
[Remove the combination meter and replace the meter/gauge assembly.]
```

STEP A

[Ignition switch OFF. Remove the combination meter. Measure the angle sensor resistance between the black/red and blue/green terminals in the black 16-pin combination meter connector. Is the resistance 4.0-6.0 k ohms?] →No.→ [Go to Step B.]

↓ Yes. Record the actual resistance reading and continue.

[Measure the angle sensor resistance between the yellow/red and blue/green terminals in the black 16-pin combination meter connector. Is the resistance 0.4-5.4 k ohms?] →No.→ [Go to Step C.]

↓ Yes. Record the actual angle sensor resistance reading to determine the appropriate test procedure.

[Remove the suspension level UP and DOWN relays.]

↓

[Was the recorded angle sensor resistance 2.5 k ohms or more?] →No.→ [Was the recorded angle sensor resistance 2.5 k ohms or less?]

↓ Yes. ↓ Yes.

[Test the actuator motor operation with a 12-volt battery as follows: Connect the positive battery (+) terminal to the UP relay connector brown/white relay box terminal and the negative (-) terminal to the DOWN relay connector green wire within 3 seconds. Did the actuator motor operate?] →No.→ [Go to Step E.]

[Test the actuator motor operation with a 12-volt battery as follows: Connect the positive battery (+) terminal to the DOWN relay connector green relay box terminal and the negative (-) terminal to the UP relay connector brown/white wire within 3 seconds. Did the actuator motor operate?]

↓ Yes. ↓ Yes.

[Go to Step D.] ← ← ←

(continued)

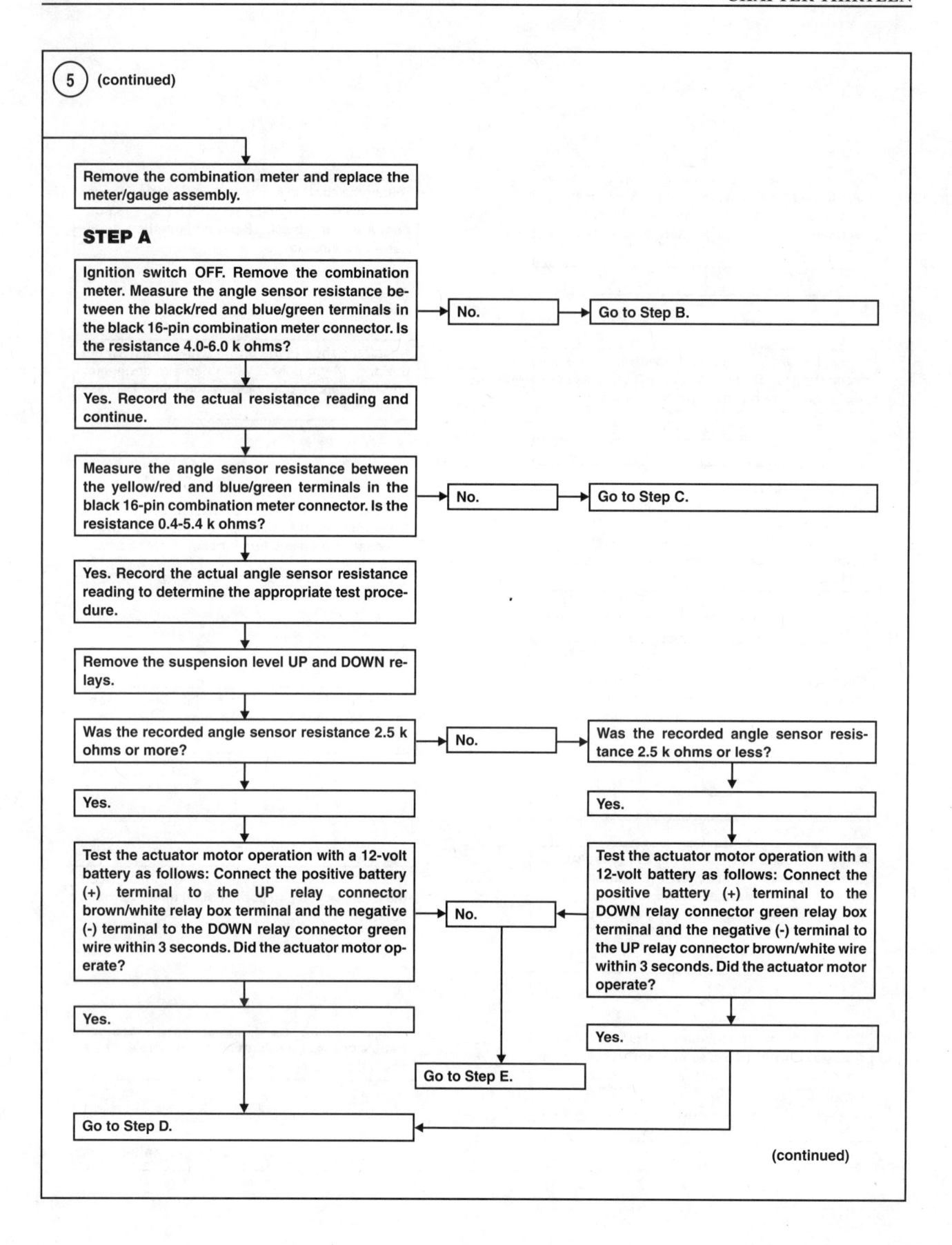

REAR SUSPENSION AND FINAL DRIVE

⑤ (continued)

STEP B

Remove the right saddlebag.

↓

Perform the following:
1. Ignition switch OFF.
2. Disconnect the gray 3-pin angle sensor connector.
3. Measure the resistance between the black/red and blue/green terminals of the gray sensor connector. Is the resistance 4.0-6.0 k ohms?

→ No. → The angle sensor is faulty. Replace the shock absorber assembly.

↓ Yes.

Check the wiring between the angle sensor and combination meter for a loose connector or wire terminal. If good, check for an open or short circuit in the black/red and blue/green wires between the angle sensor and combination meter.

STEP C

Remove the right saddlebag.

↓

Perform the following:
1. Ignition switch OFF.
2. Disconnect the gray 3-pin angle sensor connector.
3. Measure the resistance between the yellow/red and blue/green terminals of the green sensor connector. Is the resistance 4.0-6.0 k ohms?

→ No. → The angle sensor is faulty. Replace the shock absorber assembly.

↓ Yes.

Check the wiring between the angle sensor and combination meter for a loose connector or wire terminal. If good, check for an open or short circuit in the yellow/red wire between the angle sensor and combination meter.

(continued)

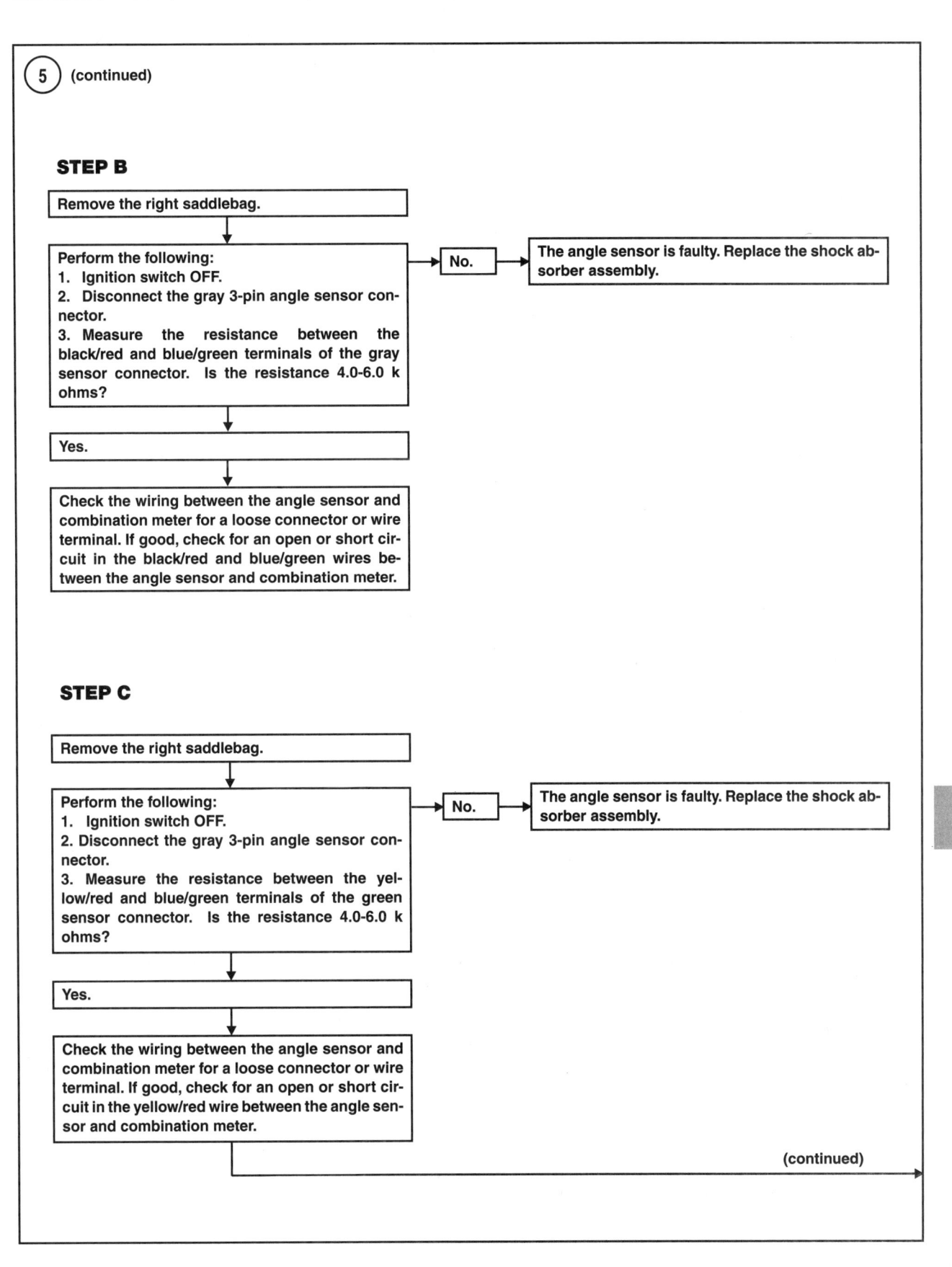

⑤ (continued)

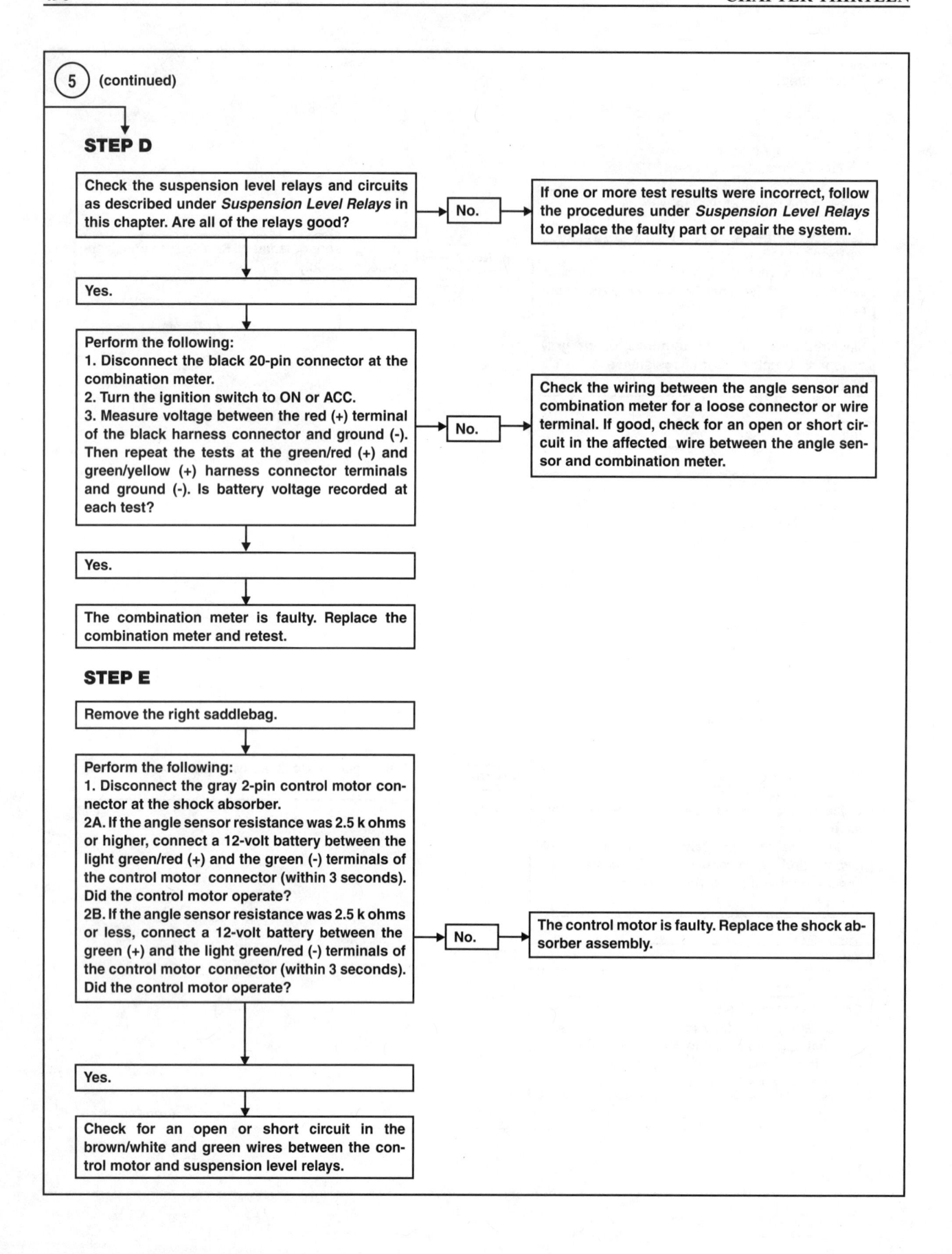

STEP D

Check the suspension level relays and circuits as described under *Suspension Level Relays* in this chapter. Are all of the relays good? → **No.** → If one or more test results were incorrect, follow the procedures under *Suspension Level Relays* to replace the faulty part or repair the system.

↓ Yes.

Perform the following:
1. Disconnect the black 20-pin connector at the combination meter.
2. Turn the ignition switch to ON or ACC.
3. Measure voltage between the red (+) terminal of the black harness connector and ground (-). Then repeat the tests at the green/red (+) and green/yellow (+) harness connector terminals and ground (-). Is battery voltage recorded at each test? → **No.** → Check the wiring between the angle sensor and combination meter for a loose connector or wire terminal. If good, check for an open or short circuit in the affected wire between the angle sensor and combination meter.

↓ Yes.

The combination meter is faulty. Replace the combination meter and retest.

STEP E

Remove the right saddlebag.

Perform the following:
1. Disconnect the gray 2-pin control motor connector at the shock absorber.
2A. If the angle sensor resistance was 2.5 k ohms or higher, connect a 12-volt battery between the light green/red (+) and the green (-) terminals of the control motor connector (within 3 seconds). Did the control motor operate?
2B. If the angle sensor resistance was 2.5 k ohms or less, connect a 12-volt battery between the green (+) and the light green/red (-) terminals of the control motor connector (within 3 seconds). Did the control motor operate? → **No.** → The control motor is faulty. Replace the shock absorber assembly.

↓ Yes.

Check for an open or short circuit in the brown/white and green wires between the control motor and suspension level relays.

REAR SUSPENSION AND FINAL DRIVE

(6)

SUSPENSION LEVEL CANNOT BE MEMORIZED WITH EITHER MEMORY SWITCH

1. Note the following:
   a. The panel switch assembly is mounted on the left side of the top shelter.
   b. The manual height switch, part of the panel switch assembly, is marked with UP and DOWN buttons.
   c. Refer to *Fairing Pockets* in Chapter Seventeen to remove and install the left fairing pocket.
   d. The gray 14-pin panel switch connector is located inside the left fairing pocket opening in the top shelter.
   e. Refer to *Switches* in Chapter Nine to test Memory Switch 1 and Memory Switch 2.
   f. Refer to *Combination Meter* in Chapter Nine to remove and install the combination meter assembly.

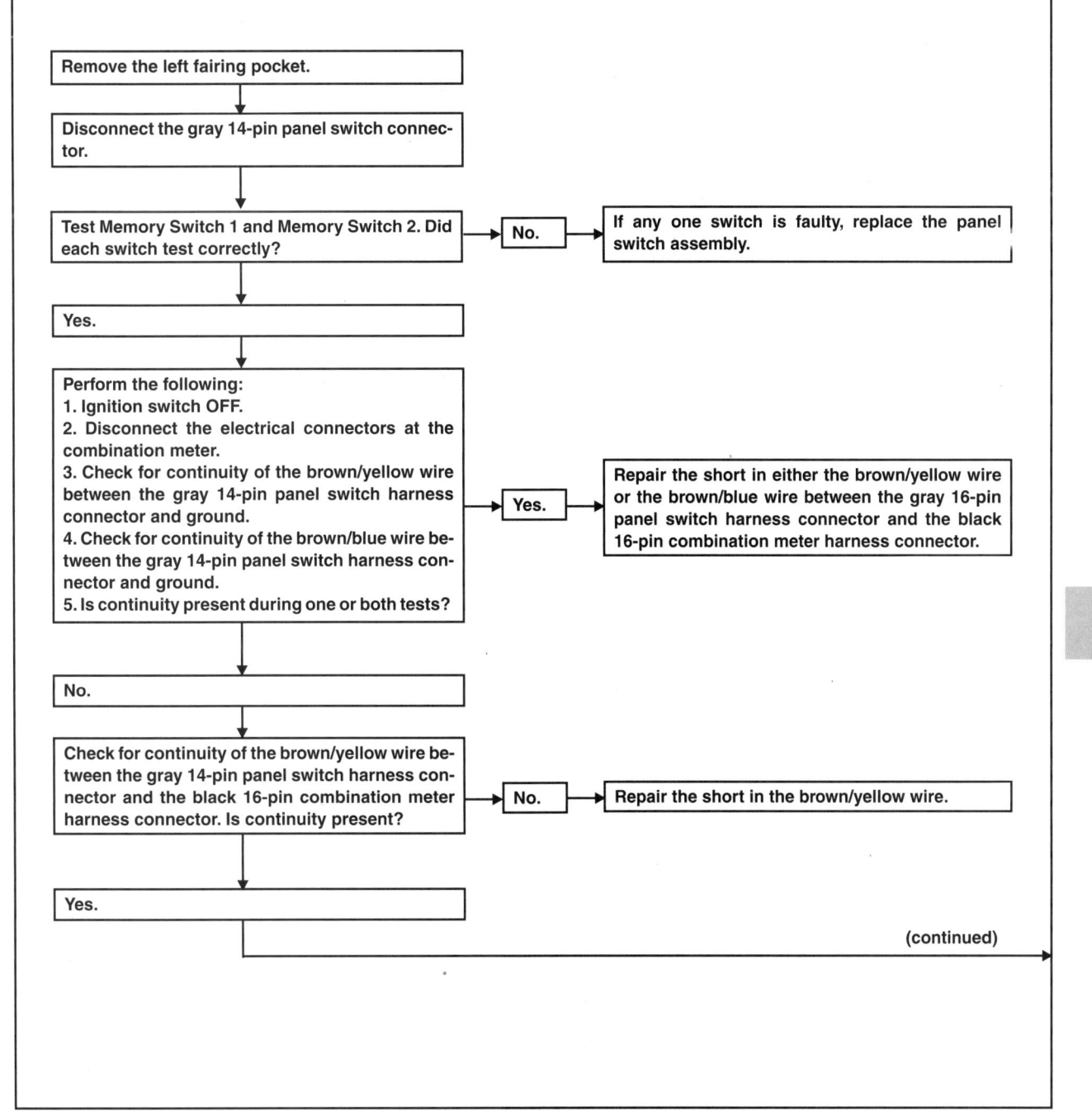

Remove the left fairing pocket.

↓

Disconnect the gray 14-pin panel switch connector.

↓

Test Memory Switch 1 and Memory Switch 2. Did each switch test correctly? → **No.** → If any one switch is faulty, replace the panel switch assembly.

↓ Yes.

Perform the following:
1. Ignition switch OFF.
2. Disconnect the electrical connectors at the combination meter.
3. Check for continuity of the brown/yellow wire between the gray 14-pin panel switch harness connector and ground.
4. Check for continuity of the brown/blue wire between the gray 14-pin panel switch harness connector and ground.
5. Is continuity present during one or both tests?

→ **Yes.** → Repair the short in either the brown/yellow wire or the brown/blue wire between the gray 16-pin panel switch harness connector and the black 16-pin combination meter harness connector.

↓ No.

Check for continuity of the brown/yellow wire between the gray 14-pin panel switch harness connector and the black 16-pin combination meter harness connector. Is continuity present? → **No.** → Repair the short in the brown/yellow wire.

↓ Yes.

(continued)

13

6 (continued)

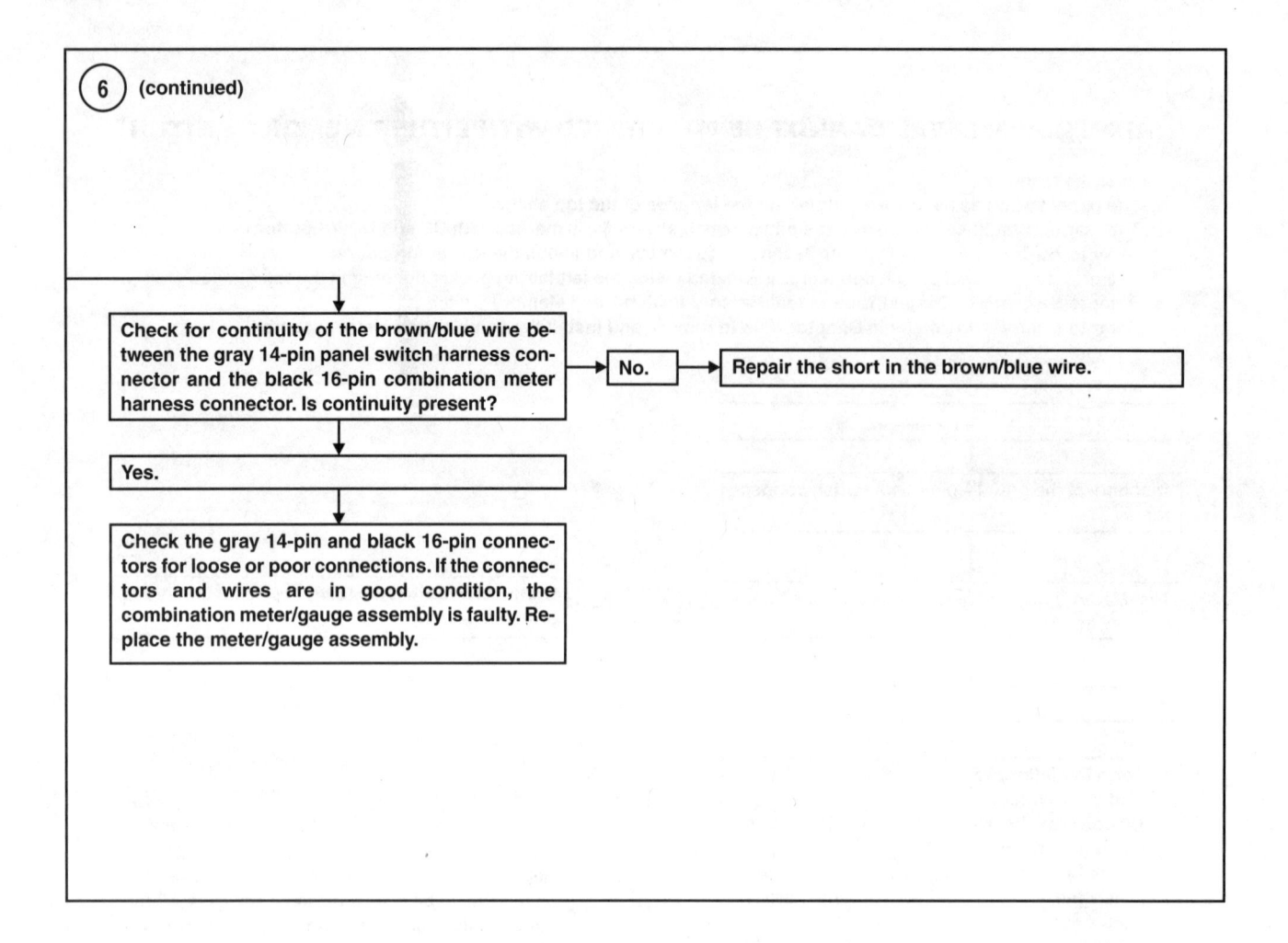

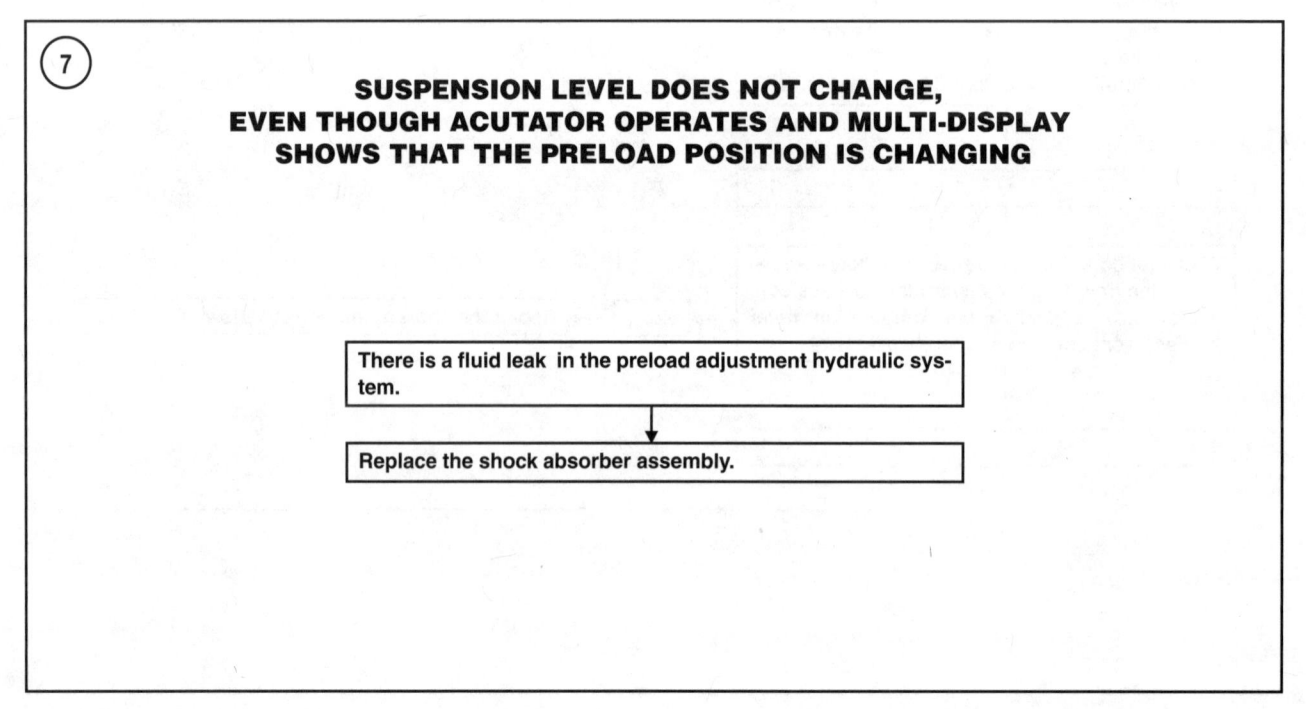

Check for continuity of the brown/blue wire between the gray 14-pin panel switch harness connector and the black 16-pin combination meter harness connector. Is continuity present? → No. → Repair the short in the brown/blue wire.

Yes.

Check the gray 14-pin and black 16-pin connectors for loose or poor connections. If the connectors and wires are in good condition, the combination meter/gauge assembly is faulty. Replace the meter/gauge assembly.

---

7 **SUSPENSION LEVEL DOES NOT CHANGE, EVEN THOUGH ACUTATOR OPERATES AND MULTI-DISPLAY SHOWS THAT THE PRELOAD POSITION IS CHANGING**

There is a fluid leak in the preload adjustment hydraulic system.

Replace the shock absorber assembly.

REAR SUSPENSION AND FINAL DRIVE

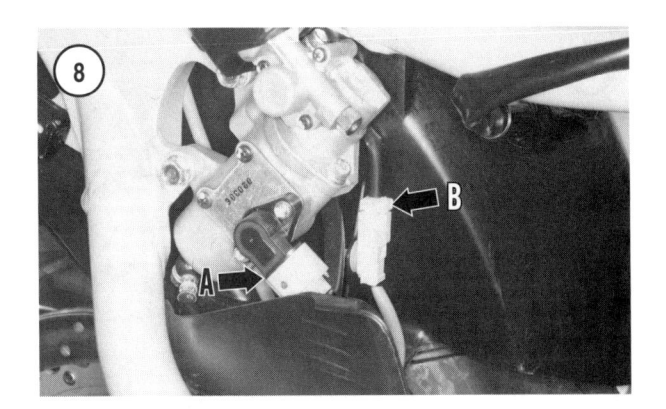

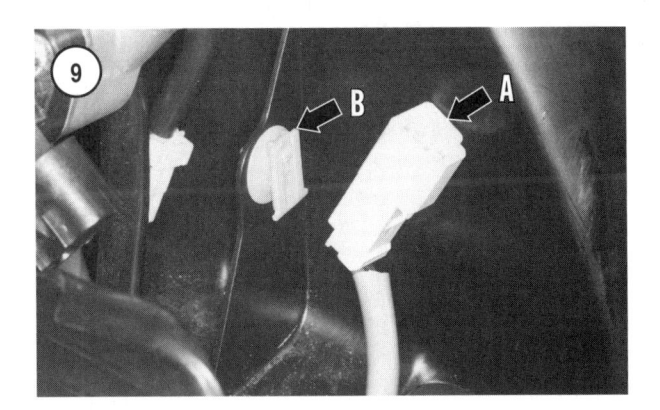

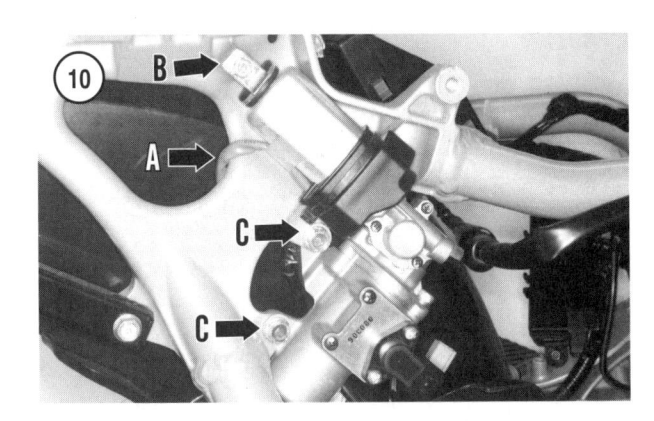

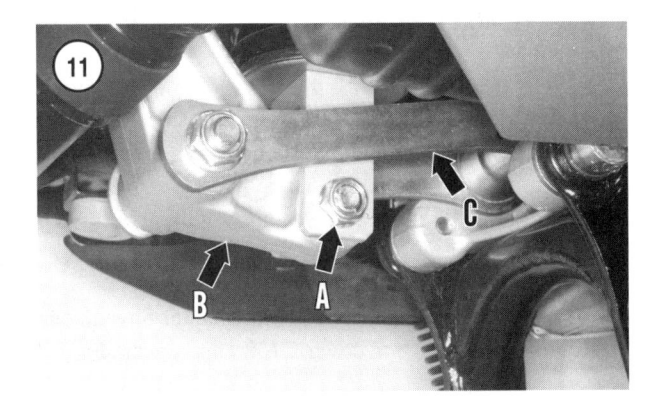

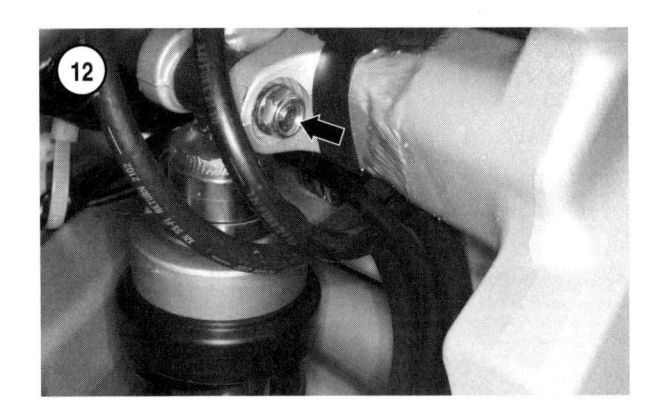

5. Record the wire harness and hose routing between the actuator and control motor.
6. Disconnect the gray 3-pin angle sensor connector (A, **Figure 8**).
7. Disconnect the green 2-pin control motor connector (B, **Figure 8**).
8. Push the control motor connector (A, **Figure 9**) up to release it from its mounting clamp (B).
9. Remove the control motor wire harness from around the frame (A, **Figure 10**).
10. Remove the 6-mm bolt (B, **Figure 10**) and the two 8 mm bolts (C).
11. Remove the shock hose from the clamp.
12. Remove the lower shock absorber mounting nut (A, **Figure 11**).
13. Place a jack underneath the shock arm (B, **Figure 11**) and raise it to remove weight from the lower shock bolt and to support the swing arm when the shock is removed. Remove the lower shock bolt.
14. Remove the upper mounting nut (**Figure 12**) and bolt and remove the shock absorber/actuator assembly.

Installation

1. Clean the shock absorber and actuator mounting bolts and nuts.
2. Position the shock in the frame with the reservoir hose bolt facing the left side. Install the upper mounting bolt (Allen head) from the left side. Install the nut (**Figure 12**) finger-tight.
3. Route the reservoir hose in front of the two fuel pump hoses.
4. Lift the swing arm and install the lower mounting bolt from the left side. Install the nut (A, **Figure 11**) finger-tight.
5. Route the reservoir hose along the right side of the frame and above the two brake pipes.
6. Route the gray 2-pin connector and its wire harness over the frame (A, **Figure 10**). Then pass it between the reservoir

and rear fender B. Make sure the main wiring harness and its two connectors are not pinched between the reservoir and rear fender B.

7. Install the three actuator mounting bolts. Tighten the two 8-mm bolts (C, **Figure 10**) to 26 N•m (19 ft.-lb.). Tighten the 6-mm bolt (B, **Figure 10**) securely.

8. Reconnect the gray 2-pin control motor connector (B, **Figure 8**), then secure the connector to the clamp on rear fender B.

9. Reconnect the gray 3-pin angle sensor connector at the actuator (A, **Figure 8**).

10. Tighten the upper and lower shock absorber mounting nuts to 42 N•m (31 ft.-lb.).

11. Reverse Steps 2-4 of *Removal*.

Inspection

Replace the shock absorber assembly (**Figure 1**) if it leaks oil or shows damage as described in this section.

> *WARNING*
> *The shock absorber housing contains high-pressure nitrogen gas. Do not tamper with or open the shock housing. Do not place it near an open flame or other extreme heat. Do not dispose of the shock assembly. Take it to a dealership where it can be deactivated and disposed of properly.*

> *CAUTION*
> *The shock absorber and actuator assembly are not serviceable. There are no replacement parts for the shock absorber or actuator assembly. Do not loosen the oil hose bolts in an attempt to separate the parts.*

1. Check the shock housing for dents, damage or oil leakage.
2. Check the banjo bolts and oil hose for leaks or other damage.
3. Check the spring for cracks or other damage. Check the spring seats for damage. The spring is not removable. If damaged, replace the shock absorber assembly.
4. Check the actuator and control motor assembly for damage. Check the wiring harness for cuts or other damage. Check both connectors for contamination or damaged pins.
5. Check the bushing (**Figure 13**) in the upper shock absorber joint for excessive wear or damage. The bushing must be a tight fit. If damaged, replace the shock absorber assembly.
6. Replace the shock mounting bolts and nuts if damaged.

SUSPENSION LEVEL RELAYS

Three relays are used to control the suspension level system. These relays are located in the main relay box (**Figure 14**) underneath the seat:
1. Main relay (11).
2. UP Relay (13).
3. DOWN relay (14).

The relays are easily unplugged or plugged into the relay terminal strip. Because only 4-terminal and 5-terminal relays are used, there is no danger of installing a relay in the wrong position. The 4-terminal relays are identical, and the 5-terminal relays are identical. Refer to *Relay Box* in Chapter Nine for more information on identifying and testing the relays.

> *NOTE*
> *A questionable relay can be quickly checked by exchanging it with a known good relay. To do this, first identify the relay as either a 4-terminal or 5-terminal type. Unplug the relay and plug in a known good relay.*

Relay and Circuit Testing

Refer to the wiring diagrams at the end of manual to identify the suspension level relay wire and fuse circuits referred to in this procedure.

1. Before testing the suspension level relays, confirm the following:
   a. Start the engine. With the transmission in neutral, confirm that the reverse system is off.
   b. Confirm that the audio system works correctly.
   c. Confirm that the multi-display works correctly.
   d. If these systems work correctly, continue with Step 2.
2. Turn the ignition switch off.
3. Remove the seat. Refer to *Seat* in Chapter Seventeen.
4. Remove the relay box cover screws and cover (**Figure 15**).

REAR SUSPENSION AND FINAL DRIVE

(14) RELAY BOX

1	2	3	4	Blank	6	7
8	9	10	11	12	13	14
15	16	17	18	19	20	21

------- 4-Terminal relays ------- | ----- 5-Terminal relays -----

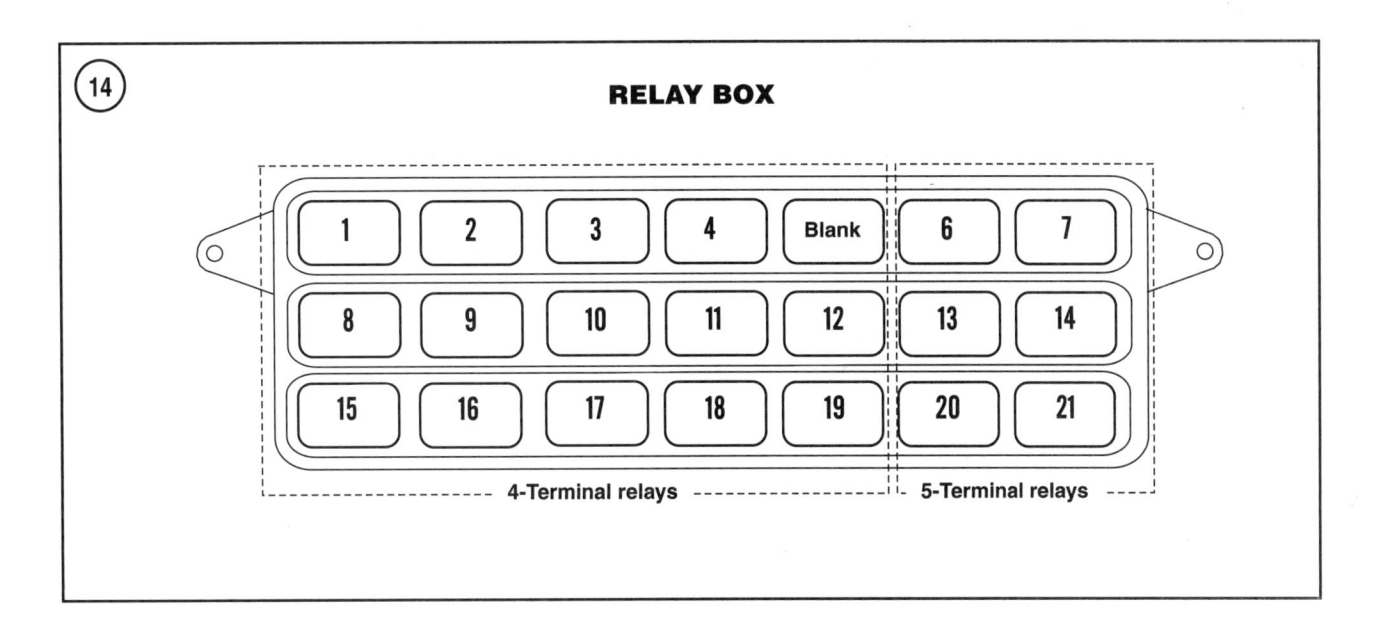

NOTE
*Refer to **Spring Preload, Adjustment** in this chapter for information on operating the manual height switch.*

5. Replace the main relay (11, **Figure 14**) with a known good 4-terminal relay. Turn the ignition switch on or to ACC and operate the manual height switch (**Figure 2**). Note the following:

   a. If the level actuator operates, the original main relay was faulty. Install a new 4-terminal relay and repeat the test.

   b. If the level actuator does not operate, turn the ignition switch off and continue with Step 6.

6. Replace the UP relay (13, **Figure 14**) with a known good 5-terminal relay. Turn the ignition switch on or to ACC and operate the manual height switch (**Figure 2**). Note the following:

   a. If the level actuator operates, the original UP relay was faulty. Install a new 5-terminal relay and repeat the test.

   b. If the level actuator does not operate, turn the ignition switch off and continue with Step 7.

7. Replace the DOWN relay (14, **Figure 14**) with a known good 5-terminal relay. Turn the ignition switch on or to ACC and operate the manual height switch (**Figure 2**). Note the following:

   a. If the level actuator operates, the original DOWN relay was faulty. Install a new 5-terminal relay and repeat the test.

   b. If the level actuator does not operate, turn the ignition switch off and continue with Step 8.

NOTE
If the three suspension level relays are working correctly, continue with Step 8 to test the individual relay circuits.

NOTE
*The meter test connections called out in Steps 8-12 are made at the relay connector terminals inside the relay box (**Figure 16**).*

8. Turn the ignition switch off and remove the suspension level main relay (11, **Figure 14**). Measure voltage between the main relay brown terminal (+) and ground (–):

   a. Battery voltage: Go to Step 9.

   b. No battery voltage: Check for an open in the brown wire between the suspension level main relay and the No. 20 fuse.

16

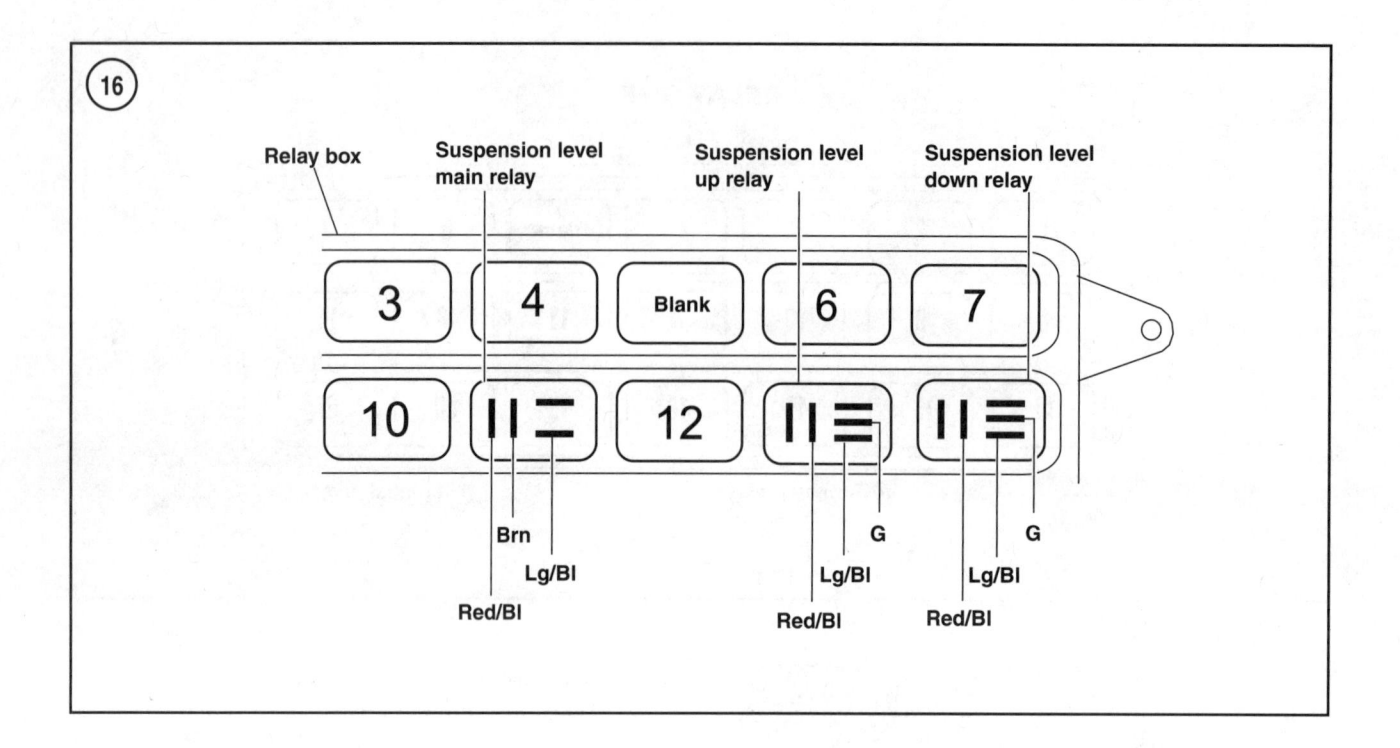

Relay box — Suspension level main relay — Suspension level up relay — Suspension level down relay

```
[ 3 ]  [ 4 ]  [Blank] [ 6 ]  [ 7 ]
[10 ]  [11≡]  [ 12 ]  [11≡]  [11≡]
        Brn             G      G
        Lg/Bl          Lg/Bl  Lg/Bl
        Red/Bl         Red/Bl Red/Bl
```

9. Turn the ignition switch on or to ACC. Measure voltage between the main relay light green/black terminal (+) and ground (–):
   a. Battery voltage: Go to Step 10.
   b. No battery voltage: Check for an open in the light green/black wire between the suspension level main relay and the ACC relay (12, **Figure 14**).

10. Turn the ignition switch off. Remove the suspension level UP (13, **Figure 14**) and DOWN (14) relays. Check continuity in the red/black wires between the suspension level main and UP relays, then between the main and DOWN relays:
    a. Continuity recorded at both tests: Go to Step 11.
    b. No continuity recorded at one or both tests: Check for an open in the red/black wire between the affected relay(s).

11. Turn the ignition switch off. Check continuity between the UP relay green wire terminal and ground. Repeat the test for the DOWN relay green wire terminal:
    a. Continuity recorded at both tests: Go to Step 12.
    b. No continuity recorded at one or both tests: Check for an open in the green wire (actuator motor ground circuit).

12. Turn the ignition switch on or to ACC. Measure voltage between the light green/black terminal (+) in the UP relay connector and ground (–). Repeat the test for the DOWN relay light green/black terminal:
    a. Battery voltage: Perform the troubleshooting procedures in **Figure 5**, Test E.
    b. No battery voltage: Check for an open in the light green/black wire between the suspension level UP and DOWN relay and the ACC relay (12, **Figure 14**).

13. Turn the ignition switch off and reinstall the relays. Recheck the suspension level adjustment operation described in this chapter under *Shock Absorber Adjustment*.

14. Install the relay box cover (**Figure 15**) and secure the mounting screws.

15. Install the seat. Refer to *Seat* in Chapter Seventeen.

SUSPENSION LINKAGE

This section describes service to the shock arm (B, **Figure 11**) and shock link (C) with the swing arm mounted on the motorcycle.

Removal/Installation

1. Support the bike on its centerstand.
2. Remove the exhaust system. Refer to *Exhaust System* in Chapter Seventeen.
3. Remove the rear wheel. Refer to *Rear Wheel* in Chapter Eleven.
4. Remove the following suspension linkage components:
   a. Shock absorber lower mounting nut and bolt (A, **Figure 17**).

REAR SUSPENSION AND FINAL DRIVE

f. Tighten the shock link-to-frame nut (D, **Figure 17**) to 64 N•m (47 ft.-lb.).

Shock Arm
Inspection/Lubrication

Replace worn or damaged parts as described in this chapter.

Refer to **Figure 19**.

1. Remove the pivot collars from the shock arm (A, **Figure 20**).
2. Pry the seals out of the shock arm (**Figure 21**).
3. Clean and dry all parts.
4. Check the shock arm for cracks and other damage.
5. Inspect the pivot collars for excessive wear, rust or damage. Check the collars for burrs or nicks that may damage the seals. Remove burrs or nicks with a fine cut file.
6. Inspect the bolts and nuts for corrosion, wear grooves or other damage. Visual damage on a bolt shoulder may indicate pivot collar wear or damage.
7. Insert and turn the pivot collars in their respective bearings (**Figure 22**). Check for any roughness or binding, indicating bearing damage.
8. Check each needle bearing (**Figure 23**) for damaged or missing needles, seizure and rust. Replace damaged bearings as described under *Shock Arm Needle Bearing Replacement* in this section.

CAUTION
Do not remove the needle bearings to inspect or lubricate them.

9. Pack the bearings with water-resistant grease.
10. Pack the lips of the new seals with the same grease and install into position on the shock arm. Install the seals (**Figure 24**)—closed side facing out—by pushing them into place by hand.
11. Apply a thin film of the same grease to the outside of each collar and install the collar into its original mounting position.

Shock Arm
Needle Bearing Replacement

A single bearing is used for the shock absorber. Replace the swing arm and shock link bearings in pairs.

1. A blind hole bearing puller equipped with an expanding collet, and a hydraulic press are required to replace the bearings:
   a. The single shock absorber side bearing (B, **Figure 20**) can be removed with a press.

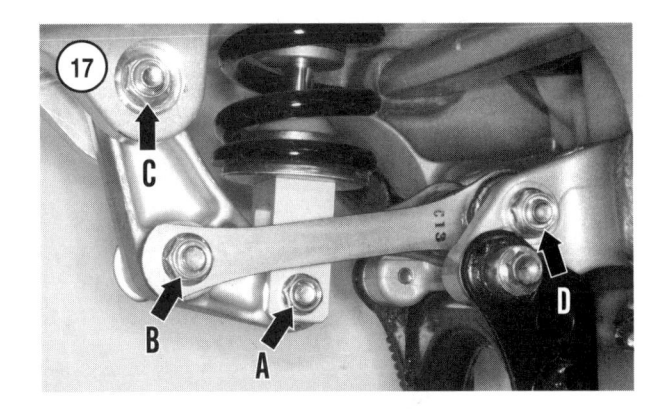

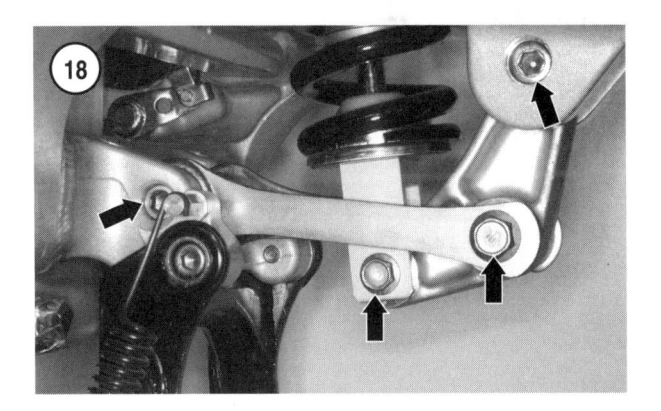

b. Shock link mounting nut and bolt (B, **Figure 17**) at the shock arm.
c. Shock arm nut and bolt (C, **Figure 17**) at the swing arm. Remove the shock arm.
d. Shock link nut and bolt (D, **Figure 17**) at the frame. Remove the shock link.

5. Inspect the shock arm and shock link seals and bearings as described in this section.
6. Clean and lubricate the shock arm and shock link mounting bolts, seals, collars and bearings as described in this section.
7. Installation is the reverse of removal. Note the following:
   a. Lubricate the shock linkage mounting bolts with grease. Do not lubricate the mounting bolt threads or nuts. These must be tightened with dry threads.
   b. Install all the mounting bolts from the left side (**Figure 18**). Install all the shock linkage assembly mounting bolts and nuts before tightening them.
   c. Tighten the lower shock absorber mounting nut (A, **Figure 17**) to 42 N•m (31 ft.-lb.).
   d. Tighten the shock arm-to-shock link mounting nut (B, **Figure 17**) to 64 N•m (47 ft.-lb.).
   e. Tighten the shock arm-to-swing arm nut (C, **Figure 17**) to 64 N•m (47 ft.-lb.).

CHAPTER THIRTEEN

REAR SUSPENSION LINKAGE

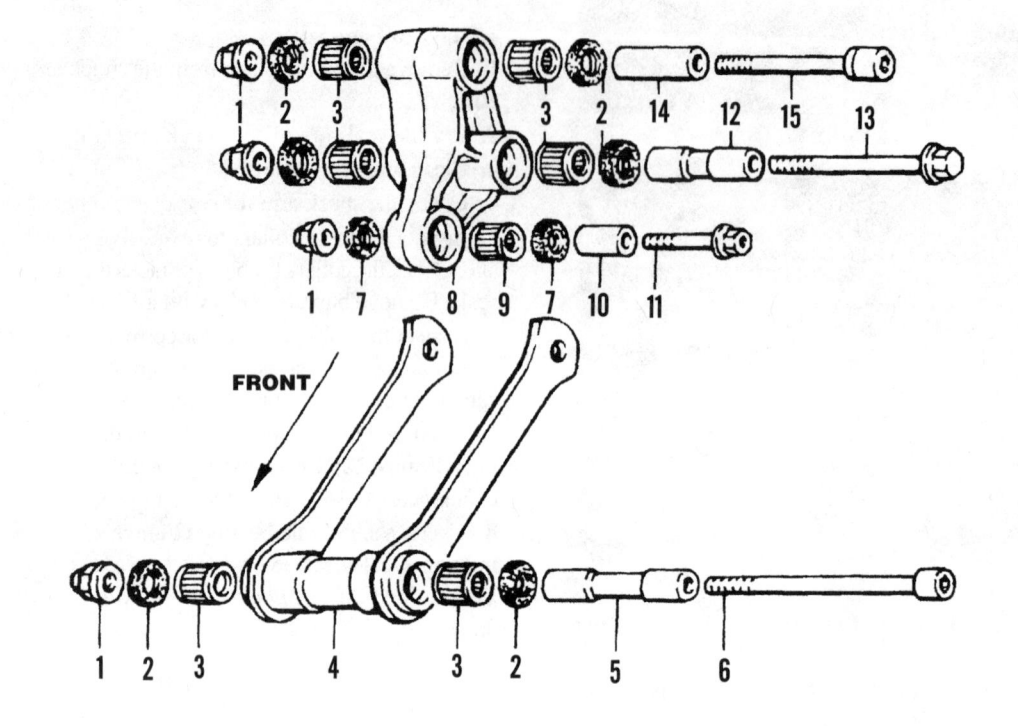

1. Nut
2. Seal
3. Bearing
4. Shock link
5. Pivot collar
6. Bolt
7. Seal
8. Shock arm
9. Bearing
10. Pivot collar
11. Bolt
12. Pivot collar
13. Bolt
14. Pivot collar
15. Bolt

REAR SUSPENSION AND FINAL DRIVE

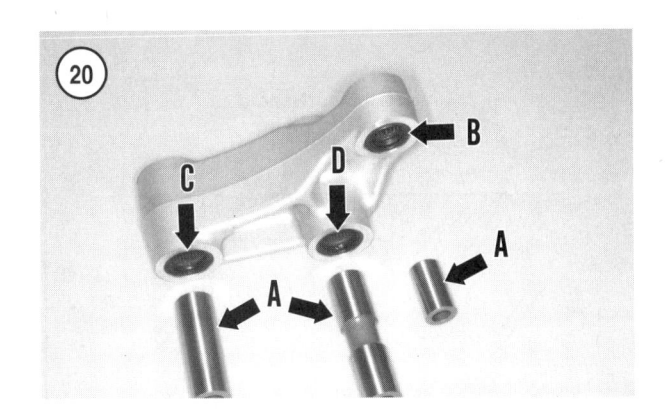

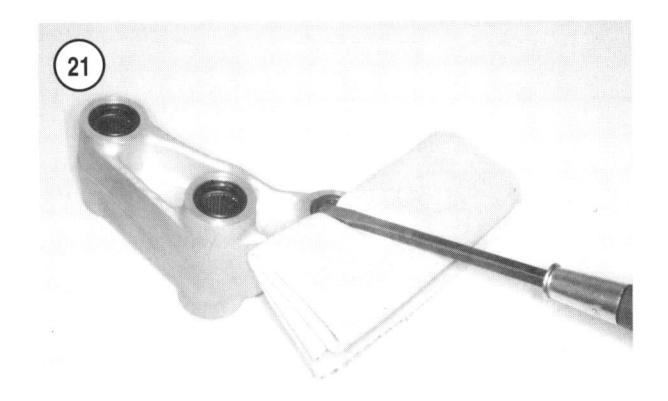

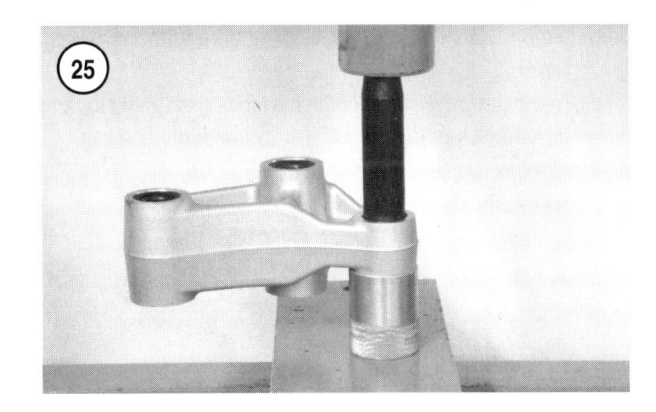

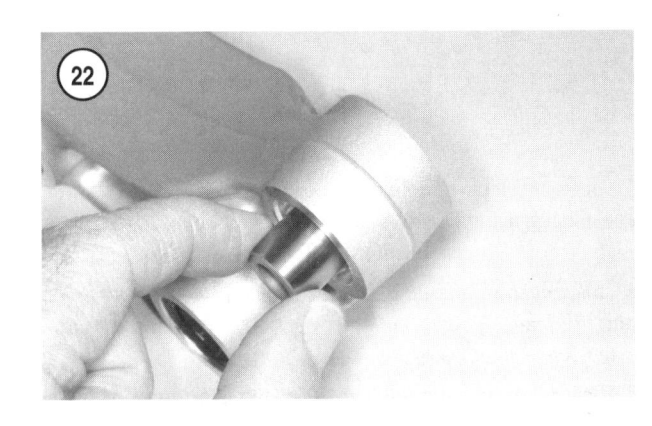

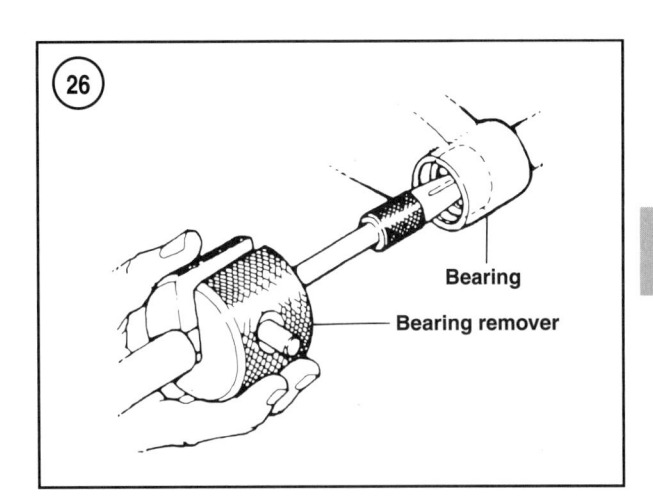

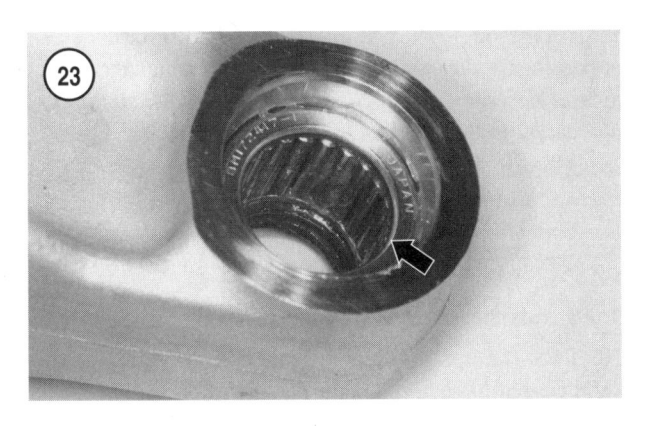

b. The swing arm (C, **Figure 20**) and shock link (D) side bearings (two bearings in each bore) must be removed with a slide hammer puller.

2. Support the shock arm in a press and press out the shock absorber side needle bearing (**Figure 25**).

3. To remove the swing arm (C, **Figure 20**) and shock link (D) bearings with a blind hole bearing puller (**Figure 26**, typical):

   a. Support the shock arm in a vise.

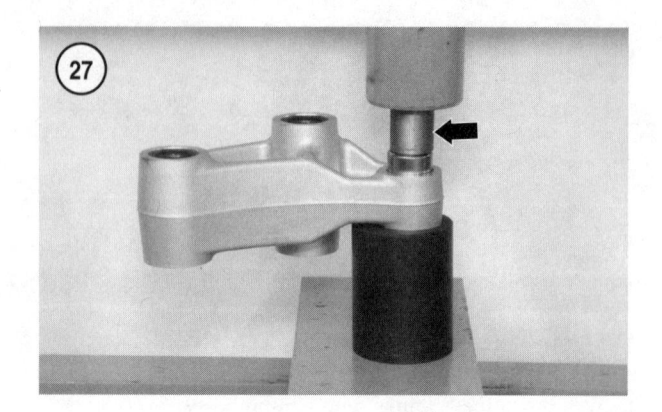

b. Heat the area around the bearing bore with a propane torch.
c. Insert the correct size collet into the bearing and lock it in place. Operate the puller to remove the bearing.
d. Reverse for the opposite bearing.
e. Allow the shock arm to cool.

NOTE
*Refer to **Bearings** in Chapter One for additional information on using heat to remove bearings.*

4. Clean and dry the shock arm.
5. Inspect each mounting bore for galling, cracks or other damage.
6. Support the shock arm in a press (**Figure 27**) and press in the bearings so their outer surface, as measured from the shock arm outer surface to the outer edge of the bearing (**Figure 28**) is as follows:
   a. Shock absorber side needle bearing (B, **Figure 20**): 5.3-5.7 mm (0.21-0.22 in.).
   b. Swing arm and shock link side needle bearings (C and D, **Figure 20**): 5.5-6.0 mm (0.22-0.24 in.).
7. Pack the new bearings with a water-resistant grease.
8. Install a collar into each bearing and pivot it by hand. Each collar should pivot smoothly. If there is any roughness or binding, the bearing was damaged during installation.

Shock Link
Inspection/Lubrication

Replace worn or damaged parts as described in this chapter.
Refer to **Figure 19**.
1. Remove the pivot collar from the shock link (**Figure 29**).
2. Pry the seals (A, **Figure 30**) out of the shock link.
3. Clean and dry all parts.
4. Check the shock link for cracks and other damage. 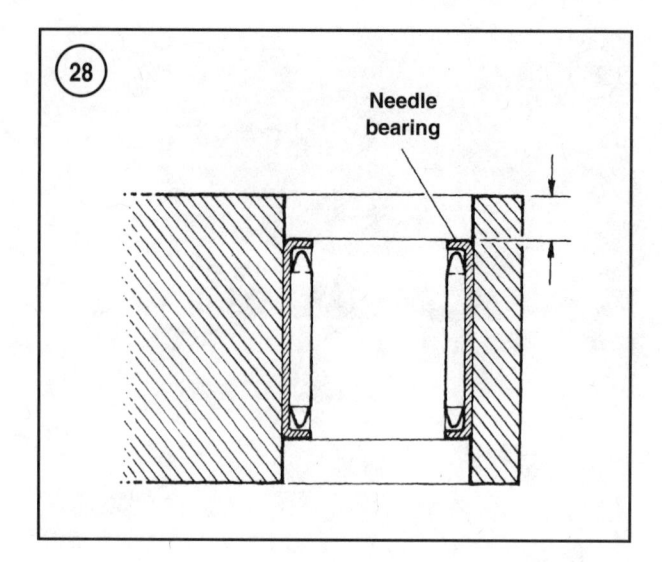 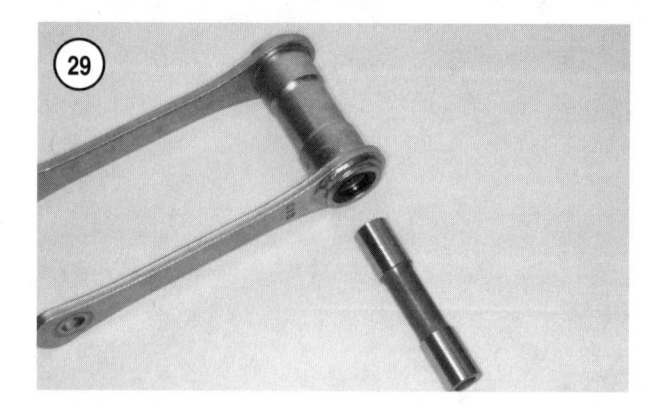
5. Inspect the pivot collar for excessive wear, rust or damage. Check the collar for burrs or nicks that may damage the seals. Remove burrs or nicks with a fine cut file.
6. Inspect the bolt and nut for corrosion, wear grooves or other damage. Visual damage on a bolt shoulder may indicate pivot collar wear or damage.
7. Insert and turn the pivot collar in the bearing. Check for any roughness or binding, indicating bearing damage.
8. Check each needle bearing (B, **Figure 30**) for damaged or missing needles, seizure and rust. Replace damaged bearings as described under *Shock Link Needle Bearing Replacement* in this section.

CAUTION
Do not remove the needle bearings for inspection or lubrication.

9. Pack the bearings with water-resistant grease.
10. Pack the lips of the new seals with the same grease and install into position on the shock link. Install the seals

REAR SUSPENSION AND FINAL DRIVE

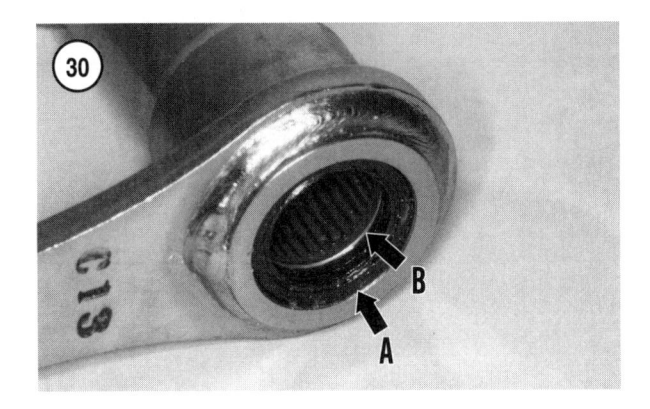

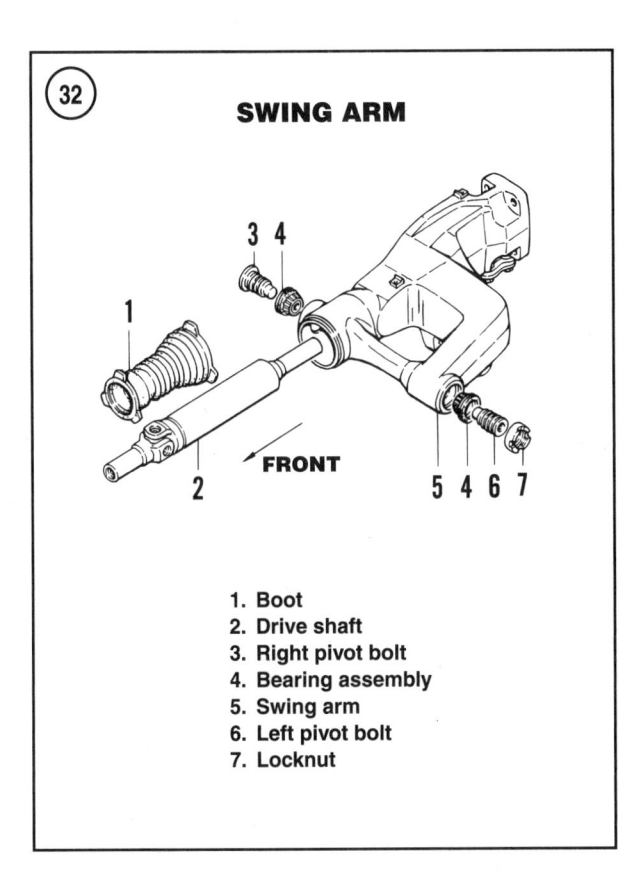

(A, **Figure 30**)—closed side facing out—by pushing them into place by hand.

11. Lubricate the collar (**Figure 29**) with grease and install into the shock link.

Shock Link
Needle Bearing Replacement

Replace both shock link bearings at the same time. Use a blind hole bearing puller equipped with an expanding collet, and a hydraulic press to replace the bearings:

1. To remove the bearings with a blind hole bearing puller (**Figure 26**, typical):
    a. Support the shock link in a vise.
    b. Heat the area around the bearing bore with a propane torch.
    c. Insert the correct size collet into the bearing and lock it in place. Operate the puller to remove the bearing.
    d. Reverse for the opposite bearing.
    e. Allow the shock link to cool.

NOTE
*Refer to **Bearings** in Chapter One for additional information on using heat to remove bearings.*

2. Clean and dry the shock link.
3. Inspect the mounting bore for galling, cracks or other damage.
4. Support the shock link in a press (**Figure 31**) and press in the bearings so their outer surface is 5.5-6.0 mm (0.22-0.24 in.) from the shock link outer surface (**Figure 28**).
5. Pack the new bearings with a water-resistant grease.
6. Install a collar into each bearing and pivot it by hand. Each collar should pivot smoothly. If there is any roughness or binding, the bearing was damaged during installation.

SWING ARM

Refer to **Figure 32**.

Tools

The following tools are required to loosen and tighten the swing arm pivot bolts and the left locknut:

1. Locknut wrench (Honda part No. 07ZMA-MCAA101 [A, **Figure 33**]).
2A. 19 mm (3/4 in.) hex socket (B, **Figure 33**).
2B. A suitable 19 mm hex socket can be assembled (C, **Figure 33**) from the following:
    a. A 5/8 in. (16 mm) spark plug socket with an external 3/4 in. (19 mm) hex shoulder (A, **Figure 34**).

SWING ARM

1. Boot
2. Drive shaft
3. Right pivot bolt
4. Bearing assembly
5. Swing arm
6. Left pivot bolt
7. Locknut

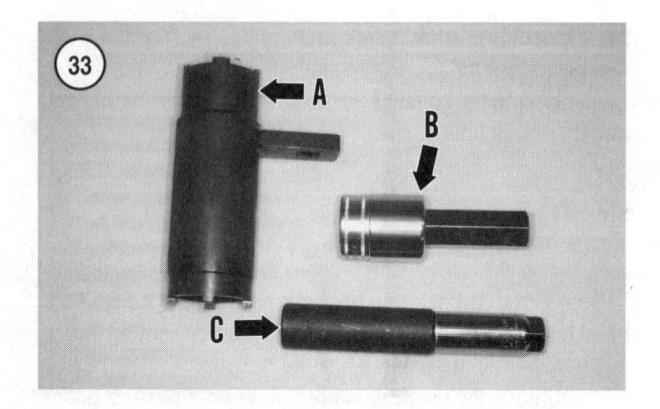

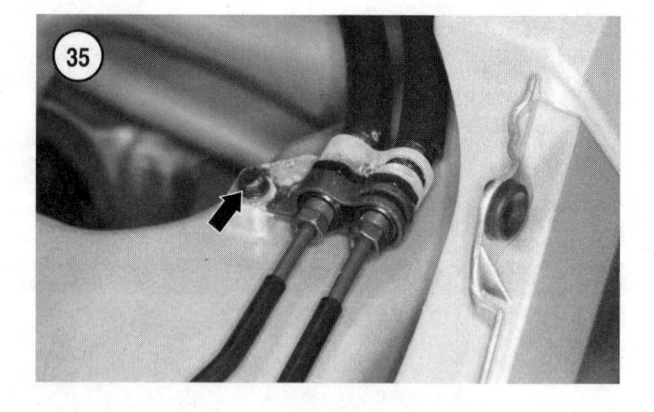

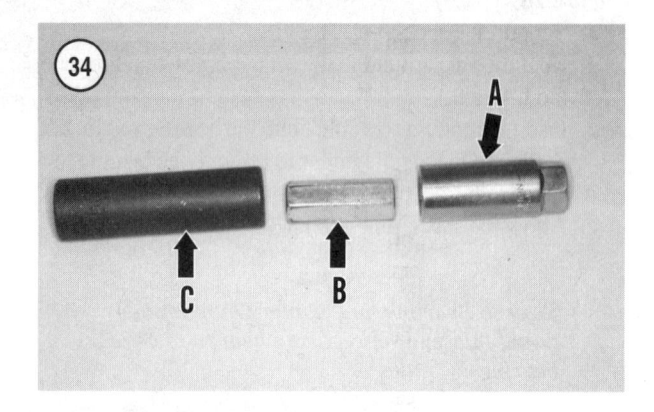

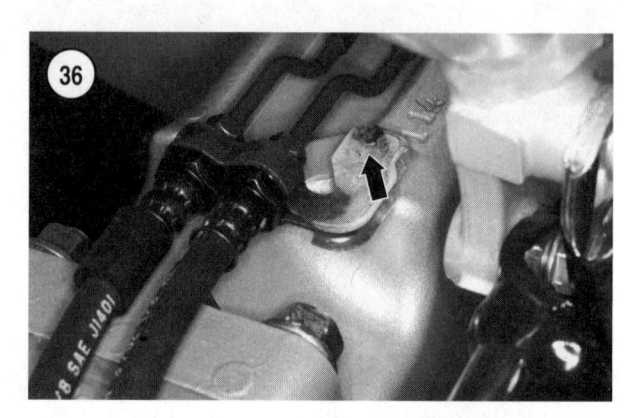

b. A 5/16-18 in. hex coupling that measures 5/8 in. across its flats (B, **Figure 34**). Hex couplings are available at most hardware stores.

c. A 5/8 in. (16 mm) socket (C, **Figure 34**).

Removal

1. Support the bike on its centerstand.
2. Remove the battery and battery tray. Refer to *Battery* in Chapter Nine.
3. Remove the right saddlebag. Refer to *Saddlebags* in Chapter Seventeen.
4. Remove the top shelter. Refer to *Top Shelter* in Chapter Seventeen.
5. Remove the fuel tank. Refer to *Fuel Tank* in Chapter Eight.
6. Remove the shock absorber as described under *Shock Absorber* in this chapter.
7. Remove the bolts and brake pipe clamps at the front (**Figure 35**) and rear (**Figure 36**) of the swing arm.
8. Remove the rear brake caliper and final drive housing as described under *Final Drive Housing* in this chapter.
9. Remove the right engine side cover as described in Chapter Seventeen under *Engine Side Covers*.

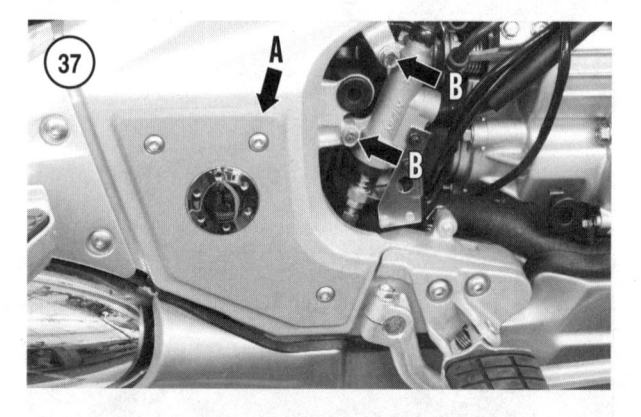

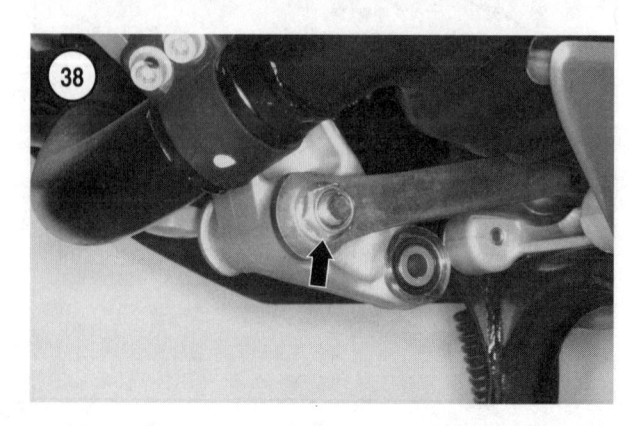

REAR SUSPENSION AND FINAL DRIVE

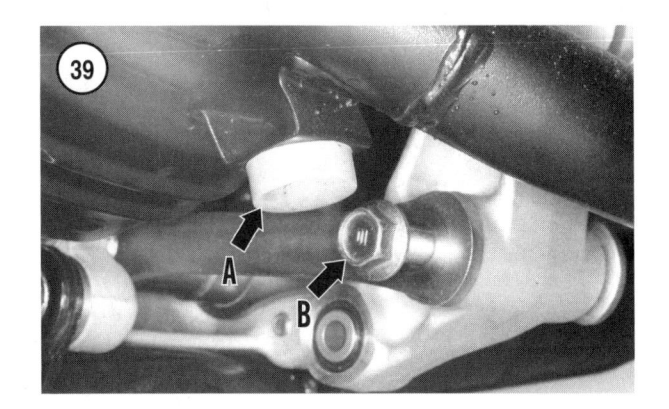

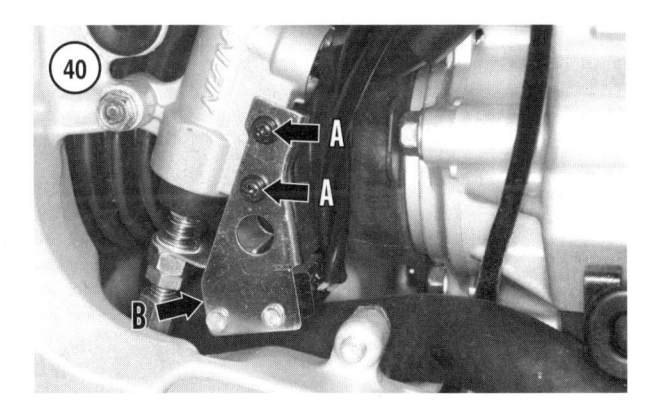

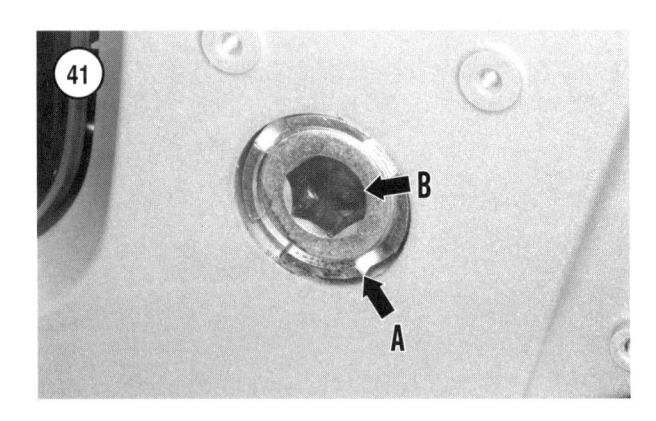

10. Remove the screws and the swing arm pivot covers (A, **Figure 37**, typical).
11. Disconnect the shock link from the shock arm as follows:
    a. Remove the shock arm-to-shock link nut (**Figure 38**).
    b. If the exhaust pipe is installed on the engine, remove the rubber centerstand stopper (A, **Figure 39**), then the bolt (B).
12. Check swing arm bearing play as follows:
    a. Grasp the rear end of the swing arm and move it up and down. The swing arm should move smoothly. If there is any binding or roughness, replace the swing arm bearings as described in this section.
    b. Now try to move the swing arm from side to side in a horizontal arc. If any movement is felt where the swing arm is mounted at the frame, replace the swing arm bearings as described in this section.
13. Remove the two screws (A, **Figure 40**) and the brake light/cruise cancel switch (B) from the rear master cylinder.
14. Disconnect the boot (1, **Figure 32**) at the engine.
15. Loosen the left pivot bolt locknut (A, **Figure 41**) with the locknut wrench (**Figure 42**).
16. Loosen and remove the left pivot bolt (B, **Figure 42**).
17. Loosen and remove the right pivot bolt (**Figure 43**).
18. Remove the swing arm/drive shaft with the shock arm attached.
19. Remove the boot (**Figure 44**).
20. Remove the drive shaft (A, **Figure 45**).
21. Inspect the swing arm and drive shaft as described in this chapter.

Installation

1. Before installing the swing arm, check the back of the engine for any loose or missing fasteners, misrouted wiring harnesses or leaking or damaged hoses.

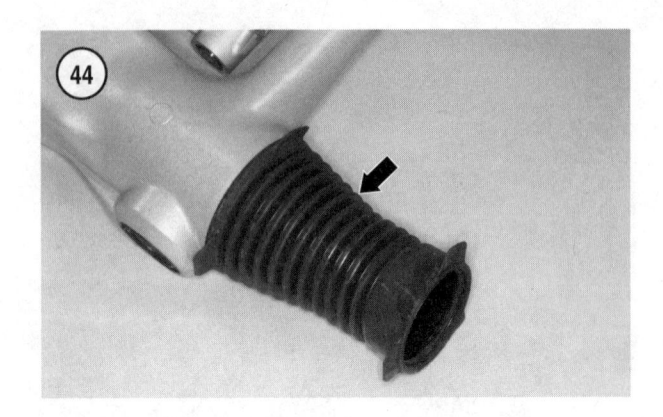

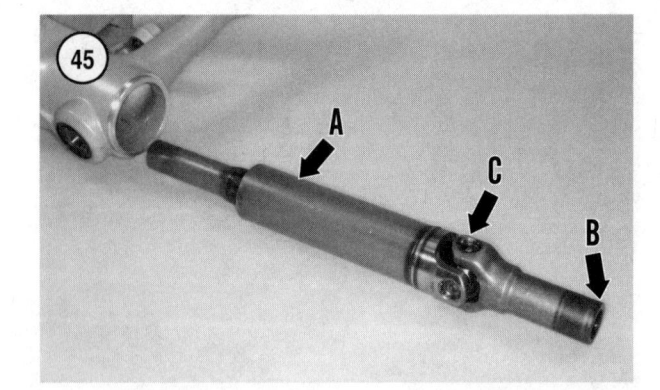

2. Clean the output shaft (**Figure 46**) of all old grease.

3. If removed, install the shock link as described under *Suspension Linkage* in this chapter.

4. If removed, install the shock arm onto the swing arm as described under *Suspension Linkage* in this chapter.

5. Clean, lubricate and install the swing arm bearings as described under *Bearing Inspection and Lubrication* in this section.

6. Clean the swing arm pivot bolt shafts and the threaded holes in the frame. There must be no grease or oil on any of these threads.

7. Lubricate the pivot bolt shaft ends (A and B, **Figure 47**) with grease and set aside until installation.

8. Remove the rear master cylinder mounting bolts (B, **Figure 37**). This provides access around the master cylinder when installing the drive shaft onto the output shaft.

9. Lubricate the front drive shaft (B, **Figure 45**) splines with 1 g (0.04 oz.) of molybdenum disulfide grease.

10. Install the drive shaft into the swing arm as shown in A, **Figure 45**.

11. Install the boot onto the swing arm (**Figure 44**).

12. Remove the joint shaft from the final drive housing and install it onto the end of the drive shaft (**Figure 48**). This will extend the length of the drive shaft to help with its installation.

13. Push the drive shaft/joint shaft forward in the swing arm.

NOTE
Shift the transmission into gear to lock the output shaft and prevent it from turning. This helps the drive shaft to engage and mesh with the output shaft in the following steps.

14. Install and position the swing arm between the frame. Position the shock arm in front of the exhaust pipe (**Figure 49**). If necessary, support the swing arm with a jack.

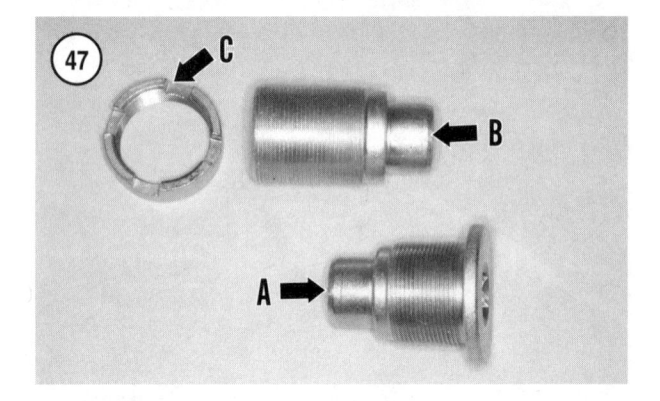

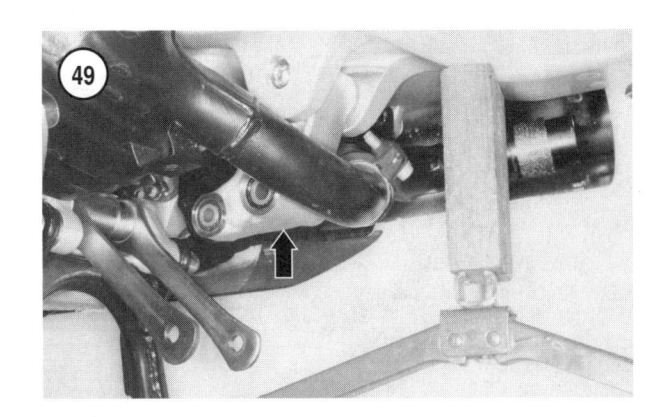

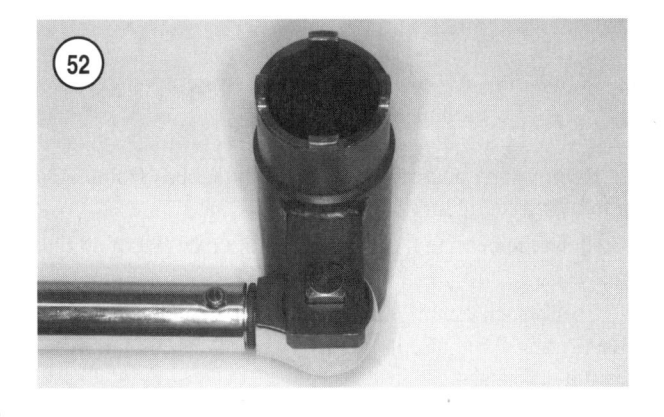

REAR SUSPENSION AND FINAL DRIVE

CAUTION
Make sure there are no wiring harnesses pinched between the swing arm and frame.

15. Push the swing arm forward and slide the drive shaft over the output shaft. Turn the joint shaft (**Figure 50**) to help with spline alignment. If necessary, pry the drive shaft upward (**Figure 51**) to aid alignment and installation.

16. Align the right swing arm bearing with the frame hole and install the right pivot bolt (A, **Figure 47**). Tighten the pivot bolt (**Figure 43**) finger-tight.

NOTE
*If it is difficult to align the right side swing arm bearing with the frame threads, the joint shaft may have dropped inside the swing arm and is pushing against the back of the swing arm. The joint shaft must extend through the end of the swing arm (**Figure 50**) for the swing arm to move forward.*

17. Install the left pivot bolt (B, **Figure 47**) and tighten by hand to make sure the swing arm alignment with the frame is correct. Then back the left pivot bolt out. This pivot bolt must be free (no pressure applied against the swing arm or bearing) when the right pivot bolt is tightened.

18. Tighten the right pivot bolt (**Figure 43**) to 108 N•m (80 ft.-lb.). Then check that the left pivot bolt can be turned by hand, indicating that it is not applying pressure against its bearing.

19. Tighten the left pivot bolt (B, **Figure 41**) to 34 N•m (25 ft.-lb.).

20. Move the swing arm up and down several times to seat the bearings. Retighten the right pivot bolt (**Figure 43**) to 108 N•m (80 ft.-lb.).

21. Install the left pivot locknut (C, **Figure 47**).

NOTE
*Mount the locknut wrench onto the torque wrench at a right-angle (**Figure 52**). This position does not extend the length of the torque wrench and the locknut can be tightened without having to recalibrate the torque setting. Refer to **Torque Adaptors** in Chapter One.*

22. Hold the left pivot bolt with a 19 mm hex socket (A, **Figure 53**) and tighten the locknut (B) to 108 N•m (80 ft.-lb.).

23. Hold the joint shaft (to prevent the drive shaft from disengaging) and move the swing arm. If there is any binding or roughness, loosen the locknut and both pivot bolts and check their installation and alignment.

24. Hold the drive shaft to prevent it from disengaging from the output shaft, then remove the joint shaft (**Figure 48**).
25. Install the final drive housing and rear brake caliper as described under *Final Drive Housing* in this chapter.
26. Install the drive boot (1, **Figure 32**) over the engine. Then check that the reserve tank hose is not pushing against the drive shaft boot. If so, the boot may contact the drive shaft and melt.
27. Install the rear master mounting bolts (B, **Figure 37**) and tighten to 12 N•m (106 in.-lb.).
28. Install and adjust the brake light/cruse switch:
   a. Use a medium strength threadlock on the switch screw threads.
   b. Install the switch (B, **Figure 40**) onto the master cylinder and secure with the mounting screws (A)—do not tighten.
   c. Turn the ignition switch on. Adjust the switch so the brake light comes on when the brake pedal moves the push rod 0.7-1.7 mm (1/32-1/16 in.)
   d. Tighten the switch screws (A, **Figure 40**) to 2 N•m (18 in.-lb.).
   e. Recheck the brake light adjustment and turn the ignition switch off.
29. Reconnect the shock link at the shock arm:
   a. Lubricate the bolt shoulder with grease. Do not lubricate the bolt threads.
   b. Install the bolt from the left side.
   c. Install the nut (**Figure 38**) and tighten to 64 N•m (47 ft.-lb.).
   d. Install the rubber centerstand stopper (A, **Figure 39**).
30. Reinstall the swing arm pivot covers (A, **Figure 37**).
31. Install the right engine side cover as described in Chapter Seventeen under *Engine Side Covers*.
32. Install the brake pipe clamps and bolts at the front (**Figure 35**) and rear (**Figure 36**) of the swing arm. Tighten the bolts to 12 N•m (106 in.-lb.).
33. Reverse Steps 1-6 of *Removal* in this section.

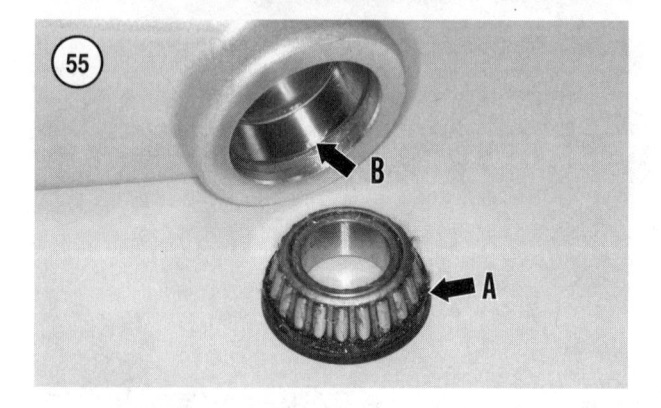

Bearing Inspection and Lubrication

The swing arm operates on tapered roller bearings and pivot bolts that thread into the frame. A shoulder on each pivot bolt rides inside the bearing.

The bearings can be removed for inspection and lubrication. A seal is permanently installed on each bearing and should not be removed. The bearing races are installed with a press fit inside the swing arm and should not be removed unless bearing replacement is required.

Damaged swing arm bearings can affect wheel alignment and handling.

NOTE
Identify the bearings so they can be reinstalled in their original position.

1. Remove the bearings (with seals attached) from each side of the swing arm (**Figure 54**).
2. Clean the bearing (A, **Figure 55**) and race (B) of all old grease.
3. Check each bearing and race for pitting, flat spots or overheating (blue discoloration). Check the seals for wear, tearing or leakage.

REAR SUSPENSION AND FINAL DRIVE

7. Install the bearing until the seal seats fully in the swing arm bore (**Figure 56**).

Bearing Race Replacement

Each bearing assembly consists of the bearing, seal, race (**Figure 57**) and grease retainer. The bearing races (B, **Figure 55**) are pressed into the swing arm. Remove the races only if the bearings must be replaced. The grease retainer behind each race is removed when the race is removed. Always replace left and right bearing sets at the same time.

NOTE
*For general information on bearing race replacement, refer to **Bearings** in Chapter One.*

Tools

The following tools (or equivalents) are required to replace the bearing races and grease retainers:
1. Slide hammer. Available from tool and automotive parts supply houses.
2. Bearing attachment (32 × 35 mm). (Honda part No. 07746-0010100) or equivalent.

NOTE
The bearing attachment will work with a slide hammer with 3/8 × 16 in. screw threads.

3. Bearing driver, 42 × 47 mm. (Honda part No. 07746-0010200) or equivalent.
4. Bearing driver handle (Honda part No. 07749-0010000) or equivalent.
5. Hydraulic press.

Procedure

1. If necessary, remove the shock arm (**Figure 58**).
2. Remove the bearing race and grease retainer from the right side:
   a. Drill a 1/2 in. hole in the right side grease retainer.
   b. Insert the end of the slide hammer shaft through the hole (from the outside) and attach the driver onto the backside of the grease retainer with the nut. See **Figure 59**.
   c. Support the swing arm in a vise with soft jaws.
   d. Operate the slide hammer to remove the race and grease retainer. Discard both parts.
3. Remove the bearing race and the grease retainer from the left side:

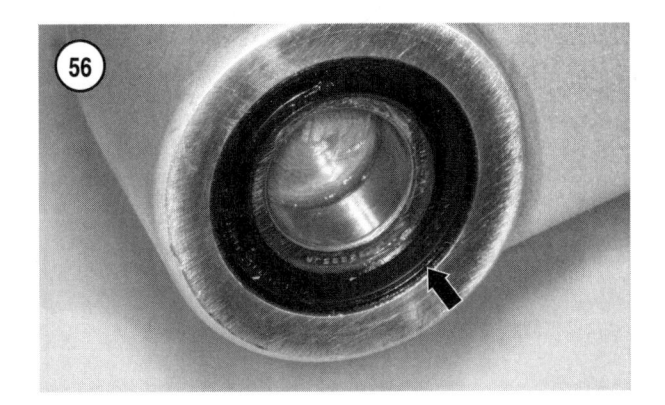

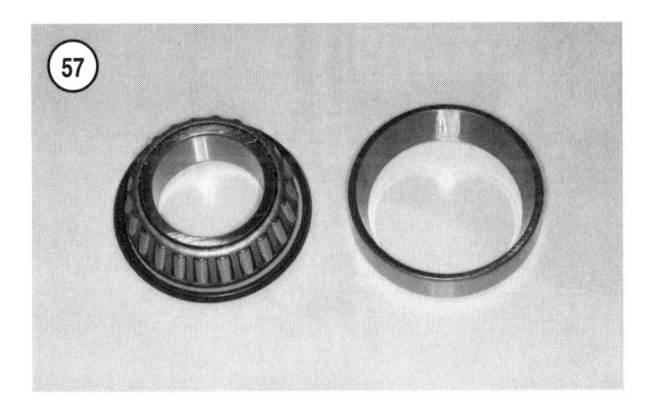

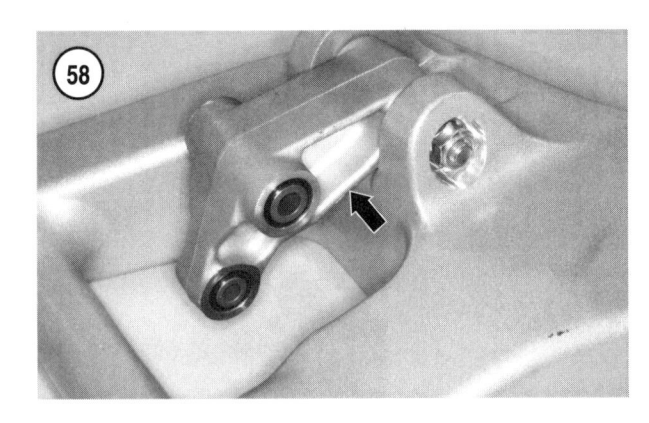

4. If the bearing, seal or race is damaged, replace the bearing as an assembly as described under *Bearing Race Replacement* in this section. The seal is integral with the bearing and cannot be replaced separately.

5. Check the grease holder (installed behind each race) for a loose fit or damage. The grease holders and bearing races are removed at the same time.

6. Lubricate the seal lips, bearing and race with grease. If the original bearing was cleaned with a degreaser, or a new bearing installed, thoroughly work the grease between the rollers.

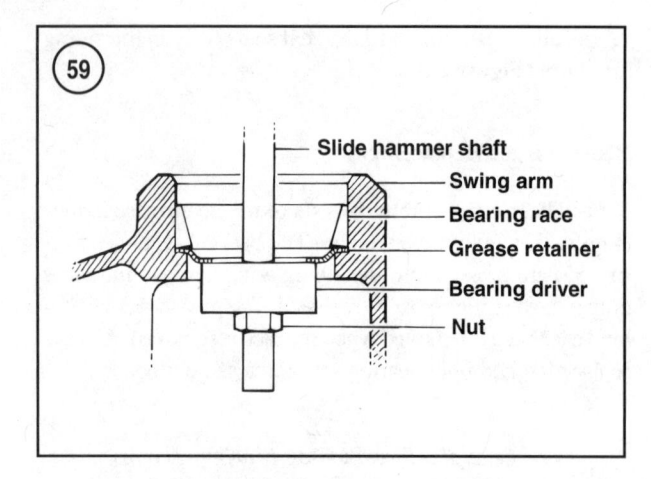

a. Install a suitable bearing driver onto a handle and insert the driver through the swing arm from the right side (**Figure 60**).
b. Position the driver against the grease retainer and press or drive out the bearing race and grease retainer. Discard both parts.
4. Clean the bearing bores in both sides of the swing arm.
5. Install the new grease retainer and race:
   a. Support the swing arm in a press.
   b. Fit the new grease retainer into the swing arm pivot (**Figure 61**).
   c. Center the new race into the bearing bore (**Figure 61**).
   d. Press the new race into the bearing bore using the 42 × 47 mm attachment and press. Make sure the race is properly seated within the bearing bore (B, **Figure 55**).
   e. Repeat for the opposite side.
6. If removed, reinstall the shock arm. Tighten the nut to 64 N•m (47 ft.-lb.).

Drive Shaft
Inspection

1. Hold the drive shaft and check that the universal joint (C, **Figure 45**) moves smoothly. If there is any roughness or noticeable play, replace the drive shaft assembly.

CAUTION
*The universal joint (C, **Figure 45**) is permanently attached to the drive shaft. Do not disassemble or repair the universal joint or drive shaft.*

2. Inspect the drive shaft and universal joint splines for wear or damage.
3. Inspect the boot for any tears or other damage.

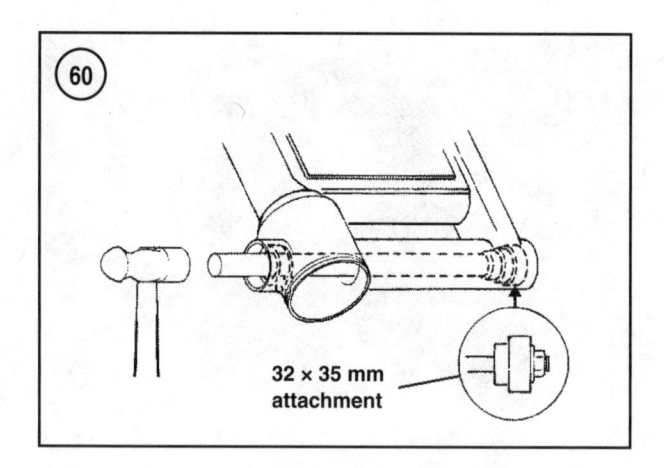

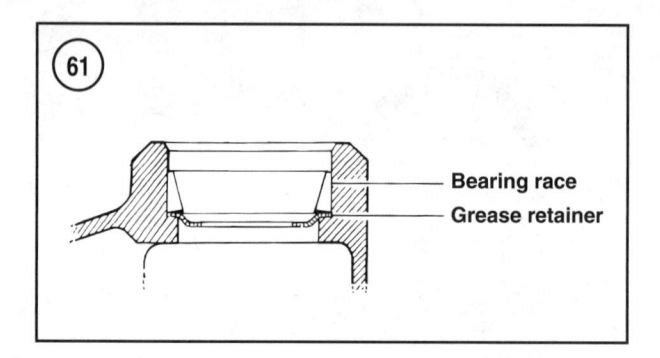

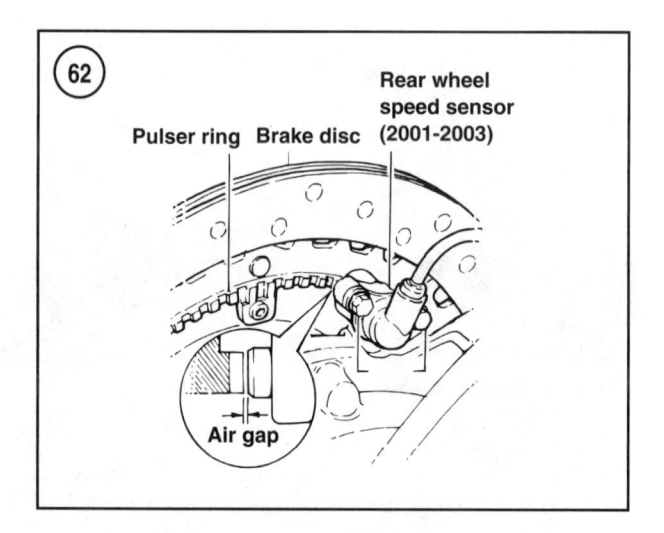

FINAL DRIVE HOUSING

Removal

1. Support the motorcycle on its centerstand.
2. Remove the right muffler:
   a. Loosen the two clamp Allen bolts.

REAR SUSPENSION AND FINAL DRIVE

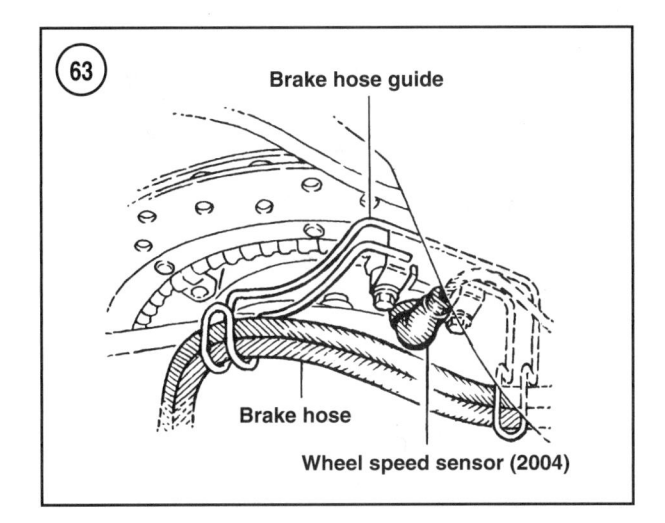

b. Remove the mounting bolt and washer and remove the muffler.

3. Remove the rear wheel. Refer to *Rear Wheel* in Chapter Eleven.

4A. On 2001-2003 GL1800A models, unbolt and remove the wheel speed sensor (**Figure 62**).

4B. On 2004 GL1800A models, unbolt and remove the brake hose guide and wheel speed sensor (**Figure 63**). Note how the wires and hoses are routed for reinstallation.

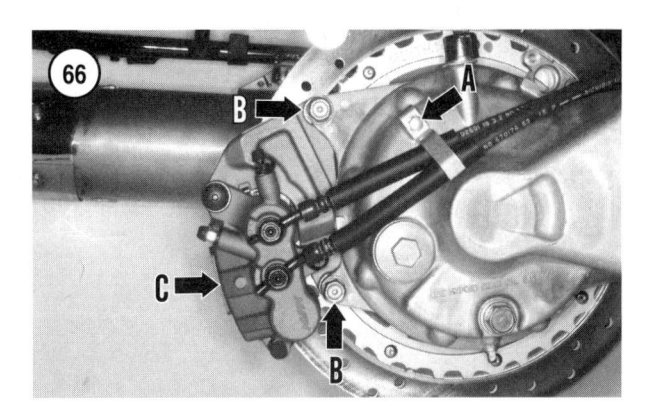

5. Loosen the two brake disc mounting screws (A, **Figure 64**) with a hand impact driver and No. 3 Phillips bit:

CAUTION
These screws use a threadlock. Engage these screws fully with the Phillips bit to avoid damaging them.

   a. Turn the brake disc so that one of the screws is at the bottom.
   b. Have an assistant apply the rear brake.

NOTE
*If the left muffler is installed on the motorcycle, mount a long extension on the impact driver when loosening the brake disc mounting screws (**Figure 65**).*

   c. Loosen, but do not remove, the two Phillips screws (A, **Figure 64**).

6. Remove the rear brake caliper and brake lines as follows:

NOTE
Do not disconnect the brake lines from the caliper.

   a. 2001-2003: Remove the bolt and hose clamp (A, **Figure 66**) at the final drive housing.
   b. Remove the two bolts (B, **Figure 66**) and rear brake caliper (C). Support the brake caliper above the swing arm with stiff wire.

7. Remove the two screws (A, **Figure 64**) and the brake disc (B).

8. Remove the nuts (A, **Figure 67**), washers and the final drive assembly. See **Figure 68**.

9. Locate and remove the two dowel pins, if necessary.

10. Secure the final drive housing and pull hard on the joint shaft (**Figure 69**) to dislodge its stopper ring from the groove in the pinion joint. See **Figure 70**.

11. Inspect the joint shaft and final drive assembly as described under *Drive Housing Inspection* in this section.

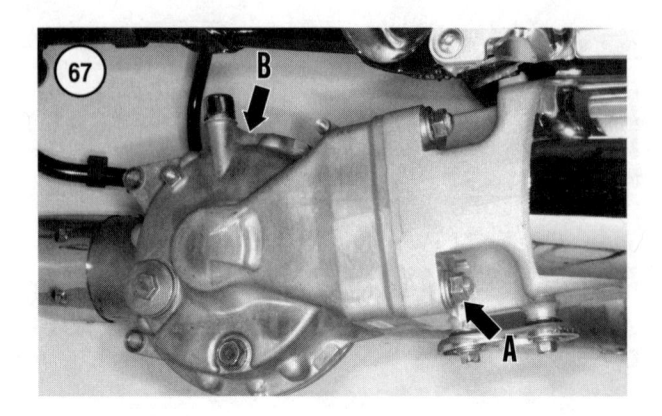

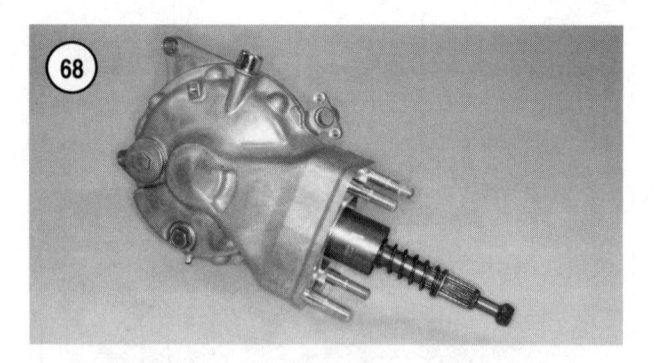

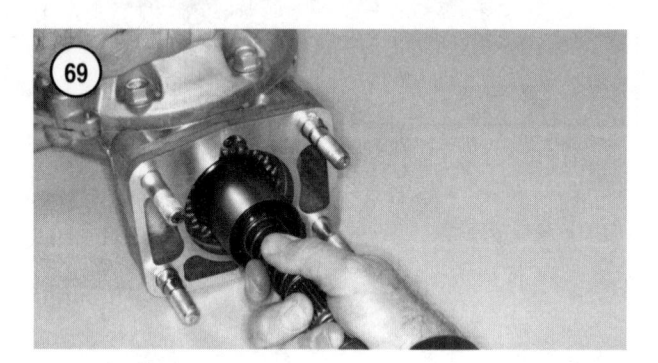

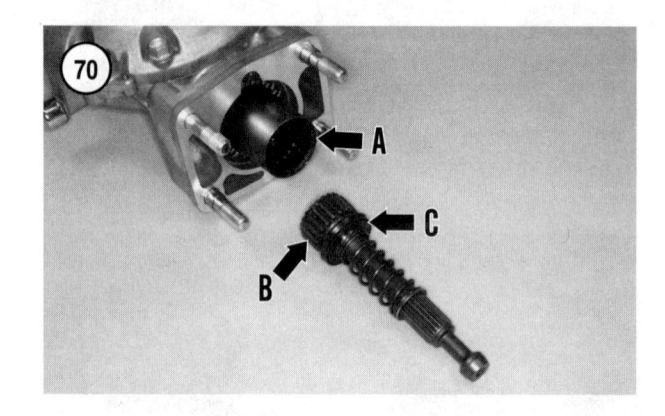

Installation

1. Install a new stopper ring and seal onto the joint shaft as described under *Joint Shaft Inspection/Overhaul* in this section.
2. Lubricate the following with molybdenum disulfide grease:
    a. Pinion joint splines (A, **Figure 70**).
    b. Joint shaft splines (B, **Figure 70**).
    c. Joint shaft seal lips (C, **Figure 70**).
3. Insert the joint shaft into the pinion joint (**Figure 69**) and push hard to seat the stopper ring on the joint shaft into the groove in the pinion joint. See **Figure 71**. Pull the joint shaft out slightly to check that the stopper ring seats in the groove completely.
4. Lubricate the joint shaft splines (A, **Figure 71**) with molybdenum disulfide grease.
5. Install the two dowel pins (B, **Figure 71**), if removed.
6. Clean the final drive housing and swing arm mating surfaces.
7. Install the final drive assembly so the end of the joint shaft enters the drive shaft. Continue to hold the final drive housing while turning the final drive gear to engage the splines on the joint shaft and drive shaft. Then push the final drive housing (B, **Figure 67**) all the way against the swing arm.
8. Install the washers and nuts (A, **Figure 67**). Tighten the final drive housing nuts to 88 N•m (65 ft.-lb.).
9. Clean the brake disc and final drive housing mating surfaces.
10. Install the brake disc (B, **Figure 64**) and secure with two new ALOC mounting screws (A). Tighten the screws to 9 N•m (80 in.-lb.).
11. Install the brake caliper (C, **Figure 66**):
    a. Secure the brake caliper with two new ALOC mounting bolts (B, **Figure 66**). Tighten the bolts to 45 N•m (33 ft.-lb.).
    b. 2001-2003: Secure the brake hoses at the final drive housing with the clamp and bolt (A, **Figure 66**). Tighten to 12 N•m (106 in.-lb.).
12A. On 2001-2003 GL1800A models, install the wheel speed sensor (**Figure 62**) and tighten the mounting bolts to 12 N•m (106 in.-lb.).
12B. On 2004 GL1800A models, install the wheel speed sensor (**Figure 63**), brake hose guide and brake hoses and tighten the mounting bolts to 12 N.m (106 in.-lb.).
13. Install the rear wheel. Refer to *Rear Wheel* in Chapter Eleven.
14. If the original oil was drained, refill the final drive housing with the recommended type and quantity gear oil. Refer to *Final Drive Oil* in Chapter Three.

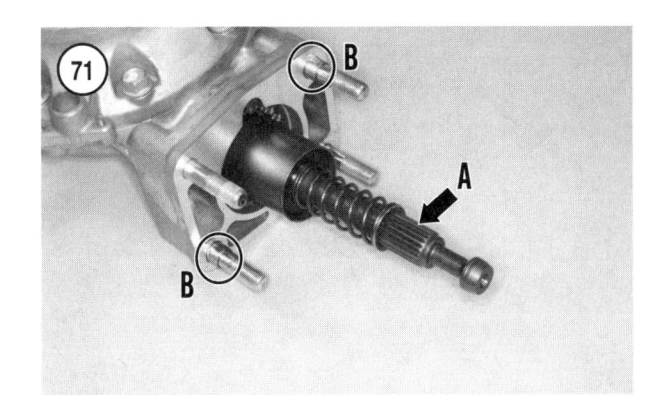

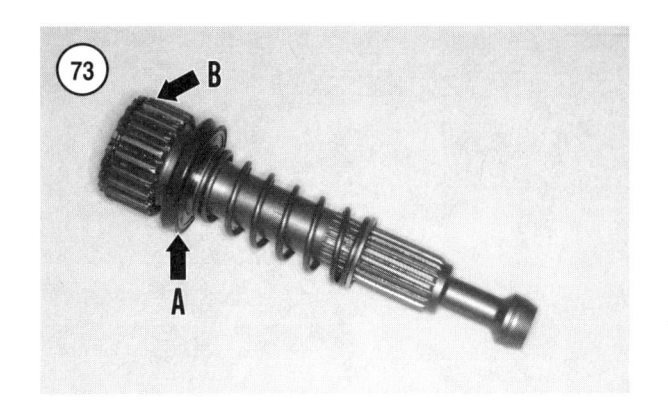

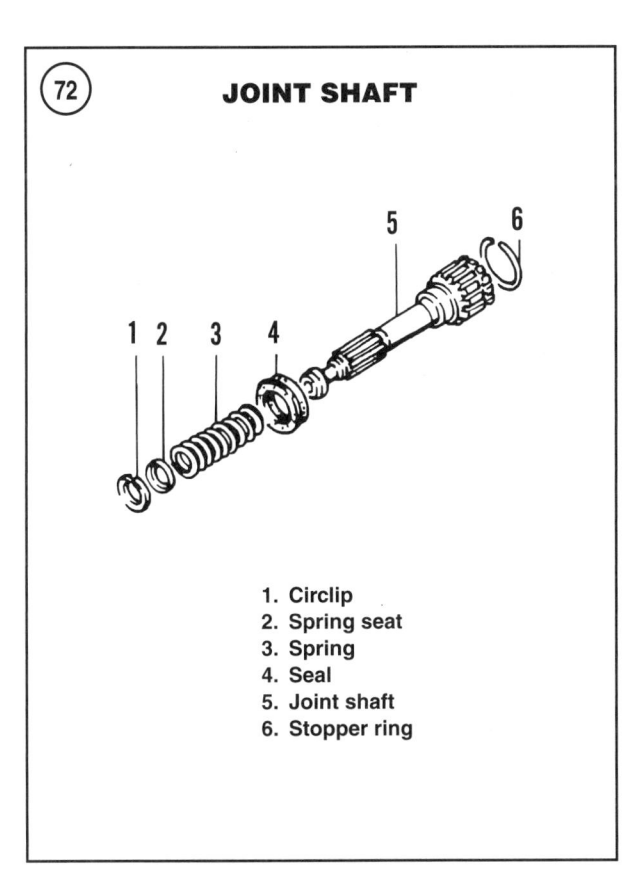

REAR SUSPENSION AND FINAL DRIVE

JOINT SHAFT

1. Circlip
2. Spring seat
3. Spring
4. Seal
5. Joint shaft
6. Stopper ring

Joint Shaft Inspection/Overhaul

The joint shaft connects the drive shaft to the final drive housing (**Figure 72**).

1. Remove the circlip and disassemble the joint shaft in the order shown in **Figure 72**.
2. Discard the circlip, seal and stopper ring.
3. Clean the spring seat, spring and joint shaft.
4. Inspect the spring seat and spring for damage.
5. Inspect the joint shaft splines for wear and damage. Inspect the circlip and stopper ring grooves for cracks and other damage.

NOTE
*If the joint shaft splines are damaged, check the drive shaft and pinion joint splines for damage. The swing arm must be removed to access and inspect the drive shaft splines. Refer to **Swing Arm** in this chapter.*

6. Lubricate the lips of a new seal with molybdenum disulfide grease. Install the seal (A, **Figure 73**) with its flat side facing toward the front (spring side).
7. Install the spring, spring seat and a new circlip. Make sure the circlip seats in the groove completely.
8. Install a new stopper ring into the joint shaft spline groove (B, **Figure 73**).

CAUTION
The stopper ring is a loose fit in the joint shaft spline groove. Do not pinch or squeeze the ring to tighten it.

Final Drive Housing Inspection

1. Inspect the housing for any external damage.
2. Turn the pinion joint (A, **Figure 74**) and check that the ring gear (B) turns smoothly. If there is any binding or

15. Install the right muffler:
   a. Install a new gasket over the right exhaust pipe, if necessary.
   b. Slide the muffler over the gasket and install its rear mounting bolt and washer.
   c. Tighten the two clamp Allen bolts to 26 N•m (20 ft.-lb.).
   d. Tighten the muffler mounting bolt securely.
16. Check rear brake operation.

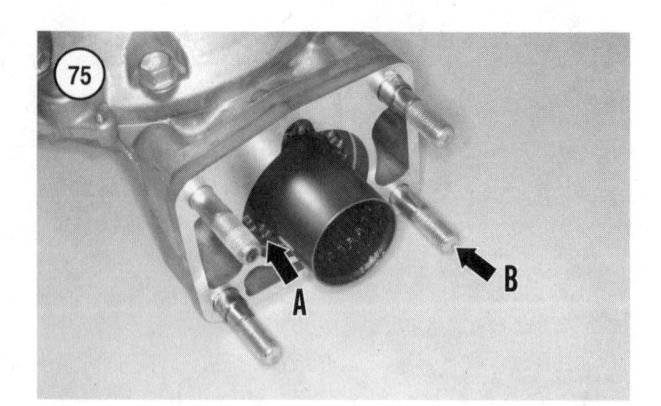

roughness, the pinion and/or ring gears may be damaged. Refer service to a Honda dealership.

3. Inspect the area around the pinion joint (A, **Figure 75**) for oil leaks. If there is an oil leak in this area, have a Honda dealership replace the pinion seal.

4. Inspect the area around the ring gear case cover (B, **Figure 74**) for oil leaks. If there is an oil leak, have a Honda dealership replace the oil seal and O-ring installed under the side flange.

5. Check for loose or damaged final drive housing studs (B, **Figure 75**). Tighten loose studs, or replace damaged studs; Refer to *Basic Service Methods* in Chapter One. Position the studs so the distance from the top of each stud to the final drive housing mating surface is 39.5-41.5 mm (1.5-1.6 in.).

6. Measure the gear assembly preload as follows:
   a. Support the final drive housing so it cannot move.
   b. Attach a socket to a beam type torque wrench that can read either N•m or in.-lb.
   c. Rotate the torque wrench (**Figure 76**) until the pinion gear shaft starts to turn and read the preload on the torque wrench. The correct preload reading is 0.2-1 N•m (1.7-8.7 in.-lb.).
   d. If the preload reading is incorrect, the bearings or ring gears in the final drive housing assembly may be incorrectly installed (if the assembly was rebuilt) or damaged. Refer service to a Honda dealership.

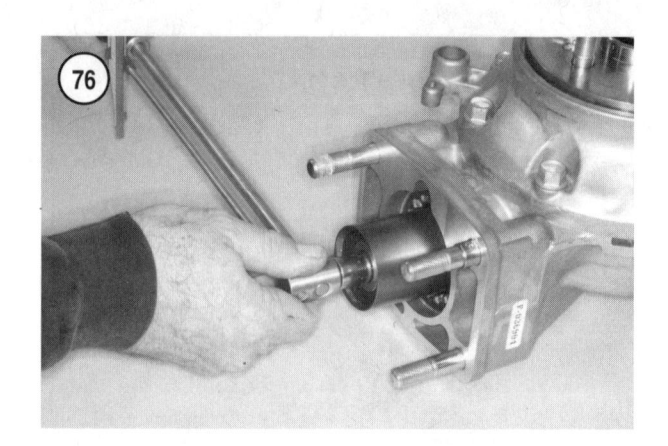

Table 1 REAR SUSPENSION SPECIFICATIONS

Rear axle travel	105 mm (4.1 in.)

Table 2 FINAL DRIVE SPECIFICATIONS

Final drive gear assembly preload	0.2-1 N•m (1.7-8.7 in.-lb.)
Final drive gear backlash	
Standard	0.05-0.15 mm (0.002-0.006 in.)
Service limit	0.30 mm (0.012)
Final drive gear oil type	SAE 80 hypoid gear oil
Final drive oil capacity	
Oil change	120 cc (4.1 U.S. oz.)
After drive gear disassembly	150 cc (5.1 U.S. oz.)

Table 3 REAR SUSPENSION AND FINAL DRIVE TORQUE SPECIFICATIONS

	N•m	in.-lb.	ft.-lb.
Brake hose clamp bolt	12	106	–
Brake pipe clamp bolts	12	106	–
Final drive housing mounting nuts	88	–	65
Final side flange screw[1]	9	80	–
Gear case cover			
10-mm bolt[2]	62	–	46
8-mm bolt[2]	25	–	18
Pinion joint nut[2]	108	–	80
Pinion retainer lock tab bolt	10	88	–
Pinion retainer[3]	147	–	108
Rear brake caliper mounting bolts[1]	45	–	33
Rear brake disc screws[1]	9	80	–
Rear brake light/cruise cancel switch			
holder screw[2]	2	18	–
Rear master cylinder mounting bolts	12	106	–
Rear pulser ring bolts (ABS)[1]	8	71	–
Rear wheel nuts	108	–	80
Shock absorber mounting nuts[4]	42	–	31
Shock arm-to-shock link nut[4]	64	–	47
Shock arm-to-swing arm nut[4]	64	–	47
Shock link-to-frame nut[4]	64	–	47
Suspension level actuator mounting			
bolt	26	–	19
Swing arm left pivot bolt	34	–	25
Swing arm left pivot locknut	108	–	80
Swing arm right pivot bolt	108	–	80
Wheel speed sensor bolts			
(2001-2003 ABS)	12	106	–
Wheel speed sensor and brake hose			
guide bolt (2004 ABS)	12	106	–

1. Install new ALOC fastener during assembly.
2. Apply a medium strength threadlock onto fastener threads.
3. Lubricate threads and seating surface with engine oil.
4. Replace U-nut with same type of fastener.

CHAPTER FOURTEEN

BRAKES

Two different brake systems are used on the models covered in this manual. The GL1800 (standard model) is equipped with a linked brake system (LBS). This system is designed to operate both the front and rear brakes when either the front brake lever or the rear brake pedal is applied. Hydraulic devices in the LBS system control brake action depending on whether the front brake, rear brake or both brakes are applied. The LBS system is a hydraulic system—no electronic controls are used.

The GL1800A model is equipped with the linked braking system (LBS) and an Anti-lock brake system (ABS). The ABS system is an electronically controlled hydraulic system designed to prevent wheel lockup during hard braking or when braking on slippery and loose road surfaces. During normal braking operations, the ABS system operates like the LBS system used on the standard GL1800 model. When the wheel is about to lock, the ABS function modulates the hydraulic pressure in the system by reducing pressure at the brake calipers. When the system senses that the wheel lock condition is reduced, full hydraulic pressure to the calipers is restored. Hydraulic pressure is regulated continuously.

Refer to Chapter Fifteen for ABS components.

Tables 1-3 (end of chapter) lists front and rear brake specifications.

LINKED BRAKE SYSTEM (LBS)

The LBS (**Figure 1**) provides simultaneous braking action to both the front and rear brakes when either the front brake lever or rear brake pedal is applied. The LBS is a fully hydraulic system. Two separate hydraulic systems provide optimum balance and operation when either the front brake lever or rear brake pedal is applied. No electronic controls are used.

Brake Calipers

The three brake calipers in the LBS system are single action, three-piston types and controlled by two separate hydraulic systems (**Figure 1**) as follows:
1. The front brake lever operates the following caliper pistons at the same time:
    a. Outer two pistons of the front right side caliper.
    b. The center piston of the front left side caliper.
    c. The center piston of the rear brake caliper.

BRAKES

LINKED BRAKE SYSTEM

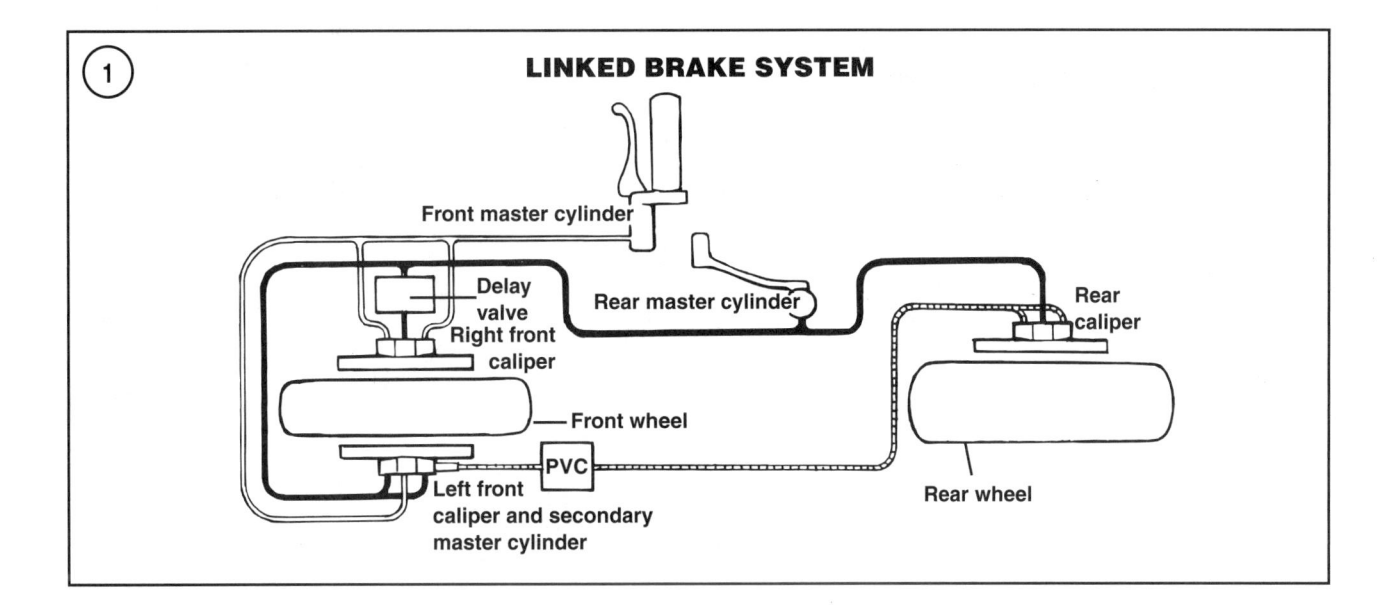

(1) Diagram showing Front master cylinder, Rear master cylinder, Delay valve, Right front caliper, Left front caliper and secondary master cylinder, PVC, Front wheel, Rear caliper, Rear wheel.

2. The rear brake pedal operates the following caliper pistons at the same time:
   a. Center piston of the rear caliper.
   b. Outer two pistons of the front left side caliper.
   c. Center piston of the front right side caliper.

Proportional Control Valve (PCV)

The PCV valve (**Figure 2**) connects the secondary master cylinder to the center piston of the rear brake caliper. The PCV regulates hydraulic pressure from the secondary master cylinder to the rear brake caliper.

Delay Valve

The delay valve (**Figure 3**) connects the rear master cylinder to both front brake calipers. The delay valve engages the front brake calipers separately to reduce fork dive when only the rear brake pedal is applied. It does this by first engaging the left front brake caliper. Then, as the rear brake pedal pressure increases, pressure is applied from the delay valve to the right front brake caliper. At a predetermined level, the pressure to both front brake calipers is equalized.

Anti-Dive Operation

The anti-dive system reduces fork compression during braking. An anti-dive valve assembly is mounted on the left front fork slider and is part of the fork damping circuit. This system is brake-activated via the secondary master cylinder. During normal fork operation, fork oil flows through the anti-dive valve. When the brake lever or brake pedal is applied, hydraulic pressure from the secondary master cylinder closes a plunger in the anti-dive valve. This restricts the flow of fork oil through the anti-dive valve and reduces fork compression.

BRAKE FLUID SELECTION

When adding brake fluid, use DOT 4 brake fluid from a sealed container. Brake fluid is glycol-based and draws moisture, greatly reducing its ability to perform correctly. Purchase brake fluid in small containers and discard any small leftover quantities.

CAUTION
Do not intermix DOT 5 (silicone-based) brake fluid, as it can cause brake system failure.

PREVENTING BRAKE FLUID DAMAGE

Many of the procedures in this chapter require handling brake fluid. Be careful not to spill any fluid, as it stains or damages most surfaces. To prevent brake fluid damage, note the following:

1. Before performing any procedure in which brake fluid could contact the motorcycle, cover the work area with a large piece of plastic. It only takes a few drops of brake fluid to cause damage.
2. Before handling brake fluid or working on the brake system, fill a small container with soap and water and keep it close to the bike while working. If brake fluid contacts the bike, clean the area and rinse it thoroughly.

3. To help control the flow of brake fluid when filling the reservoirs, punch a small hole into the seal of a new container next to the edge of the pour spout.

BRAKE SERVICE

When working on the brake system, the work area and all tools must be absolutely clean. Any tiny particles of dirt or debris in the caliper assembly or master cylinder can damage the components and prevent the system from functioning properly.

Consider the following when servicing the disc brakes.

WARNING
*Do **not** use compressed air to blow off brake parts. It may contain asbestos, which can cause lung injury and cancer. Wear a face mask that meets OHSA requirements for trapping asbestos particles, and wash hands and forearms thoroughly after completing the work. Clean brake components with water or an aerosol brake parts cleaner before working on the brake system. Dispose of all brake dust and cleaning materials properly.*

1. Do not allow disc brake fluid to contact any plastic parts or painted surfaces, as damage will result.
2. Always keep the master cylinder reservoir and spare cans of brake fluid closed to prevent dust or moisture from entering. This contaminates the brake fluid and can cause brake failure.
3. Handle the brake components carefully when servicing them. Use only DOT 4 brake fluid or isopropyl alcohol to wash rubber parts in the brake system. Never allow any petroleum-based cleaner to contact any of the rubber parts. These chemicals cause the rubber to swell, requiring their replacement.
4. Do not allow any grease or oil to contact the brake pads.
5. When cleaning the brake components, wear rubber gloves to keep brake fluid off skin.
6. When loosening any brake hose banjo bolt or brake pipe nut, the brake system is open. The system must be bled to remove air bubbles. Also, if the brake feels spongy, this usually means there are air bubbles in the system. Bleed the brakes, as described in this chapter.

CAUTION
Never reuse brake fluid (fluid expelled during brake bleeding). Contaminated brake fluid can cause brake failure. Dispose of used brake fluid in an environmentally safe manner.

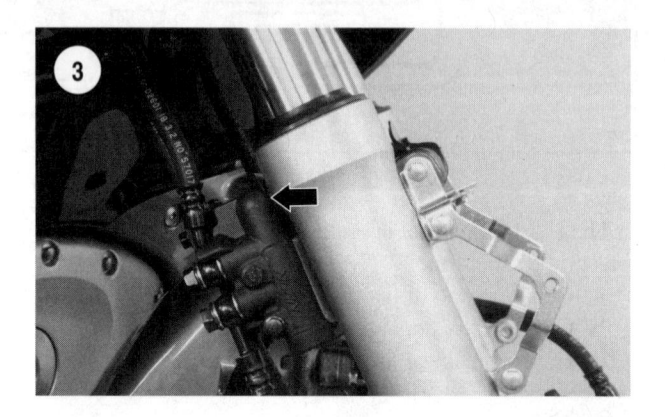

BRAKE PADS

There is no recommended mileage interval for changing the brake pads. Pad wear depends greatly on riding habits and the condition of the brake system. As the brake pads wear, the brake fluid level drops in the reservoir and automatically adjusts for wear. Replace the brake pads if

BRAKES

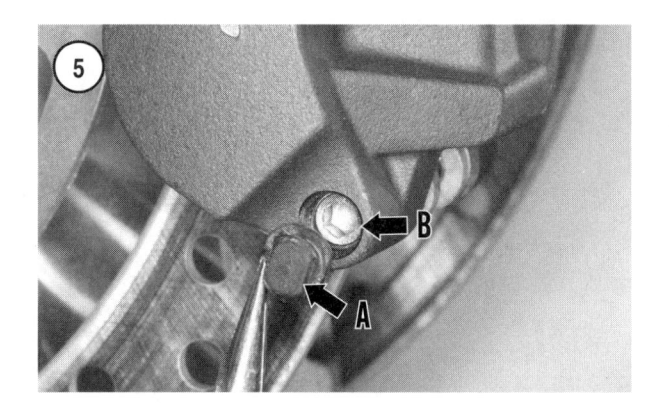

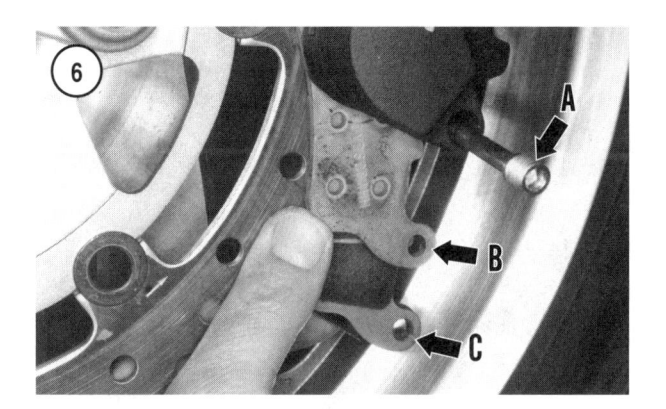

the pad wear indicator is worn to the edge of the brake pad. See **Figure 4**, typical.

CAUTION
Because the wear indicator is very close to the metal backing plate, check the brake pads more frequently when the wear limit line approaches the disc. If pad wear is uneven, the backing plate may contact the disc and cause damage.

The brake pads can be replaced with the brake calipers mounted on the bike.

Front Brake Pad Replacement

Always replace the front brake pads in sets by servicing both front calipers at the same time. Never use one new brake pad with a used brake pad in a caliper. Never replace the brake pads in one brake caliper without replacing them in the other caliper. To do so causes an unbalanced braking condition.

1. Read the information listed under *Brake Service* in this chapter.

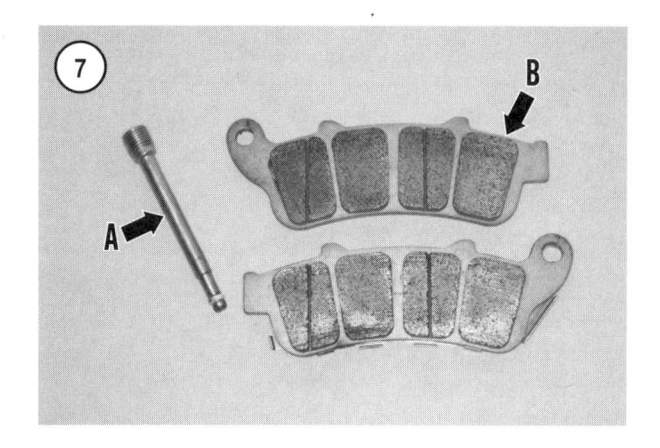

2. Remove the front master cylinder cover and use a large syringe to remove and discard about 50 percent of the fluid from the reservoir. This step prevents the master cylinder from overflowing when the caliper pistons are compressed for reinstallation. Do not drain the entire reservoir or air will enter the system.

CAUTION
Do not allow the master cylinder reservoir to overflow when performing Step 3. Brake fluid damages most surfaces it contacts.

3. Hold the caliper housing from the outside and push it toward its brake disc (**Figure 4**). This pushes the pistons into the caliper to make room for the new brake pads.

NOTE
The pistons should move smoothly when compressing them in Step 3. If not, check the caliper for sticking pistons or damaged caliper bores, pistons and seals. Repair requires overhauling the brake caliper assembly. Also note that road debris collected on the exposed part of the pistons can tear the seals as the pistons pass through them.

4. Remove the pad pin plug (A, **Figure 5**) and loosen the pad pin (B).
5. Remove the pad pin (A, **Figure 6**) and both brake pads (B and C).
6. Make sure the pad spring is in good condition and installed inside the caliper. Replace the pad spring if it appears weak or damaged.
7. Inspect the pad pin (A, **Figure 7**) for excessive wear, corrosion or damage. Remove corrosion and dirt from the pad pin surface. A dirty or damaged pad pin surface prevents the brake pads from sliding properly and causes brake drag and overheating of the brake disc.

14

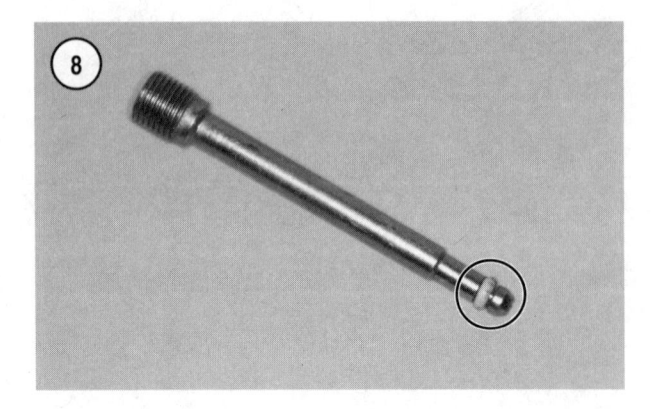

8. Replace the stopper ring in the end of the pad pin (**Figure 8**) if damaged.
9. Inspect the brake pads (B, **Figure 7**) as follows:
   a. Inspect the friction material for light surface dirt, grease and oil contamination. Remove light contamination with sandpaper. If the contamination has penetrated the surface, replace the brake pads.
   b. Inspect the brake pads for excessive wear or damage. Replace the brake pads when the friction material is worn to the wear indicator line (**Figure 3**).
   c. Inspect the brake pads for uneven wear. If one pad has worn more than the other, it may be binding or the caliper is not sliding properly.
   d. Inspect the shim (**Figure 9**) on the backside of the inner pad for corrosion, tightness and damage.

NOTE
*If brake fluid is leaking from around the pistons, refer to **Brake Calipers** in this chapter.*

10. Service the brake disc as follows:
    a. Use brake cleaner and a fine-grade emery cloth to remove road debris and brake pad residue from the brake disc. Clean both sides of the disc.

CAUTION
Cleaning the brake disc is especially important if changing brake pad compounds. Many compounds are not compatible with each other.

    b. Check the brake disc for wear as described in this chapter.

WARNING
Use just enough grease to lubricate the O-ring in the next step. Excess grease could contaminate the brake pads.

11. Lubricate the stopper ring on the end of the pad pin (**Figure 8**) with silicone brake grease.

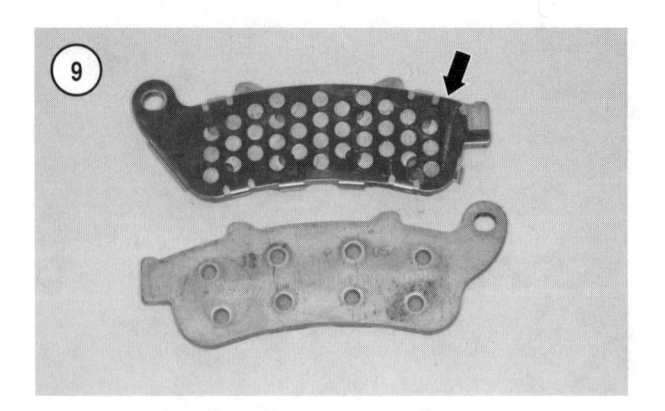

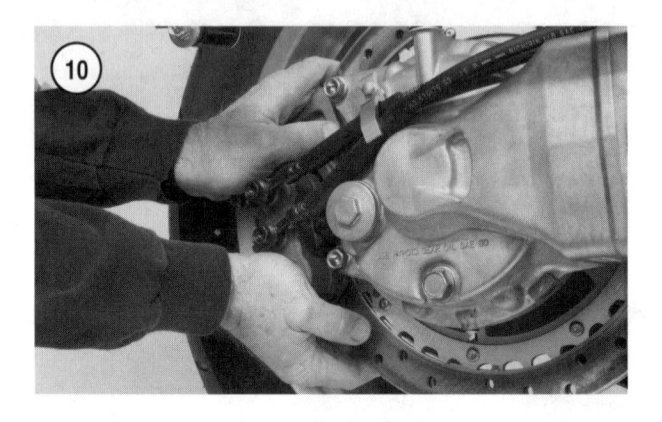

12. Install the brake pads into the caliper so the friction material on both brake pads faces toward the brake disc. The pad with the shim is the inner pad (C, **Figure 6**). Insert the extended arm on the end of each pad into the pad retainer in the caliper bracket.

13. Push both brake pads against the pad spring and install the pad pin (B, **Figure 5**) through the brake caliper and brake pad holes.

14. Tighten the pad pin (B, **Figure 5**) to 18 N•m (13 ft.-lb.).

15. Install the pad pin plug (A, **Figure 5**).

16. Repeat for the opposite brake assembly.

17. Operate the front brake lever and rear brake pedal to seat the pads against the disc, then check the brake fluid level in the reservoir. If necessary, add new DOT 4 brake fluid (Chapter Three).

WARNING
Do not ride the motorcycle until the front and rear brakes operate correctly with full hydraulic advantage.

BRAKES

from overflowing when the caliper pistons are compressed for reinstallation. Do not drain the entire reservoir or air will enter the system.

CAUTION
Do not allow the master cylinder reservoir to overflow when performing Step 3. Brake fluid damages most surfaces it contacts.

3. Hold the caliper housing from the outside and push it toward its brake disc (**Figure 10**). This pushes the pistons into the caliper to make room for the new brake pads.

NOTE
The pistons should move smoothly when compressing them in Step 3. If not, check the caliper for sticking pistons or damaged caliper bores, pistons and seals. Repair requires overhauling the brake caliper assembly. Also note that road debris collected on the exposed part of the pistons can tear the seals as the pistons pass through them.

4. Remove the pad pin plug (A, **Figure 11**) and loosen the pad pin (B).
5. Remove the pad pin (A, **Figure 12**) and both brake pads (B).
6. Make sure the pad spring is in good condition and installed inside the caliper. Replace the pad spring if it appears weak or damaged.
7. Inspect the pad pin (A, **Figure 7**) for excessive wear, corrosion or damage. Remove corrosion and dirt from the pad pin surface. A dirty or damaged pad pin surface prevents the brake pads from sliding properly and causes brake drag and overheating of the brake disc.
8. Replace the stopper ring in the end of the pad pin (**Figure 8**) if damaged.
9. Inspect the brake pads (B, **Figure 7**) as follows:
   a. Inspect the friction material for light surface dirt, grease and oil contamination. Remove light contamination with sandpaper. If the contamination has penetrated the surface, replace the brake pads.
   b. Inspect the brake pads for excessive wear or damage. Replace the brake pads when the friction material is worn to the wear indicator line (**Figure 3**).
   c. Inspect the brake pads for uneven wear. If one pad has worn more than the other, it may be binding or the caliper is not sliding properly.
   d. Inspect the shim (**Figure 13**) on the backside of each pad for corrosion, tightness and damage.

NOTE
*If brake fluid is leaking from around the pistons, refer to **Brake Calipers** in this chapter.*

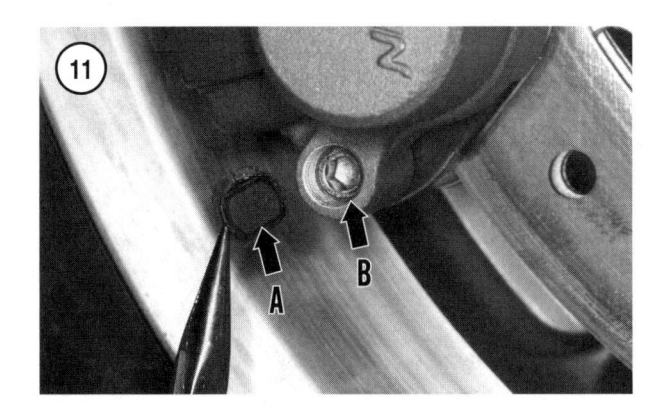

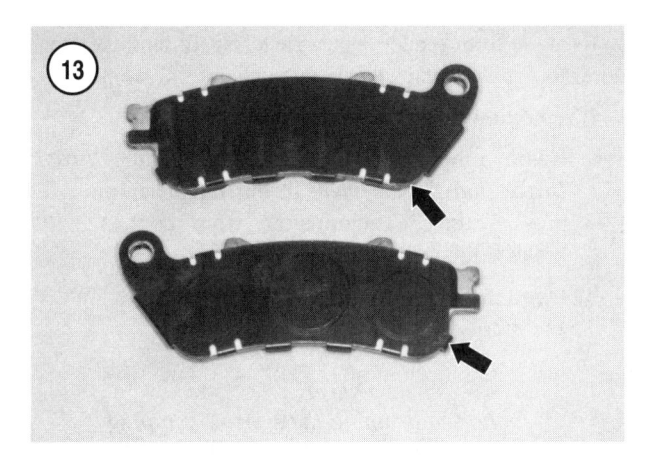

Rear Brake Pad Replacement

Always replace the rear brake pads in pairs. Never use one new brake pad with a used brake pad in the caliper.

1. Read the prior information listed under *Brake Service* in this chapter.
2. Remove the rear master cylinder cover and use a large syringe to remove and discard about 50 percent of the fluid from the reservoir. This prevents the master cylinder

10. Service the brake disc as follows:
   a. Use brake cleaner and a fine-grade emery cloth to remove road debris and brake pad residue from the brake disc. Clean both sides of the disc.

   CAUTION
   Cleaning the brake disc is especially important if changing brake pad compounds. Many compounds are not compatible with each other.

   b. Check the brake disc for wear as described in this chapter.

WARNING
Use just enough grease to lubricate the stopper ring in the next step. Excess grease could contaminate the brake pads.

11. Lubricate the stopper ring on the end of the pad pin (**Figure 8**) with silicone brake grease.
12. Install the brake pads into the caliper (B, **Figure 12**) so the friction material on both brake pads faces toward the brake disc. Insert the extended arm on the end of each pad into the pad retainer in the caliper bracket (**Figure 14**).
13. Push both brake pads against the pad spring and install the pad pin (A, **Figure 12**) through the brake caliper and brake pad holes.
14. Tighten the pad pin (B, **Figure 11**) to 18 N•m (13 ft.-lb.).
15. Install the pad pin plug (A, **Figure 11**).
16. Operate the front brake lever and rear brake pedal to seat the pads against the disc, then check the brake fluid level in the reservoir. If necessary, add new DOT 4 brake fluid (Chapter Three).

WARNING
Do not ride the motorcycle until the front and rear brakes operate correctly with full hydraulic advantage.

BRAKE CALIPERS

This section describes removal/installation and overhaul of the front and rear brake calipers.

The brake calipers can be removed without having to disconnect their brake hoses. Disconnect the brake hoses if the caliper requires overhaul.

Left Brake Caliper
Removal

1. Support the motorcycle on its centerstand.

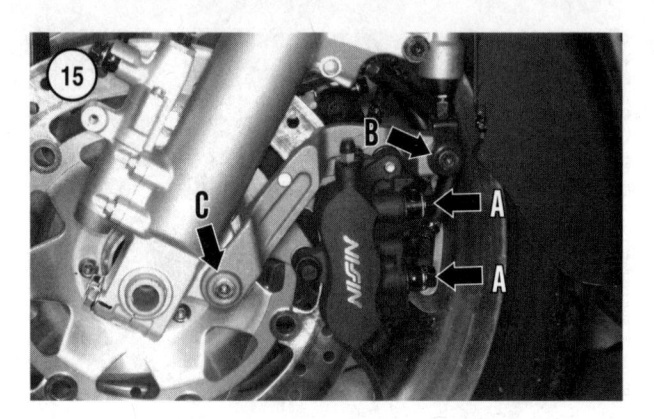

2. Remove front fender A and the left front fender cover. Refer to *Front Fender Assembly* in Chapter Seventeen.

3. If removing the caliper from the motorcycle:
   a. Remove the brake pads as described under *Front Brake Pad Replacement* in this chapter. This will prevent their contamination from contact with brake fluid.
   b. Drain the brake fluid from the front master cylinder as described in this chapter.

NOTE
An option to draining the system is to plug the bolt hole openings on the caliper and the banjo fitting on the brake hoses.

4. Remove the two brake hose banjo bolts and washers at the caliper (A, **Figure 15**). Plug the hoses to prevent leakage and hose contamination.

5. Remove the secondary master cylinder joint bolt (B, **Figure 15**) and the caliper pivot bolt (C), then remove the brake caliper.

6. If the brake hoses were not disconnected at the caliper, insert a spacer block between the brake pads and support the caliper with a wire hook.

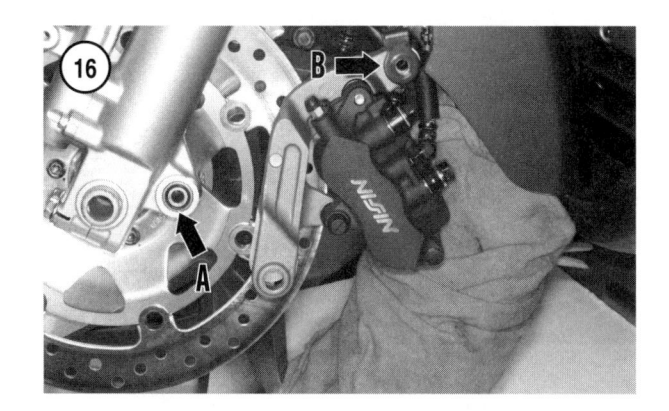

BRAKES

5. Install the caliper assembly over the brake disc. If the pads are installed in the caliper, be careful not to damage their leading edge.

6. Install a new secondary master cylinder joint bolt (B, **Figure 15**) and a new left front brake caliper pivot bolt (C):

   a. Tighten the left brake caliper secondary master cylinder joint bolt (B, **Figure 15**) to 25 N•m (19 ft.-lb.).

   b. Tighten the left front brake caliper pivot bolt (C, **Figure 15**) to 31 N•m (23 ft.-lb.).

NOTE
Figure 17 shows the alignment of the brake hoses on the caliper.

7. Place a new washer on each side of each brake hose (A, **Figure 15**). Then thread the banjo bolts into the caliper and tighten to 34 N•m (25 ft.-lb.).

8. If removed, install the brake pads as described in this chapter.

9. Bleed the brake system as described in this chapter.

10. Operate the front brake lever and rear brake pedal to seat the pads against the brake disc.

11. Install the left front fender cover and front fender A. Refer to *Front Fender Assembly* in Chapter Seventeen.

NOTE
The spacer block prevents the pistons from being forced out of the caliper if the front or rear brakes are applied while the brake caliper is removed from the brake disc.

7. Remove and clean the collar (A, **Figure 16**) and bearing. If the collar appears worn, inspect the bearings for damage. To replace the bearings and/or seals, refer to *Brake Caliper Pivot Bearing Replacement (Left Fork Tube)* in Chapter Twelve.

8. Service the brake caliper as described in this chapter.

Left Brake Caliper
Installation

1. Lubricate the left caliper pivot bearings and collar with grease. Install the collar (A, **Figure 16**) into the bearings.

2. Install the left brake caliper onto the caliper bracket, if removed.

3. Lubricate the upper pivot collar with grease and install it in the caliper bracket (B, **Figure 16**).

4. Remove the spacer block from between the brake pads, if used.

Right Brake Caliper
Removal

1. Support the motorcycle on its centerstand.

2. Remove front fender A and the left front fender cover. Refer to *Front Fender Assembly* in Chapter Seventeen.

3. If the caliper is going to be removed from the motorcycle:

   a. Remove the brake pads as described under *Front Brake Pad Replacement* in this chapter. This will prevent their contamination from contact with brake fluid.

   b. Drain the brake fluid from the front master cylinder as described in this chapter.

NOTE
An option to draining the system is to plug the bolt hole openings on the caliper and the banjo fitting on the brake hoses.

4. Remove the two brake hose banjo bolts and washers at the caliper (A, **Figure 18**). Plug the hoses to prevent leakage and hose contamination.

5A. On GL1800 models, remove the two mounting bolts (B, **Figure 18**) and brake caliper.

5B. On GL1800A models, remove the wheel speed sensor wire harness and brake caliper as follows (**Figure 19**):

a. Remove the bolt and upper clamp from the slider.
b. Remove the two bolts, clamp and wheel speed sensor from the caliper bracket.
c. Remove the two mounting bolts and the brake caliper.

6. If the brake hoses were not disconnected at the caliper, insert a spacer block between the brake pads and support the caliper with a wire hook.

NOTE
The spacer block prevents the pistons from being forced out of the caliper if the front or rear brakes are applied while the brake caliper is removed from the brake disc.

7. Service the brake caliper as described in this chapter.

Right Brake Caliper
Installation

1. Install the right brake caliper onto the caliper bracket, if removed.
2. Remove the spacer block from between the brake pads, if used.
3. Install the caliper assembly over the brake disc. If the pads are installed in the caliper, be careful not to damage their leading edge.
4A. On GL1800 models, install two new right front brake caliper mounting bolts (B, **Figure 18**) and tighten to 31 N•m (23 ft.-lb.).
4B. On GL1800A models, install the brake caliper and route the wheel speed sensor wiring harness as follows (**Figure 19**):
   a. Install the wheel speed sensor onto the slider and tighten the mounting bolts to 12 N•m (106 in.-lb.). Secure the wire harness clamp with the upper mounting bolt.
   b. Route the wire harness along the slider and secure with the upper clamp. Tighten the bolt securely.
5. Place a new washer on each side of each brake hose (A, **Figure 18**). Then thread the banjo bolts into the caliper and tighten to 34 N•m (25 ft.-lb.). Make sure the brake hose stop seats against the caliper as shown in **Figure 18**.
6. If removed, install the brake pads as described in this chapter.
7. Bleed the brake system as described in this chapter.
8. Operate the front brake lever and rear brake pedal to seat the pads against the brake disc.
9. Install the left front fender cover and front fender A. Refer to *Front Fender Assembly* in Chapter Seventeen.

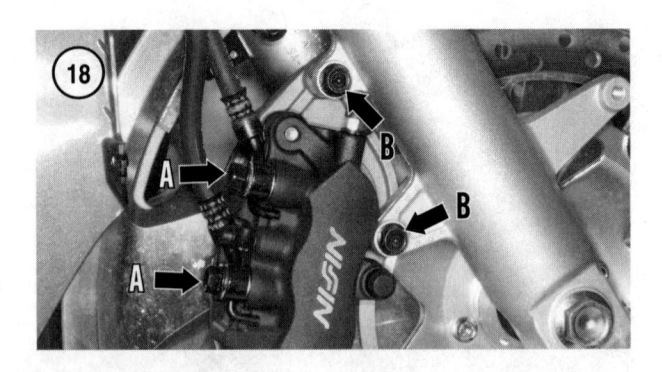

Rear Brake Caliper
Removal

1. Support the motorcycle on its centerstand.
2. Remove the rear wheel (Chapter Eleven).
3. If the caliper is going to be removed from the motorcycle:
   a. Remove the brake pads as described under *Rear Brake Pad Replacement* in this chapter. This will prevent their contamination from contact with brake fluid.
   b. Drain the brake fluid from the rear master cylinder as described in this chapter.

NOTE
An option to draining the system is to plug the bolt hole openings on the caliper and the banjo fitting on the brake hoses.

4. Remove the two brake hose banjo bolts and washers at the caliper (A, **Figure 20**). Plug the hoses to prevent leakage and hose contamination.
5. Remove the two mounting bolts (B, **Figure 20**) and brake caliper.
6. If the brake hoses were not disconnected at the caliper, insert a spacer block between the brake pads and support the caliper with a wire hook.

NOTE
The spacer block prevents the pistons from being forced out of the caliper if the front or rear brakes are applied while the brake caliper is removed from the brake disc.

7. Service the brake caliper as described in this chapter.

Rear Brake Caliper
Installation

1. Install the brake caliper onto the caliper bracket, if removed.

BRAKES

19

- White tape
- 3-way joint
- Banjo bolts
- Clamp (ABS)
- Brake hoses
- Stopper
- Clamp (ABS)
- Delay valve
- Bolt (10-mm head)
- Bolt (8-mm head)
- Speed sensor wire (ABS)
- Speed sensor clamp bolt (ABS)
- Clamp (ABS)
- Caliper mounting bolt
- Caliper mounting bolt
- Clamp (ABS)
- Wheel speed sensor (ABS)

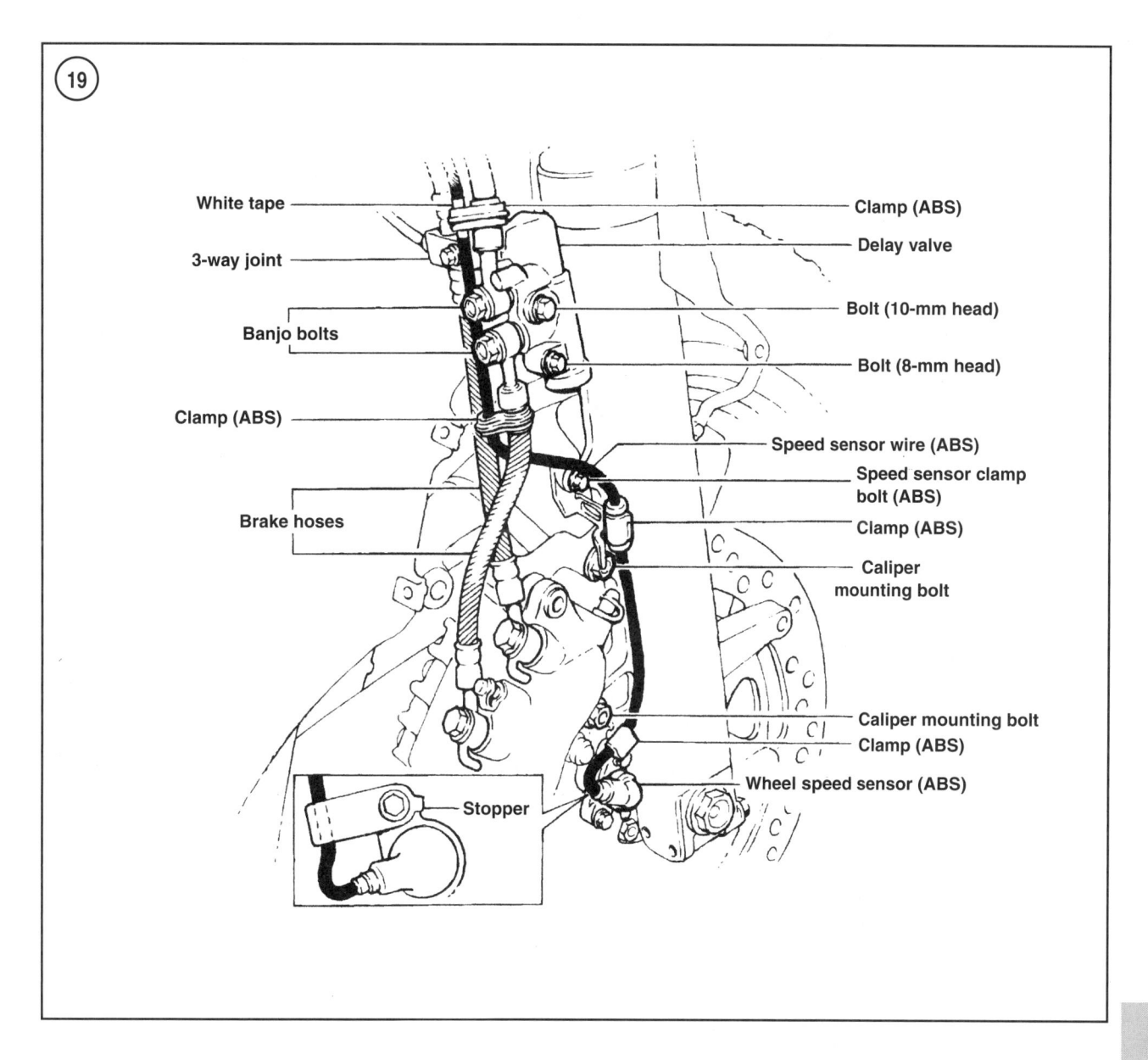

2. Remove the spacer block from between the brake pads, if used.

3. Install the caliper assembly over the brake disc. If the pads are installed in the caliper, be careful not to damage their leading edge.

4. Install two new rear brake caliper mounting bolts (B, **Figure 20**) and tighten to 45 N•m (33 ft.-lb.).

5. Place a new washer on each side of each brake hose (A, **Figure 20**). Then thread the banjo bolts into the caliper and tighten to 34 N•m (25 ft.-lb.). Make sure each brake hose stop seats against the caliper as shown in **Figure 20**.

6. If removed, install the brake pads as described in this chapter.

7. Bleed the brake system as described in this chapter.

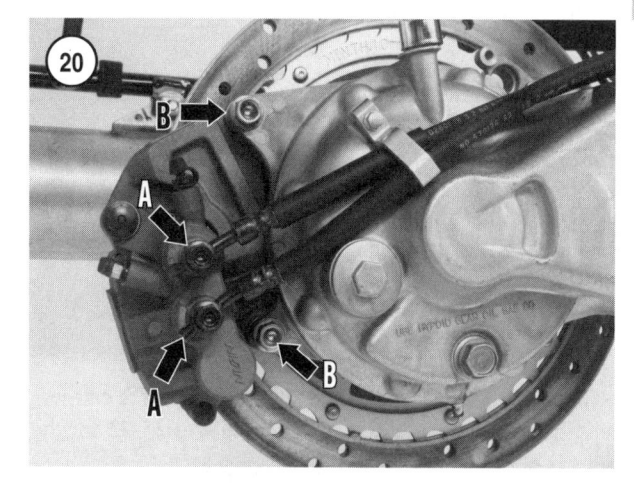

LEFT FRONT BRAKE CALIPER

1. Brake pads
2. Pad spring
3. Bolt
4. Caliper half
5. Piston (22 mm)
6. Piston (27 mm)
7. Dust seal
8. Piston seal
9. Dust seal
10. Piston seal
11. Boot
12. Housing shaft
13. Bleed screw
14. Cover
15. Housing
16. Stopper ring
17. Pad pin
18. Pad pin plug
19. Bolt
20. Bracket shaft
21. Boot
22. Caliper bracket
23. Pad retainer
24. Bushing

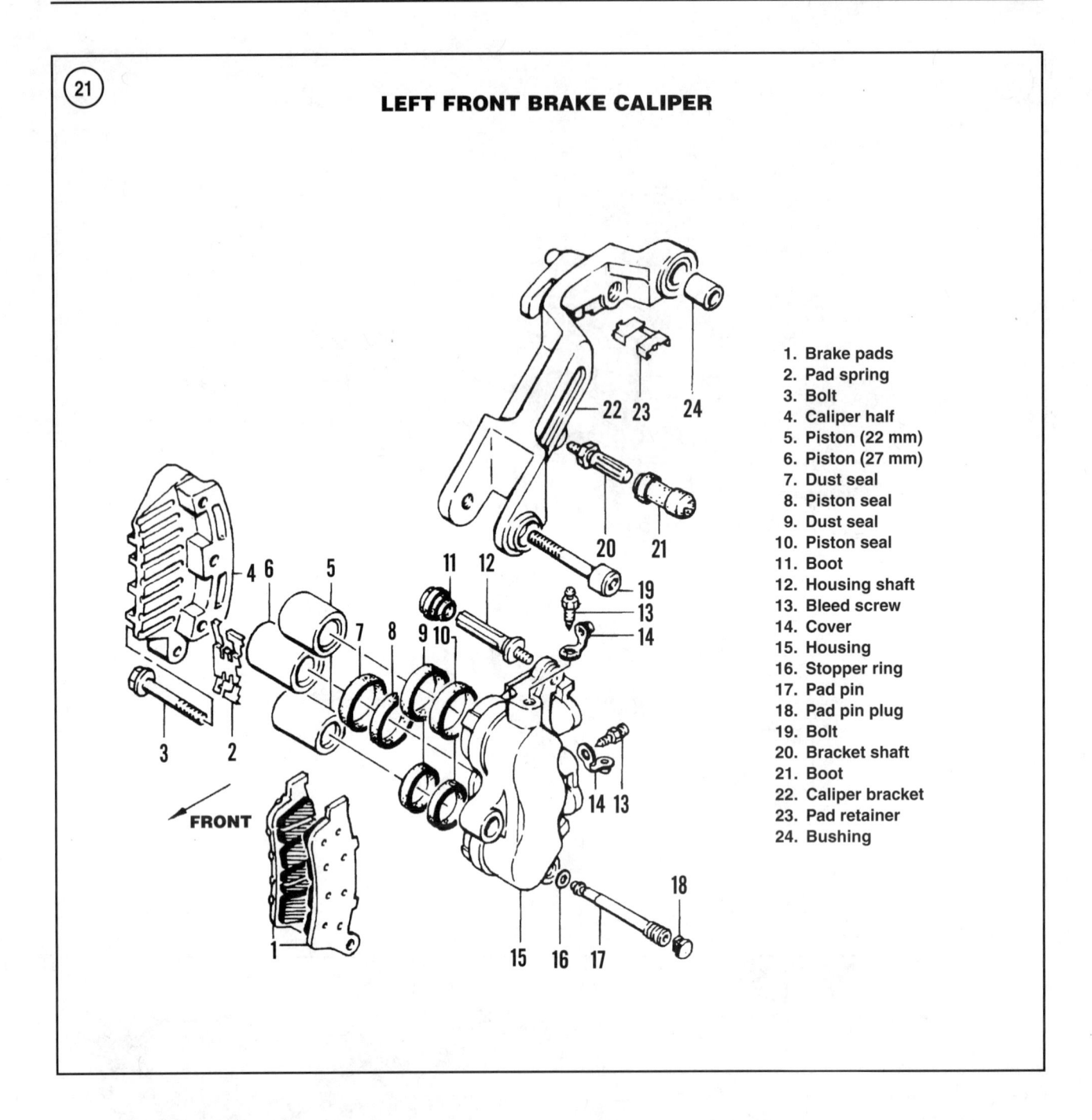

8. Install the rear wheel (Chapter Eleven).
9. Operate the front brake lever and rear brake pedal to seat the pads against the brake disc.

BRAKE CALIPER OVERHAUL

This procedure is applicable to both front brake calipers and the rear brake caliper. Refer to **Figures 21-23** for the caliper being worked on.

Disassembly

Brake caliper service involves separating the caliper housing halves and removing the pistons and seals. Use an air compressor during the disassembly procedure.

1. Remove the brake pads as described in this chapter.
2. Remove the brake caliper as described in this chapter.
3. Remove the pad spring.
4. Remove the caliper bracket and caliper pin boots.

BRAKES

(22) RIGHT FRONT BRAKE CALIPER

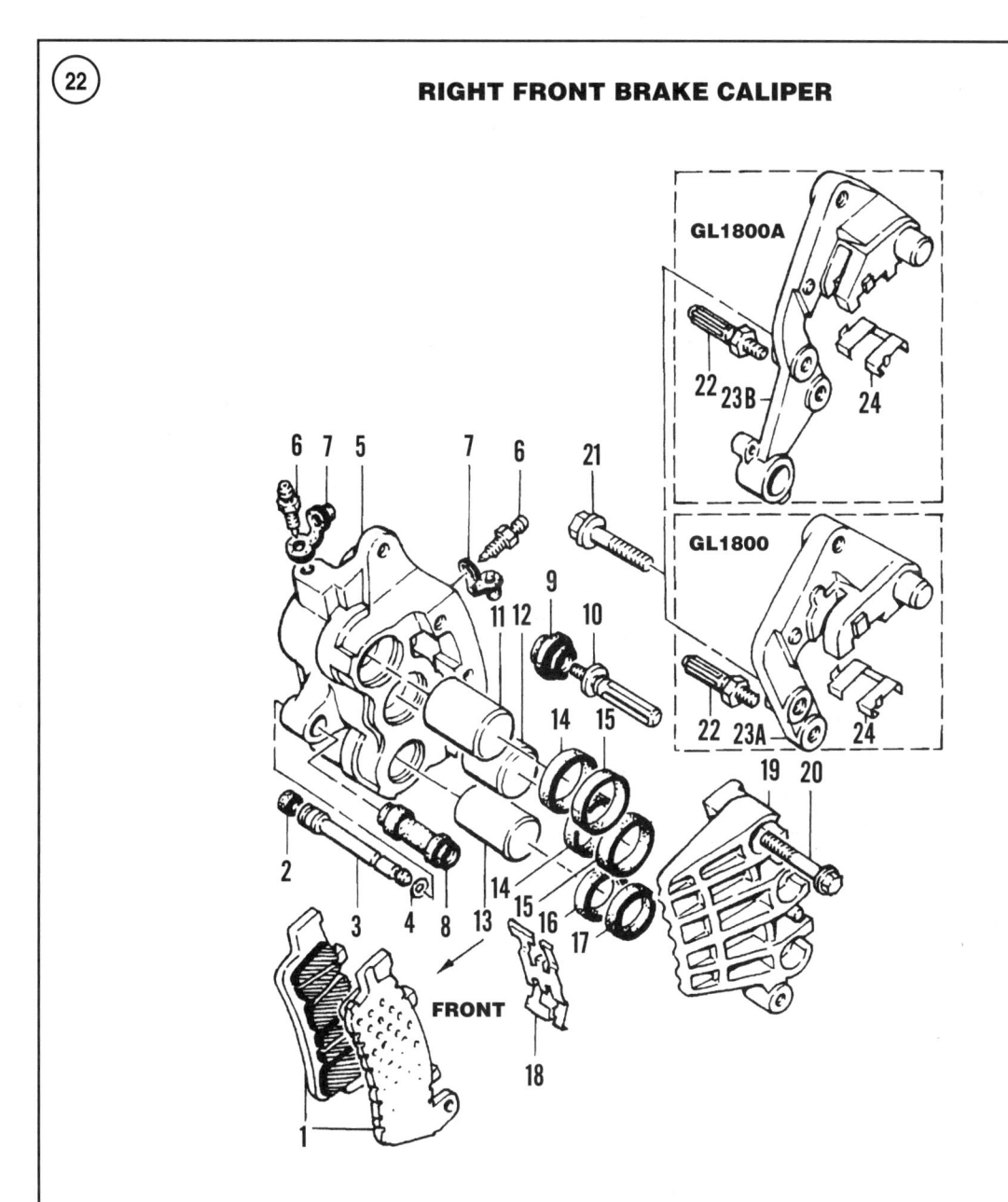

1. Brake pads
2. Pad pin plug
3. Pad pin
4. Stopper ring
5. Housing
6. Bleed screw
7. Cover
8. Boot
9. Boot
10. Housing shaft
11. Piston (25 mm; long)
12. Piston (25 mm; short)
13. Piston (22 mm)
14. Piston seal
15. Dust seal
16. Piston seal
17. Dust seal
18. Pad spring
19. Caliper half
20. Bolt
21. Bolt
22. Bracket shaft
23A. Caliper bracket
23B. Caliper bracket
24. Pad retainer

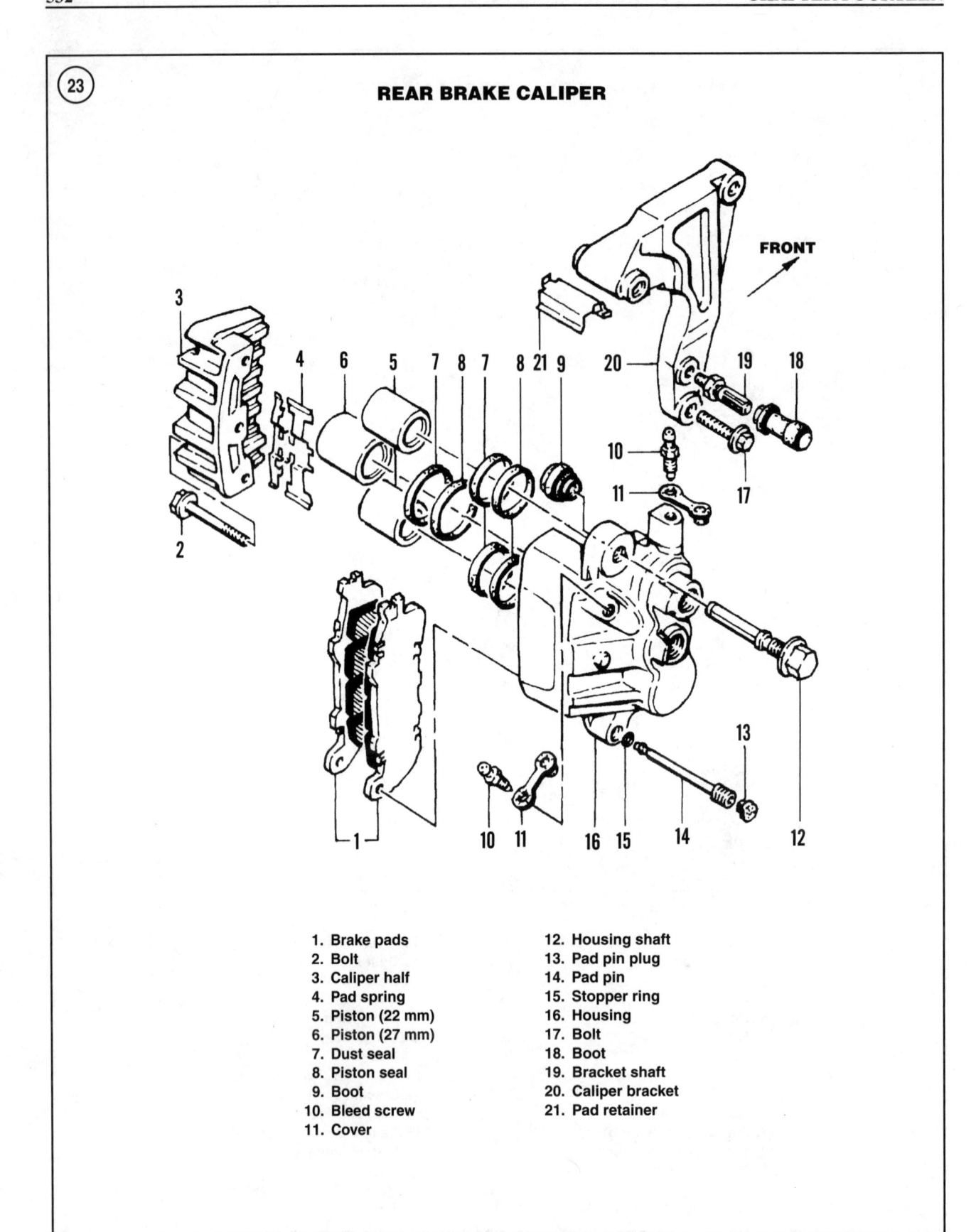

REAR BRAKE CALIPER

23

1. Brake pads
2. Bolt
3. Caliper half
4. Pad spring
5. Piston (22 mm)
6. Piston (27 mm)
7. Dust seal
8. Piston seal
9. Boot
10. Bleed screw
11. Cover
12. Housing shaft
13. Pad pin plug
14. Pad pin
15. Stopper ring
16. Housing
17. Bolt
18. Boot
19. Bracket shaft
20. Caliper bracket
21. Pad retainer

BRAKES

24

Air hose

25

Piston seals
Dust seals
Pistons

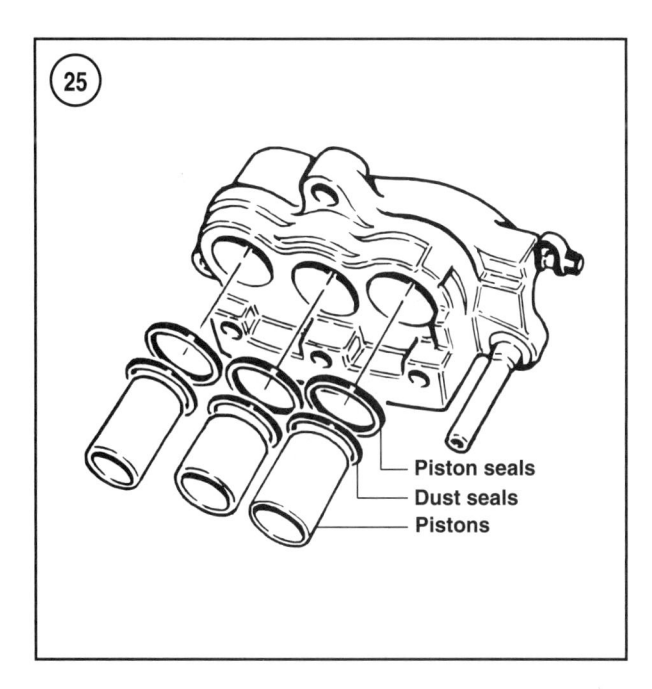

5. Remove the brake caliper assembly bolts and separate the brake caliper halves. Discard the bolts.
6. Remove the pistons as follows:

WARNING
Compressed air forces the pistons out of the caliper under considerable force. Do not cushion the pistons by hand, as injury could result.

   a. Depending on the caliper, different size and/or length pistons are used. Mark the pistons so they can be installed in their original bores.
   b. Support the caliper on a wooden block with the piston side facing down. Place a thick towel between the pistons and workbench. Make sure there is enough space underneath the caliper to remove the pistons completely.
   c. Direct compressed air through the brake hose port (**Figure 24**, typical) to remove the pistons.
7. To remove the dust and piston seals (**Figure 25**):
   a. Different size dust and pistons seals are used. Identify the seals after removing them to identify the replacement seals.
   b. Use a small tool to carefully pry the dust and piston seals from the grooves in the cylinder bore. Repeat for each bore.
8. Remove the bleed valve and its cover from the caliper.

Inspection

When measuring the brake caliper components, compare the actual measurements to the specifications in **Table 1** (front calipers) or **Table 2** (rear caliper). Replace worn or damaged parts as described in this section.

1. Clean and dry the caliper assembly as follows:
   a. Handle the brake components carefully when servicing them.
   b. Remove all threadlocking compound residue from the brake caliper housing halves and their mounting bolt threads.
   c. Use only DOT 4 brake fluid or isopropyl alcohol to wash rubber parts in the brake system. Never allow any petroleum-based cleaner to contact the rubber parts. These chemicals cause the rubber to swell, requiring their replacement.
   d. Clean the dust and piston seal grooves carefully to avoid damaging the caliper bore. Use a small pick or brush to clean the grooves. If a hard varnish residue has built up in the grooves, soak the caliper housing halves in solvent to help soften the residue. Then wash the caliper halves in soapy water and rinse completely.
   e. If alcohol or solvent was used to clean the caliper, blow dry with compressed air.
   f. Check the fluid passages to make sure they are clean and dry.
   g. After cleaning the parts, place them on a clean lint-free cloth until reassembly.

CAUTION
Do not get any oil or grease onto any of the brake caliper components. These chemicals cause the rubber parts in the brake system to swell, permanently damaging them.

2. Check the caliper housing mating surfaces for burrs or other damage.

3. Check each cylinder bore for corrosion, deep scratches and other wear marks. Do not hone the cylinder bores.

NOTE
*Each caliper uses different bore diameters. Refer to the upper, center and lower brake caliper inside diameter and the corresponding piston outside diameter specifications in **Table 1** or **Table 2** when measuring the parts in the following steps.*

4. Measure each caliper cylinder bore inside diameter.
5. Inspect the pistons for pitting, corrosion, cracks or other damage.
6. Measure each piston outside diameter (**Figure 26**).
7. Clean the bleed screws with compressed air. Check the screw threads for damage. Replace the covers if missing or damaged.
8. Clean the banjo bolts with compressed air.
9. Inspect the caliper shaft and boot assembly as follows:

NOTE
*The brake calipers are of a floating design. Shafts mounted on the caliper housing (**Figure 27**, typical) and caliper bracket allow the brake caliper to slide or float during piston movement. Rubber boots installed over each shaft help to control caliper movement by preventing excessive shaft vibration and play and to prevent dirt from damaging the shaft operating surfaces. Grooved or damaged shafts will prevent or bind caliper movement. This condition causes brake pads to wear unevenly, causing brake drag and overheating of the brake disc and brake fluid. The shafts and rubber boots are an important part of the brake caliper and must be maintained to provide proper brake operation.*

   a. Inspect the rubber boots (**Figures 21-23**) for age deterioration and damage.
   b. Inspect the caliper housing and caliper bracket shafts (**Figures 21-23**) for excessive wear, uneven wear (steps) and other damage. Replace damaged shafts as required.

10. When reinstalling or replacing damaged shafts, apply a medium strength threadlock onto the shaft threads and tighten as follows:
   a. Front brake caliper housing shaft: 23 N•m (17 ft.-lb.).
   b. Front brake caliper bracket shaft: 13 N•m (115 in.-lb.).
   c. Rear brake caliper housing shaft: 27 N•m (20 ft.-lb.).
   d. Rear brake caliper bracket shaft: 23 N•m (17 ft.-lb.).

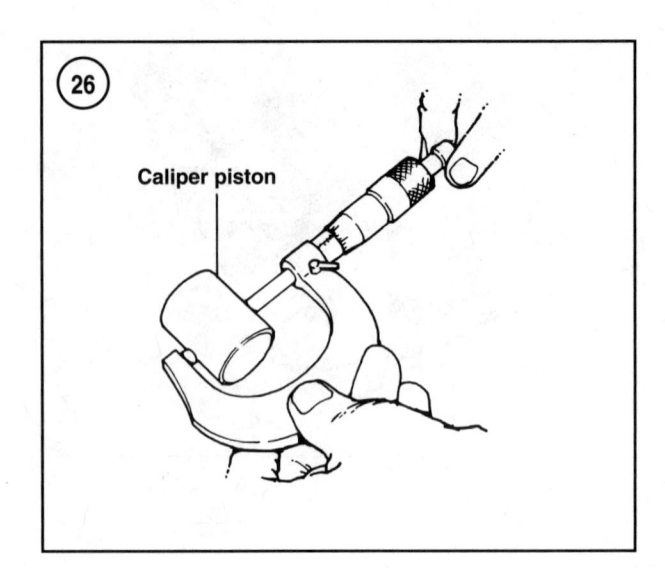

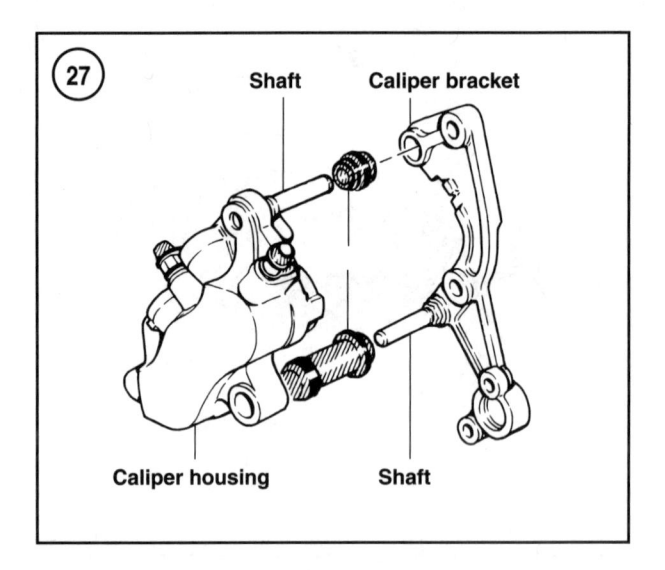

Assembly

Use new DOT 4 brake fluid when lubricating the parts in the following steps.

1. Install the bleed screws and covers into the caliper. Tighten finger-tight.

NOTE
*Compare the new seals with the original parts (identified during removal) to determine their size and bore locations. Refer to the brake caliper cylinder bore upper, center and lower measurements in **Table 1** or **Table 2**.*

BRAKES

2. Soak the *new* piston and dust seals in brake fluid.

3. Lubricate the pistons and cylinder bores with brake fluid.

NOTE
The piston seals are thicker than the dust seals (Figure 25).

4. Install a *new* piston seal into each cylinder bore rear groove.

5. Install a *new* dust seal into each cylinder bore front groove.

NOTE
Check that each seal fits squarely in its groove.

6. Install each piston into its respective caliper bore with its open side facing out. To prevent the pistons from damaging the seals, turn them into the bore by hand. Install the pistons until they bottom out.

7. Assemble the caliper housing halves. Install new brake caliper assembly bolts and tighten finger-tight. Then check the caliper housing mating surfaces. Make sure the surfaces are flush all the way around the caliper halves.

8. Tighten the brake caliper assembly bolts to 32 N•m (24 ft.-lb.).

9. Install the pad spring.

10. Apply silicone brake grease to the inside of the rubber boots and along the shafts. Install the small boot onto the caliper bracket, and the large boot through the caliper housing. See **Figure 27**, typical.

11. Install the brake pad shim onto the caliper bracket.

12. Install the caliper bracket over the caliper housing.

13. Install the brake caliper and brake pads as described in this chapter.

FRONT MASTER CYLINDER

Read the information listed under *Brake Service* in this chapter before servicing the front master cylinder.

Removal/Installation

1. Support the bike on its centerstand.

2. Cover the top shelter and fairing to prevent damage from brake fluid contact.

CAUTION
Wash brake fluid off any surface immediately, as it damages the finish. Use soapy water and rinse completely.

3. Remove the master cylinder cover assembly. Empty the brake fluid reservoir with a syringe. Reinstall the parts.

NOTE
The front brake light switch and front cruise cancel switch are enclosed in the same switch housing mounted on the bottom of the front master cylinder. The two upper (smaller) terminals plug into the brake light switch. The two lower (larger) terminals plug into the cruise cancel switch. While the connectors can be disconnected with the switch mounted on the master cylinder handlebar, it is easier to remove the switch and leave the connectors attached to the switch.

4. Remove the screw (A, **Figure 28**) and switch housing (B). Remove the switch and connectors from beneath the housing (**Figure 29**).

5. Remove the banjo bolt (A, **Figure 30**) and washers securing the brake hose to the master cylinder. Plug the brake hose and master cylinder bolt opening to prevent leakage.

6. Remove the holder cap, bolts (**Figure 31**), holder and master cylinder.

7. If necessary, service the master cylinder as described in this chapter.

8. Clean the handlebar, master cylinder and clamp mating surfaces.

9. Installation is the reverse of removal. Note the following:

   a. Mount the master cylinder onto the handlebar and align the upper master cylinder and clamp mating surfaces with the punch mark on the handlebar.

   b. Install the holder and mounting bolts (**Figure 31**). Tighten the upper mounting bolt, then the lower mounting bolt to 12 N•m (106 in.-lb.).

   c. Secure the brake hose to the master cylinder with the banjo bolt (A, **Figure 30**) and two *new* washers—install a washer on each side of the brake hose. Position the brake hose arm against the master cylinder bracket as shown in B, **Figure 30** and tighten the banjo bolt to 34 N•m (25 ft.-lb.).

   d. Bleed the brake system as described in this chapter.

   e. Turn the ignition switch on and make sure the rear brake light comes on when operating the front brake lever.

Disassembly

Refer to **Figure 32**.

1. Remove the master cylinder as described in this chapter.
2. Remove the screw and the switch (11, **Figure 32**), if necessary.
3. Remove the nut and pivot bolt, then remove the front brake lever and its adjuster assembly.
4. Remove the master cylinder cover and diaphragm assembly. Drain the reservoir and discard the brake fluid.
5. Remove the protector (18, **Figure 32**) from the reservoir.
6. Remove the dust cover (**Figure 33**) from the groove in the end of the piston.

NOTE
If brake fluid is leaking from the piston bore, the piston cups are worn or damaged. Replace the piston assembly.

NOTE
To aid in the removal and installation of the master cylinder snap ring, thread a bolt and nut into the brake hose port, and secure the bolt in a vise.

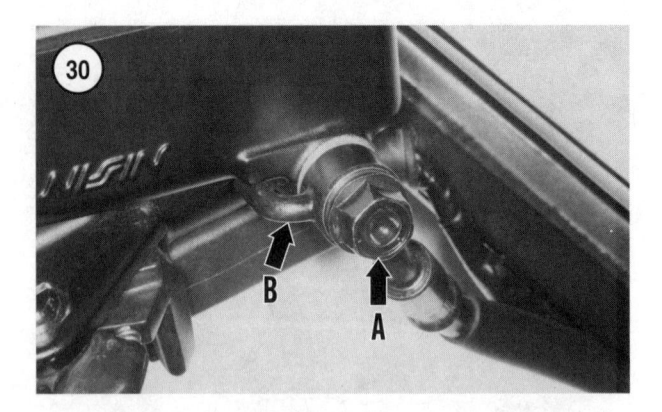

7. Compress the piston and remove the snap ring (5, **Figure 32**) from the groove in the master cylinder.
8. Remove the piston assembly (**Figure 34**) from the master cylinder bore. Do not remove the secondary cup from the piston.

Inspection

When measuring the front master cylinder components, compare the actual measurements to the specifications in **Table 1**. Replace worn or damaged parts as described in this section.

1. Clean and dry the master cylinder assembly as follows:

   a. Handle the brake components carefully when servicing them.

   b. Use only DOT 4 brake fluid or isopropyl alcohol to wash rubber parts in the brake system. Never allow any petroleum-based cleaner to contact the rubber parts. These chemicals cause the rubber to swell, requiring their replacement.

   c. Clean the master cylinder snap ring groove carefully. Use a small pick or brush to clean the groove. If a hard varnish residue has built up in the groove, soak the master cylinder in solvent to help soften

BRAKES

32 **FRONT MASTER CYLINDER**

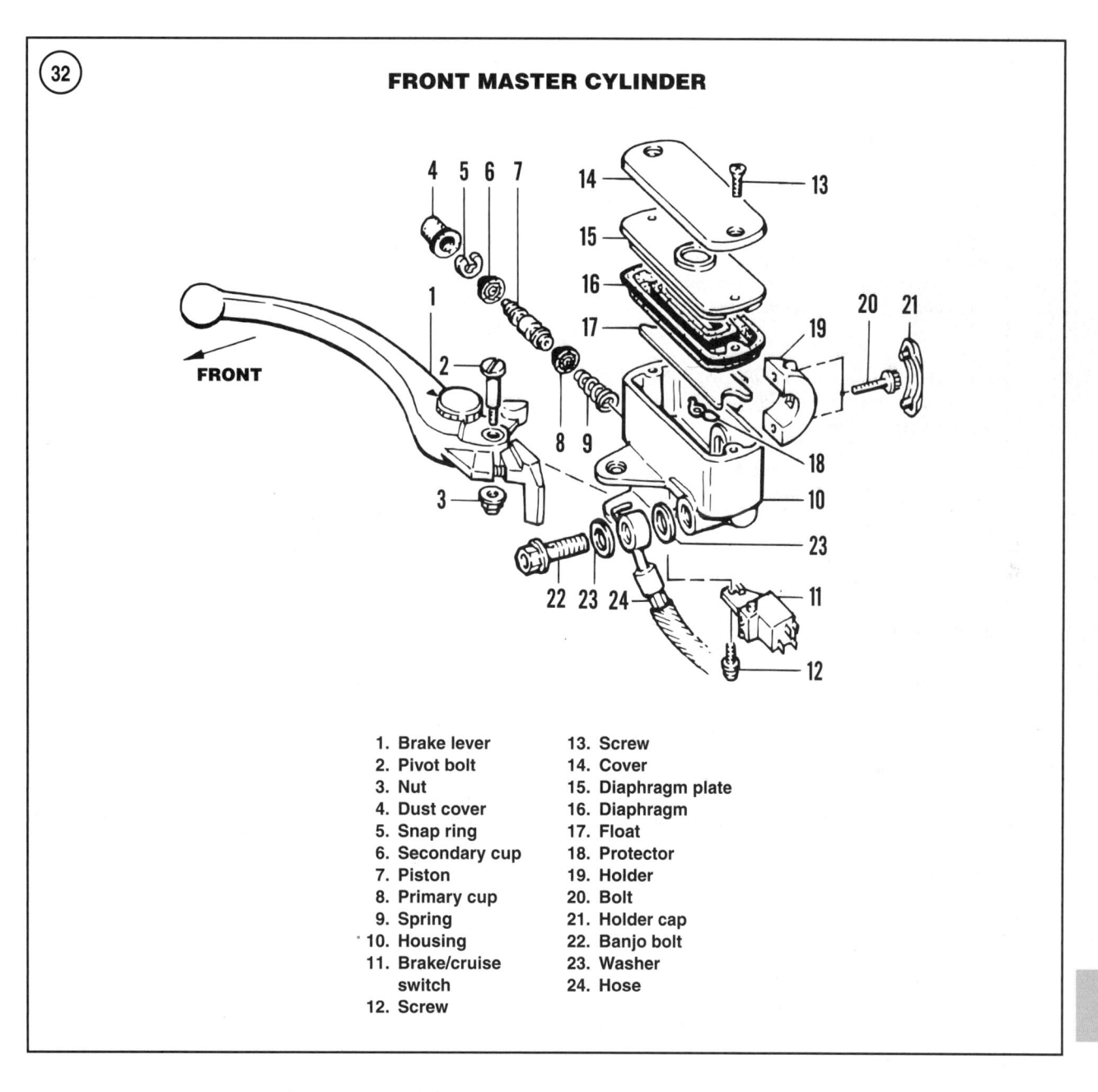

1. Brake lever
2. Pivot bolt
3. Nut
4. Dust cover
5. Snap ring
6. Secondary cup
7. Piston
8. Primary cup
9. Spring
10. Housing
11. Brake/cruise switch
12. Screw
13. Screw
14. Cover
15. Diaphragm plate
16. Diaphragm
17. Float
18. Protector
19. Holder
20. Bolt
21. Holder cap
22. Banjo bolt
23. Washer
24. Hose

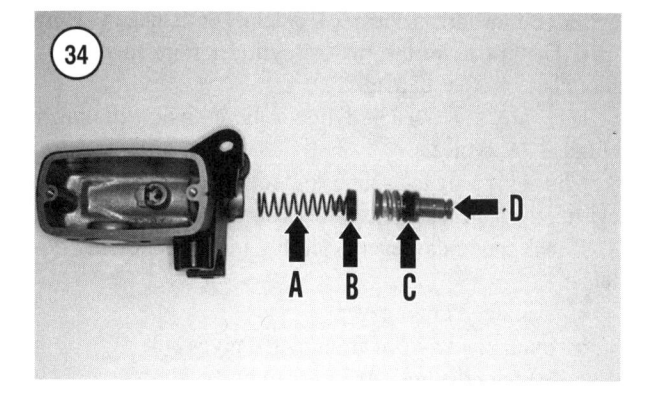

the residue. Then wash in soapy water and rinse completely.
  d. Blow the master cylinder dry with compressed air.
  e. Place cleaned parts on a clean lint-free cloth until reassembly.

WARNING
Do not get any oil or grease onto any of the components. These chemicals cause the rubber parts in the brake system to swell, permanently damaging them.

CAUTION
Do not remove the primary and secondary cups from the piston assembly for cleaning or inspection purposes.

2. Check the piston assembly for the following defects:
   a. Broken, distorted or collapsed piston return spring (A, **Figure 34**).
   b. Worn, cracked, damaged or swollen primary (B, **Figure 34**) and secondary cups (C).
   c. Scratched or damaged piston (D, **Figure 34**).
   d. Corroded, weak or damaged snap ring.
   e. Worn or damaged dust cover.
   f. If any of these parts are worn or damaged, replace the piston assembly.
3. Measure the piston outside diameter (**Figure 35**).
4. To assemble a new piston assembly, perform the following:
   a. If replacing the piston, install the new secondary cup onto the piston. Use the original piston assembly (**Figure 34**) as a reference when installing the new cup onto the piston.
   b. Before installing the new piston cups, lubricate them with brake fluid.
   c. Clean the new piston in brake fluid.
   d. Install the primary cup (B, **Figure 34**) onto the spring (B). Install the secondary cup (C, **Figure 34**) onto the piston.
5. Inspect the master cylinder bore. Replace the master cylinder if its bore is corroded, cracked or damaged in any way. Do not hone the master cylinder bore to remove scratches or other damage.
6. Measure the master cylinder bore inside diameter (**Figure 36**, typical).
7. Check for plugged supply and relief ports in the master cylinder. Clean with compressed air.
8. Check the brake lever assembly for the following defects:
   a. Damaged brake lever.
   b. Excessively worn or damaged pivot bolt.
   c. Damaged adjuster arm assembly.

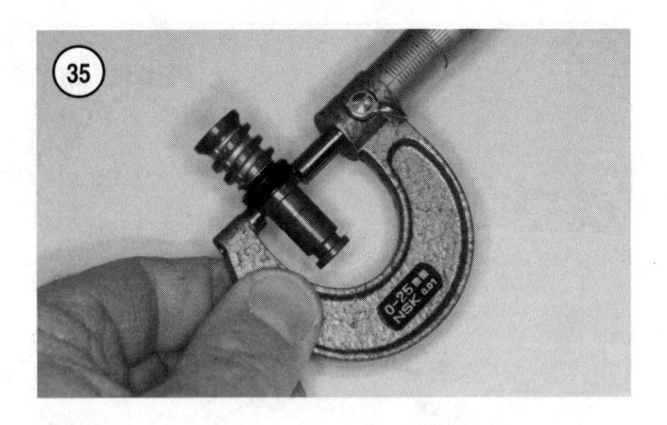

Assembly

1. If installing a new piston assembly, assemble it as described under *Inspection* in this section.
2. Lubricate the piston assembly and cylinder bore with DOT 4 brake fluid.

CAUTION
Do not allow the piston cups to tear or turn inside out when installing them into the master cylinder bore. Both cups are larger than the bore.

3. Insert the piston assembly—spring end first—into the master cylinder bore (A, **Figure 34**).
4. Compress the piston assembly and install a new snap ring (A, **Figure 37**)—flat side facing out—into the bore groove.

CAUTION
The snap ring must seat in the master cylinder groove completely. Push and release the piston a few times to make sure it moves smoothly and that the snap ring does not pop out.

BRAKES

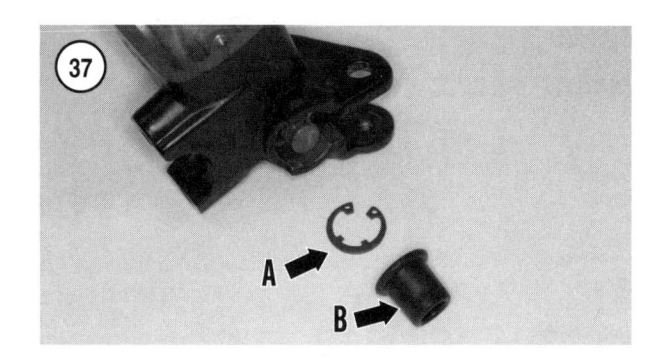

5. Slide the dust cover (B, **Figure 37**) over the piston. Seat the large end against the snap ring and the outer lip end into the groove in the end of the piston (D, **Figure 34**).
6. Install the brake lever assembly as follows:
   a. Lubricate the pivot bolt with silicone brake grease.
   b. Install the brake lever and adjuster assembly.
   c. Install and tighten the brake lever pivot bolt (2, **Figure 32**) to 1 N•m (8.8 in.-lb.). Check that the brake lever moves freely. If there is any binding or roughness, remove the pivot bolt and brake lever and inspect the parts.
   d. Hold the pivot bolt, then install and tighten the brake lever pivot nut (3, **Figure 32**) to 6 N•m (53 in.-lb.). Check that the brake lever moves freely.
7. Install the protector into the port channel in the reservoir (**Figure 38**).

NOTE
The protector prevents brake fluid from exiting the reservoir when the brake lever is operated with the cover removed during brake bleeding.

8. Install the master cylinder and front brake light/front cruise cancel switch (11, **Figure 32**) and secure with the screw.

REAR MASTER CYLINDER AND REAR BRAKE PEDAL

Read the information listed under *Brake Service* in this chapter before servicing the rear master cylinder.

Removal

Refer to **Figure 39**.
1. Support the motorcycle on its centerstand.
2. Remove the right engine side cover. Refer to *Engine Side Covers* in Chapter Seventeen.

CAUTION
Wipe up any spilled brake fluid immediately, as it damages the finish of most plastic and metal surfaces. Use soapy water and rinse thoroughly.

3. Drain the rear brake pedal line as described in this chapter.
4. Remove the two screws (A, **Figure 40**) and the switch housing (B) from the master cylinder.
5. Remove the snap ring (14, **Figure 39**) and disconnect the hose joint from the master cylinder. Remove the O-ring.
6. Remove the banjo bolt (6, **Figure 39**) and washers. Plug the hoses to prevent brake fluid from leaking out.
7. Remove the bolts and rider footpeg.
8. Remove the pinch bolt (4, **Figure 39**) and the brake pedal (3).
9. Disconnect the spring (17, **Figure 39**).
10. Remove the master cylinder mounting bolts (5, **Figure 39**).
11. Push the pivot shaft (18, **Figure 39**) through the frame, then remove the master cylinder/pivot shaft assembly.

Installation

Refer to **Figure 39**.
1. Replace the O-ring (10, **Figure 39**) and snap ring (14) if necessary.
2. Lubricate the pivot shaft with grease.
3. Install the master cylinder/pivot shaft assembly into the frame. Push the water hose aside when installing the assembly and insert the pivot shaft through the frame.
4. Install the rear master cylinder mounting bolts (5, **Figure 39**) and tighten to 12 N•m (106 in.-lb.).
5. Hook the return spring onto the pivot shaft and frame pin.

540 CHAPTER FOURTEEN

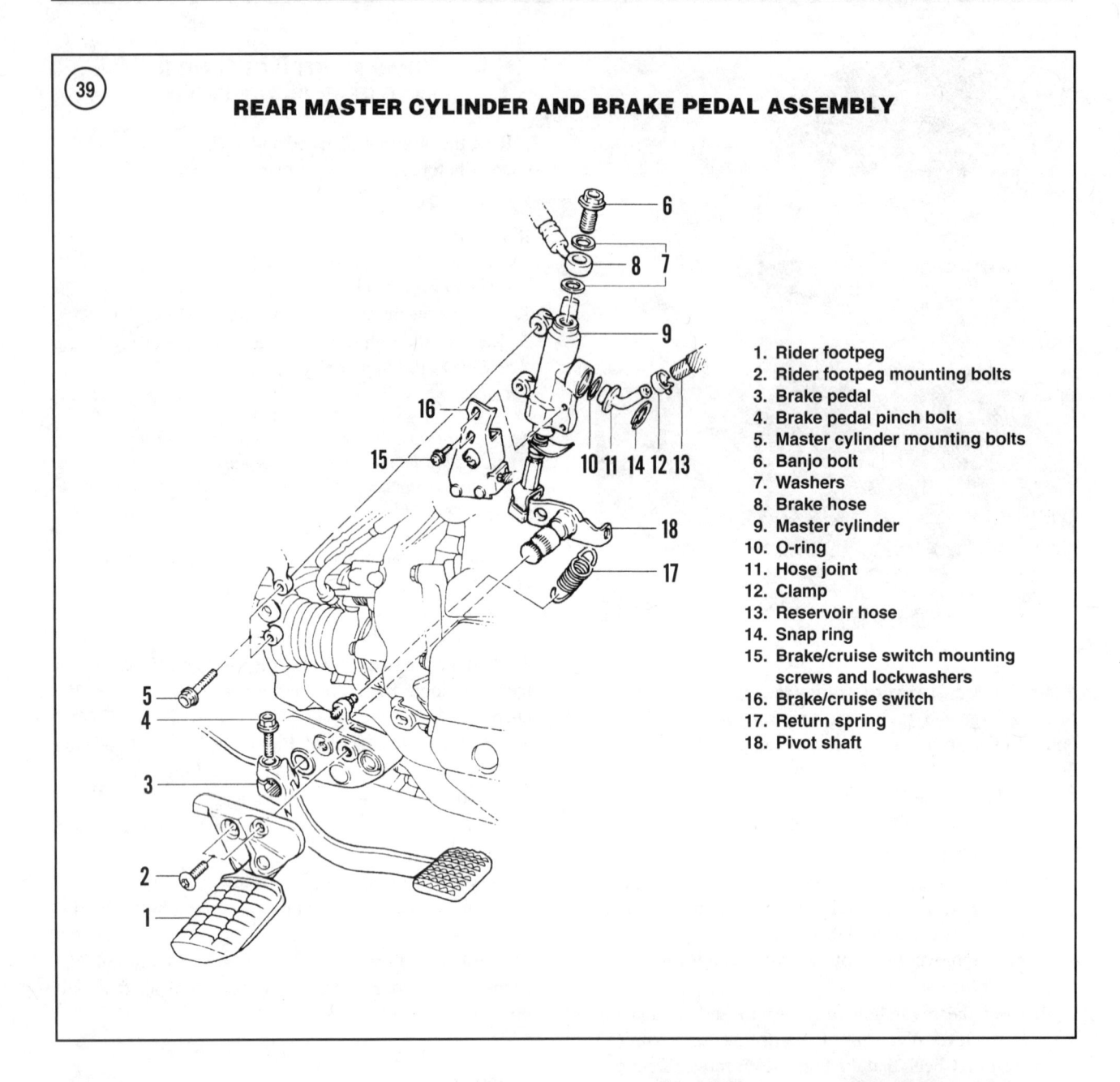

39 REAR MASTER CYLINDER AND BRAKE PEDAL ASSEMBLY

1. Rider footpeg
2. Rider footpeg mounting bolts
3. Brake pedal
4. Brake pedal pinch bolt
5. Master cylinder mounting bolts
6. Banjo bolt
7. Washers
8. Brake hose
9. Master cylinder
10. O-ring
11. Hose joint
12. Clamp
13. Reservoir hose
14. Snap ring
15. Brake/cruise switch mounting screws and lockwashers
16. Brake/cruise switch
17. Return spring
18. Pivot shaft

6. Lubricate the O-ring (10, **Figure 39**) with brake fluid and install it onto the hose joint.

7. Install the hose joint into the master cylinder and secure it with a new snap ring.

8. Secure the brake hose to the master cylinder with the banjo bolt (6, **Figure 39**) and washers. Install a washer on each side of the hose. Tighten the banjo bolt to 34 N•m (25 ft.-lb.).

9. Install the brake pedal onto the pedal pivot shaft—align the notch in the brake pedal with the punch mark on the pivot shaft. Install the pinch bolt and tighten to 26 N•m (19 ft.-lb.).

BRAKES

41 REAR MASTER CYLINDER

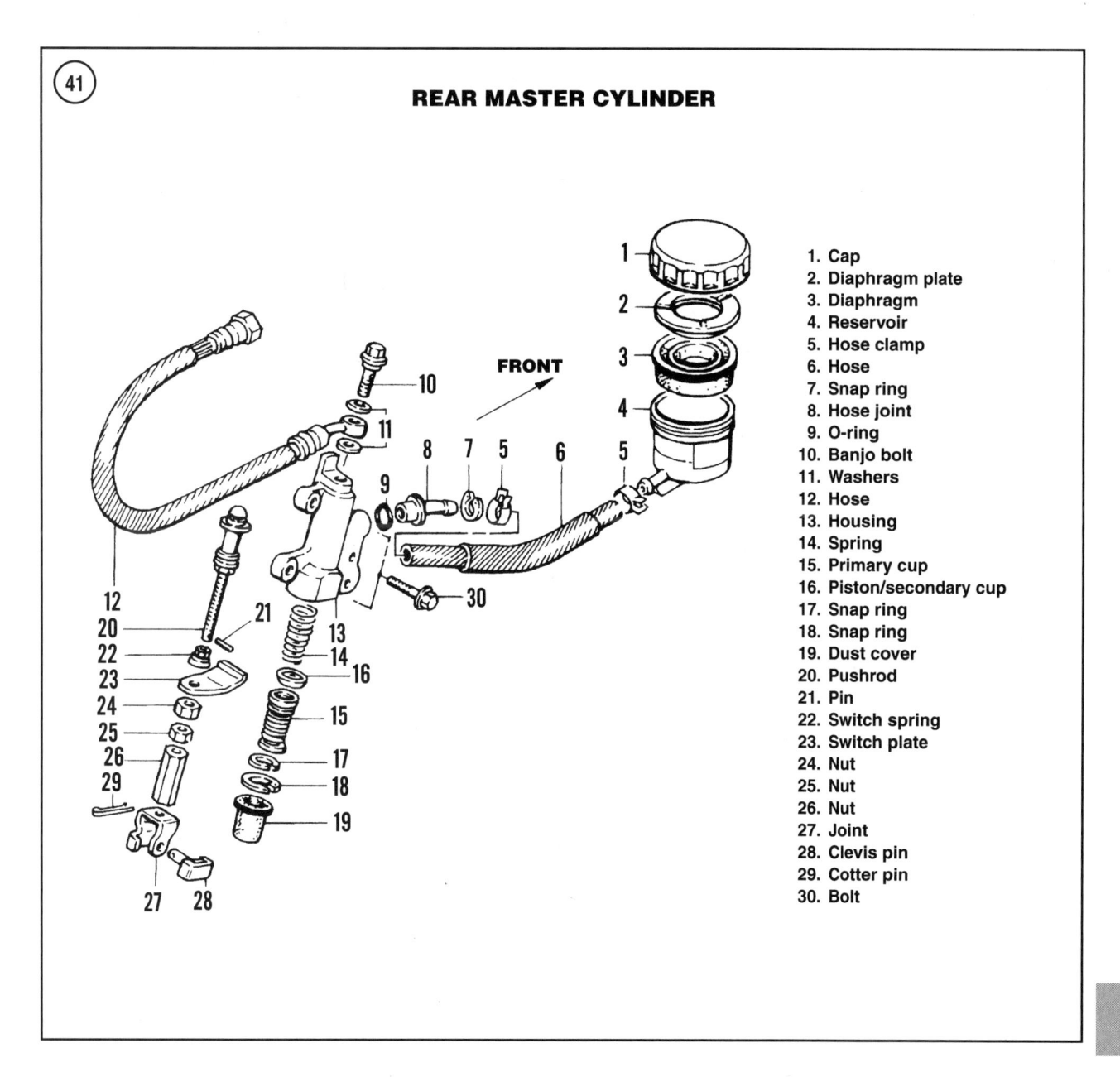

1. Cap
2. Diaphragm plate
3. Diaphragm
4. Reservoir
5. Hose clamp
6. Hose
7. Snap ring
8. Hose joint
9. O-ring
10. Banjo bolt
11. Washers
12. Hose
13. Housing
14. Spring
15. Primary cup
16. Piston/secondary cup
17. Snap ring
18. Snap ring
19. Dust cover
20. Pushrod
21. Pin
22. Switch spring
23. Switch plate
24. Nut
25. Nut
26. Nut
27. Joint
28. Clevis pin
29. Cotter pin
30. Bolt

10. Install the rider footpeg and tighten the mounting bolts (2, **Figure 39**) to 26 N•m (19 ft.-lb.).

11. Install and adjust the brake light/cruse switch:

   a. Use a medium strength threadlock on the switch screw threads.

   b. Install the switch (B, **Figure 40**) onto the master cylinder and secure with the mounting screws (A)—do not tighten.

   c. Turn the ignition switch on. Adjust the switch so the brake light comes on when the brake pedal moves the push rod 0.7-1.7 mm (1/32-1/16 in.).

   d. Tighten the switch screws (A, **Figure 40**) to 2 N•m (17 in.-lb.).

   e. Recheck the brake light adjustment and turn the ignition switch off.

12. Bleed the brake system as described in this chapter.
13. Install the engine cover.
14. Turn the ignition switch on and check that the rear brake light comes on when pressing the rear brake pedal.

Disassembly

Refer to **Figure 41**.

1. Remove the cotter pin and disconnect the pivot shaft from the joint.
2. Slide the dust cover out of the master cylinder bore.

WARNING
If brake fluid is leaking from the master cylinder bore, the piston cups are excessively worn or damaged. Replace the piston assembly during reassembly.

NOTE
To aid in the removal/installation of the piston snap ring, thread a bolt and nut into the master cylinder, and secure the bolt in a vise.

3. Compress the piston and remove the snap ring (**Figure 42**) from the groove in the master cylinder, then remove the pushrod/piston assembly (**Figure 43**).

Inspection

When measuring the master cylinder components, compare the actual measurements to the specifications in **Table 2**. Replace worn or damaged parts as described in this section.

1. Clean and dry the master cylinder assembly as follows:
    a. Handle the brake components carefully when servicing them.
    b. Use only DOT 4 brake fluid or isopropyl alcohol to wash rubber parts (dust cover, piston seals and O-ring) in the brake system. Never allow any petroleum-based cleaner to contact the rubber parts. These chemicals cause the rubber to swell, requiring their replacement.
    c. Clean the master cylinder snap ring groove carefully. Use a small pick or brush to clean the groove. If a hard varnish residue has built up in the groove, soak the master cylinder in solvent to help soften the residue. Then wash in soapy water and rinse completely.
    d. Blow the master cylinder dry with compressed air.
    e. After cleaning the parts, place them on a clean lint-free cloth until reassembly.

CAUTION
Do not get any oil or grease onto any of the components. These chemicals cause the rubber parts in the brake system to swell, permanently damaging them.

CAUTION
Do not remove the primary and secondary cups from the piston assembly for cleaning or inspection purposes.

2. Check the piston assembly (**Figure 43**) for the following defects:
    a. Broken, distorted or collapsed piston spring.
    b. Worn, cracked, damaged or swollen primary and secondary cups.
    c. Scratched, scored or damaged piston.
    d. Corroded, weak or damaged snap rings.
    e. Worn or damaged dust cover.
    f. If any of these parts are worn or damaged, replace the piston assembly.
3. Measure the piston outside diameter at the point indicated in **Figure 43**.

NOTE
To install a new piston, separate the pushrod from the piston as described in Step 9.

4. To assemble a new piston assembly, perform the following:
    a. When replacing the piston, install the new secondary cup onto the piston. Use the original piston assembly as a reference when installing the new cup onto the piston.
    b. Before installing the new piston cups, lubricate them with brake fluid.
    c. Clean the new piston in brake fluid.
    d. Install the primary cup onto the return spring.
5. Inspect the master cylinder bore. Replace the master cylinder if its bore is pitted, corroded, cracked or damaged in any way. Do not hone the master cylinder bore to remove scratches or other damage.

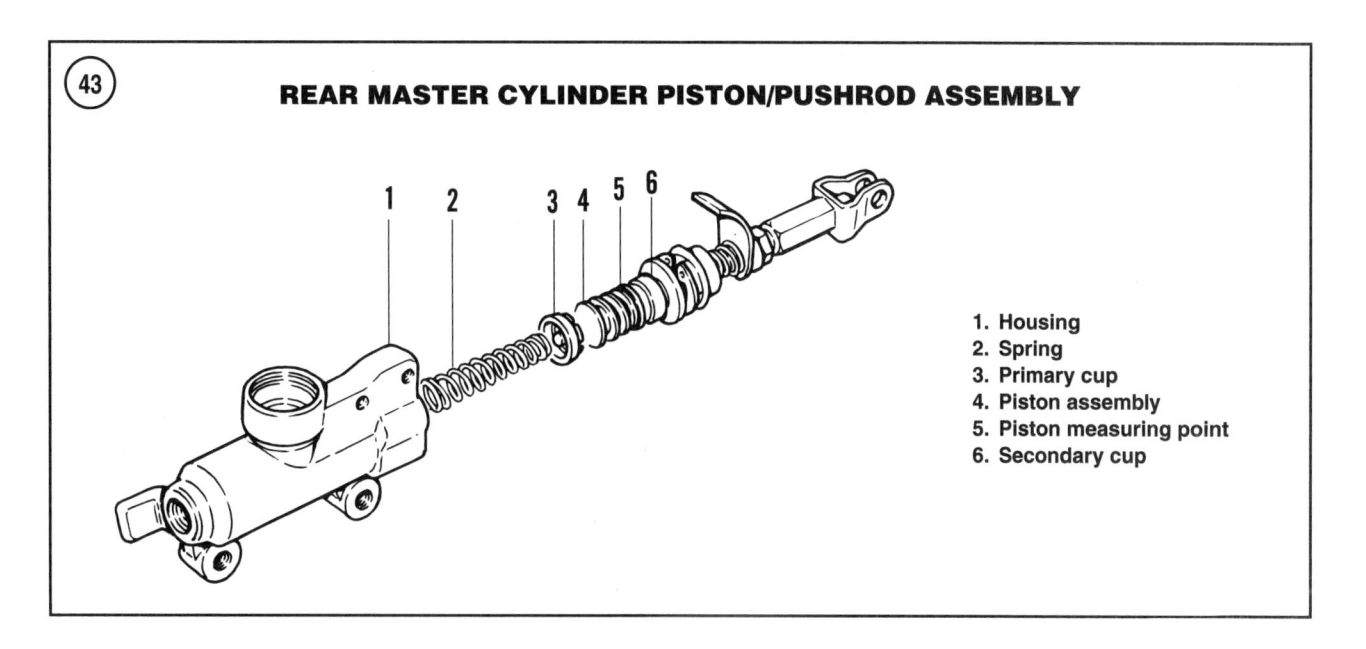

BRAKES

Figure 43: REAR MASTER CYLINDER PISTON/PUSHROD ASSEMBLY

1. Housing
2. Spring
3. Primary cup
4. Piston assembly
5. Piston measuring point
6. Secondary cup

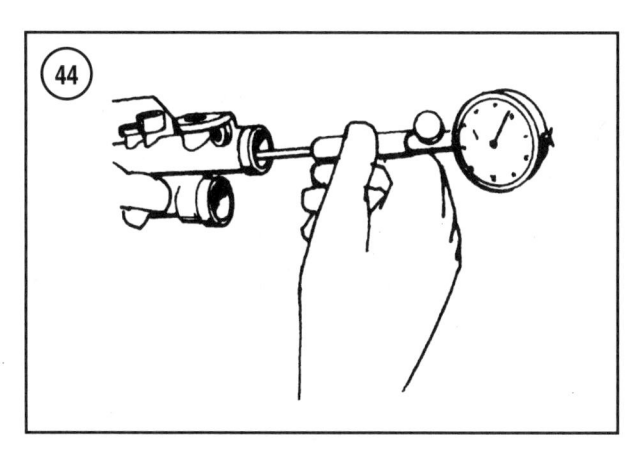

6. Measure the master cylinder bore inside diameter (**Figure 44**, typical).
7. Check for plugged supply and relief ports in the master cylinder. Clean with compressed air.
8. Check the pushrod assembly (**Figure 43**) for the following defects:
   a. Corroded or damaged pushrod.
   b. Damaged pushrod joint.
   c. Weak or damaged snap rings.
9. To service the pushrod, perform the following:
   a. Remove the snap ring and separate the piston from the pushrod.
   b. Loosen the locknuts as required to replace components installed on the pushrod.
   c. Replace worn or damaged parts.
   d. Use new snap rings to assemble the pushrod in the order shown in **Figure 41**. Do not tighten the locknuts. The pushrod length must be adjusted after reassembling the master cylinder.

Assembly

1. If installing a new piston assembly, assemble it onto the pushrod as described under *Inspection* in this section.
2. Lubricate the piston assembly and cylinder bore with DOT 4 brake fluid.
3. Install the spring—large end first—into the cylinder as shown in **Figure 43**. The primary cup installed on the small spring end must be facing out, toward the bore opening.

CAUTION
Do not allow the piston cups to tear or turn inside out when installing them into the master cylinder bore. Both cups are larger than the bore.

4. Insert the piston/pushrod assembly into the master cylinder bore as shown in **Figure 43**.
5. Push the pushrod to compress the piston assembly and position the washer below the snap ring groove, then install the snap ring into the master cylinder groove.

CAUTION
The snap ring must seat in the master cylinder groove completely. Push and release the piston a few times to make sure it moves smoothly and that the snap ring does not pop out.

45 Rear master cylinder — Locknut — Joint — 106 mm (4.2 in.) Pushrod length

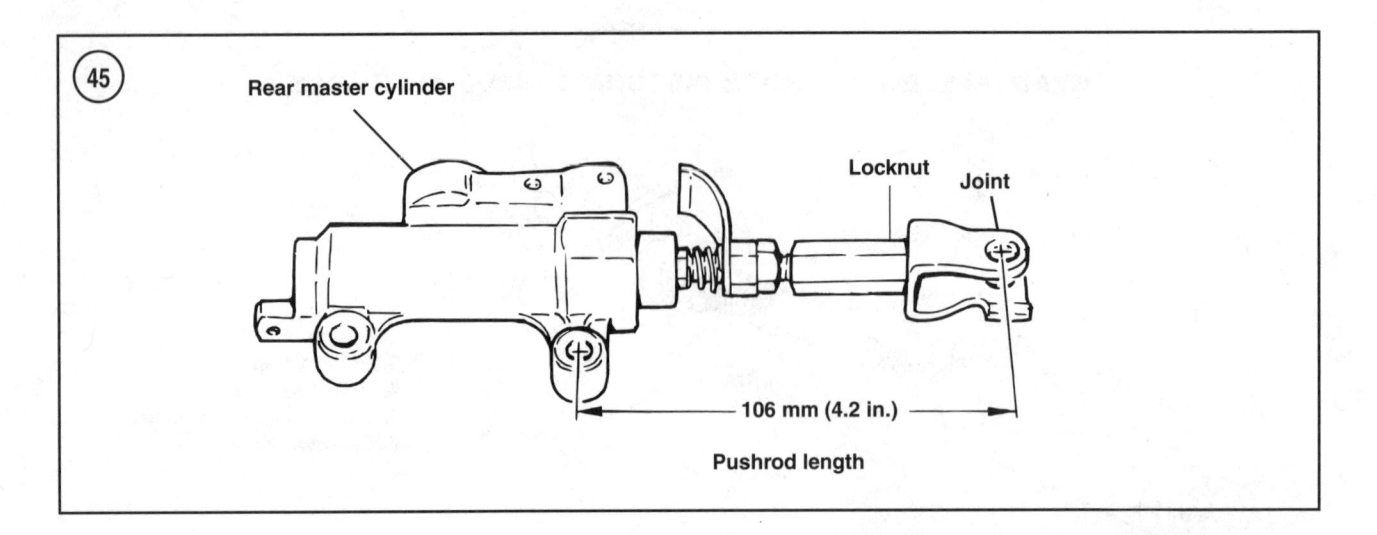

6. Lubricate the boot groove in the pushrod with silicone brake grease. Then slide the boot down the pushrod and seat it against the washer. Seat the outer end of the boot onto the pushrod.
7. If loosened, tighten the switch plate locknut (25, **Figure 41**) to 18 N•m (13 ft.-lb.).
8. Measure the pushrod length as shown in **Figure 45**. If incorrect, loosen the locknut and turn the joint as required. Tighten the locknut to 18 N•m (13 ft.-lb.).
9. Secure the pivot arm onto the joint with the clevis pin and a *new* cotter pin (**Figure 39**).
10. Install the master cylinder as described in this chapter.

SECONDARY MASTER CYLINDER

The secondary master cylinder is mounted on the left fork slider.

Removal/Installation

1. Remove front fender B. Refer to *Front Fender Assembly* in Chapter Seventeen.
2. Drain the rear brake pedal line as described in this chapter.
3. Remove the upper brake hose banjo bolt and washers (A, **Figure 46**) from the secondary master cylinder. Plug the hose end to prevent leakage.
4. Remove the brake caliper secondary master cylinder joint mounting bolt (B, **Figure 46**).
5. Remove the secondary master cylinder mounting bolts (C, **Figure 46**).
6. Remove the front brake hose banjo bolt and washers (**Figure 47**) from the secondary master cylinder. Plug the hose end to prevent leakage.

7. If necessary, remove the collar from the caliper bracket.

8. Installation is the reverse of removal. Note the following:

   a. Install new banjo bolt washers.

   b. Tighten the front banjo bolt (**Figure 47**) to 34 N•m (25 ft.-lb.).

BRAKES

Figure 48: SECONDARY MASTER CYLINDER

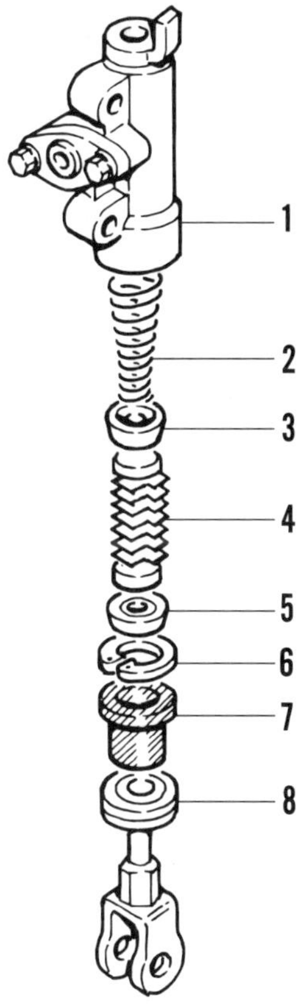

1. Housing
2. Spring
3. Primary cup
4. Piston
5. Secondary cup
6. Snap ring
7. Dust cover
8. Pushrod assembly
9. Locknut
10. Joint

c. Install new secondary master cylinder mounting bolts (C, **Figure 46**) and tighten to 31 N•m (23 ft.-lb.).
d. Install a new left front brake caliper-to-secondary master cylinder joint bolt (B, **Figure 46**) and tighten to 25 N•m (19 ft.-lb.).
e. Tighten the upper banjo bolt (A, **Figure 46**) to 34 N•m (25 ft.-lb.).
f. Bleed the brake system as described in this chapter.

Disassembly

Refer to **Figure 48**.

1. Slide the boot up the pushrod and away from the snap ring.
2. Compress the pushrod to release tension against the snap ring and remove the snap ring from the groove in the master cylinder.
3. Remove the pushrod.

NOTE
If brake fluid is leaking from the piston bore, the piston cups are worn or damaged. Replace the piston assembly.

4. Remove the piston assembly from the master cylinder bore. Do not remove the secondary cup from the piston.

Inspection

When measuring the secondary master cylinder components, compare the actual measurements to the specifications in **Table 2**. Replace worn or damaged parts as described in this section.

1. Clean and dry the master cylinder assembly as follows:
   a. Handle the brake components carefully when servicing them.
   b. Use only DOT 4 brake fluid or isopropyl alcohol to wash rubber parts (dust cover and piston seals) in the brake system. Never allow any petroleum-based cleaner to contact the rubber parts. These chemicals cause the rubber to swell, requiring their replacement.
   c. Clean the master cylinder snap ring groove carefully. Use a small pick or brush to clean the groove. If a hard varnish residue has built up in the groove, soak the master cylinder in solvent to help soften the residue. Then wash in soapy water and rinse completely.
   d. Blow the master cylinder dry with compressed air.
   e. Place cleaned parts on a clean lint-free cloth until reassembly.

CAUTION
Do not get any oil or grease onto any of the components. These chemicals cause the rubber parts in the brake system to swell, permanently damaging them.

CAUTION
Do not remove the secondary cup from the piston assembly for cleaning or inspection purposes.

2. Check the piston assembly (**Figure 49**) for the following defects:
   a. Broken, distorted or collapsed piston return spring (A, **Figure 49**).
   b. Worn, cracked, damaged or swollen primary (B, **Figure 49**) and secondary cups (C).
   c. Scratched, scored or damaged piston (D, **Figure 49**).
   d. Corroded, weak or damaged snap ring.
   e. Worn or damaged dust cover.
   f. If any of these parts are worn or damaged, replace the piston assembly.
3. Measure the piston outside diameter (**Figure 50**).
4. To assemble a new piston assembly, perform the following:
   a. Before installing the new piston cups, lubricate them with brake fluid.
   b. Clean the new piston in brake fluid.
   c. Install the secondary (C, **Figure 49**) cup onto the piston. Refer to the old piston and cup as a guide.
   d. Install the new primary cup onto the spring (B, **Figure 49**).
5. Inspect the master cylinder bore. Replace the master cylinder if its bore is corroded, cracked or damaged in any way. Do not hone the master cylinder bore to remove scratches or other damage.
6. Measure the master cylinder bore inside diameter (**Figure 44**, typical).
7. Inspect the pushrod and replace if damaged.

Assembly

1. If installing a new piston assembly, assemble it as described under *Inspection* in this section.
2. Lubricate the piston assembly and cylinder bore with DOT 4 brake fluid.
3. Install the spring and piston assembly into the master cylinder bore (**Figure 48**).

CAUTION
Do not allow the piston cups to tear or turn inside out when installing them into the master cylinder bore. Both cups are larger than the bore.

4. Lubricate the end of the pushrod (where it contacts the piston) with silicone brake grease.
5. Install the pushrod into the master cylinder bore and use it to compress the piston assembly. Then install the snap ring into the master cylinder bore groove.

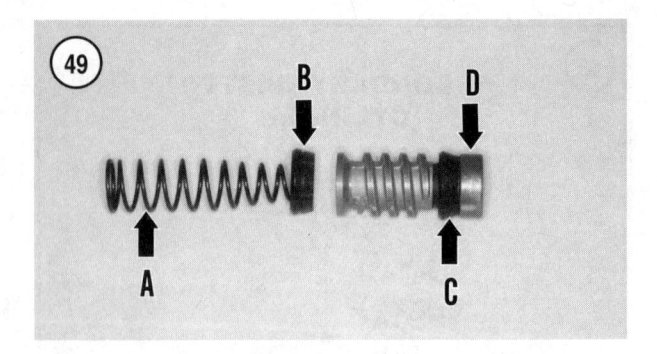

CAUTION
The snap ring must seat in the master cylinder groove completely. Push and release the pushrod a few times to make sure it moves smoothly and that the snap ring does not pop out.

6. Slide the dust cover down the pushrod and install its end into the master cylinder bore and bottom it against the snap ring.
7. Turn the joint to adjust the pushrod to the length shown in **Figure 51**. Then tighten the master cylinder pushrod locknut to 18 N•m (13 ft.-lb.).
8. Install the secondary master cylinder as described in this chapter.

DELAY VALVE

The delay valve is mounted on the right fork slider.

Removal/Installation

Refer to **Figure 52**.
1. Remove right front fender cover. Refer to *Front Fender Assembly* in Chapter Seventeen.
2. Drain the rear brake pedal line as described in this chapter.

BRAKES

51

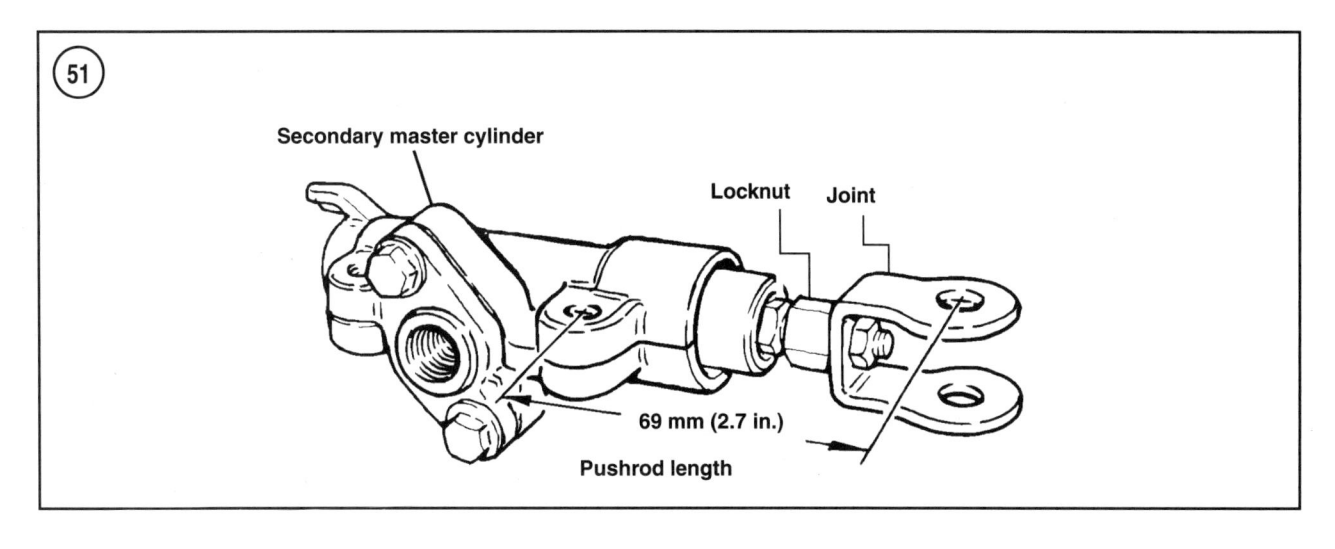

Secondary master cylinder — Locknut — Joint

69 mm (2.7 in.) Pushrod length

52

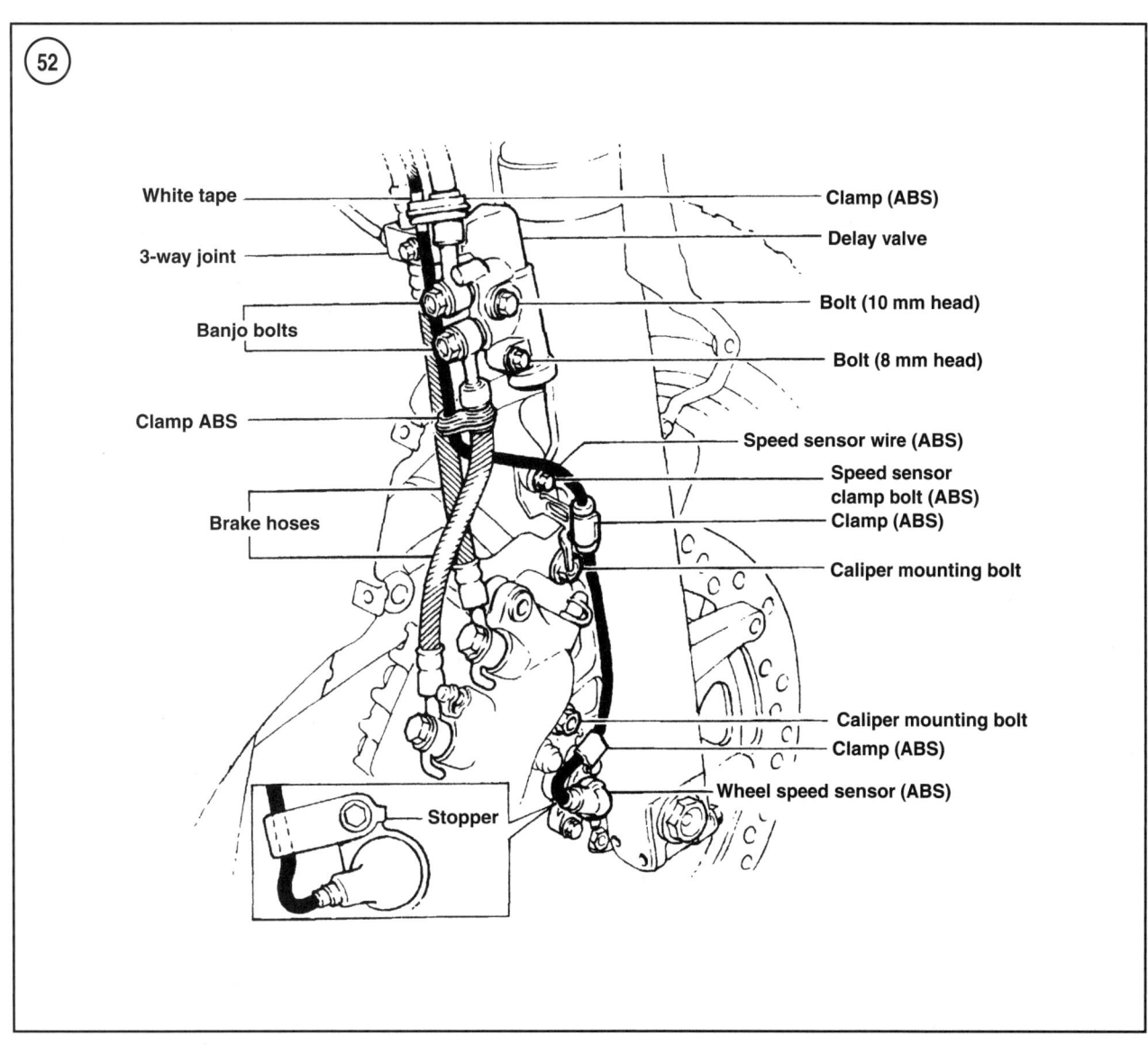

- White tape
- 3-way joint
- Banjo bolts
- Clamp ABS
- Brake hoses
- Stopper
- Clamp (ABS)
- Delay valve
- Bolt (10 mm head)
- Bolt (8 mm head)
- Speed sensor wire (ABS)
- Speed sensor clamp bolt (ABS)
- Clamp (ABS)
- Caliper mounting bolt
- Caliper mounting bolt
- Clamp (ABS)
- Wheel speed sensor (ABS)

3. On GL1800A models, remove the upper and lower ABS and brake hose clamps.
4. Remove the brake hose banjo bolts and washers. Plug the hoses to prevent leakage.
5. Remove the bolt and 3-way joint.
6. Loosen the brake pipe joint nut and disconnect the brake pipe at the delay valve.
7. Remove the delay valve mounting bolts and delay valve.
8. Installation is the reverse of removal. Note the following:
   a. Install new banjo bolt washers.
   b. On GL1800A models, route the brake hoses and wheel speed sensor wire as shown in **Figure 52**.
   c. Tighten the delay valve mounting bolts to 12 N•m (106 in.-lb.).
   d. Lubricate the brake pipe joint threads with DOT 4 brake fluid. Insert the brake pipe into the delay valve and tighten the brake pipe joint nut to 17 N•m (12 ft.-lb.).
   e. Tighten the 3-way joint bolt to 12 N•m (106 in.-lb.).
   f. Connect the brake hoses to the delay valve using *new* washers. Place a washer on each side of the brake hose fittings. Tighten the banjo bolts to 34 N•m (25 ft.-lb.).
   g. On GL1800A models, secure the wheel speed sensor wire with the two clamps. Center the upper clamp between the white tape marks on the speed sensor wire.
   h. Bleed the brake system as described in this chapter.

PROPORTIONAL CONTROL VALVE

The proportional control valve (PCV) is mounted on the frame behind the left radiator.

Removal/Installation

1. Remove the left radiator. Refer to *Left Radiator* in Chapter Ten.
2. Drain the rear brake pedal line as described in this chapter.
3. Loosen the brake pipe joint nuts and disconnect the brake pipes (A, **Figure 53**) from the PCV.
4. Remove the bolts (B, **Figure 53**) and the PCV.
5. Installation is the reverse of removal. Note the following:
   a. Tighten the PCV mounting bolts (B, **Figure 53**) to 12 N•m (106 in.-lb.).
   b. Identify the two brake pipes in **Figure 53** so they are not installed incorrectly.
   c. Lubricate the joint nuts with DOT 4 brake fluid. Reconnect the brake pipes and tighten the nuts to 17 N•m (12 ft.-lbs).
   d. Bleed the brake system as described in this chapter.

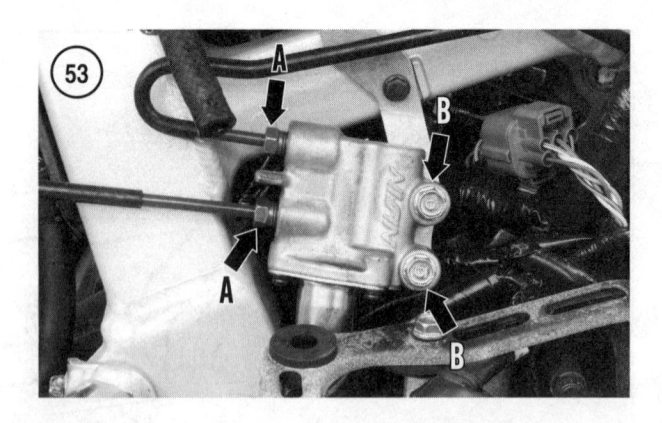

BRAKE HOSE AND BRAKE PIPE REPLACEMENT

Check the rubber brake hoses and metal brake pipes at the brake inspection intervals listed in Chapter Three. Replace the brake hoses if they are worn or damaged, or if they have bulges or signs of chafing. Replace the brake pipes if they are cracked or leaking.

To replace a brake hose or brake pipe, perform the following:
1. Drain the brake system as described in this chapter.
2. Use a plastic drop cloth to cover areas that could be damaged by spilled brake fluid.
3. When removing a brake hose or brake pipe, note the following:
   a. Record the hose or pipe routing on a piece of paper.
   b. Remove any bolts or brackets securing the brake hose or brake pipe to the frame or suspension component.

BRAKES

55

c. Before removing the banjo bolts, note how the end of the brake hose is installed or indexed against the part it is threaded into. The hoses must be installed facing in their original position.

4. Replace damaged banjo bolts.

5. Reverse these steps to install the new brake hoses, while noting the following:

   a. Compare the new and old hoses to make sure they are the same.
   b. Clean the *new* washers, banjo bolts and hose ends to remove any contamination.
   c. Referring to the notes made during removal, route the brake hose along its original path.
   d. Install a *new* banjo bolt washer (**Figure 54**) on each side of the brake hose.
   e. Tighten the banjo bolts to 34 N•m (25 ft.-lb.).

6. Reverse these steps to install new brake pipes, while noting the following:

   a. Prefabricated steel brake pipes are available from Honda. These pipes are equipped with joint nuts and preflared ends.

   WARNING
   Copper tubing cracks and corrodes easily. Never use this tubing in a hydraulic brake system as it can cause leaks and brake system failure.

   b. Blow the new brake pipes out with compressed air before installing them.
   c. Clean the pipe ends and joint nuts to remove any contamination.
   d. Do not bend the brake pipes or try to force them into position during installation. This creases the metal and causes leaks.
   e. If there is any chance of the tubing ends becoming contaminated with grease or debris when installing them, cover the ends with a small plastic bag.
   f. Install the brake pipes following their original mounting position.
   g. Brake pipes can be damaged from vibration and heat. Install the brake pipes in their original positions while using the original mounting fasteners and heat shields.
   h. When installing a brake pipe, thread both brake pipe joint nuts into their mating joints a few turns only. Do not tighten one nut fully before attempting to install and tighten the other nut. Also make sure there is no stress on the brake pipe.
   i. Lubricate the brake pipe joint nuts with DOT 4 brake fluid before installing them.
   j. Tighten the brake pipe joint nuts to 17 N•m (13 ft.-lb.).

7. After installing new hoses and pipes around the steering and front fork area, turn the handlebars from side to side to make sure the hose or pipe does not rub against any part or pull away from its brake unit.

8. Bleed the brake system as described in this chapter.

BRAKE DISC

This section applies to the front and rear brake discs.

Inspection

The front and rear brake discs can be inspected while installed on the motorcycle. Small marks on the disc are not important, but deep scratches or other marks may reduce braking effectiveness and increase brake pad wear. If these grooves are evident and the brake pads are wearing rapidly, replace the brake disc.

Refer to **Table 1** and **Table 2** for brake disc specifications.

1. Support the motorcycle with the wheel (front or rear) off the ground.

2. Measure the disc thickness at several locations around the disc (**Figure 55**). Replace the disc if its thickness at any point is less than the minimum allowable specification stamped on the disc or less than the service limit in **Table 1** or **Table 2**.

3. Position a dial indicator stem against the brake disc (**Figure 56**). Zero the dial gauge and slowly turn the wheel and measure runout. If the disc runout is excessive:

   a. Check for loose or missing fasteners.
   b. Remove the front wheel and check the wheel bearings.

c. Remove the rear wheel and rear brake caliper and inspect the disc mounting on the final drive housing.

4. Clean the disc of any rust or corrosion and wipe clean with brake cleaner. Never use an oil-based solvent that may leave an oil residue on the disc.

Front Brake Disc
Removal/Installation

1. Remove the front wheel (Chapter Ten).
2. On GL1800A models, remove the bolts and front pulser ring from the right side of the hub.
3. Remove the bolts securing the brake disc to the wheel and remove the disc (**Figure 57**).
4. Perform any necessary service to the front hub (wheel bearing or tire replacement) before installing the brake discs.
5. Clean the brake disc threaded holes in the front hub.
6. Clean the brake disc mounting surface on the front hub.
7. Install the brake disc with its directional arrow facing toward the wheel's normal rotating direction.
8. Install *new* brake disc mounting bolts and tighten to 20 N•m (15 ft.-lb.).
9. On GL1800A models, install the pulser ring onto the right hub. Install new bolts and tighten in a crisscross pattern in two or three steps to 8 N•m (71 in.-lb.).
10. Clean the disc of any rust or corrosion and spray clean with brake cleaner. Never use an oil-based solvent that may leave an oil residue on the disc.
11. Install the front wheel (Chapter Ten).

Rear Brake Disc
Removal

1. Support the motorcycle on its centerstand. 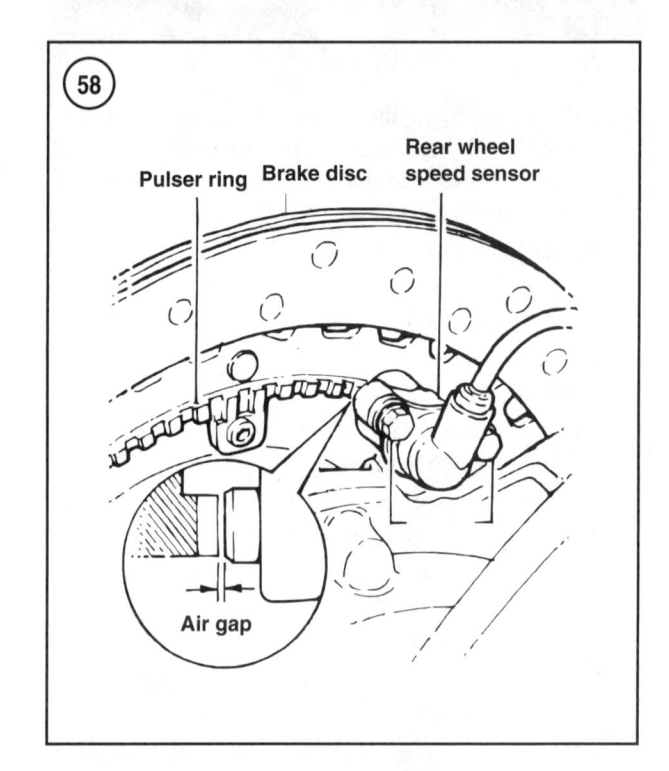
2. Remove the rear wheel. Refer to *Rear Wheel* in Chapter Eleven.
3. On GL1800A models, remove the wheel speed sensor (**Figure 58**, typical). Note how the sensor wire (and brake hose guide on 2004 models) is routed and route it along the same path during installation. Refer to *Wheel Speed Sensors* in Chapter Fifteen.
4. Loosen the two brake disc mounting screws (A, **Figure 59**) with a hand impact driver and No. 3 Phillips bit.

CAUTION
These screws use a threadlock. Engage these screws fully with the Phillips bit to avoid damaging them.

BRAKES

Figure 59 (photo showing rear brake disc with labels A, A, and B)

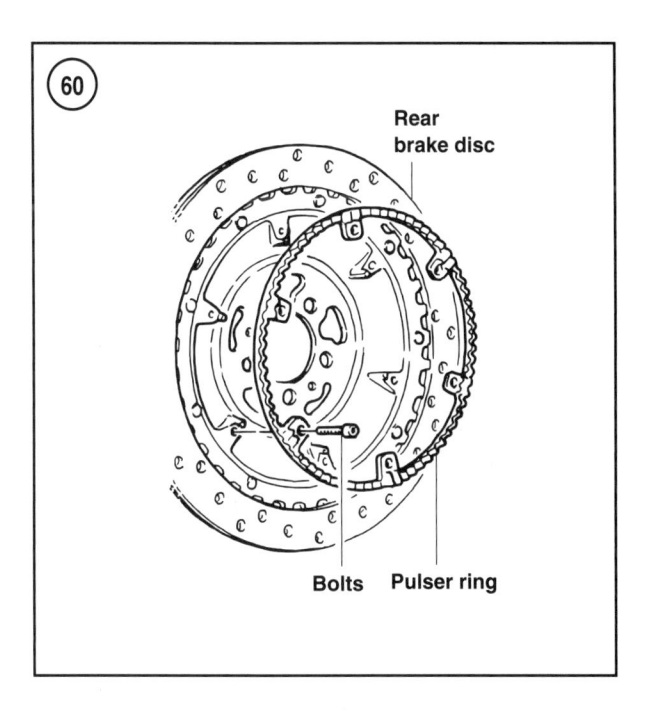

Figure 60 — Rear brake disc, Bolts, Pulser ring

NOTE
If the left muffler is installed on the motorcycle, mount a long extension on the impact driver when loosening the brake disc mounting screws.

5. Remove the rear brake caliper as described under *Rear Brake Caliper* in this chapter.

6. Remove the two screws (A, **Figure 59**) and the brake disc (B).

7. On GL1800A models, remove the six mounting bolts and the pulser ring (**Figure 60**), if necessary.

Installation

1. Clean the brake disc and final drive housing mating surfaces.

2. On GL1800A models, clean the pulser ring, if removed. Secure the pulser ring (**Figure 60**) onto the brake disc with new mounting bolts and tighten to 8 N•m (71 in.-lb.).

3. Install the brake disc (B, **Figure 59**) onto the final drive housing. Make sure it seats flush against the housing. Secure with two new ALOC mounting screws (A, **Figure 59**) and tighten to 9 N•m (80 in.-lb.).

4. Install the brake caliper as described in this chapter.

5. Install the wheel speed sensor and tighten its mounting bolts (**Figure 58**, typical) to 12 N•m (106 in.-lb.).

6. Install the rear wheel. Refer to *Rear Wheel* in Chapter Eleven.

BRAKE BLEEDING SERVICE TIPS AND TOOLS

Bleeding the brakes removes air from the brake system. Air in the brake system increases brake lever or pedal travel while causing it to feel spongy and less responsive. Under extreme braking (heat) conditions, it can cause complete loss of the brake.

The brake system can be bled manually or with the use of a vacuum pump. Both methods are described in this section

When adding brake fluid during the bleeding process, use DOT 4 brake fluid. Do not reuse brake fluid drained from the system or use DOT 5 (silicone based) brake fluid. Brake fluid damages most surfaces, so wipe up any spills immediately with soapy water and rinse completely.

NOTE
When bleeding the brakes, check the fluid level in the front and rear master cylinders frequently to prevent them from running dry, especially when using a vacuum pump. If air enters the system rebleed the system.

General Bleeding Tips

When bleeding the brakes, note the following:

1. Clean the bleed screws and the area around the screws of all dirt and debris. Make sure the passageway in the end of the screw is open and clear.

2. Use a box-end wrench to open and close the bleed screws. This prevents damage to the hex-head.

3. Replace bleed screws with damaged hex-heads. These are difficult to loosen and cannot be tightened fully.

61 Vacuum pump / Reservoir / Box-end wrench

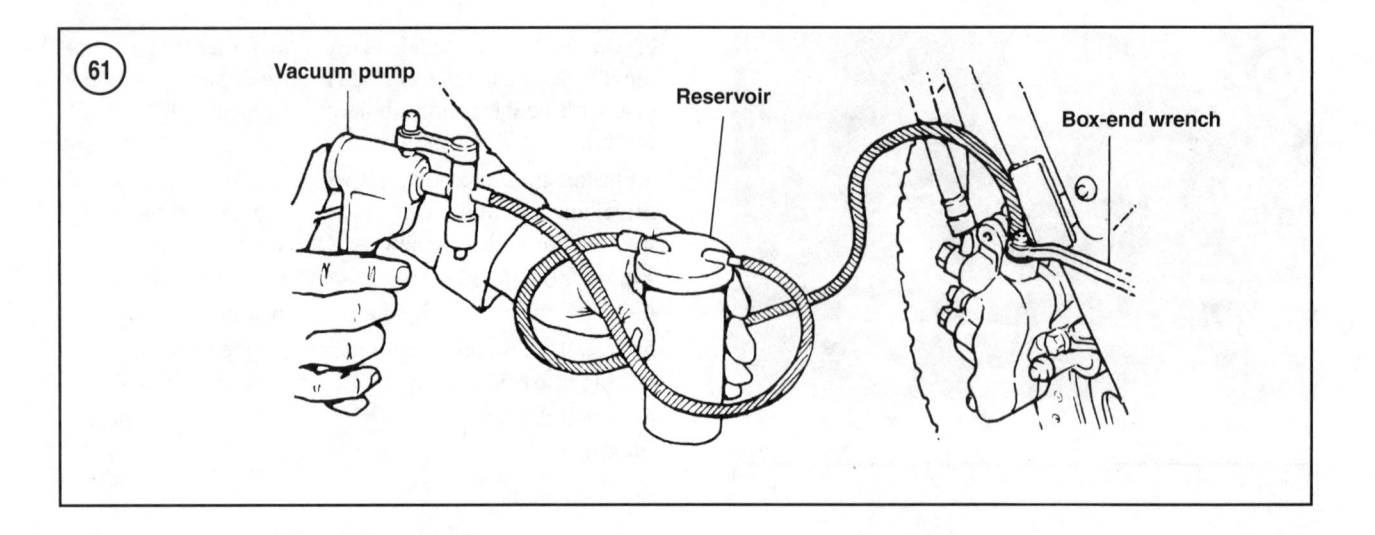

4. Install the box-end wrench on the bleed screw before installing the catch hose. This allows operation of the wrench without having to disconnect the hose.

5. Use a clear catch hose to allow visual inspection of the brake fluid as it leaves the caliper or brake unit. Air bubbles visible in the catch hose indicate there still may be air trapped in the brake system.

6. Depending on the play of the bleed screw when it is loosened, it is possible to see air exiting through the catch hose even through there is no air in the brake system. A loose or damaged catch hose also causes air leaks. In both cases, air is being introduced into the bleed system at the bleed screw threads and catch hose connection and not from within the brake system itself. This condition can be misleading and cause excessive brake bleeding when there is no air in the system.

7. Open the bleed screw just enough to allow fluid to pass through the screw and into the catch bottle. The farther the bleed screw is opened, the looser the valve becomes. This allows air to be drawn into the system from around the screw threads.

8. If air is suspected of entering from around the bleed screw threads, pack the area around the bleed screw with silicone brake grease.

WARNING
Do not force grease into the caliper past the bleed screw threads. This can block the bleed screw passageway and contaminate the brake fluid.

9. If the system is difficult to bleed, tap the banjo bolt on the master cylinder a few times. It is not uncommon for air bubbles to become trapped in the hose connection where the brake fluid exits the master cylinder. When a number of bubbles appear in the master cylinder reservoir after tapping the banjo bolt, it means air was trapped in this area. Also, tap the other bolt and line connection points at the calipers and other brake units.

Vacuum Tools and Bleeding

Vacuum bleeding can be accomplished by using either a hand-operated or compressed air tool. The tools described below can be used by one person to drain and bleed the GL1800/A brake systems.

Hand-operated vacuum pump

This is a one-person procedure where a hand-operated vacuum pump, two lengths of clear hose and a separate reservoir are used to drain and bleed the brake system (**Figure 61**).

1. Connect the catch hose between the bleed screw and catch bottle. Connect the other hose between the catch bottle and vacuum pump. See the tool manufacturer's instructions for additional information.

2. Secure the vacuum pump to the motorcycle with a length of stiff wire so it is possible to check and refill the master cylinder reservoirs without having to disconnect the catch hose.

3. Operate the vacuum pump to create a vacuum in the catch hose connected to the bleed screw. Then open the bleed screw with a wrench to allow brake fluid to be drawn through the master cylinder, brake hoses and lines. Close the bleed screw before the brake fluid stops flowing from the system (no more vacuum in line) or before the master cylinder reservoir runs empty.

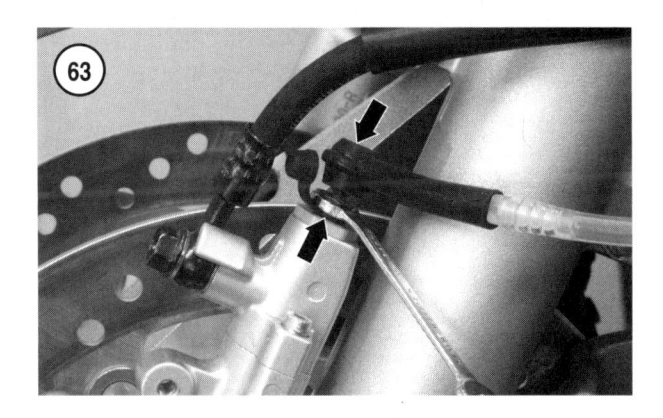

BRAKES

Figure 62

Figure 63

Figure 64 Catch hose, Box-end wrench, Catch bottle

4. Repeat Step 3 until the brake fluid running through the vacuum hose is a clear and solid stream without air bubbles.

NOTE
When using a vacuum pump, observe the brake fluid level in the reservoir frequently as it will drop quite rapidly.

Compressed air vacuum pump

This tool (A, **Figure 62**, typical) uses compressed air to create a powerful vacuum to drain and bleed brake systems. An air compressor (80-120 psi) is required. The handle (B, **Figure 62**) on top of the cover allows the user to control the amount of brake fluid removed from the system.

1. Assemble the tool, following the manufacturer's instructions.
2. Connect a box-end wrench onto the bleed screw. Connect the vacuum hose onto the bleed screw (**Figure 63**).
3. Connect a compressed air source (C, **Figure 62**) to the vacuum tool.
4. Depress the lever on top of the pump, then open the bleed screw slightly. As long as the lever is depressed, a vacuum is created in the canister and brake fluid will evacuate from the line. Releasing the lever stops the vacuum. Because this tool will drain the master cylinder rapidly, use an assistant to refill the master cylinder during the procedure. Otherwise, release the lever and tighten the bleed screw. Then refill the master cylinder and continue the procedure.

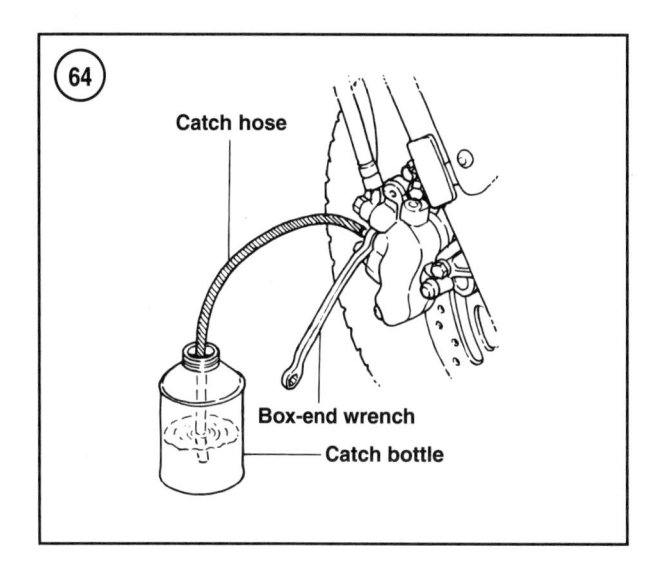

NOTE
Always close the bleed screw before releasing the lever on top of the pump.

5. Continue until the brake fluid running through the vacuum hose is a clear and solid stream without air bubbles.

NOTE
*Air drawn in around the bleed screw will cause bubbles to form in the vacuum hose. While this is normal, it is misleading as it always appears there is air in the system, even when the system has been bled completely. Block off the bleed valve with silicone brake grease as described in **General Bleeding Tips** (Step 8).*

Manual Tools and Bleeding

This is a one-person procedure that requires a reservoir bottle, length of clear hose (catch hose), wrench and DOT 4 brake fluid (**Figure 64**).

1. Connect the catch hose to the bleed valve on the brake caliper. Submerge the other end of the hose into the bottle partially filled with DOT 4 brake fluid. This prevents air from being drawn into the catch hose and back into the brake system.

65

Right front caliper
Front master cylinder
Left front caliper
Lines being bled in black

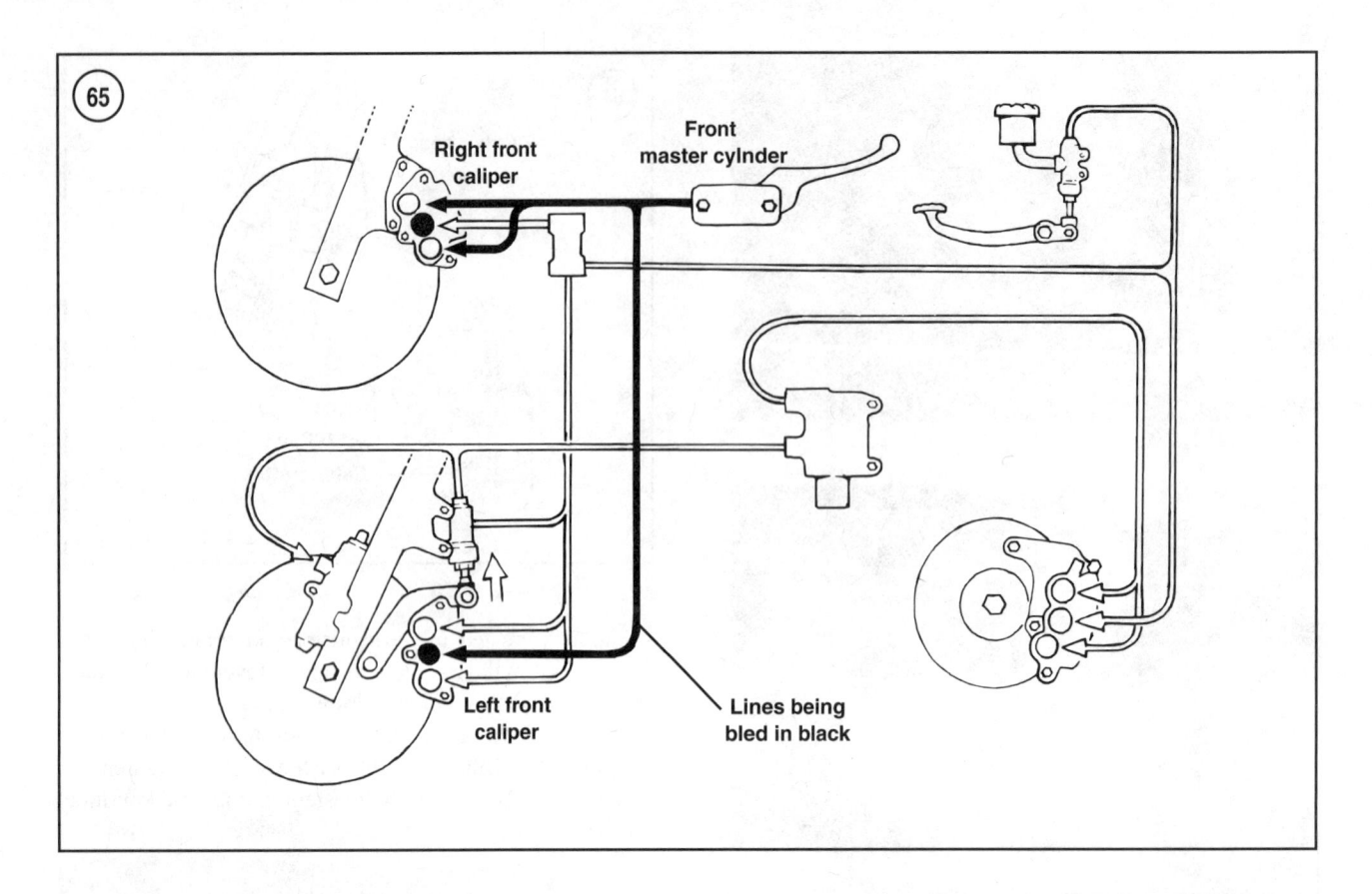

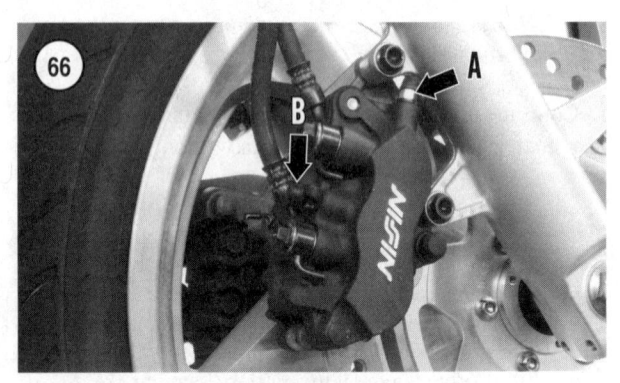

66

67

2. Apply the front brake lever or rear brake pedal until it stops and holds in this position.

3. Open the bleed screw with a wrench and let the lever or pedal move to the limit of its travel, then close the bleed screw.

NOTE
*When bleeding the brakes manually, make sure to close the bleed screw **before** releasing the brake lever or pedal. This prevents air from being drawn back into the system on the lever's or pedal's return stroke if the bleed screw was left open.*

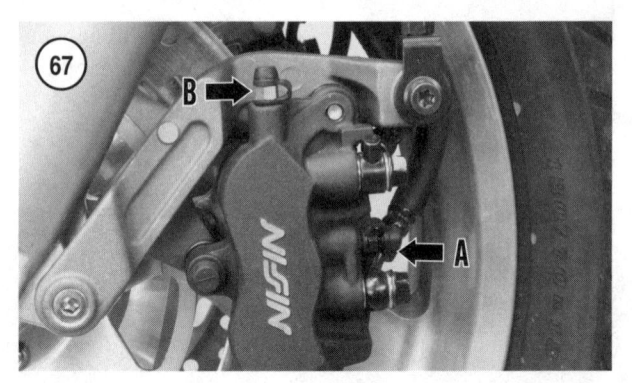

4. Release the lever or pedal slowly, then repeat Step 2 and Step 3 until the brake fluid running through the vacuum hose is a clear and solid stream without air bubbles.

BRAKE FLUID DRAINING

Before disconnecting a brake hose from a brake component, drain the brake fluid as described in this section. Doing so reduces the amount of brake fluid that can spill out when disconnecting the brake hoses and lines from the system.

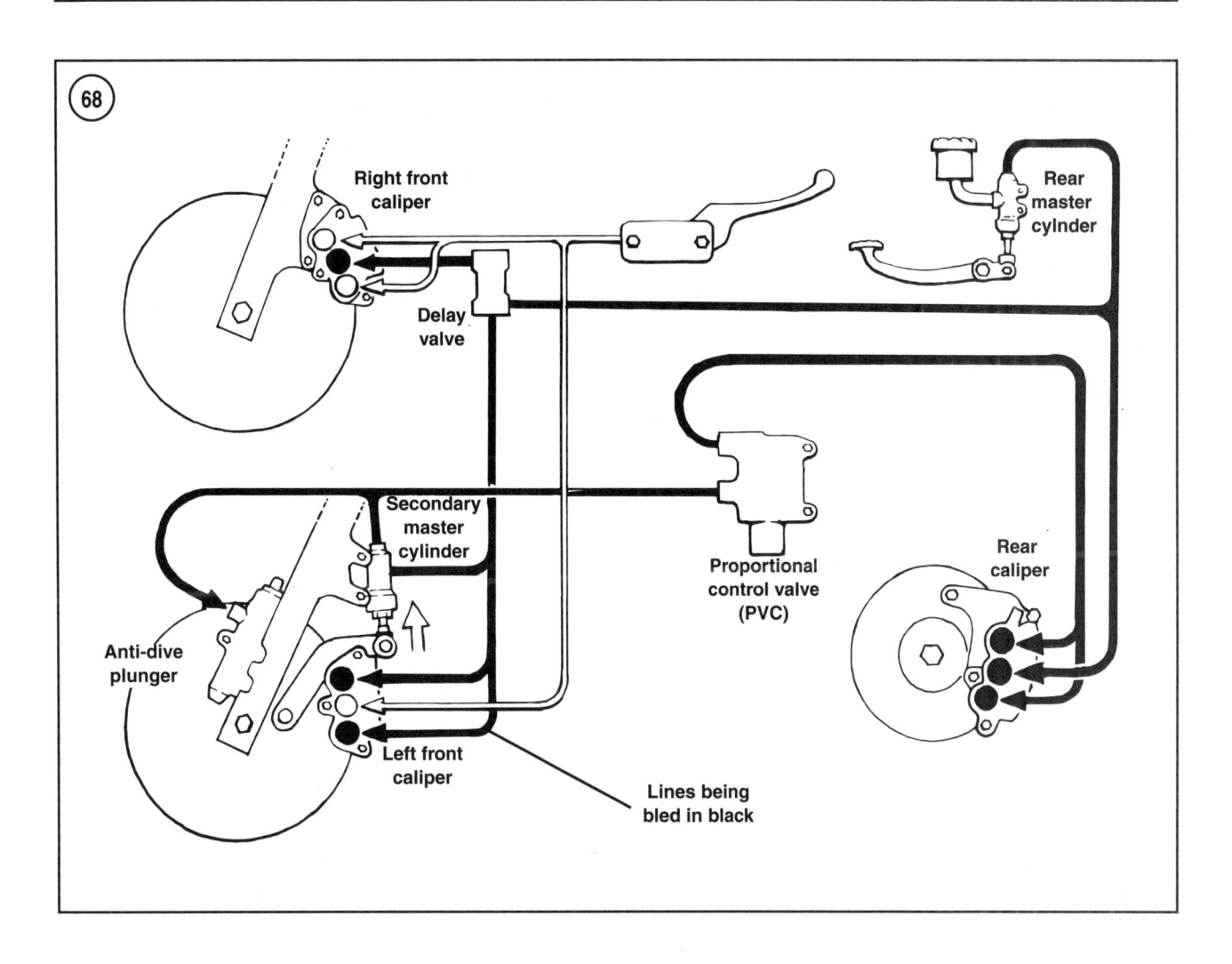

BRAKES

68

- Right front caliper
- Delay valve
- Secondary master cylinder
- Anti-dive plunger
- Left front caliper
- Rear master cylinder
- Proportional control valve (PVC)
- Rear caliper
- Lines being bled in black

Front Brake Lever Line

Figure 65 identifies the brake lever lines that will be drained in this procedure.

1. Read the information listed under *Brake Bleeding Service Tips and Tools* in this chapter for the selection, installation and operation of a brake bleeder.

2. Support the bike on its centerstand.

3. Remove front fender A and both fender covers. Refer to *Front Fender Assembly* in Chapter Seventeen.

4. Turn the handlebars to level the front master cylinder and remove the cover assembly.

5. Connect a brake bleeder to the right front brake caliper upper bleed screw (A, **Figure 66**). Open the bleed screw and operate the brake bleeder until brake fluid stops flowing. Tighten the bleed screw.

6. Repeat Step 5 at the left front caliper lower bleed screw (A, **Figure 67**).

7. Disconnect the bleeder tool.

Rear Brake Pedal Line

Figure 68 identifies the brake pedal lines that will be drained in this procedure.

1. Read the information listed under *Brake Bleeding Service Tips and Tools* in this chapter for the selection, installation and operation of a brake bleeder.

2. Support the bike on its centerstand.

3. Remove front fender A and both fender covers. Refer to *Front Fender Assembly* in Chapter Seventeen.

4. Remove the right engine side cover. Refer to *Engine Side Covers* in Chapter Seventeen.

5. Remove the rear reservoir cap and diaphragm.

6. Connect a brake bleeder to the left front brake caliper upper bleed valve (B, **Figure 67**). Open the bleed valve and operate the brake bleeder until brake fluid stops flowing. Tighten the bleed valve.

7. Repeat Step 6 at the following bleed valves.

   a. Right front brake caliper lower bleed valve (B, **Figure 66**).

b. Anti-drive plunger bleed valve (**Figure 69**).
c. Both bleed screws at the rear caliper (A and B, **Figure 70**).
8. Disconnect the bleeder tool.

FLUSHING THE BRAKE SYSTEM

When flushing the brake system, use DOT 4 brake fluid as a flushing fluid. Flushing consists of pulling new brake fluid through the system until the new fluid appears at the caliper and without any air bubbles. To flush the brake system, follow one of the bleeding procedures described under *Bleeding The System* in this chapter.

CAUTION
Never reuse old brake fluid. Properly discard all brake fluid flushed from the system.

BLEEDING THE SYSTEM

Because of the interconnection and operation of the linked braking system, filling the brake pipes and bleeding air from the system can become an elaborate and time-consuming procedure. To reduce frustration and service time, perform the brake fluid filling and bleeding procedures in the order given.

1. Read the information listed under *Brake Bleeding Service Tips and Tools* in this chapter.

CAUTION
Cover all parts that could become contaminated by the accidental spilling of brake fluid. Wash any spilled brake fluid from any surface immediately because the brake fluid damages the finish. Use soapy water and rinse completely.

2. Support the bike on its centerstand.
3. Check that all the brake system banjo bolts and hose fittings are tight.
4. Connect a brake bleeder to the bleed valve specified in the procedure. Refer to *Brake Bleeding Service Tips and Tools* for a description of bleeding tools and how to connect them to the brake system.

NOTE
When using a vacuum pump in the following sections, observe the brake fluid level in the reservoir as it will drop quite rapidly. This is especially true for the rear reservoir because it does not hold as much brake fluid as the front reservoir. Stop often and check the brake fluid level. Maintain the level at 10 mm (.40 in.) from the top of the reservoir to prevent air from being drawn into the system.

Filling/Bleeding the Front Brake Lever Line

Figure 65 shows the front brake lever brake pipes that will be filled and bled in this procedure. These lines are separate from the rear brake pedal lines.

1. Read the information listed under *Brake Bleeding Service Tips and Tools* in this chapter for the selection, installation and operation of a brake bleeder.
2. Support the motorcycle on its centerstand.
3. Remove front fender A and both fender covers. Refer to *Front Fender Assembly* in Chapter Seventeen.
4. Turn the handlebars to level the front master cylinder and remove cover assembly. Fill the reservoir to about 10 mm (.40 in.) from the top.

NOTE
If the brake lever system was drained of all brake fluid, start with Step 5 and Step 6 to pull brake fluid through the system. Do not be concerned about the air bubbles exiting the system at this time. When the brake lines are full of fluid, repeat Step 5 and Step 6 to remove all air from the lines.

5. Connect a brake bleeder to the right front brake caliper upper bleed screw (A, **Figure 66**). Open the bleed screw and operate the brake bleeder to bleed the brake line. Repeat until the brake fluid exiting the catch hose is clear and free of air.
6. Repeat Step 5 at the left front caliper lower bleed screw (A, **Figure 67**).
7. Repeat Step 5 and Step 6 as necessary until the front brake lever feels firm when applying it and there are no air bubbles in the catch hose.

BRAKES

70

B
A

NOTE
*If the system is difficult to bleed, refer to **General Bleeding Tips** in this chapter.*

NOTE
*If the brake lever feels firm but there are air bubbles in the catch line, air may be leaking around the bleed valve threads. Refer to **General Bleeding Tips** in this chapter.*

8. After bleeding each of the bleed screws and their separate systems so no air bubbles appear in the catch hose, test the feel of the brake lever. It should be firm and should offer the same resistance each time it is operated. If the brake lever feels spongy, air is trapped in the system and the bleeding procedure must be continued.

9. Tighten each bleed screw to 6 N•m (53 in.-lb.).

10. If necessary, add DOT 4 brake fluid to correct the level in the master cylinder reservoir. It must be above the level line.

11. Test ride the bike slowly at first to make sure the brakes are operating correctly.

WARNING
Do not ride the bike until the brakes and the brake light are working properly.

Filling/Bleeding the Rear Brake Pedal Line

Due to the number and length of the brake pipes in this system, approximately 500 ml (16.9 U.S. oz./14.1 Imp. oz.) of DOT 4 brake fluid is required.

1. Read the information listed under *Brake Bleeding Service Tips and Tools* in this chapter for the selection, installation and operation of a brake bleeder.
2. Support the bike on its centerstand.
3. Remove front fender A and both fender covers. Refer to *Front Fender Assembly* in Chapter Seventeen.
4. Remove the right engine side cover. Refer to *Engine Side Covers* in Chapter Seventeen.

5. Remove the rear reservoir cap and diaphragm.

NOTE
If the brake pedal system was drained of all brake fluid, start with Step 6 and Step 7 to pull brake fluid through the system. Do not be concerned about the air bubbles exiting the system at this time. When the brake lines are full of fluid, repeat Step 6 and Step 7 to remove all air from the lines.

6. Connect a brake bleeder to the left front brake caliper upper bleed screw (B, **Figure 67**).

Open the bleed screw and operate the brake bleeder to bleed the brake line. Repeat until the brake fluid exiting the catch hose is clear and free of air.

7. Repeat Step 6 at the following bleed screws in order:
    a. Right front brake caliper lower bleed screw (B, **Figure 66**).
    b. Rear brake caliper lower bleed screw (A, **Figure 70**).
    c. Anti-dive plunger bleed screw (**Figure 69**).
    d. Rear brake caliper upper bleed screw (B, **Figure 70**).

NOTE
*If the system is difficult to bleed, refer to **General Bleeding Tips** in this chapter.*

NOTE
*If the brake lever feels firm but there are air bubbles in the catch line, air may be leaking around the bleed valve threads. Refer to **General Bleeding Tips** in this chapter.*

8. After bleeding each of the bleed screws and their separate systems so no air bubbles appear in the catch hose, test the feel of the brake pedal. It should be firm and should offer the same resistance each time it is operated. If the brake pedal feels spongy, air is trapped in the system and the bleeding procedure must be continued.

9. Tighten each bleed screw to 6 N•m (53 in.-lb.).

10. If necessary, add DOT 4 brake fluid to correct the level in the master cylinder reservoir. It must be above the level line.

11. Test ride the bike slowly at first to make sure the brakes are operating correctly.

WARNING
Do not ride the bike until the brakes and the brake light are working properly.

Table 1 FRONT BRAKE SERVICE SPECIFICATIONS

	New mm (in.)	Service limit mm (in.)
Brake disc runout	–	0.30 (0.012)
Brake disc thickness	4.5 (0.18)	3.5 (0.14)
Brake caliper cylinder inside diameter		
Left caliper		
Upper	22.650-22.700 (0.8917-0.8937)	22.710 (0.8941)
Center	27.000-27.050 (1.0630-1.0650)	27.060 (1.0654)
Lower	22.650-22.700 (0.8917-0.8937)	22.710 (0.8941)
Right caliper		
Upper	25.400-25.450 (1.0000-1.0020)	25.460 (1.0024)
Center	25.400-25.450 (1.0000-1.0020)	25.460 (1.0024)
Lower	22.650-22.700 (0.8917-0.8937)	22.710 (0.8941)
Brake caliper piston outside diameter		
Left caliper		
Upper	22.585-22.618 (0.8892-0.8905)	22.560 (0.8882)
Center	26.935-26.968 (1.0604-1.0617)	26.910 (1.0590)
Lower	22.585-22.618 (0.8892-0.8905)	22.560 (0.8882)
Right caliper		
Upper	25.335-25.368 (0.9974-0.9987)	25.310 (0.9960)
Middle	25.335-25.368 (0.9974-0.9987)	25.310 (0.9960)
Lower	22.585-22.618 (0.8892-0.8905)	22.560 (0.8882)
Master cylinder bore inside diameter	14.000-14.043 (0.5512-0.5529)	14.055 (0.5533)
Master cylinder piston outside diameter	13.957-13.984 (0.5495-0.5506)	13.945 (0.5490)

Table 2 REAR BRAKE SERVICE SPECIFICATIONS

	New mm (in.)	Service limit mm (in.)
Brake disc runout	–	0.30 (0.012)
Brake disc thickness	11.0 (0.43)	10.0 (0.39)
Brake caliper cylinder inside diameter		
Upper	22.650-22.700 (0.8917-0.8937)	22.710 (0.8941)
Center	27.000-27.050 (1.0630-1.0650)	27.060 (1.0654)
Lower	22.650-22.700 (0.8917-0.8937)	22.710 (0.8941)
Brake caliper piston outside diameter		
Upper	22.585-22.618 (0.8892-0.8905)	22.560 (0.8882)
Center	26.935-26.968 (1.0604-1.0617)	26.910 (1.0590)
Lower	22.585-22.618 (0.8892-0.8905)	22.560 (0.8882)
Master cylinder bore inside diameter	17.460-17.503 (0.6874-0.6891)	17.515 (0.6896)
Master cylinder piston outside diameter	17.417-17.444 (0.6857-0.6868)	17.405 (0.6852)
Secondary master cylinder bore inside diameter	14.000-14.043 (0.5512-0.5529)	14.055 (0.5533)
Secondary master cylinder piston outside diameter	13.957-13.984 (0.5495-0.5506)	13.945 (0.5490)

BRAKES

Table 3 BRAKE TORQUE SPECIFICATIONS

	N•m	in.-lb.	ft.-lb.
Banjo bolts	34	–	25
Bleed screws	6	53	–
Brake hose bracket bolts	12	106	–
Brake hose clamp bolt	12	106	–
Brake hose 3-way joint bolt	12	106	–
Brake lever pivot bolt	1	8.8	–
Brake lever pivot nut	6	53	–
Brake pipe joint nuts[1]	17	–	12
Delay valve mounting bolt	12	106	–
Front brake caliper assembly bolts[2]	32	–	24
Front brake caliper bracket shaft[3]	13	115	–
Front brake caliper housing shaft[3]	23	–	17
Front brake disc bolts[2]	20	–	15
Front brake light/cruise cancel switch screws	1	8.8	–
Front master cylinder holder bolts	12	106	–
Front master cylinder reservoir cap screw	2	18	–
Left front brake caliper pivot bolt[2]	31	–	23
Left front brake caliper secondary master cylinder joint bolt[2]	25	–	18
Pad pin	18	–	13
Proportional control valve mounting bolts	12	106	–
Pulser ring mounting bolts (ABS)[2]	8	71	–
Rear brake caliper assembly bolts[2]	32	–	24
Rear brake caliper bracket shaft[3]	23	–	17
Rear brake caliper mounting bolts[2]	45	–	33
Rear brake caliper housing shaft[3]	27	–	20
Rear brake disc screws[2]	9	80	–
Rear brake cruise switch holder[3]	2	18	–
Rear brake pedal pinch bolt	26	–	19
Rear and secondary master cylinder pushrod locknut	18	–	13
Rear master cylinder mounting bolts	12	106	–
Rear master cylinder reservoir mounting bolt	12	106	–
Rear master cylinder switch plate locknut	18	–	13
Rider footpeg mounting bolts	26	–	19
Right front brake caliper mounting bolts[2]	31	–	23
Secondary master cylinder mounting bolts[2]	31	–	23
Wheel speed sensor mounting bolt (ABS)	12	106	–

1. Lubricate threads with brake fluid.
2. Install new ALOC fastener during assembly.
3. Apply a medium strength threadlock onto fastener threads.

CHAPTER FIFTEEN

ANTI-LOCK BRAKE SYSTEM

GL1800A models are equipped with an anti-lock brake system (ABS) that prevents wheel lockup during hard braking or when braking on slippery surfaces. During operation, the ABS rapidly pumps the brake(s), which interrupts the flow of hydraulic fluid to the brake caliper(s) of the wheel approaching lockup.

The ABS system includes a front and rear modulator (each with a control motor), front and rear wheel speed sensors, front and rear wheel pulser rings, ABS indicator and the ABS electronic control unit (ECU). See **Figure 1**.

The ABS control unit monitors the rotational speed of the front and rear wheels. When it determines that a wheel is approaching lockup, the ABS control unit modulates the control piston, which opens and closes the cut-off valve. Hydraulic pressure from the master cylinder to the caliper is momentarily interrupted and then reapplied. The ABS control unit repeats this cycle and rapidly pumps the brake(s) until secure braking is restored.

The linked brake system (LBS) and ABS are hydraulically independent. If the ABS system fails or shuts down for any reason, it is disabled, but the linked brake system will still operate normally. In addition, problems that occur in the non-ABS part of the brake system can effect the ABS system and cause it to shut down.

This chapter describes troubleshooting and replacement procedures for the ABS components. Replace any part that is damaged. During assembly, install new fasteners when instructed to do so and tighten the fasteners to the correct specification. **Tables 1-3** are located at the end of this chapter.

Before proceeding any further, read *Preventing Brake Fluid Damage* and *Brake Service* in Chapter Fourteen. These sections contain pertinent service information that must be considered when troubleshooting the ABS system.

WARNING
When working on any part of the ABS system, the work area and all tools must be absolutely clean. ABS components can be damaged by even tiny particles of grit that enter the system.

If there is any doubt about your ability to correctly and safely troubleshoot or service the ABS system, take the job to a Honda dealership.

ABS SERVICE PRECAUTIONS

Before troubleshooting or servicing the ABS, note the following:
1. Handle the ABS components carefully. The ECU can be damaged if dropped.
2. Turn the ignition switch off before disconnecting or reconnecting ABS harness connectors. If current is flowing

ANTI-LOCK BRAKE SYSTEM

ABS COMPONENTS AND CONNECTORS

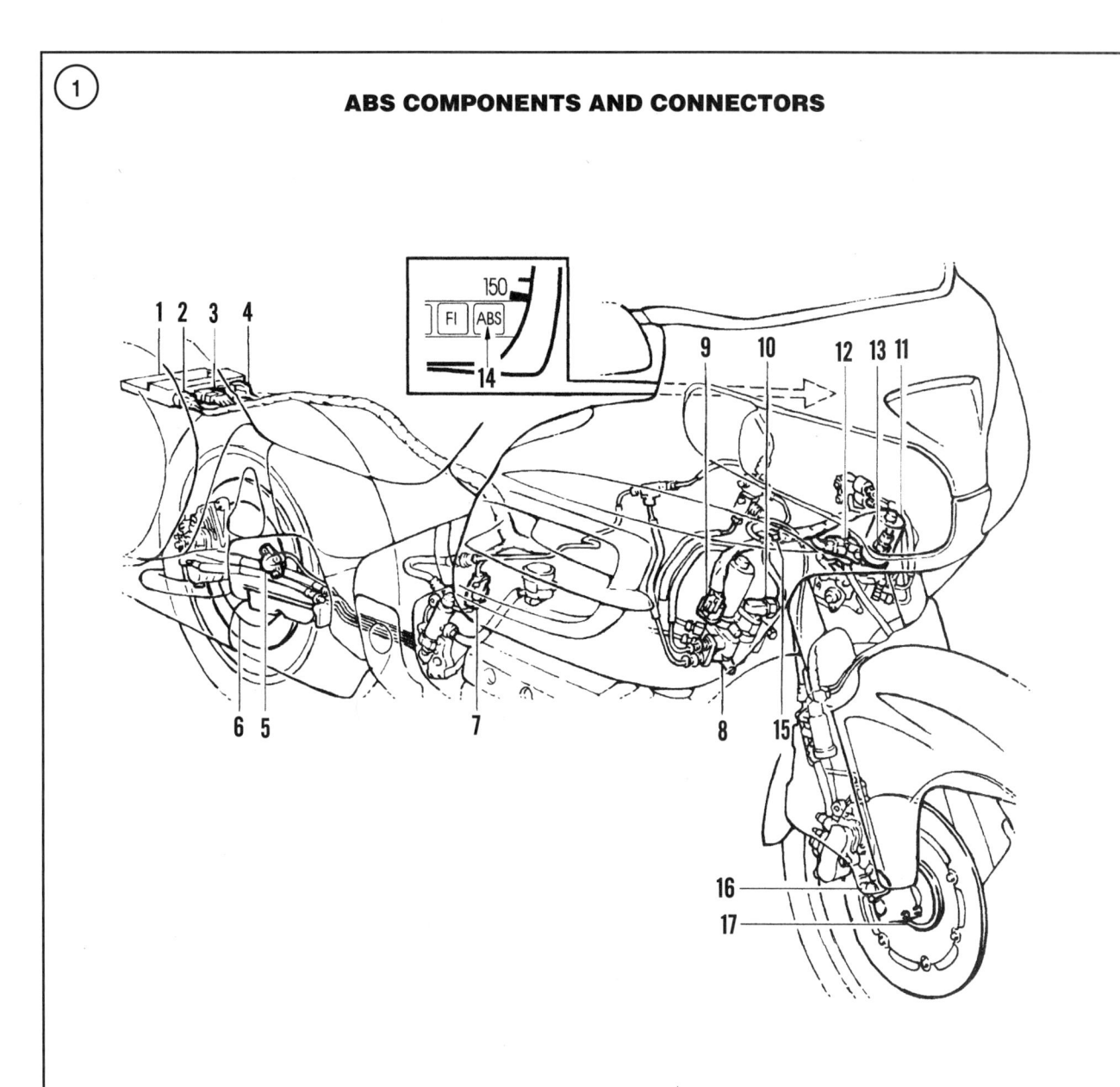

1. ABS electronic control unit (ECU)
2. ECU black 5-pin connector
3. ECU black 12-pin connector
4. ECU brown 5-pin connector
5. Rear wheel speed sensor
6. Rear pulser ring
7. Rear wheel speed sensor gray 2-pin connector
8. Front ABS modulator
9. Front ABS modulator control motor black 2-pin connector
10. Front ABS modulator crank angle gray 3-pin connector
11. Rear ABS modulator
12. Rear ABS modulator crank angle sensor gray 3-pin connector
13. Rear ABS modulator control motor black 2-pin connector
14. ABS indicator
15. Front wheel speed sensor orange 2-pin connector
16. Front wheel speed sensor
17. Front pulser ring

when a connector is disconnected, a voltage spike can damage the ECU.

3. While the ECU is constantly monitoring the ABS, it does not recognize problems in other parts of the brake system unless they affect the ABS. For example, worn brake pads and low brake fluid level. Inspect non-ABS components at the intervals specified in Chapter Three, or more often when towing a trailer or when operating under severe riding conditions.

4. Be careful not to damage the ABS wiring harness or connectors during service procedures. Note the routing of the wire harness and the type, number and position of fasteners and clamps used to secure the wiring to the frame and other components.

5. The metal brake lines can be creased and permanently damaged from improper handling. Remove, position and secure these lines carefully.

6. The ABS modulators are not designed to be serviced. Replace the modulators as a unit when faulty.

7. Use OEM fasteners (or equivalents) when replacement is required during service.

ABS TROUBLESHOOTING

When the engine is started, the ECU performs self-diagnosis and checks the operating conditions of the ABS components. Self-diagnosis starts when the ignition switch is turned on and ends when the motorcycle speed reaches 10 km/h (6 mph). If the system is operational, the ABS indicator (**Figure 2**) turns off. This check must be observed each time the motorcycle is ridden.

If a problem is detected, the ABS stores a diagnostic trouble code (DTC) and the ABS indicator light will flash or stay on. A DTC is also set in the ECU memory. When the ABS indicator is flashing or stays on, the ABS function is disabled. However, even when the ABS is disabled, the LBS system (both front and rear brakes) still operates normally. DTC(s) can be retrieved by performing the DTC retrieval procedure described in this section.

Self-Diagnosis Check

Perform the following to initiate self-diagnosis.
1. Turn the ignition switch on.
2. Check that the ABS indicator (**Figure 2**) turns on.

NOTE
*If the ABS indicator does not turn on when the ignition switch is turned on, perform the **Power/Ground Circuit Test** under **Combination Meter** in Chapter Nine.*

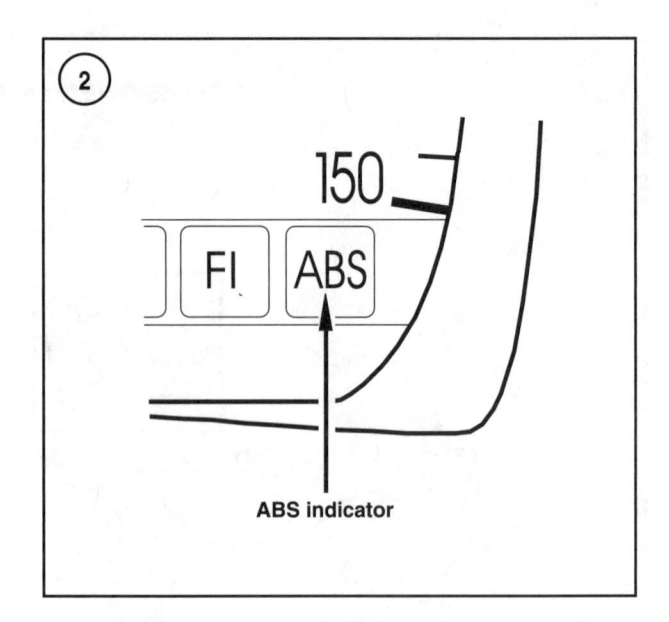

ABS indicator

3. Start the engine.
4. Ride the motorcycle until the speed reaches approximately 10 km/h (6 mph).
5. Observe the ABS indicator (**Figure 2**):
   a. ABS is normal if the ABS indicator turns off.
   b. If the ABS indicator remains on or flashes, a malfunction has been detected by the ECU and the ABS system is turned off. Perform the *Pre-Inspection* described in this section.

Pre-Inspection

When the ABS indicator flashes or stays on, make the following general checks before testing the ABS:

1. The electrical components in the ABS require a fully charged battery. When troubleshooting the ABS, make sure the battery is fully charged. When in doubt, test the battery. Refer to *Battery* in Chapter Nine. If the charging system is suspect, test the charging system output. Refer to *Charging System* in Chapter Nine.

2. Check the overall condition of the linked braking system. This includes brake pad wear, tightness of the component fasteners, and brake fluid levels in both master cylinder reservoirs. Check all banjo bolts and metal brake lines for tightness. Check for any brake fluid leaks.

3. Inspect the entire ABS wire harness, starting at the ECU (**Figure 1**). Check for chaffing and other apparent damage. Particularly check the front and rear wheel speed sensors, wiring harness and connectors.

4. Because electrical components in the ABS operate on low voltage, they are sensitive to any increase in resistance in the circuit. Visually inspect the ABS circuit for

ANTI-LOCK BRAKE SYSTEM

FUSE BOX ③

ACC Terminal

5, 6, 17, 7, 18, 8, 19, 9, 20, 10, 21, 11, 22, 12, 23, 1, 13, 24, 2, 14, 25, 3, 15, 26, 4, 16, 27

Main fuse B 100A

Speed limiter fuse 70A

Diagnostic Trouble Codes

Diagnostic trouble codes (DTCs) are retrieved from the ECU by triggering a series of timed flashes across the ABS indicator (**Figure 2**). The number of flashes displayed by the ABS indicator indicates the stored DTC. See **Table 1** for a description of each DTC.

Note the following before retrieving DTCs:

1. The DTC set in the ECU memory remains in memory until the problem is repaired and the DTC is erased. Turning the ignition switch off during or after retrieving the DTC does not erase it. However, to view the DTC again after turning the ignition switch off and then on, the retrieval procedure must be repeated. See *Retrieving DTCs*.
2. The ECU can store two DTCs. The most current problem is displayed first.
3. Record each DTC in the order displayed.
4. After troubleshooting and repairing the ABS, erase the DTC and perform the self-diagnosis. If the ABS indicator does not flash or stay on after completing the self-diagnosis, the problem has been repaired.

Retrieving DTC(s)

NOTE
Do not start the engine during this procedure.

1. Turn the ignition switch off.
2. Remove the No. 3 and No. 4 ABS motor fuses (**Figure 3**) and inspect them. Refer to *Fuses* in Chapter Nine. Note the following:
    a. If the fuses are not blown, do not install either fuse. Go to Step 3.
    b. If one or both fuses are blown, refer to *DTC Flow Charts* in this section and follow the Code 4 or Code 5 troubleshooting procedure, depending on which fuse is blown.
3. Turn the ignition switch on and watch the ABS indicator (**Figure 2**). It will come on for five seconds and then blink off.
4. Immediately—within three seconds of the indicator going off, install *either* the No. 3 or No. 4 ABS motor fuse (**Figure 3**).
5. The ABS diagnostic system now enters the start mode—the ABS indicator (**Figure 2**) will light and remain on for 3 seconds, blink off for 1 second, and then flash the DTC. Count the number of flashes. The system repeats this pattern, beginning at the start code until the ignition switch is turned off or until the DTC is cleared. The display pattern for one code is:
    a. 3 seconds on (start code).

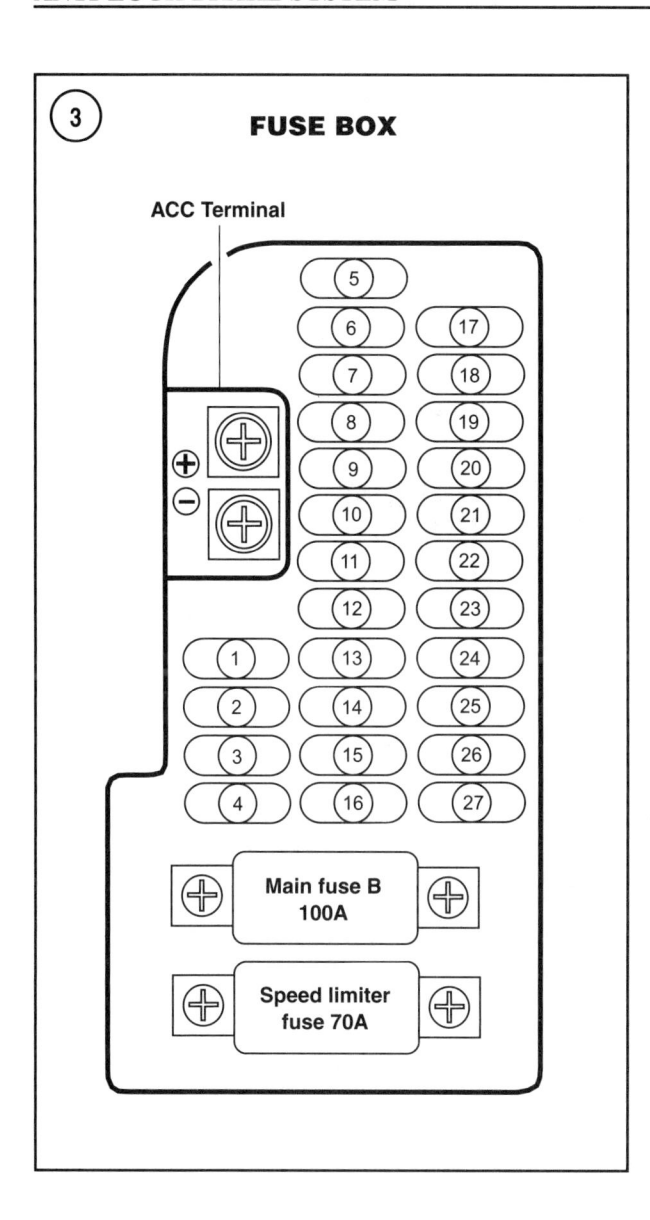

any loose or damaged connectors. Then check the voltage drop across the suspect connectors. Refer to Chapter Two for typical voltage drop testing. A voltage drop exceeding 0.5 volt indicates excessive resistance and a problem in the circuit.

5. If the ABS indicator flashed or stayed on after reinstalling a wheel, check the wheel speed sensor and pulser ring for damage. Check the wheel sensor air gap as described under *Wheel Speed Sensor* in this chapter. Check the pulser ring for debris or chipped or damaged teeth.

6. After performing these general checks, retrieve the DTC(s), and then clear them. Refer to *Diagnostic Trouble Codes* in this section. If the ABS indicator flashes or remains on, retrieve the DTC and troubleshoot the system by referring to *DTC Flow Charts* in this section.

DTC PATTERN FOR ONE CODE

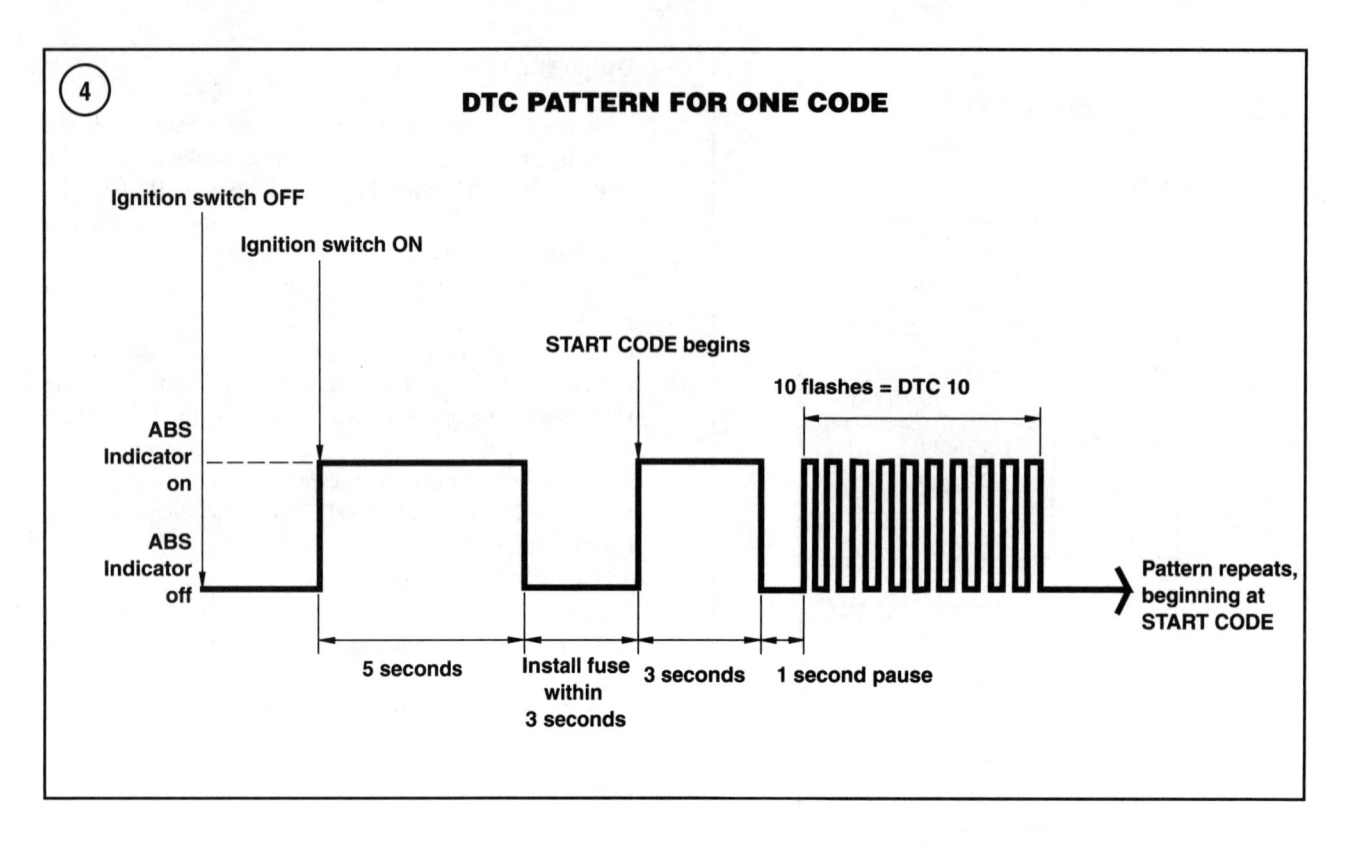

④

- Ignition switch OFF
- Ignition switch ON
- START CODE begins
- 10 flashes = DTC 10
- ABS Indicator on / ABS Indicator off
- 5 seconds
- Install fuse within 3 seconds
- 3 seconds
- 1 second pause
- Pattern repeats, beginning at START CODE

DTC PATTERN FOR TWO CODES

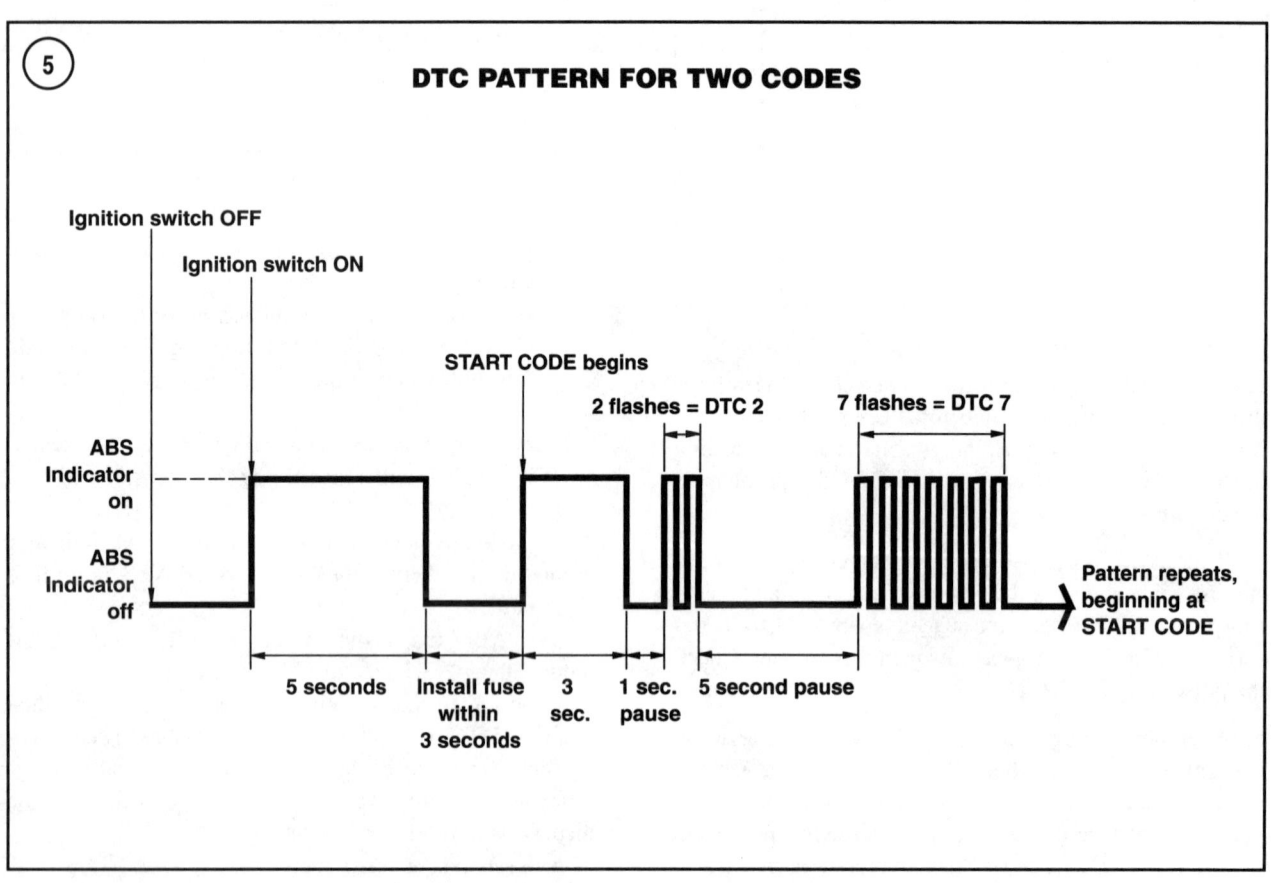

⑤

- Ignition switch OFF
- Ignition switch ON
- START CODE begins
- 2 flashes = DTC 2
- 7 flashes = DTC 7
- ABS Indicator on / ABS Indicator off
- 5 seconds
- Install fuse within 3 seconds
- 3 sec.
- 1 sec. pause
- 5 second pause
- Pattern repeats, beginning at START CODE

ANTI-LOCK BRAKE SYSTEM

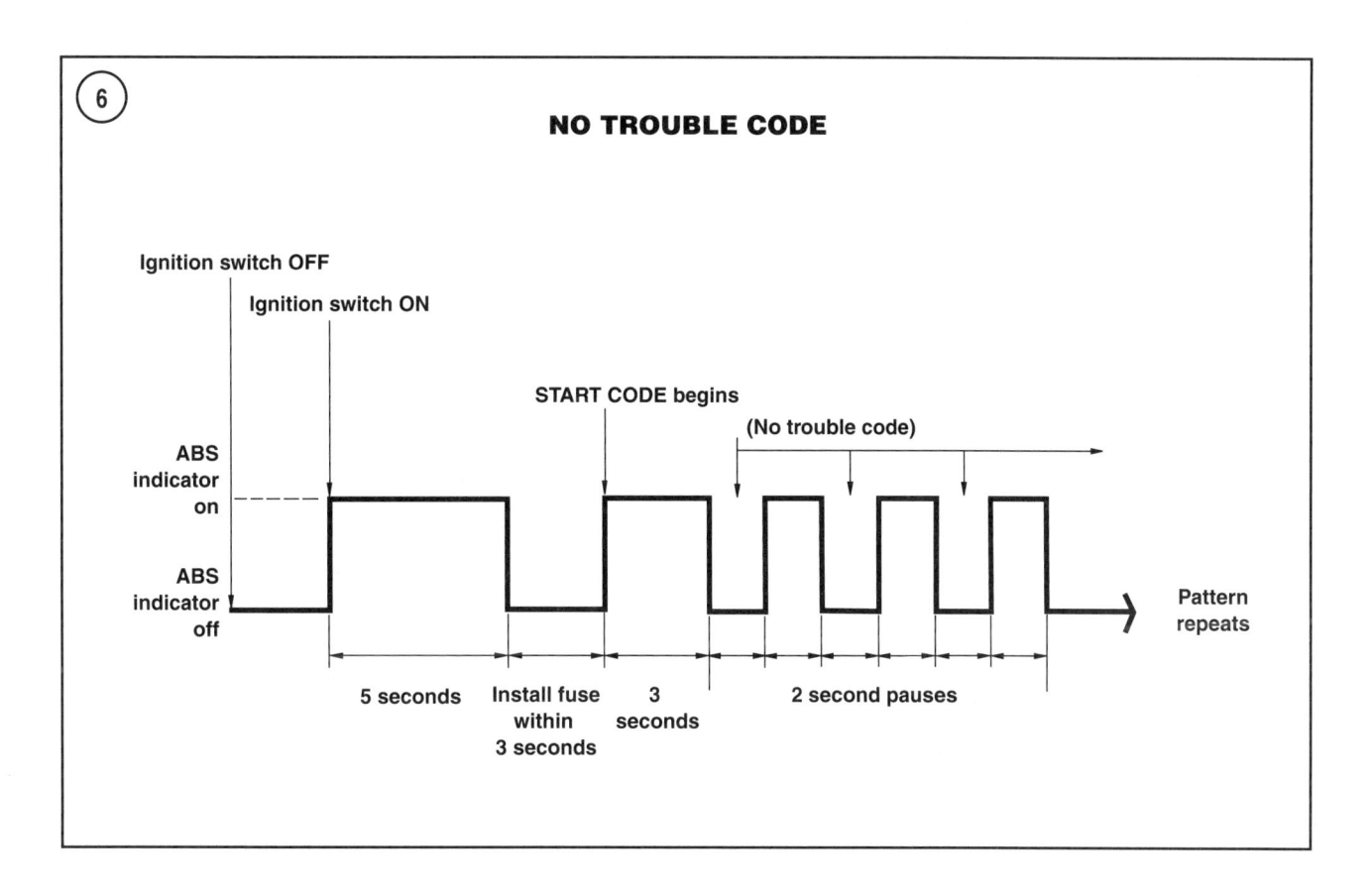

6

NO TROUBLE CODE

Figure 6: Timing diagram showing ABS indicator on/off pattern: Ignition switch OFF, then Ignition switch ON, ABS indicator on for 5 seconds, off (Install fuse within 3 seconds), START CODE begins (3 seconds on), then No trouble code flashes with 2 second pauses, pattern repeats.

  b. 1 second off.
  c. DTC is flashed.
  d. The pattern repeats.

NOTE
Figure 4 *shows a graphic representation of retrieving DTC 10.*

6. The ECU can store up to two codes. When there are two codes, they are separated by a 5-second pause. The display pattern for two codes is:
  a. 3 seconds on (start code).
  b. 1 second off.
  c. Latest-stored DTC is flashed.
  d. 5 seconds off.
  e. Second DTC is flashed.
  f. The pattern repeats.

NOTE
Figure 5 *shows a graphic representation of retrieving two DTC 2 and 7.*

NOTE
When two codes are stored, the first one displayed is the one that occurred most recently. The earlier DTC is displayed second. Troubleshoot the first code displayed first.

Correcting this problem may also correct the source of the earlier DTC.

7. If there are no codes stored in the ECU, a normal code is displayed. This consists of the ABS indicator flashing on and off every 2 seconds. **Figure 6** shows a graphic representation of a normal code.

8. After retrieving and writing down any codes, refer to the ABS diagnostic trouble code chart in **Figure 7**. The chart lists the DTC number, the problem area and specifies a flow chart to follow to determine whether the problem is in the wiring harness or caused by a damaged component. Refer to *DTC Flow Charts* in this section.

9. After repairing the problem, erase the trouble code(s) and perform the self-diagnosis to confirm the problem has been repaired.

Clearing DTCs

1. Retrieve the DTC as previously described, and write down its DTC number.

2. To clear a DTC, reinstall the remaining ABS motor fuse (No. 3 or No. 4) while the ABS indicator is flashing the DTC.

⑦ DIAGNOSTIC TROUBLE CODES

DTC	Problem	DTC flowchart*
2 and 8	Faulty front wheel speed sensor system and ECU	Figure 8
3 and 9	Faulty rear wheel speed sensor system and ECU	Figure 9
4	Faulty front modulator control motor system	Figure 10
5	Faulty rear modulator control motor system	Figure 11
6	Faulty front modulator crank angle sensor system	Figure 12
7	Faulty rear modulator crank angle sensor system	Figure 13
10	Faulty ECU (front relay circuit)	Figure 14
11	Faulty ECU (rear relay circuit)	Figure 15
12 and 13	Faulty ECU (front and rear motor driver circuit)	Figure 16
14	Faulty power circuit	Figure 17
None	ABS indicator stays ON	Figure 18

*Refer to *Electrical Component Replacement* in Chapter Nine when component replacement is called for in a flowchart.

3. The DTC is cleared when the ABS indicator stays on (does not flash).

DTC Flow Charts

To troubleshoot the ABS system, perform the self-diagnosis as described in this section and retrieve any DTC(s). Find the DTC in **Figure 7**, turn to the indicated diagnostic flow chart (**Figures 8-18**), and perform the relevant procedures until the problem is resolved.

When using the flow charts, note the following:

1. Refer to the wiring diagrams at the end of this manual to identify the connectors. Refer to **Figure 1** to locate the individual ABS components.

2. Unless otherwise specified, perform the test procedures with the ignition switch off.

3. Do not disconnect or reconnect electrical connectors when the ignition switch is on. A voltage spike may damage the ECU.

4. Do not perform tests with a battery charger attached to the battery or circuit. Test and charge the battery as described in Chapter Nine.

5. After troubleshooting, clear any DTC and perform self-diagnosis to confirm that the problem has been corrected.

6. Refer to *Wheel Speed Sensors* in this chapter to adjust the air gap and service the wheel speed sensor as described in the chart.

ANTI-LOCK BRAKE SYSTEM

(8)

CODE 2 and 8: FAULTY FRONT WHEEL SPEED SENSOR SYSTEM AND ECU

1. The following riding conditions can produce a false DTC:
   a. Incorrect tire size or inflation pressure. Check the tire size and tire pressure before troubleshooting.
   b. Riding continuously on a bumpy road.
   c. If the rear wheel turned for more than 30 seconds with the engine running and the front wheel was not turning.
   d. Powerful radio waves (electromagnetic interference) disrupted the ECU. This is normally a temporary failure. To check, erase the DTC and perform the self-diagnosis check. If the ABS indicator goes off, this was a temporary failure and the ABS is working normally.

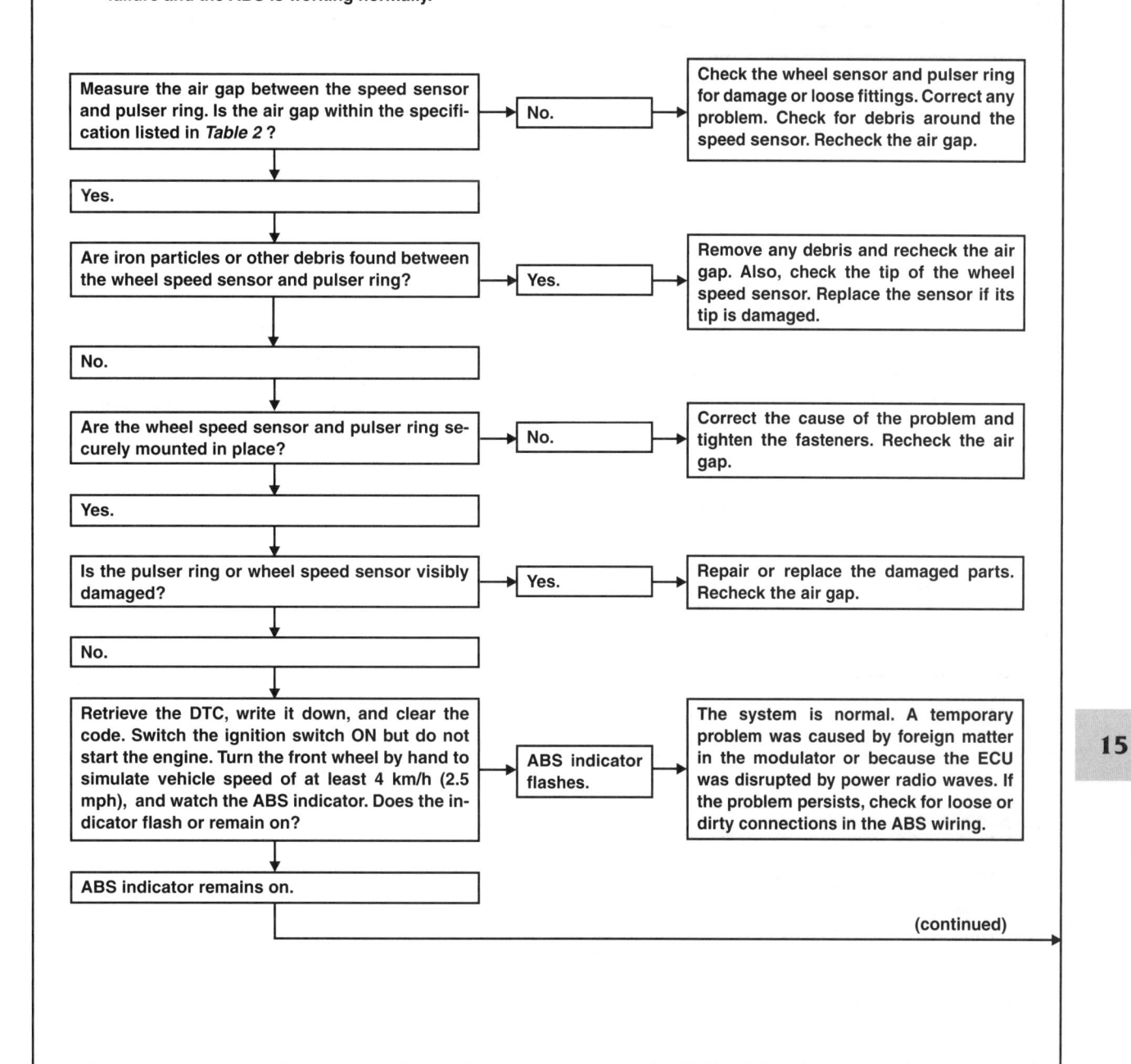

Measure the air gap between the speed sensor and pulser ring. Is the air gap within the specification listed in *Table 2*? → **No.** → Check the wheel sensor and pulser ring for damage or loose fittings. Correct any problem. Check for debris around the speed sensor. Recheck the air gap.

↓ Yes.

Are iron particles or other debris found between the wheel speed sensor and pulser ring? → **Yes.** → Remove any debris and recheck the air gap. Also, check the tip of the wheel speed sensor. Replace the sensor if its tip is damaged.

↓ No.

Are the wheel speed sensor and pulser ring securely mounted in place? → **No.** → Correct the cause of the problem and tighten the fasteners. Recheck the air gap.

↓ Yes.

Is the pulser ring or wheel speed sensor visibly damaged? → **Yes.** → Repair or replace the damaged parts. Recheck the air gap.

↓ No.

Retrieve the DTC, write it down, and clear the code. Switch the ignition switch ON but do not start the engine. Turn the front wheel by hand to simulate vehicle speed of at least 4 km/h (2.5 mph), and watch the ABS indicator. Does the indicator flash or remain on? → **ABS indicator flashes.** → The system is normal. A temporary problem was caused by foreign matter in the modulator or because the ECU was disrupted by power radio waves. If the problem persists, check for loose or dirty connections in the ABS wiring.

↓ ABS indicator remains on.

(continued)

15

⑧ (continued)

Turn the ignition switch OFF. Disconnect the black 12-pin harness connector from the ECU. Check for continuity between the black/white and green/orange terminals of harness connector and ground. Is continuity present?

→ **Yes.** → There should be no continuity. Disconnect the orange 2-pin harness connector from the front wheel speed sensor. Check for continuity between the black/white or green/orange terminal of harness connector and ground. Is continuity present?

- **No.** → The speed sensor is faulty. Replace the speed sensor and retest.
- **Yes.** → Repair the short in the black/white or green/orange wires between the speed sensor and ECU harness connectors.

No.

Ignition switch OFF. Disconnect the orange 2-pin harness connector from the front wheel speed sensor. Short the black/white and green/orange terminals of the harness connector with a jumper wire. Check for continuity between the black/white and green/orange terminals in the ECU 12-pin harness connector. Is continuity present?

→ **No.** → Repair the open in either the black/white or green/orange wires located between the front wheel sensor and ECU harness connectors.

Yes.

Perform the following:
1. Replace the front wheel speed sensor with a new unit.
2. Reconnect the ECU 12-pin harness connector and the wheel speed sensor 2-pin harness connector.
3. Perform the self-diagnosis and check the ABS indicator. Does the indicator flash or turn off?

→ **The ABS indicator flashes.** → The ECU is faulty.

ABS indicator turns off.

The original wheel speed sensor was faulty. The system is now operating normally.

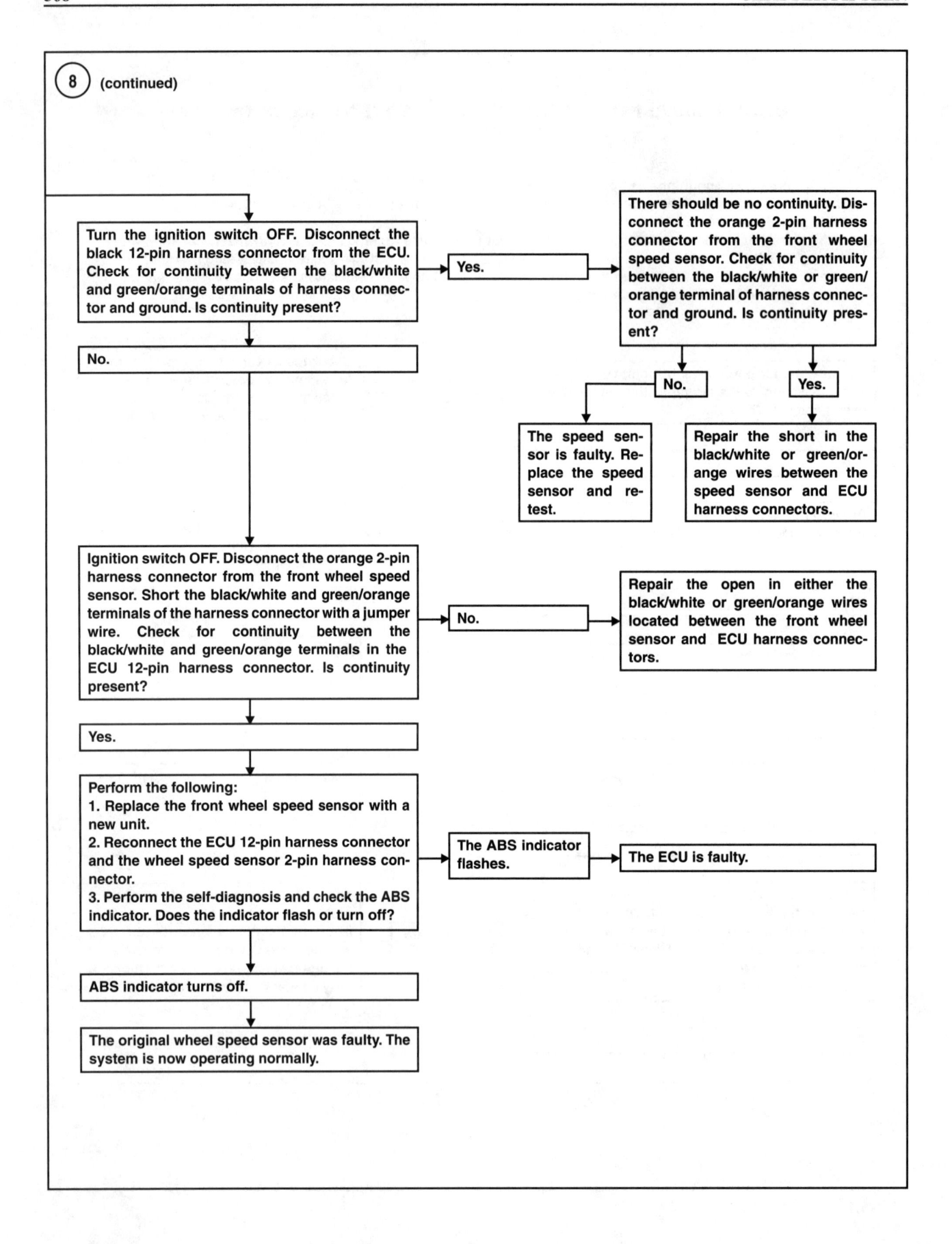

ANTI-LOCK BRAKE SYSTEM

(9)

CODE 3 and 9: FAULTY REAR WHEEL SPEED SENSOR SYSTEM AND ECU

1. The following riding conditions can produce a false DTC:
   a. Incorrect tire size or inflation pressure. Check the tire size and tire pressure before troubleshooting.
   b. Riding continuously on a bumpy road.
   c. If the rear wheel turned for more than 30 seconds with the engine running and the front wheel was not turning.
   d. Powerful radio waves (electromagnetic interference) disrupted the ECU. This is normally a temporary failure. To check, erase the DTC and perform the self-diagnosis check. If the ABS indicator goes off, this was a temporary failure and the ABS is working normally.

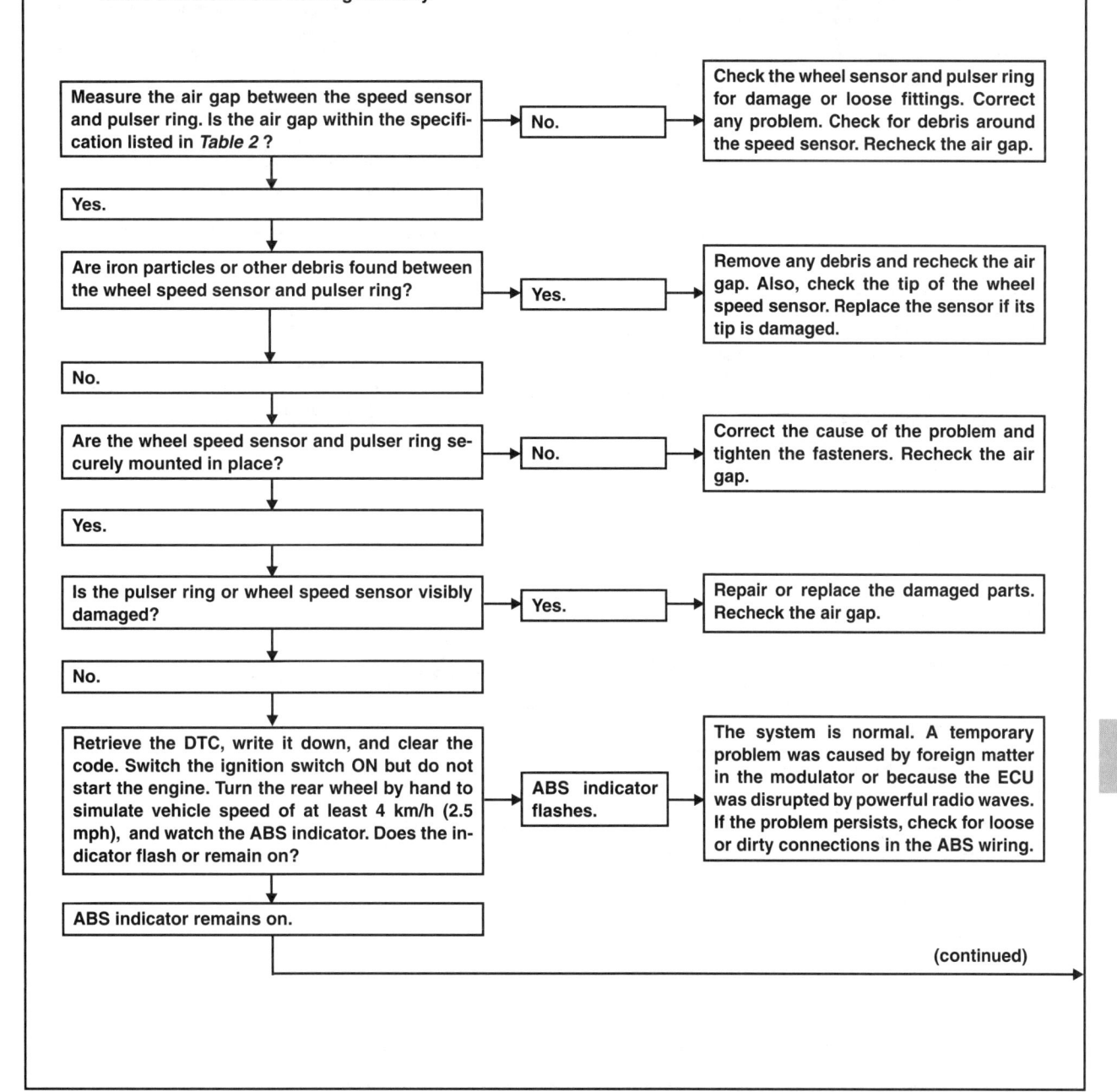

Measure the air gap between the speed sensor and pulser ring. Is the air gap within the specification listed in *Table 2*? → **No.** → Check the wheel sensor and pulser ring for damage or loose fittings. Correct any problem. Check for debris around the speed sensor. Recheck the air gap.

↓ Yes.

Are iron particles or other debris found between the wheel speed sensor and pulser ring? → **Yes.** → Remove any debris and recheck the air gap. Also, check the tip of the wheel speed sensor. Replace the sensor if its tip is damaged.

↓ No.

Are the wheel speed sensor and pulser ring securely mounted in place? → **No.** → Correct the cause of the problem and tighten the fasteners. Recheck the air gap.

↓ Yes.

Is the pulser ring or wheel speed sensor visibly damaged? → **Yes.** → Repair or replace the damaged parts. Recheck the air gap.

↓ No.

Retrieve the DTC, write it down, and clear the code. Switch the ignition switch ON but do not start the engine. Turn the rear wheel by hand to simulate vehicle speed of at least 4 km/h (2.5 mph), and watch the ABS indicator. Does the indicator flash or remain on? → **ABS indicator flashes.** → The system is normal. A temporary problem was caused by foreign matter in the modulator or because the ECU was disrupted by powerful radio waves. If the problem persists, check for loose or dirty connections in the ABS wiring.

↓ ABS indicator remains on.

(continued)

15

9 (continued)

```
                    ┌──────────────────────────────────────┐
                    │ Turn the ignition switch OFF.        │                           ┌──────────────────────────────────────┐
                    │ Disconnect the black 12-pin harness  │                           │ There should be no continuity. Dis-  │
                    │ connector from the ECU. Check for    │         Yes.              │ connect the orange 2-pin harness     │
                    │ continuity between the black/orange  │ ────────────────────────► │ connector from the rear wheel        │
                    │ and blue/yellow terminals of harness │                           │ speed sensor. Check for continuity   │
                    │ connector and ground. Is continuity  │                           │ between the black/orange and         │
                    │ present?                             │                           │ blue/yellow terminal of harness      │
                    └──────────────────────────────────────┘                           │ connector and ground. Is continu-    │
                                    │                                                  │ ity present?                         │
                                    │ No.                                              └──────────────────────────────────────┘
                                    ▼                                                          │                    │
                                                                                            No.│                    │Yes.
                                                                                               ▼                    ▼
                                                                                  ┌──────────────────┐  ┌──────────────────────────┐
                                                                                  │ The speed sensor │  │ Repair the short in the  │
                                                                                  │ is faulty. Replace│ │ black/orange or blue/yel-│
                                                                                  │ the speed sensor │  │ low wires between the    │
                                                                                  │ and retest.      │  │ speed sensor and ECU     │
                                                                                  └──────────────────┘  │ harness connectors.      │
                                                                                                        └──────────────────────────┘

   ┌──────────────────────────────────────────┐
   │ Ignition switch OFF. Disconnect the       │
   │ orange 2-pin harness connector from the  │                              ┌──────────────────────────────────────┐
   │ rear wheel speed sensor. Short the       │                              │ Repair the open in either the        │
   │ black/orange and blue/yellow terminals   │         No.                  │ black/orange or blue/yellow wires    │
   │ of the harness connector with a jumper   │ ────────────────────────────►│ located between the rear wheel       │
   │ wire. Check for continuity between the   │                              │ sensor and ECU harness connec-       │
   │ black/orange and blue/yellow terminals   │                              │ tors.                                │
   │ in the ECU 12-pin harness connector.     │                              └──────────────────────────────────────┘
   │ Is continuity present?                   │
   └──────────────────────────────────────────┘
                │
                │ Yes.
                ▼
   ┌──────────────────────────────────────────┐
   │ Perform the following:                    │
   │ 1. Replace the rear wheel speed sensor    │                 ┌──────────────────┐      ┌─────────────────────┐
   │    with a new unit.                       │                 │ The ABS indicator│      │ The ECU is faulty.  │
   │ 2. Reconnect the ECU 12-pin harness       │ ──────────────► │ flashes.         │ ────►│                     │
   │    connector and the wheel speed sensor   │                 └──────────────────┘      └─────────────────────┘
   │    2-pin harness connector.               │
   │ 3. Perform the self-diagnosis and check   │
   │    the ABS indicator. Does the indicator  │
   │    flash or turn off?                     │
   └──────────────────────────────────────────┘
                │
                │ ABS indicator turns off.
                ▼
   ┌──────────────────────────────────────────┐
   │ The original wheel speed sensor was       │
   │ faulty. The system is now operating       │
   │ normally.                                 │
   └──────────────────────────────────────────┘
```

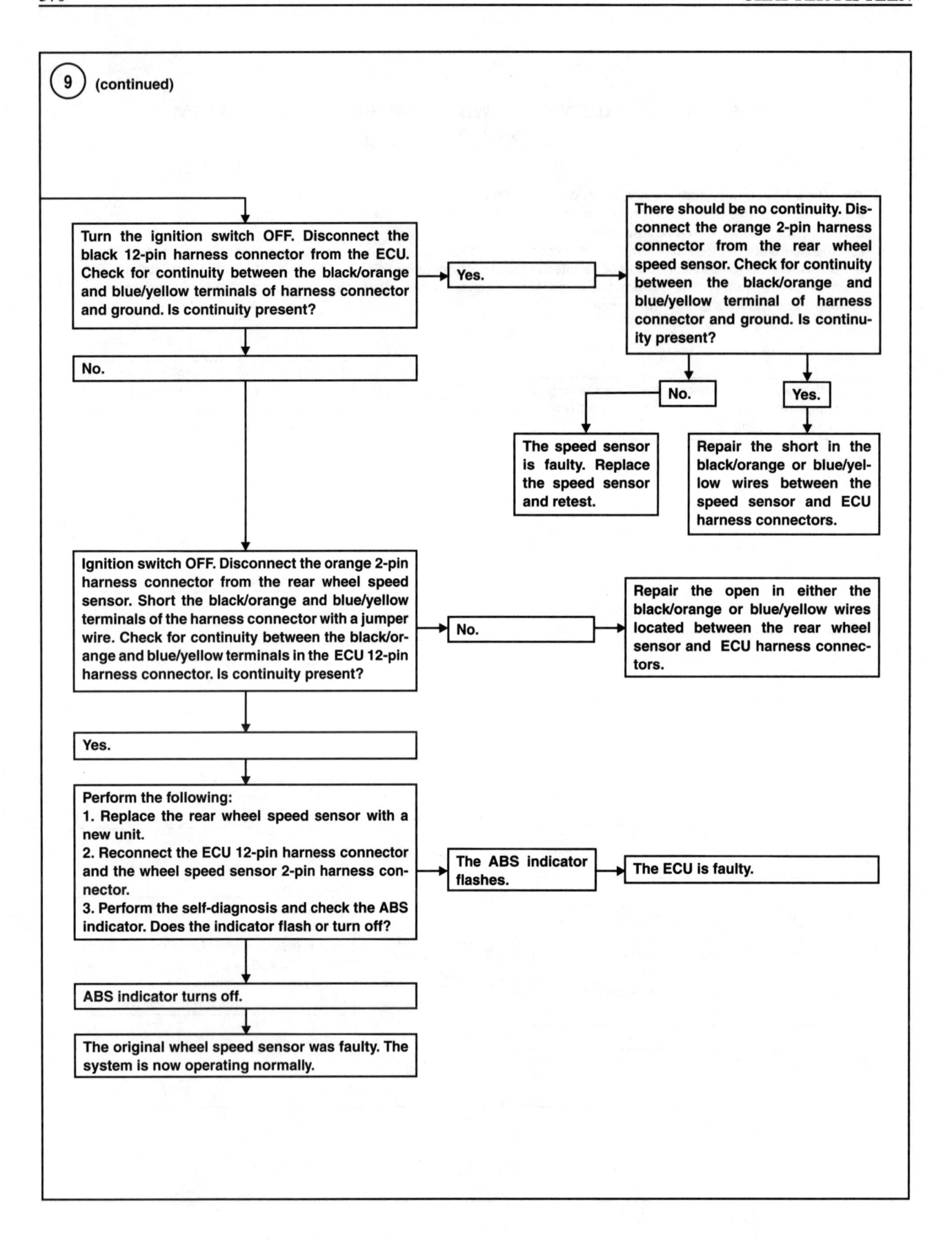

ANTI-LOCK BRAKE SYSTEM

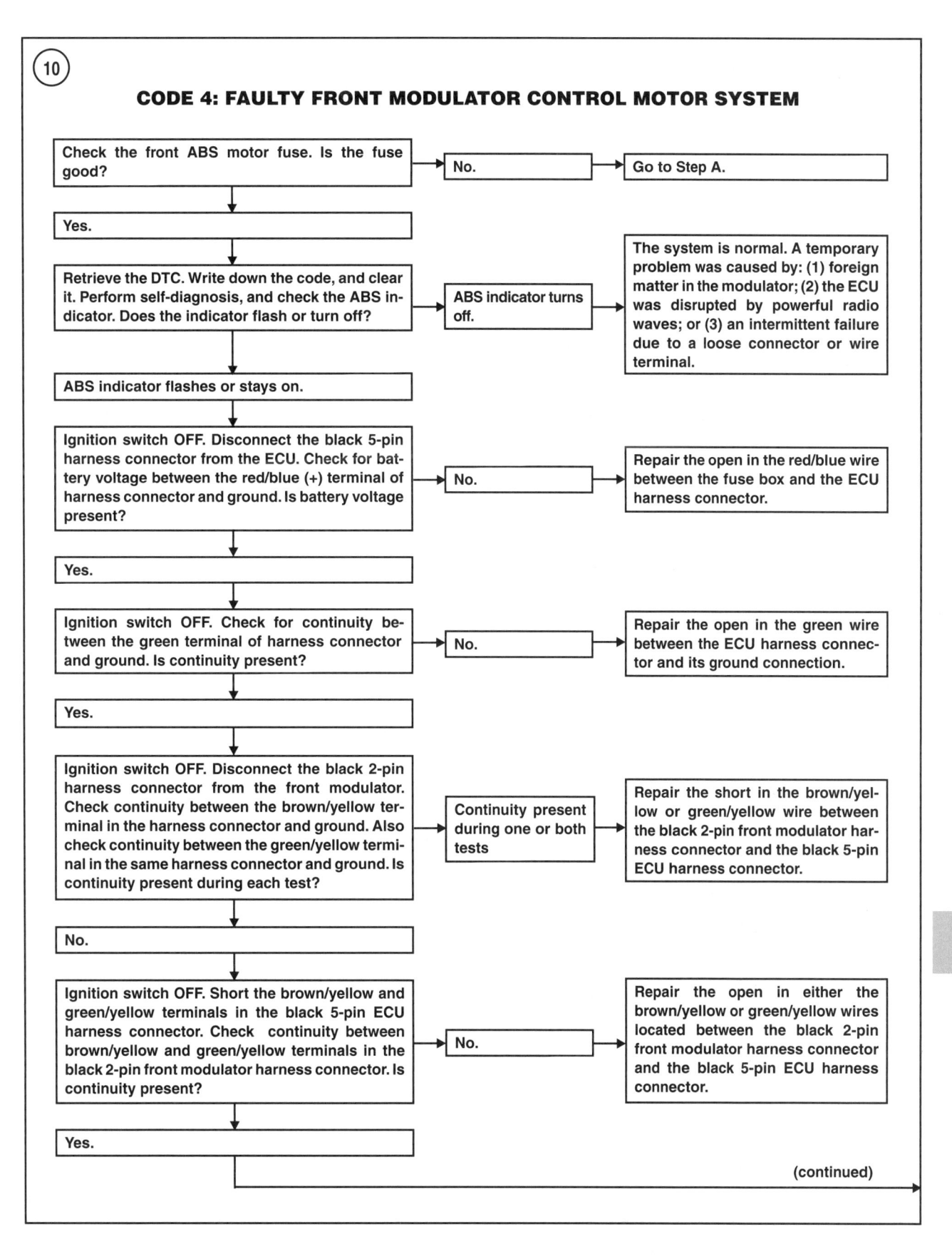

(10)

CODE 4: FAULTY FRONT MODULATOR CONTROL MOTOR SYSTEM

Check the front ABS motor fuse. Is the fuse good? → **No.** → Go to Step A.

↓ Yes.

Retrieve the DTC. Write down the code, and clear it. Perform self-diagnosis, and check the ABS indicator. Does the indicator flash or turn off? → **ABS indicator turns off.** → The system is normal. A temporary problem was caused by: (1) foreign matter in the modulator; (2) the ECU was disrupted by powerful radio waves; or (3) an intermittent failure due to a loose connector or wire terminal.

↓ ABS indicator flashes or stays on.

Ignition switch OFF. Disconnect the black 5-pin harness connector from the ECU. Check for battery voltage between the red/blue (+) terminal of harness connector and ground. Is battery voltage present? → **No.** → Repair the open in the red/blue wire between the fuse box and the ECU harness connector.

↓ Yes.

Ignition switch OFF. Check for continuity between the green terminal of harness connector and ground. Is continuity present? → **No.** → Repair the open in the green wire between the ECU harness connector and its ground connection.

↓ Yes.

Ignition switch OFF. Disconnect the black 2-pin harness connector from the front modulator. Check continuity between the brown/yellow terminal in the harness connector and ground. Also check continuity between the green/yellow terminal in the same harness connector and ground. Is continuity present during each test? → **Continuity present during one or both tests** → Repair the short in the brown/yellow or green/yellow wire between the black 2-pin front modulator harness connector and the black 5-pin ECU harness connector.

↓ No.

Ignition switch OFF. Short the brown/yellow and green/yellow terminals in the black 5-pin ECU harness connector. Check continuity between brown/yellow and green/yellow terminals in the black 2-pin front modulator harness connector. Is continuity present? → **No.** → Repair the open in either the brown/yellow or green/yellow wires located between the black 2-pin front modulator harness connector and the black 5-pin ECU harness connector.

↓ Yes.

(continued)

(10) (continued)

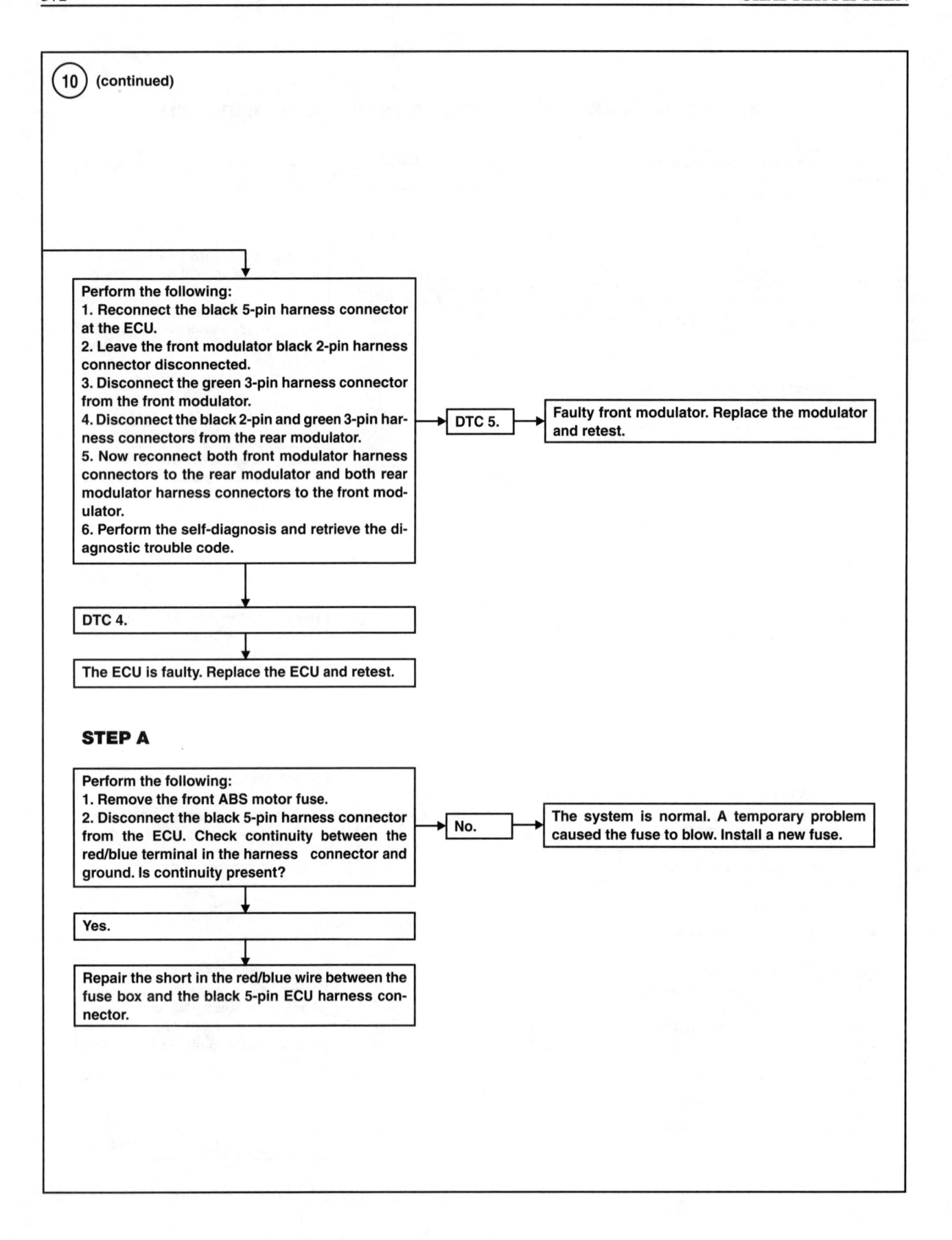

Perform the following:
1. Reconnect the black 5-pin harness connector at the ECU.
2. Leave the front modulator black 2-pin harness connector disconnected.
3. Disconnect the green 3-pin harness connector from the front modulator.
4. Disconnect the black 2-pin and green 3-pin harness connectors from the rear modulator.
5. Now reconnect both front modulator harness connectors to the rear modulator and both rear modulator harness connectors to the front modulator.
6. Perform the self-diagnosis and retrieve the diagnostic trouble code.

→ DTC 5. → Faulty front modulator. Replace the modulator and retest.

↓

DTC 4.

The ECU is faulty. Replace the ECU and retest.

STEP A

Perform the following:
1. Remove the front ABS motor fuse.
2. Disconnect the black 5-pin harness connector from the ECU. Check continuity between the red/blue terminal in the harness connector and ground. Is continuity present?

→ No. → The system is normal. A temporary problem caused the fuse to blow. Install a new fuse.

↓

Yes.

Repair the short in the red/blue wire between the fuse box and the black 5-pin ECU harness connector.

ANTI-LOCK BRAKE SYSTEM

(11)

CODE 5: FAULTY REAR MODULATOR CONTROL MOTOR SYSTEM

Check the rear ABS motor fuse. Is the fuse good? → No. → Go to Step A.

↓ Yes.

Retrieve the DTC. Write down the DTC, and clear it. Perform self-diagnosis, and check the ABS indicator. Does the indicator flash or turn off? → ABS indicator turns off. → The system is normal. A temporary problem was caused by: (1) foreign matter in the modulator; (2) the ECU was disrupted by powerful radio waves; or (3) an intermittent failure due to a loose connector or wire terminal.

↓ ABS indicator flashes or stays on.

Ignition switch OFF. Disconnect the brown 5-pin harness connector from the ECU. Check for battery voltage between the red/green (+) terminal of harness connector and ground (-). Is battery voltage present? → No. → Repair the open in the red/green wire between the fuse box and the ECU harness connector.

↓ Yes.

Ignition switch OFF. Check continuity between the green terminal of harness connector and ground. Is continuity present? → No. → Repair the open in the green wire between the ECU harness connector and its ground connection.

↓ Yes.

Ignition switch OFF. Disconnect the black 2-pin harness connector from the rear modulator. Check continuity between the brown/light green terminal in the harness connector and ground. Also check continuity between the red/black terminal in the same harness connector and ground. Is continuity present during each test? → Continuity present during one or both tests. → Repair the short in the brown/light green or red/black wire between the black 2-pin rear modulator harness connector and the brown 5-pin ECU harness connector.

↓ No.

Ignition switch OFF. Short the brown/light green and red/black terminals in the brown 5-pin ECU harness connector. Check continuity between brown/light green and red/black terminals in the black 2-pin rear modulator harness connector. Is continuity present? → No. → Repair the open in either the brown/light green or red/black wires located between the black 2-pin rear modulator harness connector and the brown 5-pin ECU harness connector.

↓ Yes.

(continued)

(11) (continued)

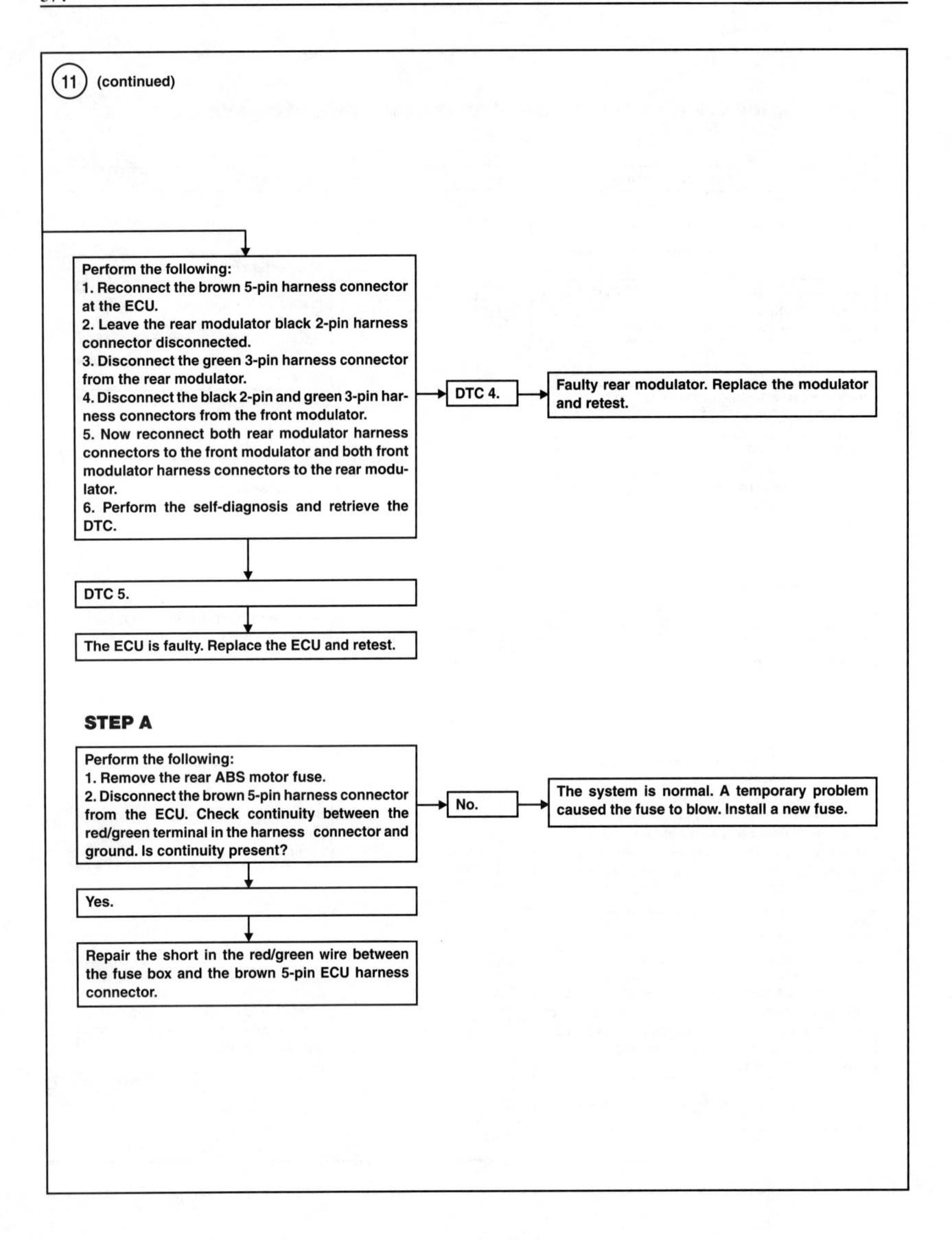

Perform the following:
1. Reconnect the brown 5-pin harness connector at the ECU.
2. Leave the rear modulator black 2-pin harness connector disconnected.
3. Disconnect the green 3-pin harness connector from the rear modulator.
4. Disconnect the black 2-pin and green 3-pin harness connectors from the front modulator.
5. Now reconnect both rear modulator harness connectors to the front modulator and both front modulator harness connectors to the rear modulator.
6. Perform the self-diagnosis and retrieve the DTC.

→ DTC 4. → Faulty rear modulator. Replace the modulator and retest.

DTC 5.

The ECU is faulty. Replace the ECU and retest.

STEP A

Perform the following:
1. Remove the rear ABS motor fuse.
2. Disconnect the brown 5-pin harness connector from the ECU. Check continuity between the red/green terminal in the harness connector and ground. Is continuity present?

→ No. → The system is normal. A temporary problem caused the fuse to blow. Install a new fuse.

Yes.

Repair the short in the red/green wire between the fuse box and the brown 5-pin ECU harness connector.

ANTI-LOCK BRAKE SYSTEM

⑫ CODE 6: FAULTY FRONT MODULATOR CRANK ANGLE SENSOR SYSTEM

STEP A

```
Turn the ignition switch ON, and check the ABS indicator. Does the indicator flash or does it remain on.
   │
   ├──► ABS indicator remains on. ──► Go to Step B.
   ▼
ABS indicator flashes.
   │
   ▼
Disconnect the green 3-pin harness connector from the front modulator. Turn the ignition switch ON and measure voltage between the orange/green (+) and pink/black (-) terminals of harness connector. Is the voltage 4.5-5.5 volts?
   │
   ├──► No. ──► Go to Step C.
   ▼
Yes.
   │
   ▼
Ignition switch OFF. Disconnect the black 12-pin harness connector from the ECU. Check for continuity between the white/red terminal of harness connector and ground. Is continuity present?
   │
   ├──► Yes. ──► Repair the short in the white/red wire between the 12-pin ECU and 3-pin front modulator harness connectors.
   ▼
No.
   │
   ▼
Ground the white/red terminal in the 12-pin ECU harness connector with a jumper wire. Check for continuity between the white/red terminal of the front modulator green 3-pin harness connector and ground. Is continuity present?
   │
   ├──► No. ──► Repair the open in the white/red wire between the 12-pin ECU and 3-pin front modulator harness connectors.
   ▼
Yes.
   │
   ▼ (continued)
```

⑫ (continued)

Perform the following:
1. Reconnect the black 12-pin harness connector at the ECU.
2. Leave the front modulator green 3-pin harness connector disconnected.
3. Disconnect the black 2-pin harness connector from the front modulator.
4. Disconnect the black 2-pin and green 3-pin harness connectors from the rear modulator.
5. Now reconnect both front modulator harness connectors to the rear modulator and both rear modulator harness connectors to the front modulator.
6. Perform the self-diagnosis and retrieve the DTC.

→ DTC 7. → The front modulator is faulty. Replace the modulator and retest.

↓

DTC 6.

↓

The ECU is faulty. Replace the ECU and retest.

STEP B

Retrieve the DTC. Write down the code, and clear it. Perform self-diagnosis, and check the ABS indicator. Does the indicator flash or turn off?

→ ABS indicator turns off. → The system is normal. A temporary problem was caused by: (1) foreign matter in the modulator; (2) because the ECU was disrupted by powerful radio waves; or (3) an intermittent failure due to a loose connector or wire terminal.

↓

ABS indicator flashes or stays on.

Go to Step A and follow steps under ABS indicator flashes to continue troubleshooting.

(continued)

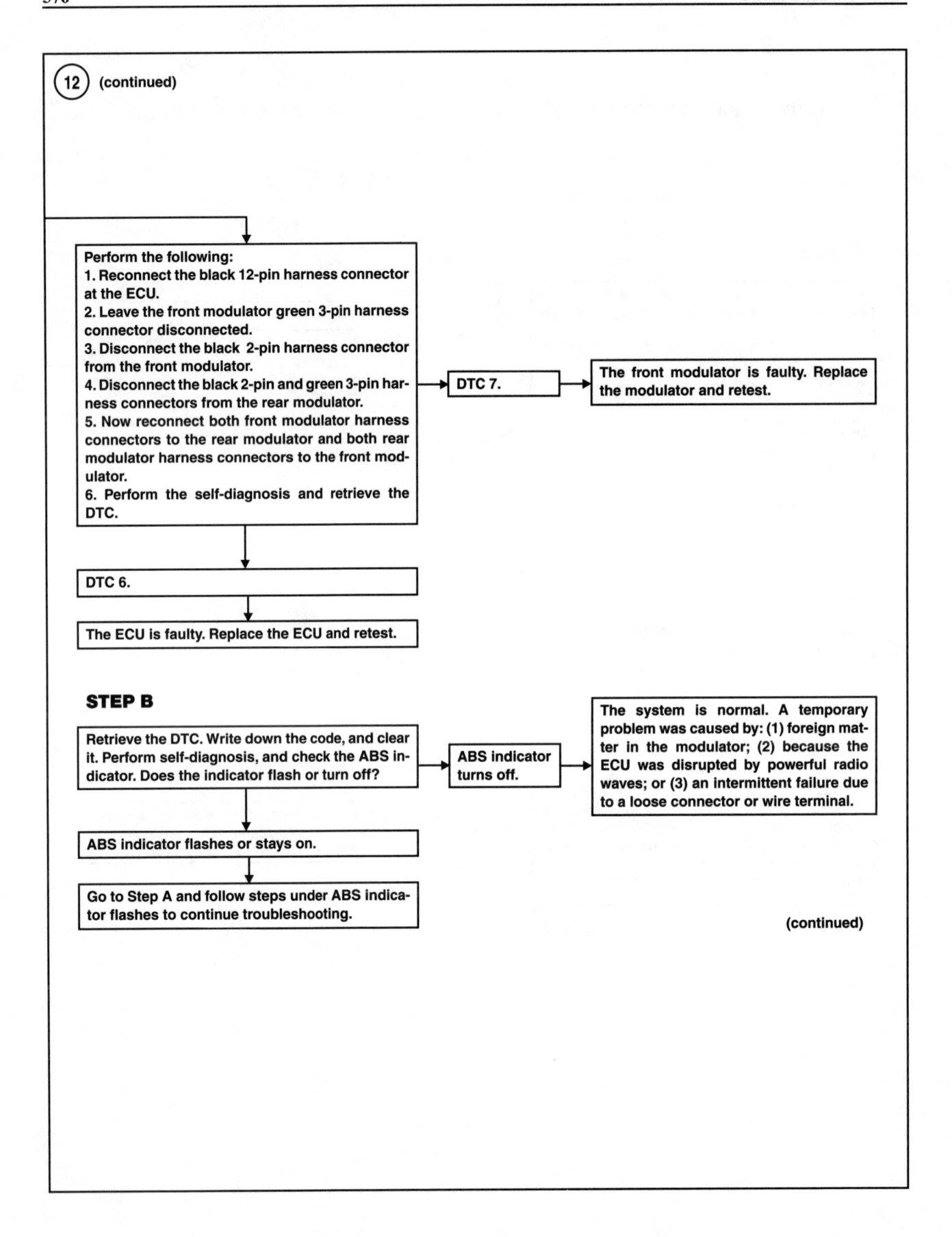

ANTI-LOCK BRAKE SYSTEM

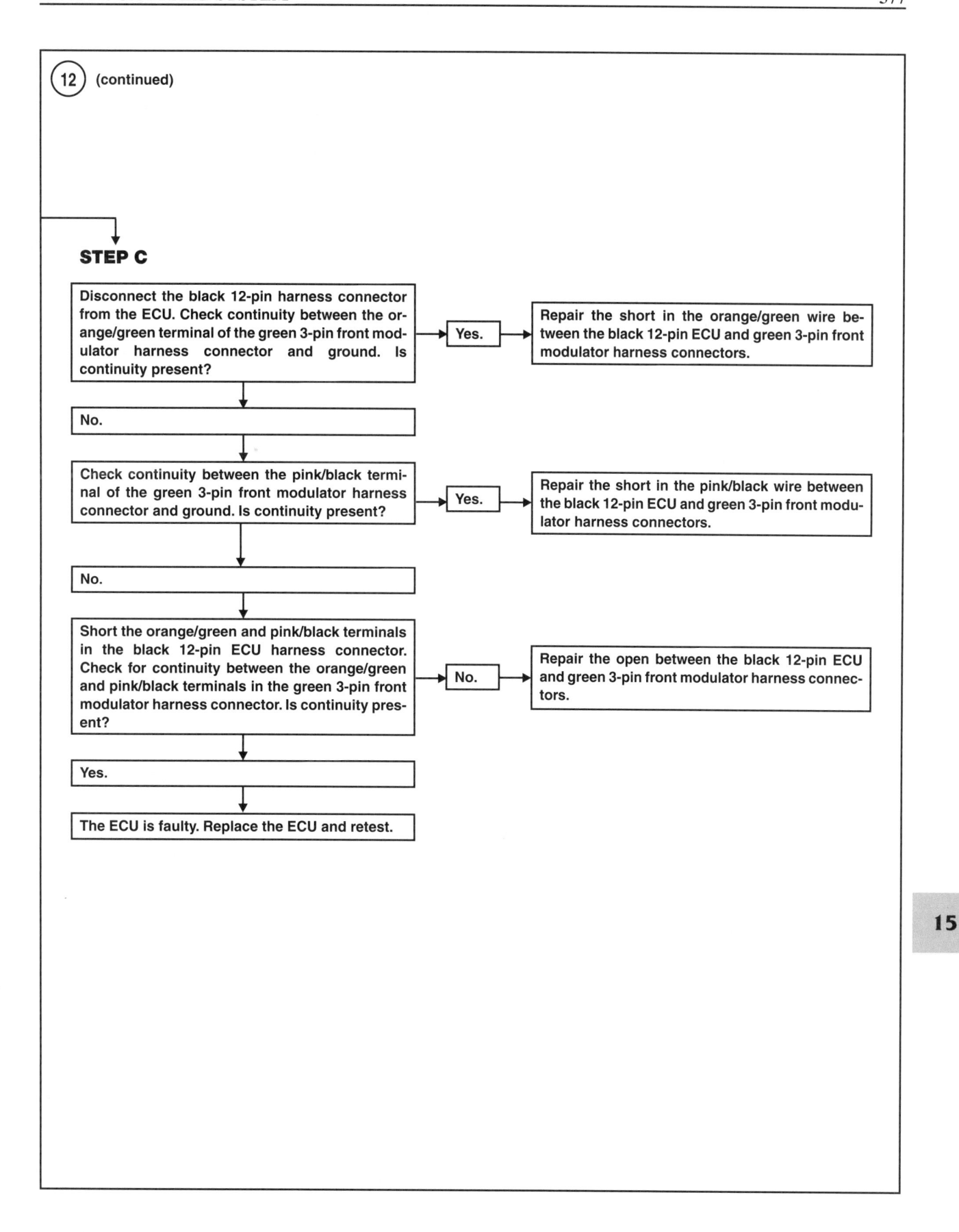

(12) (continued)

STEP C

Disconnect the black 12-pin harness connector from the ECU. Check continuity between the orange/green terminal of the green 3-pin front modulator harness connector and ground. Is continuity present?

→ Yes. → Repair the short in the orange/green wire between the black 12-pin ECU and green 3-pin front modulator harness connectors.

↓ No.

Check continuity between the pink/black terminal of the green 3-pin front modulator harness connector and ground. Is continuity present?

→ Yes. → Repair the short in the pink/black wire between the black 12-pin ECU and green 3-pin front modulator harness connectors.

↓ No.

Short the orange/green and pink/black terminals in the black 12-pin ECU harness connector. Check for continuity between the orange/green and pink/black terminals in the green 3-pin front modulator harness connector. Is continuity present?

→ No. → Repair the open between the black 12-pin ECU and green 3-pin front modulator harness connectors.

↓ Yes.

The ECU is faulty. Replace the ECU and retest.

⑬ CODE 7: FAULTY REAR MODULATOR CRANK ANGLE SENSOR SYSTEM

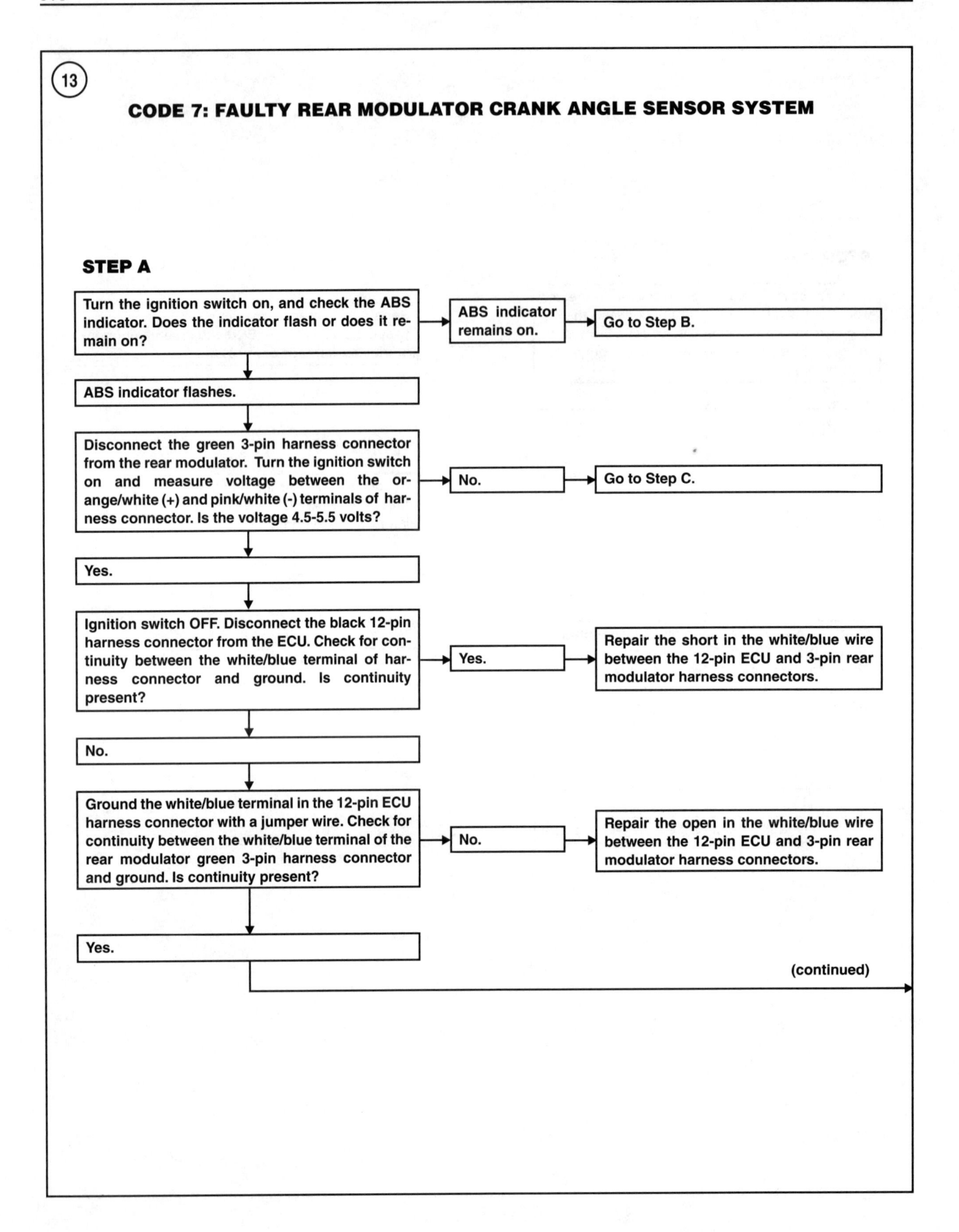

STEP A

Turn the ignition switch on, and check the ABS indicator. Does the indicator flash or does it remain on? → ABS indicator remains on. → Go to Step B.

↓ ABS indicator flashes.

Disconnect the green 3-pin harness connector from the rear modulator. Turn the ignition switch on and measure voltage between the orange/white (+) and pink/white (-) terminals of harness connector. Is the voltage 4.5-5.5 volts? → No. → Go to Step C.

↓ Yes.

Ignition switch OFF. Disconnect the black 12-pin harness connector from the ECU. Check for continuity between the white/blue terminal of harness connector and ground. Is continuity present? → Yes. → Repair the short in the white/blue wire between the 12-pin ECU and 3-pin rear modulator harness connectors.

↓ No.

Ground the white/blue terminal in the 12-pin ECU harness connector with a jumper wire. Check for continuity between the white/blue terminal of the rear modulator green 3-pin harness connector and ground. Is continuity present? → No. → Repair the open in the white/blue wire between the 12-pin ECU and 3-pin rear modulator harness connectors.

↓ Yes.

(continued)

ANTI-LOCK BRAKE SYSTEM

(13) (continued)

Perform the following:
1. Reconnect the black 12-pin harness connector at the ECU.
2. Leave the rear modulator green 3-pin harness connector disconnected.
3. Disconnect the black 2-pin harness connector from the rear modulator.
4. Disconnect the black 2-pin and green 3-pin harness connectors from the front modulator.
5. Now reconnect both rear modulator harness connectors to the front modulator and both front modulator harness connectors to the rear modulator.
6. Perform the self-diagnosis and retrieve the DTC.

→ DTC 6. → The rear modulator is faulty. Replace the modulator and retest.

↓

DTC 7.

The ECU is faulty. Replace the ECU and retest.

STEP B

Retrieve the DTC. Write down the code, and clear it. Perform self-diagnosis, and check the ABS indicator. Does the indicator flash or turn off?

→ ABS indicator turns off. → The system is normal. A temporary problem was caused by: (1) foreign matter in the modulator; (2) the ECU was disrupted by powerful radio waves; or (3) an intermittent failure due to a loose connector or wire terminal.

↓

ABS indicator flashes or stays on.

↓

Return to Step A and follow steps under ABS indicator flashes to continue troubleshooting.

(continued)

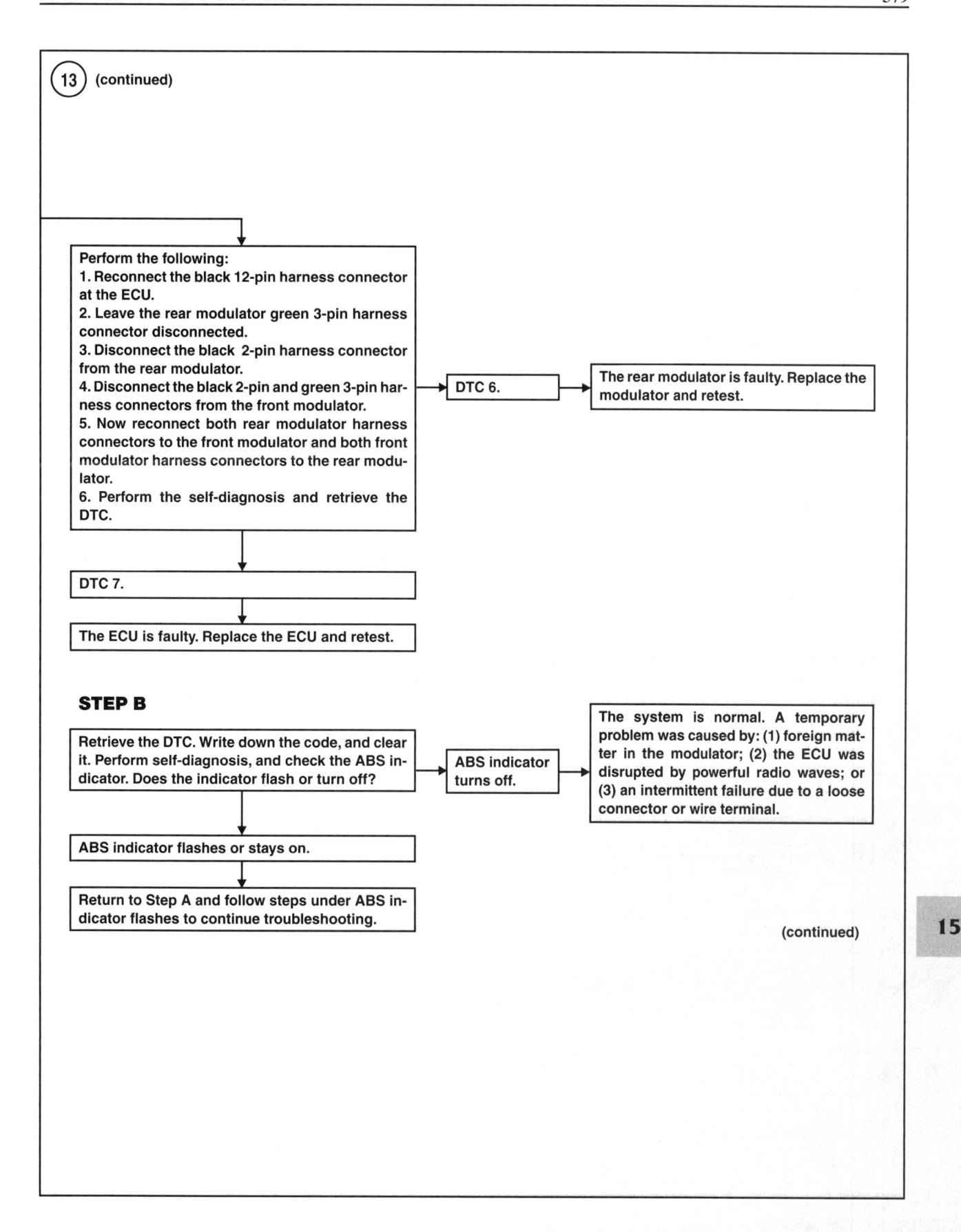

(13) (continued)

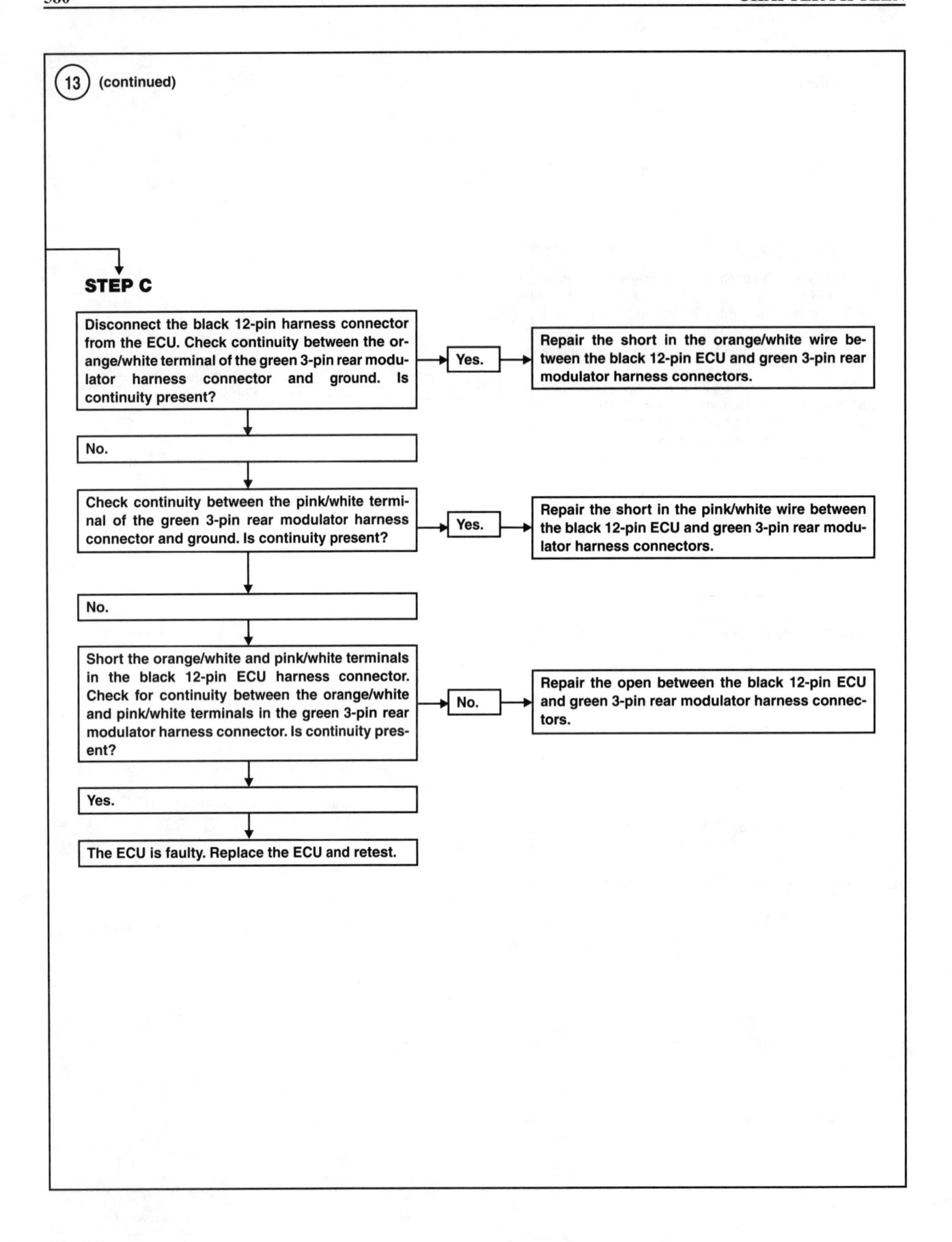

STEP C

Disconnect the black 12-pin harness connector from the ECU. Check continuity between the orange/white terminal of the green 3-pin rear modulator harness connector and ground. Is continuity present? → **Yes.** → Repair the short in the orange/white wire between the black 12-pin ECU and green 3-pin rear modulator harness connectors.

↓ No.

Check continuity between the pink/white terminal of the green 3-pin rear modulator harness connector and ground. Is continuity present? → **Yes.** → Repair the short in the pink/white wire between the black 12-pin ECU and green 3-pin rear modulator harness connectors.

↓ No.

Short the orange/white and pink/white terminals in the black 12-pin ECU harness connector. Check for continuity between the orange/white and pink/white terminals in the green 3-pin rear modulator harness connector. Is continuity present? → **No.** → Repair the open between the black 12-pin ECU and green 3-pin rear modulator harness connectors.

↓ Yes.

The ECU is faulty. Replace the ECU and retest.

ANTI-LOCK BRAKE SYSTEM

CODE 10: FAULTY ECU (FRONT RELAY CIRCUIT)

An ECU diagnostic trouble code (DTC) can be set when the ECU has been disrupted by an extremely strong electromagnetic field (radio waves, high-tension wires, etc.). This is typically a temporary problem. If electromagnetic interference is the source of this DTC, normal operation usually returns once the DTC is cleared.

STEP A

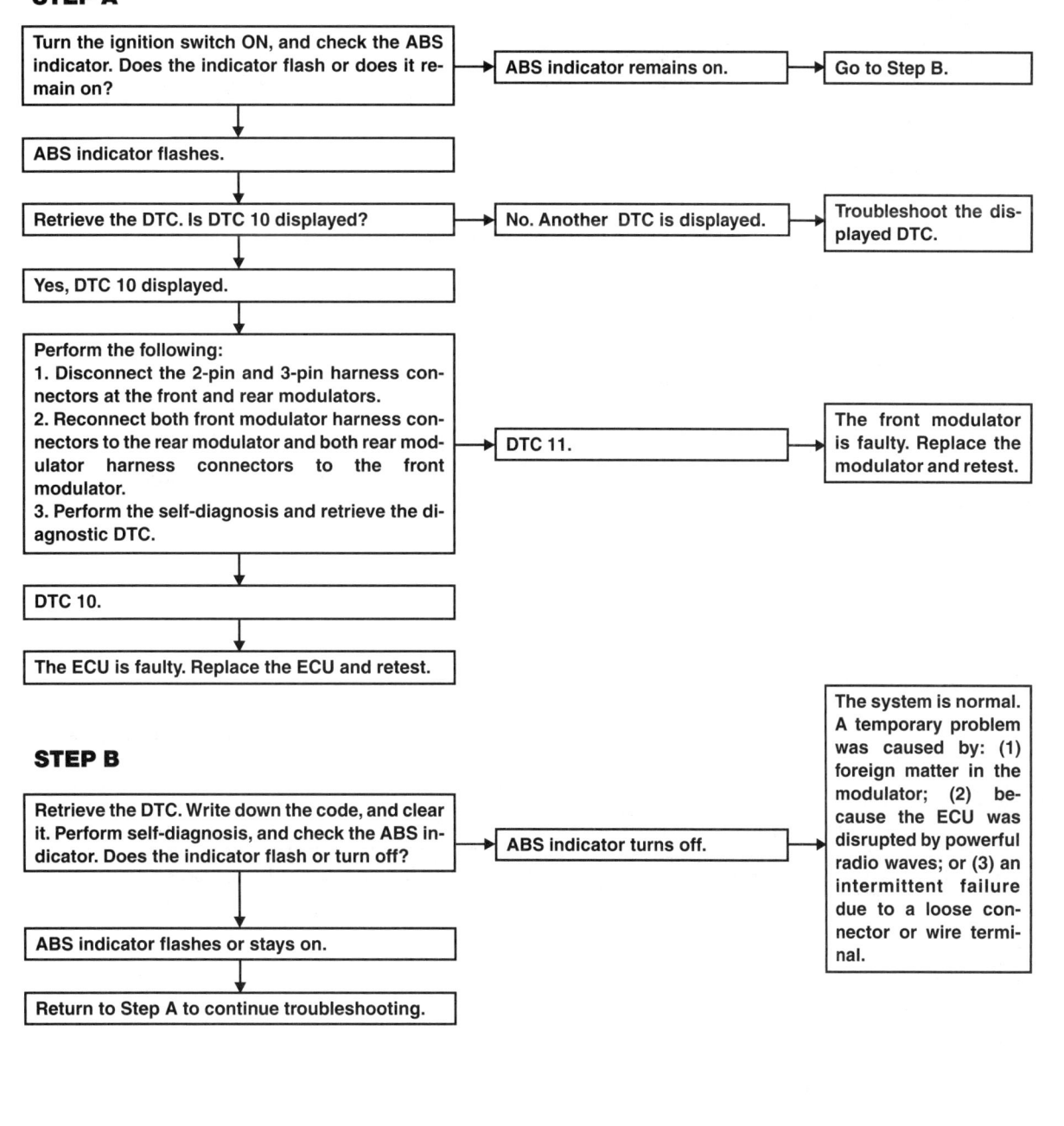

Turn the ignition switch ON, and check the ABS indicator. Does the indicator flash or does it remain on?
→ ABS indicator remains on. → Go to Step B.

↓ ABS indicator flashes.

Retrieve the DTC. Is DTC 10 displayed?
→ No. Another DTC is displayed. → Troubleshoot the displayed DTC.

↓ Yes, DTC 10 displayed.

Perform the following:
1. Disconnect the 2-pin and 3-pin harness connectors at the front and rear modulators.
2. Reconnect both front modulator harness connectors to the rear modulator and both rear modulator harness connectors to the front modulator.
3. Perform the self-diagnosis and retrieve the diagnostic DTC.

→ DTC 11. → The front modulator is faulty. Replace the modulator and retest.

↓ DTC 10.

The ECU is faulty. Replace the ECU and retest.

STEP B

Retrieve the DTC. Write down the code, and clear it. Perform self-diagnosis, and check the ABS indicator. Does the indicator flash or turn off?
→ ABS indicator turns off. → The system is normal. A temporary problem was caused by: (1) foreign matter in the modulator; (2) because the ECU was disrupted by powerful radio waves; or (3) an intermittent failure due to a loose connector or wire terminal.

↓ ABS indicator flashes or stays on.

Return to Step A to continue troubleshooting.

CODE 11: FAULTY ECU (REAR RELAY CIRCUIT)

An ECU diagnostic trouble codes (DTC) can be set when the ECU has been disrupted by an extremely strong electromagnetic field (radio waves, high-tension wires, etc.). This is typically a temporary problem. If electromagnetic interference is the source of this DTC, normal operation usually returns once the DTC is cleared.

STEP A

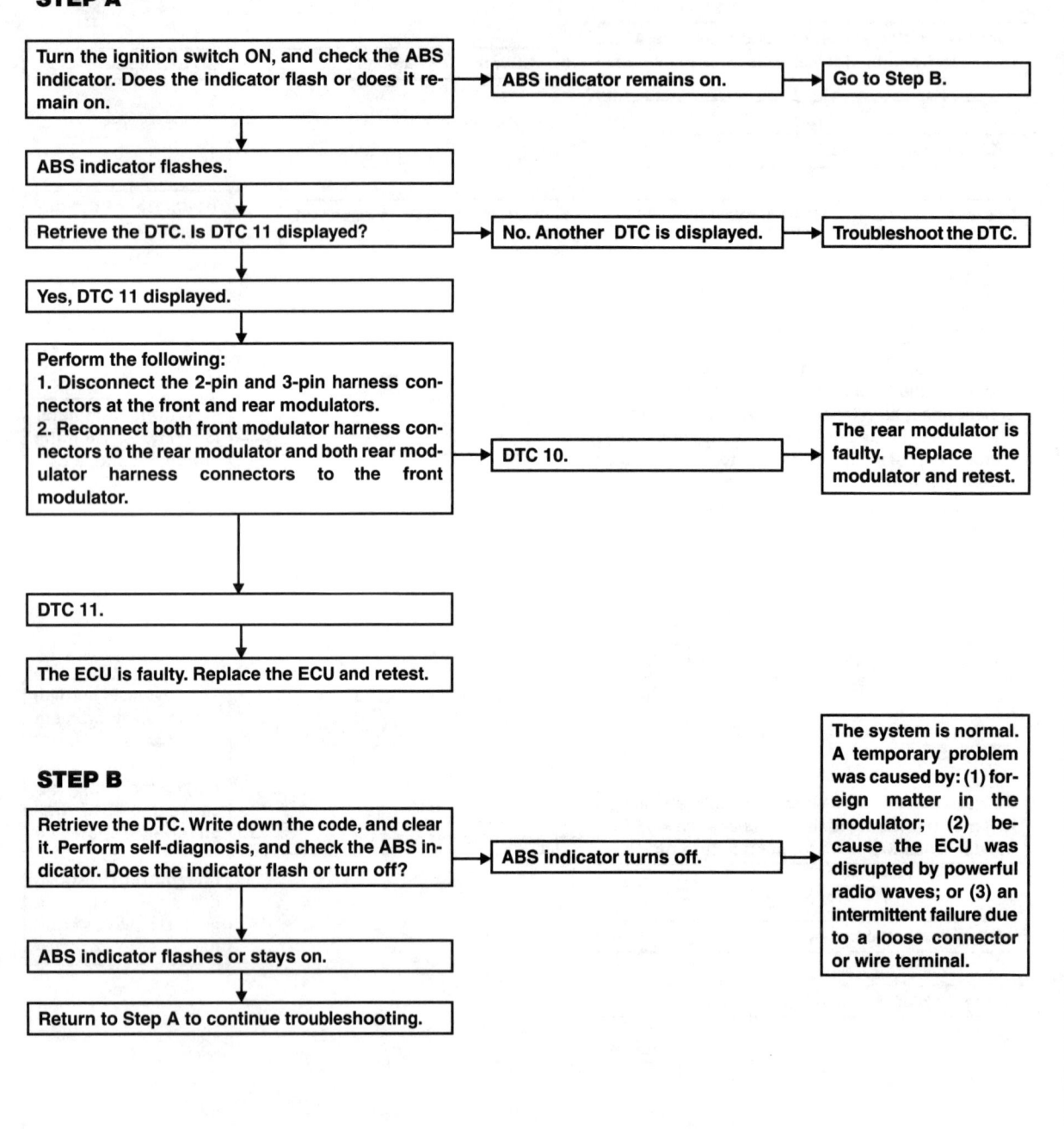

Turn the ignition switch ON, and check the ABS indicator. Does the indicator flash or does it remain on.
- ABS indicator remains on. → Go to Step B.
- ABS indicator flashes.

Retrieve the DTC. Is DTC 11 displayed?
- No. Another DTC is displayed. → Troubleshoot the DTC.
- Yes, DTC 11 displayed.

Perform the following:
1. Disconnect the 2-pin and 3-pin harness connectors at the front and rear modulators.
2. Reconnect both front modulator harness connectors to the rear modulator and both rear modulator harness connectors to the front modulator.

- DTC 10. → The rear modulator is faulty. Replace the modulator and retest.
- DTC 11. → The ECU is faulty. Replace the ECU and retest.

STEP B

Retrieve the DTC. Write down the code, and clear it. Perform self-diagnosis, and check the ABS indicator. Does the indicator flash or turn off?
- ABS indicator turns off. → The system is normal. A temporary problem was caused by: (1) foreign matter in the modulator; (2) because the ECU was disrupted by powerful radio waves; or (3) an intermittent failure due to a loose connector or wire terminal.
- ABS indicator flashes or stays on. → Return to Step A to continue troubleshooting.

ANTI-LOCK BRAKE SYSTEM

⑯

CODE 12 and 13: FAULTY ECU
(FRONT AND REAR MOTOR DRIVER CIRCUIT)

An ECU diagnostic trouble code (DTC) can be set when the ECU has been disrupted by an extremely strong electromagnetic field (radio waves, high-tension wires, etc.). This is typically a temporary problem. If electromagnetic interference is the source of this DTC, normal operation usually returns once the DTC is cleared.

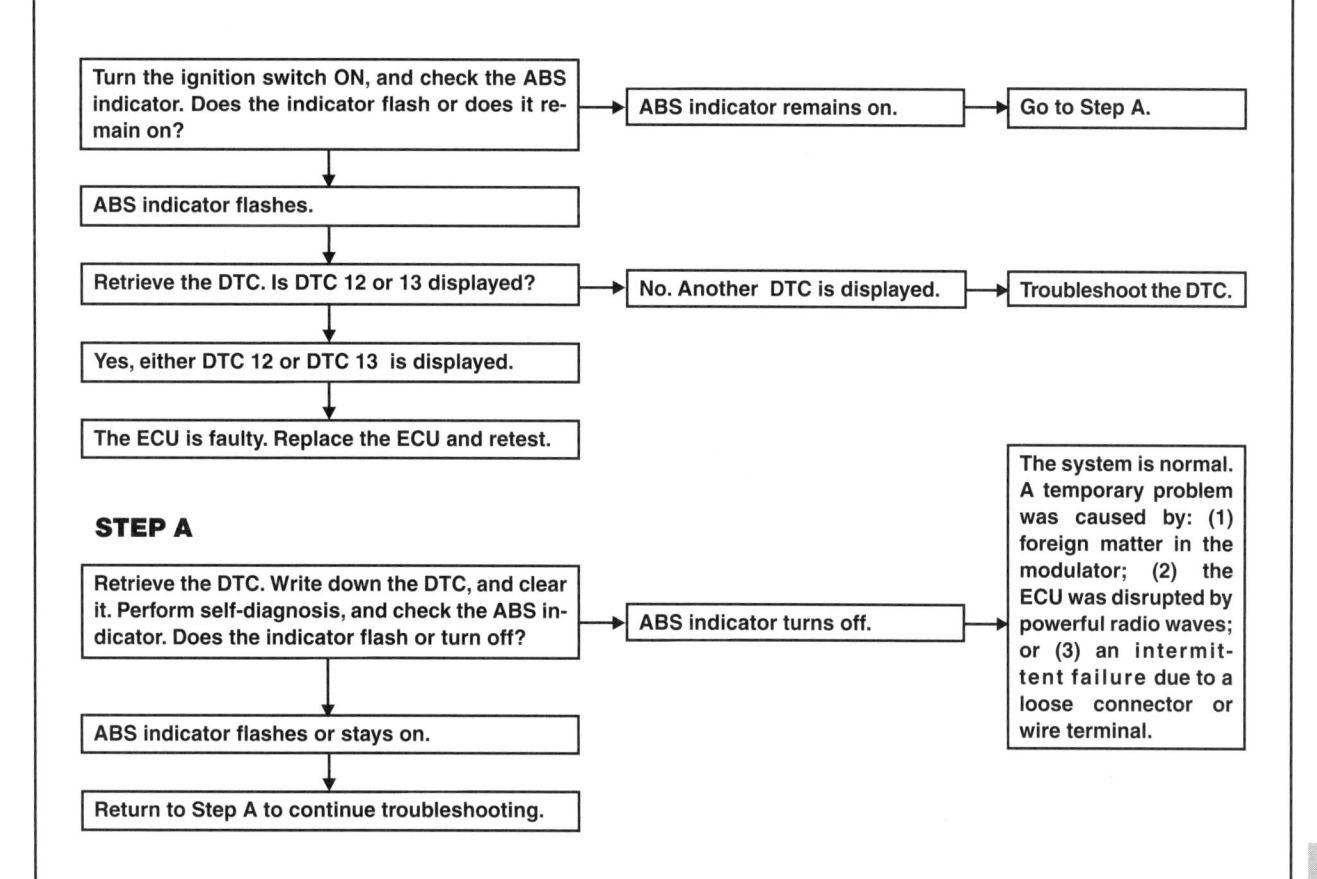

Turn the ignition switch ON, and check the ABS indicator. Does the indicator flash or does it remain on? → ABS indicator remains on. → Go to Step A.

↓

ABS indicator flashes.

↓

Retrieve the DTC. Is DTC 12 or 13 displayed? → No. Another DTC is displayed. → Troubleshoot the DTC.

↓

Yes, either DTC 12 or DTC 13 is displayed.

↓

The ECU is faulty. Replace the ECU and retest.

STEP A

Retrieve the DTC. Write down the DTC, and clear it. Perform self-diagnosis, and check the ABS indicator. Does the indicator flash or turn off? → ABS indicator turns off. → The system is normal. A temporary problem was caused by: (1) foreign matter in the modulator; (2) the ECU was disrupted by powerful radio waves; or (3) an intermittent failure due to a loose connector or wire terminal.

↓

ABS indicator flashes or stays on.

↓

Return to Step A to continue troubleshooting.

⑰

CODE 14: FAULTY POWER CIRCUIT

1. Note the following:
   a. Check the idle speed as described in Chapter Three.
   b. Low battery voltage can cause a malfunction with the ABS power circuit and set DTC 14.
   c. Low battery voltage can result when operating the motorcycle with accessories that place an excessive strain on the charging system, or if the ignition switch was left on (engine not running) for an extended period of time. If the battery is weak, charge the battery, clear the DTC, and perform the self-diagnosis check.

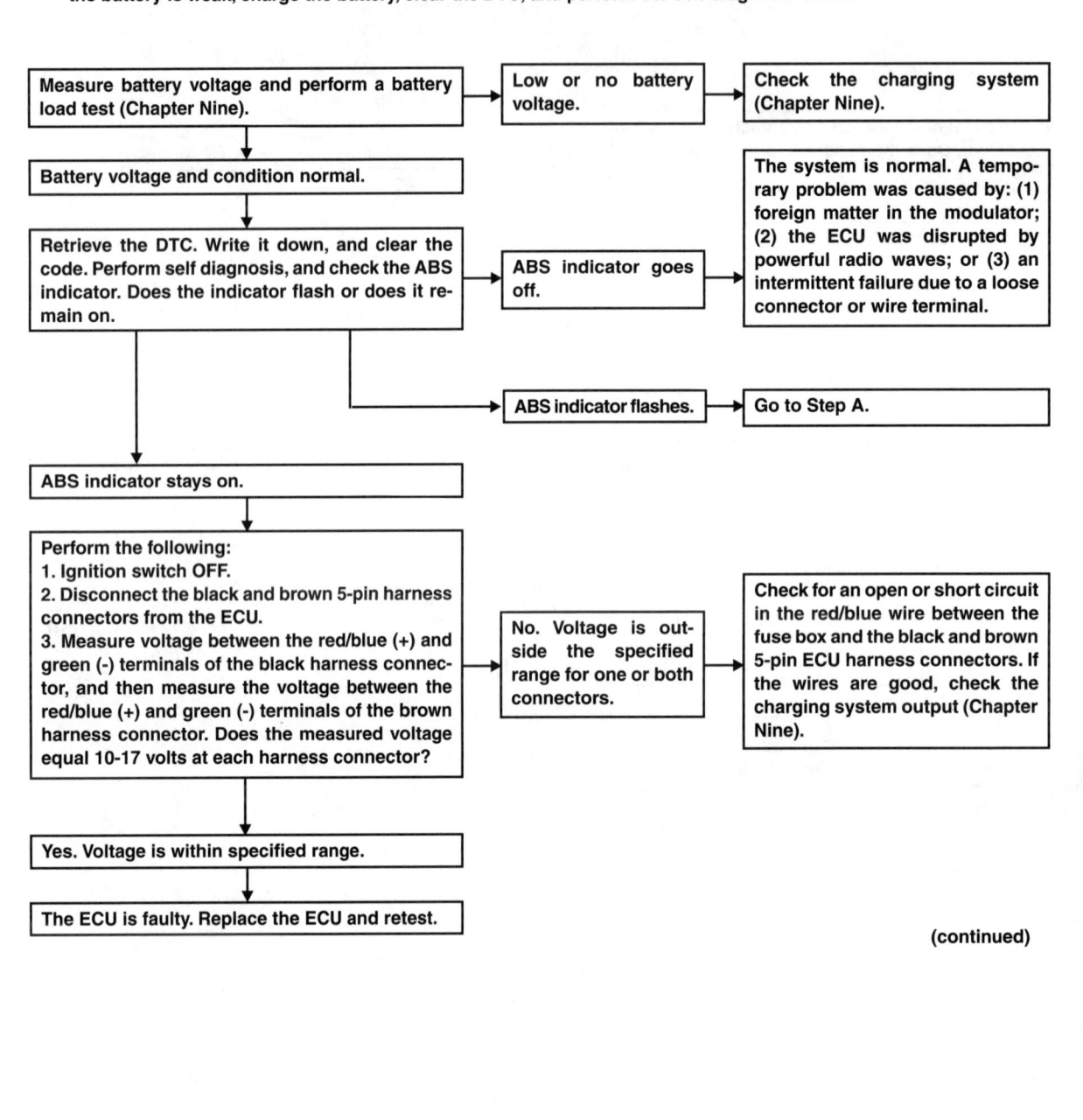

Measure battery voltage and perform a battery load test (Chapter Nine). → Low or no battery voltage. → Check the charging system (Chapter Nine).

↓

Battery voltage and condition normal.

↓

Retrieve the DTC. Write it down, and clear the code. Perform self diagnosis, and check the ABS indicator. Does the indicator flash or does it remain on. → ABS indicator goes off. → The system is normal. A temporary problem was caused by: (1) foreign matter in the modulator; (2) the ECU was disrupted by powerful radio waves; or (3) an intermittent failure due to a loose connector or wire terminal.

→ ABS indicator flashes. → Go to Step A.

↓

ABS indicator stays on.

↓

Perform the following:
1. Ignition switch OFF.
2. Disconnect the black and brown 5-pin harness connectors from the ECU.
3. Measure voltage between the red/blue (+) and green (-) terminals of the black harness connector, and then measure the voltage between the red/blue (+) and green (-) terminals of the brown harness connector. Does the measured voltage equal 10-17 volts at each harness connector?
→ No. Voltage is outside the specified range for one or both connectors. → Check for an open or short circuit in the red/blue wire between the fuse box and the black and brown 5-pin ECU harness connectors. If the wires are good, check the charging system output (Chapter Nine).

↓

Yes. Voltage is within specified range.

↓

The ECU is faulty. Replace the ECU and retest.

(continued)

ANTI-LOCK BRAKE SYSTEM

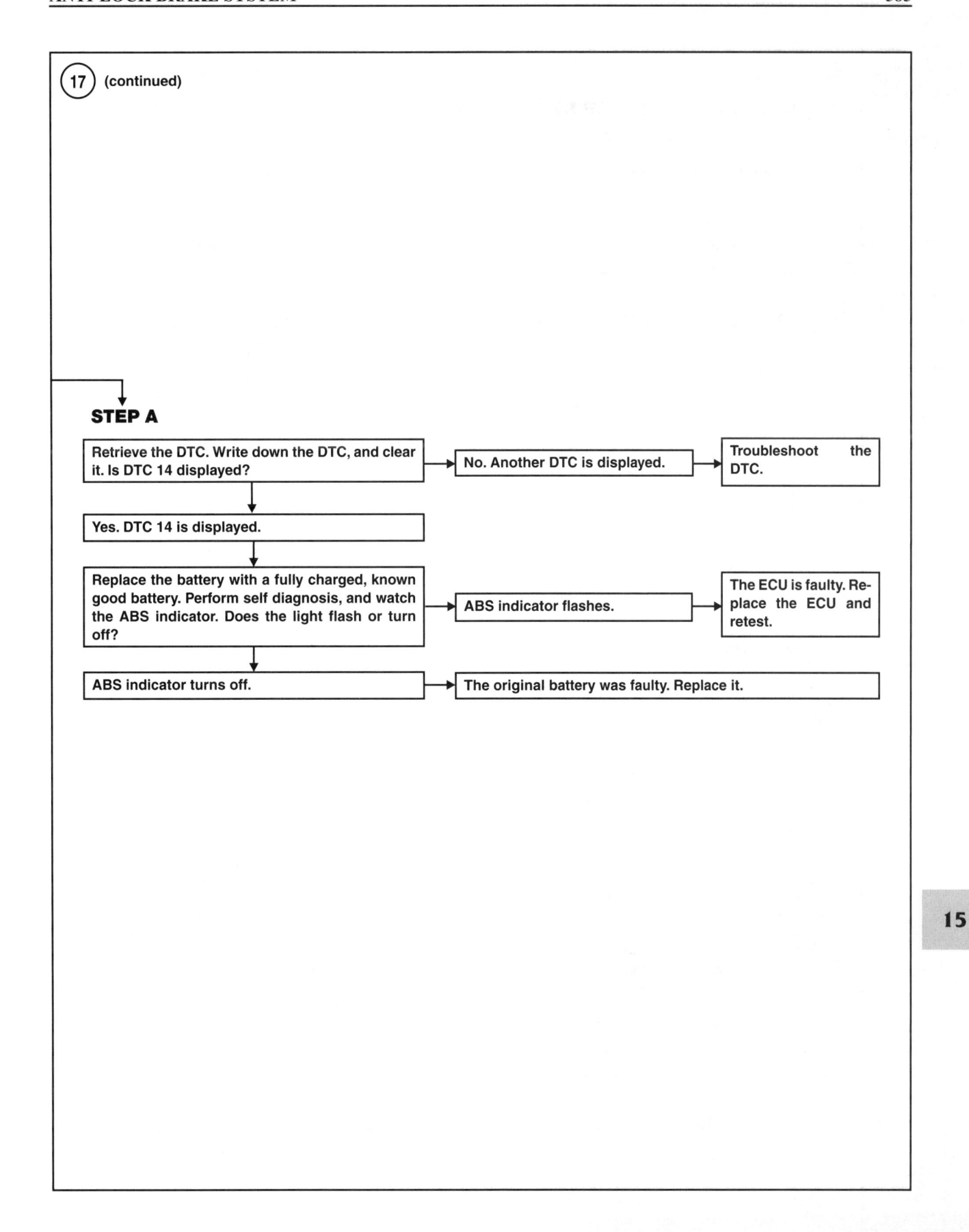

17 (continued)

STEP A

| Retrieve the DTC. Write down the DTC, and clear it. Is DTC 14 displayed? | → No. Another DTC is displayed. → Troubleshoot the DTC. |

↓ Yes. DTC 14 is displayed.

| Replace the battery with a fully charged, known good battery. Perform self diagnosis, and watch the ABS indicator. Does the light flash or turn off? | → ABS indicator flashes. → The ECU is faulty. Replace the ECU and retest. |

↓ ABS indicator turns off. → The original battery was faulty. Replace it.

(18)

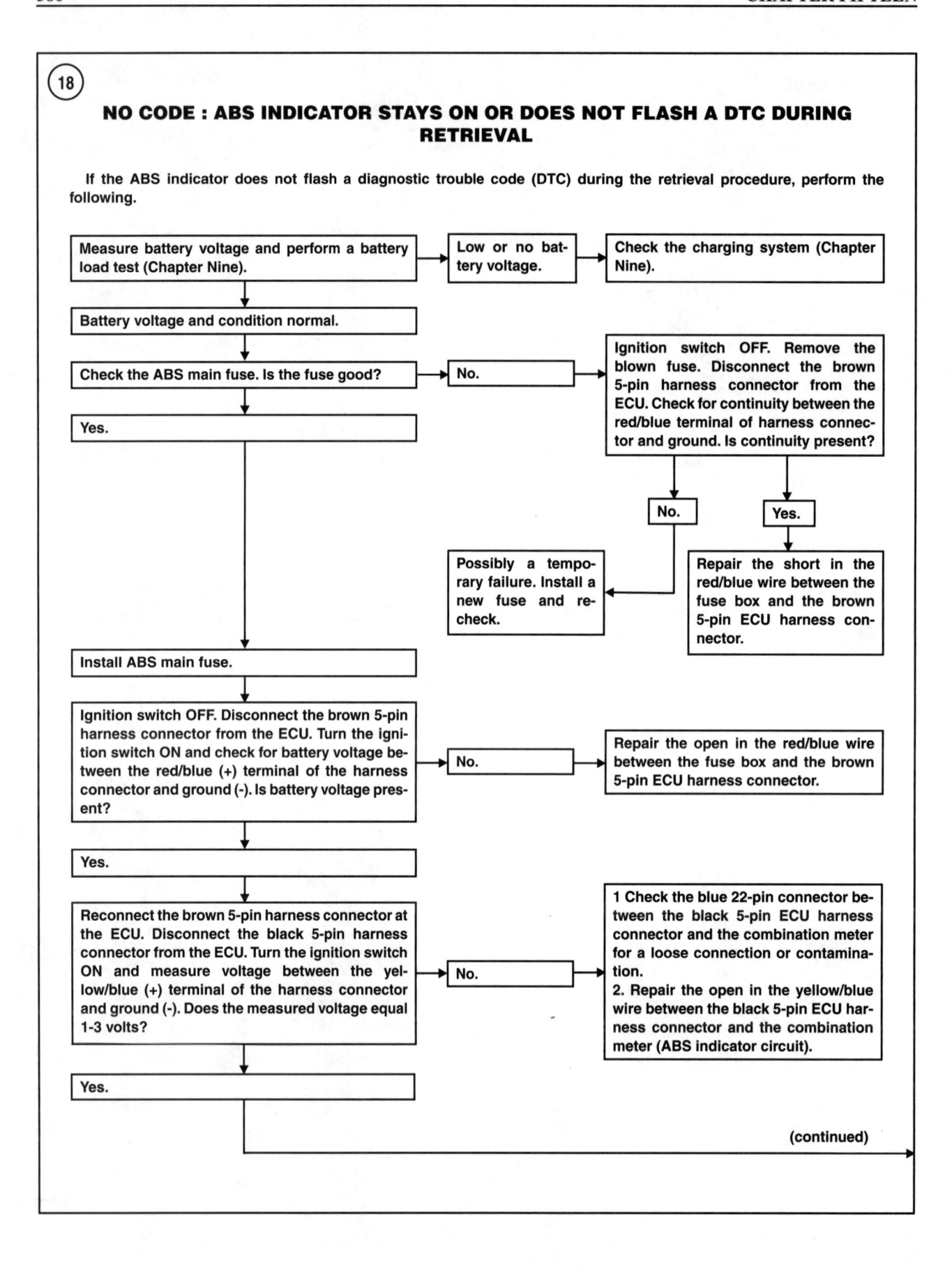

NO CODE : ABS INDICATOR STAYS ON OR DOES NOT FLASH A DTC DURING RETRIEVAL

If the ABS indicator does not flash a diagnostic trouble code (DTC) during the retrieval procedure, perform the following.

Measure battery voltage and perform a battery load test (Chapter Nine). → Low or no battery voltage. → Check the charging system (Chapter Nine).

↓

Battery voltage and condition normal.

↓

Check the ABS main fuse. Is the fuse good? → No. → Ignition switch OFF. Remove the blown fuse. Disconnect the brown 5-pin harness connector from the ECU. Check for continuity between the red/blue terminal of harness connector and ground. Is continuity present?

↓ Yes.

 — No. → Possibly a temporary failure. Install a new fuse and recheck.

 — Yes. → Repair the short in the red/blue wire between the fuse box and the brown 5-pin ECU harness connector.

Install ABS main fuse.

↓

Ignition switch OFF. Disconnect the brown 5-pin harness connector from the ECU. Turn the ignition switch ON and check for battery voltage between the red/blue (+) terminal of the harness connector and ground (-). Is battery voltage present? → No. → Repair the open in the red/blue wire between the fuse box and the brown 5-pin ECU harness connector.

↓ Yes.

Reconnect the brown 5-pin harness connector at the ECU. Disconnect the black 5-pin harness connector from the ECU. Turn the ignition switch ON and measure voltage between the yellow/blue (+) terminal of the harness connector and ground (-). Does the measured voltage equal 1-3 volts? → No. → 1 Check the blue 22-pin connector between the black 5-pin ECU harness connector and the combination meter for a loose connection or contamination.
2. Repair the open in the yellow/blue wire between the black 5-pin ECU harness connector and the combination meter (ABS indicator circuit).

↓ Yes.

(continued)

ANTI-LOCK BRAKE SYSTEM

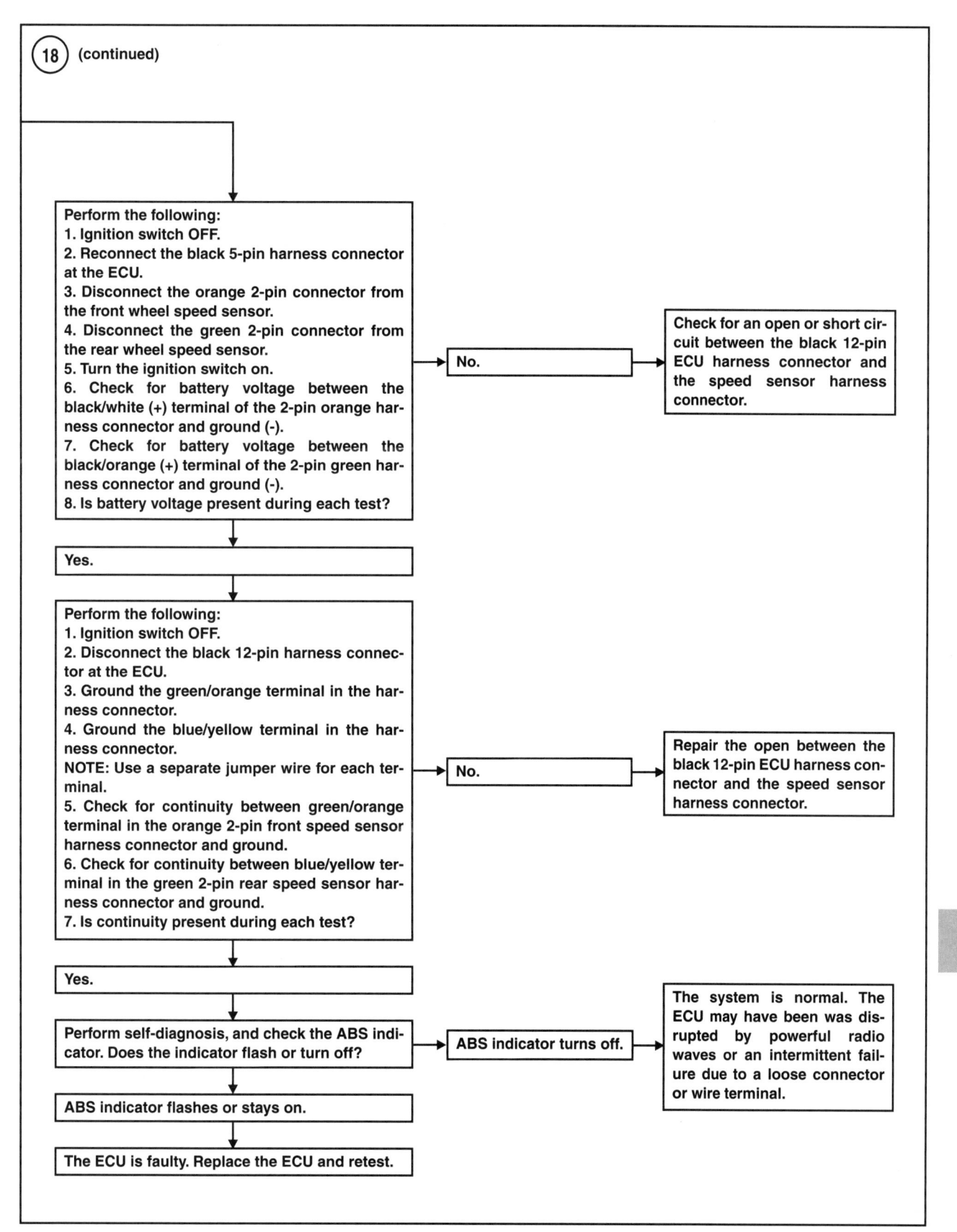

(18) (continued)

Perform the following:
1. Ignition switch OFF.
2. Reconnect the black 5-pin harness connector at the ECU.
3. Disconnect the orange 2-pin connector from the front wheel speed sensor.
4. Disconnect the green 2-pin connector from the rear wheel speed sensor.
5. Turn the ignition switch on.
6. Check for battery voltage between the black/white (+) terminal of the 2-pin orange harness connector and ground (-).
7. Check for battery voltage between the black/orange (+) terminal of the 2-pin green harness connector and ground (-).
8. Is battery voltage present during each test?

→ No. → Check for an open or short circuit between the black 12-pin ECU harness connector and the speed sensor harness connector.

Yes.

Perform the following:
1. Ignition switch OFF.
2. Disconnect the black 12-pin harness connector at the ECU.
3. Ground the green/orange terminal in the harness connector.
4. Ground the blue/yellow terminal in the harness connector.
NOTE: Use a separate jumper wire for each terminal.
5. Check for continuity between green/orange terminal in the orange 2-pin front speed sensor harness connector and ground.
6. Check for continuity between blue/yellow terminal in the green 2-pin rear speed sensor harness connector and ground.
7. Is continuity present during each test?

→ No. → Repair the open between the black 12-pin ECU harness connector and the speed sensor harness connector.

Yes.

Perform self-diagnosis, and check the ABS indicator. Does the indicator flash or turn off?

→ ABS indicator turns off. → The system is normal. The ECU may have been was disrupted by powerful radio waves or an intermittent failure due to a loose connector or wire terminal.

ABS indicator flashes or stays on.

The ECU is faulty. Replace the ECU and retest.

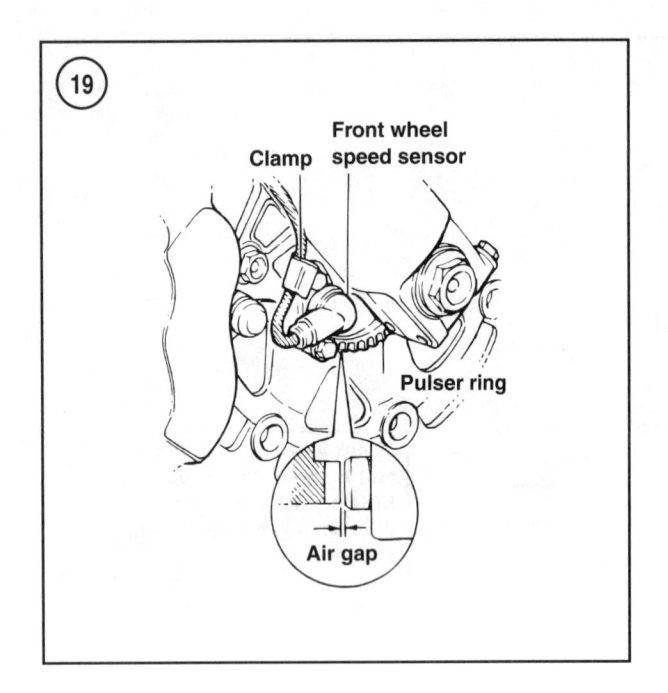

Figure 19: Front wheel speed sensor, Clamp, Pulser ring, Air gap

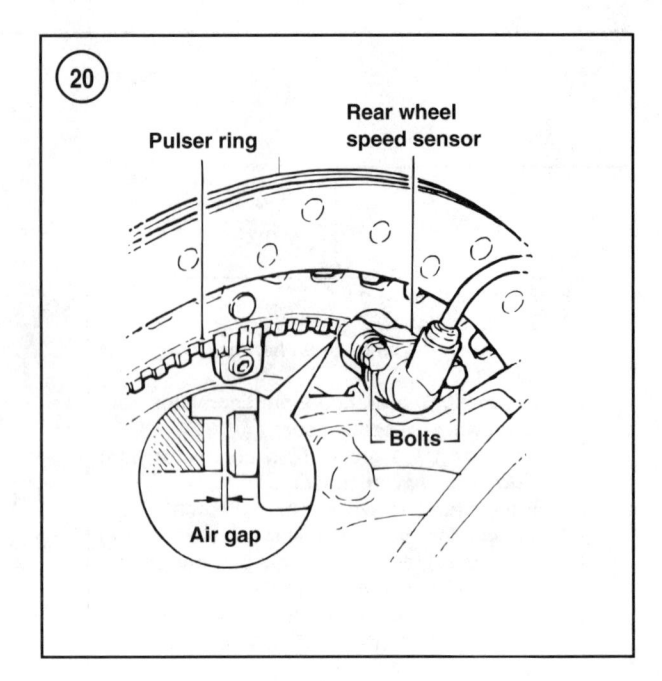

Figure 20: Pulser ring, Rear wheel speed sensor, Bolts, Air gap

WHEEL SPEED SENSORS

The wheel speed sensors and pulser rings work as magnetic triggering units to signal the ECU on the speed of the wheels. The wheel speed sensors are permanent magnet-type sensors and are non-adjustable.

The front wheel speed sensor mounts on the right fork tube. The rear wheel speed sensor mounts on the final drive unit. Each sensor aligns with a pulser ring. The front pulser ring is mounted onto the front wheel hub. The rear pulser ring is mounted onto the rear brake disc.

As the wheel turns, projections on the pulser ring pass across the wheel speed sensor and disturb the magnetic field provided by the speed sensor magnet. This causes the magnetic field to turn off and on and create small voltage signals or pulses at the speed sensor. Because the frequency of the voltage signals varies proportionally to the speed of the pulser ring (wheel), the ECU uses these signals to calculate wheel speed and to determine whether the wheels are about to lock.

Air Gap Inspection

The air gap is a fixed distance between the wheel speed sensor and pulser ring to ensure proper signal pickup from the sensor to the ECU. The air gap is non-adjustable.

NOTE
An incorrect air gap measurement can set a diagnostic trouble code (DTC) in the ECU.

1. Place the motorcycle on its centerstand.

2A. Front wheel—Support the motorcycle with the front wheel off the ground. Refer to *Bike Lift* in Chapter Eleven.

2B. Rear wheel—Remove rear fender A. Refer to *Rear Fender A* in Chapter Seventeen.

3. Spin the wheel slowly by hand. Inspect the pulser ring for any chipped or damaged projections, road tar or other debris that could affect the measurement.

4. Measure the air gap between the wheel sensor and the pulser ring with a non-magnetic feeler gauge. See **Figure 19** (front) or **Figure 20** (rear). Rotate the wheel and measure the air gap all the way around the pulser ring circumference. The standard air gap measurement for the front and rear wheels is 0.4-1.2 mm (0.02-0.05 in.).

5. The air gap is not adjustable. If the air gap is incorrect, perform the following:

    a. Check the speed sensor for loose or damaged mounting bolts.

    b. Check for a damaged speed sensor mounting bracket.

    c. Check the front fork for damage.

WARNING
Do not shim the speed sensor to correct the air gap. An incorrect measurement indicates that the wheel speed sensor, or a related component, is improperly installed or damaged.

ANTI-LOCK BRAKE SYSTEM

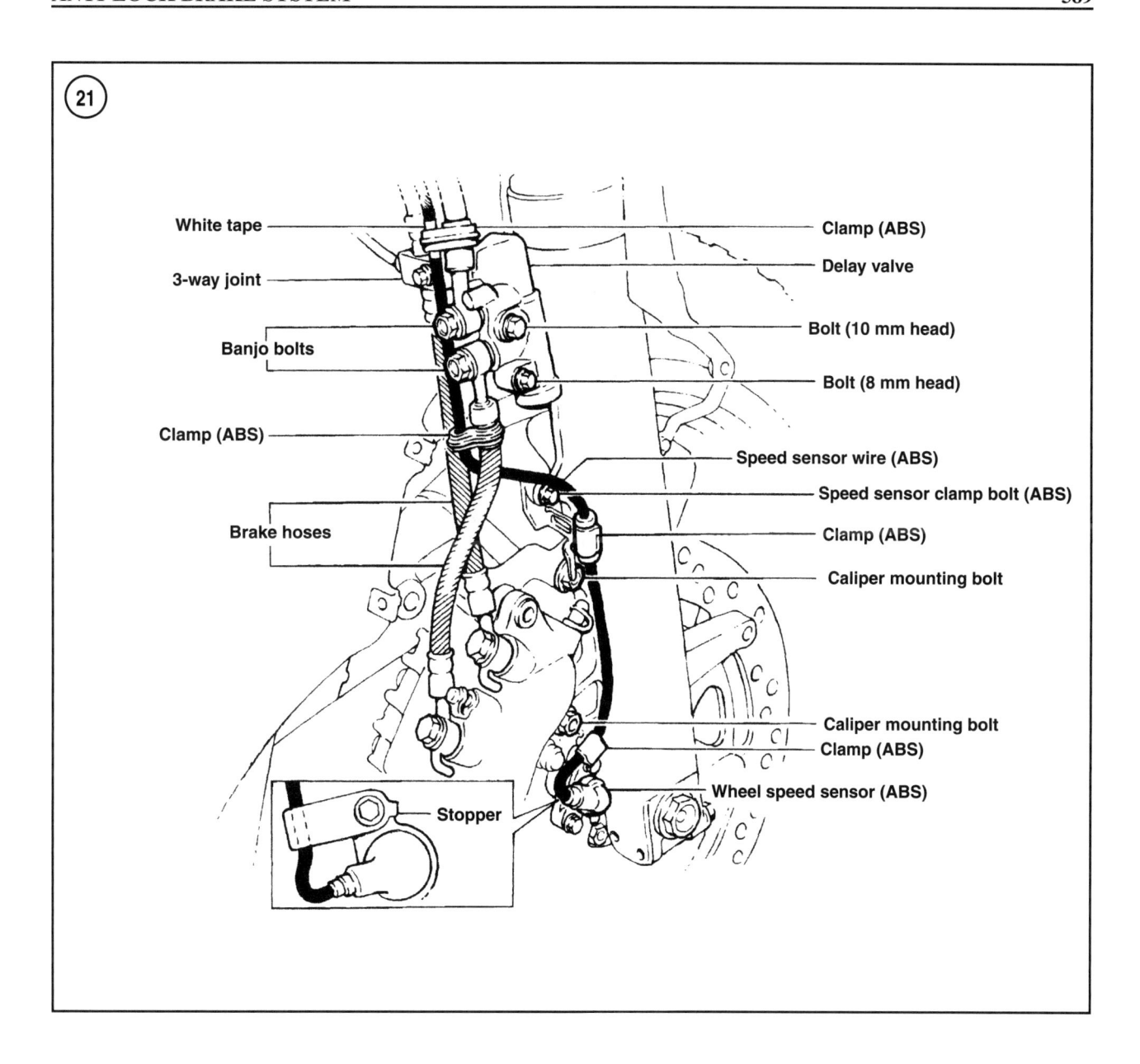

Figure 21 labels:
- White tape
- 3-way joint
- Banjo bolts
- Clamp (ABS)
- Brake hoses
- Stopper
- Clamp (ABS)
- Delay valve
- Bolt (10 mm head)
- Bolt (8 mm head)
- Speed sensor wire (ABS)
- Speed sensor clamp bolt (ABS)
- Clamp (ABS)
- Caliper mounting bolt
- Caliper mounting bolt
- Clamp (ABS)
- Wheel speed sensor (ABS)

Front Wheel Speed Sensor Removal/Installation

1. Turn the ignition switch off.

2. Remove the following as described in Chapter Seventeen:
   a. Right front fender cover.
   b. Center inner fairing.

3. Note how the wheel speed sensor wire is routed along the fork tube and frame, and release the wire from any clamp that secures it in place (**Figure 21**).

4. Disconnect the wheel speed sensor orange 2-pin connector (**Figure 22**).

5. Remove the sensor mounting bolts and remove the sensor (**Figure 19**).

6. Installation is the reverse of removal, plus the following:

   a. Route the sensor wire along the same path noted during removal. Note how the stopper on the lower wire clamp (**Figure 21**) seats against the wheel speed sensor.

   b. Tighten the wheel speed sensor mounting bolts to 12 N•m (106 in.-lb.).

   c. Measure the air gap as described in this section.

ABS MODULATOR ASSEMBLY

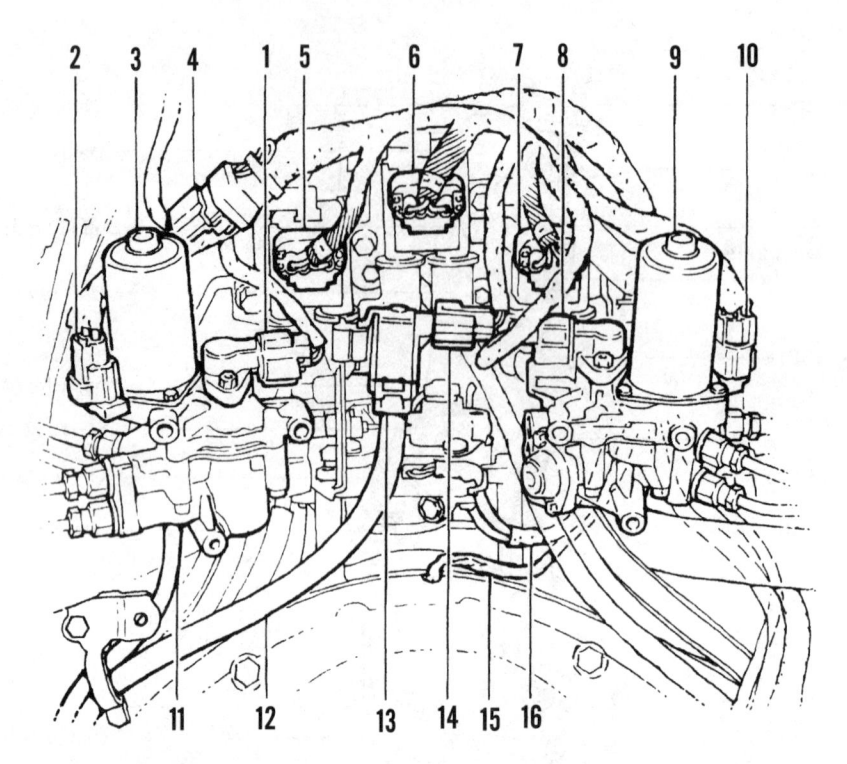

㉒

1. Front ABS modulator crank angle gray 3-pin connector
2. Front ABS modulator control motor black 2-pin connector
3. Front ABS modulator
4. Front wheel speed sensor orange 2-pin connector
5. No. 1/2 ignition coil and connector
6. No. 3/4 ignition coil and connector
7. No. 5/6 ignition coil and connector
8. Rear ABS modulator crank angle sensor gray 3-pin connector
9. Rear ABS modulator
10. Rear ABS modulator control motor black 2-pin connector
11. Fuel tank-to-EVAP canister No. 1 tube
12. No. 4 hose
13. Evaporative emission (EVAP) control solenoid valve
14. EVAP black 2-pin connector
15. Ignition pulse generator wire
16. Reverse sift actuator wire

Rear Wheel Speed Sensor Removal/Installation

1. Turn the ignition switch off.

2. Remove the right saddlebag. Refer to *Saddlebags* in Chapter Seventeen.

3. Remove the fuel tank. Refer to *Fuel Tank* in Chapter Eight.

4. Note how the wheel speed sensor wire is routed along the frame, and release the wire from any clamp that secures it in place.

5. Disconnect the wheel speed sensor gray 2-pin connector located near the thermostat housing.

6. On 2001-2003 models, unbolt and remove the wheel speed sensor (Figure 20, typical). On 2004, unbolt and remove the brake hose guide and wheel speed sensor. Note how the wires and hoses are routed for reinstallation.

7. Installation is the reverse of removal, plus the following:

   a. Route the sensor wire along the same path noted during removal. On 2004 models, route the brake hoses in the brake hose guide.

ANTI-LOCK BRAKE SYSTEM

REAR PULSER RING

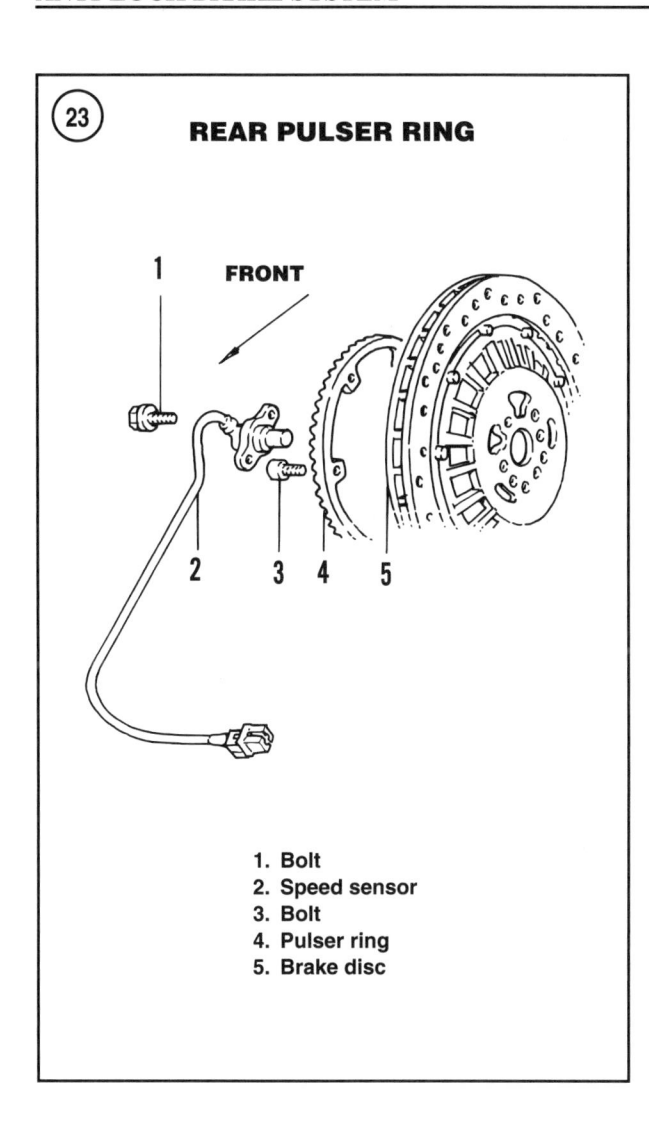

1. Bolt
2. Speed sensor
3. Bolt
4. Pulser ring
5. Brake disc

b. Tighten the wheel speed sensor mounting bolts to 12 N•m (106 in.-lb.).
c. Measure the air gap as described in this section.

PULSER RING

Front Pulser Ring Removal/Installation

1. Remove the front wheel. Refer to *Front Wheel* in Chapter Twelve.
2. Remove the mounting bolts and remove the pulser ring (**Figure 19**) from the wheel hub.
3. Inspect the pulser ring for any chipped or damaged projections, road tar or other debris.
4. Make sure the pulser ring and hub mounting surfaces are clean.
5. Installation is the reverse of removal. Note the following:

   a. Mount the pulser ring using new mounting bolts.
   b. Tighten the pulser ring mounting bolts in two or three steps to 8 N•m (71 in.-lb.).
   c. Check the air gap. Refer to *Wheel Speed Sensor* in this chapter.

Rear Pulser Ring Removal/Installation

1. Remove the rear wheel. Refer to *Rear Wheel* in Chapter Twelve.
2. Remove the mounting bolts and remove the pulser ring (4, **Figure 23**) from the brake disc.
3. Inspect the pulser ring for any chipped or damaged projections, road tar or other debris.
4. Make sure the pulser ring and hub mounting surfaces are clean.
5. Installation is the reverse of removal. Note the following:

   a. Mount the pulser ring using new mounting bolts.
   b. Tighten the pulser ring mounting bolts in two or three steps to 8 N•m (71 in.-lb.).
   c. Check the air gap. Refer to *Wheel Speed Sensor* in this chapter.

ABS MODULATORS

Separate front and rear ABS modulators are used to control or modulate brake fluid pressure at the brake calipers.

Removal/Installation (Front and Rear Modulators)

The front and rear modulators are removed at the same time and then separated.

Refer to **Figure 22** and **Figure 24**.

1. Remove the center inner fairing. Refer to *Front Lower Fairings* in Chapter Seventeen.
2. Drain the front and rear brake system as described in Chapter Fourteen.
3. Remove the ignition coils. Refer to *Ignition Coils* in Chapter Nine.

NOTE
Note the harness routing and identify the 2-pin and 3-pin connectors before disconnecting them.

4. Disconnect the control motor black 2-pin connector from each modulator.

592 CHAPTER FIFTEEN

㉔ ABS MODULATORS

1. Brake pipe (black)
2. Brake pipe (white)
3. Brake pipe (green)
4. Brake pipe (yellow)
5. Front ABS modulator control motor black 2-pin connector
6. Front ABS modulator
7. Front ABS modulator crank angle gray 3-pin connector
8. Rear ABS modulator crank angle sensor gray 3-pin connector
9. ABS modulator mounting bracket
10. Rear ABS modulator
11. Brake pipe (white)
12. Brake pipe (blue)
13. Rear ABS modulator control motor black 2-pin connector
14. Brake pipe (yellow)
15. Brake pipe (green)

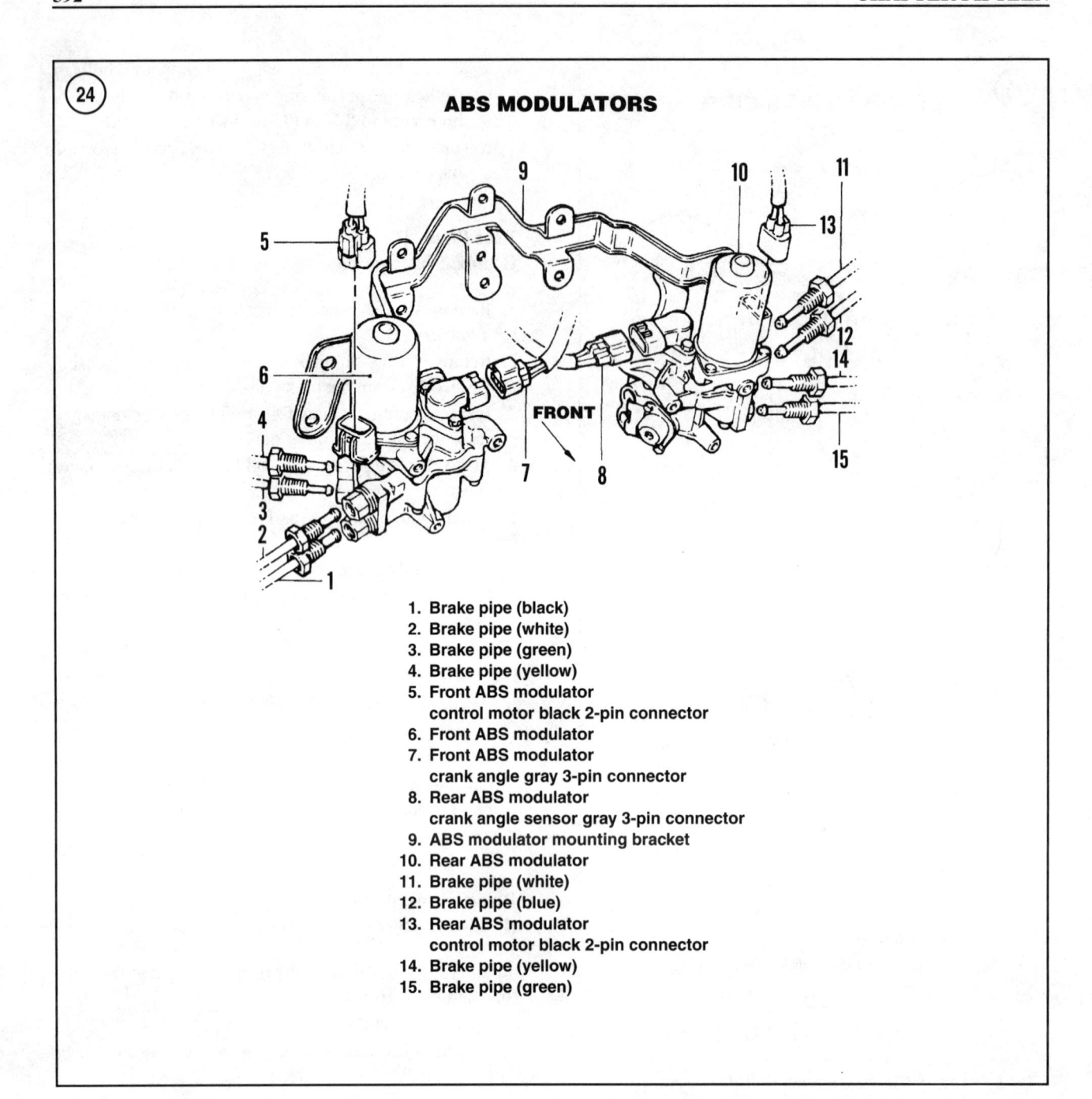

5. Disconnect the crank angle sensor gray 3-pin connector from each modulator.

WARNING
*The four brake pipes and their related ports on the front and rear modulators are identified by matching color marks on the ABS modulators with color marks on the brake pipes (**Figure 25**). Make sure the identification marks and color marks are legible before disconnecting the brake pipes. If necessary, make new reference marks on each brake pipe and its port. Each brake pipe must be connected to the correct port during assembly.*

CAUTION
The metal brake pipes are easily damaged. Handle them carefully when disconnecting them from the modulators.

ANTI-LOCK BRAKE SYSTEM

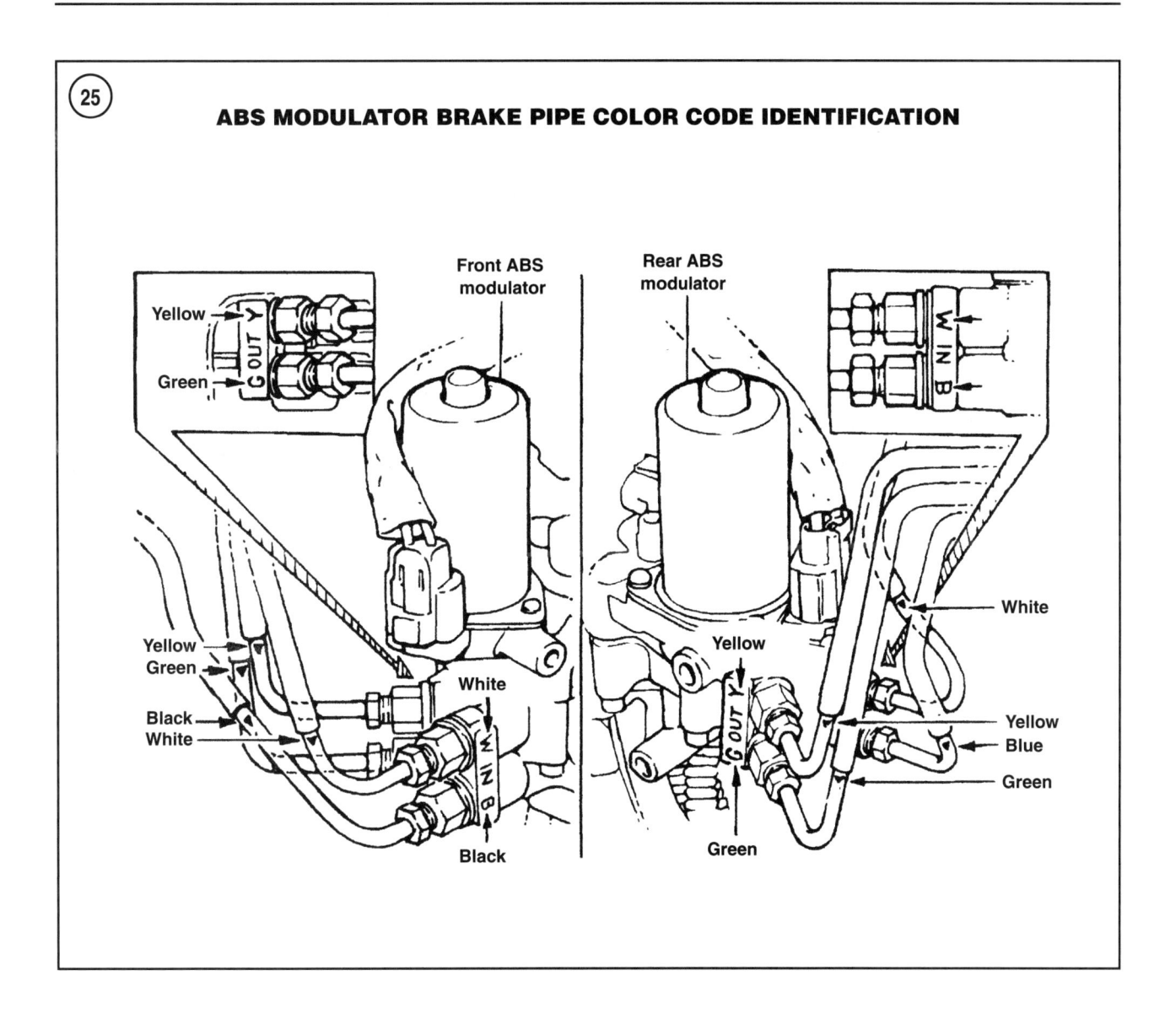

ABS MODULATOR BRAKE PIPE COLOR CODE IDENTIFICATION

(Figure 25: Front ABS modulator — Yellow, Green, Yellow, Green, Black, White, White, Black. Rear ABS modulator — Yellow, White, Yellow, Blue, Green, Green.)

6. Disconnect each brake pipe from its port on the front and rear modulators. Seal a plastic bag over the end of each pipe to prevent leakage and pipe contamination.

7. Remove the spark plug wire clip from the mounting stay.

CAUTION
Brake fluid will leak from the modulators when removing them.

8. Remove the bolts and remove the modulators and mounting bracket from the frame.

9. If necessary, unbolt and remove the front and rear modulators from the mounting bracket (**Figure 24**).

10. Store the modulators in a plastic bag to prevent contamination.

11. Clean the fasteners and check for damage.

12. Check the brake pipes for creasing, fluid leaks and other damage.

13. If removed, install the modulators onto the stay. Tighten the mounting bolts to 12 N•m (106 in.-lb.). See **Figure 24** to identify the front and rear modulators.

14. Install the modulators and stay, making sure not to bend or damage the metal pipes. Install and tighten the stay mounting bolts to 12 N•m (106 in.-lb.).

15. Wipe the brake pipe ends and joint nut threads with a clean rag.

16. Install each brake pipe onto the proper port on the modulator. Make sure the color mark on the brake pipe matches the identification mark on the port (**Figure 25**).

17. Lubricate the brake pipe joint nut threads with DOT 4 brake fluid. Thread the brake pipe joint nuts into the modulator ports and tighten to 17 N•m (12 ft.-lb.).

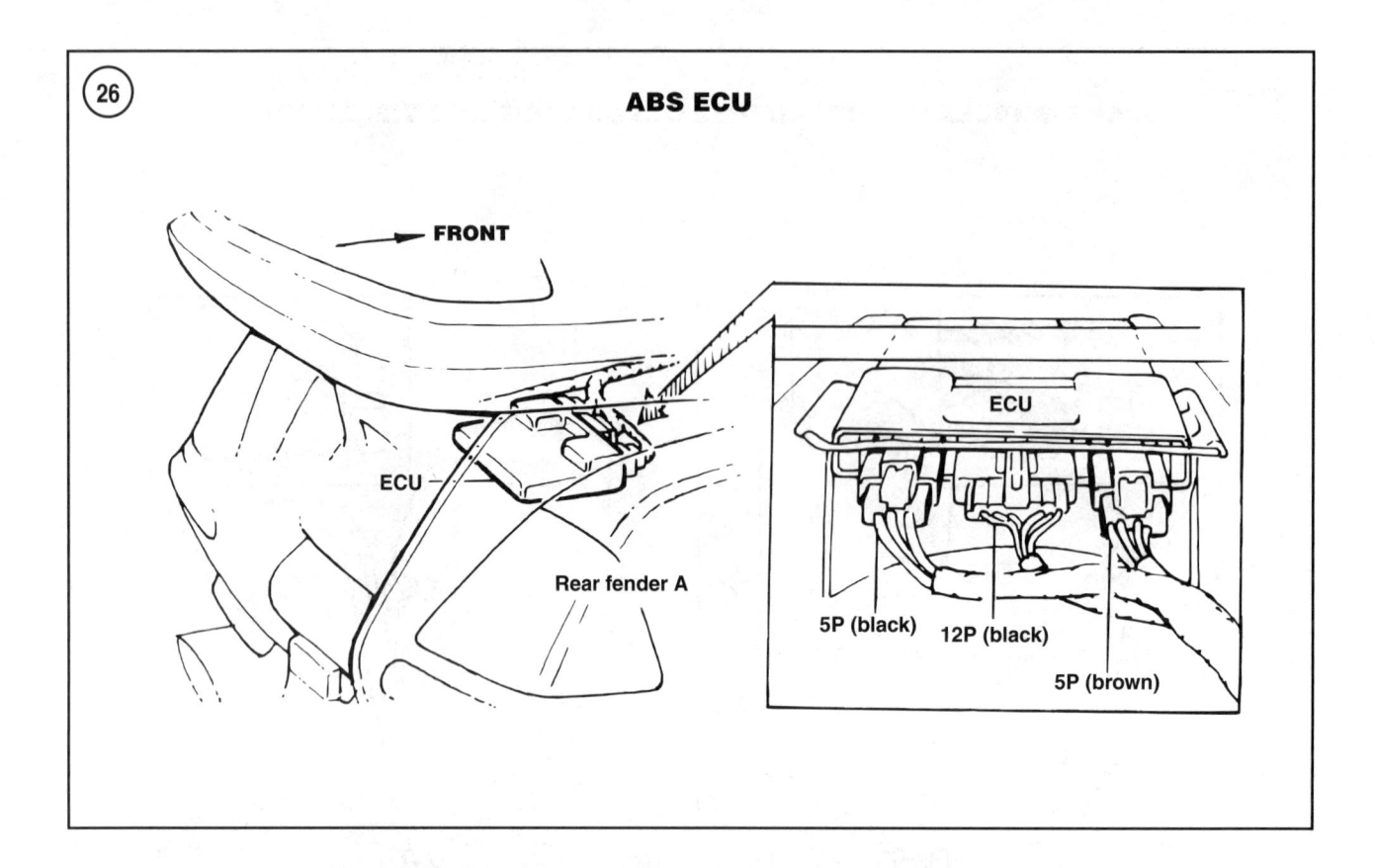

Figure 26 ABS ECU
- FRONT
- ECU
- Rear fender A
- 5P (black)
- 12P (black)
- 5P (brown)

18. Reconnect the connectors to the modulators.
19. Fill and bleed the brake system as described in Chapter Fourteen. Check the brake pipes for fluid leaks when bleeding the system.
20. Reverse Steps 1-3 to complete installation.
21. Start the engine and perform the self-diagnosis as described in this chapter.

ABS ELECTRONIC CONTROL UNIT (ECU)

The ABS electronic control unit (ECU) is mounted on rear fender B and secured with a rubber mounting band (**Figure 26**).

Removal/Installation

1. Read *ABS General Information* in this chapter before removing or installing the ECU.
2. Start the engine and perform the self-diagnosis to check for any stored diagnostic trouble codes in the ECU. Refer to *ABS Troubleshooting* in this chapter.
3. Disconnect the negative cable from the battery. Refer to *Battery* in Chapter Nine.

NOTE
If troubleshooting the ABS system, visually inspect the wiring harness and connectors at the ECU before disconnecting or removing parts. Loose or contaminated connectors or problems with the wiring harness are common causes of ABS problems.

4. Remove rear fender A. Refer to *Rear Fender A* in Chapter Seventeen.
5. Disconnect the black 5-pin, black 12-pin and brown 5-pin harness connectors at the ECU (**Figure 26**).
6. Unhook the mounting band and remove the ECU from rear fender B. Handle the ECU carefully.

CAUTION
Store the ECU in a safe place if it is to be reinstalled.

7. Make sure the harness connectors are clean. Check for bent or damaged contacts.
8. Replace the mounting band if weak or damaged.
9. Check the mounting band hooks on rear fender B for cracks or damage.
10. Installation is the reverse of removal.
11. Start the engine and perform the self-diagnosis as described in this chapter.

Table 1 ABS DIAGNOSTIC TROUBLE CODES

DTC	Probematic part/system	Probably faulty part/system
2	Front wheel speed sensor system	Front wheel speed sensor. Front pulser ring. Wire harness. ECU. Front tire. Front wheel. Riding conditions.
3	Rear wheel speed sensor system	Rear wheel speed sensor. Rear pulser ring. Wire harness. ECU. Rear tire. Rear wheel. Riding conditions.
4	Front control motor system	Blown No. 3 fuse (front ABS motor fuse). Front modulator control motor. Front modulator crank angle sensor. Wiring harness. ECU.
5	Rear control motor system	Blown No. 4 fuse (rear ABS motor fuse). Rear modulator control motor. Rear modulator crank angle sensor. Wiring harness. ECU.
6	Front modulator crank angle system	Front crank angle sensor. Wire harness. ECU.
7	Rear modulator crank angle system	Rear crank angle sensor. Wire harness. ECU.
8	Front control circuit in ECU	Blown No. 3 fuse (front ABS motor fuse). Front modulator control motor. Front crank angle sensor. Front wheel speed sensor. Front pulser ring. Wire harness. ECU. Front tire. Front wheel. Riding conditions.
9	Rear control circuit in ECU	Blown No. 4 fuse (rear ABS motor fuse). Rear modulator control motor. Rear crank angle sensor. Rear wheel speed sensor. Rear pulser ring. Wire harness. ECU. Rear tire. Rear wheel. Riding conditions.

(continued)

Table 1 ABS DIAGNOSTIC TROUBLE CODES (continued)

DTC	Probematic part/system	Probably faulty part/system
10	Front relay circuit in ECU	Blown No. 3 fuse (front ABS motor fuse). Front modulator control motor. Front crank angle sensor. Wire harness. ECU.
11	Rear relay circuit in ECU	Blown No. 4 fuse (rear ABS motor fuse). Rear modulator control motor. Rear crank angle sensor. Wire harness. ECU.
12	Front motor driver circuit in ECU	Blown No. 3 fuse (front ABS motor fuse). Front modulator control motor. Wire harness. ECU.
13	Rear motor driver circuit in ECU	Blown No. 4 fuse (rear ABS motor fuse). Rear modulator control motor. Wire harness. ECU.
14	Power circuit	Power circuit (charging). Wire harness. ECU.
None	Problems in ABS circuit not detected by ECU	ABS main circuit. Power circuit (charging). Wire harness. ECU. ABS indicator.

Table 2 WHEEL SPEED SENSOR

Wheel speed sensor air gap	0.4-1.2 mm (0.02-0.05 in.)

Table 3 ANTI-LOCK BRAKE SYSTEM TORQUE SPECIFICATIONS

	N•m	in.-lb.	ft.-lb.
ABS modulator brake pipe joint bolt	34	–	25
ABS modulator mounting bolts	12	106	–
Brake pipe joint nuts[1]	17	–	12
Ignition coil/modulator mounting bracket bolts	12	106	–
Pulser ring mounting bolts[2]	8	71	–
Wheel speed sensor mounting bolts (2001-2003)	12	106	–
Wheel speed sensor and brake hose guide bolts (2004)	12	106	–

1. Apply brake fluid to threads.
2. Install new ALOC fasteners during assembly

CHAPTER SIXTEEN

CRUISE CONTROL SYSTEM

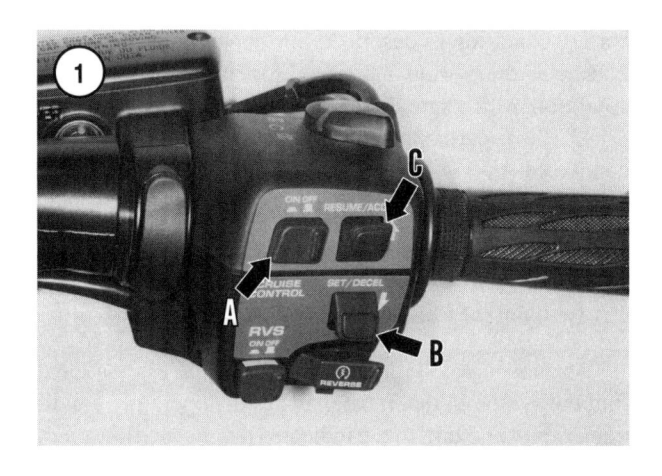

All models are equipped with a cruise control system. The system allows speed to be set between 30-100 mph (48-161 km/h) when the transmission is in fourth or fifth (overdrive) gear.

This chapter describes troubleshooting and replacement procedures for the cruise control components.

If there is any doubt about your ability to correctly and safely troubleshoot or service the cruise control system, take the job to a Honda dealership.

CRUISE CONTROL SYSTEM COMPONENTS

The cruise control system consists of the following components:
1. Cruise actuator relays.
2. Cruise cancel switch (front brake).
3. Cruise cancel switch (rear brake).
4. Cruise cancel switch (throttle).
5. Cruise cancel switch (clutch).
6. Cruise control actuator cable.
7. Cruise/reverse control module.
8. Cruise actuator.
9. Gauge assembly.
10. Gear position switch.
11. Cruise control switch assembly.

CRUISE CONTROL TROUBLESHOOTING

Before troubleshooting or servicing the cruise control system, note the following:
1. The cruise control operational switches are mounted on the right handlebar switch assembly (**Figure 1**):
    a. Cruise control main switch (A).
    b. SET/DECEL switch (B).
    c. RESUME/ACCEL switch (C).

② CRUISE CONTROL TROUBLESHOOTING CHART

Symptom or condition	Figure number
Cruise control will not set when cruise main switch is turned on	Figure 3
Cruise ON indicator does not light	Figure 4
Cruise ON indicator flashes when cruise main switch is on	Figure 5
Cruise ON indicator flashes when speed reaches 16 mph (25 km/h) when cruise main switch is turned on	Figure 6

2. Be careful not to damage the cruise control wiring harness or connectors during service procedures. Note the routing of the wire harness and the type, number and position of fasteners used to secure the wiring to the frame and other components.
3. Use original equipment fasteners (or equivalents) when replacement is required during service.
4. For the cruise control to operate, the following conditions must be maintained:
   a. The engine stop switch must be set to RUN.
   b. The cruise control switch must be set to ON.
   c. Front and rear brake must be released.
   d. Clutch must be fully engaged.
   e. Throttle must not be closed.
   f. Motorcycle speed must be maintained within 30-100 mph (48-161 k/mh).
   g. Transmission must be in fourth or fifth gear.
5. The cruise control will disengage when any of the following conditions are performed:
   a. When either brake is applied.
   b. The clutch lever is applied (releasing the clutch).
   c. The throttle is closed.
   d. The cruise control switch is turned OFF.

Visual Inspection

Intermittent problems, which are the most difficult to troubleshoot, are often caused by poor electrical connections or damaged wiring. These types of problems can be found with a thorough visual inspection. However, this requires removing the body panels to access the connectors.

1. Make sure the battery is fully charged. Check that the battery connections are clean and free of corrosion. Check battery condition as described in Chapter Nine.
2. Check the wiring harnesses for proper routing and condition. Make sure they are properly secured.
3. All electrical connectors must be clean and properly connected.
4. Check that the ignition/cruise fuse (fuse No. 19) is in good condition.

Cruise Control Flow Charts

To troubleshoot the cruise control, refer to **Figure 2** and find the symptom that most matches the operating condition or fault and turn to the indicated diagnostic flow chart (**Figures 3-6**). Perform the relevant procedures until the problem is resolved.

When using the flow charts, note the following:
1. Refer to the wiring diagrams at the end of this manual to identify the connectors.
2. Unless otherwise specified, perform the test procedures with the ignition switch off.
3. Do not disconnect or reconnect electrical connectors when the ignition switch is on. A voltage spike may damage the cruise/reverse control module.
4. Do not perform tests with a battery charger attached to the battery or circuit. Test and charge the battery as described in Chapter Nine.

CRUISE CONTROL SYSTEM

(3)

CRUISE CONTROL WILL NOT SET WHEN CRUISE MAIN SWITCH IS TURNED ON

1. Begin troubleshooting by making sure the following conditions are met:
   a. The cruise ON indicator comes on when the cruise main switch is turned on.
   b. The speedometer and tachometer are working correctly.
   c. The cruise actuator cable adjustment is correct.
2. Note the following:
   a. Refer to *Cruise/Reverse Control Module* in this chapter for additional information when performing tests at the black and gray 26-pin cruise/reverse control module harness connectors.
   b. Refer to *Switches* in Chapter Nine for information on testing and replacing the gear position switch and the cruise control switches mounted in the right-side handlebar switch housing. The right-side handlebar switch 18-pin harness connector called out in this procedure changed colors between 2003 and 2004 model years—2001-2003: green connector; 2004: blue connector.
   c. Refer to *Switches* in this chapter to test and replace the clutch cruise cancel switch.
   d. Remove the top shelter to access the cruise actuator. Refer to *Top Shelter* in Chapter Seventeen.
   e. Refer to *Cruise Actuator* in this chapter for additional testing and replacement of the cruise actuator.
   f. A separate 12-volt battery and test leads are required to check the cruise actuator.

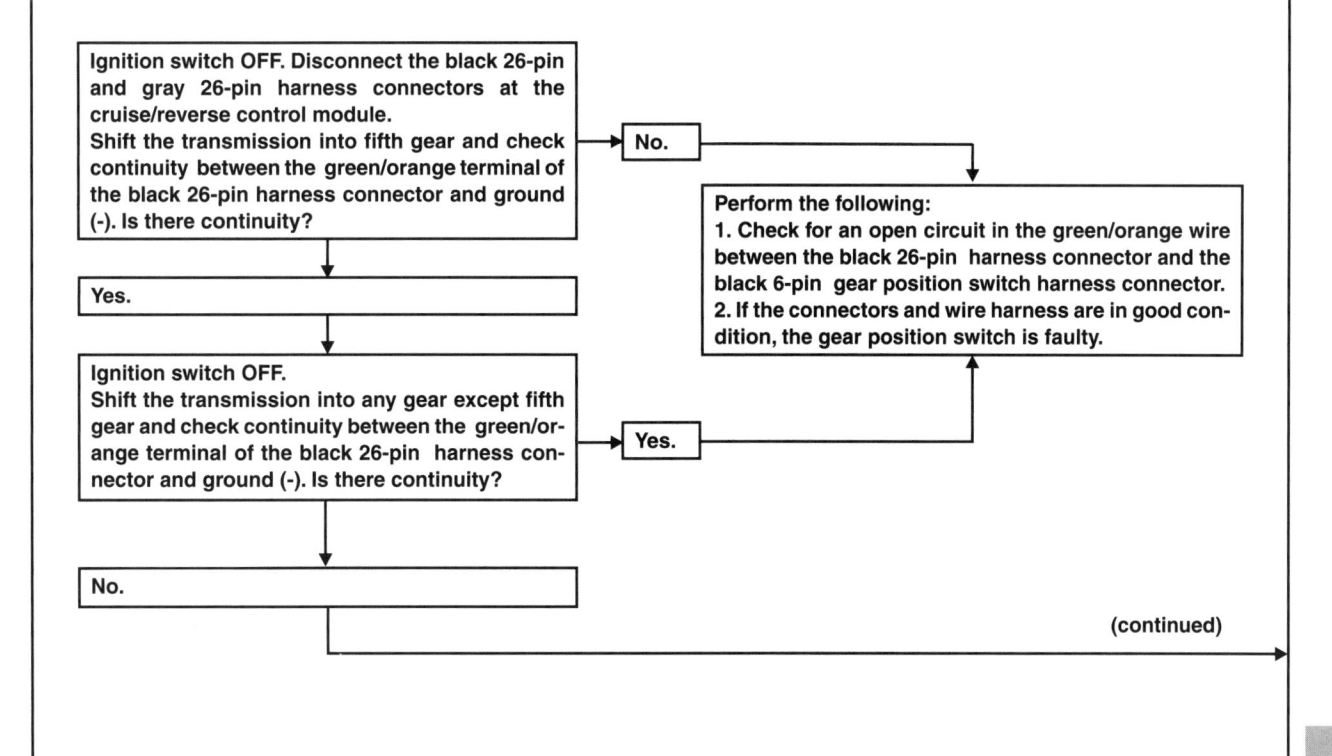

Ignition switch OFF. Disconnect the black 26-pin and gray 26-pin harness connectors at the cruise/reverse control module.
Shift the transmission into fifth gear and check continuity between the green/orange terminal of the black 26-pin harness connector and ground (-). Is there continuity?

→ No. →

Perform the following:
1. Check for an open circuit in the green/orange wire between the black 26-pin harness connector and the black 6-pin gear position switch harness connector.
2. If the connectors and wire harness are in good condition, the gear position switch is faulty.

↓ Yes.

Ignition switch OFF.
Shift the transmission into any gear except fifth gear and check continuity between the green/orange terminal of the black 26-pin harness connector and ground (-). Is there continuity?

→ Yes. ↑

↓ No.

(continued)

③ (continued)

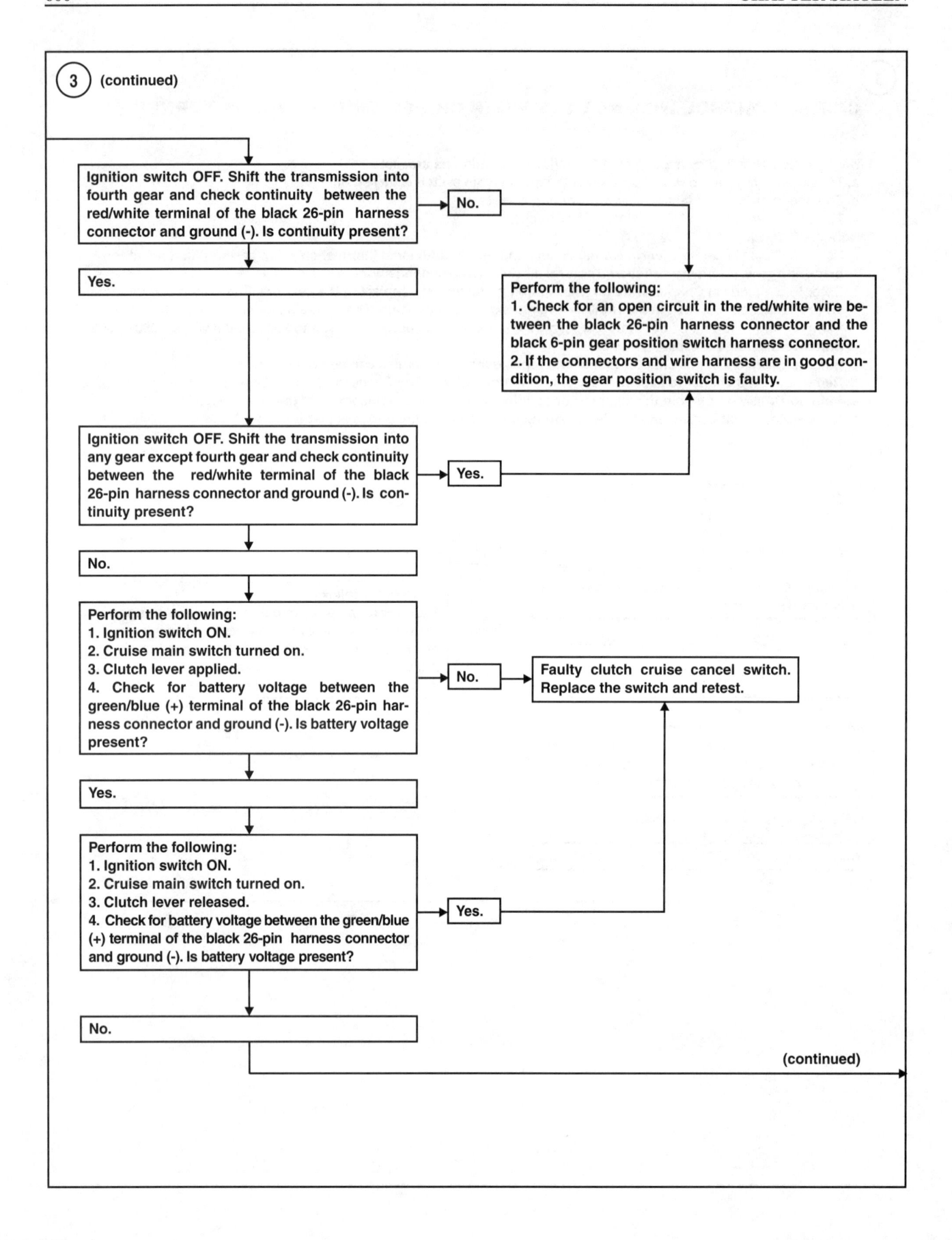

Ignition switch OFF. Shift the transmission into fourth gear and check continuity between the red/white terminal of the black 26-pin harness connector and ground (-). Is continuity present?

→ **No.** → **Perform the following:**
1. Check for an open circuit in the red/white wire between the black 26-pin harness connector and the black 6-pin gear position switch harness connector.
2. If the connectors and wire harness are in good condition, the gear position switch is faulty.

↓ **Yes.**

Ignition switch OFF. Shift the transmission into any gear except fourth gear and check continuity between the red/white terminal of the black 26-pin harness connector and ground (-). Is continuity present?

→ **Yes.** → (to the box above)

↓ **No.**

Perform the following:
1. Ignition switch ON.
2. Cruise main switch turned on.
3. Clutch lever applied.
4. Check for battery voltage between the green/blue (+) terminal of the black 26-pin harness connector and ground (-). Is battery voltage present?

→ **No.** → **Faulty clutch cruise cancel switch. Replace the switch and retest.**

↓ **Yes.**

Perform the following:
1. Ignition switch ON.
2. Cruise main switch turned on.
3. Clutch lever released.
4. Check for battery voltage between the green/blue (+) terminal of the black 26-pin harness connector and ground (-). Is battery voltage present?

→ **Yes.** → (to Faulty clutch cruise cancel switch box)

↓ **No.**

→ **(continued)**

CRUISE CONTROL SYSTEM

③ (continued)

Perform the following:
1. Ignition switch ON.
2. Cruise main switch turned on.
3. SET/DECEL switch free.
4. Check for battery voltage between the white/yellow (+) terminal of the black 26-pin harness connector and ground (-). Is battery voltage present?

→ **No.** → Perform the following:
1. Check for an open circuit in the white/yellow wire between the black 26-pin harness connector and the 18-pin right handlebar switch harness connector.
2. If the connectors and wire harness are in good condition, the SET/DECEL switch is faulty. Replace the right handlebar switch and retest.

Perform the following:
1. Ignition switch ON.
2. Cruise main switch turned on.
3. Push the SET/DECEL switch.
4. Check for battery voltage between the white/yellow (+) terminal of the black 26-pin harness connector and ground (-). Is battery voltage present?

→ **Yes.** ↑

No.

Perform the following:
1. Ignition switch ON.
2. Cruise main switch turned on.
3. Push the SET/DECEL switch.
4. Turn the rear wheel by hand while checking voltage between the white/black (+) terminal of the black 26-pin harness connector and ground (-). Does the battery voltage vary between 0-5 volts?

→ **No.** → Check for an open circuit in the white/black wire between the black 26-pin harness connector and the white 3-pin speed sensor harness connector.

Yes.

Ignition switch ON. Operate the starter button to turn the engine over while checking the voltage between the yellow/green (+) terminal of the black 26-pin harness connector and ground (-). Is the voltage 10.5 volts or higher?

→ **No.** → Check for an open circuit in the yellow/green wire between the black 26-pin harness connector and the gray 22-pin engine control module (ECM) harness connector. (Refer to the wiring diagrams for the ECM harness connector in this circuit.).

Yes.

Ignition switch OFF. Measure resistance between the brown/black terminal of the gray 26-pin harness connector and ground (-). Is the resistance 35-45 ohms?

→ **No.** → Perform the following:
1. Check for an open or short circuit in the brown/black wire between the gray 26-pin harness connector and the black 6-pin cruise actuator harness connector.
2. Check for an open circuit in the green wire between the black 6-pin cruise actuator harness connector and ground.
3. If the wires and connectors in Step 1 and Step 2 are in good condition, the cruise actuator clutch is faulty. Replace the cruise actuator and retest.

Yes.

(continued)

③ (continued)

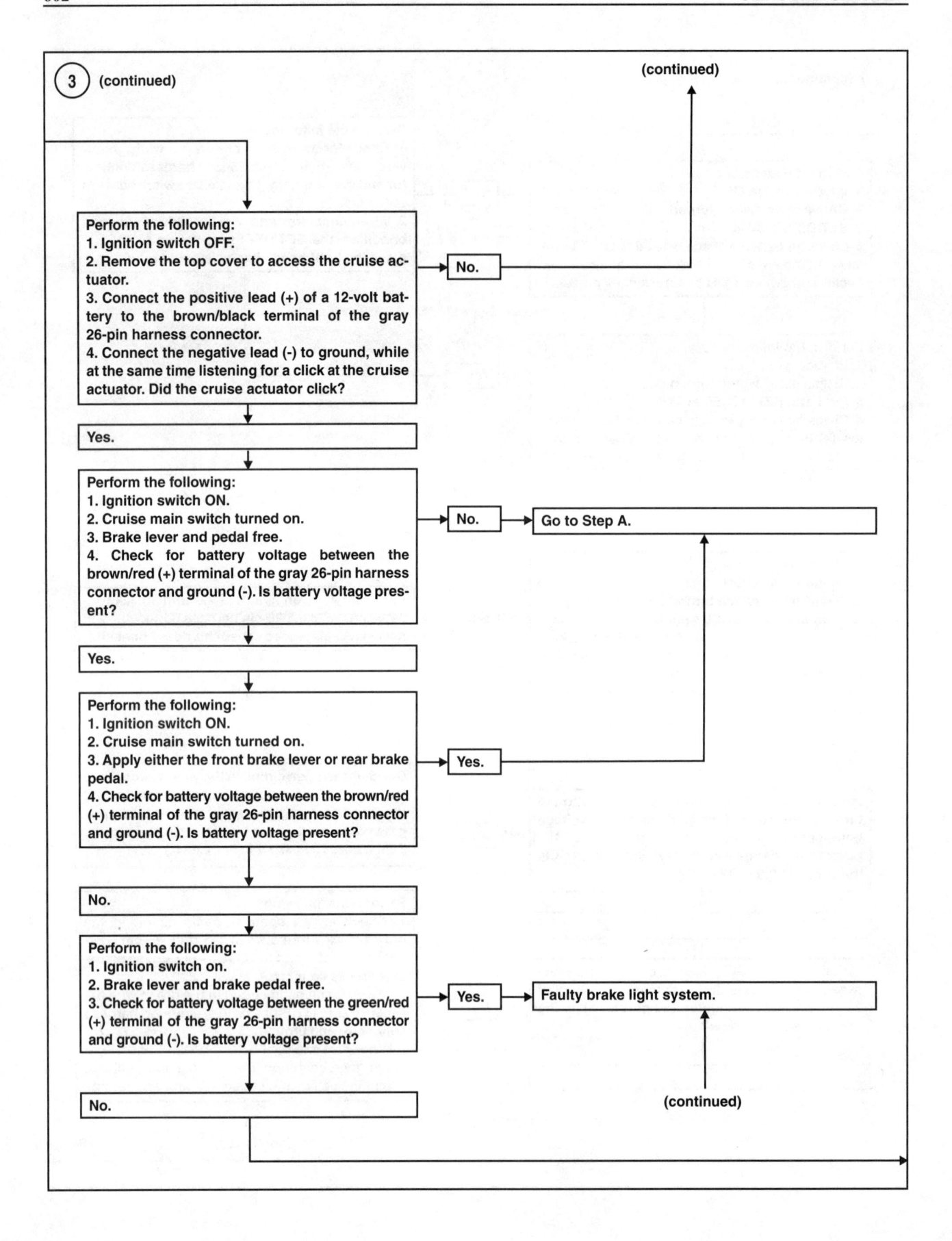

Perform the following:
1. Ignition switch OFF.
2. Remove the top cover to access the cruise actuator.
3. Connect the positive lead (+) of a 12-volt battery to the brown/black terminal of the gray 26-pin harness connector.
4. Connect the negative lead (-) to ground, while at the same time listening for a click at the cruise actuator. Did the cruise actuator click? → **No.** → (continued)

↓ Yes.

Perform the following:
1. Ignition switch ON.
2. Cruise main switch turned on.
3. Brake lever and pedal free.
4. Check for battery voltage between the brown/red (+) terminal of the gray 26-pin harness connector and ground (-). Is battery voltage present? → **No.** → Go to Step A.

↓ Yes.

Perform the following:
1. Ignition switch ON.
2. Cruise main switch turned on.
3. Apply either the front brake lever or rear brake pedal.
4. Check for battery voltage between the brown/red (+) terminal of the gray 26-pin harness connector and ground (-). Is battery voltage present? → **Yes.** → (to Go to Step A)

↓ No.

Perform the following:
1. Ignition switch on.
2. Brake lever and brake pedal free.
3. Check for battery voltage between the green/red (+) terminal of the gray 26-pin harness connector and ground (-). Is battery voltage present? → **Yes.** → Faulty brake light system.

↓ No. → (continued)

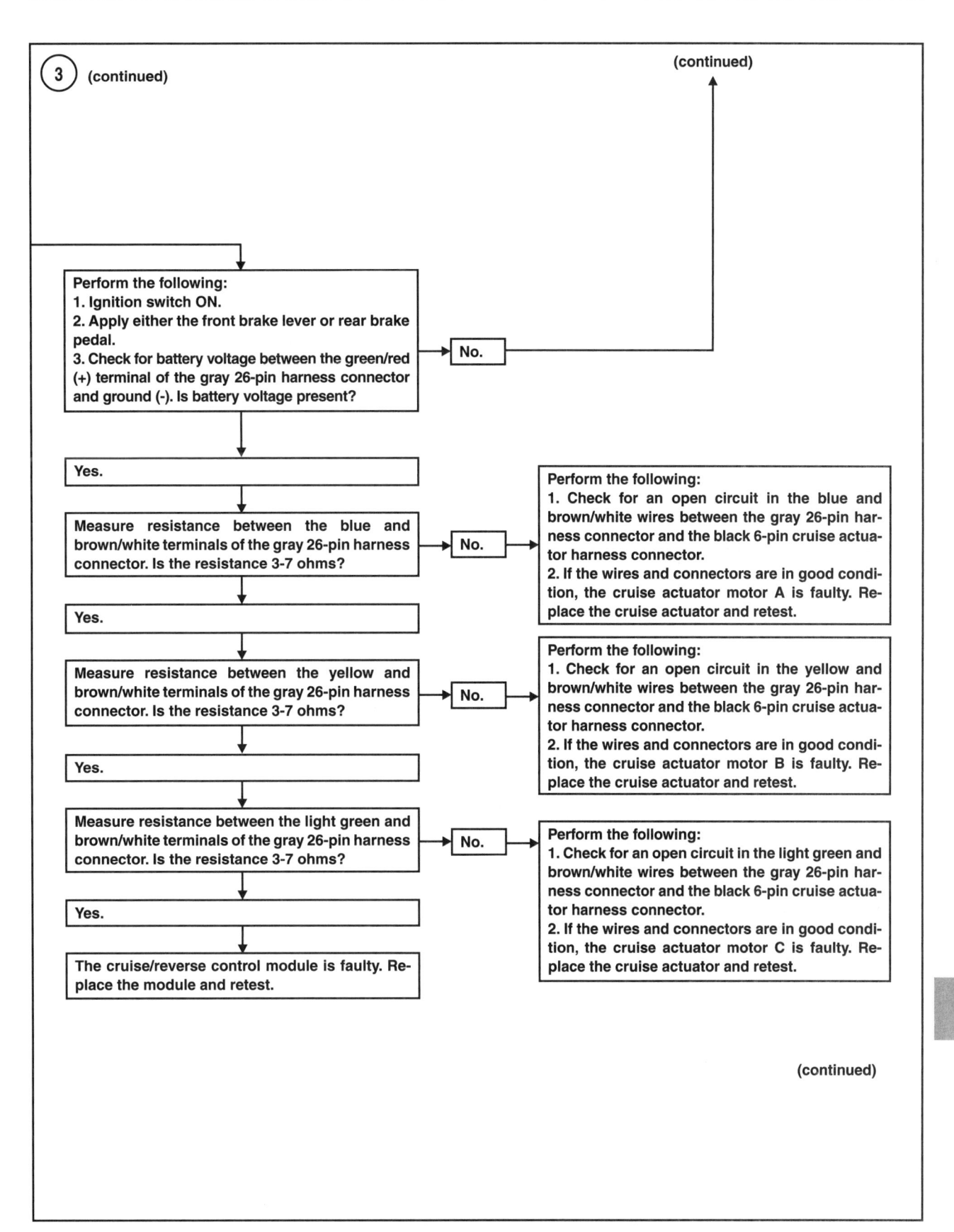

CRUISE CONTROL SYSTEM

③ (continued)   **(continued)**

Perform the following:
1. Ignition switch ON.
2. Apply either the front brake lever or rear brake pedal.
3. Check for battery voltage between the green/red (+) terminal of the gray 26-pin harness connector and ground (-). Is battery voltage present?

→ **No.** → (continued)

↓ **Yes.**

Measure resistance between the blue and brown/white terminals of the gray 26-pin harness connector. Is the resistance 3-7 ohms?

→ **No.** → Perform the following:
1. Check for an open circuit in the blue and brown/white wires between the gray 26-pin harness connector and the black 6-pin cruise actuator harness connector.
2. If the wires and connectors are in good condition, the cruise actuator motor A is faulty. Replace the cruise actuator and retest.

↓ **Yes.**

Measure resistance between the yellow and brown/white terminals of the gray 26-pin harness connector. Is the resistance 3-7 ohms?

→ **No.** → Perform the following:
1. Check for an open circuit in the yellow and brown/white wires between the gray 26-pin harness connector and the black 6-pin cruise actuator harness connector.
2. If the wires and connectors are in good condition, the cruise actuator motor B is faulty. Replace the cruise actuator and retest.

↓ **Yes.**

Measure resistance between the light green and brown/white terminals of the gray 26-pin harness connector. Is the resistance 3-7 ohms?

→ **No.** → Perform the following:
1. Check for an open circuit in the light green and brown/white wires between the gray 26-pin harness connector and the black 6-pin cruise actuator harness connector.
2. If the wires and connectors are in good condition, the cruise actuator motor C is faulty. Replace the cruise actuator and retest.

↓ **Yes.**

The cruise/reverse control module is faulty. Replace the module and retest.

(continued)

③ (continued)

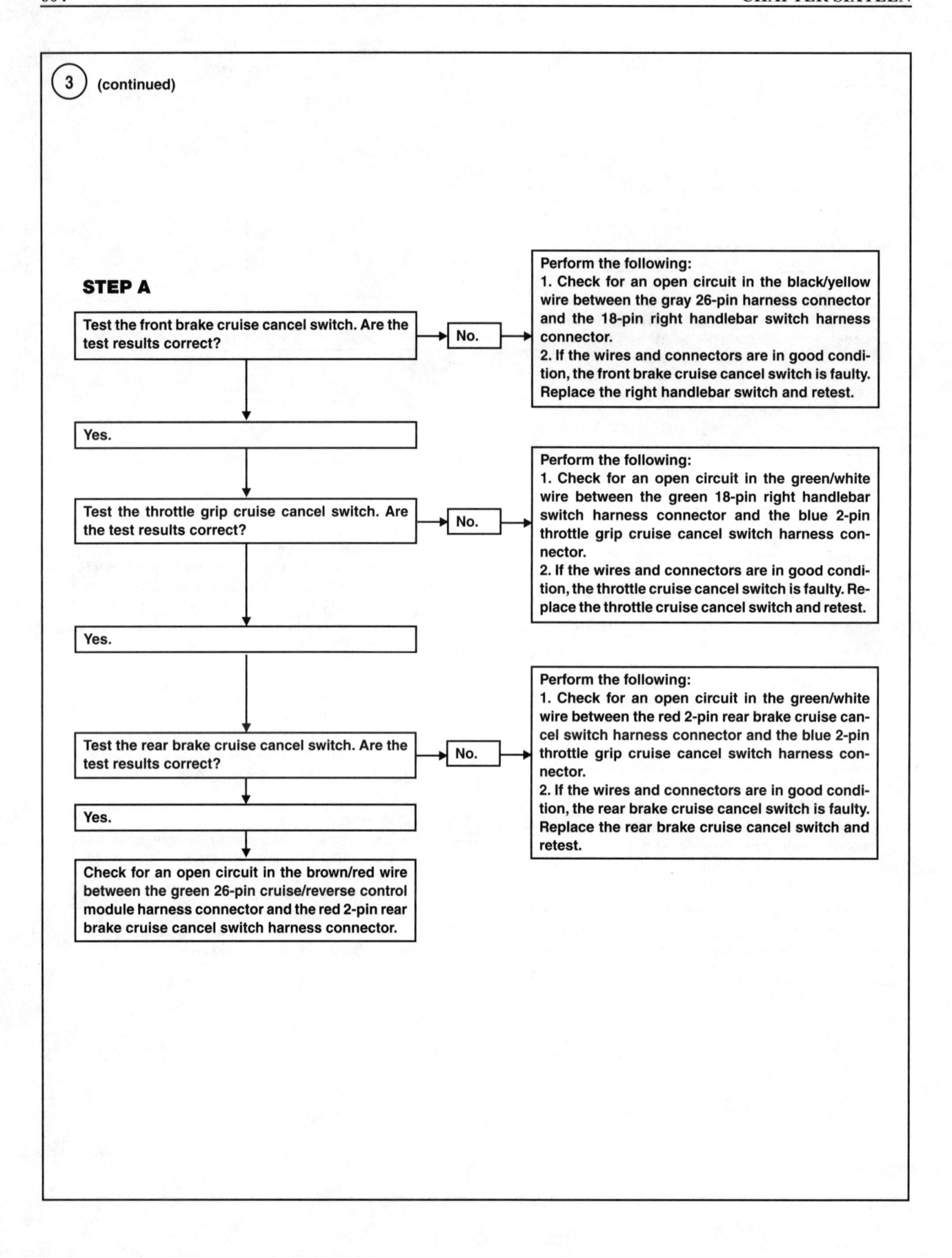

STEP A

Test the front brake cruise cancel switch. Are the test results correct? → **No.** →
Perform the following:
1. Check for an open circuit in the black/yellow wire between the gray 26-pin harness connector and the 18-pin right handlebar switch harness connector.
2. If the wires and connectors are in good condition, the front brake cruise cancel switch is faulty. Replace the right handlebar switch and retest.

↓ Yes.

Test the throttle grip cruise cancel switch. Are the test results correct? → **No.** →
Perform the following:
1. Check for an open circuit in the green/white wire between the green 18-pin right handlebar switch harness connector and the blue 2-pin throttle grip cruise cancel switch harness connector.
2. If the wires and connectors are in good condition, the throttle cruise cancel switch is faulty. Replace the throttle cruise cancel switch and retest.

↓ Yes.

Test the rear brake cruise cancel switch. Are the test results correct? → **No.** →
Perform the following:
1. Check for an open circuit in the green/white wire between the red 2-pin rear brake cruise cancel switch harness connector and the blue 2-pin throttle grip cruise cancel switch harness connector.
2. If the wires and connectors are in good condition, the rear brake cruise cancel switch is faulty. Replace the rear brake cruise cancel switch and retest.

↓ Yes.

Check for an open circuit in the brown/red wire between the green 26-pin cruise/reverse control module harness connector and the red 2-pin rear brake cruise cancel switch harness connector.

CRUISE CONTROL SYSTEM

④

CRUISE ON INDICATOR DOES NOT LIGHT

1. The cruise ON indicator should light when the following conditions are met:
   a. Cruise main switch is pushed to ON.
   b. Ignition switch is turned on.
   c. Engine stop switch is turned to RUN.
2. Note the following:
   a. Refer to *Fuses* in Chapter Nine to inspect the No. 19 ignition cruise fuse.
   b. Refer to *Cruise/Reverse Control Module* in this chapter for additional information when checking voltage at the black 26-pin cruise/reverse control module harness connector.
   c. The cruise control switch assembly consists of three separate switches. These switches are mounted in the right-side handlebar switch housing. Refer to *Switches* in Chapter Nine to test the main cruise control switch and to access the right handlebar switch 18-pin harness connector. (2001-2003: green; 2004: blue).
   d. Refer to *Combination Meter* in Chapter Nine to access the combination meter and its harness connectors.

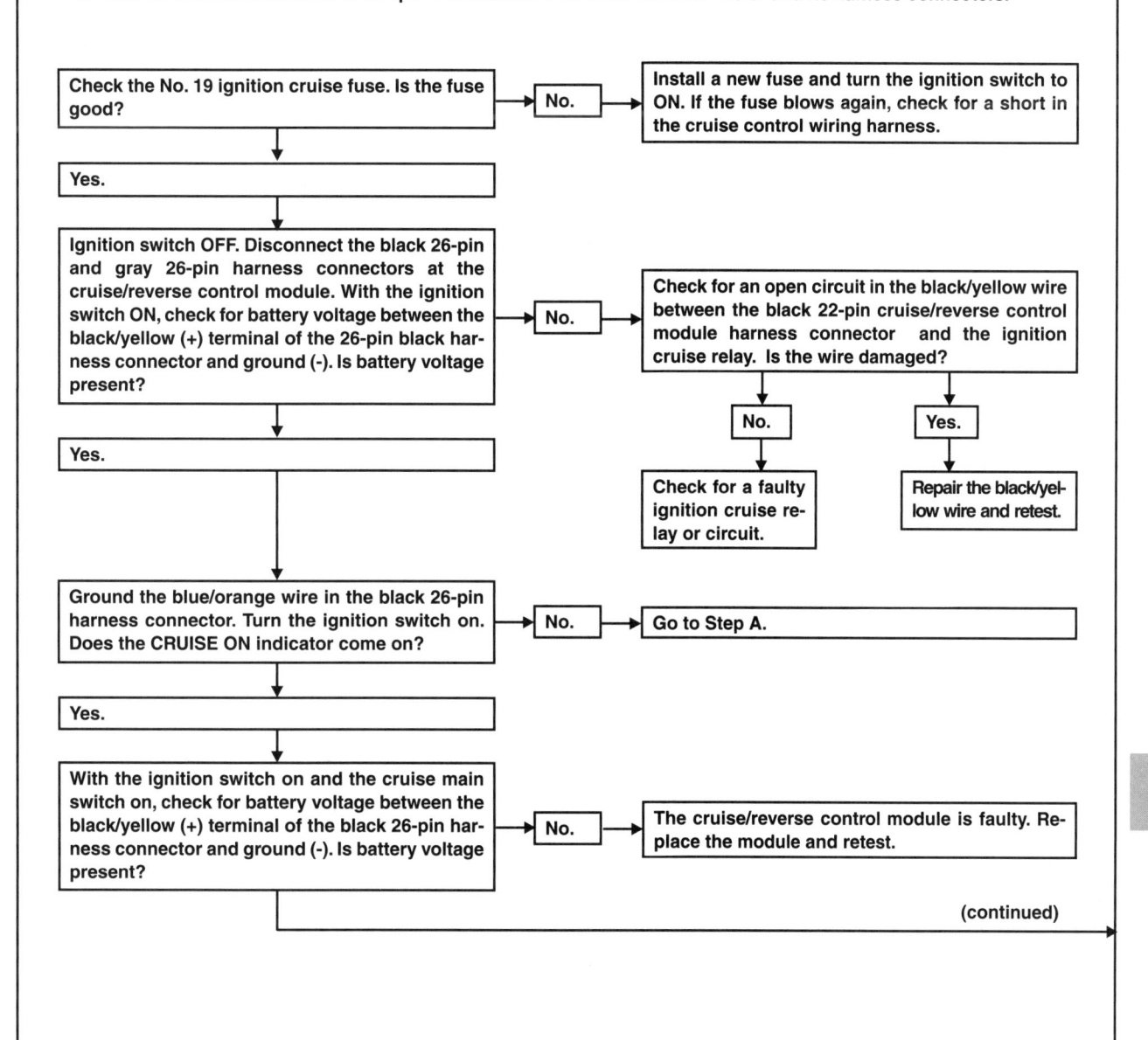

Check the No. 19 ignition cruise fuse. Is the fuse good? → **No.** → Install a new fuse and turn the ignition switch to ON. If the fuse blows again, check for a short in the cruise control wiring harness.

↓ Yes.

Ignition switch OFF. Disconnect the black 26-pin and gray 26-pin harness connectors at the cruise/reverse control module. With the ignition switch ON, check for battery voltage between the black/yellow (+) terminal of the 26-pin black harness connector and ground (-). Is battery voltage present? → **No.** → Check for an open circuit in the black/yellow wire between the black 22-pin cruise/reverse control module harness connector and the ignition cruise relay. Is the wire damaged?

  - **No.** → Check for a faulty ignition cruise relay or circuit.
  - **Yes.** → Repair the black/yellow wire and retest.

↓ Yes.

Ground the blue/orange wire in the black 26-pin harness connector. Turn the ignition switch on. Does the CRUISE ON indicator come on? → **No.** → Go to Step A.

↓ Yes.

With the ignition switch on and the cruise main switch on, check for battery voltage between the black/yellow (+) terminal of the black 26-pin harness connector and ground (-). Is battery voltage present? → **No.** → The cruise/reverse control module is faulty. Replace the module and retest.

(continued)

④ (continued)

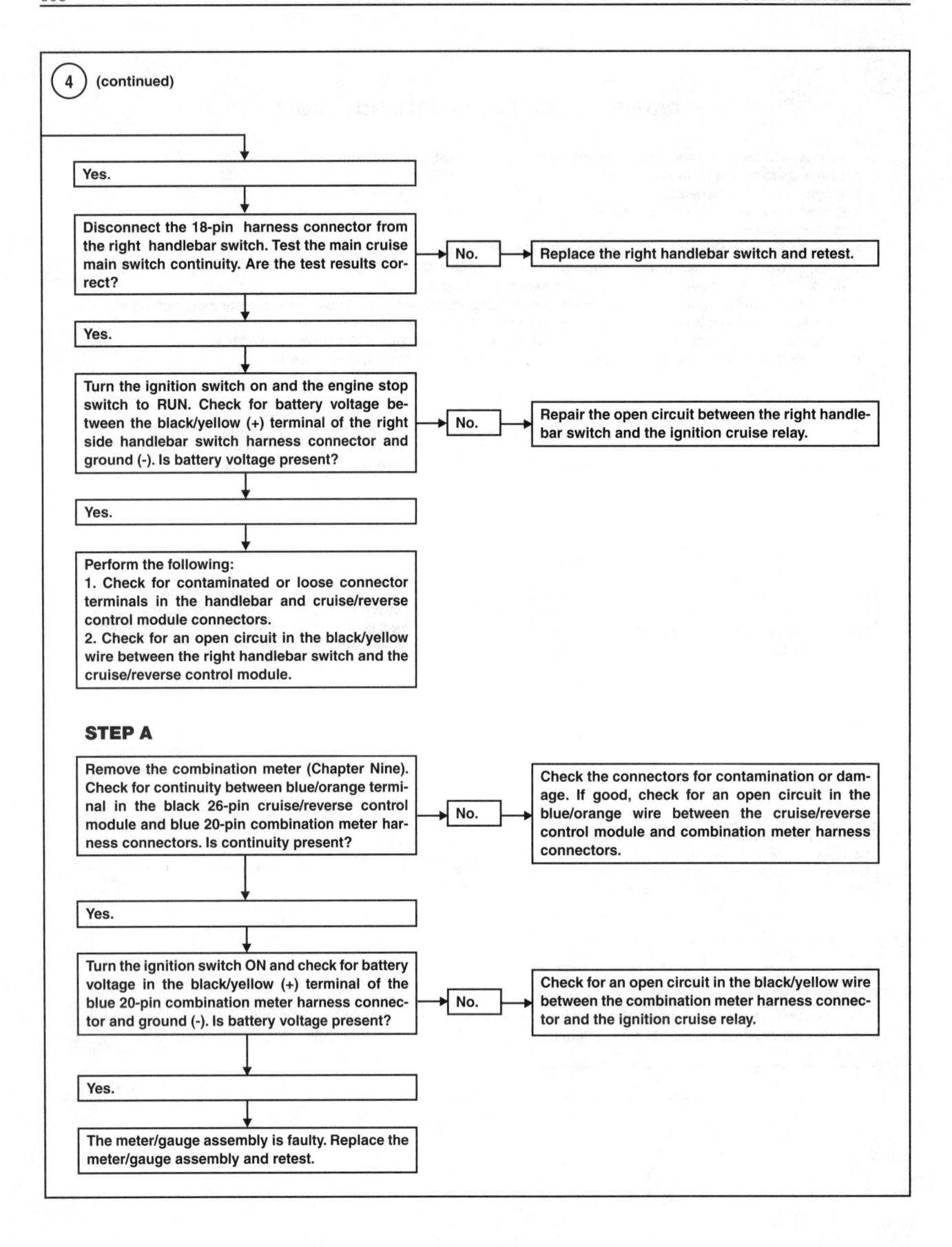

↓

Yes.

↓

Disconnect the 18-pin harness connector from the right handlebar switch. Test the main cruise main switch continuity. Are the test results correct? → **No.** → Replace the right handlebar switch and retest.

↓

Yes.

↓

Turn the ignition switch on and the engine stop switch to RUN. Check for battery voltage between the black/yellow (+) terminal of the right side handlebar switch harness connector and ground (-). Is battery voltage present? → **No.** → Repair the open circuit between the right handlebar switch and the ignition cruise relay.

↓

Yes.

↓

Perform the following:
1. Check for contaminated or loose connector terminals in the handlebar and cruise/reverse control module connectors.
2. Check for an open circuit in the black/yellow wire between the right handlebar switch and the cruise/reverse control module.

STEP A

Remove the combination meter (Chapter Nine). Check for continuity between blue/orange terminal in the black 26-pin cruise/reverse control module and blue 20-pin combination meter harness connectors. Is continuity present? → **No.** → Check the connectors for contamination or damage. If good, check for an open circuit in the blue/orange wire between the cruise/reverse control module and combination meter harness connectors.

↓

Yes.

↓

Turn the ignition switch ON and check for battery voltage in the black/yellow (+) terminal of the blue 20-pin combination meter harness connector and ground (-). Is battery voltage present? → **No.** → Check for an open circuit in the black/yellow wire between the combination meter harness connector and the ignition cruise relay.

↓

Yes.

↓

The meter/gauge assembly is faulty. Replace the meter/gauge assembly and retest.

CRUISE CONTROL SYSTEM

⑤ CRUISE ON INDICATOR FLASHES WHEN CRUISE MAIN SWITCH IS ON

1. If the cruise actuator is faulty, the cruise control system will operate in a fail-safe mode—when the cruise main switch is pushed to ON, the ignition switch is turned to ON, and the engine stop switch to RUN, the cruise ON indicator will flash.
2. The cruise ON indicator should light when the following conditions are met:
   a. Cruise main switch is pushed to ON.
   b. Ignition switch is turned on.
   c. Engine stop switch is turned to RUN.
3. Note the following:
   a. Refer to *Cruise/Reverse Control Module* in this chapter when checking voltage at the gray 26-pin cruise/reverse control module harness connector.
   b. Refer to *Ignition Cruise Relays* in this chapter to test and replace the cruise actuator relay.

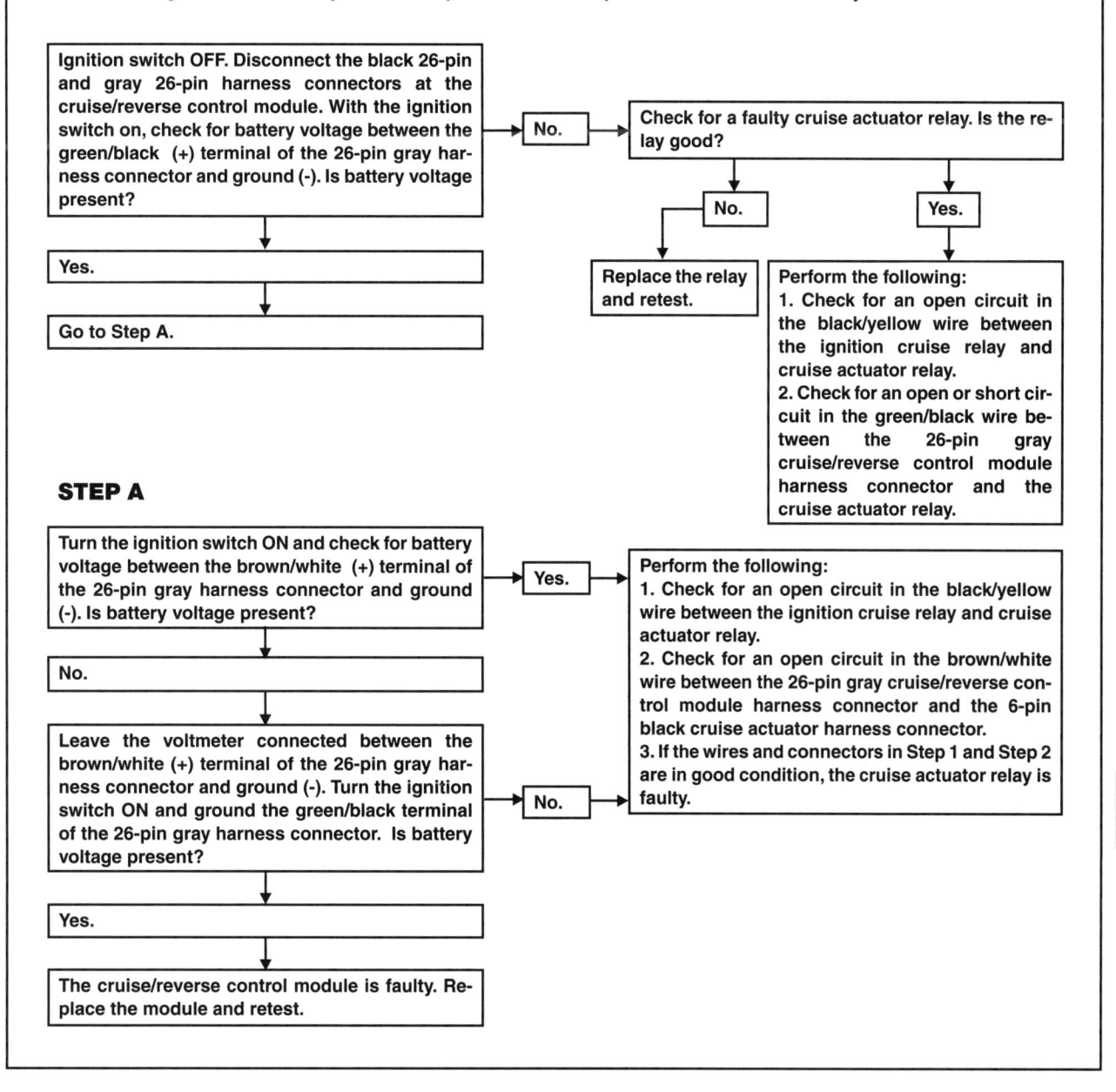

Ignition switch OFF. Disconnect the black 26-pin and gray 26-pin harness connectors at the cruise/reverse control module. With the ignition switch on, check for battery voltage between the green/black (+) terminal of the 26-pin gray harness connector and ground (-). Is battery voltage present?

→ **No.** → **Check for a faulty cruise actuator relay. Is the relay good?**
  - **No.** → Replace the relay and retest.
  - **Yes.** → Perform the following:
    1. Check for an open circuit in the black/yellow wire between the ignition cruise relay and cruise actuator relay.
    2. Check for an open or short circuit in the green/black wire between the 26-pin gray cruise/reverse control module harness connector and the cruise actuator relay.

↓ **Yes.**

Go to Step A.

STEP A

Turn the ignition switch ON and check for battery voltage between the brown/white (+) terminal of the 26-pin gray harness connector and ground (-). Is battery voltage present?

→ **Yes.** → Perform the following:
1. Check for an open circuit in the black/yellow wire between the ignition cruise relay and cruise actuator relay.
2. Check for an open circuit in the brown/white wire between the 26-pin gray cruise/reverse control module harness connector and the 6-pin black cruise actuator harness connector.
3. If the wires and connectors in Step 1 and Step 2 are in good condition, the cruise actuator relay is faulty.

↓ **No.**

Leave the voltmeter connected between the brown/white (+) terminal of the 26-pin gray harness connector and ground (-). Turn the ignition switch ON and ground the green/black terminal of the 26-pin gray harness connector. Is battery voltage present?

→ **No.** → (same perform the following as above)

↓ **Yes.**

The cruise/reverse control module is faulty. Replace the module and retest.

6

CRUISE ON INDICATOR FLASHES WHEN SPEED REACHES 16 MPH (25 KM/H) WHEN CRUISE MAIN SWITCH IS TURNED ON

1. If the cruise actuator is faulty, the cruise control system will operate in a fail-safe mode—when the cruise main switch is pushed to ON, the ignition switch is turned to ON, and the engine stop switch to RUN, the cruise ON indicator will flash.
2. Note the following:
   a. Refer to *Cruise/Reverse Control Module* in this chapter when performing tests at the gray 26-pin cruise/reverse control module harness connector.
   b. Remove the top shelter to access the cruise actuator. Refer to *Top Shelter* in Chapter Seventeen.
   c. A separate 12-volt battery and test leads are required to check the cruise actuator.
   d. Refer to *Cruise Actuator* in this chapter for additional testing and replacement of the cruise actuator.

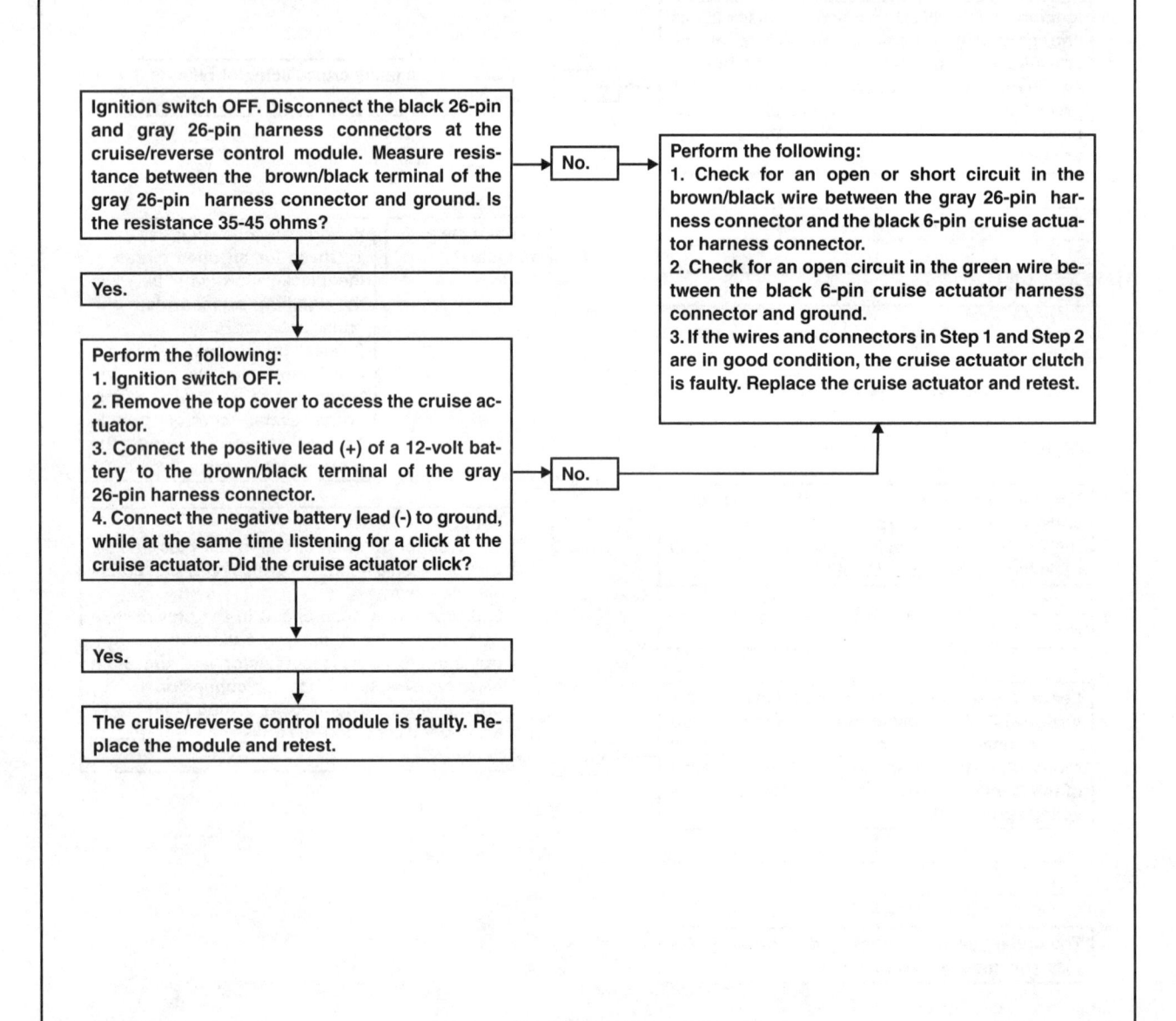

Ignition switch OFF. Disconnect the black 26-pin and gray 26-pin harness connectors at the cruise/reverse control module. Measure resistance between the brown/black terminal of the gray 26-pin harness connector and ground. Is the resistance 35-45 ohms?

→ **No.** → Perform the following:
1. Check for an open or short circuit in the brown/black wire between the gray 26-pin harness connector and the black 6-pin cruise actuator harness connector.
2. Check for an open circuit in the green wire between the black 6-pin cruise actuator harness connector and ground.
3. If the wires and connectors in Step 1 and Step 2 are in good condition, the cruise actuator clutch is faulty. Replace the cruise actuator and retest.

↓ **Yes.**

Perform the following:
1. Ignition switch OFF.
2. Remove the top cover to access the cruise actuator.
3. Connect the positive lead (+) of a 12-volt battery to the brown/black terminal of the gray 26-pin harness connector.
4. Connect the negative battery lead (-) to ground, while at the same time listening for a click at the cruise actuator. Did the cruise actuator click?

→ **No.** → (see box above)

↓ **Yes.**

The cruise/reverse control module is faulty. Replace the module and retest.

CRUISE CONTROL SYSTEM

age tester. Refer to Chapter Nine for peak voltage test procedures.

7. When having to shift the reverse actuator from neutral into reverse:
   a. Turn the ignition switch off. Reconnect the control module connectors.
   b. Shift transmission into neutral, then turn the ignition switch on.
   c. Push the reverse shift switch to on. The reverse position switch turns to reverse and the reverse indicator lamp turns on.
   d. Turn the ignition switch off and disconnect the control module connectors. Perform the test as described in **Figure 8**.

8. When having to shift the reverse actuator from reverse into neutral:
   a. Turn the ignition switch off. Reconnect the control module connectors.
   b. Turn the ignition switch on.
   c. Turn the reverse switch off. The reverse position switch moves to the neutral position and the neutral indicator turns on.
   d. Turn the ignition switch off and disconnect the control module connectors. Perform the test as described in **Figure 8**.

9. Reverse Steps 1-3 to complete installation.

CRUISE/REVERSE CONTROL MODULE

The cruise control module is an integral part of the reverse control module.

Main Wiring Harness Check

This section tests the cruise/reverse control module harness connectors and individual wires.

1. Turn the ignition switch off.
2. Remove the top shelter. Refer to *Top Shelter* in Chapter Seventeen.
3. Disconnect the black 26-pin (A, **Figure 7**) and gray 26-pin (B) harness connectors from the cruise/reverse control module. Press the locking tab on the bottom of each connector and pull on the connector to disconnect.
4. Inspect the connector and pins to make sure they are making good contact.
5. Check the connectors for any bent or corroded pins. Clean and repair any pins as required, and recheck the system. If the pins are in good condition, continue with Step 6.
6. Perform the input tests described in **Figure 8** at the black 26-pin and gray 26-pin harness side connectors. Note the following:
   a. Unless otherwise specified, perform the test procedures with the ignition switch off.
   b. When an incorrect reading is obtained, check the connector and related wiring harness for loose or contaminated terminals.
   c. When testing the engine speed pulse circuit (connector No. 21), measure voltage using a peak volt-

Control Module Replacement

1. Turn the ignition switch off.
2. Remove the top shelter. See Chapter Seventeen under *Top Shelter*.
3. Disconnect the black 26-pin (A, **Figure 7**) and gray 26-pin (B) harness connectors from the cruise/reverse control module. Press the locking tab on the bottom of each connector and pull on the connector to disconnect.
4. Carefully lift the holder clamps (**Figure 9**) and remove the cruise/reverse control module. See **Figure 10**.
5. Installation is the reverse of these steps.

CRUISE CONTROL RELAYS

The cruise actuator relay (2, **Figure 11**) and ignition cruise relay (6) control the cruise control system. These relays are located in the main relay box underneath the seat.

The relays are easily unplugged or plugged into the relay terminal strip. Because only 4-terminal and 5-terminal relays are used, there is no danger of installing a relay in the wrong position. The 4-terminal relays are identical, and the 5-terminal relays are identical. Refer to *Relay Box*

⑧ CRUISE/REVERSE CONTROL MODULE PIN TESTING

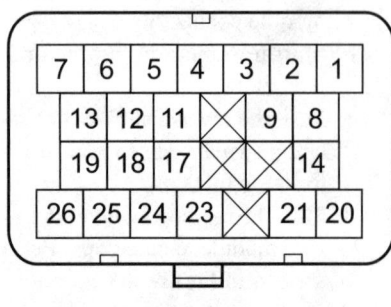

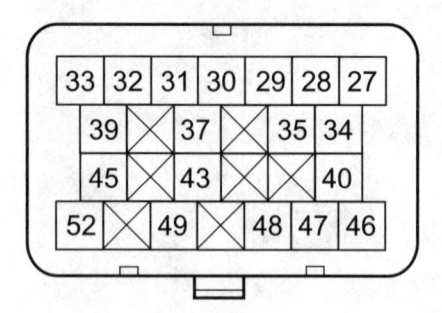

```
 7  6  5  4  3  2  1         33 32 31 30 29 28 27
13 12 11  X  9  8             39  X 37  X 35 34
19 18 17  X  X 14             45  X 43  X  X 40
26 25 24 23  X 21 20          52  X 49  X 48 47 46
```

Black 26-pin   Grey 26-pin

Terminal No./Circuit	Wire Color	Test Conditions	Results
1. Reverse shift motor (+) circuit.	Control module pink and pink wires and reverse shift relays 2 and 3 connectors.	Remove reverse shift relays 2 and 3. Check continuity between the same wire color terminals.	Continuity.
2. Gear position switch (overdrive).	Green/orange and ground.	1. Transmission shifted into any gear except fifth. 2. Transmission shifted into fifth gear.	1. No continuity. 2. Continuity.
3. Gear position switch (fourth gear).	Red/white and ground.	1. Transmission shifted into fourth gear. 2. Transmission shifted into any gear except fourth.	1. Continuity. 2. No continuity.
4. Clutch cruise cancel switch.	Green/blue (+) and ground (-).	1. Ignition switch ON; cruise main switch ON; clutch lever applied. 2. Ignition switch ON; cruise main switch ON; clutch lever released.	1. Battery voltage. 2. No battery voltage.
5. RESUME/ACCEL switch.	White/blue (+) and ground (-).	1. Ignition switch ON; cruise main switch ON; RESUME/ACCEL switch pushed. 2. Ignition switch ON; cruise main switch ON; RESUME/ACCEL switch free.	1. Battery voltage. 2. No battery voltage.

(continued)

CRUISE CONTROL SYSTEM

⑧ (continued)

Terminal No./Circuit	Wire Color	Test Conditions	Results
6. SET/DECEL switch.	White/yellow (+) and ground (-).	1. Ignition switch ON; cruise main switch ON; SET/DECEL switch pushed. 2. Ignition switch ON; cruise main switch ON; SET/CECEL switch free.	1. Battery voltage. 2. No battery voltage.
7. Speed sensor.	White/black (+) and ground (-).	Ignition switch ON. Turn rear wheel slowly.	0 to 5 volts (pulse reading).
8. Reverse shift motor ground circuit.	Green and ground.	–	Continuity.
9. Cruise main switch.	Black/yellow (+) and ground (-).	Ignition switch ON. Cruise main switch ON.	Battery voltage.
11. Sidestand switch.	Green/white and ground.	1. Sidestand up. 2. Sidestand down.	1. Continuity. 2. No continuity.
12. Starter/reverse motor.	Pink/white and starter/reverse motor cable terminals of starter relay switch B.	–	Continuity.
13. CRUISE ON indicator.	Blue/orange.	Ground blue/orange with jumper wire and turn ignition switch on.	CRUISE ON indicator should light.
14. Cruise power input.	Black/yellow (+) and ground (-).	Ignition switch ON.	Battery voltage.
17. Reverse switch.	Green/orange and ground.	1. Reverse shift actuator in NEUTRAL. 2. Reverse shift actuator in REVERSE.	1. Continuity. 2. No continuity.
18. Starter/reverse switch.	Yellow/red (+) and ground (-).	Ignition switch ON; engine stop switch in RUN; starter/reverse switch pushed.	Battery voltage.
19. Speed limiter fuse.	Yellow and ground.	–	Continuity.
20. Ground.	Green and ground.	–	Continuity.
21. Engine speed pulse.	Yellow/green (+) and ground (-).	Turn engine over with starter motor.	Voltage reading 10.5 volts or higher. (Measure with peak voltage tester).
23. Reverse switch.	Yellow/white (+) and ground (-).	Ignition switch ON. Reverse switch ON.	Battery voltage.
24. Reverse indicator.	White/red (+) and ground (-).	Ground white/red with jumper wire with ignition switch ON.	Reverse indicator should light.
25. Reverse position switch.	White/blue (+) and ground (-).	Reverse shift actuator in reverse position. Ignition switch ON.	Battery voltage.

(continued)

⑧ (continued)

Terminal No./Circuit	Wire Color	Test Conditions	Results
26. CRUISE SET indicator.	Blue/white.	Ground blue/white wire with jumper wire. Ignition switch ON.	CRUISE SET indicator should light.
27. Reverse shift relay 3.	1. Yellow/white and yellow/white of the control module and reverse shift relay 3. 2. Yellow/white and ground.	Remove reverse shift relay 3. Check for continuity between same color terminals.	1. Continuity. 2. No continuity.
28. Power control relay 1.	Orange (+) and ground (-).	Reverse shift actuator in reverse. Ignition switch ON.	Battery voltage.
29. Speed limiter relay.	Gray (+) and ground (-).	Reverse shift actuator in reverse. Ignition switch on.	Battery voltage.
30. Cruise actuator clutch.	Brown/black and ground.	1. – 2. Connect 12-volt battery to brown/black (+) and ground (-).	1. 35-45 ohms. 2. Actuator clutch should click.
31. Cruise cancel switch.	Brown/red (+) and ground (-).	1. Ignition switch ON; cruise main switch ON; brake lever and brake pedal free. 2. Ignition switch ON; cruise main switch ON; brake lever and brake pedal applied.	1. Battery voltage.
32. Cruise actuator relay output.	Brown/white (+) and ground (-).	1. Ignition switch ON. 2. Ground No. 50 green/black terminal with jumper wire. Ignition switch ON.	1. 35-45 ohms. 2. Actuator clutch should click.
33. Cruise actuator motor A.	Blue and No. 32 brown/white.	–	3-7 ohms.
34. Reverse shift relay 2.	1. Blue/white and blue/white terminals at the control module and reverse shift relay 2. 2. Blue/white and ground.	1. Remove reverse shift relay 2 and check for continuity between same color terminals. 2. –	1. Continuity. 2. No continuity.
35. Brake light.	Green/red (+) and ground (-).	1. Ignition switch ON. Brake lever and brake pedal free. 2. Ignition switch ON. Brake lever and brake pedal applied.	1. No voltage. 2. Voltage.

(continued)

CRUISE CONTROL SYSTEM

⑧ (continued)

Terminal No./Circuit	Wire Color	Test Conditions	Results
37. Reverse regulator	Light blue and light blue of the reverse regulator and control module terminators.	Remove the reverse regulator and check for continuity between these same color wire terminals.	Continuity.
39. Cruise actuator motor B.	Yellow and No. 32 Brown/white.	–	3-7 ohms.
40. Reverse shift relay 1.	Blue/black (+) and ground (-).	Ignition switch ON.	Battery voltage.
43. Reverse relay.	Yellow/red (+) and ground (-).	Ignition switch ON.	Battery voltage.
45. Cruise actuator motor B.	Light green and No. 32 brown/white.	–	3-7 ohms.
46. Power control relay 2.	White (+) and ground (-).	Reverse shift actuator in reverse. Ignition switch ON.	Battery voltage.
47. Neutral switch.	Light green/red and ground.	Transmission in NEUTRAL.	Continuity.
48. Oil pressure switch.	Green and ground.	1. Engine off. 2. Engine running at idle speed.	1. Continuity. 2. No continuity.
50. Cruise actuator relay coil.	Green/black (+) and ground (-).	Ignition switch ON.	Battery voltage.
52. Ground.	Green and ground.	–	Continuity.

*Starter/reverse and cruise control tests are grouped by number as follows:
Starter/reverse system: 1, 8, 11, 12, 17, 18, 19, 20, 23, 24, 25, 27, 28, 29, 34, 37, 40, 43, 46, 47, 48 and 52.
Cruise system: 2, 3, 4, 5, 6, 7, 9, 13, 14, 20, 21, 26, 30, 31, 32, 33, 35, 39, 45, 50 and 52.

RELAY BOX

1	2	3	4	Blank	5	6
8	9	10	11	12	13	14
15	16	17	18	19	20	21

------- 4-Terminal relays ------- | ---- 5-Terminal relays ----

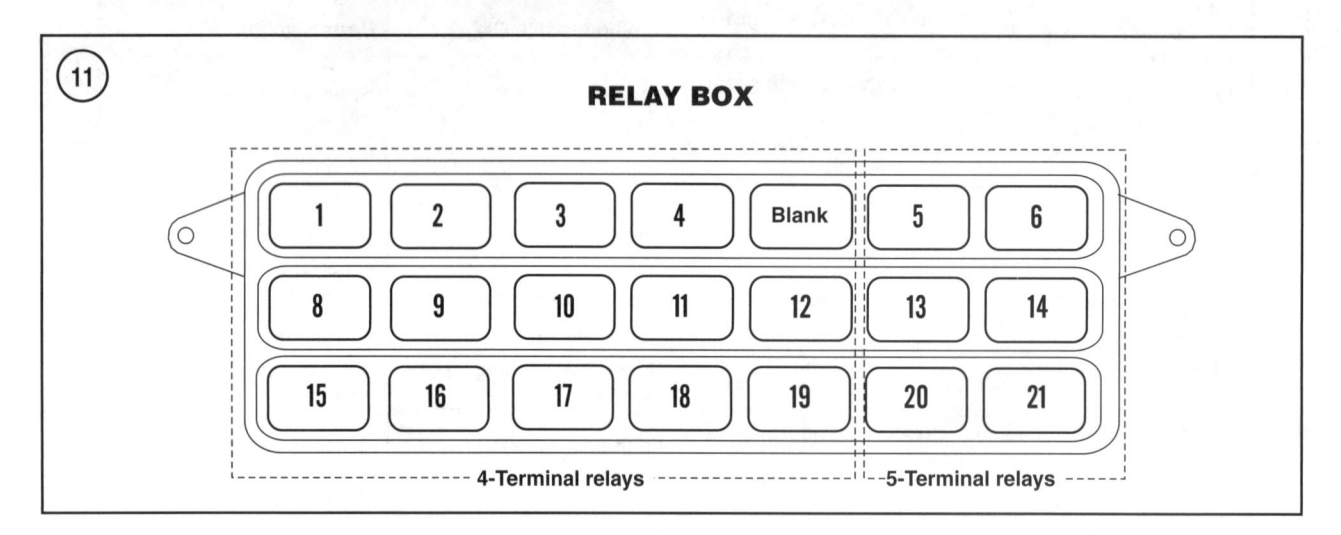

in Chapter Nine for more information on identifying and testing the relays.

NOTE
A questionable relay can be quickly checked by exchanging it with a known good relay. To do this, first identify the relay as either a 4-terminal or 5-terminal type. Unplug the relay and plug in a known good relay.

Relay and Circuit Testing

Refer to the wiring diagrams at the end of the manual to identify the cruise control wire and fuse circuits referred to in this procedure.

1. Turn the ignition switch off.
2. Remove the seat. Refer to *Seat* in Chapter Seventeen.
3. Remove the relay box cover screws and cover (**Figure 12**).

NOTE
*The meter test connections called out in Steps 4-8 are made at the relay connector terminals inside the relay box (**Figure 13**).*

4. Turn the ignition switch off. Check continuity between the cruise actuator relay black/yellow No. 1 terminal and

CRUISE CONTROL SYSTEM

13

Relay box — 2 Cruise actuator relay — Ignition cruise relay — Grn

| 1 | | 3 | 4 | Blank | | 7 |

Blk/yel (terminal No. 1) — Blk/yel (terminal No. 3) — Blk/yel (terminal No. 1) — Yel — Blk/yel (terminal No. 3)

14

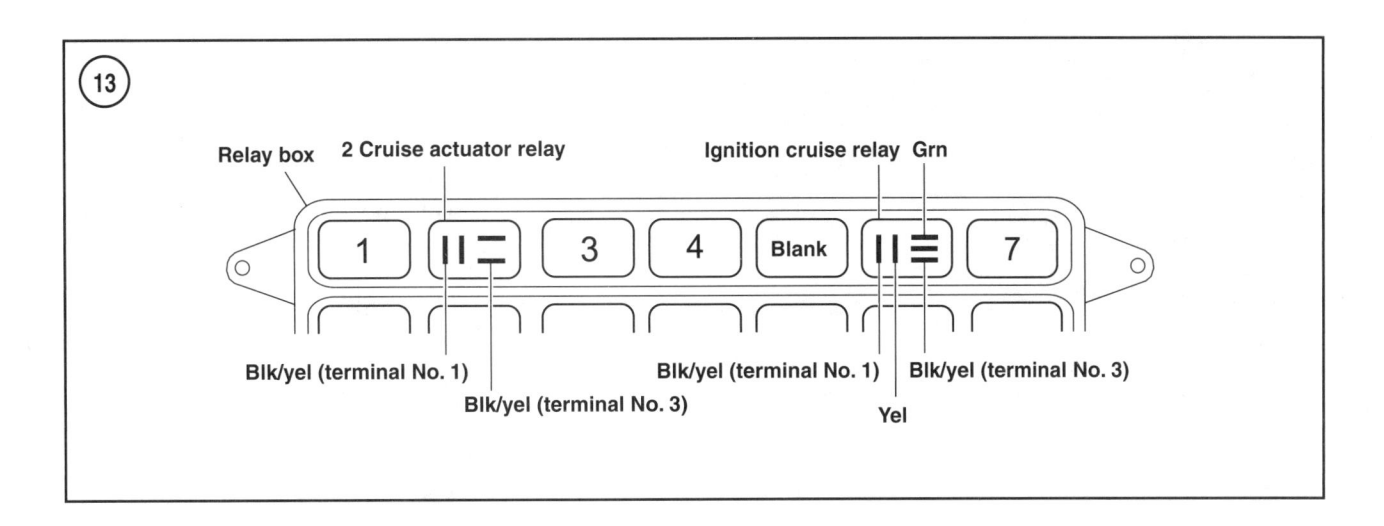

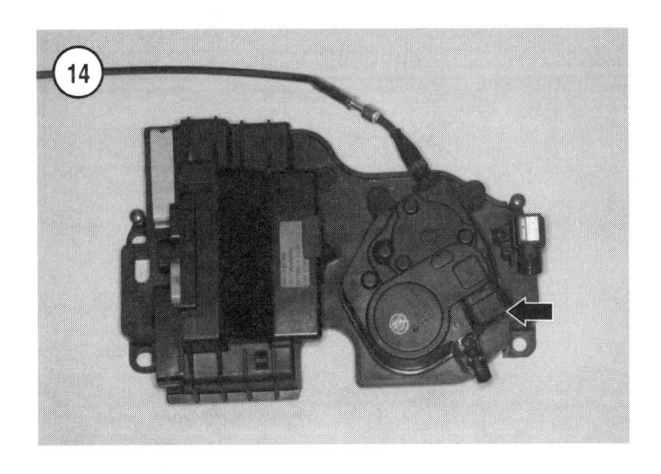

the ignition cruise relay black/yellow No. 1 terminal. Then repeat the test between the cruise actuator relay black/yellow No. 3 terminal and the ignition cruise relay black/yellow No. 3 terminal:

a. Continuity recorded at both tests: Go to Step 5.

b. No continuity recorded at one or both tests: Check for an open circuit in the black/yellow wires.

5. Replace the ignition cruise relay (6, **Figure 11**) with a known good 5-terminal relay. Turn the ignition switch on and measure voltage between the cruise actuator black/yellow No. 3 terminal (+) and ground (–):

a. Battery voltage: Replace the original ignition cruise relay with a new relay.

b. No battery voltage: Go to Step 6.

6. Turn the ignition switch off. Remove the ignition cruise relay. Check continuity between the ignition cruise relay green terminal and ground:

a. Continuity: Go to Step 7.

b. No continuity: Check for an open circuit in the green wire between the relay box and the cruise actuator black 6-pin connector.

7. Turn the ignition switch on. Measure voltage between the ignition cruise relay yellow terminal (+) and ground (–):

a. Battery voltage: Go to Step 8.

b. No battery voltage: Check for an open circuit in the yellow wire between the ignition cruise relay and the fuse box (at the No. 19 fuse).

8. Turn the ignition switch on. Measure voltage between the ignition cruise relay black/yellow No. 3 terminal (+) and ground (–):

a. Battery voltage: The relay circuit is good. Check for loose or dirty contacts.

b. No battery voltage: Check for an open circuit in the black/yellow wire between PGM-FI ignition relay and the ignition cruise relay.

9. Turn the ignition switch off and reinstall the relays.

10. Install the relay box cover (**Figure 12**) and mounting screws.

11. Install the seat. Refer to *Seat* in Chapter Seventeen.

CRUISE ACTUATOR

The cruise actuator is mounted onto the control module holder. **Figure 14** shows the module removed for clarity. The cruise actuator cable is permanently fixed to the cruise actuator. Do not attempt to remove it.

Testing

1. Remove the top shelter. Refer to *Top Shelter* in Chapter Seventeen.

2. Disconnect the black 6-pin cruise actuator connector (**Figure 15**).
3. Connect a 12-volt battery to the cruise actuator No. 6 terminal (+) and to the No. 3 terminal (–) (**Figure 16**). The cruise actuator should click. Disconnect the battery leads.
4. Measure the resistance between the No. 1 and No. 2 terminals and then between the No. 1 and No. 5 terminals (**Figure 16**). The standard resistance reading for both tests is 3-5 ohms.
5. If the cruise actuator failed either test, replace it as described in this section.
6. If the cruise actuator passed both tests, reverse Step 1 and Step 2 to complete installation.

Replacement

1. Remove the top shelter. Refer to *Top Shelter* in Chapter Seventeen.
2. Remove the air filter housing. Refer to *Air Filter Housing* in Chapter Eight.
3. Disconnect the actuator cable from the cable stay, then disconnect it from the actuator drum on the throttle body (**Figure 17**).
4. Turn the control module holder (**Figure 14**) over and remove the three cruise actuator mounting screws (**Figure 18**) and cruise actuator.
5. Install the new cruise actuator and secure with the three mounting screws.
6. Connect the new cruise actuator cable onto the actuator drum and against the throttle body cable stay (**Figure 17**). Position the cable so the distance from the throttle body cable stay to the end of the cable cap is 34 mm (1.34 in.) as shown in **Figure 19**. Tighten the locknuts and recheck the adjustment.
7. Reinstall the air filter housing.
8. Perform the *Cruise Actuator Cable Adjustment* in this section.
9. Install the top shelter.

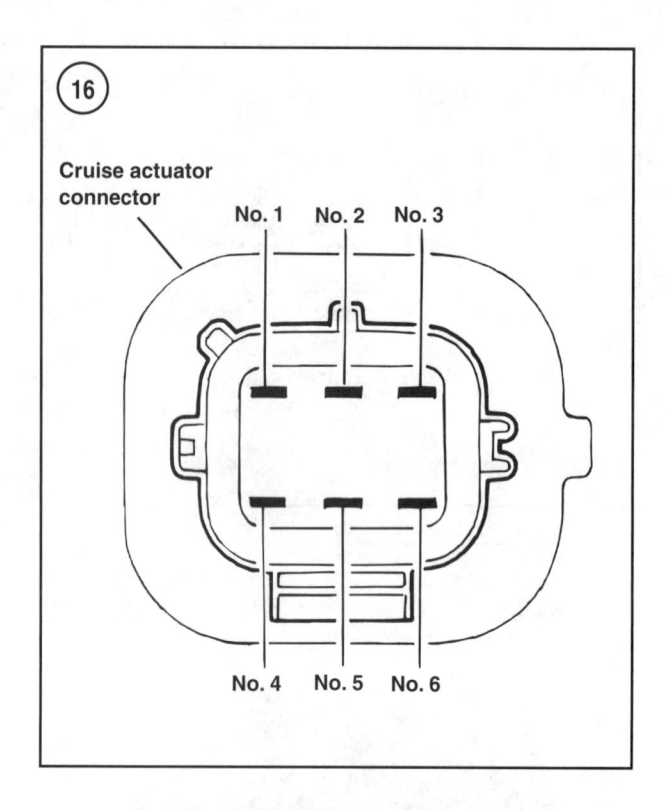

CRUISE CONTROL SYSTEM

Figure 19: Throttle body cable stay, 34 mm (1.34 in.), Cable cap end, Cruise actuator cable, Locknuts

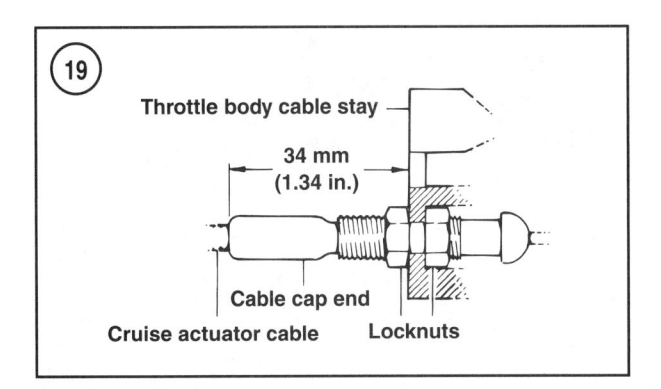

Figure 21: Actuator drum, Throttle drum

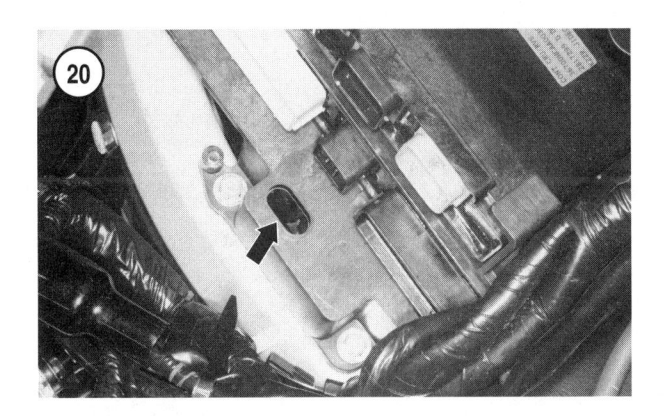

3. Locate the inspection hole (**Figure 20**) in the front part of the control module holder.

4. Direct a flashlight through the inspection hole. The point on the actuator drum should align with the groove in the throttle drum (**Figure 21**). If the alignment is incorrect, loosen the locknut and turn the cruise actuator cable adjuster (**Figure 22**) as required to make the alignment. Tighten the locknut and recheck the alignment.

5. Install the top shelter.

CRUISE CONTROL SWITCHES

Cruise Control Switch Assembly

The cruise control switch assembly is mounted in the right-side switch housing (**Figure 1**):
1. Cruise control main switch (A).
2. SET/DECEL switch (B).
3. RESUME/ACCEL switch (C).

Refer to *Handlebar Switches* in Chapter Nine under to test these switches.

Clutch Cancel Switch Testing/Replacement

The clutch cancel switch (A, **Figure 23**) is mounted underneath the clutch switch.

1. Disconnect the connectors (B, **Figure 23**).
2. Check for continuity between the switch terminals. There should be no continuity with the clutch lever released and continuity with the clutch lever applied. Replace the switch if faulty.
3. Turn the ignition switch on and push the cruise control main switch (A, **Figure 1**). Measure voltage between the black/yellow wire terminal (+) and ground (−). There should be battery voltage. Turn the ignition switch off.
4. Remove the screw and switch (**Figure 24**).
5. Installation is the reverse of removal.

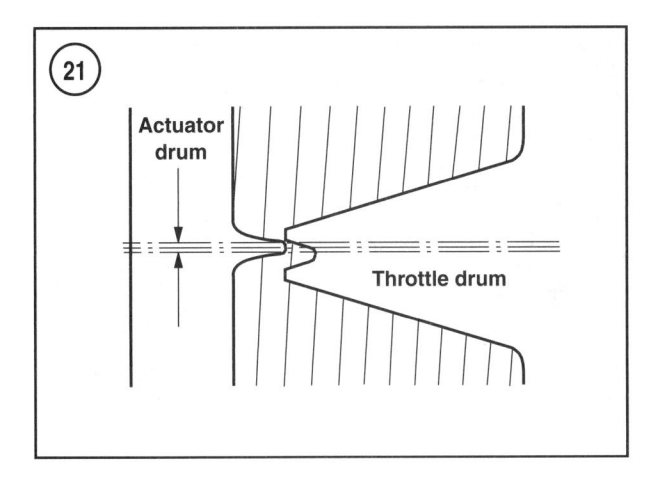

Cable Adjustment

1. Support the motorcycle on its centerstand.

2. Remove the top shelter. Refer to *Top Shelter* in Chapter Seventeen.

NOTE
Figure 20 shows the control module holder with the module connectors disconnected to better illustrate the step.

6. Reposition the switch cover.

Front Brake Cancel Switch Testing/Replacement

The front brake cancel switch and front brake light switch are enclosed in the same switch housing mounted on the bottom of the front master cylinder. The two lower (larger) terminals (A, **Figure 25**) plug into the cruise cancel switch. The two upper (smaller) terminals (B, **Figure 25**) plug into the brake light switch.

1. Disconnect the connectors from the lower switch terminals (**Figure 26** and **Figure 27**).
2. Check for continuity between the switch terminals. There should be continuity with the brake lever released and no continuity with the brake lever applied. Replace the switch if faulty.
3. Turn the ignition switch on and push the cruise control main switch (A, **Figure 1**). Measure voltage between the black/blue wire terminal (+) and ground (–). There should be battery voltage. Turn the ignition switch off.
4. Remove the screw and switch.
5. Installation is the reverse of removal.
6. Make sure all connectors are plugged tightly into the switch.
7. Reposition the switch cover.

Throttle Cancel Switch

Testing

1. Remove the air filter housing. Refer to *Air Filter Housing* in Chapter Eight.
2. Disconnect the throttle cancel switch blue 2-pin connector located in the connector pouch (A, **Figure 28**) in front of the throttle body.
3. Check continuity between the two connector terminals. There should be continuity with the throttle grip in

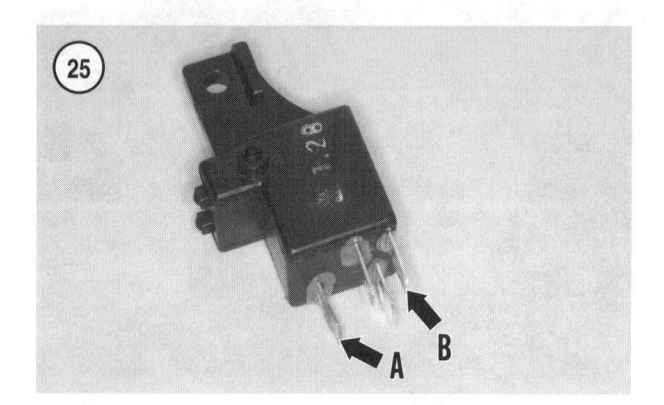

CRUISE CONTROL SYSTEM

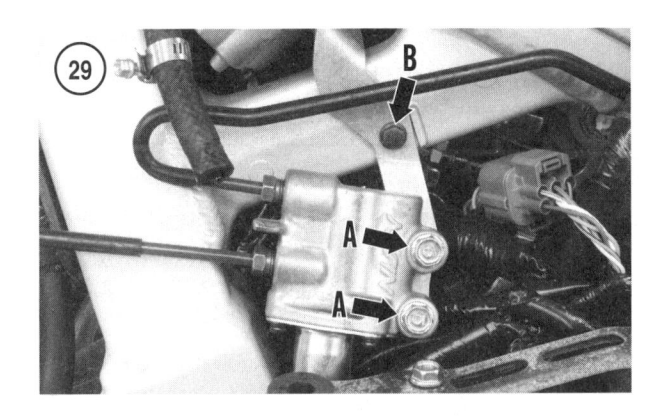

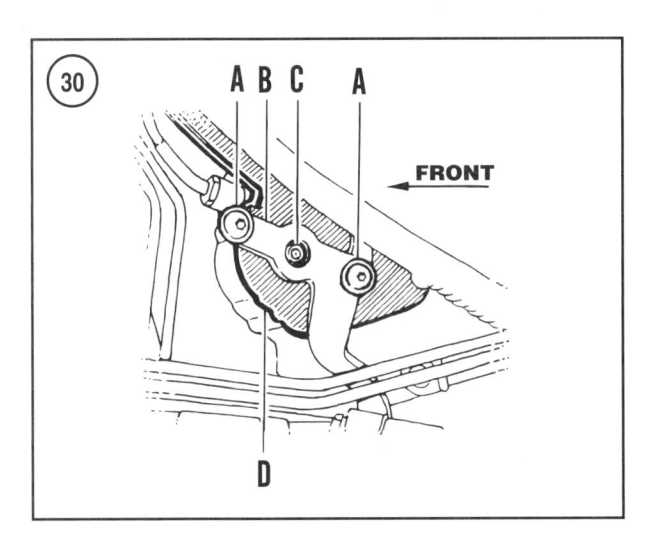

any position. Open and close the throttle to check for continuity in different positions.

4. Open the throttle grip to any position. Hold the throttle drum (B, **Figure 28**) to prevent it from closing and release the throttle grip. There should be no continuity.

5. Release the throttle drum. Operate the throttle grip several times to ensure the throttle cable is positioned correctly.

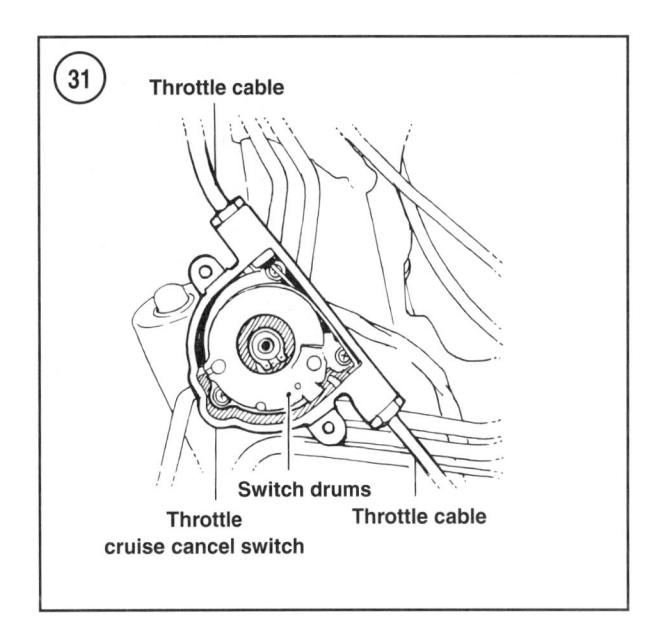

6. Turn the ignition switch on and push the cruise control main switch (A, **Figure 1**). Measure voltage between the green/white wire (without the brown tube terminal) (+) and ground (–). There should be battery voltage. Turn the ignition switch off.

7. If necessary, replace the switch as described in the following procedure.

8. Reverse Steps 1 and Step 2 to complete installation.

Replacement

1. Remove the air filter housing. Refer to *Air Filter Housing* in Chapter Eight.

2. Remove the left radiator. Refer to *Radiators* in Chapter Ten.

3. Disconnect the throttle cruise cancel switch blue 2-pin connector located in the connector pouch (A, **Figure 28**) in front of the throttle body.

4. Loosen the locknuts and disconnect the return side throttle cable (C, **Figure 28**) from the throttle drum.

5. Remove the two mounting bolts (A, **Figure 29**) and remove the PCV valve from its mounting bracket. Remove the bolt (B, **Figure 29**) and the mounting bracket.

6. Remove the following (**Figure 30**):
   a. Two Allen bolts (A).
   b. Mounting bracket (B).
   c. Center bolt (C).
   d. Switch cover (D).

7. Disconnect the throttle cables from the switch drums (**Figure 31**).

8. Discard the old throttle cancel switch.

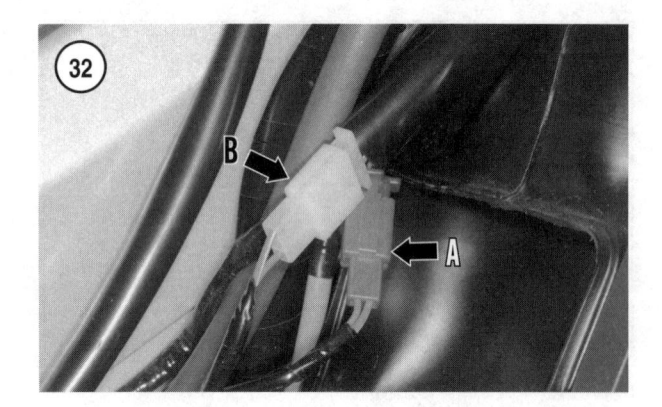

9. Clean the switch housing of all old grease and cable lubricant.
10. Lubricate the sliding parts of the switch with grease.
11. Installation is the reverse of removal. Note the following:
    a. With all the hardware removed to access this switch, now would be a good time to lubricate the throttle cables. Refer to *Throttle Cable Lubrication* in Chapter Three.
    b. Tighten the PCV mounting bolts (A, **Figure 29**) to 12 N•m (106 in.-lb.).
    c. Check and adjust the throttle cable free play before installing the top shelter. Refer to *Throttle Cable Operation and Adjustment* in Chapter Three.

Rear Brake Cancel Switch Testing/Replacement

The rear brake light switch and rear brake cancel switch are enclosed in the same switch housing.
1. Remove the top shelter. Refer to *Top Shelter* in Chapter Seventeen.
2. Disconnect the rear brake cruise cancel switch red 2-pin connector (A, **Figure 32**). The white connector (B, **Figure 32**) is used for the brake light part of the switch.
3. Check for continuity between the switch terminals. There should be no continuity with the brake pedal released and continuity with the brake pedal applied. Replace the switch if faulty.
4. Turn the ignition switch on and push the cruise control main switch (A, **Figure 1**). Measure voltage between the green/white wire terminal (+) and ground (–). There should be battery voltage. Turn the ignition switch OFF.
5. To replace and adjust the switch:
    a. Remove the right engine side cover. Refer to *Engine Side Covers* in Chapter Seventeen.
    b. Disconnect the switch connectors (A and B, **Figure 32**).
    c. Remove the screws and switch (**Figure 33**).
    d. Use a medium strength threadlock on the switch screw threads.
    e. Install the switch onto the master cylinder and secure with the mounting screws (**Figure 33**)—do not tighten.
    f. Turn the ignition switch on. Adjust the switch so that the brake light comes on when the brake pedal moves the push rod 0.7-1.7 mm (1/32-1/16 in.).
    g. Tighten the switch screws (**Figure 33**) to 2 N•m (17 in.-lb.).
    h. Recheck the brake light adjustment and turn the ignition switch OFF.
6. Installation is the reverse of removal.

WARNING
Do not ride the motorcycle until the rear brake light works correctly.

CHAPTER SEVENTEEN

BODY COMPONENTS AND EXHAUST SYSTEM

This chapter contains procedures for the fairing components and exhaust system. Procedures on how to use and troubleshoot the remote transmitter are also described.

Whenever you remove a body or frame member, reinstall all mounting hardware (i.e. small brackets, bolts, nuts, rubber bushings, metal collars, etc.) onto the removed part so they will not be misplaced. Parts and the way they attach to the frame may differ slightly from the those used in the service procedures described in this chapter.

The plastic fairing and frame cover parts are expensive to replace. After removing each part from the bike, wrap it in a blanket or towel, and store it in an area where it will not be damaged.

When installing body and frame parts, tighten the fasteners to the specifications provided. **Table 1** is at the end of this chapter.

COLOR LABEL

The color label is mounted on the back of the fuel fill cover lid. Refer to the color code when ordering color-coded parts. Refer to Chapter One for frame and engine serial number locations.

ENGINE SIDE COVERS

Removal/Installation

1. A small plastic spoiler (**Figure 1**) is mounted at the rear of each injection cover. Squeeze the spoiler to release its upper and lower tabs and remove it.

NOTE
*The two dots in **Figure 2** and **Figure 3** show where the bosses are attached to the engine side covers.*

2. Pull out on the bottom and rear side of the engine side cover to disconnect its two bosses from the grommets.
3. Disconnect the engine side cover from the tab on the injector over, then remove the engine side cover.
4. Check for dislodged grommets on the frame brackets (**Figure 4**, typical) and on each cover.
5. Inspect the covers for damage. The insulator (**Figure 5**) installed on the left cover can be replaced separately.

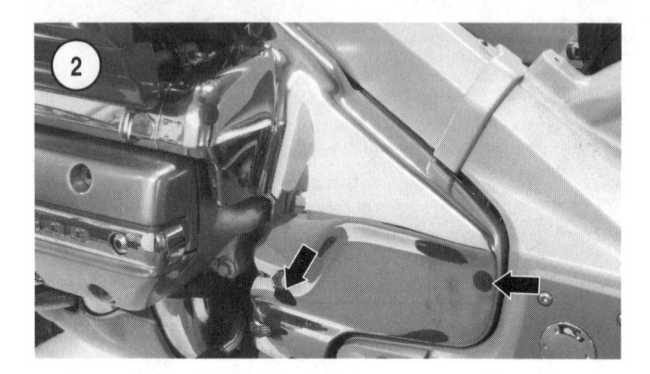

6. Install by reversing these steps.

SIDE COVERS

When removing the side cover, it is important to apply pressure evenly at the mounting bosses fixed to the back of the side cover to prevent them from snapping off.

Removal/Installation

NOTE
*The four white dots in **Figure 6** and **Figure 7** show where the mounting bosses are attached to the side cover.*

1. Grasp the side cover and pull it to release its four bosses from the grommets mounted on the frame.
2. Check for missing or damaged grommets.
3. Install the side cover by aligning the bosses with the grommets, then push the cover into position.

SEAT

The seat (A, **Figure 8**) is removed as a complete assembly. The passenger seat back (B, **Figure 8**) is removed separately.

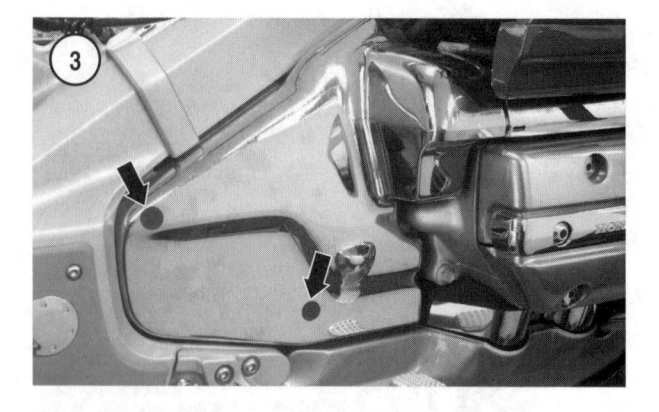

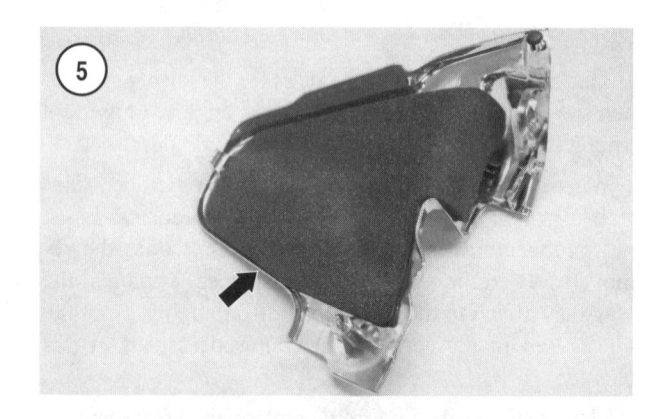

BODY COMPONENTS AND EXHAUST SYSTEM

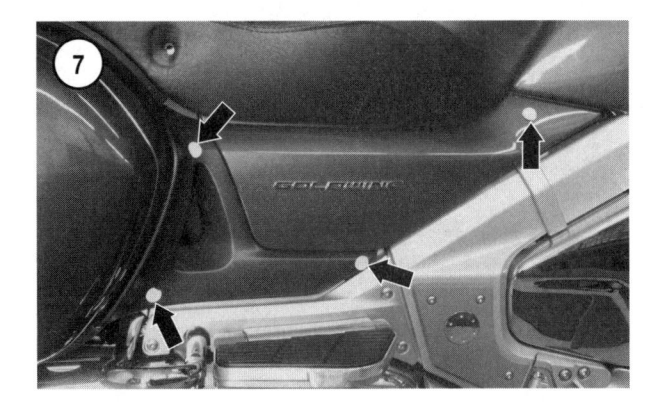

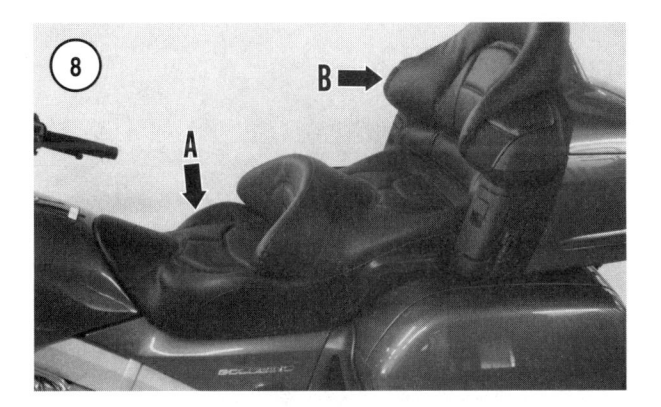

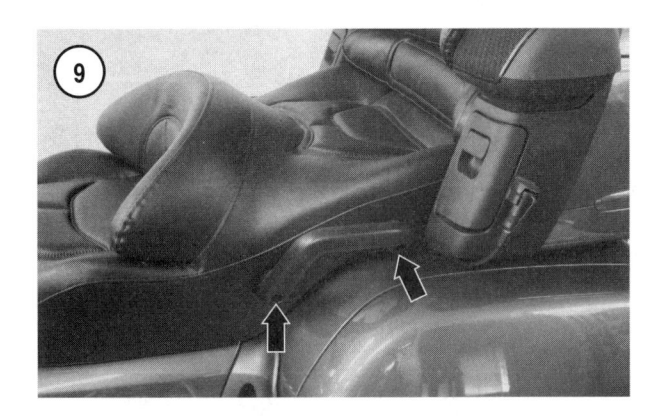

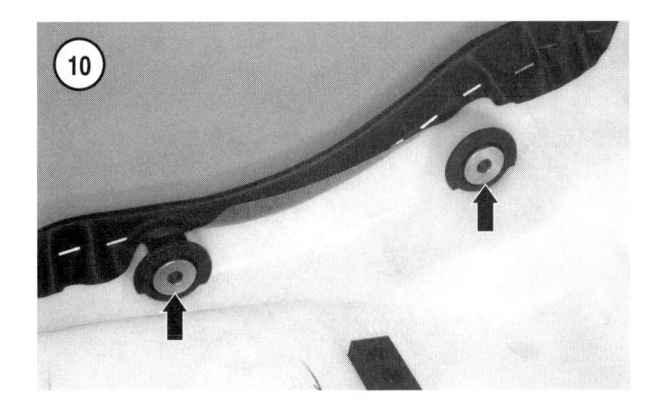

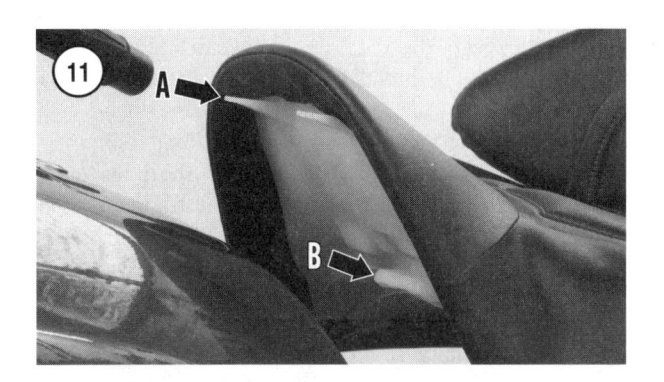

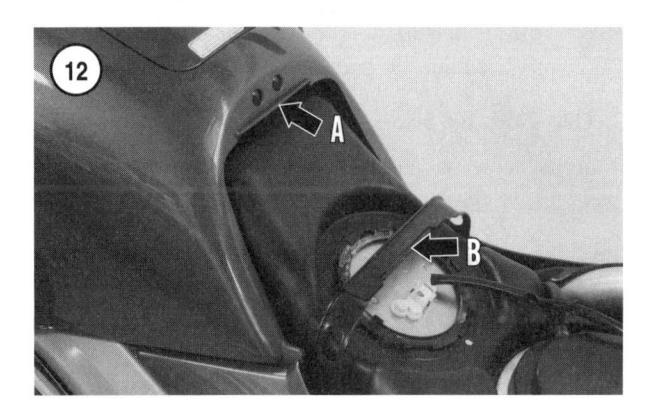

Removal/Installation

NOTE
The left side passenger hand grip is used when moving the bike on its centerstand. Place the bike on its centerstand before removing the seat.

1. Support the bike on its centerstand.
2. Remove the bolts and passenger hand grips (**Figure 9**).
3. Raise the rear of the seat (A, **Figure 8**), then slide it rearward and remove it.
4. Replace missing or damaged seat dampers (four total) mounted on the bottom of the seat. Then check that the dampers seat flush in the seat (**Figure 10**). They can dislodge during seat removal and may interfere with seat installation.
5. Install the seat by placing it on top of the frame, then pushing it rearward as far as possible. If necessary, pull the rear seat flap (movable part of seat) out to gain additional room. Then grasp the front part of the seat and push the seat rearward again to provide room for the upper seat tab to clear the shelter stay. Lower the seat so the upper seat tabs (A, **Figure 11**) fit under the top shelter (A, **Figure 12**) and the two lower seat tabs (B, **Figure 11**) seat under the seat bracket (B, **Figure 12**).

6. Push the rear portion of the seat down to align the holes in the seat with the threaded holes in the frame. Working on one side at a time, position the passenger hand grip (**Figure 9**) along the seat and install the rear mounting bolt first, then the front bolt. Repeat on the other side. When all four bolts are threaded into the frame, tighten them securely.

> *NOTE*
> *If a bolt is difficult to install, check for a dislodged damper/collar assembly.*

> *NOTE*
> *Make sure the hand grip mounting bolts are threaded into the frame.*

7. Check that the seat is secure.

PASSENGER SEAT BACK

The passenger seat back (B, **Figure 8**) is mounted on the front of the trunk.

Removal/Installation

The seat back can be removed with the seat mounted on the motorcycle.

1. Open the trunk.

> *NOTE*
> *Cover the gap between the trunk and trunk lid to prevent the seat back fasteners from falling beneath the seat.*

2. Remove the two Phillips screws and washers (**Figure 13**).
3. Close the trunk lid and slide the seat up to remove it.
4. Installation is the reverse of removal. Align the two upper brackets (A, **Figure 14**) on the seat back with the mating notches in the trunk (B). Install the seat back, then slide down so the two screw bosses fit into the holes in the trunk.

FRONT FENDER ASSEMBLY

Front Fender A and Fender Covers Removal/Installation

1. Support the motorcycle on its centerstand.
2. Remove the four mounting bolts, two rubber washers and front fender A (**Figure 15**).
3. Remove the two mounting bolts and the front fender cover (**Figure 16**). Repeat for the opposite side.

BODY COMPONENTS AND EXHAUST SYSTEM

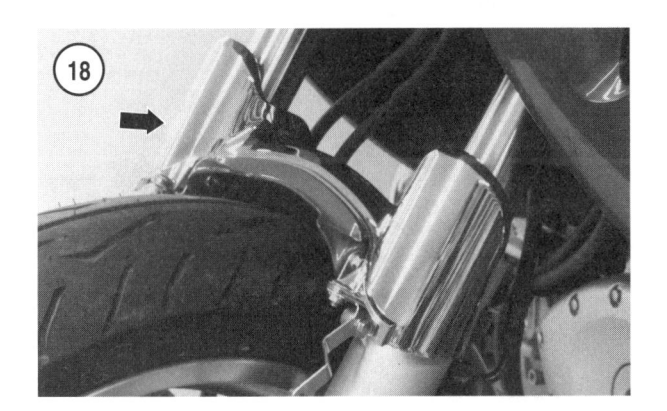

4. Installation is the reverse of removal. The rubber washers and long mounting bolts are installed in the front fender A lower mounting holes.

FRONT FENDER TOP COVER

Removal/Installation

1. Remove front fender A and both front fender covers as described in this chapter.

2. Remove the trim clip (**Figure 17**) from the top cover.

3. Remove the mounting bolts and the top cover (**Figure 18**).

4. Installation is the reverse of removal.

FRONT FENDER B

Removal/Installation

1. Remove the following as described in this section:

   a. Front fender A and both front fender covers.

   b. Front fender top cover.

2. Remove the mounting bolt and brake pipe bracket mounting bolt (**Figure 19**) from the left side.

3. Remove the fender mounting bolts (**Figure 20**) from the right side.

4. Slide the fender (**Figure 21**) behind the fork tubes and remove it.

NOTE
If the front lower fairing and center fairings were previously removed, it will be easier to install them before installing front fender B.

5. Installation is the reverse of removal.

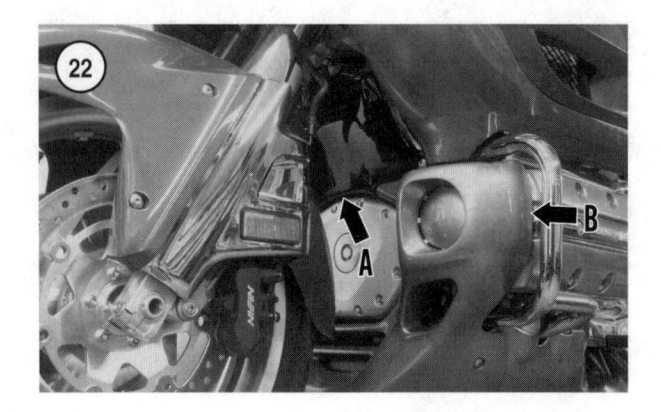

FRONT LOWER FAIRINGS

The front lower fairing is a subassembly consisting of a center inner fairing and front lower fairing. The center inner fairing must be removed first. Because the front wheel must be turned to one side when removing and installing these fairings, remove these parts before mounting the motorcycle on a lift with a front wheel vise.

Removal

1. To remove the center inner fairing (A, **Figure 22**), remove the six trim clips and the two mounting bolts. Turn the front wheel so it faces forward, then pull the center inner fairing forward and away from the front fairing. Turn the front wheel as required to provide access and remove the center inner fairing.
2. To remove the front lower fairing (B, **Figure 22**), remove the two mounting bolts.

Installation

NOTE
If both the center inner fairing and front lower fairing were removed, it is easier to position the center inner fairing first.

1. Install the center inner fairing (A, **Figure 22**) into position. Position the top of the center inner fairing behind the turn cancel switch screw and bracket assembly (**Figure 23**). Do not install any fasteners at this time.
2. Install the front lower fairing as follows:
   a. Make sure the two nut clamps are installed on the front lower fairing.
   b. Install the front lower fairing (B, **Figure 22**). Insert its upper edge behind the center inner fairing and behind the exposed tab on each inner fairing (where the two outer mounting bolts are installed).

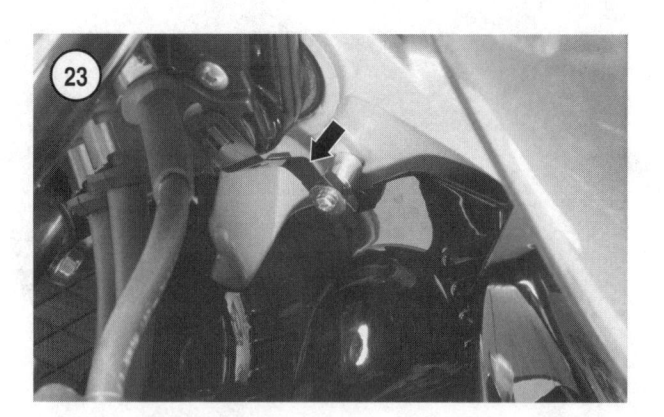

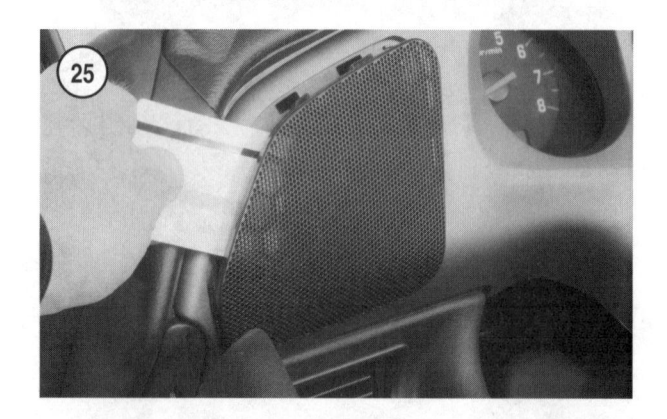

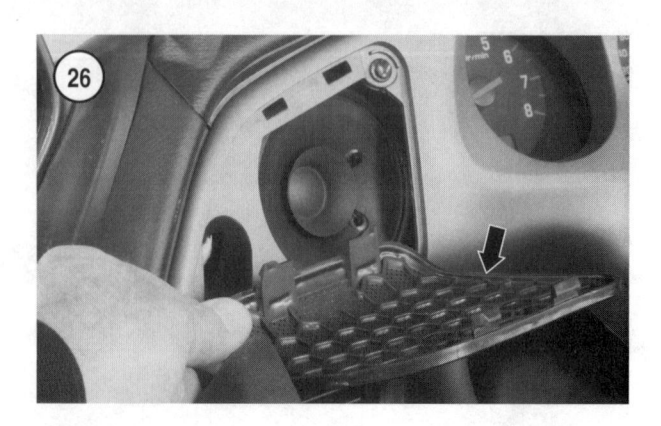

BODY COMPONENTS AND EXHAUST SYSTEM

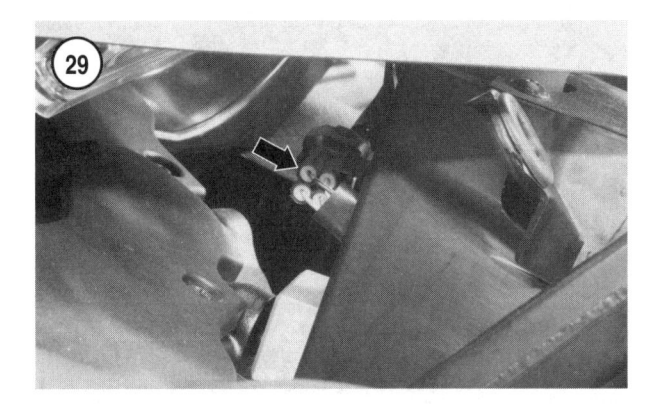

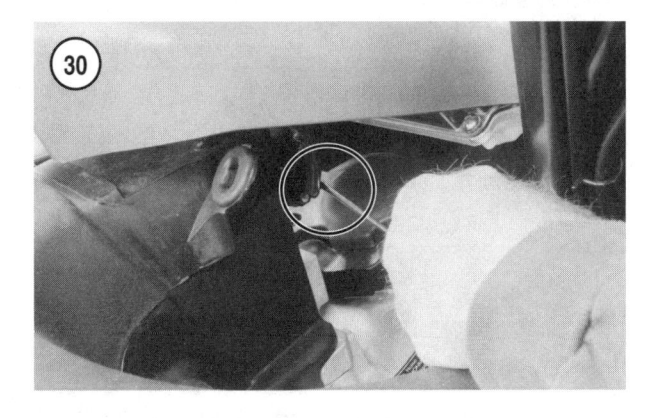

c. Install the remaining front lower fairing Allen bolts finger-tight.
3. Install all the trim clips.
4. Tighten the front lower fairing Allen bolts securely.

METER PANEL

The meter panel (**Figure 24**) surrounds the gauges, multi-display, ignition switch and speakers. Speaker covers are installed in the meter panel to protect the speakers.

Removal/Installation

1. Support the bike on its centerstand.

NOTE
Four clip tabs (mounted on the backside of the speaker covers) secure the covers to the meter panel.

2. Insert a thin, plastic card into the groove between the speaker cover and meter panel (**Figure 25**). Pry the card against the speaker cover to move the cover in slightly and release its clip tabs. Remove the speaker cover (**Figure 26**). Repeat for the other side.
3. Remove the trim clip (A, **Figure 27**) and mounting bolt (B) from the speaker cavity. Repeat for the other side.

NOTE
*The white dots in **Figure 28** show where the alignment pins in the meter panel are located.*

4. Pull the meter panel rearward to disconnect its pins (**Figure 28**) from the grommets.

NOTE
*The black 4-pin multi-display connector is mounted on the front side of the ignition switch (**Figure 29**).*

5. Working from the right side, locate the black 4-pin connector. Locate the tab at the front of the connector, then push the tab forward (**Figure 30**) to unlock the connectors and disconnect them.
6. Remove the meter panel.
7. Inspect for missing or damaged grommets.
8. Installation is the reverse of these steps, plus the following:
   a. Fit the windshield lever lock rubber cover into its mating tabs and brackets (**Figure 31**).
   b. Align the four alignment pins with the holes.
   c. Align the two tabs on the bottom of the meter panel with the mating slots.

MIRRORS AND FRONT TURN SIGNALS

Removal/Installation

1. Remove the meter panel as described under *Meter Panel* in this chapter.

   NOTE
   *When pulling the mirror boot (**Figure 32**) away from the front fairing, note how the tabs on the boot seat into grooves to hold the boot in place.*

2. Pull the mirror boot (**Figure 33**) away from the front fairing and fold over the mirror housing.
3. Remove the mirror cover (A, **Figure 34**):
   a. Remove the trim clip (B, **Figure 34**) and screw (C) that hold the mirror cover to the front fairing.
   b. Pull/disconnect the windshield adjuster boot (**Figure 35**) away from the mirror cover.

   NOTE
   *The two white dots in **Figure 36** show where the mirror cover mounting tabs are located.*

   CAUTION
   If the mirror cover's upper mounting tab is a tight fit, it can be difficult to free it from the slot in the upper fairing. Protect the fairing with a blanket should the mirror cover break free when removing it under pressure.

   c. Lift the bottom of the mirror cover (A, **Figure 34**) to disconnect its mounting tab from the slot in the front fairing, then pull the mirror cover straight down to disconnect its upper mounting tab from the fairing.

   NOTE
   Reposition the mirror housing to access the three mounting screws.

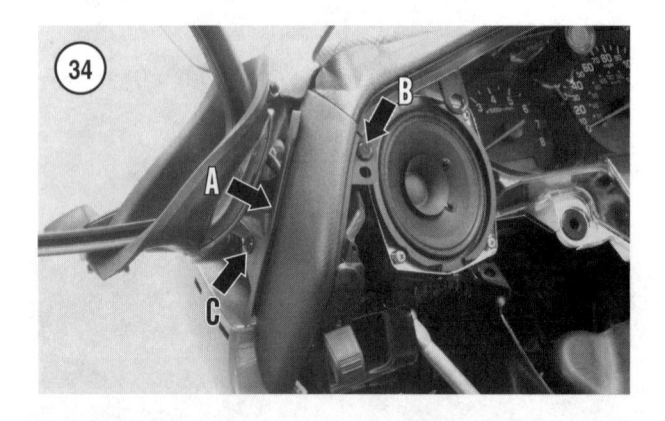

BODY COMPONENTS AND EXHAUST SYSTEM

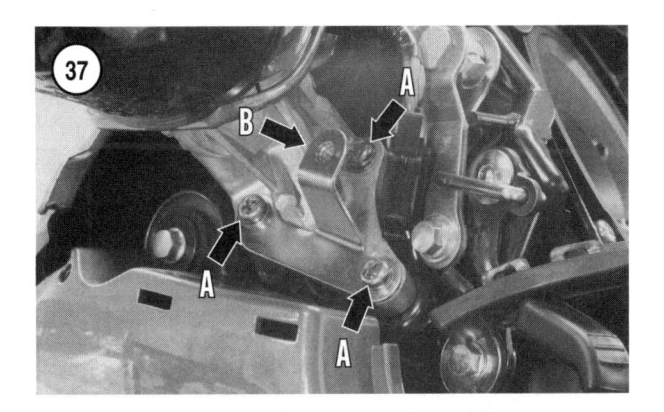

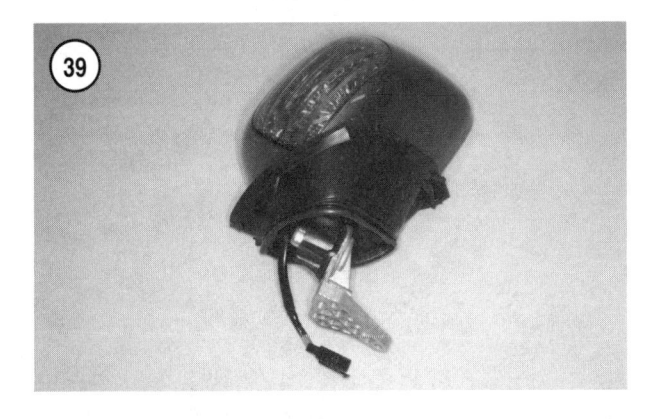

4. Remove the three screws (A, **Figure 37**), bracket (B) and mirror from the mounting bracket.

5. Disconnect the turn signal 3-pin connector (**Figure 38**) and remove the mirror assembly. See **Figure 39**.

6. Installation is the reverse of these steps, plus the following:
   a. Install the mirror boot and windshield height lever boot tabs.
   b. Install the trim clip straight through the mirror cover so its lower end rests outside the speaker frame. If the trim clip is installed at an inward angle, its tip may contact the speaker and damage its paper element.
   c. Check the turn signal light operation and mirror adjustment before riding the motorcycle.

WINDSHIELD ASSEMBLY

A properly maintained and adjusted windshield is an important part of riding comfort. This section describes basic maintenance, adjustment and replacement procedures for the windshield and its mounting assembly.

Windshield Height Adjustment

A 6-position manual ratchet adjuster allows the rider to vary the windshield height by more than 102 mm (4 in.). Optional windshields are also available from Honda in two different heights.

NOTE
The suggested height for a windshield is just below eye level. While some riders prefer to look through a windshield, wearing glasses and/or using a helmet shield can affect visibility. To arrive at an ideal windshield height, wear normal riding gear (pants, helmet and glasses) when adjusting it. Then test ride the bike to check the windshield's performance in regards to visibility, noise and wind buffeting.

NOTE
The windshield should move under light resistance. If the windshield is difficult to move after unlocking it, binds on either side or in the middle, one or more of the rubber windshield screen holders has probably pulled away from the windshield holder. Remove the windshield and holder plate to reposition or replace the rubber windshield holder(s).

1. To raise the windshield:

a. Lift both adjust levers (**Figure 40**) to unlock the windshield.
b. Grasp the windshield on both sides and raise it to the desired position. Each time a click is heard, the windshield changed position. After hearing a click, lower the windshield a few millimeters to remove backlash and seat the adjuster.
c. Align the horizontal marks on the windshield with the upper edge of the instrument panel.
d. Lower both adjusters (**Figure 40**) to lock the windshield.

2. To lower the windshield and reset the ratchet mechanism:
   a. Raise both adjust levers (**Figure 40**) to unlock the windshield.
   b. Grasp the windshield on both sides and raise it until the upper mark on the windshield (O) aligns with the upper edge of the instrument panel.

NOTE
The windshield cannot be lowered until the upper mark on the windshield (O) is brought to align with the upper edge of the instrument panel. **Figure 41** *shows the windshield marks.*

   c. Lower the windshield all the way to reset the ratchet mechanism.
   d. Raise the windshield to the desired position. Align the horizontal marks on the windshield with the upper edge of the instrument panel.
   e. Lower both adjusters (**Figure 40**) to lock the windshield.

Windshield Garnish
Removal/Installation

The windshield garnish (**Figure 42**) is a decorative cover that hides and protects the lower part of the windshield, the adjuster mechanism and mounting fasteners.

1. Remove the left and right side mirrors as described under *Mirrors and Front Turn Signals* in this chapter.
2. Remove the bolt, washer and rubber washer from each side (**Figure 43**).

NOTE
The windshield garnish will still fit tightly against the windshield and fairing after removing the two mounting bolts. Four pins installed in the backside of the garnish fit into rubber grommets. Disconnect the garnish from the outer grommets first (one at a

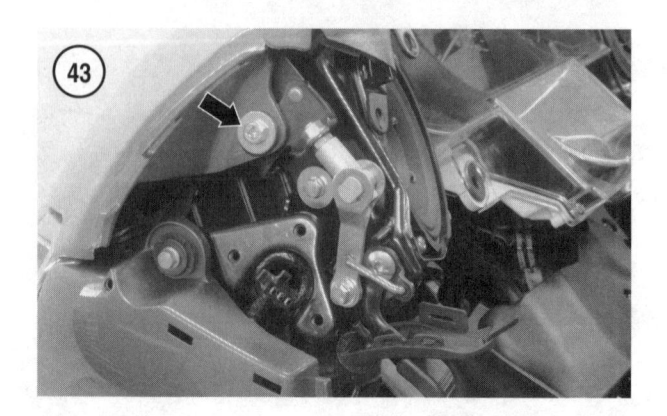

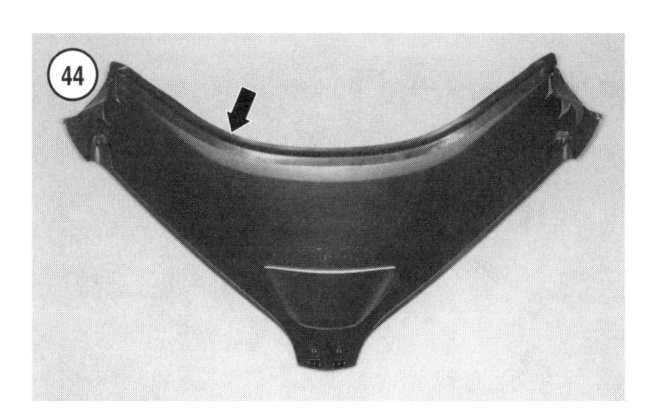

BODY COMPONENTS AND EXHAUST SYSTEM

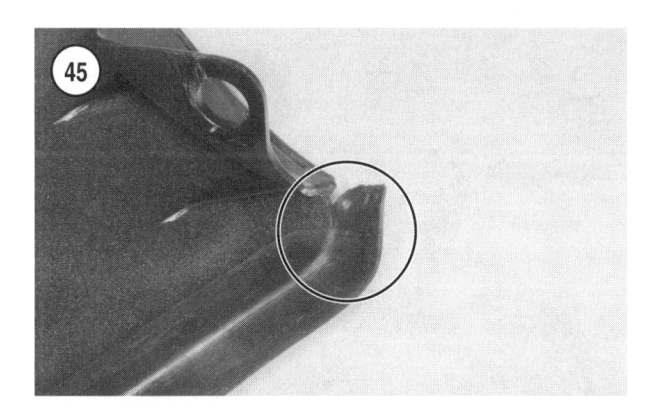

time), then from the middle grommets (Steps 3-5).

NOTE
*The white dots in **Figure 42** show where the two center and left side pins are positioned on the backside of the garnish. The right side pin is mounted symmetrically with the left pin.*

3. Pull the windshield garnish at its side (to release its mounting hole from the stud), while at the same time lifting the end of the garnish to disconnect the outer pin from the grommet.
4. Repeat Step 3 for the opposite side.
5. Pull the front, lower part of the garnish straight out to release its pins from the front grommets, and remove the garnish.
6. Check for missing or damaged grommets.
7. Check the seal (**Figure 44**) for cracks, deterioration or other damage. To replace the seal:
   a. Pull the old seal off the garnish.
   b. Pull the adhesive strip (if used) off the new seal.
   c. Start at one end and install the new seal over the top edge of the garnish. Both ends of the seal should overlap the garnish as shown in **Figure 45**.
8. Installation is the reverse of these steps.

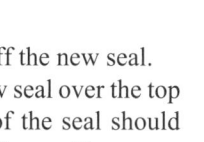

Windshield Holder and Windshield Removal/Installation

1. Remove the windshield garnish as described in this section.
2. Remove the nut from each side of the windshield holder (**Figure 46**, typical).
3. Turn the garnish cover as shown in A, **Figure 47** to uncover the mounting bolt.
4. Remove the bolt (B, **Figure 47**) and the windshield holder (C).
5A. On 2001-2002 models, remove the windshield as follows:
   a. Remove the four windshield screws (2, **Figure 48**).
   b. Remove the two bolts (3, **Figure 48**) from the middle of the windshield, and remove the windshield.
5B. On 2003-on models, remove the windshield as follows:
   a. Remove the screw, plastic washer and holder (A, **Figure 49**) from each side of the windshield.
   b. Remove the two center bolts (B, **Figure 49**) and remove the windshield.
6. Install the windshield by reversing these steps. Tighten each windshield screw and bolt securely. Do not overtighten.

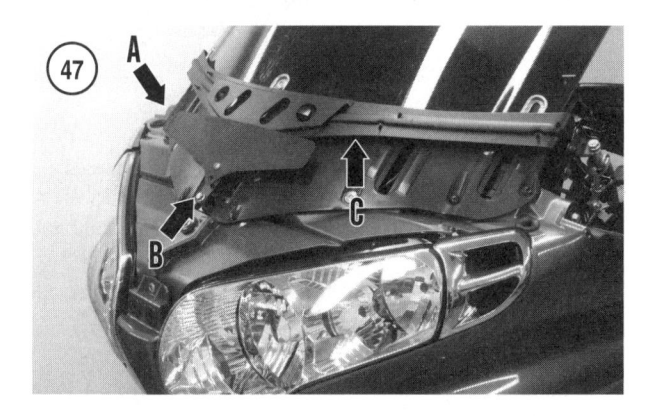

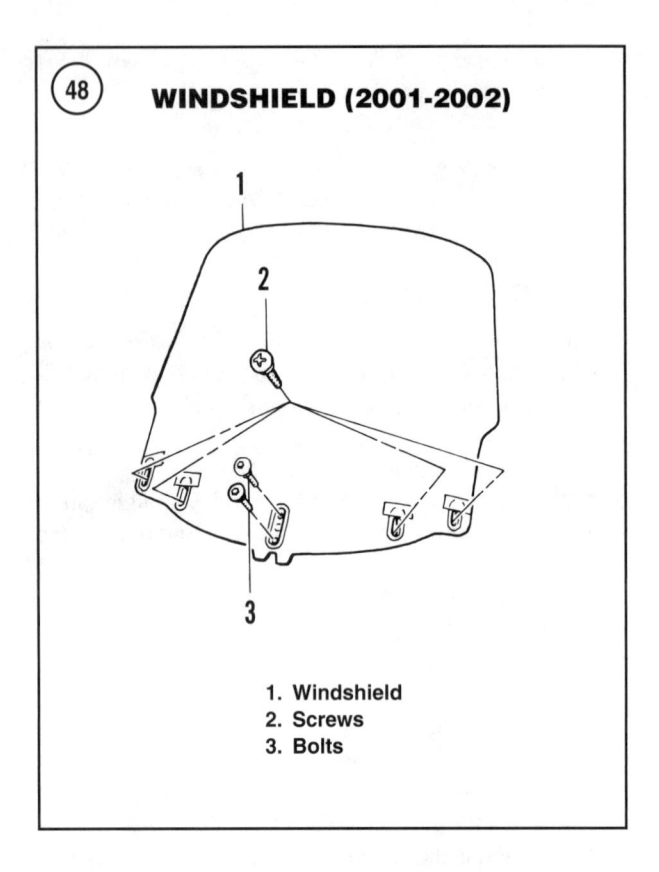

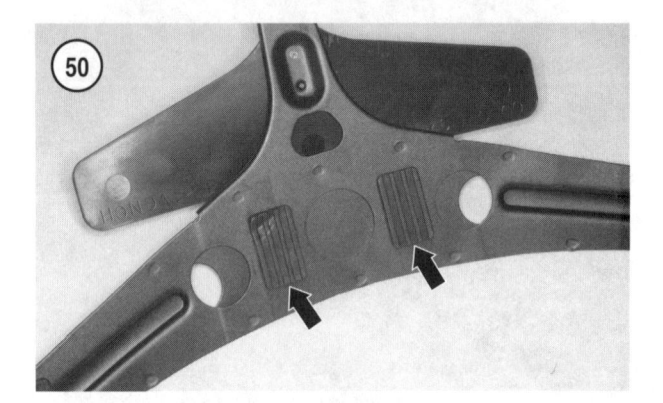

48 WINDSHIELD (2001-2002)

1. Windshield
2. Screws
3. Bolts

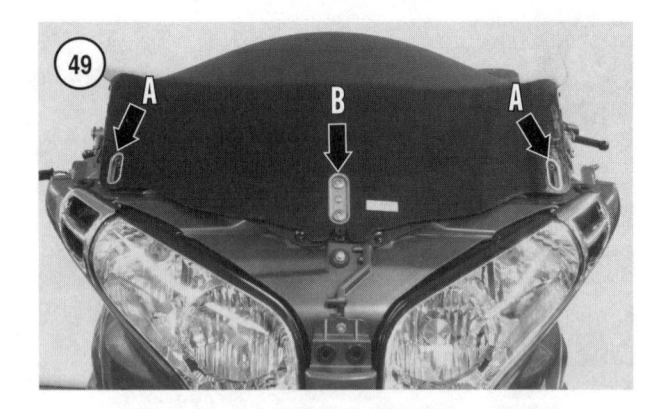

7. Note the following when installing the windshield holder:
   a. Make sure the windshield lock arms are in the unlock position.
   b. Make sure the rubber dampers (**Figure 50**, typical) mounted beneath the windshield holder are positioned correctly.
   c. Install the windshield holder and its center bolt. Check that the windshield holder locating tabs are hooked under the fairing bracket tabs (**Figure 51**).
   d. Tighten the two nuts and center bolt securely.
   e. Turn the cover down.
   f. Operate the windshield locks to make sure they lock and unlock the windshield correctly.

Windshield Locks
Troubleshooting

If the windshield locks are difficult to move to the lock position after removing and installing the windshield holder, check that the windshield holder locating tabs are hooked under the fairing bracket tabs (**Figure 51**). If the windshield holder tabs are hooked incorrectly over the fairing bracket tabs, the windshield lock must also force the holder tab down when being moved to the lock position.

Windshield Lock Mechanism
Removal/Installation

The windshield lock mechanisms are mounted on the front fairing mounting bracket. See **Figure 52** (left) and **Figure 53** (right).

1. Remove the windshield holder and windshield as described in this chapter.
2. Remove the pivot and mounting bolts and remove the lock mechanism.

BODY COMPONENTS AND EXHAUST SYSTEM

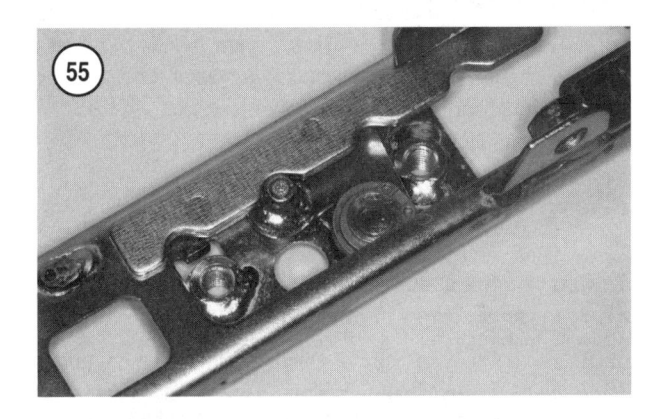

3. Installation is the reverse of these steps. Operate the lock to make sure the windshield is locked firmly in position.

Windshield Ratchet Mechanism Removal/Installation

The windshield ratchet mechanism is mounted inside a track welded onto the upper fairing mounting bracket (**Figure 54**).

1. Remove the front fairing. See *Front Fairing* in this chapter.
2. Remove the mounting bolts, then lift and remove the ratchet mechanism from the track.
3. Refer to **Figure 55** to check the return spring. Remove the snap ring and replace the return spring if damaged.
4. Check the adjustment detents in the guide for excessive wear or damage.
5. Installation is the reverse of these steps. Tighten the mounting bolts securely.
6. Check the ratchet mechanism operation as follows:

NOTE
This check can also be performed with the windshield mounted in place.

   a. Position the ratchet arm (**Figure 54**) at the bottom of its guide.
   b. Raise the ratchet arm in its guide. The ratchet should click (five times) as it engages each detent.
   c. After the ratchet arm clicks five times, raise it to the top of its guide. The ratchet should release itself (no tension on ratchet) after the fifth detent position. The ratchet guide is now free of all tension.
   d. Lower the ratchet until it stops. Then push the ratchet guide to the bottom of its guide where it will reset itself (tension on ratchet).

Meter Panel Visor Removal/Installation

The meter panel visor (**Figure 56**) mounts across the front fairing and over the instrument panel. The visor supports the windshield and windshield holder and shields the instrument panel. Vertical slots in the meter panel visor provide windshield adjustment.

NOTE
Different length bolts and screws secure the meter panel visor to the front fairing. Identify these fasteners when removing them.

17

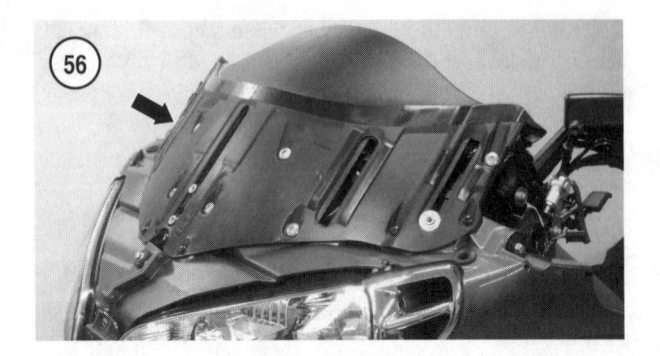

1. Remove the trim clips, bolts and screws securing the meter panel visor to the front fairing.

2. Remove the metal panel visor.

3. The movable windshield nuts can be replaced by removing the E-clip, washer and the nut (**Figure 57**). Reverse to install.

NOTE
The cover placed across the top of the visor that shields the instrument panel is an integral part of the visor assembly. Do not remove it.

4. Installation is the reverse of these steps.

Center Fresh Air Visor (2004)

Removal/Installation

1. Remove the nuts, washers and bolts (**Figure 58**), then release the tab to remove the grille, grommets and visor from the windshield.

2. Refer to **Figure 59** to remove the lever assembly from the visor.

3. Install the lever assembly as follows:
   a. Install the lid on the visor. Then install the click plate by aligning its mounting hole with the boss on the visor.
   b. Install the spring and ball into the lever hole (**Figure 59**). Align the slot in the lever with the pin on the lid and install the lever. Install the lever cover and screw and tighten to 4 N•m (35 in.-lb.).

4. Install the grommets into the visor holes. Then align the tab on the grille with the slot in the visor and install the assembly onto the windshield. Install the washers, bolts and nuts. Tighten the bolts to 1 N•m (8.8 in.-lb.).

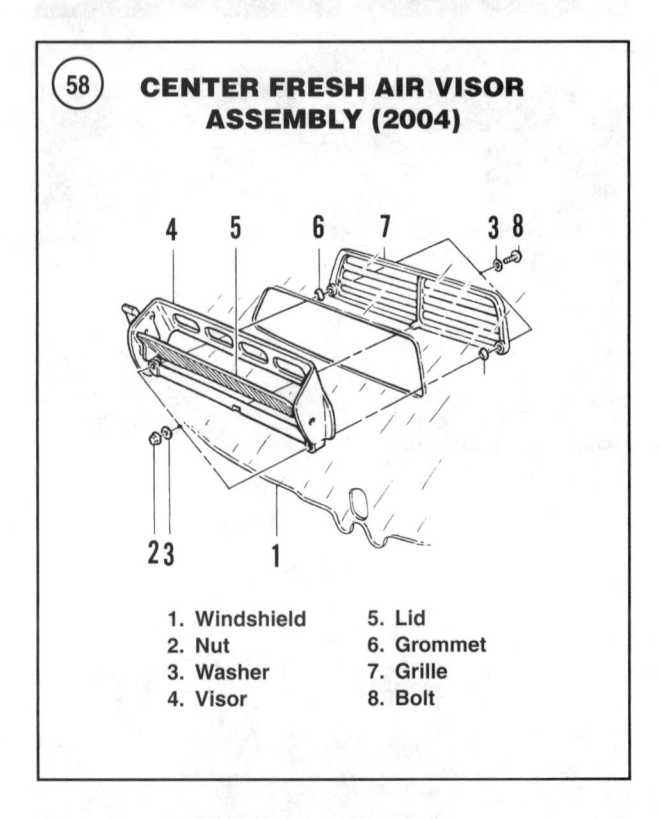

CENTER FRESH AIR VISOR ASSEMBLY (2004)

1. Windshield
2. Nut
3. Washer
4. Visor
5. Lid
6. Grommet
7. Grille
8. Bolt

INNER FAIRINGS

The inner fairing assembly consists of the front fairing garnish and inner fairings. Both mount below the headlights alongside the front fairing to cover and protect the components mounted inside the front fairing.

Removal/Installation

1. Remove the center inner fairing as described under *Front Lower Fairings* in this chapter.

2. Remove the windshield garnish as described under *Windshield Assembly* in this chapter.

BODY COMPONENTS AND EXHAUST SYSTEM

VISOR (2004)

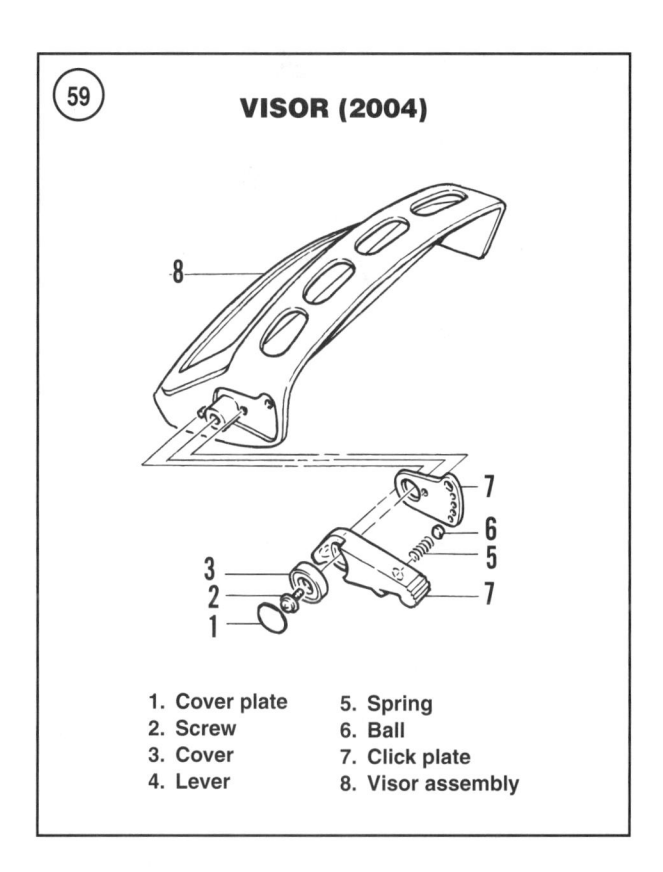

1. Cover plate
2. Screw
3. Cover
4. Lever
5. Spring
6. Ball
7. Click plate
8. Visor assembly

3. Remove the bolt and screws securing the front fairing garnish (**Figure 60**) to the lower fairing, and remove the front fairing garnish.

4. Remove the screw, trim clips and inner fairing (**Figure 61**). Repeat for the opposite side.

5. Installation is the reverse of these steps. Note the following.

6. If the top shelter inner covers were removed, install them now so the trim clips, used to secure the inner fairing to the inner covers can be installed.

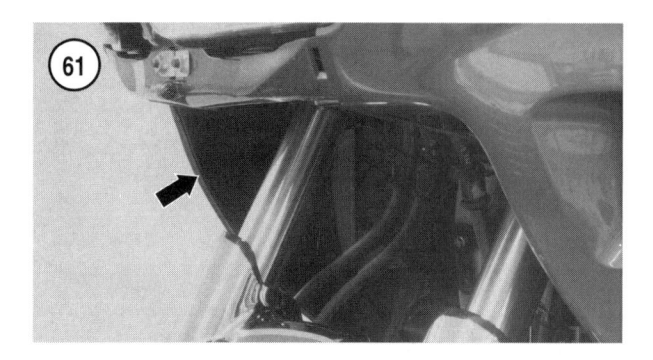

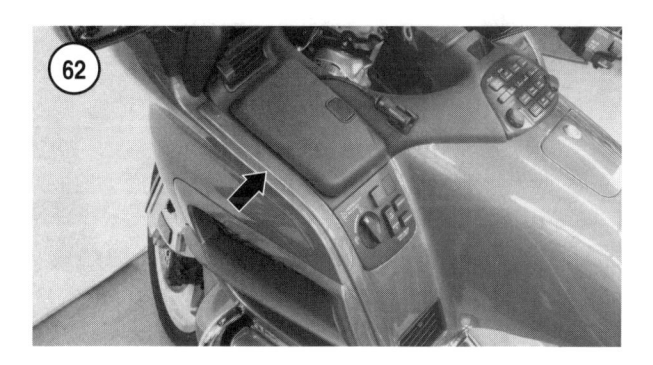

FRONT FAIRING AND FAIRING MOLDING

The front fairing is a subassembly consisting of a separate left and right cowl assembly. The headlight housing, open air temperature sensor and bank angle sensor are mounted inside the front fairing. Remove the front fairing as an assembly. Then, if necessary, remove and service the headlight assembly and cowl parts separately.

This section describes service for the fairing molding and front fairing assembly.

Fairing Molding Removal/Installation

The fairing molding (**Figure 62**) conceals the gap between the front fairing and top shelter assembly. Remove the fairing molding before removing the front fairing.

NOTE
Figure 63 shows a fairing molding removed to illustrate its mounting tabs (A).

1. Grasp or pry the fairing molding at its top edge, then carefully lift it to release its front mounting tab (**Figure 64**). When the first tab is free, continue to pull the fairing molding rearward to release the other mounting tabs in order (top to bottom).

2. Repeat for the opposite side.

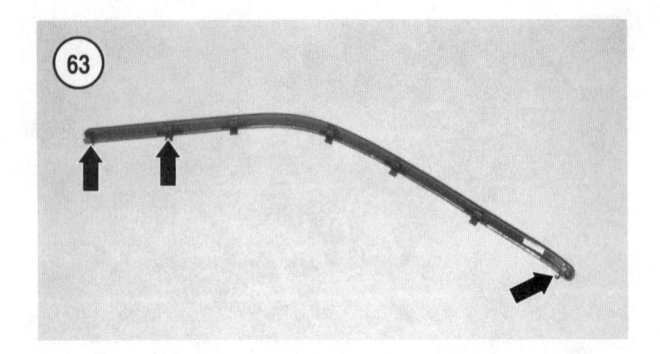

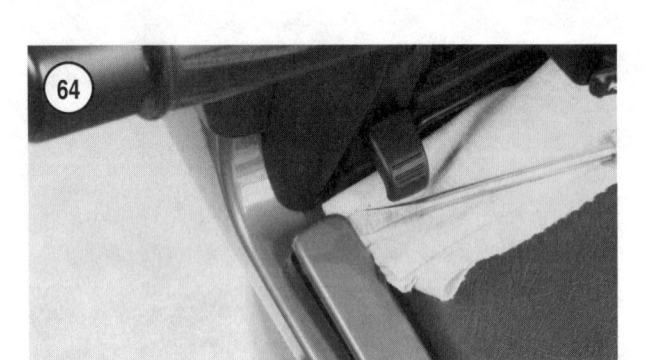

3. Check for any damaged molding lock clips.

NOTE
*The fairing moldings are identified with an L or R. The upper molding end uses one center lock clip (B, **Figure 63**). The bottom end uses two side clips (C, **Figure 63**).*

4. Install the fairing molding by snapping the lower end into the fairing hole, then work upward by installing the lock clips in order. To install the top clip, pull the fairing molding back slightly to arch its top, then align and install the upper lock clip.

**Front Fairing
Removal/Installation**

1. Disconnect the negative battery cable as described in Chapter Nine under *Battery*.
2. Remove the meter panel as described in this chapter.
3. Remove the left and right side fairing molding as described in this section.

*NOTE
Different length fasteners secure the front fairing to the frame. Identify the fasteners for installation.*

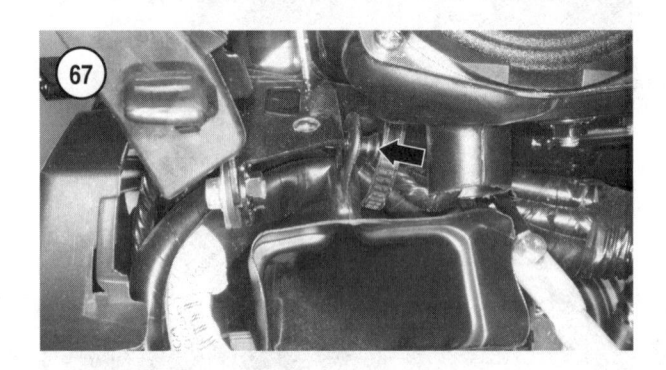

4. Remove the following components as described in this chapter:
   a. Mirrors.
   b. Windshield garnish.
   c. Windshield holder.
   d. Windshield.
   e. Meter panel visor.
   f. Center inner fairing.
   g. Front fairing garnish.
   h. Left and right side inner fairings.
5. Disconnect the six headlight connectors:
   a. Two high beam connectors.
   b. Two low beam connectors.
   c. Two headlight adjusting motor connectors (**Figure 65**, typical).

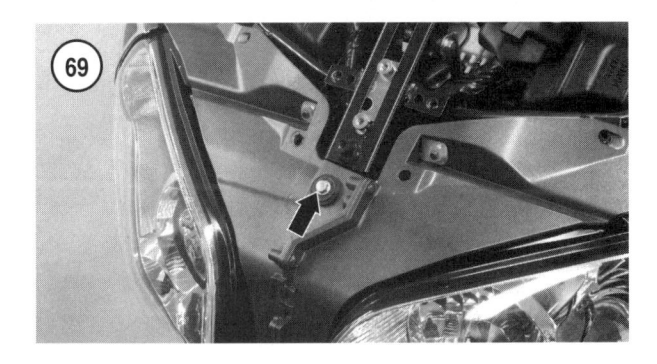

BODY COMPONENTS AND EXHAUST SYSTEM

6. Place a thick blanket across the front fender to protect it when removing and installing the front fairing.

7. If the top shelter is installed on the motorcycle, remove the Allen bolts (**Figure 66**) from the rear side of the front fairing.

8. Remove the push pins (**Figure 67**) securing the left and right side air vents to the fairing mounting bracket.

CAUTION
Three flange bolts are now securing the front fairing. Have someone assist with the removal of the front fairing in the following steps.

9. While a helper holds the front fairing, remove the three flange bolts and collars from the front fairing (**Figure 68** and **Figure 69**). Have the helper move the fairing outward to uncover the open air temperature sensor (A, **Figure 70**) and bank angle sensor (B) connectors (located on right side of fairing), and disconnect them. Remove the front fairing (**Figure 71**) from the motorcycle. See **Figure 72**.

10. Clean the electrical connectors with contact cleaner.

11. Check the area uncovered by the front fairing for any disconnected connectors, loose harness clamps, or loose or missing fasteners.

12. Make sure the headlight covers fit snugly over the bulb holders.

13. Check the installation of the frame grommets. Reposition or secure if necessary.

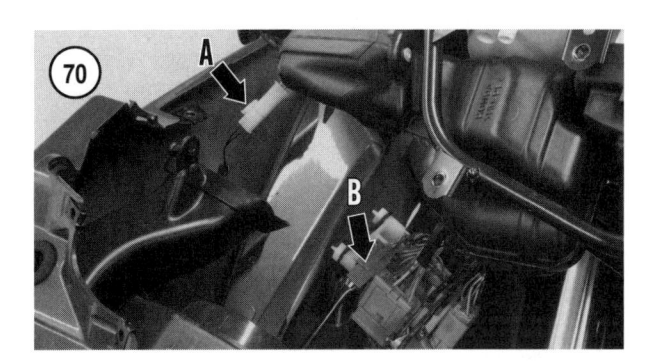

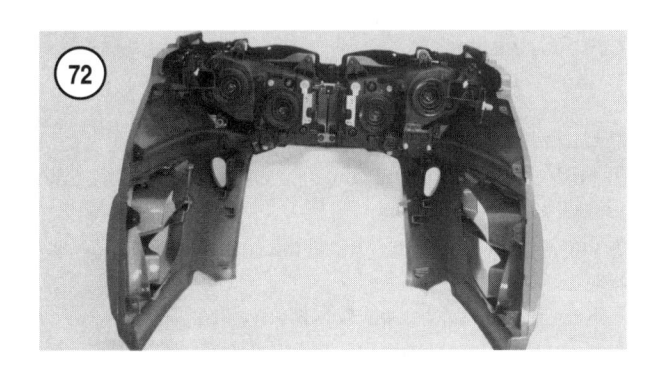

14. While your helper guides the front fairing into position, reconnect the open air temperature sensor (A, **Figure 70**) and bank angle sensor (B) connectors. Then install the front fairing by inserting its two bosses (**Figure 73**) into the frame grommets.

15. Guide the left intake air vent through the wiring harness when installing the fairing into position.

16. Install the front mounting bolt (**Figure 69**).

17. Check that the open air temperature sensor and bank angle sensor connectors and wiring harnesses are routed correctly on the right side of the fairing.

18. Locate and position the turn signal 3-pin connectors toward the outside of the fairing.

19. Install the two push pin connectors (**Figure 67**) through the air vent and fairing mounting bracket holes.

20. Reconnect the six headlight connectors.

21. Install the remaining fairing mounting bracket bolts (**Figure 68**). Tighten the three bolts securely.

22. After completing assembly, check the headlight operation, both HI and LO beams.

23. If the top shelter is installed on the motorcycle, install the Allen bolts (**Figure 66**) at the rear side of the front fairing.

24. Reverse Steps 1-4 to complete installation.

FAIRING POCKETS

Left Fairing Pocket
Removal/Installation

1. Open the fairing pocket lid.

2. Push the center part of the trim clips (**Figure 74**) in to unlock and remove them.

3. On 2002-on models, remove the two screws and washers.

4. Remove the fairing pocket (**Figure 75**).

5. Installation is the reverse of these steps.

Right Fairing Pocket
Removal/Installation

1. Use the ignition key to open the fairing pocket lid. Turn the key clockwise to unlock.

2. Push the center part of the trim clips (**Figure 76**) in to unlock and remove them.

3. On 2002-on models, remove the two screws and washers.

4. Remove the opener cable from the groove in the side of the fairing pocket (A, **Figure 77**). Then disconnect the cable end (B) from the lock mechanism and remove the fairing pocket.

5. Installation is the reverse of these steps. Close the lid to lock it.

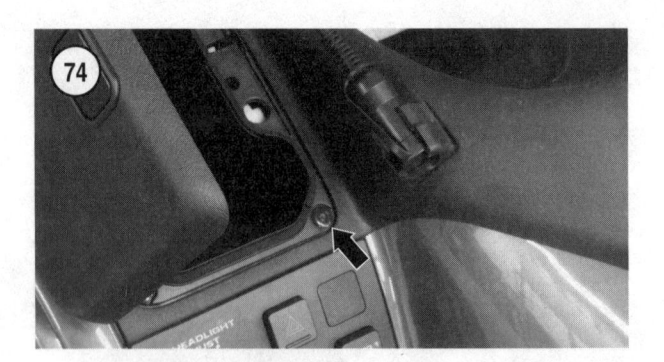

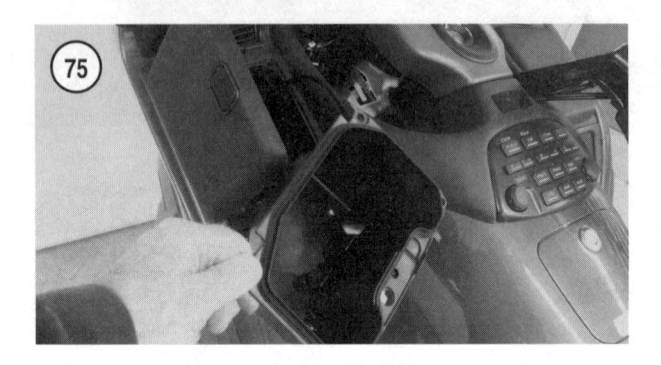

TOP SHELTER

The top shelter assembly (**Figure 78**) consists of the top shelter and the left and right side fairing pockets. The top shelter houses all the audio controls, the headlight beam adjuster and rear spring pre-load adjuster.

Removal/Installation

1. Remove the seat as described under *Seat* in this chapter.

2. Remove the left and right side covers as described under *Side Covers* in this chapter.

3. Remove the meter panel as described under *Meter Panel* in this chapter.

4. Remove the left and right side fairing molding as described under *Front Fairing and Fairing Molding* in this chapter.

5. Remove the left and right side fairing pockets as described in this chapter.

6. Remove the headset connector (**Figure 79**) from the holder on the left side of the top shelter.

NOTE
*Different length fasteners secure the top shelter (**Figure 80**) to the motorcycle. Identify the fasteners for installation.*

7. Remove the top shelter mounting bolts.

BODY COMPONENTS AND EXHAUST SYSTEM

8. Remove the nuts from the rear, bottom side of the top shelter. One nut is used on each side.

9. Release both windshield lever locks.

10. Spread the bottom sides of the top shelter and slip it over the two studs.

11. Lift the left side of the top shelter and disconnect the gray left panel switch connector (**Figure 81**) as follows:

   a. Remove the plastic band from around the wire harness.

   NOTE
   The gray connector uses two lock tabs. One connector unlocks the connector halves. The other one unlocks the connector when it is mounted over a flat mounting bracket (not used).

   b. Press the connector lock tab and disconnect the connectors.

   NOTE
   ***Figure 82** shows the antenna connector (A), and the left (B) and right (C) side audio connectors with the top shelter removed for clarity.*

12. Lift the rear of the top shelter and locate the antenna connector (**Figure 80**). Pull the antenna connector straight out to disconnect it from the audio unit.

13. Lift the front of the top shelter to expose the two audio unit connectors.

   a. Disconnect the left connector by pressing its lock tab (on left side of connector.).

   b. Disconnect the right connector by pressing its lock tab (on right side of connector).

14. Lift the top shelter and remove it from the motorcycle.

15. Identify the following connectors (**Figure 82**) before installing the top shelter:

   a. Antenna connector (A).
   b. Speaker connector (B).
   c. Audio connectors (C).

16. Check the area uncovered by the top shelter for any disconnected connectors, loose harness clamps, or loose or missing fasteners.

17. Unlock the windshield adjusters.

18. Place the top shelter into position.

19. Locate the headset connector and pass it through the left fairing pocket opening.

20. Locate the panel switch connector inside the left fairing pocket opening. Do not connect it.

21. Locate and position the antenna connector. Do not connect it.

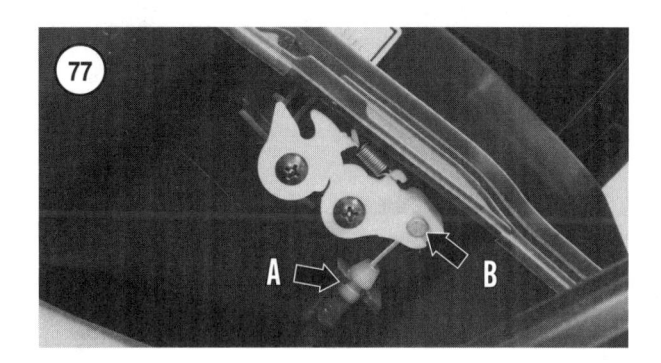

22. Lift the top shelter and connect the two audio unit connectors. Match the harness connectors when connecting them.

23. Push the antenna connector into the audio unit.

24. Align the top shelter mounting bolt holes and position the top shelter into position.

NOTE
If the top shelter does not drop into position, use the key to open the fuel tank cover. Then check to see if the fuel tank drain pan is interfering with the audio unit.

25. Working through the left fairing pocket opening, reconnect the panel switch connector. Then secure the connector and wiring harness to the main wiring harness with the plastic clamp. Secure the connector so it does not contact the radiator.

26. Slowly work to align the top shelter with the fairing mounting holes, meter panel mounting holes, the two rear studs and the top shelter inner cover.

27. Refer to **Figure 80** to identify and install the top shelter mounting bolts. Tighten each bolt securely.

28. Fit the windshield lock lever through the rubber boot (**Figure 83**).

29. Reverse Steps 1-6 to complete installation.

78

79

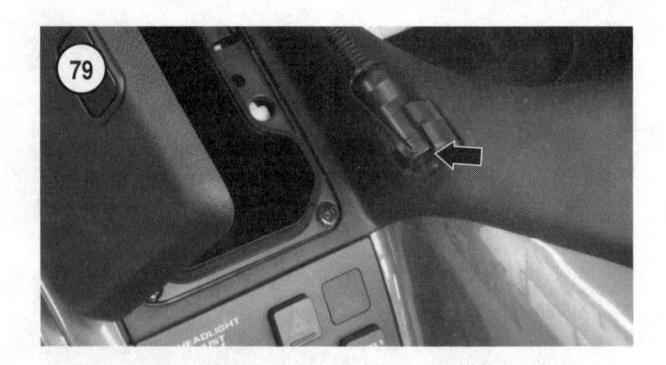

80 TOP SHELTER

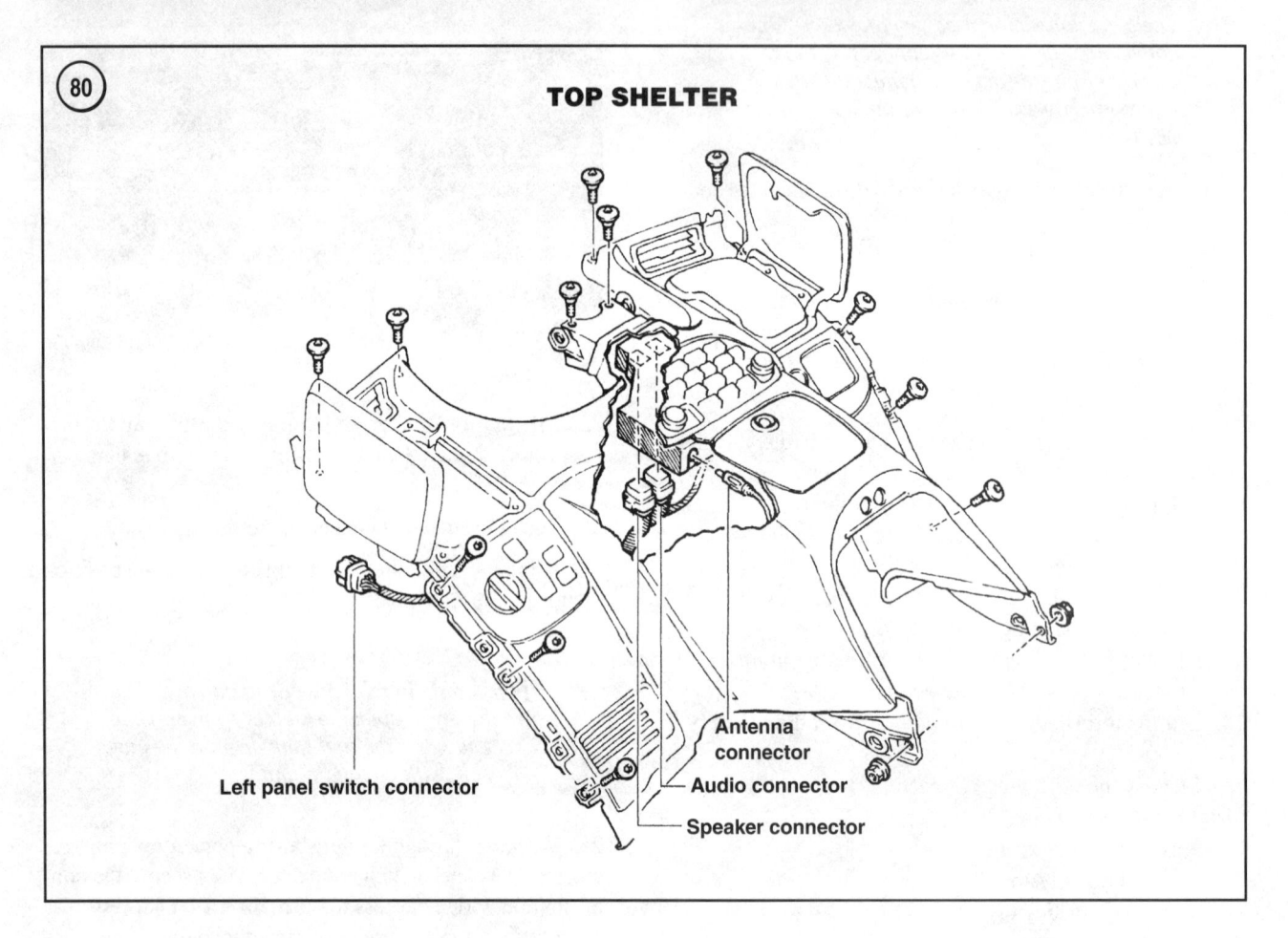

Left panel switch connector
Antenna connector
Audio connector
Speaker connector

REAR FENDER A

Rear fender A (**Figure 84**) is installed underneath the trunk and across the area separating the left and right saddlebags. The license plate mounts onto rear fender A.

Removal/Installation

1. Remove the license plate from the holder on rear fender A.

2. Remove the four mounting bolts and the center flange bolt from rear fender A.

NOTE
*The white dots in **Figure 84** show where the fender tabs are located.*

3. Release the tabs that secure rear fender A to the saddlebags, and remove rear fender A (**Figure 84**).

4. Installation is the reverse of these steps.

BODY COMPONENTS AND EXHAUST SYSTEM

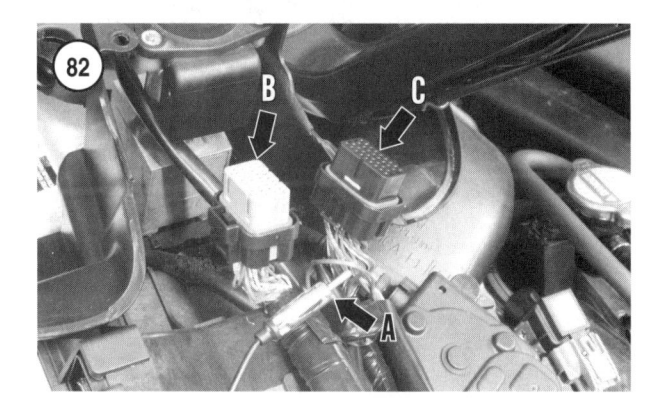

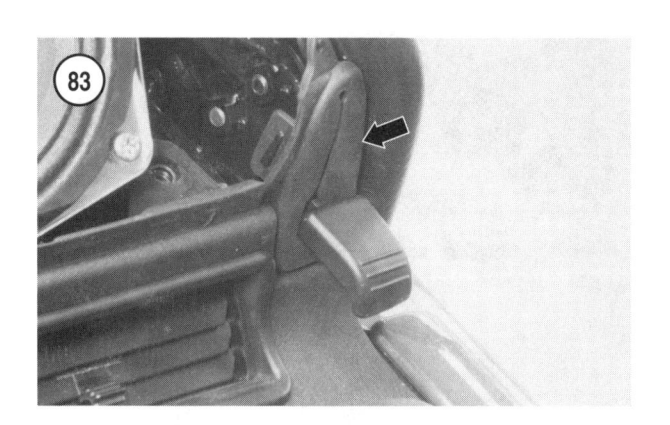

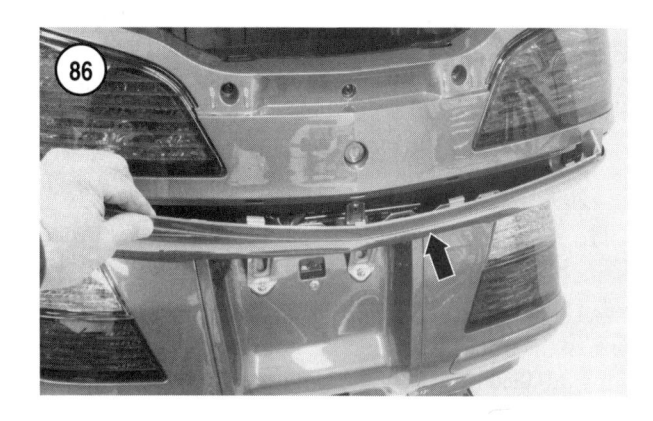

TRUNK LOWER COVER AND MOLDING STRIPS

The trunk lower cover (A, **Figure 85**) mounts around the bottom of the trunk. The side (B, **Figure 85**) and center (C) molding strips cover the gap between the trunk and trunk lower cover.

Removal/Installation

1. Open the trunk lid.

2. Remove the four self-tapping screws from inside the trunk (two each side) and remove the left and right side moldings (B, **Figure 85**).

3. Remove the two self-tapping screws from outside the trunk and remove the center molding (**Figure 86**).

4. Remove the seven self-tapping screws from the trunk lower cover (four inside the trunk and three outside).

5. Lightly pull the trunk lower cover forward to release the four screw studs from the trunk, then release the three compartment lid handles from the notch in the lower cover and remove the cover (**Figure 87**).

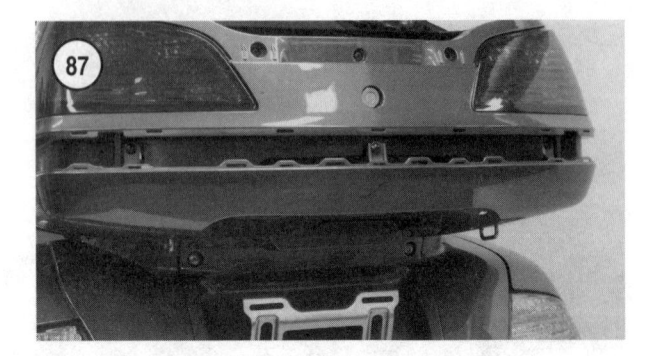

6. Installation is the reverse of these steps.

TRUNK

The trunk is mounted across the back of the motorcycle and over the saddlebags. The trunk has a storage capacity of 66 liters (61 liters if the optional CD changer is installed). Cargo weight in the trunk must not exceed 9.0 kg (20 lbs.).

Use the ignition key or the remote transmitter to lock and unlock the trunk and saddlebags. To use the key, insert the ignition key and turn it clockwise to unlock the compartment. Turn the key counterclockwise to lock the compartment. To use the remote transmitter, refer to *Remote Transmitter* in this chapter. To open the trunk after it is unlocked, pull its middle latch lever down. Close the trunk by pressing its lid down evenly with both hands.

NOTE
If the trunk and saddlebag open indicator is lit and flashing OPEN after the ignition switch is turned on, the trunk lid (and/or a saddlebag lid) is not properly closed.

A separate storage compartment is mounted in the bottom of the trunk. To open the storage box lid, push the lid forward, then raise it.

This section describes service procedures for the trunk lid, trunk lock cover and trunk.

Trunk Lid
Removal/Installation

1. Remove the seat as described under *Seat* in this chapter.
2. Open the trunk lid.
3. Disconnect the trunk control unit 12-pin connector (**Figure 88**).
4. Note the routing path of the trunk lid wiring harness before removing it (**Figure 89**).
5. Disconnect the bands securing the wire harness and remove the wire harness from the trunk lid (**Figure 90**).
6. Remove the bolts (**Figure 90** and **Figure 91**) and the trunk lid. See **Figure 92**.

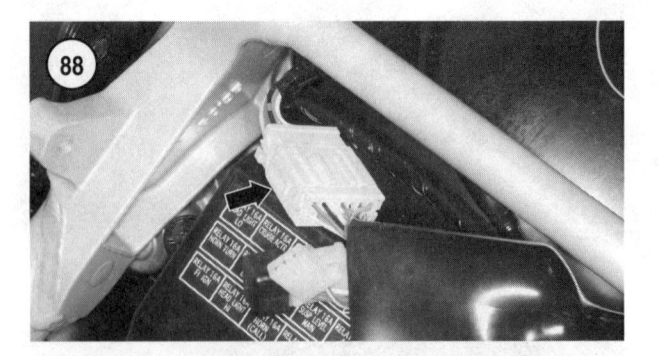

BODY COMPONENTS AND EXHAUST SYSTEM

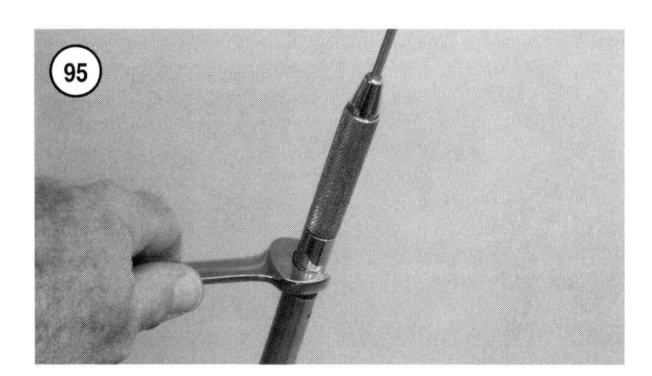

Trunk Lock Cover
Removal/Installation

The trunk lock cover is mounted inside the rear part of the trunk and protects the lock mechanism from damage.
1. Open the trunk lid.
2. Remove the two self-tapping screws and the three screws, and remove the trunk lock cover (**Figure 93**).
3. Installation is the reverse of these steps.

Trunk
Removal/Installation

1. Remove the following as described in this chapter:
    a. Seat.
    b. Rear fender A.
    c. Trunk lower cover.
    d. Trunk lock cover.
2. Detach the passenger headset connector from its holder on the left side of the trunk (**Figure 94**).
3. Loosen the nut and remove the antenna (**Figure 95**) from the base.

NOTE
Draw a diagram of the saddlebag opener cables, wiring harness and connectors before disconnecting them in the following steps.

4. Disconnect the left and right side saddlebag opener cables. These enter the trunk from the bottom side. Pry the plastic cable holders out of the metal brackets. See **Figure 96** (left) and **Figure 97** (right).
5. Disconnect the following electrical connectors:
    a. Brown 2-pin license plate light connector (**Figure 98**).
    b. Gray 8-pin connector (**Figure 99**).
6. Remove the three Acorn nuts (**Figure 100**) and pull the taillight lens assembly away from the trunk. Disconnect the two connectors (**Figure 101**) and remove the lens as-

7. Installation is the reverse of these steps, plus the following:

    a. Route the wiring harness along its original path inside the trunk lid and secure it with the bands. Make sure the wiring harness is not pinched when the trunk lid is closed.

    b. Clean and reconnect the 12-pin connector.

    c. Close the trunk. Check that the gap between the lid and compartment is even all around.

    d. Operate the remote transmitter to check the operation of the trunk lock.

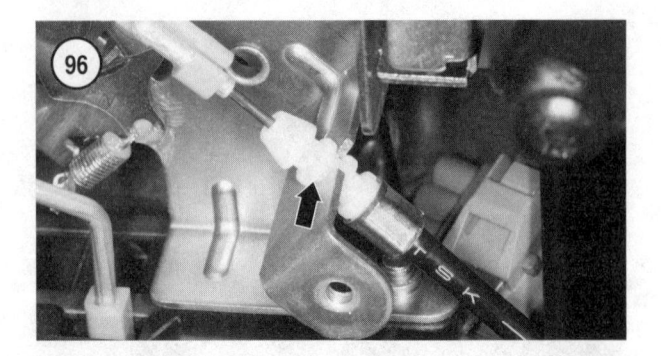

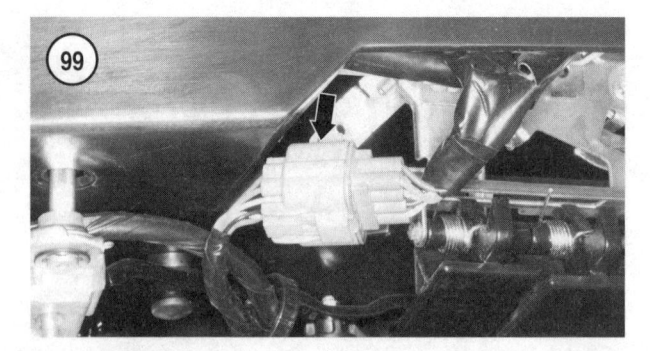

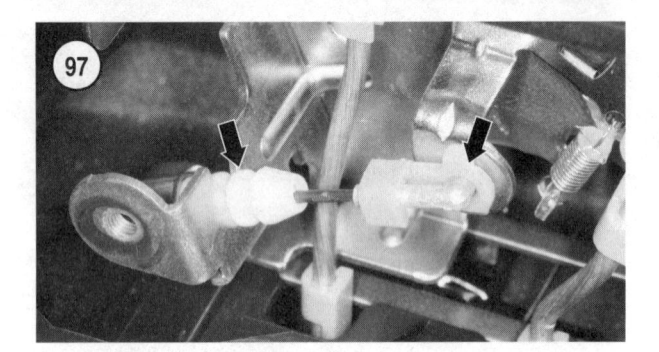

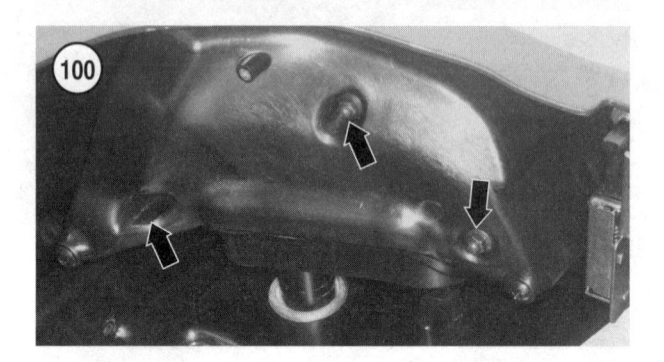

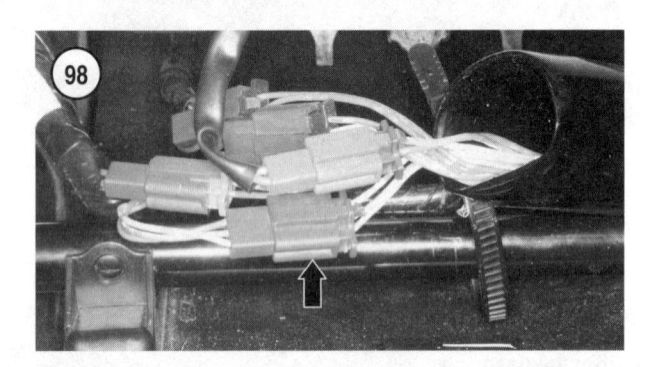

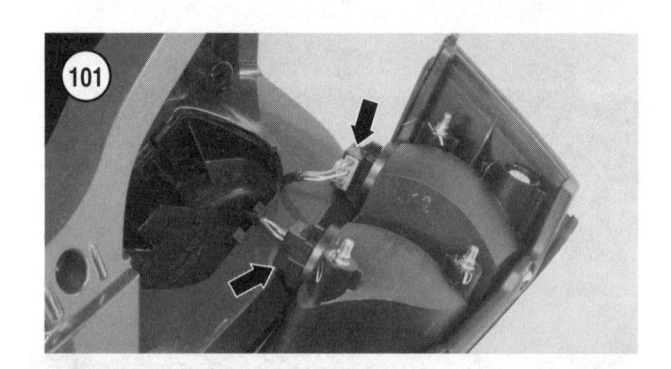

sembly. Pull the taillight connectors and harness from the trunk (both sides). Loosen the plastic band beneath the trunk and remove the taillight wiring harness. Repeat for the opposite side.

NOTE
The trunk is secured to its mounting bracket with four mounting bolts and three washers. Do not use a washer at the antenna ground strap.

7. Remove the four mounting bolts (**Figure 102**) and three washers from inside the trunk. Do not remove the trunk.

8. Disconnect the antenna connector located outside and in front of the trunk (**Figure 103**).

9. Lift the trunk slightly, and loosen the plastic band that secures the antenna cable to the frame. Pull the antenna cable through the band and away from the frame.

10. Remove the trunk.

BODY COMPONENTS AND EXHAUST SYSTEM

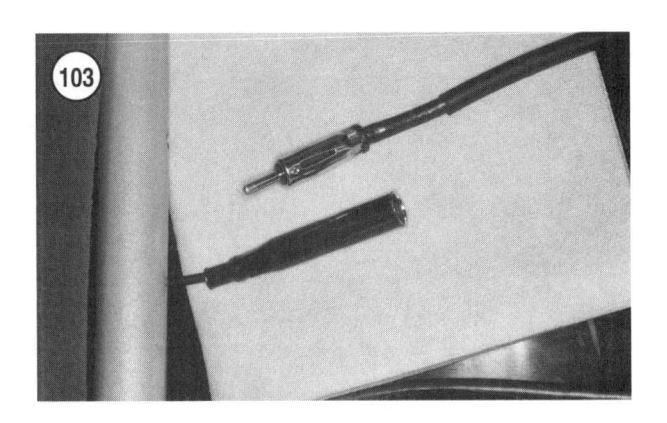

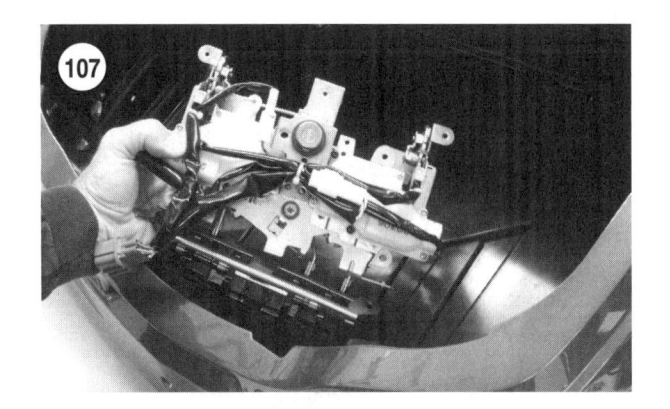

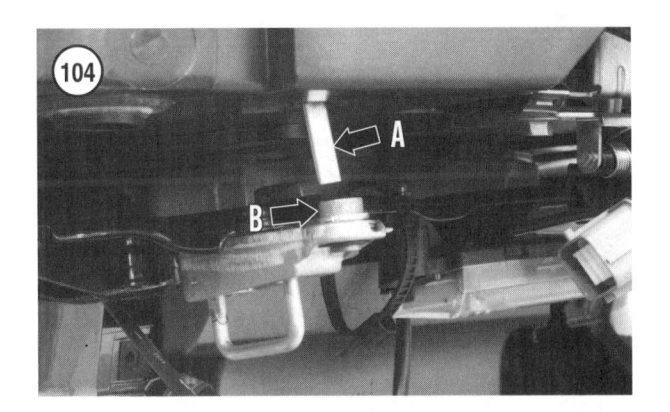

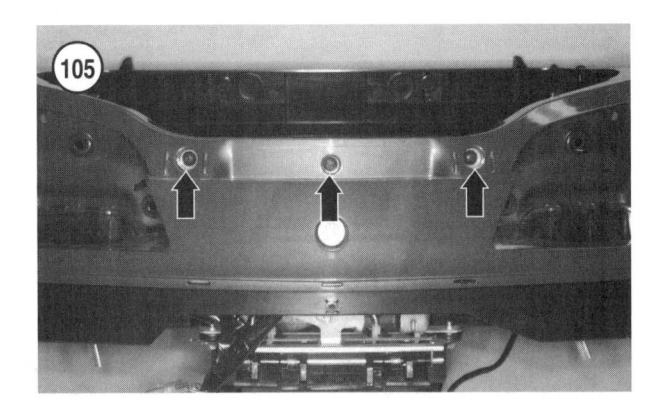

11. Installation is the reverse of these steps, plus the following:

   a. Route the wiring harnesses along their original paths. Secure the wiring harnesses with the bands.
   b. Clean the connectors with contact cleaner.
   c. Align the lock arm (A, **Figure 104**) with the tumblers (B) mounted on the frame.
   d. Check the operation of the lights, locks and remote transmitter.

Trunk Actuator

The trunk and latch handles are removed as an assembly. Do not disconnect or attempt to separate these parts.

1. Remove the three Phillips screws (**Figure 105**).
2. Remove the two bolts (**Figure 106**) beneath the lock mechanism.
3. Push the lock cylinder through the trunk while pivoting the mechanism away from the trunk. Remove the lock mechanism (**Figure 107**) from inside the trunk. See **Figure 108**.
4. Install the mechanism through the inside of the trunk. Align the three lever arms with the notches in the trunk.

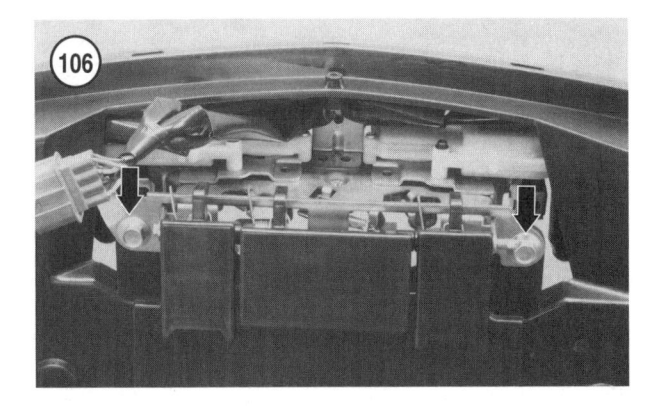

5. Route the gray 8-pin connector through the left side trunk opening.

6. Pivot the mechanism up while installing the key cylinder with the hole in the trunk. Push against the metal bracket on top of the mechanism to assist in aligning and installing the key cylinder in the trunk hole.

7. Install the shoulder bolts (**Figure 106**) and tighten finger-tight.

8. Install the three Phillips screws (**Figure 105**) and tighten securely.

9. Tighten the shoulder bolts securely.

10. Check the wire harness routing.

SADDLEBAGS

A saddlebag is mounted on each side of the motorcycle. Each saddlebag has a storage capacity of 40 liters. Cargo weight in each saddlebag must not exceed 9 kg. (20 lbs.).

Use the ignition key or the remote transmitter to lock and unlock the saddlebags. To use the key, insert the ignition key and turn it clockwise to unlock the compartment. Turn the key counterclockwise to lock the compartment. To use the remote transmitter, refer to *Remote Transmitter* in this chapter. To open a saddlebag after it is unlocked, pull its latch lever down. Close the saddlebag by pressing its lid down evenly with both hands.

NOTE
If the trunk and saddlebag open indicator is lit and flashing OPEN after the ignition switch is turned on, the trunk lid (and/or a saddlebag lid) is not properly closed.

Opening a Stuck Saddlebag

If a saddlebag is unlocked but does not open:

1. Open the trunk lid.
2. Remove the cover from the left or right access hole in the trunk floor.
3. Place a finger through the access hole and push the release rod. The saddlebag should open.

Saddlebag Removal/Installation

Follow this procedure to remove either the left or right side saddlebag.

1. Remove the following as described in this chapter:
   a. Seat.
   b. Left and right side covers.
   c. Rear fender A.
   d. Trunk lower cover.

BODY COMPONENTS AND EXHAUST SYSTEM

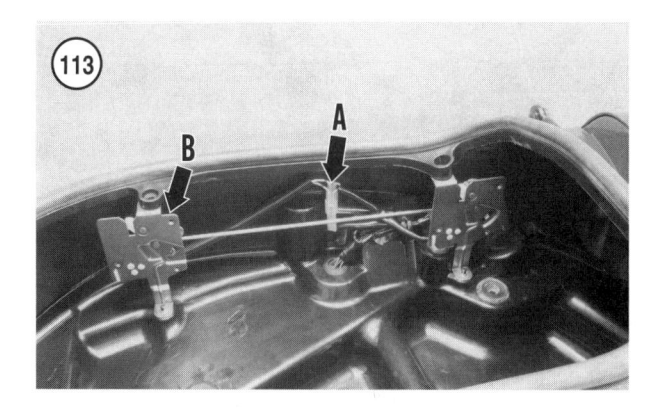

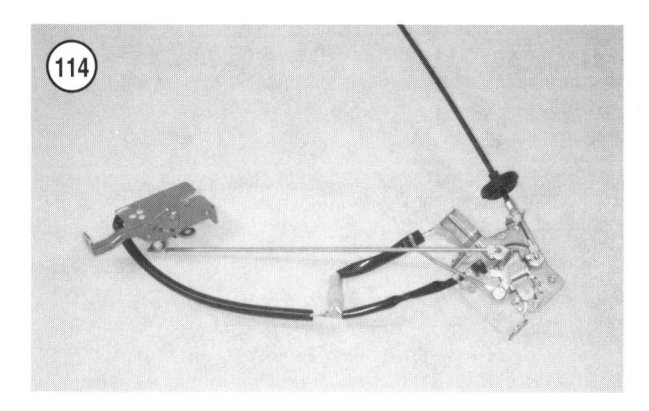

e. Trunk lock cover.

2. Open the saddlebag and empty its contents.

3. Disconnect the saddlebag opener cable. See **Figure 96** (left) or **Figure 97** (right). Note the cable routing.

4. Disconnect the two sub-wire harness connectors located underneath the seat.

5. Remove the bolts and washers that secure the saddlebag to its mounting bracket—remove the top bolts first, then the bottom bolts (**Figure 109**).

6. Close and remove the saddlebag. See **Figure 110**.

7. Remove the four collars from the mounting bolt holes in the saddlebag.

8. Installation is the reverse of these steps. Make sure the saddlebag locks and unlocks correctly with the ignition key and remote transmitter. If the saddlebag does not open, refer to *Opening a Stuck Saddlebag* in this section.

Saddlebag Opener
Removal/Installation

The saddlebag opener is the open/lock assembly for the saddlebag. It is mounted on the inside top of the saddlebag. Follow this procedure to remove either the left or right side saddlebag catch.

1. Remove the saddlebag as described in this section.

2. Remove the screws and the inner cover (**Figure 111**).

3. Perform the following at the backside of the saddlebag:

   a. Push the release cable grommet (A, **Figure 112**) inside the saddlebag.

   b. Remove the large rubber release cover (B, **Figure 112**).

4. Remove the two screws from the backside of the saddlebag and lower the opener mechanism.

5. Disconnect the white 2-pin connector (A, **Figure 113**) and remove the opener mechanism (B) from inside the saddlebag. See **Figure 114**.

6. See **Figure 115** to disconnect and replace the release cable.

7. Installation is the reverse of removal. Note the following:

   a. Make sure the side of the mechanism opposite the release cable is positioned as shown in **Figure 116**. This side can turn 180° out during installation.

   b. Align the hook in the inner cover with the release arm when installing the inner cover.

   c. Operate the manual release lever to make sure the opener mechanism operates correctly.

REAR FENDER B

Rear fender B (**Figure 117**) is mounted underneath the trunk and can be removed with the trunk and the saddlebag/trunk stay mounted on the motorcycle. A number of electrical components are mounted on rear fender B.

Removal/Installation

1. Disconnect the negative battery cable. Refer to *Battery* in Chapter Nine.
2. Remove both saddlebags as described in this chapter.
3. Remove the two screws that hold the relay box assembly to rear fender B. Do not disconnect relay connectors.
4. Remove starter relay switches A and B.
5. Remove power control relays 1 and 2.
6. Remove the speed limiter relay.
7. Remove the reverse regulator assembly from the right side of the fender.
8. Remove the ABS control unit, if so equipped.
9. Remove the bolts that hold rear fender B to the saddlebag/trunk stay. Then release the hooks and remove rear fender B.
10. Installation is the reverse of these steps.

SADDLEBAG/TRUNK STAY

The saddlebag/trunk stay (**Figure 118**) is bolted across the rear of the frame. Always check the tightening torque of the stay mounting bolts when having access to them.

This section describes service procedures for the passenger footrest under cover and the saddlebag/trunk stay.

Passenger Footrest Under Cover
Removal/Installation

This procedure can be used to remove either the left or right passenger footrest under cover.
1. Remove the side cover as described in this chapter.
2. Remove the three bolts that hold the under cover (**Figure 119**) to the frame, and remove the under cover.
3. Installation is the reverse of these steps.

Saddlebag/Trunk Stay
Removal/Installation

1. Remove the following as described in this chapter:
 a. Trunk.
 b. Both saddlebags.
 c. Both passenger footrest under covers.

2. Remove the bolts and both exhaust pipe guards.

3. Remove the two rear fender B mounting bolts at the rear of the saddlebag/trunk stay. Do not remove the two front rear fender B mounting bolts.

4. Remove the two nuts and six bolts that hold the stay to the frame, and remove the stay (**Figure 118**).

5. Installation is the reverse of these steps. Tighten the six stay mounting bolts to 26 N•m (19 ft.-lb.).

REMOTE TRANSMITTER

The remote transmitter (**Figure 120**) is an electronic remote control device used to lock and unlock the trunk and saddlebags. The transmitter is also equipped with a call function that sounds the motorcycle horn and activates the turn signal lights. Operating the call function helps to locate the motorcycle in large parking areas.

Operation

When operating and troubleshooting the remote transmitter, note the following:

BODY COMPONENTS AND EXHAUST SYSTEM

119

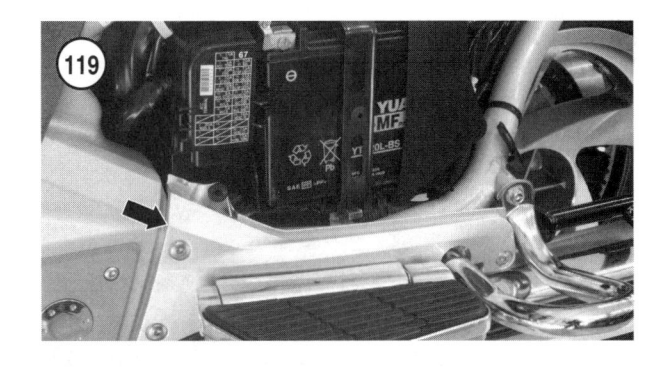

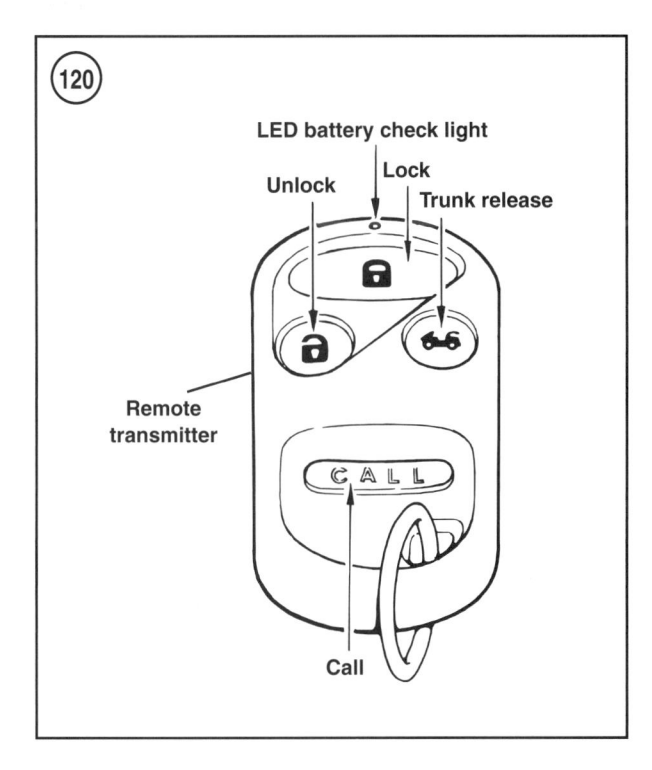

120

LED battery check light
Unlock
Lock
Trunk release
Remote transmitter
Call

1. When the ignition switch is turned on or to the ACC position, the TRUNK RELEASE and CALL functions will not work.

2. When the ignition switch is turned off or to the LOCK position, pushing the TRUNK RELEASE button will unlock the trunk and cause the turn signal lights to blink twice.

3. When the ignition key is turned on or to the ACC position, the UNLOCK and LOCK functions will work, but the turn signal lights will not blink.

4. The LOCK button does not lock the trunk if any lid is open. The turn signal lights blink ten times to warn that the trunk is not locked.

5. The trunk will automatically lock if the TRUNK RELEASE button is pushed but none of the lids are opened within 30 seconds.

6. When the CALL button is pushed for more than a 1/2 second, both horns will sound two times and the turn signal lights will blink two times.

7. With the ignition key turned off or to LOCK, pushing the LOCK button will cause the turn signal lights to blink once.

Trunk Control Unit Programming

1. Note the following before programming the control unit:
    a. Each step that is required to program the trunk control unit must be performed within 1 to 4 seconds. If an individual programming step is under or over this time limit, the control unit will lose or erase the stored codes.
    b. Three separate codes (using a separate transmitter for each code can be programmed into the trunk control unit memory at one time. If a fourth code is programmed, the first code will be erased.

2. Turn the ignition switch on. Aim the transmitter at the trunk and press the TRUNK RELEASE button (within 1-4 seconds). Then turn the ignition off (within 1-4 seconds).

3. Turn the ignition switch on. Aim the transmitter at the trunk and press the TRUNK RELEASE button (within 1-4 seconds). Then turn the ignition off (within 1-4 seconds).

4. Turn the ignition switch on. Aim the transmitter at the trunk and press the TRUNK RELEASE button (within 1-4 seconds). Then turn the ignition off (within 1-4 seconds).

5. Turn the ignition switch on. Aim the transmitter at the trunk and press the TRUNK RELEASE button (within 1-4 seconds). Check that the turn signal lights blink two times. Leave the ignition switch on.

6A. Programming one transmitter: Within 10 seconds of performing Step 4, aim the transmitter at the trunk and press the TRUNK RELEASE button. Check that the turn signal lights blink two times. Then turn the ignition switch off.

6B. Programming up to three transmitters: Within 10 seconds of performing Step 4, aim the transmitters at the trunk and press the TRUNK RELEASE button on each transmitter. Check that the turn signal lights blink two times. Then turn the ignition switch off.

7. Operate the transmitter to check that the new codes were stored correctly. If more than one code was stored into the trunk opener unit memory, check the operation of each transmitter separately.

Care and Operation of the Remote Transmitter

1. Avoid dropping the remote transmitter.
2. Protect the remote transmitter from extreme hot and cold temperatures.
3. Use a mild cleaning solution and a soft cloth to clean the transmitter housing. Do not immerse the remote transmitter in any liquid.
4. If the remote transmitter is lost, the replacement transmitter must be reprogrammed to the motorcycle as described in the prior section.
5. When the buttons must be pushed several times to operate the locks, or when the LED on the top of the transmitter (**Figure 120**) becomes dim, the battery is weak. Replace with a CR2025 battery.

Opening and Locking the Trunk and Saddlebags

To use the remote transmitter (**Figure 120**), perform the following:

NOTE
Operating the remote transmitter will lock and unlock the trunk and both saddlebags at the same time.

1. Turn the ignition switch off.
2. To lock the compartments (trunk and both saddlebags), confirm that they are fully closed. Then push the LOCK button one time. If the action is successful, the front and rear turn signals will blink one time.

NOTE
If the front and rear turn signals blink ten times after operating the lock button, one or more compartment lids were not fully closed, and the remote transmitter cannot lock any of the fully closed compartments. Check for the open compartment—the gap between the compartment lid and compartment must be even all around.

NOTE
The turn signal lights will not blink if the ignition switch is on or in the ACC position.

3. To unlock the compartments, push the UNLOCK button one time.

NOTE
If the remote transmitter is used to unlock the compartments, but a compartment lid does not opene within 30 seconds, the compartments will automatically relock.

Unlocking the Trunk Separately

The trunk lock can be unlocked separately by pressing the TRUNK RELEASE button (**Figure 120**) for approximately one second. The TRUNK RELEASE button does not operate when the ignition switch is on or in the ACC position.

Using the Call Function

Push and hold the CALL button (**Figure 120**). The horn will sound and the turn signal lights will blink two times. The CALL function does not operate when the ignition switch is on or in the ACC position.

REMOTE TRANSMITTER AND CIRCUIT TROUBLESHOOTING

If the remote transmitter does not operate properly, locate the section that best describes the problem.

Transmitter Does Not Operate the Power Trunk System

1. Check the transmitter for water damage. If good, go to Step 2.
2. Install a new battery (CR2025) and see if the transmitter now works.
   a. Yes.
   b. No. Go to Step 3.
3. Reprogram the transmitter as described in this section and see if the transmitter now works.
   a. Yes.
   b. No. Go to Step 4.
4. Open the trunk and remove the inner cover (**Figure 121**). Disconnect the white 18-pin trunk lock control unit connector (**Figure 122**).

BODY COMPONENTS AND EXHAUST SYSTEM

122 **TRUNK LOCK CONTROL UNIT**

18-pin White connector connector terminal side

Red Yel	Red	Yel			Pnk Wht	Pink	
		Grn	Brn Red			Wht	Blk

Trunk lid

Trunk lock control unit
18-pin (Wht)

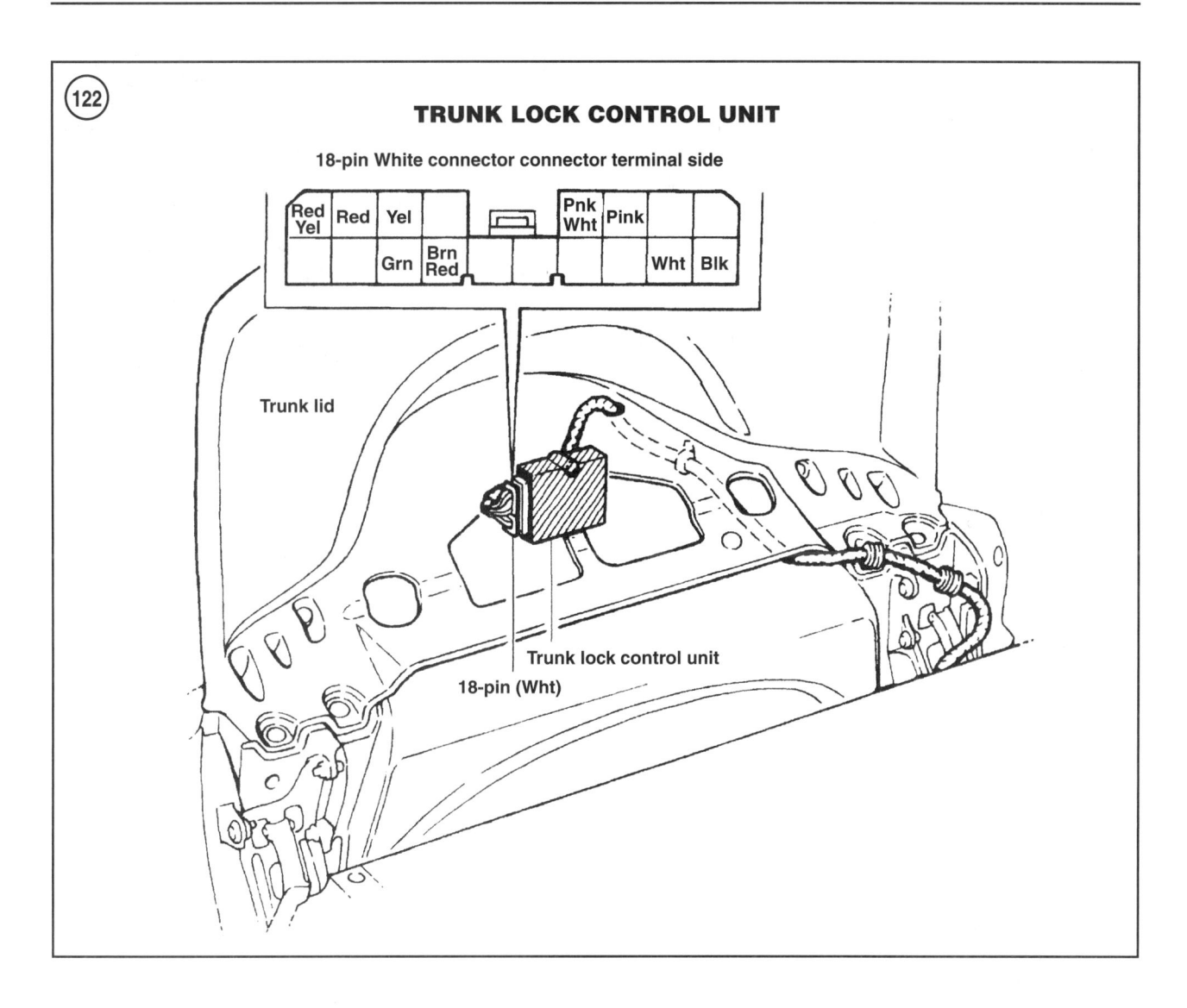

5. Check for continuity between the green terminal and ground.
   a. Yes. Go to Step 6.
   b. No. Repair the open circuit in the green wire.
6. Check for battery voltage between the red/yellow (+) connector terminal and ground (−).
   a. Yes. Replace the trunk lock control unit as described under *Trunk Lock Control Unit* in this chapter.
   b. No. Check for a blown No. 22 fuse. If the fuse is good, check the red/yellow wire for an open circuit.

Transmitter Does Not Lock or Unlock Trunk Opener

1. Open the trunk and remove the inner cover (**Figure 121**). Disconnect the white 18-pin trunk control unit connector (**Figure 122**).
2. Unlock the trunk opener with the key.
3. Briefly apply battery voltage to the yellow (+) and red (−) connector terminals. The trunk opener should lock.
   a. Yes. Go to Step 4.
   b. No. Check the red and yellow wires for an open circuit. If good, check the connector for contamination or pin damage. If good, check the lock/unlock actuator for damage.
4. Briefly apply battery voltage to the red (+) and yellow (−) connector terminals. The trunk opener should unlock.
   a. Yes. Replace the trunk lock control unit as described under *Trunk Lock Control Unit* in this chapter.
   b. No. Replace the trunk actuator as described under *Trunk* in this chapter.

Transmitter Can Unlock Trunk Opener Unit, But Can Not Unlock It

1. Check that the trunk and saddlebag lids are closed.
2. Open the trunk and remove the inner cover (**Figure 121**). Disconnect the white 18-pin trunk control unit connector (**Figure 122**).
3. Check for continuity between the brown/red terminal and ground. There should be no continuity:
    a. If there is no continuity, go to Step 4.
    b. If there is continuity, check the trunk and saddlebag open switches as described in this chapter. If the switches are good, check for a short circuit in the blue/red, white/red, brown/red and red/white wires between the open switches and the trunk control unit.
4. Unlock the trunk opener with the key.
5. Check for continuity between the white wire terminal and ground. There should be no continuity.
    a. If there is no continuity, replace the trunk lock control unit as described under *Trunk Lock Control Unit* in this chapter.
    b. If there is continuity, check for a short circuit in the white wire. If the wire is good, check the lock/unlock switch as described in this section.

Transmitter Can Lock Trunk Opener But Can Not Unlock It

1. Open the trunk and remove the inner cover (**Figure 121**). Disconnect the white 18-pin trunk control unit connector (**Figure 122**).
2. Lock the trunk with the ignition key.
3. Check for continuity between the black terminal and ground. There should be no continuity:
    a. If there is no continuity, replace the trunk lock control unit as described under *Trunk Lock Control Unit* in this chapter.
    b. If there is continuity, check for a short circuit in the black wire. If the black wire is good, check the lock/unlock switch as described in this section.

Transmitter Can Lock and Unlock Trunk Opener But the Turn Signal Lights Do Not Blink

1. Before testing, check that:
    a. The turn signal/hazard system works properly.
    b. The ignition switch is turned off or to LOCK.
    c. If both conditions are met, go to Step 2.
2. Open the trunk and remove the inner cover (**Figure 121**). Disconnect the white 18-pin trunk control unit connector (**Figure 122**).
3. Remove the hazard switch diode. Refer to *Hazard Switch Diode* in Chapter Nine.
4. Check for continuity in the pink/white wire between the hazard switch diode and trunk lock control unit.
    a. Yes. Test the hazard switch diode as described in Chapter Nine. If the diode is good, replace the trunk lock control unit as described in this chapter.
    b. No. Check the hazard switch diode and trunk lock control unit connectors for dirt or loose terminal pins. If good, check the pink/white wire for an open circuit.

Transmitter Does Not Open the Trunk Lid

Check for the following conditions:
1. See if the ignition switch is turned on or in the ACC position. If so, turn the ignition switch off and retry. If the trunk lid still does not open, go to Step 2.
2. Check for a damaged trunk lid opener actuator.
3. Check the green wire for an open circuit between the opener actuator and ground.
4. Check the blue wire for an open circuit between the open actuator and trunk control unit.
5. Check all the connectors for dirt or damage.
6. The trunk control unit may be damaged.

Saddlebag and Trunk Lid Open, But Open Indicator Does Not Work

Check for the following:
1. Damaged trunk or saddlebag open switch:
    a. Test the trunk switch as described under *Trunk Actuator Test* in this chapter.
    b. Test the saddlebag switch as described under *Saddlebag Open Switch* in this chapter.
2. Check for an open circuit in the following wires:
    a. Left saddlebag—blue/red and brown/red wires.
    b. Right saddlebag—white/red and brown/red wires.

BODY COMPONENTS AND EXHAUST SYSTEM

124

Trunk actuator 8-pin connector (connector terminal side)

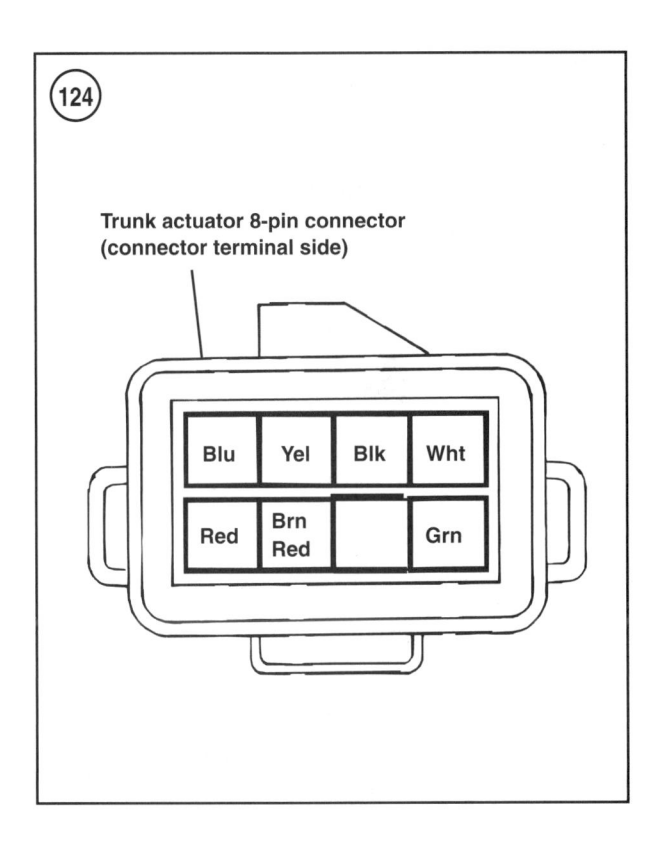

| Blu | Yel | Blk | Wht |
| Red | Brn/Red | | Grn |

c. Trunk lid—Brown/red wire.

Horns Do Not Sound When the CALL Button is Pushed

1. The horns should work when the ignition switch is turned on or to the ACC position:
    a. Yes. Go to Step 2.
    b. No. Test the horns as described in Chapter Nine under *Horns*.
2. Make sure the ignition switch is turned off or in the LOCK position when the CALL button is pushed.
    a. Yes. Go to Step 3.
    b. No. Turn the ignition switch off or to LOCK and retry the CALL button. If the horns still do not sound, go to Step 3.
3. Open the trunk and remove the inner cover (**Figure 121**). Disconnect the white 18-pin trunk control unit connector (**Figure 122**).
4. Ground the pink wire terminal with a jumper wire. Both horns should sound:
    a. Yes: Refer to *Trunk Lock Control Unit* in this chapter and replace the unit.
    b. No: Test the horn relay as described in Chapter Nine under *Relay Box*. If the horn relay is good, check for an open circuit in the pink, light green and red/yellow wires.

TRUNK LOCK CONTROL UNIT

The trunk lock control unit is mounted inside the trunk lid.

Replacement

1. Open the trunk and remove the inner cover (**Figure 121**). Disconnect the white 18-pin trunk control unit connector (**Figure 122**).
2. Remove the screws and the trunk lock control unit.
3. Installation is the reverse of these steps.

TRUNK ACTUATOR TEST

Procedure

1. Remove the trunk lower cover as described under *Trunk Lower Cover and Molding Strips* in this chapter.
2. Disconnect the trunk actuator 8-pin connector (A, **Figure 123**).
3. Perform the following tests to check the trunk actuator.
4. Installation is the reverse of removal.

Lock/unlock actuator test

1. Unlock the trunk opener with the ignition key.
2. Briefly apply battery voltage to the yellow (+) and red (−) connector terminals (**Figure 124**). The trunk should lock.
3. Briefly apply battery voltage to the red (+) and yellow (−) connector terminals (**Figure 124**). The trunk should unlock.
4. If the trunk actuator failed to operate as described in Step 2 or Step 3, replace the trunk actuator as described in this chapter.

Lock/unlock switch test

1. Check continuity between the green and white connector terminals (**Figure 124**) and note the following:
    a. There should be continuity with the actuator locked.
    b. There should be no continuity with the actuator unlocked.
2. Check continuity between the green and black connector terminals (**Figure 124**) and note the following:
    a. There should be continuity with the actuator unlocked.

b. There should be no continuity with the actuator locked.

3. If the trunk opener failed to operate as described in Step 1 or Step 2, replace the trunk actuator as described under *Trunk* in this chapter.

Trunk open switch

1. Check continuity between the green and brown/red connector terminals (**Figure 124**) as follows:
    a. There should be no continuity with the trunk lid closed.
    b. There should be continuity with the trunk lid open.
2. If the trunk opener failed to operate as described in Step 1, replace the trunk actuator as described under *Trunk* in this chapter.

Trunk lid opener actuator

1. Close the trunk lid.
2. Briefly apply battery voltage to the blue (+) and green (–) connector terminals (**Figure 124**). The trunk lid should open.
3. If the trunk lid did not open, replace the trunk actuator as described under *Trunk* in this chapter.

SADDLEBAG OPEN SWITCH TEST

1. Remove rear fender A as described in this chapter.
2. Slide the group of connectors out of the pouch (**Figure 125**).
3. Disconnect the saddlebag 2-pin and 3-pin connectors (**Figure 126**). The left saddlebag connectors are red. The right saddlebag connectors are blue.
4. Check continuity between the brown/red terminal (3-pin connector) and the green terminal (2-pin connector) as follows (**Figure 127**):
    a. There should be continuity with the saddlebag lid open.
    b. There should be no continuity with the saddlebag lid closed.
5. If the saddlebag open switch failed to operate as described in Step 4, replace the saddlebag opener assembly as described under *Saddlebags* in this chapter.
6. Installation is the reverse of removal.

OPEN SWITCH DIODE TEST

1. Remove the seat as described in this chapter.

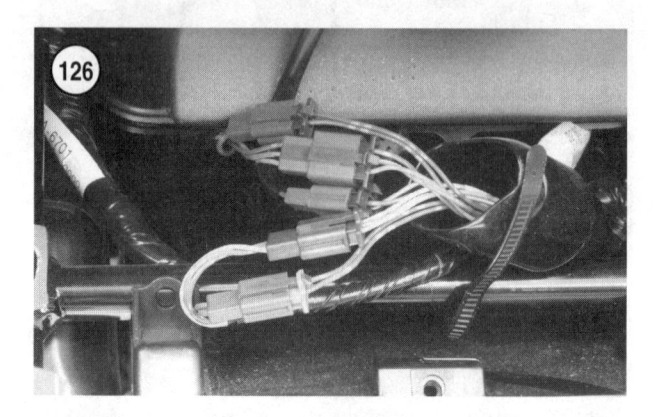

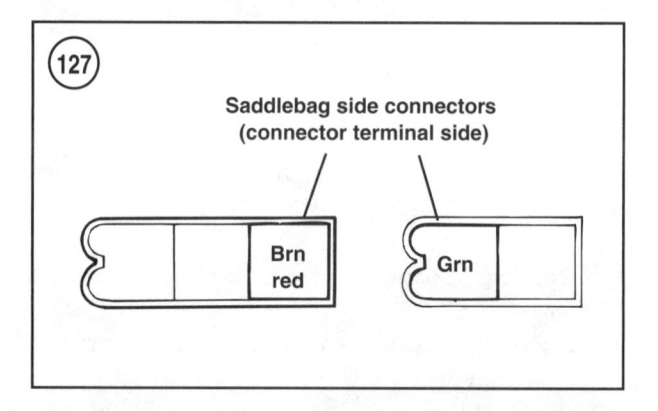

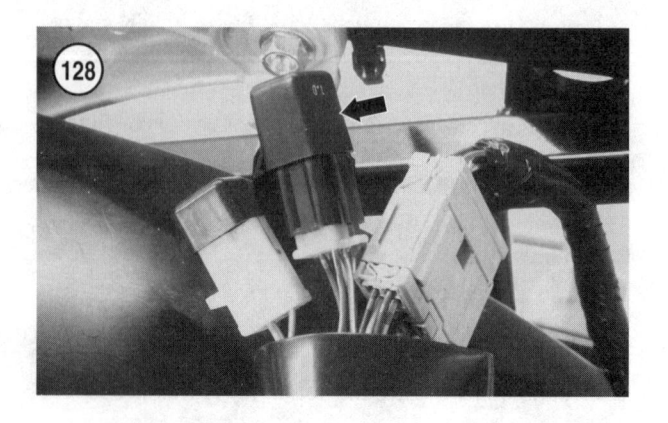

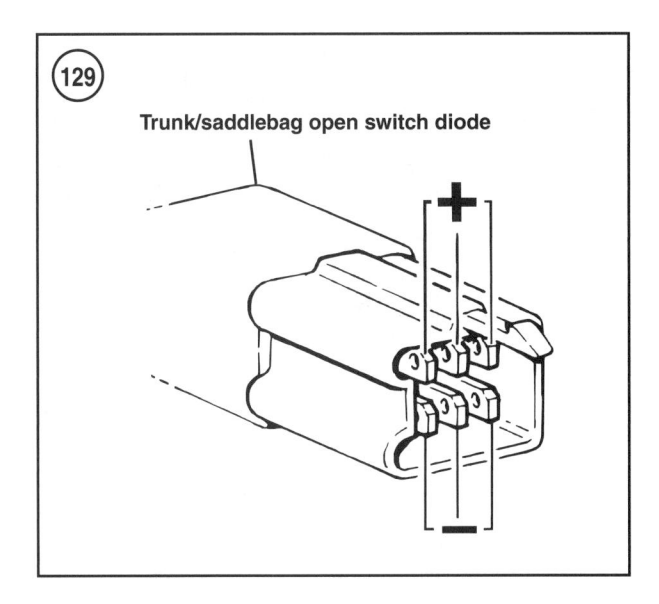

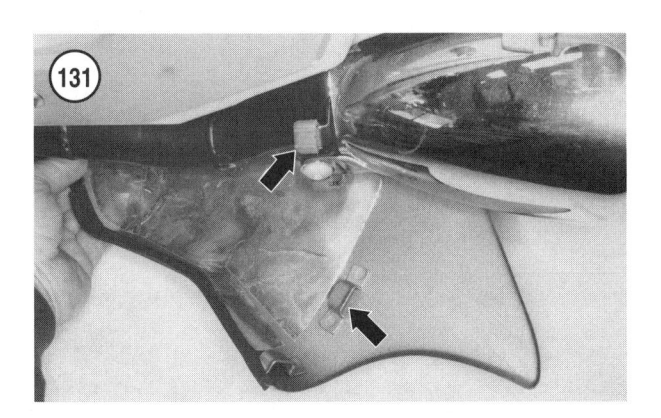

BODY COMPONENTS AND EXHAUST SYSTEM

129 Trunk/saddlebag open switch diode

check the continuity in one direction, reverse the test leads, and check the continuity in the opposite direction. Each pair should have continuity in one direction and no continuity when the test leads are reversed.

5. Replace the diode if it fails this test.

EXHAUST SYSTEM

Front Exhaust Pipe Protector Removal/Installation

1. Remove the front lower fairing as described in this chapter.

2. Remove the bolt and two nuts from the exhaust pipe protector (**Figure 130**). The fiber washers may come off or tear during removal. Install new fiber washers during installation.

3. Pull the exhaust pipe protector down and then forward and release its hooks from the mounting tabs on the front and rear cylinder exhaust pipe flange (front tabs) and the exhaust pipe (rear tabs).

4. Remove the exhaust pipe protector.

5. Installation is the reverse of removal. Note the following:

   a. Install the rubber dampers onto the front and rear tabs (if removed).
   b. Install new fiber washers.
   c. Install the exhaust pipe protector by connecting the hooks on the protector (**Figure 131**, typical) with the front and rear tabs.
   d. Tighten the exhaust pipe protector bolt and nuts to 12 N•m (106 in.-lb.).

Muffler Protector Removal/Installation

1. Remove the front exhaust pipe protector as described in this section.

2. Remove the screw and washer. Slide the muffler protector (**Figure 132**) forward to release its hooks from the mounting tabs on the muffler (**Figure 133**), and remove the protector.

3. Installation is the reverse of removal. Note the following:

   a. Install the rubber dampers onto the mounting tabs (if removed).
   b. Tighten the screw securely.

2. Locate the connector pouch next to the relay box and disconnect the trunk/saddlebag open switch diode (**Figure 128**) from its connector.

3. Set an ohmmeter to the R × 1 scale.

4. Check continuity between the + and - diode terminals (**Figure 129**). At each pair of opposite terminals (+ and -)

Muffler
Removal/Installation

The mufflers can be removed separately.
1. Support the motorcycle on the centerstand.
2. Loosen the muffler clamp bolts (A, **Figure 134**) at the exhaust pipe.
3. Remove the mounting bolt (B, **Figure 134**) and remove the muffler.
4. Replace the gasket (**Figure 135**) if leaking or damaged.
5. Installation is the reverse of removal.

Exhaust Pipe
Removal/Installation

Refer to **Figure 136**.
The exhaust pipe assembly is removed as an assembly, then the left and right sides are separated.
1. Support the motorcycle on the centerstand.
2. Remove the top shelter as described in this chapter.
3. Remove the exhaust pipe protectors as described in this section.
4. Remove the mufflers as described in this section.
5. Disconnect the left side oxygen sensor connector (**Figure 137**):
   a. Note the wire harness routing from the connector to the sensor.
   b. Lift the tab on the side of the connector to disconnect it.
   c. Remove the two clamps (at the brake fluid pipe and at the sidestand switch wire harness) that secure the oxygen sensor wire harness to the frame.

NOTE
The left oxygen sensor wire harness is routed behind the sidestand switch wire harness and above the fuel tank overflow drain hose.

6. Disconnect the right side oxygen sensor connector (**Figure 138**):
   a. Note the wire harness routing from the connector to the sensor. The wire harness is routed behind the master cylinder reservoir mounting bracket and above the lower coolant hose.
   b. Locate the oxygen sensor connector in the connector pouch beside the air intake nozzle.
   c. Lift the tab on the side of the connector to disconnect it.
7. Loosen the connecting pipe pinch bolts (**Figure 139**).
8. Loosen the rear exhaust pipe mounting bolts (each side). See **Figure 140**.

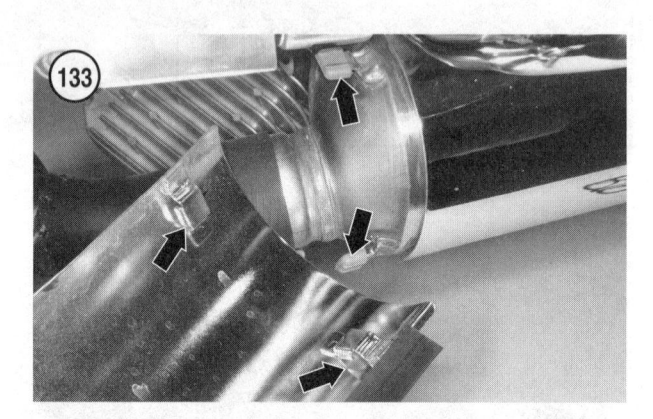

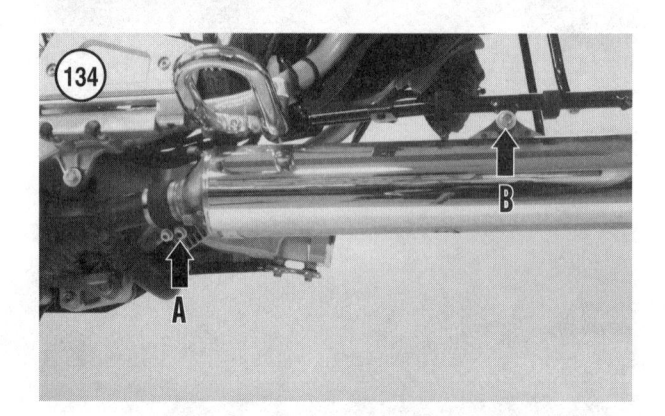

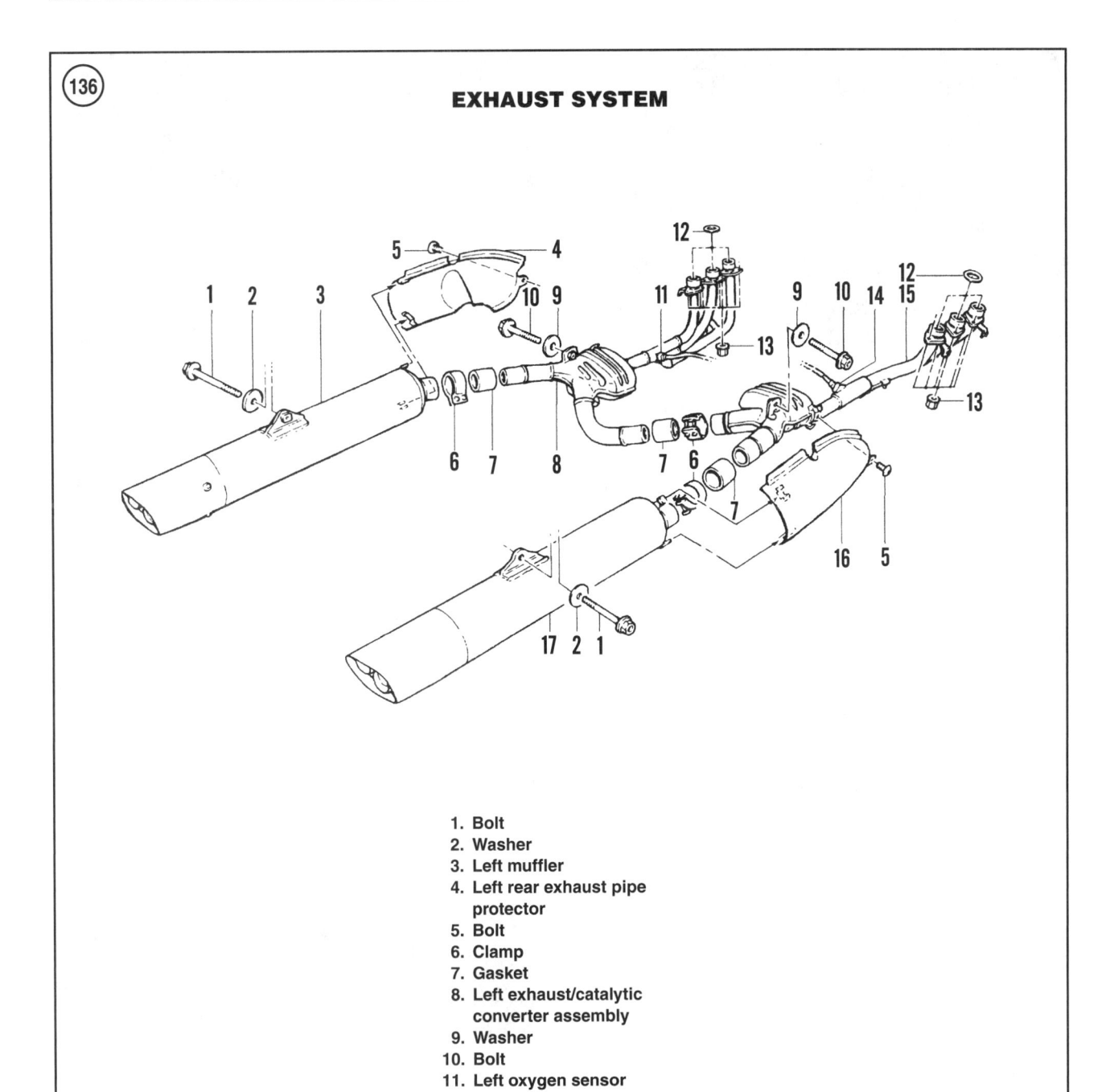

BODY COMPONENTS AND EXHAUST SYSTEM

EXHAUST SYSTEM

1. Bolt
2. Washer
3. Left muffler
4. Left rear exhaust pipe protector
5. Bolt
6. Clamp
7. Gasket
8. Left exhaust/catalytic converter assembly
9. Washer
10. Bolt
11. Left oxygen sensor
12. Gasket
13. Nut
14. Right oxygen sensor
15. Right exhaust/catalytic converter assembly
16. Right rear exhaust pipe protector
17. Right muffler

9. Remove the two lower mounting bolts (**Figure 141**) from the right front engine guard.
10. Place a small jack underneath the rear part of the exhaust pipe assembly. Position the jack so there is a small gap between the jack pad and exhaust pipe.
11. Loosen and remove the exhaust pipe nuts at each cylinder head. See **Figure 142** (left) and **Figure 143** (right).

CAUTION
Remove the exhaust pipes carefully to avoid damaging the oxygen sensors and wiring harnesses.

12. Remove the rear exhaust pipe mounting bolts (each side) and lower the exhaust pipe assembly onto the jack (**Figure 144**).
13. Carefully remove each oxygen sensor wiring harness from its path through the frame.
14. Remove the gasket (**Figure 145**) from each exhaust port.

CAUTION
The oxygen sensors angle inward on the exhaust pipe and can contact the centerstand. Remove the exhaust system as described in Step 15.

15A. If the rear wheel is installed, separate the exhaust pipes and remove them.
15B. If the rear wheel is removed, pivot one of the exhaust pipes up to provide clearance between the oxygen sensors and centerstand and remove the exhaust system.
16. **Figure 146** shows the exhaust pipe assembly.

Installation

1. If necessary, use a 6 × 1.00 metric die to repair any damaged or nicked exhaust pipe stud threads.
2. Install a new gasket into each exhaust port.

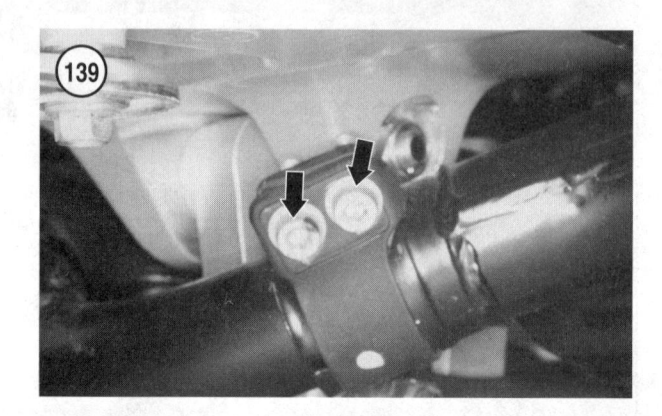

BODY COMPONENTS AND EXHAUST SYSTEM

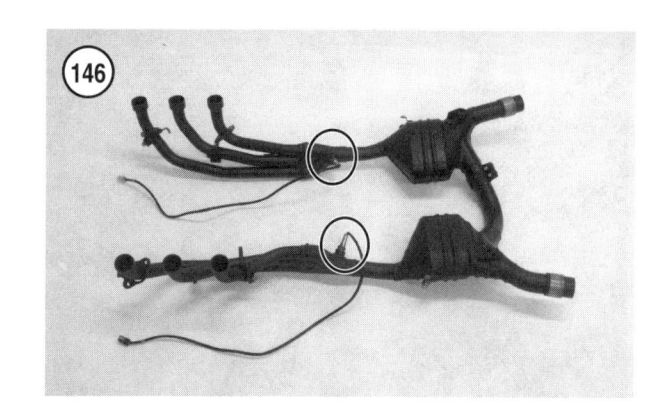

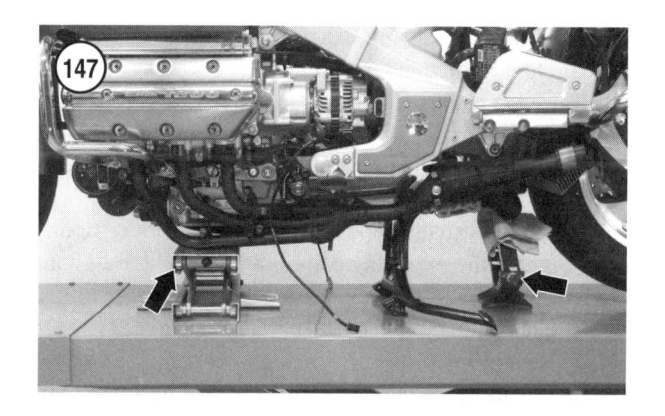

3. Place the exhaust system beneath the frame. If the rear tire is mounted on the bike, install each side separately, then connect them. Do not tighten the connecting pipe pinch bolts. Each side must pivot freely from each other to help with alignment and installation.

4. Install the oxygen sensor wiring harnesses through the frame, following their original paths.

CAUTION
When positioning and raising the exhaust system, monitor the oxygen sensors and their wiring harnesses to avoid damaging them.

5. Raise the exhaust system and support it with two jacks (**Figure 147**). Position the front exhaust pipes directly under the exhaust ports.

6. Raise the exhaust system slowly, and install the rear mounting bolts (one bolt each side). Do not force the exhaust pipe into position. Tighten the bolts finger-tight. See **Figure 140**.

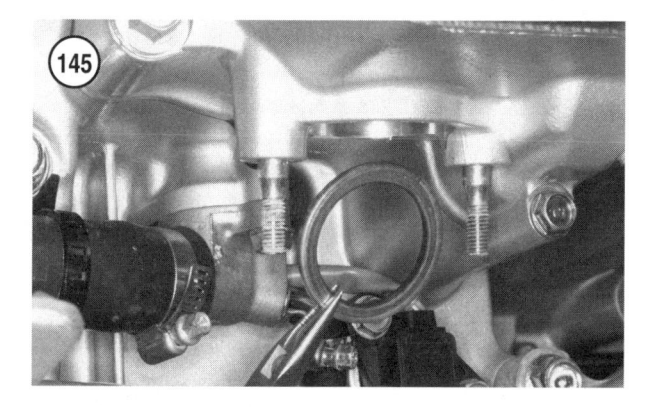

NOTE
Operate the front jack while watching and aligning the exhaust pipes with the exhaust ports. Release tension and lower the jack if the pipes bind against the cylinder head or studs.

7. Raise and install the exhaust pipes into the exhaust ports. Do not force the pipes into the ports.
8. Raise the clamps and install through the cylinder head studs. The exhaust pipe protector tabs, installed on the front and rear exhaust pipe clamps (each side), face outward (**Figure 148**, typical).
9. Install the exhaust pipe nuts and tighten finger-tight.
10. Tighten the mounting bolts and nuts in the following order:
    a. Connecting pipe pinch bolts (**Figure 139**) to 26 N•m (19 ft.-lb).
    b. Exhaust pipe nuts at cylinder head (**Figure 142** and **Figure 143**) to 12 N•m (106 in.-lb.).
    c. Exhaust pipe mounting bolts (**Figure 140**) to 26 N•m (19 ft.-lb).
11. Install the mufflers as described in this section.
12. Install the exhaust pipe protectors as described in this section.
13. Reconnect the oxygen sensor connectors (**Figure 137** and **Figure 138**).
14. Install the top shelter as described in this chapter.

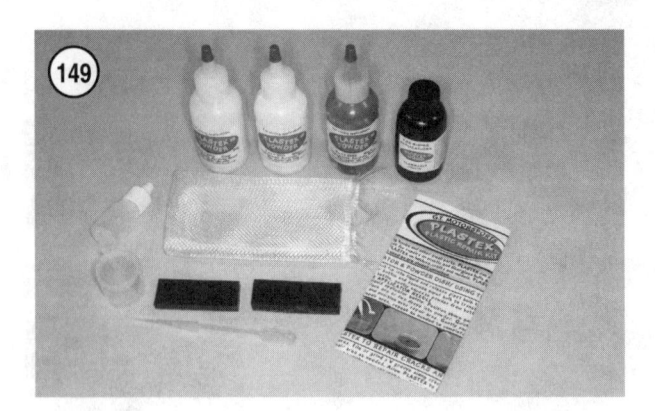

BODY REPAIR

The Gold Wing is equipped with many plastic body panels and parts. These body panels and parts are expensive to replace when damaged. A plastic tab or alignment guide can be broken when a body part is removed for service. A number of plastic repair kits are available that can be used to repair all types of plastic damage. The PLASTEX repair kit shown in **Figure 149** (available from G.T. Motorsports) can be used to repair cracks, rebuild broken tabs and fill gaps in many different types of plastic materials found on motorcycles. PLASTEX can also be used to repair stripped or damaged threads in plastic parts. PLASTEX kits are available in different sizes, from small to shop size. Because these kits are self-contained, many long distance riders pack one of the small or regular size kits to make repairs while on the road. Contact PLASTEX at *www.plastex.com*.

Table 1 EXHAUST SYSTEM TORQUE SPECIFICATIONS

	N•m	in.-lb.	ft.-lb.
Connecting pipe pinch bolts	26	–	19
Exhaust pipe nuts at cylinder heads	12	106	–
Exhaust pipe mounting bolts	26	–	19
Front exhaust pipe protector fasteners	12	106	–
Muffler clamp bolts	26	–	19
Rear exhaust pipe protector bolt	14	–	10
Saddlebag/trunk stay mounting bolts	26	–	19
Center fresh air visor (2004)	4	35	–
Center fresh air visor mounting bolt (2004)	1	8.8	–

INDEX

A

Air filter 57-60
    housing 245-246
Alternator 337-343
    and charging system
        specifications 410
Antenna 407-408
Antifreeze .... *See* Cooling System
Anti-lock brake system
    electronic control unit (ECU) 594
    modulators 591-594
    pulser ring 591
    service precautions 560-562
    specifications, torque 596
    troubleshooting 562-587
        diagnostic trouble
            codes 563-566, 595-596
        DTC flow charts 566-587
        pre-inspection 562-563
        self-diagnosis check 562
    wheel speed sensor 588-591, 596
Audio
    headset 408
    switch 407
    system 401
    unit 401-407

B

Bank angle sensor 351-352
BARO sensors 256-257, 281
Battery 327-333
    current drain (draw) 333-335
    maintenance-free voltage
        readings 410

specifications 410
Bike lift 431
Body
    color label 621
    fairings
        front, and molding 635-638
        inner 634-635
        lower front 626-627
        pockets 638
    fender
        front
            assembly 624-625
            B 625
            top cover, front 625
        rear
            A 640
            B 648
    meter panel 627
    mirrors and front turn
        signals 628-629
    open switch test 654-655
    passenger seat back 624
    preventing damage 187
    remote transmitter 648-650
        circuit troubleshooting 650-653
    repair 660
    saddlebags 646-647
        and trunk stay 648
        open switch test 654
    seat 622-624
    shelter, top 638-639
    side covers 622
        engine 621-622
    trunk 642-646
        actuator test 653-654
        lock control unit 653

    lower cover and molding
        strips 641-642
    windshield assembly 629-634
Brakes 81-83
    adjustment 83
    bleeding 556-557
    service tips and
        tools 551-554
    calipers 526-528
        overhaul 530-535
        pivot bearing 474
    disc 549-551
    fluid 87
        change 82
        draining 554-556
        inspection 81-82
        level 81-82
        preventing damage 521-522
        selection 81, 521
    flushing 556
    hose and pipe replacement ... 548-549
    inspection 81-82
    light switch
        front 396
        rear 396-397
    linked system (lbs) 520-521
    master cylinder
        front 535-539
        rear, and pedal 539-544
        secondary 544-546
    pads 522-526
    service 522
    specifications
        front 558
        rear 558
        torque 559

Brake, specifications (continued)
   switch 83, 396-397
   troubleshooting 52-55
   valve
      delay 546-548
      proportional control 548
Break-in procedure 183-184
Bulb specification 410

C

Cam chains
   and timing sprocket 101-105
   tensioner 105-106
Cam pulse generator 257, 297-300
Camshafts 91-101
Charging system 333-337
Clutch 72-73, 199-209
   cover 198-199
   cylinder
      master 191-196
      slave 196-197
   draining 188
   fluid 72-73
   specifications 210
      torque 210
   switch 395-396
   system bleeding 188-191
   troubleshooting 44-45
Combination meter 381-383
Connecting rods
   and pistons 160-173
   bearing selection 185
Coolant reserve tank 417
Coolant temperature gauge 385
Cooling fans 421-425
Cooling system 76-78
   coolant
      change 77
      level 77
      reserve tank 417
      test 77
      type 77
   fans 421-425
      relay testing 425-427
   hoses 415
   inspection 415
   pressure test 416-417
   radiators 417-421
   specifications 430
      torque 430
   temperature warning 415
   water pump
      and thermostat 427-429
      mechanical seal inspection 416
Countershaft 227-234
   specifications 236

Crankcase 150-160
   breather inspection 78-79
   cover, rear 131-133
Crankshaft 173-178
   main bearing selection 185
   specifications 184-185
Cruise control system
   actuator 615-617
   components 597
   cruise/reverse control module 609
   relays 609-614
   switches 617-620
   troubleshooting 55, 597-608
Cylinder
   head 106-111
      and valves specifications 121
      covers 89-91
   identification and firing order 57
   leakdown test, troubleshooting . 43-44

D

Delay valve 546-548
Depressurizing the fuel system . 237-240
Diagnostic flow charts 271-321
   cam pulse generator 257, 297-300
   clutch switch 320
   ECM (E-2 prom) 318
   fuel injectors 246-249, 285-296
   gear position switch 320
   IAC valve 257, 315
   sensors
      BARO 256-257, 281
      ETC 274
      IAT 360, 283
      knock 258, 311-314
      MAP 256-257, 271
      oxygen 258, 301-310
      TP 257, 276
      VSS 259, 283
Diagnostic trouble
   codes (DTC) 270, 325-326

E

ECM test harness 322-323
ECT sensor 385
Electrical system
   alternator 337-343
   antenna 407-408
   audio
      headset 408
      switch 407
      system 401
      unit 401-407
   bank angle sensor 351-352
   battery 327-333

maintenance-free voltage
      readings 410
   specifications 410
charging system 333-337
combination meter 381-383
component replacement 327
connectors 327
coolant temperature gauge 385
ECT sensor 385
engine control module 352
front turn signal 379
fuel
   gauge 385-386
   level sensor 386-387
fundamentals 19-20
fuses 400
headlight 374-376
   adjusters 376-377
      relay 377-379
      switch 377
headset junction cable 408-409
horns 398-399
ignition
   coil 343-349
   PGM-FI relay 350-351
   pulse
      generator 349-350
      rotor 350
   system troubleshooting ... 343-347
   timing 352
license plate light 380
neutral indicator 388-389
oil pressure indicator
   and switch 387-388
open air temperature sensor .. 391-392
position light relay 390-391
power control relays 366-367
relay box 399-400
reverse
   regulator 363-365
   relay 358
   resistor, assembly 365-366
   shift
      actuator 360-362
      relays 359-360
      switch 358
saddlebag combination light . 380-381
speakers 409
specifications
   alternator and charging
      system 410
   battery 410
   bulb replacement 410
   electrical system torque ... 411-412
   fuse 411
   ignition system 410
   sensor and switch test 411
   starting system 410

INDEX

speed limiter relay 367-368
speedometer 384
starter
  and reverse
    motor service 368-374
    troubleshooting 352-356
  relays 356-358
switches
  brake light
    front 83, 396
    rear 396-397
  clutch 395-396
  cruise control 395
  gear position 397-398
  handlebar 394-395
  hazard 393
    diode 393-394
  ignition 395
  left panel 393
  multi-display control 392-393
  sidestand 398
tachometer 384-385
testing, troubleshooting 46-51
trunk brake/taillight 379-380
turn signal
  cancel unit 389-390
  relay 389
wiring diagrams 667-703
Emission control systems 79
  evaporative 264-266
  labels 263
  secondary air supply system ... 263-264
  vacuum hose identification
    and fuel 237-238
Engine control module
  (ECM) 259-260
Engine
  compression test 60-61
  coolant, capacity 88
  idle speed
    adjustment 260-263
    inspection 69
  lower end 123,124-130
    break-in procedure 183-184
    connecting rod,
      bearing selection 185
    crankcase 150-160
    crankshaft 173-178
      main bearing selection 185
  level 73-74
  oil
    change 74-75
    output shaft 139-142
      and primary gears 134-139
    pressure 75-76
    pump 180-183
    strainer and pressure
      relief valve 178-179

pistons and connecting
  rods 160-173
rear crankcase cover 131-133
reverse
  shift arm 142-145
  shifter and shift drum
    lock arm 145-150
  service 123-124
specifications
  crankshaft 184-185
  oil pump 185
  output shaft 184
  piston rings and cylinder 184
  torque 122, 185-186
torque specifications 122
lubrication, troubleshooting 43
oil capacity 87
performance, troubleshooting .. 38-40
service 123-124
side covers 621-622
starting, troubleshooting 32-35
top end
  cam chains
    and timing sprocket 101-105
    tensioner 105-106
  camshafts 91-101
  cylinder head 106-111
  covers 89-91
specifications
  cylinder head and
    valves service 121
  general engine 121
  torque 122, 185-186
  valves and components 111-120
troubleshooting 41-43
ETC sensor 274
Evaporative emission
  control system 264-266
Exhaust system 655-660
  specifications, torque 660
External shift linkage 211-215

F

Fairings
  front and molding 635-638
  front, lower 626-627
  inner 634-635
  pockets 638
Fans
  cooling system 421-425
  relay testing 425-427
Fasteners 4-6, 85
Fender
  front
    assembly 624-625
    B 625
    top cover 625

rear
  A 640
  B 648
Final drive
  housing 514-518
  level relays 500-502
  linkage 502-507
  oil
    change 79
    capacity 88
  shock absorber 489-500
  specifications 518
  torque 519
  swing arm 507-514
  troubleshooting 51
Fork
  front 457-475
    assembly, right tube 469-472
    left tube 463-466
    brake caliper pivot bearing
      replacement (left tube) 474
    disassembly,
      left tube 461-463
      right tube 466-469
    inspection 472-473
    installation 458-461
    oil adjustment 474-475
    removal 457-458
    service specifications 487-488
  oil capacity 88
Front turn signals and mirrors .. 628-629
Fuel injection system
  air filter housing 245-246
  and vacuum hose
    identification 237-238
  cam pulse generator 257, 297-300
  circuit test 323-324
  component abbreviations 324
  depressurizing the system 237-240
  diagnostic trouble
    codes (DTC) 266-267 270, 325
    erasing 268-269
    flow charts 269-322
    reading 268
  engine control module (ECM) .. 259-260
  test harness 322-323
  engine idle speed adjustment ... 260-263
  flow test 241-243
  idle air control (IAC) valve .. 257, 315
  injectors 246-249, 285-296
  intake manifold 266
  malfunction indicator lamp
    (MIL) 266-324
    circuit test 323-324
  precautions 237
  pressure
    regulator 249-250
    test 240-241

Fuel injection system (continued)
- pump 252-254
- relay 254-256
- self-diagnostic check 267
- sensors
  - BARO 256-257, 281
  - ETC 274
  - intake air temperature (IAT) . 260, 283
  - knock 258, 311-314
  - MAP 256-257, 271
  - oxygen (O_2) 258, 301-310
  - throttle position (TP) 257, 276
- specifications
  - diagnostic trouble codes (DTC) 325-326
  - test 324
  - torque 324
- tank 243-245
- throttle body 250-252
- vehicle speed (VSS) 259, 283

Fuel system, troubleshooting 40-41
Fuel 56-57
- gauge 385-386
- hose inspection 69-69
- injectors 246-249, 285-296
- level sensor 386-387
- pressure regulator 249-250
- pump 252-254
- relay 254-256
- tank 243-245
- test
  - flow 241-243
  - pressure 240-241
- type 56
Fuse specifications 411
Fuses 400

G

Gear position switch 397-398
- troubleshooting 45-46
Gearshift, spindle assembly 222-224

H

Handlebars 449-457
- switches 394-395
Hazard switch 393
- diode 393-394
Headlight 374-376
- adjusters 376-377
- relay 377-379
- switch 377
Headset junction cable 408-409
Horns 398-399
- inspection 83-84
Hose identification, fuel and vacuum 237-238
Hub, front 439-443

I

Idle air control (IAC) valve 257, 315
Ignition
- coil 343-349
- cut-off system and sidestand switch 84
- PGM-FI relay 350-351
- pulse generator 349-350
- switch 395
- system
  - specifications 410
  - troubleshooting 343-347
- timing 65-66
Intake air temperature (IAT) sensor 260, 283

K

Knock sensors 258, 311-314

L

License plate light 380
Light
- brakes switch
  - front 396
  - rear 396-397
- front, turn signal 379
- inspection 83-84
- rear
  - license plate 380
  - tail/trunk brake 379-380
- saddlebag combination 380-381
Lubricants and fuel, recommended . . . 87
Lubrication
- engine
  - coolant capacity 88
  - oil
    - and filter 73-76
    - capacity 87
- final drive
  - oil 79
  - capacity 88
- fork oil capacity 88
- recommended lubricants and fuel . . . 87
- schedule 85-86
- throttle cable, lubrication 69-71

M

Mainshaft 215-222
- specifications 235
Maintenance
- air filter 57-60
- brakes 81-83
- clutch 72-73
- cooling system 76-78
- crankcase breather inspection 78-79
- emission control systems 79
- engine
  - compression test 60-61
  - coolant capacity 88
  - idle speed inspection 69
- fasteners 85
- final drive oil capacity 88
- fork oil capacity 88
- fuel 56-57
  - hose inspection 69-69
- ignition timing 65-66
- lights and horn inspection 83-84
- recommended lubricants and fuel . . . 87
- reverse operation 72
- schedule 85-86
- service torque specifications 87
- sidestand switch and ignition cut-off system 84
- spark plugs 61-65
- steering head bearing inspection . . . 84
- suspension inspection
  - front 84
  - rear 84-85
- throttle cable, operation and adjustment 71-72
- tire inflation pressure 88
- tires and wheels 79-80
- tune-up specifications 86-87
- unscheduled 85
- valve clearance 67-69
Malfunction indicator
- lamp (MIL) 266-324
- circuit test 323-324
- diagnostic trouble code 266-267
  - erasing 268-269
  - flow charts 269-322
  - reading 268
- ECM test harness 322-323
- self-diagnostic check 267
Manifold, intake 266
MAP sensors 256-257, 271
Master cylinder
- front 535-539
- rear, and brake pedal 539-544
- secondary 544-546
Meter panel 627
Mirrors and front turn signals . . 628-629
Multi-display control switch . . . 392-393

N

Neutral indicator 388-389

O

Oil
- change 74-75
- filter 73-76
- pressure, indicator and switch 387-388

INDEX

pump. 180-183
    specifications. 185
    strainer and pressure
        relief valve. 178-179
Open air temperature sensor . . . 391-392
Output shaft. 139-142
    and primary gears. 134-139
    specifications. 184
Oxygen (O_2) sensors 258, 301-310

P

Panel switch, left. 393
Passenger seat back. 624
Peak voltage. 343-344
PGM-FI ignition relay 350-351
Pistons
    and connecting rods 160-173
    rings and cylinder specifications . . 184
Position light relay. 390-391
Power control relays 366-367
Primary gears, output shaft 134-139
Proportional control valve. 548
Pulser ring. 591

R

Radiators. 417-421
Radio . . . . *See* Audio
Relay box. 399-400
Remote transmitter. 648-650
    circuit troubleshooting 650-653
Reverse
    operation 72
    regulator. 363-365
    relay 358
    resistor assembly 365-366
    shift
        actuator 360-362
        arm. 142-145
        relays 359-360
    shifter and shift drum
        lock arm 145-150
    switch 358

S

Saddlebags. 646-647
    combination light 380-381
    open switch test. 654
    trunk stay. 648
Seat 622-624
Secondary air supply system. 263-264
Sensors
    and switch
        test specifications 411
    open air temperature 391-392
Serial numbers. 3
    engine and frame. 27

Shift forks
    and drum 224-227
    and shaft specifications 235
Shift mechanism
    and transmission torque
        specifications. 236
    external linkage 211-215
    forks and drum 224-227
    gearshift spindle assembly . . . 222-224
    specifications
        fork and shaft. 235
        torque 236
Shock absorber 489-500
Side covers, engine 621-622
Sidestand switch 398
    and ignition cut-off system 84
Spark plugs 61-65
Speakers 409
Specifications
    alternator and charging system . . . 410
    anti-lock brake system torque 596
    battery 410
    brakes
        front. 558
        rear 558
        torque 559
    bulbs 410
    clutch. 210
        torque 210
    cooling system. 430
        torque 430
    countershaft. 236
    crankshaft. 184-185
    dimensions 27
    electrical system, torque. 411-412
    engine
        general. 121
        lower end, torque. 122
    exhaust system, torque 660
    final drive 518
    front fork service 487-488
    fuel injection system
        test. 324
        torque 324
    fuse 411
    ignition system
    mainshaft. 235
    oil pump 185
    output shaft 184
    piston rings and cylinder 184
    sensor and switch test 411
    shift fork and shaft 235
    shift mechanism and transmission
        torque 236
    starting system. 410
    steering and front suspension 487
    suspension
        front. 487

    fork service. 487-488
    rear. 518-519
    tire and wheel 447
    torque
        anti-lock brake system. 596
        brakes 559
        clutch. 210
        cooling system. 430
        electrical system. 411-412
        engine. 122, 185-186
        exhaust system 660
        final drive 519
        general. 29
        service 87
        steering 488
        suspension
            front. 488
            rear 519
        wheel. 448
    transmission 235
    tune up 86-87
    weight 27
    wheel
        and axle service. 447
        torque 448
Speed limiter relay. 367-368
Speedometer 384
Starter
    and reverse
        motor service 368-374
        system troubleshooting . . . 352-356
    relay
        A 356-357
        B. 357-358
Starting system specifications 410
Steering 84
    and front suspension
        troubleshooting 51-52
    bearing preload 483-484
    handlebars 449-457
    head
        and stem. 475-483
        bearing
            inspection
                races. 484-487
        specifications. 487
        torque 488
Storage. 26-27
Supplies, shop 6-8
Suspension
    front
        fork. 457-475
        inspection. 472-473
        installation 458-461
        removal 457-458
        left tube
            disassembly 461-463
            reassembly 463-466

Suspension, front, fork (continued)
 right tube
  assembly............ 469-472
  disassembly 466-469
 handlebars 449-457
 inspection 84
 oil adjustment........... 474-475
 specifications................ 487
  torque 488
 troubleshooting 51-52
 rear
  final drive housing 514-518
  inspection................ 84-85
  level relays 500-502
  linkage................. 502-507
  shock absorber 489-500
  specifications 518-519
  torque................. 518-519
Swing arm................... 507-514

T

Tachometer.................. 384-385
Tail/brake light.............. 379-380
Temperature warning system 415
Throttle
 body.................... 250-252
 cable
  lubrication 69-71
  operation and adjustment 71-72
 position (TP) sensor 257, 276
Tires
 and wheels 79-80
 changing.................. 443-447
 inflation pressure......... 79, 88, 447
 repairs 447
 safety....................... 443
 specifications................. 447
  wheel and axle service........ 447
Tools
 basic...................... 8-13
 precision measuring.......... 13-19
Top shelter 638-639
Torque specifications
 brake 558
 clutch...................... 210

electrical system........... 411-412
engine............... 122, 185-186
exhaust 660
final drive 519
fuel injection 324
service..................... 87
shift mechanism 236
steering 448
suspension
 front..................... 488
 rear 519
transmission 236
wheel...................... 448
Transmission
 assembly 215
 countershaft 227-234
 mainshaft 215-222
 shift forks and shift drum...... 224-227
 shifting check............. 234-235
 specifications................ 235
  countershaft................ 236
  mainshaft.................. 235
  shift fork and shift shaft 235
  transmission torque, and
   shift mechanism........... 236
 troubleshooting 46
Troubleshooting
 anti-lock brake system
  diagnostic trouble
   codes 563-566, 595-596
  DTC flow charts......... 566-587
  pre-inspection........... 562-563
  self-diagnosis check 562
 brakes..................... 52-55
 clutch..................... 44-45
 cruise control................. 55
 cylinder leakdown test 43-44
 electrical system testing....... 46-51
 engine..................... 41-43
  lubrication................ 43
  performance............... 38-40
  starting................. 32-38
 final drive 51
 fuel system................. 40-41
 gearshift linkage............. 45-46
 operating requirements 32

suspension and steering, front .. 51-52
transmission 46
Trunk 642-646
 actuator test 653-654
 lock control unit 653
 lower cover and
  molding strips........... 641-642
Tune-up 57
 cylinder identification and
  firing order.................. 57
 engine
  compression test.......... 60-61
  idle speed inspection 69
  ignition timing 65-66
  spark plugs................. 61-65
  valve clearance............. 67-69
Turn signal
 cancel unit 389-390
 front, light 379
 relay 389

V

Valves
 and components 111-120
 clearance 67-69
Vehicle speed sensor (VSS).... 259, 283

W

Water pump
 and thermostat 427-429
 mechanical seal inspection 416
Wheels
 and tires 79-80
 front.................... 431-436
  hub 439-443
 rear..................... 436-437
 runout and balance......... 437-439
 specifications................. 447
  torque 448
 speed sensors 588-591, 596
Windshield assembly 629-634
Wiring diagrams............. 667-703

Wiring Diagrams

STARTING AND REVERSE SYSTEMS
IGNITION, CHARGING AND
 COOLING SYSTEMS
FUEL INJECTION
ABS SYSTEM
LIGHTING SYSTEM
TURN SIGNAL AND HORN SYSTEMS

COMBINATION METER
CRUISE CONTROL SYSTEM
ACCESSORY AND TRUNK LOCK
 CONTROL SYSTEMS
AUDIO SYSTEM
SUSPENSION LEVEL CONTROL SYSTEM
FUSE BOX AND RELAY BOX

STARTING AND REVERSE SYSTEMS

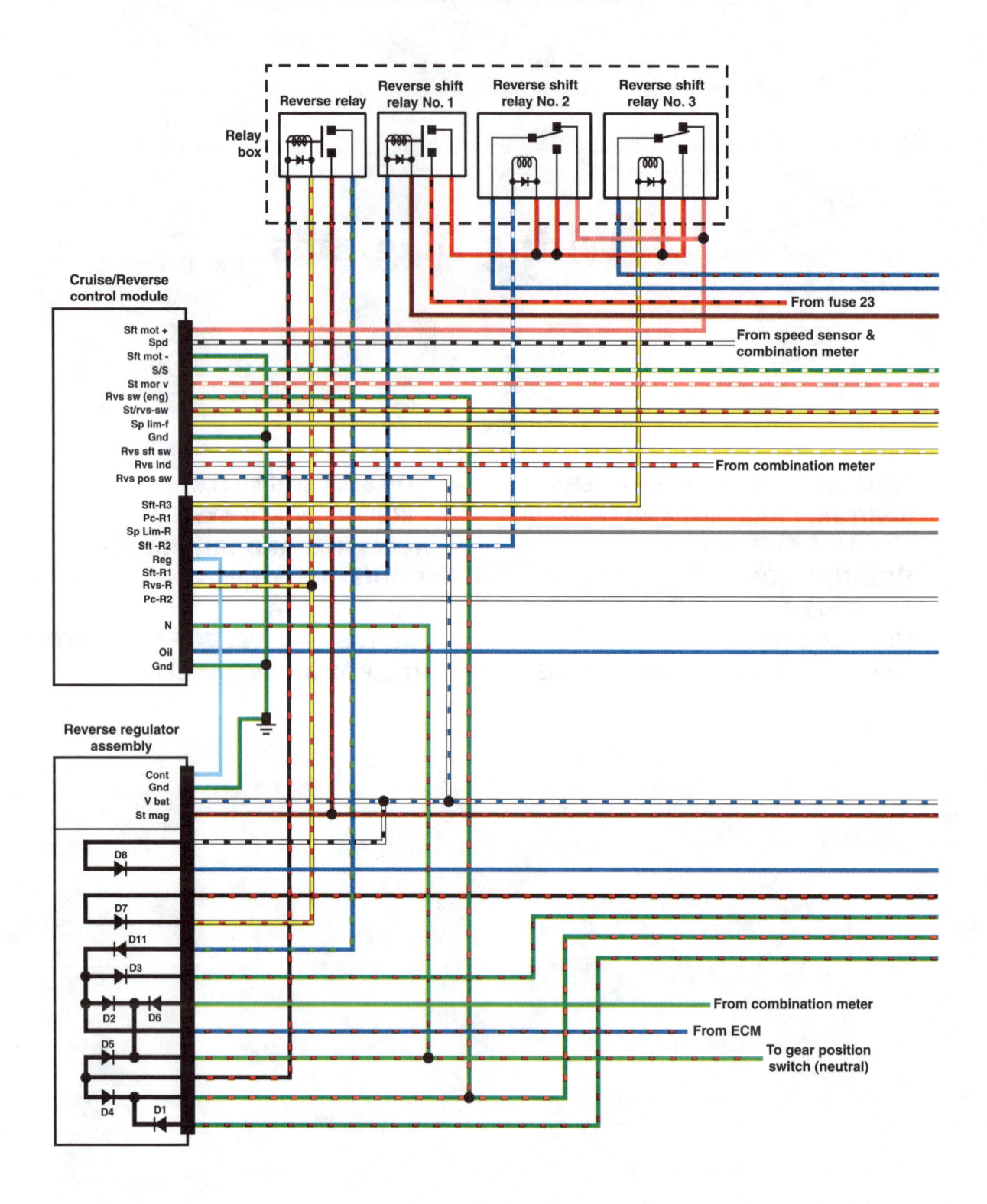

Relay box:
- Reverse relay
- Reverse shift relay No. 1
- Reverse shift relay No. 2
- Reverse shift relay No. 3

From fuse 23
From speed sensor & combination meter
From combination meter

Cruise/Reverse control module
- Sft mot +
- Spd
- Sft mot −
- S/S
- St mor v
- Rvs sw (eng)
- St/rvs-sw
- Sp lim-f
- Gnd
- Rvs sft sw
- Rvs ind
- Rvs pos sw
- Sft-R3
- Pc-R1
- Sp Lim-R
- Sft -R2
- Reg
- Sft-R1
- Rvs-R
- Pc-R2
- N
- Oil
- Gnd

Reverse regulator assembly
- Cont
- Gnd
- V bat
- St mag
- D8
- D7
- D11
- D3
- D2, D6
- D5
- D4, D1

From combination meter
From ECM
To gear position switch (neutral)

WIRING DIAGRAMS

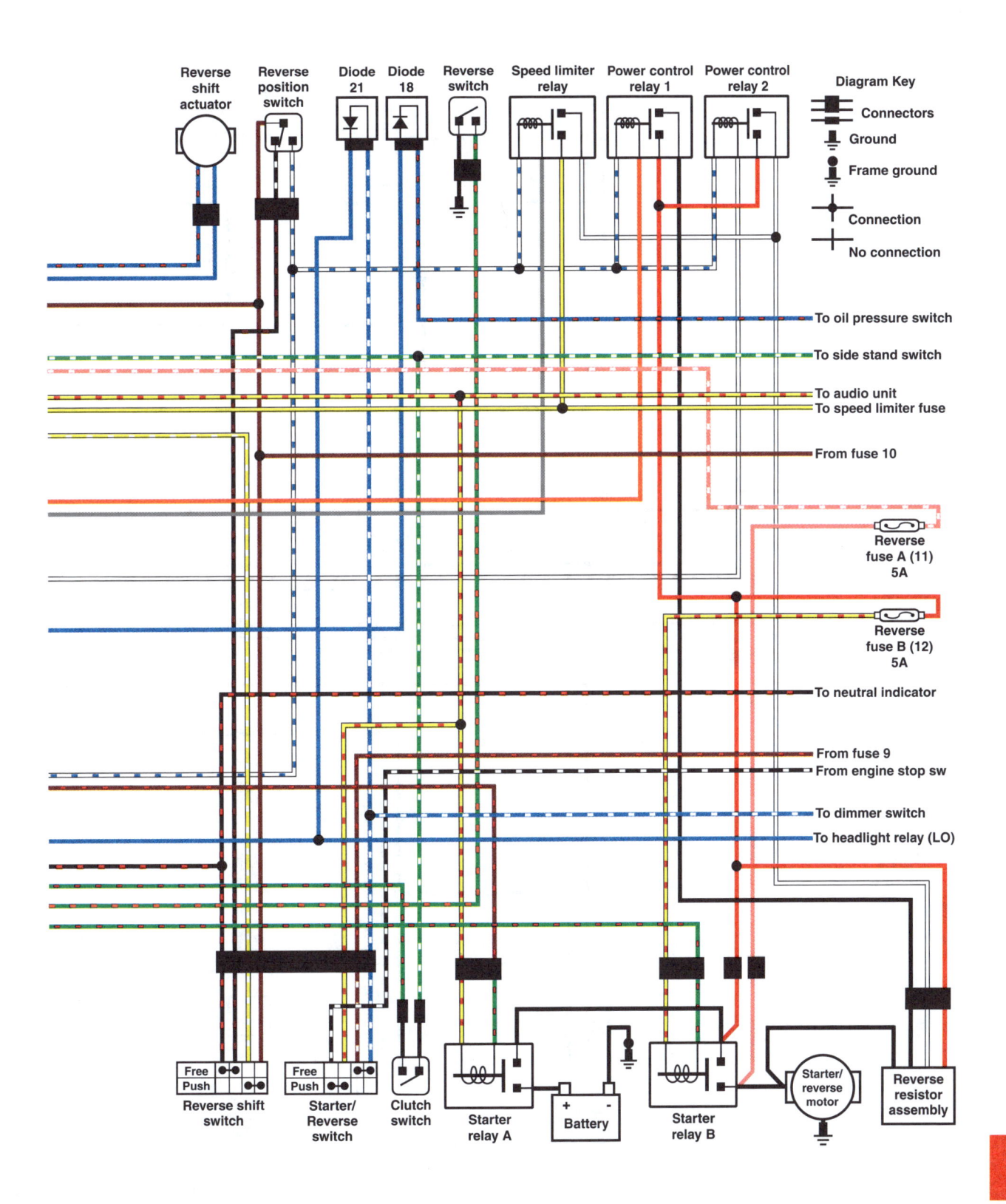

IGNITION, CHARGING AND COOLING SYSTEMS

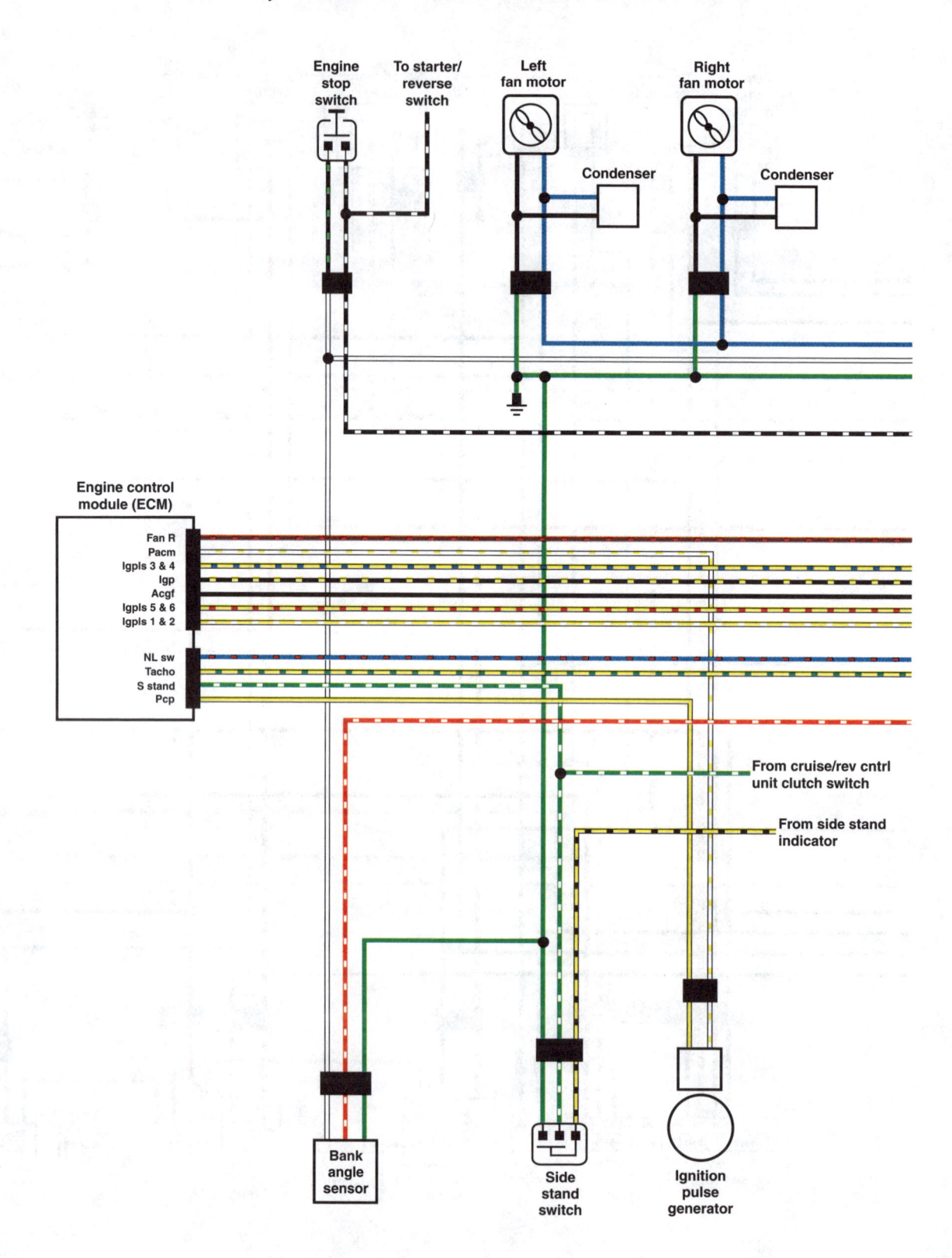

WIRING DIAGRAMS

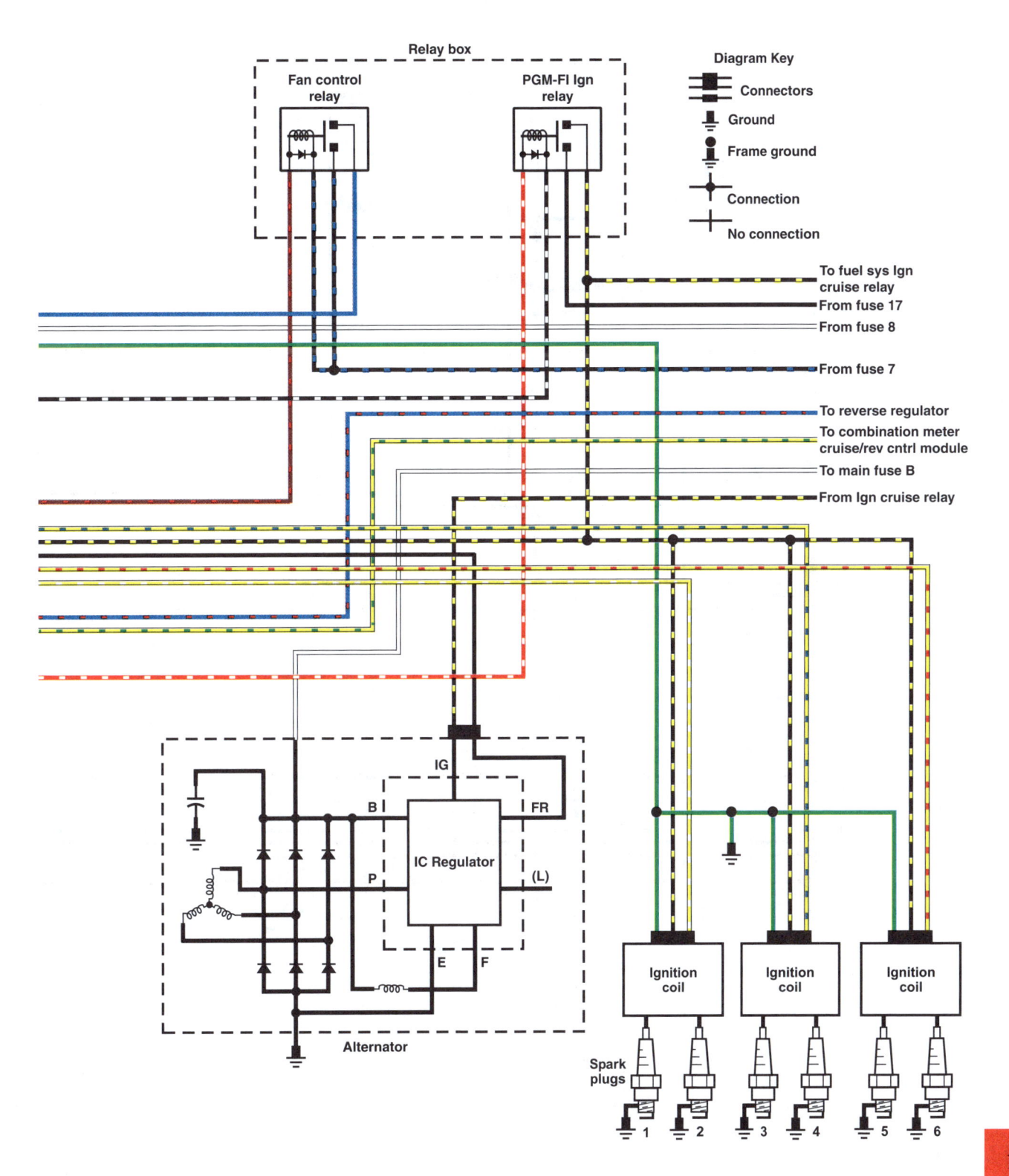

WIRING DIAGRAMS

FUEL INJECTION SYSTEM (2001-2003)

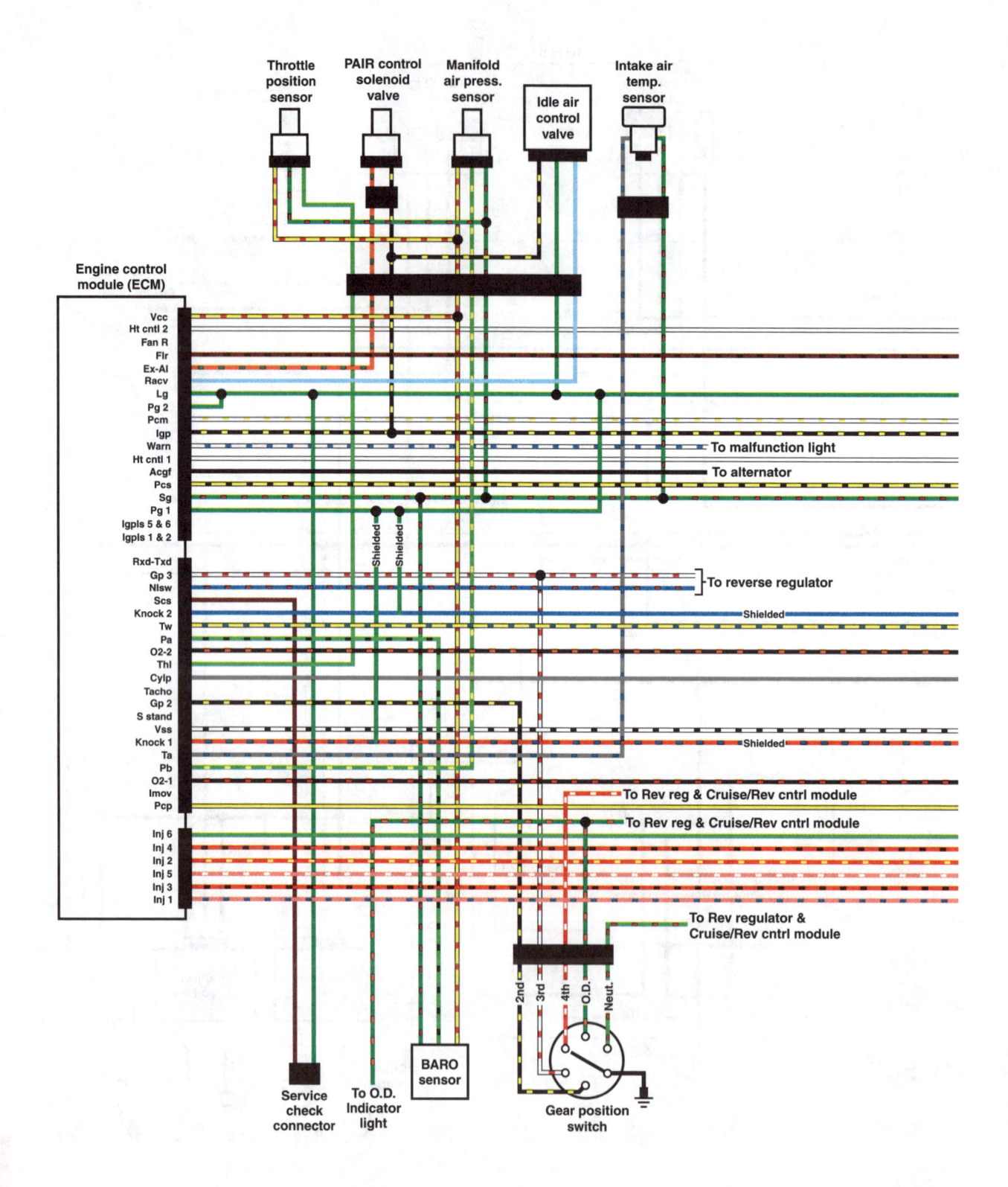

- Throttle position sensor
- PAIR control solenoid valve
- Manifold air press. sensor
- Idle air control valve
- Intake air temp. sensor

Engine control module (ECM)
- Vcc
- Ht cntl 2
- Fan R
- Flr
- Ex-Al
- Racv
- Lg
- Pg 2
- Pcm
- Igp
- Warn
- Ht cntl 1
- Acgf
- Pcs
- Sg
- Pg 1
- Igpls 5 & 6
- Igpls 1 & 2
- Rxd-Txd
- Gp 3
- Nlsw
- Scs
- Knock 2
- Tw
- Pa
- O2-2
- Thl
- Cylp
- Tacho
- Gp 2
- S stand
- Vss
- Knock 1
- Ta
- Pb
- O2-1
- Imov
- Pcp
- Inj 6
- Inj 4
- Inj 2
- Inj 5
- Inj 3
- Inj 1

- To malfunction light
- To alternator
- Shielded
- To reverse regulator
- Shielded
- Shielded
- To Rev reg & Cruise/Rev cntrl module
- To Rev reg & Cruise/Rev cntrl module
- To Rev regulator & Cruise/Rev cntrl module

- Service check connector
- To O.D. Indicator light
- BARO sensor
- Gear position switch (2nd, 3rd, 4th, O.D., Neut.)

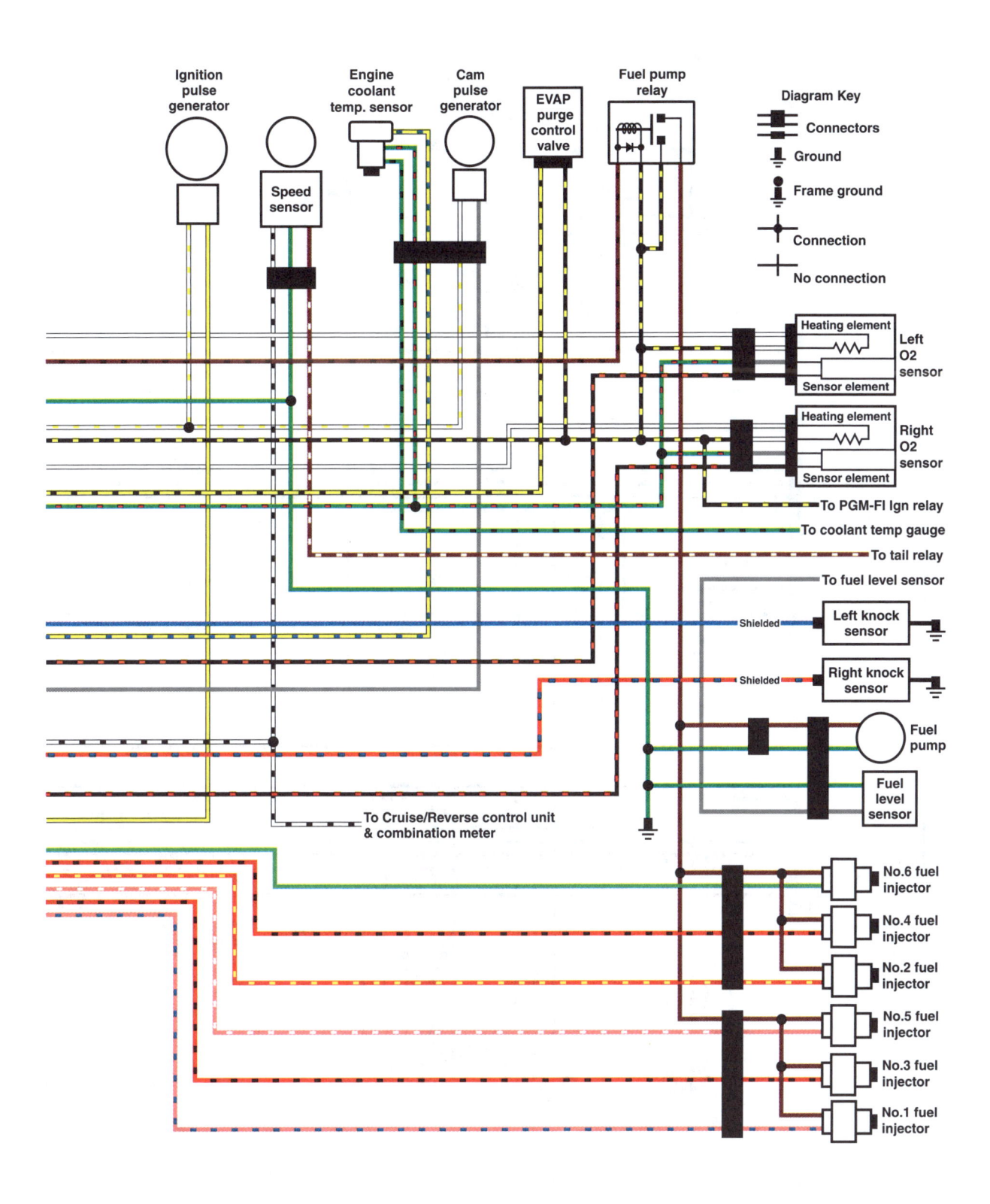

WIRING DIAGRAMS

673

WIRING DIAGRAMS

FUEL INJECTION SYSTEM (2004-2005)

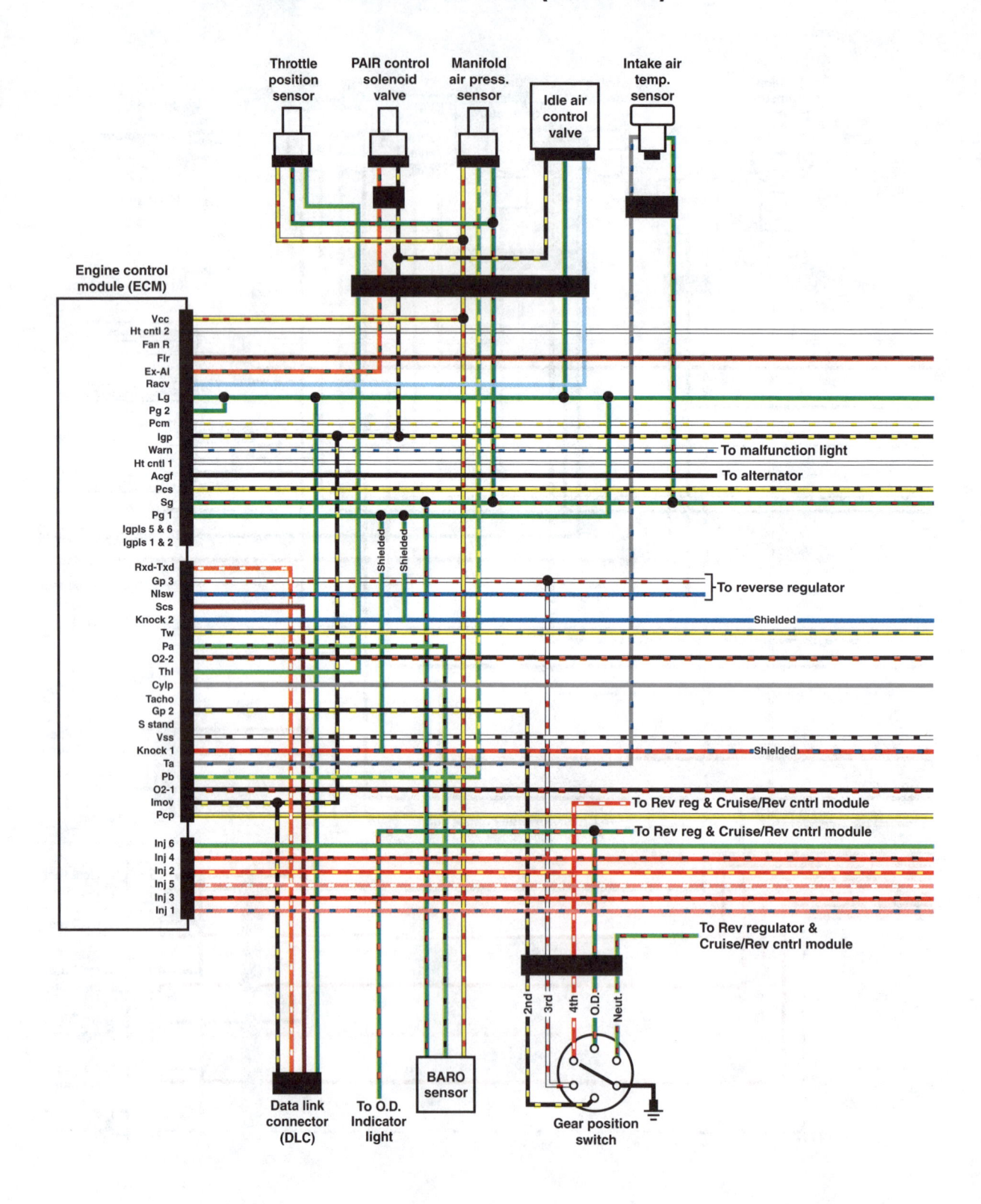

- Throttle position sensor
- PAIR control solenoid valve
- Manifold air press. sensor
- Idle air control valve
- Intake air temp. sensor

Engine control module (ECM)

- Vcc
- Ht cntl 2
- Fan R
- Flr
- Ex-Al
- Racv
- Lg
- Pg 2
- Pcm
- Igp
- Warn — To malfunction light
- Ht cntl 1
- Acgf — To alternator
- Pcs
- Sg
- Pg 1
- Igpls 5 & 6
- Igpls 1 & 2
- Rxd-Txd
- Gp 3
- Nlsw — To reverse regulator
- Scs
- Knock 2 — Shielded
- Tw
- Pa
- O2-2
- Thl
- Cylp
- Tacho
- Gp 2
- S stand
- Vss
- Knock 1 — Shielded
- Ta
- Pb
- O2-1
- Imov — To Rev reg & Cruise/Rev cntrl module
- Pcp — To Rev reg & Cruise/Rev cntrl module
- Inj 6
- Inj 4
- Inj 2
- Inj 5
- Inj 3
- Inj 1

To Rev regulator & Cruise/Rev cntrl module

Shielded (x2)

- Data link connector (DLC)
- To O.D. Indicator light
- BARO sensor
- Gear position switch (2nd, 3rd, 4th, O.D., Neut.)

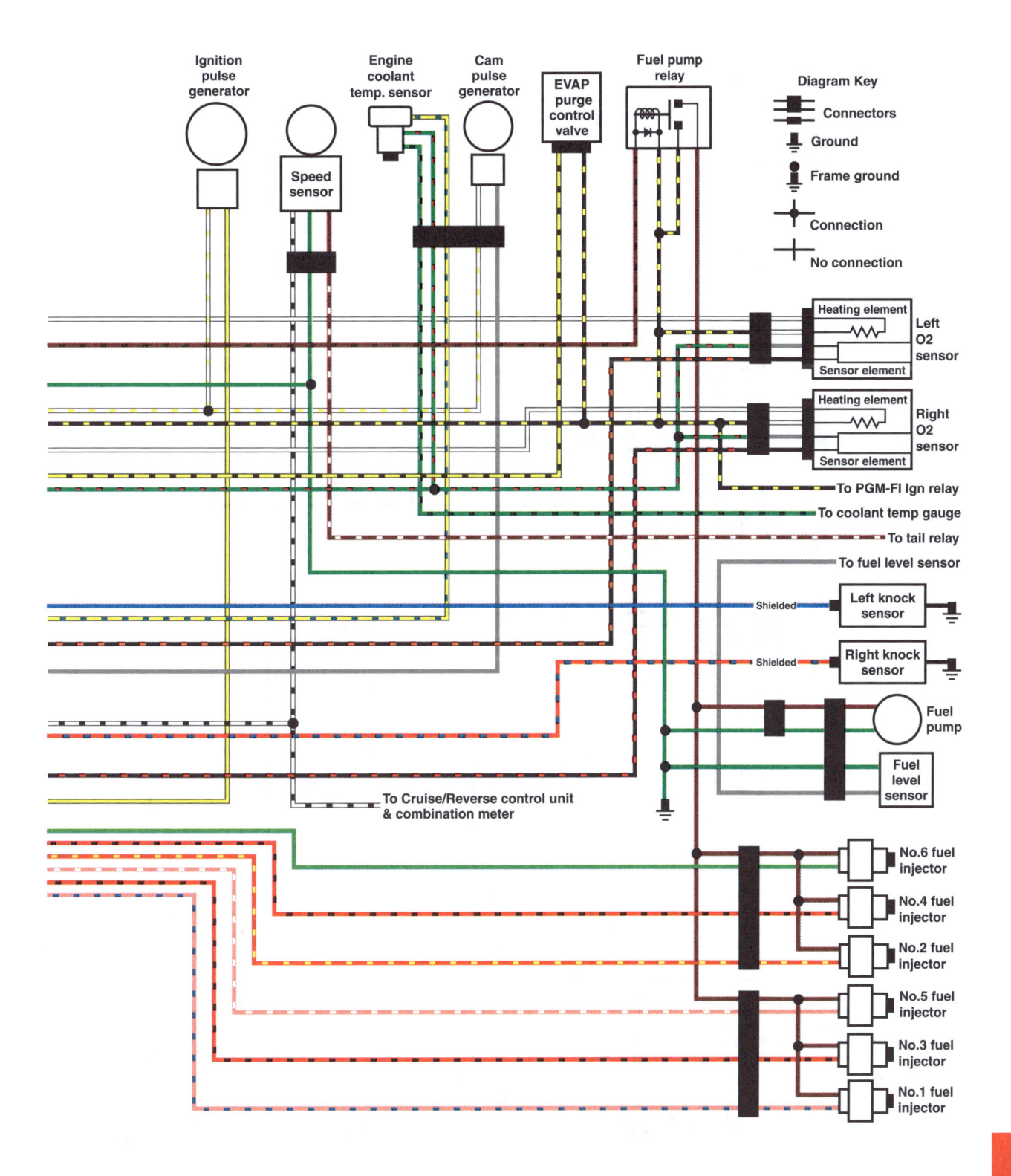

WIRING DIAGRAMS

Ignition pulse generator

Engine coolant temp. sensor

Speed sensor

Cam pulse generator

EVAP purge control valve

Fuel pump relay

Diagram Key
- Connectors
- Ground
- Frame ground
- Connection
- No connection

Left O2 sensor — Heating element / Sensor element

Right O2 sensor — Heating element / Sensor element

→ To PGM-FI Ign relay
→ To coolant temp gauge
→ To tail relay
→ To fuel level sensor

Left knock sensor (Shielded)

Right knock sensor (Shielded)

Fuel pump

Fuel level sensor

To Cruise/Reverse control unit & combination meter

- No.6 fuel injector
- No.4 fuel injector
- No.2 fuel injector
- No.5 fuel injector
- No.3 fuel injector
- No.1 fuel injector

WIRING DIAGRAMS

ABS SYSTEM (2001)

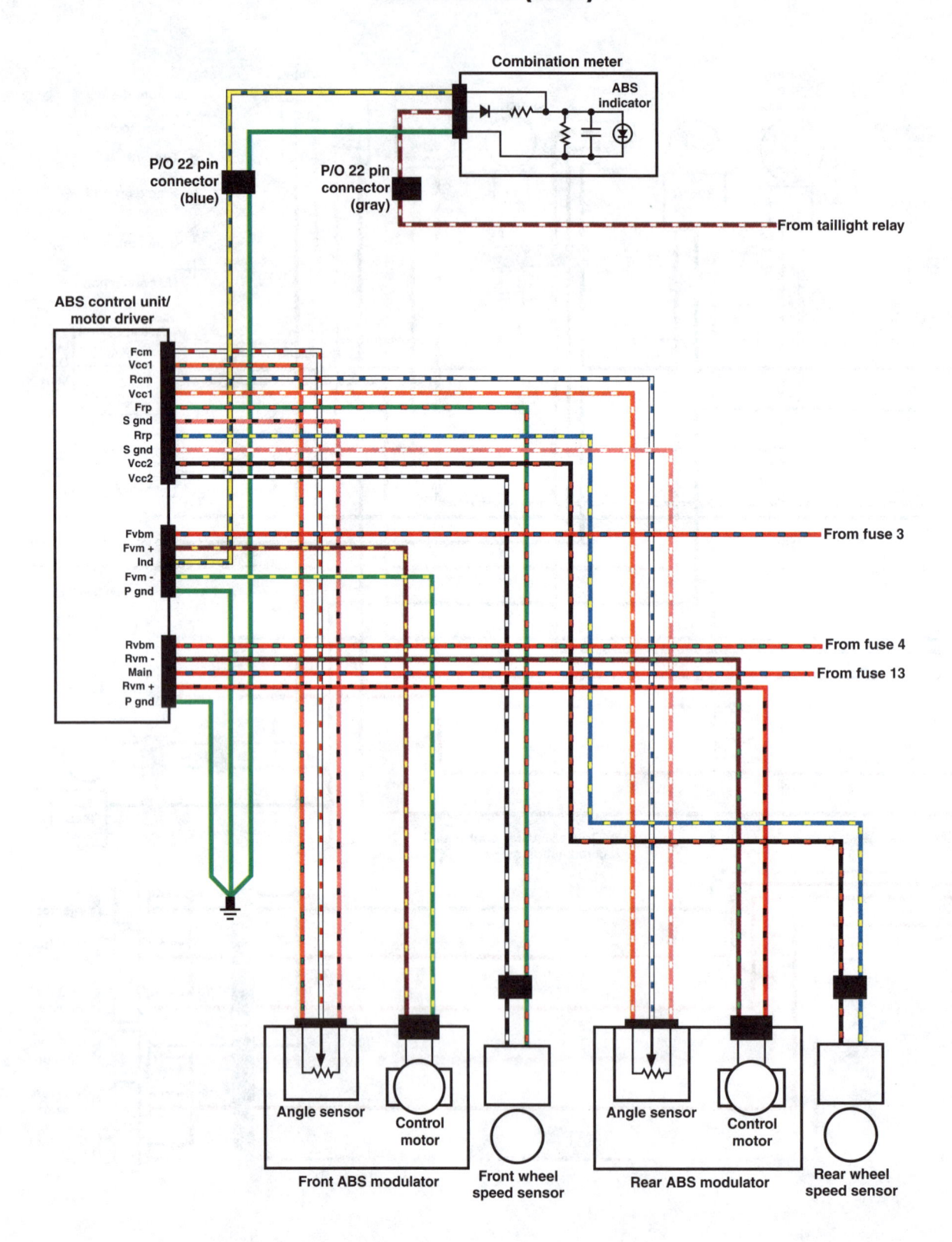

WIRING DIAGRAMS

ABS SYSTEM (2002-2005)

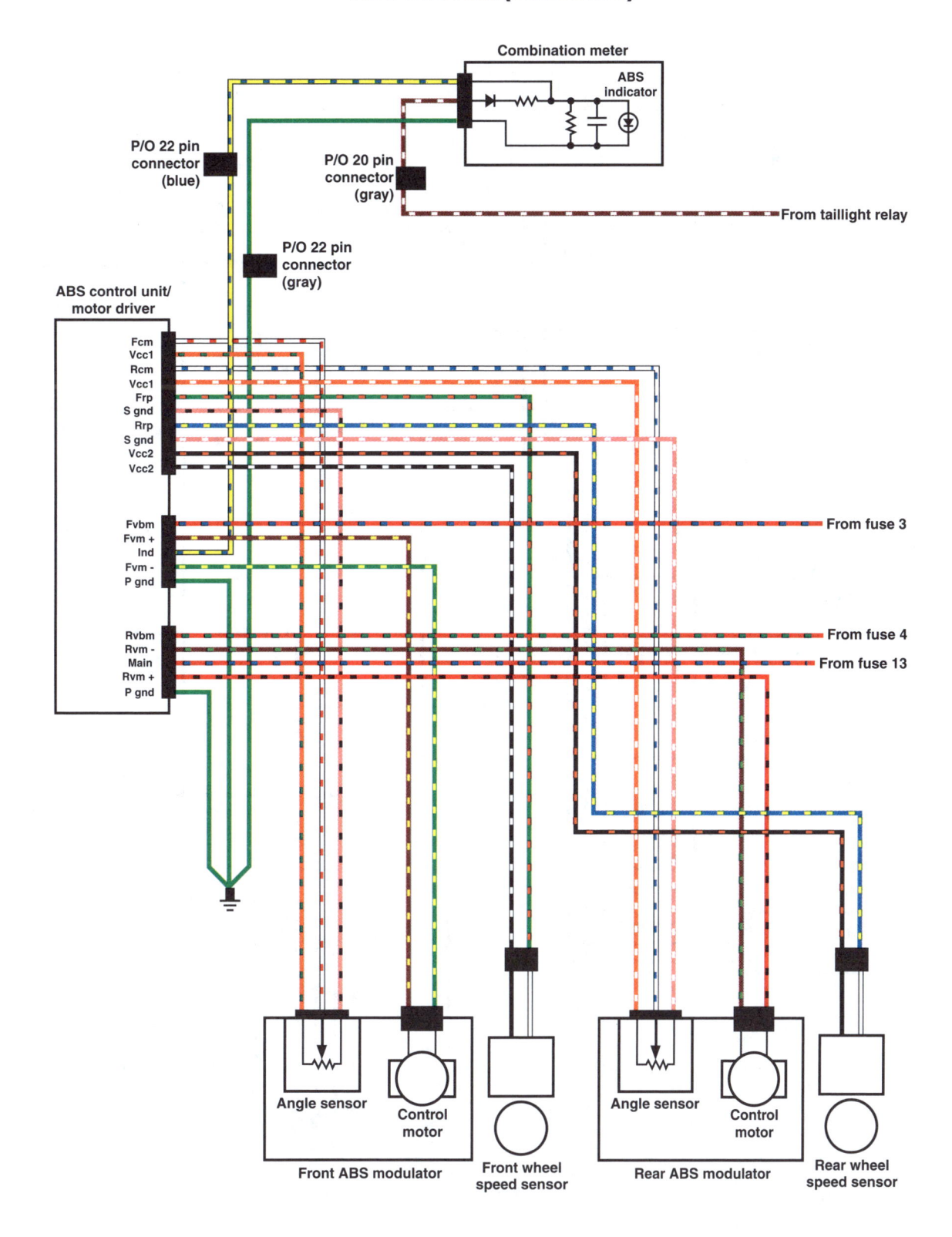

WIRING DIAGRAMS

LIGHTING SYSTEM

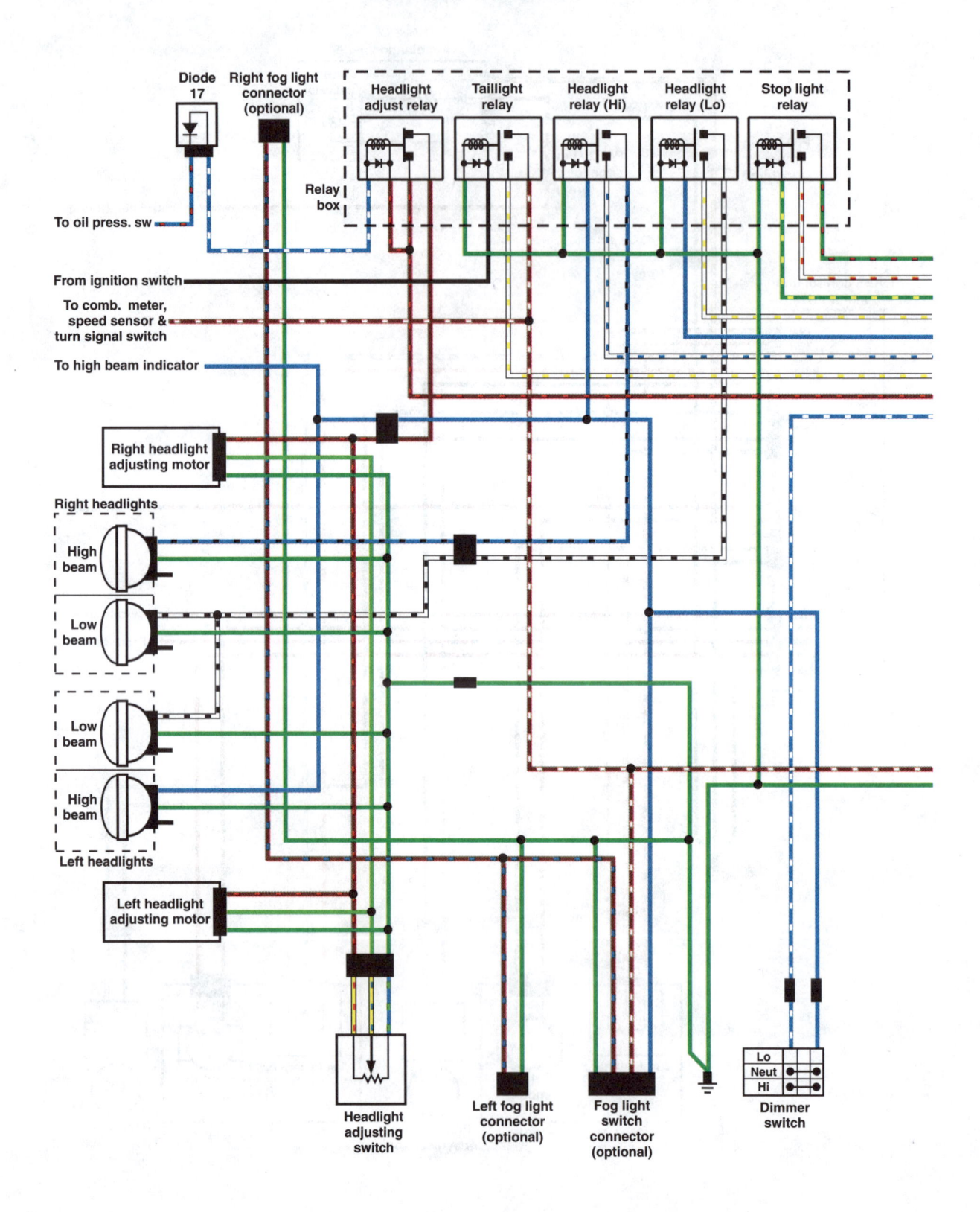

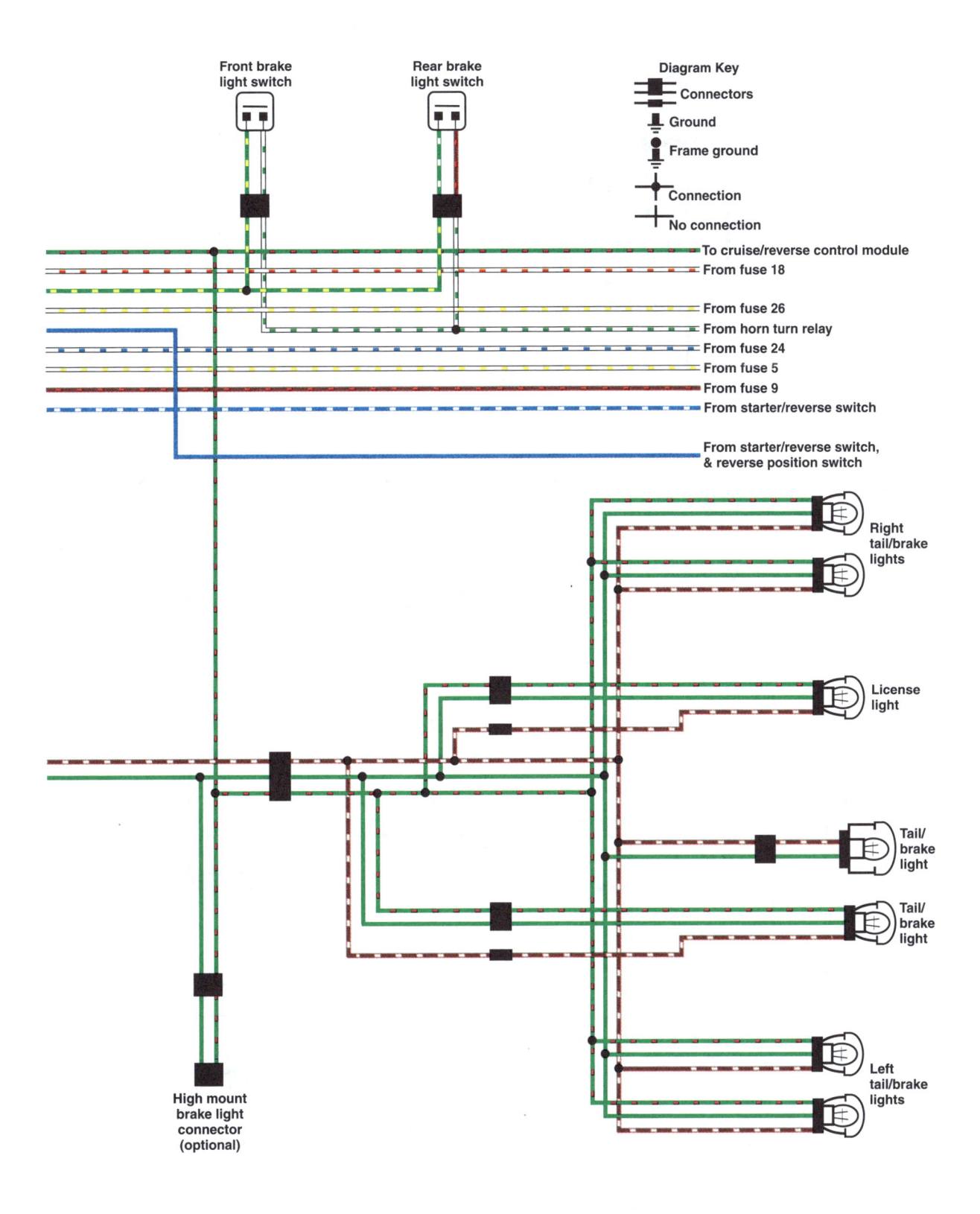

WIRING DIAGRAMS

Front brake light switch

Rear brake light switch

Diagram Key
- Connectors
- Ground
- Frame ground
- Connection
- No connection

To cruise/reverse control module
From fuse 18
From fuse 26
From horn turn relay
From fuse 24
From fuse 5
From fuse 9
From starter/reverse switch
From starter/reverse switch, & reverse position switch

Right tail/brake lights

License light

Tail/brake light

Tail/brake light

Left tail/brake lights

High mount brake light connector (optional)

WIRING DIAGRAMS

TURN SIGNAL AND HORN SYSTEMS

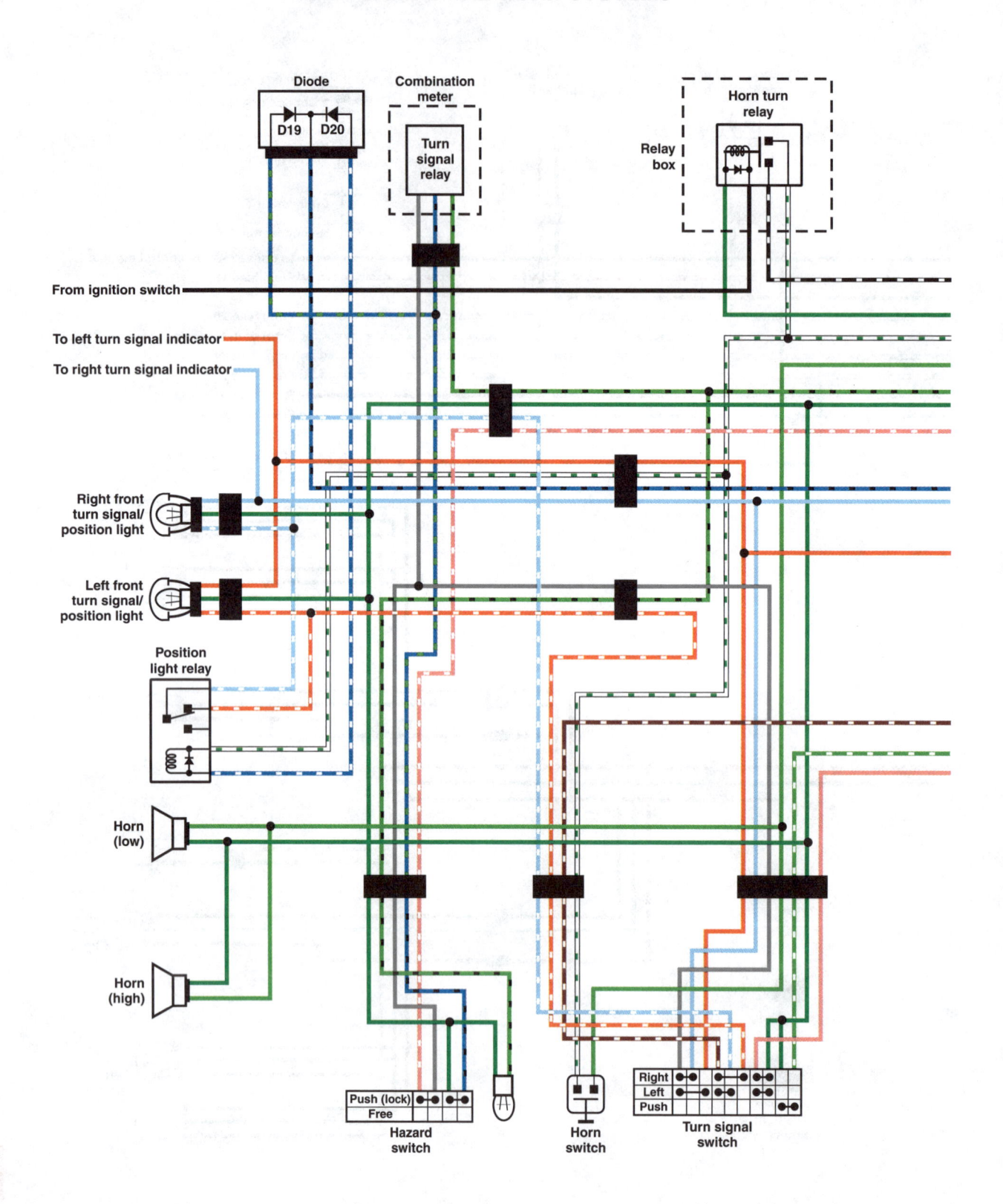

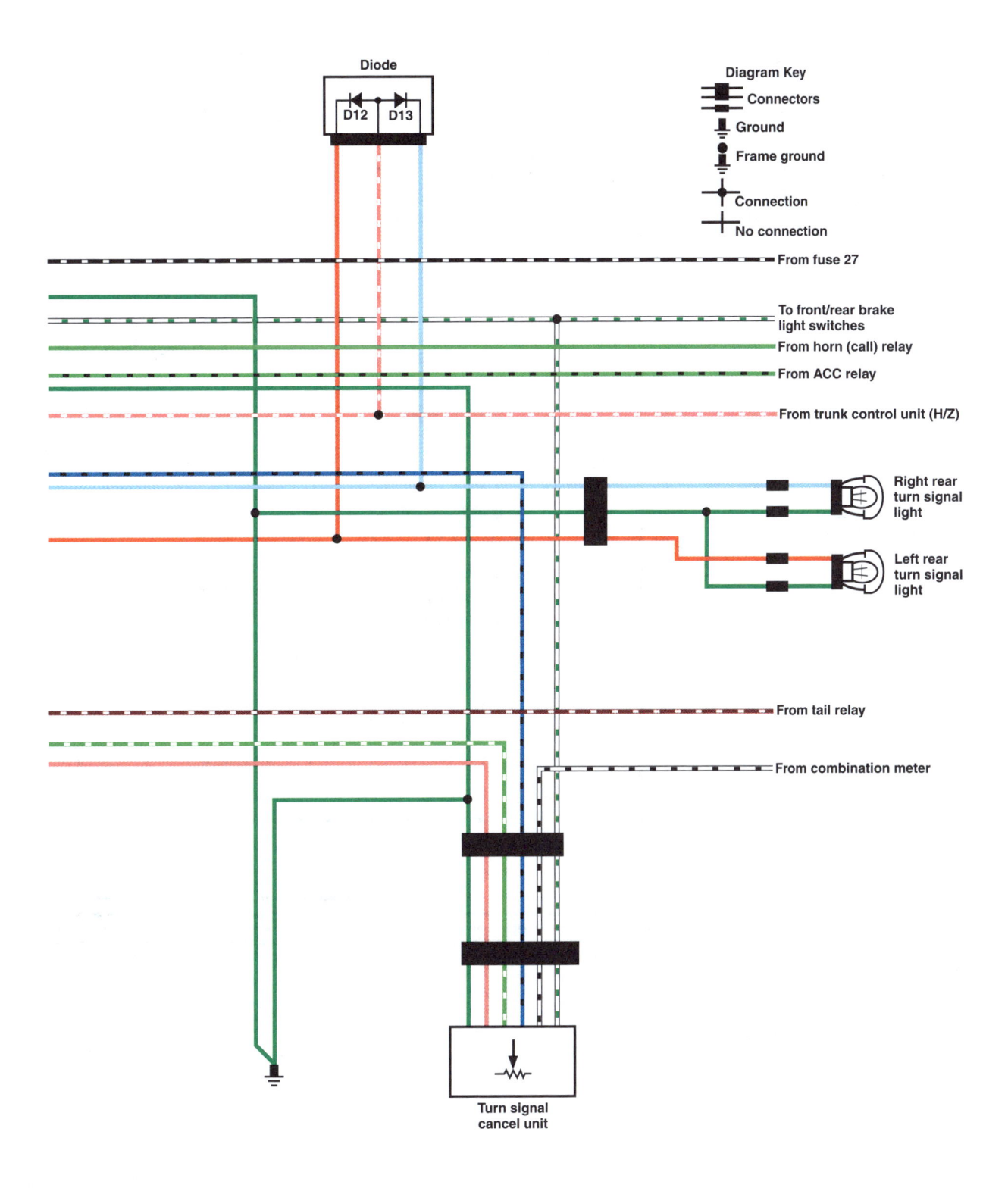

WIRING DIAGRAMS

Diode
D12 D13

Diagram Key
- Connectors
- Ground
- Frame ground
- Connection
- No connection

From fuse 27
To front/rear brake light switches
From horn (call) relay
From ACC relay
From trunk control unit (H/Z)

Right rear turn signal light
Left rear turn signal light

From tail relay
From combination meter

Turn signal cancel unit

WIRING DIAGRAMS

COMBINATION METER (2001)

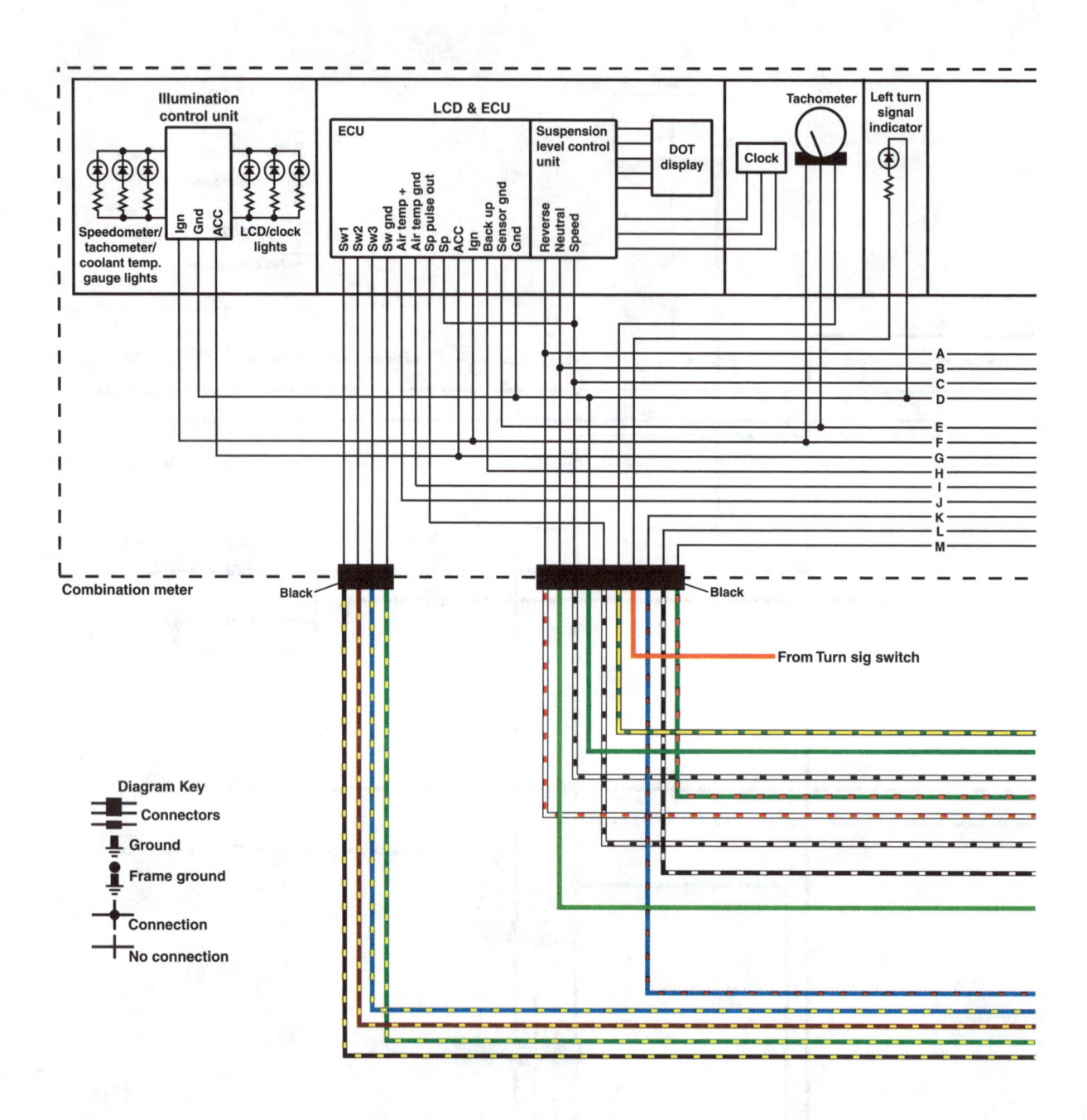

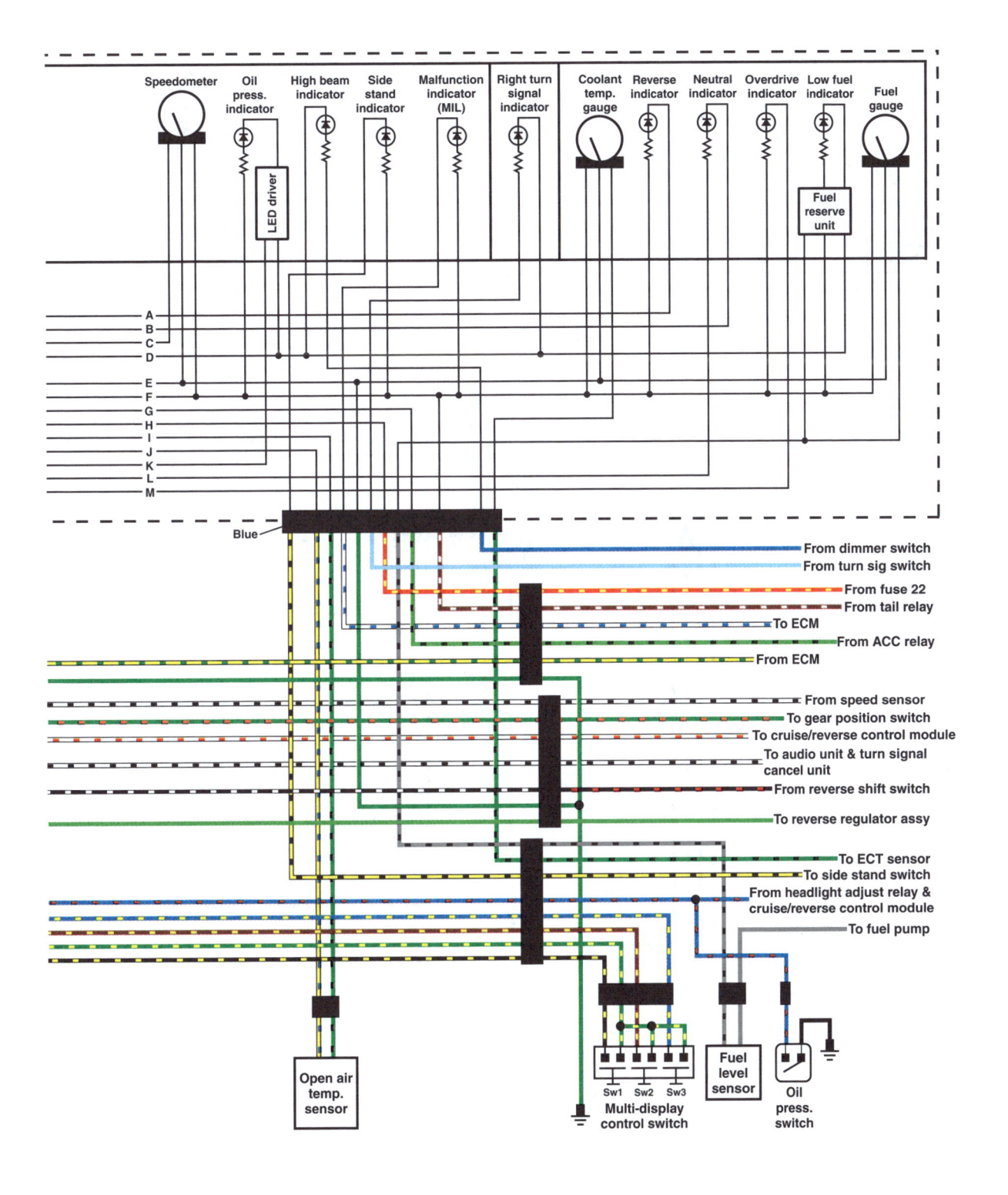

WIRING DIAGRAMS

683

Speedometer — **Oil press. indicator** — **High beam indicator** — **Side stand indicator** — **Malfunction indicator (MIL)** — **Right turn signal indicator** — **Coolant temp. gauge** — **Reverse indicator** — **Neutral indicator** — **Overdrive indicator** — **Low fuel indicator** — **Fuel gauge**

LED driver

Fuel reserve unit

A
B
C
D
E
F
G
H
I
J
K
L
M

Blue

From dimmer switch
From turn sig switch
From fuse 22
From tail relay
To ECM
From ACC relay
From ECM
From speed sensor
To gear position switch
To cruise/reverse control module
To audio unit & turn signal cancel unit
From reverse shift switch
To reverse regulator assy
To ECT sensor
To side stand switch
From headlight adjust relay & cruise/reverse control module
To fuel pump

Open air temp. sensor

Sw1 Sw2 Sw3
Multi-display control switch

Fuel level sensor

Oil press. switch

19

WIRING DIAGRAMS

COMBINATION METER (2002-2005)

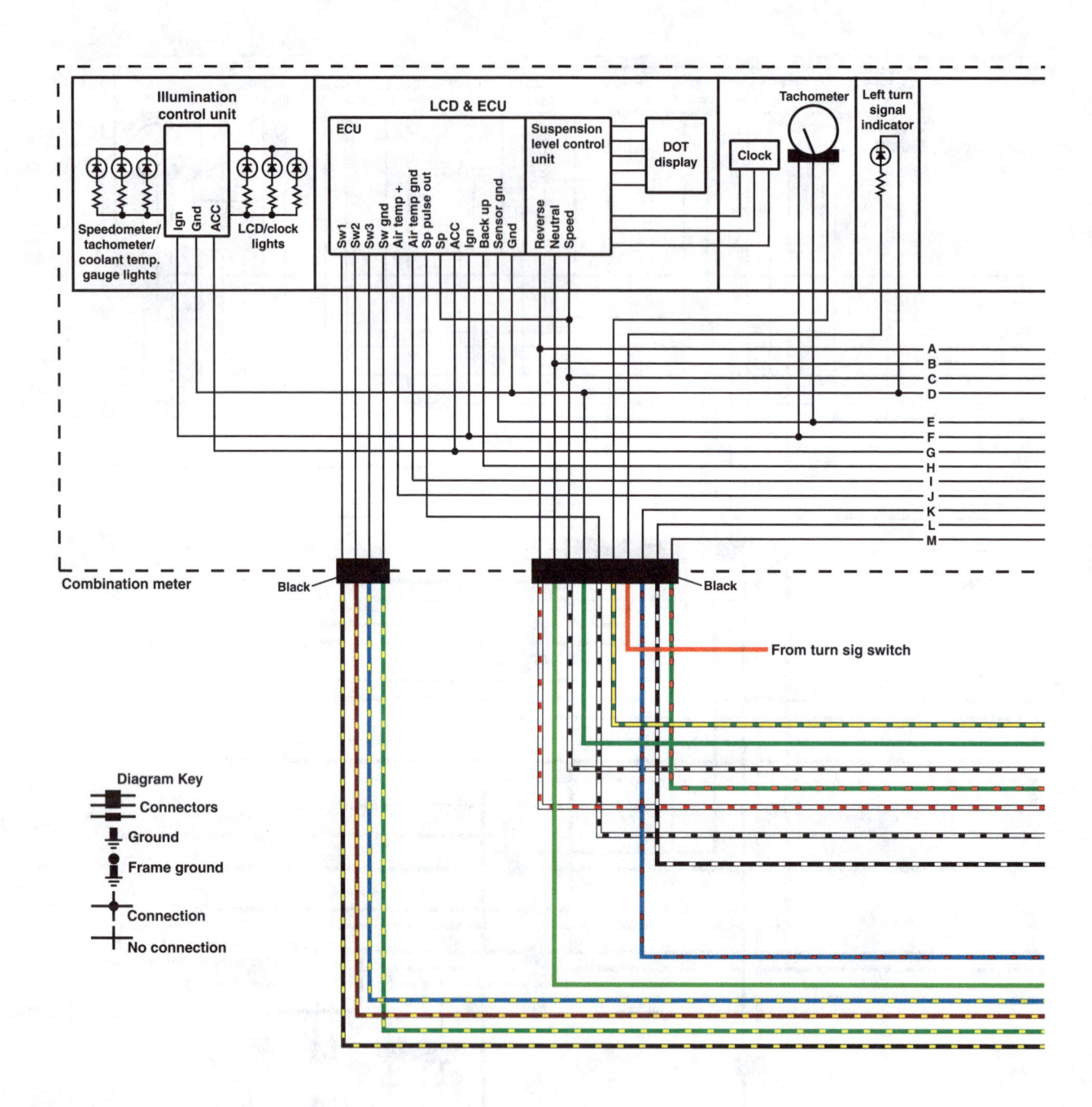

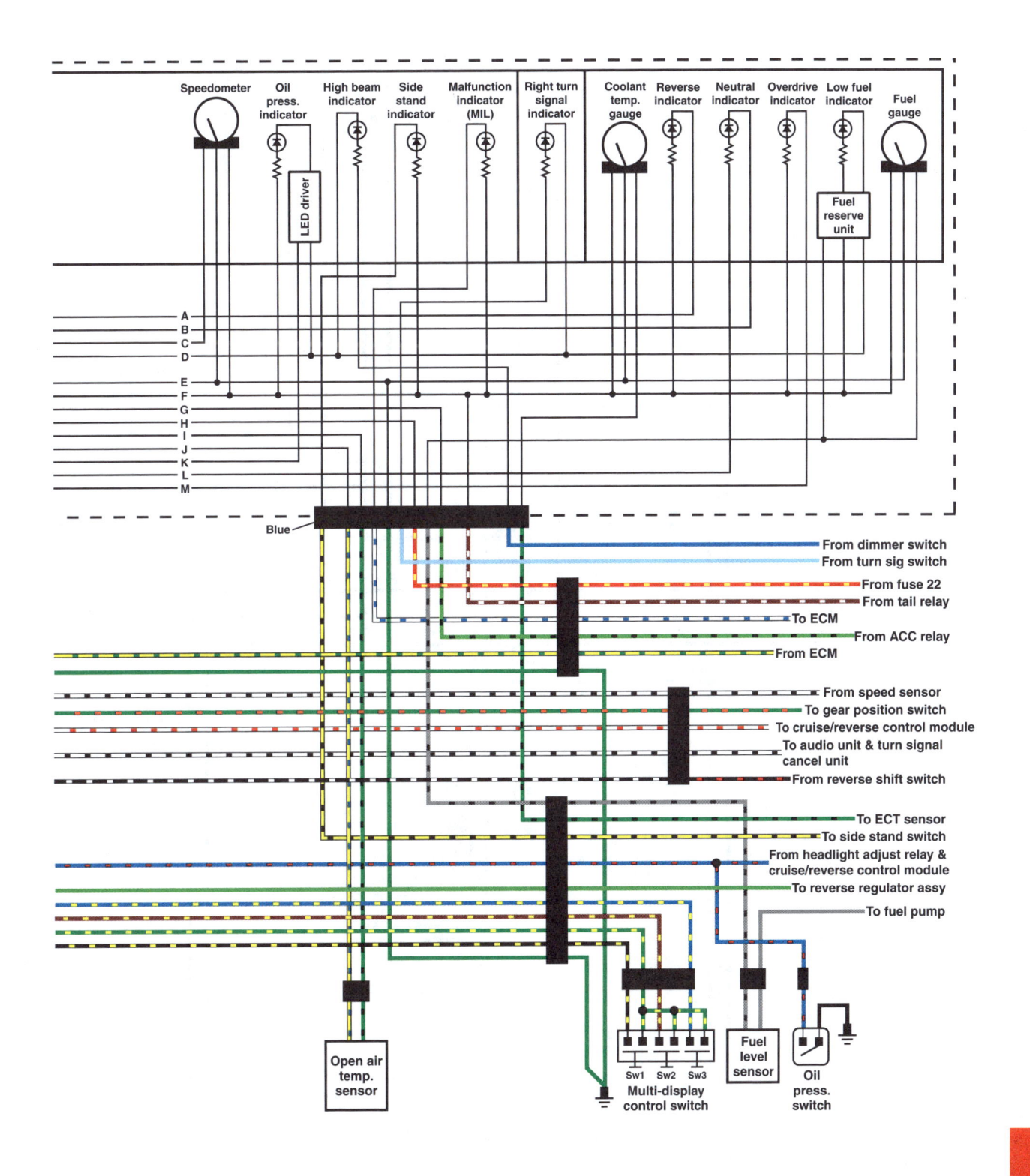

WIRING DIAGRAMS

CRUISE CONTROL SYSTEM (2001-2003)

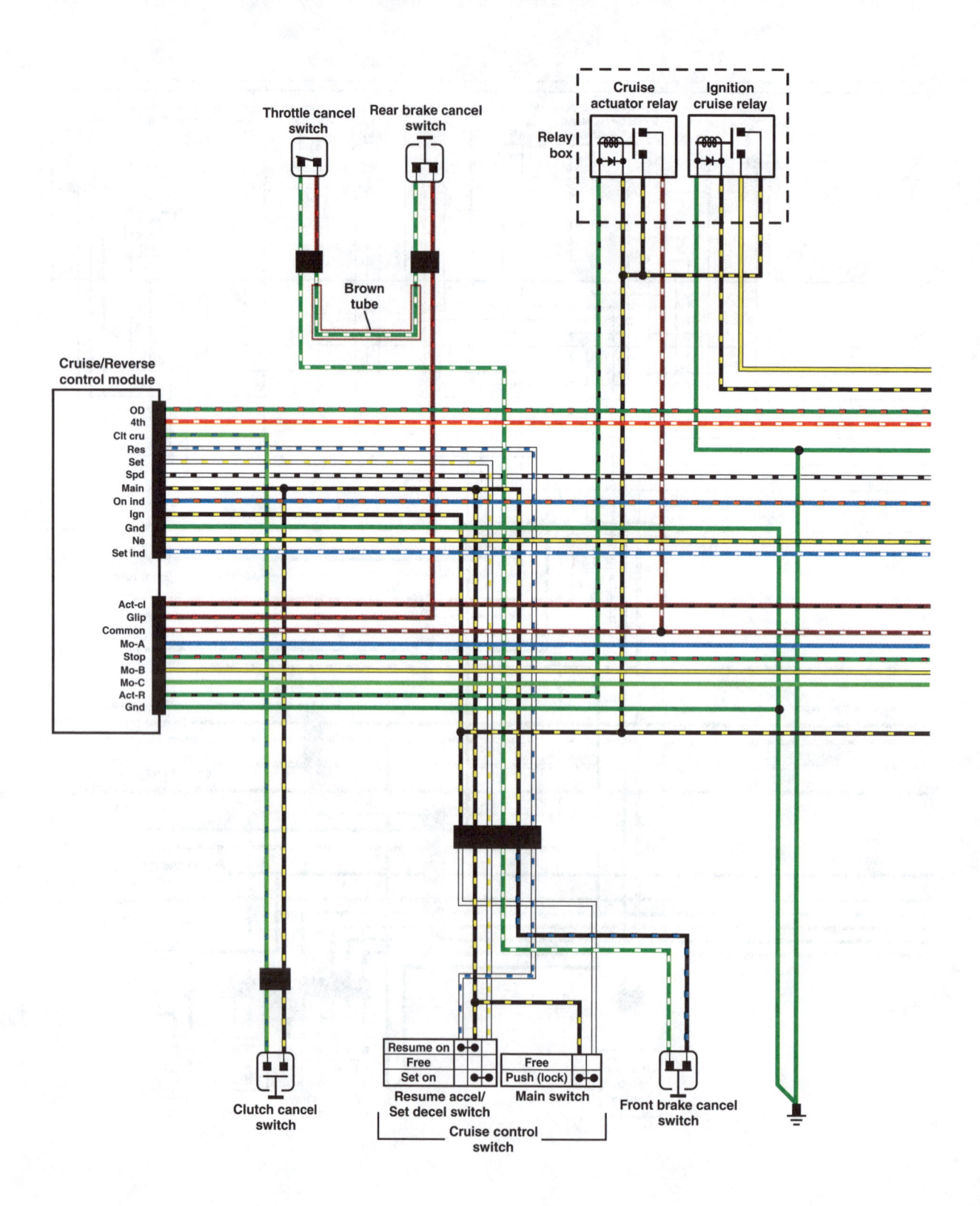

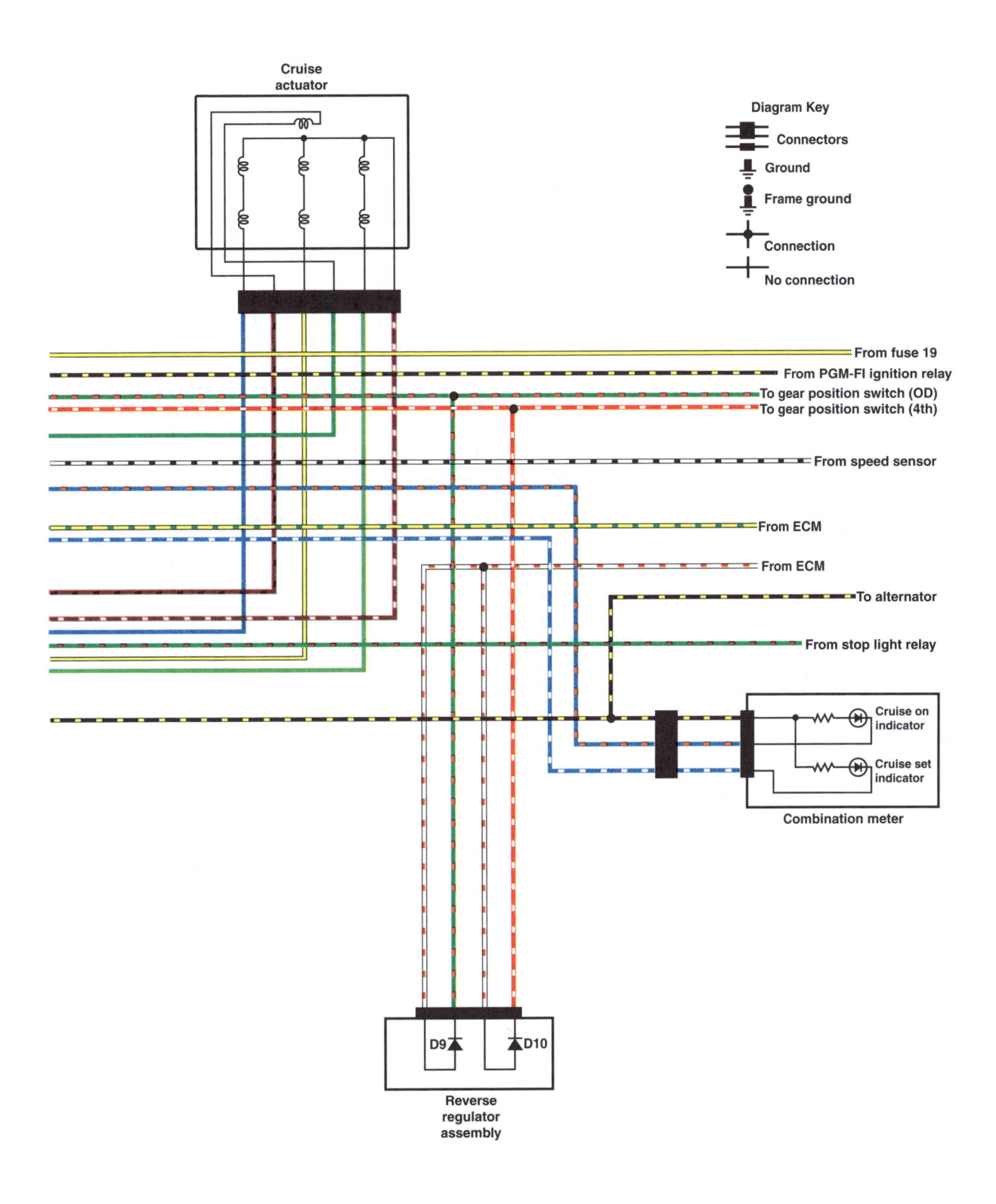

WIRING DIAGRAMS

CRUISE CONTROL SYSTEM (2004-2005)

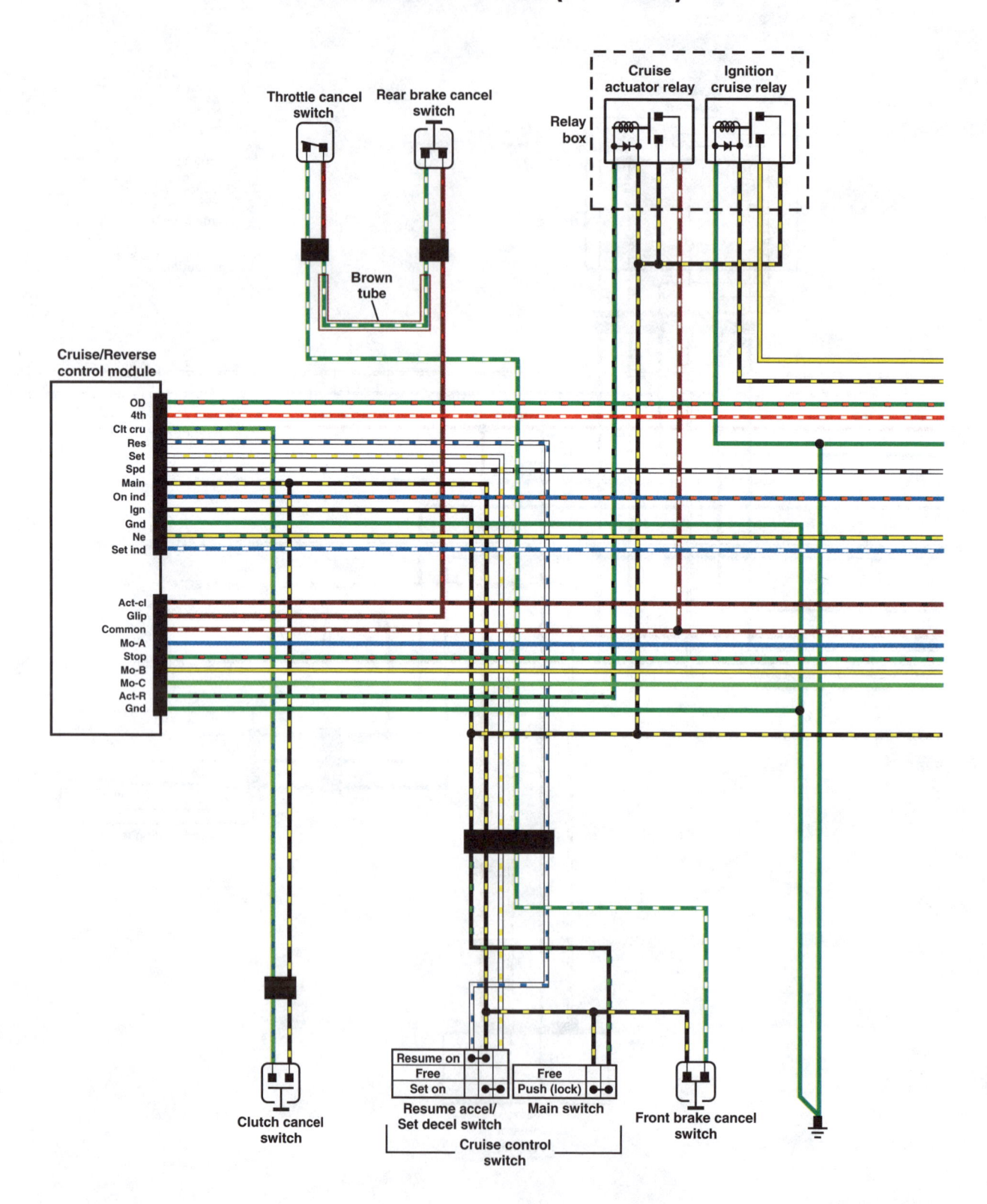

WIRING DIAGRAMS

689

Cruise actuator

Diagram Key
- Connectors
- Ground
- Frame ground
- Connection
- No connection

From fuse 19
From PGM-FI ignition relay
To gear position switch (OD)
To gear position switch (4th)
From speed sensor
From ECM
From ECM
To alternator
From stop light relay

Cruise on indicator
Cruise set indicator

Combination meter

D9 D10

Reverse regulator assembly

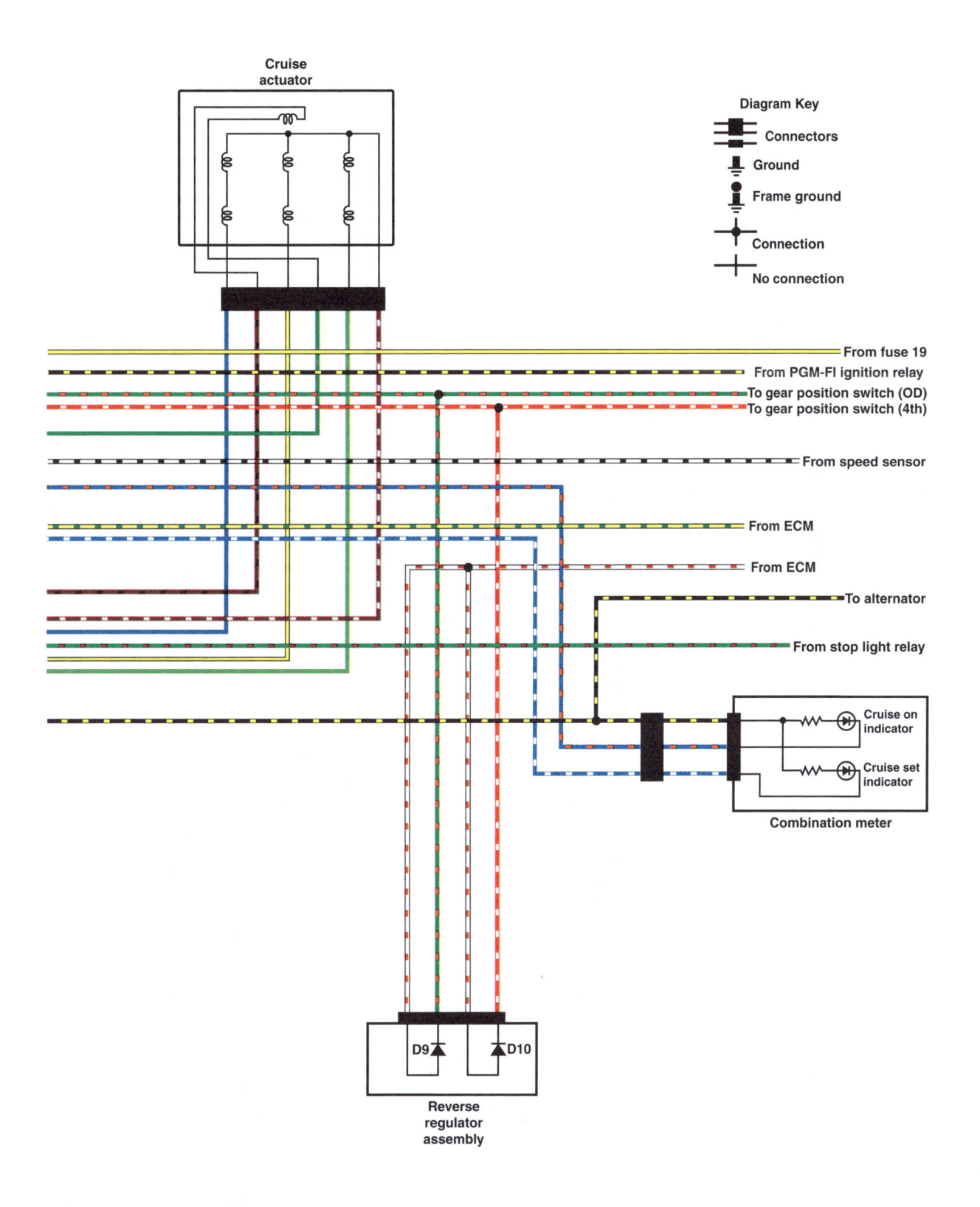

19

WIRING DIAGRAMS

ACCESSORY AND TRUNK LOCK CONTROL SYSTEMS (2001-2002)

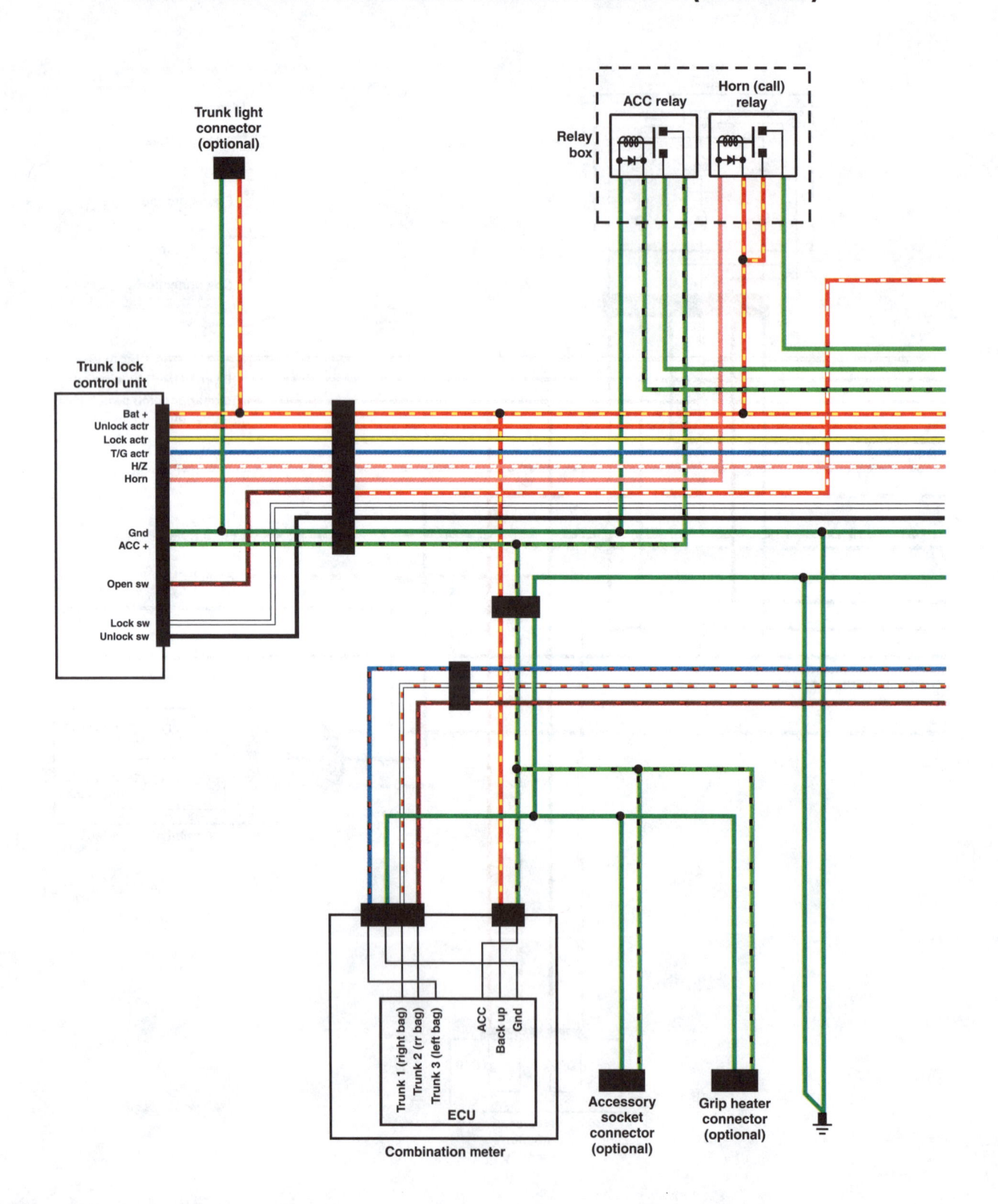

Relay box:
- ACC relay
- Horn (call) relay

Trunk light connector (optional)

Trunk lock control unit:
- Bat +
- Unlock actr
- Lock actr
- T/G actr
- H/Z
- Horn
- Gnd
- ACC +
- Open sw
- Lock sw
- Unlock sw

Combination meter / ECU:
- Trunk 1 (right bag)
- Trunk 2 (rr bag)
- Trunk 3 (left bag)
- ACC
- Back up
- Gnd

Accessory socket connector (optional)

Grip heater connector (optional)

WIRING DIAGRAMS

Diode assembly: D16, D15, D14

Accessory terminal connectors (−, +)

Fuse box: Accessory terminal fuse 5A (No. 6)

Diagram Key
- Connectors
- Ground
- Frame ground
- Connection
- No connection

To suspension control system/Audio sys/Combination meter
To horns
From fuse 21
From ignition switch (ACC)
From fuse 22
To turn signal lights

- Lock/Unlock switch
- Lock/Unlock actuator (Motor)
- Trunk actuator (Motor)
- Trunk open switch 2
- Trunk open switch 1
- Right saddlebag open switch 2
- Right saddlebag open switch 1
- Left saddlebag open switch 2
- Left saddlebag open switch 1

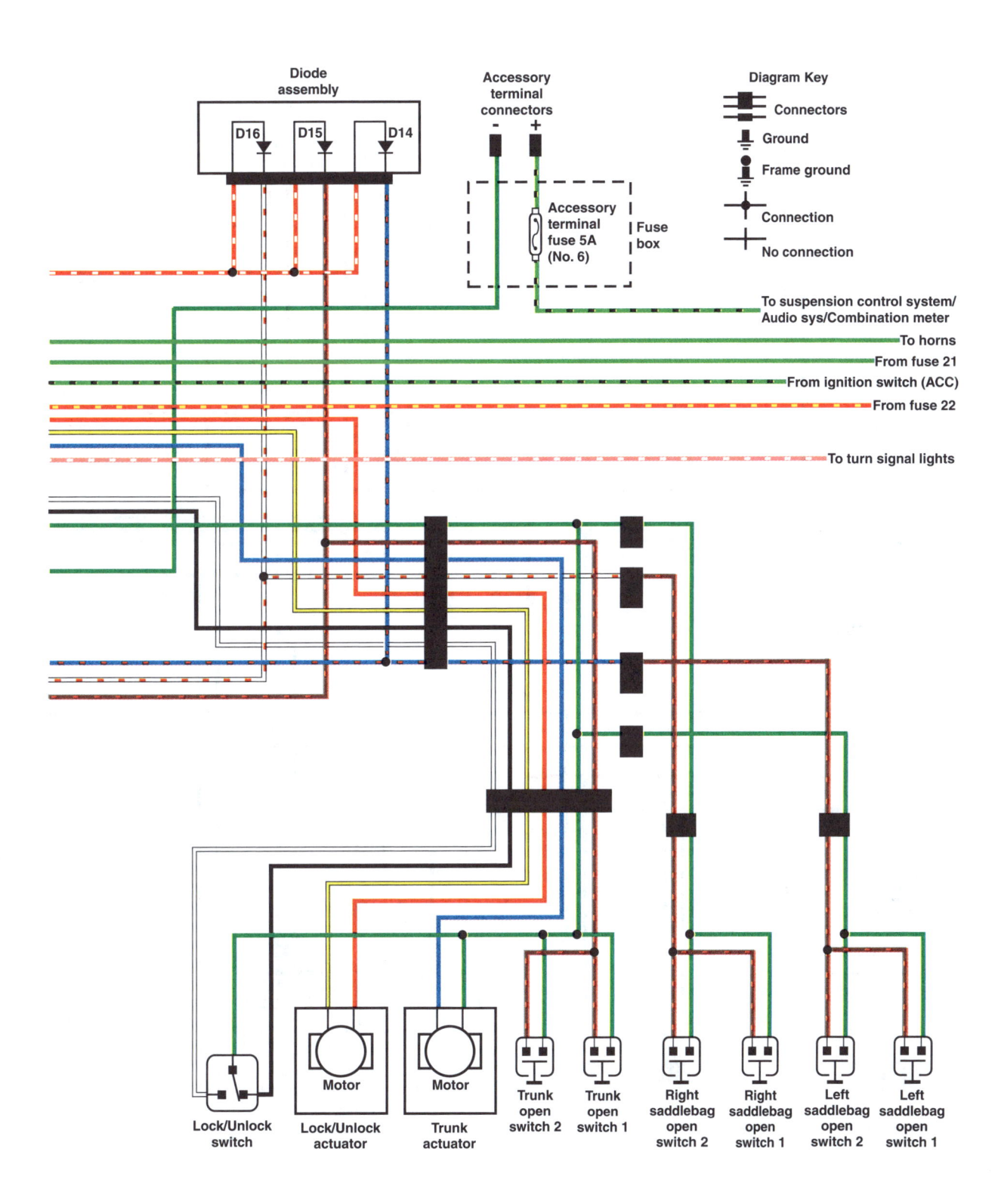

WIRING DIAGRAMS

ACCESSORY AND TRUNK LOCK CONTROL SYSTEMS (2003-2005)

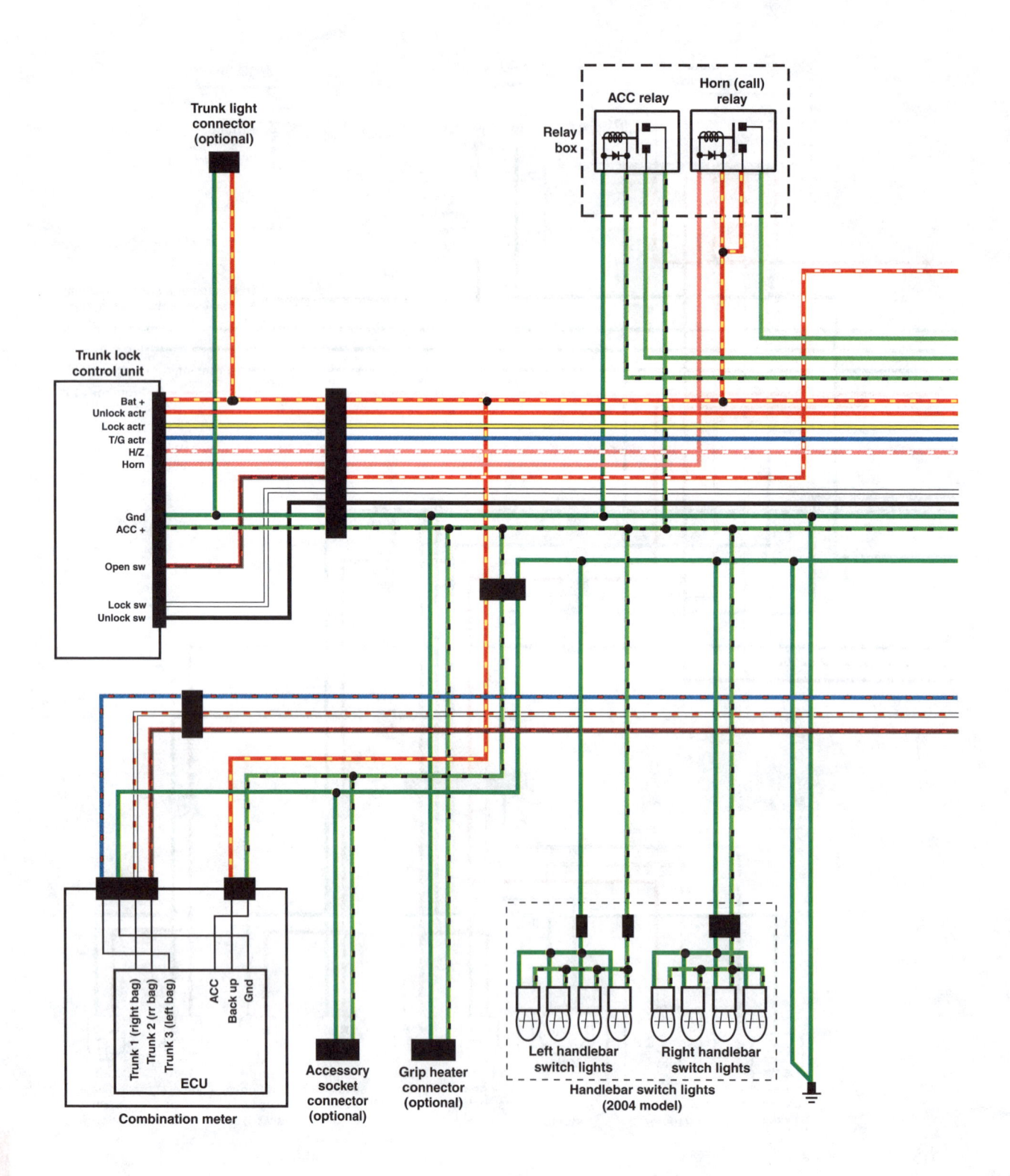

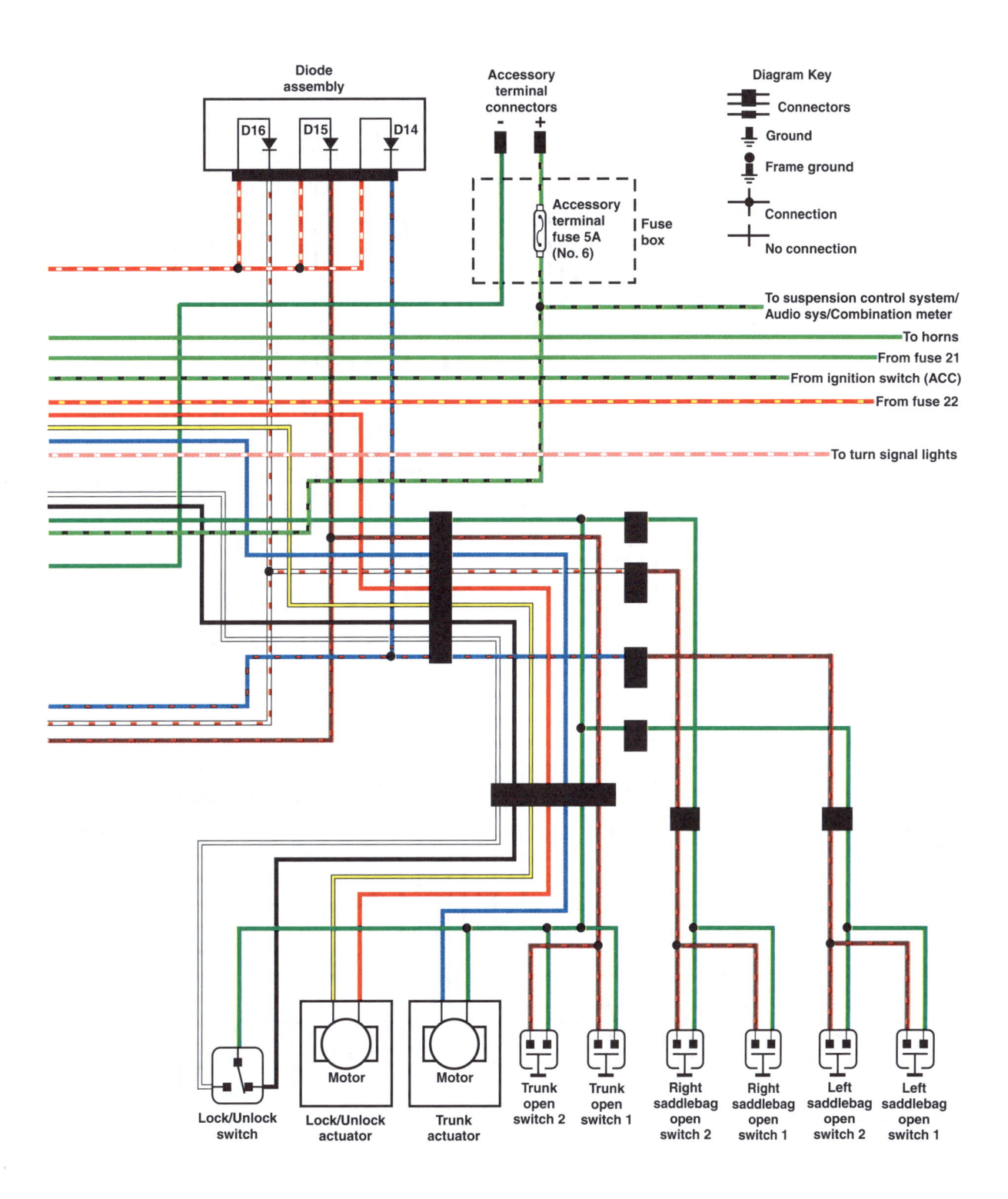

WIRING DIAGRAMS

WIRING DIAGRAMS

AUDIO SYSTEM (2001-2002)

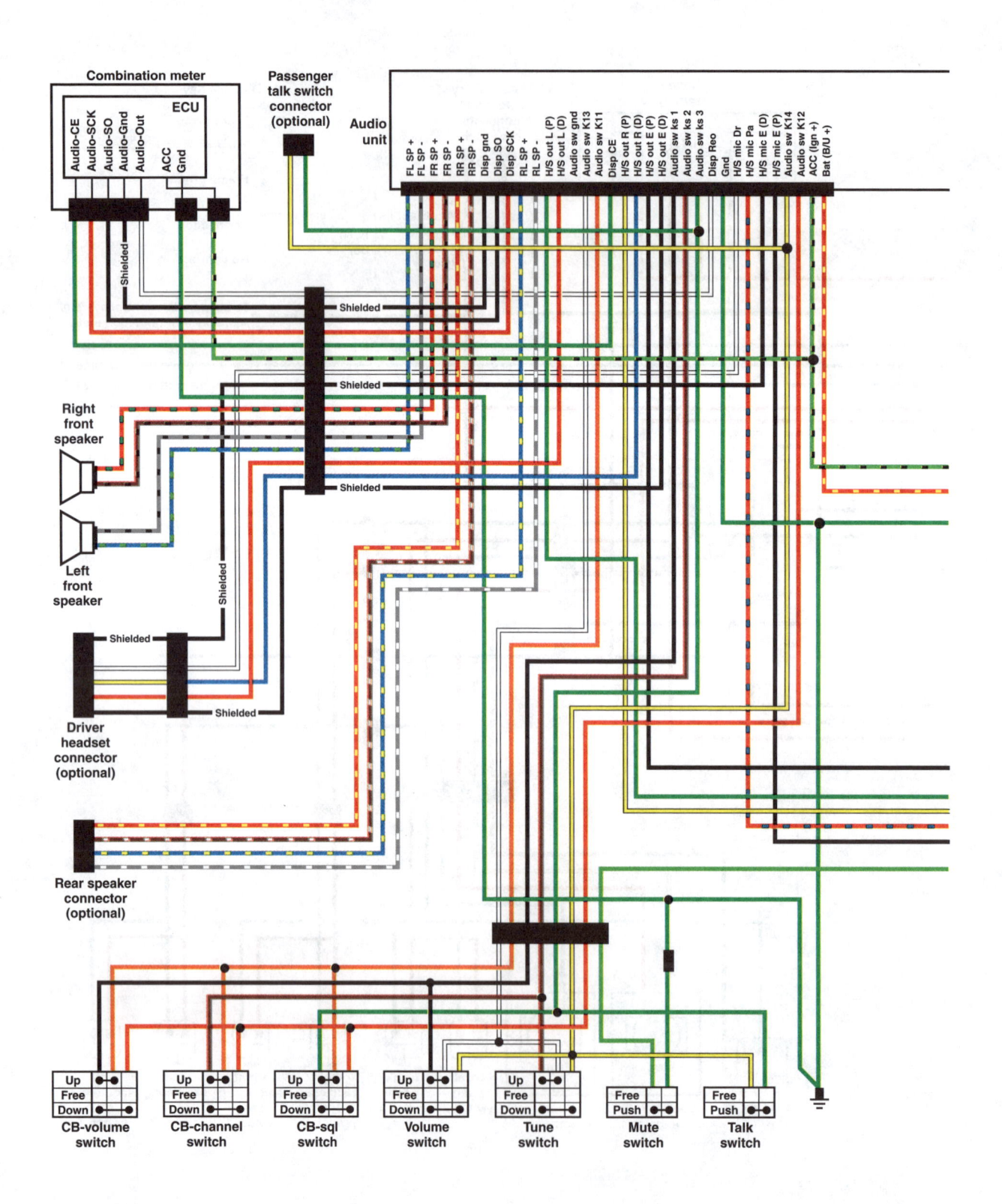

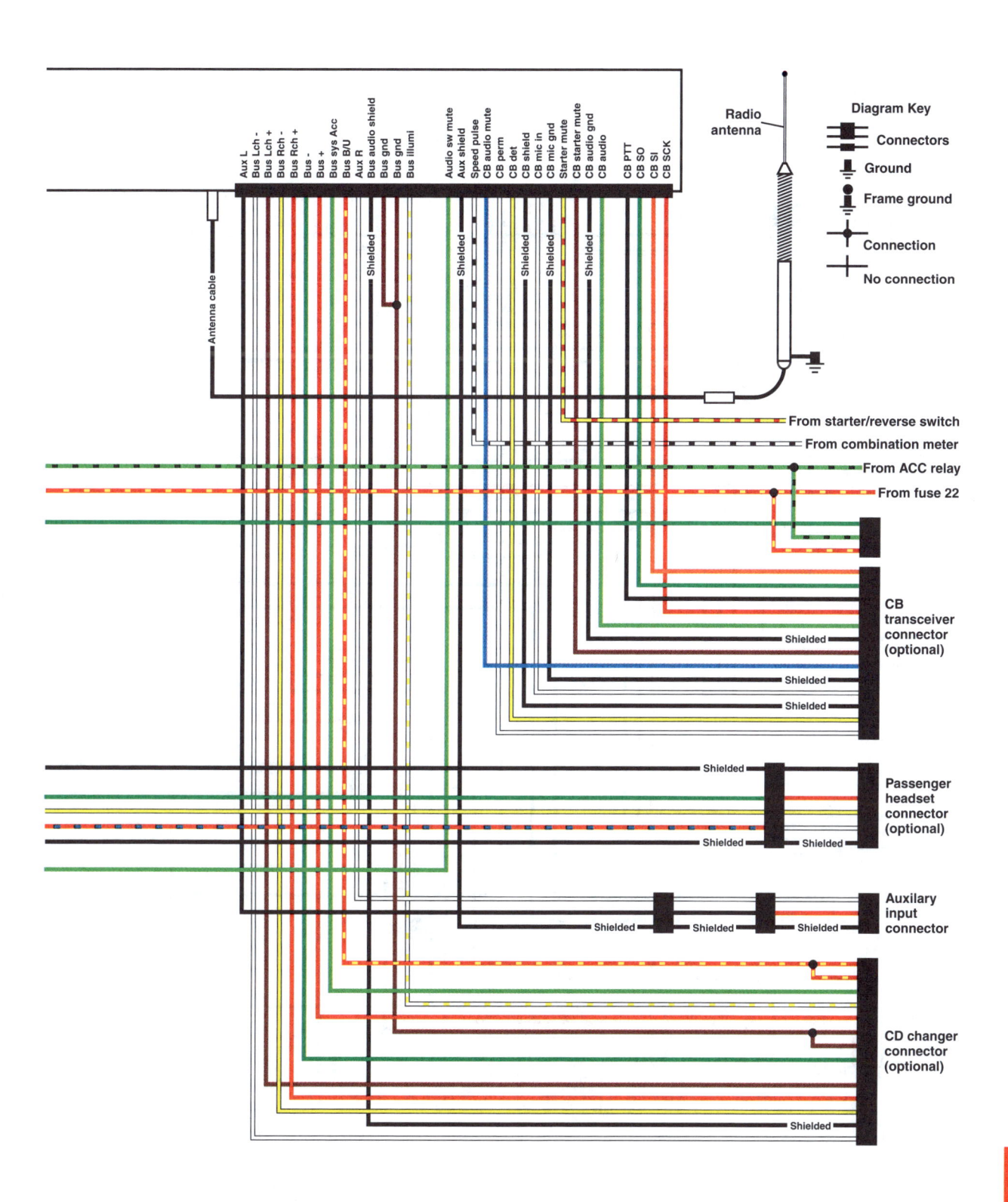

WIRING DIAGRAMS

695

Diagram Key
- Connectors
- Ground
- Frame ground
- Connection
- No connection

Radio antenna

From starter/reverse switch
From combination meter
From ACC relay
From fuse 22

CB transceiver connector (optional)

Passenger headset connector (optional)

Auxilary input connector

CD changer connector (optional)

Pin labels (left to right):
Aux L, Bus Lch −, Bus Lch +, Bus Rch −, Bus Rch +, Bus −, Bus +, Bus sys Acc, Bus B/U, Aux R, Bus audio shield, Bus gnd, Bus gnd, Bus illumi, Audio sw mute, Aux shield, Speed pulse, CB audio mute, CB perm, CB det, CB shield, CB mic in, CB mic gnd, Starter mute, CB starter mute, CB audio gnd, CB audio, CB PTT, CB SO, CB SI, CB SCK

19

WIRING DIAGRAMS

AUDIO SYSTEM (2003)

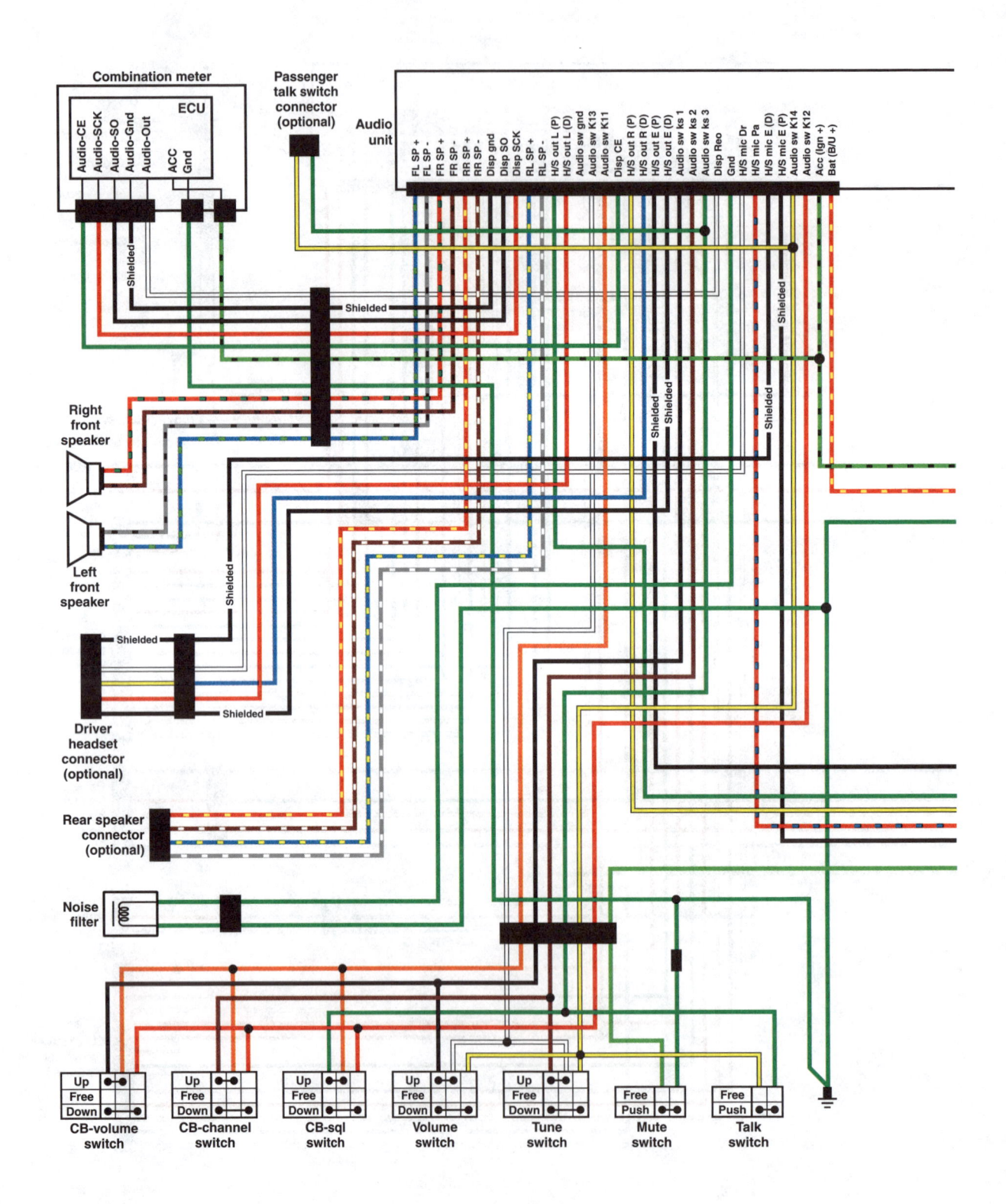

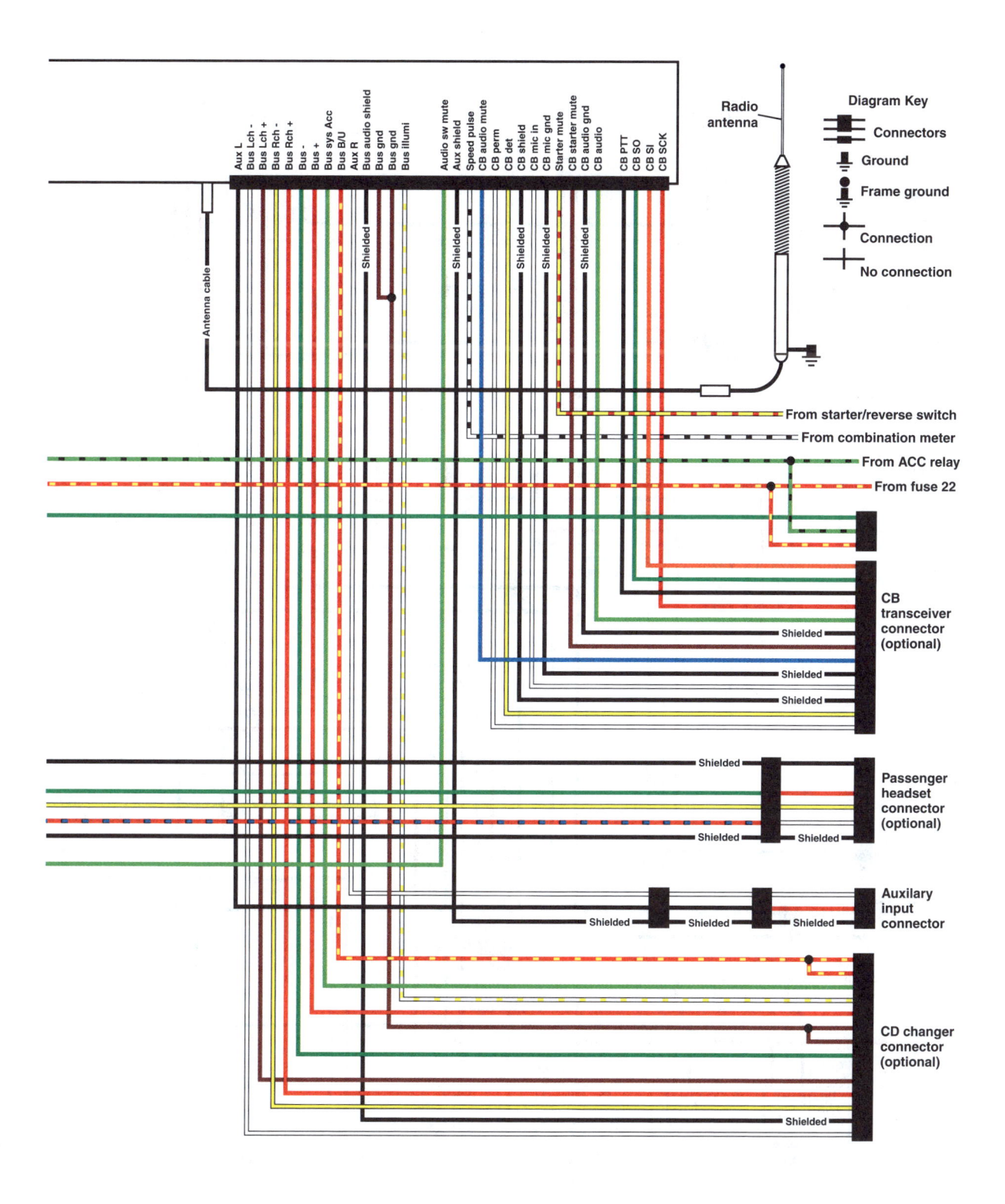

WIRING DIAGRAMS

WIRING DIAGRAMS

AUDIO SYSTEM (2004-2005)

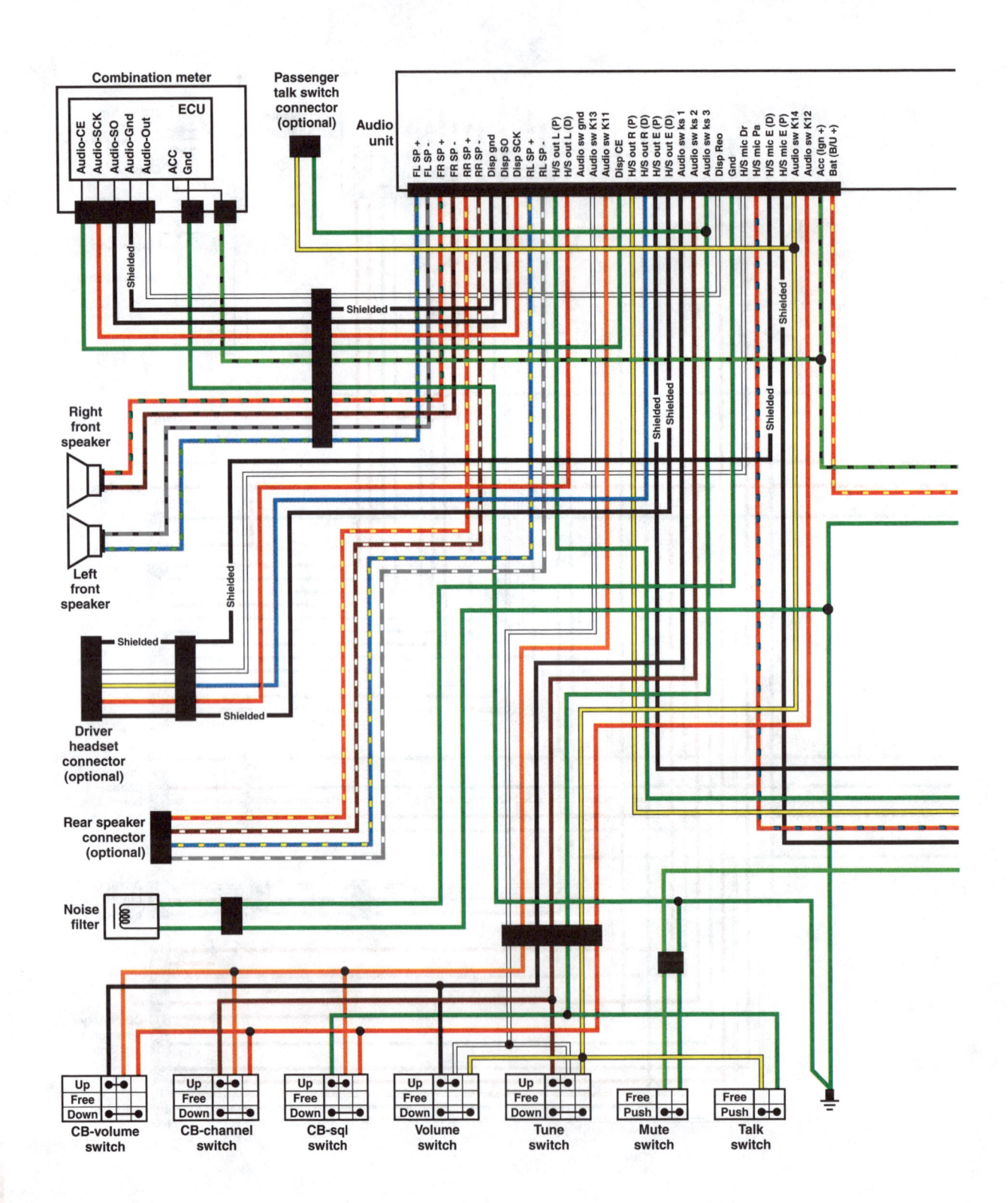

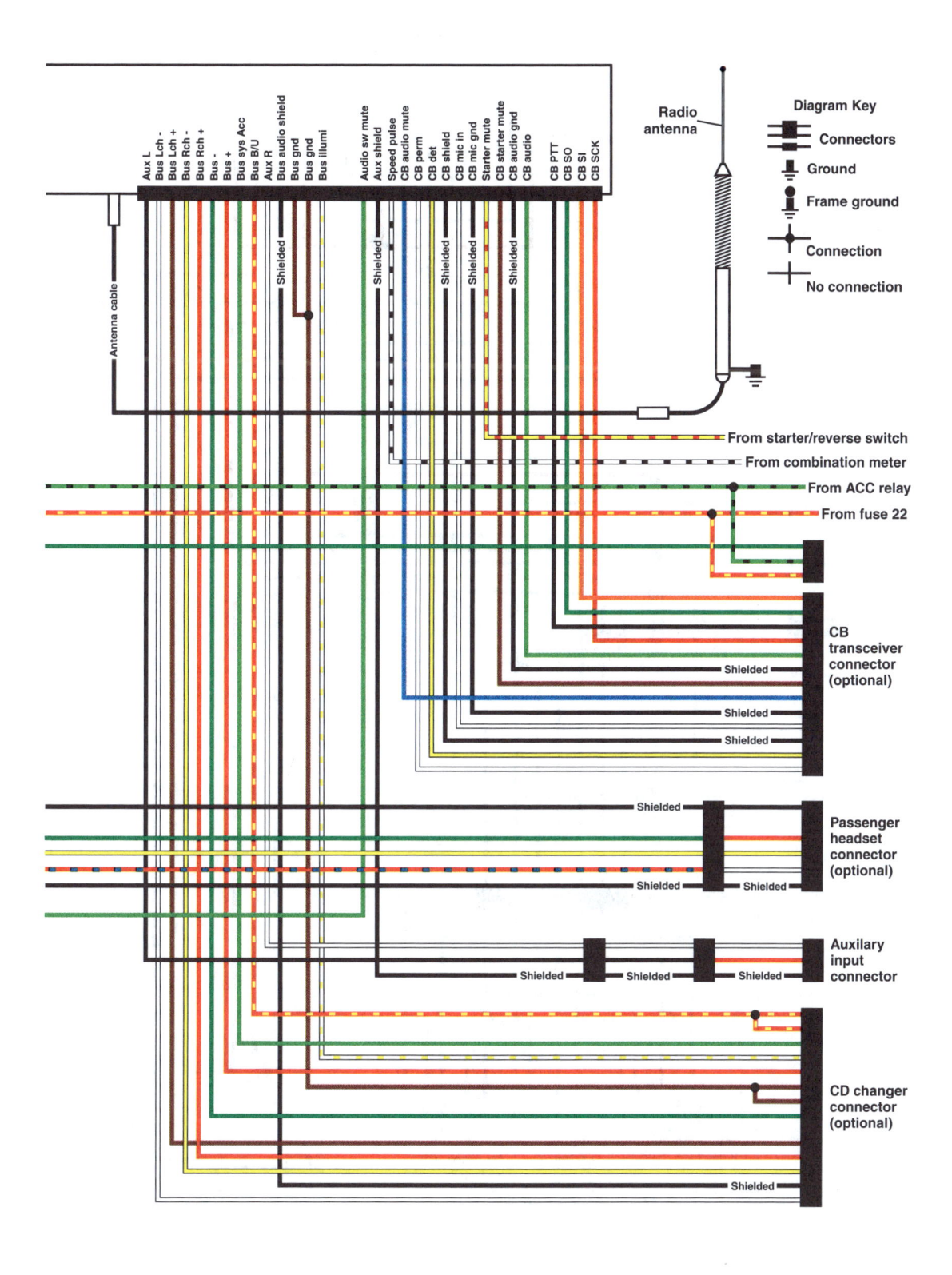

WIRING DIAGRAMS

WIRING DIAGRAMS

SUSPENSION LEVEL CONTROL SYSTEM (2001)

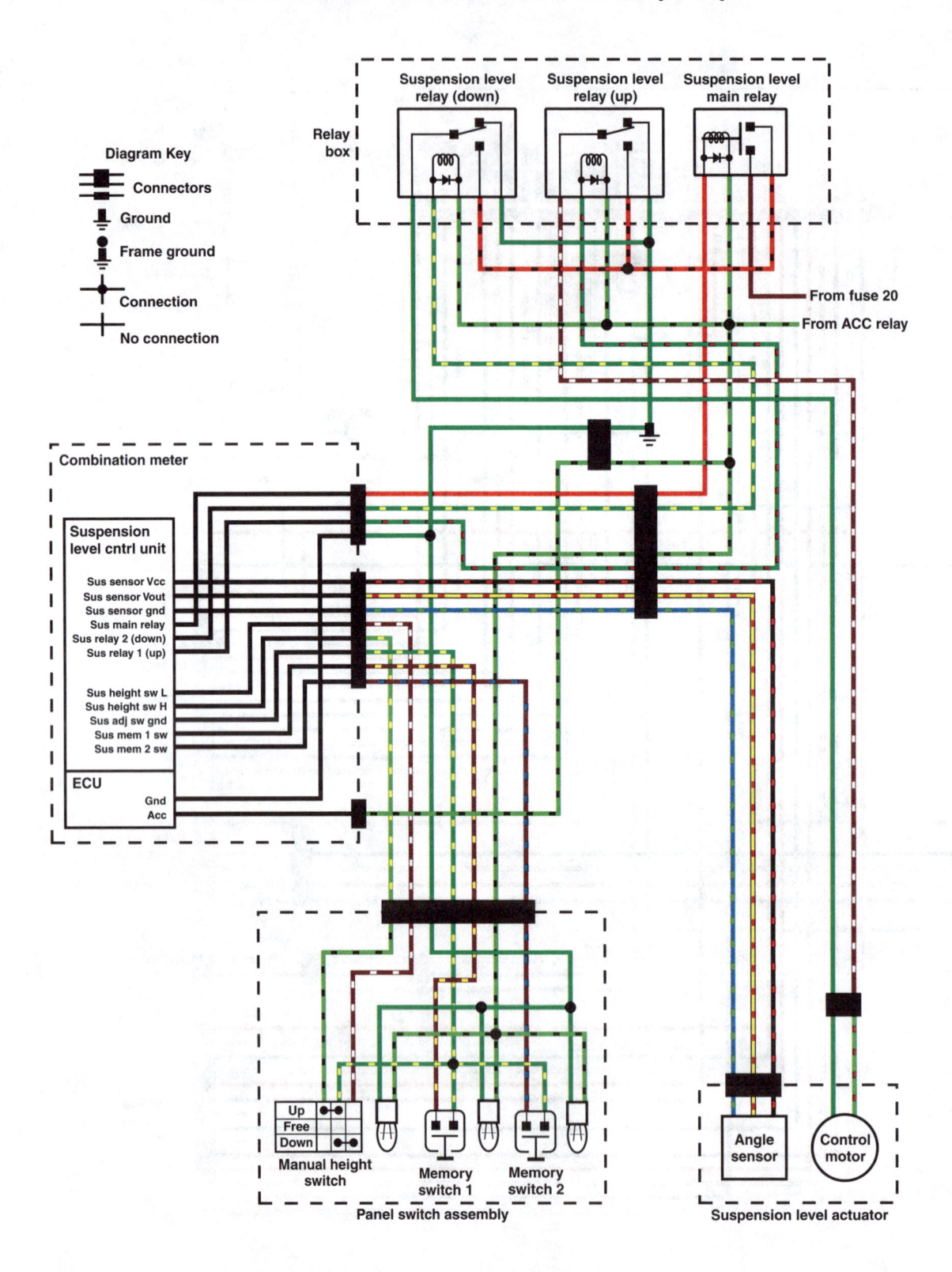

WIRING DIAGRAMS

SUSPENSION LEVEL CONTROL SYSTEM (2002-2005)

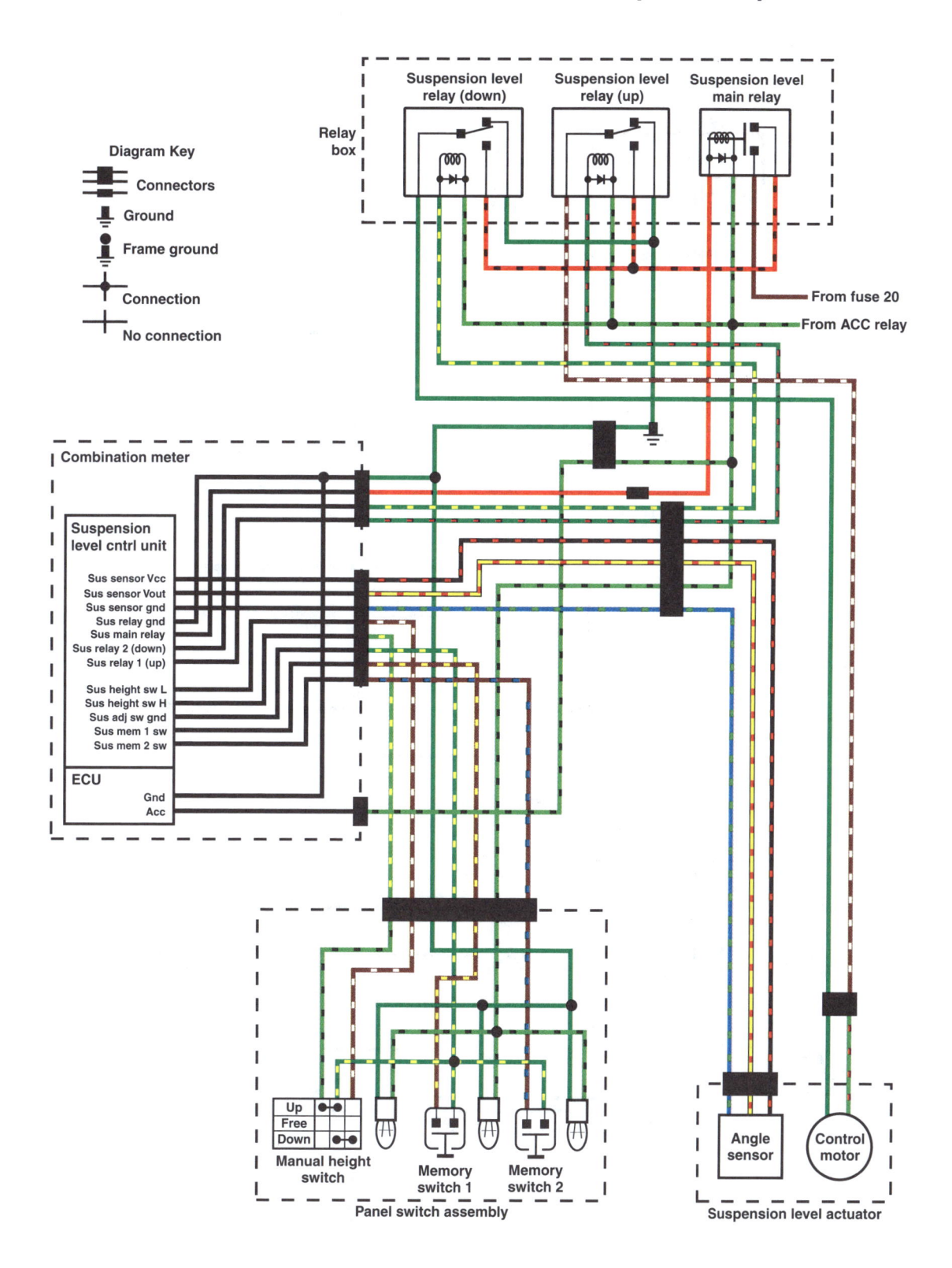

WIRING DIAGRAMS

FUSE BOX AND RELAY BOX

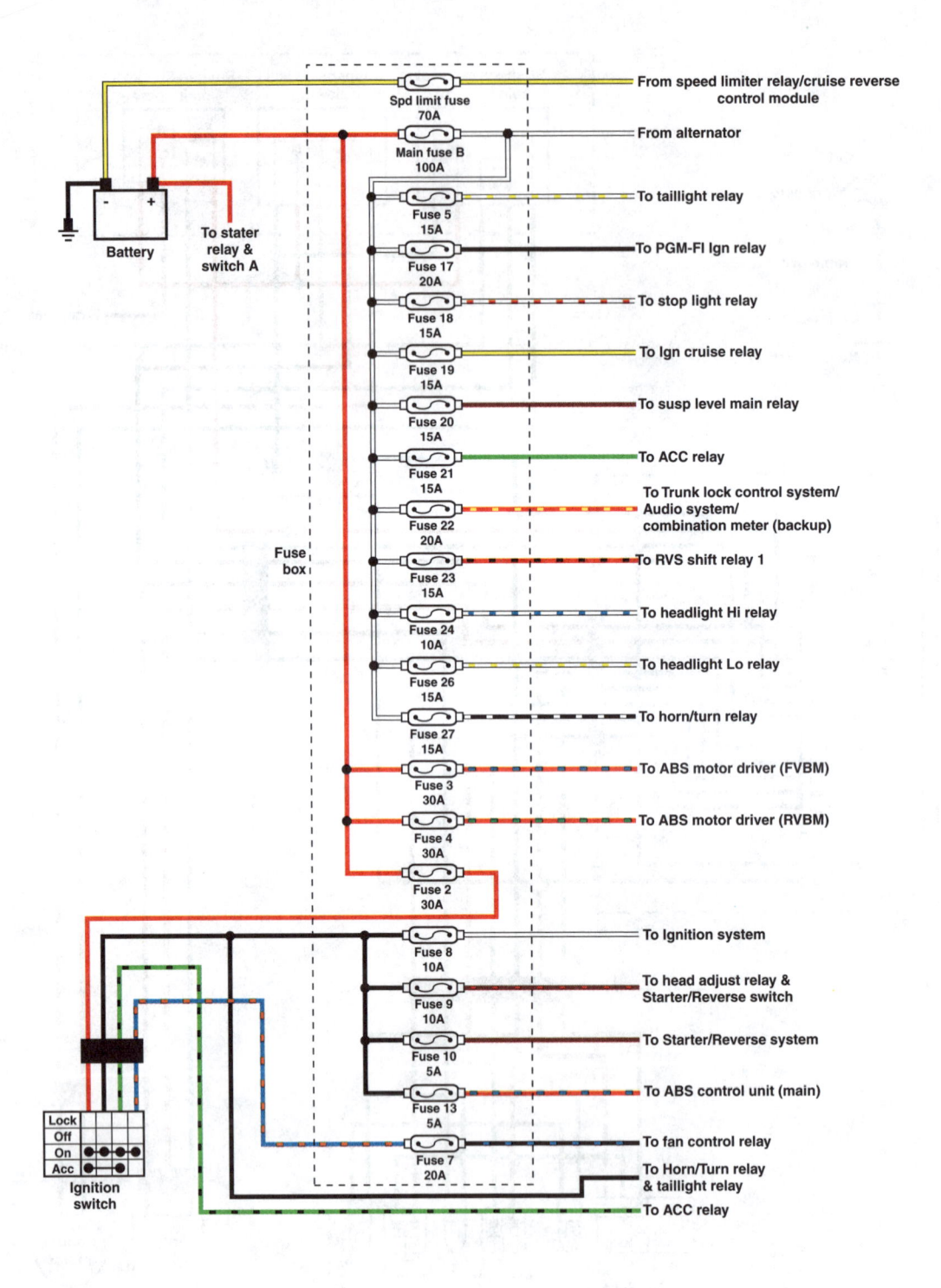

- Spd limit fuse 70A — From speed limiter relay/cruise reverse control module
- Main fuse B 100A — From alternator
- Fuse 5 15A — To taillight relay
- Fuse 17 20A — To PGM-FI Ign relay
- Fuse 18 15A — To stop light relay
- Fuse 19 15A — To Ign cruise relay
- Fuse 20 15A — To susp level main relay
- Fuse 21 15A — To ACC relay
- Fuse 22 20A — To Trunk lock control system/ Audio system/ combination meter (backup)
- Fuse 23 15A — To RVS shift relay 1
- Fuse 24 10A — To headlight Hi relay
- Fuse 26 15A — To headlight Lo relay
- Fuse 27 15A — To horn/turn relay
- Fuse 3 30A — To ABS motor driver (FVBM)
- Fuse 4 30A — To ABS motor driver (RVBM)
- Fuse 2 30A
- Fuse 8 10A — To Ignition system
- Fuse 9 10A — To head adjust relay & Starter/Reverse switch
- Fuse 10 5A — To Starter/Reverse system
- Fuse 13 5A — To ABS control unit (main)
- Fuse 7 20A — To fan control relay
- To Horn/Turn relay & taillight relay
- To ACC relay

Battery — To stater relay & switch A

Ignition switch: Lock / Off / On / Acc

WIRING DIAGRAMS

FUSE NUMBER AND AMPERAGE

1. Not used
2. Main fuse A 30A
3. ABS motor front fuse 30A
4. ABS motor rear fuse 30 A
5. Tail light fuse 30A
6. ACC terminal fuse 5A
7. Fan fuse
8. Stop switch fuse 10A
9. Headlight relay fuse 10A
10. RVS start fuse A 5A
11. RVS fuse A 5A
12. RVS fuse B 5A
13. ABS main fuse 5A
14. Not used
15. Not used
16. Not used
17. PGM-FI ignition fuse 20A
18. Stop light fuse 15A
19. Ignition cruise fuse 15A
20. Suspension level fuse 15A
21. Audio/ACC fuse 15A
22. Battery fuse 20A
23. RVS shift fuse 15A
24. Headlight HI fuse 10A
25. Not used
26. Headlight LO fuse 15A
27. Horn/turn fuse 15A

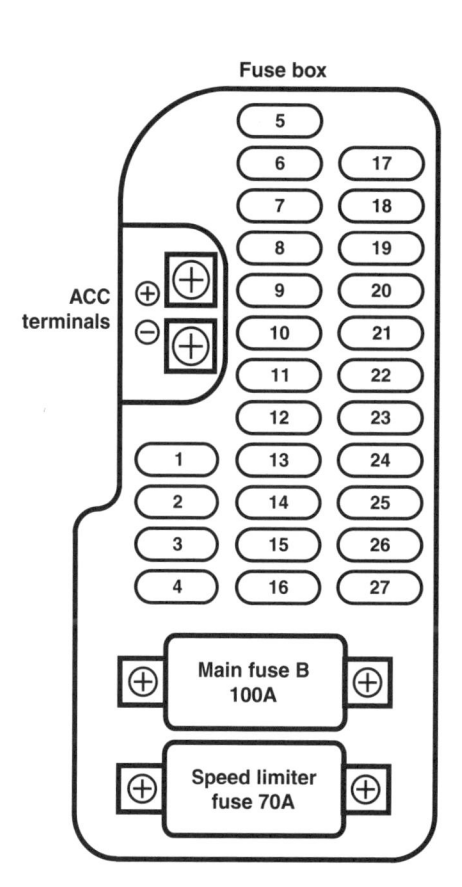

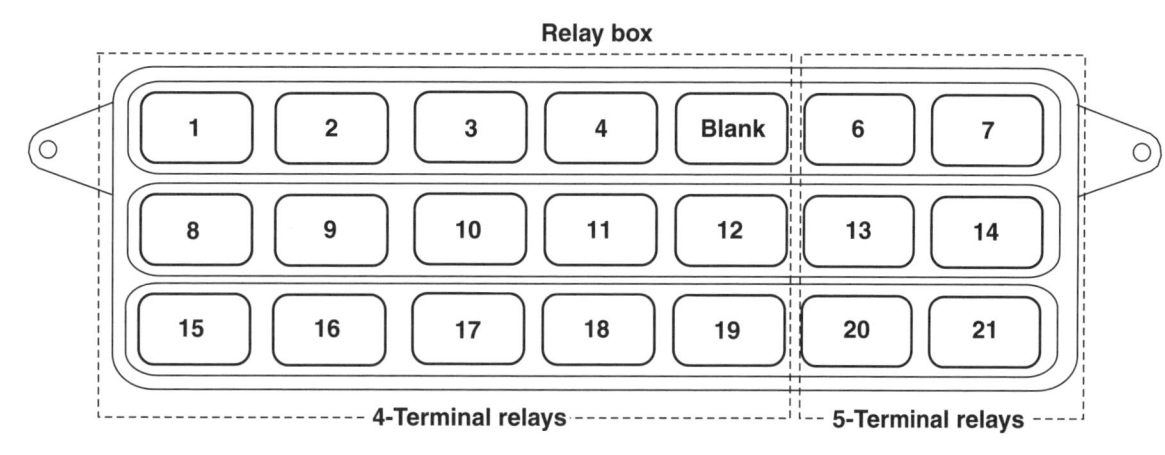

1. Headlight LO relay
2. Cruise actuator relay
3. Fan control relay
4. Fuel pump relay
5. Blank
6. Ignition cruise relay
7. Headlight adjuster relay
8. Horn turn relay
9. Stop light relay
10. Taillight relay
11. Suspension level mail relay
12. ACC relay
13. Suspension level down relay
14. Suspension level up relay
15. FI ignition relay
16. Headlight HI relay
17. Horn (call) relay
18. Reverse relay
19. Reverse shift relay 1
20. Reverse shift relay 2
21. Reverse shift relay 3

NOTES

MAINTENANCE LOG

Date	Miles	Type of Service

Check out clymer.com for our full line of powersport repair manuals.

BMW
- M308 — 500 & 600 CC Twins, 55-69
- M309 — F650, 1994-2000
- M500-3 — BMW K-Series, 85-97
- M501 — K1200RS, GT & LT, 98-05
- M502-3 — BMW R50/5-R100 GSPD, 70-96
- M503-2 — R850, R1100, R1150 and R1200C, 93-04

HARLEY-DAVIDSON
- M419 — Sportsters, 59-85
- M428 — Sportster Evolution, 86-90
- M429-4 — Sportster Evolution, 91-03
- M427 — Sportster, 04-05
- M418 — Panheads, 48-65
- M420 — Shovelheads, 66-84
- M421-3 — FLS/FXS Evolution, 84-99
- M423-2 — FLS/FXS Twin Cam, 00-05
- M422-3 — FLH/FLT/FXR Evolution, 84-99
- M430-4 — FLH/FLT Twin Cam, 99-05
- M424-2 — FXD Evolution, 91-98
- M425-3 — FXD Twin Cam, 99-05

HONDA
ATVs
- M316 — Odyssey FL250, 77-84
- M311 — ATC, TRX & Fourtrax 70-125, 70-87
- M433 — Fourtrax 90 ATV, 93-00
- M326 — ATC185 & 200, 80-86
- M347 — ATC200X & Fourtrax 200SX, 86-88
- M455 — ATC250 & Fourtrax 200/250, 84-87
- M342 — ATC250R, 81-84
- M348 — TRX250R/Fourtrax 250R & ATC250R, 85-89
- M456-3 — TRX250X 87-92; TRX300EX 93-04
- M446-2 — TRX250 Recon & ES, 97-04
- M346-3 — TRX300/Fourtrax 300 & TRX300FW/Fourtrax 4x4, 88-00
- M200 — TRX350 Rancher, 00-03
- M459-3 — TRX400 Foreman 95-03
- M454-3 — TRX400EX 99-05
- M205 — TRX450 Foreman, 98-04
- M210 — TRX500 Rubicon, 98-04

Singles
- M310-13 — 50-110cc OHC Singles, 65-99
- M319-2 — XR50R, CRF50F, XR70R & CRF70F, 97-05
- M315 — 100-350cc OHC, 69-82
- M317 — Elsinore, 125-250cc, 73-80
- M442 — CR60-125R Pro-Link, 81-88
- M431-2 — CR80R, 89-95, CR125R, 89-91
- M435 — CR80, 96-02
- M457-2 — CR125R & CR250R, 92-97
- M464 — CR125R, 1998-2002
- M443 — CR250R-500R Pro-Link, 81-87
- M432-3 — CR250R, 88-91 & CR500R, 88-01
- M437 — CR250R, 97-01
- M352 — CRF250, CRF250X & CRF450R, CRF450X, 02-05
- M312-13 — XL/XR75-100, 75-03
- M318-4 — XL/XR/TLR 125-200, 79-03
- M328-4 — XL/XR250, 78-00; XL/XR350R 83-85; XR200R, 84-85; XR250L, 91-96
- M320-2 — XR400R, 96-04
- M339-7 — XL/XR 500-650, 79-03

Twins
- M321 — 125-200cc, 65-78
- M322 — 250-350cc, 64-74
- M323 — 250-360cc Twins, 74-77
- M324-5 — Twinstar, Rebel 250 & Nighthawk 250, 78-03
- M334 — 400-450cc, 78-87
- M333 — 450 & 500cc, 65-76
- M335 — CX & GL500/650 Twins, 78-83
- M344 — VT500, 83-88
- M313 — VT700 & 750, 83-87
- M314-2 — VT750 Shadow (chain drive), 98-05
- M440 — VT1100C Shadow, 85-96
- M460-3 — VT1100C Series, 95-04

Fours
- M332 — CB350-550cc, SOHC, 71-78
- M345 — CB550 & 650, 83-85
- M336 — CB650, 79-82
- M341 — CB750 SOHC, 69-78
- M337 — CB750 DOHC, 79-82
- M436 — CB750 Nighthawk, 91-93 & 95-99
- M325 — CB900, 1000 & 1100, 80-83
- M439 — Hurricane 600, 87-90
- M441-2 — CBR600F2 & F3, 91-98
- M445 — CBR600F4, 99-06
- M434-2 — CBR900RR Fireblade, 93-99
- M329 — 500cc V-Fours, 84-86
- M438 — Honda VFR800, 98-00
- M349 — 700-1000 Interceptor, 83-85
- M458-2 — VFR700F-750F, 86-97
- M327 — 700-1100cc V-Fours, 82-88
- M340 — GL1000 & 1100, 75-83
- M504 — GL1200, 84-87
- M508 — ST1100/PAN European, 90-02

Sixes
- M505 — GL1500 Gold Wing, 88-92
- M506-2 — GL1500 Gold Wing, 93-00
- M507 — GL1800 Gold Wing, 01-04
- M462-2 — GL1500C Valkyrie, 97-03

KAWASAKI
ATVs
- M465-2 — KLF220 & KLF250 Bayou, 88-03
- M466-4 — KLF300 Bayou, 86-04
- M467 — KLF400 Bayou, 93-99
- M470 — KEF300 Lakota, 95-99
- M385 — KSF250 Mojave, 87-00

Singles
- M350-9 — Rotary Valve 80-350cc, 66-01
- M444-2 — KX60, 83-02; KX80 83-90
- M448 — KX80/85/100, 89-03
- M351 — KDX200, 83-88
- M447-3 — KX125 & KX250, 82-91 KX500, 83-04
- M472-2 — KX125, 92-00
- M473-2 — KX250, 92-00
- M474 — KLR650, 87-03

Twins
- M355 — KZ400, KZ/Z440, EN450 & EN500, 74-95
- M360-3 — EX500, GPZ500S, Ninja R, 87-02
- M356-4 — Vulcan 700 & 750, 85-04
- M354-2 — Vulcan 800 & Vulcan 800 Classic, 95-04
- M357-2 — Vulcan 1500, 87-99
- M471-2 — Vulcan Classic 1500, 96-04

Fours
- M449 — KZ500/550 & ZX550, 79-85
- M450 — KZ, Z & ZX750, 80-85
- M358 — KZ650, 77-83
- M359-3 — 900-1000cc Fours, 73-81
- M451-3 — 1000 &1100cc Fours, 81-02
- M452-3 — ZX500 & 600 Ninja, 85-97
- M453-3 — Ninja ZX900-1100 84-01
- M468-2 — Ninja ZX-6, 90-04
- M469 — ZX7 Ninja, 91-98
- M453-3 — Ninja ZX900, ZX1000 & ZX1100, 84-01
- M409 — Concours, 86-04

POLARIS
ATVs
- M496 — Polaris ATV, 85-95
- M362 — Polaris Magnum ATV, 96-98
- M363 — Scrambler 500, 4X4 97-00
- M365-2 — Sportsman/Xplorer, 96-03

SUZUKI
ATVs
- M381 — ALT/LT 125 & 185, 83-87
- M475 — LT230 & LT250, 85-90
- M380-2 — LT250R Quad Racer, 85-92
- M343 — LTF500F Quadrunner, 98-00
- M483-2 — Suzuki King Quad/Quad Runner 250, 87-98

Singles
- M371 — RM50-400 Twin Shock, 75-81
- M369 — 125-400cc 64-81
- M379 — RM125-500 Single Shock, 81-88
- M476 — DR250-350, 90-94
- M384-3 — LS650 Savage, 86-04
- M386 — RM80-250, 89-95
- M400 — RM125, 96-00
- M401 — RM250, 96-02

Twins
- M372 — GS400-450 Twins, 77-87
- M481-4 — VS700-800 Intruder, 85-04
- M482-2 — VS1400 Intruder, 87-01
- M484-3 — GS500E Twins, 89-02
- M361 — SV650, 1999-2002

Triple
- M368 — 380-750cc, 72-77

Fours
- M373 — GS550, 77-86
- M364 — GS650, 81-83
- M370 — GS750 Fours, 77-82
- M376 — GS850-1100 Shaft Drive, 79-84
- M378 — GS1100 Chain Drive, 80-81
- M383-3 — Katana 600, 88-96
- M331 — GSX-R750-1100, 86-87
- M478-2 — GSX-R600, 97-00 GSX-R750, 88-92 GSX750F Katana, 89-96
- M485 — GSX-R750, 96-99
- M377 — GSX-R1000, 01-04
- M338 — GSF600 Bandit, 95-00
- M353 — GSF1200 Bandit, 96-03

YAMAHA
ATVs
- M499 — YFM80 Badger, 85-01
- M394 — YTM/YFM200 & 225, 83-86
- M488-5 — Blaster, 88-05
- M489-2 — Timberwolf, 89-00
- M487-5 — Warrior, 87-04
- M486-5 — Banshee, 87-04
- M490-3 — Moto-4 & Big Bear, 87-04
- M493 — YFM400FW Kodiak, 93-98
- M280-2 — Raptor 660R, 01-05

Singles
- M492-2 — PW50 & PW80, BW80 Big Wheel 80, 81-02
- M410 — 80-175 Piston Port, 68-76
- M415 — 250-400cc Piston Port, 68-76
- M412 — DT & MX 100-400, 77-83
- M414 — IT125-490, 76-86
- M393 — YZ50-80 Monoshock, 78-90
- M413 — YZ100-490 Monoshock, 76-84
- M390 — YZ125-250, 85-87 YZ490, 85-90
- M391 — YZ125-250, 88-93 WR250Z, 91-93
- M497-2 — YZ125, 94-01
- M498 — YZ250, 94-98 and WR250Z, 94-97
- M406 — YZ250F & WR250F, 01-03
- M491-2 — YZ400F, YZ426F, WR400F WR426F, 98-02
- M417 — XT125-250, 80-84
- M480-3 — XT/TT 350, 85-00
- M405 — XT500 & TT500, 76-81
- M416 — XT/TT 600, 83-89

Twins
- M403 — 650cc, 70-82
- M395-10 — XV535-1100 Virago, 81-03
- M495-4 — V-Star 650, 98-05
- M281-2 — V-Star 1100, 99-05
- M282 — Road Star, 99-05

Triple
- M404 — XS750 & 850, 77-81

Fours
- M387 — XJ550, XJ600 & FJ600, 81-92
- M494 — XJ600 Seca II, 92-98
- M388 — YX600 Radian & FZ600, 86-90
- M396 — FZR600, 89-93
- M392 — FZ700-750 & Fazer, 85-87
- M411 — XS1100 Fours, 78-81
- M397 — FJ1100 & 1200, 84-93
- M375 — V-Max, 85-03
- M374 — Royal Star, 96-03
- M461 — YZF-R6, 99-04
- M398 — YZF-R1, 98-03
- M399 — FZ1, 01-04

VINTAGE MOTORCYCLES
Clymer® Collection Series
- M330 — Vintage British Street Bikes, BSA, 500-650cc Unit Twins; Norton, 750 & 850cc Commandos; Triumph, 500-750cc Twins
- M300 — Vintage Dirt Bikes, V. 1 Bultaco, 125-370cc Singles; Montesa, 123-360cc Singles; Ossa, 125-250cc Singles
- M301 — Vintage Dirt Bikes, V. 2 CZ, 125-400cc Singles; Husqvarna, 125-450cc Singles; Maico, 250-501cc Singles; Hodaka, 90-125cc Singles
- M305 — Vintage Japanese Street Bikes Honda, 250 & 305cc Twins; Kawasaki, 250-750cc Triples; Kawasaki, 900 & 1000cc Fours